W9-CRH-570

Student CD to accompany
An Introduction to Behavioral Endocrinology, Third Edition

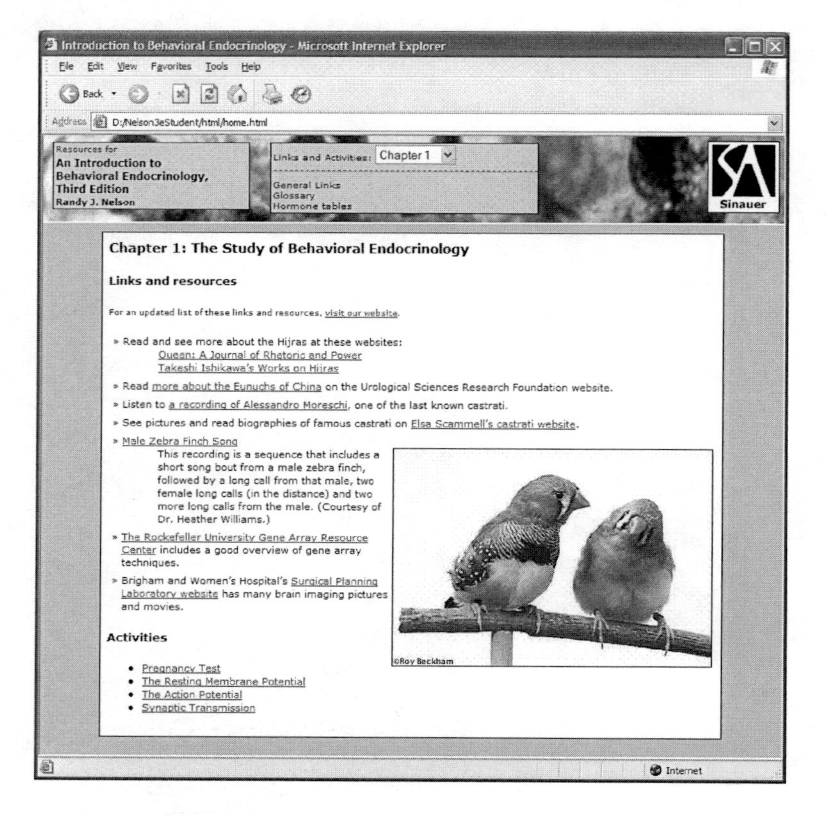

Included with each copy of *An Introduction to Behavioral Endocrinology,* Third Edition, the Student CD provides a wealth of electronic resources to help the student better understand many of the concepts and behaviors discussed in the textbook. The CD includes:

- **Animated Tutorials:** These detailed tutorials explain complex concepts and processes; each includes an introduction that provides the context for the concept to be presented, a detailed animation that explains the concept, a conclusion that summarizes what you should have learned, and a short quiz.
- **Videos:** The video segments depict interesting behaviors in a variety of organisms, helping to bring to life some of the behaviors discussed in the text.
- **Web Links:** Chapter-specific online resources include articles, photographs, activities, and audio clips that help explain and illustrate selected topics.
- **Hormone tables:** Detailed listings of hormones, their sources, and their major biological actions are provided.
- **Links:** The links are to journals, societies, and associations in the field of endocrinology; these are useful for further study or research.
- A complete **Glossary** is included for quick reference.

(Up-to-date links available online at www.sinauer.com/nelson3e-cd)

Included on the Student CD

Animated Tutorials

- Pregnancy Test
- The Resting Membrane Potential
- The Action Potential
- Synaptic Transmission
- Mitosis
- Meiosis
- Control, Regulation, and Feedback
- Complete Metamorphosis
- Hormonal Regulation of Calcium
- The Ovarian and Uterine Cycles
- Insulin and Glucose Regulation
- Hypothalamic–Pituitary–Endocrine axis
- Fertilization and Production of the Conceptus
- Development of the Male and Female Reproductive Tracts
- Development of Sexual Dimorphism
- Aromatization Hypothesis
- The Costs of Defending a Territory
- Biological Rhythms

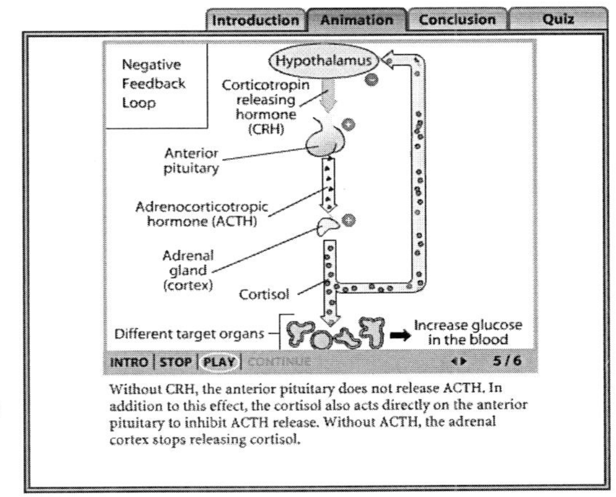

Control, Regulation, and Feedback

Videos

- Leopard Gecko Sex Determination
- Hyena Maternal Rank
- Hyena Neonatal Aggression
- Zebra Finch Courtship Song
- Rat Sexual Behavior
- Rat Mating Behavior 1
- Rat Genital Grooming
- Rat Ejaculation
- Ultrasonic Communication in Rats
- Pseudosexual Behavior in Whiptail Lizards
- Courtship in the Fish *Porichthys notatus*
- Quail Mating
- Group Mating in Rats
- Rat Mating Behavior 2
- Snake Mating Behavior
- Maternal Aggression in Rats
- Mouse Maternal Care
- Aggression in Syrian Hamsters
- Play Fighting in Syrian Hamsters
- Aggression in Rats
- Hamster Flank Marking
- Lizard Territorial Fighting (nine segments)
- Aggression in Chimpanzees
- Inter-Species Play Behavior
- Biological Clocks in Ring Doves
- Morris Water Maze
- Porsolt (Forced Swim) Test

Rat Grooming

Zebra Finch Courtship Song

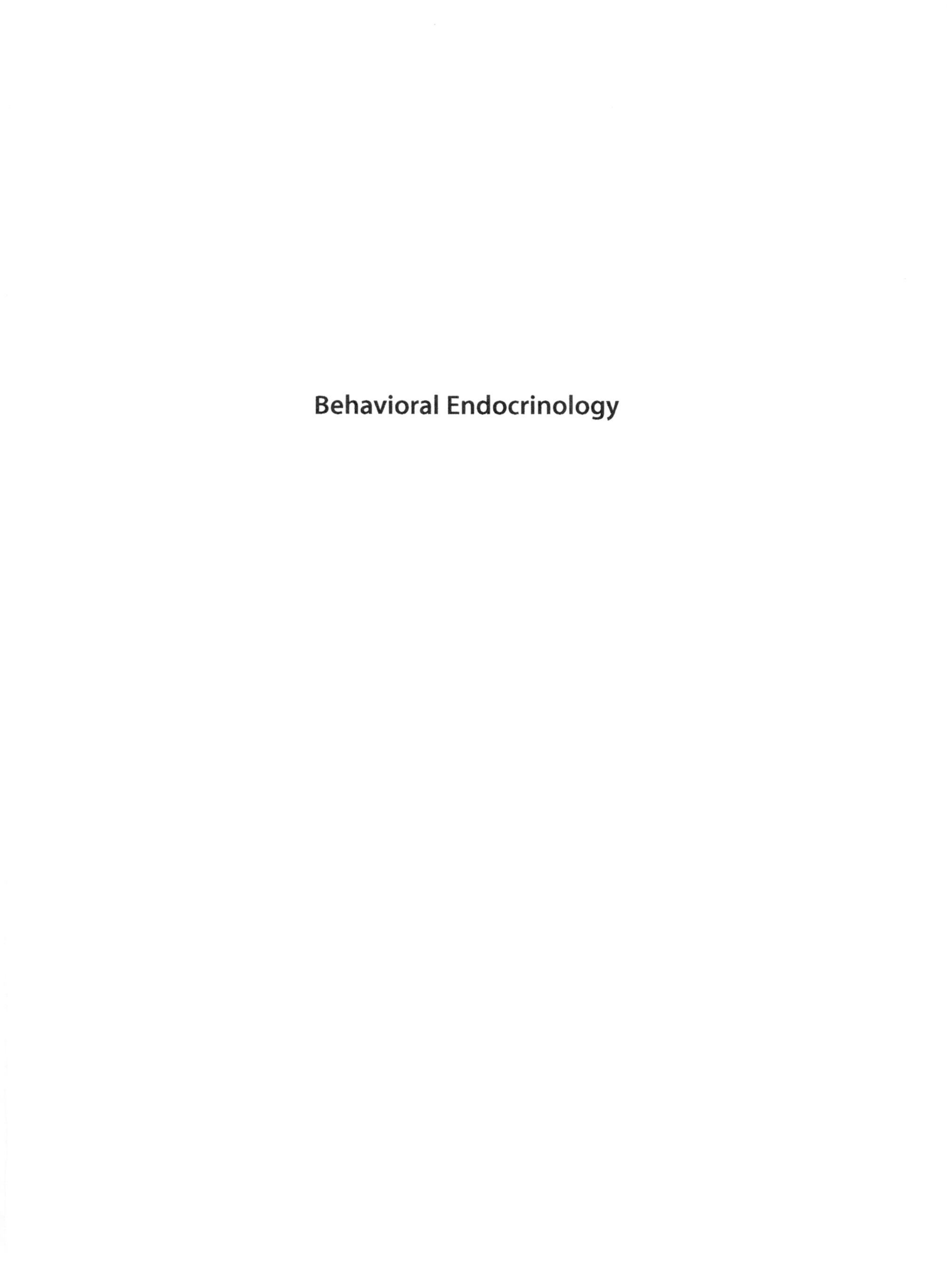

Behavioral Endocrinology

An Introduction to Behavioral Endocrinology

Third Edition

Randy J. Nelson
The Ohio State University

Sinauer Associates, Inc. Publishers
Sunderland, Massachusetts, 01375

The Cover

The cover photo is of a group of spotted hyenas (*Crocuta crocuta*) at a water hole in the Masai Mara National Reserve, Kenya. The photo was generously provided by Drs. Kay Holekamp and Laura Smale, professors at Michigan State University. Hyenas have become an iconic species in behavioral endocrinology because many aspects of their biology and behavior are sex reversed. For example, females have masculinized external genitalia. Female hyenas are also much more aggressive than males of this species. In the cover photo, a group of 56 hyenas was battling nine female lions over an African cape buffalo carcass. A few of the female hyenas took a break at the watering hole when the photo was taken. Food competition is fierce among female hyenas, which may be why females of this species evolved to be so much more aggressive than males, and the hormonal mechanisms underlying this high level of aggressiveness have led to some other unusual behavioral and morphological adaptations.

An Introduction to Behavioral Endocrinology, Third Edition

Sinauer Associates, Inc.
23 Plumtree Road
Sunderland, MA 01375
U.S.A.

FAX 413-549-1118
publish@sinauer.com, orders@sinauer.com

Library of Congress Cataloging-in-Publication Data

Nelson, Randy Joe.
An introduction to behavioral endocrinology/Randy J. Nelson.— 3rd ed.
p. cm.
Includes bibliographical references and index.
ISBN 978-0-87893-617-5 (alk. paper)
1. Psychoneuroendocrinology. I. Title.

QP356.45.N45 2005
612.4'05—dc22

2005004419

10 9 8

For Courtney, Morgan, and Justin

Contents

Chapter 5
Male Reproductive Behavior 235

Chapter 6
Female Reproductive Behavior 319

Chapter 7 Parental Behavior 387

Chapter 8 Hormones and Social Behavior 455

Preface

It is mid-January, 2005, and behavioral endocrinology has been in the news a lot this past week. Major League Baseball has finally adopted a tougher steroid-testing program that will suspend first-time offenders for 10 days and randomly test the players year-round. Only the National Hockey League maintains no steroid testing policy among major sporting leagues in the U.S. Anabolic steroids help build muscle and possibly enhance athletic performance, however, at significant health costs.

The New York Times reported that while giving a speech at a scientific conference, the president of Harvard University (who claims he was acting as a *provocateur*), suggested that one of the reasons for the lack of tenured women in academic science and engineering departments and research facilities might be that there are innate differences in the cognitive abilities of men and women. Although this was a provocative statement, and presumably uttered to promote consideration of all hypotheses for the male bias in academic science and engineering departments, it predictably caused uproar among academics and others. It is too early to predict how his statement will affect his ability to serve as president of a major U.S. university. To be fair, he is an economist by training, not a behavioral endocrinologist. There certainly are biologically-based sex differences in the brains of humans, but these small differences are highly unlikely to account for the disparity in the sex ratio in the sciences and engineering.

The field of behavioral endocrinology is a truly interdisciplinary effort. It involves the study of phenomena ranging from genetic, molecular, and cellular levels of analysis to the study of individual and social behaviors. I had several goals when I began writing a third edition of this textbook, one of which was to continue to present information about the interaction of hormones and behavior from diverse perspectives. I also wanted to pass on to readers more-detailed information about the scientists who laid the foundation for our modern studies of behavioral endocrinology. I have tried to present current hypotheses and theories in the context of their historical origins. Naturally, after five years, the book needed some general updating to remain useful, and I have endeavored to update all chapters to reflect current studies and information. Some areas, such as body mass regulation and circadian rhythm research, are moving forward with rapid advances announced weekly.

One criticism that I have received in the past, especially from colleagues teaching in Psychology departments, is that I have too much comparative work in the text. This is a criticism that I will continue to ignore. The comparative perspective is what gives behavioral endocrinology great strength, and some of the most fascinating aspects of our field come from discoveries gained on nontraditional animal models. I have continued to present this broad comparative approach in this edition. It is my hope that presenting adaptive function along with molecular and physiological mechanisms will yield greater understanding than presenting either approach alone.

I appreciate that many behavioral endocrinology students will be psychology majors. Thus, I have tried to keep the conceptual issues clear, and provide only sufficient details and examples that support the concepts. My assumption is that psychology students will have taken a course in biopsychology or neuroscience by the time they encounter this textbook, but again, I have tried to keep discussions of endocrine physiology and biochemistry to a minimum level necessary to understand the hormone–behavior interactions being discussed. Because students are likely to be familiar with the behavior of common animals such as dogs and rabbits, I have continued to use these animals as examples to help explain many concepts in this text. New to this edition of the text is a student CD that contains some wonderful videos and animations, which I hope help illustrate some of the behavioral and physiological concepts discussed.

Many topics had to be omitted or curtailed in the text. I assume that professors will use additional readings to make up for any deficiencies. Some topics covered in the text are controversial and will likely stimulate class discussions. At the end of each chapter are some questions for discussion that I hope will be potential starting points for such exchanges. A short, updated list of suggested readings is also provided at the end of each chapter where students can find reasonably current and more detailed information on the material in each chapter.

This is a very exciting time to be studying this field—either as a student or as a researcher in behavioral endocrinology. I hope that I have captured a fraction of the excitement I feel for this field and successfully conveyed it to the reader.

Acknowledgments

As I bask in the relief of completion, there are many people I wish to thank who contributed time and effort to this revision. I remain grateful to the reviewers and colleagues who provided feedback and assistance that shaped the first two editions. These individuals include Elizabeth Adkins-Regan, Gregory Ball, Jacques Balthazart, Tim Bartness, George Bentley, Dan Bernard, Staci Bilbo, Eric Bittman, Elliott Blass, Jeff Blaustein, Joan Blom, Robert Bridges, Sue Carter, Joe Casto, Silvana Chiavegatto, James Dabbs, Greg Demas, Courtney DeVries, Don Dewsbury, Debbie Drazen, Lori Flanagan-Cato, Alison Fleming, Nancy Forger, Stephen Gammie, Paul Gold, Bruce Goldman, Elizabeth Gould, David Gubernick, Tom Hahn, Joyce Hairston, Elaine Hull, Sabra Klein, Lance Kriegsfeld, Michael Leon, Vicky Luine, Margaret McCarthy, Martha McClintock, Jim McGaugh, Chris Moffatt, Celia Moore, Michael Moore, Kathie Olsen, David Olton, Deb Olster, Emilie Rissman, Mike Romero, Jay Rosenblatt, Ed Roy, Ben Sachs, Randall Sakai, Jen Sartor, Jill Schneider, Barbara Sherwin, David Shide, Rae Silver, Cheryl Sisk, Chuck Snowdon, Judith Stern, Anjali Trasy, George Wade, John Wingfield, Amy Wisniewski, Ruth Wood, Pauline Yahr, Kelly Young, and Irving Zucker.

I am grateful for assistance during the preparation of this third edition by many colleagues, who provided helpful, insightful, direct, and immensely kind comments including Elizabeth Adkins-Regan, Jessica Alexander, Jacques Balthazart, Zeynep Benderlioglu, Staci Bilbo, Jeff Blaustein, Nicky Clayton, Lique Coolen, Tara Craft, David Crews, Greg Demas, Courtney

DeVries, Debbie Drazen, Karyn Frick, Stephen Gammie, Erica Glasper, Bruce Goldman, Andrew Hotchkiss, Sabra Klein, Rosemary Knapp, Lance Kriegsfeld, Joe Lonstein, Vicky Luine, Lynn Martin, Gretchen Neigh, Vladimir Pravosudov, Brian Prendergast, Leah Pyter, Ben Sachs, Jill Schneider, Robert Spencer, Brian Trainor, Amy Wisniewski, Zach Weil, and Irving Zucker. I endeavored to incorporate virtually all of these reviewers' suggestions for changes into this new edition. Occasionally, because of stubbornness, laziness, or both traits on my part, I failed to address my friends' and colleagues' suggestions. Any remaining errors, sources of confusion, or other shortcomings in this new edition remain my sole responsibility.

I also thank Staci Bilbo, Stephanie Bowers, Greg Demas, Courtney DeVries, Erica Glasper, Lance Kriegsfeld, Leah Pyter, Brian Trainor, and Zach Weil for helping to track down various materials for the revised book manuscript. I remain very grateful to my many colleagues who kindly provided reprint or preprint copies of their papers, as well as my colleagues and friends who generously provided permission to use their graphic or photographic material in the book or accompanying CD. Special thanks to Kay Holekamp and Laura Smale, two long-time friends, for providing the beautiful image on the cover. I am especially grateful to my colleague, collaborator, and wife, Courtney DeVries, for being critical while reading chapter manuscripts, and uncritical at all other times.

I also thank the hard-working folks at Sinauer Associates. Although only my name appears on the front of the book, producing a textbook requires the hard work of many individuals. I remain grateful for the friendship and guidance provided to me during the first two editions by Pete Farley. The third edition of this book was put together by a new editorial and production team. My new book editor, Graig Donini, showed remarkable patience as several deadlines passed without text from me and he provided great help in completing the manuscript.

I am grateful to Sydney Carroll, the production editor, whose excellent eye for detail has helped improved the book. Others who deserved special thanks at Sinauer include Chris Small, production manager, Joan Gemme, compositor, Nancy Haver, line artist, and David McIntyre, photo editor. I am especially grateful for having the same outstanding copy editor for all three editions, Norma Roche. I appreciate her patient wrangling of my prose in her attempts to make me write coherently. She will probably re-write that last sentence. I also thank Jason Dirks, the media editor, for his help in designing and implementing the CD. This book is much improved because of their unrelenting hard work and uncompromising standards.

Finally, I thank the hundreds of undergraduate students who have taken my course in Behavioral Endocrinology over the past twenty years. They have provided many helpful suggestions on improving the textbook. I hope this text inspires students to enter the field of Behavioral Endocrinology, and to be as fortunate as I have been to work in such an exciting research discipline.

Supplements to accompany *An Introduction to Behavioral Endocrinology,* Third Edition

For the Student

Student CD (ISBN 0-87893-618-1)

Included with each copy of *An Introduction to Behavioral Endocrinology,* Third Edition, the Student CD provides a wealth of electronic resources to help students learn the material and to illustrate some of the behaviors and concepts discussed in the textbook. The CD includes:

- Animated Tutorials: Detailed tutorials explain complex concepts and processes; each includes an introduction, animation, conclusion, and short quiz.
- Videos: Short segments illustrate interesting behaviors in a variety of organisms.
- Web Links: These chapter-specific online resources include articles, photographs, activities, and audio clips.
- Hormone tables: A detailed listing of hormones, their sources, and their major biological actions is included.
- Links: There are links to journals, societies, and associations in the field of endocrinology.
- A complete Glossary.

For the Instructor

Instructor's Resource CD (ISBN 0-87893-619-X)

Available to qualified adopters of *An Introduction to Behavioral Endocrinology,* Third Edition, the Instructor's Resource CD includes electronic versions of all the line-art figures and tables from the textbook. Figures and tables are provided in both a high-resolution and a low-resolution JPEG format, and all have been formatted and optimized for excellent legibility and projection quality. In addition, the IRCD includes a ready-to-use PowerPoint® presentation of all the figures for each chapter of the textbook.

Test Bank (Included on the IRCD)

The new Test Bank to accompany the Third Edition includes multiple choice, short answer, and essay questions for each chapter of the textbook. Prepared by the textbook author, these questions are a helpful resource that can greatly speed the process of preparing exams and quizzes for the course.

1

The Study of Behavioral Endocrinology

The phrase "raging hormones" is a common expression that has been used to explain or excuse many types of behavior. For example, inappropriate behaviors, especially sexual behaviors, displayed by adolescent boys are often attributed to raging hormones. Women who display aggressive or assertive behaviors, especially in association with their pre-menstrual period, are often said to be affected by raging hormones. What are these hormones, and can they really take over the nervous system to direct behavior?

You may have discussed this and other questions about hormones and behavior with friends and family members. For example, does anabolic steroid abuse cause violent behavior? Does an individual's sex drive wane with aging? How does exposure to acute or chronic stress affect sexual behavior? Is the sex drive of men higher or lower than the sex drive of women? Can melatonin cure jet lag? Do seasonal cycles of depression occur in people? Why are men much more likely than women to commit violent crimes? Does postpartum depression really exist? Is homosexuality caused by hormone concentrations that are too low or too high? Is the sexual behavior of women influenced by menopause? Can leptin curb our food intake? Contrary to popular beliefs, hormones do not *cause* behavioral changes per se. Rather, hormones change the *probability* that a specific behavior will be emitted in the appropriate behavioral or social context.

Researchers in the field of **behavioral endocrinology**, the study of the interaction between hormones and behavior, attempt to address these kinds of questions in a formal, scientific manner. The study of hormones and behavior has been truly interdisciplinary: methods and techniques from one scientific discipline have been borrowed and refined by researchers in other fields. Psychologists, endocrinologists, neuroscientists, entomologists, zoologists, anatomists, physiologists, psychiatrists, and other behavioral biologists have all made contributions to the understanding of hormone–behavior interactions. This exciting commingling of scientific interests and approaches, with its ongoing synthesis of knowledge, has led to the emergence of behavioral endocrinology as a distinct and important field of study (Beach, 1975b).

Historical Roots of Behavioral Endocrinology

Psychology, as Ebbinghaus (1908) stated, has a short history but a long past, and the same can be said of behavioral endocrinology (Beach, 1974a). Although the modern era of the discipline is generally recognized to have emerged during the middle of the 20th century, with the publication of the classic book *Hormones and Behavior* (Beach, 1948), some of the relationships among the endocrine glands, their hormone products, and behavior have been implicitly recognized for centuries.

The male sex organs, or **testes**, produce and secrete a hormone called testosterone, which influences sexual behavior, aggression, territoriality, hibernation, and migration, as well as many other behaviors that differentiate males from females. The testes of mammals are usually located outside of the body cavity and can easily be damaged or removed. Thus, **castration**, the surgical removal of the testes, has historically been the most common manipulation of the endocrine system. For millennia, individuals of many species of domestic animals have been castrated to make them better to eat or easier to control, and the behavioral and physical effects of castration have been known since antiquity. Indeed, these effects were known to Aristotle, who described the effects of castration in roosters (and humans) with great detail and accuracy. For example, in *History of Animals*, written about 350 B.C., Aristotle reported that

> Birds have their testicles inside . . . Birds are castrated at the rump at the part where the two sexes unite in copulation. If you burn this area twice or thrice with hot irons, then, if the bird be full-grown, his crest grows sallow, he ceases to crow, and foregoes sexual passion; but if you cauterize the bird when young, none of these male attributes or propensities will come to him as he grows up. The case is the same for men: if you mutilate them in boyhood, the later-growing hair never comes, and the voice never changes but remains high-pitched; if they be mutilated in early manhood, the late-growths of hair quit them except the growth on the groin, and that diminishes but does not entirely depart.

Human males have been castrated for a number of reasons throughout history. For centuries, royalty employed men castrated before puberty, called **eunuchs**, to guard women from other men. For example, the Old Testament reports that these emasculated males were used to guard the women's quarters of Hebrew kings and

1.1 Eunuch of the last imperial court, photographed in China by Henri Cartier-Bresson in 1949. Note the lack of facial hair and unusually long arms.

princes (Esther 1:10). Castration in humans often has little or no effect on physical appearance or future sexual behavior when performed after the unfortunate individual attains sexual maturation; however, if human males are castrated before puberty, they will develop a characteristic physical appearance, marked by short stature and long arms (Figure 1.1), and sexual behaviors are unlikely to develop. The typical secondary male sex characters are also affected by prepubertal removal of the testes. For example, as noted by Aristotle, eunuchs never develop beards, and the pubertal change in voice does not occur. Normally during puberty, the vocal cords of males thicken in response to testosterone secreted by the testes. It is the thickened vocal cords that produce the deeper-pitched voice characteristic of males, just as the thick strings of a guitar produce deeper-pitched notes than the thin strings.

Castration was once a common practice in Europe and Asia (Box 1.1). Young boys with exceptional singing voices were castrated to prevent the pubertal changes in pitch. These singers became known as *castrati*. Although castrati were prized by church choirs for centuries, their popularity reached a peak in Europe during the 17th and 18th centuries with the development of opera, which made castrati the first superstars of the entertainment world (Heriot, 1974). The first castrato opera star, Baldassare Ferri, died in 1680 at the age of 70 with a fortune that was worth the equivalent of $3 million today. In hopes of attaining this level of wealth and fame, young boys with musical aptitude were identified early, and poor families offered their sons outright to church leaders, singing teachers, and

BOX 1.1 The Hijras of India

Over a million eunuchs, called Hijras, are estimated to be currently living in India. Traditionally, the cultural function of Indian Hijras has been to sing and dance at weddings and ceremonies associated with the birth of male children. The Hijras have become associated with fertility and form a sect within Hinduism; they worship Bahachara, a manifestation of the Hindu mother goddess. However, a significant minority of Hijras are Muslem. Although the Hijras claim the religious status of *"sannyasis"* (or celibates) because their genitalia (i.e., testes and penis) have been surgically removed, many work as prostitutes.

The Hijras are a heterogeneous population; most individuals have sexually ambiguous histories. Commonly, their genitalia were ambiguous at birth, or their sexual development was atypical during puberty (Nanda, 1990). Some Hijras engaged in homosexual sex during adolescence, and virtually all were reared as boys. The surgery is usually performed during an outdoor ritual that is accompanied by singing and dancing Hijras. Both the penis and scrotum are removed in the rapid, but nonsterile, operation. Rumors persist that young, homeless boys are taken in by the Hijras, who provide friendship and material goods, then transform them into eunuchs (e.g., Diamond, 1984). However, recent accounts of the Hijras failed to uncover any evidence of coerced or involuntary acceptance of the demasculinizing surgery (Nanda, 1990).

The most common occupation of the Hijras today is removing "bad luck." Because the Hijras consider themselves to have already suffered the very worst luck that can befall a man (despite the "elective" nature of the surgery), they will accept a little more bad luck for a fee. Thus, it is customary for new homeowners to hire Hijras to dance through all the rooms to absorb any potential bad luck. Similarly, small groups of Hijras appear uninvited at weddings to dance away any potential bad fortune for the bride and groom. Despite a certain amount of traditional charm, the arrival of the garishly dressed Hijras at a wedding is seldom a welcome sight. If they are not given a substantial fee, the Hijras will disrupt the wedding ceremony by cursing and exposing their disfigured genitalia. Not surprisingly, most families pay the fee to avoid the anatomy lesson, but these small extortions have made the Hijras rather unpopular in the cities of India.

Photo courtesy of Takeshi Ishikawa.

music academies. Thousands of boys lost their testes but never gained the celebrity or riches of the star castrati.

What did a castrato sound like? Essentially, the castrati had the range of a soprano, but the greater development of the male lungs gave their singing remarkable power. An early critic remarked, "Their timbre is as clear and piercing as that of choirboys and much more powerful; they appear to sing an octave above the natural voice of women. Their voices . . . are brilliant, light, full of sparkle, very loud, and astound with a very wide range" (Heriot, 1974).

After 200 years, the tastes of the opera-loving public changed. The rise in popularity of the female soprano voice reduced the demand for castrati, and they soon became an oddity. In 1849, the last great castrato, Giovanni Velluti, retired from opera to his villa in Venice. The last known castrato, Alessandro Moreschi, who served as the director of the Sistine Chapel Choir, as well as one of the choir's soloists, died in 1922. Before his death, he made 17 recordings that, although of poor quality by today's standards, still provide a remarkable example of the castrato voice.

Berthold's Experiment

A useful starting point for understanding research in hormones and behavior is a classic 19th-century experiment that is now considered to be the first formal study of endocrinology. This remarkable experiment conclusively demonstrated that a substance produced by the testes could travel through the bloodstream and eventually affect behavior. Professor Arnold Adolph Berthold, a Swiss–German physician and professor of physiology at the University of Göttingen (Figure 1.2), demonstrated experimentally that a product of the testes was necessary for a cockerel (an immature male chicken) to develop into a normal adult rooster.

As you probably know, roosters display several characteristic behaviors that are not typically seen among hens or immature chicks of either sex. Roosters mate with hens, they fight with other roosters, and of course, roosters crow. Moreover, roosters are larger than hens and immature birds and have distinctive plumage. On the other hand, capons, male chickens that have been castrated prior to adulthood in order to make their meat more tender, do not show many of the behavioral and physical characteristics of roosters. They do not attempt to mate with hens and are not very aggressive toward other males. Indeed, they avoid aggressive encounters, and if conditions force them to fight, they do so in a "half-hearted" manner. Finally, capons do not crow like roosters.

The behavioral and physical differences among roosters, hens, capons, and immature chickens were undoubtedly familiar to Berthold when he planned his study, which began on the second day of August, 1848, and lasted for sev-

1.2 Arnold Adolph Berthold of the University of Göttingen, who in 1849 conducted what is now recognized as the first formal experiment in endocrinology.

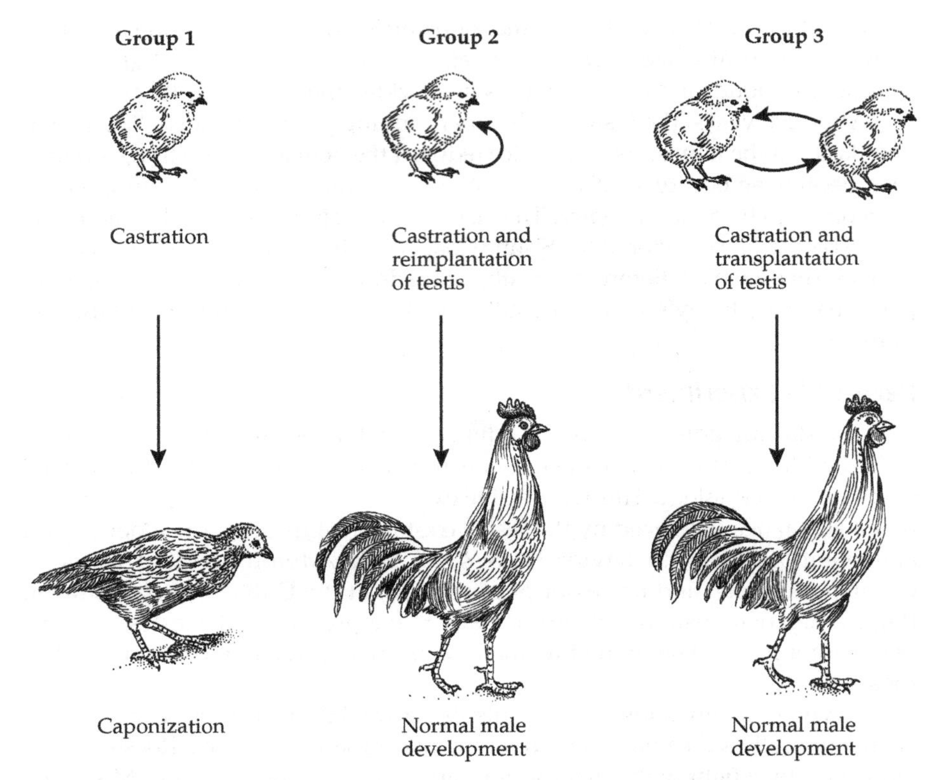

1.3 Berthold's experiment. The two birds in group 1 were castrated, and when observed several months later, were smaller than normal roosters and failed to engage in rooster-typical behaviors. The two birds in group 2 were also castrated, but one of each bird's own testes was reimplanted in its abdominal cavity. These birds looked and behaved like normal roosters when adults. The two birds in group 3 were also castrated, and one testis from each bird was transplanted into the abdomen of the other. Several months later, these birds also looked and behaved liked normal roosters. Berthold found that the reimplanted and transplanted testes in groups 2 and 3 developed vascular connections and generated sperm.

eral months (Figure 1.3; Berthold, 1849). He placed six cockerels in three experimental groups, each consisting of two birds. He removed both testes from each of the two cockerels in the first group, and as expected, these birds eventually developed as capons. They never fought with other males after castration, and they failed to crow; instead, Professor Berthold reported, they developed the "monotone voice of the capon." They avoided females and never exhibited mating behavior. Finally, these birds *looked* different from intact (uncastrated) adult males. Their bodies and heads were small, and their combs and wattles were atrophied and pale in color.

The second pair of cockerels was also castrated, but Berthold reimplanted one testis from each bird in its abdominal cavity after ensuring that all of the original vascular and neural connections had been cut. Interestingly, both birds in this group developed normal rooster behavior. According to Berthold, they "crowed lustily, often engaged in battle with each other and with other cockerels, and showed the usual reactions to hens." Their physical appearance was indistinguishable from that of other young roosters; they grew normally and possessed highly developed combs and wattles that were bright red in color.

The remaining two birds were also castrated, but after the testes were removed, Berthold placed a single testis from each bird in the other's abdominal cavity. Like the cockerels in the second experimental group, these birds also developed the "voice, sexual urge, belligerence, and growth of combs and wattles" characteristic of intact males.

After observing all six birds for several months, Berthold dissected one of the cockerels from the second group and found that the implanted testis had attached itself to the intestines, developed a vascular supply, and nearly doubled in size. Eventually, he examined all the implanted testes under a microscope and noted the presence of sperm.

Based on the results of this experiment, Berthold drew three major conclusions: (1) the testes are transplantable organs; (2) transplanted testes can function and produce sperm (Berthold drew the analogy to a tree branch that produces its own fruit after having been grafted to another tree); and (3) because the testes functioned normally after all nerves were severed, there are no specific nerves directing testicular function. To account for these findings, Berthold proposed that a secretory, blood-borne product of the transplanted testes (*productive Verhältniss der Hoden*) was responsible for the normal development of the birds in the second and third groups. It is worth noting that three of the four parameters Berthold used to formulate this hormonal hypothesis—mating, vocalization, aggression, and distinctive appearance—were behavioral.

In recent years, Berthold's experiment has been credited as the genesis of the field of endocrinology (and thus of behavioral endocrinology; Box 1.2), but his intriguing demonstration of non-neural control of behavior was apparently not embraced with great enthusiasm by his scientific contemporaries, as his paper does not seem to have been cited for nearly 60 years after its publication. Why, then, did Berthold conduct his study? His experiment was elegant in its simplicity, but unfortunately his published report was brief and had no introduction, so we cannot know for certain what questions motivated him to conduct the work (Forbes, 1949). He seems unaware of the work of John Hunter, who had reported by 1771 the successful transplantation of a testis from a rooster into a hen (with no obvious changes resulting; Forbes, 1947). We do know that Berthold had previously authored a well-known physiology textbook and had actively conducted biological research. A reading of his textbook makes it apparent that Berthold was a proponent of the pangenesis theory of inheritance. This theory, endorsed by many biologists prior to the discovery of chromosomes and genes, held that all body parts actively discharge bits and pieces of themselves into the blood system, where they are transported to the ovaries or testes and assembled into miniature offspring resembling the parents. As a consequence of this theoretical stance, Berthold had two concepts at hand when evaluating the results of his testicular

BOX 1.2 Frank A. Beach and the Origins of the Modern Era of Behavioral Endocrinology

For some time before behavioral endocrinology emerged as a recognized field, its foundations were being laid by researchers in other fields. The anatomists, physiologists, and zoologists who were doing the majority of the work on "internal secretions" prior to 1930 often used behavioral parameters in their studies. Soon thereafter, psychologists began making important contributions to the study of hormones and behavior. In the early decades of the 20th century, American psychology was undergoing a major change, both in ideology and in methodology. Led by John B. Watson, students of the "science of the mind" were casting aside introspection as a method in favor of observation and experimentation. Watson argued that only overt behavior was observable, and psychologists began describing and quantifying all types of overt behavior.

Karl S. Lashley did his graduate work under Watson at Johns Hopkins University and eventually joined the faculty at the University of Chicago. Lashley investigated the effects of removing parts of rats' brains to discover where in the brain vari-

Frank A. Beach (1911–1988)

transplantation study: (1) various parts of the body release specific agents into the blood; and (2) these agents travel through the bloodstream to particular target organs. Why Berthold did not go any further with his interesting finding is not known; he died 12 years later in 1861 without following up on his now-famous study.

What Are Hormones?

Berthold took the first step in the study of behavioral endocrinology by demonstrating that the well-known effects of the testes were due to their production of a substance that circulated in the blood. Modern studies in behavioral endocrinology have documented the effects of substances from many different glands affecting an ever greater range of behaviors.

We now refer to Berthold's "secretory blood-borne product" as a hormone, a term coined by Bayliss and Starling in 1902. **Hormones** are organic chemical mes-

ous psychological processes were carried out; he was particularly interested in finding where memories were stored. Although he never published any reports on the interaction between hormones and behavior, Lashley was clearly interested in the subject (e.g., Lashley, 1938), and several of Lashley's students became important contributors to behavioral endocrinology, including Calvin P. Stone, Josephine Ball, and Frank A. Beach.

Beach, William C. Young (see Box 3.2), and Daniel Lehrman (see Box 7.2) were especially influential during the early studies of behavioral endocrinology. Beach's dissertation at Chicago, "The Neural Basis for Innate Behavior," examined the effects of cortical tissue destruction on the maternal behavior of first-time mother rats. In 1937, Beach began working as a curator in the Department of Experimental Biology at the American Museum of Natural History in New York and began contributing to the museum's tradition of comparative behavioral experimentation. One study completed at the museum, which was a logical extension of Beach's dissertation work, is of special note: he began investigating the effects of cortical lesions on the mating behavior of male rats. He found that some brain-damaged rats continued to mate, whereas others failed to do so. Beach was concerned that his lesions were interfering indirectly with the endocrine system, so he injected the nonmating brain-injured rats with testosterone, the primary hormone secreted from the testes. The treatment evoked mating behavior in some of the lesioned rats, and this modification of behavior by hormones prompted Beach to learn more about endocrinology.

Beach audited a course in endocrinology at New York University, but was distressed by the lack of information about the behavioral effects of hormones; the professor responded to Beach's complaint by allowing him to teach one session. While preparing for the lecture, Beach discovered that no comprehensive summary of hormone–behavior interactions existed, and he prepared such a review as a term paper for the endocrinology course. Several years later, Beach expanded his paper into an influential book, *Hormones and Behavior* (Beach, 1948). The publication of this book marked the beginning of the formal study of behavioral endocrinology. Beach is credited with the genesis of this scientific discipline, and he continued to provide intellectual leadership in shaping the field for the next 40 years.

sengers produced and released by specialized glands called **endocrine glands**. Hormones are released from these glands into the bloodstream (or the tissue fluid system in invertebrates), where they may then act on target organs (or tissues) at some distance from their origin. Hormones coordinate the physiology and behavior of an animal by regulating, integrating, and controlling its bodily functions. For example, the same hormones that cause gametic (egg or sperm) maturation also promote mating behavior in many species. This dual hormonal function ensures that mating behavior occurs when animals have mature gametes available for fertilization. Another example of endocrine regulation of physiological and behavioral function is provided by the metabolic system. Several metabolic hormones work together to elevate blood glucose levels prior to awakening in anticipation of increased activity and energy demand. This "programmed" elevation of fuel availability coordinates the animal's physiology with its behavior.

Hormones are similar in function to **neurotransmitters**, the chemicals used by the nervous system in coordinating animals' activities. However, hormones

BOX 1.3 Neural Transmission versus Hormonal Communication

Although neural and hormonal communication both rely on chemical signals, there are several prominent differences between them. Communication in the nervous system is analogous to traveling on a train. You can use the train in your travel plans as long as tracks exist between your proposed origin and destination. Likewise, neural messages can travel only to destinations along existing nerve tracts. Hormonal communication, on the other hand, is like traveling in a car. You can drive to many more destinations than train travel allows because there are many more roads than railroad tracks. Likewise, hormonal messages can travel anywhere in the body via the circulatory system; any

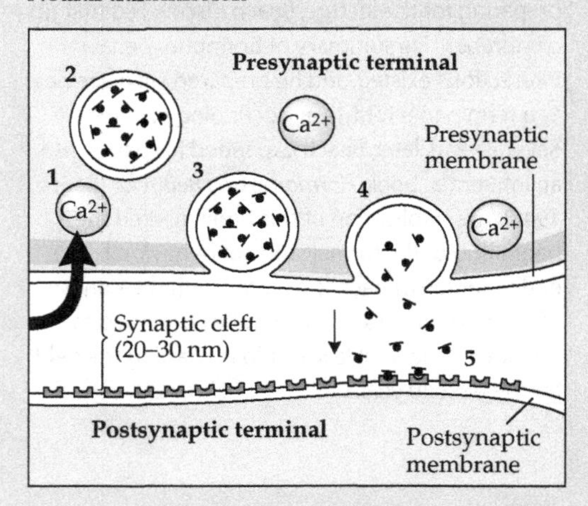

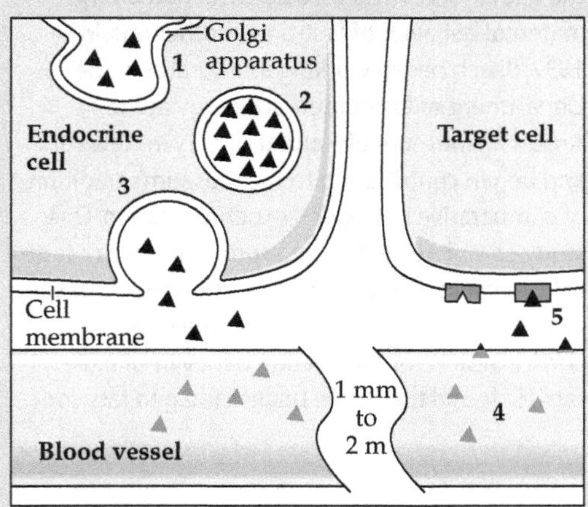

can operate over a greater distance and over a much greater temporal range than neurotransmitters (Box 1.3). Hormones are also similar to **cytokines**, chemical signals produced by cells of the immune system, and may interact with cytokines to affect behavior, especially when individuals are ill or unduly stressed.

Not all cells are influenced by each and every hormone. Rather, any given hormone can directly influence only cells that have specific **receptors** for that hormone. Cells that have these specific receptors are called **target cells** for the hormone. The interaction of a hormone with its receptor often begins a series of cellular events that eventually affect gene expression and protein synthesis. The newly synthesized proteins may activate or deactivate other genes, causing yet another cascade of cellular events. Recently, some effects of hormones on behavior have been reported that are not caused by activation of the genetic machin-

cell receiving blood is potentially able to receive a hormonal message.

Neural and hormonal communication differ in other ways as well. To illustrate them, consider the differences between digital and analog technologies. Neural messages are digital, all-or-none events that have rapid onset and offset: neural signals can take place in milliseconds. Accordingly, the nervous system mediates changes in the body that are relatively rapid. For example, the nervous system regulates immediate food intake and directs body movement. In contrast, hormonal messages are analog, graded events that may take seconds, minutes, or even hours to occur. Hormones can mediate long-term processes, such as growth, development, reproduction, and metabolism.

Hormonal and neural messages are both chemical in nature, and they are released and received by cells in a similar manner; however, there are important differences as well. As shown in the left panel of the accompanying figure, when a neural impulse arrives at a presynaptic terminal, there is an influx of calcium ions (Ca^{2+}) into the cell (1) that causes vesicles containing neural chemical messengers called neurotransmitters to move toward the presynaptic cell membrane (2). The vesicles fuse with the membrane (3) and release the neurotransmitters into the synaptic cleft (4). The neurotransmitters travel a distance of only 20–30 nanometers (30×10^{-9} m) to the membrane of the postsynaptic neuron, where they bind to receptors (5). Hormones, as shown in the right panel of the figure are manufactured in the Golgi apparatus of an endocrine cell (1). They also move toward the cell membrane in vesicles (2), which fuse with the membrane, releasing the hormone (3). However, hormones then enter the circulatory system, through which they may travel from 1 millimeter to 2 meters (4) before arriving at a cell of a target tissue, where they bind with specific receptors (5).

Another distinction between neural and hormonal communication is the degree of voluntary control that can be exerted over their functioning. In general, there is more voluntary control of neural than of hormonal signals. It is virtually impossible to will a change in your thyroid hormone levels, for example, whereas moving your limbs on command is easy.

Although these differences are significant, the division between the nervous system and the endocrine system is becoming more blurred as we learn more about how the nervous system regulates hormonal communication. A better understanding of the interface between the endocrine system and the nervous system is likely to yield important advances in the future study of the interaction between hormones and behavior.

ery; these so-called nongenomic effects of hormones on behavior will be reviewed in later chapters.

Importantly, sufficient numbers of appropriate hormone receptors must be available for a specific hormone to produce any effects. For example, if a capon had no receptors for testosterone, then implanting another testis (or giving testosterone hormone therapy) would not cause it to mate, fight, or crow. Furthermore, a common bias in behavioral endocrinology is the assumption that individual differences in the expression of a behavior reflect differences in hormone concentrations in the blood. In other words, it is assumed that roosters that crow frequently have higher blood testosterone concentrations than roosters that rarely crow. To a certain extent, this assumption is true, but such individual differences usually reflect complex influences of hormone concentrations, patterns of hormone release, numbers and locations of hormone receptors, and the efficiency of those receptors in triggering the signal transduction pathways that ultimately affect gene transcription.

Hormones commonly alter the rate of normal cellular function. Another way that hormones can affect cells is to change their morphology or size. For example, some athletes abuse anabolic steroids, which are synthetic hormones, because muscle cells grow larger after exposure to these substances. Hormones also may affect neuronal growth and development, as well as cell death throughout the nervous system. Whereas the examples we have discussed so far have all demonstrated how the presence or absence of a hormone may affect behavior, it is important to appreciate that the interactive relationship between hormones and behavior is bidirectional: hormones obviously affect behavior, but, as we will see in later chapters, behavior can also influence hormone concentrations.

The Study of Behavior

Behavioral endocrinologists are interested in how the general physiological effects of hormones may eventually alter behavior, and how behavior may influence the effects of hormones. This book will describe, both phenomenologically and functionally, how hormones affect behavior.

What is behavior? Generally, we think of behavior as "output," and because muscles are the most common output organs, or **effectors**, we tend to think of behavior as coordinated movement. Sometimes lack of movement is an important behavior, especially when stalking prey or avoiding predators, or during mating for females of many species. The excretion of scents and chemicals, changes in skin coloration, the flashing lights of fireflies, and the production of electrical signals by various species are also types of behavior.

Problems of Behavioral Research

The goals of behavioral scientists are to determine what behaviors are relevant to the question being asked, to describe those behaviors, and to interpret their function. These goals are not as simple to achieve as they may sound, and there are several pitfalls that behavioral investigators must avoid. Let us examine several of these problems in the context of a simple example: how might we begin to explain the singing of a robin (*Turdus migratorius*) that visits our backyard bird feeder each morning?

As soon as any observer begins to look at behavior, some degree of abstraction and bias is inevitable. When hens or roosters vocalize, we say that they "cluck" or "crow." However, the vocalizations of many birds are so melodious that we actually call them "singing." We may thus explain the bird's singing as a result of its being happy because we often sing when we are happy. This is obviously an anthropomorphic bias on our part. Behavioral scientists must take care not to attribute motives (hunger, fear, happiness) to animals based only on introspection, and must make an effort to observe behavior as objectively as possible.

Second, we must determine what other behaviors may be relevant to elucidating the behavior being examined. Even if an animal's behavior is videotaped continuously or observed directly for 24 hours a day, decisions must be made by the observer regarding which behaviors are meaningful and which are trivial in terms of answering the question at hand. In the case of the robin, for example, we must determine whether its presence at the bird feeder bears any relation to its singing.

Finally, to understand the causes (hormonal or otherwise) of any behavior, we must thoroughly describe that behavior (Tinbergen, 1951). When does the robin sing? What elicits its singing? Do all robins, or only some, engage in singing? As we saw in our discussion of Berthold's experiment, we must have a reasonably complete description of normal behavior before we can accurately assess the effects of any experimental manipulations on behavior.

How Is Behavior Described?

We may classify descriptions of behavior within two broad categories: (1) descriptions of action and (2) descriptions of consequence (Dewsbury, 1979). In the former, the pattern of an animal's behavioral output is described with little or no reference to the effects of the behavior on the environment. Thus, descriptions of a lion's bared teeth, a firefly's flash of light, or a songbird's singing are descriptions of action. When classifying behavior by consequence, an observer notes the effect of the behavior on the environment; the observer may describe an animal's behavior as "gathering up nest material," "depressing a lever," or "inducing a female to visit." In practice, the two types of descriptive classifications can be combined to provide a rich description of animal behavior.

The Simple System Approach

Even though we may ultimately be curious about the influence of hormones on human behavior, the unique and complex interactions of genes and environment in humans make behavioral endocrinology studies of our own species very difficult to interpret. For example, suppose we observed that people in the upper 25th percentile of body mass tend to be more socially aggressive than individuals in the lower 25th percentile of body mass. The cause of the difference in behavior might reflect differences in body size, which might result from differences in concentrations of hormones that regulate growth and development. Alternatively, the differences in body mass might reflect differences in nutrition that affect brain development and thus the development of confidence in social situations. There is also the "chicken and egg" problem to resolve; that is, do hormones affect behavior directly by affecting the brain, indirectly by affecting body size, which in turn affects brain and behavior, or does acting aggressively affect hormone concentrations, which in turn affect body development? Furthermore, humans are reared in a wide variety of environments, which complicates the assignment of causation to individual variation in behavior.

In order to untangle the contributions of various factors to hormone–behavior interactions, behavioral endocrinologists generally perform experiments on genetically identical animals in controlled environments. The development of behaviors and changes in hormone concentrations can be monitored throughout life under these conditions. Similar controlled experiments are virtually impossible to conduct on humans. Even though nonhuman animals represent "simple systems" relative to humans, it is important to appreciate that the behavioral repertoires of animals are extraordinarily varied and complex.

Most research in behavioral endocrinology involves only a few types of simple behavior. This narrow focus on only a few behavioral measures is partially a response to the enormous variation inherent in observations of complex behaviors. There are advantages and disadvantages to this approach. The advantages of using simple behaviors include ease of replication and quantification. In this way, the simple system approach parallels the reductionist approach prevalent in physiological and biochemical analyses. On the other hand, the most apparent disadvantage to studying simple behaviors is the possibility that subtle but important interactions between hormones and behavior will be neglected and overlooked. Social and other environmental factors are often absent, or significantly reduced, in the laboratory, but may also be important in attaining a complete understanding of hormone–behavior interactions. There are certainly cases in which investigators have endeavored to observe the hormonal correlates of complex behavior. But commonly, the behavioral end point studied is as simple as the presence or absence of bird song, or the occurrence of mounting behavior among male rodents.

Levels of Analysis

Once behaviors have been adequately described, we may proceed to ask about their causes (Alcock, 2001; Dewsbury, 1979; Tinbergen, 1951). For example, the zebra finch (*Taeniopygia guttata*; Figure 1.4), a native Australian songbird, is one of the

1.4 Zebra finches. These small birds have been used extensively in the study of the hormonal and neural bases of bird song. As in most songbird species, only the male zebra finch (left) sings in nature. Courtesy of Atsuko Takahashi.

species most frequently used to study bird song. The various notes zebra finches produce, the circumstances in which they sing, and even the specific muscles they use during singing have been extensively studied and described. Based on these descriptions, many researchers have begun to explore the causes of singing in male zebra finches by developing hypotheses and testing them through observation and experimentation. (As in most songbird species, females and immature zebra finches do not sing in nature.)

The generic question an animal behaviorist asks at this point in the research may be simply expressed as "What causes animal A to emit behavior X?" (Sherman, 1988), so many researchers have asked, in effect, "What causes zebra finches to sing?" You may be surprised to learn that there may be four types of correct answers to this basic question, based on four different **levels of analysis**: immediate causation, development, evolution, and adaptive function (Tinbergen, 1951).

IMMEDIATE CAUSATION The level of **immediate causation** encompasses the underlying physiological, or proximate, mechanisms responsible for a given behavior. Typically these mechanisms are mediated by the nervous and endocrine systems, which influence behavior on a moment-to-moment basis during the life of an individual. Various internal and environmental stimuli, as well as sensory and perceptual processes, are involved in the short-term regulation of behavior. Accordingly, experiments designed to address questions of immediate causation often use physiological methods such as alterations of hormone concentrations or direct manipulations of the brain. In the case of zebra finches, these kinds of experiments have revealed that elevated blood concentrations of testosterone and increased rates of neural activity in certain areas of the brain are immediate causes of singing, so a correct answer to the question posed above might be that zebra finches sing because the level of testosterone in their blood is high. This class of explanation is the one most frequently used by behavioral endocrinologists, and will be the primary focus of this book.

DEVELOPMENT The behavioral responses and repertoires of animals change throughout their lives as a result of the interaction between genes and environmental factors. Questions of **development** concern the full range of the organism's lifetime from conception to death. For example, the behavior of newborns is quite rudimentary in many species, but becomes more complex as they grow and interact with the environment. Hormonal events affecting the fetal and newborn animal can have pervasive influences later in life. Although the majority of research at the developmental level of analysis has focused on how events early in development influence animals later in life, the decay of behavioral patterns during aging is also of interest to behavioral biologists pursuing developmental questions. Possible answers to our question from the developmental perspective might be that zebra finches sing because they have undergone puberty or because they learned their songs from their fathers.

EVOLUTION **Evolutionary approaches** involve many generations of animals and address the ways that specific behaviors change during the course of natural selection. Behavioral biologists study the evolutionary bases of behavior to learn why behavior varies between closely related species as well as to understand the specific behavioral changes that occur during the evolution of new species. Behaviors rarely leave interpretable traces in the fossil record, so the study of the evolution of behavior relies on comparing existing species that vary in relatedness. An investigator working at the evolutionary level might say that zebra finches sing because they are finches, and that all finches sing because they have evolved from a common ancestral species that sang.

ADAPTIVE FUNCTION Questions of **adaptive function** are synonymous with questions of adaptive significance: they are concerned with the role that behavior plays in the adaptation of animals to their environments and with the selective forces that currently maintain behavior. At this level of analysis, it might be argued that male zebra finches sing because singing increases the likelihood that they will reproduce by attracting females to their territories or dissuading competing males from entering their territories.

Thus, there are four different types of causal explanations for a particular behavior, and there may be many correct answers to the question, "What causes male zebra finches to sing?" No one type of explanation is better or more complete than another, and in practice, the levels of questions and explanations overlap and interact in many situations. Nevertheless, it is important that researchers specify clearly the level of analysis within which they are working when they are generating hypotheses for testing. Care must be taken to avoid comparing noncompeting hypotheses at the different levels of analysis (Sherman, 1988).

For the sake of simplicity, these four levels of analysis can be grouped into sets of two, with questions of immediate causation and development grouped as "how questions" ("*How* does an animal engage in a behavior?") and questions of evolution and adaptive function as "why questions" ("*Why* does an animal engage in a particular behavior?") (Alcock, 2001). "How questions" have also been referred to as questions of *proximate causation*, and "why questions" as questions of *ultimate causation* (Wilson, 1975). To construct an exhaustive explanation of

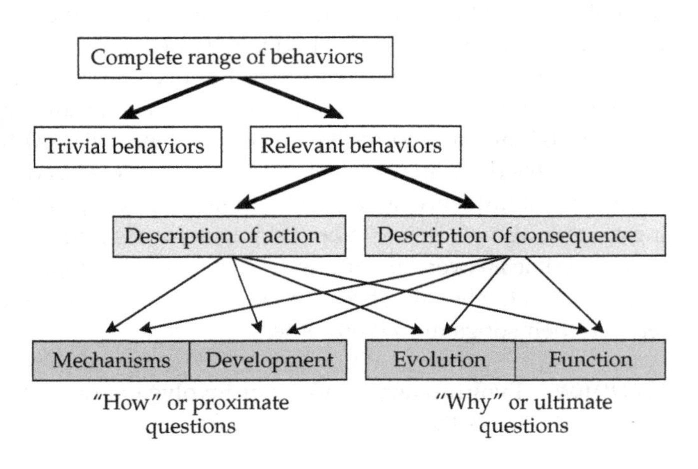

1.5 Stages of behavioral research. From the complete range of an organism's behaviors, the behavioral scientist must first determine which behaviors are relevant to the question under consideration, a process that is inherently prone to undue abstraction or bias. Descriptions of relevant behaviors may focus on the actions themselves (description of action) or on their environmental effects (description of consequence). Examination of the causes of behaviors may proceed at any of four levels of analysis that address either proximate ("how") or ultimate ("why") questions.

the causes of bird song, then, we would want to study both how birds sing and why they sing (Figure 1.5). What developmental and physiological processes occur before and during singing? What is the evolutionary history of bird song? When, phylogenetically (during evolutionary history), did singing appear among birds? What adaptive advantages do singers enjoy relative to nonsingers?

Researchers in different disciplines tend to favor particular types of questions and classes of explanations. For example, physiologists work almost exclusively at the level of immediate causation, whereas behavioral ecologists specialize in evolutionary and adaptive explanations of behavior. Behavioral endocrinologists who focus on physiology and neuroscience tend to work in laboratories, whereas behavioral endocrinologists who focus on behavioral ecology tend to work in the field. In general, laboratory data are more reliable (i.e., repeatable) than field data because the experimental conditions can be tightly controlled. However, field data tend to be more valid (i.e., more ecologically relevant) than laboratory data because they are collected in the setting where the behavior and physiology of animals evolved. The types of explanations that individual scientists pursue in conducting their research reflect their tastes and their training, but their combined efforts allow us to gain the most comprehensive understanding of animal behavior.

How Might Hormones Affect Behavior?

In terms of their behavior, one can think of animals as being made up of three interacting components: (1) input systems (sensory systems), (2) integrators (the central nervous system), and (3) output systems, or effectors (e.g., muscles) (Figure 1.6). Again, hormones do not cause behavioral changes. Rather, hormones influence these three systems so that specific stimuli are more likely to elicit certain responses in the appropriate behavioral or social context. In other words, hormones change the probability that a particular behavior will be emitted in the

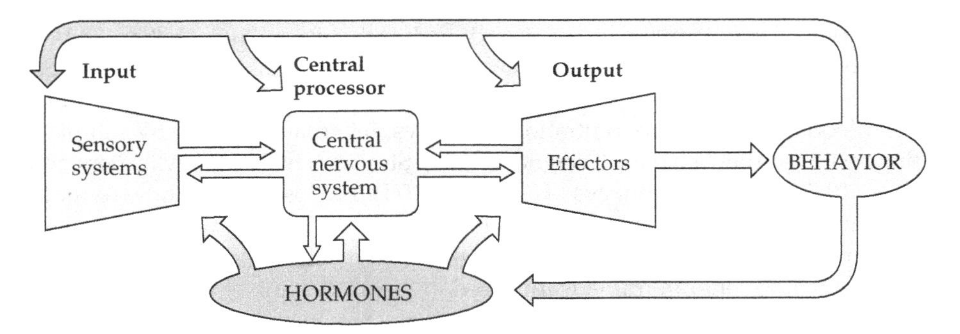

1.6 How hormones may affect behavior. Behaving animals may be thought of as being made up of three interacting components: input systems (sensory systems), central processing systems (the central nervous system), and output systems (effectors, such as muscles). Hormones may affect any or all of these three components when influencing behavior. Note that there is a bidirectional causal relationship between hormones and behavior in that an animal's behavior (or the behavior of conspecifics or predators) may affect its endocrine state.

appropriate situation. This is a critical distinction that can affect how we think of hormone–behavior relationships.

We can apply this three-component behavioral scheme by returning to our example of singing behavior in zebra finches. As previously noted, only the male zebra finch sings in nature. If the testes of adult male finches are removed, the birds stop singing, but castrated finches resume singing if the testes are reimplanted, or if they are provided with the primary testicular hormone, testosterone. Singing behavior is most frequent when blood testosterone concentrations are high. Although it is clear from these observations that testosterone is somehow involved in singing, how might the three-component framework just introduced help us to formulate hypotheses to explore testosterone's role in this behavior? By examining input systems, we could determine whether testosterone alters the birds' sensory capabilities, making the environmental cues that normally elicit singing more salient. If this were the case, females or territorial intruders might be more easily seen or heard. Testosterone also could influence the central nervous system. Neuronal architecture or the speed of neural processing could change in the presence of testosterone. Higher neural processes (e.g., motivation, attention, or perception) also might be influenced. Finally, the effector organs, muscles in this case, could be affected by the presence of testosterone. Blood testosterone concentrations might somehow affect the muscles of a songbird's syrinx (the avian vocal organ). Testosterone, therefore, could affect bird song by influencing the sensory capabilities, central nervous system, or effector organs of an individual bird. We do not understand completely how testosterone influences bird song, but in most cases, hormones can be considered to affect behavior by influencing one, two, or all three of these components, and our three-part framework can aid in the design of hypotheses and experiments to explore these issues. This conceptual scheme will provide the major organization for this book.

How Might Behavior Affect Hormones?

The zebra finch example demonstrates how hormones can affect behavior, but, as noted previously, the reciprocal relation also occurs; that is, behavior can affect hormone concentrations. For example, the sight of a territorial intruder may elevate blood testosterone concentrations in the resident male and thereby stimulate singing or fighting behavior (Wingfield, 1988). Similarly, male mice (Ginsberg and Allee, 1942) or rhesus monkeys (Rose et al., 1971) that lose a fight show reduced circulating testosterone concentrations for several days or even weeks afterward. Similar results have also been reported in humans. Testosterone concentrations are affected not only in humans involved in physical combat, but also in those involved in simulated battles. For example, testosterone concentrations were elevated in winners and reduced in losers of regional chess tournaments (Mazur et al., 1992).

People do not even have to be directly involved in a contest to have their hormones affected by the outcome. For instance, male fans of both the Brazilian and Italian soccer teams were recruited to provide saliva samples to be assayed for testosterone before and after the final game of the World Cup soccer match in 1994. Brazil and Italy were tied going into the final game, which was hard fought and tied until the final seconds, but Brazil won on a penalty kick at the last possible moment. The Brazilian fans were elated and the Italian fans were crestfallen. When the samples were assayed, 11 of 12 Brazilian fans had increased testosterone concentrations, and 9 of 9 Italian fans had decreased testosterone concentrations, compared with pre-game baseline values (Dabbs, 2000) (Figure 1.7).

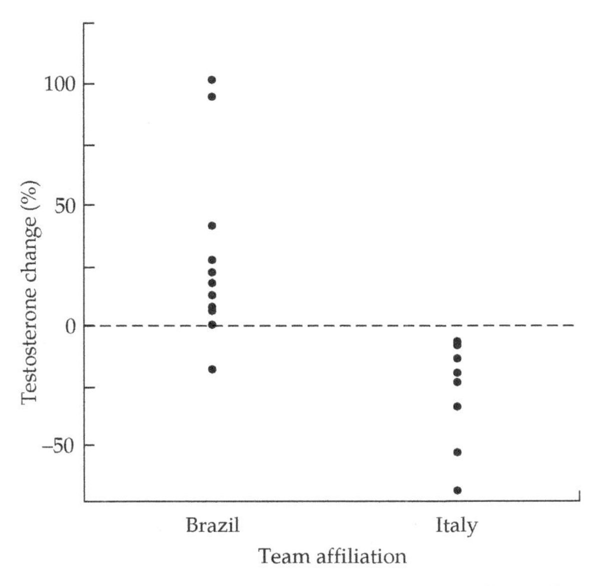

1.7 Change in the testosterone concentrations of sports fans. Testosterone concentrations of male Brazilian and Italian fans were measured in saliva samples obtained before and immediately after the final soccer match of the 1994 World Cup, which Brazil won. The graph shows the change in testosterone concentration between the two samples. After Dabbs, 2000.

Even anticipation of behavior may influence hormone concentrations. An anonymous contribution to the journal *Nature* (1970) provided an account by a gentleman whose work caused him to live in isolation on an island for days at a time. Occasionally, he returned to the mainland to pick up his mail and visit his fiancée. This man possessed a strong curiosity, and his isolated lifestyle presumably eliminated many of the usual distractions of modern life. In this peaceful milieu, he noted, while shaving, that his beard seemed thicker immediately prior to his visits to his fiancée. He began weighing the shavings, and determined that his beard thickened on days when he had sex with his fiancée. The rate of beard growth is correlated with blood concentrations of testosterone; high levels of testosterone increase the rate of beard growth, whereas low testosterone concentrations are associated with slow beard growth. Our anonymous colleague suggested that his sexual behavior, as well as his *anticipation* of sexual behavior, caused an elevation in testosterone, which in turn increased beard growth.

You probably recognize that this is not a very robust demonstration of a hormone–behavior interaction because (1) only one experimental subject was used, (2) the experimental subject was aware of the experimental conditions, which, consciously or unconsciously, may have caused him to misread the scale on critical days, and (3) other conditions, such as the drying effects of the air during his travels home, could have accounted for some of the changes in beard growth. In other words, this letter described **anecdotal evidence** of a hormone–behavior interaction—evidence that is not compelling. More recently, testosterone concentrations were measured in four heterosexual couples over a total of 22 evenings (Dabbs and Mohammed, 1992). There were two different types of evenings. On

11 evenings, the samples were obtained before and after sexual intercourse; on the remaining 11 evenings, two samples were obtained during the evening, but there was no sexual intercourse. To avoid the logistic complications of drawing blood samples, testosterone was measured in the saliva of the participants. Engaging in sexual intercourse caused testosterone concentrations to increase in both men and women. The early evening saliva samples revealed no difference in testosterone concentrations between evenings when sexual intercourse took place and evenings when it did not. These results suggest that in humans, sexual behavior increases testosterone concentrations more than high testosterone concentrations cause sexual activity (Dabbs and Mohammed, 1992). Although this is certainly a reasonable conclusion, there are alternative explanations for the results of this study. For example, perhaps physical exercise alone increases testosterone concentrations. To rule out this possibility, additional studies are required in which the level of exercise is similar between experimental groups, but differs in sexual content.

Classes of Evidence for Determining Hormone–Behavior Interactions

What sort of evidence *would* be sufficient to establish that a particular hormone affected a specific behavior or that a specific behavior changed hormone concentrations? Experiments to test hypotheses about the effects of hormones on behavior must be carefully designed, and, generally, three conditions must be satisfied by the experimental results to establish a causal link between hormones and behavior (Silver, 1978):

1. A hormonally dependent behavior should disappear when the source of the hormone is removed or the actions of the hormone are blocked.
2. After the behavior stops, restoration of the missing hormonal source or its hormone should reinstate the absent behavior.
3. Finally, hormone concentrations and the behavior in question should be covariant; that is, the behavior should be observed only when hormone concentrations are relatively high and never or rarely observed when hormone concentrations are low.

The third class of evidence has proved difficult to obtain because hormones may have a long latency of action, and because many hormones are released in a pulsatile manner. For example, if a pulse of hormone is released into the blood, and then no more is released for an hour or so, a single blood sample will not provide an accurate picture of the endocrine status of the animal under study. We might come to completely different conclusions about the effect of a hormone on behavior if we measure hormone concentrations when they are at their peak rather than when they are at their nadir. This problem can be overcome by obtaining measures in several animals and averaging across peaks and valleys, or by taking several sequential blood samples from the same animal. Another problem is that biologically effective amounts of hormones are vanishingly small and difficult to measure accurately. Effective concentrations of hormones are usually measured in micrograms (μg, 10^{-6} g), nanograms (ng, 10^{-9} g), or picograms (pg, 10^{-12} g); they are sometimes expressed as a mass percentage relative to 100 ml of blood plasma or serum (10 μg% = 10 μg/100 ml = 0.1 μg/ml). The development of techniques such as the radioimmunoassay (see next section), has increased the precision with which hormone concentrations can be measured. However, because of the difficulties associated with obtaining reliable covariant hormone–behavior measures, obtaining the first two classes of evidence usually has been considered sufficient to establish a causal link in hormone–behavior relations.

As we will see, the unique conditions of the laboratory environment may themselves cause changes in an animal's hormone concentrations and behavior that may confound the results of experiments; thus, it has become apparent that hormone–behavior relationships established in the laboratory should be verified in natural environments. The verification of hormone–behavior relationships in natural environments is not yet common, but it is a useful procedure for differentiating laboratory artifacts from true phenomena.

Common Techniques in Behavioral Endocrinology

How do we gather the evidence needed to establish hormone–behavior relationships? Because we cannot directly observe the interactions between hormones and their receptors or their intracellular consequences, we must use various indirect tools to explore these phenomena. This section describes some of the primary methods and techniques used in behavioral endocrinology. Much of the recent progress in the field has resulted from technical advances in the tools that allow us to detect, measure, and probe the functions of hormones. Therefore, familiarity with these techniques will help you to understand and assess the research to be discussed in subsequent chapters.

Ablation and Replacement

The **ablation** (removal or extirpation) of the suspected source of a hormone to determine its function is a classic technique in endocrinology. Recall that this was the method Berthold used to establish the role of the testes in the development of rooster behavior. There are four steps to this time-honored procedure:

1. A gland that is suspected to be the source of a hormone affecting a behavior is surgically removed
2. The effects of removal are observed
3. The hormone is replaced by reimplanting the removed gland, by injecting a homogenate or extract from the gland, or by injecting a purified hormone
4. A determination is made whether the observed consequences of ablation have been reversed by the replacement therapy

This technique is commonly used in endocrine research today. A traditional complementary approach to the ablation–replacement technique is the observation of behavior in individuals with diseased or congenitally dysfunctional endocrine organs. When ablation occurs in the brain, either through the actions of a researcher or through disease, the result is often called a **lesion**. Modern complementary approaches include the administration of drugs to block hormone synthesis or hormone receptor activity. More recent technologies include manipulation of genes to block hormone production or hormone receptor function (see the section on genetic manipulation).

The replacement component of this technique has been improved by technological advances, especially new recombinant DNA methods, that have made highly purified hormones readily available. Access to this virtually pure material has allowed researchers to rule out contaminants as a cause of the physiological or behavioral effects of a particular hormone. In addition, recent studies have emphasized the importance of replacing hormones in patterns and doses (physiological doses) similar to those found in nature, rather than using a single pharmacological dose (usually resulting in hormone concentrations higher than physiological levels). This has been made possible by the availability of implantable timed-release hormone capsules and minipumps that provide precisely timed infusions of purified hormones.

Bioassays

Once the existence of a hormone has been established, the next step is to identify the chemical processes involved in its actions. Classically, this has required a **bioassay**: a test of the effects of the hormone on a living animal. A living animal can serve as a reliable, quantifiable response system on which to test extracts and chemical fractions for biological activity.

A bioassay need not be conducted on the same species from which the hormone was obtained. For example, the crop sac of a pigeon (a structure that produces the "crop milk" fed to young pigeons; see Figure 7.6) can be used to measure levels of a hormone called prolactin in a rat. In incubating pigeons, prolactin stimulates growth of the epithelial cells of the crop sac to prepare it for the feeding of hatchlings; the height of the cells is correlated with the amount of prolactin in the pigeon's blood. To conduct a bioassay, researchers can inject different, known amounts of purified prolactin into pigeons, measure the resulting heights of the crop sac epithelial cells, and generate a standard dose–response curve (Figure 1.8). To measure the unknown prolactin levels of a rat, an extract from a tissue or blood sample obtained from the rat can be injected into a pigeon, and the resulting height of the epithelial cells measured and compared with the dose–response curve.

Probably the most famous endocrine bioassay was the so-called "rabbit test" (or Friedman test) for pregnancy. The rabbit test was developed by Maurice Friedman in 1929 and was the most commonly used pregnancy test in North America until the late 1950s. This procedure tested for the presence of human chorionic gonadotropin (hCG; a hormone released from the implantation site of a blastocyst that prevents menstruation) in the urine of women. The urine was injected into a rabbit, and if hCG was present in the urine, the rabbit's ovaries would form corpora lutea (ovarian endocrine structures formed following ovulation; see Chapter 2) within 48 hours. The rabbit test had several advantages over

(A)

(B)

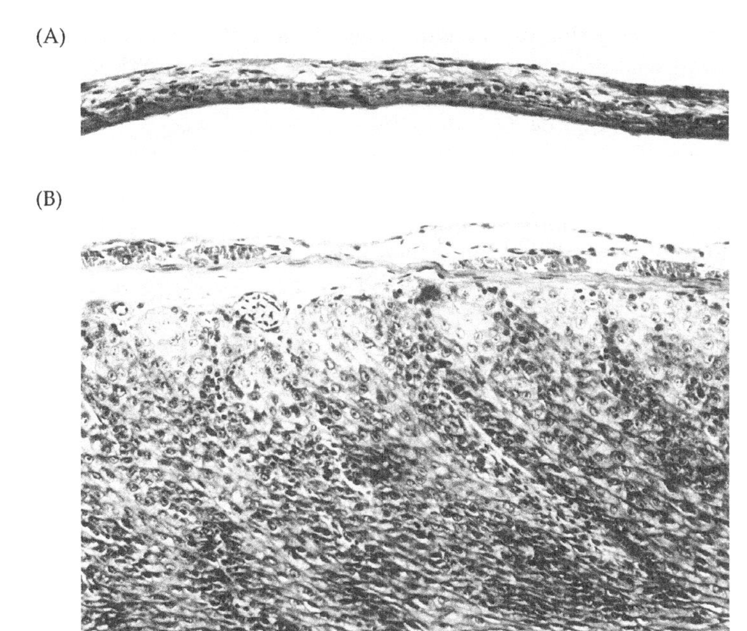

(C)

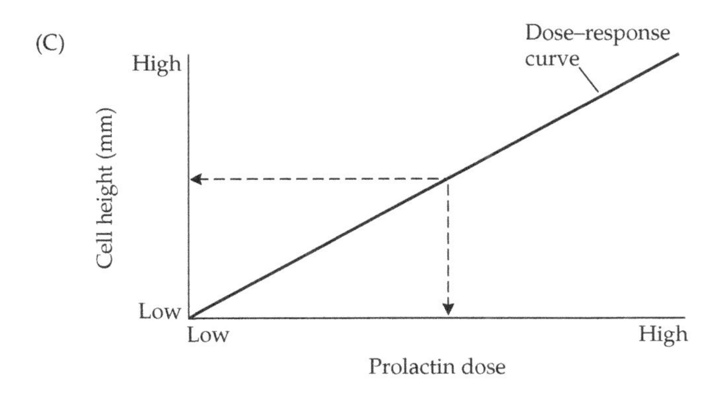

1.8 A bioassay for prolactin. (A) Photomicrograph of an untreated pigeon crop sac. (B) Prolactin injection causes a marked increase in the height of crop sac epithelial cells. (C) Different known amounts of purified prolactin can be injected into pigeons to generate a dose–response curve (solid line). The levels of prolactin in pituitary extracts from some other animal (e.g., a rat) can then be measured by injecting a pigeon with an extract sample from the animal and comparing any resulting change in epithelial cell height with the dose–response curve (dashed line). Photomicrographs courtesy of C. S. Nicoll.

the previously used Aschheim–Zondek mouse test (developed in 1928), which required six or more mice and at least 96 hours to complete. However, the rabbit test also had disadvantages. False positives could be obtained if the rabbit used was unduly stressed by the test procedure because stress can cause spontaneous corpora luteal formation in the absence of hCG.

Because biological systems are inherently subject to fluctuations induced by environmental conditions, it is essential that the conditions under which bioassays are conducted be rigidly defined. For example, in the Galli–Mainini or "frog test" for pregnancy, urine from a woman was injected into a male frog or toad. If the woman was pregnant and producing hCG, the animal would begin to produce sperm. The test provided results in 2–4 hours, and frogs were cheaper to maintain than mammals. However, there were variations in the sensitivity of the frogs: they had a greater tendency to yield "false negatives" during the summer.

The potential for contamination presents a problem in bioassays. Because the concentrations of hormones circulating in the bloodstream are so small, it is quite possible for contaminants in the tissue samples to interfere with the results. Furthermore, a bioassay is obviously only as good as the purified "standard" hormone used to calibrate the dose–response curve. For example, if the "purified" prolactin injected into pigeons in the crop sac bioassay had been contaminated by some other agent that suppressed cell growth, the resulting dose–response curve would not be accurate. Indeed, many early reports of endocrine effects on behavior and physiology were inaccurate because of contamination by hormones other than the so-called "pure" hormones. In recent years, with the advent of molecular biology tools to detect and measure biological products, the bioassay technique has been used less frequently.

BEHAVIORAL BIOASSAYS Several behavioral bioassays for hormones have been developed and used. The "water drive" of newts after injection of prolactin is one example. Prolactin causes newts to seek water; high levels of prolactin in a test sample cause a faster trip to water than low prolactin levels. A more common behavioral bioassay, which measured the behavioral effects of estrogens on the mating posture of female guinea pigs, was used for many years as the most sensitive assay for measuring circulating levels of these hormones. Again, bioassays require rigorous standardization of test conditions for accuracy and reliability. Even minute amounts of estrogen-like plant compounds (phytoestrogens) in their food are sufficient to induce female guinea pigs to display a mating posture. Consequently, laboratory personnel who used guinea pigs for estrogen behavioral bioassays had to make certain that the animals' chow did not contain alfalfa or other phytoestrogen-rich ingredient(s).

Immunoassays

Bioassays were useful because they measured a biological response to the hormone in question. In some cases, they allowed the determination of the presence or absence of a substance (as in the rabbit test), and in others they allowed quantitative measurement of specific hormones (as in the pigeon crop sac test for prolactin). However, bioassays usually required a great deal of time, labor, and the sacrifice of many animals for every assay conducted. The development of the **radioimmunoassay (RIA)** technique reduced these problems and increased the precision with which hormone concentrations could be measured. The ability to measure hormones precisely was such an important scientific advancement that one of the developers of this technique, Rosalyn Yalow, won the 1977 Nobel Prize in Physiology or Medicine. (Her close collaborator, Solomon A. Berson, died in 1972, and Nobel Prizes are not awarded posthumously.)

The concept of RIA is based on the principle of competitive binding of an antibody to its antigen. An antibody produced in response to any antigen, in this case a hormone, has a binding site that is specific for that antigen. A given amount of antibody possesses a given number of binding sites for its antigen. Antigen molecules can be "labeled" with radioactivity, and an antibody cannot discriminate between an antigen that has been radiolabeled (or "hot") and a normal, non-radioactive ("cold") antigen.

The first step in a radioimmunoassay is to inject the hormone of interest into an animal (usually a rabbit) to raise antibody; the antibody is then collected from the animal's blood and purified. To develop a standard curve, several reaction tubes are set up, each containing the same measured amount of antibody, the same measured amount of radiolabeled hormone, and different amounts of cold purified hormone of known concentration. The radiolabeled hormone and cold hormone compete for binding sites on the antibody, so the more cold hormone that is present in the tube, the less hot hormone will bind to the antibody. The quantity of hot hormone that was bound can be determined by precipitating the antibody and measuring the associated radioactivity resulting from the radiolabeled hormone that remains bound. The concentration of hormone in a sample can then be determined by subjecting it to the same procedure and comparing the results with the standard curve (Figure 1.9).

As is the case with other techniques, there are limitations to this method. First of all, RIAs require a source of highly purified hormone in order to prepare a highly specific antibody against it, so contamination of the hormone used to generate antibodies is a potential problem. In addition, because many hormones have similar chemical structures, RIAs must be tested for specificity to rule out the possibility that the antibody recognizes other antigens in addition to the one of interest. There is also the possibility that the antibody may bind not only to the intact hormone molecule, but also to a fragment of the hormone molecule that lacks biological activity. For these reasons, when the results of an RIA do not agree with those of a bioassay, the bioassay may be considered more valid even if it is less precise.

The **enzymoimmunoassay (EIA)**, like the RIA, works on the principle of competitive binding of an antibody to its antigen. The major difference between the RIA and EIA techniques is that EIAs do not require radioactive tags. Instead, the antibody is tagged with a chromogenic compound, which changes optical density (color) in response to its binding with antigen. The home pregnancy test is a familiar example of an EIA. This test, like the rabbit test, is designed to give a yes-or-no answer. However, most EIAs are developed to provide quantitative information. A standard curve is generated, as for RIAs, so that different known amounts of the hormone in question provide a gradient of color that can be read on a spectrometer. The unknown sample is then added, and the amount of hormone is interpolated by the standard curve. A similar technique is called the **enzyme-linked immunosorbent assay (ELISA)**.

Immunocytochemistry

Immunocytochemistry (**ICC**) techniques use antibodies to determine the location of a hormone or hormone receptors in the body. Antibodies linked to marker molecules, such as those of a fluorescent dye (see Polak and Van Noorden, 1997), are usually introduced into dissected tissue from an animal, where they bind with the hormone or neurotransmitter of interest. For example, if a thin slice of brain tissue is immersed in a solution of antibodies to a hormone protein linked to a fluorescent dye, and the tissue is then examined under a fluorescent microscope, concentrated spots of fluorescence will appear, indicating where the protein hormone is located (Figure 1.10). Fluorescent dyes used for ICC include fluorescein and rhodamine. Other commonly used markers are the enzyme horseradish peroxidase, for bright-field or electron microscopy; the enzyme alkaline phosphatase, for biochemical detection; and the iron-containing protein ferritin, for electron microscopy.

A common alternative ICC technique involves raising a second antibody against

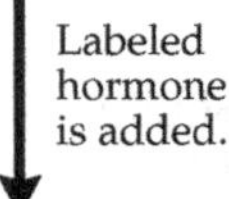

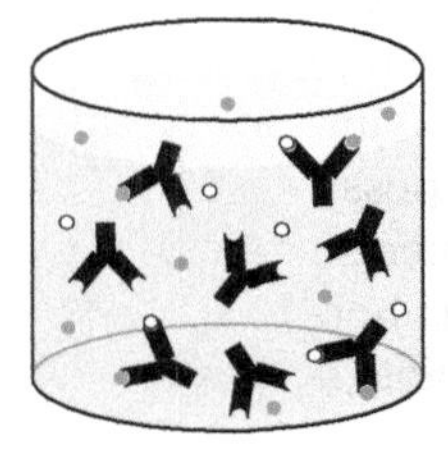

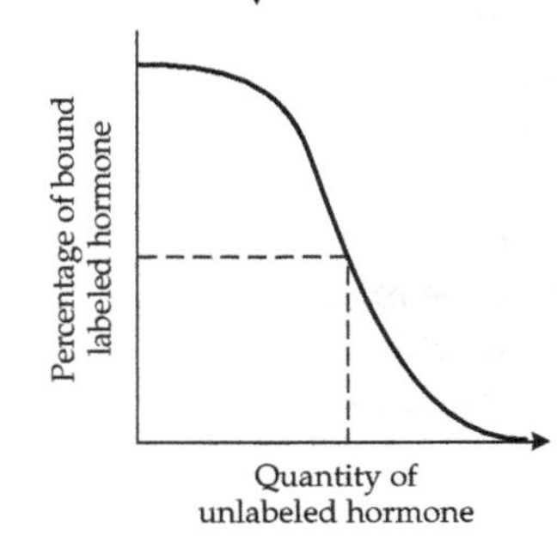

1.9 Radioimmunoassay. Purified hormone (antigen) is injected into an animal to raise antibody, which is collected and purified. Measured amounts of purified, radioactively labeled hormone are added to measured amounts of the collected antibody in several reaction tubes. The antibody binds with the radiolabeled hormone to form reversible hormone–antibody complexes. When different amounts of unlabeled hormone are added to the reaction tubes, this "cold" hormone competes with the "hot" hormone for antibody binding sites and displaces some of it. The antibody is precipitated, and the radioactivity of the bound hormone from each reaction tube is measured. In this way, a standard curve can be developed that expresses the quantity of cold hormone as a decline in radioactivity. The hormone concentration in a blood sample can then be measured by using it as cold hormone in the same procedure and comparing its effect on measured radioactivity with the standard curve.

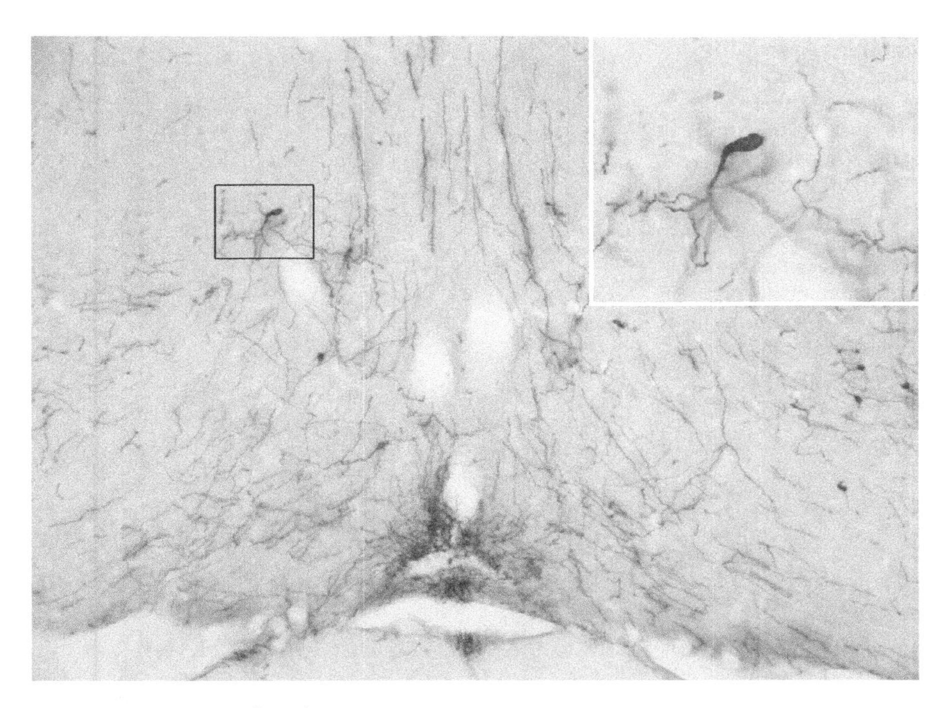

1.10 Immunocytochemistry. Antibodies to a hormone can be linked to a dye and used to determine the location of a hormone in the body. If a slice of tissue is exposed to a solution of antibody linked to such a marker, the binding of the antibody will cause those parts of the tissue containing the hormone to selectively take up the dye, making them more visible under the microscope. This figure is a low-power photomicrograph (×100) of a thin section of a rodent's brain showing immunocytochemically marked cell bodies and fibers that contain gonadotropin-releasing hormone (GnRH). The section is taken from the medial preoptic area (MPOA) at the level of the organum vasculosum of the lamina terminalis (OVLT). Box surrounds a neuron shown at high power (×400) in the inset. Note the appearance of beaded fibers characteristic of neurosecretory cells in the brain. Courtesy of Lance Kriegsfeld.

the primary antibody that recognizes the substance to be measured, then coupling the second antibody to a marker. Sometimes a secondary antibody is linked to biotin, a water-soluble vitamin, which has a strong binding affinity to avidin, a bacterial protein. If avidin is coupled to a marker molecule, it can be used in place of a second antibody. A single biotin–antibody–antigen complex may link to multiple marker molecules, which essentially amplifies the signal indicating the presence of the antigen.

Autoradiography

Many studies have demonstrated that hormone receptors are selectively concentrated in particular target tissues; estrogen receptors, for example, are concentrated in the uterus. **Autoradiography** is typically used to determine hormonal uptake and indicate receptor location. An animal can be injected with a radiola-

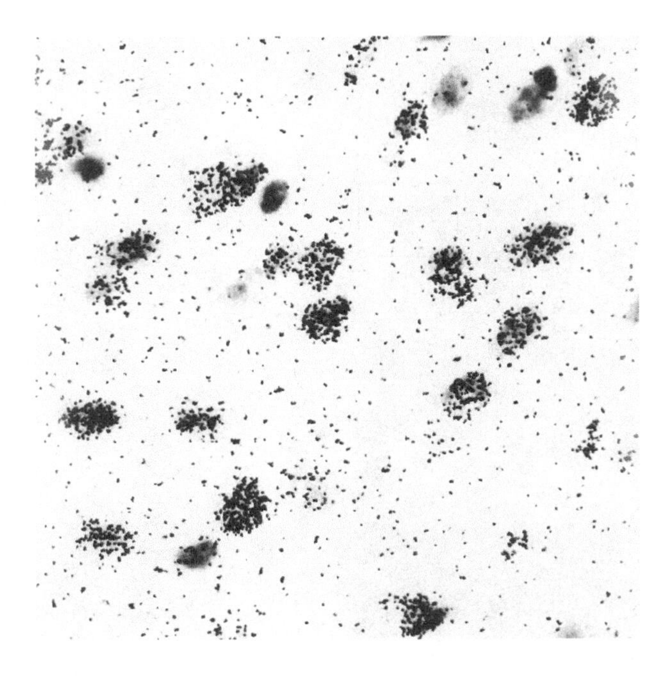

1.11 Autoradiography. An animal is injected with a radiolabeled hormone, and adjacent tissue slices are either treated with stains to reveal cellular structures or exposed to film. Those parts of the tissue that bind with the radiolabeled hormone darken the film, and when combined with the stained section, show which cell structures have taken up the hormone in the largest amounts. In this autoradiograph, cell nuclei in a monkey brain that have accumulated radiolabeled estrogen are seen as darker than the background neurons. Courtesy of Bruce McEwen.

beled hormone, or the study can be conducted entirely in vitro. Suspected target tissues are sliced into several very thin sections; adjacent sections are then subjected to different treatments. One section of the suspected target tissue is stained in the usual way to highlight various cellular structures. The next section is placed in contact with photographic film or emulsion for some period of time, and the emission of radiation from the radiolabeled hormone develops an image on the film. The areas of high radioactivity on the film can then be compared with the stained section to determine how the areas of highest hormone concentration correlate with cellular structures (Figure 1.11). This technique has been very useful in determining the sites of hormone action in nervous tissue, and consequently has increased our understanding of hormone–behavior interactions.

Blot Tests

Other techniques allow whether or not a particular protein or nucleic acid is present in a specific tissue. In the so-called **blot tests**, the tissue of interest is homogenized and the cells are lysed with detergent. The resulting homogenate is placed in gel, which is subjected to electrophoresis. **Electrophoresis** refers to the application of an electric current to a matrix or gel, which results in a gradient of molecules separating out along the current on the basis of size (smaller molecules move farther than larger molecules during a set time period). The homogenate is transferred to a membrane or filter, and the filter is then incubated with a labeled substance that can act as a tracer for the protein or nucleic acid of interest: radiolabeled complementary deoxyribonucleic acid (cDNA) for a nucleic acid assay, or an antibody that has been radiolabeled or linked to an enzyme for a protein assay. If radiolabeling is used, the filter is then put over film to locate and measure radioactivity.

In enzyme-linked protein assays, the filter is incubated with chromogenic chemicals, and standard curves reflecting different spectral densities are generated. The test used to assay DNA is called Southern blotting, after its inventor, E. M. Southern; the test used to assay RNA is called Northern blotting, and the test for proteins is called Western blotting.

Autoradiography Using In Situ Hybridization

An important tool used at the cellular level to examine gene expression is called **in situ hybridization**. This technique is used to identify cells or tissues in which messenger RNA (mRNA) molecules encoding a specific protein—for example, a hormone or a neurotransmitter—are being produced. The tissue is fixed, sliced very thinly, mounted on slides, and either dipped into emulsion or placed over film and developed with photographic chemicals. Typically, the tissue is also counterstained to identify specific cellular structures. A radiolabeled cDNA probe is introduced into the tissue. If the mRNA of interest is present in the tissue, the cDNA will form a tight association (that is, hybridize) with it. The tightly bound cDNA, and hence the mRNA, will appear as dark spots (Figure 1.12). The techniques previously described, such as blot tests, can typically determine only whether or not a particular substance is present in a specific tissue, but in situ hybridization can be used to determine whether a particular substance is *produced* in a specific tissue. Recent advances in the technique allow for the quantification of the substance being produced. Blot tests cannot match the resolution or sensitivity of in situ hybridization.

Stimulation and Recording

By using **electrical stimulation** to "turn on" specific neurons or brain centers, we can discover the effects of various endocrine treatments on the central nervous system. In this technique, a fine electrode is precisely positioned in the brain, and a weak electric current is used to stimulate neurons. This technique has been used to study the releasing and inhibiting hormones of the hypothalamus (see Chapter 2).

The electrical activity of single neurons can be monitored through the use of **single-unit recording**, which involves the placement of very small electrodes

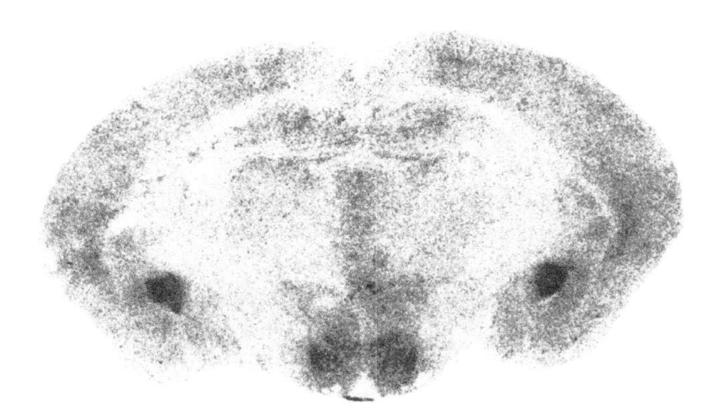

1.12 In situ audioradiography. The dark spots at the bottom of this slice of a rodent brain represent cells in the ventromedial hypothalamus (VMH) that contain mRNA for an oxytocin receptor. This tissue was treated with a specific cDNA that hybridized with the mRNA that encodes the oxytocin receptor. Courtesy of Thomas Insel and Larry Young.

in or near one neuron to record changes in its activity during and immediately after exposure to hormones. This technique can help to uncover the direct effects of various endocrine products on neural activity. Often, several neurons are recorded simultaneously and an average change in activity in these multiple units is calculated.

Pharmacological Techniques

The development of synthetic **agonists** (mimics) and **antagonists** (blockers) of hormones for medical purposes has taught us a great deal about the functioning of the endocrine system. Some specific chemical agents act to stimulate or inhibit endocrine function by affecting hormonal release; these agents are called general agonists and antagonists, respectively. Other drugs act directly on hormone receptors, either enhancing or negating the effects of the hormone under study; these drugs are referred to as receptor agonists and antagonists, respectively. Cyproterone acetate (CPA), for example, is a powerful anti-androgen (anti-testosterone) that has been used clinically as a treatment for male sex offenders (see Chapter 5). This antagonist binds to testosterone receptors but does not activate them, thereby blocking the effects of testosterone on behavior and physiology. Other examples of hormone agonists and antagonists will be presented throughout the book.

In the technique known as **cannulation**, hollow electrodes or fine tubes (cannulas) are inserted into specific areas of the brain and used to introduce substances into those sites. Davidson (1966a) used this technique to find out where in the brain testosterone acts to influence sexual behavior in rats. In Davidson's study, male rats were castrated, after which they were observed to cease mating. Testosterone was then introduced through cannulas into different areas of the brain in different rats; a control group of rats received cholesterol, the precursor of testosterone. Those animals that received testosterone in one specific location, the preoptic area of the hypothalamus, resumed sexual behavior; the rats that received cholesterol or received testosterone in other brain regions did not respond to the treatment.

Another type of cannulation involves inserting a small hollow tube into the jugular vein, carotid artery, or other blood vessel. In this way, specific hormones or pharmacological agents can later be injected directly into the animal without further disturbance, or blood samples can be obtained to correlate hormone levels with behavior. In a related technique, **anastomosis**, the blood systems of two animals are connected via cannulation tubing to see if the endocrine condition of one animal can cause a behavioral change in the other.

Microdialysis

The **microdialysis** technique is based on the principle of dialysis, in which a semipermeable membrane, which allows passage of water and small molecules, divides two fluid compartments. Developed in the 1980s, microdialysis allows assessment of responses to neurotransmitters, drugs, and hormones in a conscious animal (DeLange et al., 1997). Typically, a cannula that is divided into two compartments by a semipermeable membrane is implanted in the brain region of interest using stereotactic surgery. The end of one compartment is continuously perfused with a liquid, and molecules are exchanged with the extracellular fluid by diffusion in both directions. Hormones or drugs can be delivered through one part of the cannula while extracellular signaling molecules (e.g., neurotransmitters) can be monitored via the second compartment. Because microdialysis can be performed in awake, freely moving animals, this method is especially well-suited for the study of the interactions of hormones, neurotransmitters, and behavior. The method can both introduce and remove molecules from the brain. It is possible to sample continuously for hours or days. Typically, the samples are analyzed with high-performance liquid chromatography (HPLC) to detect such substances as amino acids, acetylcholine, biogenic amines, choline, glucose, histamine, and purines. Microdialysis is sometimes performed from the human brain for diagnostic purposes.

Brain Imaging

Several brain scanning techniques are used in behavioral endocrinology to determine brain structure and function (DeLange et al., 1997; Van Bruggen and Roberts, 2002). Comparisons can be made, for example, between the brains of men and women or among those of individuals in different hormonal conditions. One important scanning technique used to determine regional brain activation is called **positron emission tomography** (**PET**). Unlike a simple X-ray or CT scan, which reveals only anatomical details, PET scanning permits detailed measurements of real-time functioning of specific brain regions of people who are conscious and alert. PET gives a dynamic representation of the brain at work. Prior to the availability of PET scanners, changes in neurotransmitter levels or hormonal activation of specific circuits could only be inferred on the basis of autopsy data.

Before the PET scan begins, a small amount of a radioactively labeled molecule that mimics glucose or a radioactive gas such as oxygen-15 is injected into the individual. When neurons become more active, they use more glucose and oxygen, so the radioactive material is taken up at high rates by the most active neurons. This radioactive material emits positrons. When a positron collides with an electron, the collision produces two gamma rays that leave the body in opposite directions and can be detected by the PET scanner. This information about where glucose is being metabolized or oxygen is being used is then converted into a complex picture of the person's functioning brain by a computer.

A **computer-assisted tomography** (**CT**) scanner shoots fine beams of X-rays into the brain from several directions. The emitted information is fed into a computer that constructs a composite picture of the anatomical details within a "slice" through the brain of the person. **Magnetic resonance imaging** (**MRI**) does much the same thing, but uses non-ionizing radiation formed by the excitation of protons by radio-frequency energy in the presence of large magnetic fields (Van Bruggen and Roberts, 2002). MRI can be used in anatomical studies, and assessing anatomical irregularities is its main function as a medical diagnostic tool. **Functional MRI** (**fMRI**) uses a very high spatial (~1 mm) and temporal resolution to detect changes in brain activity during specific tasks or conditions. Most fMRI studies require the person to lie still in a narrow tunnel (Figure 1.13), so only

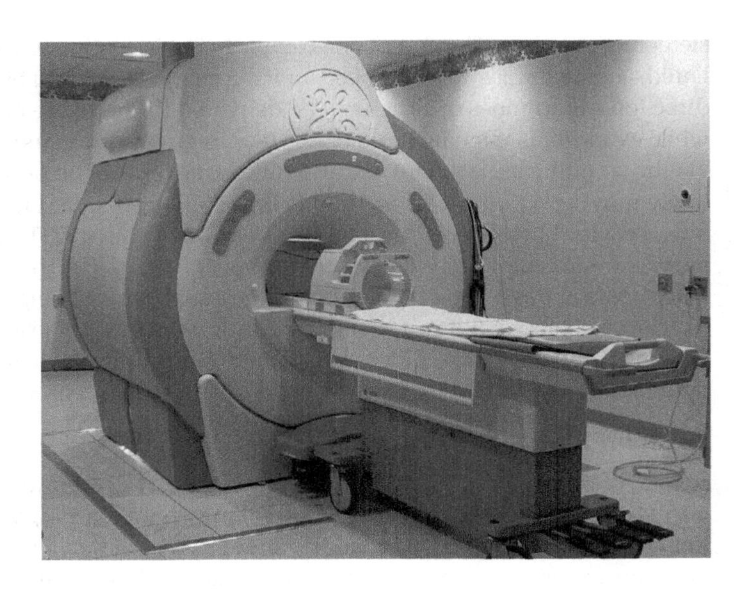

1.13 A modern MRI scanner.

cognitive or affective changes can be monitored. As noted above, when neurons become more active, they use more energy, so they require additional blood flow to deliver glucose and oxygen. The fMRI scanner detects this change in cerebral blood flow by detecting changes in the ratio of oxyhemoglobin and deoxyhemoglobin. Deoxyhemoglobin is paramagnetic (becomes magnetic in magnetic fields), whereas oxyhemoglobin is diamagnetic (does not become magnetic in magnetic fields). Thus, deoxyhemoglobin molecules act like little magnets in the large magnetic field of the MRI and dephase the signal. When brain regions increase their activity, more oxygenated blood is present than before the activation. More oxyhemoglobin results in a net decrease in paramagnetic material (deoxyhemoglobin), which leads to a net increase in signal because of reduced dephasing of the signal. A complex computer program plots all of the phase changes of the signal and applies this picture on top of a structural picture of the brain usually obtained with a CT scan (Figure 1.14).

Genetic Manipulations

With new advances in molecular biology, it is possible to perform specific genetic manipulations. In behavioral endocrinology research, common genetic manipulations include the insertion (creating a **transgenic** organism) or removal (**knockout**) of the genetic instructions encoding a hormone or the receptor for a hormone. The genetic instructions for each individual are contained in its DNA, located in the nucleus of nearly every cell. These instructions are encoded in the form of four nucleotides, adenine (A), thymine (T), cytosine (C), and guanine (G). The specific order of these four nucleotides along the "rails" of the DNA double helix forms the genetic instructions for all organisms, from those as simple as slime molds to those as complex as mice and humans. Each **gene** represents the

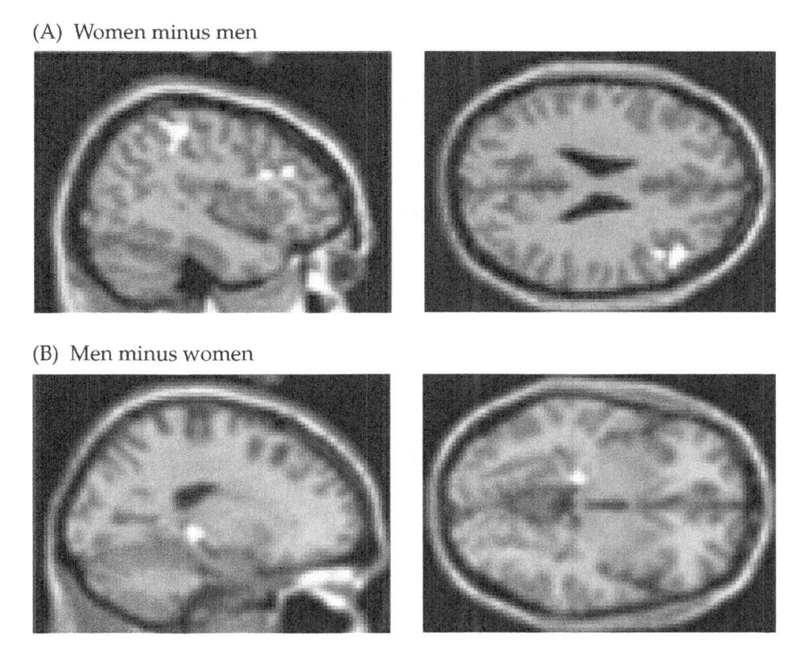

1.14 fMRI. These images are the results of fMRI scans of groups of men's and women's brains that were superimposed on CT scans of the sagittal (left) and transverse (right) planes. The bright areas represent the most active parts of the brain. Men and women were asked to navigate (via keyboard) through a spatial maze. (A) There was more activation in the right inferior parietal lobe in women than in men. (B) There was more activation in the hippocampus in men than in women. The brains of men and women show similar activation throughout the brain; for instance, when you take the female brain activation map and subtract the male brain activation map, you get the difference indicated in (A) and (B). From Grön et al., 2000.

complex instructions for the production of a specific protein in the cell. Thus, nucleotide "syntax" is critical in conveying the instructions encoded in the genes. To inactivate, or knock out, a gene, molecular biologists scramble the order of the nucleotides that make up the gene (Aguzzi et al., 1994; Soriano, 1995).

Cloning a gene is different from cloning an animal. To make Dolly, the famous cloned sheep, the researchers in Scotland first obtained ova from the ovaries of a female sheep and destroyed the cell nucleus of each one. Then a new nucleus was obtained from cells of the adult sheep that was to be cloned. The manipulated ova were stimulated to divide, and one of the resulting embryos was implanted into a surrogate mother. Thus, Dolly was genetically-identical to the individual from which she was derived. To clone a gene, the gene must first be identified, then the specific piece of DNA that contains the gene is placed into a vector. A vector is another DNA molecule (usually from a simple organism such as bacteria or yeast) that can be inserted into a viral host. This produces a new DNA molecule termed recombinant DNA. This recombinant DNA is an impor-

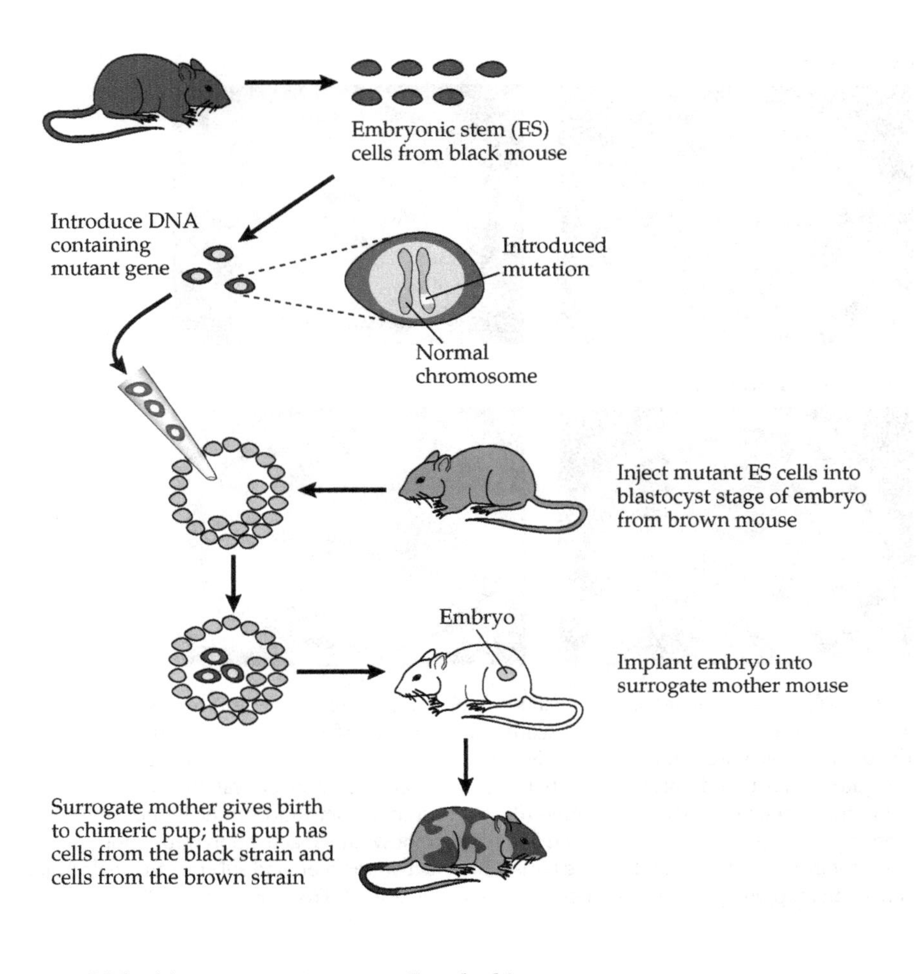
Embryonic stem (ES) cells from black mouse
Introduce DNA containing mutant gene
Introduced mutation
Normal chromosome
Inject mutant ES cells into blastocyst stage of embryo from brown mouse
Embryo
Implant embryo into surrogate mother mouse
Surrogate mother gives birth to chimeric pup; this pup has cells from the black strain and cells from the brown strain

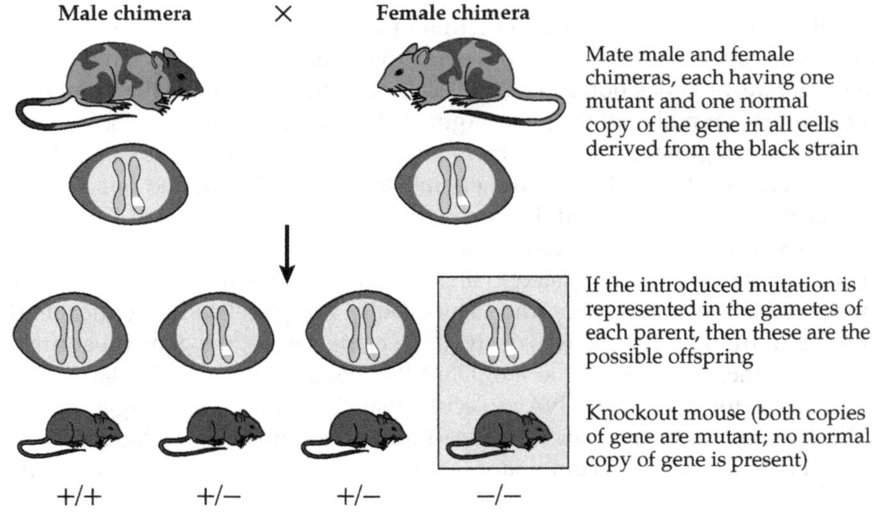
Male chimera
×
Female chimera
Mate male and female chimeras, each having one mutant and one normal copy of the gene in all cells derived from the black strain
If the introduced mutation is represented in the gametes of each parent, then these are the possible offspring
Knockout mouse (both copies of gene are mutant; no normal copy of gene is present)
+/+
+/–
+/–
–/–

◀ **1.15 Engineering knockout mice.** DNA that has been engineered to contain a mutant (inactive) copy of the gene of interest is introduced into embryonic stem cells (ES cells) from black mice that are growing in tissue culture. The black coat color serves as a genetic marker. ES cells with one mutant copy of the gene are introduced into an early mouse embryo (blastocyst), which incorporates the cells into the body of the developing mouse. Mice that are born from this manipulation are called "chimeras." These chimeric mice are mated to each other. Only chimeric mice in which the ES cells have been incorporated into the germ line (gametes) will produce offspring carrying the mutation. By simple Mendelian genetics, one in four of those offspring will be knockout mice, which contain two mutant copies of the gene. The entire process takes 6–12 months.

tant tool in understanding the function of the gene in normal and pathological states.

Among vertebrates, the identification of genes has been most successful among laboratory mice (*Mus musculus*). Consequently, mice are currently the most commonly used species in targeted gene deletion studies. Knocking out a specific gene in a mouse is an arduous task that relies on several low-probability events. First, the gene of interest must be identified, targeted, and marked precisely (Figure 1.15). This has been accomplished for an astounding number of murine (mouse) genes during the past decade (Takahashi et al., 1994). Next, a mutated form of the gene is created (i.e., a piece of DNA that contains a genetically engineered, inactive copy of the gene of interest). Mouse embryonic stem cells (ES cells) are harvested and cultured, and copies of the altered gene are introduced into the cultured cells by microinjection (Tonegawa, 1994). A very small number of the altered genes will be incorporated into the DNA of the ES cells through recombination (Bernstein and Breitman, 1989; Sedivy and Sharp, 1989). The mutant ES cells are then inserted into otherwise normal mouse embryos (blastocysts), which are implanted into surrogate mothers (Boggs, 1990; Le Mouellic et al., 1990; Steeghs et al., 1995). The ES cells are equipotential, which means that they may become incorporated into any part of the developing body. All of the cells descended from the mutant ES cells will have the altered gene; the descendants of the original blastocyst cells will have normal genes. Thus, the newborn mice will have some cells that possess a copy of the mutant gene and some cells that possess only the normal (wild-type) gene. This type of animal is called a **chimera**. If the mutated ES cells have been incorporated into the germ line (the cells destined to become the sperm and eggs), then some of the mouse's gametes will contain one heritable copy of the mutant gene. If these chimeric mice are mated to each other, then approximately half of their offspring will be heterozygous for the mutation; that is, they will possess one copy of the mutant gene. Approximately one-fourth of their offspring will be homozygous for (i.e., have two copies of) the mutant gene; the product that the gene typically encodes will be missing from these homozygous (knockout) mice (Sedivy and Sharp, 1989). These homozygous mice can then be interbred to produce pure lines of mice with the gene of interest knocked out (Galli-Taliadoros et al., 1995).

Behavioral performance can then be compared among wild-type (+/+), heterozygous (+/–), and homozygous (–/–) mice, in which the gene product is produced normally, produced at reduced levels, or completely missing, respectively. The comparison of +/+ and −/− littermates of an F_2 recombinant generation is probably the minimal acceptable control in determining behavioral effects in knockout mice (Morris and Nosten-Bertrand, 1996). The use of new inducible knockouts, in which the timing and tissue-specific placement of the targeted gene disruption can be controlled, promises to be an extremely important tool in future behavioral endocrinology research (Nelson and Chiavegatto, 2001). Similarly, the use of transgenic animals in which there is overexpression of specific genes has become an increasingly common technique in behavioral endocrinology investigations.

Gene Arrays

Another new technology that has become extremely useful in behavioral endocrinology is the **gene array** or **microarray**, which is a marriage of genomics and computer microprocessor manufacturing. Essentially, a minuscule spot of nucleic acid of known sequence (usually cDNA, although RNA is also used) is attached to a glass slide (or occasionally a nylon matrix) in a precise location, often by high-speed robotics. This identified, attached nucleic acid is called an **oligonucleotide** (Phimister, 1999). Because thousands of oligonucleotides (pieces of RNA or cDNA) can be added to the array, an experiment with a single array can provide researchers with information on thousands of genes simultaneously. The underlying principle of gene arrays is hybridization, the process by which nucleotide bases pair (i.e., A–T and G–C for DNA; A–U and G–C for RNA). A nucleic acid sample to be identified is called a **probe**. By observing which of the probes hybridize with the oligonucleotides on the array, we can identify some of the mRNAs that are present in the sample.

In behavioral endocrinology, gene arrays might be used to determine relative gene expression during the onset of a behavior, or during a developmental stage, or among individuals that vary in the frequency of a given behavior or hormonal state. For example, mRNA may be extracted from brain regions that are thought to regulate aggressive behavior. To see whether specific gene expression differs in this region between intact and castrated rats, the mRNA extracted from the brain tissue would be labeled with fluorescent dyes and added to a cDNA microarray, where it would be available for hybridization to the attached cDNA oligonucleotides. Any differences in hybridization between the samples would be indicated by changes in the color of the fluorescence readout. The relative amount of hybridization as compared to hybridization of a standard "housekeeping gene" would indicate the relative amount of gene expression in the tissue. Of course, the nucleic acid sequences of interest must be attached to the array to be detected; also, because the gene array usually provides only information about gene expression from pooled tissue samples, an additional method, such as quantitative PCR, must be conducted on individual samples.

A Case Study: Effects of Leptin on Behavior

Recently, a novel hormone was discovered that is released from **adipose** (fat) cells. This hormone was named **leptin**, a term derived from the Greek word *leptos*, which means "thin." Studies of the behavioral effects of leptin will be used here to

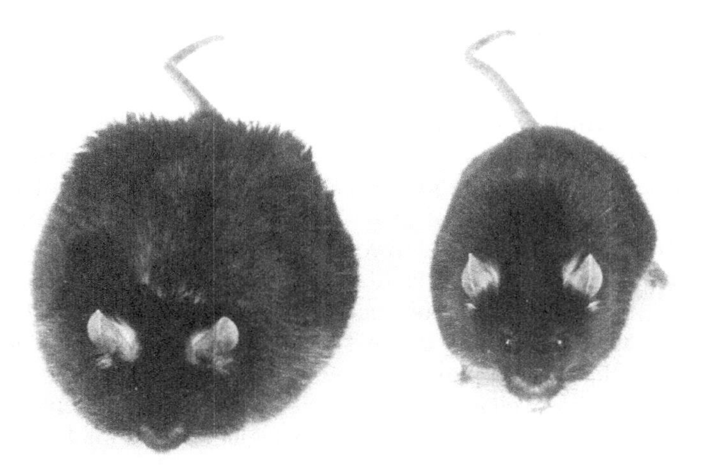

1.16 Behavioral effects of leptin. Both of these mice have defective *ob* genes. This mutation typically results in a large increase in food intake and subsequent obesity. The mouse on the right was treated daily with the protein encoded by the *ob* gene, called leptin. It weighs about 35 grams; the untreated mouse weighs about 67 grams. A mouse with a normal *ob* gene would weigh about 25 grams at this age. Photograph by John Sholtis.

demonstrate how the various techniques described above are used to understand the biological functions of hormones.

For many years, specific mutations in mice that cause extreme obesity have been recognized. Animals that are homozygous for a mutation in the *ob* gene are hyperphagic (overeat), obese, and reproductively sterile (Figure 1.16). These animals may be considered natural knockouts for the *ob* gene, which normally codes for leptin. Another gene, the *db* gene, encodes a leptin receptor; mice that are homozygous for a mutation in this gene are diabetic. Leptin is released by adipose cells into the bloodstream, where it travels to specific receptors in the central nervous system and elsewhere to regulate feeding and energy balance. Since its identification in 1994 (Zhang et al., 1994), many of the techniques described in the previous sections have been used to understand the behavioral functions of leptin.

For years, the effects of the *ob/ob* mutation on the ability to maintain normal body mass in mice have been known. However, research into the mechanisms underlying this mutation began in earnest only with the identification and sequencing of the leptin protein (Zhang et al., 1994). The *ob* gene was cloned, and a copy of the gene was inserted into bacteria; the resulting transgenic bacteria produced purified leptin. Using this purified leptin, a "replacement study" was conducted in which leptin was provided to *ob/ob* mice to determine whether replacing their missing leptin would ameliorate their hyperphagia, energy impairments, and obesity. It did.

The availability of purified leptin also allowed researchers to produce specific antibodies that could be used in developing assays to determine blood concentrations of leptin (e.g., Cohen et al., 1996). An RIA was developed to determine the blood plasma leptin concentrations of obese and diabetic humans to see whether leptin was implicated in human obesity or diabetes (McGregor et al., 1996); in this study, no connection was found. A leptin ELISA was developed for rats and mice, which was used to determine that fasting or exposure to low temperatures caused circulating leptin concentrations to fall (Hardie et al., 1996). Immunocytochemistry techniques revealed that leptin was present in both white and brown adipose tissue (Cinti et al., 1997) as well as other peripheral tissues.

Injections of leptin into the third or lateral ventricles in the brains of mice were found to reduce food intake (Campfield et al., 1995). To determine the site of action of leptin in the brain, purified leptin was radiolabeled and injected into mice. Using autoradiography, dense, specific binding of the radiolabeled leptin was found in the choroid plexus, a brain structure located in the dorsal part of the third ventricle, as well as in the lateral ventricles (Lynn et al., 1996). This tissue was then used to clone a leptin receptor (Tartaglia et al., 1995). In situ hybridization showed that the mRNA for the leptin receptor was expressed in several brain regions, including the hypothalamus (Mercer et al., 1996; Zamorano et al., 1997).

In an effort to "cure" obesity, gene replacement therapy was attempted (Chen et al., 1996; Muzzin et al., 1996). The assumption behind this effort was that mice deficient in the wild-type *ob* gene were obese because of a deficiency of leptin. If the leptin or the *ob* gene were replaced, then the individual should display normal food intake and body mass. When *ob/ob* mice were treated with a recombinant adenovirus containing the mouse *ob* gene, there was a dramatic reduction in both food intake and body mass (Muzzin et al., 1996). Similar findings were reported for rats (Chen et al., 1996). In addition to being obese, *ob/ob* mice are also sterile; treatment of *ob/ob* mice with recombinant leptin reverses the sterility (Chehab et al., 1996; Mounzih et al., 1997).

These exciting findings suggested the possibility of treating obese humans with gene therapy. However, leptin appears to have significant weight-reducing properties only in rodents. Despite screening of thousands of obese humans, only five individuals in two families have been found with mutations in the leptin gene (Montague et al., 1997; Strobel et al., 1998) (i.e., they are like the *ob/ob* mice), and three members of a single family have been found that are homozygous for mutations in the leptin receptor gene (Clement et al., 1998) (i.e., they are like the *db/db* mice). Only moderate obesity is observed among people who are heterozygous (one normal and one mutant copy) for their leptin genes (Farooqi et al., 2003). The effects of leptin treatment on human obesity have been equivocal. Treatment of a 9-year-old girl who is homozygous for a mutation in the leptin gene with daily injections of leptin for a year reduced her body mass from 94.4 kg (208 pounds) (55.9 kg was estimated to be adipose tissue [i.e., fat] to 78 kg (172 pounds) (Farooqi et al., 1999). It was reported that the leptin injections reduced her appetite and food intake. Treatment of obese individuals with normal leptin genes resulted in moderate body mass loss (–7.1 kg versus –1.3 kg for a placebo group). Because obese people tend to have higher levels of leptin than lean people, it has been proposed that obese individuals are "resistant" to leptin. Recently, it has been proposed that obesity is a disease of the blood–brain barrier (BBB). According to this hypothesis, obesity occurs in some individuals because leptin cannot cross the BBB to signal the brain that sufficient fat reserves exist (Banks, 2003). Although leptin has only been identified for about a decade, much has been learned about this hormone; however, additional research is required to sort out the role of leptin in human and nonhuman energy balance.

Taken together, all of these techniques are useful in elucidating hormone–behavior relationships; later in the book, other specific techniques will be introduced as we examine the details of specific hormone–behavior interactions. In the next chapter, many of the fruits of these techniques will be presented in the context of a general introduction to endocrine anatomy, chemistry, and physiology.

Summary

- Behavioral endocrinology is the study of the interaction between hormones and behavior. This interaction is bidirectional: hormones can affect behavior, and behavior can influence hormone concentrations.
- Hormones are chemical messengers that are released from endocrine glands and travel through the bloodstream to target cells, where they induce changes in the rate of cellular function. The endocrine system and the nervous system work together to regulate the physiology and behavior of individuals.
- Behavior is generally thought of as involving movement, but nearly any type of output can be considered behavior. A complete description of behavior is required before researchers can address questions of its causation. All behavioral biologists study a specific version of the general question, "What causes animal A to emit behavior X?"
- The four interacting levels of analysis that can be used for exploring and explaining the causes of behavior are immediate causation, development, evolution, and adaptive function.
- Hormones influence behavior by increasing the probability that a particular behavior will occur in the presence of a particular stimulus. Hormones can influence behavior by affecting an animal's sensory systems, integrators, and/or effectors or peripheral structures. Hormones can also be influenced by behavior.
- Three types of evidence are necessary to establish a causal link between hormones and behavior: (1) a behavior that depends on a particular hormone should diminish when the source of, or actions of, the hormone are removed, (2) the behavior should reappear when the hormone is reintroduced, and (3) hormone levels and the behavior in question should be covariant.
- Several techniques have been useful in advancing research in behavioral endocrinology: ablation and replacement; bioassays; modern assays that utilize the concept of competitive binding of antibodies; autoradiography; immunocytochemistry; electrical stimulation and single-unit recording; pharmacological methods; methods that make use of cannulation, including in vivo microdialysis; and gene arrays and genetic manipulations.

Questions for Discussion

1. What are some of the problems associated with attempting to determine causation in a hormone–behavior interaction? What are the best ways to address these problems?

2. An experimenter is interested in the effect of a particular hormone on aggressive behavior in butterflies. It is hypothesized that butterflies fight because of high levels of this hormone. An alternative hypothesis is that the butterflies that fight to guard territories and potential mates produce more offspring in subsequent generations. If the results of an experiment rule out the hypothesis that the hormone is required for aggressive behavior, does that necessarily mean that the alternative hypothesis is true? Why or why not?

3. Hormones cause changes in the rates of cellular processes or in cellular morphology. Suggest some ways that these hormonally induced cellular changes might theoretically produce profound changes in behavior.

4. In recent years, investigations of hormone–behavior relationships have involved increasingly elaborate and precise methodology for identifying and locating hormones. Discuss the proposition that comparable increases in sophistication, quantification, and so forth are needed on the side of behavior. Give examples of desirable and undesirable approaches.

5. In lesion studies or studies of animals with specific genes deleted, the behavioral tests study the effects of the *missing* brain region or the *missing* gene, respectively and not the effects of the brain region directly. Discuss how these conceptual shortcomings can be overcome in evaluating the results of studies using these types of procedures.

Refer to the accompanying Student CD for additional resources, including Web links, videos, animations, and additional photos.

Suggested Readings

Beach, F. A. 1948. *Hormones and Behavior*. Paul Hoeber, New York.

Beach, F. A. 1975. Behavioral endocrinology: An emerging discipline. *American Scientist*, 63: 178–187.

Brown, R. E. 1994. *An Introduction to Neuroendocrinology*. Cambridge University Press, Cambridge.

Pfaff, D. W., Arnold, A. P., Etgen, A. M., Fahrbach, S. E., and Rubin, R. T. (eds.). 2002. *Hormones, Brain, and Behavior*. Vol. 1–5. Academic Press, New York.

Pfaff, D. W., Phillips, I. M., and Rubin, R. T. 2004. *Principles of Hormone/Behavior Relations*. Academic Press, New York.

Tinbergen, N. 1951. *The Study of Instinct*. Oxford University Press, Oxford.

2

The Endocrine System

In order to understand the interaction between hormones and behavior, a basic understanding of the endocrine system is essential. A comprehensive review of endocrinology is both beyond the scope of this book and unnecessary to understand hormone–behavior interactions conceptually. But a good basic knowledge of the general principles of endocrinology, the hormones found in animals, and the organs that produce those hormones will help you in reading the discussions that follow. The goals of this chapter are to provide a solid general background in endocrinology for students who are unfamiliar with this topic; to provide a basic review for students who have previously studied endocrinology; and to serve all readers as a reference resource to aid in understanding the specific material in the following chapters. The major vertebrate endocrine organs involved in behavior and their various hormone products are described here; consequently, this chapter is rather densely packed with information. However, the individual hormones, their sources, and their physiological actions are also listed in tables in Appendix I for easy reference and review. Hormones that have well-established effects on behavior and that will be featured prominently throughout this book include the steroid hormones (androgens, estrogens, progestins, and glucocorticoids) and peptide hormones such as prolactin, the gonadotropins (luteinizing hormone and follicle-stimulating hormone), oxytocin, vasopressin, and hypothalamic releasing hormones (gonadotropin-releasing hormone and corticotropin-releasing hormone).

When you have finished reading this chapter you should be able to answer the following general questions: Where do hormones come from? Where do hormones go? What do hormones do? The bulk of this chapter is devoted to providing specific answers to these three questions, but the general answers to these questions are straightforward: Hormones are produced by glands and secreted into the blood. They travel in the blood to target tissues containing specific receptors for the hormones. By interacting with their receptors, hormones initiate biochemical events that may directly or indirectly activate genes to induce certain biological responses.*

In the first section of this chapter, the general principles of endocrine and neuroendocrine physiology are reviewed. Next, a description of the major endocrine and neuroendocrine glands and their location in humans is presented. In this context, the hormones associated with each gland will be briefly mentioned. Then hormone signal transduction, transcription, and translation is presented. Detailed descriptions of the hormones, as well as their physiological effects, are presented separately in the following section. The final sections discuss the ways that hormones are regulated and how they evolved.

Chemical Communication

The endocrine system is only one example of chemical communication, which is a ubiquitous feature of life on earth. All levels of biological organization make use of some form of chemical messages, ranging from those used to mediate intracellular processes to the substances used in communication between organs, individuals, and even populations. In all these cases, information is communicated by the release of chemical agents and their detection by receptor activation.

This chemical mediation is conducted by several interacting physiological systems. A key component of these interacting systems is the endocrine system; however, there are other systems of chemical mediation (Figure 2.1). For example, chemical mediation of intracellular events is called **intracrine** mediation. Some intracrine mediators may have changed their function over the course of evolution and now serve as hormones or pheromones (Bern, 1990). **Autocrine** cells secrete products that may feed back to affect processes in the cells that originally produced them. For example, steroid hormone-producing cells possess receptors for their own secreted products (O'Malley, 1995). Chemical mediators released by one cell that induce a biological response in an adjacent cell are called **paracrine** agents; neurons are well-known paracrine cells. Another example of paracrine processes is found during embryonic development, when cells may release chemicals that induce or influence tissue differentiation in neighboring cells. **Ectocrine** substances are released to the outside of an individual and induce a biological response in another animal; pheromones are examples of ectocrine agents.

*In some cases, hormone–receptor interactions result in nongenomic effects on cellular function. Because they are relatively fast, the so-called nongenomic effects of hormones are being intensively studied in their role in mediating behavior, although "nongenomic" is likely a misnomer as genes are certainly involved.

Intracrine mediation

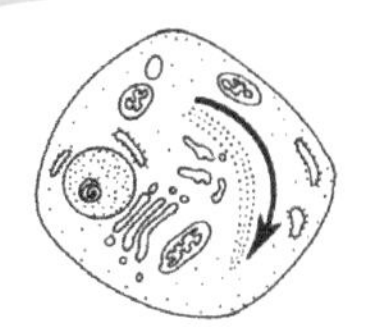

Intracrine substances regulate intracellular events.

Autocrine mediation

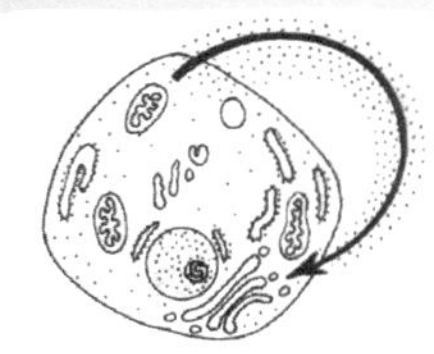

Autocrine substances feed back " to influence the same cells that secreted them.

Paracrine mediation

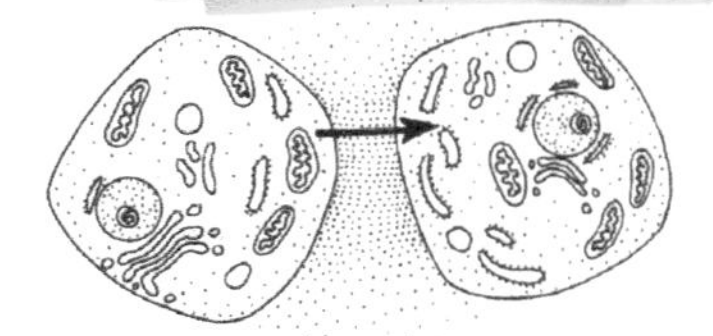

Paracrine cells secrete chemicals that affect adjacent cells.

Endocrine mediation

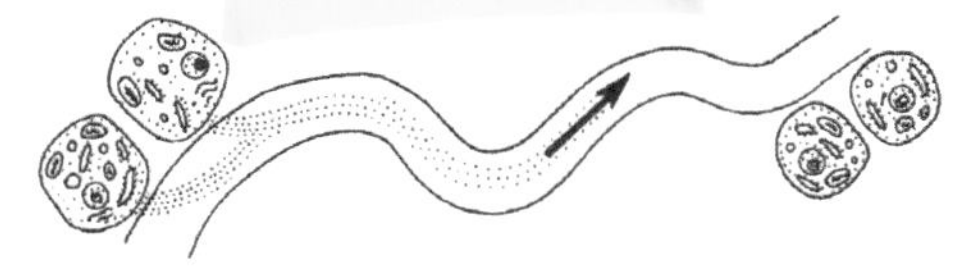

Endocrine cells secrete chemicals into the bloodstream where they may travel to distant target cells.

Ectocrine mediation

Ectocrine substances, such as pheromones, are released into the environment by individuals to communicate with others.

2.1 Systems of chemical mediation and communication. Chemical mediation occurs at all levels of biological organization, ranging from the intracrine mediation of intracellular processes to ectocrine communication, which takes place between individual animals or populations of animals.

TABLE 2.1 **Chemical communication terminology**

Chemical messenger:	Any substance that is produced by a cell that affects the function of another cell.
Cytokine:	A chemical messenger that evokes the proliferation of other cells, especially in the immune system.
Hormone:	A chemical messenger that is released into the bloodstream or tissue fluid system that affects the function of target cells some distance from the source.
Neurohormone:	A hormone produced by a neuron.
Neuropeptide:	A peptide hormone produced by a neuron.
Neurosteroid:	A steroid hormone produced by a neuron.
Neuromodulator:	A hormone that changes (modulates) the response of a neuron to some other factor.
Neurotransmitter:	A chemical messenger that acts across the neural synapse.

Source: Adapted from Hadley, 1996.

Many of these chemical communication systems have been studied independently by separate groups of scientific specialists. For example, neuroscientists have focused their studies on the chemical communication within the nervous system (via neurotransmitters), endocrinologists have mainly studied the chemical communication processes within the endocrine system (via hormones), and immunologists have focused their attention on the chemical communication within the immune system (via cytokines) (Table 2.1). Although historically these three scientific disciplines have operated mostly independently of one another, it is becoming increasingly obvious that the chemical mediators from the nervous, endocrine, and immune systems interact significantly. For example, many immune system cells have receptors for neurotransmitters, hormones, and cytokines. Similarly, many neurons have receptors for hormones and cytokines, as well as for neurotransmitters. Furthermore, the structural features of receptors and the cellular and molecular mechanisms involved in signal transduction, signal amplification, and gene transcription among the three systems are very similar. Thus, the hormone–behavior interactions that are the main focus of this book will also involve neuron–hormone–behavior interactions and immune system–hormone–behavior interactions. The separate categories of neurotransmitters, hormones, and cytokines reflect divisions within the field of biology, but probably mask the integrative nature of the various chemical messengers. In a few cases, the anatomical locations of their interactions have been identified; however, in most cases, the site(s) or mechanism(s) of integration among the nervous, endocrine, and immune systems remain to be specified.

General Features of the Endocrine System

The word *endocrine* is derived from the Greek root words *endon*, meaning "within," and *krinein*, meaning "to release," whereas the term *hormone* is based on the

Greek word *hormon*, meaning "to excite" or "to set into motion."* **Endocrinology** is the scientific study of the endocrine glands and their associated hormones. As we will see, in some cases, a special type of hormone, called a **neurohormone**, is released into the blood by a neuron, rather than by a gland. **Neuroendocrinology** is the scientific study of this transduction of a neural signal into a hormonal signal, as well as other types of relations between the nervous and endocrine systems, and will be discussed later in this chapter.

The endocrine system has several general features:

1. Endocrine glands are ductless.
2. Endocrine glands have a rich blood supply.
3. Hormones, the products of endocrine glands, are secreted into the bloodstream.
4. Hormones can travel in the blood to virtually every cell in the body, and can thus potentially interact with any cell that has appropriate receptors.
5. Hormone receptors are specific binding sites, embedded in the cell membrane or located elsewhere in the cell, that interact with a particular hormone or class of hormones.

The products of **endocrine glands** are secreted directly into the blood (Figure 2.2A). Some glands in the body, known as **exocrine glands,** have ducts, or tubes into which their products are released. The salivary, sweat, and mammary glands are well-known examples of exocrine glands (Figure 2.2B). Some glands in the body have both endocrine and exocrine structures. The pancreas, for example, contains exocrine cells that secrete digestive juices into the intestines via ducts, while the endocrine compartment of the pancreas secretes hormones, such as insulin, directly into the blood system, where they travel throughout the body to regulate energy utilization and storage. Recently, the definition of an endocrine gland has had to be reconsidered. For example, adipose tissue (fat) produces the hormone leptin, and the stomach produces a hormone called ghrelin.

Some endocrine glands, such as the thyroid gland, are among the most highly vascularized organs in the body. As noted above, hormones are released into the bloodstream, and in many instances, the agents that regulate the endocrine glands are themselves transported to the glands via the blood system. The large supply of blood speeds the transport of hormones to their target sites. Although many endocrine glands are devoid of direct neural connections, in some cases neural control of vascular flow rates can indirectly influence endocrine activity.

Recall from Chapter 1 that hormones are organic chemical messengers. Some hormones are water-soluble proteins or small peptides that are stored in the endocrine cell in secretory granules, or **vesicles,** each of which contains many hormone molecules suspended in a protein matrix. In response to a specific stimulus for secretion, the secretory vesicle fuses its membrane with the cell membrane, an opening develops, and the hormone molecules diffuse into the extracellular space.

---

*This derivation should not be taken too literally because, as we will see, hormones may have inhibitory effects as well as stimulatory ones.

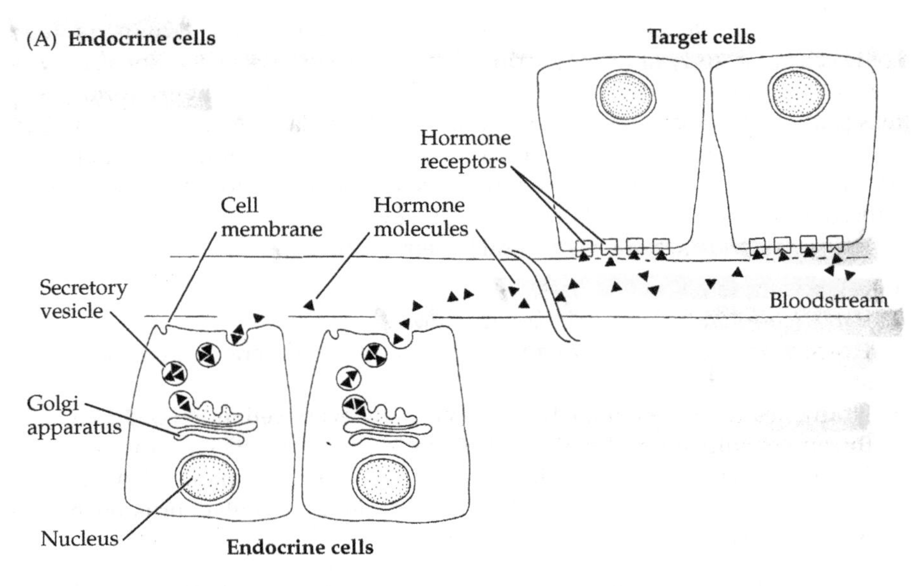

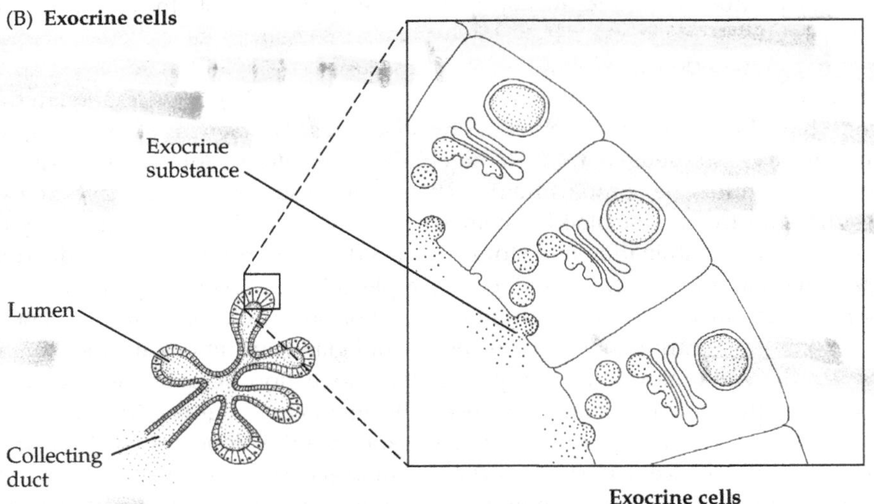

2.2 A comparison of endocrine and exocrine cells. (A) Hormones are secreted by endocrine cells into the bloodstream, where they may travel to distant target organs. (B) In contrast, exocrine cells, such as those found in the salivary, sweat, and mammary glands (and in parts of the pancreas), secrete their products into ducts that carry them to adjacent target organs or the external environment.

This process of extrusion is called **exocytosis**. The expelled hormone molecules then enter the blood system from the extracellular space (see Figure 2.2A).

Other hormones, such as steroid hormones, are lipid-soluble (i.e., fat-soluble), and because they can move easily through the cell's phospholipid membrane, they are not stored in the endocrine cells. Instead, a signal to an endocrine cell to produce steroid hormones also serves as a signal to release them into the blood as soon as they are produced by the cellular machinery.

All cells in the body, except those in the lenses of the eyes, have a direct blood supply. Thus, the blood system provides direct access to virtually every cell, and hormones can potentially interact with any cell that has appropriate receptors. Protein and peptide hormones are soluble in blood, an aqueous (watery) solution. In contrast, steroids are not very water-soluble and often form a reversible bond with a **carrier protein** while circulating in the blood. These carrier molecules have recently been discovered to serve important regulatory roles in the mediation of steroid hormone actions, although this function remains somewhat controversial.

Hormone receptors, which are either embedded in the target cell membrane or located elsewhere within the cell, interact with a particular hormone or class of hormones. The receptor is analogous to a lock, and the hormone acts as a key to the lock. Receptor proteins bind to hormones with high affinity and generally high specificity. As a result of the high affinity of hormone receptors, hormones can be very potent in their effects, despite the fact that they are found in very dilute concentrations in the blood (some as low as one-millionth of a milligram [1 picogram] per milliliter of blood plasma). However, when the blood concentration of a hormone is high, binding with receptors that are specific for other, related hormones can occur in sufficient numbers to cause a biological response (i.e., *cross-reaction*).

When sufficient receptors are not available due to a clinical condition, or because previous high concentrations of a hormone have occupied all the receptors and new ones have yet to be made, a biological response may not be sustained (see below). Such a reduction in numbers of receptors may lead to a so-called endocrine deficiency despite normal or even supernormal levels of circulating hormones. For example, a deficiency of testosterone receptors can prevent the development of male traits despite normal circulating testosterone concentrations. Conversely, elevated receptor numbers may produce clinical manifestations of endocrine excess despite a normal blood concentration of a hormone. Thus, in order to understand hormone–behavior interactions, it is sometimes necessary to characterize target tissue sensitivity (that is, the number and type of receptors possessed by the tissue in question) in addition to measuring hormone concentrations. The hormonal signal must also be terminated for proper endocrine regulation. There are several mechanisms to achieve hormone clearance from the blood. The most common method involves the uptake and degradation (metabolism) of hormones by the kidney or liver. The degraded hormones are then excreted into the urine or bile. Assessing the levels of metabolized hormones in urine or feces is a common, noninvasive way to determine hormone levels in both wild and captive animals.

The Endocrine Glands

Hormones can be divided into four classes: (1) peptides or proteins, (2) steroids, (3) monoamines, and (4) lipid-based hormones. In general, only one class of hormones is produced by a single endocrine gland, but there are some notable exceptions, as we will see below. The locations of the major endocrine organs in humans are depicted in Figure 2.3. Because the modern anatomical names for the

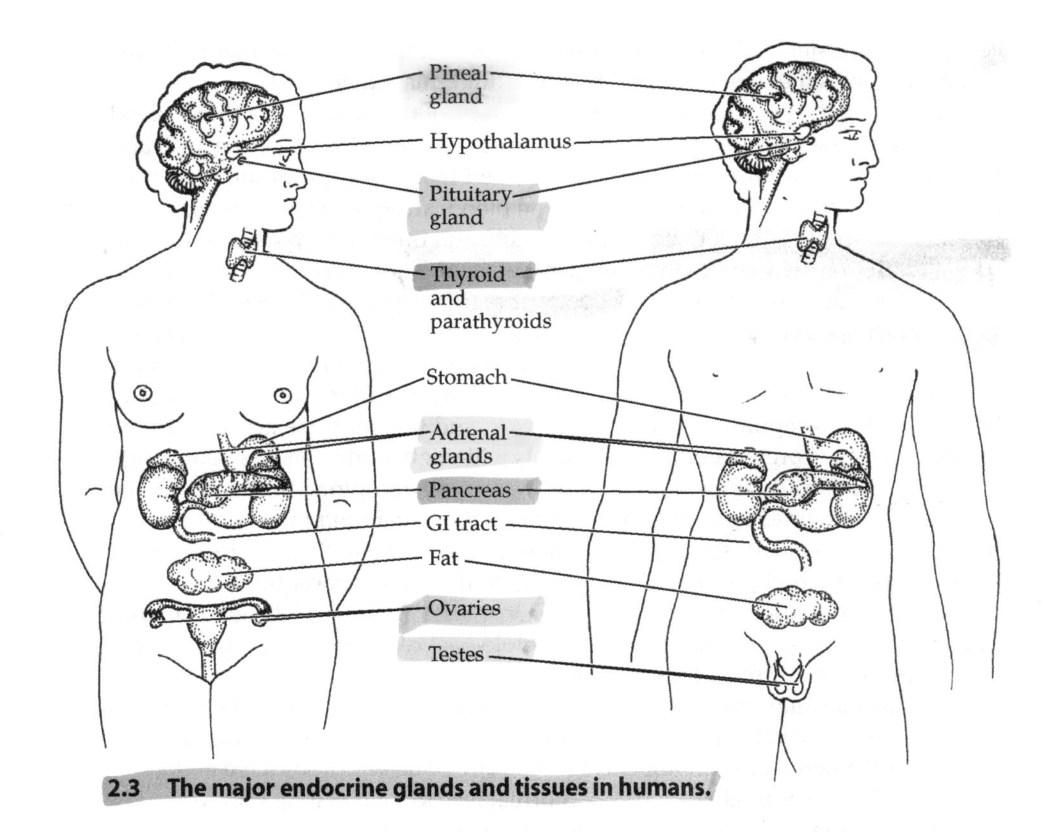

2.3 The major endocrine glands and tissues in humans.

endocrine glands typically originated from Greek or Latin terms that early anatomists used to describe the shape, location, or, occasionally, the function of a structure, the derivations for many of these names are provided as an aid to remembering and understanding the terms.

The **hypothalamus** is located directly beneath the thalamus (*hypo*, "below") at the base of the brain. The **pituitary** is located at the base of the skull in a bony depression called the sella tursica (*sella*, "saddle"; *tursica*, "Turkish"), and is connected to the base of the hypothalamus near the median eminence by a funnel-shaped stalk called the *infundibulum* ("funnel"), or pituitary stalk. The **thyroid gland** is an H-shaped organ located on the upper trachea. The **pancreas** resides within the curve of the duodenum (small intestine), behind the stomach and liver. The endocrine tissues of the gastrointestinal tract are somewhat primitively organized; small clumps of endocrine cells are scattered throughout the gut, rather than concentrated into a glandular organ. The **adrenal glands** are bilateral organs that are situated on top of the kidneys (*ad*, "toward"; *renes*, "kidneys"). The **pineal gland**, also called the *epiphysis* (*epi*, "over"; *physis*, "brain"), is located in the brain in mammals, between the telencephalon and diencephalon. The **gonads** are the reproductive organs. In sexually reproducing animals, there are two types of gonads: as noted above, males have **testes**, often located in a sac outside the abdomen called the scrotum; females have **ovaries**, located in the abdomen. A temporary endocrine organ, the **placenta**, forms in the uterus of female mammals during pregnancy.

This section describes the structure of each of these glands in detail, and, in some cases, provides brief descriptions of their endocrine products (detailed descriptions of the hormones are presented in the following section). A brief description of invertebrate endocrinology is provided in Box 2.1.

The Hypothalamus

The hypothalamus is one of the primary interfaces between the nervous system and the endocrine system. It is made up of several collections of neuronal cell bodies, or **nuclei**, at the base of the brain (Figure 2.4). The hypothalamus is a part of the brain, and it receives information via axons that form tracts, or *projections*,

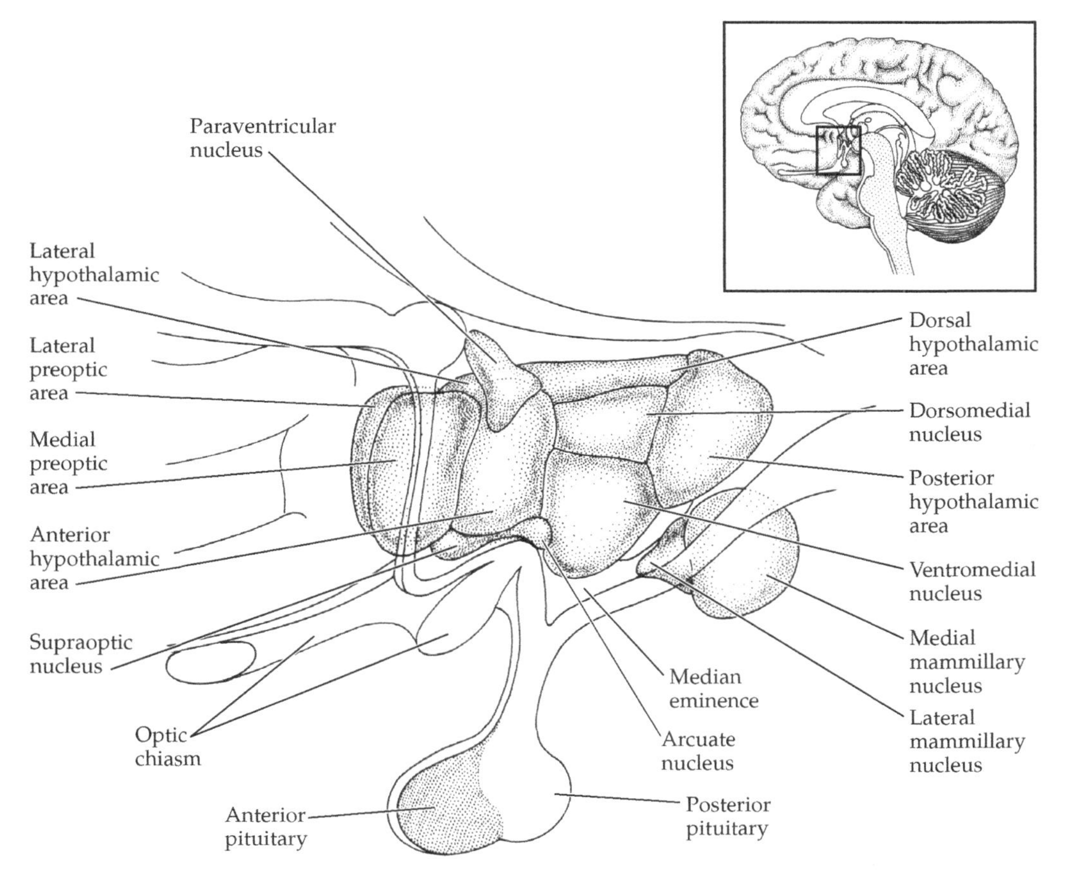

2.4 The hypothalamus is located at the base of the brain (inset), and consists of several collections of neuronal cell bodies called nuclei. The hypothalamic nuclei integrate neural information from higher brain sites and coordinate numerous physiological processes by means of specialized neurosecretory cells that produce neurohormones and secrete them into the pituitary gland via the hypothalamic–hypophyseal portal system. This schematic rendering of the hypothalamus is a composite based on anatomical studies of rodents and humans.

BOX 2.1 Invertebrate Endocrinology

More than 90% of the animal species on this planet are invertebrates. Although this book focuses on vertebrate animals, studies of invertebrates have been extremely informative to those who wish to understand the basic principles of hormone–behavior interactions. Furthermore, establishing the extent to which invertebrates and vertebrates have similar hormone–behavior interactions has provided insight into the evolution of these interactions.

Although the behavioral repertoires of invertebrates are remarkably complex and diverse, the organization of their nervous systems is relatively simple: the total number of neurons in invertebrate brains may number in the hundreds, compared with tens of billions in vertebrate brains. Many invertebrate species, such as the coelenterates (jellyfishes) and tunicates (sea squirts), have rudimentary nervous systems consisting of diffuse neural nets. Neuropeptides secreted from these neurons can affect the firing rate of neighboring neurons or diffuse throughout the tissue fluid system to affect distant non-neural tissues. In more complex invertebrates, the neurons are often organized into discrete groupings called ganglia.

A circulatory system is present among complex invertebrates such as arthropods (insects and crustaceans) and mollusks (snails); these animals also possess distinct endocrine and neuroendocrine systems. Although all arthropods have a blood–brain barrier, many other invertebrates do not, so that any neuropeptide secreted into the blood can affect any neuron (providing, of course, that the neuron in question possesses appropriate receptors).

The functions of invertebrate hormones are similar to those among vertebrate animals; that is, hormones modify body structures as well

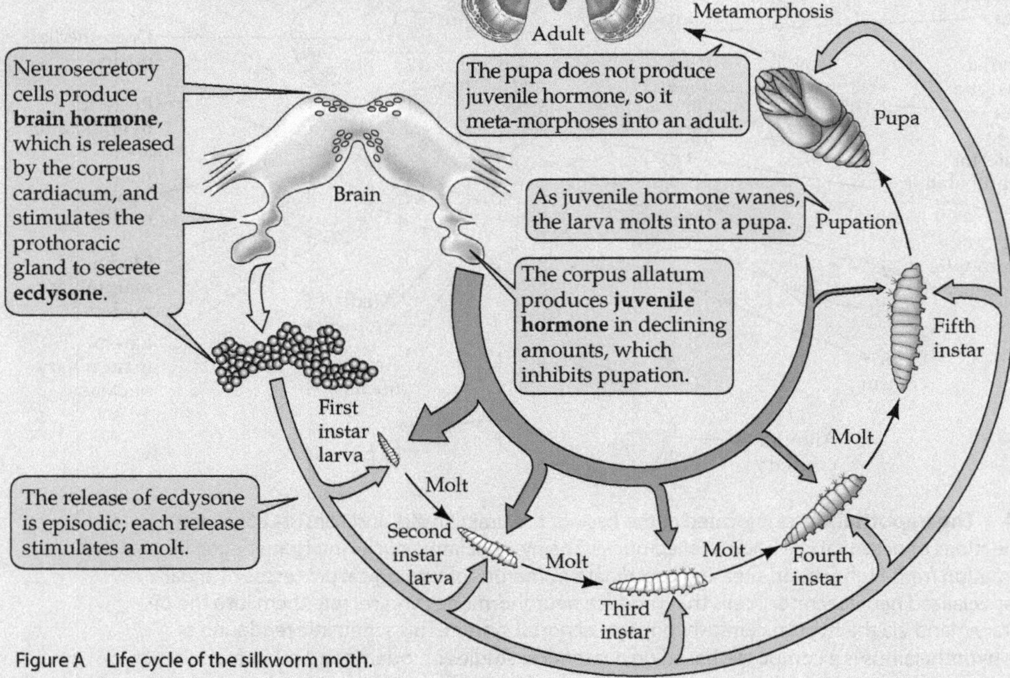

Figure A Life cycle of the silkworm moth.

as physiological and behavioral functions. Invertebrate behaviors, including eclosion (hatching), courtship, mating, parental care, ecdysis (shedding of the exoskeleton at the end of a molt), oviposition (egg laying), migration, and aggression may be affected by hormones.

The main (neuro) endocrine glands in complex invertebrates (i.e., arthropods and mollusks) include the corpora cardiaca, corpora allata, protodeal nerve, and prothoracic gland. These glands secrete steroid, peptide, monoamine, and lipid-based hormones. In the silkworm moth (*Hyalophora cecropia*), one hormone, ecdysone, triggers molting, while juvenile hormone acts to keep insects in the larval stage. As the concentration of juvenile hormone decreases over time, larvae molt into pupae (see Figure A). Pupae do not produce juvenile hormone; thus, they molt into adults.

Hormones that influence behavior tend to fall into one of four broad functional categories among invertebrates: (1) modifiers, (2) triggers, (3) primers, and (4) organizers (Truman, 1992). As in vertebrate animals, hormones function most commonly as modifiers of invertebrate behavior: in most cases, they do not "cause" behavior, but adjust the responsiveness of the nervous system so that the threshold of stimuli that elicit a particular behavior is changed. For example, during cricket courtship, females become receptive upon hearing the calling of male conspecifics. One behavioral indication that a female is receptive is that she moves toward a male cricket (or an audio speaker) that is emitting calls; this behavior is called positive phonotaxis. If the female is not mature, she ignores the calling or requires a much more intense stimulus before responding. Juvenile hormone, which is released from the corpora allata, causes maturation of the phonotaxis response. When the corpora allata are removed from immature female crickets, they never respond to males. Injection of juvenile hormone into adult female crickets that have had their corpora allata removed results in normal phonotaxis.

In some cases, hormones may directly trigger specific behaviors in invertebrate animals. One of the best-documented cases is the ecdysis, or shedding, behavior of the hawkmoth (*Manduca sexta*) (Ewer et al., 1997). Because arthropods outgrow their exoskeletons, they must shed them at the end of each molt. Ecdysis requires a series of coordinated, stereotyped muscular movements during which the individual literally crawls out of its outer shell. Until recently, this behavior was thought to be triggered by a peptide called eclosion hormone (EH). (Eclosion is the technical term for either the hatching of a larva from its egg or an adult arthropod leaving its pupal case.) However, recent work suggests that EH is not the final activator of ecdysis behavior, but rather is part of a neuropeptide cascade that triggers the behavior. The actual activator of ecdysis is a second peptide, known as crustacean cardioactive peptide; its release is controlled by EH via its up-regulation of the second messenger molecule cGMP (Gammie and Truman, 1997).

As is the case with vertebrates, invertebrate hormones may serve a priming function, preparing the nervous system to respond to subsequent hormonal signals. For example, EH can trigger the neuropeptide cascade that results in ecdysis only during a brief window of time late during each molt, but ecdysteroids, the steroid hormones that initiate molting, must first prime the system so that EH can exert its effects.

Among vertebrate animals, hormones have important effects on brain organization during development, particularly on sexual dimorphism (see Chapter 3). Although many invertebrates do not use hormones to regulate sexual differentiation of the nervous system, hormones do cause brain differentiation of invertebrate species in which there are caste systems. For example, the differential organization of the brains of worker bees and queen bees is mediated by hormones.

As the examples of the silkworm moth, cricket, and hawkmoth illustrate, hormones are important in mediating invertebrate behavior. The simplicity and diversity of invertebrate nervous systems and behaviors make this group of animals ideal for answering many questions about mechanisms underlying hormone–behavior interactions.

from various higher brain sites. This information is consolidated by the nuclei, which carry out many integrative processes, including the control of reproduction and metabolism (Griffin and Ojeda, 1988).

At the base of the hypothalamus and in the median eminence, there are modified neurons specialized for the release of chemical messengers. Although these **neurosecretory cells** function primarily as endocrine glands, they are morphologically similar to conventional neurons, having dendrites, axons, Golgi apparatuses, and neurotubules (for a review of basic features of the nervous system, see Rosenzweig et al., 2001). Their chemical messengers, called neurohormones, are released from the neuronal axon terminals in response to neuronal impulses in a manner analogous to neurotransmitters, but rather than being released into a synaptic space, they are released into blood vessels in the pituitary gland. This chemical communication system between the hypothalamus and the pituitary is one of the areas where the boundaries between the endocrine and nervous systems are blurred.

The Pituitary Gland

The word *pituitary* is derived from the Latin word for "mucous"; early anatomists erroneously believed that this gland collected waste products from the brain and excreted them via the nose. The pituitary gland was once considered the "master gland" because it mediated so many physiological processes. While it is true that the pituitary orchestrates many processes, we now know that the pituitary itself is one of the most regulated glands in the endocrine system.

The mammalian pituitary, also called the *hypophysis* (from the Greek *hypo*, "below"; *physis*, "brain") is really two distinct glands fused into one (Kannan, 1987) (Figure 2.5). The two parts have very different embryological origins. The front part, the **anterior pituitary** (also called the *pars anterior* or *adenohypophysis*; from the Greek *aden*, "gland"), develops from an embryonic structure called Rathke's pouch, which pinches off from the roof of the mouth and migrates to the final loca-

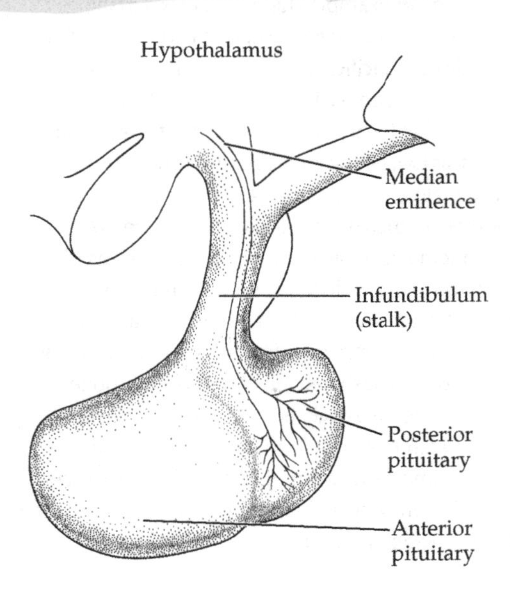

2.5 The pituitary gland has two distinct anatomical components in humans: the anterior pituitary and the posterior pituitary. The anterior and posterior pituitary have different embryological origins and different functional roles in the endocrine system.

tion of the gland. There the anterior pituitary joins the back part, or **posterior pituitary** (also called the *pars nervosa* or *neurohypophysis*), which is an outgrowth from the base of the brain. Thus, the anterior pituitary is derived from the soft tissues of the upper palate, and the posterior pituitary has a neural origin in the base of the brain. The area where the two parts of the pituitary join has a distinct anatomical organization in mammals, and is called the *pars intermedia*. Birds lack a distinct pars intermedia, and in some primitive vertebrates (for example, lampreys and hagfishes), the two parts of the pituitary remain anatomically separate.

The hypothalamus communicates with the pituitary by two methods. Neurohormones from the hypothalamus reach the anterior pituitary via the hypothalamic–hypophyseal **portal system**, a special closed blood circuit in which two beds of capillaries, one in the hypothalamus (the primary plexus) and one in the anterior pituitary (the secondary plexus), are connected by a vein that extends down the infundibulum (Figure 2.6A). The portal system ensures that blood flows primarily in one direction (although there can be some backflow), from the hypothalamus to the anterior pituitary, and also ensures that hormonal signals from the hypothalamus will be read by the pituitary rather than diluted in the general blood circulation. Evidence for hypothalamic control of the pituitary via a portal system has been documented in all but the most primitive vertebrates.

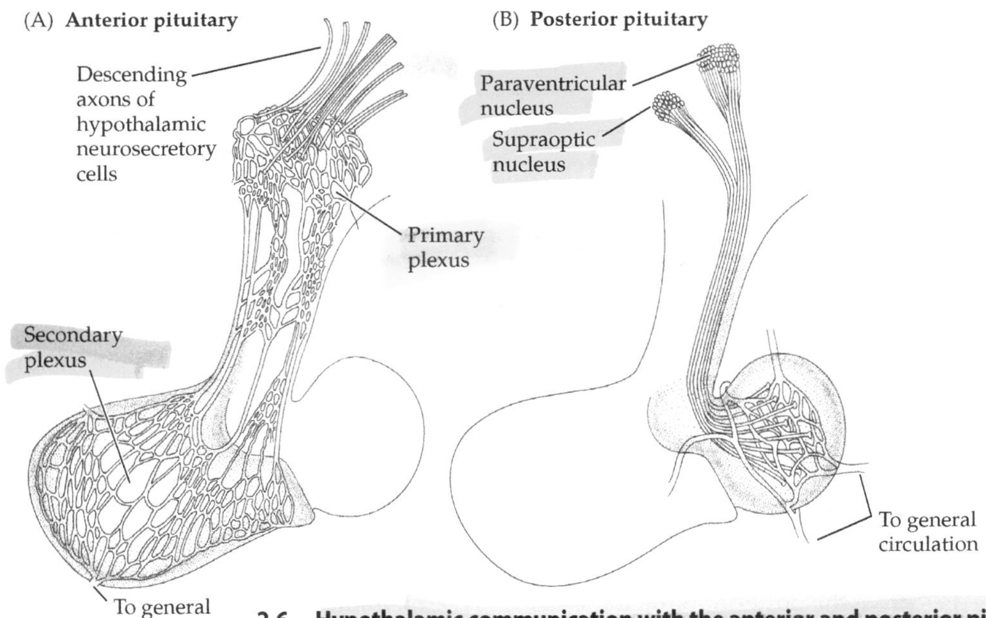

2.6 Hypothalamic communication with the anterior and posterior pituitary. (A) Hypothalamic neurosecretory cells release hormones into the primary plexus of the hypothalamic–hypophyseal portal system, a special closed blood system that connects the hypothalamus and anterior pituitary. (B) Axons from hypothalamic neurosecretory cells that produce the neurohormones oxytocin and vasopressin extend down the infundibulum and synapse on blood vessels in the posterior pituitary.

A different set of hypothalamic neurosecretory cells directly innervates the posterior pituitary. Neurohormones are secreted directly into this structure, where they enter blood vessels and the general circulation (Figure 2.6B).

THE ANTERIOR PITUITARY The hypothalamic neurohormones secreted into the portal system are called by a variety of names, including releasing factors, releasing hormones, inhibitory factors, and inhibitory hormones. These small peptides act on the anterior pituitary to stimulate or inhibit the release of the anterior pituitary's own hormones. However, for the sake of simplicity, we will refer to these hormones as **releasing hormones**.

The release of hormones from the anterior pituitary is a two-step process. When hypothalamic releasing hormones are secreted into the primary plexus of the portal system, they travel to the secondary plexus and diffuse into the anterior pituitary. In response, the anterior pituitary releases appropriate hormones of its own. These pituitary hormones are also secreted into the portal system, but due to the "one-way" directional blood flow of the system, they join the general circulation, rather than traveling to the hypothalamus. The anterior pituitary hormones are called **tropic hormones** (from the Greek *trophe*, "nourishment") because they stimulate various physiological processes, either by acting directly on target tissues or by causing other endocrine glands to release hormones.

The anterior pituitary contains three distinct types of cells, each of which produces a different group of tropic hormones, as we will see below. Growth hormone, which acts on muscle and bone cells to increase their size, is an example of an anterior pituitary tropic hormone. The anterior pituitary itself receives little neural input; however, the blood vessels in the anterior pituitary are innervated, and neural signals can increase or decrease the diameter of these vessels to alter blood flow rates and effectively alter hormone distribution.

THE POSTERIOR PITUITARY The posterior pituitary serves as a sort of reservoir for two neurohormones, oxytocin and vasopressin, which are actually manufactured in the cell bodies of the magnocellular neurons of the supraoptic and paraventricular nuclei of the hypothalamus. Axons from these cells, which function both as typical neurons that conduct neural impulses and as endocrine cells, extend down the infundibulum and terminate in the posterior pituitary. Oxytocin and vasopressin made and packaged in the Golgi apparatuses of the cell bodies are transported down the axons and stored in vesicles at the axon terminals in the posterior pituitary, from which they are released in response to a neural impulse. When the cell membrane is depolarized by a neural impulse, the hormones are released from the terminals by exocytosis and enter the bloodstream. Thus, in contrast to the two-step process in the anterior pituitary, hormones can be released from the posterior pituitary as fast as a neural impulse is conducted.

The Thyroid Gland

The thyroid gland (Figure 2.7A) consists of many sphere-shaped, colloid-filled structures called **follicles** (*follicle*, "sac") (Figure 2.7B), which produce thyroid hormones in response to an anterior pituitary tropic hormone. The thyroid is

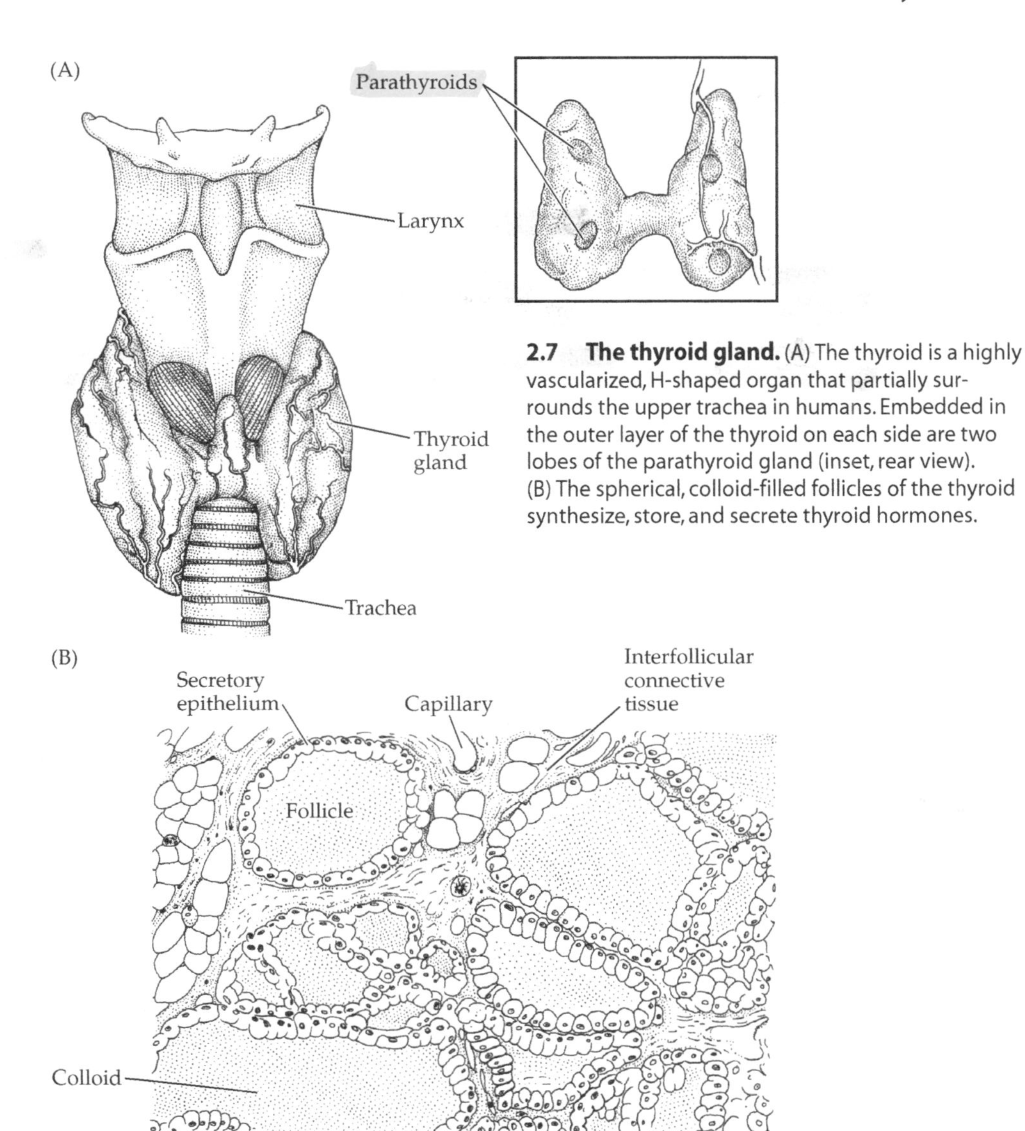

2.7 The thyroid gland. (A) The thyroid is a highly vascularized, H-shaped organ that partially surrounds the upper trachea in humans. Embedded in the outer layer of the thyroid on each side are two lobes of the parathyroid gland (inset, rear view). (B) The spherical, colloid-filled follicles of the thyroid synthesize, store, and secrete thyroid hormones.

unusual among vertebrate endocrine glands because it stores large quantities of these hormones; humans store sufficient thyroid hormone reserves for approximately 90 days. A functional explanation for this storage ability is that thyroid hormones contain iodine, which is uncommon in many human diets.

Embedded in the thyroid are other endocrine cells called **C-cells**, which secrete a protein hormone called calcitonin that is involved in calcium metabolism. Also situated in the thyroid is the **parathyroid gland** (see Figure 2.7A, inset), which usually has several parts; in humans, two lobes are present in the outer layer of the thyroid on each side. The parathyroid secretes a protein hormone, parathyroid hormone, which is also important in calcium regulation. However, the C-cells and parathyroid gland probably affect behavior only indirectly.

The Pancreas

As stated previously, the pancreas (Figure 2.8A) functions as both an endocrine and an exocrine gland. Most of the pancreas consists of exocrine cells that produce and secrete digestive juices into the intestines (Go et al., 1993), but nested throughout the exocrine tissue are islands of endocrine tissue called **islets of Langerhans** (Figure 2.8B). Within these endocrine islands are four cell types, α-cells, β-cells, δ-cells, and a few polypeptide-secreting cells, each of which secretes a different protein hormone (Figure 2.8C). For example, the β-cells secrete insulin, a hormone that is necessary for cells to utilize and store metabolic fuel.

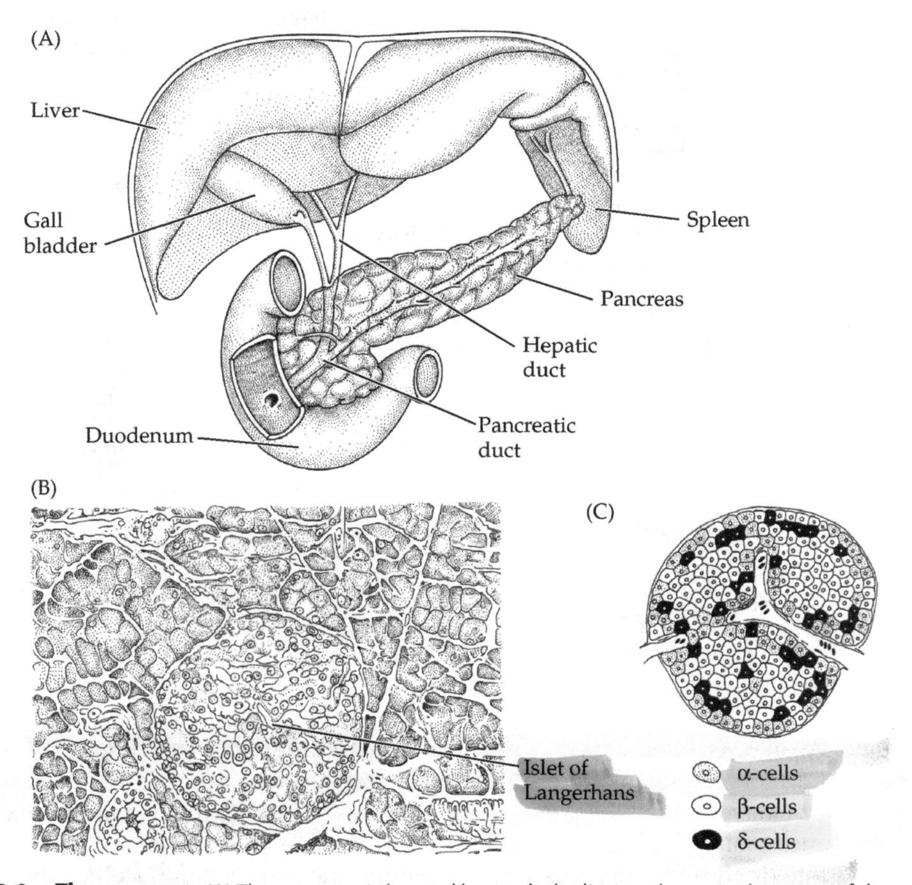

2.8 The pancreas. (A) The pancreas is located beneath the liver and rests in the curve of the duodenum, or small intestine. (B) While most of the pancreas consists of exocrine cells that secrete digestive fluids, islands of endocrine tissue called islets of Langerhans are also present. (C) The relative distribution of α-, β-, and δ-cells in a pancreatic islet; each cell type secretes a different protein hormone.

In many animals, the islets of Langerhans are innervated. In mammals, these cells are innervated by the vagus nerve, which stimulates insulin secretion when glucose concentrations in the blood increase, as they do after a meal. This mechanism allows an animal to anticipate the arrival of food in the intestine with a rapid secretion of insulin and thus promotes efficient movement of energy from the blood into the cells.

The Gastrointestinal Tract

As mentioned above, there are endocrine cells scattered throughout the gastrointestinal tract, in what is considered to be a primitive organization. The gastrointestinal hormones regulate the cells and organs in which they are produced. Such intracrine or autocrine chemical mediation is usually considered a more primitive mechanism than endocrine mediation (O'Malley, 1989). A relatively new hormone, ghrelin, has been discovered to be secreted from the stomach into the bloodstream, through which it travels to receptors in the brain to stimulate food intake and growth hormone secretion.

There are three major gastrointestinal hormones—gastrin, secretin, and cholecystokinin—all of which are protein hormones that control various aspects of digestion. In addition to their gastrointestinal roles, many of these so-called gut hormones are also found in the brain and appear to have diverse and marked behavioral effects.

The Adrenal Glands

The adrenal glands are located on top of the kidneys (Figure 2.9A). Like the pituitary gland, each adrenal gland in mammals is actually two distinct organs (James, 1992). An outer gland, the **adrenal cortex** (*cortex*, "bark"), surrounds an inner gland, the **adrenal medulla** (*medulla*, "marrow" or "innermost part") (Figure 2.9B). In many nonmammalian species, these two parts of the adrenal gland are separate.

THE ADRENAL MEDULLA The adrenal medulla is made up of **chromaffin cells**, so called because they have a high affinity for colored stains. These cells are derived from primitive neural tissue, specifically postganglionic sympathetic neurons, during embryonic development; after birth, they function as part of the autonomic nervous system. In response to neural signals, the adrenal medulla releases three monoamine hormones—epinephrine, norepinephrine, and dopamine—into the general circulation. A class of protein hormones, the enkephalins, is also released from the adrenal medulla.

THE ADRENAL CORTEX The adrenal cortex is composed of three distinct zones in mammals (Figure 2.9B; Ganong, 2003). The outer region, the **zona glomerulosa** (*glomeruli*, "small balls"), represents 10%–15% of the adrenal cortex and is characterized by whorls of epithelial cells. The middle zone, the **zona fasciculata**, is the largest zone of the adrenal cortex and makes up 75%–80% of this gland. The epithelial cells of the zona fasciculata are arranged as orderly bands (*fascicles*, "small bundles"). The **zona reticularis** is the innermost zone of the adrenal cor-

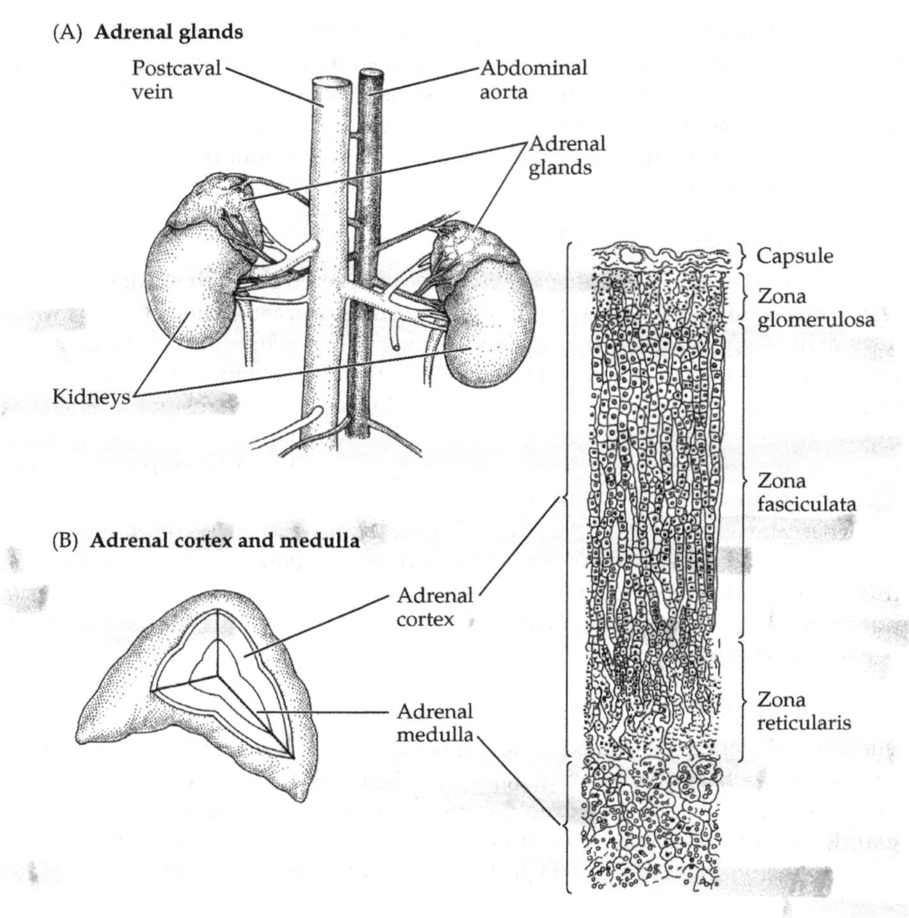

2.9 The adrenal glands. (A) The adrenal glands are located atop the kidneys. (B) Each gland consists of two distinct regions, the cortex and the medulla. The adrenal cortex also has distinct cellular zones with different functional roles: the zona glomerulosa, marked by whorls of epithelial cells; the zona fasciculata, in which the epithelial cells are organized in orderly bands; and the zona reticularis, where epithelial cells have a disorderly appearance.

tex. Here the epithelial cells are arranged in a somewhat disorderly fashion (*reticulum*, "net" or "network").

Human fetuses have very large adrenal glands that possess a fourth layer not found in adults. During fetal life, the zona glomerulosa, zona fasciculata, and zona reticularis account for only 20% of the adrenal cortex, while the other 80% of the cortex consists of the so-called *fetal zone*. The fetal zone undergoes rapid regression at birth, and its function remains unknown. There is no fetal zone per se in nonhuman animals, but an analogous layer of cortical cells appears perinatally in a number of laboratory mammals and has been named the *X zone*.

These anatomical divisions of the adrenal cortex represent functional divisions of hormone production. Although all three zones of the cortex produce steroid hormones, different types are produced in each zone. For example, aldosterone, a steroid hormone that regulates sodium levels in the blood, is produced only in the zona glomerulosa. Furthermore, each zone differs in the way its steroid production is controlled. For instance, the secretion of aldosterone from the zona glomerulosa is regulated by blood concentrations of sodium and via a blood plasma α_2-globulin-derived hormone called angiotensin II (see Chapter 8), whereas the release of steroid hormones from the zona fasciculata and zona reticularis is regulated by tropic hormones from the anterior pituitary.

The Pineal Gland

The pineal gland (Figure 2.10A) is unique among endocrine organs in the extent of its evolution, both structurally and functionally, among the vertebrates (Reiter, 1982). In all mammals examined, individual pineal cells (**pinealocytes**) function exclusively as secretory structures (Figure 2.10B; Hansen and Karasek, 1982). In nonavian and nonmammalian vertebrates, the pineal gland functions primarily as a photoreceptor organ and is often referred to as the "third eye" (Figure 2.10C; Eakin, 1973). The pineal functions secondarily as a neurosecretory organ in these animals. In birds and some reptiles, the pineal cells possess rudimentary photoreceptive structures and also function as secretory cells. The avian pineal may also serve as an important biological clock.

The primary endocrine product of the pineal gland is melatonin, which is formed from serotonin. A number of small peptide hormones have also been found in the pineal, but thus far no functional role for these substances has been documented.

The Gonads

The gonads have two functions, which are usually separately compartmentalized: the production of gametes (sperm or eggs) and the production of hormones (Knobil and Neill, 1994). The hormones produced by the gonads, primarily steroid hormones, are required for gamete development and development of the secondary sex characters. These hormones also mediate the behaviors necessary to bring sperm and eggs together. The functions of the gonads are regulated by tropic hormones from the anterior pituitary, known as gonadotropins.

THE TESTES The testes are bilateral glands, located in most mammals in an external sac called the **scrotum**, and in most other vertebrates in the abdomen. Several cell types exist in the testes. When looking through a microscope at a thin cross section slice of a testis, you can identify the **seminiferous tubules**, long, convoluted tubes in which sperm cells undergo various stages of maturation, or spermatogenesis (Figure 2.11A). The heads of nearly mature sperm are embedded in specialized cells, called **Sertoli cells**, located along the basement membrane of the tubules; these cells provide nourishment to the developing sperm and also produce a peptide hormone, inhibin, that is important in regulating one of the

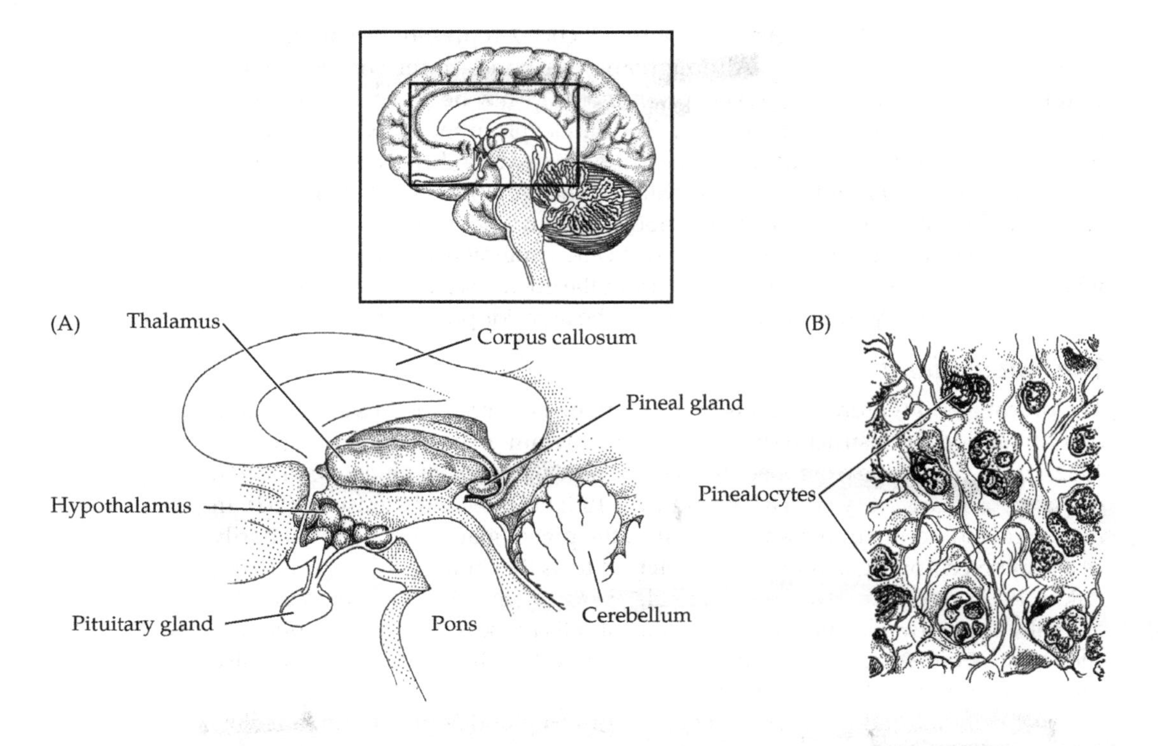

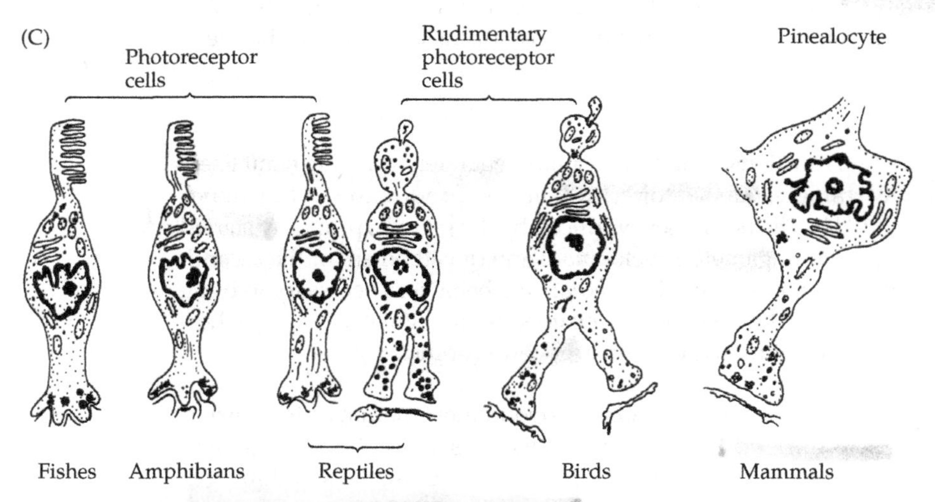

2.10 The pineal gland. (A) The pineal gland is located within the brain, between the telencephalon and diencephalon. (B) The secretory cells of the mammalian pineal, known as pinealocytes, are nested in neural tissue, produce both serotonin and melatonin, and secrete melatonin. (C) Over the course of vertebrate evolution, the primary function of pineal cells has shifted from photoreception to neurosecretion.

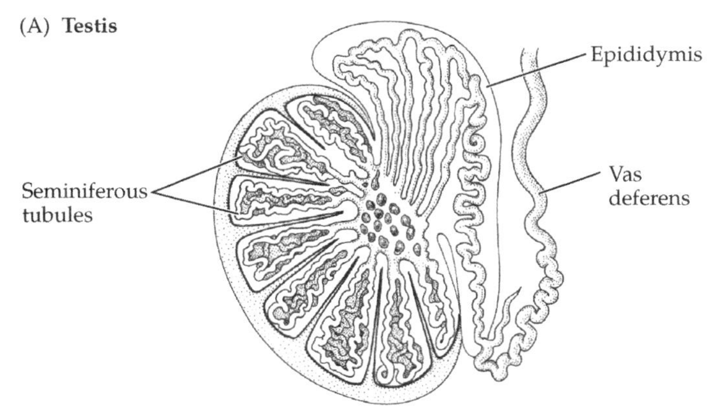

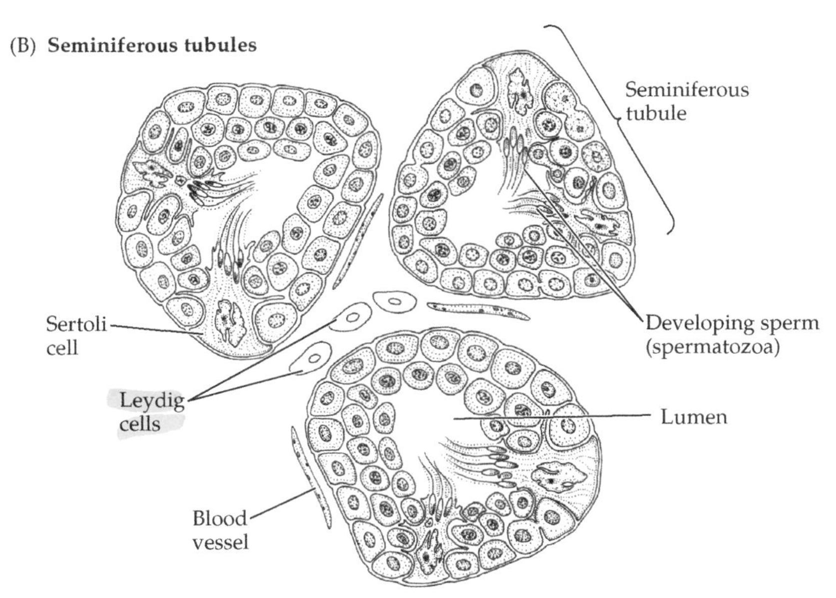

2.11 The testes. (A) A testis in cross section, showing the seminiferous tubules. (B) Seminiferous tubules in cross section, showing the heads of developing sperm (spermatozoa) embedded in Sertoli cells and the steroid-secreting Leydig cells lying between the tubules.

gonadotropins from the anterior pituitary. The primary hormone-producing cells in the testes are interspersed among the tubules and are called the **Leydig cells** or *interstitial cells* (*inter* + *stit*, "standing between") (Figure 2.11B). The Leydig cells produce steroid hormones—primarily androgens—under the influence of gonadotropins from the anterior pituitary.

THE OVARIES In mammals, the ovaries are bilateral glands located in the dorsal part of the abdominal cavity, normally below the kidneys (Figure 2.12A). Among many species of birds, only the left ovary is developed (Serra, 1983). Like the testes, the ovaries produce both gametes and hormones, and the two functions are compartmentalized. However, active ovaries exhibit cyclic changes in both

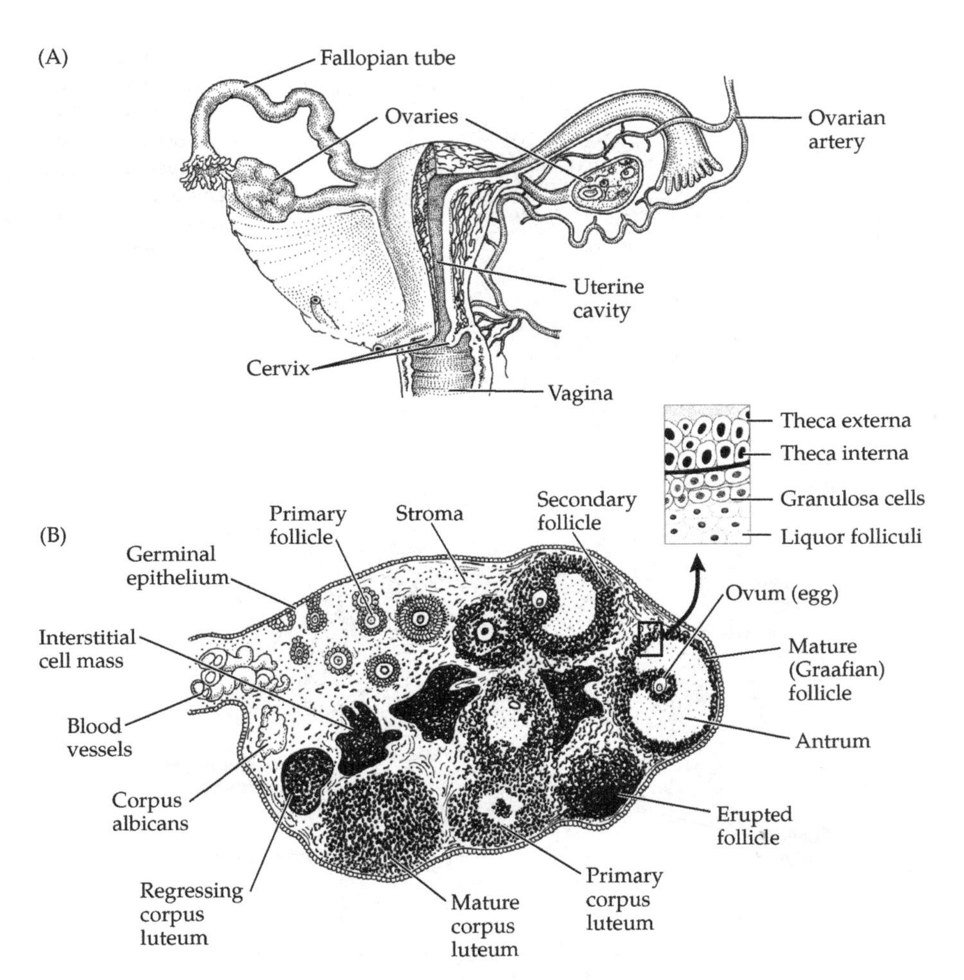

2.12 The ovaries. (A) The ovaries are almond-shaped, bilateral glands about 4–6 cm long in humans, located in the abdomen of females. (B) A schematic diagram of one magnified mammalian ovary reveals several important anatomical features and the various stages of follicular and corpora luteal development. The inset shows different types of ovarian tissue as they are organized in the mature (Graafian) follicle, including the theca externa, theca interna, and granulosa cells. Prior to ovulation, the theca interna cells and the follicular cells collaborate to synthesize estrogens. After ovulation, the cells constituting the follicle combine with the theca interna cells to form the corpus luteum, which secretes progestins.

functions, whereas the testes are tonic, or constant, in their sperm-making and secretory activities during the breeding season.

The ovary has three functional subunits: **follicles**, which each contain a developing egg, or **ovum**; **corpora lutea**, structures that develop from follicles after the ovum is released; and supporting tissue, or **stroma** (Figure 2.12B). The functional potential of ovaries can be perceived microscopically early in their development.

Within the fetal ovary are germinal epithelia that eventually develop into primordial follicles. The two ovaries of a human infant may contain half a million primary, or immature, follicles. No additional gametes are formed postnatally, and there is a continual degeneration of follicles throughout life through a process called atresia. Approximately 400 ova are ovulated by an average woman between puberty and menopause.

Each primary follicle consists of an oocyte (immature egg) surrounded by a monolayer of flattened epithelial cells called **granulosa cells**. The granulosa cells also appear to be the source of two peptide hormones, inhibin and activin, that are important in suppressing and enhancing, respectively, hormone secretion from the hypothalamus and pituitary gland. Under the influence of gonadotropins secreted by the anterior pituitary, a certain number of oocytes mature during each ovarian cycle. As an oocyte matures, the follicle tends to move deeper into the stroma, and the granulosa cells increase in number via mitotic division. Surrounding the granulosa cells, but separated from them by a basement membrane, are the **theca interna** and **theca externa** cells. These thecal cells secrete estrogens, the primary class of sex steroid hormones in females.

As the ovum continues to mature, its cell membrane becomes surrounded by an outer layer called the **zona pellucida**, and a multiple layer of epithelial cells surrounds the follicle. At this stage the maturing follicle is called a secondary follicle. A space develops between the ovum and surrounding epithelial cells. This space, called the antrum ("room"), fills with fluid prior to ovulation. As the antrum enlarges, the follicle is called a tertiary follicle. The antrum fluid is called the liquor folliculi (or follicle fluid), and is rich in steroid hormones. Just prior to ovulation, the follicle reaches its maximal size and is called a Graafian follicle. When the ovum is mature, it erupts from the Graafian follicle and travels to the mouth of the oviduct, through which it is transported to the uterus. In most mammals, fertilization normally occurs while the ovum is in the oviduct.

After the ovum is released, both the granulosa cells of the erupted follicle and the surrounding thecal cells undergo rapid mitosis, and capillaries generated from the theca vascularize the granulosa cells. In this way the follicle transforms into the corpus luteum ("yellow body"; so called because it appears orangish under a microscope). Although there is great species variation in the component tissues of the corpora lutea, in humans and many other mammals corpora lutea are derived from both granulosa cells and cells from the theca interna. The corpus luteum persists for some time on the surface of the ovary and produces another class of important sex steroid hormones, progestins. The corpus luteum eventually degenerates, leaving a scar called the corpus albicans ("white body"), which does not produce any hormones.

The stroma of the ovary consists of connective tissue and interstitial cells. The interstitial cells of the ovary, which produce sex steroid hormones, are probably analogous, if not homologous, to those of the testes.

During the follicular phase (when the primary follicle is growing), the theca interna develops receptors for an anterior pituitary hormone called luteinizing hormone (LH), and it produces androgens from cholesterol in response to LH stimulation of these receptors. The granulosa cells develop receptors for follicle-stimulating hormone (FSH), and in response to FSH from the anterior pituitary, convert androgens into estrogens. LH receptors are also expressed in the granulosa cells near the time of ovulation in response to FSH and estrogenic stimulation. Stimulation of these receptors by LH causes the granulosa cells to produce progesterone.

The Placenta

The placenta is a temporary endocrine organ that develops in the uterus during pregnancy in mammals. It forms from tissues derived from both the blastocyst and the maternal uterus. The placenta is important in maintaining nutritional, respiratory, and excretory functions for the fetus(es). It is also the source of several steroid and peptide hormones that affect both the mother and the offspring. In fact, the most common pregnancy tests measure human chorionic gonadotropin (hCG), a hormone produced by the rudimentary placenta that forms immediately after blastocyst implantation. This hormone maintains corpora luteal function (and progesterone secretion) during pregnancy and is part of the regulatory system that inhibits ovulation during pregnancy. The placenta is dislodged from the uterine wall during parturition (birth) and is expelled from the uterus at the end of pregnancy.

Cellular and Molecular Mechanisms of Hormone Action

Now that you know something about the major endocrine glands and the hormones they produce and secrete, we will examine how hormones induce reactions at their target tissues. Hormonal messages, or signals, evoke intracellular responses via **signal transduction**: the chemical hormonal "message" is transformed into intracellular events that ultimately affect cell function. The sequence of events from the time a hormone binds to its receptor to the ultimate response in a target cell is called **signal transduction**. The initiation of signal transduction pathways in neurons is the most common mechanism by which hormones influence behavior.

Hormone Receptor Types

Many different kinds of molecules constantly bombard the outer surfaces of cells throughout the body. Few molecules, however, have a greater effect on cellular function than hormones. The ability of a target cell to recognize and respond to a given hormone is mediated by the presence of specific receptor proteins in that cell, but only a small fraction of the receptors need to be activated to evoke the maximal cellular response. Receptors for steroid hormones differ in form and action from receptors for protein or peptide hormones, and we will now describe these differences in detail.

STEROID RECEPTORS Steroid and thyroid hormone receptors are located inside cells, either in the cytosol or in the nucleus. As mentioned earlier, steroids are lipid-soluble, so they can penetrate the cell membrane to bind with these intracellular receptors. When the receptors bind a specific steroid or thyroid hormone, they migrate to the nucleus (if they are not already there) to regulate gene transcription. These receptors are part of a superfamily that also includes receptors for vitamin D and retinoic acid. We will discuss steroid receptors in more detail later in the chapter.

PROTEIN AND PEPTIDE HORMONE RECEPTORS Protein and peptide hormone receptors are found embedded in the cell membrane and have at least three domains (a *domain* is a region of the receptor that has a specific recognized function): (1) an extracellular domain that specifically binds to the hormone in question to form a hormone–receptor complex (in this context, the hormone is called a **ligand**—the name for any molecule that binds to a receptor), (2) a transmembrane domain, and (3) a cytoplasmic domain (Figure 2.13). These receptors are dynamic, changing in form and position in the cell membrane, and they turn over rapidly. Though all protein hormone receptors share a basic structure, they can be divided into two functional classes: those with intrinsic enzymatic activity, and those that require an intracellular "second messenger" to exert their effects. Through a

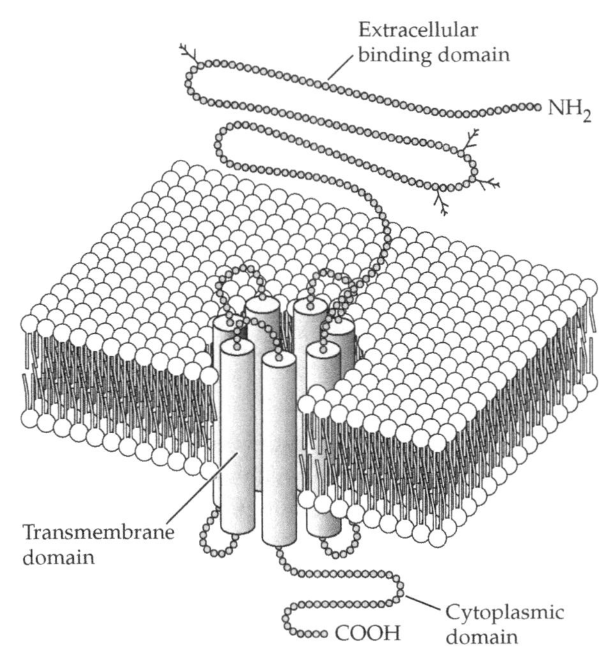

2.13 Protein and peptide hormone receptors are embedded in the phospholipid bilayer of the cell membrane. These receptors consist of three domains: an extracellular binding domain to which the peptide or protein hormone binds, a complex transmembrane domain that may traverse the membrane several times, and a cytoplasmic domain within the cell.

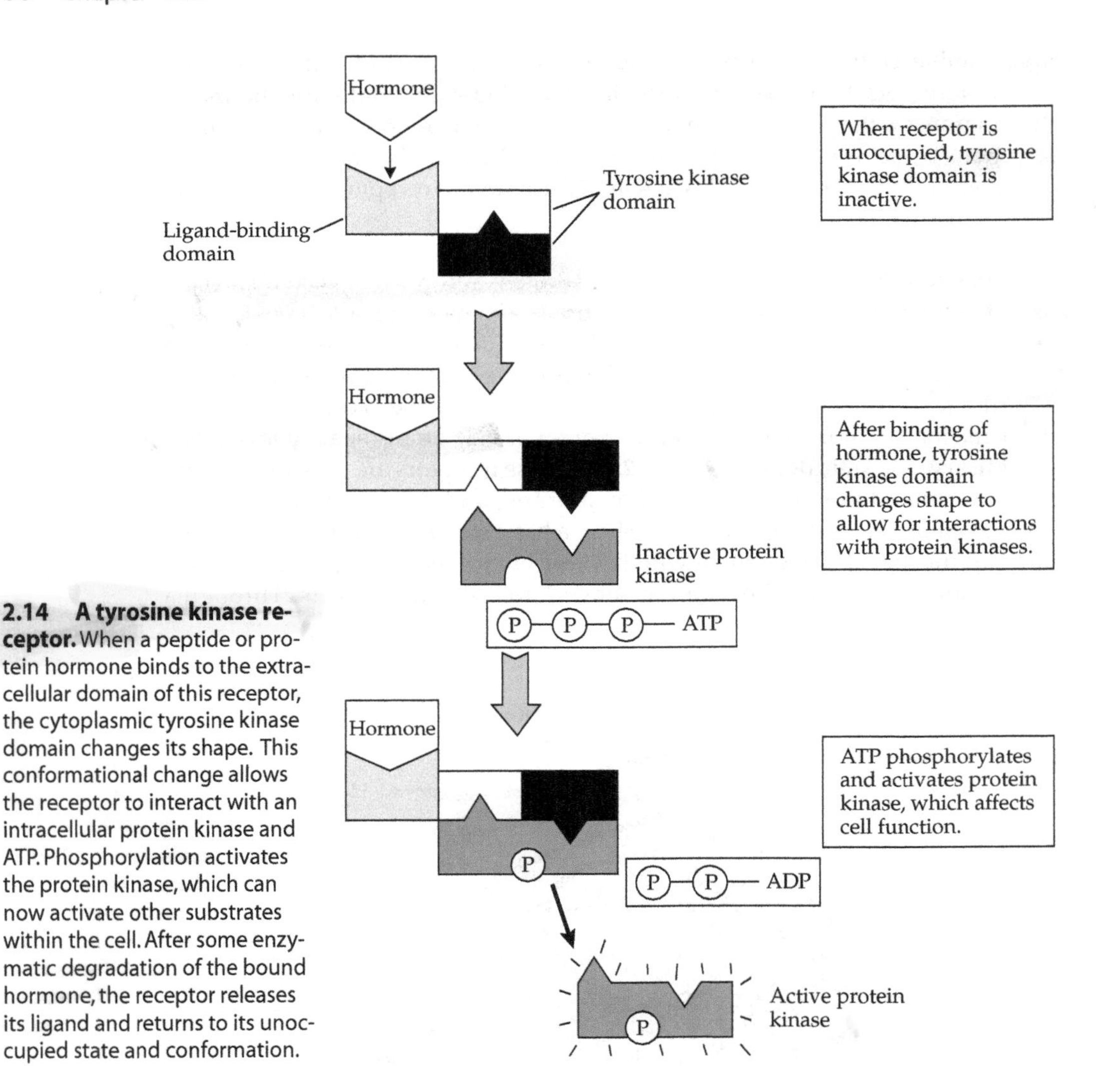

2.14 A tyrosine kinase receptor. When a peptide or protein hormone binds to the extracellular domain of this receptor, the cytoplasmic tyrosine kinase domain changes its shape. This conformational change allows the receptor to interact with an intracellular protein kinase and ATP. Phosphorylation activates the protein kinase, which can now activate other substrates within the cell. After some enzymatic degradation of the bound hormone, the receptor releases its ligand and returns to its unoccupied state and conformation.

process called **enzyme amplification**, a single protein or peptide hormone molecule triggers synthesis of thousands of enzyme molecules.

Receptors with intrinsic enzymatic activity have enzymes as part of the cytoplasmic domain. These enzymes phosphorylate (add a phosphate group to)—and thus activate—intracellular proteins. For instance, some of these receptors have a tyrosine kinase as part of their cytoplasmic domain. Tyrosine kinase is an enzyme that catalyzes the transfer of a phosphate group from ATP to a protein kinase; in turn, the activated protein kinase phosphorylates various other enzymes that cause specific changes in cellular function or membrane permeability (Figure 2.14). In addition to tyrosine kinase receptors, this group includes guanylate cyclases that produce the second messenger cGMP from GTP inside the cell (e.g., natriuretic peptide receptors), tyrosine phosphatases that dephos-

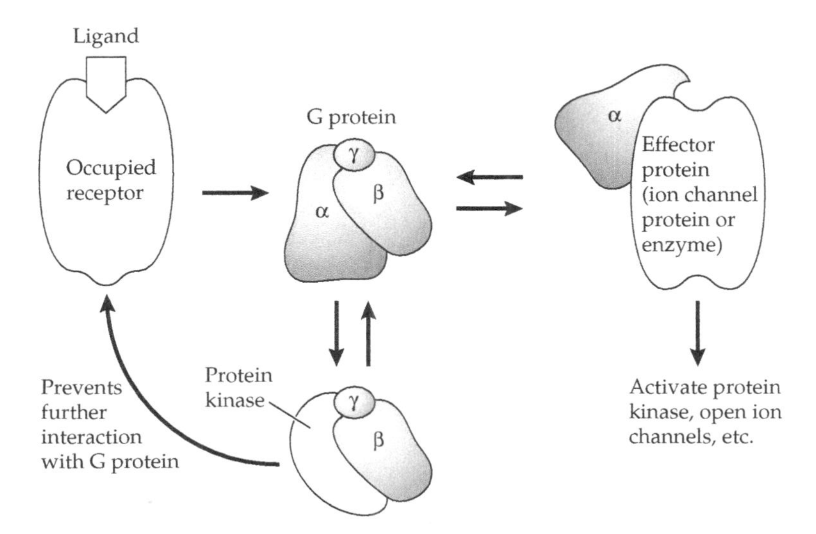

2.15 A G protein. G proteins consist of three different subunits, which often dissociate after the receptor is occupied by its hormonal ligand. The α-subunit, which has GTP-binding and hydrolyzing ability, interacts with an effector protein (either an enzyme—often adenylate cyclase—or an ion channel) to stimulate the production of cAMP. The cAMP, acting as a second messenger, causes changes in the target cell. The β- and γ-subunits form a complex that binds to a protein kinase that may feed back to influence the hormone receptor.

phorylate proteins, and serine/threonine kinases that specifically phosphorylate serine and threonine residues (e.g., receptors for activin) (Hadley, 1996).

The second class of protein and peptide hormone receptors requires a **second messenger**—a molecular "middleman" such as an enzyme or another protein—to transduce the hormonal signal. These receptors are coupled to **G proteins**, which can bind and hydrolyze GTP. There are several different types of G proteins, all of which have three different subunits, commonly labeled α, β, and γ (Figure 2.15). When a hormone binds to a G protein-coupled receptor, the hormone–receptor complex most frequently activates an intracellular enzyme, adenylate cyclase, which in turn stimulates the formation of cyclic adenosine monophosphate (cAMP) inside the cell (Figure 2.16). This cyclic nucleotide then activates specific protein kinases in the cell. When formed in response to the binding of a hormone or other ligand, cAMP is referred to as a second messenger (the hormone is the first messenger).* In some cases, G proteins open ion channels, causing an increased intracellular influx of calcium that eventually stimulates cAMP (or cGMP) production, which then activates specific kinases. G protein-

*Because G proteins often mediate between the hormonal signal (first messenger) and the cAMP (or cGMP) second messengers, students often ask why G proteins are not called second messengers, and the cAMP (and cGMP) referred to as third messengers. Although this would be technically correct, the second messengers were discovered nearly two decades before the discovery of the G proteins, and the name has persisted.

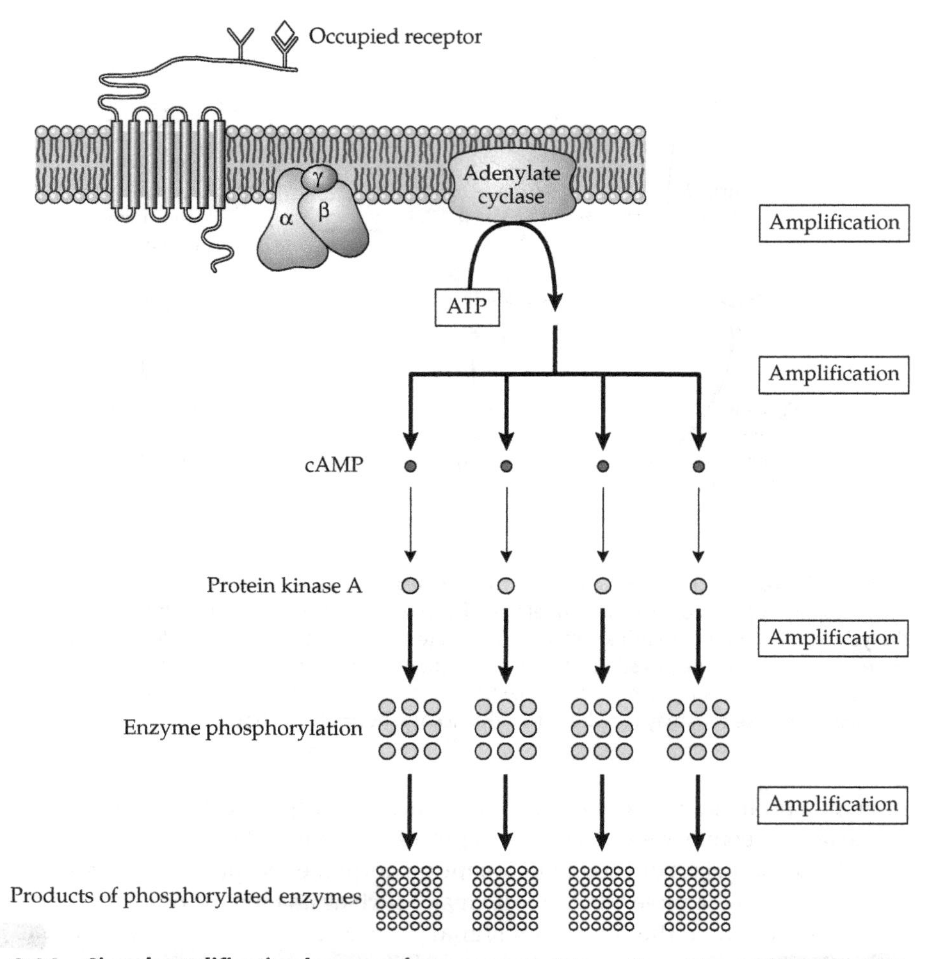

2.16 Signal amplification by second messengers. When a G protein-coupled receptor is bound to its hormonal ligand, a cascade of changes occurs that greatly amplifies the original signal. The activated receptor induces an adjacent G protein molecule, consisting of α-, β-, and γ-subunits, to activate the enzyme adenylate cyclase. Adenylate cyclase, in turn, stimulates the production of many cAMP molecules, which activate many protein kinase molecules. At each step, there is a manyfold increase in the number of molecules activated, until finally, a significant change in the cell has been evoked by a single hormone binding to the receptor.

coupled receptors all have exactly seven transmembrane domains, and are sometimes called serpentine receptors (see Figure 2.13). The G protein-coupled receptor family includes glucagon, oxytocin, and vasopressin receptors.

Different cell types have different specific cellular responses to cAMP. For example, when epinephrine binds to its receptor, the ligand–receptor complex interacts with a stimulatory G protein called G_s, which in turn interacts with adenylate cyclase. Through signal amplification, adenylate cyclase generates many cAMP molecules (see Figure 2.16; Norman and Litwack, 1987). Cell membranes also may contain another G protein, called G_i protein (i.e., inhibitory G protein); interaction with G_i protein inhibits the cAMP mechanism. We know that cAMP functions as the intracellular messenger for epinephrine because intracellular administration of cAMP duplicates all of the "hormonal" actions of epinephrine (Hadley, 1996). Once formed, cAMP can repeatedly combine with protein kinase A (PKA), an enzyme that phosphorylates (and thus activates) another enzyme called phosphorylase kinase in a variety of cells. In the liver, phosphorylase kinase then converts phosphorylase B into its active form, phosphorylase A, which breaks down glycogen into glucose that can be released into the blood to provide energy for other cells (Norris, 1996). In fat cells, cAMP activates a hormone-dependent lipase that causes hydrolysis of stored fats for energy. In heart cells or neurons, cAMP activates PKA, and may also bind to Ca^{2+}-gated ion channels, which can depolarize the cell membrane. In addition to these cytosolic mechanisms of cAMP signal transduction, cAMP may migrate to the cell nucleus and bind to a cAMP regulatory element binding protein (CREB). When bound to cAMP, CREB serves as a transcription factor that binds to DNA promoter regions in order to regulate gene transcription.

There are other signal transduction pathways that involve G proteins, but do not rely on cAMP or cGMP. One important example is the inositol triphosphate (IP_3)/diacylglycerol (DAG) pathway (Figure 2.17). As is the case with the G pro-

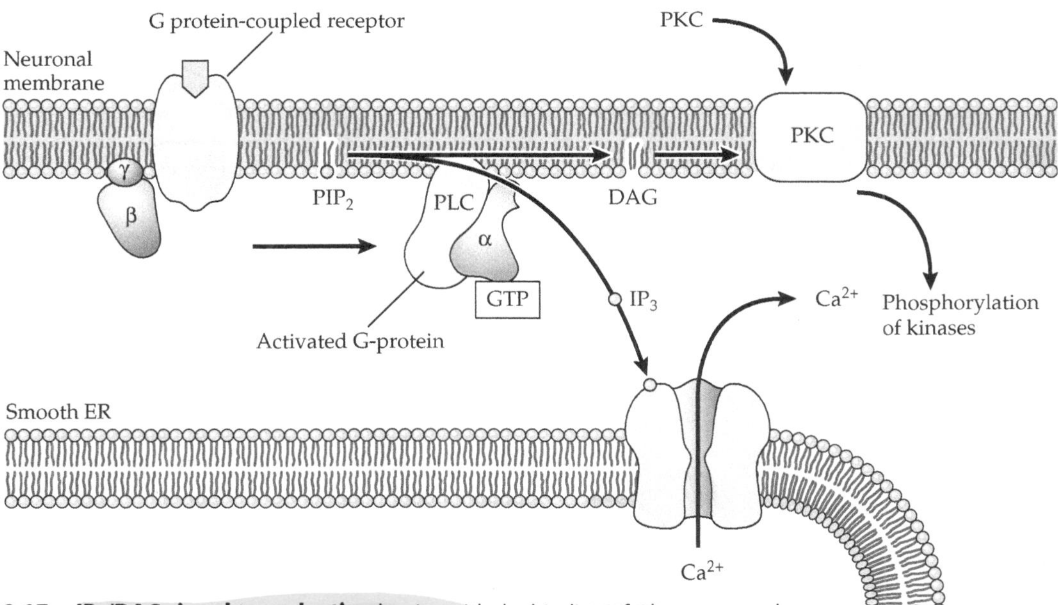

2.17 IP_3/DAG signal transduction begins with the binding of a hormone to the extracellular domain of a G protein-coupled receptor. The α-subunit of the activated G protein activates phospholipase C (PLC), which evokes the hydrolysis of PIP_2 and subsequent formation of IP_3 and DAG. IP_3 mobilizes Ca^{2+} from stores in the endoplasmic reticulum (and possibly moves Ca^{2+} into the cell). These twin signals, the rise in intracellular DAG and Ca^{2+} concentrations, evoke PKC binding to the cell membrane and activation of Ca^{2+} channels.

tein-coupled receptors we have discussed so far, the IP_3/DAG pathway begins with the binding of a hormone to the extracellular domain of a receptor and the activation of its associated G protein (see Figure 2.15). However, in this case, the dissociated α-subunit activates phospholipase C (PLC), resulting in the hydrolysis of phosphatidylinositol-4,5-biphosphate (PIP_2), which leads to the formation of IP_3 and DAG. IP_3 mobilizes Ca^{2+} from stores in the endoplasmic reticulum. These twin signals—that is, the rise in intracellular concentrations of DAG and of Ca^{2+}—evoke protein kinase C (PKC) binding to the cell membrane and activation of ion channels.

Transcription, Translation, and Post-Translational Events

The final common pathway through which hormones may affect behavior is by acting as **transcription factors**. The signal transduction pathways described above are diverse, and some, such as the cAMP/CREB pathway, ultimately affect gene transcription, the process by which the sequence of nucleotides in a single strand of DNA is transcribed into a single strand of complementary RNA (Figure 2.18). Transcription factors bind to the beginning of the DNA sequence where the gene to be transcribed is located. This binding facilitates transcription of the gene.

In order to synthesize mRNA, the two tightly twisted strands of DNA must be unraveled by agents called helicases. A gene consists of a unique linear

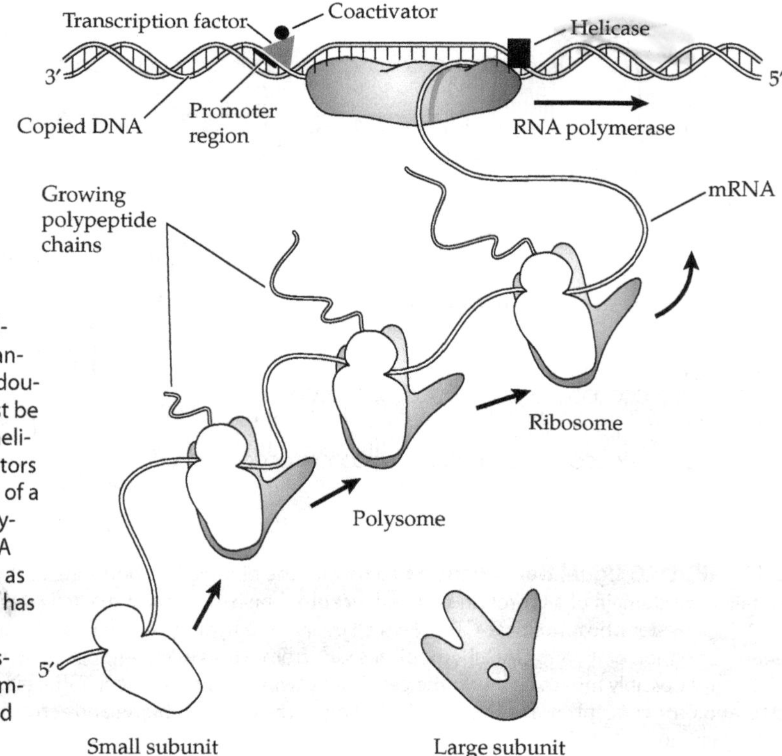

2.18 Transcription and translation can be nearly simultaneous. During gene transcription, the strands of the double-helix DNA molecule must be chemically "pried apart" by helicases. Then, transcription factors bind to the promoter region of a gene; this allows an RNA polymerase to transcribe the DNA and produce mRNA. As soon as the first part of the message has been transcribed, mRNA can leave the nucleus to be translated by the ribosomes. A complex of several ribosomes and mRNA is called a polysome.

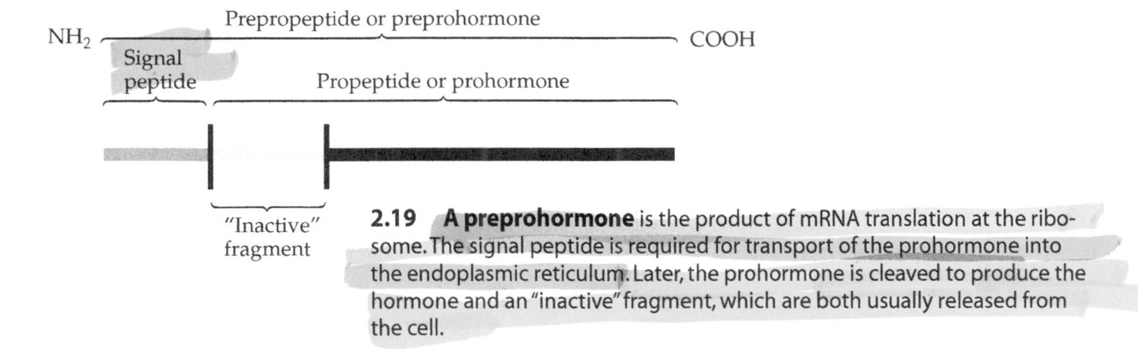

2.19 A preprohormone is the product of mRNA translation at the ribosome. The signal peptide is required for transport of the prohormone into the endoplasmic reticulum. Later, the prohormone is cleaved to produce the hormone and an "inactive" fragment, which are both usually released from the cell.

sequence of DNA. Among eukaryotic organisms, some of the nucleotide sequences within the gene are noncoding sequences, called introns, which alternate with coding sequences, called exons. There are special marker sequences denoting the start and end points of each gene. A distinct sequence of nucleotides, called a promoter or facilitory region, marks the start of the gene. The binding of a transcription factor to the promoter allows special enzymes, called RNA polymerases, to attach to the promoter and begin the process of RNA synthesis. The sequence of RNA nucleotides, determined by the sequence of nucleotides along the DNA, eventually determines the sequence of amino acids in the protein product of the gene. After transcription, enzymes clip out the intron sequences; then other enzymes splice together the remaining segments (exons) to form messenger RNA (mRNA). The mRNA leaves the cell nucleus, travels to the rough endoplasmic reticulum (RER), and serves as the template for translation into a linear sequence of amino acids, which occurs on ribosomes.

When the protein being synthesized is itself a protein hormone, the first product of translation is called a **preprohormone** (Figure 2.19). At the N-terminal end of the preprohormone is a certain amino acid sequence called the signal peptide. This peptide is separated from the preprohormone during further processing of the protein, yielding a **prohormone**, which is then packaged into vesicles before being moved to the Golgi apparatus. Typically, additional post-translational processing of the prohormone by special enzymes occurs within the rough endoplasmic reticulum, the Golgi apparatus, and the vesicles to yield the final version of the hormone (or hormones) for secretion. In some cases (e.g., thyrotropin-releasing hormone), the prohormone is cleaved into several molecules of the hormone; in other cases (e.g., pro-opiomelanocortin), several different hormones can be produced from a single prohormone, as we will see below.

The Major Vertebrate Hormones

At this point, you should know where the major endocrine glands are located, as well as their primary physiological functions and how their hormonal messages are transduced. This section describes in detail the four classes of hormones secreted by these glands: (1) **protein and peptide hormones**, (2) **steroid hormones**, (3)

monoamines, and (4) **lipid-based hormones**. It is important and useful to discriminate among the four types of hormones because they differ in several important characteristics, including their mode of release, how they move through the blood, the location of their target tissue receptors, and the manner by which the interaction of the hormone with its receptor results in a biological response. Again, all of the hormones discussed here, as well as the glands from which they originate and their primary physiological function(s), are summarized in Appendix I.

Protein and Peptide Hormones

Most vertebrate hormones are proteins. **Protein hormones**, in common with other protein molecules, are made up of individual amino acid building blocks. Protein hormones only a few amino acids in length are called **peptide hormones**, whereas larger ones are called protein or polypeptide hormones. Protein and peptide hormones include insulin, the glucagons, the neurohormones of the hypothalamus, the tropic hormones of the anterior pituitary, inhibin, calcitonin, parathyroid hormone, the gastrointestinal hormones, ghrelin, leptin, adiponectin, and the posterior pituitary hormones.

Protein and peptide hormones can be stored in endocrine cells and are released into the circulatory system by means of exocytosis (see Figure 2.2A). They are soluble in blood, and therefore do not require a carrier protein to travel to their target cells, as do steroid hormones. However, protein and peptide hormones may bind with other blood plasma proteins, which slow their metabolism by peptidases (enzymes that break down peptides) in the blood. Hormones are removed from the blood via degradation or excretion. The metabolism of a hormone is reported in terms of its **biological half-life**, which is the amount of time required to remove half of the hormone from the blood (measured by radioactive tagging). Generally, larger protein hormones have longer half-lives than smaller peptide hormones (e.g., growth hormone has 200 amino acids and a biological half-life of 20–30 minutes; thyrotropin-releasing hormone has 3 amino acids and a biological half-life of less than 5 minutes in humans) (Norris, 1996).

HYPOTHALAMIC HORMONES The peptidergic neurons of the hypothalamic median eminence secrete a number of releasing hormones and inhibiting hormones. These hypothalamic hormones are small peptides, ranging from 3 to 44 amino acids in length. The properties of releasing and inhibiting hormones are very similar to those of neurotransmitters. These hormones are best thought of as a special class of neurotransmitters that act on a variety of cells in the anterior pituitary. In fact, several of these hypothalamic hormones act as neurotransmitters elsewhere in the central nervous system.

Five hypothalamic releasing hormones—**thyrotropin-releasing hormone (TRH)**, **somatocrinin** (or **growth hormone–releasing hormone [GHRH]**), **gonadotropin-releasing hormone** (GnRH), **melanotropin-releasing hormone** (MRH), and **corticotropin-releasing hormone** (CRH)—and one inhibiting hormone (**growth hormone–inhibiting hormone**, which is also called **somatostatin**) have been isolated and characterized (Figure 2.20). **Dopamine (DA)**, a well-char-

TRH

pGLU–HIS–PRO–NH_2

GnRH

1 3 10
pGlu–His–Trp–Ser–Tyr–Gly–Leu–Arg–Pro–Gly–NH_2

Somatostatin

1 14
Ala–Gly–Cys–Lys–Asn–Phe–Phe–Trp–Lys–Thr–Phe–Thr–Ser–Cys

CRH

1
Ser–Gln–Glu–Pro–Pro–Ile–Ser–Leu–Asp–Leu–Thr–Phe–His–Leu–Leu–Arg–Glu–Val–Leu–
20 30
Glu–Met–Thr–Lys–Ala–Asp–Gln–Leu–Ala–Gln–Gln–Ala–His–Ser–Asn–Arg–Lys–Leu–Leu–
40 41
Asp–Ile–Ala–NH_2

GHRH

1 10 20
Tyr–Ala–Asp–Ala–Ile–Phe–Thr–Asn–Ser–Tyr–Arg–Lys–Val–Leu–Gly–Gln–Leu–Ser–Ala–Arg–
30
Lys–Leu–Leu–Gln–Asp–Ile–Met–Ser–Arg–Gln–Gln–Gly–Glu–Ser–Asn–Gln–Glu–Arg–Gly–
40 44
Ala–Arg–Ala–Arg–Leu–NH_2

2.20 Known amino acid sequences of releasing hormones. The sequences of TRH, GnRH, somatostatin, CRH, and GHRH have been characterized. The sequence of melanotropin-releasing hormone (MRH) has been characterized. Melanotropin inhibitory hormone (MIH) and prolactin inhibitory hormone (PIH) both appear to be dopamine. pGLU denotes pyroglutamyl; NH_2 denotes amide of the C-terminal amino acid, thus prolinamide in TRH, glycinamide in GnRH, alanylamide in CRH, or leucinamide in GHRH.

acterized neurotransmitter, and technically a monoamine, also serves as a neurohormone in the hypothalamus to inhibit the release of prolactin and melanotropin from the anterior pituitary; in these contexts, it is known, respectively, as **prolactin inhibitory hormone (PIH)** and **melanotropin inhibitory hormone (MIH)**. A recently discovered hormone called **hypocretin** (also called **orexin**) is found in cells located in the hypothalamus that project widely to the rest of the brain and spinal cord. This hormone is involved in sleep, metabolic balance, and possibly activation of the sympathetic nervous system.

Unlike steroid hormones, which are structurally identical among all vertebrates, protein and peptide hormones vary among taxonomic groups in their sequence of amino acids. Usually, the mammalian version of a hormone is isolat-

	1	2	3	4	5	6	7	8	9	10
Mammal	pGlu	His	Trp	Ser	Tyr	Gly	Leu	Arg	Pro	Gly–NH_2
Chicken I	pGlu	His	Trp	Ser	Tyr	Gly	Leu	Gln	Pro	Gly–NH_2
Catfish	pGlu	His	Trp	Ser	His	Gly	Leu	Asn	Pro	Gly–NH_2
Chicken II	pGlu	His	Trp	Ser	His	Gly	Trp	Tyr	Pro	Gly–NH_2
Dogfish	pGlu	His	Trp	Ser	His	Gly	Trp	Leu	Pro	Gly–NH_2
Salmon	pGlu	His	Trp	Ser	Tyr	Gly	Trp	Leu	Pro	Gly–NH_2
Lamprey	pGlu	His	Tyr	Ser	Leu	Glu	Trp	Lys	Pro	Gly–NH_2

2.21 Primary structures of species-specific forms of GnRH. Amino acid residues that are different from the mammalian form of GnRH are enclosed in boxes. Presumably, point mutations have occurred during evolution to yield the current forms of GnRH in different types of animals.

ed by researchers first. Because the sequence of amino acids ultimately affects the shape of the hormone molecule, protein hormones from one group of animals may not activate the receptors in another group. Injection of the mammalian version of GnRH into a frog, for example, would probably not affect its reproductive function. After conducting such an experiment, it would not be prudent to conclude that GnRH is not involved in frog reproduction (although such errors in logic have been made in endocrine investigations). Some variants of GnRH are shown in Figure 2.21. Small differences in the amino acid sequences of protein hormones exist even between species of mammals. These small variations generally do not prevent a hormone taken from one mammal from having biological activity when injected into an individual of another mammalian species. If treatment with the foreign hormone continues for several weeks, however, the recipient may mount an immune response to the hormone. This is the primary reason that protein hormones extracted from the tissues of other animals are not used for clinical treatments in humans.

ANTERIOR PITUITARY HORMONES The hormones of the anterior pituitary are protein hormones that range from 15 to about 220 amino acids in length. If the anterior pituitary is removed from an animal, sliced very thin, mounted on a slide, colored with various stains to enhance visibility, and viewed through a microscope, a characteristic pattern of cells can be observed. One type of cell stains readily with acidic stains, another type takes up basic stains, and a third type does not readily take up either acidic or basic stains. These types of cells are called acidophils, basophils, and chromatophobes, respectively. Through the use of modern techniques, particularly immunocytochemistry (ICC), we now know that specific types of hormones are made in these three cell types, and that the acidophils and basophils can be further subdivided into more cell types, each producing a single hormone.

Luteinizing hormone (LH), **follicle-stimulating hormone (FSH)**, and **thyroid-stimulating hormone (TSH)** are secreted by the basophils. They consist of 200–220 amino acids and have a molecular weight of approximately 25,000–35,000 daltons. Approximately 10%–25% of the molecular structure of each of these three hormones is made up of carbohydrates, and they are thus known collectively as

glycoproteins. Each glycoprotein is composed of two subunits, labeled α and β. These subunits have no biological activity separately; both subunits are necessary to produce a biological response. The α-subunits of LH, FSH, and TSH are identical. The β-subunit imparts the function of the molecule and also determines the species specificity. If the LH α-subunit and the TSH β-subunit are combined, the result is a biologically active TSH molecule. LH and FSH are also known as **gonadotropins** because, in response to GnRH, they stimulate steroidogenesis in the gonads as well as the development and maturation of gametes. In response to TRH from the hypothalamus, TSH, also known as **thyrotropin**, is released from the anterior pituitary and stimulates the thyroid gland to release thyroid hormones.

Growth hormone and prolactin are secreted by the acidophils and are similar in structure; both are simple proteins consisting of 190–220 amino acids. **Growth hormone** (**GH**) is released from the anterior pituitary in response to GHRH from the hypothalamus. Growth hormone shows more species specificity in its biological activity among mammals than most protein hormones. For example, rat GH has little or no biological activity in primates.* Human growth hormone (hGH) has two disulfide bonds.

Growth hormone stimulates somatic (body) growth. However, GH does not directly induce growth of the skeleton. Rather, it stimulates the production of growth-regulating substances called **somatomedins** by the liver, kidneys, and other tissues; the somatomedins cause bone to take up sulfates, leading to skeletal growth. GH and somatomedins also stimulate protein synthesis. GH promotes protein anabolism and acts as an anti-insulin. Other physiological effects of GH include fat mobilization, increased cell membrane permeability to increase the uptake of amino acids, changes in ion influxes, and increased blood sugar concentrations (hyperglycemia), as well as indirect effects on the thymus, an organ involved in immune function. Human GH also has inherent prolactin activity.

Human **prolactin** (**PRL**) has 198 amino acids and contains three disulfide bonds. The release of prolactin is stimulated by TRH from the hypothalamus. Prolactin is named for its well-known effect of promoting lactation in female mammals. However, the first function of prolactin discovered was the promotion of corpus luteum function in the ovaries of rats; thus, it was originally named luteotropic hormone. Several years passed before prolactin and luteotropic hormone were discovered to be structurally identical. The term luteotropic hormone has faded from current use; however, it reminds us that prolactin has many functions other than promoting lactation in mammals.

In fact, prolactin is a hormone that has been conserved throughout vertebrate evolution and has hundreds of different physiological functions. These functions can be broken down into five basic classes of actions related to five different physiological processes (Table 2.2). In terms of reproduction, prolactin stimulates the formation and maintenance of the corpora lutea in rats and mice, and may do so

*Lowercase letters are used with the hormone abbreviation to indicate the animal from which a particular hormone is derived. Growth hormone from rats, sheep (ovine), cattle (bovine), pigs (porcine), and mice (murine) would be abbreviated rGH, oGH, bGH, pGH, and mGH, respectively.

TABLE 2.2 Cytokines

Cytokines	Source	Biological Function
Interleukin-1α (IL-1α)	Macrophages	Activates T and B cells of the immune system; induces fever and inflammation
IL-1β	Macrophages	As IL-1α, early mediator of the inflammatory and overall immune response
IL-2	T cells	Stimulates T cell proliferation
IL-3	T cells	Induces growth of several cell types
IL-4	T cells	Promotes B cell growth and differentiation
IL-5	T cells	Stimulates B cell and eosinophil cell growth
IL-6	Macrophages; T cells	Stimulates B cells and promotes inflammation
IL-7	Stromal cells	Provokes B and T cell differentiation
IL-8	Macrophages	Mediates neutrophil attraction
IL-9	T cells	Acts as a mitogen
IL-10	T cells	Inhibits cytokine production ; anti-inflammatory
IL-11	Bone marrow	Supports hematopoeisis
IL-12	Antigen Presenting Cells (APC)	Stimulates T and Natural Killer (NK) immune cells
IL-13	T cells	Promotes B cell growth and differentiation
IL-14	Dendritic cells; T cells	Promotes B cell memory
IL-15	T cells	Stimulates T cell proliferation
Interferon-α (IFNα)	Many types of cells	Anti-viral
IFNβ	Many types of cells	Anti-viral
IFNγ	T and NK cells	Induces inflammation and activates macrophages; microbicidal
Transforming Growth Factor-β (TGFβ)	Macrophages; lymphocytes	Inhibits lymphocyte function; stimulates IL-1 production; chemoattractant for monocytes
Tumor Necrosis Factor-α (TNFα)	Macrophages	Evokes inflammation and tumor killing
TNFβ	T cells	Evokes inflammation and tumor killing; enhances phagocytosis

in other mammals as well. An example of the importance of prolactin to growth and development is seen in certain salamanders, the newts, which undergo a second metamorphosis from a terrestrial intermediate form to the aquatic sexually mature form; without prolactin, newts cannot undergo this metamorphosis. Prolactin has many osmoregulatory effects, especially in teleost (bony) fishes. Some species of minnows (such as *Fundulus heteroclitus*) are euryhaline; that is, they can migrate between seawater and fresh water. In the absence of prolactin, these minnows cannot adapt to fresh water (see Chapter 8). Prolactin is important in the development and maintenance of integumentary structures, which include the crop sac of pigeons and doves (see Figure 1.8) and the mammary glands in mammals. Finally, prolactin has synergistic actions on steroid-dependent target tissues. For example, prolactin is critical for maintaining LH receptors in the testes of some mammalian species.

Adrenocorticotropic hormone (ACTH) is made in the chromatophobes; it has 39 amino acids and weighs about 4500 daltons. Only the first 23 amino acids of the ACTH molecule are required to maintain its full physiological activity. ACTH is released in response to CRH from the hypothalamus and stimulates the adrenal cortex to secrete corticoids (especially the glucocorticoids).

ACTH comes from a much larger parent protein called **pro-opiomelanocortin (POMC)** (Figure 2.22). This large precursor molecule can be cleaved into a variety of biologically active substances. In most cases, tissue-specific enzymes are responsible for the different fates of POMC. Some of the other products of POMC include the lipotropins (which may mobilize fat), the endogenous opioids, **β-endorphin** and **met-enkephalin**, and **melanoctye-stimulating hormone (MSH)**, a pigmentation regulator.

Because the MSH molecule is produced by cleavage of the ACTH molecule after its cleavage from POMC, all ACTH includes inherent MSH activity; therefore, all ACTHs are also MSHs. MSH has little or no effect on the adrenal cortex, but functions primarily to control pigmentation in nonmammalian and nonavian vertebrates. Melanocytes are cells that contain the pigment melanin. MSH stimulates melanogenesis, the synthesis of melanin, therefore inducing a darker skin color. ACTH also results in melanocyte stimulation. Although the physiological function of MSH in mammals and birds is not well understood, this peptide does have some behavioral effects, which will be discussed in Chapter 10.

The opioids are endogenous (from within) "pain killers." They interact with opioid receptors throughout the central nervous system to ameliorate pain sensations. Exogenous opiates such as heroin and morphine interact with these same receptors. ACTH and β-endorphin are secreted simultaneously in response to CRH. These two hormones come from different parts of the POMC precursor molecule (Nakanishi et al., 1979), and both are believed to be part of the stress adaptation response. ACTH stimulates glucocorticoid secretion by the adrenal cortex, which helps an animal adjust to stressful conditions. Stressful conditions may also trigger the release of β-endorphin to reduce pain. The secretion of ACTH and β-endorphin increases after adrenalectomy, indicating that the adrenal corticoids may feed back to control their own release.

POSTERIOR PITUITARY HORMONES Two nonapeptides, oxytocin and vasopressin, are released from the posterior pituitary in mammals. These hormones are often referred to as "octapeptides" despite their having nine amino acids. However, two cysteine molecules form a cysteine–cysteine dimer that has been counted as one amino acid "unit" in the peptide sequence. The two peptides are closely related (Figure 2.23).

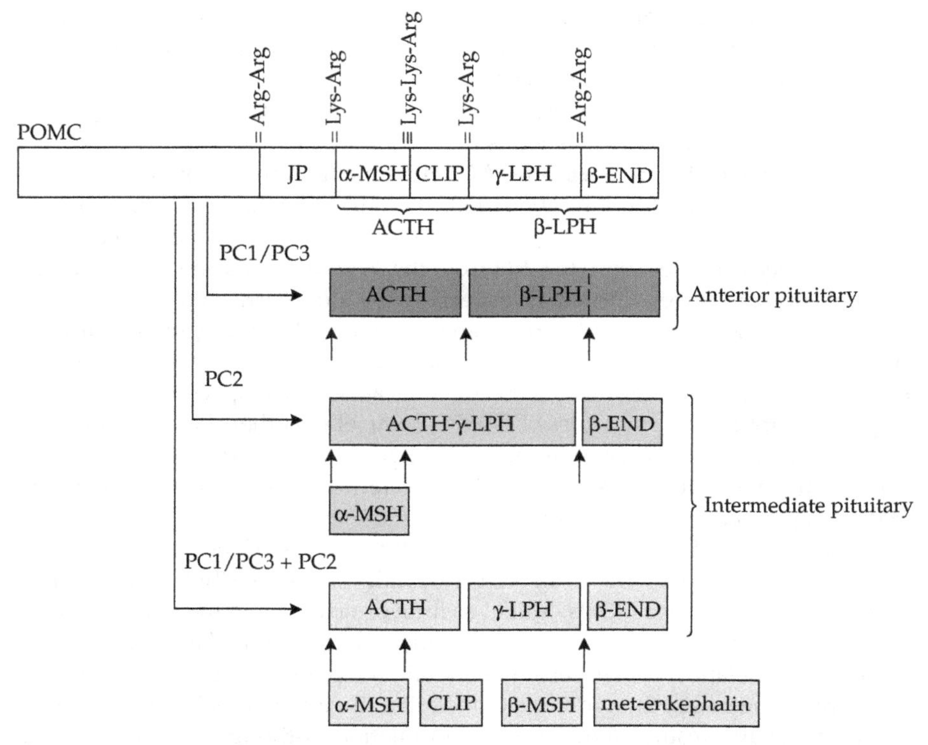

2.22 Pro-opiomelanocortin (POMC) is the precursor to a variety of biologically important substances, including ACTH, α-MSH, the lipotropins γ-LPH and β-LPH, the endogenous opioids β-endorphin and met-enkephalin, and CLIP (corticotropin-like intermediate lobe peptide). Cells in the anterior pituitary that secrete ACTH (called corticotrophs) possess proteolytic enzymes (PC1/PC3), which cause primarily ACTH to be produced from POMC. Other hormones are produced from POMC in cells between the anterior and posterior pituitary, where other proteolytic enzymes are located.

The nonapeptides are produced from prohormones that also include a carrier protein, neurophysin. Thus, for each molecule of nonapeptide produced, one molecule of neurophysin is produced. There are two types of neurophysin: the prohormone prooxyphysin is hydrolyzed to oxytocin and neurophysin I, and the prohormone propressophysin is hydrolyzed to vasopressin and neurophysin II plus a short glycopeptide (Norris, 1996) (Figure 2.24). The relationship between the carrier protein and the nonapeptide is labile. Neurophysin increases the half-lives of vasopressin and oxytocin from 3 minutes to about 30 minutes in the blood when it is bound to them, primarily by protecting these peptides from peptidases.

Oxytocin profoundly influences reproductive function in mammals. This hormone is important during birth, causing uterine contractions when the uterus is responsive to it. Oxytocin, or an artificial version of oxytocin, is often used medically to induce labor. Oxytocin cannot be used to induce abortions, however,

Oxytocin

Cys–Tyr–Ile–Gln–Asn–Cys–Pro–Leu–Gly–NH_2
1 2 3 4 5 6 7 8 9

Vasopressin

Cys–Tyr–Phe–Gln–Asn–Cys–Pro–Arg–Gly—NH_2
1 2 3 4 5 6 7 8 9

2.23 Primary structure of the nonapeptides secreted by the posterior pituitary.

because the uterus is responsive to it only when progesterone levels drop after high pregnancy-induced estrogen concentrations during the third trimester.

Oxytocin is also important in the suckling reflex. Oxytocin is released into the blood in response to sensory stimulation from the nipple and travels through the general circulation to the mammary glands (Figure 2.25A). The myoepithe-

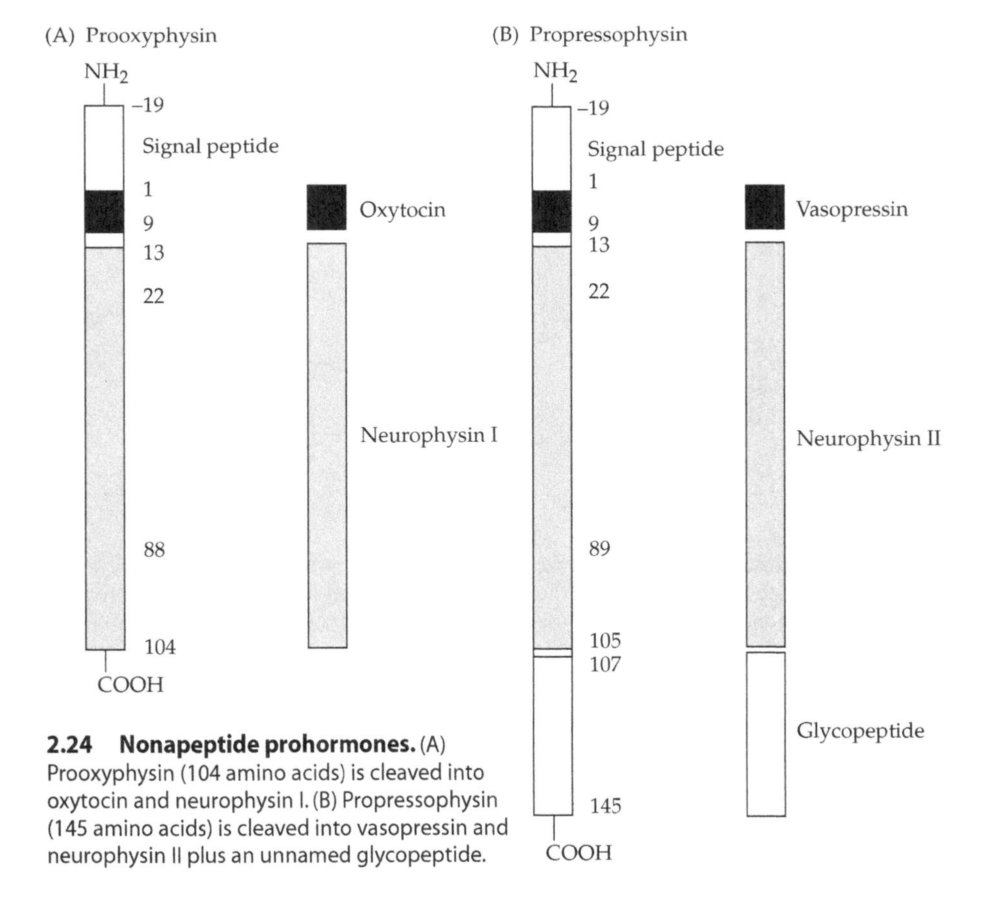

2.24 Nonapeptide prohormones. (A) Prooxyphysin (104 amino acids) is cleaved into oxytocin and neurophysin I. (B) Propressophysin (145 amino acids) is cleaved into vasopressin and neurophysin II plus an unnamed glycopeptide.

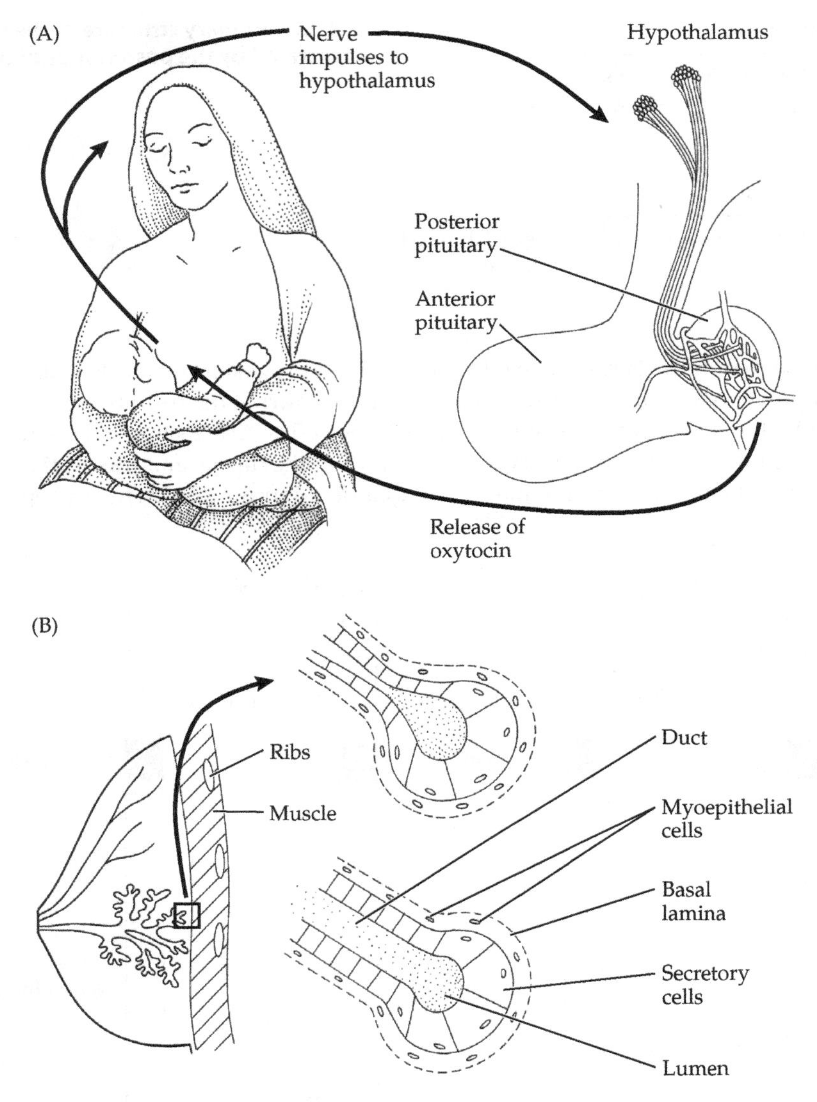

2.25 Milk letdown is mediated by oxytocin. (A) Upon stimulation of the nipple by a feeding baby, oxytocin is released from the posterior pituitary. (B) Oxytocin causes the myoepithelial cells surrounding the lumens of the mammary glands to contract. This suckling reflex is subject to classical conditioning; a lactating mother may experience milk letdown after merely hearing a crying baby.

lial cells surrounding the milk-collecting lumens of the mammary glands contract upon exposure to oxytocin, causing milk letdown (Figure 2.25B). This suckling reflex can become associated with environmental stimuli such that oxytocin is released in anticipation of the sensory stimulation arising from the nipples. Hence, the sight of the milkmaid or the sound of the milking machine may evoke milk letdown in cows, and women may respond to the cry of a hungry baby in the same way, because of prior associations between these sights and sounds and nipple stimulation.

Vasopressin, also known as **antidiuretic hormone** (**ADH**) or arginine vasopressin (AVP), is another nonapeptide found in many mammals. ADH acts to retain water in tetrapod (four-footed) vertebrates. Alcohol is a potent inhibitor of ADH secretion, thereby increasing the frequency of urination after imbibing. ADH also has pressor (hypertensive) effects in response to serious blood loss: it causes constriction of blood vessels, which helps to slow blood flow and presumably enhances the probability of survival. One of the possible serious consequences of ADH-induced pressor effects is that the arterioles of the kidney may be completely constricted, and the kidney may become permanently damaged during hemorrhage-induced shock.

THE THYROID HORMONES The thyroid hormones are derived from a single amino acid, tyrosine. Thus, they are not technically peptides, but monoamines, and they are not water-soluble, as we will see. However, the most common thyroid hormone, **thyroxine** (**T_4**), consists of two tyrosine residues and is thus considered a peptide hormone here.

The thyroid gland releases its hormones in response to TSH stimulation from the anterior pituitary. There are two biologically active thyroid hormones, both derived from a large, globular glycoprotein called thyroglobulin. The hormones are synthesized from iodinated tyrosine residues in the thyroglobulin molecule: essentially, inorganic dietary iodine reacts with water to form "active" iodine, which immediately binds to tyrosine residues in thyroglobulin. Binding of one active iodine atom to tyrosine at position 3 on the phenolic ring yields 3-monoiodotyrosine (MIT). A second iodine may attach at position 5, resulting in 3,5-diiodotyrosine (DIT) (Figure 2.26); position 3 is always more readily iodinated than position 5. MIT and DIT are joined together through an ether linkage (—O—) between the phenolic rings to form the thyronine structure called **triiodothyronine** (**T_3**), one of the secreted thyroid hormones. T_4, also known as tetraiodothyronine, is formed by the combination of two DIT molecules.

Both T_3 and T_4 are fat-soluble. Like steroid hormones, they diffuse rapidly across lipid-based cell membranes, but they need the help of a carrier protein to travel through the blood. The thyroid hormones are easily removed from the blood by the kidneys and liver if not bound to a carrier protein. About 99% of circulating thyroid hormones are reversibly bound to serum proteins. In humans, about 75% of thyroid hormones are bound to α_2-globulins, 20%–30% are bound to albumin or pre-albumin, and less than 1% are transported unattached in the blood; these proportions vary among species.

The thyroid hormones act to increase oxidation rates in tissue. They have three general effects in mammals: they affect metabolism, alter growth and differentiation, and influence reproduction. The thyroid is probably most commonly associated with its metabolic actions, especially its calorigenic, or heat-producing, effects.

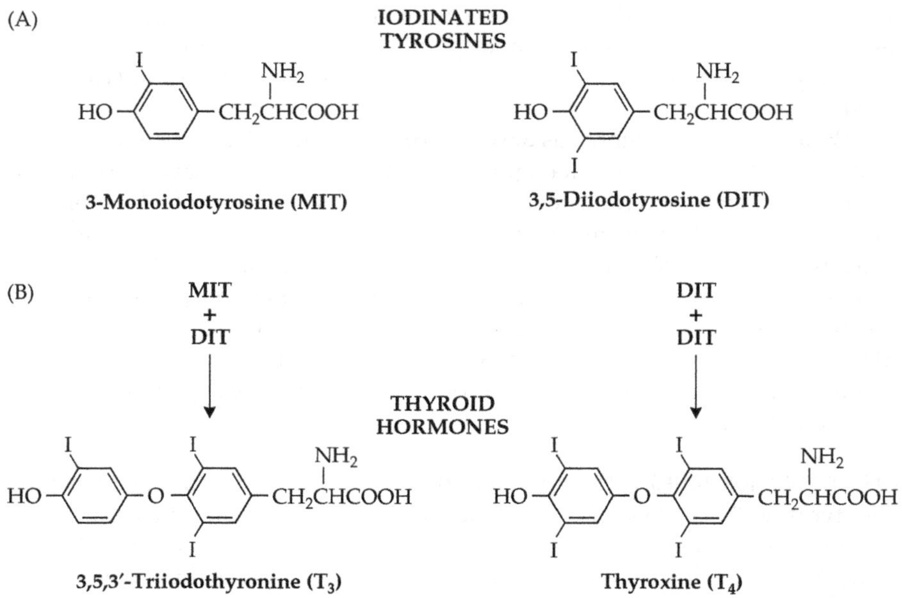

(A) IODINATED TYROSINES

3-Monoiodotyrosine (MIT)

3,5-Diiodotyrosine (DIT)

(B) MIT + DIT → THYROID HORMONES ← DIT + DIT

3,5,3′-Triiodothyronine (T_3)

Thyroxine (T_4)

2.26 Thyroid hormone synthesis. (A) The follicles of the thyroid produce a large glycoprotein molecule called thyroglobulin. Dietary iodine enters the bloodstream from the gut and is transported to the thyroid gland, where tyrosyl components of the thyroglobulin molecule become iodinated and form monoiodotyrosine (MIT) and diiodotyrosine (DIT), the precursors of the two main thyroid hormones. (B) Triiodothyronine (T_3) results from the combination of MIT and DIT, and thyroxine (T_4) is formed by the combination of two DIT molecules.

Thyroid hormones can increase the rate of glucose oxidation and thus increase the amount of metabolic heat produced in any given time. In addition, it is thought that thyroid hormones can uncouple oxidative phosphorylation, which decreases the efficiency of adenine triphosphate (ATP) synthesis, allowing the release of more heat. By analogy, if the motor of an electric fan heated up excessively when turning the fins, this would indicate that the motor was inefficient in converting electrical energy into movement. Because energy cannot be created or destroyed, the excess energy would be released as heat. Likewise, a reduction in the efficiency of ATP synthesis increases the quantity of heat released per mole of glucose oxidized. There are some data suggesting that this function may be imprecise and that the uncoupling may not occur. An alternative hypothesis to explain the ability of thyroid hormones to increase oxidation rates suggests that they stimulate extramitochondrial processes that consume nutrients, but generate fewer moles of ATP.

Thyroid hormones are probably not as important in acute responses to cold as are the adrenal corticoids, but they do play an important role in adaptation to changing temperatures. Generally, thyroid activity in mammals is greater in winter than in summer. In some nonhibernating mammals (for example, beavers and muskrats), however, thyroid activity is decreased during the winter. This is also true in hibernating species, but apparently the reduction in thyroid function does not directly cause hibernation.

The thyroid hormones also have specific effects related to carbohydrate, lipid, and protein metabolism, usually enhancing the effects of other hormones. For example, thyroid hormones can cause hyperglycemia (increased blood sugar levels), elevate lipid oxidation, and alter nitrogen balance.

Thyroid hormones are also important in growth and differentiation. Their growth-promoting actions are closely related to those of growth hormone, and they probably have priming effects (see below) on GH-sensitive target cells; that is, the effects of growth hormone probably cannot be expressed without prior or simultaneous exposure to thyroid hormones. We know that thyroid hormones mediate the secretion of growth hormone from the anterior pituitary. Thyroid hormones may also stimulate somatomedin production, thereby augmenting the actions of GH.

Thyroid hormones have both direct and indirect effects on behavior. Hypothyroidism, the insufficient production of thyroid hormone, can have a number of detrimental effects. The primary tissue affected by a lack of thyroid hormones is nervous system tissue. Hypothyroidism during development of the nervous system results in profound mental retardation, a syndrome called cretinism in humans; however, the addition of iodine to table salt has virtually eliminated this condition in industrialized countries.

Finally, the thyroid affects reproduction. Generally, sexual maturation is delayed in hypothyroid mammals. Spermatogenesis occurs in hypothyroid males, but blood levels of testosterone are reduced; as we have seen, a reduction in serum testosterone levels can greatly affect many reproductive behaviors. Ovarian cycles are irregular in hypothyroid female rats.

THE PARATHYROID AND C-CELL HORMONES The hormones produced by the parathyroid gland and by the C-cells of the thyroid are both protein hormones, and both are involved in calcium metabolism. **Parathyroid hormone (PTH)** is made up of 84 amino acids and has a molecular weight of about 9500 daltons. In terrestrial vertebrates, PTH elevates blood levels of calcium (Ca^{2+}) by increasing resorption of Ca^{2+} from the bone and absorption of Ca^{2+} from the gut via its effects on vitamin D_3. PTH also inhibits phosphate from the kidney, which reduces Ca^{2+} clearance.

Calcitonin (CT) is made up of 32 amino acids and has a molecular weight of approximately 3600 daltons. There is a disulfide bond between amino acids 1 and 7. CT is released from the C-cells of the thyroid in terrestrial vertebrates or from the ultimobranchial organ in fishes. CT acts in opposition to PTH to lower blood levels of calcium by inhibiting the release of Ca^{2+} from bone. PTH and CT are both controlled directly by blood calcium levels; there are no pituitary tropic hormones involved in their regulation. The importance of CT in mammalian calcium regulation remains an open question.

GASTROINTESTINAL HORMONES Three major gastrointestinal hormones, secretin, gastrin, and cholecystokinin (CCK), are released into the circulation and act to supplement the actions of the autonomic nervous system that mediate the process of digestion. Some of these so-called gut hormones have also been found in the brain, and several behavioral effects have been attributed to them (see Chapters 8 and 11).

Secretin is a small peptide consisting of 27 amino acids. It and other members of the secretin "family," including vasoactive intestinal polypeptide (VIP), share many amino acid sequences in common. The actions of secretin were first described by Starling and Bayliss in a series of papers that are generally considered to be the genesis of modern endocrinology (Box 2.2). The passage of food into the duodenum (small intestine) stimulates the release of secretin by the duodenal mucosa. Secretin stimulates the acinar cells of the pancreas to produce water and bicarbonate (CHO_3^-), which aid in digestion. Other actions of secretin include the stimulation of hepatic bile flow (from the liver) and pepsin secretion, as well as the inhibition of gastrointestinal tract movement and gastric acid secretion. Secretin also influences insulin release, fat cell lipolysis, and renal (kidney) function.

There are several forms of the polypeptide **gastrin**, but the C-terminal tetrapeptide amide Trp-Met-Asp-Phe-NH_2 is common to all forms. Gastrin is produced by the gastrin (or G) cells in the antral glands of the stomach. Its release is promoted by acetylcholine or by vagus nerve stimulation. At low concentrations, gastrin stimulates the secretion of water and electrolytes by the stomach, pancreas, and liver, as well as enzyme secretion by the stomach and pancreas. Hydrochloric acid in the stomach inhibits gastrin secretion via negative feedback. Gastrin also inhibits water and electrolyte absorption by the ileum. At high concentrations, gastrin stimulates the growth of gastric mucosa, release of insulin, and smooth muscle contractions of the gut, gallbladder, and uterus. Gastrin also inhibits gastric secretion at high concentrations. **Cholecystokinin (CCK)**, also called pancreozymin, is another member of the gastrin family of hormones. CCK consists of 33 amino acids and causes the exocrine pancreas to secrete digestive enzymes. CCK also causes the gallbladder to contract and release bile.

There are several additional gastrointestinal hormones, including bombesin, substance P, motilin, galanin, neurotensin, peptide YY, and neuropeptide Y. Many of these hormones have also been identified in the brain, where they appear to function as neurotransmitters or neuromodulators. Bombesin, for example, is a 14-amino acid peptide hormone that was originally isolated from frog skin (genus *Bombina*) and subsequently identified in the mammalian brain and gastrointestinal tract. A gastrin-releasing peptide (GRP), consisting of 27 amino acids, isolated from the gastrointestinal tract of hogs has virtually the same amino acid sequence as bombesin at one end of the peptide. Both GRP and bombesin stimulate gastrin release in isolated rat stomachs. Bombesin or a bombesin-like substance in the brain may be involved in feeding behavior in mammals. Other gut hormones that have been found in the brain and implicated in mammalian feeding behavior include substance P, neuropeptide Y, and galanin. When such substances are

BOX 2.2 The Discovery of Secretin

Despite Berthold's early work in 1849, the scientific study of endocrinology did not "officially" begin until the first 5 years of the 20th century, when two British physiologists, William M. Bayliss and Ernest H. Starling, described the mode of action of the hormone secretin. Working in apparent ignorance of Berthold's earlier work, Bayliss and Starling convincingly demonstrated that chemical mediation of physiological processes could occur independently from the nervous system, and that certain organs could release specific chemical agents that could travel through the blood and affect physiological processes at some distance from their source.

William M. Bayliss (1860–1924)

Ernest H. Starling (1866–1927)

Normally, the pancreas secretes digestive juices when food enters the intestine. Building on the earlier work of physiologists such as Claude Bernard and Ivan Pavlov, Bayliss and Starling denervated a loop of jejunum (intestine) in dogs. When they perfused the jejunum with weak hydrochloric acid, they observed the pancreatic secretory response. This finding demonstrated that the pancreatic secretory response does not involve the nervous system, but must be stimulated by some substance that travels through the blood. However, when Bayliss and Starling injected the original stimulant, hydrochloric acid, directly into the blood, they did not observe a pancreatic secretory response. This result indicated the presence of a mediating factor. That mediating factor was found to be secretin, a hormone that is released from the cells of the duodenal mucosa in response to acidified food entering from the stomach. The hormone travels through the blood to the exocrine cells of the pancreas to stimulate the release of pancreatic digestive juices.

Bayliss and Starling published a series of elegant studies from 1902 through 1905. For example, they demonstrated the independence of secretin release and the pancreatic secretory response from all neural influences by means of transplantation studies in which the intestinal loops containing the pancreas were moved to new locations without impairment of the secretin reaction. Their most convincing demonstration of the lack of neural involvement was obtained when the circulatory systems of two dogs were joined surgically (see below). Introduction of hydrochloric acid into the duodenum of one dog stimulated the secretion of pancreatic juices in both animals. For their work, they were nominated for a Nobel Prize in Medicine in 1913.

secreted by neurons, they are usually considered neurohormones or neuropeptides and usually interact with receptors on adjacent neurons.

STOMACH HORMONES While investigating drugs to stimulate GHRH, researchers discovered a hormone that stimulated GHRH release from the anterior pituitary. The hormone was named **ghrelin** (*ghre* = proto-Indo-European root for "grow"; *relin*, "release"). Ghrelin is a 28-amino acid peptide that is made in the stomach (Casanueva and Dieguez, 2002). When ghrelin was administered to mice to see whether it would enhance GH secretion, their food intake and fat deposition increased. Concentrations of ghrelin increased to peak levels prior to each meal (~80% increase) and fell dramatically after the meal in both mice and humans. Human participants treated with ghrelin ate about 30% more food than individuals not given the hormone. The potential of this hormone for clinical treatment of obesity is high, although obese individuals already have lower than average levels of ghrelin in their blood. A more immediate benefit of ghrelin is in the treatment of cachexia, the body wasting syndrome observed in people with many cancers and AIDS. Additional discussion of ghrelin and food intake is presented in Chapter 9.

ADIPOKINE HORMONES As noted in Chapter 1, **leptin** is a protein hormone that is secreted from adipose (fat) cells. Leptin acts on receptors in the central nervous system and at other sites to induce energy expenditure and inhibit food intake (Margetic et al., 2002). There is a relationship between body fat mass and blood leptin concentrations in some people. Leptin blood concentrations increase after a meal in concert with insulin release. If leptin is a "satiety" hormone, then it probably operates on the hypothalamus to reduce food intake, possibly by influencing the motivation to eat in some species, and by affecting the release of neuropeptides that regulate food intake (Cowley et al., 2001).

Adiponectin is a recently discovered peptide hormone that is released by fat cells (Saltiel and Kahn, 2001). Expression of adiponectin mRNA is decreased in obese mice and humans. Treatment with adiponectin lowers circulating glucose and free fatty acid levels in the blood, reduces insulin resistance, and decreases triglyceride storage in skeletal muscle in mice. Adiponectin receptors have been identified in the liver and skeletal muscles (Yamauchi et al., 2003). Mice in which the gene for adiponectin has been knocked out are insulin resistant and display diabetes. Additional research will be needed to determine whether adiponectin has any behavioral effects.

Many of the adipokine and gastrointestinal hormones we have just described operate on systems that are involved with energy balance (ingestive behaviors, activity, etc.) and also affect hypothalamic releasing hormones and tropic hormones from the anterior pituitary involved in reproductive regulation. The interaction between energy balance and reproduction will be considered in later chapters.

THE PANCREATIC HORMONES Three major peptide hormone products are secreted by the endocrine cells of the pancreas: insulin, glucagon, and somatostatin. **Insulin**

is a protein hormone that has a molecular weight of about 6000 daltons. Insulin has three disulfide bonds in most vertebrates; it is made up of two short amino acid chains, called the A- and B-chains, linked together by two disulfide bonds. Proinsulin, the precursor of insulin, possesses a 33-amino acid C-peptide in addition to the A- and B-chains (Figure 2.27). The insulin molecule has not changed in structure during the evolution of animals, and it is evident in both vertebrates and invertebrates; in other words, insulin is a highly conserved molecule. Many hormones act to increase blood glucose levels, but insulin is the only known hormone in the animal kingdom that can lower them. Consequently, anything that disrupts insulin action has a profound effect on blood glucose regulation.

The first place insulin acts after its secretion by the β-cells of the pancreas is the liver, where it promotes energy storage in the form of glycogen. Except for central nervous system tissue, all cells have receptors for insulin. When a cell's insulin receptors are activated, blood glucose is taken up into the cell and used or, in muscle or adipose cells, stored (as glycogen or fat, respectively).

Glucagon is a simple peptide that is similar to those of the secretin family. Typically, glucagon has 29 amino acids and a molecular weight of approximately 3000 daltons. Once released from the α-cells of the pancreas, glucagon travels first to the liver, where it stimulates glycogenolysis, or the breakdown of stored glycogen. Glucagon thus acts in opposition to insulin and serves to increase blood levels of glucose.

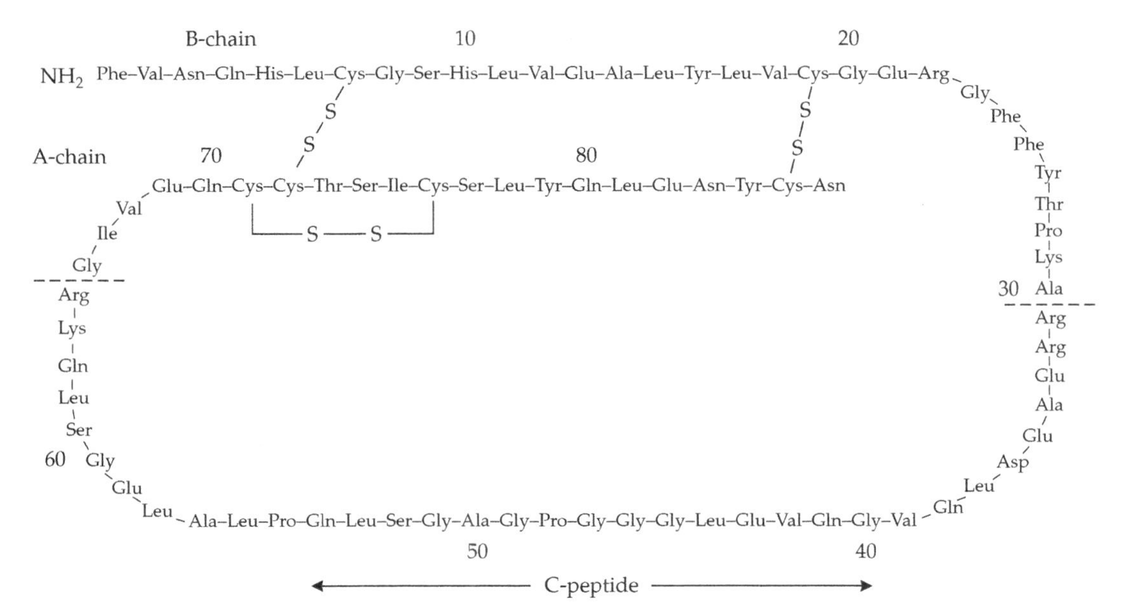

2.27 Proinsulin, the chemical precursor of insulin, consists of two short peptide chains, called the A- and B-chains, which are bound by two disulfide bonds, and a 33-amino acid C-peptide.

Somatostatin is an inhibitory hormone released from the δ-cells of the pancreas. Somatostatin consists of 14 amino acids and inhibits the release of insulin and glucagon locally in the pancreas (paracrine action). Somatostatin is also released from the hypothalamus to regulate the release of growth hormone from the anterior pituitary.

THE ADRENAL MEDULLARY HORMONES Enkephalins are released by the adrenal medulla in response to stress. A precursor protein consisting of 267 amino acids yields one leu-enkephalin and six met-enkephalins. The individual enkephalins are produced by the actions of trypsin-like enzymes on the proenkephalin precursor. As described above, enkephalins are also among the products derived from the POMC molecule and released by the anterior pituitary; however, the adrenal medullae are the major source of circulatory enkephalins. No definitive function has yet been assigned to the adrenal medullary enkephalins, so their identification as hormones remains preliminary. Because these hormones are released in response to stress, it is reasonable to suggest that they are involved in adaptation to stress, but this remains to be determined.

THE GONADAL PEPTIDE HORMONES Several peptide hormones are secreted by the gonads. Early in testicular development, a small peptide, **Müllerian inhibitory hormone** (**MIH**), inhibits development of the Müllerian duct system, the embryonic tissue that gives rise to the female accessory sex organs (see Chapter 3). The Sertoli cells in the testes and the granulosa cells in the ovaries also secrete different forms of **inhibin**, a hormone that feeds back to block the secretion of FSH from the anterior pituitary. Inhibin also actively inhibits **aromatase**, an enzyme involved in the formation of estrogens from androgens, in the granulosa cells of the ovaries. A closely related peptide, **activin**, has been discovered in both the testes and the ovaries. Its name comes from its stimulatory effects on FSH secretion. Its physiological function, if any, has yet to be determined in the testes, but it directly stimulates aromatase activity in the ovarian granulosa cells. Although additional research needs to be conducted on the inhibins and activins, these hormones are certainly candidates for regulatory effects on hormone–behavior interactions.

Another peptide hormone found in the ovaries of mammals is called **relaxin**. Relaxin, first identified in 1932, is an insulin-related peptide hormone that is produced in the corpora lutea during pregnancy. It functions to soften estrogen-primed pelvic ligaments to allow them to stretch sufficiently to permit passage of the relatively large head of a mammalian fetus through the pelvis during birth.

THE PLACENTAL HORMONES The placenta secretes a number of peptide hormones that can affect both the fetus and the mother. Some of these hormones are similar to hormones secreted elsewhere. For example, the placenta secretes prolactin and GnRH that are chemically similar to the versions released from the anterior pituitary gland and hypothalamus, respectively. Other peptide hormones are unique to the placenta, but may augment the function of hormones from other sources. For example, chorionic gonadotropin (CG), chorionic somatomammotropin (CS) (also called placental lactogen), chorionic corticotropin (CC), and chorionic thyrotropin (CT) act as "supplementary" tropic hormones to stimulate gonadal, mammary, adrenal, and thyroid functions. Recent evidence suggests that CS is important in initiating the onset of mammalian maternal behavior. Other hormones, such as POMC, β-endorphin, and α-MSH, are secreted by the placenta in vitro and appear identical to the hypothalamic versions.

The Steroid Hormones

The adrenal glands and the gonads are the most common sources of steroid hormones in vertebrates. Vertebrate steroid hormones have a characteristic chemical structure that includes three six-carbon rings plus one conjugated five-carbon ring. In the nomenclature of steroid biochemistry, substances are identified by the number of carbon atoms in their chemical structure (Figure 2.28). The precursor to all vertebrate steroid hormones is **cholesterol.** Although we mainly associate this waxy, artery-blocking substance with the bad cardiovascular consequences that can result from ingesting too much of it in our food, our bodies make substantial quantities of cholesterol from acetate in our livers. In addition to its role as a precursor to steroid hormones, cholesterol is important in many other biochemical reactions. The cholesterol molecule contains 27 carbon atoms. Thus, cholesterol is a C_{27} substance, although cholesterol itself is not a true steroid.

As noted above, steroid hormones are lipid-soluble and move easily through cell membranes. Consequently, they are never stored, but leave the cells in which they were produced almost immediately. A signal to produce steroid hormones is also a signal to release them. The response to that signal, however, can be a rather slow one: the delay between stimulus and response in biologically significant steroid production may be hours, although ACTH stimulates corticoid secretion within a few minutes and LH acts quickly to affect progesterone production during the periovulatory surge.

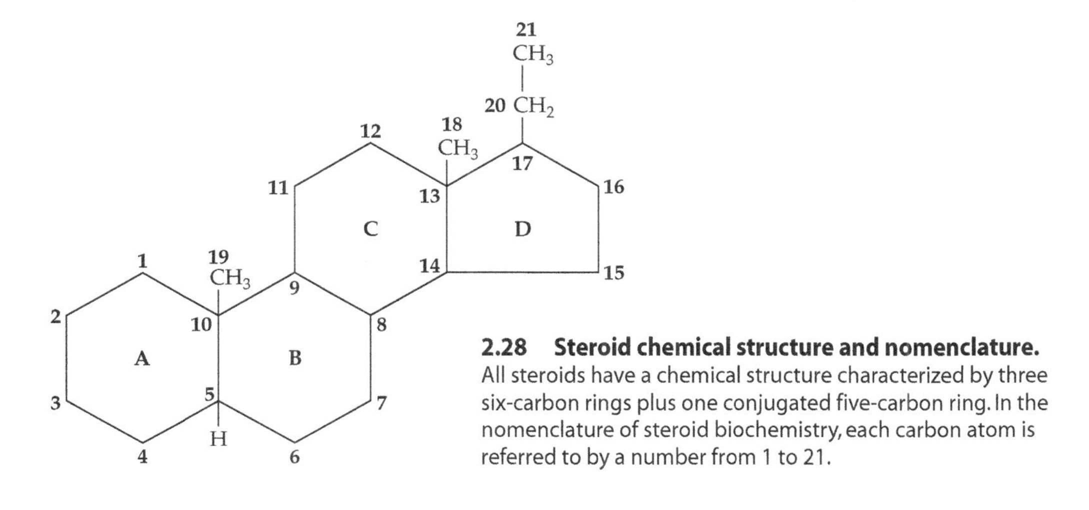

2.28 Steroid chemical structure and nomenclature. All steroids have a chemical structure characterized by three six-carbon rings plus one conjugated five-carbon ring. In the nomenclature of steroid biochemistry, each carbon atom is referred to by a number from 1 to 21.

Steroid hormones are not very soluble in water, and, as you know, blood consists mainly of water. In the circulatory system, steroid hormones must generally bind to water-soluble carrier proteins that increase the solubility of the steroids and transport them through the blood to their target tissues. These carrier proteins also protect the steroid hormones from being degraded prematurely. The target tissues have cytoplasmic receptors for steroid hormones and accumulate steroids against a concentration gradient.

Upon arrival at the target tissues, steroid hormones dissociate from their carrier proteins and diffuse through the cell membrane into the cytoplasm of the target cell, where they bind to the cytoplasmic receptors. The amino acid sequence of steroid hormone receptors is highly conserved among vertebrates. Each steroid hormone receptor comprises three major domains: the steroid hormone binds to the C-terminal domain, the central domain is involved in binding to DNA, and the N-terminal domain interacts with other DNA-binding proteins to affect transcriptional activation (Hadley, 1996). The steroid–receptor complex is transported into the cell nucleus, where it acts as a transcription factor by binding to a DNA sequence called a **hormone response element** within the promoter of a gene and stimulating or inhibiting the transcription of that gene (Figure 2.29). The precise mechanism by which the binding of a steroid–receptor complex to a specific

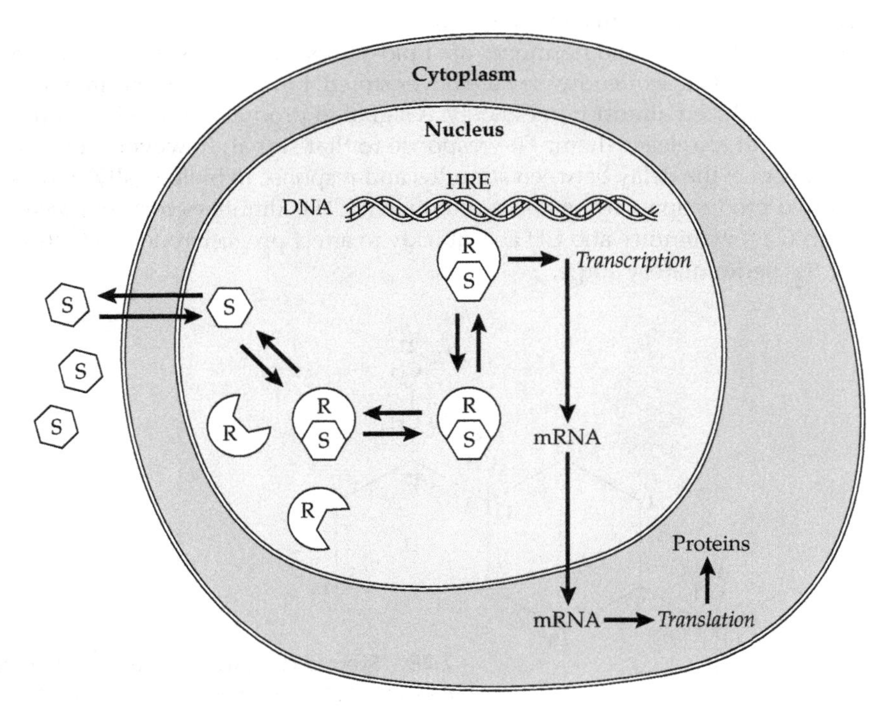

2.29 Steroid hormone receptors are generally located in the cytosol or nucleus of a cell. The steroid hormone (S) binds to its receptor (R) to form a hormone–receptor complex (transcription factor), which binds to a hormone response element (HRE) on the DNA to begin transcription of mRNA.

hormone response element evokes activation or suppression of gene transcription remains unknown, but it appears certain that co-activator proteins are often necessary (Smith et al., 1997). The transcribed mRNA migrates to the cytoplasmic rough endoplasmic reticulum, where it is translated into specific structural proteins or enzymes that produce the physiological response. Changes in the types of proteins a cell makes (i.e., the gene products) can often be observed within 30 minutes of hormone stimulation.

The actions of steroids on target tissues, therefore, are based on three factors: (1) the steroid hormone concentrations in the blood, (2) the number of available receptors in the target tissue, and (3) the availability of appropriate co-activators. Variation in any of these factors can influence the effects of steroid hormones. Blood concentrations of steroid hormones are themselves dependent on three factors: (1) the rate of steroid biosynthesis; (2) the rate of steroid inactivation by catabolism, which occurs mainly in the liver; and (3) the "tenacity" (affinity) with which the steroid hormone is bound to its plasma carrier protein. Multiple versions of steroid receptors represent a mechanism by which responsiveness to steroid hormones can be regulated. Recently, it has been determined that different "types" of steroid receptors exist; for example, three versions of the estrogen receptor (α, β, and γ) are currently recognized (Hawkins et al., 2000). The effects of environmental, social, or other extrinsic or intrinsic factors on the regulation of specific co-activators have not been studied, but represent yet another process by which individual variation in hormone–behavior interactions may be mediated (Brosens et al., 2004). The regulation of receptors and co-activators is currently a very active research area in molecular endocrinology.

THE C_{21} STEROIDS: PROGESTINS AND CORTICOIDS In response to various protein hormones from the anterior pituitary, cholesterol is converted into steroid hormones in the adrenals or gonads (Figure 2.30). The pituitary hormones induce an enzyme, called desmolase, which cleaves off the long chain of carbons from the top of the cholesterol molecule to yield **pregnenolone**, a C_{21} steroid. There are two general types of C_{21} hormones, **progestins** and **corticoids**. Pregnenolone is a progestin, and it is the obligatory precursor to all other steroid hormones. In other words, pregnenolone is a prohormone, a substance that can act as a hormone itself or can be converted into another hormone that has different endocrine properties. For example, pregnenolone can be converted into progesterone, another progestin. Progesterone is also a prohormone; it has several endocrine functions itself, and it is the major precursor for a variety of other C_{21} steroids, including the corticoids and another progestin called 17α-hydroxyprogesterone, which in turn is an important precursor of additional steroid hormone types.

Many progestins (e.g., pregnenolone and progesterone) are ubiquitous in vertebrates. Although named for their "progestational" or pregnancy-maintaining effects in rodents, progestins are found in vertebrates wherever steroidogenesis occurs. In mammals, progesterone is important in maintaining pregnancy and also in the initiation and cessation of mating behavior.

In response to hormone signals from the anterior pituitary, pregnenolone is converted by a series of enzymes in the adrenal glands to progesterone and then

Cholesterol

Pregnenolone

2.30 Biochemical pathways in steroid formation. The enzyme desmolase cleaves the chain of carbons from the top of the cholesterol molecule to form pregnenolone, a C_{21} steroid that is the obligate precursor to all other steroids.

Progesterone

11-Deoxycorticosterone

17α-Hydroxyprogesterone

Corticosterone

11-Deoxycortisol

Aldosterone

Cortisol

to the various corticoids. Like all steroid hormones, the corticoids appear to be released as they are produced. These steroids have behavioral effects associated with the maintenance of bodily functions, such as mediating salt appetite.

There are two types of corticoids: **glucocorticoids** and **mineralocorticoids**. Glucocorticoids are involved in carbohydrate metabolism and are often released in response to stressful stimuli; the two primary glucocorticoids are **corticosterone** (abbreviated B) and **cortisol** (F).* Most animals make either corticosterone or cortisol, but rarely are both glucocorticoids produced in large quantities. All reptiles and birds secrete corticosterone from their adrenals. Some mammals also secrete corticosterone; rats and mice secrete corticosterone exclusively. The primary glucocorticoid secreted by humans and other primates, however, is cortisol.†

Of the mineralocorticoids, **aldosterone** is the most important, as it is secreted by all terrestrial vertebrates and functions in ion exchange and water metabolism. Aldosterone is primarily responsible for retaining sodium ions (Na^+) and excreting potassium ions (K^+).

THE C_{19} STEROIDS: ANDROGENS Just as a specific series of enzymes in the adrenals convert pregnenolone into the corticoids, specific enzymes found primarily in the gonads convert pregnenolone into several types of C_{19} steroid hormones, called **androgens** because of their male- (*andros*) generating effects (Figure 2.31). These C_{19} steroid hormones result from the enzymatic cleavage of the ethyl group from a progestin precursor at C_{17}. **Testosterone** and **androstenedione** are biologically important androgens, as is a biochemically reduced version of testosterone, **5α- and 5β-dihydrotestosterone (DHT)**.†† These androgens are produced in the well-vascularized Leydig cells of the testes; the Sertoli cells are the source of androgen-binding proteins that carry androgens through the blood (see Figure 2.11B). Testosterone is typically reduced to DHT or converted to an estrogen in order to have a biological effect (as we will see below).

The zona reticularis of the adrenal cortex also produces a relatively weak androgen, called **dehydroepiandrosterone (DHEA)**. For the most part, the physiological functions of DHEA remain unknown, but blood concentrations of this hormone decline as individuals age. Replacement of DHEA at appropriate small doses appears to ameliorate certain effects of aging, including decreased muscle strength and weakened immune function (Yen et al., 1995). Based on these preliminary studies, alternative health care providers have advocated DHEA sup-

*The adrenal steroids were initially fractionated or isolated before their biological functions were known. Each fraction was designated compound A, compound B, and so on. Later, when identified, they retained their original letter designations, but these abbreviations are now becoming obsolete and replaced by abbreviations such as CORT.

†It is important to be aware of this distinction between the two glucocorticoids. A physician–scientist recently argued that a procedure he was performing on mice was not stressful because "the levels of cortisol were very low." However, after he was persuaded to measure corticosterone, he discovered that his procedure had indeed caused a substantial increase in this corticoid, which presumably reflected elevated stress.

††The "5" in the name indicates that the reduction occurs with the addition of a hydrogen atom at C_5; α and β refer to the orientation of the hydrogen molecule in space.

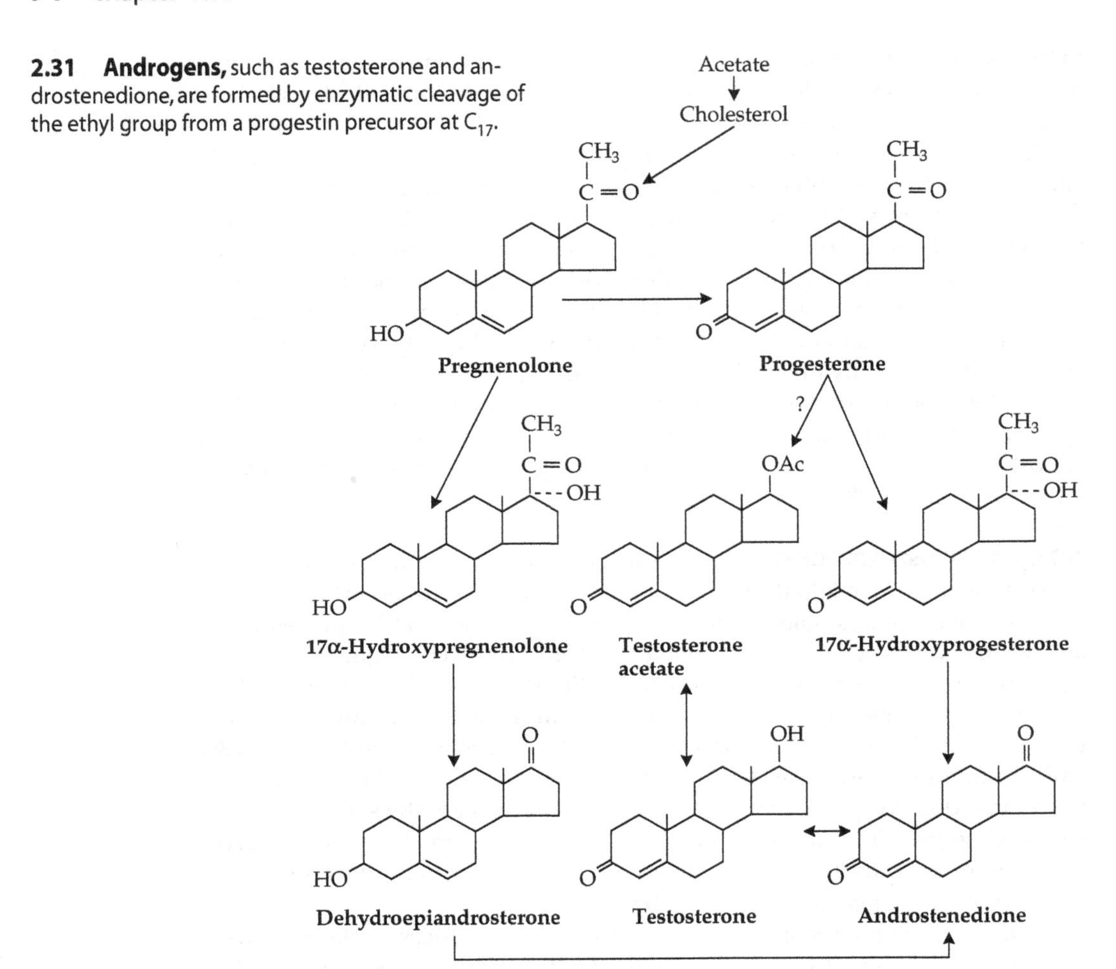

2.31 Androgens, such as testosterone and androstenedione, are formed by enzymatic cleavage of the ethyl group from a progestin precursor at C_{17}.

plementation as a treatment to reverse some of the effects of aging. However, long-term studies on the effects of DHEA therapy on human health are lacking, and it is probably prudent to await the results of such studies before indulging in self-administration of an active steroid hormone.

Androgens have many physiological and behavioral functions. These steroid hormones are necessary for spermatogenesis and the maintenance of the genital tract (for example, the vas deferens). Androgens also maintain the accessory sex organs such as the prostate, seminal vesicles, and bulbourethral glands. The male secondary sex characters are also supported by androgens: well-known examples include the pattern and density of body hair in humans, comb size in roosters, and antler growth in deer. Some species show subtle morphological effects due to androgens; for example, in mice, the salivary gland and kidney structures are affected by androgens. The liver, heart, and kidneys are generally larger in male mammals than in females with similar body masses. Of course, androgens have many effects on behavior, including courtship and copulatory behaviors, aggressive behaviors, and other social behaviors.

Metabolism is greatly affected by androgens. These hormones stimulate respiratory metabolism and are well known for their protein anabolic effects. The increase in muscle mass that results from treatment with anabolic androgens motivates many people to abuse these compounds. However, because many organs have androgen receptors, long-term problems in the liver, heart, and kidneys can result from chronic exposure to high androgen levels as hypertrophy of these organs leads to a reduction in their functional efficiency. Reproductive problems are also common among people abusing anabolic androgens. Psychological problems have also been reported for long-term androgen users. In many cases, however, a subjective sense of "invincibility" allows steroid abusers to dismiss warnings about the long-term health problems associated with androgen treatment (see Chapter 13).

THE C_{18} STEROIDS: ESTROGENS Androgens are the obligatory precursors of all **estrogens** in the same way that progestins are the obligatory precursors of all androgens. Specific enzymes, primarily present in the ovaries, convert testosterone and androstenedione to estrogens by cleaving the carbon at position 19 from these androgen precursors (Figure 2.32). This process is called **aromatization** because the removal of the carbon leaves the estrogen with a phenolic A-ring, known as an aromatic compound. Biologically significant estrogens include 17β-estradiol, estrone, and estriol.

The interstitial tissues of the ovary (see Figure 2.12B) produce steroid hormones. The theca interna cells produce progestins, and enzymes in these cells act on those progestins to produce androgens. The blood flow in the ovary moves these androgens into the granulosa cells, where enzymes rapidly convert them into estrogens. The follicles of the ovary produce increased amounts of estrogens as they mature. The corpora lutea produce progesterone in many vertebrate species, but also release estrogens, androgens, and perhaps oxytocin, a peptide hormone.

Thus, ovaries produce substantial amounts of androgens, which are normally converted immediately into estrogens. However, in some cases, excess androgens are produced and enter the general circulation before they can be converted into estrogens, affecting female physiology and behavior. Furthermore, if insufficient enzymes are present to convert androgens to estrogens in the ovaries, some androgens may be secreted into a female's blood circulation. On the other hand, if high levels of the enzymes that convert androgens to estrogens are present in the testes, then estrogens will be secreted from the testes into the blood circulation of a male. Some androgens, such as dihydrotestosterone (DHT), cannot be aromatized and therefore cannot be converted to estrogens.

Estrogens have many functions. Estrogens initiate the formation of corpora lutea. They also affect the genital tract; for example, high levels of estrogens correlate with increased uterine mass. The secondary sex characters of female mammals are influenced by estrogens. Estrogens have several metabolic functions, including effects on water metabolism: for instance, they favor the retention of

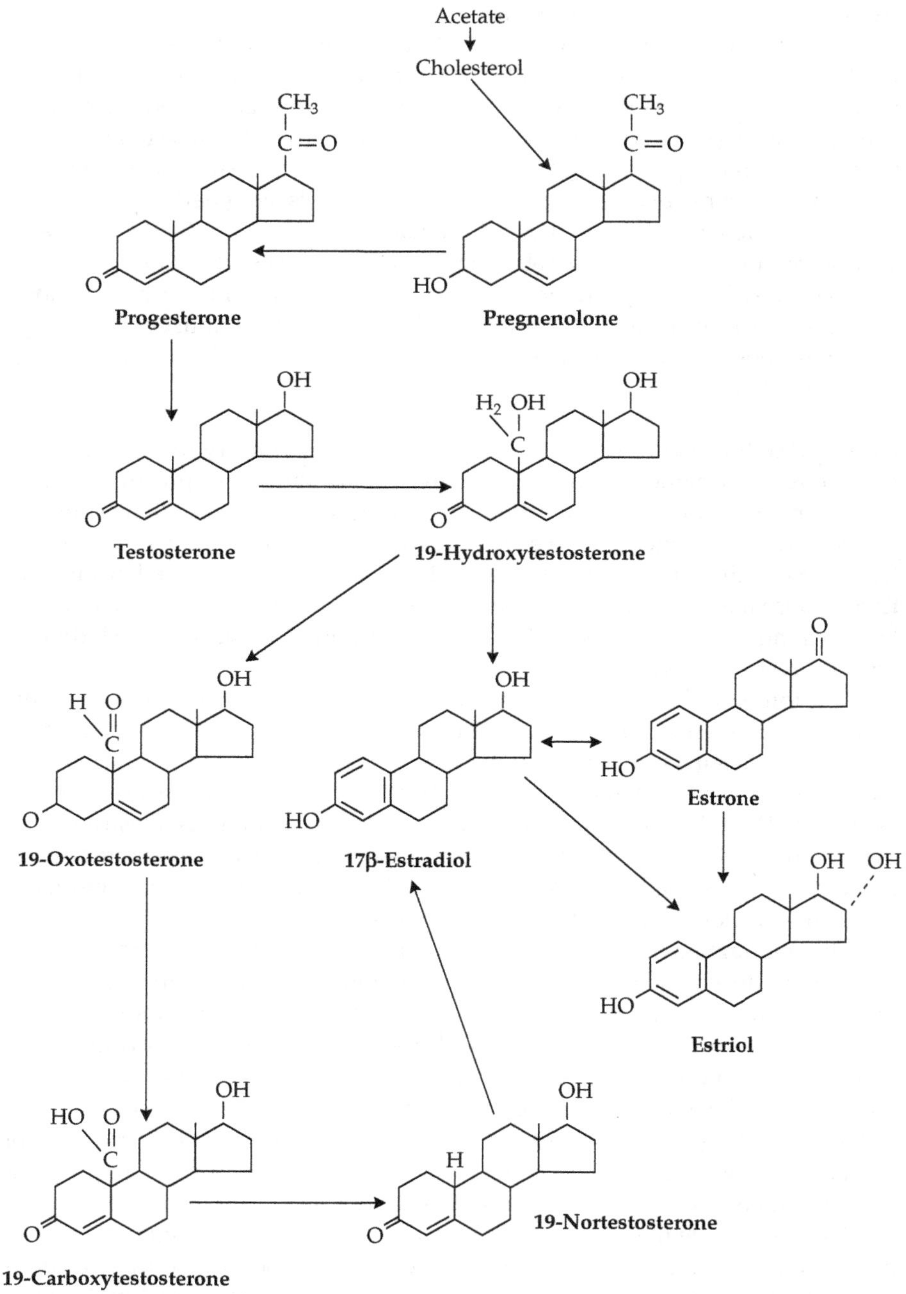
Acetate
Cholesterol
CH3
C=O
HO
Pregnenolone
CH3
C=O
O
Progesterone
OH
O
Testosterone
H2 OH
C
OH
O
19-Hydroxytestosterone
H O
C
OH
O
19-Oxotestosterone
OH
HO
17β-Estradiol
O
HO
Estrone
OH OH
HO
Estriol
HO O
C
OH
O
19-Carboxytestosterone
H
OH
O
19-Nortestosterone

2.32 Estrogens, such as 17β-estradiol, estrone, and estriol, are formed when C_{19} is cleaved from an androgen precursor in a process called aromatization.

water in humans. Estrogens are also important in calcium metabolism. More bone is made in the presence of high estrogen concentrations. During human menopause, estrogen production diminishes, and bone erodes away in a degenerative process called osteoporosis. Finally, estrogens are very important in sexual behavior and may also play a part in maternal aggression (see Chapter 7).

The placenta also produces steroid hormones. Both estrogens and progestins can be produced in significant amounts by the placenta; it also produces some androgens and corticoids. The contribution of these hormones from the placenta to the development of behavior is not well understood at this time.

NEUROSTEROIDS **Neurosteroids** are produced in both the central and peripheral nervous system, mainly in glial cells. They are synthesized de novo from cholesterol or from steroidal precursors produced elsewhere in the body. Common neurosteroids include 3 β-hydroxy-δ 5-compounds (e.g., pregnenolone and DHEA), their sulfated derivatives, and reduced steroidal metabolites (e.g., the tetrahydroderivative of progesterone, 3-α-hydroxy-5 α-pregnane-20-one (3-α,5-α-THPROG) (Frye, 2001). These compounds can function as modulators of certain neurotransmitter receptors. Progesterone also acts as a neurosteroid, and a progesterone receptor has been identified in peripheral and central glial cells (Frye, 2001). Neurosteroid concentrations in different brain regions vary according to the environment and behavioral interactions. Behaviors associated with stress, memory, sexual behavior, and aggression have all been reported to be affected by neurosteroids. Neurosteroids may also be important regulators of affect (mood) and modulate feelings of anxiety (Bitran et al., 2000).

A CAUTIONARY ASIDE Although the androgens and estrogens are commonly referred to as "sex hormones," it is important to emphasize that these hormones should not be considered "male" or "female." All male vertebrates produce some estrogens and progestins, and all female vertebrates produce androgens. The two sexes do differ in their relative concentrations of circulating androgens or estrogens, but this difference is due to the relative proportions of steroidogenic-specific enzymes in their gonads. Testes normally have more enzymes for making androgens and fewer aromatizing enzymes than do ovaries, but imbalances in the appropriate enzymes can lead to overt endocrine abnormalities. Androgens, with a few exceptions, are easily converted to estrogens. Testosterone can be considered a prohormone that is converted either to DHT or to an estrogen. Ovaries produce high concentrations of androgens, but high local concentrations of aromatizing enzymes rapidly convert those androgens to estrogens before they can enter the general circulation. Estrogens can theoretically be converted back to androgens, but this energetically expensive reaction is rare in nature.

It is also important to emphasize that, although sex steroid hormones are primarily formed in the gonads, they can also be produced in the adrenals, along with glucocorticoids and mineralocorticoids. Androgens, for example, may be synthesized in the adrenal cortex directly from corticoids if the appropriate enzymes are present. The production of the various types of steroid hormones is

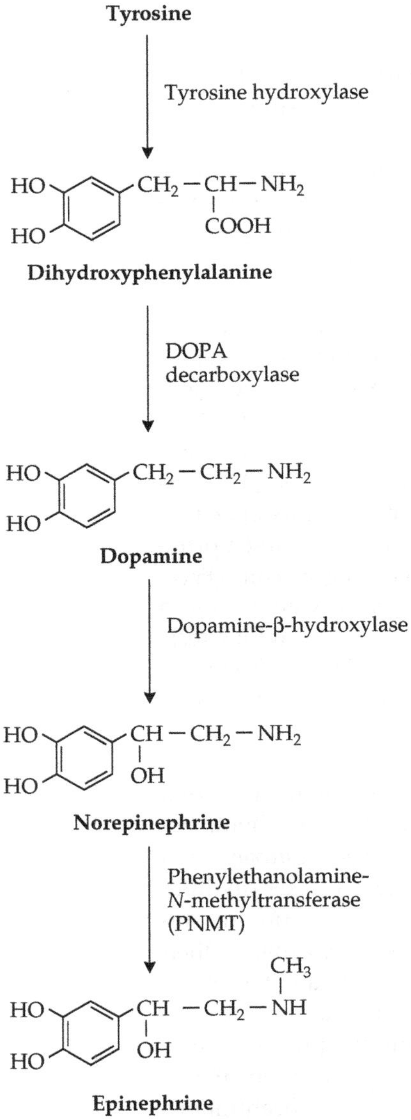

Tyrosine

Tyrosine hydroxylase

HO, HO — $CH_2-CH-NH_2$, COOH

Dihydroxyphenylalanine

DOPA decarboxylase

HO, HO — $CH_2-CH_2-NH_2$

Dopamine

Dopamine-β-hydroxylase

HO, HO — $CH-CH_2-NH_2$, OH

Norepinephrine

Phenylethanolamine-*N*-methyltransferase (PNMT)

HO, HO — $CH-CH_2-NH$, OH, CH_3

Epinephrine

2.33 Epinephrine, norepinephrine, and dopamine are catecholamines, so called because each is a catechol with an amine side group. Catecholamines are all derived from the amino acid tyrosine, which is hydroxylated to form 3,4-dihydroxyphenylalanine (DOPA). DOPA may be decarboxylated in the cytoplasm to form dopamine, which in turn may be hydroxylated to form norepinephrine. Norepinephrine can be converted to epinephrine by the addition of a methyl group to the amine.

zone-specific: the presence or absence of various enzymes in each of the three layers of the adrenal cortex determines the type of steroid hormone produced there. An aberrant gene coding for an incorrect enzyme can lead to an ovary or adrenal gland producing great quantities of androgens. The behavioral effects of such genetic errors will be discussed in Chapter 3.

The Monoamine Hormones

Monoamines are hormones that are derived from a single amino acid. There are two classes of monoamines that affect behavior: the **catecholamines** and the **indole amines**. Recall that the thyroid hormones, too, are often classified as monoamines because they are derived from a single amino acid, tyrosine.

ADRENAL MEDULLARY MONOAMINE HORMONES The adrenal medulla is the main source of two catecholamines, **epinephrine** (adrenaline) and **norepinephrine** (noradrenaline); it also releases some dopamine (Figure 2.33). Only epinephrine acts primarily as a hormone; norepinephrine and dopamine may also act as hormones, but they serve mainly as neurotransmitters. The catecholamines are derived from tyrosine and are called catecholamines because each is a catechol (that is, a dihydroxyphenol) with an amine side group. In response to sympathetic neural signals, the adrenal medulla releases these hormones into the general circulation. Typical physiological stimuli that evoke catecholamine release include stress, exercise, low temperatures, anxiety, emotionality (fight or flight responses), and hemorrhage. In humans, norepinephrine and epinephrine are released at a ratio of 1:4 (James, 1992).

The catecholamines influence both the circulatory and metabolic systems. They have five general effects: (1) increased heart rate and cardiac output; (2) vasoconstriction of the deep and superficial arteries and veins; (3) dilation of skeletal and liver blood vessels; (4) increased glycolysis; and (5) increased blood glucagon concentrations and decreased insulin secretion. All of these effects serve to prepare the body for action. Epinephrine and norepinephrine have similar physiological effects, but differ in several important respects. For instance, epinephrine elevates heart rate and affects metabolism more than norepinephrine. In contrast, norepinephrine is a more potent vasoconstrictor than epinephrine. The catecholamines affect their target cells via second messenger mechanisms.

THE PINEAL GLAND HORMONES Melatonin and serotonin are derived from the amino acid tryptophan. Melatonin is derived from serotonin in a two-step enzymatic reaction. These hormones are collectively called indole amines because the indole ring is common to both melatonin and serotonin (Figure 2.34).

Serotonin is the common name for 5-hydroxytryptamine (5-HT). Serotonin levels are high in the pineal gland during the light hours of the day, but diminish during the dark hours as serotonin is converted into melatonin. The enzyme *N*-acetyltransferase converts serotonin to *N*-acetylserotonin, which is transformed into melatonin (5-methoxy-*N*-acetyltryptamine) by the transfer of a methyl group from S-adenosylmethionine (SAM) to the 5-hydroxyl of *N*-acetylserotonin by the actions of hydroxyindole-O-methyltransferase (HIOMT). Serotonin and *N*-acetylserotonin are probably not released from the pineal gland in appreciable quantities.

Melatonin is the major hormone secreted by the pineal gland. It is highly lipid-soluble and probably leaves the cell by diffusion in a manner similar to the steroids. The melatonin receptor gene has been characterized, and codes for a 440-amino acid peptide that is found in the mammalian brain and in amphibian skin (Weaver et al., 1989; Bittman and Weaver, 1990; Dubocovich et al., 2003). Many mammals use the annual pattern of changes in day length as a cue to aid in the timing of various seasonal biological responses. Melatonin is an important hormone in the mammalian photoperiodic time-measurement mechanism (Goldman, 2001). The effects of melatonin on the seasonal organization of breeding will be described more fully in Chapter 10.

The Lipid-Based Hormones

Prostaglandins, a family of lipid-based hormones, were discovered in the 1930s. Their first known function was an ectocrine one: these compounds are found in seminal fluid and can cause uterine contraction or relaxation in the recipient. Since their initial discovery, there have been reports of a vast array of biological actions produced by the prostaglandins. However, they probably affect behavior only indirectly in vertebrates. Conversely, among many invertebrate species, prostaglandins directly mediate behavior.

All prostaglandins possess a basic 20-carbon fatty acid skeleton, called prostanoic acid, derived from essential fatty acids via cyclization and oxidation. The naturally occurring prostaglandins have been classified into four basic groups, named E, F, A, and B. These groups are distinguished by differences in the cyclopentyl group of the fatty acid skeleton. Substitutions in the side chains and the extent of saturations provide a number of compounds within each group.

Prostaglandins in the E and F series are involved in reproduction. High concentrations of prostaglandins coincide with degradation of the corpora lutea of

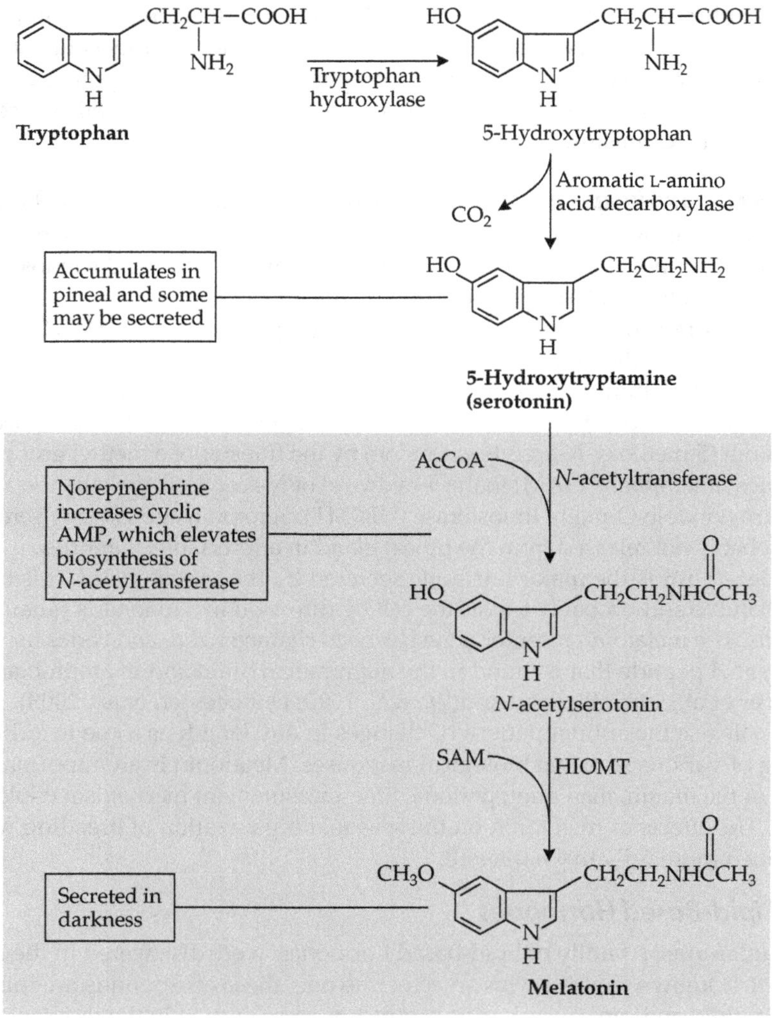

Tryptophan

Tryptophan hydroxylase

5-Hydroxytryptophan

Aromatic L-amino acid decarboxylase

CO_2

Accumulates in pineal and some may be secreted

5-Hydroxytryptamine (serotonin)

AcCoA

N-acetyltransferase

Norepinephrine increases cyclic AMP, which elevates biosynthesis of N-acetyltransferase

N-acetylserotonin

SAM

HIOMT

Secreted in darkness

Melatonin

2.34 Serotonin and melatonin are formed from tryptophan. During the light hours of the day, tryptophan is converted in a two-step enzymatic reaction to serotonin. In darkness (shaded area), increased norepinephrine secretion causes an increase in *N*-acetyltransferase, the first of two enzymes that convert serotonin to melatonin. SAM = S-adenosylmethionine; HIOMT = hydroxyindole-O-methyltransferase.

the ovary. In other systems, prostaglandins mediate a large array of actions, often opposing ones; for example, prostaglandins both stimulate and relax smooth muscle. Prostaglandins also affect cardiac functioning. In some circumstances they act as a pressor agent, whereas in other situations they have an antipressor effect.

Prostaglandins have also been implicated in influencing cyclic nucleotide formation. Prostaglandins inhibit cAMP production in adipose cells, which can be stimulated by hormones such as epinephrine. In other cell types, prostaglandins stimulate cAMP formation, and often modulate the second messenger response that mediates the actions of many peptide hormones. Prostaglandins constitute one type of the chemical messengers called eicosanoids. The prostacyclins, thromboxanes, and leukotrienes are closely related eicosanoids that may play important roles in behavioral processes.

How Hormones Are Regulated

There are two basic patterns of internal hormonal regulation: (1) hormones are regulated by the physiological by-products generated in response to their actions; and (2) hormones are regulated by the stimulatory or inhibitory effects of hormones. Within the second type of control system, there may be one, two, or three hormones in a regulatory chain, or there may be autoregulation via feedback from the circulating levels of the regulated hormone itself. Within either type of control system, feedback may be negative or positive but negative feedback is far more common (Figure 2.35).

Parathyroid hormone provides an example of the first type of regulatory mechanism. When blood levels of calcium decrease, parathyroid hormone is released. When the action of the hormone has raised the concentration of blood calcium to some optimal level, parathyroid hormone secretion stops. Calcium levels in the blood are thus maintained in an optimal range (Figure 2.36A). This simple **negative feedback** system works much the way a thermostat controls the temperature in your home.

GnRH provides an example of a hormone that is regulated by a multiple chain of negative feedback controls (Figure 2.36B). In response to environmental stimuli, some internal pacemaker, or some other intrinsic event, GnRH is released by the hypothalamus and stimulates the release of the gonadotropins from the anterior pituitary gland. The gonadotropins, in turn, stimulate steroid and gamete produc-

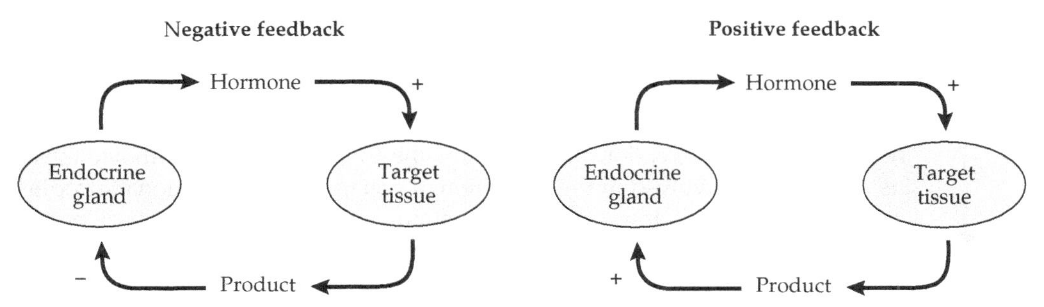

2.35 Models of negative and positive feedback. In negative feedback, the production of a product by the target tissue feeds back to the source of the hormone and causes it to stop producing the hormone. During positive feedback, the production of a product by the target tissue stimulates additional hormone production. Negative feedback is much more common than positive feedback in endocrine regulation.

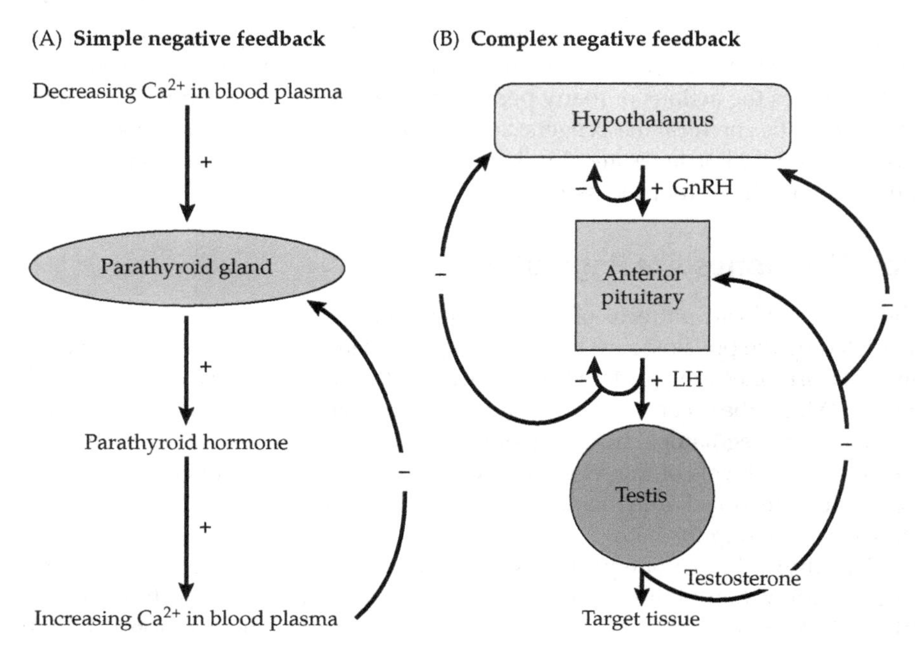

2.36 Negative feedback. (A) The parathyroid gland is regulated by a simple negative feedback mechanism. When levels of calcium in the blood decline, the parathyroid gland is stimulated to release parathyroid hormone, which causes blood calcium levels to increase. These increased calcium levels feed back to the parathyroid and inhibit the release of further parathyroid hormone until calcium levels decrease again. (B) A more complex negative feedback relationship exists among the hypothalamus, anterior pituitary, and testes. In response to certain external or endogenous stimuli, the hypothalamus releases GnRH, which stimulates the anterior pituitary to release gonadotropins such as LH. In turn, the secreted LH stimulates steroid synthesis and secretion in the testes. In addition to acting on target tissues, the steroid hormones feed back to inhibit activity in both the anterior pituitary and hypothalamus. In parallel, increasing levels of gonadotropins slow down their secretion from the anterior pituitary and GnRH secretion from the hypothalamus. Likewise, as GnRH is secreted, the hypothalamus responds to increasing levels of the hormone by slowing down its secretion.

tion in the gonads. The resulting steroid hormones feed back to turn off GnRH production in the hypothalamus, thereby shutting down gonadotropin secretion from the anterior pituitary. Peptide hormones from the gonads, such as activins and inhibins, may also be involved in the regulation of gonadotropins. The gonadotropins also feed back to shut down GnRH production, and GnRH also feeds back on the hypothalamus to regulate its own secretion (autoregulation). More detail regarding these multiple levels of control will be provided in Chapters 5 and 6.

In some cases, feedback may drive hormone concentrations away from the preprogrammed setting. This type of feedback loop is called **positive feedback** (see Figure 2.35). Positive feedback is often involved when a relatively rapid endocrine response is necessary. Hormones associated with the stress response and with ovulation are regulated through a positive feedback system. Of course, positive feedback must be tightly controlled in short-term situations; otherwise, the equilibrium within the body will be seriously disrupted, with deleterious consequences for survival. Therefore, negative feedback is the most common type of regulatory mechanism in the endocrine system.

In addition to these two general types of internal controls, the secretion of many hormones is influenced by environmental factors. For example, rodents that breed only during the summer may inhibit pituitary LH and FSH secretion when exposed to short (winter-like) day lengths. In some species, this inhibitory effect of short days occurs even in castrated animals (i.e., even in the absence of internal negative feedback actions of the gonadal hormones).

Hormones often affect the levels of their own receptors. For example, an increase in blood concentrations of prolactin stimulates the production of more prolactin receptors; this process is called receptor **up-regulation**. Similarly, high insulin concentrations reduce the number of insulin receptors in a process known as receptor **down-regulation**. Finally, hormones may regulate receptors for other hormones. Estrogens, for example, increase the number of uterine receptors for progestins. Such an effect, in which one hormone induces production of receptors for a second hormone, or otherwise brings about the conditions necessary for the second hormone to be effective, is called a **priming effect**. Priming and other permissive effects are common in behavioral endocrinology.

Hormones are often released in spurts, a process called episodic or **pulsatile secretion**. For example, in male rhesus monkeys, GnRH is released in a pulsatile pattern. Approximately every 45 minutes or so, a quantity of GnRH is released into the hypothalamic–hypophyseal portal system. Shortly thereafter, a pulse of LH is released into the general circulation. Several minutes later, a burst of testosterone is released from the testes, which suppresses further GnRH release for another 45–90 minutes. This pulsatile pattern of hormonal release has functional implications. If one chronically infuses high levels of GnRH into an animal, all of the GnRH receptors are soon occupied, and no further biological action is possible. A hormone-free recovery period during which receptors are replenished is necessary to allow a physiological response. The many factors that affect pulse frequency and amplitude will be discussed later in this book in terms of their effects on behavior.

The Evolution of Hormones

Behavioral endocrinologists are often interested in the phylogenetic relationships among hormones, endocrine gland anatomy, and behavior. Examination of these relationships provides insight into the function and evolution of hormone–behavior interactions. For example, to understand the evolution of hormonal effects on bird vocalization, it might be useful to examine the distribution of steroid hormone receptors throughout the nervous systems of different types of birds that vocalize. Because we can estimate when various bird species diverged from common ancestors, the evolution of various brain centers and their interactions with hormones can be discerned through the use of the comparative method to study diverse bird species, such as roosters, starlings, and parrots. We can learn whether analogous or homologous neural structures are necessary for vocalization in diverse birds, whether these structures serve similar functions in reptiles or amphibians, and whether these structures have retained this function in mammals. Do hormones affect vocalization in all types of birds? If not, which birds are affected by hormones? Is reptilian or amphibian vocalization affected by the same hormones or by different hormones? What about mammals? The comparative method potentially provides insight into many fascinating questions.

In most cases, the evolution of hormones has been an evolution in the function of hormones (see Table 2.2). The chemical structures of steroid hormones, for example, are virtually identical among all vertebrate animals, but the functions of these hormones have changed many times across different species. Some protein hormones have also been conserved over evolutionary time and perform diverse functions. For instance, as noted above, insulin has been found virtually unchanged in species ranging from bacteria to humans (Norman and Litwick, 1987). Point mutations have occurred during the evolution of some other protein hormones, leading to species-specific versions of those hormones (e.g., GnRH; see Figure 2.21), or to new hormones. Growth hormone and prolactin, for example, have very similar molecular structures. Assuming that mutations occur at a constant rate, the amino acid sequences of peptide hormones can be compared to determine the evolutionary time when the proteins diverged. The amount of time, in millions of years, required to change 1% of the amino acids between two proteins can be quantified. By comparing mammalian prolactin to mammalian growth hormone amino acid sequences, it was determined that these two hormones diverged in structure about 350 million years ago (Miller et al., 1983).

Substances that resemble vertebrate peptide hormones have been identified within tissues of flies, worms, protozoans, and even bacteria, but these substances probably function as tissue growth factors; that is, these chemical messengers may be released intracellularly to regulate cellular growth processes (Hadley, 1996; Norris, 1995). Cell-to-cell communication probably developed during the evolution of multicellular organisms so that processes could be synchronized among the various cells of an organism. Because receptors are necessary to mediate the effects of hormones on target tissues, there was probably simultaneous evolution of both hormones and receptors.

Studying the evolution of hormones, their receptor distributions, and behavior has provided behavioral endocrinologists with clues about how solutions to problems common to all species have evolved. During the process of natural selection, outcomes, not specific mechanisms, are selected. Consequently, vastly different mechanisms leading to similar outcomes have evolved among various species. In the following chapters, many hormone–behavior interactions that have led to increased survival and reproductive success will be described.

Summary

- Endocrinology is the study of the endocrine system. Endocrine glands are ductless glands that secrete their products, hormones, directly into the circulatory system. The anatomical locations of the major endocrine glands in humans are depicted in Figure 2.3. Tables listing the hormones and their primary physiological functions are provided in Appendix I.
- The hypothalamus is located at the base of the brain. It integrates information from many higher brain sites into blood-borne signals, serving as one of the primary interfaces between the nervous system and the endocrine system. The hypothalamus secretes small peptide hormones that mediate anterior pituitary function. It communicates with the anterior pituitary through a portal system of blood vessels and with the posterior pituitary via neuronal connections.
- The pituitary gland is really two distinct organs with different embryological origins that are fused together in most vertebrate species. The anterior pituitary produces and releases a number of tropic hormones, which regulate the production and release of other hormones and additional physiological processes. Two nonapeptide hormones are typically released from the posterior pituitary. Oxytocin regulates smooth muscle contractions during milk letdown and birth; vasopressin mediates osmoregulatory functions.
- The thyroid gland is located in the upper thorax in most vertebrate species. It secretes two hormones in response to thyroid-stimulating hormone from the anterior pituitary. The thyroid hormones, triiodotyrosine and thyroxine, elevate oxidation rates in tissue. Situated in or near the thyroid gland are two additional endocrine structures, the parathyroid gland and the C-cells of the thyroid. These structures secrete protein hormones that regulate blood levels of calcium.
- The three major peptide gastrointestinal hormones—secretin, gastrin, and cholecystokinin—are secreted from specialized endocrine cells scattered throughout the gastrointestinal tract. Ghrelin is produced in the stomach. These and other gut hormones have also been discovered elsewhere in the body, particularly in the brain, where they may have behavioral effects. In the gut, these hormones mediate the digestion process.
- Adipose (fat) cells also produce hormones, including leptin and adiponectin. Leptin acts on receptors in the central nervous system and at other sites to induce energy expenditure and inhibit food intake.
- The pancreas is both an exocrine and an endocrine organ. The endocrine compartment of the pancreas consists of islands of hormone-secreting cells nested throughout the exocrine tissue. These islets of Langerhans contain three cell types that release three different protein hormones: insulin, glucagon, and somatostatin. Insulin is the only hormone that can lower blood glucose levels.

Glucagon works in opposition to insulin to elevate blood sugar levels. Somatostatin acts as an inhibitory hormone.

- The adrenal gland consists of two distinct organs. The outer part of the gland, the adrenal cortex, is arranged in three discrete bands of cell types, which secrete three distinct types of steroid hormones: mineralocorticoids, glucocorticoids, and sex steroid hormones. The corticoids are primarily involved in mineral and carbohydrate metabolism. The adrenal medulla, or inner part of the gland, produces and releases two monoamine hormones, epinephrine and norepinephrine, which are important in the physiological response to stress.
- The pineal gland functions as a photoreceptor in some fish, amphibian, and reptilian species. The indole amine melatonin is its primary endocrine product in most vertebrates thus far examined. Melatonin has many effects on reproduction, especially in terms of the timing of puberty or seasonal breeding patterns.
- The gonads are the reproductive organs, called testes in males and ovaries in females. The gonads have two major functions: production of gametes and production of sex steroid hormones. These two functions are compartmentalized in different cell types. In the testes, spermatogenesis occurs in the seminiferous tubules. The primary steroidal products of the testes are androgens. The ovaries produce ova, which develop in follicles in association with estrogen secretion. When the ova are shed from the ovaries during ovulation, the site of rupture develops into another steroid-producing structure called the corpus luteum. The primary steroidal products of corpora lutea are progestins.
- The placenta is a temporary endocrine gland that forms in the uterus of pregnant mammals. The placenta secretes protein and peptide hormones that can affect both the mother and fetus(es).
- There are four classes of hormones: (1) proteins and peptides, (2) steroids, (3) monoamines, and (4) lipid-based hormones.
- Peptide and protein hormones are made up of amino acid chains of various lengths. Protein hormone receptors are located on the cell surface and usually involve a second messenger to mediate the physiological response to the hormone.
- Steroid hormones are derived from cholesterol. These molecules are fat-soluble and thus travel through cell membranes easily; steroid receptors are typically found inside the cell.
- Monoamine hormones are derived from a single amino acid, whereas the lipid-based hormones are derived from lipids.
- Hormones are often internally regulated by negative feedback mechanisms, either by their own concentrations or those of other hormones, or by regulatory physiological processes that involve the end-products of target tissues. Many environmental stimuli also influence hormonal secretion.

Questions for Discussion

1. Hormones control many aspects of physiology and behavior. Discuss the proposition that hormones themselves are exquisitely controlled substances.
2. How do peptide and steroid hormones differ in their production, secretion, and interaction with their receptors?
3. One focus of behavioral endocrinology has been on correlating the blood concentrations of hormones with behavior. Recently, this focus has shifted to target tissue sensitivity to hormones. To what extent are hormone receptor numbers important in understanding hormone–behavior interactions?
4. How are comparative studies useful in understanding hormone–behavior interactions? Are there any negative consequences of comparative analyses?
5. Describe the differences between the mechanisms controlling the release of LH from those controlling the release of oxytocin. Is the anatomy of the pituitary gland relevant to your answer?

Refer to the accompanying Student CD for additional resources, including Web links, videos, animations, and additional photos.

Suggested Readings

Conn, P. M., and Freeman, M. E. 2000. *Neuroendocrinology in Physiology and Medicine*. Humana Press, New York.

Ganong, W. F. 2003. *Review of Medical Physiology*. 21st edition. Lange, Los Altos, CA.

Knobil, E., and Neill, J. D. 1994. *The Physiology of Reproduction*. Raven Press, New York.

Larsen, P. R., Wilson, J. D., Schlomo, M., Foster, D. W., and Kronenberg, H. M. 2003. *Williams Textbook of Endocrinology*. Saunders, Philadelphia.

Levy, A., and Lightman, S. L. 1997. *Endocrinology*. Oxford University Press, Oxford.

Norman, A. W., and Litwack, G. 1997. *Hormones*. 2nd edition. Academic Press, New York.

Norris, D. O. 1996. *Vertebrate Endocrinology*. 3rd edition. Academic Press, New York.

3

Sex Differences in Behavior

SEX DETERMINATION AND DIFFERENTIATION

Girls and boys are different. Humans, like many other animals, are sexually dimorphic (*di*, "two"; *morph*, "type") in the size and shape of their bodies, their physiology, and their behavior. The behavior of boys and girls differs in many ways. Girls generally excel in verbal abilities relative to boys; boys are nearly twice as likely as girls to suffer from dyslexia (reading difficulties) and stuttering. Boys are generally better than girls at tasks that require visuospatial abilities. Girls engage in nurturing behaviors more frequently than boys. Over 90% of all anorexia nervosa cases involve young women. Young men are twice as likely to suffer from schizophrenia as young women. Boys are much more aggressive and generally engage in more rough-and-tumble play than girls (Hines, 2004; Ruble and Martin, 1998). Many sex differences, such as the difference in aggressiveness, persist throughout adulthood. For example, there are many more men than women serving prison sentences for violent behavior. What accounts for these sex differences?

Behavioral endocrinologists, like most people, are interested in these and other behavioral sex differences. What are the proximate causes of behavioral sex differences? This question is one version of the enduring nature-versus-nurture question. As you probably already know, all behavior results from an interaction between environmental factors, including learning and cultural influences ("nurture"), and biological factors, including genes and physiological influences ("nature"). Students of behavioral endocrinology try to discern how the behavioral differences between males and females are mediated by environmental influences, including social interactions, and the extent to which they are mediated by physiological factors, especially hormones. Trying to separate these influences can be difficult, if not impossible. Sex differences in some behaviors, such as snoring, are probably due exclusively to biological factors, whereas other behaviors, such as choosing a style of clothing or haircut, appear to reflect only cultural influences. But what about the pattern of play behavior? Is the difference in the frequency of rough-and-tumble play between boys and girls due to biological factors associated with being male or female, or due to cultural expectations and learning? If there is a combination of biological and cultural influences mediating the frequency of rough-and-tumble play, then what proportion of the variation between the sexes is due to biological factors and what proportion is due to social influences?

It is easy to speculate on these questions in a casual manner. Visit any preschool and watch 3- or 4-year-old children at play. Even when gender-neutral toys are provided, boys tend to play aggressive, rowdy games, whereas girls tend to engage in play activities in which cooperation is valued (Figure 3.1). Again, are these differences mediated by biological influences, or are they due to cultural influences? Perhaps children learn sexually dimorphic modes of behavior from TV or from their friends at nursery school—boys play with guns and trucks and girls play with dolls and tea sets. Of course, it is easy to produce examples of girls who enjoy rough-and-tumble play and boys who enjoy playing nurturing roles.

(A)

(B)

3.1 Patterns of play behavior. The play behavior of boys and girls is different from a very early age. (A) Girls tend to play games that involve groups and verbal cooperation. (B) Boys tend to play with toys that move through space, such as vehicles and balls. These patterns of play behavior probably reflect both socialization and biological differences that may involve hormones. (A) © Stock Connection/Alamy Images; (B) © Painet, Inc.

However, producing examples to support one's contentions is not the most effective way to determine the causes of a particular phenomenon. In order to ascertain the true causes underlying behavior, all biases must be eliminated from the data-gathering process. The most effective manner of obtaining information about an adequately described phenomenon is via the experimental method, primarily through the use of inductive reasoning (Platt, 1964).

Behavioral endocrinologists are particularly interested in the extent to which hormones mediate behavioral sex differences because it is known that steroid hormone concentrations differ between males and females from almost the first trimester of gestation. However, it is extremely difficult to ascertain the proximate factors underlying sex differences in human behavior experimentally. Although the critical experiments can be designed, they cannot ethically be performed on humans. Tracking hormone concentrations during sexual differentiation in utero could jeopardize the well-being of the fetus. To assess the role of hormones in rough-and-tumble play, children would have to be castrated and receive hormone replacement therapy. Other children would have to be reared in social isolation to assess the contribution of social factors to aggressive play. Obviously, such studies are not possible. However, indirect evidence has been obtained from several converging sources that can help us to understand the behavioral differences between human males and females.

Three strategies that do not require experimentation on humans have been used to explore questions about the origin of human behavioral sex differences. First, animal models have been used to study **sexual differentiation**, the developmental process of becoming male or female, which occurs before or immediately after birth or hatching. The study of nonhuman mammals has established that hormone concentrations during sexual differentiation guide the development of many physiological, morphological, and behavioral characteristics that are displayed later in life (Figure 3.2). The second strategy has involved studies of people

3.2 Rough-and-tumble play in nonhuman primates. Although it is often difficult to account for behavioral sex differences in humans, studies of nonhuman animals can often provide insights. For example, strong sex differences in rough-and-tumble play are observed in humans, as well as nonhuman primates, and these sex differences appear to be modulated by early exposure to hormones. Photo © Craig Lovell/Eagle Visions Photography/Alamy Images.

or nonhuman animals that have undergone anomalous sexual differentiation. For example, studying the behavior of girls who were exposed to a male-typical hormonal milieu in utero helps to separate the contribution of rearing conditions from the hormonal factors that may underlie behavioral sex differences. Finally, because child-rearing practices vary so greatly among different cultures, anthropologists, psychologists, and sociologists have looked for universal commonalities in the behavior of all children. Studies of sex differences in behavior that emerge consistently regardless of rearing conditions suggest that some such differences—in aggressive behavior, for example—are mediated by biological factors (Hines, 1982, 2004). Although this third approach will not be reviewed in great depth here, such behavioral surveys have indicated that several behaviors in humans may have significant biological bases that could override social effects. Of course, one could argue that human parents may universally treat their sons differently from their daughters and induce the sex differences in emitted behaviors. This argument is hard to refute because experimental controls are infrequent in long-term developmental studies of human behavior. Consequently, nonhuman animal studies have generally proven to be superior to human studies for understanding the sex differences underlying behavior because of the possibility for greater experimental control.

Sex Determination and Differentiation

What is sex? What makes you male or female? There are several different ways to answer this very basic question (Figure 3.3). First, there is **chromosomal sex**, determined by the sex chromosomes the individual receives at fertilization. In mammals, the homogametic sex is female (XX), whereas the heterogametic sex is male (XY). Chromosomal sex is fundamental during the process of sexual differentiation, and determines **gonadal sex**: the possession of either ovaries or testes. Females have ovaries, whereas males have testes. Gonadal sex is related to **gametic sex**: the ovaries of females produce large, immobile, resource-rich gametes called ova or eggs, whereas the testes of males produce small, mobile gametes called sperm. Females and males also differ on the basis of **hormonal sex**. Females of most vertebrate species tend to have high estrogen-to-

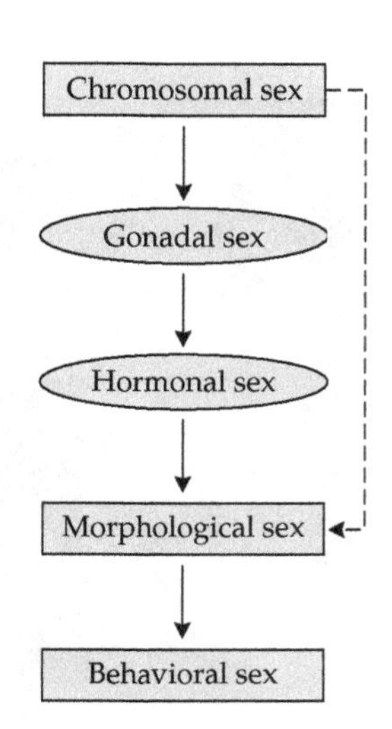

3.3 Levels of sex determination. Chromosomal sex, which is determined at conception, determines which gonads form in the embryonic individual. Gonadal sex determines the hormonal environment in which the fetus develops, and steers morphological development in a male or female direction. The resulting sex differences in the central nervous system and in some effector organs lead to the behavioral sex differences observed in later life. There is some evidence that chromosomes may also directly influence specific sexually dimorphic brain anatomy and function.

androgen ratios of circulating steroid hormone concentrations; males have the opposite pattern. **Morphological sex** refers to the differences in body type between males and females. Male mammals are typically larger than females and possess different external genitalia. In addition, males of many species differ from females in coloration, the presence of horns, antlers, and other ornamentation, and body shape. **Behavioral sex** can be discriminated on the basis of male-typical and female-typical behaviors. For example, females of many species often care for young, whereas males, especially male mammals, rarely provide parental care (see Chapter 7). Male and female birds of many species share equally in parental care, but they display other types of behavioral sex differences. For example, males of many avian species sing, but their female conspecifics do not.

Additional categories of sex classification exist among humans. For example, **gender identity** reflects the sex, or gender, that individuals feel themselves to be; gender identity usually corresponds to the expression of a **gender role**, a culturally based summary of sex-specific behaviors. **Sexual orientation** or **sexual preference** can also categorize males and females: males generally prefer female sexual partners, whereas females generally prefer male sexual partners. Finally, there is **legal sex**. You are recognized by governmental agencies as male or female because of an "M" or "F" on your birth certificate, driver's license, or some other official document. If individuals sporting "M" identification documents enter a women's rest room, they can be arrested in most parts of North America, even if they are exhibiting female-typical behavior (for example, wearing a silk evening gown). Sex differences occur at each of these systematic levels.

Ultimate Causes of Sex Differences

Most of the animals with which you are familiar engage in sexual reproduction. In fact, most animal species currently on the planet engage in sexual reproduction. However, a survey of the entire animal kingdom reveals that a number of animals, especially insect species, breed asexually. There are several forms of asexual reproduction. Among vertebrate animals, the process of asexual reproduction is called **parthenogenesis**. In parthenogenetic vertebrates, there is only one sex: female. Parthenogenetic females produce genetically identical eggs that develop into female offspring that are all genetically identical to their mother.

From the Darwinian perspective, reproductive success reflects the amount of genetic material an individual contributes to subsequent generations. Thus, asexual reproduction may seem like an extremely efficient system. This mode of breeding must have some distinct disadvantages, however, if so many animals breed sexually. One hypothesis about why asexual reproduction is rare among vertebrates is that asexual individuals, because they have such similar genotypes, provide little variation on which natural selection can act. Thus, if the conditions of its environment change, an asexual species could become extinct. Sexual reproduction, through the separation of chromosome pairs into haploid gametes and their recombination in each offspring, provides genetic variability and hence evolutionary flexibility (Howard and Lively, 1994). Those animals and plants that have the option of either sexual or asexual reproduction usually opt for sexual reproduction when the environment becomes unstable. By analogy, would you rather have 100 lottery tickets with the same number or 50 tickets, each with a slightly different number? Another problem that can burden asexual species is the likelihood that pathogens may become specialized to exploit a single genotype (Ewald, 1994). Because pathogens reproduce faster than their hosts, they can rapidly evolve ways to override the immunological defenses of their hosts (Judson, 1997; Ladle et al., 1993). By shuffling their genetic material, host species reduce the odds that pathogens will be preadapted to exploit their offspring. Why sex evolved remains a question that is far from settled among evolutionary biologists.

Sexually reproducing species have two sexes: male and female. Why do the two sexes differ in appearance and behavior? In other words, what are the ultimate causes of the behavioral differences between males and females? We cannot travel back in time to watch the slow evolution of sexual dimorphism in a given species and note the ecological or social factors associated with the development of sex differences. But we can address these "why questions" by examining present-day species that show little or no sexual dimorphism and comparing them with species that are sexually dimorphic. When these comparisons are made, a striking relationship between sexual dimorphism and mating system becomes obvious. Species that are **monogamous** (have a single mating partner) display

BOX 3.1 Behavioral Sex Role Reversals

In virtually every mammalian and avian species studied, males are more aggressive than their female conspecifics. They also tend to weigh more, and are (particularly among mammals) more likely to establish and defend breeding territories and less likely to provide parental care to the young. There are several notable exceptions to this dogma, however, and the study of these unusual cases is useful for shedding light on the more typical situation as well as for understanding how reliable the "generalities" of the interaction between hormones and behavior are.

Female spotted hyenas (*Crocuta crocuta*) are socially dominant: in addition to eliciting submissive postures and vocalizations from males, they are also allowed first access to food (Frank, 1986; Kruuk, 1972). In addition to their masculinized behavior, these females have masculinized external genitalia. Indeed, the species was once thought to be hermaphroditic because all individuals, even nursing mothers, possess scrota and penis-like structures. The photographs in this box show the external genitalia of male (left) and female (right)

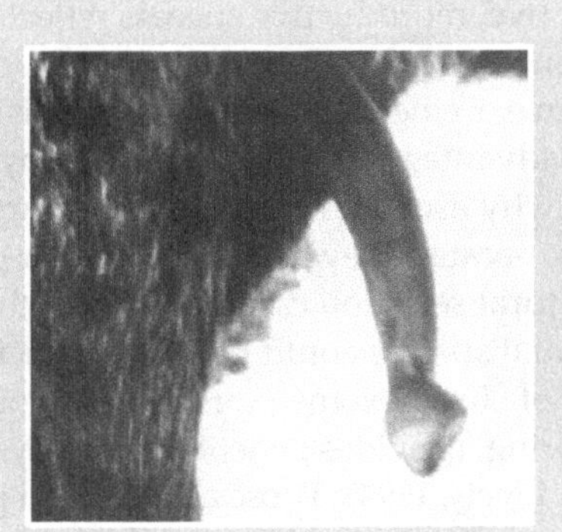

Photographs courtesy of Stephen Glickman.

less sexual dimorphism than species that are **polygamous** (have multiple mating partners) (Trivers, 1972). In polygamous species, in which members of one sex compete to mate with the other sex, sexual selection may act to favor certain behavioral or morphological traits in the competing sex (Andersson, 1994).

Sexual selection is a subcategory of natural selection (Andersson, 1994; Trivers, 1972). Generally, the rule among animal species is that males compete and females choose. For instance, female deer choose to mate with bucks that control the largest feeding areas (intersexual selection). Bucks obtain feeding areas prior to the breeding season by fighting among themselves and establishing a social hierarchy (intrasexual selection). The largest bucks usually have the largest antlers, and they typically win the most fights, move to the top of the hierarchy, and control the best feeding areas. Thus, they are able to mate with the most females and pass on their genes to the most offspring (Lincoln et al., 1972). Consequently, females indirectly select for males with large body and antler size, as well as for aggressive behavior. Because males typically compete for females, they are usually bigger, more colorful, and more aggressive than females. In species in which the sex roles are reversed and females compete with one another for males (Box 3.1), the females are typically larger and more colorful than their male conspecifics (Trivers, 1972). Thus, sexual selection favors

hyenas. The vaginal labia in spotted hyenas are fused to form a fat-filled pseudoscrotum. The clitoris develops into a pseudopenis through which the urogenital tract passes. Females urinate, copulate, and deliver young through the urogenital tract. The pseudopenis possesses full erectile function, and both males and females routinely display erections as part of their social interactions (Glickman et al., 1987). Erectile function in both sexes is observed from infancy.

The extreme masculinization of female behavior and morphology appears to be organized by high concentrations of ovarian androstenedione in pregnant females, which is converted to testosterone by the placenta (Glickman et al., 1987, 1992). However, exposure to anti-androgens in utero does not block the development of a pseudopenis in female hyenas (Drea et al., 1998). It is possible that some other factor stimulates the androgen receptors, leading to external masculinization, but the androgen receptors themselves appear relatively unaltered (Catalano et al., 2002). Adult males exhibit higher androgen concentrations than adult females. Thus, adult concentrations of androgens do not appear to account for the difference in social dominance (see Chapter 10).

Spotted sandpipers (*Actitis macularia*) are birds that display atypical sex roles. Females establish and defend a feeding territory and compete with other females for access to males (Fivizzani and Oring, 1986). Males incubate the eggs and brood the young with little or no assistance from the females. Prior to incubation, blood plasma concentrations of testosterone and dihydrotestosterone are substantially higher in males than in females. As incubation proceeds, testosterone levels in males plummet 25-fold. Mated females exhibit testosterone concentrations that are 7-fold higher than those of unmated females (Fivizzani and Oring, 1986). These hormone profiles are similar to those of other avian species in which males assist with care of the young. Thus, the reversal of sex roles in spotted sandpipers appears to be unrelated to adult hormonal effects.

sexual dimorphisms, and probably amplifies them over time (Zahavi and Zahavi, 1997).

Although monogamy is asserted to be the most common mating system practiced by humans (*Homo sapiens*) in Western societies, examination of many societies suggests that humans are mildly to strongly polygynous; that is, each male may mate with more than one female (Alexander et al., 1979; Daly and Wilson, 1983; Martin and May, 1981; Murdock, 1967). Approximately 700 out of 850 human societies described in an extensive ethnographic catalog practiced polygyny (Murdock, 1967). Furthermore, as noted above, the sex differences observed among humans are atypical for monogamous animals, but are common in polygynous and promiscuous species. Human sexual dimorphisms include increased body size and strength (Alexander et al., 1979), later puberty (Bronson, 1988; Tanner, 1962), and higher mortality rates (Shapiro et al., 1968) in men than in women. Importantly, cross-cultural studies of humans have revealed behavioral sexual dimorphisms that are also consistent with those of other polygynous species, including increased courtship activity (Daly and Wilson, 1983), higher levels of aggression (Moyer, 1976), and reduced parental care (Clutton-Brock, 1991) in males than in females. Thus, the morphological, physiological, and behavioral differences between men and women may reflect the evolutionary history of the species' most common mating pattern, namely, polygyny.

Again, we cannot go back in time to test these ultimate hypotheses, but we can use them to make predictions about current behaviors, and we can conduct animal experiments designed to test our predictions. For example, males of polygynous species typically move around over larger areas than conspecific females or males of monogamous species, presumably because wandering around increases their chances of finding additional mates. One might predict that females and males of monogamous species will have poorer direction-finding skills or spatial aptitude in general than males of these wide-ranging polygynous species. This prediction appears to be true when tested. Males of polygynous species, including rats, mice, meadow voles (a small field rodent; *Microtus pennsylvanicus*), and humans, are superior in spatial aptitudes to females of the same species (see the next section). Moreover, male meadow voles are superior in spatial abilities (navigating a maze) to male pine voles (*Microtus pinetorum*), a monogamous species of the same genus (Gaulin and FitzGerald, 1986).

The ultimate level of explanation is useful for guiding predictions about the proximate mechanisms that cause males and females to behave differently. The remainder of this chapter will review the proximate processes of sex determination and sexual differentiation. In other words, it will look at how sexually dimorphic behaviors arise.

Proximate Causes of Sex Differences

Because behavioral sex differences are so ubiquitous among animals, behavioral endocrinologists have focused significant effort on studying this phenomenon. One very important concept developed to guide this effort is the so-called **organizational/activational hypothesis**. This fundamental hypothesis on how hormones guide behavioral sex differences will be presented within its historical context. Basically, the organizational/activational hypothesis suggests that behavioral sex differences result from (1) differential exposure to hormones that act early in development to organize the neural circuitry underlying sexually dimorphic behaviors and (2) differential exposure to sex steroid hormones later in life that activate the neural circuitry previously organized. The workings of this process in several animal models commonly used to study behavioral sex differences will be described in detail in the next section. Sex differences in brain structures, how these differences arise, and how these differences might mediate behavioral differences between males and females will also be reviewed in this chapter, as well as in Chapter 4.

A number of clinical syndromes exist among humans and nonhuman animals that cause anomalies in the process of sexual differentiation. Many of these syndromes are congenital; others are the result of endocrine treatment during gestation. The behavioral consequences resulting from these developmental irregularities in humans, as well as from animal studies in which these clinical syndromes are simulated, will be presented in this chapter. Issues of gender, sex role, and sexual orientation among humans, as well as cognitive sex differences, will be explored in Chapter 4.

There are real and substantial differences between males and females. Some of these sex differences may be apparent at birth; others appear later in development. Our goal is to understand how behavioral sex differences are mediated. But in order to do this, it is necessary to understand sexual differentiation in general. If we limit our discussion to mammals and birds, then basically, chromosomal sex determines gonadal sex, and all subsequent sexual differentiation is normally the result of differential exposure to gonadal steroid hormones. Thus, gonadal sex determines hormonal sex, which in turn influences morphological sex. Morphological differences in the central nervous system, as well as in some effector organs, lead to behavioral sex differences (see Figure 3.3). To understand the specifics of how sexually dimorphic behaviors arise, a discussion of embryology is necessary. This discussion will yield a rationale for the principles underlying behavioral sex differences.

Mammalian Sexual Differentiation

The primary step in the process of mammalian sexual differentiation occurs at fertilization. An ovum can be fertilized by a sperm bearing either an X or a Y chromosome. This event, called **sex determination**, has far-reaching consequences for the differentiation of the embryonic gonads, as well as subsequent behavioral differences.

Each individual embryonic mammal, whether XX or XY, exhibits a thickened ridge of tissue on the ventromedial surface of each mesonephros (protokidney), known as the **germinal ridge** (Figure 3.4; Jost, 1979). At this stage, this primordial gonad is said to be *indifferent* or *bipotential*. In most mammals studied to date, whether the germinal ridge will develop into a testis or an ovary is determined by the cellular expression of the **testis determination factor** (**TDF**)—a protein encoded

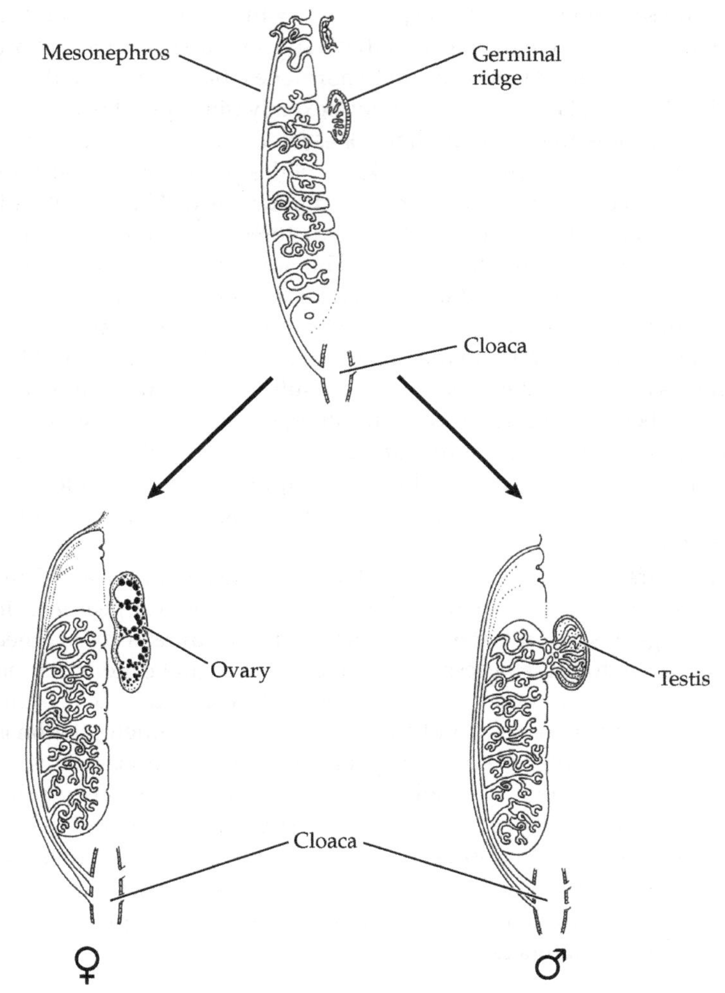

3.4 Gonadal development of the human embryo. Each individual, regardless of chromosomal sex, develops a thickening called the germinal ridge on the ventromedial surface of each mesonephros (protokidney). The germinal ridge is a bipotential primordial gonad; it can become either an ovary or a testis. If the *SRY* gene is expressed and the testis determination factor protein is present, the medulla of the germinal ridge develops, forming a testis. If the *SRY* gene is not expressed, the cortex of the germinal ridge develops, and an ovary is formed.

by a gene known as ***SRY*** (*s*ex-determining *r*egion on the *Y* chromosome) (Berta et al., 1990). When the *SRY* gene is expressed in the indifferent gonads, the testis determination factor, binds to a specific hormone response element in the promoter region of other genes that it regulates (Harley and Goodfellow, 1995). When the protein products of these genes are produced, the middle (medulla) of the germinal ridge develops, and a testis forms (Berta et al., 1990). If the *SRY* gene is not expressed, then the outer part of the germinal ridge (the cortex) develops, and an ovary is formed. It is possible for the *SRY* gene to be expressed in one gonad but not in the other, which leads to unilateral differentiation: in other words, a testis develops on one side while an ovary develops on the other side, suggesting that the testis determination factor functions only locally and is not a blood-borne agent. Partial expression of the *SRY* gene can lead to incomplete gonadal differentiation, yielding an ovotestis. Mice that are chromosomal XY males but do not have the *SRY* gene develop ovaries; similarly, XX transgenic mice into which an *SRY* gene has been inserted develop testes (Goodfellow and Lovell-Badge, 1993; McElreavey et al., 1995). In non-mammalian vertebrates, *SRY* is not a testis-determining gene.

Hormonal secretions from the developing gonads determine whether the individual develops in a male or female manner. The mammalian embryonic testes produce androgens, as well as peptide hormones, that steer the development of the body, central nervous system, and subsequent behavior in a male direction. The embryonic ovaries of mammals are virtually quiescent and do not secrete high concentrations of hormones (but see Arnold and Breedlove, 1985; Döhler et al., 1982). In the presence of ovaries, or in the complete absence of any gonads, morphological, neural, and later, behavioral development follows a female pathway. (As we will see, however, low estrogen concentrations are probably required for normal female neural and behavioral development.) Thus, the prevailing hypothesis about the initiation of sexual differentiation indicates that the gonads are differentiated by genetic influences (*SRY*), and that all other sexual differentiation reflects hormonal mediation.

It is possible, however, that some sex differences in brain and behavior might be mediated directly by gene expression in nongonadal tissue (Arnold, 2002), without the involvement of hormones. For example, the *SRY* gene is transcribed in the hypothalamus and midbrain of adult male mice (Lahr et al., 1995). Indeed, sexually dimorphic transcription of over 50 genes was observed in the brains of mice 10.5 days post-conception, before the gonads had developed (Dewing et al., 2003). Because only genetic males possess a Y chromosome, any gene located on the Y chromosome that is involved in neural development is a candidate to mediate sex differences in brain and behavior directly, without invoking hormones. Mice can be genetically engineered to differ in their sex chromosomes (i.e., XX or XY) but possess the same type of gonads (i.e., ovaries or testes). Such mice show subtle differences in certain brain cell types but most sex differences in these mice appear to rely on hormonal signals regardless of genotype. For example, mice with ovaries, regardless of sex chromosomes, have fewer cells expressing tyrosine hydroxylase in the anteroventral periventricular nucleus of the preoptic area than mice with testes (Arnold et al., 2004). Sex chromosomes also appear to influence the density of vasopressin-containing fibers in the lateral septum, a part of the limbic system located just below the lateral ventricles (DeVries et al., 2002). Additional work is required to determine the importance of all of these factors in the development and expression of sexually dimorphic behaviors.

In contrast to the single, bipotential primordial gonad, dual **anlagen** (primordia) for the accessory sex organs are present early during ontogeny (Figure 3.5). The

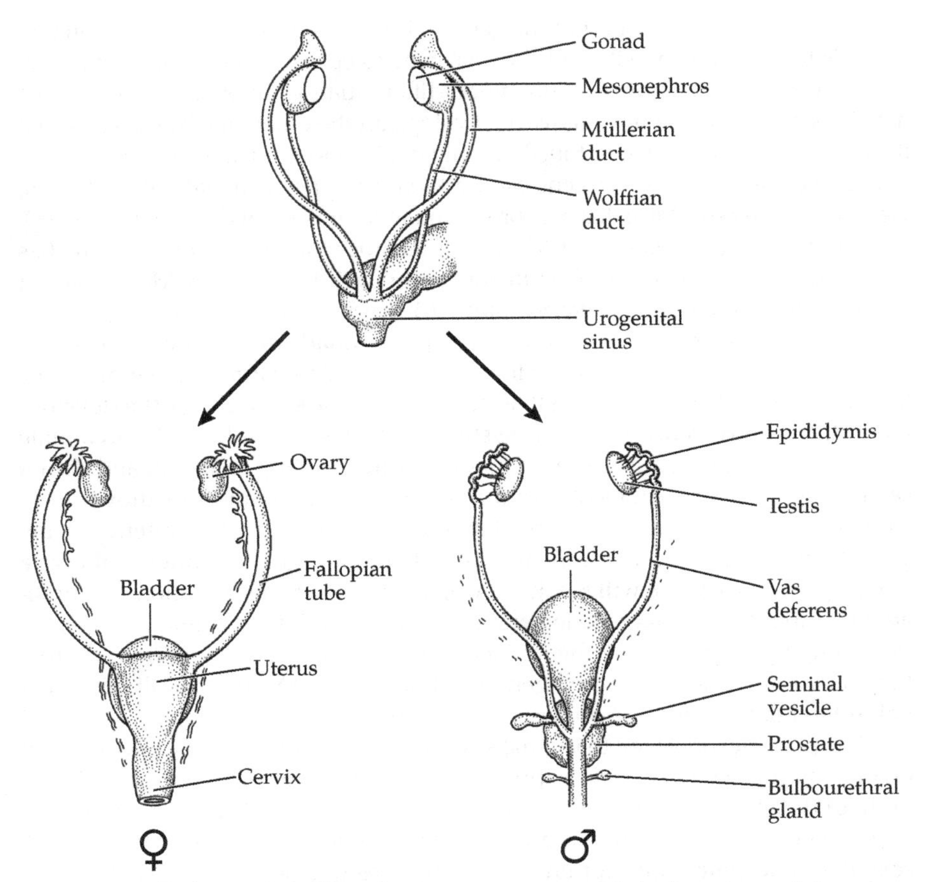

3.5 The Müllerian and Wolffian duct systems are normally both present early in embryonic development. In the absence of testicular hormones, the Müllerian duct system develops into the fallopian tubes, uterus, and, the Wolffian ducts regress. This is the normal course of events in female mammals. When testicular hormones are present, as is normally the case in male mammals, the Wolffian ducts eventually develop into the seminal vesicles and vas deferens, and the Müllerian ducts regress. Because anlagen for each of these duct systems are initially present in the embryo, it is possible in some circumstances for both systems to develop in a single individual.

accessory sex organs connect the gonads to the outside environment. The **Müllerian duct system** develops into the female accessory sex organs: the fallopian tubes, uterus, and cervix. The **Wolffian duct system** develops into the male accessory sex organs, which connect the testes to the outside environment via the penis. Later-developing components of the Wolffian duct system include the seminal vesicles and vas deferens. During normal sexual differentiation, the Wolffian duct system develops in males, whereas the Müllerian duct system regresses. Conversely, in females, the Müllerian duct system develops, whereas the Wolffian duct system regresses. In humans, this process occurs during the first trimester of gestation.

In the presence of ovaries or in the complete absence of gonads, normal development of the Müllerian ducts is accompanied by complete regression of the Wolffian duct system. Again, no hormones are necessary to permit normal female development among mammals. Male development of the accessory sex organs requires two products from the embryonic testes: testosterone and a peptide hormone called **Müllerian inhibitory hormone** (**MIH**). Testosterone is necessary to stimulate Wolffian duct development. Müllerian inhibitory hormone, as its name implies, causes the regression of the Müllerian duct system. If testosterone is absent early in development, then Wolffian duct development fails to occur. If MIH is not secreted at the proper time, then the Müllerian duct system develops. Because of the twin anlagen, it is possible for both systems to develop in a single individual.

In order for normal female morphological development of the accessory sex organs to occur, the embryo must become feminized (Müllerian duct development), as well as demasculinized (Wolffian duct regression). Normal male development requires both masculinization (Wolffian duct development) and defeminization (Müllerian duct regression). In other words, sexual differentiation of the accessory sex organs proceeds along two continua: (1) a masculinization–demasculinization scale and (2) a feminization–defeminization scale (Figure 3.6). **Masculinization** is the induction of male traits; **feminization** is the induction of female traits. **Demasculinization** is the removal of the potential for male traits, whereas **defeminization** is the removal of the potential for female traits. This nomenclature will be important for describing the development of behavioral sex differences as well as morphological ones.

The difference in external genitalia is the most obvious difference between the sexes at birth and has been used for generations to assign the sex of humans. During embryological development, the urogenital sinus is surrounded on both sides by long, thickened urogenital ridges, which are flanked by two flaps of skin called the **genital folds** (Figure 3.7). In front of (anteroventral to) the urogenital opening, the two ridges meet to form a median outgrowth called the **genital tubercle**. The genital tubercle and folds are common to both sexes and develop into the external genitalia. Female humans, as well as many other species of mammals, possess a clitoris and vaginal labia, which develop from the genital tubercle

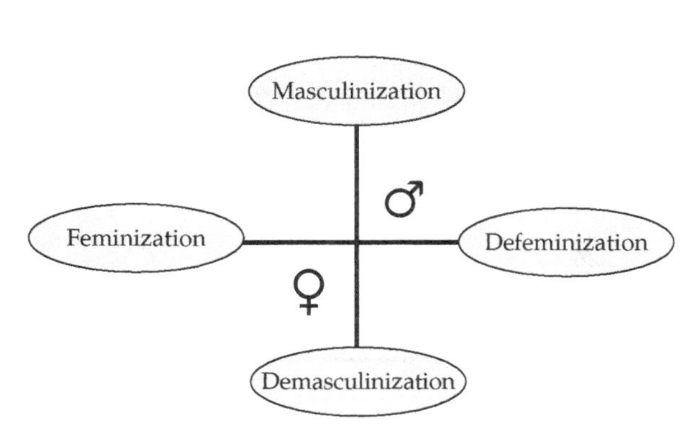

3.6 Normal development of the accessory sex organs occurs along two dimensions. Normal females must become feminized (Müllerian duct development) as well as demasculinized (Wolffian duct regression). Normal males must become masculinized (Wolffian duct development) as well as defeminized (Müllerian duct regression). Thus, sexual dimorphism in accessory sex organs, as well as in many behaviors, requires development along two separate continua, a masculinization–demasculinization continuum and a feminization–defeminization continuum. Male-typical behavior should be masculinized and defeminized.

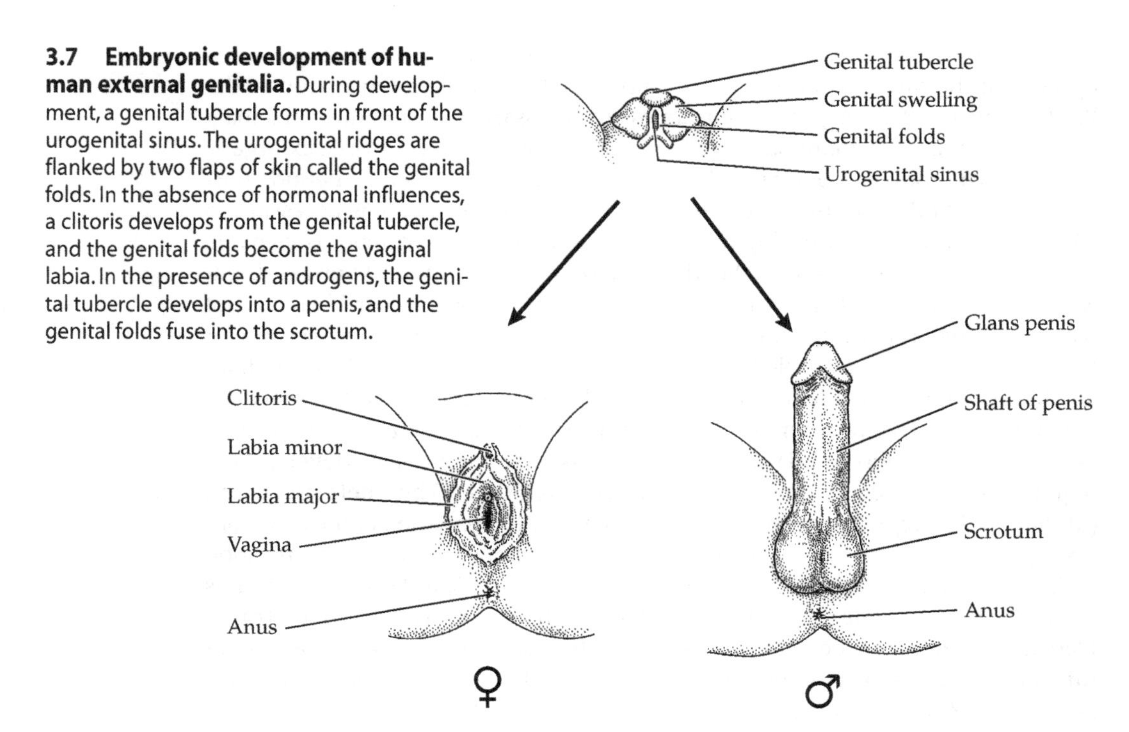

3.7 Embryonic development of human external genitalia. During development, a genital tubercle forms in front of the urogenital sinus. The urogenital ridges are flanked by two flaps of skin called the genital folds. In the absence of hormonal influences, a clitoris develops from the genital tubercle, and the genital folds become the vaginal labia. In the presence of androgens, the genital tubercle develops into a penis, and the genital folds fuse into the scrotum.

and genital folds, respectively. Males possess a penis, which develops from the genital tubercle, and a scrotal sac, which results from the fusing of the genital folds and eventually contains the testes. Consequently, development of one type of genitalia occurs at the expense of the other type. In other words, a single continuum of feminine to masculine external genital development exists (Figure 3.8).

Androgens are responsible for the differentiation of the external genitalia. In the absence of sex steroid hormones, a clitoris develops from the genital tubercle and the vaginal labia develop from the genital folds. In the presence of androgens, the urethral groove fuses, the genital tubercle develops into a penis, and the genital folds fuse into the scrotum (Wilson et al., 1981). One of the androgenic metabolites of testosterone, 5α-dihydrotestosterone (DHT), is critical for this process of genital fusing. Testosterone is converted to DHT by an enzyme called **5α-reductase**, which is locally abundant in the embryonic genital skin of both females and males (Wilson et al., 1981). If unusually high concentrations of androgens are available to a female fetus, they are converted to DHT in the genital skin,

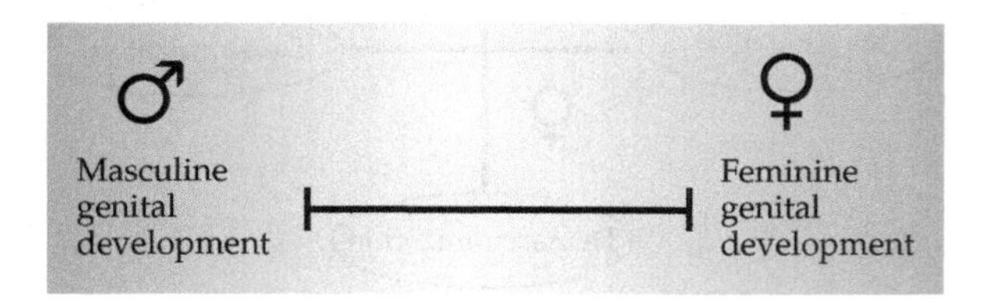

3.8 Normal development of the external genitalia proceeds along a single masculinization–feminization continuum. Because there is a single anlage for female and male external genitalia, the development of one type of genitalia precludes the development of the other.

and the development of male-typical external genitalia proceeds. Males that congenitally lack 5α-reductase undergo incomplete differentiation of the external genitalia and may be considered female at birth (see Figure 3.13). Females that lack 5α-reductase suffer no grave consequences, because normal female genital development occurs in the absence of sex steroid hormones.

Anomalous Mammalian Sexual Differentiation

The process of mammalian sexual differentiation is summarized in Figure 3.9 (MacLaughlin and Donahoe, 2004). This process is complex, and as the case of

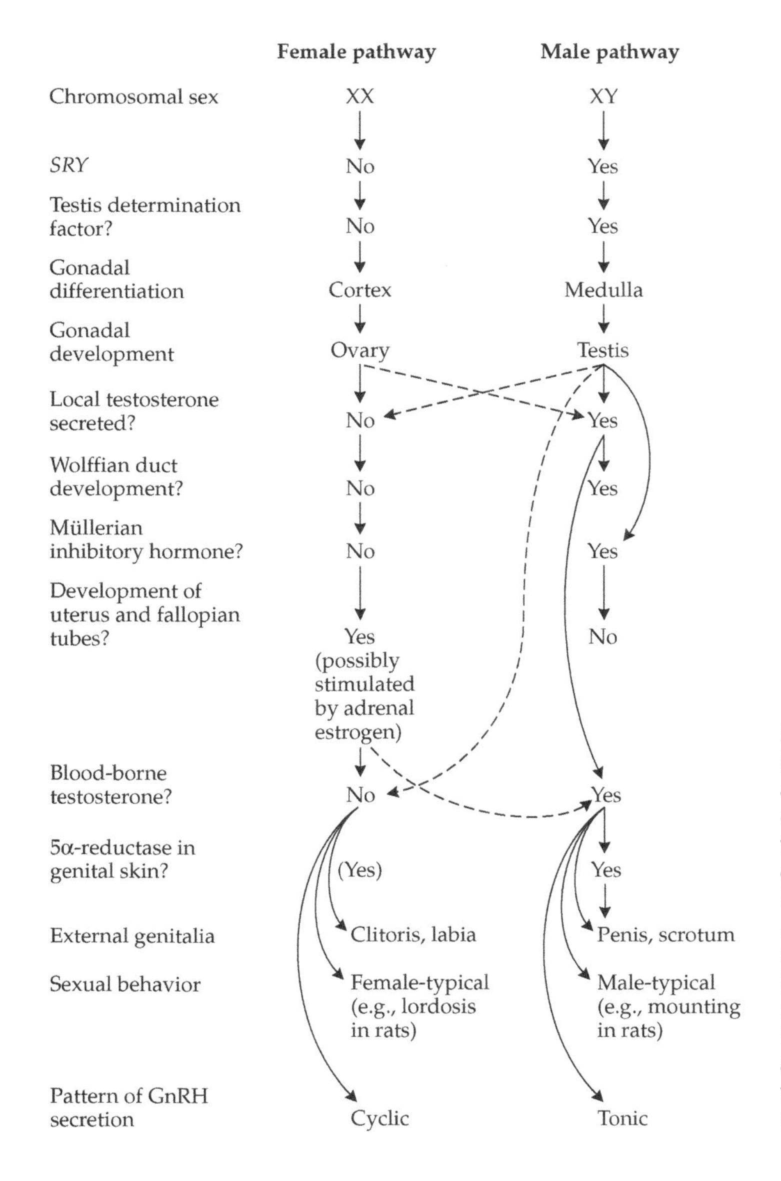

3.9 Sexual differentiation in humans is complex, and there are several stages where errors can occur. Chromosomal sex determination takes place when an egg is fertilized by a Y- or X-bearing sperm. The normal developmental pathways for females and males are indicated by the solid lines; common developmental errors are indicated by the dashed lines, which show where individuals can be "shunted" onto the developmental pathway of the opposite chromosomal sex.

5α-reductase deficiency shows, the potential for error is present. People and nonhuman animals that have undergone such anomalous sexual differentiation are of interest to behavioral endocrinologists because they can serve as "experiments of nature," and, as such, they can provide important information about the workings of the normal process of sexual differentiation.

Errors in sexual differentiation have been recognized by humans since early in their history. One very old idea was that male babies came from the right testis and females came from the left. **Hermaphrodites**, individuals who possess both ovaries and testes, were thought to arise when both testes simultaneously contributed to the offspring. All human babies born with ambiguous genital development used to be called hermaphrodites (Ellis, 1945), but they are now considered to be **pseudohermaphrodites**, because only one set of gonads is usually present. True hermaphrodites, with two sets of gonads, are extremely rare.

ANOMALIES IN FEMALES The female organization is basic among mammals. Because females are the so-called "neutral" or "default" sex, neither ovaries nor hormones are necessary for female development of the body prior to puberty. It is becoming increasingly clear, however, that low concentrations of estrogens are critical for normal female development of the brain (Döhler and Hancke, 1978; Döhler et al., 1982; Toran-Allerand, 1984). It has been hypothesized that females are the ancestral sex and that males are the derived sex (Crews, 1993).

One out of every 3000 live births in humans exhibits **Turner syndrome**, characterized by a congenital lack of, or damage to, the second X (or a Y) chromosome (XO) (Figure 3.10; Zinn et al., 1993). Individuals with Turner syndrome are unambiguously sexed as girls at birth. The gonads rarely develop completely, but they are clearly recognizable as ovaries. The dysgenic ovaries fail to produce steroid hormones, however, so girls with Turner syndrome must be treated with sex steroid hormones in their mid-teens to induce puberty. The consequences of missing a second sex chromosome are widespread and are not limited to sexual differentiation. Additional endocrine problems associated with Turner syndrome are reflected in the slow growth rates of afflicted girls. Both neural and non-neural tissues are also affected; hearing loss and mental retardation, as well as kidney dysfunction and webbing of the neck, are observed in some indi-

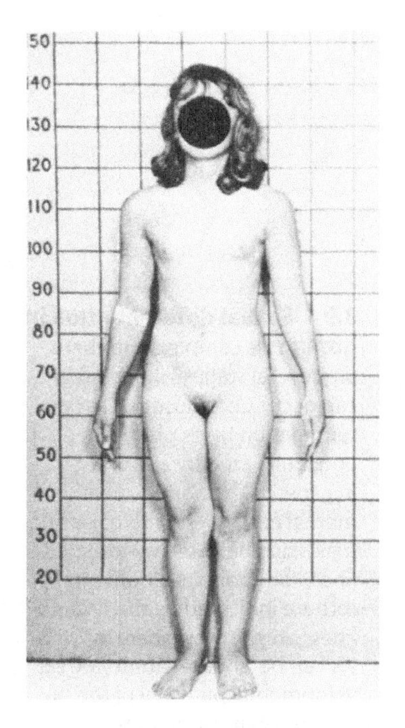

3.10 Turner syndrome results from a congenital absence of one of the X chromosomes. Without hormone therapy, these individuals do not undergo puberty. This 21-year-old XO woman is shown prior to estrogen therapy. She exhibits the short stature associated with the syndrome; however, she does not display other possible symptoms of Turner syndrome, such as webbed skin on the neck. Courtesy of John Money.

viduals born with Turner syndrome. These individuals are of interest to behavioral endocrinologists because they are not exposed to steroid hormones prenatally or postnatally until the age of 16 or 17, when exogenous hormonal treatments begin.

Recently, an intriguing study was conducted that examined the role of gene imprinting on behavior in girls with Turner syndrome. Because females normally inherit one X chromosome from their mother and another X chromosome from their father, one of the X chromosomes in each cell is usually disabled so that normal transcription can occur. The process of inactivating either maternal or paternal genes is called **gene imprinting**. Girls with Turner syndrome have only a single X chromosome, which is either maternal (X^m) or paternal (X^p). X^mO girls tend to suffer from neurodevelopmental disorders of social cognition (e.g., autism) more than X^pO girls (Skuse et al., 1997). Because the X chromosome of all males necessarily comes from the mother, these results suggest that the vulnerability of males to cognition and social adjustment problems is not necessarily due to the presence of specific genes on the Y chromosome, but could rather be due to genes on the maternal X chromosome (which presumably act independently of sex steroid hormones). Alternatively, it is possible that pieces of the Y chromosome might be expressed in neural tissue to cause these behavioral difficulties, because many individuals with Turner syndrome are so-called genetic mosaics (Henn and Zang, 1997), possessing cells with differing genetic contents. This issue will probably go unresolved for some time because it is virtually impossible to observe the chromosomal makeup of brain cells in living people (Skuse and Jacobs, 1997).

The most common reason for anomalous sexual differentiation in human females is prenatal exposure to androgens, either from exogenous or endogenous sources. Exogenous androgen exposure, or other steroid hormones that activate androgen receptors, most commonly occurs during treatment of the pregnant mother with steroid hormones to maintain the pregnancy. Exposure to the artificial steroid hormones diethylstilbestrol (DES) and medroxyprogesterone acetate (MPA), for instance, often causes masculinization of reproductive function and subsequent behavior in the exposed children (see Chapter 4). Endogenous androgens are most likely to come from one of two sources: the ovaries or the adrenal glands. For example, both males and females afflicted with **congenital adrenal hyperplasia (CAH)** possess fetal adrenal glands that produce high concentrations of androgens instead of cortisol. CAH does not cause sexual development problems in genetic males, but results in moderate to severe masculinization of the genitalia in afflicted females (Figure 3.11). The clitoris can be enlarged into a penis-sized structure, and the labia majora may be fused into a scrotum-like organ. The anomalous genitalia can be corrected at birth by surgery, and the endocrine disorder can be treated by lifelong cortisol treatments.

ANOMALIES IN MALES There are more steps that require physiological intervention in male than in female sexual differentiation, and consequently, more clinical problems in male sexual development can occur. For example, some XY individuals look and act like males throughout their lives, but during surgery for abdominal cramps, a uterus and small fallopian tubes are discovered. These indi-

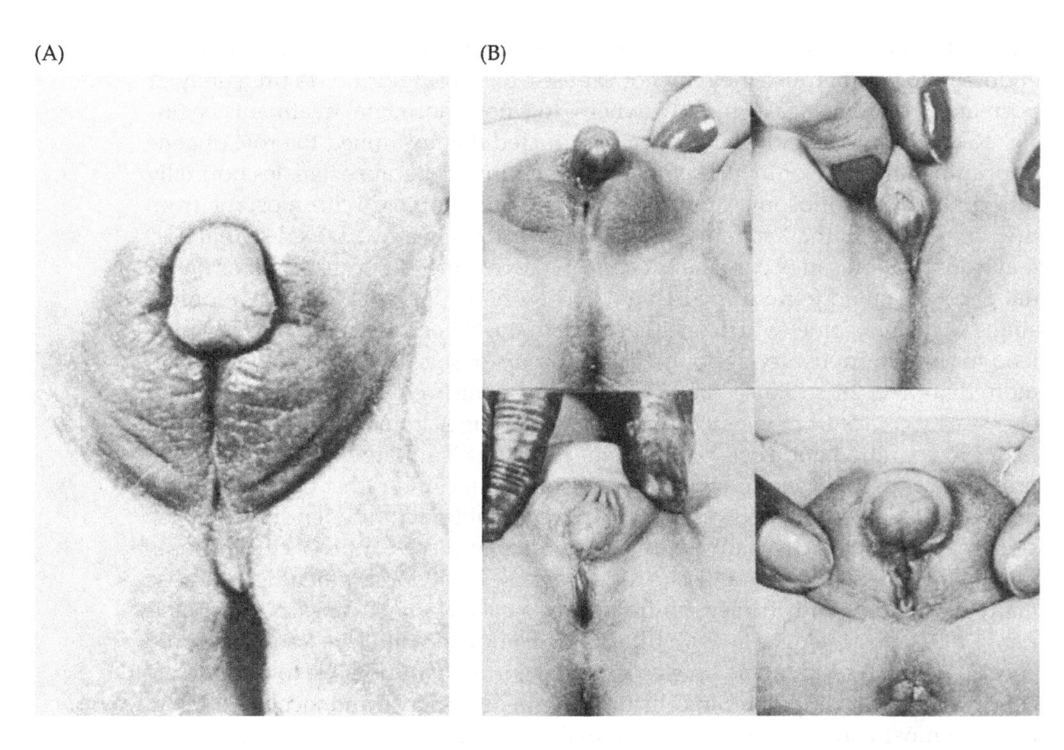

3.11 Partial masculinization of the external genitalia in genetic females can be caused by prenatal exposure to sufficient progestins to activate androgen receptors or by congenital adrenal hyperplasia (CAH). (A) The mother of this individual was treated with progestins to prevent miscarriage after suffering symptoms of premature labor. Note the enlarged clitoris and fusing of the vaginal labia majora. (B) Congenital adrenal hyperplasia, a condition in which the adrenal glands secrete high levels of androgens, can cause varying degrees of masculinization of the external genitalia in genetic females, including clitoral enlargement and fusing of the urethral groove. Courtesy of John Money.

viduals lack either Müllerian inhibitory hormone or the receptors that respond to this hormone; all other components of their sexual differentiation are male-typical. This sort of error in sexual differentiation does not have important consequences for male behavior, but in other cases anomalous sexual differentiation produces dramatic behavioral effects.

Anomalous sexual differentiation occurs in XY individuals afflicted with a condition known as **testicular feminization mutation** (**TFM**) in rodents and as **androgen insensitivity syndrome** (**AIS**) in humans. AIS can be complete (CAIS) or partial (PAIS). The tissues of individuals with CAIS do not possess functional androgen receptors. A genetic mutation on the X chromosome involving a single base pair substitution in the gene for the androgen receptor causes this insensitivity to androgens in rats (Yarbrough et al., 1990). In humans, there are several mutations that cause androgen insensitivity (Brinkmann et al., 1996). Genetic XX females with such mutations have a second X chromosome that contains the normal gene for androgen receptors, so they suffer no ill effects. However, in genetic XY males, because a Y chromosome is present, the *SRY* gene is activated, and testicular development proceeds, accompanied by significant prenatal and postnatal androgen secretion. The testes also produce Müllerian inhibitory hormone, which causes the regression of the Müllerian duct system. However, the Wolffian duct system is not stimulated to develop, even in the presence of androgens. XY individuals born with CAIS have perfectly normal-appearing female external genitalia and are sexed and reared as girls. The condition is usually discovered during adolescence when menstruation fails to occur. Upon examination, the vagina is found to be reduced in length (blind vagina). Because the Müllerian ducts failed to develop in these individuals, there are no uteri or fallopian tubes, and unfortunately, they are sterile. The testes are usually removed surgically and estrogen treatment provided, although this treatment has been questioned by some individuals who have received it. Genetic male XY individuals with CAIS display normal female body shapes and regard themselves unequivocally as female (Wisniewski et al., 2000) (Figure 3.12).

Behavioral endocrinologists are interested in CAIS (TFM) because this genetic defect has been discovered in several nonhuman species, including rats, mice, cattle, and chimpanzees (Olsen, 1992), allowing experimental study of its physiological mechanisms. In addition, individuals with CAIS have normal receptors for, and normal concentrations of, the nonandrogenic hormones (such as estrogens) involved in sexual differentiation, and thus provide researchers with the ability to separate the contributions of these hormones from those of androgens to typical behavioral sexual differentiation.

As noted above, another genetic dysfunction that leads to anomalous sexual differentiation is 5α-reductase deficiency. Genetic males (XY) with 5α-reductase deficiency are born with ambiguous genitalia and small, undescended testes. They are usually considered females at birth and

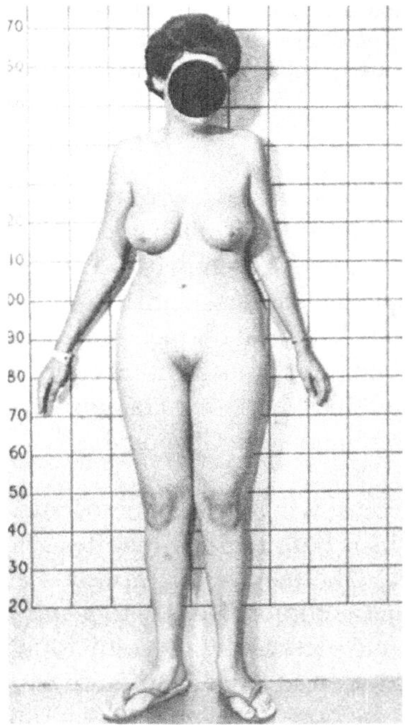

3.12 Complete androgen insensitivity syndrome (CAIS) is a condition caused by an absence of functional androgen receptors in genetic male (XY) individuals. Because individuals with CAIS completely lack androgen receptors, development of secondary sexual characteristics at puberty proceeds in a feminine direction. Although individuals with CAIS exhibit a normal female body morphology, the gonads are testes, and the accessory sex organs do not develop normally, resulting in sterility. Individuals may also display partial androgen insensitivity syndrome (PAIS). Courtesy of John Money.

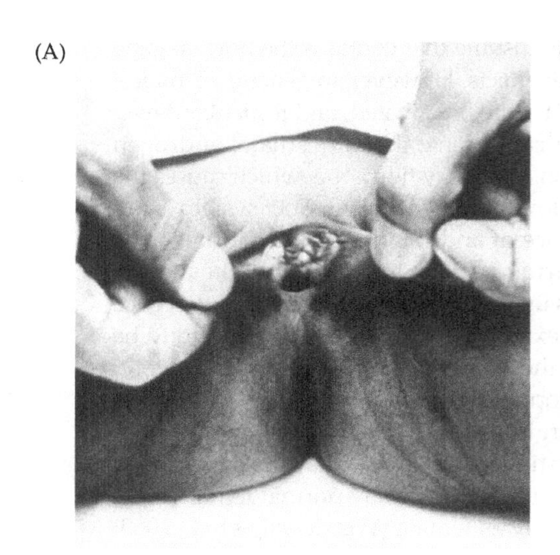

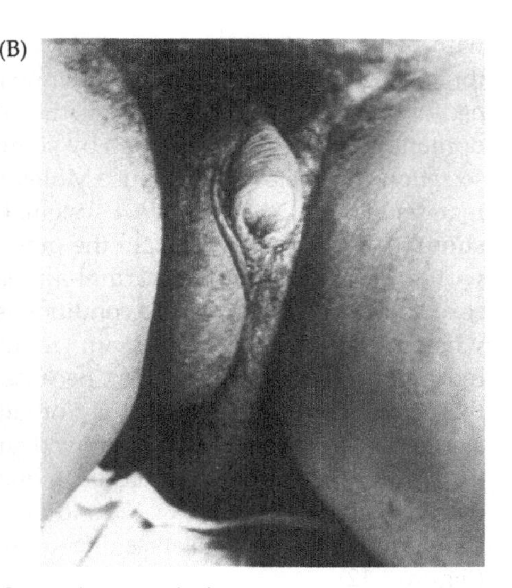

3.13 5α-reductase deficiency. (A) A lack of 5α-reductase, which converts testosterone into DHT, results in incomplete genital masculinization at birth. Note the incomplete fusing of the urogenital groove, the lack of penile development, and the presence of hypospadias, the opening of the urethra on the underside of the penis. Most individuals with 5α-reductase deficiency are sexed as girls at birth. (B) The same individual at puberty. Although DHT is still not present, high blood concentrations of other androgens at puberty can activate DHT receptors and cause delayed and partial masculinization of the external genitalia. Despite pubertal penile development, hypospadias remains, which contributes to infertility. Because these androgen-induced changes in the external genitalia are accompanied by a change to a male-typical body type, some individuals with 5α-reductase deficiency who have been raised as females may take on a male gender identity and gender role at puberty. Courtesy of Julianne Imperato-McGinley.

reared as females. At puberty, testosterone masculinizes the body, which develops male-typical musculature and axillary hair growth, and the genitalia develop to resemble a male-typical penis and scrotum (Figure 3.13). The urethra usually opens at the base of the penis; this condition, called **hypospadias**, substantially reduces fertility. Because these individuals are exposed to male-typical hormones pre- and postnatally, but are typically reared as females until puberty, behavioral endocrinologists have studied them in order to understand the contribution of hormones versus rearing to human behavioral sexual differentiation (see Chapter 4).

TRISOMIC ANOMALIES Occasionally, babies are born with extra chromosomes. Approximately one infant in every 600 live births is born with **Klinefelter syndrome** (Figure 3.14A; Nielsen and Wohlert, 1990). These individuals possess an extra X chromosome (XXY). The presence of the Y chromosome is sufficient for the *SRY* gene to be activated and masculinization to occur, and these individuals are sexed as males at birth. Although the testes develop sufficiently to cause masculinization, these individuals are usually sterile because of reduced sperm pro-

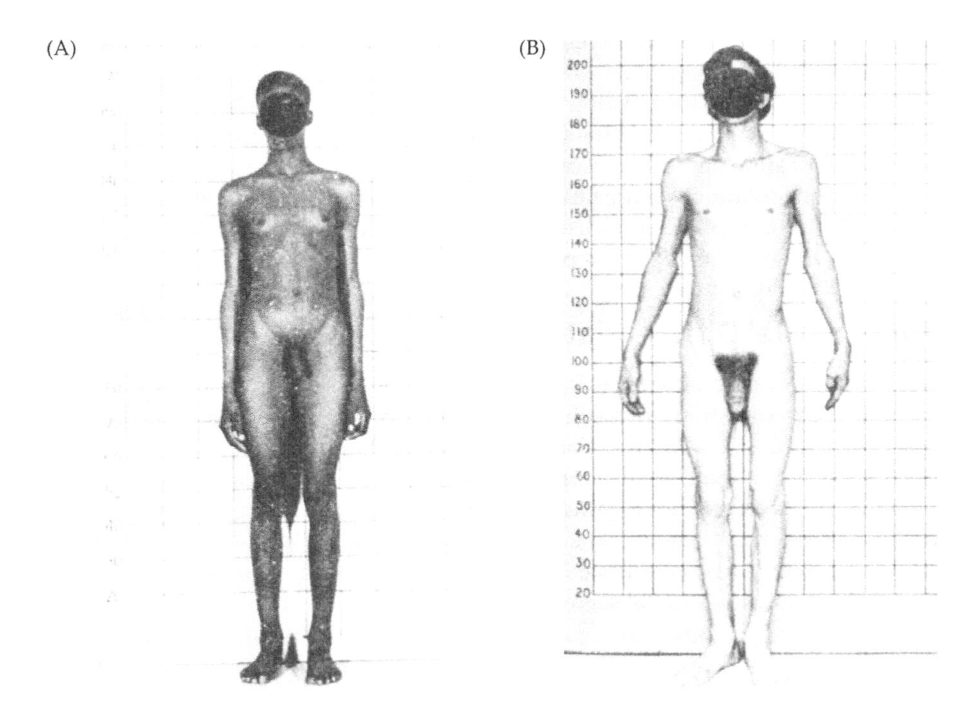

3.14 Trisomic genetic anomalies. (A) This 18-year-old XXY male shows the underdeveloped external genitalia, gynecomastia (breast development), and disproportionately long limbs characteristic of Klinefelter syndrome. (B) This young adult male with the XYY genotype is well over 2 meters tall. Like those with Klinefelter syndrome, XYY individuals are usually sterile. Other characteristic traits of the syndrome are above-average height and below-average intelligence. Courtesy of John Money.

duction. Learning disabilities are also commonly observed among individuals with Klinefelter syndrome.

Approximately one infant in every 850 live births possesses an XYY genotype (Figure 3.14B; Nielsen and Wohlert, 1990). These individuals are considered male at birth, but in common with XXY individuals, XYY men may be sterile. The XYY genotype was once thought to result in increased aggressiveness because a survey of prison populations revealed that a disproportionately large number of these individuals were represented (Hook, 1973; Jacobs et al., 1965). It was proposed that the Y chromosome was responsible for aggressiveness in normal males; therefore, it was hypothesized that individuals with two Y chromosomes displayed increased aggression, which led to their imprisonment. Further analyses revealed that individuals with XYY genotypes were not necessarily more aggressive than XY males, but that they possessed two other traits that increased the likelihood of criminal prosecution: they were less intelligent than average and they were much taller than average. Thus, these individuals may not engage in more criminal activities than average, but apparently they tend to be apprehended more readily (Witkin et al., 1976).

In sum, possession of a single Y chromosome sets off a chain reaction that leads to the differentiation of the testes. Hormonal secretions from the testes lead to masculinization and defeminization. In the absence of these hormonal events, feminization and demasculinization occur. In the following sections, the similarities and differences in the process of sexual differentiation among mammals, birds, and other animals will be presented.

Avian Sexual Differentiation

Studies of nonmammalian species offer intriguing insights into our own species by demonstrating alternative solutions to common problems of adaptation. The process of avian sexual differentiation is similar to that in mammals, but with several interesting differences. Among birds, females are the heterogametic sex (WZ), whereas males are the homogametic sex (ZZ). In the case of copulatory behavior, males are the default sex: in the absence of gonadal secretions, the masculine developmental track is followed, and the feminine pattern of development is attained by active hormonal secretion (Figure 3.15; Balthazart and Adkins-Regan

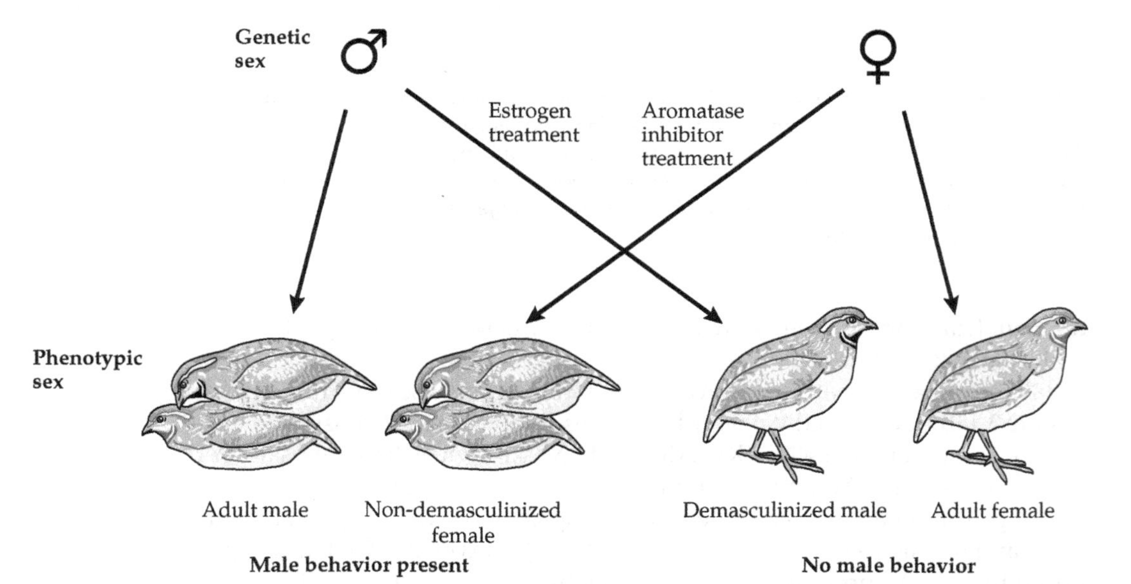

3.15 Development of female copulatory behavior requires active hormonal secretion. The genetic sex of birds in the egg is either male or female. If only the presence or absence of male copulatory behavior is considered, then males generally show masculine patterns and females generally show feminine patterns of copulatory behavior. If estrogen is given to a genetic male bird in the egg, he is demasculinized. If a genetic female bird is given an aromatase inhibitor in the egg, preventing the formation of estrogens, she is not demasculinized, and shows male copulatory behavior as an adult. After Balthazart and Ball, 1995.

2002). In other avian systems (i.e., the mechanisms underlying singing behavior), testicular hormones exert a masculinizing effect, whereas the absence of hormonal stimulation leads to the feminine pattern of no song (but, as noted below, it is still not clear what the mechanisms are for sexual differentiation of the neuronal system underlying bird song). For yet other aspects of sexual differentiation, the default condition in birds is neither male nor female; rather, ovarian or testicular hormones must exert feminizing or masculinizing effects, respectively, during normal sexual differentiation (Adkins-Regan, 1987).

At the present time, no sex-determining gene has been identified in birds. The presence of the W sex chromosome appears to cause the primordial gonad to secrete estrogens. The estrogens stimulate proliferation of the germinal epithelium in the left gonad, leading to cortical (ovarian) development. Only the left ovary of birds usually develops (Nalbandov, 1976). If chicken (*Gallus domesticus*) eggs containing ZZ chromosomal males are injected with estrogens, gonadal tissue develops into ovarian tissue. Similarly, when the embryonic gonads of one sex are grafted into an embryo of the other sex, or when embryonic gonads are grown together in culture dishes, ovaries are observed to feminize testicular development, but the presence of testes does not affect ovarian development (Haffen and Wolff, 1977).

Normal sexual differentiation of the avian Müllerian duct system provides an example in which neither the male nor the female pattern of development can be regarded as neutral. Like mammals, avian embryos of both sexes initially possess both Müllerian and Wolffian duct primordia. The Müllerian duct system develops in female chickens to form the oviduct and the shell gland (Figure 3.16), and regresses in male chickens during days 8–13 of incubation. Only the left duct

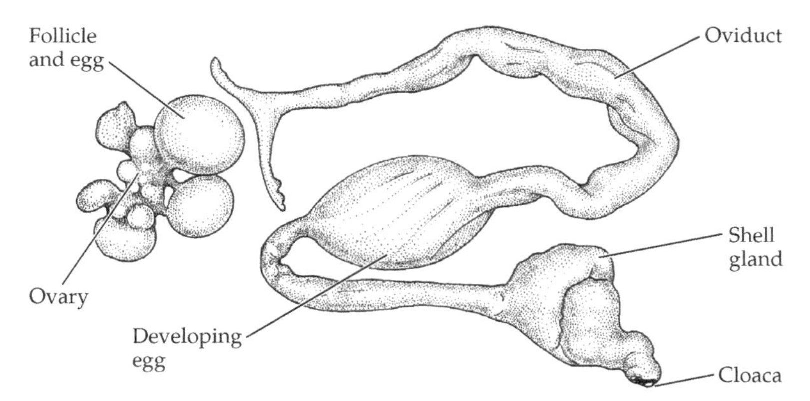

3.16 Ovary and oviduct of a chicken. Eggs (ova) are shown in the ovary at different stages of preovulatory development. Chickens ovulate virtually every day, and the egg travels down the oviduct to the shell gland, where a calcium-rich shell is laid down prior to oviposition (egg laying). Other birds have not been bred to be as prolific in their egg-laying behavior as chickens; typically most, but not all, songbirds and seabirds lay only 1–4 eggs per clutch and rear 1–2 clutches per year. After Nalbandov, 1976.

develops in female chickens; the right duct regresses during days 9–18 of incubation. Thus, feminization of the accessory sex organs requires development of only the left components of the Müllerian duct system. If embryonic chickens or ducks are gonadectomized by focal X-irradiation prior to day 8 of incubation, then both duct systems persist; in other words, neither defeminization (regression of the Müllerian duct system) nor feminization (development of the left components of the Müllerian duct system) occurs in the absence of gonadal influences (Wolff and Wolff, 1951).

Sexually dimorphic avian behavior also reflects this variation in hormonal influences. In some cases, as in copulatory behavior, the masculine behavior pattern is the default condition, whereas the feminine pattern is produced by hormonal activation that causes demasculinization. In the case of other behaviors, testicular hormones exert a defeminizing or masculinizing effect, whereas the absence of hormonal stimulation leads to the feminine pattern.

Alternative Reproductive Tactics and Male Polymorphism

There are other situations in which the default condition is neither male nor female, but rather, ovarian or testicular hormones must exert feminizing or masculinizing effects, respectively, during normal sexual differentiation of behavior. The previous sections have assumed that there are two distinct sexes: males and females. Within some species, however, there are multiple "types" (or **morphs**) of the same sex. Different male morphs often look very distinct, and they often exhibit quite dissimilar behavioral patterns, as well as alternative mating tactics. In some cases, a subset of males within a population resemble females, whereas the alternative male morph differs from females in appearance and behavior; in other cases, there are several different male morphs (**polymorphisms**) within a population, all of which differ from females in appearance and behavior (Moore et al., 1998).

Different male morphs often pursue different tactics to achieve reproductive success. Some morphs are territorial and exclude other males from (and attempt to attract females to) their defended territories. Other morphs are "satellite" or "sneaker" males. Satellite males usually loiter near the edges of a territorial male's property and attempt to mate with females that are attracted to his territory. Sneaker males often resemble females; they "sneak" into a territorial male's property along with visiting females, then attempt clandestine matings with the females before the resident male discovers the deception.

Males that change phenotypes, either permanently or temporarily, during their lives are said to have *plastic* phenotypes (Moore, 1991). For example, male spring peepers (frogs of the genera *Pseudacris*, *Rana*, and *Hyla*) display one of two alternative mating tactics: they either vocalize to attract females, or remain silent near calling males and attempt to intercept females that are moving toward the callers. Calling is energetically expensive, and males can switch between these behavioral strategies on different nights (Lance and Wells, 1993).

Male morphs that permanently differentiate into a single phenotype for life are said to have *fixed* phenotypes (Moore, 1991). For example, male ruffs (*Philomachus pugnax*; a type of sandpiper) become either darkly colored and highly territorial or lightly colored with the behavioral characteristics of satellite males (Lank et al., 1995). Several species of fish, including sunfishes, swordtails, platyfishes, and midshipmen, display male polymorphism in appearance and mating tactics (Bass, 1996). One of the most interesting examples of alternative male phenotypes comes from the midshipman fishes.

There are two types of male plainfin midshipmen (*Porichthys notatus*) (Figure 3.17). Type I males build nests and attract females to their nests by generating an advertisement call, which sounds like a series of long-duration "hums." If you walk along a Pacific beach anywhere from Northern California to the Canadian border on a summer night, you are likely to hear the "songs" of Type I midshipman males calling females to their nests (Bass, 1996). A responding female inspects the nest, and if it meets her specifications, she deposits her eggs. The Type I male fertilizes the eggs, then maintains a vigilant defense of the clutch and the nest site against rivals. While guarding the eggs, he emits an aggressive call (or grunt) to warn away potential intruders (Bass, 1995). Type I males are nearly eight times larger than Type II males, which physically resemble females (Bass, 1995). Type II males never build nests, guard eggs, or vocalize to attract females;

3.17 Different male morphs of the plainfin midshipman fish. A nest containing a Type I male (center), a female (right), and Type II males (far right and lower left). Also shown on the left are newly hatched fry attached to a small rock by an adhesive disk at the base of the yolk sac. During the breeding season, male plainfin midshipmen (*Porichthys notatus*) build nests under rocky shelters along the intertidal and subtidal zones of the western coast of the United States and Canada. Photograph by Margaret A. Marchaterre, courtesy of Andrew Bass.

rather, they have evolved an alternative reproductive tactic as sneaker/satellite males. They swim into nest sites guarded by Type I males and release sperm. Presumably, the Type I males do not challenge them because they resemble females.

Type II midshipman males undergo sexual maturation nearly a year earlier than Type I males, and this precocious puberty differentiates the two male phenotypes. However, it has not yet been established whether the difference in the timing of puberty is the primary determining factor in the developmental pathways leading to the different male morphs. Both types of males (and females) have detectable testosterone concentrations in their blood, with Type II males displaying the highest concentrations, followed by females, then Type I males (Brantley et al., 1993). The primary teleost fish androgen, however, is 11-ketotestosterone. The 11-ketotestosterone concentrations in Type I males are five times higher than their testosterone concentrations; neither Type II males nor females have detectable 11-ketotestosterone levels (Bass, 1996). Only females have any detectable circulating estrogens. Presumably, these different patterns of steroid hormones activate receptors in various neural tissues involved in growth, reproductive behavior, and vocalization. Thus, one reasonable hypothesis is that 11-ketotestosterone mediates nest building and vocal courtship behavior in Type I males.

In several lizard species, males also exhibit alternative behavioral mating strategies and body types. For example, male tree lizards (*Urosaurus ornatus*) display striking polymorphisms in the color of a body structure called a "dewlap" (Moore et al., 1998). The dewlap is a flap of skin attached to the throats of lizards that the males extend during courtship or aggressive encounters. The dewlaps of male tree lizards may have solid colors (typically orange, yellow, blue, or violet) or have a central spot surrounded by a contrasting background color (e.g., an orange spot surrounded by blue) (Hews et al., 1997). Nine different color morphs have been identified among tree lizards in Arizona, but most populations consist of two or three different male morphs. One well-studied population contains two types of males that are present in nearly equal proportions: (1) territorial males that have an orange dewlap sporting a blue spot, and (2) nonterritorial males with a solid orange dewlap (Thompson and Moore, 1992). The nonterritorial males display one of two dominant behavioral patterns: (1) sedentary satellite behavior or (2) nomadic behavior. The polymorphisms in dewlap color and territoriality appear to be permanently organized early during post-hatching development (Moore et al., 1998). If immature males are exposed to high testosterone and progesterone concentrations, then territorial, orange-blue males result; if they are not exposed to high testosterone and progesterone concentrations, they develop into nonterritorial males with orange dewlaps (Figure 3.18). However, the nonterritorial tactics of the orange morphs appear to be affected by environmental conditions. Stressful ambient conditions (e.g., drought) cause sustained corticosterone secretion, which promotes nomadic, wandering behavior (Moore et al., 1998). Thus, the two male phenotypes are permanently fixed by early hormonal conditions, whereas the flexibility of tactic switching (i.e., sedentary versus nomadic) among males with orange dewlaps is mediated by the temporary effects of corticosterone or testosterone concentrations during adulthood.

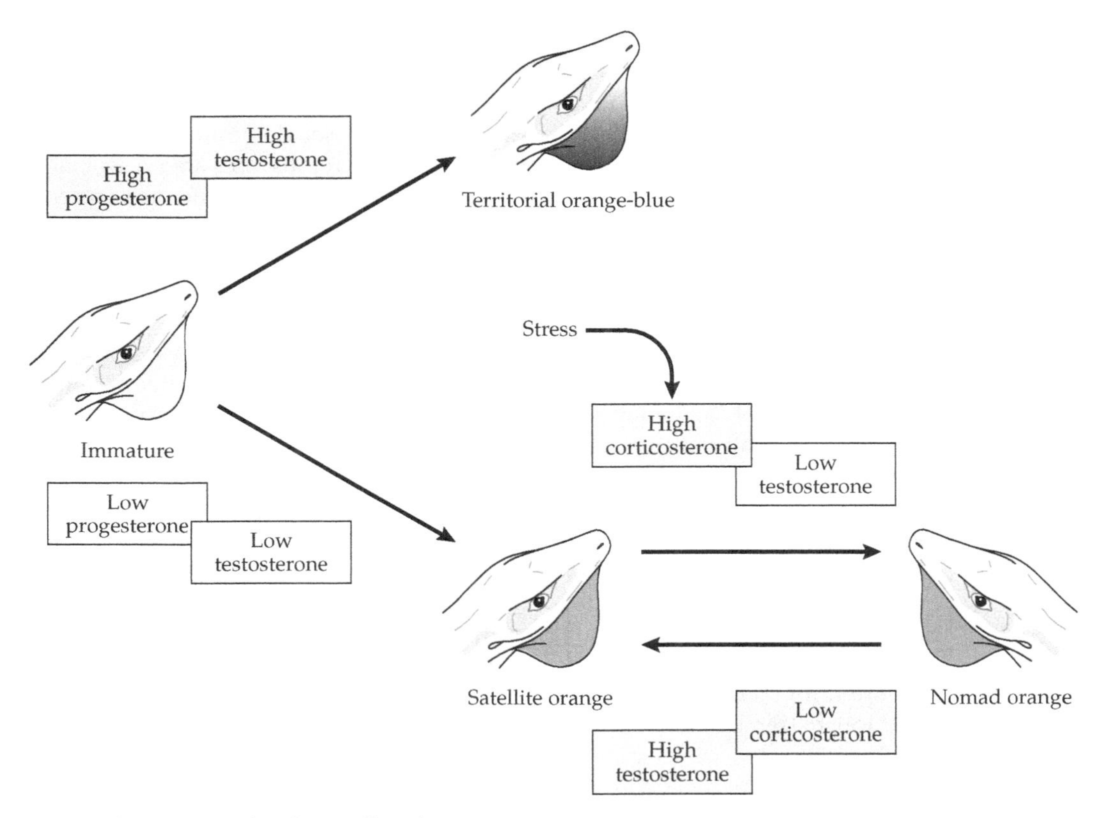

3.18 Three types of male tree lizards coexist in some Arizona populations: (1) territorial males, (2) sedentary nonterritorial males, and (3) nomadic nonterritorial males. The territorial males develop an orange dewlap with a blue spot; this morph results from early post-hatching exposure to high testosterone and progesterone concentrations. In the absence of high testosterone and progesterone concentrations, males develop a solid orange dewlap and become nonterritorial. In adulthood, if stressful environmental conditions result in high corticosterone concentrations, and thus lowered testosterone concentrations, nonterritorial males become nomadic; if low corticosterone concentrations, and thus higher testosterone concentrations, are present, nonterritorial males become sedentary. Territorial orange-blue males remain territorial regardless of environmental conditions and corticosterone concentrations. After Moore et al., 1998.

To date, alternative reproductive tactics and multiple morphs have not been described for females. Whether this is a true sex difference or simply represents a failure to discover these alternatives among females remains unknown.

Environmental Sex Determination in Reptiles and Fishes

In mammals and birds, sex is determined by the presence or absence of sex-specific genes. If homologous sex chromosomes are present, then female mammals or male birds develop; if heterologous sex chromosomes are present, then male mammals or female birds develop. Sex determination is dependent only on genotypic differences. Mammals and birds evolved from reptilian ancestral forms about 325 million and 150 million years ago, respectively. Some modern reptilian species, like mammals and birds, display genotypic sex determination. In some of these species males develop if homologous sex chromosomes are present, and in other species females develop if homologous sex chromosomes are present. However, in some lizards, most turtles, and probably all crocodilian species, environmental factors, not genotypic information, direct sex determination (Crews et al., 1988a; Korpelainen, 1990). Indeed, sex chromosomes are generally absent in these species (Crews and Bull, 1987).

The environmental factor most commonly involved in reptilian sex determination is temperature (Bull, 1980; Korpelainen, 1990), although water potential (concentration of salts) may also mediate sex determination in some species (for example, the painted turtle, *Chrysemys picta*: Gutzke and Paukstis, 1983). The embryos of a species with **temperature-dependent sex determination** are bipotential and can develop into a male or a female solely on the basis of the temperature at which the egg incubates (Figure 3.19). Relatively high temperatures (>30°C) produce males, and cool temperatures (<28°C) produce females in some lizard species, as well as in alligators (Crews et al., 1988a). The opposite pattern of temperature effects on sex determination is observed among many turtle species. Yet another pattern of temperature-dependent sex determination has been reported for snapping turtles and crocodiles: females are produced at the high and low extremes, whereas males are produced at intermediate temperatures (Bull, 1983).

Temperature-dependent sex determination could allow the mother to determine the sex of her offspring by varying the temperature of the nest in which her eggs are incubated. Although this is an extremely exciting possibility, there is no evidence thus far that sex ratio is manipulated by parental care (Clutton-Brock, 1991), nor is there any evidence that temperature exerts its effects by means of differential mortality in either sex (Crews et al., 1988a). Rather, incubation temperature determines whether the primordial gonad develops into a testis or an ovary. The primordial gonad in reptiles consists of distinct cortical and medullary tissue. If the cortical portion of the gonad develops, then an ovary develops; if the medullary portion of the gonad develops, then a testis develops. The physiological mechanisms by which temperature determines which portion of the gonad develops are beginning to be unraveled (Crews, 2003b).

As in birds, steroid hormones influence gonadal differentiation, and steroid treatment can override temperature effects on gonadal development. That is, if eggs are incubated at male-producing temperatures but treated with estrogens, females develop; conversely, if eggs are incubated at female-producing temperatures but treated with androgens, males develop. The mechanism by which temperature affects steroidogenesis remains unknown. Presumably, the enzymatic conversions necessary for steroid metabolism are temperature-sensitive. It is possible that temperature controls the activity of aromatase (which converts testosterone to estradiol) or reductase (which converts testosterone to dihydro-

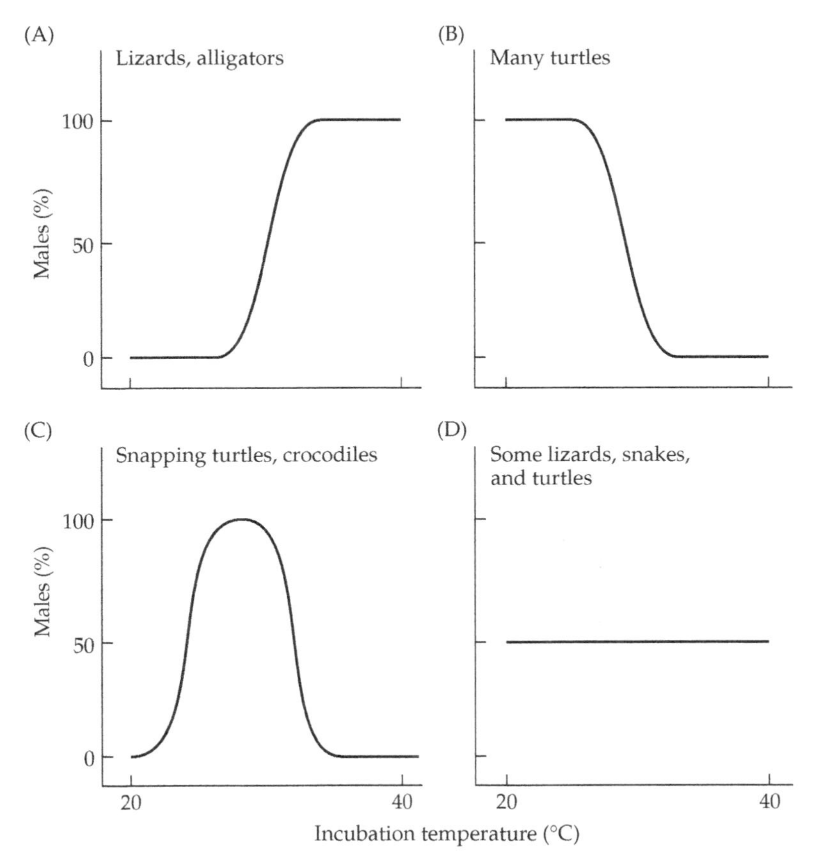

3.19 Temperature-dependent sex determination in reptiles. (A) Exposure to relatively high ambient temperatures during incubation results in male-typical development among the offspring of some lizards and alligators. (B) In many turtle species, high temperatures during incubation result in female-typical offspring development. (C) Moderate temperatures result in male offspring in snapping turtles and crocodiles, while exposure of eggs to either high or low temperatures results in female offspring in these species. (D) Sex determination in some reptilian species is unaffected by incubation temperatures. Data from Crews et al., 1988a.

testosterone) in reptilian gonads (Crews and Bergeron, 1994; Rhen and Lang, 1994). High temperatures might inactivate enzymes necessary for converting androgens to estrogens in some species, whereas low temperatures could have the same effect in other species. Thus, the development of the reptilian gonads into testes or ovaries may start a cascade of events that results in subsequent male-typical or female-typical behavior, respectively, in adult reptiles.

Alternatively, aromatase activity in the brain may differ between the putative sexes. For example, at the start of the temperature-sensitive period in red-eared slider turtles (*Trachemys scripta*), the pattern of aromatase activity in the gonads and adrenals was similar between the sexes, but in the brain, turtles destined to become females displayed significantly higher concentrations of aromatase than turtles destined to become males. By the time of hatching, this aromatase activity had waned (Willingham et al., 2000). These findings suggest that the brain is the primary site of aromatase response to temperature, which affects estradiol concentrations, which in turn may affect the neuroendocrine events regulating sex steroid hormone production (Willingham et al., 2000).

What happens if reptile eggs are maintained at intermediate temperatures? Based on the outcome of one study of leopard geckos (*Eublepharis macularius*), intermediate temperatures can have interesting consequences for adult sexual behavior (Gutzke and Crews, 1988). In this species, temperatures of 32°C produce mainly male geckos, but a few female offspring are also produced; temperatures of 26°C produce only females. When females incubated at these two temperatures, as well as at a third, intermediate temperature (29°C), were compared to adults in mating tests, female geckos from all three groups elicited equivalent amounts of courtship from males suggesting that they were considered equally feminine by males. Females from the low-temperature group responded to male courtship with female-typical behavior. However, females from the high-temperature group responded to courtship with male-typical behavior. Females incubated at intermediate temperatures displayed adult sexual behavior intermediate between the two extremes, showing some female-typical and some male-typical courtship behaviors. Interestingly, incubation at high temperatures increased adult levels of aggressive behavior in both sexes (Sakata and Crews, 2004) (Figure 3.20).

As mammals and birds, animals with temperature-dependent sex determination remain the same sex throughout their lives (Crews et al., 1988a). However, some other animals, especially fishes, can switch sexes during their lifetimes! Several species of fish are hermaphrodites (Demski, 1987). **Simultaneous hermaphrodites** possess **ovotestes** that produce both eggs and sperm, and they alternate between two behavioral roles in providing eggs or sperm during spawning. There are also two types of **sequential hermaphrodites**: animals that begin life as

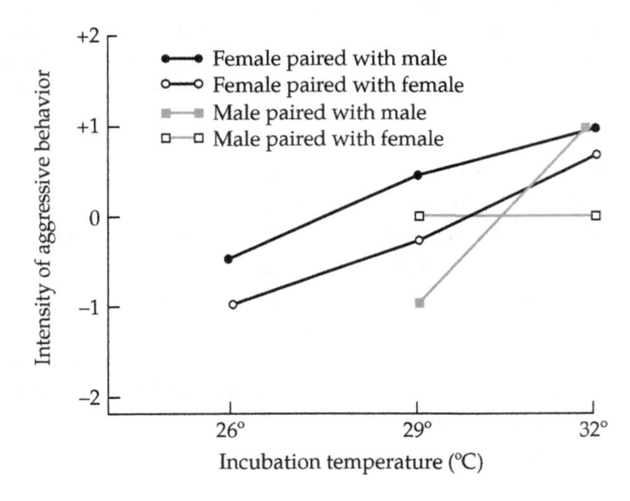

3.20 Aggression in adult female geckos can be influenced by incubation temperature. Relatively high temperatures (32°C) produce few female offspring, and relatively low temperatures (26°C) produce only female offspring in geckos. Females produced at low incubation temperatures display low levels of aggression as adults, compared with females incubated at high temperatures, which display male-like aggression as adults. −2 = fleeing; −1 = low posture; 0 = no response; +1 = high posture; +2 = attack. Data from Gutzke and Crews, 1988.

3.21 Sequential hermaphroditism in the striped parrotfish (*Scarus iserti*). (A) Young females of this species have bold stripes, and defend territories along the edges of reefs. (B) As females mature and grow in size, they change sex and adopt male-typical coloration and behavior, joining other males in defending large territories encompassing several "harems" of young females. Courtesy of Robert W. Warner.

one sex, then change to the other sex as adults in response to social, environmental, or genotypic factors (Demski, 1987; Perry and Grober, 2003). Several species of sequential hermaphroditic reef fish have been identified, including wrasses, sea basses, gobies, anemonefishes, and parrotfishes (Bass, 1996). **Protogynous** types begin life as females, then change into males; **protandrous** types are first males, then change into females. Such sequential sex changes involve a change in gonadal type and function, a change in the dominant gonadal steroid secreted, a change in morphology (body type), and a change in behavior (Figure 3.21). Clownfish (*Amphiprion percula*), as depicted in the popular movie *Finding Nemo*, defend an anemone home. The typical family unit consists of a mating pair and one to three juveniles. Unlike the movie, however, if the mother disappears, the father undergoes a dramatic transformation to become the egg-laying female, and the oldest juvenile grows rapidly to become the male mate (Buston, 2003). In other words, these fish are protandrous.

Although many species undergo sex changes, the specific trigger varies. In some cases, it appears to reflect an attainment of a certain body size. For example, a meta-analysis of roughly 80 species of sex-switching fishes, crustaceans, mollusks, echinoderms, and polychaete worms concluded that, on average, these animals change sex when they attain 72% of the maximal size for their species (Allsop and West, 2003). Over 90% of the variance in the timing of adult sex changes could be accounted for by this body mass set point rule. Although such a high degree of consistency regarding the average timing of adult sex changes may indicate that similar life history events underlie the decision to switch sexes, specific social factors show variation in their timing (Perry and Grober, 2003). Presumably, models of "ultimate" factors driving the decision to switch sexes are based on "proximate" factors underlying the process.

Stoplight parrotfish (*Sparisoma viride*) are coral reef fish that begin life as females and change into males (Cardwell and Liley, 1991a,b); in other words, these fish are protogynous. In many cases, the sex change is accompanied by a striking change in body coloration, from the female-typical "initial phase" coloration to the male-typical "terminal phase" coloration. Some individuals, however, undergo the sex change from female to male without the change in coloration. These female mimics compete with terminal-phase males by using a "sneaker" mating tactic. Terminal-phase males establish territories from which they exclude other males, and in which they court females. If a courted female releases her eggs, a female mimic may enter the territory undetected and fertilize some of the eggs. This male morph with its alternative reproductive tactic is maintained in the population because of its occasional success in attaining fertilizations.

The sex change in parrotfish is accompanied by changes in the pattern of circulating steroid hormones. Females have undetectable concentrations of the primary fish androgen 11-ketotestosterone, moderate concentrations of testosterone, and high concentrations of 17β-estradiol. During the transition from initial to terminal coloration phases, concentrations of 11-ketotestosterone rise dramatically, and estrogen concentrations decline. Males have high concentrations of 11-ketotestosterone and testosterone, but low concentrations of estrogens. Injection of adult females with 11-ketotestosterone causes a precocious change in gonadal, gametic, and behavioral sex. Female mimics, which maintain the initial-phase coloration, do not display the increase in blood concentrations of 11-ketotestosterone.

The Effects of Hormones on Sexually Dimorphic Behaviors

As we have seen, the initial step in sex determination, the development of a testis or an ovary, triggers a cascade of events that results in sexually dimorphic behaviors (see Figure 3.9). In order to understand the interaction between hormones and behavior, behavioral endocrinologists have studied several animal models that show distinct behavioral sex differences. By studying the development, neural and hormonal bases, and molecular mechanisms of a simple behavior that is displayed by one sex but not by the other, researchers hope to elucidate general principles that can be expanded from these simple systems to explain more complex behavioral sex differences, such as those displayed by humans.

The Organizational/Activational Hypothesis

In 1959, W. C. Young (Box 3.2) and his colleagues published a classic study on the effects of prenatal and early postnatal androgen treatment on female reproductive behavior (Phoenix et al., 1959). This experiment rivals Berthold's study with roosters (see Chapter 1) in theoretical importance to behavioral endocrinology. Because of the profound effect of Young's study on the shaping of the field, it will be described here in detail.

Females and males of many species show sexually dimorphic mating postures. Female rodents stand immobile with arched backs and allow males to mount them. This mating posture is called **lordosis** (see Chapter 6). Males, on the other hand, mount females. Females rarely exhibit **mounting** behavior, the masculine behavior pattern, and males typically do not assume the lordosis posture, the feminine behavior pattern, when mounted by other males.*

Mating behavior in both sexes is under the control of gonadal steroid hormones. For example, castration of adult males stops mounting behavior, and testosterone replacement therapy restores mounting behavior to its original levels. Logically, we could conclude that testosterone causes mounting behavior. However, injection of adult females with testosterone does not increase their mounting behavior to male-typical levels. Females somehow lose the potential to exhibit male-typical behavior during development; in other words, the behavior of females is demasculinized. Similarly, castration of adult males and subsequent treatment with estrogens and progestins does not lead to the appearance of female-typical mating behavior. Males become defeminized during development; that is, they lose the potential to display female-typical behavior.

Professor Young and his research group sought to understand how these behavioral differences in hormonal response were mediated. They reasoned that hormonal events early in development must be responsible for the induction of feminine and masculine behavioral patterns in general, as well as for the shifting of the probabilities of masculine and feminine mating behaviors in females and males in the presence of the appropriate stimulus. In the experiments designed to address this hypothesis, testosterone propionate was administered to pregnant guinea pigs throughout most of their 69-day gestation period. Some pregnant females were given large doses of the hormone. At birth, some of the female offspring from these pregnancies possessed external genitalia that were indistinguishable from those of their brothers or of typical males. These females were labeled hermaphrodites by the researchers.† Another group of females was exposed to smaller doses of the androgen prenatally; these females had no visible abnormalities of the external genitalia and were referred to as unmodified females. The researchers waited for the guinea pigs to mature to examine their behavior in the second phase of the study.

In adulthood, both groups of androgen-exposed females, as well as androgen-exposed male and control male and female guinea pigs, were gonadectomized and then injected with estrogen and progesterone to stimulate female sexual behavior. Each of these animals was paired with a normal male guinea pig.

---

*It should be noted here that this yes/no dichotomy is presented for clarity's sake; in reality, males of some strains of rats show lordosis regularly, and females of some rat strains display frequent mounting behavior. A sex "difference" is often one of frequency, behavioral threshold, or the quality of the behavior.

†Actually, these females were pseudohermaphrodites because they all possessed two ovaries, but recall that it had become common in medicine to refer to individuals with ambiguous external genitalia as hermaphrodites (Ellis, 1945).

BOX 3.2 *William C. Young*

William Caldwell Young was an important pioneer in behavioral endocrinology. Young received his undergraduate degree at Amherst College, where he worked with Professor Harold H. Plough. He published his senior thesis on the effects of high temperature on fertility in *Drosophila* (Young and Plough, 1926). Young attended the University of Chicago for his graduate training and worked in the laboratory of Carl R. Moore in the zoology department.

Carl Moore had earned his doctorate at the University of Chicago working in the laboratory of Frank Lillie. Moore had published a series of 11 papers with the generic title "On the Physiological Properties of the Gonads as Controllers of Somatic and Psychic Characteristics." He noted in his studies that, although castration of male guinea pigs resulted in rapid and nearly uniform degeneration of the epididymis and seminal vesicles, the timing of the cessation of sexual behavior varied enormously among individuals. Moore interpreted these results to suggest that reproductive behavior was "utterly capricious, unordered by hormonal events, and unrelated to variables of significance to reproductive biology" (Goy, 1967).

William C. Young (1899–1965)

Young produced a classic study of epididymal sperm transport for his dissertation project and was awarded his doctorate in 1927. During the course of these studies Young became interested in the hormonal control of female mating behavior, but his mentor, Professor Moore, actively discouraged behavioral research because his own experi-

Some time later, all of the animals were injected with androgens to stimulate male sexual behavior and paired with a typical conspecific female that was in mating condition. The results were impressive. Androgens given to guinea pigs early in life (1) decreased the ability of both experimental groups of females to display lordosis in adulthood, (2) enhanced the ability of both experimental groups of females to display mounting behavior in response to testosterone therapy, but (3) caused no deleterious effect on mounting behavior or other masculine behavioral patterns in males similarly treated.

Not only were the data of considerable interest, but the conceptual framework in which the authors chose to cast them was rich and wide. Professor Young and his colleagues made the following speculations (Phoenix et al., 1959; Feder, 1981):

1. A clear distinction can be made between the prenatal actions of hormones in causing differentiation, or *organization*, of neural substrates for behavior and

ence suggested that the variation inherent in behavioral studies was daunting.

Young joined the biology department at Brown University in 1928, and submitted a grant application to continue his work on the reproductive morphology and physiology of guinea pigs. While waiting for the funding agency's decision, he had "nothing better to do than investigate the inexpensive problem of the morphological changes associated with female sexual behavior in the guinea pig" (Goy, 1967). By the time he received the news that his application was not to be funded, Young and his research group had already started a productive line of research that used behavior as an indirect measure of ovarian activity (Dempsey, 1968). In 1939, Young moved to the Yerkes Laboratories of Primate Biology in Orange Park, Florida, where he interacted with R. M. Yerkes and other psychologists during his 4-year tenure. His research began to include behavior as a primary object of study.

Apparently, Young felt isolated in a nonacademic research setting (Goy, 1967), and left the Yerkes laboratory in 1943 to teach at Cedar Crest College. He then accepted an offer to set up an endocrinology laboratory in the anatomy department at the University of Kansas. This laboratory was very active, and studies of hormonal control of reproductive function, including mating behavior, were conducted there by a stellar team of graduate students and postdoctoral fellows, including M. Diamond, H. Feder, A. Gerall, R. Goy, J. Grunt, C. Phoenix, W. Riss, E. Valenstein, and others. The famous study of the organizational effects of prenatal androgens on guinea pig mating behavior was conducted in Young's Kansas laboratory (Phoenix et al., 1959). This study has been one of the cornerstones of modern behavioral endocrinology research. While at Kansas, Young also edited the third edition of *Sex and Internal Secretions*, published in 1961, which was the "bible" for reproductive biologists at that time, and continues to be an important sourcebook 45 years later.

In 1963, Young moved to the Oregon Regional Primate Center, where he became chairman of the reproductive physiology department. There, his organizational/activational hypothesis of behavioral sexual dimorphism was examined in rhesus monkeys (Goy and Phoenix, 1972; Goy and Resko, 1972). William C. Young is one of the most important figures in behavioral endocrinology because of his many significant research contributions, his theoretical contributions, which were profoundly important in framing research problems, and the graduate and postdoctoral students trained in his laboratories who continue to make important contributions to the field (Beach, 1981).

the actions of hormones in adulthood in causing *activation* of these substrates. This is a very important point, and provides the basis for the organizational/activational hypothesis of sexually dimorphic behaviors. During development, the sex steroid hormones organize or establish the components of the nervous system that will be needed for subsequent male- or female-typical behaviors. In the adult, these same hormones activate, modulate, or inhibit the function of these existing neural circuits.

2. Critical periods of perinatal development exist during which an animal is maximally susceptible to the organizing effects of steroids on neural tissue. (We now know that the timing of the organizational effects of steroids may be broader in scope than originally formulated by Young, Phoenix, and their collaborators.)

3. Organization of the neural tissues mediating mating behavior is in some ways analogous to the differentiation of the accessory sex organs. That is, the nervous systems of males are normally masculinized and defeminized during development, whereas the nervous systems of females are normally feminized and demasculinized.
4. The organizing effects of steroid hormones on prenatal neural tissues are subtle and reflect alterations in function rather than in structure, because even females whose external genitalia are not physically masculinized demonstrate behavioral masculinization and defeminization. (As you will soon learn, however, it has recently been established that changes in neural structures are associated with sexually dimorphic behaviors.)
5. The idea that prenatal hormones act to organize the neural substrates for behavior has possible implications for the study of behavior in primates, including humans.

This study profoundly affected the way behavioral endocrinologists conceptualized the interactions of hormones and sexually dimorphic behaviors. This groundbreaking study did not arise out of thin air, however, but was built on three separate, but related, lines of research, involving (1) embryological studies showing that prenatal androgen exposure results in masculinized external genitalia, (2) studies of the effects of blood-borne substances in masculinizing the female twins of male cattle, and (3) endocrinological studies investigating the control of ovulation.

PRENATAL ANDROGENS AND MASCULINIZED GENITALIA Anatomists had studied the sexual differentiation of the body and had discovered what many of us have observed: the most striking sexual dimorphism in most species is that of the external genitalia. In one early study, pregnant rats were injected with androgens throughout their pregnancy (Hamilton and Gardner, 1937). Female rats whose mothers had received androgens during pregnancy were born with masculinized external genitalia: in other words, they developed penes. This study, as well as others, contributed to the development of the organizational/activational hypothesis by demonstrating that early endocrine treatments could masculinize or feminize anatomical and functional features.

FREEMARTINS The idea of hormonal regulation of the process of sexual differentiation was originally given impetus by studies of a phenomenon known as freemartinism (Lillie, 1917). A **freemartin** is a sterile female twin of a male found in cattle, sheep, goats, and pigs. Freemartins described in cattle have the following characteristics: (1) the freemartin calf is a sexually abnormal chromosomal female (XX) born as a twin to an apparently normal male; (2) about 90% of female cattle born as twins to males are freemartins; (3) the external genitalia of the freemartin are of the female type, and there is mammary gland development; (4) there is variable development of the internal genital tracts; and (5) the gonads are atrophic but resemble testes more than ovaries. To account for freemartinism, a hormon-

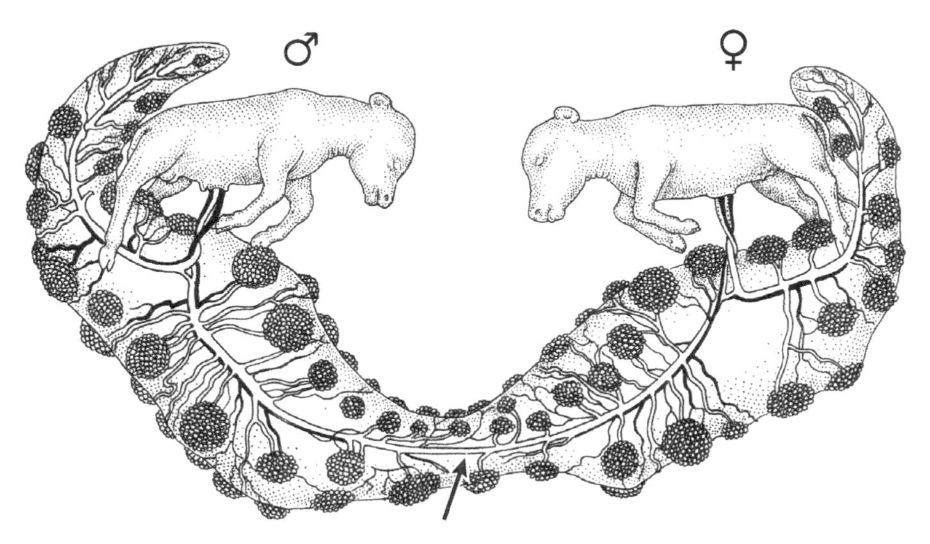

3.22 Freemartins are sterile female twins of a male found in cattle, sheep, goats, and pigs. It was hypothesized early in the 20th century that masculinizing factors travel from the male twin to the female twin through the vascular connections of the placenta because of a vascular fusion, or anastomosis (shown by arrow), and affect the internal anatomy of the female. However, attempts to produce freemartins by administering androgens to pregnant cows have not been successful, and the mechanism underlying freemartinism remains unknown. After Lillie, 1917.

al theory was proposed, suggesting that androgens produced by the male fetus are able to reach the female due to fusion of some of the placental blood vessels that supply the twins. Evidence to support this idea included the findings that such vascular connections (anastomoses) do occur in cattle and that in the 10% or so of cases in which the female twin is fertile, the placental anastomoses are obliterated (Figure 3.22). That the male produced a blood-borne factor was deduced by the observation that the male twin seemed normal and the finding, based on histological techniques, that the prenatal testis was active prior to activation of the female sibling's ovary.

However, administration of androgens to pregnant cows never results in freemartins. Female fetuses of androgen-treated mothers have ovaries and internal genitalia of the female type (predominantly), but masculinized external genitalia. Thus one of the oldest questions about sexual differentiation is still unsolved. Apparently, an exchange of cells between twins during early pregnancy, or passage from the male to the female fetus of a nonsteroidal factor that favors testicular development, mediates the freemartin phenomenon.

CONTROL OF OVULATION The third group of studies that laid the groundwork for the organizational/activational hypothesis examined the hormonal control of ovulation. Because studies of this process have continued to shed light on the mechanisms of sexual differentiation, they will be described in detail.

Female mammals display cyclic gonadal function, whereas males have more or less tonic reproductive function: that is, ovulation, the expulsion of a ripened egg from the ovary, occurs cyclically, whereas sperm production is constant throughout the breeding season. The reproductive behavior of the sexes follows these fundamentally different patterns of gamete production, with females displaying cycles of mating behavior and males displaying more or less continuous willingness to mate.

What causes these patterns? Gonadal function is driven by gonadotropins secreted from the anterior pituitary (see Chapter 2), but what drives the gonadotropins? Gonadotropin-releasing hormone (GnRH) secreted from the median eminence of the hypothalamus determines the pattern of release of luteinizing hormone (LH), a gonadotropin, from the anterior pituitary. In both male and female rats, this process begins when a series of LH pulses, 20–40 minutes apart (60 minutes in humans), are released into the circulation in response to pulses of GnRH from the hypothalamus; these pulses of LH often precede pulsatile releases of gonadal steroid hormones (Figure 3.23). Thus, the hypothalamus provides a pulsatile delivery of GnRH in both males and females, which subsequently drives the anterior pituitary, which eventually drives gonadal function. As circulating concentrations of LH and sex steroid hormones rise, they feed back to suppress further GnRH secretion (negative feedback). Similarly, the injection of steroid hormones suppresses GnRH and gonadotropin secretion in a dose-dependent manner.

In females, there is a steady increase in the amount of estrogen produced by the ovaries against the backdrop of continuous pulses of GnRH secretion and negative feedback endocrine regulation. Somehow, females escape from rigid nega-

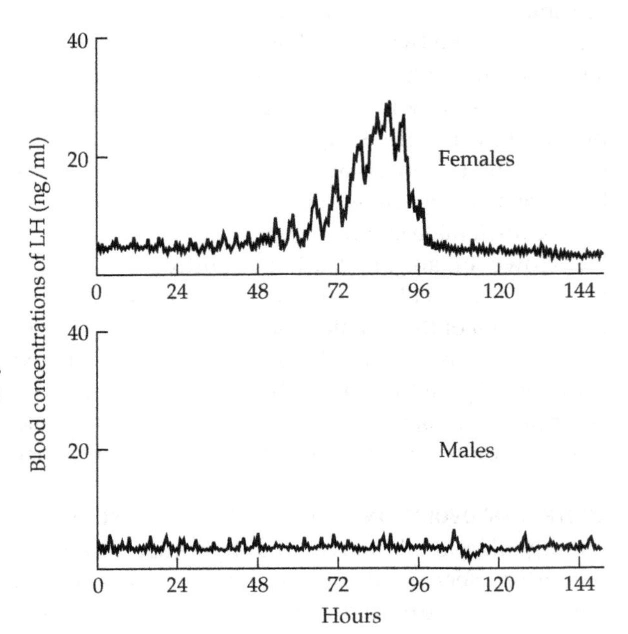

3.23 LH profiles of female and male rats. In both male and female rats, LH is released in a pulsatile fashion from the anterior pituitary in response to pulses of GnRH from the hypothalamus. In females, the pulse frequency and amplitude increase around the time of ovulation, as negative feedback mechanisms are temporarily overwhelmed by increasing blood estrogen concentrations and positive feedback mechanisms are engaged.

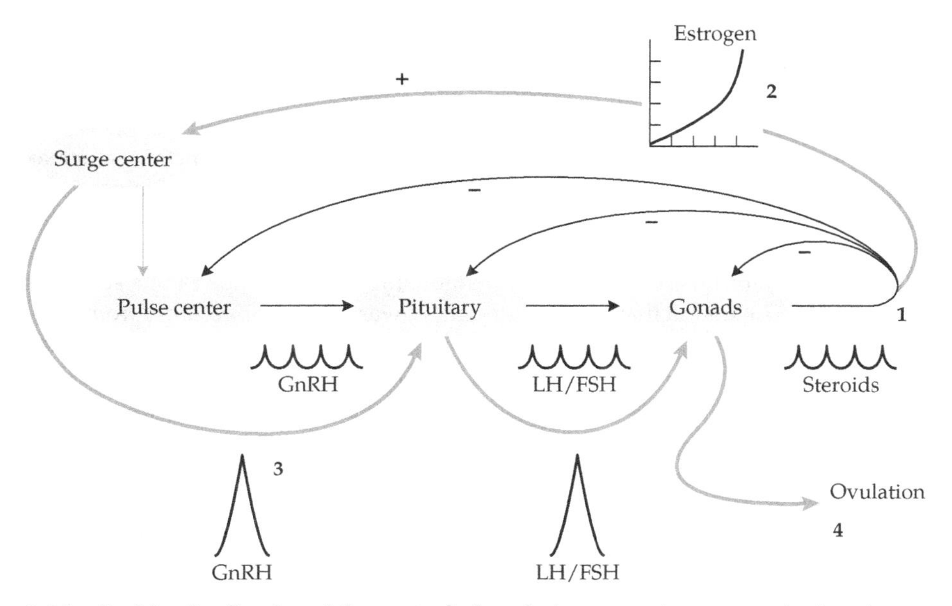

3.24 Positive feedback and the control of ovulation. Steroid secretion in both males and females is usually regulated by negative feedback (black lines), whereby increasing gonadal steroid concentrations feed back to the gonads, anterior pituitary, and hypothalamus to slow secretion of GnRH, gonadotropins, and gonadal steroids. However, on a cyclic basis, females escape this rigid negative feedback control (1; gray lines), and blood concentrations of estrogen steadily increase (2). Through a positive feedback mechanism, the surge center responds to the rising estrogen concentrations by releasing a surge of GnRH, which stimulates the anterior pituitary to release corresponding surges of LH and FSH (3), which stimulate the gonads and cause ovulation (4). After ovulation, the negative feedback mechanisms are once again engaged, and LH and FSH secretion resume a pulsatile pattern.

tive feedback control. As estrogen accumulates in the blood, a surge of GnRH is released from a putative "surge center" in the hypothalamus. The surge of GnRH causes a surge of LH and follicle-stimulating hormone (FSH) secretion, which in turn stimulates ovulation via the positive feedback mechanism illustrated in Figure 3.24. This pattern can be demonstrated in the following manner. Injecting an ovariectomized rat with low doses of estrogen causes a reduction of GnRH and LH concentrations via negative feedback; injecting an ovariectomized rat with high doses of estrogen, however, causes an increase in GnRH and LH concentrations by the process of positive feedback. Injecting a castrated male rat with low or high amounts of estrogen also causes reduction of GnRH and LH concentrations due to negative feedback (Hovath et al., 1997). Injections of pharmacologically high doses of estrogens also fail to evoke a positive feedback response in male primates (Karsch et al., 1973). In monkeys, the preovulatory gonadotropin surge does not require elevated GnRH from the hypothalamus; the positive effect of estrogens on LH and FSH secretion in primates may reflect elevated numbers of GnRH receptors on LH- and FSH-producing cells (Karsch, 1987).

What is the basis of the sex difference in the regulation of endocrine function? Initially, it was proposed that the pituitary gland itself differs between the sexes. Females were thought to possess pituitaries that display cyclic function, whereas males were believed to possess pituitaries that display tonic function (Pfeiffer, 1936, 1937). According to this logic, a male does not normally ovulate because his pituitary does not display cyclic function—and more obviously, because he does not have an ovary. But if an ovary is transplanted into an adult male rat, he still does not ovulate. The notion that the pituitary gland was masculinized arose from pituitary gland transplant studies. In an extremely delicate operation, the pituitary gland can be surgically removed from a normal female rat and transplanted into a hypophysectomized female rat. If the transplanted pituitary gland is able to establish a blood supply, it can support ovulation in the recipient female. However, if the recipient of the transplanted female pituitary is a male with a transplanted ovary, ovulation is not observed. If female rats have testicular tissue implanted under the skin within a few days of birth, they also fail to sustain ovulation, either from their own ovaries or from transplanted ovaries. Similarly, the pituitary glands of male rats that are castrated within a few days of birth are able to support ovulation with an implanted ovary. On the basis of these results, it was reasonably concluded, first, that ovulation does not require neural input, because transplanted ovaries that establish vascular, but not neural, connections can ovulate, and second, that exposure to androgens early in life results in a masculine pattern of pituitary function that is incapable of sustaining ovulation, whereas lack of perinatal androgen exposure results in a feminine pattern of ovulation-supporting pituitary function (Pfeiffer, 1936). Conceptually, the argument that control of ovulation was sexually differentiated was correct, but the endocrine structure identified as sexually differentiated was incorrect.

A series of studies on the physiological control of ovulation in rabbits demonstrated that pituitary function is in fact controlled by the hypothalamus (Harris, 1937, 1948, 1955). These studies also demonstrated that the "male" pituitary can, in fact, sustain ovulation. If the pituitary is removed from a female rat and replaced by a pituitary from an adult male, then ovulation occurs, provided that vascular connections with the hypothalamus are established (Harris and Jacobsohn, 1952). Thus, the sexual dimorphism in the ability to support ovulation is not at the level of the pituitary, but rather, reflects changes induced by early androgen exposure upstream in the hypothalamus.

Many studies of the process of sexual differentiation have used rats rather than guinea pigs because investigators do not have to make endocrine manipulations via the mother, but can influence the sexual differentiation of each individual rat directly after birth. The masculinizing actions of the guinea pig fetal testis are completed approximately halfway through the 69-day gestation period, or about 35 days post-conception (Goy et al., 1964). Guinea pigs are **precocial** animals, which means that the young are born fully furred, able to walk about, and able to regulate their body temperature. In contrast, rats are **altricial** animals: their gestation period is only about 21 days, and they are born in a very immature state, blind and hairless, and unable to walk or thermoregulate. Sexual differentiation in rats occurs during the first 10 days or so after birth, again about 35 days post-conception. This developmental pattern allows researchers to make endocrine manipulations in individual newborn rats. In rats, there is a critical period for sexual differentiation, or more properly, a period of maximal susceptibility to the effects of the hormones that induce behavioral sexual dimorphisms. If male rats are castrated at 1 day of age and implanted with an ovary as adults, ovarian cycles that are identical to female cycles are observed. However, if male rats are castrated at 20 days of age and implanted with an ovary as adults, no cycles in ovarian function are observed (Pfeiffer, 1935, 1936). Further studies have indicated that castration after 3 or 4 days of age also fails to preserve the potential for cyclic ovarian function in adult males. The secretion of testosterone by the male during the first several hours of life destroys forever the potential for the cycling that is seen in females.

The blood testosterone concentrations of male rat pups approach adult concentrations by about 6 hours after birth. This elevation in blood testosterone concentrations has been attributed to two factors: release of the infant's testes from negative feedback suppression by placental gonadotropins and retarded steroid clearance rates as steroid metabolism shifts from maternal systems to the infant's own liver function (Baum et al., 1988). This perinatal surge of testosterone appears to destroy the neural circuitry involved in the release of surges of GnRH from the hypothalamus, and thus permanently eliminates the potential for feminine positive feedback effects. In the absence of high concentrations of gonadal steroids, the neural circuitry involved in GnRH surge generation is spared. Male human babies also display this surge in blood testosterone concentrations at birth (Forest and Cathiard, 1975). However, they retain the ability to show feminine positive feedback responses when injected with estrogens. In other words, the presence of these positive feedback mechanisms is not a sexually dimorphic trait in humans, or in other primates examined (Hodges and Hearne, 1978; Karsch et al., 1973; Norman and Spies, 1986; see Chapter 4).

Where in the brain does the rodent sex difference in positive feedback occur? Recall that both males and females release LH in pulsatile bursts (see Figure 3.23). This pulsatile release of LH follows pulsatile bursts of GnRH in both sexes. In addition to these pulses of GnRH, females also show large, cyclic surges of GnRH that are not seen in males. The pulse generator in both males and females appears to be located in the arcuate nuclei and the ventromedial nuclei of the hypothalamus, which are often collectively considered to be part of the medial basal hypothalamus. However, most of the GnRH cells in females are located in the rostral aspects of the preoptic area, that is, in front of the optic chiasm (Chappel, 1985). The surge generator is located in the region of the preoptic area and suprachiasmatic nuclei (Figure 3.25; Pierce, 1988). As you will learn in Chapter 8, there are biological timekeepers located in these regions that control ovulation. A daily "minisurge" of GnRH is elicited by one of these clocks, and when this minisurge coincides with high estrogen concentrations, females' GnRH cells release large and rapid pulses of

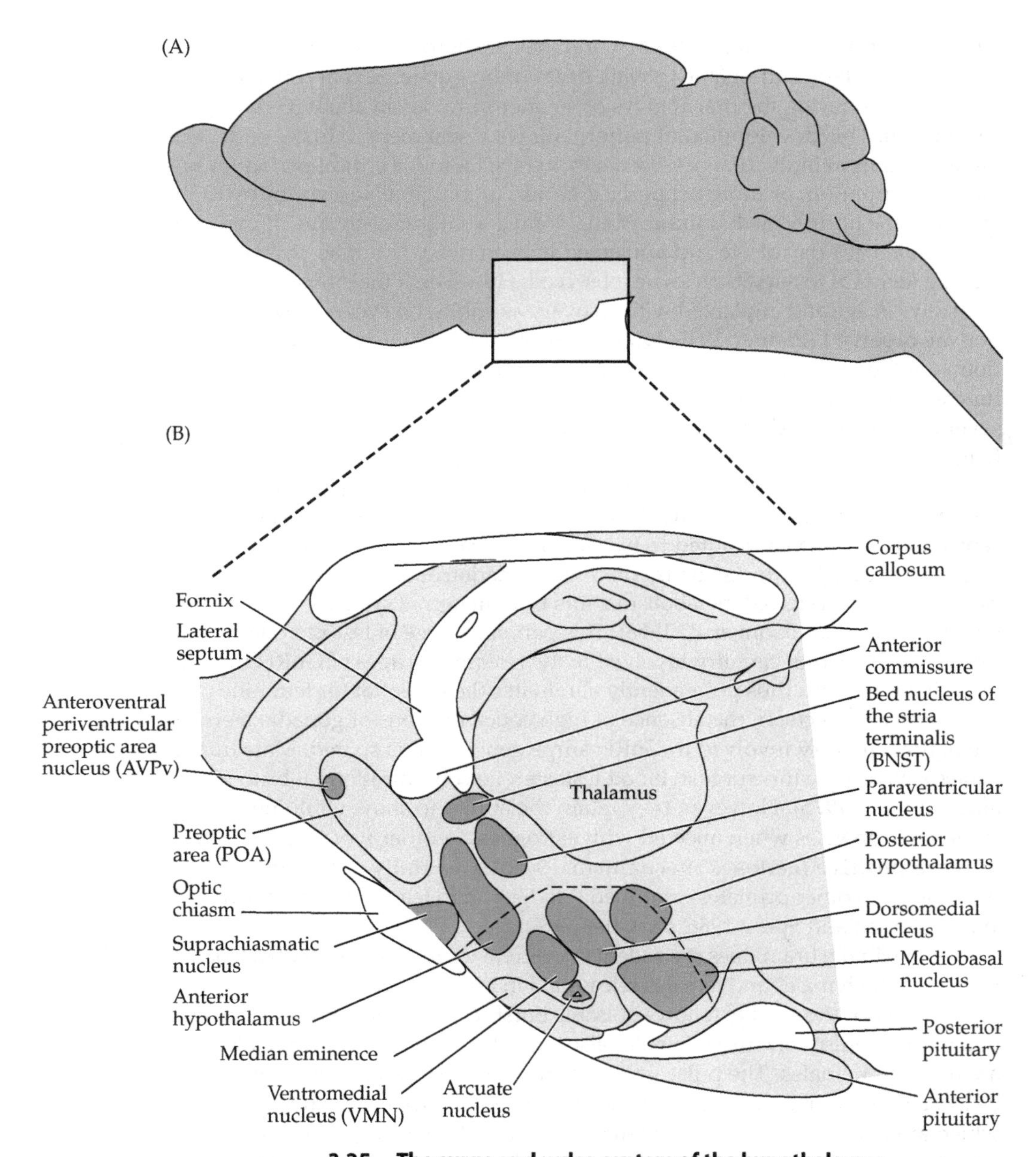

3.25 The surge and pulse centers of the hypothalamus. (A) Schematic sagittal view of a rat brain, showing the hypothalamic nuclei involved in the regulation of gonadotropin pulses and surges. (B) Pulsatile gonadotropin secretion remains normal after isolation of the dorsomedial and medial basal hypothalamus in females (below the dotted line), but the periovulatory surges observed in normal females no longer occur, indicating that the female surge center is located in the nuclei adjacent to the preoptic area (POA) of the hypothalamus.

GnRH, resulting in a GnRH and LH surge. Destruction or surgical isolation of these timekeeping cells prevents the surge of GnRH in response to estrogen treatment, but these procedures do not affect negative feedback; negative feedback in both males and females appears to be regulated in the medial basal hypothalamus.

Early exposure to androgens has been hypothesized to destroy the neural connections between the surge generator and the pulse generator. As we will see, in some cases, early androgen treatment promotes death of neurons; in other cases, early androgen exposure spares neurons from death. Another hypothesis states that androgenization may diminish pubertal estrogen binding to receptors in the surge generator tissues so that surges are not observed (Gerall and Givon, 1992). Either hypothesis is consistent with the observation that no amount of estrogen treatment in adulthood can evoke a GnRH surge in males or in females treated with androgens perinatally: the underlying neural machinery is either not present or nonfunctional.

Sexual Differentiation and Behavior

As Young and his colleagues demonstrated, the potential for feminine or masculine behavior, like other aspects of sexual differentiation, appears to be organized by early exposure to hormones. Subsequent studies of rats have confirmed this view (Figure 3.26). Female rats that are ovariectomized as adults and injected with androgens do not exhibit male-typical mounting behavior in the presence of other females that are sexually receptive. However, if female rats are injected with androgens prior to 10 days of age and injected again with androgens as adults,

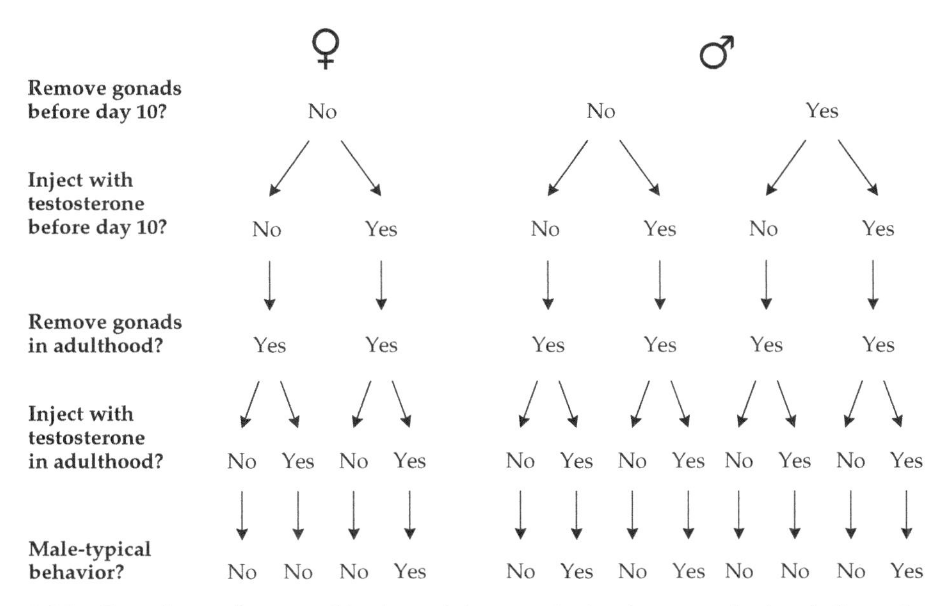

3.26 Experimental protocol for determining organizational versus activational effects of androgens on sexually dimorphic mating behavior in rodents.

they do display male-typical mounting behavior in the presence of receptive females. Similarly, if male rats are castrated at 20 days of age and injected with estrogens and progestins as adults, they do not display feminine mating postures. But if male rats are castrated at 1 day of age and injected with estrogens and progestins as adults, female sexual behavior is observed; that is, these males exhibit lordosis when mounted by male conspecifics (Grady et al., 1965). Thus, exposure to androgens prior to day 20 of life permanently organizes the brain to permit the later expression of masculine behavior.

In addition to mating behavior, many adult behavioral sex differences in rats (for example, in aggressive behavior, taste preferences, and parental behavior) are organized by steroid hormones perinatally and activated by steroid hormones in adulthood. The presence of such adult sexually dimorphic behaviors can be manipulated in a predictable way in infancy. Masculinization and defeminization of subsequent behaviors occur when rats are exposed to androgens prior to 10 days of age; feminization and demasculinization of subsequent behaviors occur when rats are not exposed to androgens prior to 10 days of age. Different sexually dimorphic behaviors do vary slightly in their organizational requirements. Small variations in the timing and dosage of androgen exposure do not influence all sexually dimorphic behaviors in rats equally; however, most behaviors are masculinized by androgen exposure on or before day 10 (Grady et al., 1965). Masculine behaviors are activated in adulthood by androgens and feminine behaviors are activated in adulthood by estrogens and progestins.

It is not testosterone or any other androgen per se, however, that is responsible for the masculinization of neural organization. As we saw in Chapter 2, testosterone is a prohormone for both an estrogen (17β-estradiol) and an androgen that cannot be converted into an estrogen (DHT). Thus, additional studies are always required to determine whether the effects of testosterone are mediated via estrogen or androgen receptors. There is now overwhelming evidence that testosterone is aromatized to an estrogen in the brain (Figure 3.27) and that many of its

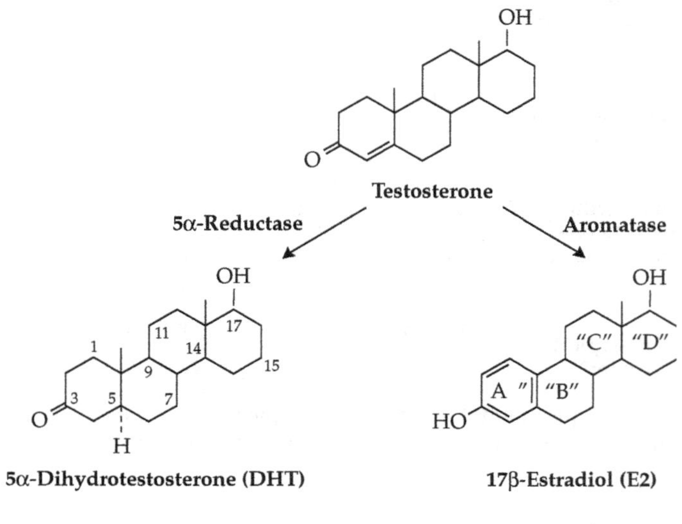

OH
O
Testosterone
5α-Reductase
Aromatase
OH
11
17
1
14
15
9
3
5
7
O
H
5α-Dihydrotestosterone (DHT)
OH
"C"
"D"
"A"
"B"
HO
17β-Estradiol (E2)

3.27 Testosterone is a prohormone for an estrogen (estradiol) and for an androgen (dihydrotestosterone) that cannot be converted into an estrogen. Thus, additional studies are always required to determine whether the effects of testosterone are mediated via estrogen or androgen receptors.

masculinizing effects on behavior are dependent on this cellular conversion (Gorski, 1993; Reddy et al., 1974). Indeed, an injection of estrogens before 10 days of age masculinizes later sexual behavior among rats even more effectively than an injection of androgens (Booth, 1977; Feder and Whalen, 1965).

You may be wondering, if estrogens cause masculinization of the nervous system, then why aren't all females masculinized? The ovaries of pregnant rats produce high concentrations of estrogens during gestation. Estrogens, like other fat-soluble steroids, could easily pass the blood–placental barrier, enter the fetal circulation, and theoretically masculinize female fetuses. However, perinatal rats produce large quantities of a protein called α-fetoprotein, which actively binds and sequesters circulating estrogens; the bound estrogens are removed through the placenta and metabolized by the maternal liver. This protein does not bind androgens effectively, however, so testosterone from the gonads of a male fetus reaches its brain, where it is aromatized to estrogen and masculinizes behavior. The α-fetoprotein protects female fetuses from estrogenic steroids, and because females normally have low circulating concentrations of androgens, masculinization does not occur. However, some artificial estrogens do not bind very well to α-fetoprotein (for example, DES), and these artificial estrogens, as we saw above, can have long-term ill effects on female health, as well as masculinizing effects on female morphology and behavior.

The Role of Steroid Receptors

Recently, molecular biological techniques have been applied to the problem of sexual differentiation in rat brains. The results of these studies demonstrate the importance of neural estrogen receptors in masculinization and defeminization (McCarthy et al., 1993a). Recall that the DNA in the nucleus of each cell contains specific coding instructions for proteins. These instructions are carried by messenger ribonucleic acid (mRNA) from the nucleus to wherever protein production occurs in the cytosol. The mRNA sequence that codes for the protein—in this case, the protein that makes up a neural estrogen receptor—provides the message with the instructions for making the receptor. Once the mRNA sequence for a certain protein is known, a piece of DNA can be generated, consisting of 10–30 bases, that is complementary to the sense mRNA. This piece of DNA, technically called an oligodeoxynucleotide, is referred to as "antisense" because it consists of a sequence of bases that is the mirror image of the "sense" message. When an animal is treated with this antisense RNA, it enters the cells, binds to the mRNA, and prevents the production of the receptor that is normally encoded by the mRNA.

In a study that demonstrated the role of estrogen receptors in sexual differentiation in the rat brain, 3-day-old female rats either received hypothalamic infusions of antisense RNA for estrogen receptors, or were subjected to one of two control procedures. Control rat pups were infused with either a scrambled series of 15 bases ("nonsense") dissolved in sesame oil, or the oil vehicle alone (McCarthy et al., 1993b). The females were then injected either with a low dose of testosterone propionate (50 μg), which usually produces slight masculinization in females, or with the oil vehicle. When the females reached adulthood, they were ovariectomized and injected with estrogen and progesterone. One hundred percent of the females that were not subjected to early testosterone treatment showed lordosis (feminine mating behavior) as adults (McCarthy et al., 1993b). Fewer than 20% of the females that received testosterone and received oil infused into the hypothalamus at 3 days of age exhibited lordosis as adults. Similarly, about 20% of the females that received testosterone and received "nonsense" DNA infused into the hypothalamus at 3 days of age exhibited lordosis as adults. In other words, these females were defeminized by the early testosterone treatment. However, early infusion of the estrogen receptor antisense RNA into the hypothalamus blocked many of the effects of early testosterone treatment; approximately 80% of these females showed lordosis behavior after estrogen and progesterone priming as adults. These results indicate that the organizational effects of early androgen exposure are partially dependent on the synthesis of estrogen receptors in the brain.

Mice with targeted deletions (knockouts) of the α estrogen receptor gene have been produced (Korach, 1994). Because these mice, referred to as αERKO mice, develop with the α estrogen receptor missing throughout life, one might predict that the males would not display normal masculine mating behavior. Male αERKO mice displayed normal mounting behaviors, reduced levels of intromissions, and no ejaculations (Ogawa et al., 1997, 2000). Male αERKO mice were also less aggressive than normal wild-type males. Because many so-called estrogen-dependent processes remained functional in the αERKO mice, the researchers reevaluated their hypotheses about the role of estrogen receptors in the development of sexually dimorphic behaviors. It was soon discovered that an alternative form of the estrogen receptor (termed the β estrogen receptor) persists in αERKO mice (see Chapter 5). Male mice in which the gene for the β estrogen receptor was disabled (βERKO mice) retained virtually all components of masculine sexual behavior (Ogawa et al., 1999). Still other mice were generated that lacked both the α and β estrogen receptor genes (αβERKO mice). Male αβERKO mice failed to display any components of masculine sexual behavior, including mounting behavior or ultrasonic vocalizations (Ogawa et al., 2000). These double knockout mice also displayed the reduced aggressive behaviors of αERKO knockout mice. Supporting a redundancy in function, these results suggest that either one of the estrogen receptors is sufficient to maintain mounting behavior in male mice. In contrast, offensive aggression specifically requires the presence of the α estrogen receptor gene.

Recent studies, however, suggest that the contribution of functional steroid receptors to sexually dimorphic behaviors is even more complicated. If the masculinizing effects of testosterone are exerted through its conversion by brain aromatase into estrogens that activate one of the estrogen receptors, then deletion of androgen receptors should not affect attainment of sexually dimorphic behaviors. However, male androgen receptor knockout (ARKO) mice showed marked reductions in male-typical reproductive and aggressive behaviors (Sato et al., 2004). Female ARKO mice displayed normal feminine sexual behaviors. Although treatment with the nonaromatizable androgen DHT failed to normalize male mating behaviors in male ARKO mice, DHT partially restored impaired male aggressive behaviors, suggesting the possibility of a second androgen receptor subtype. Interestingly, DHT treatment restored male-typical behaviors in αERKO mice (Sato et al., 2004). These results suggest that different patterns of sexually dimorphic behavior require different patterns of function by steroid receptor genes (Garey et al., 2003).

Conclusions

From the studies of rodents described above, three main principles of sexual differentiation of behavior can be distilled: (1) gonadal steroid hormones have organizing effects on behavior; (2) the organizing effects of steroid hormones are relatively constrained to a particular time during development; (3) an asymmetry exists in the effects of testes and of ovaries on the organization of behavior (but see Arnold and Matthews, 1988; Breedlove, 1992). The early effects of hormones are considered to cause permanent, irreversible behavioral characteristics (Arnold and Breedlove, 1985; Phoenix et al., 1959); early androgen treatment, for example, causes relatively irreversible and permanent masculinization of rodent copulatory behavior.

These early organizational effects can be contrasted with other reversible behavioral influences called activational effects (Phoenix et al., 1959). Androgens provided to rodents in adulthood, for example, activate male copulatory behavior by acting on the structures organized earlier by these same hormones. The activational effects of hormones on adult behavior are temporary and usually wane soon after the hormone is metabolized. The organizational and activational effects of hormones may actually be quite similar when considered at the molecular level. The permanency of the effects of early exposure to androgens may reflect events unique to a limited critical period of development. For example, in rats less than 10 days of age, gonadal steroid hormones may prevent neuronal death or promote neuronal survival during a period of programmed cell death (**apoptosis**) that is limited either temporally or developmentally (Gorski, 1993).

The notion of a critical period for organizational effects arises from the observation that these effects occur only during a specific temporal window of development. The mating behavior of rats, for example, can be masculinized by physiological concentrations of androgens only prior to about day 10. Masculinization can be induced between days 10 and 20 with pharmacological doses of androgens, but at some point (about day 25) no amount of androgen treatment can cause masculinization. This critical period reflects developmental processes that can be altered by steroid hormones only during particular stages of ontogeny; some of these developmental processes will be described in the next chapter. Increasing numbers of examples of organizational effects on adult behaviors that occur later in life (for example, postpuberty) have been reported in recent years (Adkins-Regan et al., 1989; Bloch and Gorski, 1988; Cherry et al., 1990; Commins and Yahr, 1984; DeVries et al., 1985). The classical perspective of organizational/activational effects of hormones has recently been modified to incorporate new evidence that organizational effects of steroid hormones can occur outside of the established perinatal critical period; i.e., during puberty (Sisk et al., 2004). Indeed, several critical periods during which hormones may change brain organization may exist during development (Sisk and Foster, 2004).

Finally, the asymmetry of gonadal effects on behavioral organization is striking. In mammals, in the absence of gonads or steroid hormones, female-typical behavior patterns emerge; that is, behavior is feminized and demasculinized.

Exposure to androgens causes the organization of male-typical behaviors; that is, behavior is masculinized and defeminized. This observation has led to the idea that females are the neutral or default sex in mammals, but recent studies indicate that hormones may be necessary for normal development of feminine behavior. Studies of avian sexual behavioral differentiation suggest that male birds are the default or neutral sex (see Box 4.1). To some extent, the asymmetry of testicular and ovarian effects on behavioral sexual differentiation in birds and mammals reflects the differential sensitivity (i.e., receptor availability) of the underlying neural tissue to estrogens and androgens. However, in both cases, estrogens ultimately act at the cellular level to masculinize and defeminize sexual behaviors.

More recently, adjustments to the organizational/activational hypothesis have been suggested (Wallen and Baum, 2002). For example, altricial mammalian species appear to rely more on aromatization for masculinization than do precocial species. An emerging issue is the importance of hormones in demasculinization of female brain and behavior and in defeminization of male brain and behavior in becoming a female or male, respectively (Wallen and Baum, 2002).

Environmental Influences on Mammalian Sexual Differentiation of the Nervous System

As noted above, the majority of behavioral sexual differentiation occurs postnatally in altricial rodents such as rats, mice, and hamsters. However, small and very subtle developmental events that occur in the egg (Box 3.3) or in utero can also result in significant behavioral effects in adulthood. In addition to the effects of the intrauterine environment, differences in maternal care and exposure to environmental chemicals can also affect sexual differentiation.

EFFECTS OF THE INTRAUTERINE ENVIRONMENT Although much of the process of sexual differentiation occurs postnatally, some sexual differentiation processes begin in utero for rats. For example, the position in the uterine horn where a female rat gestates can have important effects on subsequent adult physiology and behavior. Fetal rodents lie adjacent to one another in the uterine horns much like peas in a pod (Figure 3.28). Because the sexes are arranged randomly, a female fetus may be located between two sisters (a position designated 0-M for "no males"), between a sister and a brother (1-M), or between two brothers (2-M) (Clemens et al., 1978; vom Saal, 1979). This is a fine natural preparation for investigating whether variations in the intrauterine environment have postnatal effects on the physiology and behavior of rodents. In these studies, rat pups are delivered by cesarean section and their intrauterine positions noted.

Only a summary comparing two positions, 0-M and 2-M, will be presented here. Ultimately, there is no difference in reproductive capacity between 0-M and 2-M female mice in the laboratory: 0-M and 2-M females both become pregnant and produce equivalent numbers of offspring. However, interfemale aggression is higher in 2-M females than in 0-M females. Males spend more time with 0-M females than with 2-M females; that is, 0-M females are more attractive to males than 2-M females. The reproductive cycles of 0-M females are more easily inhibit-

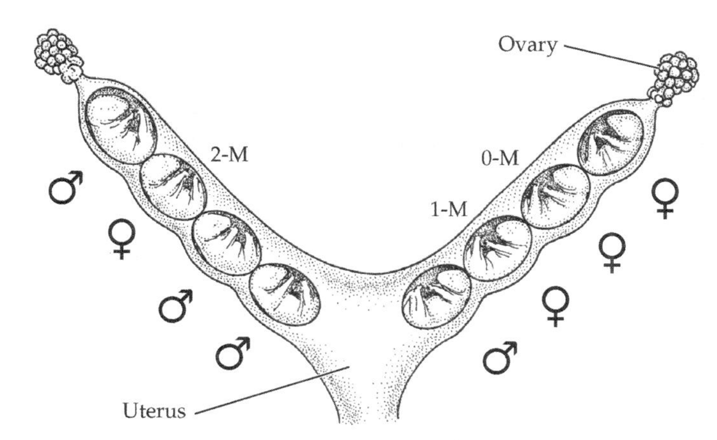

3.28 Rat pups gestating in utero are packed into the two uterine horns like peas in a pod. The sex of any given pup's neighbors in utero is entirely random. Thus, a female may gestate between two brothers (a position signified as 2-M), between a brother and a sister (1-M), or between two sisters (0-M). A female's uterine position in relation to male siblings can exert subtle influences on several behaviors later in life because it determines the degree of her exposure to androgens in utero.

ed by exposure to other adult females than are those of 2-M females (see Chapter 6). Finally, 2-M females have longer ovarian cycles than 0-M females. Thus, intrauterine position can affect several physiological and behavioral characteristics of adult female rodents. In nature, enhanced aggressiveness or lengthened reproductive cycles could significantly affect fitness.

Maternal stress can also affect subsequent adult reproductive behavior. When pregnant rats are stressed by restraining them under bright lights several times per day, their male fetal offspring produce less androgen than control males (Ward and Weisz, 1980). In adulthood, the mating behavior of these male rats is adversely affected (Grisham et al., 1991; Ward and Reed, 1985), and certain parts of the nervous system are more female- than male-typical (see Chapters 4 and 11). In addition to changes in copulatory behaviors, male rats that are prenatally stressed show reduced infanticidal behaviors, increased parental behaviors, and reduced aggressiveness (Ward, 1992). The duration of rough-and-tumble play in prenatally stressed males is comparable to that in females (Figure 3.29) (Ward and Stehm, 1991).

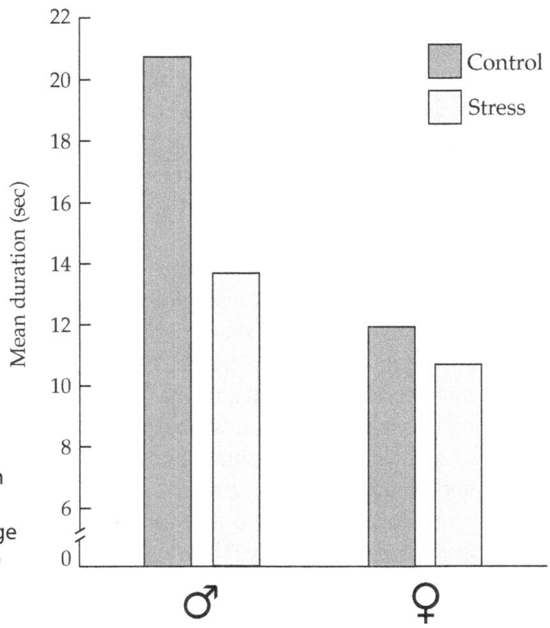

3.29 Rough-and-tumble play behavior is demasculinized in males by stress in utero. When pregnant rats were stressed by exposure to bright lights during the last trimester of gestation, the average duration of rough-and-tumble play (at 31 days of age) in their male offspring was similar to that in females. After Ward and Stehm, 1991.

BOX 3.3 Epigenetic Effects on Sexual Dimorphism: Direct Maternal Provisioning of Steroids to Offspring

Maternal hormones can be transferred directly to offspring via the egg yolk (Adkins-Regan, 1981; Schwabl, 1996). The extent to which maternal provisioning of eggs with steroids can influence sexually dimorphic behaviors (or even cause sex reversals in the developing young) remains controversial (Pike and Petrie, 2003). However, it is now well established that the amounts of steroids deposited in eggs vary, and that this variation reflects factors ranging from environmental to social to maternal conditions (Petrie et al., 2001).

Testosterone concentrations in canary (*Serinus canaria*) eggs increase as a function of laying order: regardless of chromosomal sex, eggs that are laid last in a clutch are provisioned with more testosterone than eggs laid first (Schwabl, 1993). This variation in testosterone was found to have functional effects on the offspring. The social rank of hatchlings was positively correlated with testosterone concentrations in the egg, suggesting that differential androgen levels in the egg affected subsequent behaviors. These results were the first to demonstrate that mother birds could provision their offspring with hormones to alter subsequent behavior. This epigenetic mechanism (factors that modulated gene expression) was hypothesized to account for some of the variation in offspring behavior (Schwabl, 1993).

Why would females sequester varying amounts of hormones in their eggs? Androgens speed growth and development of the offspring, and they increase begging behaviors (Schwabl, 1996). One adaptive functional reason that mothers would differentially supply eggs with sex steroid hormones would be to allow them to gain the upper hand in parent–offspring conflicts (Trivers, 1974). There is an inherent conflict of interests between parents and offspring. Because parents share an average of 50% of their genes with each offspring, all the offspring have equal value to them, and parents attempt to equalize the resources that they provide to the offspring. However, each offspring values itself twice as much as it values its siblings (100% of genes shared with self vs. 50% shared with siblings and only 25% shared with half-siblings). Thus, it is in the interest of each offspring to compete with its siblings for parental resources and to get as large a share of those resources for itself as it can. As a result, offspring will always demand more resources than their parents are willing to provide (Trivers, 1974).

Provisioning eggs with androgens can reduce the conflict between the chicks that hatch first and their siblings that hatch later by enhancing the growth and development of the latter. Otherwise, in clutches of eggs that hatch asynchronously, later-hatched siblings would be at a disadvantage in relation to their older nestmates in begging for food. To test this hypothesis, first-laid black-headed gull (*Larus ridibundus*) eggs were used to create artificial clutches of three eggs each that varied in androgen content (Eising et al., 2001). In some clutches, all of the eggs received a control treatment with an oil vehicle, whereas other clutches were experimentally manipulated so that eggs had three concentrations of testosterone: constant low, low supplement, or high supplement, and then the eggs were returned to the nest. Testosterone-treated eggs hatched an average of 0.5 days earlier than oil-treated eggs. Androgen treatment did not affect the survival of the chicks, but did enhance their growth (Eising et al., 2001). Androgen-treated, last-hatched gull chicks with experimental high androgen supplementation had higher body mass and longer legs than last-hatched gulls treated with oil. Moreover, having two younger androgen-treated siblings retarded the growth and development of the first-hatched (non-supplemented) gulls. Thus, it appears that differential transfer of steroid hormones by the mother could "level" the competition among siblings in the nest.

The advanced time of hatching in androgen-treated chicks may reflect, in part, accelerated development of the hatching muscle (musculus complexus) (Lipar and Ketterson, 2000). This muscle is important in breaking open the shell during hatching and for dorsal flexion of the neck during begging. Yolk androgen concentrations also increased with laying order in red-winged blackbirds (*Agelaius phoeniceus*), regardless of sex. The size of the hatching muscle also increased as a function of the order of laying, and this increase was positively correlated with testosterone concentrations in the egg. Direct injections of testosterone into eggs increased the size of the hatching muscle, whereas treatment with a testosterone antagonist, flutamide, decreased the relative mass of this muscle (Lipar and Ketterson, 2000).

In addition to the order of egg laying, social factors can influence maternal contribution of steroids to the eggs. The attractiveness of male zebra finches (*Taeniopygia guttata*) can easily be manipulated by providing leg bands of different colors. Females preferentially mate with red-banded males and avoid green-banded ones (Gil et al., 1999). Manipulation of color bands has a greater effect on male attractiveness than any other measured male characteristic thus far discovered (Cuthill et al., 1997). In one study, male zebra finches were randomly fitted with either red or green leg bands (Gil et al., 1999). Females were randomly assigned to the "attractive" red-banded males or the "undesirable" green-banded males. Their eggs were removed immediately after laying for hormone analyses and replaced with "dummy" eggs. At the end of laying, the dummy eggs were removed, and the females were provided with a new male with a different-colored leg band. At the end of the study, it was discovered that females mated with the "attractive" red-banded males placed more testosterone and DHT in their eggs than the same females when paired with "undesirable" green-banded males (Gil et al., 1999).

The social rank of leghorn hens affected the amount of testosterone in their eggs. Overall, the eggs of low-ranked hens had about the same concentrations of androgens in male as in female embryos (Muller et al., 2002). However, dominant hens allocated significantly more androgens to male than to female eggs. In contrast, subordinate females transferred more androgens to female than to male eggs. The investigators attribute this pattern to the benefits of high reproductive fitness in high-status roosters and the low reproductive fitness of low-status roosters. Low-status males achieve virtually no reproductive success, but females, regardless of their social status, attain nearly identical reproductive success. Thus, dominant females would benefit by producing sons, whereas subordinate females would benefit by producing daughters.

It remains controversial whether maternal allocations of steroid hormones can reverse the sex of offspring after fertilization and sex determination has occurred (e.g., Pike and Petrie, 2003). Although it is theoretically possible, no compelling evidence currently exists to suggest that maternal post-fertilization manipulations can alter offspring sex ratio by hormonally reversing sex determination to alter sexual differentiation in birds. However, in reptiles, it appears that maternal allocation of steroid hormones can influence offspring sex. In eggs from painted turtles (*Chrysemys picta*), seasonal changes in yolk steroid concentrations appear to influence the sex of the offspring. The hatchling sex ratio at 28°C shifted from 72% male to 76% female from spring to summer, which corresponded to a seasonal change in the ratio of egg estradiol to testosterone concentrations (Bowden et al., 2000). The ability of a female to control the sex of her offspring may be an important adaptive function to maintain environmental sex determination in some species.

EFFECTS OF MATERNAL CARE Sex steroid hormones can affect the way in which maternal care is delivered, which in turn can further affect sexual differentiation of behavior. Mother rats lick the anogenital regions of their newborn pups. The primary function of licking the pups is to stimulate the elimination of wastes, but newborn males require maternal licking if they are to develop normal adult mating behavior. Mother rats spend more time licking male pups than licking female pups (C. L. Moore, 1984, 1986; Moore and Morelli, 1979), apparently because mothers prefer the chemosensory cues associated with male pup urine. Testosterone contributes to this preference, because female rat pups injected with testosterone are licked as often as males. Mother rats that are made anosmic (unable to smell) do not lick their male offspring at higher rates, and these males display altered patterns of copulation in adulthood (see Box 4.4). Differential parental treatment of male and female infants has been reported for several primate species as well, including humans (Stewart, 1988). These differential behavioral interactions, as well as maternal stress, can have significant modifying effects on the neural tissues that may underlie sexually dimorphic behaviors (see Chapter 4).

ENVIRONMENTAL ENDOCRINE DISRUPTORS A number of commercial chemical agents, including many pesticides, some types of polychlorinated biphenyls (PCBs), and combustion products from plastics, can have pronounced estrogenic effects on animals (Raloff, 1994a,b; Ottinger and vom Saal, 2002). Although these substances are not steroids, they apparently mimic estrogens and bind to estrogen receptors. Other **endocrine-disrupting chemicals (EDCs)** mimic the effects of androgens or thyroid hormones. Some of these chemicals can spread widely throughout the environment and persist there for a long time. One of the first demonstrations that these chemical agents could affect reproduction was the link made in the 1950s between the now-banned pesticide DDT and the thinning of eggshells in many avian species. As you probably know, animals at the tops of food chains ingest all of the toxic chemicals that have accumulated in the bodies of the animals they consume (a process called *bioaccumulation*). Birds that consume fish, reptiles, and insects are particularly at risk for contamination by bioaccumulation. Many bird species, including bald eagles, were put at risk of extinction by DDT, which caused them to lay abnormally thin-shelled eggs that were inadvertently crushed by incubating parents.

DDT had other dramatic effects on reproduction in birds. One California gull population (*Larus californicus*) on Santa Barbara Island, off the coast of California, was heavily contaminated with DDT. When eggs did manage to hatch, the morphology of the offspring was markedly atypical. As we have seen, female birds normally develop only the left ovary and oviduct, but young females on Santa Barbara Island showed some development of the right reproductive tract as well. Young males also developed oviducts, and their right gonad often developed as an ovotestis (Fry and Toone, 1981). The behavior of adults was also atypical. Two parent gulls are required to obtain sufficient food to rear offspring successfully, but on Santa Barbara Island, most males were not engaging in reproductive behavior. Breeding males had become scarce in the population, which led to females predominating on the island; apparently many females mated with the scarce reproductively competent males, then shared nests in female–female pairs. This rise in so-called "lesbian gulls" was thought to reflect the estrogenic effects of DDT on male mating behavior, because when eggs were experimentally treated with estradiol or DDT, males hatching from those eggs demonstrated the same feminized structures and reduced sex drive as the males on Santa Barbara Island (Fry and Toone, 1981). Since the banning of DDT in 1972, the gull sex ratio on Santa Barbara Island has been slowly returning to normal levels, and the incidence of morphological and behavioral abnormalities is waning.

On the other side of the North American continent, biologists have been concerned with the dwindling numbers of the Florida panther (*Felis concolor coryi*). One reason for the panthers' decline is human encroachment into their habitats, but these panthers also seem to exhibit increasingly high rates of **cryptorchidism**, a condition in which the testes remain in the abdominal cavity after birth (Raloff, 1994a). Normally, the testes develop near the kidneys and descend into the scrotum through the inguinal canals (small openings in the pelvis) around the time of birth. The mammalian scrotal sac has evolved a number of adaptations to maintain the testes at temperatures less than the core body temperature. Mammalian sperm are very sensitive to high temperatures and may be damaged by temperatures above 37°C. Failure of the testes to descend into the scrotum is associated with infertility because a high proportion of the sperm are damaged.

Only about 50 Florida panthers are estimated to remain in the wild, and the high incidence of cryptorchidism was thought to be due to a genetic bottleneck, or reduction in genetic variation, in this dwindling population. Recent data do not rule out that possibility, but they also suggest that environmental estrogenic pesticides may be responsible for the increase in the incidence of cryptorchidism. When blood samples from Florida panthers were analyzed for steroid hormone concentrations, investigators discovered that several cryptorchid males had unusual hormone profiles showing higher circulating levels of estrogens than androgens, and that at least one female had higher circulating levels of androgens than estrogens. Investigators suspect that steroidogenic agents (compounds that stimulate steroid production) or steroid mimics in the environment are responsible for these unusual steroid levels. Consequently, approximately a hundred wildlife preserves in the southeastern United States that are managed by the U.S. Fish and Wildlife Service are now forbidden from using any known estrogenic chemicals, including pesticides and herbicides.

One of the most common herbicides in use in the world is atrazine (2-chloro-4-ethytlamino-6-isopropylamine-1,3,5-triazine). Atrazine is the most common herbicide used in the U.S., and has been detected in virtually every U.S. waterway examined. Although it is generally considered to be safe because of its relatively short half-life in the environment and lack of bioaccumulation, it can affect reproductive function in adult rats, though only at extremely high doses. Atrazine does not appear to affect adult frogs. However, laboratory studies have indicated that sexual development in frogs (*Xenopus laevis*, *Rana pipiens*, and *Hyla regilla*) is sig-

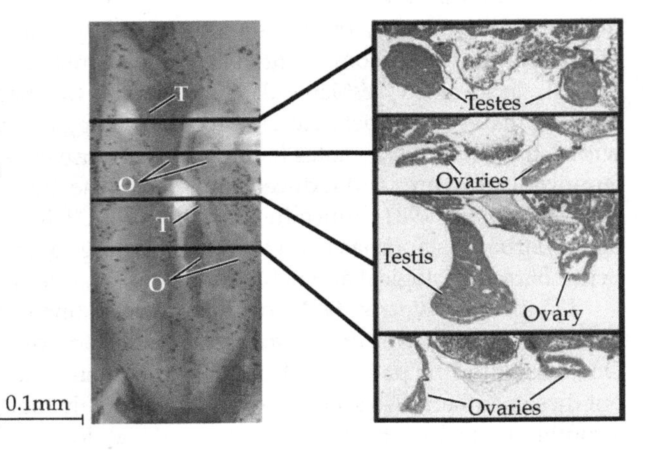

3.30 Atrazine exposure causes testicular malformations in frogs. The most severely malformed testes in tadpoles treated with various concentrations of atrazine failed to produce any sperm, but appeared to function as ovaries and produced ova. From Hayes et al., 2002b.

nificantly impaired by modest, ecologically relevant levels of atrazine (Figure 3.30) (Hayes et al., 2002a). At study sites in the western and midwestern United States, 10%–92% of male frogs were observed to have testicular abnormalities (Hayes et al., 2002b). The most severely malformed testes displayed ova production (Hayes et al., 2002b). Female malformations were not noted. Some of these testicular malformations could be induced in laboratory studies using various concentrations of atrazine (Hayes et al., 2002b, 2003). To date, these results have not been replicated and are controversial. Of course, given the complex brew of chemicals that constitutes many waterways in agricultural and urban regions, additional chemicals could be interacting with atrazine to evoke effects on reproductive development. It has been suggested that endocrine-disrupting chemicals may be contributing to the worldwide reduction in amphibian populations. Other endocrine disruptors may be causing sex reversal in male Chinook salmon (*Oncorhynchus tshawytcha*) as they develop, contributing to the rapid and extraordinary population decline of this species (Nagler et al., 2001).

A new problem that has been identified in polluted rivers and streams is the biotransformation of pollutants into potent hormones. Female eastern mosquitofish (*Gambusia holbrooki*) are being masculinized in a polluted river in Florida. To determine the source of the masculinizing chemical, water was collected from this river and analyzed with sophisticated chemical tools. It was discovered that the mixture of pollutants in the Fenholloway River was producing trace amounts of an environmental androgen (Jenkins et al., 2001). What was surprising was that this androgen was identified as androstenedione, a relatively mild anabolic androgenic steroid that has been linked to performance enhancement among U.S. Major League Baseball players. In the sediment of the river, even higher concentrations of androstenedione, as well as progesterone, were detected (Jenkins et al., 2003). These two hormones have relatively low androgenic activities and are unlikely to account for the masculinization of the female mosquitofish. Additional analyses of the water are likely to reveal additional androgenic substances.

Are steroidogenic agents affecting human reproduction? Probably. Thus far, disruptions in breeding attributed to environmental hormones have been reported for severely contaminated populations of amphibians, reptiles, birds, and mammals throughout the world. Humans, residing atop their food chain, may be ingesting far more of these steroidogenic agents than they might suspect. As with the Florida panthers, the rate of cryptorchidism in baby boys is apparently increasing (Carlsen et al., 1992); the incidence in the 1950s of about 1.6% of live births had nearly doubled by the 1970s to 2.9% in Western Europe. The incidence of hypospadias, an abnormality in which the urethra opens on the underside of the penis, has also more than doubled in Great Britain between 1964 and 1983. The age of puberty in girls is also declining among industrialized societies. Even more disturbing, the incidence of testicular, prostate, ovarian, mammary, and cervical cancer, cancers possibly associated with endocrine factors, is increasing sharply in several European and North American countries: the rate of testicular cancer has nearly tripled during the past half century, and the incidence appears to be increasing in industrialized countries (Giwercman and Skakkebaek, 1992; Petitti and Porterfield, 1992).

Plastics and pesticides contain endocrine-disrupting chemicals that are known to interfere with mammalian development (Gray et al., 2000). For example, exposure of developing rodents to EDCs at doses likely to be experienced by humans advances puberty and alters their reproductive function (Howdeshell et al., 1999; Figure 3.31). The reproductive effects that have become apparent in humans over the last century, such as the increased incidence of genital abnormalities in boys and earlier puberty in girls, are consistent with those seen in animals after exposure to high doses of EDCs.

(A)

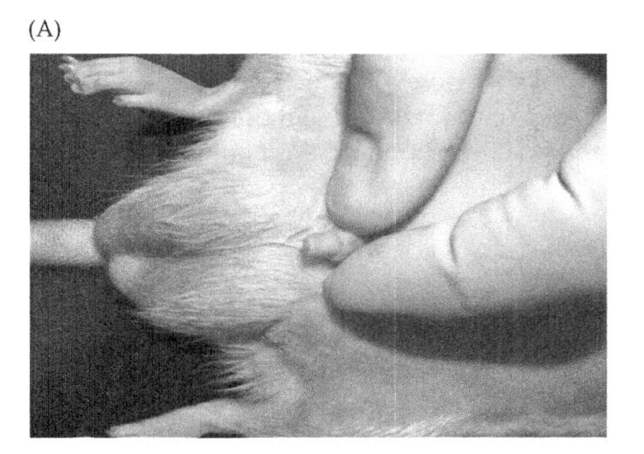

(B)

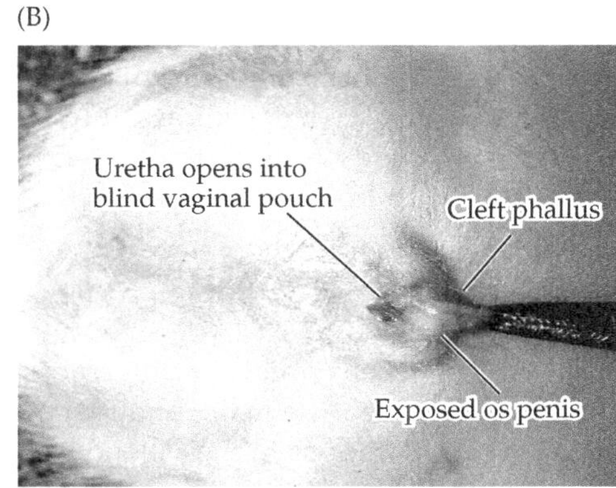

3.31 External genitalia of adult male rats that were exposed prenatally to (A) corn oil or (B) vinclozolin, an anti-androgen. Exposure to vinclozolin early in development resulted in an abnormal sexual phenotype and multiple reproductive malformations, including hypospadias, cleft phallus, female-like anogenital distance, ectopic (undescended) testes, vaginal pouches, and small or absent accessory sex glands. Photo from L. E. Gray, U.S. Environmental Protection Agency.

Human fertility also seems to be waning in recent years. Although undescended testes can be moved to their correct location in the scrotum by surgical means, boys born with this condition often suffer fertility difficulties later in life. Across human populations, sperm counts appear to be declining. In 1940, the average density of human sperm was 113 million per millimeter of semen; in 1990, this figure had dropped to 66 million per millimeter in the United States and Western Europe. Researchers also estimate that the volume of semen produced by men has dropped about 20% during the past 50 years, reducing sperm count per ejaculation even further (Carlsen et al., 1992). More recently, between 1981 and 1991, the proportion of normal sperm decreased significantly, from 56.4% to 26.9%, with a parallel increase in the incidence of partial and complete arrest of spermatogenesis (Pajarinen et al., 1997). During this period, the size of seminiferous tubules decreased, the amount of fibrotic tissue in the testes increased, and testicular mass decreased significantly. These alterations in testicular characteristics over time could not be explained by changes in body mass index or in tobacco, alcohol, or other drug use. The authors of this study suggested that deteriorating spermatogenesis might be one important factor in the explanation of the declining sperm counts observed worldwide. Sperm count and fertility are affected by estrogens.

In addition to mimics of steroid hormones, several other endocrine-disrupting chemicals have been discovered in a wide variety of common chemical compounds. These chemicals include substances that interfere with androgens, thyroid hormones, growth factors, and prolactin. Although an important book, *Our Stolen Future* (Colborn et al., 1997), has motivated the U.S. federal granting agencies to begin funding systematic studies on the toxicology and developmental effects of endocrine disruptors, the effects of these chemicals on behavior remain, for the most part, unknown. Certainly, the extent to which human behavior has been affected by environmental steroidogenic agents remains unknown. Are aggressive, parental, or sexual behaviors in humans altered by the more than 500 measurable chemicals in our bodies that were not there prior to the 20th century? No one knows. It seems prudent to curtail the use of these substances until their effects on the reproduction and health of humans and nonhuman animals are understood.

Summary

- Behavioral sex differences are common in humans and in nonhuman animals. Because males and females differ in the ratio of androgen to estrogen concentrations in their blood plasma, behavioral endocrinologists have been particularly interested in the extent to which behavioral sex differences are mediated by hormones.
- Individuals of asexual vertebrate species produce only one sex: females. Production of offspring by asexual reproduction is very efficient, but such species risk extinction if environmental conditions change drastically because there is no genetic variation among the offspring. The recombination of genetic mate-

rial during sexual reproduction produces genetic variation. Sexual species produce two sexes: females and males.

- The ultimate cause of sex differences appears to reflect sexual selection, a subcategory of natural selection. Animals with polygynous mating systems display more sexual dimorphism than monogamous animals. Humans, which are mildly to moderately polygynous, display several sexual dimorphisms, including larger body size, retarded puberty, increased courtship activity, higher levels of aggression, and reduced parental care in men as compared with women.
- The process of becoming female or male is called sexual differentiation. The primary step in sexual differentiation occurs at fertilization. In mammals, the ovum (which always contains an X chromosome) can be fertilized by a sperm bearing either a Y or an X chromosome; this process is called sex determination. The chromosomal sex of homogametic mammals (XX) is female; the chromosomal sex of heterogametic mammals (XY) is male. Chromosomal sex determines gonadal sex. All subsequent sexual differentiation is normally the result of differential exposure to gonadal steroid hormones. Thus, gonadal sex determines hormonal sex, which regulates morphological sex. Morphological differences in the central nervous system, as well as in some effector organs, lead to behavioral sex differences.
- The gonads and the external genitalia of both sexes each develop from a single anlage; these organs differ between males and females along a continuum of masculinity and femininity. In contrast, there are dual anlagen for the accessory sex organs. Normal female development of these organs requires feminization (Müllerian duct development) as well as demasculinization (Wolffian duct regression). Normal male development requires masculinization (Wolffian duct development), as well as defeminization (Müllerian duct regression). Thus, sexual differentiation of the accessory sex organs proceeds along two continua: a masculinization–demasculinization scale and a feminization–defeminization scale. Masculinization is the induction of male traits. Feminization is the induction of female traits. Demasculinization is the removal of the potential for male traits, whereas defeminization is the removal of the potential for female traits. This nomenclature is also appropriate for describing behavioral sexual differentiation.
- The process of sexual differentiation is complex, and the potential for errors is present. Perinatal exposure to androgens is the most common cause of anomalous sexual differentiation among females. The source of androgens may be endogenous (as in congenital adrenal hyperplasia) or exogenous (exposure to androgenic substances such as DES or MPA). Turner syndrome results when the second X chromosome is missing or damaged; these individuals possess dysgenic ovaries and are not exposed to steroid hormones until puberty.
- In mammals, females are the "neutral" sex; additional steps are required for male differentiation, and more steps bring more possibilities for errors in differentiation. Some examples of male anomalous sexual differentiation include 5α-

reductase deficiency (XY individuals are born with ambiguous external genitalia because of a lack of DHT and are reared as females, but masculinization occurs during puberty) and androgen insensitivity syndrome or TFM (XY individuals lack receptors for androgens and develop as females). By studying individuals in which the process of sexual differentiation is atypical, behavioral endocrinologists glean hints about the process of typical sexual differentiation.

- In birds, the homogametic sex is male (ZZ), whereas the heterogametic sex is female (WZ). Males appear to be the "neutral" sex among birds in most aspects of sexual differentiation. The embryonic ovary has the ability to feminize the embryonic testis.
- In some species, alternative mating strategies among males are associated with differences in body type and behavior, which are often regulated by changes in hormone secretion.
- Some reptilian species are like birds in that the male is homogametic; other species resemble mammals in that the female is homogametic. In yet other reptilian species, sex chromosomes are absent, and sex determination occurs in response to incubation temperature. Substantial variation in this process exists; high ambient temperatures evoke male development in some species, but female development in others. Some fish species possess the ability to undergo sex changes in adulthood in response to social or physiological cues.
- Gonadal steroid hormones have organizing effects on behavior. The organizing effects of steroid hormones are generally restricted to the early stages of development. An asymmetry exists in the effects of testes and ovaries on the organization of behavior in mammals. Hormone exposure early in life has organizational effects on subsequent rodent mating behavior: early androgen treatment, for example, causes relatively irreversible and permanent masculinization of rodent copulatory behavior. These early hormonal effects can be contrasted with the reversible behavioral influences of steroid hormones provided in adulthood, which are called activational effects.
- Females undergo cycles of reproductive physiology and behavior that correspond to the cyclic release of eggs during the breeding season; males produce sperm at relatively constant rates throughout the breeding season and exhibit relatively constant reproductive behavior. The cyclic nature of female reproductive function is driven by the cyclic release of GnRH from surge generators located in the anterior hypothalamus. Both males and females display pulsatile release of GnRH from the medial basal hypothalamus. Exposure to androgens early in development abolishes forever the potential to generate GnRH surges, either by destroying the neural connections between the pulse and surge generators or by reducing estrogen receptor availability in the surge generator tissues.
- Environmental factors can influence sexual dimorphism. These factors include temperature, intrauterine position and conditions, and chemicals that mimic hormones. Because the combination of these factors is usually complex, deter-

mining their individual contribution to atypical sexual characteristics is complicated. Another potential complication contributing to sexually dimorphic characters in birds and some reptiles is variation in maternal contribution of hormones to the egg.

- Surprisingly, mice with targeted deletions (knockouts) of the α estrogen receptor gene (αERKO) display normal mounting behaviors, reduced levels of intromissions, but no ejaculations, and are less aggressive than normal wild-type (WT) males. Male mice that had the gene for a second estrogen receptor (β estrogen receptor) disabled (βERKO) retain virtually all components of sexual behaviors. Mice lacking both ER-α and ER-β genes fail to display any components of sexual or aggressive behaviors.

Questions for Discussion

1. What are some behavioral sex differences that you have noticed between boys and girls? What causes girls and boys to choose different toys? Do you think that the sex differences you have noted arise from biological causes or are learned? How would you go about establishing your opinions as fact?
2. "Men are too aggressive to write and produce cartoon programs appropriate for children." Support and refute this statement based on sound scientific principles.
3. Why is it inappropriate to refer to androgens as "male" hormones and estrogens as "female" hormones?
4. Discuss the early hormone–anatomy correlates that were the conceptual forerunners of the organizational/activational hypothesis proposed by Young and colleagues.
5. Discuss the following proposition: The process of developing sexually dimorphic behaviors is conceptually similar to the development of sexually dimorphic accessory sex organs, and dissimilar from the development of sexually dimorphic external genitalia.
6. Is it inevitable that certain occupations (e.g., science and engineering) will be dominated by men, whereas others (e.g., child care, elementary school teaching) will be dominated by women?

Refer to the accompanying Student CD for additional resources, including Web links, videos, animations, and additional photos.

Suggested Readings

Arnold, A. P. 2002. Concepts of genetic and hormonal induction of vertebrate sexual differentiation in the twentieth century, with special reference to the brain. In D. W. Pfaff, A. P. Arnold, A. M. Etgen, S. E. Fahrbach, and R. T. Rubin (eds.), *Hormones, Brain, and Behavior*, Volume 4, pp. 105–135. Academic Press, New York.

Balthazart, J. and Adkins-Regan, E. 2002. Sexual differentiation of brain and behavior in birds. In D. W. Pfaff, A. P. Arnold, A. M. Etgen, S. E. Fahrbach, and R. T. Rubin (eds.), *Hormones, Brain, and Behavior*, Volume 4, pp. 223–301. Academic Press, New York.

Bass, A. H. 1996. Shaping brain sexuality. *American Scientist*, 84: 352–361.

Crews D. 2003. Sex determination: Where environment and genetics meet. *Evolution and Development*, 5: 50–55.

Hines, M. 2004. *Brain Gender*. Oxford University Press, Oxford.

Phoenix, C. H., Goy, R. W., Gerall, A. A., and Young, W. C. 1959. Organizing action of prenatally administered testosterone propionate on the tissues mediating mating behavior in the female guinea pig. *Endocrinology*, 65: 369–382.

Sisk, L. L., and Foster, D. L. 2004. The neural basis of puberty and adolescence. *Nature Neuroscience*, 7: 1040–1047.

Wallen, K., and Baum, M. J. 2002. Masculinization and defeminization in altricial and precocial mammals: Comparative aspects of steroid hormone action. In D. W. Pfaff, A. P. Arnold, A. M. Etgen, S. E. Fahrbach, and R. T. Rubin (eds.), *Hormones, Brain, and Behavior*, Volume 4, pp. 385–423. Academic Press, New York.

4

Sex Differences in Behavior

ANIMAL MODELS AND HUMANS

Boys generally prefer toys such as trucks and balls, and girls generally prefer toys such as dolls. Although it is doubtful that there are genes that encode a preference for toy trucks on the Y chromosome, it is possible that hormones might shape the development of a child's brain to prefer certain types of toys or styles of play behavior. It is commonly believed that children learn which types of toys and which styles of play are appropriate to their gender. How can we understand and separate the contribution of physiological mechanisms from that of learning to understand sex differences in human behaviors? To untangle these issues, behavioral endocrinologists often use animal models.

In contrast to humans, in which sex differences are usually only a matter of degree (often slight), in some animals, members of only one sex exhibit a particular behavior. Studies of such behaviors are particularly valuable for understanding the interaction among behavior, hormones, and the nervous system. For example, a study with vervet monkeys (*Cercopithecus aethiops*) calls into question the primacy of learning in the establishment of toy preferences (Alexander and Hines, 2002). Female vervet monkeys preferred girl-typical toys, such as dolls or cooking pots, whereas male vervet monkeys preferred boy-typical toys, such as cars or balls (Figure 4.1). There were no sex differences in preference for gender-neutral toys, such as picture books or stuffed animals. Presumably, monkeys have no prior concept of "boy" or "girl" toys. What, then, underlies the sex difference in toy preference? It is possible that certain attributes of toys (or objects) appeal to either males

(A)

(B)

4.1 Sex differences in monkey play. (A) Female and (B) male vervet monkeys prefer dolls and toy cars, respectively, as do their human counterparts. Courtesy of Melissa Hines.

or females (Alexander, 2003). Toys that appeal to boys or male vervet monkeys—in this case, a ball or toy car—are objects that can be moved actively through space, toys that can be incorporated into active, rough-and-tumble play (Alexander, 2003). The appeal of toys that girls or female vervet monkeys prefer appears to be based on color. Pink and red (the colors of the doll and pot) may evoke attention to infants (Alexander, 2003).

Human society may reinforce the stereotypical responses to gender-typical toys. Are the sex differences in toy preferences or play activity, for example, the inevitable consequences of the differential endocrine environments of boys and girls, or are these differences imposed by cultural practices and beliefs? Or are these differences due to some combination of endocrine and cultural factors? The goal of this chapter is to examine the factors that cause male and female brains to develop along distinct pathways underlying behavioral sexual dimorphisms.

When W. C. Young and his colleagues proposed the organizational/activational hypothesis of hormonal sexual differentiation (see Chapter 3), they drew an analogy between the effects of early testosterone exposure on the development of the accessory sex organs and its effects on the neural tissues underlying mating behavior. Because early exposure to hormones was found to change the probability of specific behaviors, and because all behavior is mediated by the nervous system, it was reasonable to suggest that hormones caused changes in the nervous system. The authors warned, however, that, unlike the embryologists who had found pronounced structural changes in the reproductive organs in response to perinatal androgen exposure, neurologists and psychologists interested in the effects of androgens on neural tissues were unlikely to find alterations so drastic. Instead, a more subtle change, reflected in function rather than in visible structure, would be expected (Phoenix et al., 1959). Therefore, most endocrinologists in the 1950s and 1960s believed that structural changes in the central nervous system mediated behavioral differences between the two sexes, but assumed that any underlying changes in brain tissue would be beyond the technical abilities of the day to observe. Despite the lack of any empirical data showing direct evidence of changes in neural tissue that mediate sexually dimorphic behaviors, hypothetical models of neural modulation of behavior in response to early hormone exposure flourished. Because the evidence of hormonally induced neural changes was indirect, other researchers resisted the temptation to speculate on mechanisms beyond their data. They explained changes in the adult copulatory ability of neonatally castrated rats, for example, by sensory deficiencies that resulted from inadequate penile development due to androgen deprivation (Beach, 1971). Thus, the effects of perinatal steroid hormone manipulations on subsequent adult behavior were explained as indirect effects on genital (non-neural) morphology (Box 4.1).

In the early 1970s, however, researchers discovered subtle, microscopic sex differences in the neural connections of the hypothalamus, and soon thereafter macroscopic sex differences in brain organization were reported. It was shown that these sex differences in brain organization could be reversed by early hormonal manipulation. Even the staunchest critics of the organizational/activational hypothesis were eventually persuaded that early hormone treatment changed neural tissues underlying sexually dimorphic behaviors, and that these behaviors could be activated by the effects of gonadal steroid hormones on these same tissues in adulthood (e.g., Beach, 1975). It is now known, however, that hormonal effects on the nervous system are not necessarily direct and thus, part of Beach's explanations remains relevant. Hormones may also act on non-neural tissues such as penile or syrinx development, and especially those that potentially change sensory input to affect neural development (reviewed in Simerly, 2002; DeVries and Simerly, 2002).

Neural Bases of Mammalian Sex Differences

The first clear-cut, though minuscule, sexual dimorphism found in the brain was observed during electron microscopic examination of synapses in the **medial preoptic area (MPOA)**, an area just anterior to the hypothalamus in rats (Raisman and Field, 1973a) (see Figure 3.25). Because previous studies had shown that lesions or stimulation of the MPOA led to alterations in sexual behavior (see Chapter 5), the possibility of sex differences in the neural organization of this area was particularly intriguing. The researchers carefully counted and categorized the synapses in the MPOA that survived after axons projecting from another area of the brain, the stria terminalis, had been cut. They discovered that females had more synapses on dendritic spines and fewer synapses on dendritic shafts than males, whereas males had more synapses on dendritic shafts and fewer synapses on dendritic spines than females (Raisman and Field, 1973a). Dendritic spines are small outgrowths along the dendritic shafts that give dendrites possessing them a "rough" appearance. These spines have become the focus of much research, especially on learning and memory, because their number and function may vary in response to experience (Figure 4.2). In male rats castrated on day 1 of life, the female pattern of synapses was observed (Raisman and Field, 1973b). Similarly, in females injected with testosterone prior to 4 days of age, the male pattern of synaptic organization was observed. The functional meaning of these data remains unknown; in other words, how the observed sex differences in synaptic organization mediate behavior, or even if they mediate behavior, remains unresolved. However, this observation began to fulfill the prediction made years

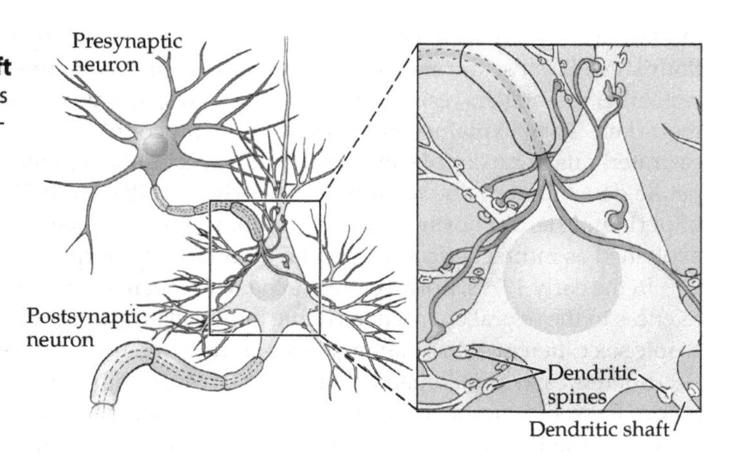

4.2 Synapses may form either on dendritic spines or on the shaft of a dendrite. In the MPOA, male rats tend to have more synapses from certain neurons on dendritic shafts and fewer synapses on dendritic spines than females; females have more synapses on dendritic spines and fewer synapses on dendritic shafts than males. Hormone manipulations immediately postpartum can affect this pattern of neuronal connectivity.

BOX 4.1 *The Organization of Avian Copulatory Behavior*

Birds have been useful models for the study of neural sexual differentiation because they allow researchers to avoid one confounding factor that is present in the study of mammals: treatments that masculinize behavior in mammals usually masculinize the external genitalia as well. For this reason, researchers found it inappropriate to assume that sex differences in behavior reflect changes in the nervous system. In other words, the sex differences in behavior might simply be due to the changes in external genital morphology (Beach, 1971). One way to resolve the controversy about the contributions of neural and non-neural changes to the development of sex differences in behavior is to discover structural changes in the nervous system that correspond to behavioral differences. Although sex differences in neural structure, neurotransmitter systems, and neuronal numbers have been reported in the mammalian nervous system (Breedlove, 1992), these morphological differences have not been linked directly to the regulation of sexually dimorphic behaviors (Balthazart et al., 1996).

Another approach to this issue has involved studies of hormone–behavior interactions during the development of birds. Males and females of most avian species do not differ in external genital morphology: both sexes have a single cloacal opening. Thus, neural and genital morphological changes are not confounded in birds as they are in mammals (Adkins-Regan, 1987). Studies of avian sexual differentiation of behavior have provided the most convincing evidence that early hormone exposure changes the neural structures that control adult sexually dimorphic behaviors.

Japanese quail (*Coturnix japonica*, see Figure 5.36) is the avian species most commonly studied to understand the hormonal bases of mating

before that a "more subtle change reflected in function rather than visible structure" was involved in mediating sex differences in behavior (Phoenix et al., 1959).

Since this early report, more obvious structural sex differences have been observed in mammalian brains. Because of its prominent role in the mediation of mating behavior in several species, the **preoptic area (POA)*** of the hypothalamus has received special attention. In rats, for example, there is a collection of cell bodies (a nucleus) in the medial preoptic area that is five to seven times larger in males than in females (Gorski et al., 1978). The difference is so prominent that rat brains can be accurately classified as either male or female by examining thin

*There has been some confusion regarding the nomenclature of the preoptic area. Many sources consider the preoptic area (or preoptic region) to be part of the hypothalamus. Other sources note that the preoptic area derives from different cell types during embryology and they exclude the preoptic area from the hypothalamus. To add to the confusion, there are substantial species differences in the developmental-based division of the preoptic area from the hypothalamus. In this text, we will generally use the nomenclature of preoptic area of the hypothalamus. This is a location anterior to the hypothalamus. There are also paired, specific clusters of neural cell bodies (soma) called nuclei that are called the preoptic nuclei in most mammals. Medial simply means the middle part of the preoptic area.

behavior. Mating behavior is sexually dimorphic in quail. Males strut and crow before and after copulatory mounting (Adkins, 1978). Castration of males reduces these male-typical behaviors, and testosterone therapy restores them (Adkins, 1975). Females never display copulatory mounting behaviors, and they crow and strut less frequently than males, even when they are injected in adulthood with androgens (Adkins, 1975; Balthazart et al., 1983). Generally, estrogenic metabolites of androgens activate copulatory behavior in adult males, whereas androgenic metabolites of testosterone activate the strutting and crowing behaviors (Adkins, 1978; Adkins-Regan, 1996; Balthazart et al., 1985).

If eggs containing male quail embryos are injected with either testosterone or estradiol, the males exhibit fewer male-typical copulatory behaviors as adults than untreated males (Adkins-Regan, 1987). In other words, the males are demasculinized. These demasculinizing effects of testosterone or estradiol occur only prior to day 12 of the 17-day incubation period. If eggs containing female quail embryos are injected with androgens or estradiol, adult female behavior is relatively unaffected (Adkins-Regan, 1987). In other words, the females are neither defeminized nor masculinized by the endocrine manipulations. Treatment of female quail in the egg prior to day 9 of incubation with an anti-estrogen (a substance that binds to estrogen receptors and prevents natural estrogens from binding) produced masculinization of adult copulatory behavior (Adkins, 1976). In other words, anti-estrogens prevented the demasculinization that is produced by endogenous estrogens (Balthazart and Foidart, 1993).

Male and female quail do not differ in their concentrations of circulating androgens immediately post-hatching, but females do exhibit higher circulating estrogen concentrations during the last few days before hatching occurs (Balthazart and Foidart, 1993). Taken together, these results suggest that the estrogens secreted from the ovaries normally demasculinize the copulatory behavior of female quail. Male quail apparently do not secrete sufficient quantities of sex steroid hormones to demasculinize their own behavior.

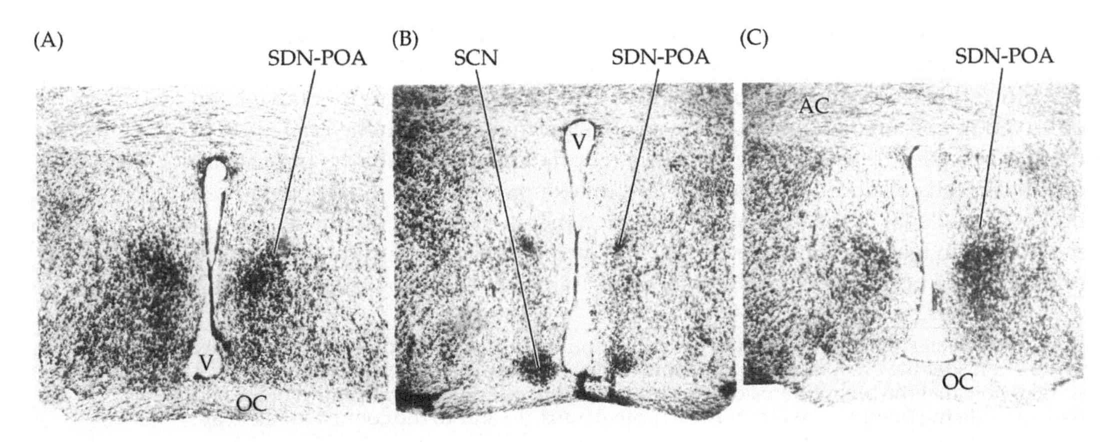

4.3 The sexually dimorphic nuclei of the preoptic area (SDN-POA) in rats are shown in cross section in micrographs of the brains of (A) a male, (B) a female, and (C) a female treated with testosterone as a newborn. Note that the SDN-POA of the male are substantially larger than those of the untreated female, but are equal in size to those of the testosterone-treated female. OC = optic chiasm; SCN = suprachiasmatic nucleus; V = third ventricle. Courtesy of Roger Gorski.

slices dissected through the hypothalamus with the naked eye (Figure 4.3). This region of the preoptic area has been termed the **sexually dimorphic nucleus of the preoptic area (SDN-POA)** (Gorski et al., 1978). The SDN-POA of females treated early in life with androgens can approach the size of the masculine SDN-POA, whereas the size of the feminine SDN-POA can be attained in males by neonatal castration (Gorski, 1984). Presumably, a sex difference in the size of a nucleus or other brain region could reflect differences in cell number or in cell size between males and females. The sex difference in the size of the SDN-POA is due to a decreased number of cells in females and castrated males; androgens or their estrogenic metabolites seem either to augment neurogenesis (Jacobson and Gorski, 1981) or reduce cell death in this region (Forger, 1998). Rats with complete androgen insensitivity syndrome (testicular feminization), which have deficient androgen receptors but normal estrogen receptors, display female-typical genitalia, but male-typical SDN-POA (Jacobson, 1980).

The functional significance of the sex difference in the size of the SDN-POA is not really known. Lesions to the entire POA disrupt normal mating behavior in a wide variety of species, including rodents and primates (Hart and Leedy, 1985). If only the SDN-POA is lesioned in females, they continue to display normal reproductive cycles (Arendash and Gorski, 1983). If only the SDN-POA is lesioned in males, they also continue to display normal mating behavior, or in some cases, exhibit temporary and mild copulatory dysfunction (Arendash and Gorski, 1983; DeJonge et al., 1989). Similar results have been documented with other species. For example, the sexually dimorphic region of the preoptic area/anterior hypothalamus is not necessary for male copulatory behavior in ferrets (Baum et al., 1990; Cherry and Baum, 1990). In rats, the cell-dense central core of the medial preoptic nucleus is the most sexually dimorphic part of this complex; males display more neurons than females, and these neurons are rich in androgen receptors (Madeira et al., 1999). In the absence of androgens, these neurons undergo programmed cell death (apoptosis) (Dodson and Gorski, 1993).

Sex differences have been reported in several other brain structures of rodents. For example, the volume of the medial amygdala and the medial posterior area of an associated part of the limbic system called the **bed nucleus of the stria terminalis** (**BNST**) are about 20% larger in males than in females (del Abril et al., 1987; Mizukami et al., 1983). The amygdala is an almond-shaped structure in the brain that is often involved in aggression or reproductive behavior. These two brain regions may be part of a sexually dimorphic neural circuit that includes the SDN-POA as well as the rostral aspects of the preoptic area of the hypothalamus such as the **anteroventral periventricular nucleus** (**AVPv**) (Simerly and Swanson, 1986). In contrast to the cell groups mentioned thus far, the AVPv is larger in females, or in males castrated before 10 days of age, than in intact males, and it is the source of a sexually dimorphic projection to the arcuate nucleus (Simerly, 2002). Recall that the AVPv is important in the secretion of GnRH (see Chapter 3); because the regulation of GnRH differs substantially between the sexes, it is a reasonable brain region in which to expect sex differences.

Several sexual dimorphisms in the central nervous systems of humans have also been described (Table 4.1). One of the earliest such differences noted was that

TABLE 4.1 Structural sex differences in the central nervous system of humans

Brain region	Difference
Hypothalamus	
SDN-POA	♂ > ♀
INAH-3	♂ > ♀
Bed nucleus of the stria terminalis (BNST)	♂ > ♀
Suprachiasmatic nuclei (SCN)	♀ > ♂[a]
Spinal cord	
Onuf's nucleus (no. of motoneurons)	♂ > ♀
Structures associated with language	
Planum temporale	♀ > ♂[b]
Dorsolateral prefrontal cortex	♀ > ♂
Superior temporal gyrus	♀ > ♂
Structures connecting the hemispheres	
Corpus callosum (posterior portion)	♀ > ♂[c]
Anterior commissure	♀ > ♂
Massa intermedia of thalamus	♀ > ♂

Source: Forger, 1998.
[a]More elongated in ♀
[b]Left and right more symmetrical in size in ♀
[c]More bulbous in ♀

women have smaller brains than men. Initially, this observation was considered evidence supporting the intellectual inferiority of women and justifying the lack of educational opportunities afforded them in the late 18th and early 19th centuries. It was eventually noted, however, that when the larger male body mass is taken into consideration, the relative brain sizes of women and men are equivalent. Recent studies using a new three-dimensional MRI device found more folding (i.e., gyrification and fissuration) of the brain surface in women than in men (Luders et al., 2004). This increased complexity of folds provides more cortical surface area in the frontal and parietal lobes, which may compensate for the smaller female brain size and account for some of the behavioral differences between males and females that are described below.

As in other mammals, a sexually dimorphic nucleus was discovered in the POA of humans (Swaab and Fliers, 1985). Because this nucleus resembles the SDN-POA of rats and is larger in males than in females, it was also named the SDN-POA. There has been some controversy, however, about the boundaries of the human SDN-POA and the consistency of the observed sex differences in its volume (see, for example, Allen et al., 1989). In other studies, the nuclei of the human POA have been subdivided into four smaller regions, called the **interstitial nuclei of the anterior hypothalamus (INAH)**, abbreviated INAH-1, INAH-2, and so on (Allen et al., 1989). Under this neuroanatomical classification scheme, INAH-1 was considered to be equivalent to the SDN-POA, and, in contrast to the previous report, no sex differences were observed in this tiny brain region. The volumes of INAH-2 and INAH-3, however, were reported to be larger in men than in women (Figure 4.4; Allen et al., 1989). In another study, a sex difference in nuclear volume was reported in INAH-3, but not in INAH-2 (LeVay, 1991; Byne et al., 2001).

Because brain tissue from humans is obtained at different times after death and is fixed with a variety of methods, it is not particularly surprising that contradictory results have been obtained. The report by LeVay (1991) generated enormous debate because, in addition to the observation that INAH-3 was smaller in women than in men, INAH-3 was reported to be smaller in homosexual men than in heterosexual men (LeVay, 1991). Although sexual orientation has not traditionally been part of medical records, the sexual preference of the homosexual men in LeVay's sample was included in their health records because the vast majority of these men had died from complications resulting from AIDS. Their brains were compared with those of men whose sexual orientation was unknown, but assumed to be heterosexual, and who had died primarily from other causes. However, not all of the obvious potential confounding variables—namely, HIV infection, age, body size (people infected with HIV often have reduced body weight), brain size, and testosterone concentrations (males infected with HIV often show end-stage reductions in plasma testosterone concentrations)—have been ruled out as causative factors underlying the differences in INAH-3 size between the heterosexual and homosexual men (Byne and Parsons, 1993; Byne et al., 2001; Swaab et al., 2001). These findings have fueled the debate on whether sexual orientation is mediated primarily by biological or by environmental factors, a debate that will be considered below.

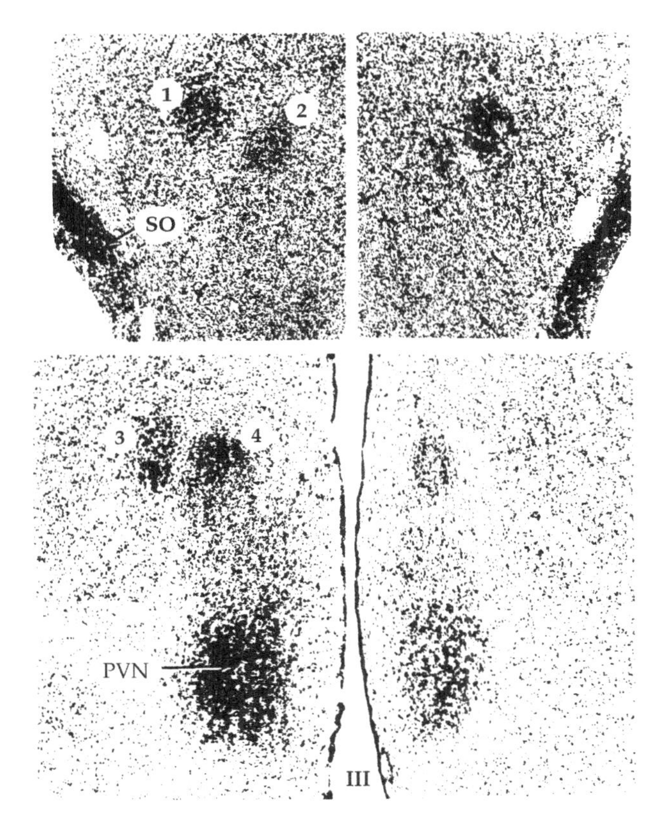

4.4 The interstitial nuclei of the anterior hypothalamus (INAH). Photomicrographs of the human anterior hypothalamus in cross section, showing the four INAH in a male (left; 1–4) and the corresponding areas in a female (right). Precise quantitative procedures (Allen et al., 1989) have shown that the volumes of INAH-2 and INAH-3 are greater in men. The lower photomicrographs are taken from a region caudal to the area depicted in the upper photomicrographs. SO = supraoptic nucleus; PVN = paraventricular nucleus; III = third ventricle. Courtesy of Roger Gorski.

Thus, INAH-3 size is correlated with sexual orientation in adults, but this observation does not provide a genetic basis for homosexuality, nor does it suggest any "mechanism" underlying homosexuality. What are the possible mechanisms? It is possible that homosexual men are genetically determined to have a smaller INAH-3 than heterosexual men. Or the differences in INAH-3 size between homosexual and heterosexual men could be due to organizational influences of hormones. It is also possible that homosexual behavior decreases INAH-3 size. Studies in African cichlid fish (*Haplochromis burtoni*) provide an excellent example of how behavior feeds back to influence brain structures (White and Fernald, 1997). There are two types of adult male cichlids: males with and those without territories. Territorial males are brightly colored, with blue or yellow basic body coloration. In contrast, nonterritorial males are cryptically colored, making them difficult to distinguish from the background and from females, which are similarly camouflaged (Figure 4.5). When males become territorial, the GnRH neurons in the POA increase in size, whereas males that lose their territories show a reduction in the size of their POA neurons (Fernald, 2002; Insel and Fernald, 2004). These data indicate that social experience can provoke neuronal size changes.

Despite the importance of sexual dimorphisms in the size of brain structures as gross indicators of sex differences in brain function and behavior, these structural

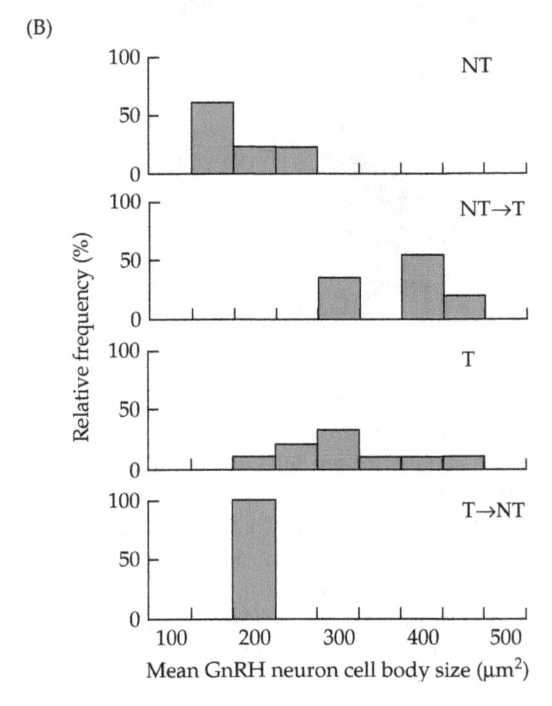

4.5 Cichlid fish show changes in neuronal cell size in response to social conditions. (A) Photo of African cichlid fish (*Haplochromis burtoni*). The top fish has markings typical of a non-territorial male, which are less flashy (more cryptic) than the markings of a territorial male. A territorial male is pictured below. Note the striking spots on the fins and the dark forehead. (B) GnRH cell size in the POA of fish in each of four social categories: nonterritorial (NT), moved from nonterritorial to territorial (NT→T), territorial (T), and moved from territorial to nonterritorial (T→NT). (A) courtesy of Russell Fernald, 2002.

differences have not been directly linked to behavior, especially in mammals. The critical questions regarding, for example, what features in the brain are necessary and sufficient to account for sex differences in the ability of sex steroid hormones to activate adult sexual behavior remain, for the most part, unanswered (DeVries and Simerly, 2002). In addition to studies of nuclear volume, behavioral endocrinologists have examined sex differences in neurochemistry, neural connectivity, enzyme activation, hormone and neurotransmitter receptor distribution, and gene expression in the brains of males and females in an attempt to answer this question.

Molecular Sex Differences in the Brain

One possible explanation of how adult hormone treatment elicits different sexual behaviors in males and in females involves sex differences in the number and distribution of neural estrogen and androgen receptors. In most early studies, no differences in steroid hormone binding sites were reported; however, recent studies using more sophisticated techniques have revealed some modest sex differences (Brown et al., 1992). For instance, sex differences in the distribution of corticosteroid receptors have been reported in rat brains. Three days after both gonads and adrenal glands had been removed, mineralocorticoid binding in males was higher than in females throughout the hippocampus, although glucocorticoid binding in the brain was equivalent between the sexes (MacLusky et al., 1996). These results may reflect sex differences in neural stress responses.

In addition, and more to the point, there are sex differences in the distribution of androgen and estrogen receptors throughout the nervous system (MacLusky et al., 1996). The highest density of sex steroid hormone receptors is in the hypothalamus and parts of the limbic system that connect to the hypothalamus (DeVries and Simerly, 2002). There is significant overlap among the distributions of neurons that express progesterone receptors, androgen receptors, and the two types of estrogen receptors (Figure 4.6). Androgen receptor binding and mRNA expression appear to be higher in the medial amygdaloid nucleus, BNST, preoptic periventricular nucleus, and ventromedial nucleus of the hypothalamus (VMN) in male than in female rats (DeVries and Simerly, 2002). Estrogen and progesterone receptor binding and mRNA expression appear to be higher in the preoptic periventricular nucleus, medial preoptic nucleus, and VMN in female than in male rats (DeVries and Simerly, 2002). Sex steroid receptor gene expression is regulated, in part, by circulating steroid hormone concentrations, although this regulation may be region-specific. For example, testosterone up-regulates androgen receptors in the medial amygdala, but down-regulates those receptors in the medial preoptic nucleus (Simerly, 1993; Burgess and Handa, 1993).

Several studies that reveal sex differences in the distribution or regulation of neurotransmitters or their receptors have been reported. Only indirect links, however, have been established between sex differences in neurotransmitter distribution and in the ability of steroid hormones to evoke adult sexual behavior. For example, in the AVPv, which is larger in female than in male rodents, dopamine-containing neurons are more plentiful in females than in males. The number of neurons expressing tyrosine hydroxylase (a convenient marker of dopaminergic neurons) is equivalent in male and female rats between birth and postnatal day 2. In the presence of testosterone or estrogens, however, tyrosine hydroxylase mRNA decreases in males, leading to dramatic sex differences by postnatal day 10 (reviewed in DeVries and Simerly, 2002). Mice with an androgen receptor mutation display this sex difference, whereas male mice genetically engineered to lack the α estrogen receptor do not show this decrease in the number of dopaminergic neurons; thus, the sex difference appears to be a result of the organizational actions of estrogens.

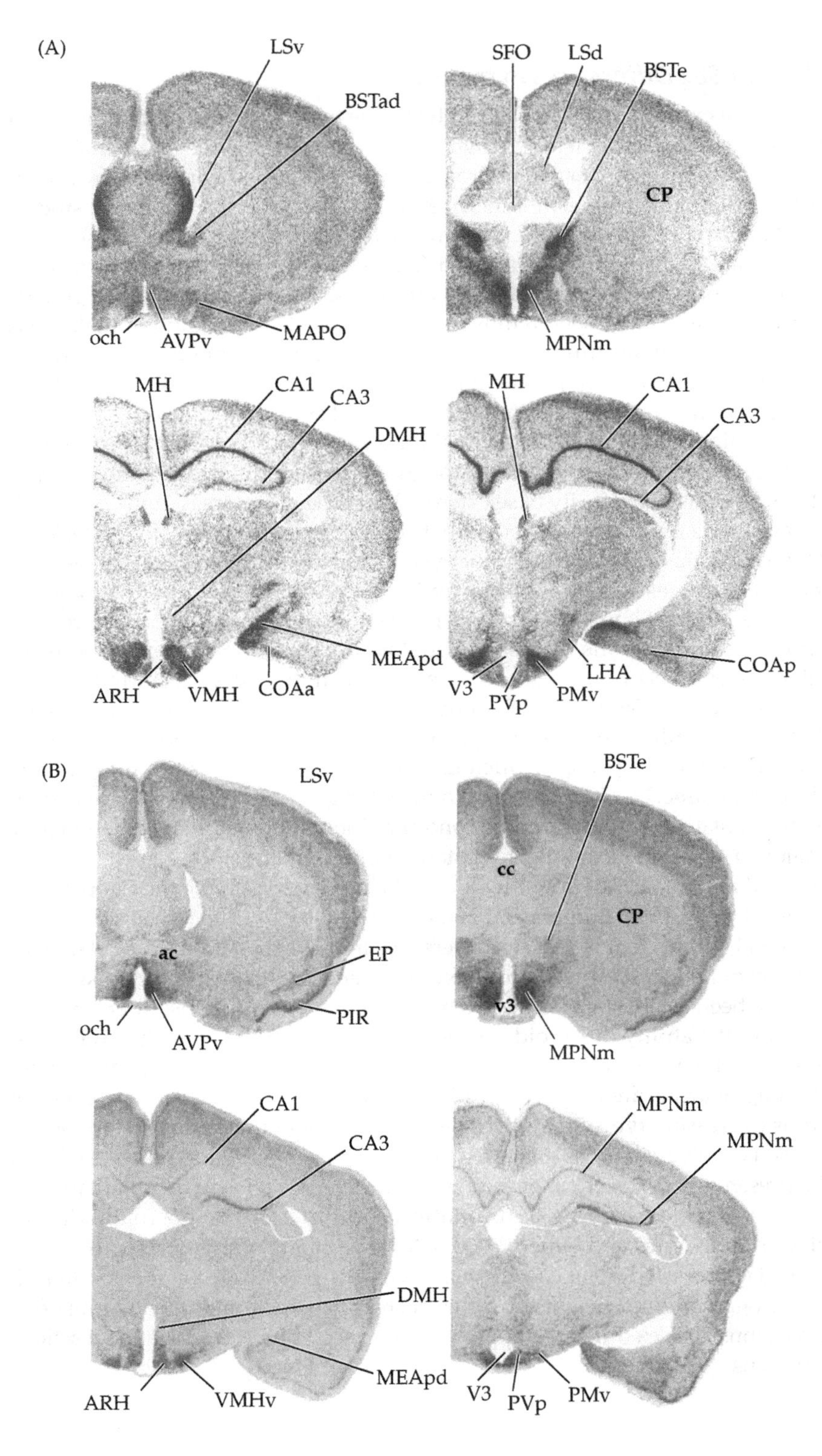
(A)
LSv
BSTad
och
AVPv
MAPO
SFO
LSd
BSTe
CP
MPNm
MH
CA1
CA3
DMH
MEApd
ARH
VMH
COAa
MH
CA1
CA3
LHA
COAp
V3
PVp
PMv
(B)
LSv
ac
EP
PIR
och
AVPv
BSTe
cc
CP
v3
MPNm
CA1
CA3
DMH
MEApd
ARH
VMHv
MPNm
MPNm
V3
PVp
PMv

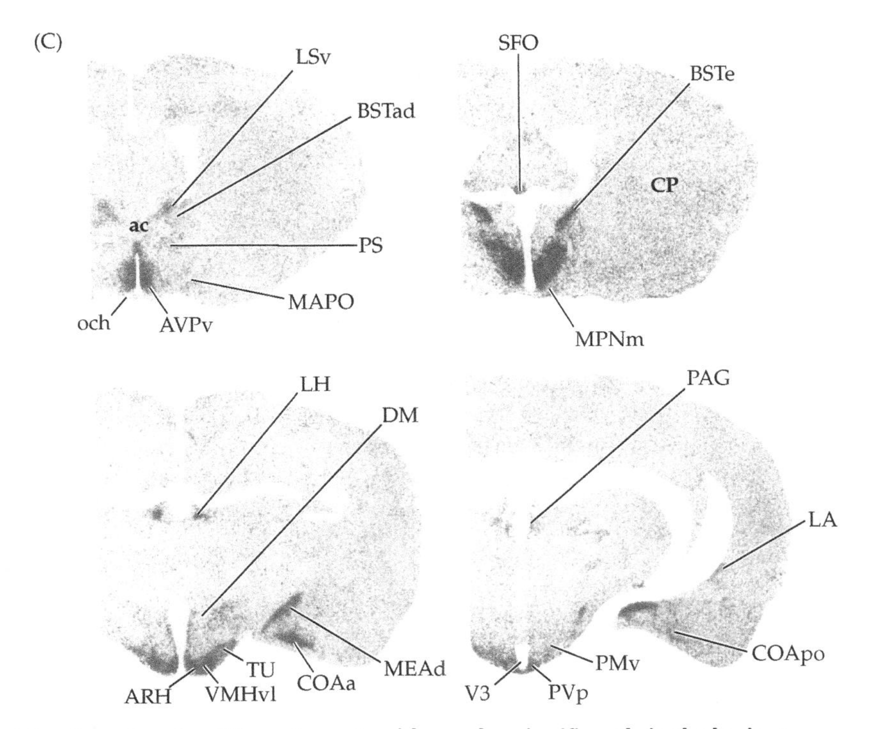

4.6 Receptors for different sex steroids overlap significantly in the brain. Cross sections of rat brain through different brain regions show dark spots that indicate (A) androgen receptors, (B) α estrogen receptors, and (C) progestin receptors. ac = anterior commissure; AVPv = anteroventral periventricular nucleus; ARH = arcuate nucleus; BSTad = anterodorsal nucleus of BNST; BSTe = encapsulated nucleus of BNST; CA1/ CA3 = fields of hippocampus; cc = corpus collosum; COa = cortical nucleus of the amygdala (anterior) ; COApo = cortical nucleus of the amygdala (posterior); CP = caudoputamen; DM = dorsomedial hypothalamic nucleus; EP = endopiriform nucleus; LA = lateral nucleus of the amygdala; LH = lateral habenula; LSv = lateral septal nucleus (later part); MAPO = magnocellular preoptic nucleus; och = optic chiasm; PAG = periaqueductal gray; PIR = piriform cortex; PS = parastrial nucleus; PVp = posterior periventricular nucleus; SFO = subfornical organ; TU = tuberal nucleus; V3 = third ventricle. Courtesy of Simerly, 2002.

The innervation of several brain regions, including the lateral septum, by neurons that release vasopressin shows marked sex differences in rats (DeVries et al., 1983; van Leeuwen et al., 1985). In addition, male rats have two to three times more vasopressin-expressing neurons than females in the BNST and medial amygdala (DeVries and Simerly, 2002). These regions of the brain are involved in sexual behaviors in both males and females, and the vasopressin projections of the BNST and medial amygdala may be indirectly involved in these behaviors. Despite correlations between sexual behavior and vasopressin expression in response to the presence and absence of sex steroid hormones, the functional meaning of sex differences in vasopressinergic neuronal number and projections remains obscure (DeVries, 1995).

In contrast to the unknown role of vasopressin in rats, a nanopeptide related to vasopressin, **arginine vasotocin (AVT)**, has been shown to modulate several aspects of sexually dimorphic behavior in bullfrogs (*Rana catesbeiana*) (Boyd, 1997). Only male bullfrogs give the so-called "mate call," and only females respond to mate calls by moving toward the source of the call (*phonotaxis*). Males respond to other males' mate calls by emitting their own mate calls. Both sexes give "release calls" when mounted inappropriately. Injection of males, but not females, with AVT increases the rate of vocalizing. Injection of AVT causes females to be attracted to the source of a mate call (a male frog or a speaker). Six separate populations of AVT-releasing neurons have been found in bullfrog brains (Figure 4.7), and receptors for AVT are located in brain regions that are linked to important reproductive behaviors, including vocalization, phonotaxis, and locomotor activity. There are sex differences in the numbers of AVT receptors in the amygdala, hypothalamus, pretrigeminal nucleus, and dorsolateral nucleus. More importantly, steroid hormones affect AVT receptor availability. Estradiol modulates AVT receptor numbers in the amygdala of both sexes, and both estradiol and DHT affect

(A)

(B)

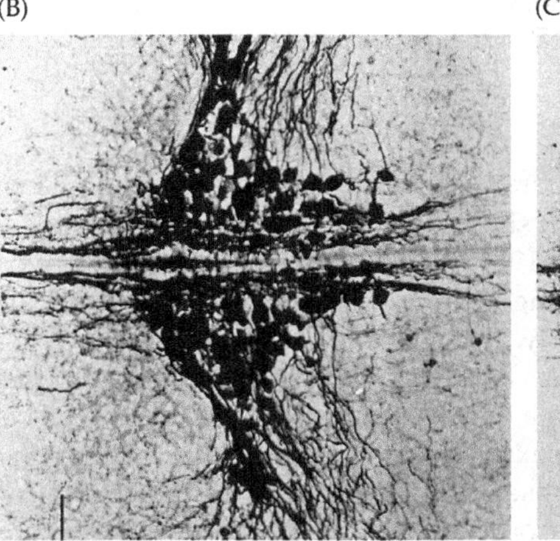

(C)

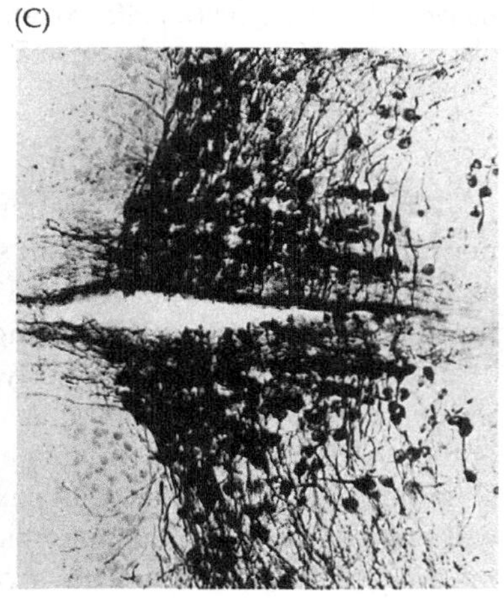

4.7 Sex differences in the bullfrog brain mediate calling behavior. (A) A male bullfrog calling. This behavior attracts potential mates and wards off competing males. (B, C) Immunocytochemistry reveals greater numbers of neuronal cell bodies and fibers containing arginine vasotocin (AVT) in the preoptic area of a male bullfrog brain (C) than in the same region of a female brain (B). (A) courtesy of Sunny K. Boyd; (B), (C) from Boyd et al., 1992.

AVT receptor numbers in the pretrigeminal nucleus of males. Thus, in bullfrogs, AVT seems to function as a neurotransmitter for reproductively relevant behaviors, and AVT activity is modulated by steroid hormones.

Aggression in mammals is often mediated by the neurotransmitter serotonin. There are at least 14 different types of serotonin receptors in mammals, and activation of one of three or four different receptors modulates aggressive behavior. In one study, using positron emission tomography (PET) and a selective radiotracer that binds only to the type 2 serotonin receptor, the distribution of this receptor was mapped in living, age-matched men and women. Significantly higher numbers of type 2 serotonin receptors were found in the brains of the men (Biver et al., 1996). More recently, a similar PET study on living people reported significantly higher type 1A serotonin receptor binding potential and greater type 1A serotonin receptor distribution in women than in men (Parsey et al., 2002); these data confirmed previous postmortem binding data. The mechanisms by which sex steroids influence serotonin or vasopressin receptor numbers and distribution have not yet been discovered.

As noted in Chapter 2, testosterone serves as a prohormone for the production of either estradiol or dihydrotestosterone (DHT) in many mammals and birds. One strategy for understanding the role of testosterone in evoking sex differences in neural tissue is to identify neurons containing testosterone receptors as well as enzymes that convert testosterone to other hormones (Hutchison and Beyer, 1994; Schlinger, 1997). Neurons that contain testosterone receptors and aromatase, 5α-, or 5β-reductase are likely candidates for part of a circuit mediating sexually dimorphic behavior(s) (Pinckard et al., 2000; Roselli et al., 1996, 1997). There are substantial sex differences in aromatase activity in several brain regions of rats, including the BNST, the medial preoptic nucleus, and the VMN; aromatase activity was two to four times higher in males than in females in these areas (Roselli et al., 1997). Aromatase activity appears to be regulated by androgen receptors in the preoptic area and the hypothalamus; testosterone and DHT, but not estrogens, maintain aromatase activity in these regions in castrated rats. There are no sex differences in aromatase activity or aromatase mRNA in the amygdala, and neither aromatase levels nor aromatase mRNA expression were affected by steroid hormones in that area. These results suggest that adult sex differences in responsiveness to androgens may reflect differences in enzymatic gene expression in neural circuits that are involved in regulating sexually dimorphic behaviors (Balthazart and Ball, 1998; Celotti et al., 1997; DeVries and Simerly, 2002; Pinckard et al., 2000; Roselli et al., 1997; Schlinger et al., 2001; see the next section).

Animal Models for Sexually Dimorphic Behaviors

Because of the inherent difficulties in studying the mechanisms underlying human behavioral sex differences, behavioral endocrinologists have relied on animal models to understand the genesis and mechanisms of behavioral sex differences. Several of these animal models are described below.

Bird Song

Birds provide the best evidence that behavioral sex differences are the result of hormonally induced structural changes in the brain (Balthazart and Adkins-Regan, 2002; Nottebohm and Arnold, 1976). In contrast to mammals, in which structural differences in neural tissues have been only indirectly linked to behavior, structural differences in avian brains have been directly linked to a sexually dimorphic behavior: bird song. A subset of birds belonging to the order Passeriformes (the oscines, or songbirds) produce complex vocalizations that, for the most part, are learned during development. These complex, learned vocalizations are referred to as song.

In most species of songbirds, song production is sexually dimorphic. Males usually sing more than females, although this sex difference is not constant across species. Male zebra finches (*Taeniopygia guttata*), like many male songbirds, sing in order to attract females and ward off competing males. Female zebra finches never sing, even after testosterone treatment in adulthood (Adkins-Regan and Ascenzi, 1987). In contrast, there appears to be no sex difference in the singing behavior of bay wrens (*Thryothorus nigricapillus*), a tropical duetting species in which males and females participate equally in producing two-bird song (Brenowitz and Arnold, 1985, 1986; Brenowitz et al., 1985; Brenowitz, 1997). Other species, such as canaries (*Serinus canaria*), fall somewhere between these two extremes; that is, females sing, but they sing less than males (Figure 4.8). These sex differences in behavior reflect sex differences in the neural centers of the brain that control singing (Schlinger, 1998; Schlinger et al., 2001; Balthazart and Adkins-Regan, 2002).

NEURAL BASES OF SEX DIFFERENCES IN BIRD SONG In contrast to the early predictions of subtle, hormonally induced changes in brain structure (Phoenix et al., 1959), several brain regions in songbirds exhibit substantial sex differences in size. Some of the sex differences in the nervous systems of songbirds are listed in Table 4.2.

4.8 Singing in female songbirds falls along a broad continuum: females of some species, such as zebra finches, never sing in nature; females of other species, such as bay wrens, sing as frequently as male conspecifics. Most species are intermediate between these two extremes. For example, female canaries sing, but not as frequently as males, and their songs are generally less complex than those of males. After Brenowitz, 1997.

TABLE 4.2 Sexual dimorphism in songbird brains

Brain Structure	Species
Volumes of song system nuclei lMAN, DM, and nXIIts	
lMAN volume (♂ > ♀)	Zebra finch, starling, dark-eyed junco
DM volume (♂ > ♀)	Zebra finch
nXIIts volume (♂ > ♀)	Zebra finch, canary, red-winged blackbird
Number of neurons	
HVC and RA (♂ > ♀)	Zebra finch, bush shrike
RA neurons (♂ slightly > ♀)	White-browned robin chat, bay wren, buff-breasted wren
lMAN (♂ > ♀)	Zebra finch
Size of neurons	
Neuronal somata in HVC (♂ > ♀)	Zebra finch, Carolina wren, bush shrike
Neuronal somata in RA (♂ > ♀)	Zebra finch, Carolina wren
Neuronal somata in RA (♂ = ♀)	White-browed robin chat, bay wren, buff-breasted wren
Neuronal somata in lMAN (♂ > ♀)	Zebra finch
Dendritic fields in some HVC neurons (♂ > ♀)	Canary
Dendritic fields in RA (♂ > ♀)	Zebra finch and canary
Dendritic field sizes in RA (♂ = ♀)	Buff-breasted wren
Dendritic fields in lMAN (♂ > ♀)	Zebra finch
Number of neurons with sex steroid receptors	
Cells with androgen receptors in HVC and lMAN (♂ > ♀)	Zebra finch
Cells with androgen and estrogen receptors in HVC (♂ > ♀)	Canary
Cells with androgen receptors in HVC and lMAN (♂ = ♀)	Bay wren, rufous-and-white wren
Connectivity	
HVC neurons projecting to RA (♂ > ♀)	Zebra finch
HVC neurons projecting to area X (♂ > ♀)	Zebra finch
lMAN neurons projecting to RA (♂ > ♀)	Zebra finch
Neurochemistry	
RA activity induced by $GABA_A$ receptor antagonist (♂ < ♀)	Zebra finch
Ascending catecholaminergic and enkephalinergic projections to area X (♂ > ♀)	Zebra finch
Acetylcholinesterase staining in area X (♂ > ♀)	Zebra finch

Source: Balthazart and Adkins-Regan, 2002.

Two major brain circuits, the efferent motor pathway and the auditory transmission pathway, have been implicated in the learning and production of bird song (Arnold and Jordan, 1988). These pathways were first described in canaries and zebra finches (Nottebohm and Arnold, 1976; Nottebohm et al., 1982) and have been subsequently identified in several oscine species. Sexual dimorphisms in the sizes of nuclei in these pathways seem to parallel sex differences in singing behavior. For example, there are large differences in the size of several song control nuclei between male and female zebra finches (Bottjer et al., 1985; Nottebohm and Arnold, 1976), but these dimorphisms are less extreme in canaries, and they are undetectable in bay wrens (Brenowitz, 1991, 1997; Brenowitz and Arnold, 1986; Brenowitz et al., 1985).

The efferent motor pathway is necessary for the production of song in adult birds. This pathway consists of the interconnected brain areas that control neural output to the syrinx (the avian vocal production organ). The locations of the song control nuclei in this pathway for a "generic" songbird brain are illustrated in Figure 4.9. The "high vocal center" (HVC) sends an efferent projection to the robust nucleus of the archistriatum (RA). A projection from RA travels to the tracheosyringeal division of the nucleus of the hypoglossal nerve (nXIIts), either directly or indirectly via the dorsomedial (DM) portion of the nucleus intercollicularis (ICo). From nXIIts, motor signals travel by way of the tracheosyringeal nerve to the syrinx. Lesions of the HVC and RA result in song production deficits in adulthood (Nottebohm et al., 1976, 1982).

Information from HVC can also reach RA through a more circuitous route (Bottjer et al., 1989; Okuhata and Saito, 1987). This second route, the auditory transmission pathway, has been called the "recursive loop" (Nottebohm et al.,

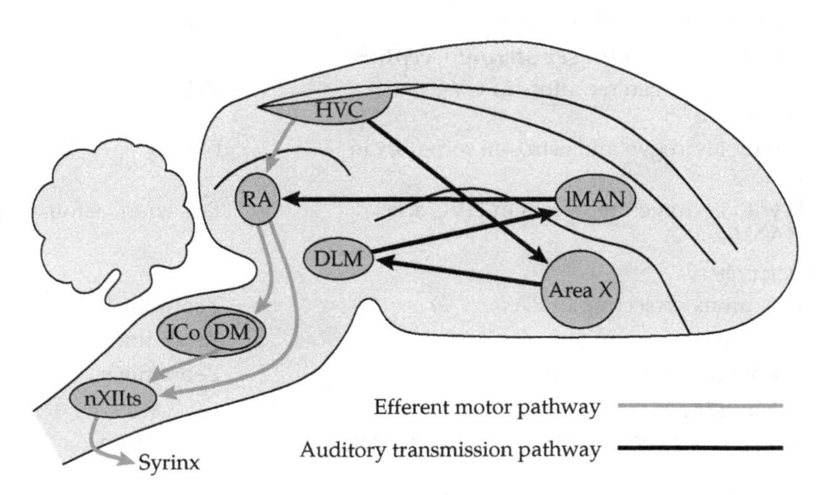

4.9 The neural basis of bird song, shown in a "generic" songbird brain. In species in which large sex differences in singing behavior are observed, the nuclei involved in these brain circuits are larger in males than in females. Courtesy of Gregory Ball, Daniel Bernard, and Joseph Casto.

1990). Information in the recursive loop travels from HVC to area X (a forebrain nucleus located in the parolfactory lobe), then to the medial nucleus of the dorsolateral thalamus (DLM), to the lateral portion of the magnocellular nucleus of the anterior neostriatum (lMAN), and finally to RA. This recursive pathway has been implicated in the process of song learning.

The song learning process occurs in two stages. During the *sensory stage*, a young bird hears and memorizes the song(s) of adult birds. The sensory stage is followed by a *sensory–motor stage*, in which the bird tries to reproduce the song that is stored in its memory. Learning at this stage proceeds by trial and error. The sensory–motor stage ends when the bird can reproduce full adult song. At this point song is considered *crystallized* and is relatively impervious to change. Damage to the recursive loop in zebra finches, especially area X and lMAN, before species-typical song is crystallized results in song abnormalities (Bottjer et al., 1984; Mooney, 1999; Scharff and Nottebohm, 1991; Schlinger, 1998).

HORMONAL INFLUENCES ON SONG CONTROL CENTERS Because sexual dimorphisms in the size of their song control areas are so conspicuous, zebra finches have proved to be an excellent model species in which to study the effects of sex steroid hormones on the neural structures involved in song (Arnold et al., 1987). The HVC and RA of male zebra finches are three to six times larger than those of female conspecifics, and one song control region, area X, cannot even be discerned in females (Nottebohm, 1991; Nottebohm and Arnold, 1976). The larger size of these nuclei in males is due to the fact that the neurons in these nuclei are larger, more numerous, and farther apart (Arnold and Gorski, 1984; Arnold and Matthews, 1988). Although castration of adult zebra finches leads to a reduction in singing, it does not reduce the size of the brain nuclei controlling song production. Similarly, androgen treatment of adult female zebra finches does not induce changes either in singing or in the size of the song control regions. Thus, activational effects of steroid hormones do not account for the sex differences in singing behavior or in brain nucleus size in zebra finches.

The sexual dimorphisms in brain nucleus size and in the subsequent singing behavior of zebra finches are organized early in development. Adult singing is activated by testosterone acting on the male-typical neural structure (Figure 4.10; Gurney and Konishi, 1980). Both male and female zebra finches have been studied, but only the results of endocrine manipulations of females will be presented here. Females treated with DHT soon after hatching and given no endocrine therapy as adults exhibited increased numbers of neurons in RA and HVC, but these nuclei were approximately the same size as those of untreated females. These females did not sing. If females were injected with DHT post-hatching, then treated with testosterone or DHT in adulthood, the RA and HVC were larger and had more neurons than those of typical females, but were smaller and had fewer neurons than those of typical males. These females also failed to sing as adults. Females that were injected with estradiol post-hatching but not treated with hormones as adults did produce some song. Their RA and HVC were larger in size and had more and larger neurons than those of untreated females, but these brain

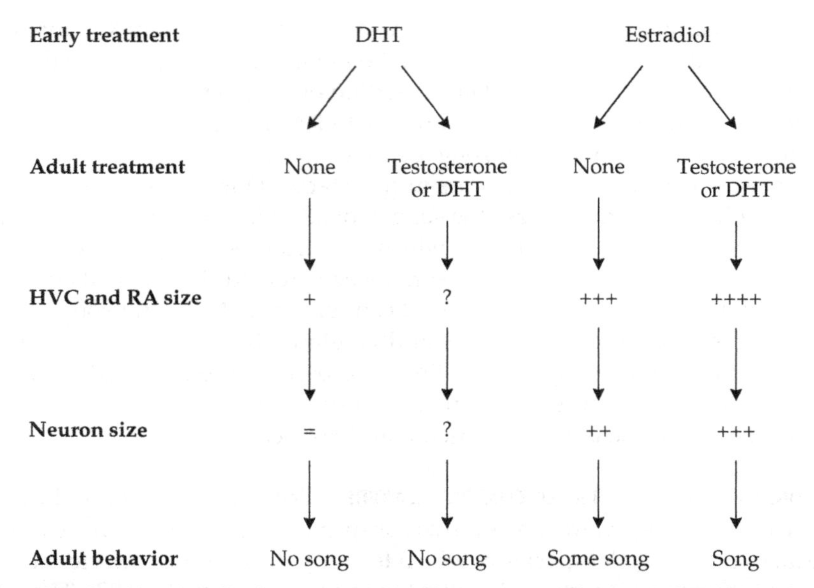

4.10 Singing in zebra finches is organized by estrogens but activated by androgens. Only females that are injected with estradiol after hatching and injected with either DHT or testosterone in adulthood show neural development comparable to that of males and display singing behavior. Thus, early exposure to estradiol is necessary for androgens to have the song-activating effects seen in adult male birds. The question marks indicate that these brain measurements were not obtained for this group. The plus signs indicate the degree of increase in neuron number and size compared with normal females. After Gurney and Konishi, 1980.

regions were still smaller and had fewer neurons than those of untreated males. If females were injected post-hatching with estradiol, then treated with either testosterone or DHT in adulthood, their RA and HVC approached the size of those of untreated males, and they sang!

The failure of hormonal manipulations to produce females indistinguishable from males in the size of their neural structures and behavior has provoked the hypothesis that genes may have direct effects on sexual differentiation in birds (Wade, 2001; Arnold, 2004; see Chapter 3). In any case, early exposure to estradiol certainly increases sensitivity to androgens in adulthood. This pattern of hormonally mediated organizational effects in birds is very different from that underlying other avian behaviors, such as quail copulatory behavior (see Box 4.1). The reasons underlying this difference remain unknown. In addition, the source of the early estrogen in normal male development is uncertain. Presumably, the gonads produce androgens or estrogens that normally masculinize male zebra finches. However, castration fails to prevent masculinization of male zebra finches (Adkins-Regan and Ascenzi, 1990). Furthermore, the presence of functional testicular tissue does not masculinize the development of the song system in genetically female zebra finches (Wade and Arnold, 1996).

Significant evidence suggests that the organizational/activational hypothesis that was developed to explain mammalian sexual differentiation accounts for the sexual differentiation of the zebra finch song system: (1) female hatchlings treated with estradiol undergo substantial masculinization of song neural circuitry and singing behavior, as we have just seen; (2) male nestlings have significant circulating estradiol concentrations; and (3) some aspects of male song circuitry are demasculinized by treatment with a 5α-reductase inhibitor (Balthazart and Adkins-Regan, 2002). There is also some compelling evidence that the organizational/activational hypothesis falls short of explaining the sexual differentiation of zebra finch song. For example, (1) effective masculinizing doses of estradiol are high, often toxic, and some females are not masculinized even by very high doses of estradiol; (2) males castrated as hatchlings sing normally; (3) the song control nuclei have few estrogen receptors during early development, (4) the sex differences in neuron size and number in the HVC are present by post-hatching day 9, despite the lack of estrogen receptors at that age; (5) treatment of hatchling males with anti-estrogens, anti-androgens, or aromatase inhibitors fails to demasculinize the song system or singing behavior; and (6) females hatched from eggs treated with Fadrozole, an aromatase inhibitor, have testes or ovotestes, but still possess female-typical song control systems and do not sing (Balthazart and Adkins-Regan, 2002). Taken together, these findings suggest that although hormones can masculinize the song control nuclei of female zebra finches, nonhormonal factors, perhaps specific activation of genes on the Y chromosome, normally induce masculinization of the zebra finch song system (Wade, 2001; Arnold, 2002).

PHOTOPERIOD EFFECTS Canaries also display sex differences in song production. Female canaries sing much less frequently than males, and their song is much less complex than that of males. However, the frequency of female singing can be increased to nearly male levels by injecting adults with androgens, which also causes a dramatic increase in the size of the adult female canary HVC and RA (Nottebohm, 1980a,b, 1989). Male canaries sing more frequently in spring than in winter, and they appear to lose components of their songs after each breeding season and incorporate new components each spring (Leitner et al., 2001). Their song production is mediated by day length, or **photoperiod** (the number of hours of light per day). In spring, the photoperiod increases, the testes grow and secrete androgens, the frequency of singing increases, the song repertoire enlarges, and, based on laboratory studies, it was assumed that the HVC and the RA double in size. In autumn, the photoperiod decreases, the testes regress in size and androgen production virtually stops, the frequency of singing decreases, the song repertoire shrinks, and the HVC and RA regress in size as well (Nottebohm, 1989). Treatment with testosterone in autumn mimics spring hormonal conditions and supports song production. The seasonal plasticity in behavior is therefore assumed to reflect seasonal changes in brain morphology induced by hormonal changes (Tramontin and Brenowitz, 2000).

Some mild controversy exists about these seasonal changes in brain nuclear volume, related to the way the boundaries of the nuclei are defined. Only sections of the brain stained with Nissl stain appear to display the seasonal change in nuclear size (Gahr, 1990); when brains are treated with other types of histological stains, they fail to display seasonal changes in brain nuclear volume. The functional meaning of seasonal variation in Nissl-stained brain tissue remains unspecified. Nissl stain is used to visualize cell bodies and dense areas of the cell (usually the cell nucleolus), but variation in the degree to which Nissl stain is taken up by cells can lead to differences in measured sizes of cell nuclei that are more apparent than real. In one study, alternate slices of the brain were either stained with Nissl stain or treated for autoradiographic detection of neurotransmitter receptors. Autoradiography for α_2-adrenergic receptors showed increases in the volume of the HVC in starlings with elevated testosterone concentrations as compared with starlings with low testosterone concentrations, and these results corresponded exactly with those from the alternate Nissl-stained slices (Bernard and Ball, 1993).

These seasonal changes in brain nuclear volume are not observed to the same degree in other species of seasonally breeding birds. For example, white-crowned sparrows (*Zonotrichia leucophrys*) display seasonal cycles of androgen secretion and song production, but do not exhibit a seasonal cycle of changes in song system nuclear volume (Baker et al., 1984). Studies of canaries in their natural habitat verified the seasonal changes in song complexity in the absence of seasonal changes in the gross anatomy or ultrastructure of HVC or RA (Leitner et al., 2001). The source of the seasonal plasticity in song complexity in these birds remains unspecified.

Estrogens appear to be necessary to activate the neural machinery underlying the song system in birds. The testes of birds primarily produce androgens, which enter the circulation. The androgens enter neurons containing aromatase, which converts them to estrogens. Aromatase is generally localized in neurons lying near other neurons with estrogen receptors in the hypothalamus and preoptic area of songbird brains, as well as in limbic structures and in the structures constituting the neural circuits controlling bird song (Balthazart and Ball, 1998; Schlinger, 1998). Indeed, the brain is the primary source of estrogens, which activate masculine behaviors in many bird species (Schlinger et al., 2001; Arnold and Schlinger, 1992).

Courtship Behavior of the Plainfin Midshipman Fish

As described in Chapter 3, there are two very different types of male plainfin midshipman fish (see Figure 3.17). The so-called Type I males are large and olive-gray, and they attract females to their nests with a persistent "humming" call, which has earned these fish the nickname "canary bird fish." The call is produced by a pair of sonic muscles that vibrate the swim bladder like the skin of a drum (Figure 4.11A,B). Type I males also build nests and guard them. These males have high testosterone and 11-ketotestosterone concentrations. Their sonic muscles and the motoneurons that innervate them are large, and the motoneurons exhibit a high discharge frequency matching the rhythm of the sonic muscles' pacemaker cells (about 20% higher than females or Type II males) (Bass, 1996). Importantly, the

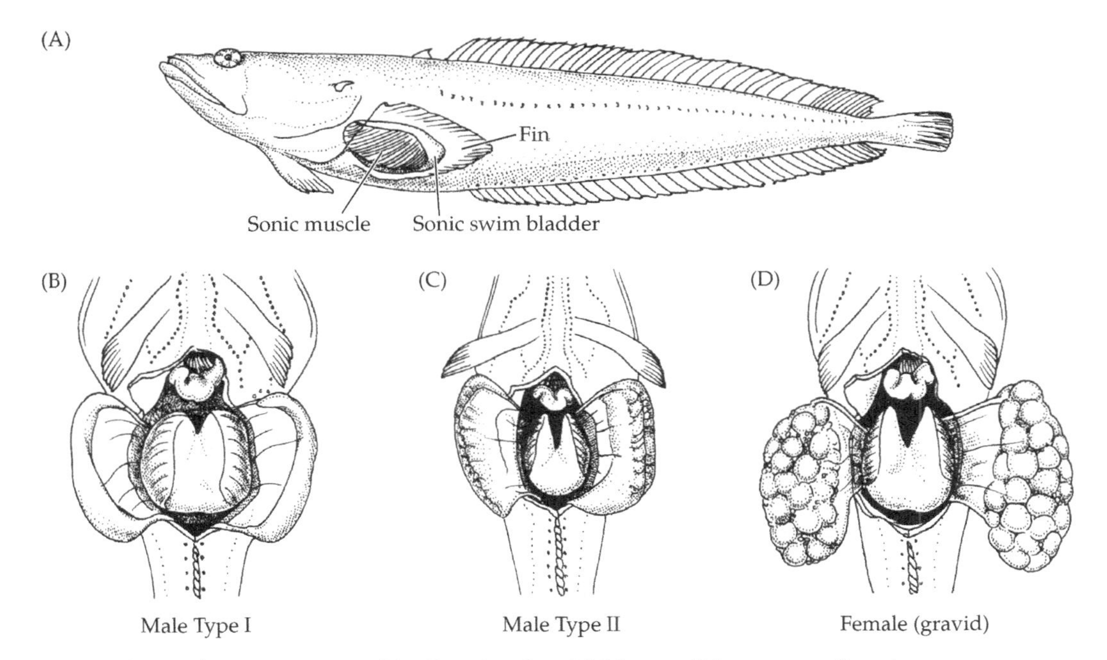

4.11 The sonic organs are used by Type I male midshipman fish to attract females to their nests. (A) The sonic organs comprise a pair of sonic muscles attached to the swim bladder; contraction of the sonic muscles vibrates the swim bladder like the skin of a drum. (B–D) The sonic organs of (B) Type I males are well developed compared with those of (C) Type II males or (D) females. The ratio of sonic muscle mass to total body mass is 6 times greater in Type I males than in Type II males. In contrast, the gonad-to-body mass ratio is 9 and 20 times higher in Type II males and gravid females, respectively, than in Type I males. After Bass, 1996.

duration of the vocalizations is modulated by steroids, and the sonic organs of these fish respond rapidly to steroids, suggesting a nongenomic mechanism of action (Remage-Healey and Bass, 2004).

In contrast, Type II males do not build nests, guard eggs, or vocalize to attract females. Rather, Type II males are characterized as "sneaker" males. They follow females into Type I males' nests and release sperm when the females release their eggs. Because Type II males resemble females in size and color, the Type I males do not exclude them from their nests effectively (Bass, 1996). The sonic system musculature and innervation of Type II males is small and resembles that of females (Figure 4.11C,D). The primary circulating steroid hormone in Type II males is testosterone, and that in females is estradiol. Females are physically larger than Type II males, but they also fail to engage in much vocal behavior.

The steroid hormone 11-ketotestosterone is produced only by Type I males. This hormone is particularly effective in stimulating development of the sonic musculature and neural circuitry. The endocrine cascade leading to puberty occurs 3 to 4 months earlier in Type II males and in females than in Type I males, precluding development of the sonic musculature and neural circuitry (Bass, 1996). The existence of two distinct male forms within a population exhibiting marked differences in behavior provides a terrific opportunity to study hormone–behavior interactions.

Communication requires a sender and a receiver. Hormones not only affect the sender, but can also affect the perceptual mechanisms of the receiver that are necessary to respond to the sender's signals (Sisneros et al., 2004). For many species, the peak sensitivity for each sensory modality is closely correlated with biologically important signals. For example, in the túngara frog (*Physalaemus pustulosus*), there is a match between the emphasized frequencies in the mating call and the tuning of the frog's two inner ear organs (Ryan, 1985). One can imagine that it could be adaptive to adjust the peak sensitivity of the sensory organs to reproductive stimuli during the breeding season, and perhaps to predator stimuli outside of the breeding season. Such a system has been described in the midshipman fish. Nonreproductive female midshipman fish treated with either testosterone or 17β-estradiol increase their sensitivity to the frequency of male vocalizations (Sisneros et al., 2004). In other words, steroid hormones secreted during the breeding season mediate the seasonal plasticity of the perceptual mechanisms. This sensory plasticity provides an adaptable mechanism that enhances coupling between sender and receiver in vocal communication (Sisneros et al., 2004).

Urinary Posture in Canines

The difference in urinary posture between male and female dogs is a well-known sexually dimorphic behavior. The difference in body posture during urination is observed not only among domestic dogs, but also among wild canids (Martins and Valle, 1948). Although this might seem at first to be a silly topic for research, this sexually dimorphic behavior has provided endocrinologists with a useful measure in studies of sexual differentiation. If you were challenged to assign the sex of a dog 100 meters away, your task would be immediately simplified if you saw the dog raise one of its rear legs and urinate. If you have observed many dogs, then you are also in a position to state that the male dog has undergone puberty, because only male dogs that are sexually mature regularly display the raised-leg urinary posture. Females very rarely raise a leg when urinating, although adult males occasionally assume the squatting posture characteristic of female dogs. The important features of this behavior are that (1) it is sexually dimorphic; (2) the sexual dimorphism appears around the time of puberty; and (3) it is not a reproductive behavior per se. Behavioral endocrinologists have formalized these observations during controlled studies to elucidate the role of sex hormones in mediating this sexual dimorphism (Figure 4.12).

In a study that followed in the conceptual footsteps of the classic experiment conducted on guinea pigs by Young, Phoenix, and co-workers (Phoenix et al., 1959), beagle puppies were subjected to several different early hormonal environments (Beach, 1974b). Beagles have a gestation period of about 58–63 days, and sexual differentiation begins late in gestation and continues for 10–15 days postnatally. To study hormonal effects on behavioral differentiation, some females were

4.12 Urination postures of domestic dogs exhibit a clear sex difference that has been exploited in studies of sexual differentiation. Adult male dogs usually raise a rear leg and orient the stream of urine toward some target to mark it; adult females usually adopt a squatting posture. Puppies of both sexes usually assume a squatting position during urination, although male dogs begin to show a "lean-forward" urinary posture around the time of puberty. After Beach, 1974b.

injected with testosterone 1–3 days after birth. The timing of the treatment for this experimental group was equivalent to that of the androgen treatment given to newborn rats to masculinize their mating behavior. Females of a second experimental group were exposed to testosterone in utero by injecting their mothers with this steroid hormone late in pregnancy. This experimental treatment was reminiscent of the early endocrine treatment of guinea pigs, a species for which sexual differentiation occurs prenatally. A third experimental group simulated more closely what normally occurs for male dogs: female pups were exposed to testosterone in utero as well as immediately postpartum. Additional experimental groups included normal females that were ovariectomized in adulthood, males castrated at birth, and males castrated in adulthood.

With no additional hormone replacement therapy later in life, postpubertal female dogs that were exposed to testosterone both in utero and postpartum displayed the male urinary posture about 50% of the time. Many of their reproductive behaviors were also masculinized and defeminized. Their external genitalia, but not their internal sex organs, were masculinized; these females had a penis through which the urogenital opening passed (Figure 4.13). Females that were exposed to testosterone for shorter durations and control females rarely displayed the male-typical urinary posture. Males, even if castrated in infancy, began shifting from the juvenile squatting posture to the male-typical posture at about 4–6 months of age, the usual time of puberty (Beach, 1974b; Ranson and Beach, 1985). These experiments show that the sex difference in canine urinary posture is organized by sex steroid hormones, but does not require their presence for activation of the behavior.

(A)

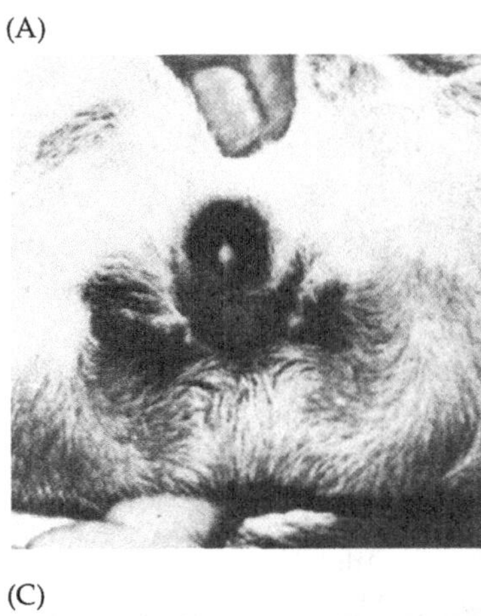

(B)

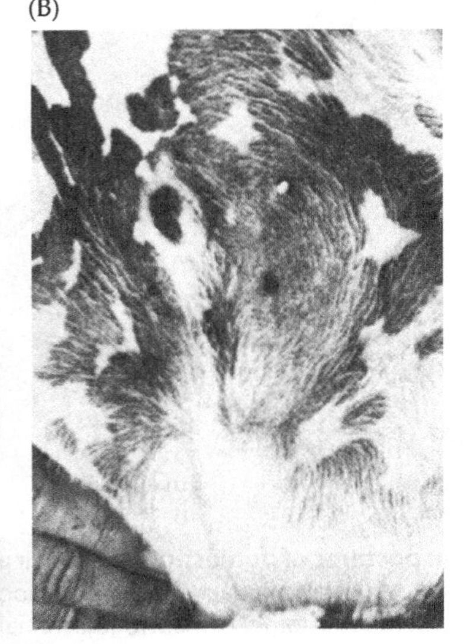

(C)

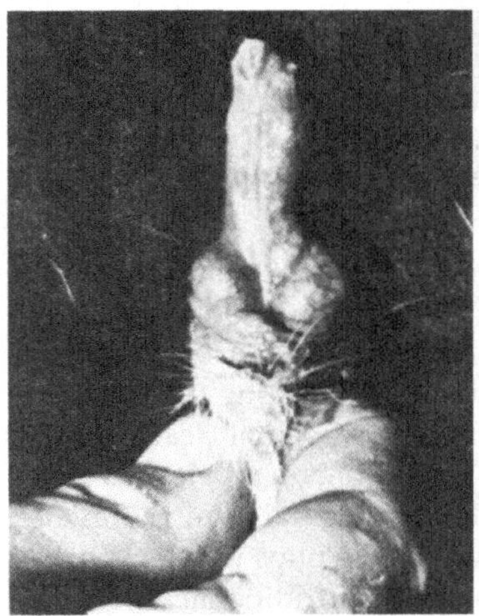

4.13 Modified external genitalia of female dogs resulting from androgen exposure. (A) Appearance of the external genitalia of a normal, untreated female. (B) A female treated with androgens in utero shows labial swelling and clitoral enlargement. (C) A female treated with androgens in utero and immediately after birth shows a clitoris formed into a structure that is indistinguishable from a penis; this pseudopenis displays erectile function. Despite this masculinization of the external genitalia, these females have structurally normal ovaries and derivatives of the Müllerian duct system, and they show estrous bleeding through the pseudopenis if treated with estrogen in adulthood. From Beach et al., 1983.

An interesting facet of the hormonal mediation of urinary posture in dogs is the way this behavior develops (Ranson and Beach, 1985). The developmental pattern of this behavior indicates that the male puppy's perception of his world changes as he experiences puberty. Initially, male and female puppies display similar squatting urinary postures. Females do not exhibit place preference for urination around the time of puberty, but as males mature, they begin to spend much time sniffing and exploring vertical objects in their environment. Objects with certain dimensions are preferred. As males become sexually mature, they lift a rear leg and direct the stream of urine onto these vertical objects (Ranson and Beach, 1985). These behaviors are the antecedents of the territorial marking behaviors that males will engage in throughout their adult lives. The process by which certain vertical objects become worthy of investigation and marking for males or perinatally androgenized females, but not typical females, remains unspecified, but it appears that hormones affect the perceptual focus (or attention) of young dogs so that they attend preferentially to objects of certain dimensions when marking.

Courtship Behavior of Electric Fishes

Another sexual dimorphism exists in the electrical discharge patterns of weakly electric fishes. Strongly electric fishes, such as the freshwater electric eel (*Electrophorus*) and saltwater electric rays (*Torpedo*), can generate hundreds of volts of electricity, which they use as an offensive weapon when hunting for prey and as a defensive weapon when threatened. In contrast, weakly electric fishes generate only a few volts, and use their low-voltage electrical signals for orientation, as well as during courtship. The two main groups of weakly electric fishes (the mormyriforms of Africa and the gymnotiforms of South America) both dwell among the vegetation along the murky bottoms of freshwater ponds and lakes. Because it is virtually impossible to study these creatures in the turbid water in which they live, the scenario that follows is derived from observations in laboratory tanks.

The electric organs of electric fishes are specialized effectors, called *electrocytes*, that have evolved from modified neurons, muscle cells, or muscle endplates. In weakly electric fishes, the electrocytes are arranged in an electrical discharge organ located in the tail (Figure 4.14A). Receptors for incoming electrical signals are scattered along the body, but are concentrated along the midline. The fish "see" or orient in murky water by discharging electrical signals and detecting any distortion in the pattern of their emitted signals (Carr, 1993). Their electroreceptors also detect electrical signals emitted by conspecifics within a short range (70–100 cm) (Hopkins, 1984).

Courtship occurs in nature during the breeding season, which is closely linked to the seasonal rains. During the breeding season, if a mature male and female locate each other, courtship begins in earnest. In some species, males emit a rapid train of electrical "chirps" (Hagedorn, 1986). These "chirps" are actually silent electrical discharges, but are named for the noise they make when the electrical signal is transduced to sound via a loudspeaker. This signal, or even a recording of this signal, is sufficiently potent to induce gravid (ready to spawn) females to release their eggs.

Although both sexes produce similar low-voltage signals during electrolocation, males usually produce much more electrical activity than females during courtship. Males, females, and sexually immature juveniles can be discriminated by the frequency pattern of their electric organ discharges (EOD; reviewed in Landsman et al., 1990). The EOD pattern in *Gnathonemus*, for instance, is sexually dimorphic. As shown in Figure 4.14B, Fourier transformations of a typical male and female EOD in *Gnathonemus petersii* show clear differences in the peak power spectral frequency (PPSF). Information about species, sex, and sexual maturation is transmitted to other fish by the electrical signal. For example, individuals of the species *Sternopygus macrurus* display a species-characteristic wave pattern in their electrical discharge. Sexually mature males discharge a wave pattern of 50 to 90 Hz (cycles per second), sexually mature females display patterns of 100 to 150 Hz,

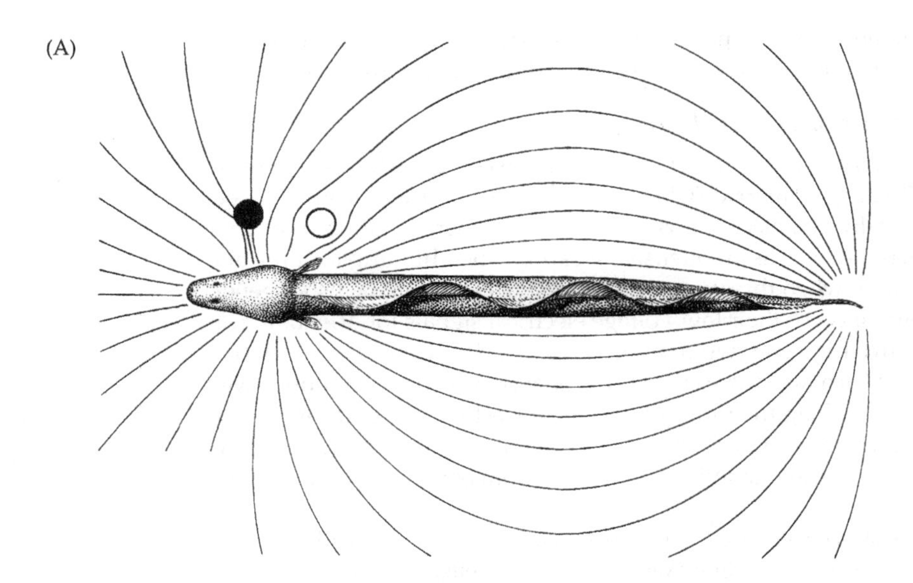

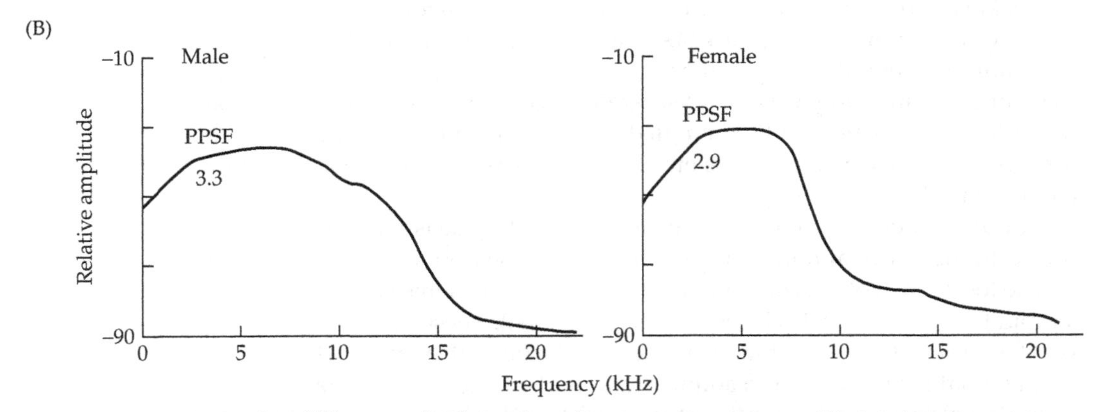

4.14 Sex differences in electrolocation. (A) Animate objects (filled circle) are good conductors of electricity and elevate receptor activity; inanimate objects (open circle) depress receptor activity. This process is called electrolocation and is conceptually similar to the echolocation used by bats, moths, and other animals for orientation. (B) Fourier transformations of a typical male and female EOD in *Gnathonemus petersii* show clear differences in the peak power spectral frequency (PPSF) a statistical analysis that permits visualization of sound waves. Information about species, sex, and sexual maturation is transmitted to other fish by the electrical signal. (B) after Landsman, 1991.

and sexually immature juveniles display intermediate between those of the adults (Hopkins, 1972). In many species, the sex difference in electrical discharge patterns is apparent only during the breeding season (Hopkins, 1988). The fish continuously discharge their electrical signals as they navigate through their habitat, so information relevant to sexual reproduction is simultaneously broadcast.

Sex steroid hormones appear to be important in mediating electrical signaling during courtship, as well as in influencing the frequency of the EOD. Treatment of females or juvenile fish of either sex with androgens, especially naturally occurring fish androgens (for example, testosterone and 11-ketotestosterone), produces electrical discharge patterns similar to those of adult males in breeding condition, whereas castration of adult males shifts the pattern of electrical discharge to one resembling that of conspecific females (Bass, 1986). The finding that hormone treatments can change adult female behavior shows that the sex difference in electrical discharge pattern is due exclusively to activational effects of androgens. Sex steroid hormones also play activational roles in mediating the onset of electrical signaling (Landsman and Moller, 1988).

Sex steroid hormones also appear to affect the sensory system mediating reception of electrical signals (Bass and Hopkins, 1984). The electroreceptors of an individual fish are maximally sensitive, or "tuned," to its own unique pattern of electric organ discharges. This "tuning" obviously aids in electrolocation. Androgens, as well as shifting of the EOD frequency, shift the receptor tuning to the new EOD. Androgens shift the tuning curve—that is, the sensitivity of the receptors—in a masculine direction in females (Bass and Hopkins, 1984). These results are especially intriguing because this model may yield evidence about how steroid hormones affect sensory receptors directly, and thus may provide a neural basis for understanding the sexual dimorphism in sensory abilities observed among many species (Hopkins, 1988). Studies of phenomena such as the development of marking in dogs, female responses to bird song, and changes in tuning in midshipman fish and weakly electric fishes could lead to important insights about how hormones change perception.

Rough-and-Tumble Play in Primates

In rhesus monkeys and many other mammalian species, males engage in much more play behavior than their female peers throughout development (Hinde, 1966; Meaney, 1988; Pellis et al., 1997). A number of other play-associated social behaviors are sexually dimorphic in rhesus monkeys. Males, using specific gestures, initiate play more often than females. Males also engage in more threat behaviors than females. A larger proportion of male play behavior involves simulated fighting or rough-and-tumble play as compared with the play behavior of females. Finally, males engage in pursuit play at higher rates than females (Goy and Phoenix, 1972). In order to understand these sex differences in play behavior, as well as other components of primate behavior, a series of studies was conducted on two populations of rhesus monkeys (Goy, 1978; Goy and Phoenix, 1972) and one population of marmosets (Abbott, 1984) treated prenatally with steroid hormones. The studies of rhesus monkeys (Goy, 1978) will be outlined here.

When pregnant rhesus monkeys were injected with androgens, the external genitalia of their female offspring were masculinized. When the behavior of these pseudohermaphroditic females was compared with that of females and males resulting from nonmanipulated pregnancies, it was found that males engaged in much more threat, play initiation, rough-and-tumble play, and pursuit play

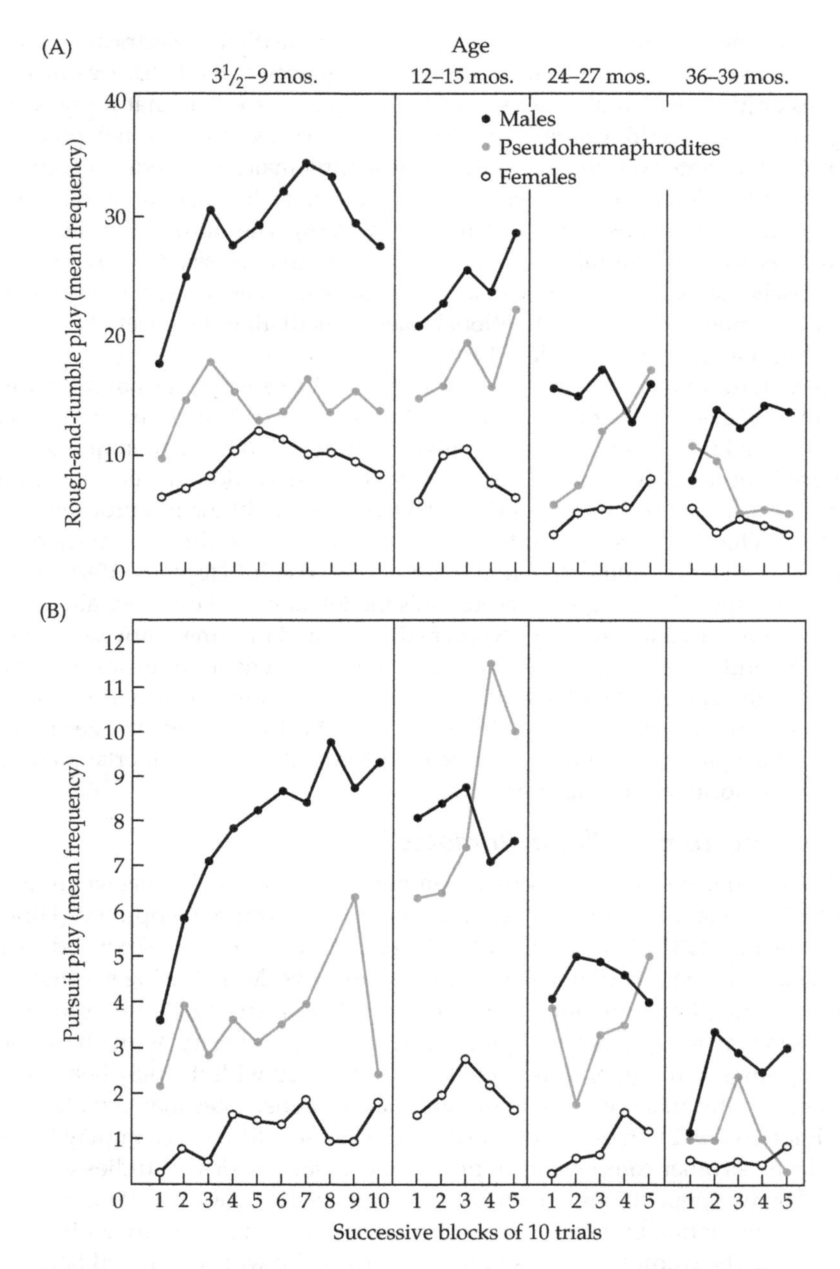

4.15 The frequency of rough-and-tumble and pursuit play is affected by early exposure to androgens. Rates of (A) rough-and-tumble play and (B) pursuit play at different ages are shown for normal males, normal females, and pseudohermaphroditic females treated in utero with androgens. Males engage in these play behaviors more frequently than normal females, but pseudohermaphrodites are intermediate between males and females, indicating that play behavior is masculinized by early androgen treatment. Data from Goy and Phoenix, 1972.

behavior than normal females (Figure 4.15). The pseudohermaphrodites engaged in these behaviors at frequencies that were between the observed rates for males and females (until the age of 4) (Goy and Phoenix, 1972). This sex difference in behavior is not dependent on activational effects of hormones; in other words, the sex difference is organized prenatally. This sex difference is also not dependent on the presence of altered genitalia. By changing the timing of the androgen treatment, juvenile play behavior could be affected independently of any hormonal effects on the external genitals (Goy et al., 1988). Castration or other postnatal endocrine manipulation did not affect the amount of threat, play initiation, rough-and-tumble play, or pursuit play behavior. Interestingly, in order for these sexually dimorphic behaviors to be expressed at all, appropriate social stimulation had to be present during development. Rhesus monkeys that were reared in social isolation with access to only surrogate mothers did not develop the appropriate patterns of play. These results suggest that organization by hormones interacts with environmental factors to produce normal play behavior.

As noted in the introduction to this chapter, human boys and girls also differ in the amount of rough-and-tumble play in which they engage, and there is some evidence suggesting that this difference, too, is organized prenatally by steroid hormones. In one series of studies, aggression was examined in children who were exposed prenatally to a synthetic progestin (MPA) that had been provided to their mothers to prevent premature labor (Reinisch, 1981). This progestin reportedly has androgenic effects, but does not usually cause masculinization of the external genitalia at low therapeutic doses; in some instances, prenatal exposure to MPA enhances feminine traits. Seventeen female and eight male patients were compared with their unexposed, same-sex siblings, using their responses to paper-and-pencil tests as well as questionnaire data obtained from friends and family members (Ehrhardt et al., 1977). Girls exposed to prenatal progestins were considered to be ultrafeminine for some traits. They were very interested in clothes and playing child-rearing games, and were less often characterized as tomboys by themselves or by their peers than were unaffected girls. Despite this so-called enhanced feminization, girls exposed to MPA engaged in higher-energy play activities and excelled at athletic endeavors as compared with control females. In another study, both males and females exposed to MPA scored higher on aggression on paper-and-pencil tests than their unexposed siblings (Reinisch, 1981). However, although paper-and-pencil tests of aggressive behavior provide interesting data, direct observations of aggressive behavior provide more valid results (see Chapter 9).

Girls born with congenital adrenal hyperplasia, which results in exposure to high androgen concentrations early in development (see Chapter 3), displayed elevated rates of energetic play, as well as initiation of fighting and rough-and-tumble play (Ehrhardt and Money, 1967; Ehrhardt et al., 1968a). In a later study, girls with CAH were found to be masculinized and defeminized in several ways; compared with their sisters, they played more with boy-typical toys, were more likely to use aggression when provoked, and showed less interest in infants than typical girls (Berenbaum et al., 2000). Sex-atypical behavior appeared to be significantly associated with degree of inferred prenatal, but not postnatal, excessive androgen exposure. Considered together, the evidence suggests that gonadal steroid hormones may contribute to behavioral sex differences, especially in play behavior, among humans and other primates (Collaer and Hines, 1995).

Sex Differences in Human Behavior

We have seen that many sex differences in behavior arise because the neural substrates of sexually dimorphic behaviors are organized by hormones early in development and the behaviors are activated by hormones later in development (Breedlove et al., 2002). The mating behavior of male rodents, for example, is both organized and activated by steroid hormones. Other sex differences, such as rates of rough-and-tumble play in rhesus monkeys, arise only from organizational effects of steroid hormones. Still other behavioral sex differences, such as the electrical discharge patterns of electric fishes, may result only from activational effects of hormones. However, some sex differences in behavior, such as the learning of a sex role among humans, may be unrelated to hormones (Figure 4.16). Behavioral endocrinologists aim to disentangle the contribution of hormones from the contribution of the environment to determine the cause of such behavioral sex differences.

As noted in Chapter 3, researchers cannot ethically manipulate hormone concentrations in humans to observe their effects on brain and behavior. Consequently, the effects of hormones on human brains and behavior must be inferred by observing the outcomes of so-called "experiments of nature," situations in which individuals have been exposed to atypical levels of hormones during some developmental period of their lives. Early exposure to MPA provides one example, and several other such conditions were described in Chapter 3. The primary goal of studying the effects of unusual hormone exposures on human brains and behavior is to understand the effects of more typical endocrine events. As we review studies on human gender identity, gender role, and sexual orientation, it is important to appreciate that at first glance, these studies provide powerful evidence that gender identity and gender role are learned. But they do not rule out the potential effects of early hormone exposure. When a child is born with ambiguous external genitalia, for example, the decision for surgical correction in a male or female direction is not based on the whim of the surgeon; if there is substantial genital development, then a medical decision is made to modify the genitals in the male direction and rear the child as a boy; if there is little genital development, then surgical modifications are

Activational effects	Organizational effects: Yes	Organizational effects: No
Yes	Rodent sexual behavior	Electrical discharge patterns of electric fishes
No	Primate rough-and-tumble play behavior	Human gender role learning

4.16 Contributions of activational and organizational effects of hormones to behavior. Sexually dimorphic behaviors may be organized and activated, organized but not activated, activated but not organized, or not influenced by hormones.

made in the female direction, and the child is reared as a girl (Box 4.2). The extent of genital development probably covaries with the extent of androgen exposure, so again, the sex of rearing is confounded by early hormone exposure.

Sexual differentiation in humans occurs early during gestation, near the end of the first trimester and throughout the first few weeks of the second trimester of pregnancy. Using the models developed in rodents, dogs, and rhesus monkeys, we can predict that exposure to high concentrations of sex steroid hormones during this time will masculinize and defeminize human brains and behavior, whereas low concentrations of steroid hormones early in development will tend to feminize and demasculinize brains and behavior. Two basic clinical populations have been examined in this regard: (1) individuals with congenital disorders that expose them to unusually high or low hormone concentrations during development, such as CAH, Turner syndrome, 5α-reductase deficiency, and CAIS (or PAIS); and (2) individuals whose mothers were treated with steroid hormones during pregnancy for medical reasons (see Chapter 3). Exposure to prescribed hormones almost always occurs prenatally and may involve natural or artificial estrogens (such as DES) as well as progestins (such as MPA).

Compared with the animal studies previously described, these clinical studies have many inherent flaws. In animal studies, individuals from inbred strains of genetically identical animals can all be treated with equal amounts of steroid hormones at the same time during development and examined under standard test conditions. The human clinical studies, on the other hand, are summaries of individual cases that may differ in the severity and timing of the disorder, or in the type, dose, and timing of the exposure to prescribed steroid hormones. Often the effects of prescribed hormones are confounded with the effects of the medical condition (for example, toxemia) that prompted treatment of the mother in the first place. Furthermore, parental expectations of children may be affected (see Box 4.2).

It is important to keep in mind that there is more variation *within* each sex than *between* the sexes for virtually all human behaviors (Figure 4.17). That is,

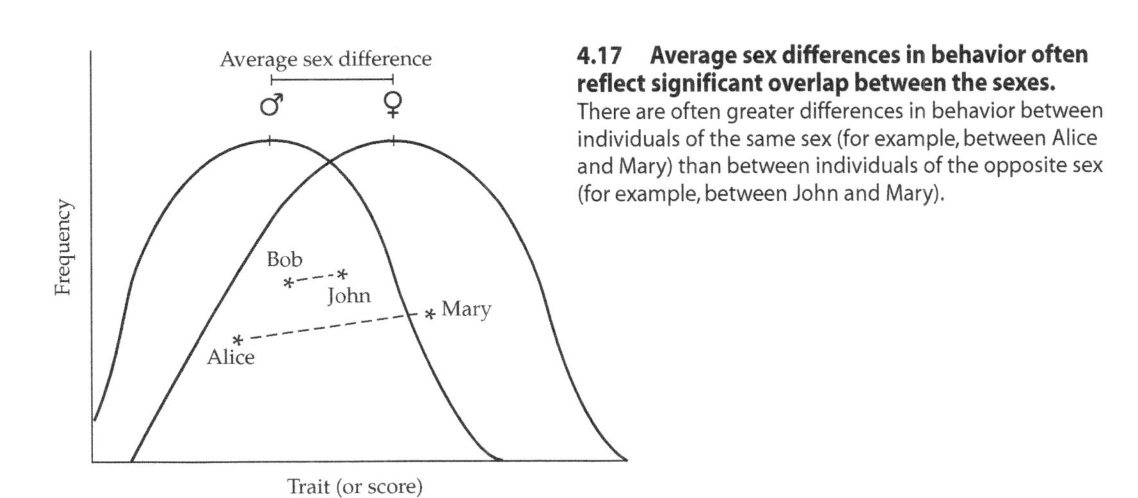

4.17 Average sex differences in behavior often reflect significant overlap between the sexes. There are often greater differences in behavior between individuals of the same sex (for example, between Alice and Mary) than between individuals of the opposite sex (for example, between John and Mary).

BOX 4.2 Ambiguous Genitalia: Which Course of Treatment?

As we saw in Chapter 3, there are occasional developmental anomalies in sexual differentiation that result in ambiguous external genitalia at birth. Prior to the 20th century, pseudohermaphrodites received no surgical treatment as newborns, and some earned their living displaying their bodies in sideshows. The accompanying figures show one such "bearded lady," Marie-Madeleine Lefort, in youth and in old age. Lefort, who possessed both masculine and feminine traits, worked in carnivals throughout Europe in the 1800s (Dreger, 1998). It is now standard practice to surgically alter ambiguous genitalia shortly after birth, but the decisions involved in such procedures are far from simple.

Most studies, including those described in this book, include large numbers of subjects in an effort to ensure that their findings will be generalizable to many individuals. In medical settings, physicians often base treatment decisions on guidelines that emerge from clinical trials that sometimes include thousands of patients. Sometimes this is impractical, however, because particular conditions occur only rarely. Physicians study rare cases in depth, and sometimes publish their findings as "case studies" or "case histories," which preserve patients' confidentiality and sometimes present hypotheses about the causes of a condition and the effectiveness of a given treatment.

Pseudohermaphroditism is fairly rare, and when a baby is born with ambiguous genitalia, physicians and parents must make treatment decisions based on a body of knowledge that has been gleaned from only a small number of cases. Although it is relatively easy to debate this issue from the comfort of the classroom, in the clinical setting a decision must be made quickly and in an emotionally charged atmosphere.

What are the options? One can determine the newborn's genetic sex and make surgical (and later endocrine) alterations so that the external genitalia match the genetic sex. This strategy assumes that genetic sex is supreme in determining psychosexual attributes such as gender identity or gender role. In contrast, one might assume that prenatal hormonal effects on the brain are reflected in genital development. According to this view, if the genitalia appear more female- than male-typical, then corrective surgery that enhances the female genital characteristics should be performed and the individual should be raised as a female, and vice versa if the genitalia appear more male- than female-typical. These courses of action would be pursued regardless of chromosomal or gonadal sex, and based solely on genital appearance—an "anatomy is destiny" approach, which assumes that environmental (i.e., rearing) factors are more important than early biological factors in determining psychosexual development.

So imagine you are the physician in charge in the maternity ward, and a beautiful baby has been born with atypical genitalia. The parents are asking you what they should do. Until recently, virtually all physicians opted for the easiest surgical procedure and rearing the child in a sex-specific manner that matched the altered genitalia, regardless of chromosomal sex. These decisions were based on the classic and influential work of Dr. John Money and his colleagues at The Johns Hopkins University. Money's work on individuals born with ambiguous genitalia (reviewed in Money and Ehrhardt, 1972; follow-up studies are discussed in Money et al., 1986, and Money and Norman, 1987) had convinced him that babies are born neutral with respect to psychosexual development: if a child is raised unambiguously as one sex or the other, that child will develop a gender identity to match the sex assignment. Of course, for the sex of rearing to be unambiguous, the genitals must match the

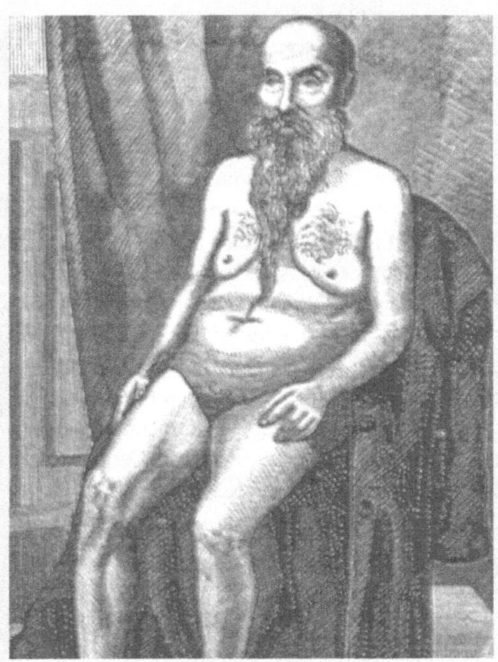

Courtesy of the Wangensteen Historical Library, University of Minnesota, Minneapolis, MN.

assigned sex. One classic case, now known as the "John/Joan" case, provided some of the strongest possible evidence for Money's position. But that case has received a great deal of attention recently (including a feature in *Rolling Stone* magazine [Colapinto, 1997]) because the conclusions drawn in Money's original case study (Money and Ehrhardt, 1972) differ sharply from those of a follow-up study reported nearly 25 years later (Diamond and Sigmundson, 1997).

The original case study described a little boy, given the pseudonym "John," who had his penis severely damaged during circumcision at 8 months of age; essentially, the penis was burned almost completely off. John was brought to The Johns Hopkins Hospital several months later for a consultation, and it was decided that he should be reared as a girl (called "Joan" in the literature). A complete penilectomy was performed; at the time, reconstructive penile surgery had not yet been (and still has not been) perfected. The fact that John was genetically male, and had been born with normal male genitalia, made his case a particularly strong clinical test of the "sex-of-rearing" hypothesis. He also had a twin brother who could serve as a sort of "baseline control" when assessing the treatment's success. "John" became "Joan" at 17 months of age, which was, according to Dr. Money, within an acceptable window of "early re-assignment." A series of follow-up reports indicated that Joan had successfully adopted a female gender identity, although she clearly displayed many "tomboyish" traits such as "abundant physical energy" and a "high level of activity," and was "often the dominant one in a girls' group" (Money and Ehrhardt, 1972). Around the time of puberty, Joan and her family decided to end her participation in the study, and no more follow-up examinations were conducted by the Johns Hopkins research group.

Continued on following page

BOX 4.2 *continued*

Diamond and Sigmundson (1997) presented a startling report that, 25 years after the original report, Joan was living as a man who had had reconstructive surgery with a penile prosthesis, was married to a woman, and was a father to adopted children. In other words, his gender role and identity were firmly male. Now known again in the literature as "John," he also reported that he had experienced substantial mental anguish during childhood and adolescence, and a growing sense that he was not a girl. These feelings culminated in a decision at age 14 to reverse his gender role and identity. The authors of this follow-up study claim that this outcome shows that biological factors are extremely powerful in psychosexual development. Essentially, Diamond and Sigmundson's report is the latest installment in a long series of doctrinal debates on this topic between Drs. Money and Diamond about the primacy of environmental versus biological factors in psychosexual development (Zucker, 1996). Tragically, both twin brothers committed suicide, suggesting that John struggled, even with life as a male.

Clearly, there is a strong interaction between biological and environmental factors in determining psychosexual development, and the truth is likely to lie somewhere between the two extreme positions. Unfortunately, this is not particularly helpful to physicians and parents agonizing over how to help babies born with ambiguous genitalia attain the best quality of life. Those of us who are used to conducting experimental studies never rely on a single incident or an "*N* of 1" (Favreau, 1993) to draw a firm conclusion. Additional studies on human psychosexual development are being conducted to help make sound clinical decisions in cases of anomalous sexual differentiation.

The two options for treatment presented above suggest an acceptance of the division of humanity into only two sexes and an assumption that making individuals fit into one of these two "normal" categories is a desirable goal of the medical profession. A third option is more daring, and has rarely been chosen in the past 40 years: that is, do nothing, in the belief that intersexuality—another term for pseudohermaphroditism—is not an "error," but simply unusual. Recently, vocal spokespersons for intersex individuals have expressed hostility toward surgical treatments. The imposition of "normality" by surgical means is often seen in hindsight as a judgmental assertion of abnormality, and intersex individuals demand acceptance as born. This position was declared most elegantly by an anonymous author who noted, "I was born whole and beautiful, but different. The error was not in my body, nor in my sex organs, but in the determination of the culture, carried out by physicians with my parents' permission, to erase my intersexuality. 'Sex error' is no less stigmatizing than 'defect' or 'deficiency.' Our path to healing lies in embracing our intersexual selves, not in labeling our bodies as having committed some 'error.'"

there is considerable overlap between males and females for most behaviors. With a sufficiently large sample size, relatively small differences in the performance or the frequency of a given behavior can be statistically significant, and these differences are reported in the scientific and popular press as "significant sex differences." Because there is often more variation within each sex than between the sexes, however, it is appropriate to talk about the "effect size," a measure of the variation between sexes corrected for the variation within each sex (Collaer and Hines, 1988; Forger, 1998). The effect size is "moderate" when the average difference between the sexes is about half the size of the standard deviation of the

TABLE 4.3 **Effect size for behavioral sex differences in humans**

Behavior	**Approximate effect size**[a]
Aggression	Moderate (♂ > ♀)
Rough-and-tumble play	Moderate–large (♂ > ♀)
Childhood activity levels	Moderate–large (♂ > ♀)
Overall verbal abilities	Negligible–small (♀ > ♂)
Speech production	Small (♀ > ♂)
Verbal fluency	Moderate (♀ > ♂)
Perceptual speed	Moderate (♀ > ♂)
3-D visual rotation	Large (♂ > ♀)
2-D visual rotation	Small (♂ > ♀)
Spatial perception	Small–moderate (♂ > ♀)
Overall quantitative abilities	Small–moderate (♂ > ♀)
Quantitative problem solving	Small–moderate (♂ > ♀)

Source: Collaer and Hines, 1988; data from J. Cohen, 1977; Kimura, 1992; Forger, 1998.

[a]Effect sizes of ≥0.8 standard deviations are considered large, of ~0.5 standard deviations, moderate; of ~0.2, small; and of < 0.2, negligible.

scores within each sex. For example, the sex difference in height between 16-year-old male and female humans has a moderate effect size (Table 4.3) (Kimura, 1992; Collaer and Hines, 1988).We will normally be discussing only slight behavioral sex differences as we discuss humans; this is very different from the research strategy using animal models in which only one sex displays a particular behavior. These caveats should be borne in mind as we discuss what these studies have revealed.

Gender Role

The most prominent behavioral sex differences between men and women are observed in gender role, gender identity, and sexual orientation/sexual preference. Most, but certainly not all, women assume a feminine gender role, perceive themselves as female, and are sexually attracted to men. Similarly, most, but not all, men assume a masculine gender role, perceive themselves as male, and are sexually attracted to women. Estimates of the number of individuals who do not fall into these categories vary.

Every society has a set of expectations for the behavior of males and females. Children learn these expectations through their play activities and through shaping by their parents, caregivers, siblings, and peers. Young males may be scolded because "boys do not cry," or a young girl may be disciplined more harshly for fighting at school than her brother. The assumption of culturally based behavioral patterns that are specific to one's own sex is the adoption of a **gender role**. Gender roles are learned early in life, and it is difficult to reverse this early learning. For example, among the Nuer people of western Africa, the men tend cattle and the women cook meals (Evans-Pritchard, 1963). Men never cook in this society, and women never work with livestock. For these two tasks, the gender roles are distinct: cooking is considered a feminine gender role, whereas tending cattle is considered a masculine gender role. In North American society, gender roles are less distinguished by occupation than they were 30 years ago, but they still exist. Because gender roles vary substantially among human cultures and over time in the same culture, it has sometimes been assumed that gender roles are entirely mediated by environmental factors. Nonetheless, subtle influences of early endocrine conditions on subsequent gender roles can be demonstrated.

Girls born with CAH, for example, display varying degrees of masculinization at birth. Typically, the endocrine dysfunction can be managed by exogenous hormone treatment and the masculinized genitalia can be surgically corrected. However, these girls display some degree of behavioral masculinization; as described by their siblings and parents, as well as in self-reports, they

1. engage in increased physical activity with higher levels of rough-and-tumble play;
2. are characterized as "tomboys";
3. prefer male playmates;
4. prefer male-typical as compared to female-typical toys;
5. engage in fewer games of simulated maternal behaviors;
6. display reduced interest in infant care;
7. are less likely to be interested in hairstyles, makeup, or jewelry;
8. engage in fewer fantasies about marriage and maternity, as compared with their unaffected sisters (Ehrhardt and Baker, 1974; Ehrhardt et al., 1968a,b).

Remember that no such study can be conducted in an "unbiased" way, because the parents, and often the child, know that there was a hormonal dysfunction that required surgical modification of the genitalia. This knowledge may lead to subtle ambiguities in perception of the individual's gender and could result in exaggerated reports of masculine behavior.

Researchers addressed the issue of parental bias by secretly observing unsupervised play sessions among girls with CAH and their unaffected sisters and first cousins (Hines and Kaufman, 1994; Hines, 2004). The play behavior of CAH girls was significantly masculinized. In other words, the CAH girls played with toys typically preferred by boys. Given a choice of transportation and construction toys, books, board games, dolls, or kitchen supplies, CAH girls preferred to play with the toys generally preferred by boys and avoided by their unaffected female relatives. Of course, this preference for masculine toys could still reflect the biases of the parents or other environmental influences. However, one might expect that parents would be as likely to encourage feminine preferences in toys for their CAH daughters as for their unaffected daughters.

In the draw-a-person test (a psychological test during which a child is asked to draw a person; drawing a person of the same sex is considered to indicate satisfaction with gender role), both females and males with CAH drew the sex-appropriate picture (Hines, 1982). Incidentally, boys exposed to high prenatal androgen concentrations engaged in higher physical activity levels and more rough-and-tumble play, and were more likely to excel in aggressive sports, than their unaffected brothers (Money and Ehrhardt, 1972). This was typically viewed as a good outcome by their parents. Although one may quibble about the relevance of the gender role identifiers used in this study 25 years ago to discriminate boys from girls (for example, preference for dolls as play objects), a subset of the results is strikingly similar to the behavioral data obtained with early androgen treatment of female rhesus monkeys (Goy and Phoenix, 1972) and more recent data in vervet monkeys (Alexander and Hines, 2002).

Imperato-McGinley and her colleagues (Imperato-McGinley et al., 1974, 1979) described a group of Central American males in whom an adolescent gender role change appears to be common. In the Santo Domingo region of the Dominican Republic live several families with a high incidence of 5α-reductase deficiency. The affected individuals are XY chromosomal males that lack 5α-reductase, the enzyme that converts testosterone to DHT (Savage et al., 1980). These individuals have ambiguous external genitalia and small, undescended testes at birth. They are generally reared as girls until puberty, when testosterone masculinizes their bodies and genitalia (see Figure 3.13). Most individuals with this condition successfully switch gender role from female to male at puberty. Of the twenty-one individuals studied in the original reports, only two retained the female gender role, and one of these was known to frequent a local female prostitute. Because these individuals were exposed to testosterone during development and puberty, the authors of the study argued that hormones are more powerful than rearing in determining gender role.

Before this interpretation of the supremacy of hormonal factors over rearing conditions can be accepted, however, the process of change and the particular developmental and cultural backgrounds of these individuals must be carefully examined. In the Dominican Republic, these individuals have been given the colorful Spanish name *guevedoces*, which literally translates into "eggs at 12." In Papua New Guinea, the frequency of individuals with 5α-reductase deficiency is also relatively high, and this population has also been the subject of study (Herdt and Davidson, 1988). Individuals in the New Guinea study population with 5α-reductase deficiency also apparently undergo successful change in gender identity and develop a heterosexual male sexual orientation after puberty. The pidgin nickname for these individuals is "turnim-man" (Herdt and Davidson, 1988). The assignment of a special nickname for this syndrome in both cultures suggests that the nature of 5α-reductase deficiency is well known, and it is possible that rearing was not unambiguously female for these individuals. Furthermore, these individuals may have been motivated to switch gender roles because of the greater prestige and power afforded to males in both societies.

As we have seen, early androgen exposure seems to result in male-like patterns of play. These findings occur cross-culturally and among different species (Goy and Phoenix, 1972; Meany, 1988; Hines, 1982; Pellis et al., 1997). However, virtually all other aspects of gender role agree with the sex of rearing. For example, boys (Meyer-Bahlburg et al., 1977) and girls (Ehrhardt et al., 1977) that were exposed in utero to MPA, an artificial progestin with androgenic effects, were completely satisfied with the male and female gender role, respectively.

Gender Identity

The process by which individuals come to view themselves as either male or female reflects the development of **gender identity**. Gender identity is generally considered by clinicians to be irreversibly established by 2 years of age (Green, 1987; Money, 1988; Money and Ehrhardt, 1972; Wisniewski et al., 2001; but see Diamond, 1996; Diamond and Sigmundson, 1997 for opposing views) (see also Box 4.2). To what extent is gender identity the result of parental influences and cultural expectations, and to what extent is it the result of perinatal hormonal influences?

Until recently, it was generally accepted that the sex of rearing is the dominant feature mediating gender identity; that is, that the gender identity of children reflects the sex of rearing regardless of early endocrine environment. For example, girls born with CAH often display clitoral enlargement to the extent that visual assignment of sex at birth is difficult. However, if the problem is corrected surgically and there is no ambiguity about assignment of sex on the part of the parents, these girls generally develop normal female gender identity (Ehrhardt and Meyer-Bahlburg, 1981).

A potential problem for the notion that gender identity is primarily the result of parental influences is raised by the studies of *guevedoces* in the Dominican Republic (Imperato-McGinley et al., 1974, 1979). As described above, these genetic male individuals were reared as girls and had established a female gender identity, but changed their gender identity from female to male when their bodies became masculinized at puberty (Imperato-McGinley et al., 1974, 1979). Change of gender identity was observed even among a few individuals who had married men in their early teens, and then remarried women after puberty. These observations suggest that biological factors—namely, the presence or absence of androgens—mediate gender identity regardless of the sex of rearing. However, this conclusion must be tempered by the same two considerations that pertain to using these individuals to draw conclusions about gender role. First, as noted above, it is not at all clear that the sex of rearing was unambiguously female (Rubin et al., 1981). Because the 5α-reductase deficiency syndrome was so well known in this village, parents may have known that an unusual metamorphosis was possible at puberty, and they may have treated the affected individuals differently from genetic females, preparing them for a gender identity change. Furthermore, the society in this region of the world is strictly sexually segregated, and opportunities for work, education, and so forth are limited for women. Consequently, these individuals may have been motivated by social factors to change their gender identity from female to male. In other cases, changing gender identity is a long and difficult psychological process (Green, 1987). Thus, the extent to which gender identity is mediated by early hormone exposure or parental influences remains uncertain.

In an effort to control for degree of masculinization at birth, two studies were conducted in which chromosomally male infants born with ambiguous genitalia were matched for genital phenotype. Some were raised male and some were raised female; these choices were made according to who the attending physician was, the religious/cultural background of family, and so forth. The children in the first study had congenital micropenis, the other study was of boys with truly ambiguous genitalia resulting from a variety of causes. Gender identity was consistent with sex of rearing in most cases, and in the relatively rare cases in which dissatisfaction with sex of rearing occurred, it occurred as often in those raised male as in those raised female (Wisniewski et al., 2001; Migeon et al., 2002).

Sexual Orientation/Sexual Preference

In most cases, there is a marked sex difference in erotic attraction among humans: males are generally attracted to females and females are generally attracted to males. But this is not always true. The process of developing an erotic preference for same-sex partners is as complex as the process of developing an erotic preference for opposite-sex partners (Gorman, 1994). Very little is known about how heterosexuals form sexual attractions, and even less is understood about how homosexuals form sexual attractions. This section has a dual title because the terms used in discussions of this issue connote subtle beliefs about causality. The term **sexual orientation** suggests that homosexuality is the result of biological factors. The term **sexual preference** indicates to some that homosexuality is a lifestyle, a choice of same-sex erotic partners that is learned. The two terms will be used interchangeably here because, as with heterosexuality, both environmental and biological factors are likely to be involved in the development of homosexuality. This statement is based on the observation that monozygotic twins (with identical genes) do not always display the same sexual orientation (for example, see Bailey and Pillard, 1991; Buhrich et al., 1991).

Because erotic attractions typically begin at puberty, sex steroid hormones have been assumed to be involved in their development (McClintock and Herdt, 1996). One common, but largely unsubstantiated, hypothesis about the cause of homosexuality is that the hormone concentrations of homosexuals are different from those of heterosexuals; that is, heterosexual men and homosexual women are attracted to women because of high androgen concentrations (or low estrogen concentrations) and heterosexual women and homosexual men are attracted to men because of low androgen concentrations (or high estrogen concentrations). However, many studies have failed to find any consistent evidence that adult homosexual men or women differ from their heterosexual counterparts in blood concentrations of androgens or estrogens. Some studies have reported lower blood androgen values in homosexual than in heterosexual men, other studies have reported no differences, and a few studies have indicated that homosexual men have higher blood androgen levels than heterosexual men (reviewed in Bancroft, 1984; Meyer-Bahlburg, 1984). Similarly, there has been no consistent indication that hormonal imbalances underlie *transsexuality*, the feeling in an anatomically typical individual that he or she is actually a member of the other sex (Gooren, 1990a).

Some studies have suggested a difference in adult neuroendocrine regulatory mechanisms between heterosexual and homosexual men. For example, the positive feedback response to estrogen treatment was reported to be greater in homosexual than in heterosexual men (Gladue et al., 1984). As we saw in Chapter 3, low doses of estrogens usually have a negative feedback effect on the secretion of gonadotropins in both sexes (Lacroix et al., 1979; Tsai and Yen, 1971). However, treatment with high doses of estrogens evokes a surge of LH and FSH release in women after the initial suppression (a positive feedback response) (Tsai and Yen, 1971). These studies generated substantial interest because the results suggested that the brains, or at least the control mechanisms of gonadotropin secretion, of homosexual men are feminized relative to those of heterosexual men. These results seemed to be in accord with the previous studies of feminization and demasculinization of the neuroendocrine regulatory mechanisms associated with ovulation in rats (see Chapter 3). However, the sex difference in steroid effects on gonadotropin release is obvious in rats, but not in rhesus monkeys (for example, see Karsch et al., 1973). Male rats never show positive feedback effects of estrogen treatment unless they are castrated at birth and do not receive steroid hormone replacement treatment. Other studies of humans have noted a positive feedback effect in men under various experimental conditions (Barbarino et al., 1983; Kastin et al., 1972b; Kuhn and Reiter, 1976). Several subsequent experiments have failed to replicate the original finding that homosexual and heterosexual men differ in their neuroendocrine responses to estrogen treatment (for example, Gooren, 1986; Gooren et al., 1984; Hendricks et al., 1989). Thus, it appears that homosexual and heterosexual men cannot be reliably discriminated by their gonadotropin response to estrogen treatment.

If hormones are implicated in mediating human sexual preference, then exposure to hormones early in development may be involved. Again, it is not possible to manipulate early hormonal environments, or even to obtain blood samples from a human fetus without substantial risk to its well-being, and animal models are generally of limited use in understanding sexual orientation (Adkins-Regan, 1988) (Box 4.3). Consequently, experiments of nature must be examined to see whether changes in sexual orientation arise after unusual hormonal exposures. Studies of this type show that individual sexual preference does not necessarily reflect early endocrine events. For example, some XY individuals exhibit partial androgen insensitivity and are born with ambiguous genitalia. If they are surgically corrected and reared as girls, then these individuals are generally sexually attracted to men; if they are surgically corrected and reared as boys, then these individuals are generally sexually attracted to women (Money and Ogunro, 1974; Migeon et al., 2002). Similarly, if XX individuals with CAH are surgically corrected and reared as boys, they develop a male-typical sexual orientation; females with CAH who are surgically corrected and reared as girls typically develop a sexual attraction to men (Money and Dalery, 1976; but see below).

An example of hormones apparently reigning supreme in affecting sexual preference is observed among XY individuals with 5α-reductase deficiency. Despite being reared as girls, and in some cases married to men at an early age, most of these individuals switch gender roles and gender identities after puberty (Imperato-McGinley et al., 1974, 1979). Furthermore, the vast majority of individuals in this population display an erotic orientation toward women. Again, the sex of rearing may not have been unambiguously female (Rubin et al., 1981).

More recent studies have reported a slightly higher than typical rate of bisexual and homosexual fantasy, or less sexual activity with a partner (Wisniewski et al., 2004), but again no significant differences in the incidence of bisexual or homosexual behaviors, among women with CAH whose genitalia were masculinized in utero by adrenal androgens (Money et al., 1984; Mulaikal et al., 1987; but see Migeon and Wisniewski, 1998). DES, a synthetic estrogen used to avoid miscarriages, has been reported to cause significant masculinizing and defeminizing effects in nonhuman animals (Hines, 1991), but the extent of behavioral effects of DES in humans remains somewhat controversial. For example, prenatal exposure to DES in humans may slightly increase the incidence of bisexual and homosexual activities to about 25% from a 15%–20% incidence in nonexposed women. Despite this finding, it is important to note that the majority (about 75%) of the women in this study who were exposed to an unusual prenatal endocrine milieu reported exclusively heterosexual behavior (Ehrhardt et al., 1985). Taken together, these studies indicate that prenatal sex steroid hormones may not play a primary role in sexual orientation. However, the incidence of homosexuality and bisexuality among the general population is not really known. The "increase in incidence" to about 25% is based on classic data that have recently been called into question (Kinsey et al., 1948). An alternative perspective would suggest that the early hormonal environment is involved in sexual orientation, but the effects of elevated endogenous adrenal androgen levels or exogenous steroid treatment may not mimic the fetal conditions of high local hormone concentrations or the specific timing necessary for producing an effect.

The differences in neural organization recently discovered among heterosexual men, heterosexual women, and homosexual men are intriguing. But, as we pointed out earlier in this chapter, the differences in brain morphology observed between homosexual men and heterosexual men could result from, rather than cause, homosexual behavior (LeVay, 1991; Breedlove, 1997). A recent study in which male rats were paired with individual ovariectomized females provides support for such a hypothesis (Breedlove, 1997). Some males were paired with ovariectomized females that were hormonally primed, and they engaged in frequent mating behavior. Other males were paired with females that did not receive hormone replacement therapy, and they never mated. The motoneurons in the spinal nucleus of the bulbocavernosus (SNB) mediate penile erection in rats (Box 4.4). Microscopic sections of SNB tissue from these rats were stained with a Nissl stain, the same histological stain used in human postmortem studies to highlight cell bodies (LeVay, 1991). Adult male rats that had engaged in sexual behavior had smaller neuronal cell bodies and nuclei in the SNB than did noncopulators (Breedlove, 1997). These results indicate that adult sexual behavior can affect neural morphology. Taken together, the evidence suggests that sexual orientation or sexual preference is the result of a complex interaction between biology and environment, but there is no strong evidence that prenatal steroid hormone exposure directly orchestrates sexual orientation. The development of homosexual erotic attraction remains as mysterious as the development of heterosexual erotic attraction (Gorman, 1994).

BOX 4.3 Hormonal Influences on Mate Choice

Although there are enormous species differences in the type and extent of sexual dimorphism observed, nearly all animal species exhibit the following sex difference: females are generally attracted to males, and males are generally attracted to females (Adkins-Regan, 1998). Despite the fact that this sex difference in sexual attraction is a critically important component of individual fitness, little is known about the mechanisms that account for it. Studies at the ultimate level of analysis have investigated species preferences and individual partner preferences, but surprisingly few data have been collected that address the problem of sexual partner preferences (Adkins-Regan, 1998). In most cases, mate preferences have been experimentally determined using a two-choice test in which one individual (the "test" individual) can choose between mates A (e.g., a male conspecific) and B (e.g., a female conspecific); usually individuals A and B are tethered in some way so that the test animal can move freely between them. The animal with which the test individual spends the most time with is assumed to be the animal it prefers as a mate. Recent evidence suggests that hormones can affect sexual preferences, but possibly via an indirect path.

The activational role of hormones in mate preference or mate choice remains ambiguous: when given a choice between a male and a female, gonadectomized test males usually lose their preference to spend time with females. Similarly, gonadectomized test females usually lose their preference to spend time with males. Hormone replacement therapy typically restores the heterosexual preference. Importantly, however, no endocrine treatment in adulthood has been shown to reverse sexual preferences. Thus, it seems that the activational effects of hormones act on sexual motivation, rather than on sexual preference.

Some interesting studies on sheep, however, have revealed that different hormone concentrations might account for differences in sexual preference (Perkins et al., 1995). Researchers at a USDA sheep breeding facility discovered that up to 16% of male sheep never mate with females during the breeding season. Approximately 6% of these males never show any sexual activity, but another 10% prefer males over females; that is, they are homosexual males. Homosexual ewes have not been identified. Homosexual rams display a level of estrogen binding in the amygdala (an almond-shaped structure in the brain that is often involved in aggression or reproductive behaviors) that is similar to that of females and much less than that of heterosexual rams. The level of aromatase in the preoptic area of the brain is also lower in homosexual

Sex Differences in Cognitive Abilities

Sex differences in human brain size have been reported for years (Crichton-Browne, 1880; Swaab and Hofman, 1984), and, as described above, specific differences between male and female humans in the size or shape of certain brain regions have also been reported. These differences in brain structure are subtle, but can be readily discerned in most cases. Male and female humans certainly behave differently, and if human behavior reflects some manifestation of central nervous system activity or organization, then differences in brain activity or structure are to be expected. But because male and female humans overlap substantially in their behavioral

than in heterosexual rams. Furthermore, homosexual rams have lower circulating testosterone concentrations than heterosexual rams (Perkins et al., 1995). Taken together, these results indicate that steroid hormone concentrations, receptor availability, and converting enzymes differ between male sheep that prefer females and those that prefer males. It is not known whether the difference in the brain is caused by behavior or vice versa. No studies have been conducted on humans to determine whether there is a similar difference in hormone binding between male heterosexual and homosexual individuals.

Recently, studies of the contribution of the organizational effects of hormones to partner preferences were conducted in zebra finches (*Taeniopygia guttata*), which are monogamous (Adkins-Regan et al., 1997; Adkins-Regan, 1998). Female zebra finches were exposed either to early treatment with estradiol (during the first 2 weeks post-hatching) or to no treatment, then housed for the next 100 days in an all-female or a mixed-sex aviary (Mansukhani et al., 1996). These females were then given testosterone in adulthood and tested for "social" partner and sexual partner preferences. Females reared in all-female groups were more likely to prefer other females in a two-choice test, regardless of their early hormone treatment; however, females that were both exposed to early estradiol treatment and reared in an all-female group not only preferred other females, but also formed pairs with females in the aviary. Perhaps females must interact with adult males to learn to choose them later in adulthood. In any case, these data suggest that, in contrast to adult hormone manipulations, steroid hormone manipulations early in life can reverse sexual partner preference.

A recent finding demonstrated the importance of the interaction between environment and genes in the sexual behavior of fruit flies (*Drosophila*) (Zhang and Odenwald, 1995). These flies were genetically manipulated so that a tryptophan/guanine transmembrane transporter gene was activated. The activation of this single gene caused male recipients to vigorously court other males (Zhang and Odenwald, 1995). Mutations that removed this gene appeared to reverse this behavior. Notably, when wild-type flies lacking the mutation were courted by homosexual males, they altered their behavior and sexual preference and actively participated in the male–male courtship behaviors (Zhang and Odenwald, 1995). Thus, in *Drosophila*, both genetic and environmental factors can be shown to affect male sexual preference.

Taken together, these results suggest that hormones interact with environmental conditions to elicit homosexual preferences. This may also be the case for humans, but again, it is difficult to conduct tightly controlled studies with humans.

repertoires (see Figure 4.17), any sex differences in the brain are expected to be subtle. Logically, as was the case for bird song, sex differences in human brain morphology should be found in regions where behaviors that are clearly sexually dimorphic are processed. Thus, an important strategy for studying human behavioral sex differences is to study behaviors that are sexually dimorphic and to map these behavioral differences onto sex differences in brain organization. The roles of the dimorphic brain regions in mediating behavior can then be assessed by examining patients that sustain damage to the designated areas.

It should be emphasized here that there is no direct evidence that the sex differences in brain morphology thus far discovered mediate human behavioral sex

BOX 4.4 The Spinal Nucleus of the Bulbocavernosus Muscle

One way to overcome the problems inherent in understanding the functional significance of neural sex differences is to study a sexually dimorphic system with a very simple function. One such mammalian system that has been studied extensively is located in the spinal cord of rats. The spinal motoneurons that innervate the muscles attached to the base of the penis are located in a particular position in the spinal cord, and collectively are referred to as the **spinal nucleus of the bulbocavernosus (SNB)** (Breedlove, 1992). These neurons control the striated bulbocavernosus and levator ani muscles that are responsible for penile erection in rodents, as well as the external anal sphincter (Figure A). The SNB is larger and has more neurons in males than in females because the SNB, and the muscles that control erectile function, are diminished in size or completely absent in adult female rats (Breedlove and Arnold, 1983a,b). Humans also have the bulbocavernosus muscle, but the sex difference is much less pronounced in humans than in rats (Forger and Breedlove, 1986). In men, the bulbocavernosus muscle has a function similar to that in rats, contracting rhythmically during erection and ejaculation. Women retain a bulbocavernosus muscle, albeit smaller and modified in form. The motoneurons of Onuf's nucleus, the human homologue of the rat SNB, are more numerous in men than in women (Forger and Breedlove, 1986).

During fetal development, female rats have bulbocavernosus and levator ani muscles similar in size to those of males, and these muscles, which connect to the base of the clitoris in females, are innervated by the SNB (Rand and Breedlove, 1987). However, these muscles atrophy and the SNB motoneurons die around the time of birth. Androgen treatment spares the muscles and sec-

Figure A

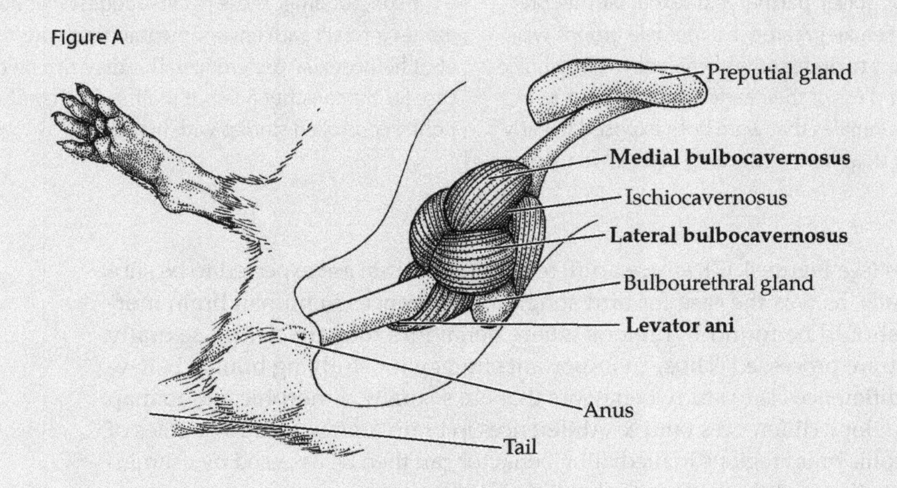

Figure B Photo courtesy of Nancy Forger.

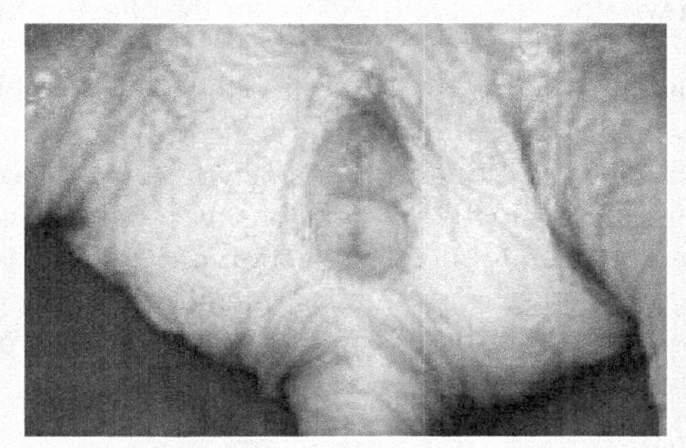

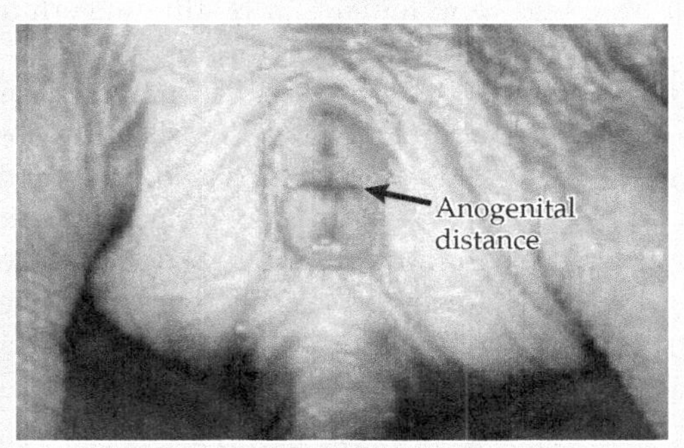

ondarily spares the SNB. Once the muscles have regressed, androgen treatment cannot restore the muscle cells, so the critical period of androgen effectiveness in sparing motoneurons in the SNB is defined by the rate of muscular atrophy. The muscle cells, but not the motoneurons of the SNB, possess androgen receptors during the time of fetal sexual differentiation (Fishman et al., 1990). Thus, androgens act on the neural tissue of the SNB indirectly by preventing apoptosis in the muscles directly. Androgens, but not estrogenic metabolites of androgens, prevent cell death. This assertion is based on the observation that genetic XY males with testicular feminization mutation, who lack functional androgen receptors but retain functional estrogen receptors as well as aromatase activity, exhibit feminine SNB and associated musculature (Breedlove and Arnold, 1981).

Environmental effects can also influence the size of the SNB. Mother rats spend more time licking and attending to the perianal regions of male than of female pups. Anosmic mothers reduced the time spent licking their male offspring, and these males exhibited a 10% reduction in the number of motoneurons in the SNB (Moore et al., 1992; Michel and Moore, 1995). This reduction in SNB size may contribute to the adult deficits in male mating behavior observed in these males. In utero stress and other factors can also feminize the SNB of male rats (Ward, 1992).

The sex difference in the SNB has been detected in all rodent species examined—except one (Peroulakis et al., 2002). Naked mole-rats (*Heterocephalus glaber*) are hystricomorph rodents found in eastern Africa. In the wild, these animals live entirely underground in colonies of 60–90 individuals. They subsist on buried tubers. Their reproductive biology is unusual for mammals because each colony has a single breeding female, called the queen, and one to three breeding males. All other individuals in the colony, called subordinates, are reproductively inactive. This social organization is common among eusocial insects such as ants, bees, and termites, but extremely rare among vertebrates. Subordinate males have very low testosterone concentrations, and they have never been observed to engage in mating behavior. Indeed, no behavioral sex differences have ever been observed among the subordinate individuals of this species (Lacey and Sherman, 1991). The external genital anatomy of subordinate male and female mole-rats is indistinguishable (Figure B), and there are no sex differences in the perineal muscles or the perineal motoneurons (Peroulakis et al., 2002).

differences. Similarly, there is no direct evidence that they do not. A further complication exists, as we saw above in the case of sexual orientation: if brain structural differences that correspond to behavioral differences do exist between males and females, it is still not certain whether they are the cause or the effect of the behavioral dimorphisms.

With these disclaimers in mind, there are several sex differences in cognitive abilities that are consistently observed cross-culturally and hence are good candidates to be mediated primarily by physiological factors; namely, sex steroid hormones. Corresponding sexual dimorphisms in the brain regions that are involved in these cognitive processes have also been found. This section presents a brief review of sex differences in perception and sensory abilities, lateralization of cognitive function, verbal skills, and visuospatial abilities among humans (reviewed in Velle, 1987).

Perception and Sensory Abilities

Sensation and perception are possibly the most basic cognitive functions. **Sensation** is the initial processing of sensory information as it enters the nervous system through the sensory receptors. **Perception** is the transduction of this sensory information into biologically meaningful information. Sex differences have been widely reported in both sensation and perception. There are many subtle and a few marked differences between males and females in sensory and perceptual abilities.

Three general aspects of sensory and perceptual function have been the subject of research: sensitivity, discrimination, and preference. To study sensitivity, experimental subjects are usually presented first with no stimuli, then stimuli are presented and increased in small incremental steps until the subject reports a sensory experience. For example, a subject may be placed in a dark room and asked to stare at a wall and push a button whenever they perceive a pulse of light. The experimenter may initially flash only one or two photons of light, but continues to provide stronger stimuli until the subject reliably reports the presence of the light. To study discrimination, sensory stimuli that are closely related are presented to the subject and are made increasingly different by the experimenter until the subject reports the presence of more than one stimulus. For example, two equimolar sucrose solutions may be provided to the subject, who should not report any difference in taste. The experimenter then covertly increases the concentration of sucrose in one of the cups very slightly until the subject reports a difference between the two solutions. Preference can be determined in a straightforward manner. For instance, a subject might be water-deprived for 24 hours, then provided with cups of saline solution of varying concentrations and asked to sample all of the cups. The experimenter simply records the volume consumed from each cup to determine the preferred concentration.

OLFACTION The olfactory sensory system (the sense of smell) transduces information about airborne chemicals into central nervous system signals. There are important sex differences in olfactory sensitivity and odor identification ability. Women are approximately 1000 times more sensitive to musklike odors (for example, pentadecanolide and oxohexadecanolide) than men. The increased sensitivity of women to these odors begins at puberty and appears to be estrogen-dependent (Koelega and Koster, 1974).

In another study, olfactory stimuli (eugenol, phenyl ethyl alcohol, or phenyl ethyl alcohol alternating with hydrogen sulfide) were delivered to the nostrils of young men and women while they were subjected to functional magnetic resonance imaging (fMRI) (Yousem et al., 1999). The women's group-averaged brain activation maps showed up to eight times more activated voxels (units of brain volume activation) than the men's for specific regions of the brain (frontal and perisylvian regions). Generally, more women than men showed activation. The functional meaning of these results remains unknown.

Olfactory information can also affect mood, and these mood effects may differ between the sexes (Jacob et al., 2001, 2002). For example, Δ4,16-androstadien-3-one, but not other musky odors such as androstenol or muscone, affects psychological state, reducing negative mood and increasing positive mood (Jacob et al., 2001, 2002). Women tended to experience an immediate increase in positive mood when in the presence of a male tester, whereas the responses of men were unaffected by the sex of the experimenter (Jacob et al., 2001). It is not apparent whether these differences are organized or activated by hormones.

For other odors, however, such as amyl acetate, a sex difference appears before puberty; girls are far superior to boys at detecting this odor at low concentrations. Periovulatory women, as well as women in early pregnancy, display enhanced olfactory sensitivity. Menstruating women and women tested late in pregnancy exhibit lowered olfactory sensitivities. Women are better than men at all ages at identifying odors, and this sex difference has been demonstrated in cross-cultural studies (Doty et al., 1984, 1985). These consistent and reliable sex differences in olfactory acuity, as well as the variation in olfactory sensitivity observed across the menstrual cycle, during pregnancy, and in individuals with clinical endocrine disorders, strongly suggest that sex steroid hormones influence olfactory sensitivity. For example, a study of women with irregular menstrual cycles showed that these women exhibit reduced olfactory sensitivity compared with women with regular menstrual cycles; fully 20% of the subjects were completely unresponsive to odors, but many of them did not recognize that they possessed a sensory impairment (Marshall and Henkin, 1971). The mechanism by which hormones affect olfactory sensitivity remains unspecified in most people.

An olfactory deficit in one clinical population indicates that olfaction and hormones may be only indirectly related in these individuals. Men with **Kallmann syndrome** possess small testes, are infertile, and are anosmic (cannot detect odors). Their olfactory deficit results from a congenital lack of olfactory bulb development during early ontogeny. Embryological studies have indicated that the hypothalamic neurons that secrete GnRH originate in the olfactory bulbs, then migrate to the hypothalamus (Figure 4.18; Schwanzel-Fukuda and Pfaff, 1989). In Kallmann syndrome, the lack of olfactory bulb development interferes with the

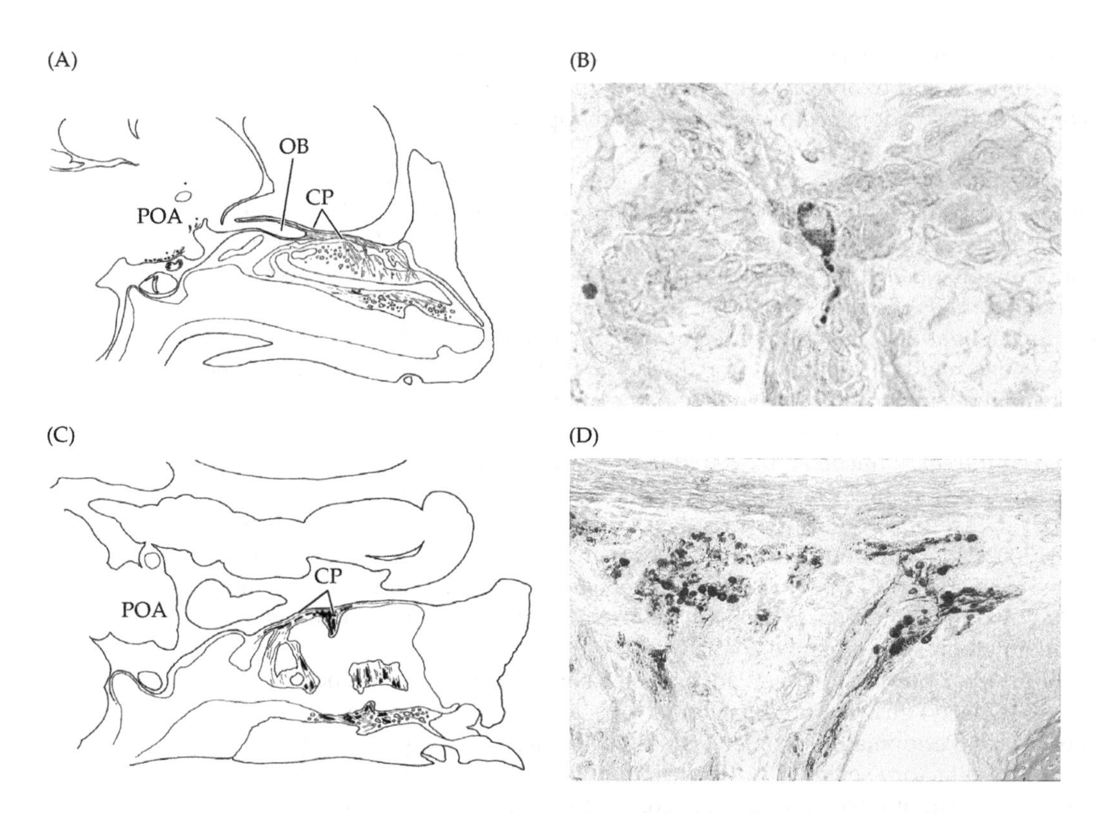

4.18 Congenital absence of the olfactory bulbs in Kallmann syndrome interferes with the migration of GnRH neurons to their proper destination in the hypothalamus. (A) Microprojection drawing of a sagittal section through the brain and nasal region of a normal 19-week-old human fetus, showing olfactory bulb (OB) development and a normal distribution of GnRH-immunoreactive cells in the nose and preoptic area (POA) of the brain (black dots). (B) Photomicrograph of the same section at higher magnification, showing a single GnRH-immunoreactive cell in the nose tissue, in a ganglion of the terminal nerve. (C) Microprojection drawing of the same section from another fetus, also 19 weeks old, with Kallmann syndrome. Note that the olfactory bulbs are absent and that no GnRH immunoreactivity is visible in the brain. GnRH-positive cells are seen as dots and heavy black lines in the nose, between the meninges, and on the dorsal surface of the cribriform plate (CP), the bony shelf on which the olfactory bulbs are normally located. (D) Photomicrograph of this brain section, again at higher magnification, showing thick clumps and clusters of GnRH-immunoreactive cells in the nose. Courtesy of Marlene Schwanzel-Fukuda.

migration of the GnRH neurons which originate in the olfactory bulb, but normally migrate to the hypothalamus, so that reproductive difficulties ensue (Schwanzel-Fukuda et al., 1989). Thus, the sensory deficit is not affected directly by hormones in this case, but rather by a congenital condition that blocks normal olfactory bulb development as well as normal endocrine function.

TASTE The gustatory sense (taste) provides information to the nervous system about chemicals in solution. On tests of taste perception, women, on average, display more sensitivity than men (Doty, 1997). Typically, a drop of a representative solution is placed on the four areas of the tongue that have specific receptors for sweet, sour, salty, and bitter solutions. Solutions commonly used include sucrose, citric acid, sodium chloride, and quinine sulfate (Fikentscher et al., 1977). Women are superior to men in naming tastes and in discriminating tastes, especially bitter tastes (Doty, 1978). The physiological bases for these sex differences in the sensation and perception of taste in humans are unknown. Differential sex steroid hormone concentrations are suspected to be involved because the sex differences in taste perception arise after puberty, are exaggerated during pregnancy and the follicular phase of the menstrual cycle, and are diminished somewhat after menopause (Doty, 1978; Fleming, 1988; Fleming and Pliner, 1983; Kuga et al., 2002).

Sex differences in taste preferences have also been reported in many species of laboratory animals. In the case of rats, the effects of hormones in mediating this sex difference are well known (Krecek, 1973; Valenstein et al., 1967; Wade, 1972; Zucker, 1969). Females display estrogen-dependent preferences for sweet tastes (for example, sugar or saccharin), organized by the lack of perinatal androgens and activated by estrogens in adulthood.

AUDITION Sound pressure changes are perceived by the auditory sensory system (hearing). There is ample evidence that women are more sensitive than men to sound. Women hear pure tones at lower thresholds than men; this is particularly true at higher frequencies and occurs at all ages (Corso, 1959). Similarly, tolerance for white noise (noise composed of all sound frequencies; an example of white noise is the hiss of an FM radio not tuned to a station) is significantly lower in females than in males. In children ranging in age from 5 to 12, the average noise tolerance level was about 82 decibels (dB) for boys and 73 dB for girls (Elliot, 1971); among college students ranging in age from 18 to 26, the average noise tolerance level was about 83 dB for men and 76 dB for women (McGuinness, 1972). (Remember that decibels are measured on a log scale, so a difference of 10 dB represents a doubling of loudness.)

In one interesting study of babies 12–14 weeks of age, the infants were trained by operant conditioning (rewarded for the appropriate behavior) to maintain visual fixation on a white circle (Watson, 1969). Both sexes could perform the task equally well under the right circumstances. In addition to demonstrating that human babies were capable of learning at a much earlier age than previously believed, this study also demonstrated a sex difference in reward contingencies. Girls could maintain visual orientation when a low-frequency tone was provided as the reward, but not when a visual stimulus was presented as the reinforcement. Baby boys performed the task better when the reward was visual. This study suggests that at a very young age, females react to auditory information preferentially, whereas males respond better to visual information. One might conclude from this study that physiological factors must mediate these effects in such young children; however, this conclusion must be tempered by the observation that parents interact with baby girls differently from baby boys, and that this difference in interaction begins as soon as the sex of the child is known. Whatever the cause of the reinforcement preference, this auditory propensity in girls may be related to the increased verbal abilities displayed by girls relative to boys (see below).

In addition to hearing sounds, the ear also produces faint, echolike noises generated by the cochlea of the inner ear, called click-evoked otoacoustic emissions (EOAEs) (McFadden, 2002). These emissions are present from birth and remain relatively stable throughout life (in the absence of ear damage). They can be measured by inserting a very small speaker/microphone into the outer ear, sending a series of clicks into the ear, and recording the responses. Women generally have louder EOAEs than men, a sex difference that is present from birth, suggesting that this phenomenon is organized by early hormone exposure. In support of this notion, earlier studies reported that women with twin brothers have more male-like "quiet ears" (McFadden and Loehlin, 1995), presumably in response to exposure to prenatal androgens. A recent study reported that homosexual and bisexual women produce weaker EOAEs than heterosexual women (McFadden and Pasanen, 1998). EOAEs did not differ among male homosexual, heterosexual, and bisexual men. Importantly, there are few data suggesting that androgens from a male twin affect the brain and behavior of a female twin, or that there is an increase in the prevalence of homosexuality among females with twin brothers. Furthermore, because there was significant overlap in the EOAEs between heterosexual and homosexual/bisexual women, it is not possible to predict homosexual behavior simply by measuring female EOAEs. However, oral contraceptives appear to masculinize EOAEs in women (McFadden, 2000). EOAEs may be useful in guiding the course of treatment for children born with ambiguous genitalia. As noted previously, treatment is often directed by the degree of masculinization of the external genitalia, which may or may not correspond to the extent of masculinization of the central nervous system. It may be possible to assess the degree of brain masculinization at a very early age with little risk or expense because EOAEs are already routinely measured to assess hearing in infants (McFadden, 1999).

VISION The visual system transduces light energy into the electrochemical signals of the nervous system. As we have just seen, baby boys find visual information more rewarding than auditory information. With few exceptions, visual acuity is better in men than in women; that is, men see better than women. Sex differences also exist in tolerance of light intensity; on average, women tolerate higher levels of light intensity than men. Conversely, when subjects with good visual acuity are tested, women undergo dark adaptation more quickly than men (McGuinness, 1976). Visual perception, especially in relation to visuospatial tasks, is markedly better among males than females at all ages.

In rats, males have about 20% more neurons than females in the primary visual cortex (Nunez et al., 2000, 2001). In one study, females were implanted with DHT or estradiol capsules on postnatal day 1. Females exposed to DHT showed the male pattern of developmental cell death in this brain region, whereas females exposed to early estrogen treatment were indistinguishable from normal females in apoptosis (Nunez et al., 2000). These results indicate that perinatal androgens inhibit cell death in the primary visual cortex of rats.

Lateralization of Cognitive Function

The two cerebral hemispheres of the brain are somewhat specialized for processing different types of cognitive tasks (Wisniewski, 1998). In right-handed individuals, generally the right side of the brain specializes in spatial processing, whereas the left hemisphere is best at processing verbal information (Hines, 1982; Kimura and Harshman, 1984). This tendency for cognitive skills to be concentrated in one hemisphere is called cerebral specialization or **lateralization**. Males and females differ in the extent to which cognitive function exhibits cerebral lateralization: males tend to be more lateralized than females. That is, different cognitive functions tend to be more confined to separate hemispheres in males, whereas female cognitive function is generally more evenly distributed between both hemispheres.

Cerebral lateralization can be measured in several ways. Researchers take advantage of the fact that auditory information enters one ear and visual information enters one visual field, but then is processed by the contralateral hemisphere. For instance, auditory signals can be presented via headphones to one ear or the other. Commonly, a brief tone or a word is embedded in white noise, and the subject must indicate as quickly as possible when the signal is present. A variation of this technique involves presenting two signals (words) simultaneously to the two ears; again, the subject must react as quickly as possible. The discrepancy in response speed or accuracy between the two ears reflects the degree of lateralization. A similar type of technique using a special apparatus called a tachistoscope to present visual stimuli to only one visual field has been employed to assess lateralization of visual processing.

People typically can detect sounds better with the right ear; that is, auditory processing is better in the left than in the right hemisphere. This facility has been referred to as the right-ear advantage. Perhaps not surprisingly, the processing of language and speech for most people occurs predominantly in the left hemisphere. However, men are much slower, and make more mistakes, in responding to auditory information arriving at the left ear than women, and women show less discrepancy than men in both response speed and response accuracy between the two ears and between the left and right visual fields. In other words, the brains of women are less lateralized than those of men for auditory and visual information processing. Additional support for this view comes from the clinical literature on verbal deficits caused by cerebral hemorrhages (strokes) (McGlone, 1980). In general, men who have suffered a stroke on the left side of the brain exhibit verbal deficits, but women with cerebral vascular damage on the left side may or may not exhibit a verbal deficit. If men suffer a stroke on the right side of the brain, their visuospatial abilities are affected. In many cases, women who suffer strokes on the right side do not exhibit any obvious decrement in cognitive or behavioral abilities. Thus, the masculine pattern of cognitive processing involves distinct lateralization of function, whereas the feminine pattern exhibits more equal distribution of cognitive function between the two hemispheres.

Prenatal steroid hormones can affect lateralization patterns. Women with Turner syndrome do not secrete steroid hormones prenatally and do not exhibit lateralization of auditory information (Gordon and Galatzer, 1980; Netley, 1977, 1983). These women may have other atypical features, however, such as retarded mental development, which may affect cognitive processing tasks independently of hormones. In contrast, women exposed to DES in utero displayed the male-typical pattern of increased lateralization of auditory and visual information processing as compared with their unexposed sisters (Hines, 1991).

Sex differences have also been reported in the lateralization of human brain morphology. The left hemisphere is slightly larger and weighs more than the right hemisphere of the human brain, but the weight difference between the two hemispheres is less pronounced in females than in males. The two hemispheres of the brain communicate with each other through several fiber pathways; one of these links is the corpus callosum. Sex differences in the size and shape of the corpus callosum, detected by magnetic resonance imaging and other techniques, suggest a neuroanatomical correlate that could account for the sex differences in lateralization of cognitive function (Figure 4.19), but there are many conflicting reports that fail to find a morphological sex difference in the corpus callosum (Allen et al., 1991; Crichton-Browne, 1880; deLacoste-Utamsing and Holloway, 1982; deLacoste et al., 1986; Wada, 1976).

Verbal Skills

There is no sex difference in scores on modern standardized tests of intelligence (Hines, 1982). That is, boys and girls do not differ predictably in IQ. However, some subsets of the skills and abilities measured on intelligence tests show consistent, though small, sex differences in performance. On average, females excel at verbal tasks, perceptual skills, fine motor skills, and mathematical calculations, whereas males outperform females on targeted directed motor skills (e.g., guiding or interception of a projectile), quantitative tasks, and visuospatial abilities, including map reading, sense of direction, and mathematical reasoning (Bryant, 1982; Gladue et al., 1990; Hines, 1982; Kimura, 1992; Velle, 1987) (Figure 4.20).

Females generally display better language comprehension, faster language acquisition (both first and additional languages), and better spelling, verbal fluency, and grammar skills than males (Hines, 1991; Netley, 1983). Many of these sex differences in verbal performance arise after puberty, which makes them likely candidates for mediation by hormonal effects (Maccoby and Jacklin, 1974); however, social and other environmental factors cannot be ruled out. No consistent effect of prenatal endocrine dysfunctions on subsequent verbal abilities has been demonstrated (reviewed in Hines, 1982).

Several studies have suggested a neuroanatomical basis for these sex differences in performance. In most people, the planum temporale, a flattened area of the temporal lobe, is larger in the left cerebral hemisphere than in the right. Based on earlier case studies of individuals with brain damage restricted to the planum

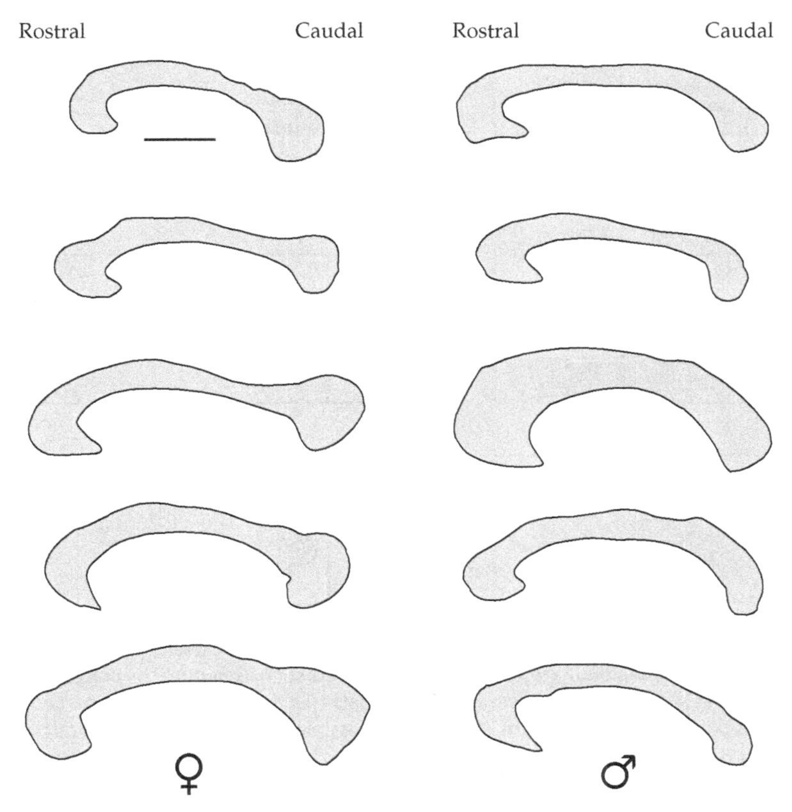

4.19 A possible sex difference in the corpus callosum is shown in these drawings based on midline cross sections of human brains. In females, the caudal portion of the corpus callosum is more bulbous than in males. It has been suggested that some sex differences in the lateralization of cognitive function might be accounted for by differential numbers of axons in the caudal corpus callosum (splenium). Because this section of the corpus callosum connects the occipital lobes, which process visual information, sex differences in visual information processing have also been attributed to sexual dimorphism in this structure. These attributions are controversial. (Scale bar = 2 cm.) After deLacoste-Utamsing and Holloway, 1982.

temporale, it has been concluded that this brain region is involved in speech. The asymmetry between the right and left planum temporale is less marked in women than in men (Wada et al., 1975). In other words, its size is less lateralized in females. In another study, the volume of the superior temporal cortex and the cortical volume fraction of Broca's area, which is also involved in speech, was about 20% larger in female than in male brains (Harasty et al., 1997), a result that corresponds well with the findings on sex differences in verbal function.

With the advent of new imaging tools such as functional magnetic resonance imaging (fMRI), it is now possible to view the living, awake brain and monitor specific brain region activity during the performance of a particular task. One

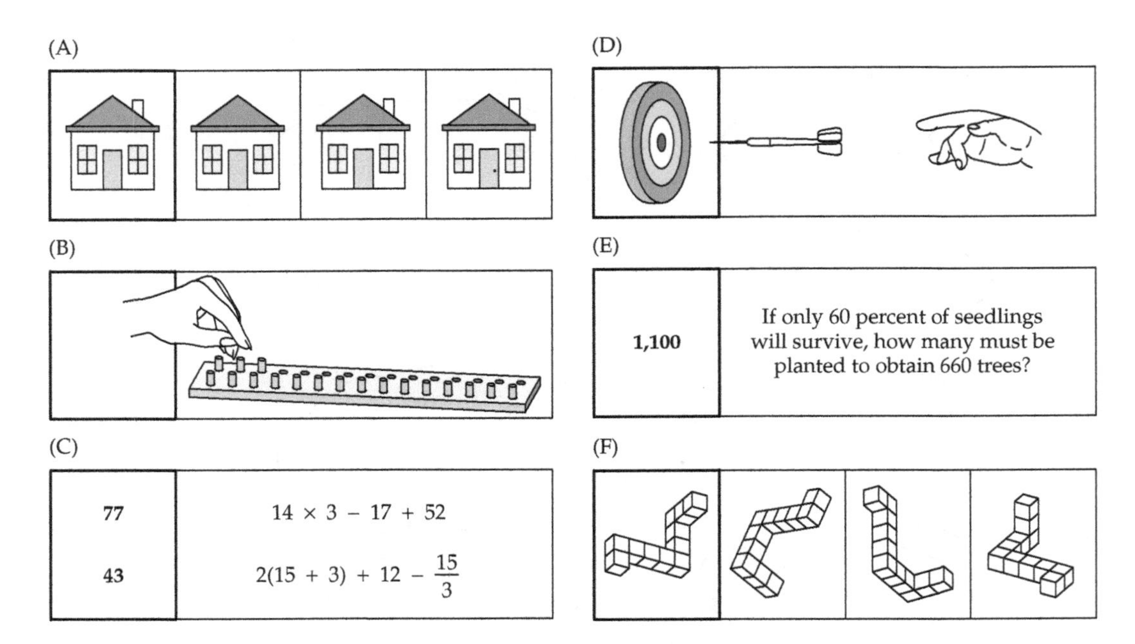

4.20 Performance on certain tasks favor one sex over the other. Specifically, women tend to outperform men on (A) tests of perceptual speed, such as finding the match to the house on the far left as quickly as possible; (B) manual tasks that require fine motor coordination, such as placing pegs on a board as quickly as possible; and (C) mathematical calculations. Men, on the other hand, are more accurate than women in (D) target-directed motor skills such as catching a ball or throwing a dart, regardless of previous sports experience. Men also outperform women in (E) mathematical reasoning and in (F) mentally rotating an object. After Kimura, 1992.

recent study using fMRI revealed that men and women use different parts of their brains to solve similar phonological tasks. Nineteen males and nineteen females were asked to report whether or not two nonsense words rhymed. Among males, brain activation during this rhyming task was concentrated in the left inferior frontal gyrus region. In contrast, both the left and right inferior frontal gyri were activated in females performing the rhyming task, and the pattern of activation was more diffuse than for males (Shaywitz et al., 1995) (Figure 4.21). These striking data provide evidence that the brains of men and women operate differently when processing certain features of language.

Mathematical Reasoning and Visuospatial Abilities

Sex differences in performance on mathematical reasoning tasks have been consistently observed within the general population for decades. In general, males outperform females on tests of mathematical reasoning ability, such as the mathematics section of the Scholastic Aptitude Test (SAT-M) (Benbow and Stanley, 1983; Gouchie and Kimura, 1991). This sex difference increases dramatically at

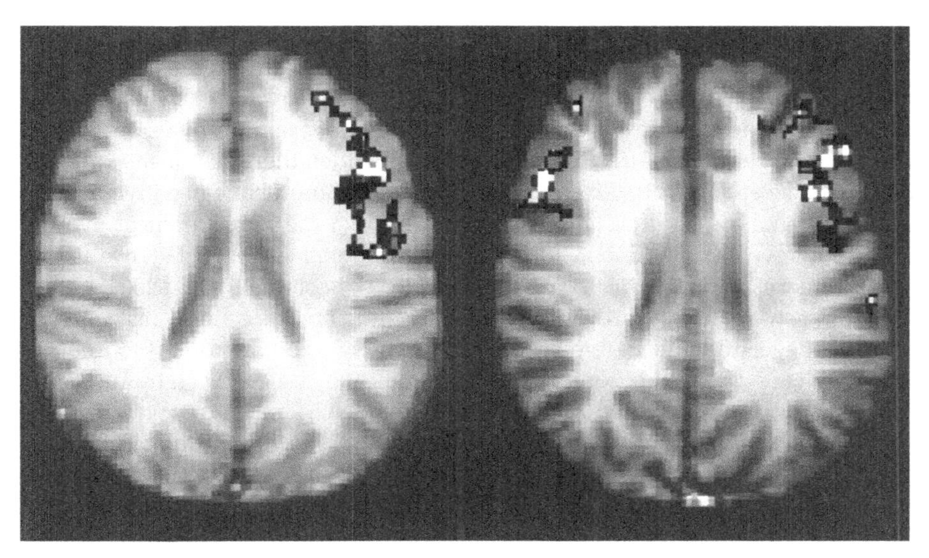

4.21 Men and women use different parts of their brains to perform similar language tasks. Among males (left), functional MRI (fMRI) reveals that brain activation during rhyming tasks is restricted to the left inferior frontal gyrus. In females (right), both the left and right inferior frontal gyri are activated, and the pattern of activation is more diffuse. From Shaywitz et al., 1995; courtesy of NMR Research/Yale Medical School.

the high end of performance: the ratio of seventh-grade boys to girls receiving SAT-M scores of 500 or above is 2.1:1; for scores of 600 or above, the ratio is 4.1:1; and for scores of 700 or above, the ratio is 12.9:1 (Benbow and Stanley, 1983). The activational effects of hormones are not likely to mediate this sex difference because it is observed prior to the onset of puberty (Wisniewski, 1998).

On average, males perform visuospatial tasks better than females (McGee, 1979). Visuospatial tasks include mental rotation of objects, map reading, mental visualization of relationships among objects in space, solving maze problems, and shape recognition (see Figure 4.20) (Hines, 1991). The difference in visuospatial abilities between boys and girls is small or absent prior to puberty, but present after puberty, suggesting that hormonal influences could be involved.

Visualization of space, objects, or relationships among objects in space is a common test of visuospatial abilities. For example, an individual may be asked to go through the alphabet mentally as fast as possible and count the number of letters containing the sound "ee," including the letter e (Coltheart et al., 1975). This task is a test of verbal ability, and females are faster than males. In a visuospatial test, a person is asked to go through the alphabet mentally as quickly as possible and count all of the uppercase letters with a curve in their form (Coltheart et al., 1975). Males perform better than females at this task. However, in a reaction-time fMRI study of a visuomotor response task (such as tracking), no sex differences in brain volume activation were reported in the right visual, left visual, left primary motor, left supplementary motor, and left anterior cingulate areas. The authors concluded, based on these results, that sex seems to have little influence on brain activation when performance on a simple reaction-time task is compared between men and women (Mikhelashvili-Browner et al., 2003). However, if visual processing in an emotional context is assessed with fMRI—for example, by using pictures from the International Affective Picture System—then there are clear sex differences (Wrase et al., 2003). Given that many visual stimuli have negative or positive emotional content, it seems prudent to consider the sex differences in brain activation revealed by these fMRI studies of emotional content when evaluating sex differences in brain function (Wrase et al., 2003; Sabatinelli et al., 2004).

Similarly, males perform better at map reading and directional tasks. In one study, college students were placed in a windowless room and told to imagine that they were facing the front of a familiar building, the college administration building, for instance. With this imaginary orientation in space, they were requested to point to other familiar landmarks, such as the stadium or the clock tower. When the degrees of error in the students' estimated orientation reports were computed, it was found that men made fewer errors in estimation of direction than women (Bryant, 1982, 1991). Of course, college men may have more experience in navigating through space than college women. Boys may have been allowed to range farther from home than girls as they were growing up, or allowed to drive more frequently than girls. Alternatively, this sex difference may be due to activational effects of hormones in adulthood.

In a recent study, some postmenopausal women were treated with estradiol, and others with methylated testosterone, for 3 months. The women receiving testosterone supplements performed better than the estrogen-treated women in a map memory test (Wisniewski et al., 2002). In another study of navigation, men and women were asked to navigate a virtual maze while undergoing fMRI (Grön et al., 2000) (Figure 4.22). Performing this task activated the medial occipital gyri, lateral and medial parietal regions, right hippocampus, and posterior cingulated and parahippocampal gyri in both sexes. In addition to the common brain regions activated, males also activated the left hippocampus, whereas females also activated the right parietal and right prefrontal cortex (Grön et al., 2000; see Figure 1.13).

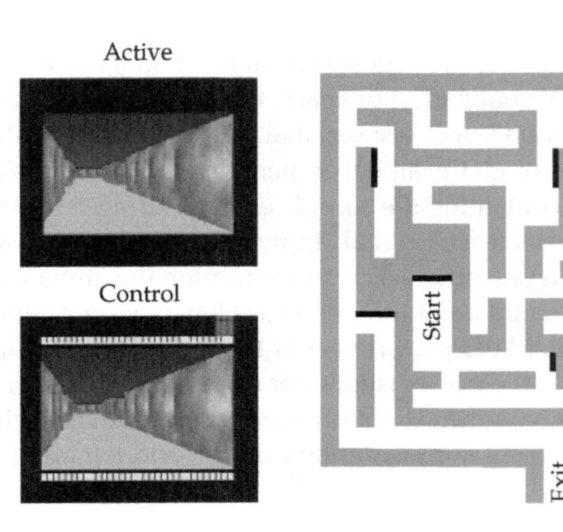

4.22 Visuospatial navigation can be studied using fMRI and a virtual maze. Because fMRI requires subjects to lie very still, researchers developed a virtual maze on a computer screen that people could navigate while remaining in the fMRI device From Grön et al., 2000.

ANIMAL MODELS OF VISUOSPATIAL SKILLS Because it is so difficult to account for different experiential effects in humans, animal models, in which environmental experiences can be somewhat equalized, have been employed to understand sex differences in visuospatial abilities. There is no perfect animal model for human visuospatial skills. However, animals can be taught to run mazes, a task that incorporates visuospatial (or perhaps olfactory–spatial) skills. A consistent sex difference in maze-running ability is observed among many rodent species (reviewed in Gaulin and FitzGerald, 1986; Haaren et al., 1990): males learn mazes faster and make fewer errors than females (see Chapter 12; McNemar and Stone, 1932). This sex difference in maze learning is thought to be mediated by hormonal effects on the hippocampus, a structure in the brain that is important for spatial learning (McEwen, 2001). Recall from Chapter 3 that male meadow voles learn mazes better than female conspecifics, whereas there is no difference in maze-learning abilities between male and female pine voles (Gaulin and FitzGerald, 1986). The hippocampus of male meadow voles is approximately 10% larger than the hippocampus of females, but hippocampal size does not differ between male and female pine voles (Jacobs et al., 1990). Whether hippocampal size drives the difference in behavior or whether sex-related differences in experience in spatial habitats drive hippocampal size has recently been investigated; the results will be discussed in Chapter 12.

CLINICAL MODELS OF VISUOSPATIAL SKILLS Sex differences in human cognitive behavior have been documented, but how do we gather evidence to determine the contribution of hormones to these differences in human behavior? Indirect evidence from people with atypical endocrine function suggests that early hormone exposure influences visuospatial abilities among humans. Individuals with Turner syndrome display a marked deficit in visuospatial abilities (Hines, 1982; Netley, 1983). Because these individuals possess dysgenic ovaries and produce no steroid hormones, these results are consistent with the idea that early exposure to androgens (or androgen metabolites) promotes subsequent visuospatial abilities. In some studies of females with CAH, who experience high levels of early androgen exposure, there have been reports of enhanced visuospatial abilities, but these reports have not consistently shown a clear trend (Hines, 1991; Hines et al., 2003) (Box 4.5). Similarly, exogenous prenatal hormone exposure does not consistently affect visuospatial abilities. Except for one brief report (Wada, 1976), there have not been reports of sex differences in brain regions that process visuospatial information corresponding to those that have been found in brain regions that process verbal information (but see Wrase et al., 2003; Sabatinelli et al., 2004).

The organizational/activational hypothesis suggests that the human central nervous system is ordered in a male or female manner early in prenatal development by the presence or absence, respectively, of gonadal steroid hormones, and that behavioral sex differences arise when these prenatally organized neural circuits are activated by sex steroids secreted at puberty. Thus, sex differences in behavior reflect differential prenatal exposure to steroid hormones. If the time frame of the organizational/activational hypothesis is expanded, however, an alternative view emerges. In a very interesting study, individuals that underwent puberty early were compared with individuals that underwent puberty late

BOX 4.5 Hormones, Sex Differences, and Art

In addition to the sex difference in rough-and-tumble play, boys and girls differ in other aspects of their play behavior. For example, boys tend to play with vehicles and building blocks, whereas females tend to favor dolls. Free-style drawings done by boys and by girls also differ on a number of parameters, including color and motifs. Girls tend to draw flowers and people in "warm" colors, whereas boys tend to draw vehicles and other moving objects and use "cold" colors (Iijima et al., 2001) (see the graph and the table). In addition, girls tend to line up their motifs at ground level, whereas boys tend to draw from an overhead (bird's-eye) perspective and to pile objects on top of one another. Do boys and girls perceive the world differently? Or do they express their similar perceptions from different perspectives?

Motifs in Children's Drawings

Motif	Boy (%)	Girl (%)
Moving objects (vehicle, train, aircraft, etc.)	92.4***	4.6
Person	26.5	96.6***
Flower	7.2	57.0***
Butterfly	3.2	23.4***
Sun	50.8	76.5***
Mountain	14.5**	3.1
House and building	17.7	33.5**
Tree	9.6	23.4*
Ground	42.7	57.8*
Cloud	25.0	32.8
Sky	41.9	49.2

*** $P < 0.001$
** $P < 0.005$
* $P < 0.05$

The two pictures on the left in the group of children's drawings were drawn by a 6-year-old boy (A) and a 5-year-old boy (B). Note the vehicular motifs and the overhead view (A) and the piling of objects (B). The two drawings in the center were drawn by a 5-year-old girl (C) and a 6-year-old girl (D). Note that the motifs are lined up at ground level. The two drawings on the right were drawn by a 5-year-old girl (E) and a 7-year-old girl (F) with CAH. Note the vehicular motifs and the piling of

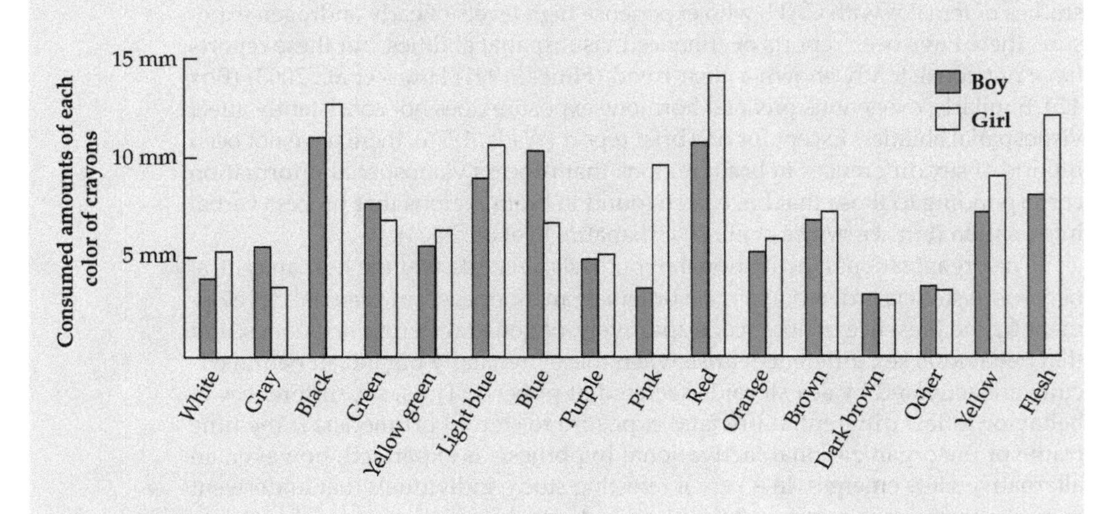

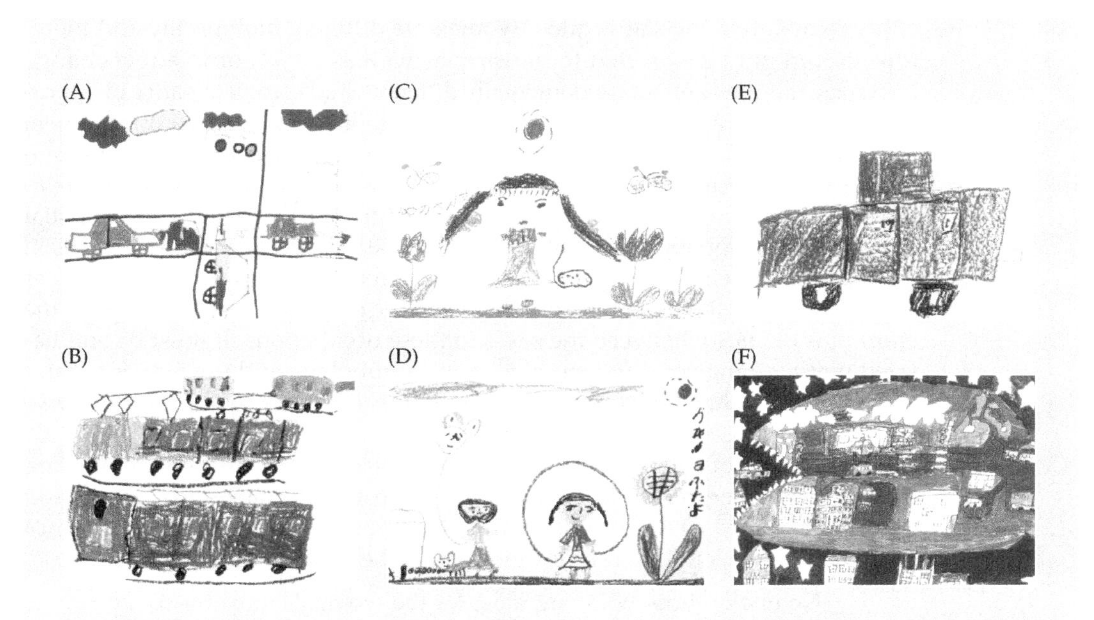

Photographs courtesy of Dr. Yasumasa Arai

objects. When a number of these aspects of children's drawings were calculated, the "feminine index" of pictures drawn by CAH girls was much lower than that of pictures drawn by unaffected girls, whereas the "masculine index" was much higher than that for unaffected girls and comparable to that of unaffected boys (Iijima et al., 2001). All of the CAH girls in this study had plastic surgery to make their external genitalia female-typical, had a female gender identity, and were being reared as girls. However, the CAH females appeared to display bisexual traits in their drawings.

(Waber, 1976). Regardless of sex, early-maturing adolescents performed better than late-maturing adolescents on tests of verbal abilities, and late-maturing individuals outperformed early-maturing individuals on tests of spatial abilities. Because females typically undergo puberty at an earlier age than males, the differences in cognitive function between males and females could reflect differential brain organization mediated by peripubertal sex steroid hormones coincident with differential rates of brain maturation (Waber, 1976).

Conclusions

A final note about sex differences in human cognitive performance. This area of research is highly politicized, both within and outside the scientific community. Some people see the discovery of sex differences in brain function, organization, or morphology as evidence of biological differences between the sexes in ability. Others view sex differences in the human brain as reflecting a social history of differential treatment of the two sexes, and thus mutable. The data can be used either to maintain the status quo ("women are different biologically and therefore should not be expected to perform as well as men") or to effect change ("women and men are not fundamentally different, and complete parity in opportunity should be a societal goal") (Harris, 1980). These issues are critical when, for example, the United States Employment Service indicates that successful candidates for technical–scientific occupations—jobs in diverse fields including architecture, engineering, chemistry, physics, and mathematics—require visuospatial aptitudes in the top tenth percentile (Harris, 1980). If career choices are made on the basis of these requirements, then people may sort themselves into career tracks because of their test scores. Thus, scientific data can be used to explain the enormous disparity between the sexes in these occupations. It must be emphasized that the range of differences in brain morphology and function is greater within each sex than between the two sexes (Breedlove, 1994; Crichton-Browne, 1880; McGlone, 1980) (see Figure 4.17).

Are studies of subtle differences in the human brain worthwhile? How are the data to be interpreted? These are difficult questions that require sensitive and carefully reasoned answers. A quote from Mrs. Fawcett, a leader of the women's movement in the late 19th century, nicely puts the debate into perspective:

> No one of those who care most for the women's movement cares one jot to prove or to maintain that men's brains and women's brains are exactly alike or exactly equal. All we ask is that the social and legal status of women should be such as to foster, not to suppress, any gift of art, literature, learning, or goodness with which women may be endowed. (Quoted in Romanes, 1887)

Summary

- The organizational and activational actions of steroid hormones cause changes in neural tissue. Sexual dimorphisms in brain structure have been reported for a number of neural regions. Some of these dimorphisms are subtle, such as the differences in synapse patterns in the MPOA, but others are marked, such as the difference in the size of the rat SDN-POA. Other sex differences have been reported in brain structure and in neurotransmitter secretion, receptor numbers, and receptor distribution patterns. In all mammalian cases, these neural sexual dimorphisms can be altered in a predictable fashion by early steroid hormone treatment. Sexual dimorphisms in the brain can be affected by circulating hormones in adults as well. However, the functional significance of sexually dimorphic brain structures in mammals is still under investigation.
- A strong relationship among hormones, brain structure, and sexually dimorphic behavior has been established in studies of bird song. The neural pathways underlying the production of bird song involve several interconnected nuclei, some of which show sexual dimorphism. The HVC and RA of male zebra finches, which sing, are three to six times larger than those of female conspecifics, which do not sing. Bird song is organized and activated by sex steroid hormones in some species, and only activated by hormones in other species.

The sexual dimorphisms in brain nucleus size and subsequent singing behavior of zebra finches appear to be organized early in development. In contrast, some female canaries will sing if treated in adulthood with testosterone. Adult androgen treatment increases the size of the song control regions of the brain. Bird species in which both sexes sing exhibit no sex differences in the size of the neural structures making up the song control regions.

- Communication requires a sender and a receiver. Hormones affect both the sender, and the perceptional mechanisms in the receiving animal. For many species, the peak sensitivity for each sensory modality is closely correlated to biologically-important signals. In túngara frogs (*Physalaemus pustulosus*) there is a match between the emphasized frequencies in the mating call and the tuning of the frog's two inner ear organs. It is likely adaptive to adjust the peak sensitivity of the sensory organs to reproductive stimuli during the breeding season, and perhaps predator stimuli outside of the breeding season. Such a system has been described in midshipman fish. Nonreproductive female midshipman fish treated with either testosterone or 17β-estradiol increase their sensitivity of the frequency content of male vocalizations. Such sensory plasticity provides an adaptable mechanism that enhances coupling between sender and receiver in vocal communication.
- The urinary posture of canines is sexually dimorphic. The male-typical urinary posture is organized by perinatal androgen exposure. Hormone exposure during puberty is not necessary to activate the male-typical urinary posture.
- Weakly electric fishes orient in the murky water of their habitat by discharging electrical signals and detecting any distortion in the pattern of their emitted signals, a process called electrolocation. During courtship, their electroreceptors detect signals emitted by conspecifics within a short range (70–100 cm). Sex steroid hormones play activational roles in mediating the onset, as well as the pattern, of electrical signaling. Treatment of females or juvenile fish of either sex with androgens, especially naturally occurring fish androgens, produces electrical discharge patterns similar to those of adult males in breeding condition. Castration of adult males shifts the pattern of electrical discharges to resemble that of conspecific females.
- In rhesus monkeys and many other mammalian species, including humans, males engage in much more play behavior than their female peers throughout development. A larger proportion of male than female play behavior involves simulated fighting or rough-and-tumble play. The sex difference in play behavior is organized prenatally, and castration or other postnatal endocrine manipulation does not affect the amount of threat, play initiation, rough-and-tumble play, or pursuit play among monkeys.
- Three common sex differences observed in humans are gender role, gender identity, and sexual orientation/preference. Gender role is the sum of culturally based behavior patterns that are specific to one sex or the other. Gender role appears to be primarily learned; however, there is currently great controversy about this topic. Individuals with early endocrine dysfunction may adopt an atypical gender role. Similarly, gender identity, whether individuals view

themselves as male or as female, appears to be primarily learned. Studies of individuals with 5α-reductase deficiency who change gender role and gender identity at puberty suggest that hormones, rather than rearing, could determine gender role and/or gender identity.

- Very little is known about how heterosexuals develop erotic attractions, and even less is known about how homosexuals develop erotic attractions. Because erotic attractions typically begin at the time of puberty, sex steroid hormones have been assumed to be involved. A common hypothesis about the cause of homosexuality is that the hormone levels of homosexuals are different from those of heterosexuals. However, many studies have failed to find any evidence that adult homosexual men or women differ from their heterosexual counterparts in blood concentrations of androgens or estrogens. Evidence for effects of early exposure to steroids on sexual preference is also inconclusive.
- Sex differences in human brain size have been reported for years. More recently, sex differences in specific brain structures have been discovered. Sex differences in a number of cognitive functions have also been reported. Females are more sensitive to auditory information, whereas males are more sensitive to visual information. Females are also more sensitive than males to taste and olfactory input.
- Women display less lateralization of cognitive functions than men. On average, females generally excel in verbal, perceptual, and fine motor skills, whereas males outperform females on quantitative and visuospatial tasks, including map reading and direction finding. Although reliable sex differences can be documented, these differences in ability are slight; there is more variation within each sex than between the sexes in most cognitive abilities.

Questions for Discussion

1. When this book was published, both the Ladies' Professional Golf Association (LPGA) and the Ladies' European Tour (LET) stipulated that players must have been born female to be eligible for membership. Is this a fair or reasonable rule? Why or why not?
2. Imagine that you discovered that the brains of architects were different from those of nonarchitects; specifically, that the "drawstraightem nuclei" of the right occipital lobe were enlarged in architects as compared with nonarchitects. Would you argue that architects were destined to be architects because of their brain organization, or that their experience as architects changed their brains? How would you resolve this issue? Expand your arguments to include brain differences between homosexuals and heterosexuals.
3. What are the advantages and disadvantages of studying animal models to understand human behavioral sexual dimorphisms? Discuss the advantages and disadvantages of three different model systems. What characteristics of an an-

imal model would provide an ideal experimental system for understanding human behavioral sexual dimorphisms?

4. Using organ discharge patterns in electric fishes and urinary behavior in dogs, discuss the proposition that hormones affect perception. Can you conceive of some possible ways this might occur using a proximate level of explanation? Can you conceive of some possible reasons why this might occur using an ultimate level of explanation?
5. Are studies of individuals with endocrine dysfunction a reasonable way to understand typical sexual dimorphisms in human behavior? Why or why not?
6. Some individuals are born with genitalia that are neither definitely male nor female. Physicians and parents often must make choices about treatment and make a sex assignment at birth. Some adult individuals who were born with ambiguous genitalia claim that, in retrospect, they would have been happier "unaltered" than forced into one particular sexual category. Keeping these views in mind, what should physicians do or advise when individuals are born with atypical genitalia?

Refer to the accompanying Student CD for additional resources, including Web links, videos, animations, and additional photos.

Suggested Readings

Balthazart, J., and Adkins-Regan, E. 2002. Sexual differentiation of brain and behavior in birds. In D. W. Pfaff, A. P. Arnold, A. M. Etgen, S. E. Fahrbach, and R. T. Rubin (eds.), *Hormones, Brain and Behavior*, Volume 4, pp. 223–301. Academic Press, New York.

Breedlove, S. M., Jordan, C. L., and Kelley, D. B. 2002. What neuromuscular systems tell us about hormones and behavior. In D. W. Pfaff, A. P. Arnold, A. M. Etgen, S. E. Fahrbach, and R. T. Rubin (eds.), *Hormones, Brain and Behavior*, Volume 4, pp. 193–221. Academic Press, New York.

DeVries, G. J., and Simerly, R. B. 2002. Anatomy, development, and function of sexually dimorphic neural circuits in the mammalian brain. In D. W. Pfaff, A. P. Arnold, A. M. Etgen, S. E. Fahrbach, and R. T. Rubin (eds.), *Hormones, Brain and Behavior*, Volume 4, pp. 137–191. Academic Press, New York.

Hines, M. 2003. *Brain Gender*. Oxford University Press, New York.

LeVay, S. 1993. *The Sexual Brain*. MIT Press, Cambridge, MA.

Migeon, C. J., and Wisiewski, A. B. 2003. Human sex differentiation and its abnormalities. *Best Practices in Research and Clinical Obstetrics and Gynaecology*, 17:1–18.

5

Male Reproductive Behavior

Males of many species, including salmon, marsupial mice, and praying mantises, face certain death in order to mate. Stories abound in the natural history literature describing the heroic journeys of males of these and other species, who overcome obstacles and dangers just to gain the opportunity to fight other males for the *possibility* of mating. Male vipers, for example, engage in a deadly battle in which only one animal survives; the survivor courts the attending female that observes the contest. Although their competition is less overtly lethal, male elephant seals also compete vigorously for the opportunity to mate. They migrate hundreds of kilometers to the northern California coast in order to set up breeding territories, and fight ferociously with their neighbors for the best real estate. Beachfront property is the best territory to control because females prefer locations close to the water and tend to mate with males occupying these prime spots. Females presumably prefer beachfront locations because their pups will be safer from terrestrial predators along the edge of the water. Once the territories are established, the high-ranking males are clustered around the water's edge, surrounded by lower-ranking males. As females come ashore, the lower-ranking males continually challenge the dominant males and try to mate with the females (Figure 5.1).

Why do male elephant seals expend so much energy fighting among themselves in order to secure a piece of the beach valued by females? Males occupying desirable waterfront territories mate with many more females than do males locat-

5.1 A dispute over territory between two male elephant seals. Waterfront areas are hotly contested among males of this species because females prefer to have their pups near the water—which provides an easy escape route from terrestrial predators for themselves and their offspring—and thus prefer to mate with males that control these territories. Male elephant seals are highly aggressive: they have been known to accidentally crush their own offspring in an attempt to keep other males away from their territories. Photo © Richard R. Hansen/Photo Researchers, Inc.

ed in peripheral areas because females prefer to mate with the males located in territories that provide the best place to deliver and nurse their pups (LeBoeuf, 1974). Although females directly choose the areas in which they want to rear their young, by doing so they also indirectly choose males possessing a constellation of genes that promotes the ability to fight other males and win. The males' proximity to the water provides a reliable indication of these dominance traits. By mating with dominant "alpha" males, females increase the probability that their sons will possess the genes associated with reproductive success, and thus enhance their own fitness. Males are driven to fight for prime territories because if they fail to secure territories that are preferred by females, then they are unlikely to mate, but if they do secure prime territories, they have the chance to sire many offspring. Thus, the ultimate cause of the males' strong motivation to mate is to increase their reproductive fitness.

Males and females differ in their potential rates of reproduction. Females produce relatively few, large, resource-rich gametes. The number of offspring a female can produce in her lifetime is limited to the number of eggs she can produce or the number of young she can rear. Females generally invest much more than males in each offspring. A male, on the other hand, may produce billions of gametes in his lifetime, and he invests relatively little energy in each one (see Chapter 7). A female elephant seal, for example, can produce and raise only one pup per year, but a male can potentially father many pups per year. Thus, the difference in potential reproductive success forms the basis of the different mating strategies of the two sexes (Box 5.1). Females improve their fitness by choosing the best male possible with whom to combine their genes; males can often best achieve reproductive success by seeking to mate with as many females as possible. In other words, females have generally evolved to be "choosy" and males have generally evolved to be "ardent." The proximate bases of sexual behavior reflect these ultimate, evolutionary factors.

BOX 5.1 Battle of the "Sexes"

As noted in the previous two chapters, most animals are either male or female; that is, they possess either testes or ovaries. Males and females typically have conflicting interests as mating partners, and individuals have evolved behaviors to protect and promote their sex-specific self-interest. Hermaphrodites possess both types of gonads and produce both eggs and sperm. Yet, even among hermaphrodites, conflicts of interest involving mating arise. For example, hermaphroditic animals often participate in elaborate courtship displays to determine which individual will donate sperm and which individual will donate ova during a particular encounter. In some circumstances, it is advantageous to donate sperm; in other cases, it is advantageous to provide ova.

One dramatic case in which it appears advantageous to donate sperm involves hermaphroditic flatworms (*Cotylea*) that typically mate in pairs. These flatworms rear up and literally fence with their penes (as shown in the accompanying figure). Each flatworm thrusts and parries, trying to stab an exposed area of its mating partner's body while trying to avoid getting stabbed itself (Michiels and Newman, 1998). An individual that successfully pierces the body of its mating partner injects sperm that travel to the partner's ovaries to inseminate its ova. These intense "fencing" bouts may last 20–60 minutes before one individual succeeds in inseminating the other. Apparently, the costs of producing ova, and of mending the stab wounds, are sufficiently high to favor competition among these hermaphroditic flatworms to donate sperm.

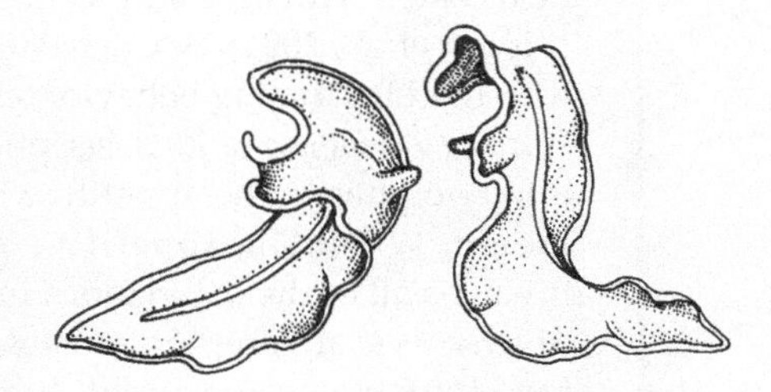

After Michiels and Newman, 1998

These dueling flatworms exemplify the complexity of so-called "simple" organisms. In contrast to sexually reproducing animals, individual hermaphroditic flatworms may have identical, but incompatible, interests as mating partners; thus, as you might predict, cooperation between mating partners should be rare. Furthermore, the mechanisms underlying aggression and reproductive motivation are probably linked.

The Proximate Bases of Male Sexual Behavior

The physiological components of male sexual behavior are the topic of this chapter. More specifically, the interaction between the endocrine and nervous systems in mediating male mating behavior will be presented. Most of what we know about the physiological mechanisms underlying sexual behavior has been gained through studies of nonhuman animals, especially rodents; examples of these studies will be described below.

As described in Chapter 3, **sex** is a division of gamete types, and **sexual behavior** is behavior that has evolved to bring the two gamete types together. Male sexual behavior is defined as all the behaviors necessary and sufficient to deliver male gametes (sperm) to female gametes (ova or eggs). The hormones involved in gamete maturation—namely, the sex steroid hormones—have been co-opted over evolutionary time to regulate the web of behaviors necessary to bring the two sexes together for successful gametic union (Crews, 1984). **Sex drive**, the motivational force that propels individuals to seek sexual union, is a very powerful engine underlying the behavior of all animals, including humans. Sex drive is primitive. Although the nervous system of the nematode worm *Caenorhabditis elegans* comprises only a few hundred neurons (White et al., 1986), these worms display complex patterns of sexual behavior (Lipton et al., 2004). Males engage in mate-searching behavior that resembles the motivated behaviors of vertebrates. When males are isolated from potential mating partners, they will leave an area with plentiful water and food and wander about their environment (in the laboratory, a petri dish coated with agar and seeded *E. coli* bacteria), presumably in search of a mate. (Lipton et al., 2004). Sex drive tends to be strong in males of all species, and the threshold for mating behavior tends to be low.

For purposes of its description here, all male sexual behavior can be divided into two phases: the appetitive phase and the consummatory phase (Figure 5.2; Hinde, 1970). The **appetitive phase** is roughly equivalent to courtship and involves all of the behaviors a male uses to gain access to a female. Behaviors as diverse as searching for females, fighting for territory, advertising, or providing food for females may occur during the appetitive phase. Courtship functions as a communication opportunity during which information about species, readiness to mate, resources, and genetic endowments is shared. The appetitive phase lasts much longer than the second phase, called the **consummatory phase**, during which copulation occurs. Although separating mating behavior into these two phases is likely to divide a unified behavioral program artificially, this dichoto-

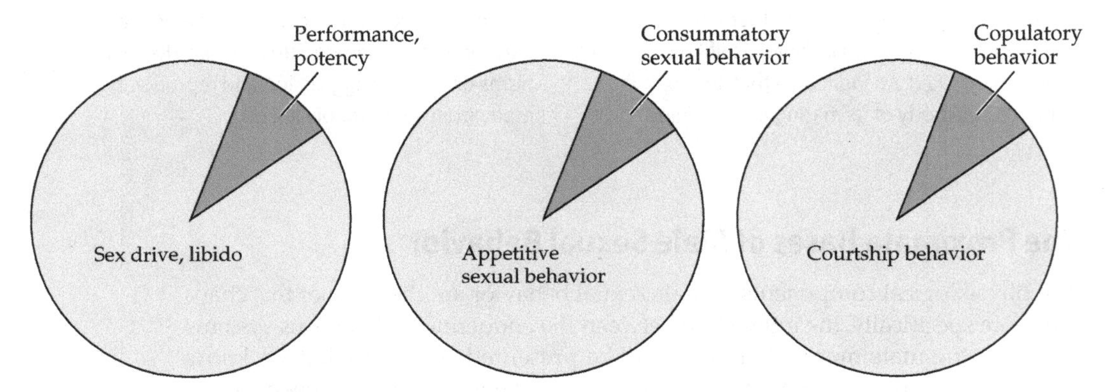

5.2 Male sexual behavior can be operationally divided into two phases: the appetitive phase (seeking sexual encounters) and the consummatory phase (engaging in copulation). The behavioral components of the appetitive phase are variously referred to as sex drive, libido, or courtship behavior; those of the consummatory phase are known as performance, potency, or copulatory behavior. Males expend much more time and energy seeking copulation than actually copulating. This division of male sexual behavior into two phases is useful for heuristic purposes as well as in medical diagnoses because hormones can affect each component differently.

my remains valuable for understanding the hormonal and neural bases of male reproductive behavior. Thus, the distinction between seeking sex (sex drive, motivation, or libido) and engaging successfully in a mating act (performance, potency) can be useful for heuristic purposes, as well as in medical diagnoses (Meisel and Sachs, 1994; but see Sachs, 1995b). Hormones affect both sex drive and sexual performance, but often to differing degrees.

Male sex drive is expressed overtly after puberty as the testes become active. The coincidence between the onset of sexual interest and puberty suggests that the testes influence sexual motivation. The observation that removal of the testes typically reduces the frequency of all male mating behaviors provides further evidence that the testes are important in the regulation of male sexual behavior (see Chapter 1). Post-castration treatment with androgens, the principal class of sex steroid hormones produced in the testes, restores reproductive behaviors to precastration levels. Thus, it is reasonable to conclude that androgens cause sexual behavior.

However, the statement that androgens cause sexual behavior requires some qualification. First, recall that hormones change the *probability* that a particular behavior will be exhibited *in a specific context*. Second, the effects of experience interact significantly with those of hormones to affect behavior. These two caveats are important to keep in mind when discussing the effects of hormones on behavior in general and on male sexual behavior specifically. One outdated model of hormone–behavior interactions arose from physiological models—the so-called *hydraulic models*—once proposed by ethologists to explain innate, reflexive patterns of behavior. According to these simplistic models, hormonal effects on behavior were analogous to a faucet: for example, if the "androgen spigot" was turned on, then male mating behavior or aggression ensued. But hormone–behavior interactions are far more complex than suggested by this model. About half the readers of this sentence have significant concentrations of androgens in their blood, yet they are reading, not copulating. How can we say, then, that androgens cause sexual behavior? It is more accurate to state that androgens, as well as other sex steroid hormones, appear to affect the likelihood of mating behaviors by reducing the threshold for these behaviors in the presence of the appropriate stimuli in the appropriate social context. All other things being equal, a male with high blood androgen concentrations is more likely to copulate with a conspecific female than a male with low androgen concentrations. In other words, in terms of a stimulus–response paradigm of sexual behavior, hormones facilitate the male response, probably by affecting the perception and processing of stimuli associated with a female. As we will see, androgens have many direct effects on the nervous system, such as regulating neurotransmitter levels as well as affecting the availability of neurotransmitter receptors and other proteins that affect neurotransmitter function (Hull et al., 2002).

Males of different species vary in the frequency of their copulatory behavior, and vast species differences exist in the extent to which male reproductive behavior relies on hormones. Similarly, enormous individual differences exist within any given species in both the frequency of sexual behavior and the extent to which sexual behavior is mediated by hormones. For example, the regulation of rodent sexual behavior is highly dependent on hormones, but in primates, the control of sexual behavior is less dependent on hormones and more dependent on socialization and learning. Compared with rats, then, primates are relatively less reliant on hormones to regulate sexual behavior. However, primates are not unique in this regard, and there is no evidence that species located "higher" on the phylogenetic scale have been emancipated from direct hormonal control of mating behavior as compared with so-called "lower" species. Examination of the regulation of mating behavior in nonmammalian animals also reveals varying degrees of hormonal dependence. An example of a reptile species (the red-sided garter snake) in which reproductive behavior is completely divorced from hormonal control will be presented to help counter the belief that humans are unique in this regard.

This chapter examines the endocrine mediation of sexual behavior among males, and it will address many of these comparative issues. The chapter will also describe the ways in which hormones interact with the nervous system in males. Animal models, mainly rodents, have been helpful for understanding the physiological mechanisms underlying sexual behavior in men, both appetitive and consummatory. The results of research based on these models will be presented in detail below. But first, the history of research into the control of male sexual behavior will be briefly reviewed.

Historical Origins of Research on Male Sexual Behavior

Castration virtually eliminates mating behavior among males of many species, a fact that has been known since the earliest recorded history (see Chapter 1). As noted in the 12th verse of the 19th chapter of the Gospel of Matthew in the New Testament, "There are eunuchs who are born so from their mother's womb; and there are eunuchs who are made so by men; and there are eunuchs who have made themselves so for the sake of the kingdom of heaven. Let him accept it who can." Some people have suggested that this passage indicates that Jesus believed that castration could clear the path to heaven. In the late 1990s, members of the San Diego cult Heaven's Gate apparently believed this to be true. Eight of the 18 males, including the cult leader, who committed suicide together prior to their imaginary journey to a waiting "Mother Ship" reportedly had performed self-castration. Many religious sects throughout history have believed that possession of functional testes blocked rational and/or pious thought.

The English word *testicle* is derived from the Latin *testiculus*, a diminutive, but synonymous, form of the Latin term *testis*, which means "witness." The verb *testify* has a similar derivation. Witnesses in ancient Roman courts often covered their testes with their right hands and swore "on their virility" that they were about to state the truth (that is, provide testimony). This practice illustrates that a relationship among the testes, virility, and mating behavior has been well established for many centuries, but many misconceptions have arisen regarding the nature of the interaction between the testes and sexual behavior.

One common misconception about the regulation of male sex drive that persisted through the late 1800s was that distention, or swelling, of the seminal vesicles activates male sexual behavior (Carter, 1974). Neural stretch receptors were recent discoveries in the 1880s, and it was proposed by several researchers of the time that these receptors, activated by bulging seminal vesicles, induced copulation. Variations of the idea that male sex drive was the result of "pressure" sensations from the accessory sex organs that had to be relieved lasted well into the 20th century (for example, Ball, 1934a; Nissen, 1929). Anecdotally, many men report "pressure" in their testes or seminal vesicles when they are sexually aroused. A major piece of scientific evidence supporting the "distended seminal vesicle" theory came from work on frogs in mating condition (Tarchanoff, 1887, cited in Steinach, 1894). Male frogs were reported to continue copulating despite removal of major portions of their anatomy, including their testes; after physical separation from their partners, castrated frogs would immediately re-mount the female. However, draining the seminal vesicles led to rapid separation of the pairs and a loss of subsequent sexual activity for the male. Thus, neural impulses from the full seminal vesicles were thought to generate sexual behavior, or more commonly, sex drive.

Eugen Steinach, apparently influenced by the writing of Sigmund Freud, postulated the existence of an incipient sexuality that existed prior to puberty and was triggered by nervous impulses from the swelling reproductive glands. He believed that once the process was initiated, psychic influences or sensory input could send the mechanisms underlying reproductive behavior into action. He attempted to reduce mating behavior in rats by removing their seminal vesicles; however, this procedure had little effect on their mating behavior (Steinach, 1894). As we shall see later, some of Steinach's views on the regulation of sexual behavior, especially the importance of sensory information, remain part of current hypotheses.

Rather than confirming the pressure hypothesis, subsequent research supported the idea that hormones from the testes generate male mating behavior. But even the presence or absence of gonads is not the only mediating factor in male sexual behavior. As mentioned above, sexual experience also affects post-castration responses. Virtually all sexually inexperienced male rats exhibit mating behavior when given the opportunity to interact with estrous females soon after castration, but their interest in females wanes in about 2 weeks. Males with prior sexual experience, on the other hand, vary in how long they maintain reproductive behavior after castration; some males lose interest in females rapidly, whereas others continue to mate for weeks (reviewed in Larsson, 2003). The contribution of experience to the interaction between hormones and sexual behavior remains unspecified.

Steinach and his early 20th-century contemporaries were aware of the effect of sexual experience in determining the rate at which post-castrational reductions in male sexual behavior would occur in domesticated animals. For instance, tomcats with sexual experience prior to neutering may continue to engage in copulatory behavior for months after the testes are removed, usually to the chagrin of pet owners (Dunbar, 1975). Even Aristotle noted in his *Historia Animalium* that castrated bulls occasionally continued to copulate with cows. It was also well known during the 19th century that men who lost their testes through disease or accident maintained sexual behavior for some time, often indefinitely, after the injury. According to Pfüger (1877, cited in Steinach, 1894), one way in which the difference between human sexual behavior and that of other animals could be explained was on the basis of humans' unique mental faculties; that is, their "psychic qualities, the vigor of men's fantasies, and their powerful memories."

Certainly, "psychic" qualities must be called forth as an explanation for the claims made by Charles Edouard Brown-Séquard at the Société de Biologie de Paris in 1899 (Brown-Séquard, 1899). Ten years earlier, Brown-Séquard, a prominent researcher, had published findings in the *Archives de Physiologie Normale et Pathologique,* claiming that injections of endocrine extracts had astounding rejuvenating effects on several physical parameters, including sexual vigor. At the Société in 1899, Brown-Séquard sought to demonstrate his findings by injecting himself with an aqueous solution of homogenized dog and guinea pig testes. His claim that the injections had amazing restorative effects prompted sales of endocrine extract "treatments" (Figure 5.3). However, androgens were later discovered to be lipid-soluble steroid hormones; therefore, water-based injections are unlikely to have produced the behavioral effects reported by Brown-Séquard because they would have been unlikely to contain many androgen molecules. The

(B)

PROFESSOR

BROWN SEQUARD'S

METHOD.

EXTRACTS OF ANIMAL ORGANS.

Testicle Extract,
Grey Matter Extract,
Tyroid Gland Extract, &c., &c.

Concentrated Solutions at 30%.

These preparations, completely aseptic, are mailed to any distance on receipt of a money order. Directions sent with the fluids.

Price for 25 Injections, $2.50.
Syringe Specially Gauged, (3 cubic c.,) $2.50.

Used in the Hospitals of Paris, New York, Boston, etc.
Circular Sent on Application.

New York Biological and Vaccinal Institute,

Laboratory of Bovine Vaccine and of Biological Products.

GEO. G. RAMBAUD, Chemist and Bacteriologist, Superintendent.

PASTEUR INSTITUTE BUILDING, NEW YORK CITY.

Advertisement, New York Therapeutic Review, *1893.*

5.3 An early misunderstanding of the effects of hormones on male sexual behavior. (A) Charles Edouard Brown-Séquard claimed that injections of endocrine extracts could restore sexual vigor. (Photogravure by Heliog Dujardin.) (B) A 19th-century advertisement for extracts based on Brown-Séquard's fallacious claims. Hoping to capitalize on Brown-Séquard's findings, charlatans sold many such "rejuvenating" preparations. (A) courtesy of the National Library of Medicine.

improvements in stamina that he noted were probably the result of a "placebo effect" produced by his belief in the treatment. Although his final studies on himself overshadowed a career of bona fide accomplishments in endocrine research, the furor created by Brown-Séquard's demonstration initiated medical interest in the sex steroid hormones, for which he must be given credit.

As noted in Chapter 3, a masculine or feminine behavior pattern is not determined simply by the presence or absence of a particular behavior, but is often a matter of the frequency of, or the threshold for, a certain behavior. Quantitative measures of male sexual behavior are therefore an important means of describing it. One of the earliest attempts at quantification of mammalian sexual behavior was made by Calvin P. Stone, a psychologist, who determined the age at which male rats first exhibited copulatory ability (Stone, 1922). He tested males for mating behavior every day from 21 to 60 days of age, and discovered that the average age at which first copulation was observed was about 50 days (Stone, 1924). Stone also determined that developing male rats maintained copulatory ability for about 14 days after castration (Stone, 1927), and reported that substantial variation exists in the maintenance of intermale aggression and mating behavior after castration.

In addition to measuring sexual performance, psychologists developed ways of measuring the strength of the sex drive (for example, Tsai, 1925; Warner, 1927). Given a choice, they asked, will a rat prefer to satiate its hunger, thirst, or sex drive? What levels of deprivation are necessary for one biological need to override another? How much work will an individual perform to attain sex? And how much pain will an individual bear to reduce its sex drive? The sex drives of individuals of many species were quantified by a battery of motivational tests (for example, Nissen, 1929; Stone et al., 1935; and see below). (Remember that "motivation" is a psychological hypothetical construct; like "learning" or "attention," motivation cannot be measured directly. Only performance on a test designed to assess such a hypothetical construct indirectly can be measured.)

Early in the 20th century, the ablation–replacement technique (see Chapter 1) was used to determine the behavioral effects of sex steroid hormones. For instance, a series of studies suggested that ovarian grafts in castrated young male rats or guinea pigs modified their subsequent behavior and physiology so that they "became" females (Steinach, 1910, 1913). In guinea pigs, growth of the mammary glands and milk secretion were reported, but these findings were never replicated completely. On the other hand, grafts of testicular tissue in young females influenced their development so that they became masculinized as they matured. These latter findings have been replicated many times in several species. A secreted product from the interstitial cells of the transplanted gonad was proposed as the controlling factor in these transsexual changes because the secondary sex characters of the opposite sex were not manifested unless the implanted gonad developed vascular connections (Steinach, 1940). We now know that the organizing properties of steroid hormones from the testes cause the appearance of male-like characters in early-treated females (see Chapter 3).

An obvious refinement in experiments addressing endocrine effects on male sexual behavior was the discovery of the active agents in the testes that were responsible for the observed behavioral effects. After the estrus-inducing hormone from the ovarian follicle was discovered to be lipid-soluble, the search for the *andros*-generating hormone began. Injections of an extract of bull testes dissolved in oil were found to produce a rapid regeneration of capons' combs (McGee et al., 1928). Pure crystalline hormone from testicular tissue was soon isolated and named *testosterone* (David et al., 1935). A year later, preparation of synthetic testosterone from cholesterol became feasible, and large amounts of the hormone rapidly became available for clinical and experimental studies. Several investigators demonstrated the restoration of sexual behavior in castrated male rats after injections of testosterone propionate, a stable, injectable form of testosterone (Moore and Price, 1938; Shapiro, 1937; Stone, 1938a,b, 1939). For example, a dose–response experiment revealed that 50–75 µg of testosterone propionate per day was necessary to maintain adult male mating behavior after castration (Figure 5.4; Beach and Holz-Tucker, 1949).

Many early researchers who attempted to link hormones and behavior came to approach their work, both theoretically and methodologically, with a "one hormone equals one behavior" philosophy (reviewed in Beach, 1948). Differences in behavior were thought to represent differences in the underlying hormone that controlled the behavior in question. However, we now know that hormonal effects on behavior rarely reflect this type of unitary relationship. For example, male copulatory behavior can be restored in castrated males by injections of either estrogens or androgens (Ball, 1937; Beach, 1942c). Furthermore, massive doses of testosterone propionate can cause either feminine or masculine mating behaviors in castrated males. These results indicate that the action of any particular sex

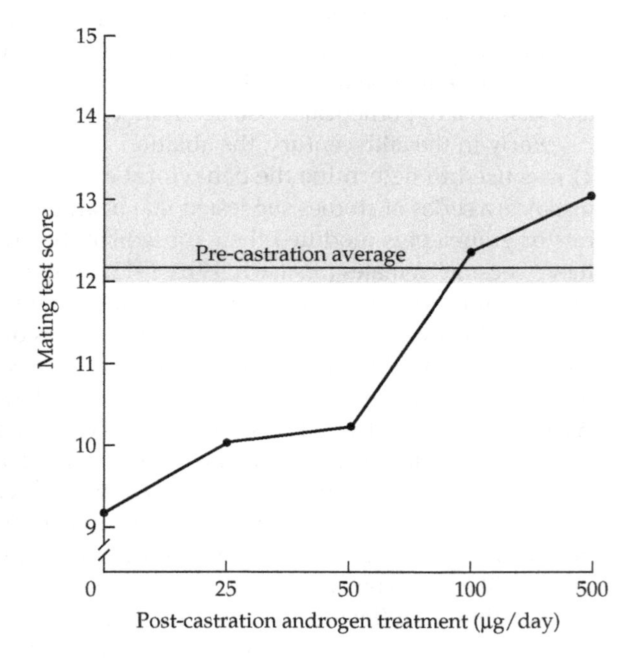

5.4 Testosterone treatment maintains sexual behavior after castration. Mating behavior declines after castration in male rats, but can be maintained by testosterone treatment. The maintenance of sexual behavior by testosterone is dose-dependent: in this study, only rats receiving testosterone doses of 100 µg/day or greater displayed sexual behavior at pre-castration levels. After Beach and Holz-Tucker, 1949.

steroid hormone cannot be behavior-specific; every individual probably possesses the behavioral repertoire of both sexes, but one set of responses appears to have a lower threshold for expression in the presence of a particular hormone(s) (Beach, 1948). It has been shown that aromatization of androgens into estrogens occurs in the nervous system, which accounts for many of these puzzling results in non-primates. Indeed, in many species, the brain produces the most estrogens of any organ in the body. Nonetheless, the empirical observation that both estrogens and androgens can support mating behavior in castrated males led eventually to an appreciation and understanding of the neural bases of reproductive behaviors and to the formulation of modern theories of hormone–behavior interactions. Current views about the hormonal regulation of behavior focus on the substrate or target tissues of the hormones, and research is now aimed at elucidating how differences in behavior reflect differences in target tissue sensitivity.

Male Sexual Behavior in Rodents

As with other areas of behavioral endocrinology research, most work on the effects of hormones on male reproductive behavior has been performed on laboratory rats (*Rattus norvegicus*), and many current theoretical considerations have arisen from these studies. Work with animal models has also provided useful information that is applicable to human clinical conditions (Ågmo and Ellingsen, 2003). Consequently, many of the concepts we discuss will be illustrated with examples from studies of rat mating behavior. In order to understand what these studies have revealed about the hormonal and neural bases underlying male reproductive behavior, a thorough description of rat mating behavior is necessary.

Mating Behavior: A Description

In rats and many other rodent species, there are three easily distinguishable behavioral components of the consummatory phase of male mating behavior: (1) mounting, (2) intromission, and (3) ejaculation. In most standard mating tests, one male and one female are introduced into a glass enclosure (usually an empty aquarium). Prior to mounting, several precopulatory male behaviors are observed; these behaviors are relatively easy to quantify and provide a useful measure of male sexual motivation. *Mount latency*, for instance, is the time from the introduction of the male and female until the first mount. Sexual motivation is considered high if mount latencies are low, and vice versa. Sexual motivation can be tested by placing some sort of obstacle between the male and the female and observing how long it takes the male to overcome the obstacle. The so-called Columbia University obstruction test, using a physical barrier or an electrified floor, was often used in research of this sort (for example, Stone et al., 1935; Warner, 1927) (Figure 5.5). In other obstruction tests, the male must run on a treadmill that is moving away from the female, climb a ladder, or press a bar. Presumably, high sexual motivation can be inferred if the male passes the obstruction at a high speed or endures great pain to reach the female. Unmotivated males will not work or suffer to obtain access to a female (Lopez et al., 1999).

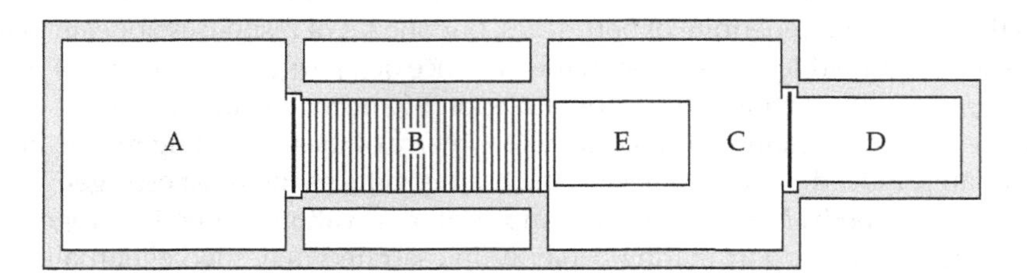

5.5 The Columbia University obstruction test was used to measure sexual motivation. In this version, a rat is placed in the entrance compartment (A), and the time it takes to cross the electrified floor (B) to gain access to another animal in the holding compartment (D) is operationally defined as a measure of sexual motivation. When males require little time to cross the grid, they are considered highly motivated. The door to the holding compartment (C) opens when the rat reaches the release plate (E) on the floor. After Warden et al., 1935.

Soon after introduction of the pair, the male begins to investigate the female, sniffing her mouth and anogenital region (Stone, 1922). Moments later, he attempts to mount her. If she is in estrus, or mating condition, the female exhibits the lordosis posture, which facilitates intromission (Figure 5.6A). **Mounting** is operationally defined as the male assuming a copulatory position, but not inserting his penis into the female's vagina. In some cases, males may mount the side or head of the female, and occasionally males may mount other males. In other cases, males may not mount at all. During time-limited tests (1 hour, for example), some males consistently fail to initiate sexual behavior by mounting. Males may fail to mount a female because of low sex drive, decreased penile sensitivity, or reduced erectile potential. One measure used to describe sexual behavior is the *intermount interval* (IMI), or the average time between successive mounts.

Intromission can be defined as the penis entering the vagina during a mount. In rats and mice, intromission is associated with thrusting motions of the hindquarters (Figure 5.6B), but in many species, thrusting is not observed. For example, male guinea pigs and rabbits ejaculate during the first intromission and complete copulation within a few seconds without thrusting. Counting the number of intromissions prior to each ejaculation provides a useful measure of reproductive behavior for males that exhibit multiple intromissions; however, the utility of this measure is limited because interindividual differences are common. Another common measure of male sexual behavior is the *inter-intromission interval* (III), the average time between successive intromissions. It is also possible to measure the copulatory motor patterns and make detailed analyses of animals at different ages or in different hormonal conditions (Morali et al., 2003).

Ejaculation is the forceful expulsion of semen from the male's body via the urethra. Ejaculation is behaviorally defined in rats as the culmination of vigorous intravaginal thrusting accompanied by the arching of the male's spine and often the lifting of his forepaws off the female prior to withdrawal (Figure 5.6C). During ejaculation, a sperm plug is often deposited in the vagina. In some rodent species, this plug effectively blocks intromission by other males until the sperm have had

(A)

(B)

(C)

5.6 Sexual behavior in rats. A male rat first investigates the anogenital region of a female, and if she is in estrus, he will mount her, clasping her hindquarters with his forepaws. This tactile stimulation causes her to display the lordosis posture, arching her back and deflecting her tail (A). Lordosis facilitates intromission, or insertion of the male's penis into the female's vagina, accompanied by thrusting of his hindquarters (B). After several seconds, the male dismounts, grooms himself, and soon remounts. After several intromissions, the male ejaculates, forcefully expelling semen into the female's vagina (C). Courtesy of Lique Coolen.

the opportunity to fertilize the estrous female. In laboratory tests, the number of intromissions prior to ejaculation, as well as the ejaculatory latency, the time from the first intromission to ejaculation, is also typically recorded. The mating potential of any given male is determined by the number of ejaculations during a time-limited test or by the number of ejaculations prior to his attaining sexual satiety (Sachs and Meisel, 1988). Most male laboratory rats can ejaculate 5–8 times during an unlimited-time mating test (Ågmo, 1997).

Male rats often "sing" after an ejaculation; a special ultrasonic detector is required to hear these vocalizations, which are in the 20–22 kHz range. Observers can detect singing by noting the rapid shallow respirations that correspond with the ultrasonic vocalizations. After ejaculation, male rats usually become sexually inactive and rather lethargic. The male may groom his genitals, then lie down and sleep. He virtually ignores the female. This postejaculatory sequence of behaviors is not due to ejaculation per se, because artificial electroejaculation does not induce these behaviors.

The time between an ejaculation and the onset of the next copulatory series is called the *postejaculatory interval* (PEI). The PEI is less than 30 seconds in Syrian hamsters (*Mesocricetus auratus*) (Bunnell et al., 1976), about 5–15 minutes in laboratory rats, and may last for hours or days in some other species (Dewsbury, 1972; Money, 1961; Fernandez-Guasti and Rodriguez-Manzo, 2003). The PEI is generally considered to be composed of two separate periods, an absolute refractory phase and a relative refractory phase (Beach and Holz-Tucker, 1949). Males are completely nonresponsive to sexual, mildly painful, and other stimuli during the absolute refractory phase. A new or very potent sexual stimulus may elicit responsiveness in a male rat during the relative refractory phase (see below). A male rat with a PEI greater than 90 minutes is usually considered sexually "exhausted" or satiated (Ågmo, 1997).

Male copulatory behavior in mammals can be classified into several categories based on the presence or absence of the following features (Dewsbury, 1972) (Figure 5.7; Table 5.1): (1) copulatory lock, (2) intravaginal thrusting, (3) mul-

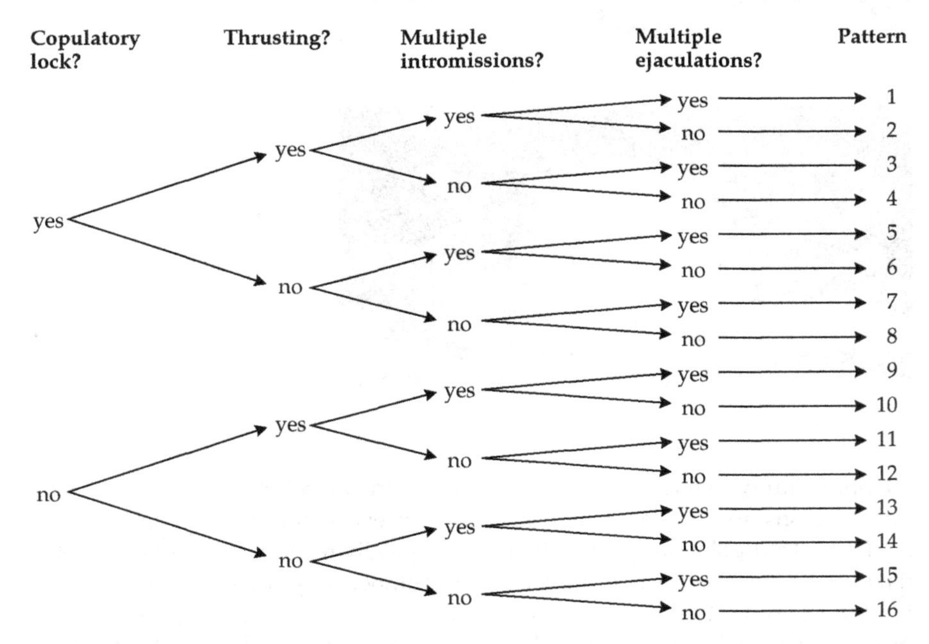

5.7 Male copulatory patterns in mammals can be classified by the presence or absence of four basic features during a mating bout: copulatory lock, thrusting, multiple intromissions, and multiple ejaculations. Species can be grouped according to the sixteen different mating patterns that emerge from this classification scheme. After Dewsbury, 1972.

TABLE 5.1 Examples of copulatory patterns in mammals

Common name	Taxonomic name	Lock?	Thrust?	Multiple intromissions?	Multiple ejaculations?	Pattern[a]
Dog	*Canis familiaris*	Yes	Yes	No	Yes	3
Wolf	*Canis lupus*	Yes	Yes	No	Yes	3
Golden mouse	*Ochrotomys nuttalli*	Yes	No	No	Yes	7
House mouse	*Mus musculus*	No	Yes	Yes	Yes	9
Montane vole	*Microtus montanus*	No	Yes	Yes	Yes	9
Rhesus monkey	*Macaca mulatta*	No	Yes	No	Yes	11
Bonnet macaque	*Macaca rudiata*	No	Yes	No	Yes	11
Meadow vole	*Microtus pennsylvanicus*	No	Yes	No	Yes	11
Norway rat	*Rattus norvegicus*	No	No	Yes	Yes	13
Mongolian gerbil	*Meriones unguiculatus*	No	No	Yes	Yes	13
Bison	*Bison bison*	No	No	No	Yes	15
Black-tailed deer	*Odocoileus hemionus*	No	No	No	No	16

Source: Dewsbury, 1972.
[a]See Figure 5.7.

tiple intromissions, and (4) multiple ejaculations. Animals that show copulatory lock, such as dogs, have a penis that swells after ejaculation, which facilitates sperm transport to the female. The lock may remain in place for several minutes, during which time it is virtually impossible for the pair to disengage (Figure 5.8). Males of some species, such as guinea pigs, do not show pelvic thrusting and ejaculate with a single intromission. Male rats, on the other hand, engage in about 20 mounts and 10 to 15 intromissions prior to the first ejaculation, but these values vary among rat strains; additional intromissions are usually required for each subsequent ejaculation (Beach and Jordan, 1956).

The various patterns of male copulatory behavior have been related to the ecology of particular species, and several broad generalizations can be made (Dewsbury, 1972). For example, predator species are more likely to lock than prey species; prey species cannot risk being immobilized for long periods, and there are few prey species that require lengthy sperm transport periods (see Table 5.1). Animals that mate at night are more likely to engage in copulatory locking than diurnal creatures, presumably because of a similar lack of predation pressures. Virtually all rodent species that lock (for example, golden mice, *Ochrotomys nuttalli*, and southern grasshopper mice, *Onychomys torridus*) are nocturnal.

Hormonal Correlates of Male Mating Behavior

As we have seen, testosterone is necessary for the maintenance of mating behavior in male rats. Castration leads to a reduction in sexual responsiveness; both motivation and performance wane. For most rodent species, sexual behavior is markedly reduced immediately after castration. Sex drive also declines rapidly

5.8 Sexual behavior in the dog. A male dog first investigates the anogenital region of a female (top). If she is in estrus (mating condition), she will allow the male to mount her (middle left). The hindquarters of the male thrust with increasing intensity until ejaculation occurs some seconds later. After ejaculation, the penis remains "locked" in the vagina, and the male steps over the female as he dismounts (middle right). They remain in this copulatory lock for several minutes to facilitate sperm transport (bottom). After Beach and LeBoeuf, 1967.

after the testes are removed. Castrated males will not investigate females, nor will they work or suffer pain to reach them. If castrated rats are subjected to daily mating tests, the timing of the disappearance of components of sexual behavior follows a characteristic pattern. The effects of castration are observed within days. Males first begin to take longer to initiate mounting and intromissions (Hull et al., 2002; Meisel and Sachs, 1994). Another early effect of castration is that fewer intromissions occur prior to ejaculation. This finding may seem paradoxical: fewer intromissions before ejaculation might appear to reflect increased reproductive performance. However, a male that normally has eight intromissions prior to ejaculation is not necessarily a more efficient or effective copulator than one that typ-

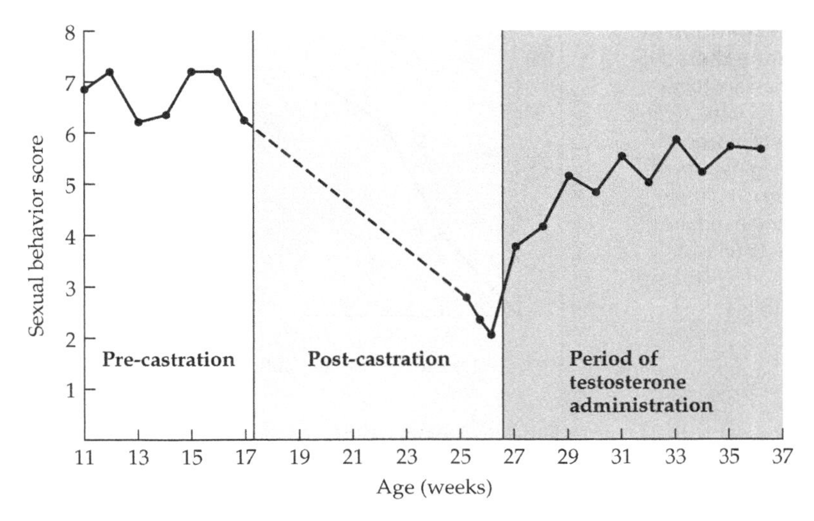

5.9 Sexual behavior can be restored by testosterone therapy. After castration, guinea pigs displayed marked declines in sexual behavior, but testosterone replacement therapy gradually restored sexual behavior to pre-castration levels. After Valenstein and Young, 1955.

ically has fifteen intromissions before ejaculation. Fewer intromissions may be less likely to provide the female with sufficient stimulation to induce a progestational state (see Chapter 6), resulting in a failure of blastocyst implantation (Wilson et al., 1965). By a week or two post-castration, rats cannot mate to ejaculation. The inability to ejaculate is soon followed by a decline in the number of mounts with intromissions, and finally the male no longer mounts females.

The effects of castration on male rodent reproductive behavior can be reversed by testosterone treatment (Figure 5.9). The restoration of copulatory behaviors after sustained androgen therapy mirrors the disappearance of those behaviors: first mounting recurs, followed by intromissions, then ejaculations. Of course, neither intromissions nor ejaculations can occur in the absence of mounting, and ejaculations will not be observed in the absence of intromissions, yet it is theoretically possible that all the behaviors could be restored simultaneously. However, they generally reappear sequentially over the course of several days (Larsson, 1979). This rigid sequential ordering of behavior suggests, although it does not prove, that mounting, intromission, and ejaculation behaviors have different sensitivities to testicular hormones.

If androgen replacement therapy is initiated immediately after castration (maintenance treatment), then lower amounts of hormone are required to maintain reproductive behavior than are necessary to restore sexual behavior some time later, after it has stopped (restoration treatment) (Figure 5.10; Davidson, 1966b). Similarly, male hamsters arousing from hibernation may require a large pulse of androgens to activate their reproductive systems (Berndtson and Desjardins, 1974). Continuous exposure of brain and sensory neural tissues to androgens apparently preserves their responsiveness to these hormones.

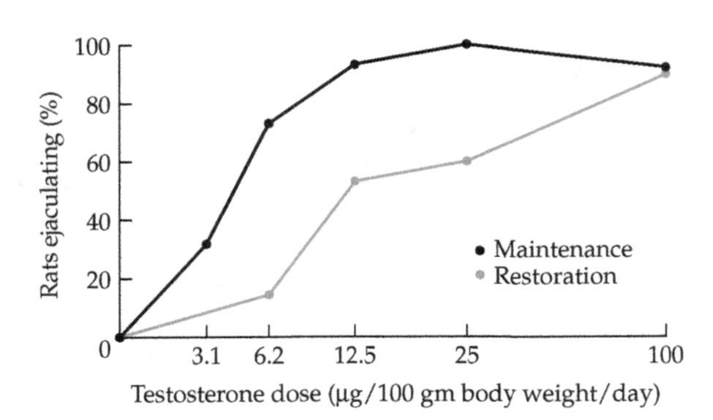

5.10 Maintenance versus restoration of sexual behavior by testosterone therapy. The amount of testosterone necessary to restore full sexual behavior is greater if the treatment begins after all sexual behavior stops (restoration regimen) than if it begins immediately after castration (maintenance regimen). Presumably, the restoration regimen requires higher doses because steroid receptors decrease in number if not maintained by circulating androgens. After Davidson, 1966a.

Two androgens, testosterone and androstenedione, a weakly androgenic precursor of testosterone, can maintain mating behavior in castrated rodents (Sachs and Meisel, 1988). Dihydrotestosterone (DHT), another product of testosterone, does not prevent the post-castration decline in reproductive behavior, regardless of whether it is provided at the time of castration or post-castration (Feder, 1971). In contrast to DHT, estradiol, an estrogen, is very effective in activating mating behavior in castrated male rodents (Davidson, 1969; Södersten, 1973). As you know, both testosterone and androstenedione can be aromatized to estradiol and other estrogens, but DHT cannot be converted into an estrogen (see Chapter 2). Taken together, these findings suggest that testosterone and androstenedione produce their behavioral effects after first being converted into estrogens (Larsson, 2003). Further support for this hypothesis arises from the observation that injecting castrated rats with specific estrogen receptor blocking agents renders subsequent androgen therapy ineffective in sustaining copulatory behavior (Beyer et al., 1976). DHT is not completely without effect; estradiol plus DHT treatment restores the mating behavior of castrated rats to the level of gonadally intact individuals (Feder et al., 1974). DHT, as we shall see, appears to be important for maintaining penile tactile sensitivity. In general, it appears that estradiol affects the central nervous system to promote mating behavior, and that DHT affects neurons in the periphery to maintain tactile sensory feedback. Thus, testosterone from the testes appears to function primarily as a prohormone providing estrogens to the central nervous system and DHT to the periphery to regulate sexual behavior.

Penile responses can be tested by placing a rat or mouse on its back and retracting the penile sheath to the base of the glans penis (Sachs, 1995a). The rat penis contains two principal erectile tissues, the corpora cavernosa and the corpus spongiosum (Box 5.2). The pressure of the sheath causes the erection reflex of the glans, which is due to the engorgement of the corpus spongiosum with venous blood. An erection of the penile shaft is depicted in Figure 5.11. If the pressure on the penis continues, penile reflexes called "flips" are noted. Flips are due to the action of the corpora cavernosa and the striated penile muscles. Intense glans erections, called cups, are observed after prolonged penile stimulation (Figure 5.12). These three penile responses are similar in form to those observed both

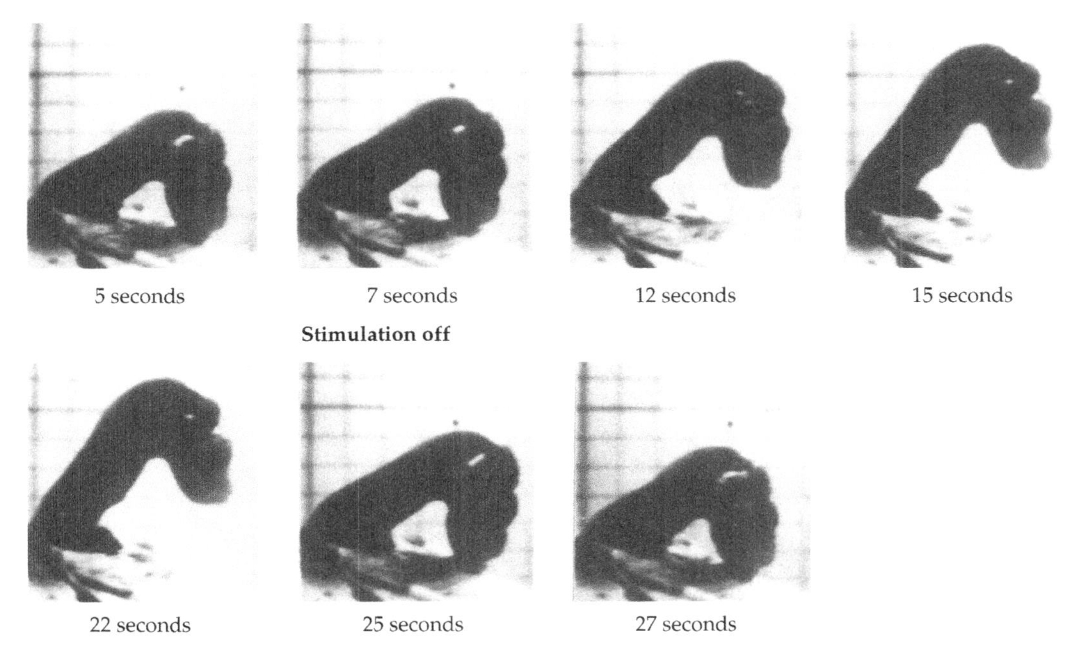

5.11 Erection in rats can be induced by bilateral electrical stimulation of the cavernosus nerve, as seen in these sequential video images of a 22-second stimulation period. Tumescence begins between 5 and 7 seconds after the initiation of stimulation; detumescence is brisk and occurs within 5 seconds of cessation of the electrical current. Courtesy of David Quinlan.

during normal sexual behavior (*in copula*) and during artificial stimulation (*ex copula*) (Meisel et al., 1984; Sachs, 1995b). Developmental studies of male rats show a remarkable coincidence between the average age of onset of erections and mounts (40.0 versus 40.8 days, respectively), flips and intromissions (44.0 versus 43.8 days, respectively), and cups and ejaculations (about 47.5 days in both cases) (Sachs and Meisel, 1979). This coincidence in development between penile reflexes and copulatory behaviors suggests a functional relationship, as well as separate underlying mechanisms and possibly different sensitivities to hormonal regulation for each of the three components of mating behavior.

If animals fail to mate after castration, ascertaining whether sexual performance, sex drive, or both functions have been affected by the surgery is difficult. One way to differentiate between the effects of hormones on sex drive and on mating performance is to isolate the brain—presumably the source of sex drive—from the spinal cord. We know that the entire erectile repertoire is programmed in the spinal cord, because appropriate stimulation causes a rat with a spinal cord severed from the brain to show erections, as well as the penile reflexes underlying intromission, thrusting, and ejaculation (Meisel and Sachs, 1994). The term "sex drive" implies that the brain is "driving" behavior through *excitatory* mes-

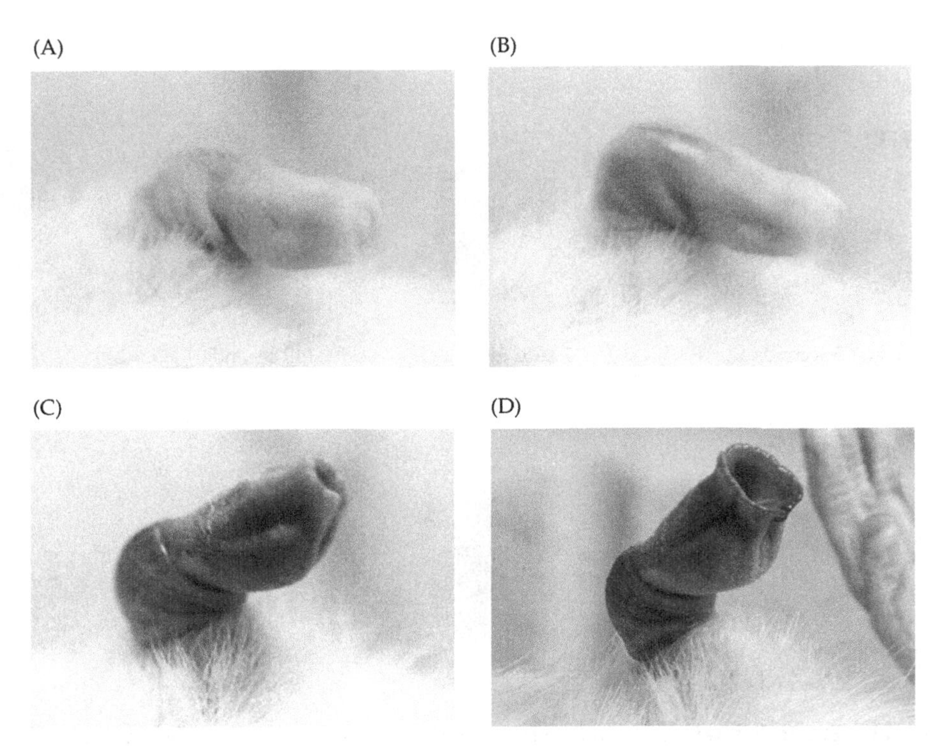

5.12 Reflexive erections in rats can be induced by retraction of the penile shaft. In A–C, the glans is directed toward the tail, the normal orientation. In D, the glans is oriented toward the head, the position necessary for intromission. (A) The quiescent rat penis. (B) Tumescence and slight elevation of the penile shaft without glans erection. (C) Glans erection. (D) Intense glans erection ("cup") and further elevation of the penile shaft. Courtesy of Ben Sachs.

sages to the periphery via the spinal cord. Although the brain does send some excitatory signals (see Box 5.2), it acts primarily to *inhibit* the spinal mechanisms of erection (Beach, 1967; Sachs and Bitran, 1990). In spinally transected rodents, the stimuli required to induce an erection are much less than the stimuli necessary for inducing erections in intact males. As we shall see, clinical data suggest an inhibitory influence of the human brain on erectile function as well.

If spinally transected rats are also castrated, their penile reflexes begin to wane after 24 hours, and disappear after 12 days (Hart et al., 1983). The loss of penile responses after castration follows an ordered pattern similar to the degradation of post-castration mating behavior: cups, then flips, and finally erections disappear (Davidson et al., 1978). (As discussed previously, however, the order of decline and restoration of copulatory behaviors is somewhat constrained by the nature of the chaining of these processes.) These results suggest that androgens are necessary, but not sufficient, for erections. Appropriate penile stimulation is necessary for penile erection, and androgens reduce the amount of stimuli required for a penile response to be observed.

BOX 5.2 Anatomy of the Penis

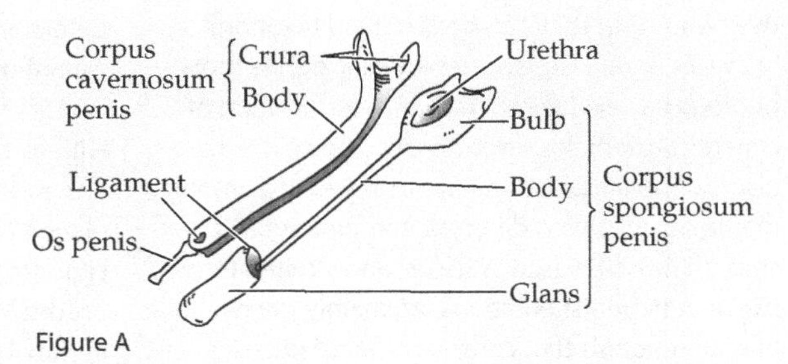

Figure A

The mammalian penis has two important structural components, the paired corpora cavernosa and the corpus spongiosum (see Figure A). (Both components are critical erectile tissues comprising smooth muscle and blood vessels and are usually associated with a set of striated muscles. The corpora cavernosa and the corpus spongiosum usually function together during tumescence and detumescence, but they may become functionally dissociated.

In humans, the corpora cavernosa are twin "tunnels" of highly vascular tissue that occupy most of the penile body. A muscle called the ischiocavernosus is connected to the base (crura) of each corpus cavernosum. The glans and the central tissue of the penis make up the corpus spongiosum. Some species have a bone, called the os penis, located within the corpus spongiosum. In rats, the bone is present only in the glans; in other species, such as dogs, the os penis is found in the penile body. The base of the corpus spongiosum is connected to the bulbospongiosus striated muscle in primates, which is called the bulbocavernosus muscle in other species, including rodents.

There is great diversity in penile morphology across species. Some species possess such distinc-

Continued on following page

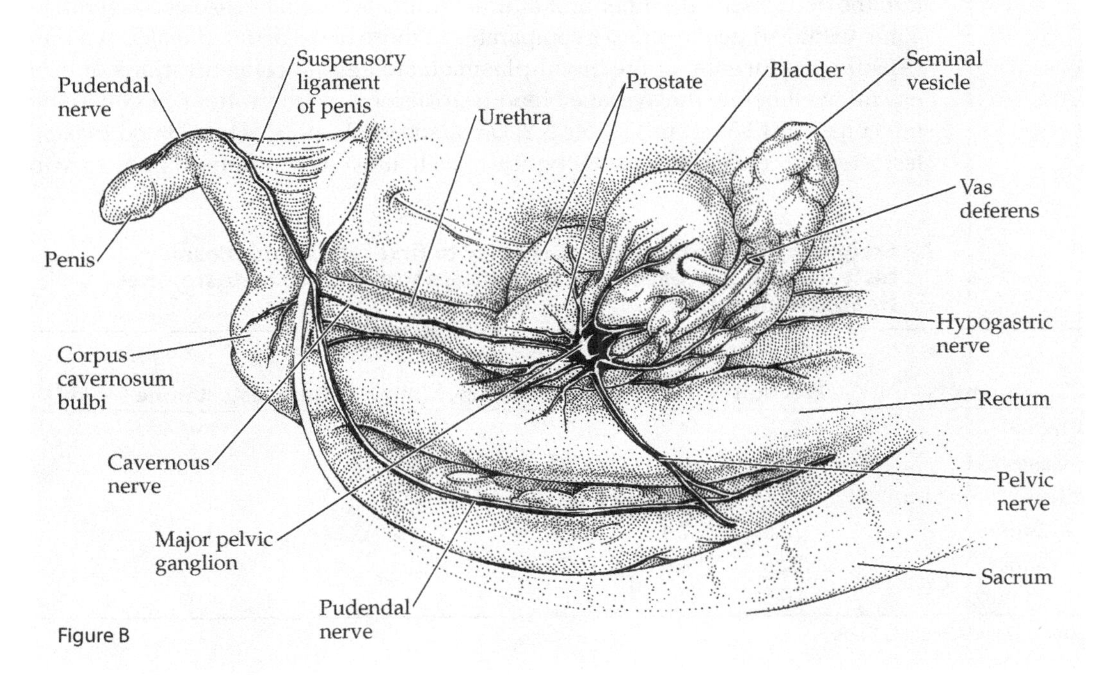

Figure B

BOX 5.2 *continued*

tive penes that they can be classified taxonomically by this organ. This species-specific penile morphology has evolved to maximize fertilization of conspecific females. Enormously diverse mechanisms of penile erection have also evolved among mammals, involving different combinations of smooth muscle, vascular tissue, and striated muscles. In rats, penile erections are mainly the result of neural signals that contract striated muscles, causing erection of the glans penis (see Figure B). In humans, penile erections are the result mainly of vascular changes that cause the corpora cavernosa to fill with blood, resulting in tumescence of the penile body (Tanagho et al., 1988).

The innervation of the penis and the details of how penile erection in humans is mediated by the nervous system have been established, but the details are incomplete (McConnell et al., 1982; Tanagho et al., 1988). It has been known for over a century that it is possible to induce penile erections with electrical stimulation of certain pelvic nerves in rats and dogs (Eckhardt, 1863). Direct electrical stimulation of the rat cavernosus nerve results in tumescence of the penile shaft—because of corpus cavernosum blood engorgement—but not tumescence of the glans (Burgers et al., 1991; Quinlan et al., 1989). The latter may be the result of a somatic response produced by the erect corpora cavernosa; contraction of the bulbocavernosus muscle may compress the erect shaft against the os penis in the glans, resulting in straightening of the glans.

Normal penile reflexes are maintained or restored by testosterone. In one study, castrated rats were implanted with Silastic capsules of testosterone that varied in length (from 2 to 18 mm). The hormone seeps out of the capsules at a constant rate; therefore, longer capsules result in more hormone being released into the body tissue fluid per unit of time. Animals bearing capsules longer than 6 mm exhibited penile reflexes comparable to those of uncastrated males. A 6 mm capsule produced average blood plasma testosterone concentrations of 0.79 ng/ml, well below the average blood testosterone concentrations of gonadally intact males (1.95 ng/ml) (Table 5.2; Davidson et al., 1978). Thus, blood plasma testosterone concentrations are typically well above the minimum necessary to

TABLE 5.2 Frequency of sexual reflexes, latency to first reflex, and plasma testosterone concentrations in intact, castrated, and testosterone-treated male rats

	Sexual reflex				
Treatment	**Erection**	**Flip**	**Cup**	**Latency (min)**	**Plasma testosterone (ng/ml)**
Intact	22.2	4.2	1.3	5.5	1.95
Castrated	12.3	0.5	0.04	9.9	<0.2
Testosterone implant					
2 mm	18.4	2.4	0.04	6.5	0.40
6 mm	22.5	6.0	0.6	5.4	0.79
18 mm	20.8	3.5	1.0	6.5	1.09

Source: Davidson et al., 1978.

maintain copulatory behaviors or penile reflexes, so a slight decrease, even 30%, from normal blood androgen concentrations should not be expected to influence mating behavior. Even an average reduction of blood testosterone concentrations of 50% would probably not affect behavior.

Testosterone per se also does not seem to regulate penile responses directly. Rather, testosterone serves as a precursor to DHT, which directly regulates penile responses. DHT maintains or restores penile reflexes to pre-castration levels in both spinally transected and spinally intact rats; estradiol, another testosterone metabolite, does not seem to affect penile reflexes (Gray et al., 1980; Hart, 1979; Meisel et al., 1984). Castrated rats implanted with Silastic capsules of estradiol maintained mating behavior at a level comparable to gonadally intact animals; however, the rate at which penile reflexes decreased in estrogen-treated males was similar to that in untreated castrated individuals (Meisel et al., 1984). Thus, it appears that testosterone, after being aromatized to estradiol in neural tissue, mediates copulatory behavior, whereas testosterone, after conversion to DHT, mediates penile reflexes and sensitivity to tactile feedback (Meisel and Sachs, 1994). Again, testosterone appears to act as a prohormone, from which steroids affecting central nervous system processing and peripheral sensory receptor sensitivities are produced to ensure successful copulation.

Brain Mechanisms of Male Mating Behavior

Behavior is mediated by the central nervous system. The results of the experiments discussed above suggest that hormones produce their effects on male reproductive behavior by acting on the central nervous system tissues controlling reproductive behaviors. But what are the neural systems underlying reproductive behavior? In order to answer this question, a number of neural manipulations have been employed. Historically, brain lesioning techniques have been used to find out where in the brain sexual behavior is regulated. The logic behind lesioning techniques is that removal of a critical component of the neural mechanisms underlying sexual behavior should result in disruptions of sexual behavior. Initially, rather large lesions were performed to study sexual behavior. Perhaps not surprisingly, removing the entire neocortex of rats diminished their sexual behavior; removal of the frontal cortex also effectively disrupted rat copulation (Larsson, 1962; Lashley, 1938). More recently, researchers have tried to limit the scope of brain lesions to locate more precisely the brain areas that regulate sexual behavior.

THE PREOPTIC AREA The region of the brain anterior to the hypothalamus, especially the preoptic area (POA) (Figure 5.13), appears to be critical for integrating environmental, physiological, and psychological information prior to and during successful copulation (Crews and Silver, 1985; Sachs and Meisel, 1988). The POA contains several nuclei from which axons project to other brain regions. Some studies have focused on the region of the POA along both sides of the midline of the brain, called the medial preoptic area (MPOA). In virtually all species studied to date, lesions of the POA in adult males eliminate sexual performance,

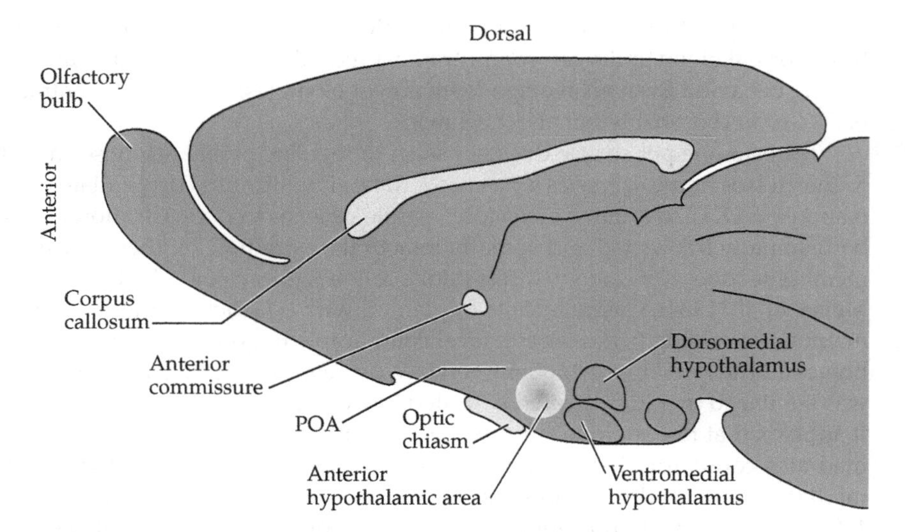

5.13 Regions that are essential to the control of sexual performance in male rats, seen in a schematic sagittal section of the brain. The preoptic area (POA) appears to be especially important for the integration of environmental, physiological, and psychological information prior to and during successful copulation; POA lesions reduce or eliminate male sexual behavior in virtually every vertebrate species examined. The POA apparently regulates endocrine function by interacting with hypothalamic nuclei and mediates parasympathetic functions associated with male copulation. After Pfaff, 1980.

although sexual motivation appears unaffected (but see Paredes, 2003): male rats with POA lesions fail to mount females even when tested 8 months after surgery, although they will press bars or run mazes to gain access to females (Ginton and Merari, 1977; Heimer and Larsson, 1966). Neither testosterone treatment nor access to multiple females compensates for POA lesions.

What does a POA lesion accomplish? It destroys cell bodies that send out axons to many other parts of the brain. Many of these neurons use dopamine as their neurotransmitter, so postsynaptic dopaminergic neurons that receive messages from these POA neurons are deprived of input (Hull et al., 1997). Treating POA-lesioned male rats with lisuride, a chemical that mimics dopamine, transiently activates copulation in these animals, with many of them copulating to ejaculation (Hansen et al., 1982). Presumably, the dopaminergic neurons destroyed by a POA lesion are part of a neural circuit that integrates and regulates copulatory behavior. In the absence of dopamine–receptor interactions "downstream" from the POA, copulatory behavior is not observed. By replacing the dopamine with an agonist that can interact with and excite neurons "downstream" from the POA, copulatory behavior is restored. If this hypothesis is true, then drugs that increase dopamine synthesis or stimulate postsynaptic dopamine receptor sites should also facilitate copulatory behavior among intact male rats, and this has been observed (Ahlenius and Larsson, 1984; Napoli-Farris et al., 1984; Paglietti et al., 1978; Sachs, 1995b). Similarly, one might predict that drugs that suppress dopaminergic activity would reduce male rat sexual behaviors, and this has also been observed (for example, Ahlenius and Larsson, 1984; Napoli-Farris et al., 1984). The claim that dopaminergic axons are mainly responsible for mediating male copulatory behavior in rats is strengthened by the observation that no other pharmacological treatment, including GnRH or naloxone, reinstates copulatory behavior in POA-lesioned rats. Taken together, these results indicate that the dopaminergic neurons destroyed by POA lesions are necessary for normal copulatory behaviors in male rats (Mas, 1995).

It appears that activation of μ opioid receptors in the medial preoptic area occurs after male sexual behavior. When male rats were allowed to mate to ejaculation, μ opioid receptors were activated and internalized in MPOA neurons within 30 minutes, and this process continued for 6 hours post-copulation (Coolen et al., 2004). Prior treatment of rats with naloxone, an opioid antagonist, prevented the internalization of μ opioid receptors after copulation. These results support the hypothesis that male sexual behavior evokes secretion of endogenous opioids, and that the MPOA is part of the brain circuitry mediating the rewarding properties of sexual behavior (Coolen et al., 2004).

Social history can dramatically affect the outcome of POA lesions in young rats. Juvenile male rats with POA lesions reared in social isolation never copulate as adults (Twiggs et al., 1978), but, similar lesions in juvenile rats have virtually no effect on adult copulatory behavior if the lesioned rats are reared in heterosexual groups (Twiggs et al., 1978). Exactly what component of group living ameliorates the effects of POA lesions remains unspecified. A reasonable hypothesis is that social interactions somehow elevate dopamine levels. This result is intriguing because it shows the importance of social conditions in mediating rodent brain plasticity and reproductive function after a substantial neural insult, and because social conditions also play a major role in the development of normal sexual behavior among primates (Money, 1988).

Remote cues from females result in so-called non-contact erections in rats (Sachs et al., 1994). These non-contact erections are analogous to psychogenic erections in humans that occur in response to visual, auditory, chemosensory, or imaginative stimuli (Meisel and Sachs, 1994). Thus, studies of non-contact erections should help us to trace the neural circuits involved in sexual arousal or motivation prior to copulation. Lesions of the medial amygdala (see below) inhibit non-contact erections; however, lesions of the bed nucleus of the stria terminalis (BNST) or paraventricular nucleus of the hypothalamus (PVN) cause only mild impairments, and lesions of the medial preoptic area have no obvious effect (Liu et al., 1997b).

THE CHEMOSENSORY SYSTEM Olfaction is critical for successful expression of male reproductive behavior among many rodent species. Chemosensory cues are also critical in mediating many other social interactions. Experimental blocking of the sense of smell usually results in social and reproductive behavioral deficits.

The **olfactory bulbs** are located at the front of the brain (see Figure 5.13) and are made up of two anatomically distinct regions, the main olfactory bulbs and the accessory olfactory bulbs (Figure 5.14). The olfactory neurons are bipolar cells

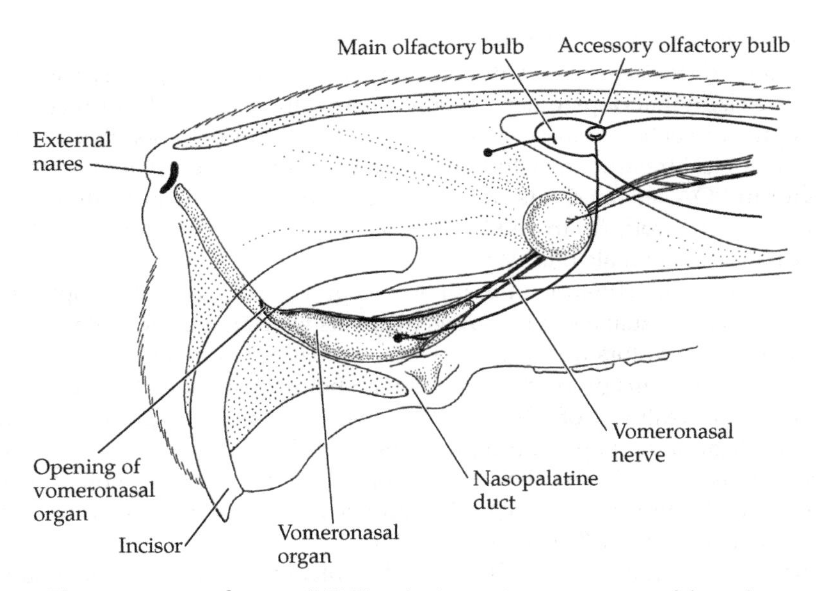

5.14 The vomeronasal organ (VNO) and other major components of the rodent chemosensory system, seen in a lateral view of the guinea pig snout. This structure may be absent in most primate species, but is essential to normal rodent reproductive behavior. When male rodents engage in anogenital investigation of females prior to mounting, vaginal chemosignals are maneuvered to the roof of the mouth, then "pumped" through the nasopalatine duct into the opening of the VNO. Other signals may enter via the external nares (nostrils) and may be pulled down to the VNO from the floor of the nares. Neural information moves from the VNO to the accessory olfactory bulbs via the vomeronasal nerve; thus, information originally obtained from activation of VNO receptors is processed by the accessory olfactory bulbs. Information from the olfactory receptors is processed separately in the main olfactory bulbs. After Wysocki, 1979.

located in the olfactory neuroepithelia at the rear of the nasal cavity. Their axons form the first cranial nerve and terminate in the main olfactory bulbs (Scalia and Winans, 1976). In many mammalian species, a portion of the olfactory neuroepithelium is discretely organized into a **vomeronasal organ** (also called Jacobson's organ), an encapsulated structure on each side of the nasal septum located near the floor of the nasal cavity (see Figure 5.14; Moulton, 1967). Rodents have highly developed vomeronasal organs, but this structure is regressed or absent in some primate species (for example, great apes and humans) and aquatic mammals. The neurons of the vomeronasal organ connect to the accessory olfactory bulb via the vomeronasal nerves (Alberts, 1974; Moulton, 1967; Wysocki, 1979).

Historically, several methods have been used to impair the chemosensory system in order to ascertain its behavioral and physiological role (reviewed in Alberts, 1974; Murphy, 1976). The most common such technique is olfactory bulbectomy (surgical removal of the olfactory bulbs), but this method has some drawbacks. Removal of the olfactory bulbs destroys approximately 4% of the entire central nervous system in rodents (Cain, 1974). Consequently, substantial neural damage can occur, and separation of the effects of such damage from the effects of sensory loss after bulbectomy is not always possible. Furthermore, the olfactory bulbs integrate nonsensory functions as well as sensory functions (Alberts, 1974), and the proportion of damage to sensory versus nonsensory functions caused by olfactory bulbectomy may vary among species. Finally, the procedure is irreversible.

Several acute, somewhat reversible, methods of impairing olfaction have been used to reduce the neural damage associated with olfactory bulbectomy. One frequently used technique is intranasal infusion of zinc sulfate ($ZnSO_4$) (Alberts, 1974; Alberts and Galef, 1971), which temporarily destroys the olfactory neuroepithelia, rendering the animal anosmic (unable to smell). The chemosensory receptors in the nasal cavity regenerate after several weeks. Two problems with using $ZnSO_4$ are uncertainty about its side effects and about the extent of the olfactory impairment (Murphy, 1976). Anesthesia of the nasal epithelia is another method used to produce acute olfactory impairment (for example, Doty and Anisko, 1973).

The results of olfactory bulbectomy have varied among studies and among species. Surgical ablation of the olfactory bulbs of sexually naive rats had no discernible effect on subsequent mating behavior in one early study (Stone, 1922). In contrast, a later study revealed that many bulbectomized rats were sexually impaired; some stopped mating completely after the surgery (Beach, 1942a). Sexually naive rats exhibited profound behavioral deficits after bulbectomy in other studies (Beach, 1942a; Bermant and Taylor, 1969). Further analyses of the impairments in mating behavior following olfactory bulbectomy revealed that male rats failed to achieve ejaculation after a series of intromissions or simply did not initiate copulation at all, suggesting that both sexual performance and motivation were impaired by olfactory bulbectomy (Larsson, 1969). Olfactory bulbectomy completely eliminated sexual behavior in male house mice (*Mus musculus*) (Rowe and Smith, 1972; Whitten, 1956a), but removal of the olfactory bulbs had variable effects on the copulatory behaviors of male guinea pigs (*Cavia porcellus*) (Beauchamp et al., 1977). Local anesthesia of the nasal mucosa (Doty and Anisko, 1973) and olfactory bulbectomy both virtually eliminated sexual behavior in male Syrian hamsters (*Mesocricetus auratus*) (Murphy and Schneider, 1970; Winans and Powers, 1974).

Copulation can be activated in bulbectomized rats by techniques that increase the general level of arousal in the animals. Thus, a mildly painful tail pinch or an electric shock to the flank will stimulate bulbectomized rats to copulate to ejaculation. Such arousal "therapy" is only temporarily restorative; additional arousing stimuli must be administered prior to subsequent mating sessions conducted a few days later (Meisel et al., 1980). These findings suggest that olfactory bulb tissues, or neurons that are connected to the bulbs, are part of a neural circuit involved in male sexual motivation among rodents.

Inputs from both the main olfactory neurons and the vomeronasal organ are apparently necessary for rodent mating behavior (Guillamón and Segovia, 1997). Olfactory bulbectomy tends to destroy both the main olfactory bulbs and the accessory bulbs, so differentiation of the contributions of the olfactory and vomeronasal inputs requires manipulations of the respective sensory receptors. While treatment with $ZnSO_4$ alone had no effect, destruction of the vomeronasal nerve alone stopped mating behavior in about one-third of the male hamsters tested (Powers and Winans, 1975). Thus, the vomeronasal organ has an important, but not a critical, role in normal reproductive behavioral function in these animals. In contrast, ablation of the vomeronasal organ of male house mice and pine voles (*Microtus pinetorum*) eliminated the surges of LH associated with the presence of females (see below) and stopped mating behavior (Lepri et al., 1985; Lepri and Wysocki, 1987; Wysocki et al., 1983). Because the deficits in mouse and vole mating behavior following either olfactory bulbectomy or vomeronasal organ ablation are similar, it appears that the vomeronasal organ/accessory olfactory bulb system is equally or more important in mediating mating behavior than the olfactory neuroepithelia/main olfactory bulb system.

In mice and rats, two large chemical families of chemosensory receptors (the V1Rs and V2Rs) have been characterized in the distinct VNO regions of the olfactory neuroepithelia (Dulac and Torello, 2003). The VNO neurons express specific receptors, which appear to respond to either male or female urine; other neurons do not discriminate between male and female urine, suggesting that other attributes of the chemostimuli are encoded by these cells (Dulac and Torello, 2003). Recently, it was suggested that a functional vomeronasal organ is necessary for males to discriminate between males and females during mating (Stowers et al., 2002). Although mice with their vomeronasal organs surgically ablated no longer preferred urinary stimuli from estrous females, these males as well as males with intact vomeronasal organs, preferred to mount an estrous female rather than a castrated male (Pankevich, et al., 2004). Thus, the vomeronasal organ does not seem necessary for discrimination of sexual partners, but appears to be important for detection of nonvolatile components of chemosensory stimuli such as urine that may be important in prolonging contact with appropriate mating partners (Pankevich et al., 2004).

THE ROLE OF THE AMYGDALA Projections from the accessory and main olfactory bulbs travel to the **amygdala** (from the Greek for "almond"), an almond-shaped structure located in each temporal lobe of the brain, which is critical for the integration of sensory information important in sexual behavior. Two regions of the amygdala have been studied in rodents: the basal and lateral collections of neuronal cell bodies (the basolateral nuclei) and the cortical and medial nuclei (corticomedial nuclei) (Figure 5.15). Removal of the basolateral nuclei of the amygdala generally does not affect the reproductive behavior of male rodents, but does reduce sexual motivation (Everitt, 1990). Lesions of the corticomedial nuclei, on the other hand, increase the ejaculation latencies of rats (Giantonio et al., 1970) and completely abolish copulation in male hamsters (Lehman and Winans, 1982). Information from the amygdala is relayed to the MPOA via the stria terminalis and the ventral amygdalofugal pathway. Predictably, lesions of these relay structures produce reproductive deficits similar in nature to corticomedial amygdala lesions (Figure 5.16; Giantonio et al., 1970).

Exposure to the chemosignals in the vaginal secretions of female hamsters produces sex-specific behaviors in the recipient animal: females mark over the scent, whereas males initiate copulatory behavior aimed at the source of the scent (Swann and Fiber, 1997). Although these behaviors are markedly different, the responses of both sexes to the chemosignals of female hamsters involve the main

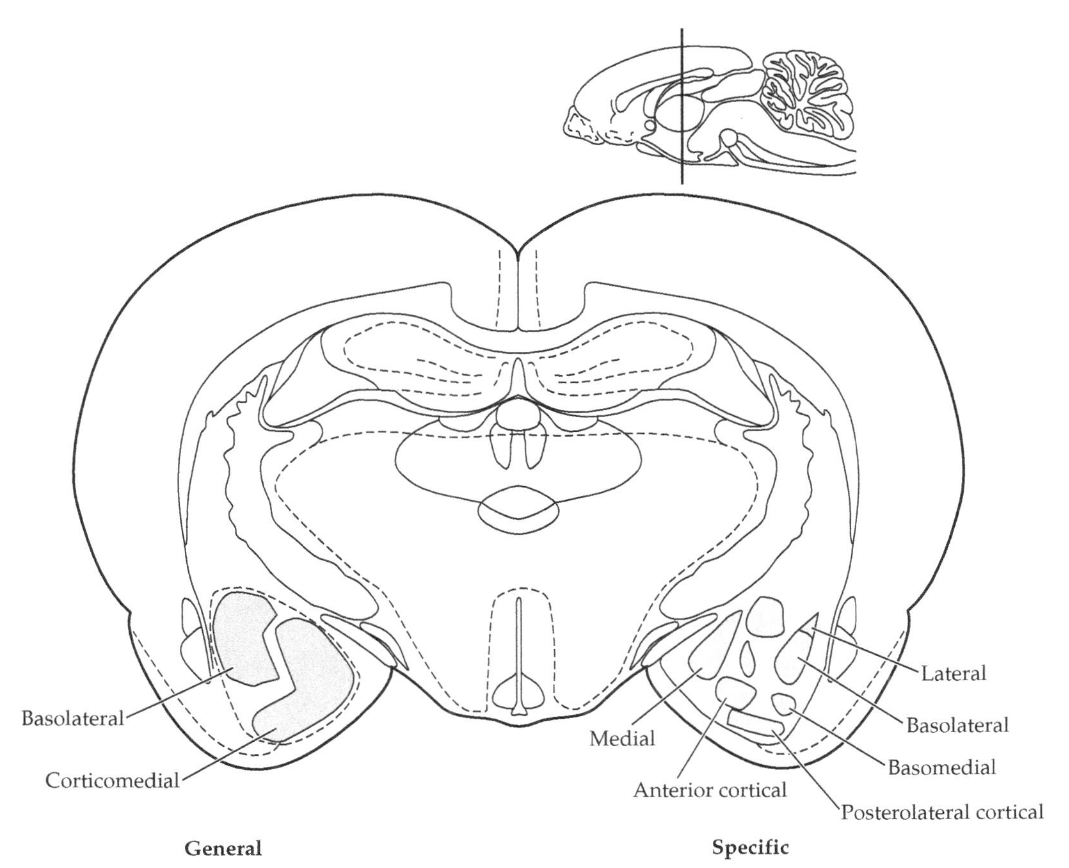

5.15 The amygdala, an almond-shaped structure located in each temporal lobe, seen in a schematic drawing of a coronal section of the rat brain. Two general amygdaloid regions, the basolateral and corticomedial nuclei (left), have been studied extensively because they receive neural input from the main olfactory bulbs. Destruction of the corticomedial nuclei, but not the basolateral nuclei, severely affects male copulatory behavior in rodents. The specific nuclei of the basolateral and corticomedial amygdala that receive olfactory input are depicted at the right.

olfactory system. In females, the neural circuitry involved includes the medial nucleus of the amygdala and the posterior medial subdivision of the bed nucleus of the stria terminalis. In addition to these two neural components, the magnocellular subdivision of the medial preoptic area is activated in males exposed to female vaginal secretions (Swann and Fiber, 1997). The integrity of this brain region is necessary for normal mating behavior in male hamsters (Swann et al., 2003). The bed nucleus of the stria terminalis of males can be activated by vaginal secretions only if plasma testosterone concentrations are sufficient (i.e., only in gonadally intact males) (Swann and Fiber, 1997).

ELECTRICAL STIMULATION AND RECORDING STUDIES In addition to lesion studies, researchers have employed electrical stimulation and recording studies to locate

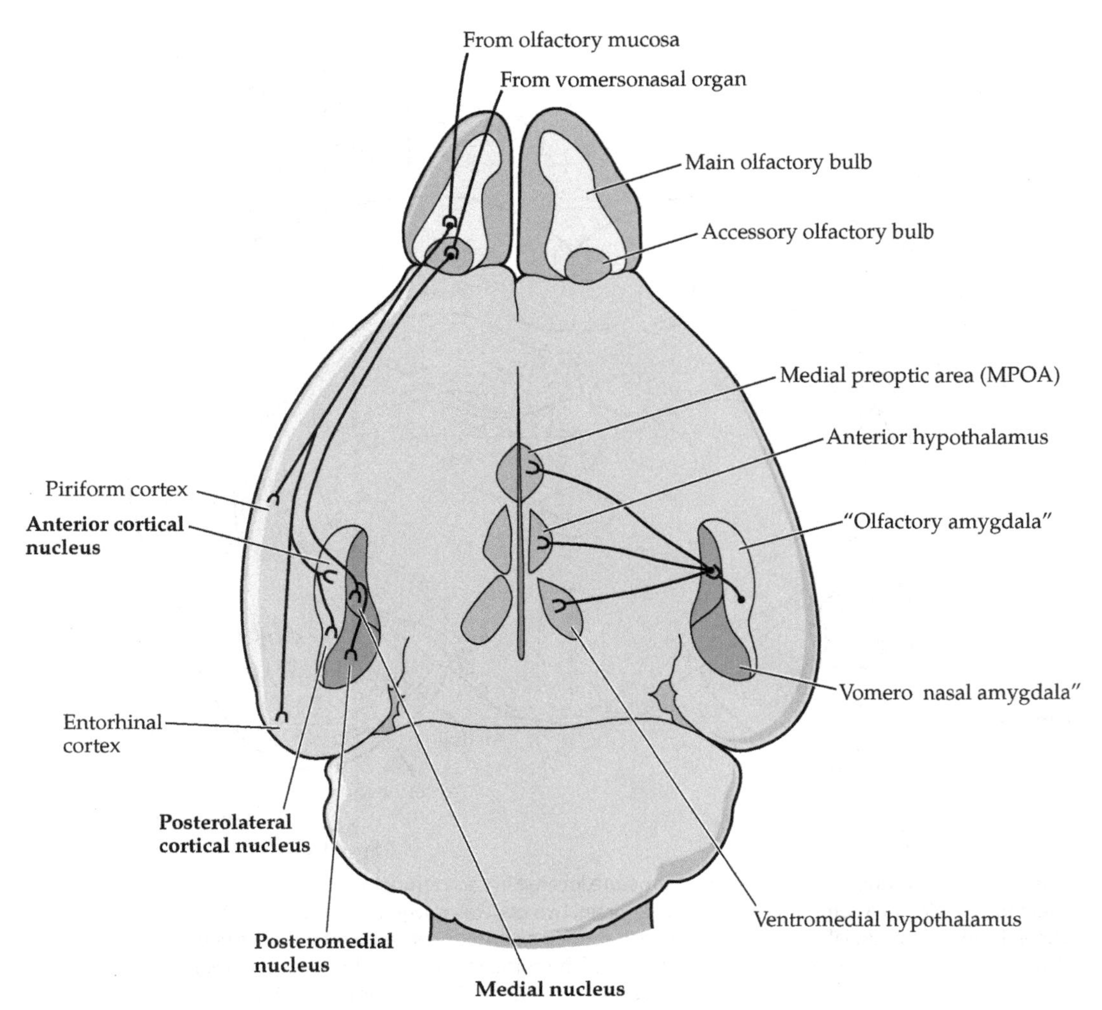

5.16 Neural pathways in the rat olfactory system shown in a schematic horizontal section. Axons from the olfactory mucosa (shown at left) synapse in the main olfactory bulb, where sensory information is sent to the piriform and entorhinal cortex and the anterior and posterolateral cortical nuclei of the amygdala; axons from the vomeronasal organ (VNO) synapse in the accessory olfactory bulb, which sends axons to the medial and posteromedial nuclei of the amygdala. (Right) Because of these projections, the cortical portion of the amygdala can be considered the "olfactory amygdala," and the medial portion the "vomeronasal amygdala." The olfactory amygdala innervates the vomeronasal amygdala, from which signals are sent to central structures including the medial preoptic area (MPOA), the anterior hypothalamus, and the ventromedial hypothalamus. Courtesy of Ruth Wood.

the neural circuitry underlying male reproductive behavior. Generally, electrical stimulation of a brain site produces a behavioral response that is opposite from the effect of destroying the area. Recording of the electrical activity in the brain in response to specific stimuli can also reveal what parts of the brain are involved in a behavioral response.

Recall that POA lesions, especially MPOA lesions, disrupt copulatory behaviors. Electrical stimulation of the MPOA accelerates ejaculation in male rats (Malsbury, 1971; Van Dis and Larsson, 1971). The MPOA appears to be crucial for integrating important external and internal information. Prior to making a successful mating response, a male rat must orchestrate and organize a multitude of stimuli. The environmental context, including the time of day, and stimuli associated with the estrous female, as well as the endocrine state of the male and his memories of previous sexual encounters, if any, must be processed and integrated in the MPOA. Lesions of the MPOA not only disrupt sexual behavior, but also interrupt several other motivated behaviors, including maternal behavior (see Chapter 7), locomotory behavior (King, 1979), drinking (Mogenson et al., 1980; Rolls and Rolls, 1982), and thermoregulatory behaviors (Satinoff and Prosser, 1988; Szymusiak and Satinoff, 1982).

Male rats must integrate auditory, olfactory, and tactile sensory cues in order to mate successfully. The MPOA appears to be crucial for processing this sensory information (Hull et al., 1997; Melis and Argiolas, 1995). Early studies found that adult male rats with no previous sexual experience required at least two of these three sensory inputs, and it was thought that it did not matter which two sensory channels were available. Rats with previous sexual experience were thought to require only one source of sensory information to engage in successful copulation (Beach, 1942b; Stone, 1923). Current views suggest that olfaction and somatosensation (tactile cues) are critical for the expression of appropriate copulatory behaviors, although the specific somatosensory requirements for male copulatory behaviors remain unknown (Stern, 1990).

As the male's brain processes incoming olfactory sensory information, a sufficiently powerful stimulus evokes a neural response from the MPOA, which results in the appropriate motor output, as well as a burst of GnRH release from the hypothalamus, which begins an endocrine cascade resulting in elevated testosterone secretion (Purvis and Haynes, 1974). Chemosensory stimuli from an estrous female induce electrical activity in the MPOA as well as in the olfactory bulbs of a male rat. Castration of the male does not affect the electrical activity in the olfactory bulbs; in other words, the chemosensory cues associated with the female continue to be processed at this early level of sensory input into the brain. However, the electrical activity of the MPOA, several synapses downstream from the olfactory bulbs, is no longer evoked by estrous female odors. The neurally coded sensory signal from the bulbs no longer influences the output of the MPOA of castrated males (Pfaff and Pfaffmann, 1969). Testosterone replacement therapy appears to amplify the chemosensory signal so that the MPOA again responds with neural activity (Figure 5.17).

IMPLANT STUDIES As we unravel the basic neural circuitry underlying sexual behavior, we should presumably discover where exactly hormones interact with the central nervous system to mediate mating behavior. Brain implant studies

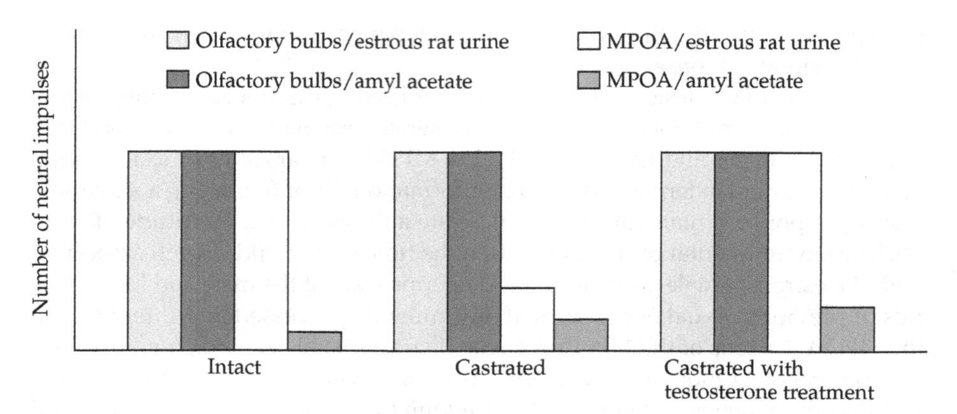

5.17 Castration reduces neural responsiveness in the MPOA. Electrical recording in the olfactory bulbs and MPOA of intact male rats reveals that the urine of estrous females causes increased neural activity in both regions. After castration, estrous urine continues to increase neural activity in the olfactory bulbs, but not in the MPOA. Testosterone replacement therapy restores MPOA responsiveness to estrous urine to levels seen in intact rats. After Pfaff and Pfaffmann, 1969.

have provided some clarification of these questions. Implants of crystalline testosterone into the MPOA of castrated male rats facilitated copulation in 100% of the animals; conversely, no castrated rats with MPOA implants of crystalline cholesterol (a hormonally inert precursor of testosterone) mated (Davidson, 1966a). Implantation of testosterone into the MPOA of castrated male rats and simultaneous treatment with an androgen aromatization inhibitor resulted in mating frequencies much lower than those observed in castrated males with MPOA implants of estrogen (Christensen and Clemens, 1975). These results again suggest that testosterone exerts its effects on mating behavior via aromatization to estrogen. Recall that systemic injections of low doses of DHT did not affect the reproductive behavior of castrated male rats. However, DHT injections paired with implants of estradiol into the MPOA elicited full copulatory behavior in castrated males, providing additional evidence that estradiol mediates central mechanisms of mating behavior, whereas DHT is important in maintaining peripheral sensitivity in castrated males (Davis and Barfield, 1979). Nevertheless, androgens must be able to interact with central androgen receptors to initiate male sexual behavior. Male sexual behavior was inhibited in males that received intracranial implants of hydroxyflutamide, an androgen receptor blocker, into the preoptic area or hypothalamus, but not the amygdala or the septal region, which are parts of the limbic system involved in motivation (McGinnis et al., 1996).

AUTORADIOGRAPHIC AND IMMUNOCYTOCHEMICAL STUDIES To gain additional information about where sex steroid hormones might exert their influence on behavior, autoradiographic studies have been performed to map the distribution of sex steroid receptors in the central nervous system. After an injection of radiolabeled testosterone, evidence of receptor binding is found in several specific regions of the

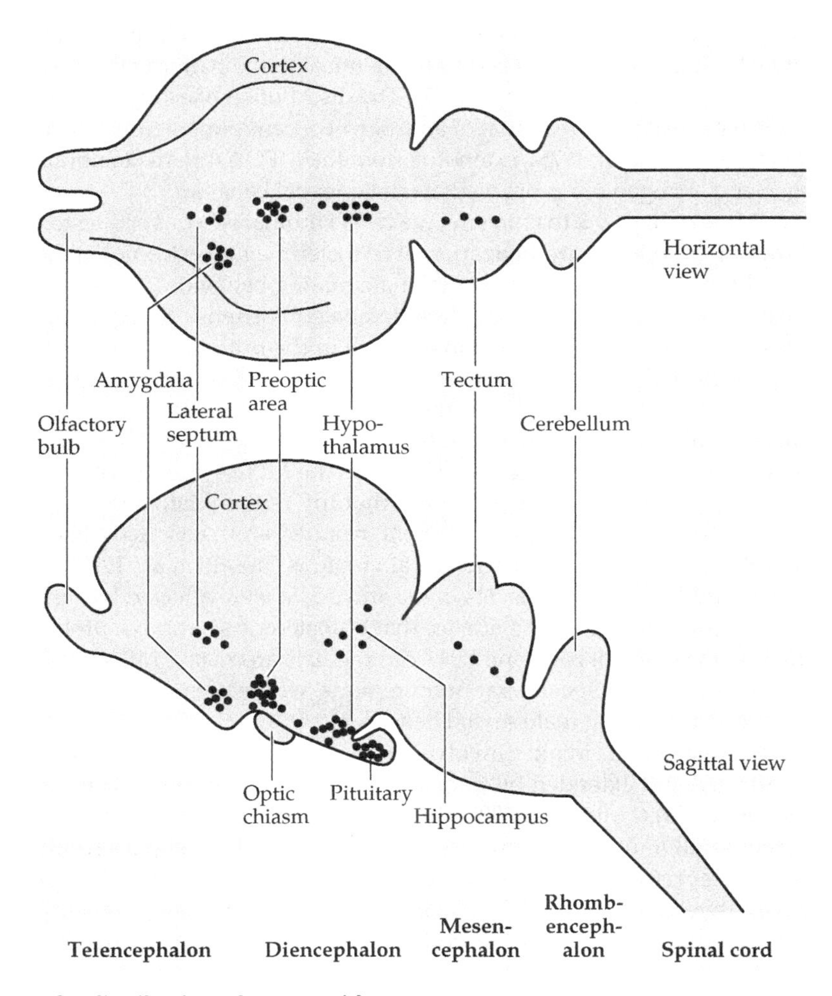

5.18 The distribution of sex steroid receptors depicted in a hypothetical generalized vertebrate brain. Note that most steroid receptors are clustered in the preoptic area, the lateral septum, the amygdala, the hypothalamus, the hippocampus, and the pituitary. After Morrell and Pfaff, 1978.

rat brain involved in sexual behavior: (1) the MPOA, (2) the bed nucleus of the stria terminalis (see Figure 5.19), and (3) the corticomedial nuclei of the amygdala (Figure 5.18; Sar and Stumpf, 1977). Other brain regions, such as the ventromedial nuclei (VMN) and arcuate nuclei (AN) of the hypothalamus, were also heavily labeled, but lesion studies suggest that these nuclei are not directly involved in the control of male mammalian mating behaviors. That is, males perform normally in mating tests after lesions of the VMN and AN.

Because testosterone can be converted into other androgens or into estrogens inside of neurons, binding studies that specifically examined estradiol and DHT receptors have also been conducted to ascertain the separate roles of these steroid hormones in specific neural target tissues. The distribution of cells that concentrate labeled DHT in the rat brain is essentially identical to the pattern of testosterone-concentrating cells (Sar and Stumpf, 1977). The distribution of estrogen-concentrating cells is more extensive than that of androgen-concentrating neurons (Commins and Yahr, 1985; Sheridan, 1978), extending from the MPOA into the forebrain. Both androgen and estrogen receptors mediate male sexual behavior.

Recall from Chapter 2 that an enzyme called aromatase converts testosterone into estradiol. In rodents, aromatization of testosterone into estradiol in the brain is required for the expression of normal male mating behavior (Christensen and Clemens, 1975; Clancy et al., 1995). A radiolabeled antisense RNA probe for rat aromatase mRNA can be used to show where in the brain aromatase mRNA is present (Roselli et al., 1997). Aromatase mRNA has been detected in many areas of the rat brain; the MPOA, VMN, and the medial and cortical nuclei of the amygdala are especially rich in aromatase activity (Roselli et al., 1997). Castrating male rats caused aromatase mRNA concentrations in the MPOA and hypothalamus to drop after 7 days; androgen replacement therapy (either testosterone or dihydrotestosterone), but not estrogen treatment, restored aromatase activity and aromatase mRNA concentrations in those brain regions (Roselli et al., 1997). Neither aromatase activity nor mRNA levels in the amygdala were affected by castration. These findings provide further evidence that testosterone serves as a prohormone in mediating male sexual behavior. DHT can regulate aromatase mRNA transcription and/or stability in specific rat brain regions, whereas estradiol can activate neural circuits regulating male sexual behavior.

Much of the neural circuitry involved in mediating male mating behavior has been confirmed and extended by tracking the activation of so-called **immediate early genes (IEGs)** (Hull et al., 2002). In neurons, these genes are activated early during the signal transduction process whereby extracellular signals result in the expression of specific genes. The nature and function of the "activation" of IEGs remain controversial (Hull et al., 2002); nevertheless, the presence of their protein products is thought to indicate the initial activation of the genetic machinery of neurons. The protein products of IEGs, such as the *fos*, *jun*, and *egr-1* families, can be detected by immunocytochemical methods. Analysis of IEG protein products in neurons has confirmed that copulatory stimuli activate neurons in several steroid-concentrating brain regions, including the MPOA, lateral septum, BNST, PVN, VMN, medial amygdala, as well as ventral premammillary nuclei, central tegmental field, mesencephalic central gray region, and perpendicular nuclei (Pfaus and Heeb, 1997) (Figure 5.19; also see Chapter 10). Although devoid of intracellular sex steroid hormone receptors, the ventral and dorsal striatum and the cortex also display significant activation after sexual behavior. Note that the MPOA, BNST, and corticomedial amygdala are regions of the brain that other methods have shown to be critical for regulating male sexual behavior. Although many nonspecific stimuli can "activate" neurons and increase IEG expression, it is a reasonable strategy to use IEG expression as one of several tools to identify neural circuits involved in sexual behavior. A putative neural circuit of male sexual behavior in rodents in depicted in Figure 5.20.

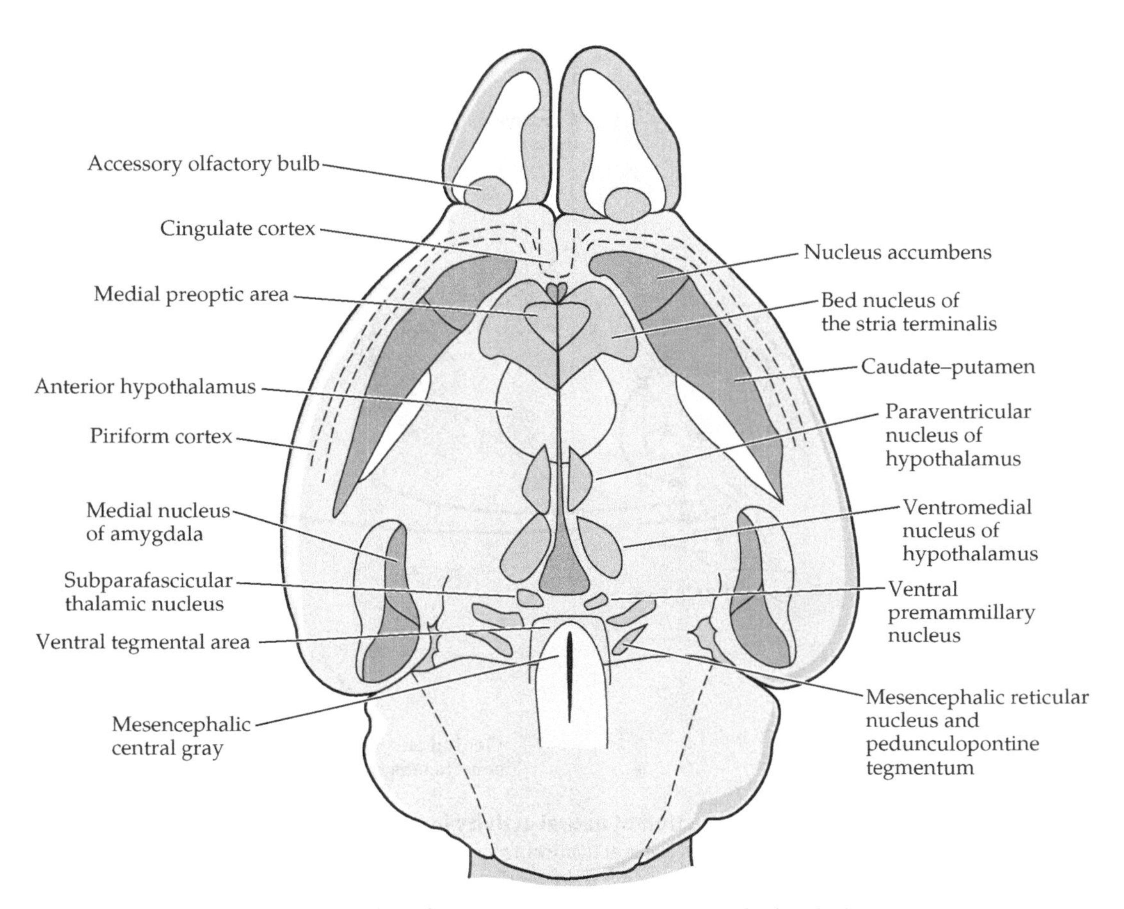

5.19 Brain regions in rodents that show *fos* activation after sexual stimulation. This schematic illustration of a horizontal section of rat brain depicts the various regions that are activated during male sexual behavior. After Pfaus and Heeb, 1997.

In one specific example, IEG expression has been used to identify the neural circuitry involved with ejaculation (Hull et al., 2002). The medial part of the paracellular subparafascicular nucleus (SPFp), which is located in the posterior thalamus, is activated during ejaculation in rats (Coolen et al., 2003a). This region of the brain is ideally located to serve as a processing center for sensory stimuli because olfactory and other sensory information is relayed through the thalamus. In addition, sensory information arrives in the thalamus from the spinal cord. Tract tracing studies revealed that the SPFp receives input from a cluster of neurons in the lumbar spinothalamic region of the spinal cord (Coolen et al., 2003b). This cluster of neurons is activated only during ejaculation, but not during other components of male sexual behavior (Truitt and Coolen, 2002). When these cells were lesioned with a drug that targeted only these specific neurons, rats failed to ejaculate, but all other components of their sexual behavior remained intact (Truitt et al., 2003).

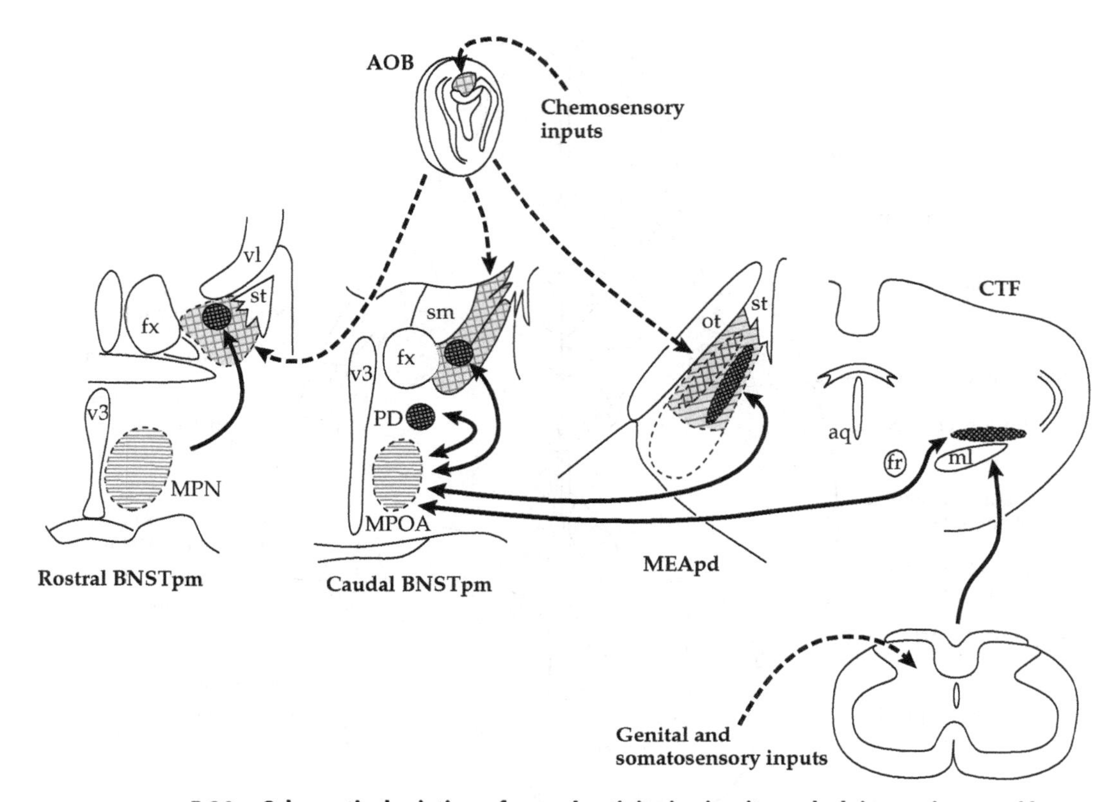

5.20 Schematic depiction of neural activity in circuits underlying male sexual behavior, as indicated by *fos* gene activation. *fos* activation by chemosensory cues is indicated by diagonal stripes from upper left to lower right. *fos* activation caused by ejaculation is depicted by solid shading. *fos* activation by consummatory behaviors are indicated by diagonal stripes from lower left to upper right. aq = aqueduct; AOB = accessory olfactory bulb; BNSTpm = posteromedial bed nucleus of the stria terminalis; CGF = central tegmental field; Fr = fasciculus retroflexus; fx = fornix; LSSC = lumbosacral spinal cord; MEApd = posterodorsal medial amygdala; MPN = medial preoptic nucleus; ml = medial lemniscus; MPOA = medial preoptic area; ot = optic tract; PD = posterodorsal preoptic nucleus; sm = stria medularis; st = stria terminalis; v3 = third ventricle; vl = lateral ventricle. From Hull et al., 2002.

THE ROLE OF NEUROTRANSMITTERS Presumably, if hormones are to influence behavior, they must affect neurotransmitter function. In the case of male sexual behavior, dopamine seems to play a central role (Melis and Argiolas, 1995; see above). Although many other neurotransmitters are involved, a discussion of the effects of hormones on every neurotransmitter system is beyond the scope of this book. We will look at dopamine as an example of how hormones and their receptors function to affect neurotransmission.

Some neurotransmitters act to stimulate sexual behavior, whereas other neurotransmitters act to inhibit it (Table 5.3). Dopamine appears to facilitate male sexual behavior by removing tonic inhibition (Chevalier and Deniau, 1990) of brain

TABLE 5.3 **Effects of various neurotransmitters on male sexual behavior**

Neurotransmitter	Effect on copulation	Effect on penile erection
Norepinephrine	α_1 receptor activity ↑ copulation α_2 receptor activity ↓ copulation β_1 receptor activity has no effect β_2 receptor activity ↑ copulation	α_1 receptor activity ↓ reflexive erection α_2 receptor activity ↑ reflexive erection β activity inhibits reflexive erection (β receptor subtype not specified)
Dopamine	Presynaptic activity ↓ copulation	Postsynaptic activity ↑ reflexive erection, but ↓ spontaneous erection
Serotonin (5-HT)	5-HT_{1A} activity ↑ copulation $5\text{-HT}_{1B/1C}$ activity ↓ copulation 5-HT_2 activity ↓ copulation	5-HT_{1A} activity ↓ erection 5-HT_{1C} activity ↑ spontaneous erection 5-HT_2 activity ↓ spontaneous erection
γ-Aminobutyric acid (GABA)	$\text{GABA}_{A/B}$ activity ↓ copulation	$\text{GABA}_{A/B}$ activity ↑ spontaneous erection
Acetylcholine	Inconclusive	Inconclusive
Endorphins	Activity ↓ copulation	Activity ↓ erection
Neuropeptide Y	Activity ↓ copulation	No effect
Oxytocin	Activity ↑ copulation	Activity ↑ spontaneous erection

Source: Meisel and Sachs, 1994.
↑ = facilitates; ↓ = decreases.

regions that mediate sensorimotor abilities (Hull et al., 1997). Three major integrative dopaminergic systems regulate sexual motivation, genital responses, and body postures during copulation in male rats (Hull et al., 1997): the nigrostriatal tract, the mesolimbic tract, and the medial preoptic area (Putnam et al., 2001). Sensory input from an estrous female before or during copulation evokes the release of dopamine in each of these three tracts (Mas, 1995; Hull et al., 1995). The nigrostriatal tract is the largest dopaminergic system and mediates the initiation of movement. The nigrostriatal tract is damaged in humans with Parkinson's disease, which is characterized by tremors, slow movements, and impairments in the initiation of movement. In rats, the nigrostriatal tract appears to be involved in the muscular movements associated with mounting females (Robbins and Everitt, 1992). The mesolimbic tract is important in reward and appetitive behaviors such as brain self-stimulation, drug addiction, and food, alcohol, and water intake, as well as sexual behavior (Hull et al., 1997; Balfour et al., 2004). The mesolimbic tract terminates in the nucleus accumbens, and blocking or stimulating dopamine receptors in this region decreases or restores, respectively, behaviors associated with sexual motivation (Pfaus and Phillips, 1991; Everitt, 1990); however, activation of the nucleus accumbens may only enhance generalized appetitive behavior (Hull et al., 2002).

As noted above, lesions of the MPOA impair sexual performance, but spare sexual motivation, in male rats (Everitt, 1990). However, more recent studies suggest that the MPOA contributes to sexual motivation (Hull et al., 1995, 1997; Paredes, 2003). The MPOA is critical for male sexual behavior in all vertebrate animals thus far studied (Meisel and Sachs, 1994). This structure receives input from and sends output to virtually every sensory modality (Hull et al., 1997, 2002). This reciprocity in connections provides a means for the MPOA to modulate sensory processes, and for sensory information to affect the integration of sexual motivation. Many dopaminergic neurons within or connected to the MPOA possess receptors for sex steroid hormones (Simerly and Swanson, 1986). Similarly, steroid hormone receptors are embedded within the membranes of many neurons constituting nondopaminergic neurotransmitter systems. Testosterone replacement therapy in sufficient doses to restore copulation in castrated male rats also increased medial preoptic dopamine secretion (Putnam et al., 2003).

Microdialysis has revealed a consistent pattern of increased dopamine concentrations in the MPOA of male rats in the presence of an estrous female housed behind a perforated barrier (Hull et al., 1997). A variety of other stimuli, including access to highly palatable food or a male conspecific, did not affect MPOA dopamine secretion. Castration attenuated the female-induced elevation of dopamine in the MPOA (Du et al., 1998). However, castrated males that did show an elevation of dopamine in the MPOA copulated with females, regardless of their testosterone concentrations (Hull et al., 1995) (Figure 5.21). Dopamine appears to integrate sensorimotor information in the MPOA, resulting in facilitation of male sexual behavior (Hull et al., 2002).

The mesolimbic tract is important in controlling both pathological behaviors such as drug addition and biologically relevant rewarding motivated behaviors (Balfour et al., 2004). The mesolimbic tract consists primarily of dopaminergic neurons that project from the ventral tegmental area (VTA) in the hindbrain to the nucleus accumbens in the forebrain. Local interneurons that secrete GABA inhibit firing in these dopaminergic neurons; these interneurons are, in turn, modulated by activation of μ opioid receptors (Balfour et al., 2004). The μ opioid receptors were observed to be internalized in neurons in the VTA after copulation or after exposure to sex-related environmental cues. These stimuli also activated dopaminergic neurons in these brain regions (Balfour et al., 2004).

Dopamine also seems to facilitate sexual behavior in primates (Hull et al., 1997; Melis and Argiolas, 1995). Treatment of male rhesus monkeys (*Macaca mulatta*) with a dopamine agonist, apomorphine, resulted in dose-dependent enhancements of sexual responses toward females that males could see, hear, and smell, but not touch. For instance, low doses (25–100 μg/kg) of apomorphine caused yawning, a behavior often observed in sexually aroused male rhesus, whereas moderate doses (50–200 μg/kg) caused penile erections and masturbation, occasionally to ejaculation (Pomerantz, 1990). Interestingly, the males required the presence of a female in order to show these sexual behaviors. These results are reminiscent of the effects of testosterone in facilitating male sexual behaviors in the presence of the appropriate stimuli, and they also suggest that hormones mediate male primate behavior by acting via dopaminergic pathways. Apomorphine also induces penile erections in men, even those with erectile dysfunction (Lal et al., 1984, 1987).

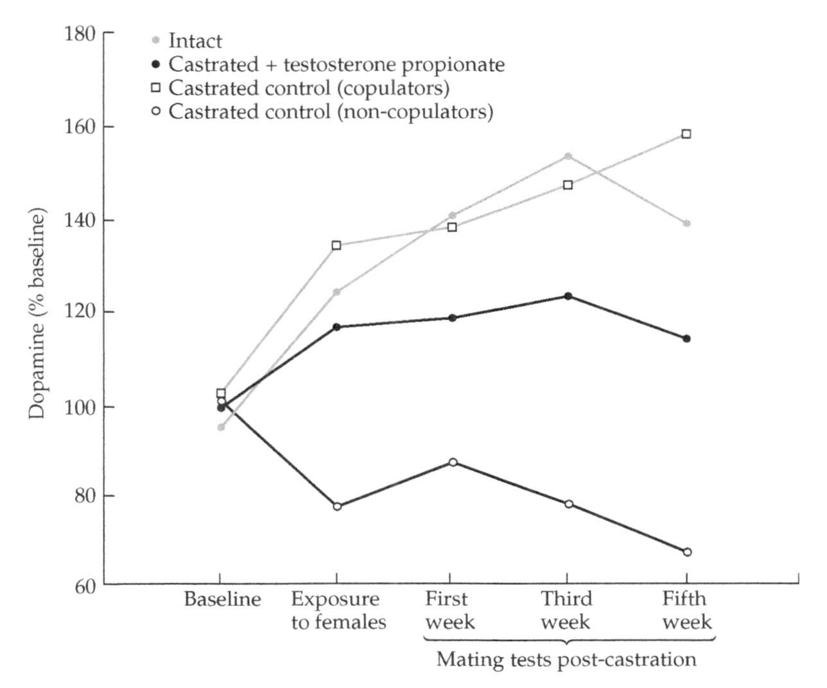

5.21 Extracellular dopamine in the MPOA is elevated by cues from the female. Dopamine concentrations in the MPOA of male rats were measured by microdialysis in the presence of an estrous female rat housed behind a perforated barrier. In the absence of the precopulatory rise in dopamine, males failed to copulate when the barrier was removed. In contrast, males that showed an increase of approximately 50% in extracellular dopamine in the MPOA copulated with females, regardless of testosterone concentrations. After Hull et al., 1995.

GENE KNOCKOUT STUDIES As described in Chapter 1, new advances in molecular biology have made it possible to perform specific genetic manipulations. An increasingly common genetic technique used in behavioral endocrinology studies is the removal (knockout) of the genetic code for a specific hormone or its receptor. Virtually all of the knockout studies of hormone–behavior interactions to date have been performed on mice (*Mus musculus*) (Nelson, 1997). Thus far, male sexual behavior in mice has been examined in several knockout varieties. For example, mice that have had the gene encoding the μ opioid receptor deleted display reduced mating activity (Tian et al., 1997). The μ opioid receptor mediates the pain-attenuating features of endogenous and exogenous opioids such as morphine and heroin. The "pleasurable" aspects of copulation may require the μ opioid receptor (van Furth et al., 1995).

Nitric oxide (NO) is well established as a neurotransmitter (Nelson et al., 1997) (Box 5.3). NO is formed from arginine by an enzyme called nitric oxide synthase (NOS). There are three isoforms of NOS. Male mice that lack the gene that codes for the neuronal isoform of NOS ($nNOS^{-/-}$) displayed persistent mounting behav-

BOX 5.3 Erectile Dysfunction, Nitric Oxide, and Viagra

Approximately 50% of men over the age of 40 have some degree of erectile dysfunction (ED). Because penile erections are mainly vascular events, medical problems such as diabetes, high blood pressure, high cholesterol, and heart disease, all of which affect blood flow, are frequently underlying causes. Therefore, it is important for individuals experiencing erectile dysfunction to be screened for these conditions. Furthermore, if these conditions are not treated, there could be a progression of erectile dysfunction.

Androgens are essential for the expression of normal sex drive (libido) in men, but their role in the maintenance of the erectile response in humans is controversial. Castration in rats causes both a loss of penile reflexes and a marked reduction in the erectile response to electrical stimulation of the cavernosal nerve. Both of these effects can be reversed by testosterone or DHT, but not estradiol, replacement (Lugg et al., 1995).

However, nitric oxide (NO) has emerged as a key mediator of erection. The physiological mechanism of penile erection involves the release of NO in the corpora cavernosa (the paired chambers in the penis where blood is trapped) during sexual stimulation (Burnett, 1995). Nitrates such as nitroglycerin are broken down to nitric oxide, a very transient and reactive free radical that not only acts as a vasodilator, but is vital in the regulation of mitochondrial respiration in both skeletal muscle and other organ systems. Nitric oxide is normally produced by the vascular endothelial tissues to facilitate blood flow, stimulate cellular respiration, and help prevent plaque formation on interior vessel walls. NO also appears to function as a neurotransmitter. However, NO is a very toxic free radical that can cause substantial tissue damage in high concentrations, especially in the brain. In a stroke, for example, large amounts of nitric oxide are released from nerve cells to cause damage to surrounding tissues.

In the penis, NO is released from both neuronal and endothelial sources (Nelson et al., 1997). NO combines with the enzyme guanylate cyclase, which converts GTP to cyclic guanosine monophosphate (cGMP). The resulting elevation of cGMP concentrations in the penis produces smooth muscle relaxation in the corpora cavernosa, allowing the inflow of blood.

Recall that the loss of penile reflexes in rats following castration can be reversed by testosterone or DHT replacement. Production of the synthetic enzyme (nitric oxide synthase) that produces NO and a by-product, citrulline, from arginine is also dependent on DHT (Lugg et al., 1995; see also Box 8.1). DHT production wanes in men after the age of 40, so the increased prevalence of erectile dysfunction at this age might reflect the interaction between DHT and NO.

In March 1998, the Pfizer Pharmaceutical Company released the drug Viagra (the generic name is sildenafil citrate; see Figure A) for the treatment of erectile dysfunction. Sildenafil acts by potentiating

ior, even toward anestrous females that refused to cooperate with their mating attempts (Nelson et al., 1995). Testosterone concentrations were not elevated in these $nNOS^{-/-}$ mice. Pharmacological inhibition of NO decreases dopamine secretion in the MPOA, which could facilitate male sexual behavior (Lorrain et al., 1996; Dominguez et al., 2004). The reason for these conflicting effects of genetic versus pharmacological inhibition of NOS on sexual behavior remains unspecified.

Viagra (sildenafil citrate)

Figure A Viagra (sildenafil citrate)

the effects of NO on the mechanism of penile erection; it enhances the effect of NO on cGMP by inhibiting the enzyme phosphodiesterase type 5 (PDE5), which breaks down cGMP in the corpora cavernosa. Individuals can take sildenafil 30–60 minutes prior to sexual activity. When sexual activity stimulates NO release, inhibition of PDE5 by sildenafil causes increased levels of cGMP in the corpora cavernosa, resulting in the desired smooth muscle relaxation and inflow of blood to the corpora cavernosa. Thus, sildenafil does not cause an erection per se, but enhances or potentiates an erection. Current evidence suggests that sildenafil does not affect sexual motivation.

Sildenafil at recommended doses (50–100 mg) has no effect on penile erection in the absence of sexual stimulation. At appropriate doses and in the presence of sexual stimulation, however, sildenafil is remarkably effective. One study showed a covariant relationship in sildenafil dosage and the number of patients reporting improvement in erectile function: with a 25 mg dose of sildenafil, 63% of patients reported improvement, while 82% reported improvement with a 100 mg dose (see Figure B). Although anecdotal information suggests that sildenafil also enhances female sexual performance, no controlled studies of females have been conducted at this time.

Shortly after its release, Viagra became the fastest-selling drug in history. At $10 per pill, current annual sales of Viagra are approximately $1.5 billion. However, it is estimated that only 13% of men suffering from erectile dysfunction currently seek medication. With the recent release of two new drugs to treat ED, Levitra (vardenafil) by Bayer and GlaxoSmithKline and Cialis (tadalafil) from Lilly and Icos, we can expect the marketing of these drugs to the remaining 87% (estimated at 30 million men in the United States) to be fierce.

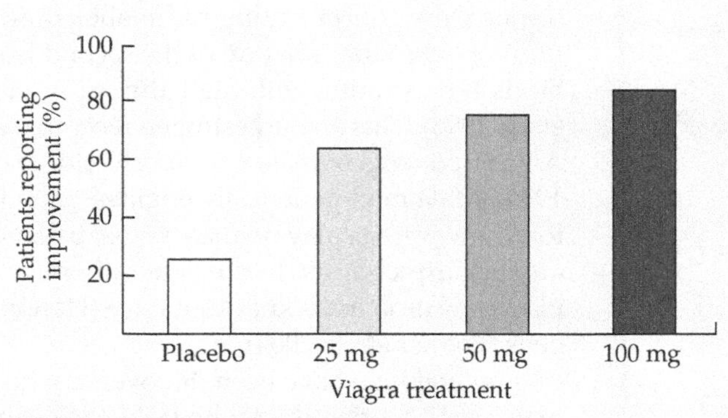

Figure B Effectiveness of sildenafil

Another gas, carbon monoxide (CO), has also been implicated as a neurotransmitter (Dawson and Snyder, 1994). Mice with targeted deletions of the gene encoding heme oxygenase-2 ($HO2^{-/-}$), an enzyme that produces endogenous carbon monoxide in neurons, displayed altered mating behavior. $HO2^{-/-}$ mice showed markedly diminished ejaculatory behavior both in mating tests and when ejaculations were elicited by neural stimulation *ex copula* (Burnett et al., 1998). In many cases, the effects of a particular missing gene on sexual behavior are investigated only when a breeding colony of knockout mice either fails to produce offspring or increases at a reduced rate. In these cases, the investigators did not anticipate the impairments of male sexual behavior (Burnett et al., 1998; Nelson et al., 1995; Tian et al., 1997).

In contrast to the serendipitous, unexpected, and somewhat subtle results of the targeted deletion of the genes encoding the μ opioid receptor, nNOS, and HO2, the deletion of genes encoding androgen or estrogen receptors would be expected to result in predictable, severe defects in male mating behavior. However, as we saw in Chapter 3, male mice in which the α estrogen receptor has been knocked out (αERKO mice) show only subtle deficits in male sexual behavior, mainly reduced ejaculations (Eddy et al., 1996; Ogawa et al., 1995); otherwise, they show mounting and intromission behaviors comparable to those of normal, wild-type (WT) mice (Ogawa et al., 1995). If male WT and αERKO mice are castrated and implanted with Silastic capsules filled with testosterone, so that their blood concentrations of testosterone are similar, then a deficit in the rate and onset of intromissions is observed in the αERKO as compared with the WT mice (Rissman et al., 1997b).

Recall from Chapter 3 that, based on the results of early injections of estrogen receptor antisense RNA, estrogen receptors appear to be necessary for the organization of masculinization and defeminization of behavior (McCarthy et al., 1993a). How do we reconcile the results of these studies with the observation that male αERKO mouse mating behavior is essentially unaffected (Ogawa et al., 1995)? In vivo autoradiography studies indicated that neurons in αERKO mouse brains were concentrating radiolabeled estrogens (Shughrue et al., 1996). This finding led to the isolation of the second isoform of the estrogen receptor, which binds 17β-estradiol with high affinity, in rat prostate and ovarian tissue (Kuiper et al., 1996). This second estrogen receptor was designated ERβ (Shughrue, 1998). All components of male sexual behavior are intact in βERKO mice (Ogawa et al., 1999). Male mice genetically engineered to lack both the α and β estrogen receptors, however, display no male sexual behaviors (Ogawa et al., 2000). These results are in sharp contrast to the generally intact, but reduced, mating behavior displayed by aromatase knockout mice (Honda et al., 1998), which produce no estrogens (Bakker et al., 2004).

Several men have been discovered who lack the gene for aromatase, and one man has been identified who lacks the gene for ERα (Simpson and Davis, 2000). A man lacking aromatase activity because of a mutation in an aromatase gene was given a series of hormone treatments, during which time he maintained a diary of his sexual behavior and thoughts (Carani et al., 1999). Psychosexual and sexual behavioral evaluations were performed before and during testosterone treatment or three doses of estradiol treatments. The gender identity interview and the sexological interview indicated that this individual was clearly male, and that his orientation was heterosexual. Significant modification of the patient's sexual behavior, including increased libido, frequency of sexual intercourse, masturbation, and erotic fantasies, occurred only during the estradiol treatments. Treatment with estradiol also reduced his scores on the Beck Depression Inventory (showing improved mood) and the Spielberger Trait Anxiety Inventory (STAI) (Carani et al., 1999). The authors concluded that estrogens do not affect gender identity and sexual orientation in men, but may influence male sexual activity (Carani et al., 1999).

Conclusions

How do steroid hormones affect male mating behavior? Taken together, the studies on rodents that have been described suggest that steroid hormones activate certain genes in neurons. Recall that neurons, like other cells, have a nucleus, and within that nucleus are genes located along the chromosomes. Hormones turn on DNA transcription within neurons to produce proteins (see Chapter 2). Generally, steroid hormones activate certain genes whose products increase DNA transcription and protein production by other genes. The result of this genomic expression may be an increase or a decrease in the number of hormone receptors, or a change in the presence or amounts of enzymes that affect neurotransmitter or neurohormone production, neurotransmitter receptor production, or even recycling of neurotransmitters. For example, dopamine may not be recycled quickly in some circumstances; increased dopamine concentrations in the MPOA stimulate male sexual behavior (Bitran et al., 1988; Hull et al., 1997, 2002). Thus, a genomic signal initiated by steroid hormones that reduced dopamine recycling rates, thereby increasing dopamine concentrations, could stimulate male copulatory behavior in the appropriate social context. New protein synthesis in response to steroidal influences on neuronal genes may allow new neuronal connections to be made, facilitating sensory input associated with estrous females or even memories of prior sexual encounters (for example, see Pfaff and Pfaffmann, 1969; Stern, 1990). For instance, implants of testosterone into the MPOA of rats stimulate dendritic branching and other structural changes there (Meisel and Sachs, 1994). These effects are presumably due to testosterone's effects on genes that code for specific structural proteins.

Mating behaviors are very complex, and many endocrine, neural, and environmental stimuli must interact to produce successful copulation. Further discoveries of the physiological mechanisms by which hormones affect male mating behavior await our understanding of precisely how a change in protein synthesis in a brain region becomes amplified into a behavioral response.

Social Influences on Male Mating Behavior

As our discussions of the endocrine and neural mechanisms underlying mating behavior in male rodents have shown, hormones affect behavior by changing the thresholds at which specific behaviors are displayed in response to particular stimuli. But many environmental factors, including social cues, can modulate hormone–behavior interactions. One environmental stimulus that greatly affects males is the presence of females. Females affect both sexual motivation (appetitive behavior) and performance (consummatory behavior) among males.

If a male rat is placed in a box with an estrous female, he will mate to satiation; that is, he may ejaculate seven or eight times over the course of several hours. He is operationally defined as satiated when no mating behavior is observed for 45 minutes or longer. However, if a new female is introduced into the mating arena, the so-called satiated male often immediately resumes copulation. This phenomenon has been termed the "Coolidge effect," in honor of an anecdote involving the former President and First Lady of the United States (Bermant et al., 1968).*

The Coolidge effect—the enhanced mating performance of males with novel females—is a striking phenomenon in some species, including rats and cattle, but in general it is not a very robust effect among mammals. Among humans, there are many anecdotal reports of enormous individual differences in the stimulatory effects of novel females on males' copulatory performances, but not very many convincing data. There is a complete absence of this phenomenon in some other species. Males of some monogamous rodent species, for example, do not show the Coolidge effect; for a male prairie vole (*Microtus ochrogaster*), the opportunity to mate with a novel female results instead in resumption of copulation with his mate (Getz et al., 1987; Gray and Dewsbury, 1973). The endocrine bases, if any, of the Coolidge effect remain unspecified. However, its neurobiological bases may involve an augmentation of dopamine release in the nucleus accumbens (mesolimbic tract) during copulation (Fiorino et al., 1997).

In order to study the effects of hormones on reproductive behavior in the laboratory, the physical space in which animals interact is often simplified. Typically, the mating behavior of rats is assessed in a small, empty aquarium. The rich complexity of the natural habitat, including space, odors, and escape paths, is eliminated to control as many variables as possible during the experiment. This simplification of the physical space in which behavior is examined may make experiments easier to analyze, but it also removes much of the rich environmental context that may play important roles in the regulation of behavior.

Natural sexual behavior may be unintentionally constrained by simplified testing environments. The stimuli associated with copulation normally induce luteal function and subsequent progesterone production in female rats (see Chapter 6). The timing of the vaginal stimulation caused by copulation may be important for inducing luteal function; thus the interval between, and the pacing of, intromissions varies on a somewhat species-specific basis. The optimal pattern of intromissions for a species has been called the "vaginal code" (Diamond, 1970). Under natural conditions, female rats control the pacing of reproductive activities to match closely the physical stimuli necessary for optimal reproductive efficiency. Female rats can adapt to the typical mating arena provided in most behavioral endocrinology studies, in which the pace of mating is faster than the

---

*According to the story, President Coolidge and his wife were visiting a farm in the Midwest and were given separate tours by the owners. Both President and Mrs. Coolidge noted during their tours that only one rooster was associated with the large flock of hens. Mrs. Coolidge asked the farmer how many times per day the rooster engaged in romance. "Several times a day," the farmer replied. "Please relay that information to the President," responded the First Lady, apparently impressed by the rooster's performance. Later, during his tour, President Coolidge was given this same information about the copulatory prowess of the rooster. The President pressed further, "Same hen each time?" "Oh no," replied the farmer, "A different hen each time." "Please relay that information to Mrs. Coolidge."

physiological optimum. However, they may produce less progesterone during these behavioral assessments, and thus produce fewer offspring, than females mating in naturalistic settings. Providing a seminatural environment to rats results in a slower pace of mating behaviors (McClintock, 1987). The reason for this reduction in mating pace is not certain, but the animals appear to interact with the environment and also to move farther away from one another during mating. The running and chasing observed during rodent mating bouts conducted in seminatural enclosures is often qualitatively and quantitatively different from what is observed during typical laboratory mating tests. In many species for which both laboratory and field hormonal data have been collected, hormone concentrations vary substantially between the two situations. In any case, it is clear that the females control the pacing of mounts, intromissions, and ejaculations when in larger enclosures, whereas males control the pacing of sexual behavior in the traditional small enclosures (Larsson, 2003).

Not only the physical environment, but also the social environment of rodents is artificially simplified in laboratory studies of hormonal effects on reproductive behavior. Typical studies with laboratory rats pair a single male with a single female and record the resulting behavior. However, rats in nature do not necessarily mate in pairs; several males and females may mate in a group. Sperm competition studies suggest that the male having the most ejaculations with a female, or the last ejaculation during her series of copulations, usually sires the most offspring. Presumably, strategies have evolved for males to compete for copulations and to time their copulatory acts to maximize their chances of fertilizing a female. One interesting advantage for rats mating in groups (a *panogamous* mating system) is that because each sex can take turns mating with different partners, each sex can mate at a different "optimal" pace for fulfilling their respective neuroendocrine stimulus requirements (Figure 5.22; McClintock, 1984; see also Chapter 6).

The stimulus value of the female is another variable that may influence the outcome of behavioral endocrine investigations of male sexual behavior. Males have different mating responses to females in naturally occurring estrus than to females that are in artificially primed estrus (Hardy and DeBold, 1971). Males also respond differentially to females brought into estrus with different hormonal treatments. For example, male rats with corticomedial amygdala lesions responded slowly to estrogen-injected females in one intriguing study. However, there was no difference in the rate of copulation between lesioned and intact males when the female was brought into estrus with both estrogen and progesterone (Perkins et al., 1980). Perhaps the additional hormonal treatment provided females with enhanced stimulus value that overcame the effects of the brain lesion. Additional studies are required to understand the effect of the hormonal condition of estrous females on male mating behaviors.

The presence of female rodents, especially novel females, induces an elevation in blood plasma testosterone concentrations in male rodents during mating (Bronson and Desjardins, 1982a; Purvis and Haynes, 1974). Prior to mating, sensory cues associated with females can cause a rapid increase in circulating LH and testosterone concentrations in sexually experienced male mice (Batty, 1978). This

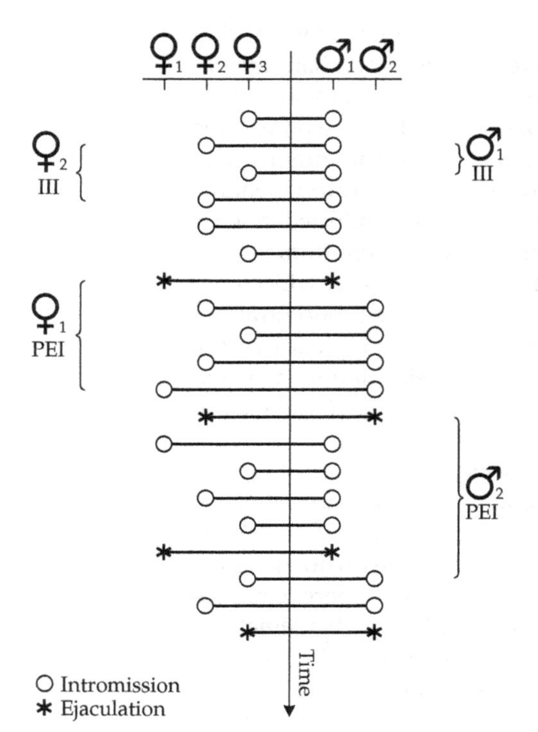

5.22 Copulatory sequence of rats mating in groups. In this example, three estrous females (left) are copulating with two males (right). A mating event is signified by either an open circle (intromission) or a star (ejaculation); the female and male participating in a given event are indicated by a horizontal line. The first mating event occurs when male 1 intromits with female 3; then female 2 receives an intromission from male 1, followed by male 1 intromitting female 3, then female 2 again. Thus, the inter-intromission interval (III) is different for female 2 than for male 1; in a more typical paired mating test, the III would be identical for each partner. Similarly, the postejaculatory interval (PEI) for female 1 is different than the PEI for male 2, but would be identical in a standard paired mating situation. The time line indicates the order and relative temporal relationship of mating events, but does not reflect actual temporal intervals. After McClintock, 1984.

response can be classically conditioned to previously neutral stimuli (Graham and Desjardins, 1980). A natural stimulus, normally referred to as the unconditioned stimulus, normally causes some reflexive response, called an unconditioned response. In Pavlov's famous studies, for example, food in the mouth, the unconditioned stimulus, naturally caused a dog to increase salivation, the unconditioned response. If a previously neutral stimulus, one that normally does not cause the biological response (in this case, a ringing bell), is paired repeatedly with the unconditioned stimulus, then the neutral stimulus, now called the conditioned stimulus, can cause the conditioned response (salivation) in the absence of the unconditioned stimulus (food).

Female presence causes a rise in blood LH and testosterone concentrations in male house mice. Exposure to the odor of wintergreen normally has no endocrine effects in mice. If female exposure is repeatedly paired with wintergreen odor, however, eventually the odor will evoke increased plasma LH and testosterone concentrations in male mice in the absence of a female. In other words, the males will "learn" to increase their hormonal concentrations whenever they experience the odor of wintergreen (Figure 5.23). The sights and sounds associated with estrous females may promote endocrine changes prior to the onset of mating because of previous associations between these stimuli and mating behavior. The role of learning in mating and pre-mating elevations in reproductive hormones, and thus in mediating copulatory behaviors, is not yet well

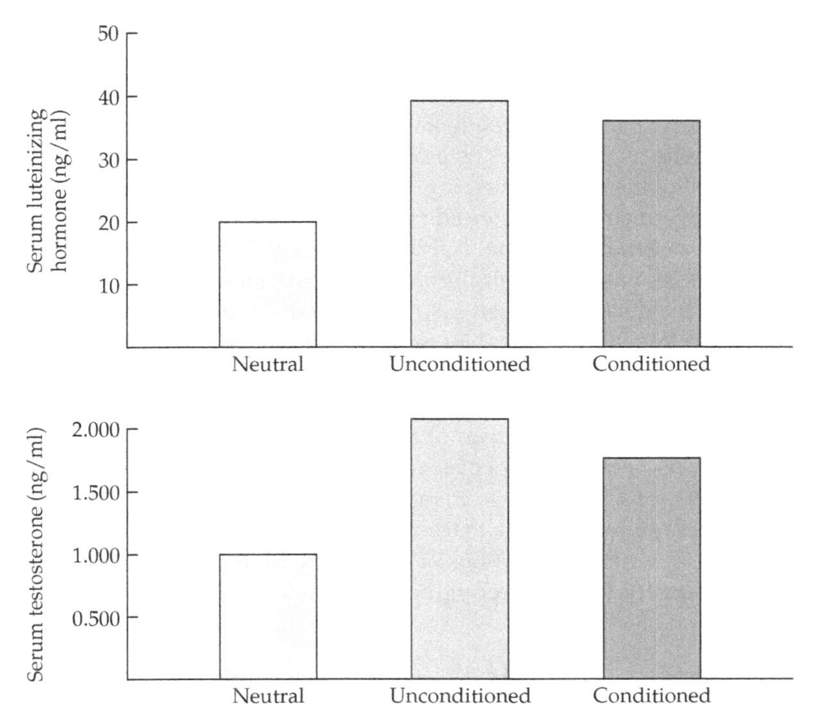

5.23 LH secretion can be modified by classical conditioning. Upon the first presentation of a neutral stimulus—in this case, the odor of wintergreen—to male house mice of the CF-1 strain, there was no change in serum hormone concentrations (Neutral). The mice were then repeatedly exposed to the wintergreen odor paired with the presentation of a female. After training, conditioned males showed increased LH and testosterone secretion in response to the wintergreen odor alone (Conditioned), at levels close to those shown by unconditioned males in response to a female (Unconditioned). After Graham and Desjardins, 1980.

understood, but this is one possible mechanism by which experience may interact with the endocrine system to preserve hormone-dependent behaviors in the absence of the hormone.

Also not well understood is the observation, initially reported by Steinach (1936) and later verified (Folman and Drori, 1966), that socially isolated castrated males display a more rapid post-castration reduction in reproductive organ mass than do heterosexually reared males. Social isolation is stressful for many mammalian species, and stress could mediate the reproductive effects reported in these studies (see Chapter 11). Again, social influences on reproductive physiology, morphology, and behavior are potent, but are unfortunately often ignored in the study of hormone–behavior interactions.

Individual Differences in Male Mating Behavior

Vast individual differences exist in the amount of sex drive and sexual performance observed among individuals of many species. That is, the sexual behavior of males falls along a continuum between hyposexual and hypersexual activity. If male rats are tested in typical time-limited mating tests, some will mate several times within the time constraints, but others will not mate at all. An early hypothesis proposed to account for individual differences in mating behavior was that animals with high sex drive had higher circulating testosterone concentrations than animals with low sex drive. This hypothesis was tested initially on guinea pigs prior to the availability of direct assays of circulating steroid hormones (Grunt and Young, 1952, 1953). Males were pre-screened for sexual activity and categorized as high, medium, or low sex drive males based on their number of ejaculations during a time-limited mating test. All the males were then castrated, and their copulatory behavior was tested during the following weeks (Figure 5.24). Low sex drive males stopped mating first, followed by the medium and high sex drive animals. After 16 weeks, all the males had completely stopped displaying mating behavior. They were then injected with low (50 µg/day) doses of testosterone propionate, and each male returned to his pre-castration level of sexual behavior. A similar study examining high and low rates of sexual activity in rats yielded similar results (Larsson, 1966). The results of these studies indicate that animals do not differ in their mating behavior because of different concentrations of hormones. More likely, differences exist in the target tissues that mediate reproductive behaviors.

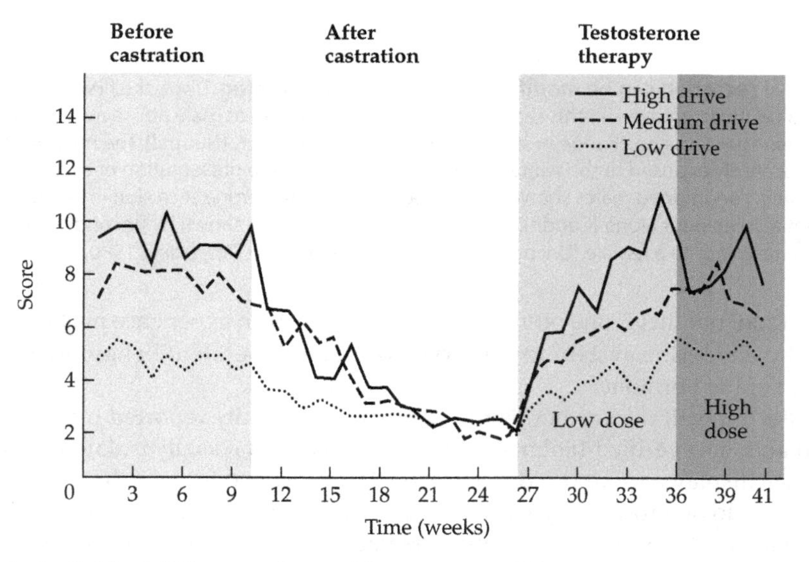

5.24 Individual differences in sex drive are retained following castration and restoration therapy. Guinea pigs that were classified as having high, medium, or low sex drives based on behavioral assessments were castrated. After castration, the sexual behavior score eventually dropped to an equal baseline for all three groups. Both low-dose and high-dose testosterone therapy restored copulatory behavior in the castrated males, and the differences in sexual behavior among the three groups were still observed. This study demonstrated that differences in blood androgen concentrations do not correspond closely to individual differences in sexual behavior, which are more likely due to variation in target tissue sensitivity to androgens. After Grunt and Young, 1952.

With technological advances in hormone measurement, it is possible to address the question of hormone concentrations directly. When such a study was conducted on laboratory rats, its outcome was in agreement with the earlier study by Grunt and Young on guinea pigs: differences in blood hormone concentrations did not account for the differences observed in the frequency of sexual behaviors. Males were identified as copulators or noncopulators, and blood plasma testosterone was assayed for all the animals. All the males in this study had testosterone concentrations between 2 and 3 ng/ml of plasma. The known copulators were then castrated, and testosterone-filled Silastic capsules of different lengths were implanted in them, resulting in different blood concentrations of testosterone. Normal sexual behavior was restored in most of the males, even those with very low testosterone concentrations. Seventy-two percent of the males ejaculated normally with only a 2 mm capsule, even though their plasma testosterone concentrations were much lower than normal values (Figure 5.25; Damassa et al., 1977).

This study demonstrates that very low concentrations of circulating testosterone can maintain sexual behavior. Why, then, do males have circulating andro-

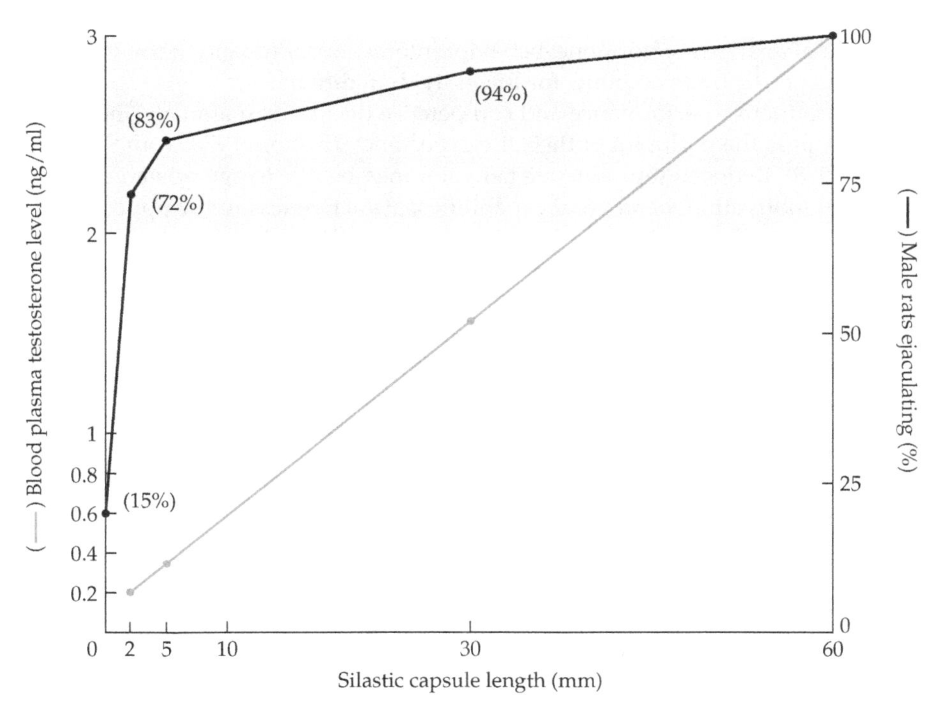

5.25 Clinically low testosterone concentrations can restore mating behavior in castrated rats. Rats that were known copulators and had plasma testosterone concentrations of 2–3 ng/ml (shaded area) were castrated. Most of the rats exhibited normal ejaculatory behavior after implantation with a 2 mm Silastic testosterone capsule, even though the capsule treatment resulted in plasma testosterone concentrations that were much lower than those found in intact rats. After Damassa et al., 1977.

gen concentrations an order of magnitude higher than necessary to maintain sexual behavior? Probably because local testis androgen concentrations must be very high to support spermatogenesis, and because mating behavior has evolved to depend on the diluted concentrations of androgens in the general circulation. Thus, large variations in circulating concentrations of testosterone (of even an order of magnitude) presumably do not affect behavior. Relatively high androgen concentrations are also important in aggression and in maintenance of other male traits. Indeed, high androgen levels support secondary sex characters that are often the criteria by which females choose mates (Andersson, 1994).

Differences among genetic strains in reproductive behavioral responsiveness to hormones have also been noted. In one study, for example, several strains of mice were examined for male copulatory behavior, and a strain was eventually discovered in which males continued to copulate for 2 years after castration (McGill, 1962, 1977). Usually mice, like rats, stop mating soon after castration, so genotype appears to be one of the most salient factors underlying the dependency of male sexual behavior on hormones in mice. But how does the genotype code for behavioral responsiveness to hormones? The answer to this specific question, and the answer to the general question of how the genotype codes for any behavior, must be obtained in order to make the link between molecular biology and organismal analyses of hormone–behavior interactions. Probably, a constellation of strain-specific traits accounts for the individual differences.

Reproductive performance and competence decline as mammals, including humans, pass the midpoint of their life expectancy (Bishop, 1970; Vom Saal and Finch, 1988). Reduced reproductive behavior may be due to age-related changes in sexual motivation, perceptual capabilities, attractiveness to the opposite sex, physical prowess, or some combination of these factors. Reductions in the frequency of sexual behavior may or may not correlate with changes in reproductive physiology or morphology. Aged mice (30 months old) have reduced gonadal mass, reduced rates of spermatogenesis, and low average blood plasma LH and testosterone concentrations compared with 6-month-old mice (Bronson and Desjardins, 1982b). Numerous syndromes can be described in which various physiological systems change, leading to decreased reproductive competence. But the most interesting question vis-à-vis issues of aging is why some healthy male mice can mate at 30 months of age and others cannot, despite equivalent blood concentrations of testosterone.

We currently do not understand completely the physiological mechanisms underlying individual differences in reproductive activity. Uptake and metabolism of testosterone in the MPOA does not differ between sexually active and inactive rats. However, estrogen receptor numbers in the MPOA are significantly lower in sexually nonresponsive male rats than in copulating males (Clark et al., 1985). Perhaps protein synthesis is differentially affected by hormones, causing changes in the synthesis of hormones, receptors, or structural proteins necessary for dendritic spines. Rats with inherited insensitivity to androgens lack specific androgen receptors, but have normal estrogen receptors; testosterone causes less of an increase in their sexual behavior than in normal rats (Beach and Buehler,

1977). With the exception of receptor numbers in the MPOA, we cannot yet point to any structural difference between the brains of copulators and those of noncopulators. Further analyses are necessary to understand individual, strain, and species differences in reproductive drive and performance.

Individual differences in adult sexual behavior may reflect different experiences during development. For example, if pregnant rats are stressed by exposure to bright lights and consumption of alcohol, their male offspring fail to ejaculate as adults (I. L. Ward et al., 1996). Although their adult blood LH and testosterone concentrations are in the normal range, these males can ejaculate only with pharmacological testosterone treatment. The effects of prenatal stress and alcohol exposure are maximal if they coincide with days 18 and 19 of gestation, a time when males normally secrete testosterone.

Male Sexual Behavior in Primates

Despite the importance of sexual motivation in human behavior, the physiological components of human sex drive and performance have not been well studied, and with only a few notable exceptions, scientific descriptions of human sexual behavior are rare. The theoretical bases of sexual behavior across humans and nonhumans are rarely linked; investigators in each field have developed separate scientific literatures and rarely interact (Pfaus, 1996). The best descriptions of the power of sexual motivation among men are often provided by playwrights, poets, and novelists. In literature, however, descriptions of sex drive and performance are masked or presented under the guise of burning passions, emotional imperatives, or even a desperate desire to marry. Shakespeare's Romeo, for example, is a classic tragic character because of his nearly obsessive desire for Juliet. Despite a number of obstructions, some social, others truly physical, Romeo is highly motivated to interact with Juliet, and we in the audience share his intense emotions as he risks everything for her. Their story is a great tragedy because the audience can see, although Romeo is blinded to it, the trajectory of his life as he attempts to overcome the obstructions in his path and obtain his goal of being with Juliet. Romeo's powerful sex drive, couched in terms of passion, overruns his life, and from the beginning of the story, we uneasily sense the sad end of the play. But to what extent does art reflect life, or in this case, hormones and behavior?

The Strength of the Sex Drive in Human Males

In their classic book, *Patterns of Sexual Behavior*, Clellan Ford and Frank Beach (1951) combined information about sexual behavior from three major sources: (1) anthropological data on human sexual behavior from non-Western societies, (2) the Kinsey studies on the sexual habits and attitudes of married Americans (Kinsey et al., 1948), and (3) data about sexual behavior in nonhuman animals. This synthesis of knowledge about sexual behavior laid the foundation for much of the subsequent research in human sexual behavior. One striking observation in this book is how powerful the motivation to engage in sexual activities truly is among humans as well as nonhuman animals. For example, young people in every culture examined often participate in premarital sexual activities, despite threats in some societies of extremely severe disgrace and punishment if discovered. This is particularly true among sexually restrictive societies, including North America. In another sexually restrictive society that inhabited the Pacific Gilbert Islands, a girl's chastity prior to marriage was required. If a sexual liaison was discovered, both parties were immediately put to death (Ford and Beach, 1951). Examples of instantaneous death sentences for unmarried couples who engage in sexual intercourse have been reported from many societies, both Western and non-Western. To engage in sexual activities, men often risk injury or death from a woman's disapproving relatives or husband. The lack of sexual restraint in the face of such drastic and immediate threats provides a striking and compelling indication of the power of the motivation driving the sexual behavior of men.

If instant death sentences are ineffective in controlling sex drive, then policymakers and educators may be naive in their belief that adolescents will curtail their sexual activities because of the vague possibility of an unwanted pregnancy or contracting a disease like AIDS that may not have any ill effects for years. To support this assertion, recall that 80 years ago, the possibility of contracting a horrible venereal disease, namely, syphilis, did not really reduce sexual behavior. In fact, the "Roaring Twenties" were a time of sexual liberation in both North America and Europe (Quétel, 1990). There was no cure for syphilis prior to the development of antibiotics, and this disease often led eventually to blindness and severe neurological dysfunction prior to a painful death. Treatments of the day included cellular poisons such as mercury and arsenic, which, like AZT treatment of AIDS today, had debilitating side effects. Until antibiotics became generally available, sexual abstinence or the use of condoms provided the best protection against syphilis. However, many men and women have reported that during sexual arousal they were so motivated to engage in sex that the possible consequences of the act were forgotten.

Understanding the physiological bases of the human sex drive would seem very important if we hope to reduce the number of unwanted pregnancies or prevent the spread of sexually transmitted diseases like AIDS. There is something different and possibly unique about sexual motivation, as compared with other motivated behaviors, that impairs decision-making processes. Do hormones contribute to the "clouding of logic" many individuals experience during sexual arousal? In addition, an understanding of the physiological mechanisms underlying typical human male sexual behavior is obviously required before we can understand undesirable behaviors such as rape or sexual molestation. But, as noted above, basic research on human sexual behavior is rare, and the physiological mechanisms of the human sex drive remain obscure. The majority of human sex research has focused on performance, primarily emphasizing the physiology of erectile function (Sachs and Meisel, 1988; Tanagho et al., 1988). With very few exceptions, funding sources have been reluctant to provide money for basic research on human sexual behavior. Sex research currently being conducted on animal models may help us to understand the physiological bases of human sexual behavior as well as the basic processes and mechanisms underlying all reproductive behavior. Animal research, mainly on rats, provides most of our knowledge of the effects of hormone–behavior interactions. However, basic data on human behavior and physiology would be useful in making informed public policies. Given the fundamental importance of understanding human sexual behavior, remarkably little research is presently being conducted.

Human Male Sexual Behavior: A Description

Before we describe what is known about the physiological mechanisms underlying human sexual behavior, a brief description of human sexual behavior will be provided (see also Chapter 6). Scientific descriptions of human male sexual behavior are available, the best known of which come from the so-called Kinsey reports of the late 1940s (Kinsey et al., 1948; Masters and Johnson, 1966; Money, 1988), but their utility is somewhat limited. As you might have guessed, human sex researchers, with clipboards and stopwatches in hand, have not placed large numbers of couples in glass-enclosed rooms and recorded the ensuing sexual behaviors. Yet, the sorts of descriptive analyses that have been conducted on rodents are necessary before issues of causation can be addressed in humans.

Like rodent mating behavior, sexual behavior in men can be divided into two components: (1) sex drive (appetitive behavior), which is sometimes called sexual motivation or libido, and (2) sexual performance (consummatory behavior), or potency. As we discovered in the discussion of rodent mating behavior, both components are required for normal sexual behavior, and one or both components can fail. Hormones appear to be crucial for maintaining both behavioral components, although much more research is needed to verify this claim.

There is no universal sexual position for humans. In contrast, if you see two dogs copulating at a distance, you can easily assign the sex of each participant with virtual certainty. If she is in estrus, the female stands relatively motionless, deflects her tail, and allows the male to mount her. The male mounts her from behind, clasps his forepaws around her sides, and repeatedly thrusts his hindquarters as he intromits (see Figure 5.8). This mating position, with the male mounting the female from the rear, is common among nearly all mammals, and few nonhuman examples of other mating positions exist. Aristotle reported in his *Historia Animalium* that hedgehogs (a European relative of porcupines) mated belly to belly. However, later observations proved Aristotle wrong. His notion was probably due to his disbelief that these beasts could manage to mate and avoid each other's sharp quills in any other fashion. But hedgehogs mate in the same position as dogs; that is, the male mounts the female from the rear and intromits (Reed, 1946). Many humans also copulate in this position, with the woman allowing the man to enter from behind. This position is not the dominant sexual position in any culture sampled, probably because of the lack of clitoral stimulation it affords (Ford and Beach, 1951; Kinsey et al., 1953). Copulation from behind is most commonly practiced lying side by side when the woman is pregnant or the couple is attempting discretion in crowded sleeping quarters.

Far from having a stereotyped mating posture, humans copulate in a variety of positions (Figure 5.26). Generally, one position for copulation is preferred and dominates in most societies, but other positions are also usually practiced. Most

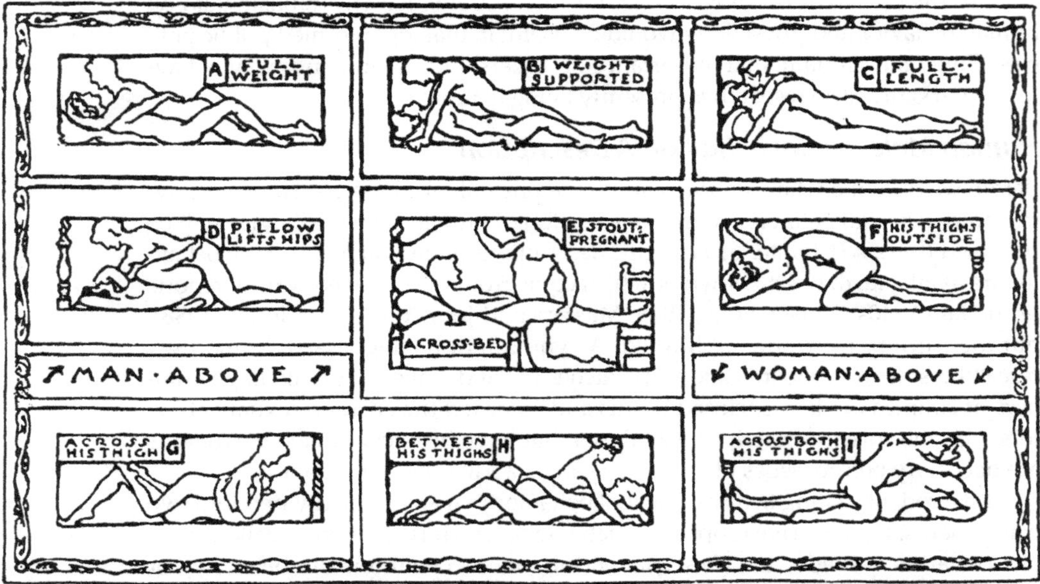

5.26 Human copulation is not constrained to stereotyped positions. This illustration from a 1940s manual used by North American physicians in marriage counseling depicts several possible coital postures, and well-known Eastern works such as the Kama Sutra describe many others. From Dickinson, 1949.

humans mate in face-to-face positions.* The most common sexual position among humans of all cultures appears to be some version of the woman sitting or squatting on the supine man. Of course, the most common sexual position among Europeans and North Americans is the face-to-face position with the man above the woman, sometimes derisively called the "missionary position."

The point of this shopping list of human sexual positions is that the mere description of human sexual behavior indicates that, in contrast to rodents, stereotyped mating sequences are not observed in humans. There are some basic constraints, but within these constraints many sexual positions exist among humans. Stereotyped mating postures like lordosis do not exist in women, and male copulatory behavior in men cannot be partitioned easily into mounting, intromissions, and ejaculations. The lack of stereotyped behaviors suggests that hormonal regulation of the muscle patterns underlying human sexual behavior is unlikely. Therefore, the hormonal regulation of human copulatory behavior must differ fundamentally from that of rodent mating behaviors, although hormonal control of human male sexual motivation and penile erection may be similar to that in rodents.

Nonhuman Primate Male Sexual Behavior: A Description

In contrast to humans, the mating posture of nearly all adult nonhuman primate species is virtually the same—and it is the position least preferred by women. Female nonhuman primates turn their backs to the male and bend at the waist to reveal their genitalia. The vaginae of all nonhuman primates are located more posteriorly than the vaginal opening of women, and thus intromission from the front is nearly impossible. Typically, the male clasps the female around her waist and intromits as he remains standing (Figure 5.27). Nonhuman primates have occasionally been reported to employ unusual mating positions in zoos (usually a male sitting while a female backs into his erect penis), but nearly all observations of mating positions among natural populations of nonhuman primates have indicated that copulation occurs only with male entry from the rear. A few exceptions to this pattern have been observed. One great ape, the orangutan (*Pongo pygmaeus*), has been observed in nature mating face-to-face while hanging upside down by the toes (Mitani, 1985)! Thus, the mating patterns of nonhuman primate males are less stereotyped than those of rodents or carnivores, but more programmed than those of humans.

The primary exception is provided by bonobos (*Pan paniscus*), also known as pygmy chimpanzees. These animals are very closely related to humans (about 98% gene homology), and primatologists have been impressed by their social interactions. Unlike the male-dominated, highly "political" and confrontational nature of common chimpanzee (*Pan troglodytes*) social life, bonobo society is best characterized by the phrase "make love, not war" (de Waal and Lanting, 1997). Bonobos use sexual interactions to resolve a wide range of social conflicts. Impor-

*The Hebrew word for "knowing" someone and "copulating" with someone is the same; hence the term, "knowing someone in the Biblical sense." The reason for the two meanings of the word apparently arises from the observation that humans copulate face to face and, unlike other animals, "know" their partner.

5.27 The typical mating posture of nonhuman primates is exhibited by copulating rhesus monkeys. In this posture, the female faces away from the male and bends from the waist to reveal her genitalia. The male usually clasps the female around the waist and intromits from the rear as he remains standing. The vaginae of nonhuman primates are located more posteriorly than the vaginal opening of women, making penile entry from the front virtually impossible in these species. Courtesy of Doug Meikle.

tantly, bonobos also have been observed to engage in numerous sexual postures, including the face-to-face position (Figure 5.28). In fact, most of the copulations observed in captivity (70.1% at the San Diego Zoo) (de Waal, 1987) are face to face. Face-to-face positions are also observed in the wild, but not as often as reported in captivity (29.1%) (Kano, 1992). One hypothesis for this difference is that the

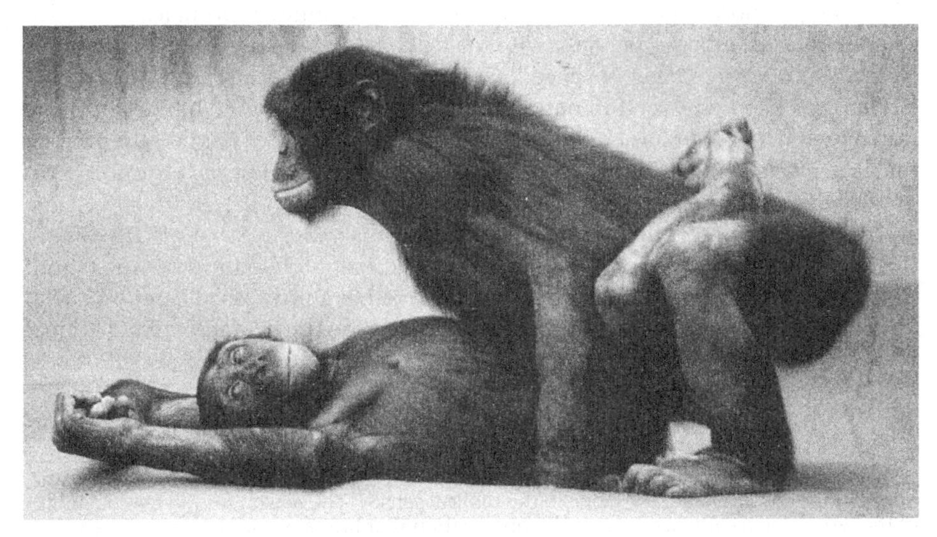

5.28 Face-to-face mating in bonobos. Courtesy of Frans de Waal.

observers' presence in the wild disturbs bonobos and induces them to climb trees, where face-to-face copulation is more difficult. Importantly, the female genitalia of bonobos are situated more frontally than in other nonhuman primates, facilitating face-to-face copulations.

Hormonal Correlates of Primate Male Sexual Behavior

Prepubescent boys do not engage in sexual activities outside of the context of play. After puberty, sexual behaviors are expressed. The average age of onset of masturbation among North American boys is about 13.5 years (Bancroft, 1978). The average age of first intercourse is 17. Frequency of sexual behavior usually peaks during the twenties and slowly declines thereafter throughout life (Figure 5.29). The general pattern of average androgen concentrations found in men of different ages corresponds to these average levels of sexual activity. Blood plasma concentrations of testosterone peak during the mid- and late teen years and remain high throughout much of the twenties. Testosterone concentrations typically diminish as men age, usually showing a sharp decline in the sixth or seventh decade of life. This decline in blood testosterone concentrations mirrors the pattern of male fertility and sperm counts (Figure 5.30).

All of these data suggest, but do not prove, that human sexual behavior is influenced by androgens. As you probably already have surmised, there is vast individual variation in the age of onset of human sexual activities, as well as in the lifelong frequency of sexual behaviors (Figure 5.31). Individual variation in the role of the testes in sexual behavior is also large among human males. The effects of castration on men's sexual behaviors vary from absolutely "no loss of sexual capacities and responsiveness" to a "decrease or total loss" (Money, 1961). In one study of men who were castrated for "treatment" or punishment of sex crimes, there appeared to be three classes of behavioral responses. More than half of the men stopped exhibiting sexual behavior shortly after castration. In other words, they responded to castration like a rodent, with a rapid cessation of sexual behavior. The frequency of post-castration sexual behavior decreased more gradually in about one-quarter of the men; in some cases many years elapsed before sexual behavior had completely waned. In these men, castration appeared to hasten the

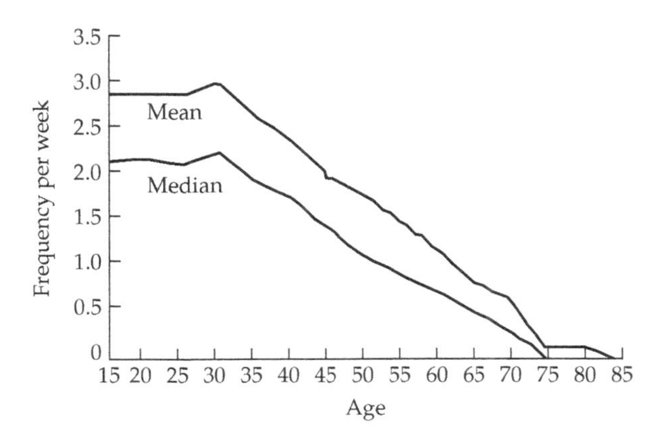

5.29 The frequency of male sexual behavior changes with age. The frequency of total sexual outlet, including heterosexual, homosexual, and autoerotic activities, peaks in the early 30s and declines thereafter, according to the Kinsey data. Although more recent surveys suggest that the peak frequency has shifted to the mid-20s, it is remarkable that human males continue to engage in sexual behaviors well into their 80s, when plasma levels of testosterone may have fallen substantially. After Kinsey et al., 1948.

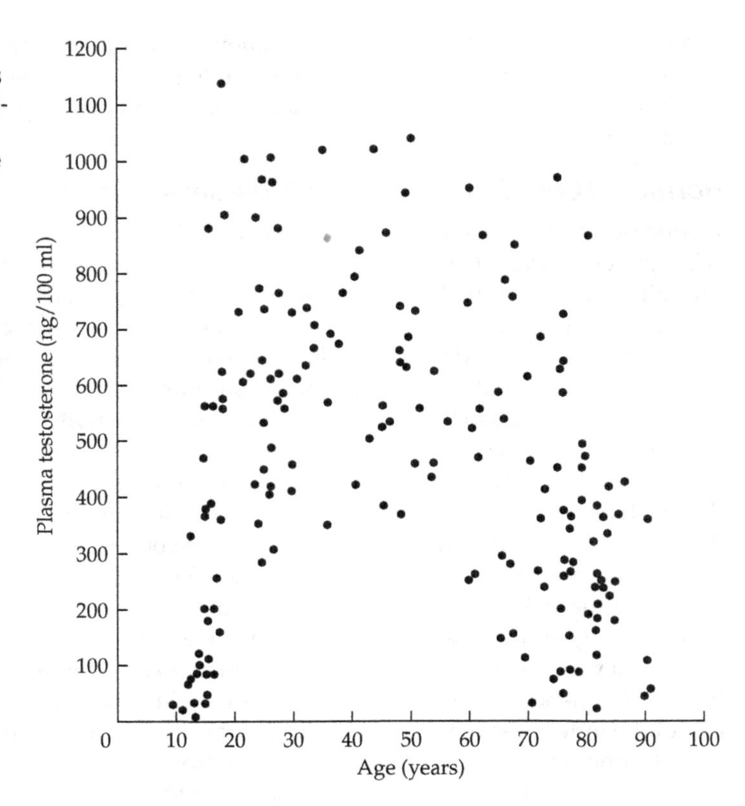

5.30 Plasma testosterone concentrations in human males change with age. There is great variation among men in plasma concentrations of testosterone, but the general trend is for testosterone levels to increase during the teens and 20s, stay generally stable until the 60s, and then gradually decrease to prepubertal levels into the 90s. Note, however, that 90-year-old individuals may have plasma testosterone levels comparable to those of many teenage males. After Vermeulen et al., 1972.

normal age-related decline in sexual behavior. The frequency of sexual behavior was unchanged after castration for about 10% of the men (Heim and Hursch, 1979). The age at castration accounted for some of the variation in behavioral response; older men appeared to show the greatest reduction in the frequency of post-castration sexual behavior. However, the variation in behavioral responsiveness to castration remains largely unexplained for all vertebrates thus far studied.

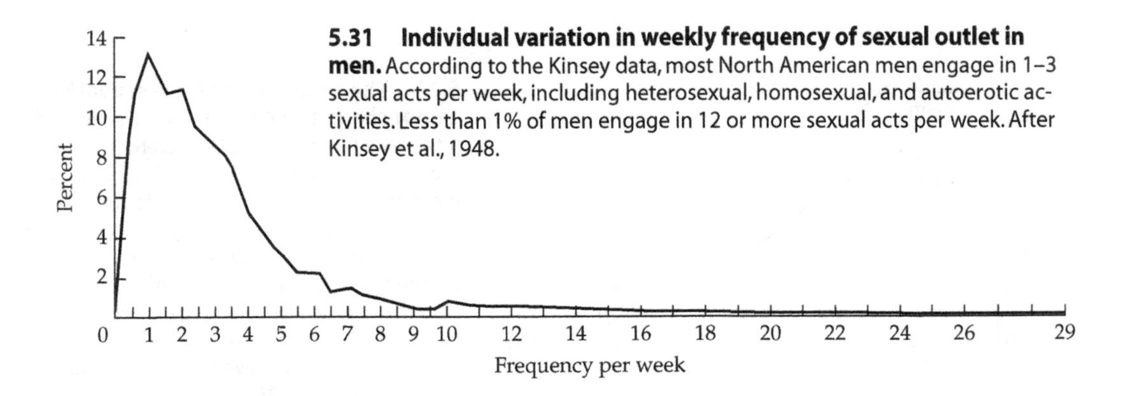

5.31 Individual variation in weekly frequency of sexual outlet in men. According to the Kinsey data, most North American men engage in 1–3 sexual acts per week, including heterosexual, homosexual, and autoerotic activities. Less than 1% of men engage in 12 or more sexual acts per week. After Kinsey et al., 1948.

Very few studies have directly examined the role of hormones in the sexual behavior of men (Wang and Swerdloff, 1997). One of the best experiments was a double-blind study performed on hypogonadal men at the Stanford Medical Center. These men met the clinical definition of "hypogonadal" because their blood plasma testosterone values were less than 3 ng/ml (Davidson et al., 1982). In the Stanford study, six hypogonadal patients were injected with one of three different substances: (1) 100 mg of a long-lasting androgen called testosterone enanthate dissolved in oil, (2) 400 mg of testosterone enanthate in oil, or (3) the oil vehicle alone.* Each patient received each of the three treatments in a different order; 6 weeks elapsed between treatments. The patients kept daily logs of their erotic thoughts, penile erections, and sexual activities during the treatments. The low and high doses of testosterone resulted in blood concentrations of about 6 and 15 ng/ml, respectively; the oil treatment did not change plasma testosterone concentrations from the pre-study average of approximately 1.5 ng/ml. The researchers found that the incidence of erections depended on the dose and timing of androgen treatment. Peak numbers of erections occurred about 1 week after the onset of the high dose of testosterone enanthate (Davidson et al., 1982). Injections of oil alone did not influence sexual thoughts or behaviors. Low doses of testosterone enanthate caused intermediate behavioral effects (Figure 5.32).

Other studies have demonstrated a dose-dependent effect of androgens on sexual fantasies and subsequent sexual arousal resulting from those erotic thoughts (O'Carroll et al., 1985). Interestingly, chronic monitoring of penile girth revealed that erections occurred normally among hypogonadal men in response to sexual fantasies or viewing of erotic films (LaFerla et al., 1978). However, these men did not have spontaneous and nocturnal erections. This study suggests that, in contrast to rats, androgens in men appear to be unnecessary to increase the probability of a sexual response in the presence of the appropriate stimuli.

Does testosterone affect sexual behavior in humans directly, or does it serve as a prohormone, as it does in rodents? The answer is not clear. Testosterone and DHT were equally effective in ameliorating the reduction of sexual activity in hypogonadal men. Neither blocking of estrogen receptors in normal men nor treating them with an aromatase inhibitor, a drug that blocks the conversion of androgens to estrogens, influenced their sexual behavior (Gooren, 1985). The results of additional studies have hinted that estrogens may influence sexual behavior in men receiving these steroids for treatment of a variety of diseases, but there has been no conclusive demonstration of an effect of estrogen on sexual behavior in normal human males (reviewed in Meisel and Sachs, 1994).

---

*This study is referred to as "double-blind" because neither the patients nor the investigators knew which treatment the patients were receiving. Recall that Brown-Séquard knew that he was injecting an aqueous solution of testicular tissue into himself, so his beliefs about its potential effects probably influenced his reports of increased abilities. If someone else had injected Brown-Séquard either with an androgen or with the inert vehicle in which the steroid hormone would have been dissolved, then the study would have been considered *single-blind*. That is, the patient would be ignorant of the treatment, but the investigator would be informed. In *double-blind* studies, neither the subjects nor the researchers are informed, so neither can unconsciously influence the outcome.

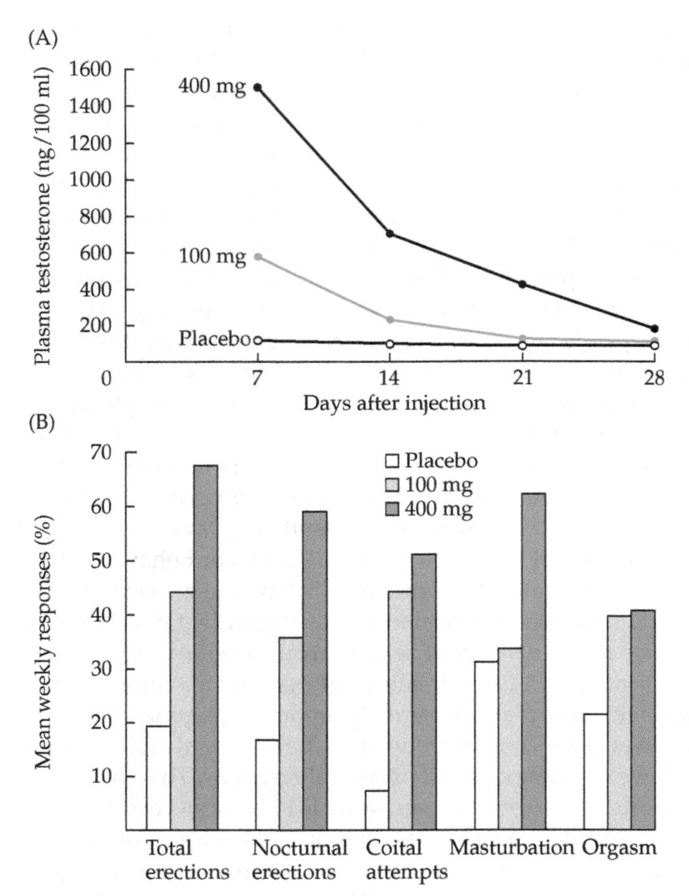

5.32 Effects of testosterone treatment on hypogonadal men. (A) Men with low concentrations of blood plasma testosterone (<3 ng/ml) received three injections, 6 weeks apart: a placebo (oil vehicle alone) and low (100 mg) and high (400 mg) doses of testosterone enanthate, both of which temporarily increased plasma testosterone concentrations in a dose-dependent manner. (B) The patients' self-reports indicated that the incidence of erections and certain sexual behaviors also increased in a dose-dependent manner. These data are plotted as percent change from baseline rates of these events. After Davidson et al., 1979.

TREATMENT OF PARAPHILIAS Despite a lack of data, assertions of a strong link between blood androgen concentrations and human sexual behavior are common, as is the belief that removal of the source of these androgens will diminish human sexual behavior. This belief has found expression in some legal systems that sentence some sex offenders to "chemical castrations" in lieu of prison sentences (for example, State of Michigan v. Roger Gauntlett, 1984). The two most common chemical compounds used for these purposes are medroxyprogesterone acetate (MPA) and cyproterone acetate (CPA). Both substances are used for treatment of sex offenders in Western Europe, but only MPA is currently approved for treatment of aberrant sexual behavior in North America. They are primarily used to treat *paraphiliac* disorders, in which individuals exhibit atypical sexual behaviors such as heterosexual and homosexual pedophilia, sexual sadism, self-mutilation, self-strangulation (asphyxiophilia), lust homicide, voyeurism, exhibitionism, fetishism, transvestitism, and frotteurism (rubbing against persons or objects for sexual stimulation) (Lehne, 1988).

CPA is a potent anti-androgen that interferes with the binding of androgen to its receptors. MPA reduces plasma concentrations of testosterone in two ways. First, it reduces LH secretion, and decreased LH secretion leads to reduced secretion of testicular androgens. Second, MPA raises the level of specific enzymes in the liver that metabolize testosterone, accelerating its clearance from the blood (Gordon et al., 1970). Blood plasma testosterone concentrations are usually less than 1 ng/ml after 60 days of MPA administration (Meyer et al., 1985).

Many studies have suggested that MPA treatment is effective in reducing paraphiliac activities (Cooper, 1986). Most of these studies were based on self-reports from patients receiving MPA treatment, either voluntary or court-ordered; in other words, former sex offenders were asked whether they continued to engage in their inappropriate sexual behaviors, and, perhaps predictably, most reported in the surveys that they did not. Such studies typically did not use control groups. In one controlled, double-blind study, penile engorgement (measured by a device, called a plethysmograph, that monitors pressure changes) in response to erotic slides was not affected by MPA treatment, despite a 68% reduction in plasma testosterone values. Physiological arousal was similar in patients treated with MPA and with a placebo. However, the patient self-ratings of arousal in response to the erotic slides revealed a reduction in self-perceived arousal in response to erotic visual stimuli (Langevin et al., 1979). Most clinicians agree that MPA treatment does not actually "cure" paraphiliac behaviors, but rather reduces the frequency of paraphiliac fantasies. This reduction in fantasies seems to allow the individual to benefit from counseling therapy (Money and Bennett, 1981).

One study reported that blood plasma LH concentrations were higher in men viewing erotic films than in men viewing nature films (LaFerla et al., 1978). The elevated LH concentrations were positively correlated with self-reports of sexual arousal; that is, men who reported that they were highly aroused by the erotic movies had higher LH concentrations than men who reported less arousal.

In another study, rapists and nonrapists watched pornographic movie scenes, and both became sexually aroused when viewing depictions of consensual sex. Similarly, both groups of men became sexually aroused when watching scenes of simulated rape in which the women acted as if they were involuntarily experiencing pleasure. However, the rapists, but not the nonrapists, became sexually aroused when viewing rape scenes in which the women appeared to be experiencing displeasure, pain, or suffering. Thus, rapists and nonrapists do not appear to differ in what activates sexual arousal, but differ in what normally terminates sexual arousal (Abel et al., 1977; Malamuth et al., 1980). Rapists and nonrapists do not differ in their average blood concentrations of androgens, despite a tendency of violent men housed in prison to have higher average blood values of testosterone than nonviolent prisoners (see Chapter 8).

Rapists are reported to be insensitive not only to the suffering of rape victims, but to the feelings of all women (Malamuth et al., 1980). Sexually aggressive men possessing these insensitive attitudes toward women may believe that neutral, or even negative, behaviors exhibited by women actually represent receptive or proceptive behaviors (see Chapter 6). Rapists often report that they believe their victims derive pleasure from a sexual assault (Malamuth et al., 1980). Although castration of rapists has been proposed as a punishment for sexual assaults, recall that this surgery is ineffective in curbing sexual activities in nearly half of the normal men who undergo it. The extent to which androgens permit males to discriminate between receptive and nonreceptive females has not been investigated in humans. Perhaps rapists differ from other men fundamentally, because of hormones, learning, or both, in how their brains process information about sexually receptive behavior in others.

NONHUMAN PRIMATE STUDIES Investigations of nonhuman primates have not yielded much more specific information about the hormonal regulation of sexual behavior in humans. This lack of progress is due, in part, to the fact that most of the primate species investigated, unlike humans, show large seasonal variations in reproductive function. Free-ranging rhesus monkeys, for example, exhibit seasonal covariation between sexual behavior and androgen-dependent secondary sex characters, such as scrotal sac color (Vandenbergh, 1969). Furthermore, large, inadequately explained inter-experimental variations have been reported in the extent of hormonal influences on male nonhuman primate sexual behavior.

Generally, free-ranging castrated adult male rhesus monkeys exhibit fewer sexual behaviors than intact males (Wilson and Vessey, 1968). The effects of castration and testosterone replacement therapy on the sexual behavior of adult male rhesus monkeys have also been examined in captive animals. Prior to castration, all of the males exhibited at least one ejaculation during a mating test. Half of the animals ejaculated when tested 6 months after castration, and 30% ejaculated 1 year post-castration. Two of ten males failed to ejaculate after castration. Daily intramuscular injections of testosterone propionate (1 mg/kg) restored mating behaviors to pre-castration levels (Phoenix et al., 1973). This study demonstrates that the hormonal control of sexual behavior in nonhuman primates is similar to the regulation of human sexual behavior; that is, enormous individual variation exists in the extent to which males are dependent on steroid hormones to initiate mating behaviors. Other studies on nonhuman primate male sexual behavior have revealed the same general outcome (reviewed in Eberhart, 1988).

Brain Mechanisms of Primate Male Sexual Behavior

As stated previously, rodents have been the primary animal model used in studies of reproductive hormone–behavior relationships. However, rodents differ from humans in many important and fundamental ways. Therefore, experiments that are considered unethical to perform on humans have been performed on nonhuman primates, animals that presumably share more characteristics with humans than do rodents. Studies of the brain mechanisms underlying primate sexual behavior have relied almost exclusively on rhesus monkeys (Eberhart, 1988; Michael et al., 1992; Michael and Zumpe, 1996).

Lesions of the MPOA severely disrupt copulation in male rhesus monkeys. MPOA lesions in primates produce mating deficits that are superficially similar to those observed in MPOA-lesioned rodents, but additional descriptive analyses of the lesioned males revealed that they continue to masturbate, maintain erections, and ejaculate at the same rate as they did prior to the surgery (Slimp et al., 1978). Clearly, destruction of the MPOA in male rhesus monkeys does not abolish the physiological mechanisms underlying sexual arousal, or even possibly sexual performance, but these lesions severely interfere with their ability to copulate with estrous females. As in rodents, the primate MPOA integrates internal and external information relevant to mating behavior. Destruction of the MPOA disables this sensory integration and thus apparently blocks the generation of appropriate sexual behavior in the presence of the appropriate stimuli.

Electrical stimulation of the MPOA causes penile erection in socially isolated male rhesus monkeys (MacLean and Ploog, 1962). Remote electrical stimulation of different brain sites of males living alone and in heterosexual social groups produced various behavioral responses: (1) erections, (2) mounting, and (3) mounting with erections. Of 59 sites within the diencephalic and telencephalic regions of the brain, stimulation of 19 sites, including several hypothalamic areas and the MPOA, evoked erections, but did not affect, or in some instances actually decreased, the number of mounts. Stimulation of 8 sites produced mounting behavior in the absence of erections, whereas 9 sites produced both erections in social isolation and mounts with erections in social groups (Perachio, 1978; Perachio et al., 1973, 1979). Some of the stimulation sites that caused erections also produced behaviors not normally associated with sex, including eating and urination. This finding suggests that no "sex centers" were being activated by the electrical stimulation, but that possibly some general level of arousal was increased.

Chemosensory cues are not as important for successful mating in primates as they are in rodents. Removal of the olfactory bulbs has no effect on rhesus monkey mating behavior (Goldfoot et al., 1978). Lesions that interrupt the function of the amygdala have produced mixed results. The most consistent series of studies on the influence of the amygdala on sexual behavior arose from studies of a phenomenon called the Klüver–Bucy syndrome in monkeys. Scientists became interested in the influences of the amygdala and temporal lobe on sexual behavior after reports of so-called hypersexuality in primates in which the temporal lobes had been surgically removed (Klüver and Bucy, 1939). Removal of the temporal lobes removes both the temporal lobe cortices and the underlying limbic structures, including the amygdala. Postoperatively, these monkeys became very docile and fearless. Inappropriate behaviors, including ingestion of inedible material and indiscriminate mounting behavior, were observed. Removal of only the temporal cortex did not produce the Klüver–Bucy syndrome, suggesting that deeper, limbic structures were involved in the hypersexuality (Eberhart, 1988).

Autoradiographic studies of steroid hormone binding in male primate brains have revealed that cells concentrating androgens, estrogens, and progestins are most abundant around the ventricles. That is, the hypothalamus, the amygdala, and the hippocampus, a bilateral forebrain limbic structure of the temporal lobe, show significant numbers of steroid hormone receptors (Eberhart, 1988; Michael et al., 1995). Thus, the pattern of steroid receptors in the male primate brain closely resembles the pattern found in rodents. In fact, the distribution of sex steroid hormone receptors is a rather conserved feature throughout the evolution of vertebrate brains (Balthazart and Ball, 1992) (see Figure 5.18). However, the location of steroid-metabolizing enzymes in neural tissues appears to have changed substantially during the evolution of vertebrates.

Because these initial studies of primates generally agreed with the results of steroid receptor and enzyme labeling studies in rodents, further data on adult primates were not obtained for several years, and subtle differences between primates and rodents may have been missed. For instance, the data for DHT binding came from a single male rhesus monkey (Sheridan et al., 1982), and the data for testosterone binding came from only two males (Bonsall et al., 1985). Although the economy of using the fewest possible animals should be applauded, the danger of this approach is that researchers may accept that steroid binding patterns are identical in disparate groups of animals when the brain regions regulating sexual behavior are only similar. More recent studies have confirmed the patterns of steroid hormone binding and have mapped out the development of these binding sites (for example, Bonsall and Michael, 1992; Michael et al., 1992).

Electrode recording during testosterone treatment of gonadally intact male rhesus monkeys revealed slow, synchronized electrical activity representing patterns of neural firing among groups of neurons in the anterior and posterior hypothalamus, as well as in the mammillary bodies, structures located just behind and below the hypothalamus. These wave forms appeared to spread to the amygdala and hippocampal regions during the recording session. Synchronization of neural activity in these brain regions was intensified and accelerated after castration (Mangat et al., 1978a,b). In other studies, androgens appeared to change the speed of neural transmission, as well as the pattern of neural firing (Eberhart, 1988).

Studies of primates using fMRI have revealed interesting similarities and differences between primates and rodents in the brain regions involved in sexual behavior. For example, when male common marmosets (*Callithrix jacchus*), restrained in an fMRI machine, were exposed to odors from either ovulating or non-ovulating females, brain activity increased in the POA and anterior hypothalamus (Ferris et al., 2001). In addition to these brain regions, odors from periovulatory females significantly increased activity patterns in several cortical areas, including the striatum, hippocampus, septum, periaqueductal gray, and cerebellum—brain regions not only involved with sexual behavior, but also involved in emotional processing and reward (Ferris et al., 2004) (Figure 5.33).

When positron emission tomography (PET) was used to scan the brains of men while they were brought to orgasm by their female partners, the ventral tegmental area (VTA)—a cluster of cells known to play a key role in reward and

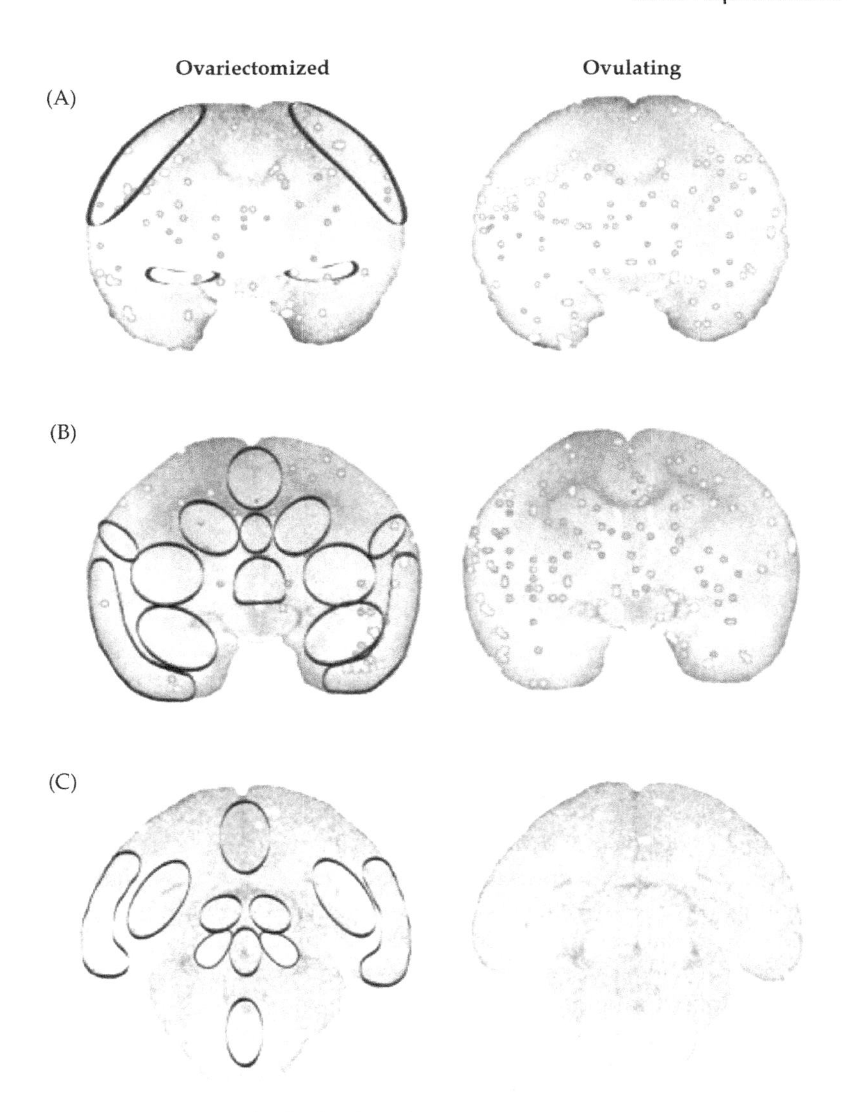

5.33 Activation patterns in the brains of common marmosets exposed to odors from either ovariectomized or ovulating females. Positive activation is indicated by circles in cross sections obtained by fMRI at the level of (A) the prefrontal cortex, (B) the amygdala, and (C) the midbrain. Courtesy of Craig Ferris.

euphoria—was activated (Figure 5.34). Recall that this brain region is important in rodent ejaculation (Balfour et al., 2004); it is also part of the reward system activated during a heroin high. Some other brain regions that are strongly activated in the brains of rodents at ejaculation are also activated in humans. In particular, activation was noted in the posterior thalamus, including the paracellular sub-

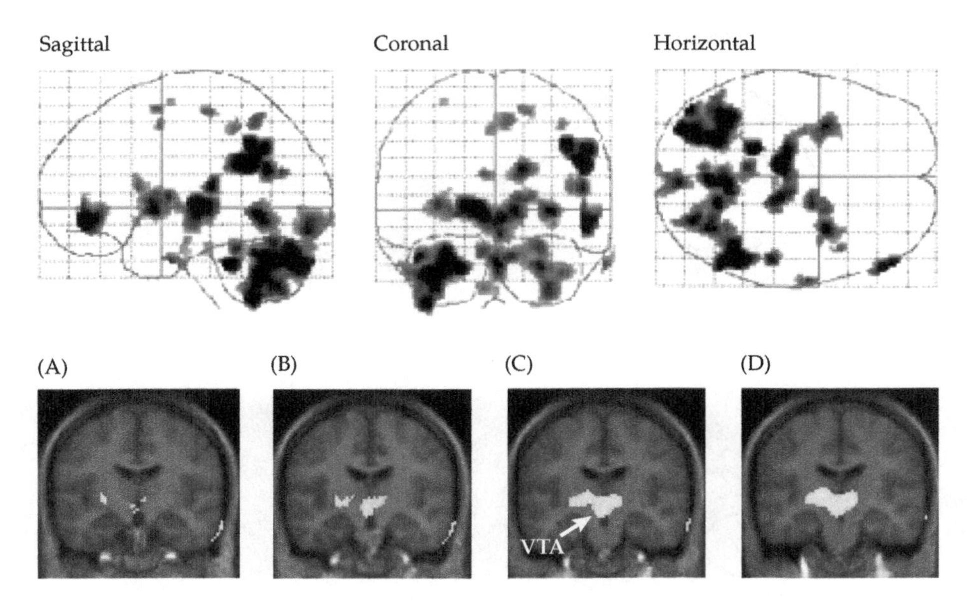

5.34 Activation patterns of the brains of men undergoing ejaculation. The top panel shows external views in the sagittal, coronal, and horizontal planes; the most active areas are darkest. The bottom panel shows cross-sectional views (A–D); the most active areas are brightest. Note that the VTA is activated during ejaculation. Courtesy of G. Holstege.

parafascicular nucleus (SPFp), further confirming the importance of this thalamic structure and its spinal inputs for ejaculation. However, PET scanning failed to demonstrate ejaculation-related activation in other areas in humans, including the medial amygdala and bed nucleus of the stria terminals (Holstege et al., 2003). Finally, activation was not observed in the human preoptic area, which shows robust activation during male rodent sexual behavior, although this activation is not specifically related to ejaculation (Coolen et al., 2004).

Social Influences on Primate Male Sexual Behavior

As we have seen, behavior can feed back to affect hormone concentrations in rodents. Interactions with women can also affect hormone–behavior interactions in men. The man we met in Chapter 1, presumably a little bored, maintained daily records of the weight of the beard trimmings in his electric razor (Anonymous, 1970). He noticed that his beard trimmings were heavier immediately prior to and during his visits to his fiancée than at other times. He attributed the increased beard growth to elevated androgen concentrations induced by sexual anticipation and sexual activity. This demonstration has many flaws as a scientific study, but the general concept that behavior can feed back to affect hormone concentrations has been observed experimentally in several other contexts. A few studies have directly measured androgens prior to, during, or immediately after sexual behavior in men. In one such study, blood plasma concentrations of testosterone increased immediately prior to and after ejaculation during copulation (Fox et al., 1972). In another study, testosterone, as well as DHT and androstenedione, was elevated following masturbation, suggesting that ejaculation per se stimulates steroid hormone production in men (Purvis et al., 1976). Many other studies, however, have found no hormonal changes after masturbation or coital ejaculation (for example, Lee et al., 1974; Lincoln, 1974; Stearns et al., 1973).

The endocrine effects of visual erotic stimuli in men were investigated in another study. Fifteen normal adult males viewed either erotic or sexually neutral films. Changes in penile girth, self-reports of arousal, and blood concentrations of testosterone, LH, prolactin, cortisol, ACTH, and β-endorphin were recorded. Sexual arousal, as indicated by both self-assessment and erectile function, was greater in all the men when viewing the erotic than when viewing the neutral films. However, no hormonal changes corresponding to the sexually aroused state were observed (Carani et al., 1990).

Individual Variation and Aging

Men in their sixties start to show decreases in circulating blood concentrations of androgens (see Figure 5.30). However, approximately one-half of healthy men over 80 years of age have plasma testosterone concentrations that fall into the normal range for men aged 25–59 (Tserotas and Merino, 1998). The mechanisms underlying the enormous variation in the sexual behavior of aging men are unknown. A small proportion of aging males with sexual dysfunction have hormonal deficiencies; however, most do not. If an elderly man complains of impotence, and his blood plasma testosterone concentrations are below 1 ng/ml, any of three possible neuroendocrine problems may be responsible: (1) the hypothalamus may not release sufficient GnRH to start the endocrine cascade that normally stimulates testosterone production, (2) the anterior pituitary may not release sufficient gonadotropins, or (3) the Leydig cells may not produce sufficient androgens in response to gonadotropin stimulation. In most cases, men over the age of 60 exhibiting hypogonadal concentrations of androgens also have elevated plasma concentrations of LH, thus indicating a failure of the Leydig cells to produce sufficient steroid hormones to feed back and maintain normal blood concentrations of LH.

As we have seen, there is great individual variation among men in their dependence on steroid hormones for the maintenance of sexual behavior. It has been suggested that male animals are less dependent on hormones for sexual behavior the "higher they are on the phylogenetic scale" (Beach, 1947). If experiments on different species in which proven copulators are castrated and later examined for mating behavior are compared, the following observations can be made: (1) most rodents and birds show a decrease in sexual behavior immediately after castration, (2) cats also exhibit a striking decline in sexual behavior after castration, and (3) the post-castration sexual behavior of monkeys and humans is very dependent on prior experience. Dogs typically show no behavioral effects on mating during the first 15 weeks post-castration. But again, the effect of gonadectomy is very dependent on experience even in so-called lower animals.

For example, a sexually experienced cat or dog may continue to copulate for a long time after being neutered (Dunbar, 1975; Rosenblatt and Aronson, 1958). Adrenalectomized, castrated dogs may continue to display complete copulatory behavior for years after the surgery, indicating a lack of dependence on steroid hormones for mediation of sexual behavior (Beach, 1970). The extent of individual differences in post-castration mating behavior is great regardless of species; that is, the variation within species is generally as large as that between species in the reliance of sexual behavior on hormones (Hart, 1974).

There is no correlation between blood testosterone concentrations and frequency of sexual behaviors among human males (W. A. Brown et al., 1978). In other words, variation in human masculine behaviors does not correlate with differences in blood androgen concentrations. Furthermore, as described in Chapter 4, there is no reliable difference in blood androgen concentrations between homosexual and heterosexual men, and it is unlikely that differences in male sexual orientation are due to blood androgen concentrations.

Peptide Hormones and Male Sexual Behavior

Steroids are the predominant hormones involved in the endocrine regulation of sexual behavior, but other hormones, particularly peptide hormones, appear to modulate their effects (Box 5.4). For example, GnRH has subtle effects on ejaculation latencies in rats (Moss et al., 1975; Myers and Baum, 1980). Elevated prolactin concentrations interfere with male reproductive behavior. Other peptides that have been reported to influence male sexual behavior include vasopressin, oxytocin, CRH, and the neurohormones cholecystokinin (CCK), vasoactive intestinal polypeptide (VIP), galanin, and neuropeptide Y (Hull et al., 2002).

The most consistent reports of peptidergic influences on male reproductive behavior have involved endorphins. Providing male rats with naloxone, an antagonist to endorphins, typically enhances reproductive behavioral performance. When gonadally-intact noncopulators, for example, were given either naloxone or saline injections, nearly three-quarters of the naloxone-treated animals, but none of the saline-treated animals, mated to ejaculation in subsequent tests (Gessa et al., 1979). Thus, readiness to mate appears to be inversely correlated with the amount of endorphin receptors available. Recall that sexually inactive male rats are nonreactive to many stimuli, and that a tail pinch is very effective in initiating copulatory behaviors (Meisel et al., 1980; Pfaus and Gorzalka, 1987). Endorphins and other opioids appear to affect the dopaminergic systems involved in the release of GnRH and thus subsequently suppress blood concentrations of LH and androgens (Pfaus and Gorzalka, 1987). Opioids, from endogenous or exogenous sources, tend to inhibit male sexual behavior in both humans and nonhuman animals (reviewed in Pfaus and Gorzalka, 1987). Chronic heroin use, for example, suppresses plasma testosterone and LH concentrations. Chronic heroin use among human males throughout puberty does not affect later testosterone or LH blood concentrations after heroin use is discontinued (Mendelson and Mello, 1982). This and other studies suggest that a tolerance for opiates, in terms of sexual behavior, does not develop.

BOX 5.4 *Sodefrin, a Female-Attracting Pheromone in Newts*

Recent evidence suggests that amphibians are decreasing in numbers throughout the world. Some suggest that environmental endocrine disruptors are involved, whereas others hypothesize that other types of toxins are affecting reproductive function in these animals. A fungus or other pathogen may also be involved in the declining numbers of amphibians. Whatever the cause of the decline in amphibian populations, it serves as an early warning signal of environmental degradation for humans. We must understand the reproductive physiology and behavior of amphibians in order to understand how they, and we, might be affected by environmental agents.

Figure A Courtesy of Sakae Kikuyama

If given a choice, female newts and salamanders (urodele amphibians) appear to be attracted to water in which a male conspecific was recently courting a female. During courtship, male red-bellied newts (*Cynops pyrrhogaster*) fan the water with their tails. This action tends to move water from the male's cloacal opening to the snout of the female. Females follow these males and keep their snouts in contact with the male's tail (Figure A). Upon close inspection, it was noted that courting males extend small tubules from the cloaca in the presence of females (Kikuyama et al., 1997). These tubules are connected to the abdominal glands. Males without abdominal glands are not successful at courting females (Malacarne et al., 1984). Electrophysiological recordings of the olfactory nerve showed that the neuroepithelial cells of female crested newts (*Triturus cristatus*) became more active when exposed to male odors or extracts of the male abdominal glands (Cedrini and Fasolo, 1970). Taken together, this evidence suggested that male urodeles were emitting a chemical signal that attracted conspecific females.

In an elegant series of experiments, a team of researchers in Japan has isolated, characterized, localized, and quantified the female-attracting agent (pheromone) in male red-bellied newts (Kikuyama et al., 1995). Using reverse-phase high-pressure liquid chromatography, they isolated a decapeptide from the abdominal glands of males. This decapeptide, which had no sequence homology with any known peptide, was named sodefrin (Kikuyama et al., 1995).*

When small blocks of sponge were impregnated with sodefrin, the minimal effective concentration for attracting females (in 3000 ml of water) was determined to be between 0.1 (picomolar) pM and 1.0 pM. A synthetic version was produced, and its effectiveness in attracting females was similar to that of the native peptide (Kikuyama et al., 1997).

*The name was derived from two Japanese words prominent in a well-known poem. Nukada (Figure B) was a famous and beautiful poet who married an emperor. One day, she went out with the emperor to the outskirts of the city, but the emperor's brother, who was once her lover, solicited her by waving (*furu*) his sleeves (*sode*). She responded to his unwanted advances with a poem that loosely translates, "Don't be so bold as to wave your sleeves to me. People will say that we are still in love" (Sakae Kikuyama, personal communication).

Continued on following page

BOX 5.4 *continued*

Figure B Courtesy of Sakae Kikuyama

This peptide appeared to act through the female's olfactory system because females whose nostrils were plugged with cotton and petroleum jelly were not attracted to the sodefrin.

An antibody to sodefrin was made in rabbits, and immunohistochemical and radioimmunoassay techniques were used to locate and measure, respectively, sodefrin in male red-bellied newts (Kikuyama et al., 1997). Sodefrin was located primarily in secretory granules in the apical portions of epithelial cells in the abdominal glands. Castration and hypophysectomy reduced sodefrin concentrations substantially, and testosterone and prolactin replacement therapy prevented this reduction. Testosterone appears to play a more critical role than prolactin in mediating sodefrin concentrations (Toyoda et al., 1994).

This pheromone seems to show species specificity. Sodefrin from red-bellied newts did not serve as an attractant to female sword-tailed newts (*Cynops ensicauda*), a congeneric (belonging to the same genus) species (Kikuyama et al., 1997), and it appears that another pheromone with a different amino acid sequence may act as a chemical attractant in congeneric species. The sodefrin studies were the first to identify an amphibian pheromone, and the first to discover a peptide pheromone among vertebrate animals (Toyada et al., 2004). Because of their diversity and their potential for small changes in amino acid sequence, other peptides could easily serve as chemical signals conveying species-specific messages.

Male Reproductive Behavior in Birds

Reproductive behavior in male birds, both appetitive and consummatory, is highly dependent on androgens (reviewed in Ball and Balthazart, 2002). Males of most bird species engage in some sort of ritualized courtship behavior, and castration eliminates this behavior. Since Berthold's findings regarding castration in roosters (see Chapter 1), many species of birds have been studied. In virtually all species, mating behavior wanes rapidly after castration, and treatment of castrated male birds with androgens restores courtship and mating behaviors (reviewed in Silver et al., 1979; Crews and Silver, 1985; Ball and Balthazart, 2002). Either testosterone or DHT restores most mating behaviors to pre-castration levels in male birds (Adkins and Adler, 1972; Adkins and Nock, 1976; Beach and Inman, 1965; Cheng and Lehrman, 1973; Hutchison, 1970; Silver, 1977). DHT, however, does not restore copulation; rather, DHT restores precopulatory displays. Male pigeons that congenitally lack testes exhibit little sexual behavior as compared with normal males (Riddle, 1924; 1925).

Birds have some specific features as animal models for the study of hormone–behavior interactions. Blood concentrations of androgens in birds covary with their breeding season; that is, androgen concentrations are typically highest at the beginning of the breeding season and lowest during the early winter (Wingfield and Farner, 1980). These changes in steroid concentrations are linked to seasonal fluctuations in day length (photoperiod; see Chapter 10). Additionally, unlike female rodents, which provide rather consistent behavioral stimuli to males, female birds do not exhibit such rigid consistency. Consequently, the behavior of the male may be less easy to separate from the behavior or responsiveness of the female.

The earliest, and arguably the best, descriptive studies linking testicular steroid hormones with avian reproductive behavior were conducted on ring doves (*Streptopelia risoria*) (Lehrman, 1965). Ring doves, however, may not be an ideal avian model because several features of the endocrine control of behavior may be unique to the species. One such unusual feature is the lack of photoperiodic effects on the reproductive behavior of male ring doves. Second, ring doves possess a crop sac, a highly specialized adaptation that requires unique hormonal stimulation (see Chapter 1). Most breeding behaviors in the ring dove have been linked to changes in hormone concentrations (Figure 5.35). However, there is a gradual shift from the aggressive courtship behavior observed early in the courtship sequence to the nest-oriented courtship behavior observed later that does not reflect a change in hormones, but probably reflects testosterone-induced changes in aromatase activity. That is, testosterone directly mediates the aggressive courtship behaviors, but it must be converted in the brain to estrogen to mediate the nest-oriented courtship behaviors (Hutchison and Steimer, 1983).

Japanese quail (*Coturnix japonica*) have also been used as a model species to study the hormonal control of male avian sexual behavior (Balthazart, 1989; Balthazart and Ball, 1992). These birds provide a useful animal model for investigating hormone–reproductive behavior interactions because they have been selectively bred to mate quickly and their other reproductive behaviors have waned; for example, in the laboratory, these birds no longer incubate their eggs. Consequently, hormonal effects on their sexual behavior can be separated from effects on other reproductive behaviors. When a sexually mature pair of quail is introduced into a neutral mating arena, copulation usually is initiated within 10 seconds. The male grabs the neck of the female, mounts her, and brings his cloacal area into contact with hers (Figure 5.36). The male typically ejaculates during this brief cloacal contact period. After copulation, the male struts around the arena, often with a wing lowered toward the female. He may vocalize or crow. Castration eliminates all these behaviors within days, and testosterone therapy restores copulation to pre-castration levels. DHT is ineffective in restoring copulatory behavior in castrated quail, but activates strutting and crowing (Balthazart, 1989).

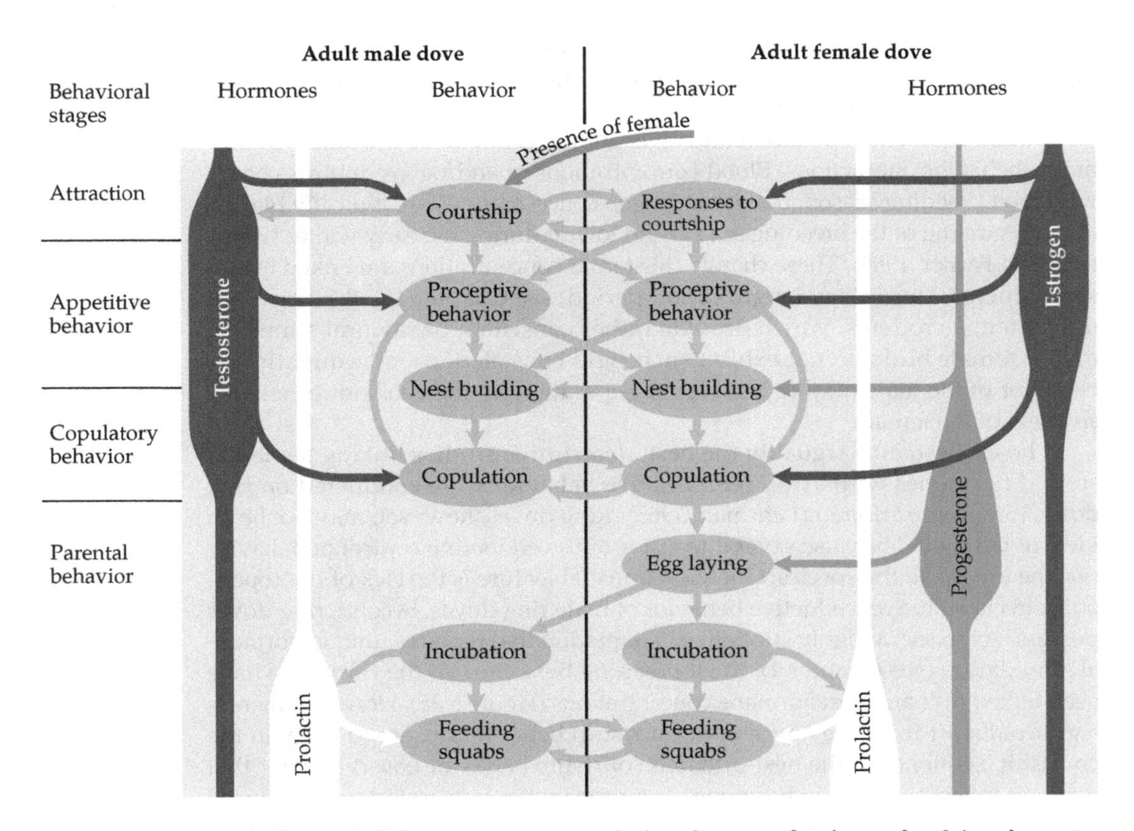

5.35 Changes in hormone secretion during the reproductive cycle of ring doves follow a consistent pattern. In females, ovarian growth and estrogen secretion coincide with courtship behavior; ovulation and rising progesterone levels induce incubation behavior and suppress bowing and cooing courtship behaviors. Prolactin induces the formation of the crop sac and brood patch and also stimulates parental behaviors. In males, testosterone stimulates courtship behavior, and prolactin appears to be involved in incubation and other parental behavior. Testosterone levels decrease and courtship behavior is suppressed during incubation and paternal care. Progesterone does not appear to be involved in male ring dove reproductive behavior. As soon as one cycle is completed in the laboratory setting, a new cycle begins with the onset of courtship behavior. After Rosenzweig and Leiman, 1989.

Studies of Japanese quail have found that testosterone must be converted to estrogen in the POA before it produces an effect on copulatory behavior (Adkins-Regan, 1996). Blocking aromatase activity or estrogen binding prevents androgens from restoring copulatory behavior to pre-castration rates (Adkins-Regan, 1987; Balthazart, 1989). When immunocytochemical techniques are used, the preoptic area, especially the preoptic medial nucleus (POM), which is analogous to the MPOA in mammals, stains positively for estrogen receptors, as well as for cells containing the enzyme aromatase (Balthazart and Ball, 1993; Panzica et al., 1996), indicating that androgens are converted to estrogens in these or possibly

(A)

(B)

(C)

(D)

5.36 Copulation in Japanese quail may take only 10 to 15 seconds. In a typical sequence, the male (A) struts around, (B) crows, then rushes at the female, (C) mounts her, and (D) positions his cloaca against hers. Because female birds do not exhibit rigid consistency in their behavior from one mating test to the next, as do female rodents, it is difficult to separate factors associated with male behavior from those resulting from the behavior or responsiveness of the female. Courtesy of Jacques Balthazart.

adjacent cells (Figure 5.37). Other sites where this aromatization occurs include the nucleus striae terminalis, nucleus taeniae, infundibular hypothalamic nucleus, ventromedial nuclei, and mesencephalic substantia grisea (comparable to the periaqueductal gray in mammals) (Adkins-Regan, 1996) Taken together, these data suggest that estrogens might influence behavior by affecting neurotransmitter

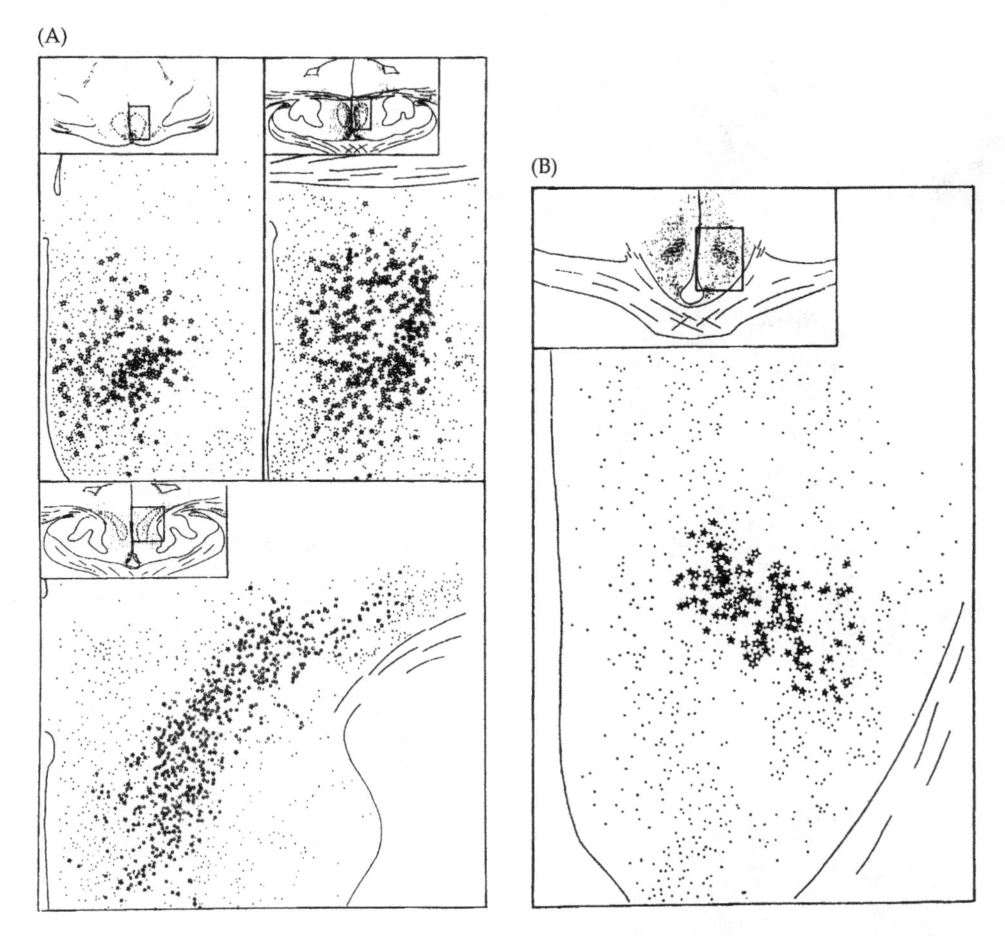

5.37 Aromatase and estrogen receptors in the quail brain shown in camera lucida drawings of sections double-labeled for aromatase and estrogen. Each panel shows an enlarged area of the brain, the location of which is indicated by the inset. (A) Sections of the preoptic area (POA), where only a small proportion of cells expressed both aromatase and estrogen receptor activity. Top left, middle part of the medial preoptic nucleus (POM); top right, POM at the level of the anterior commissure; bottom, caudal POA and septum at the level of the nucleus striae terminalis. (B) One section of the caudal hypothalamus at the level of the nucleus inferioris hypothalami. Solid circles = cells labeled for estrogen receptors; open stars = cells labeled for aromatase; solid stars = cells double-labeled for estrogen receptors and aromatase. From Balthazart et al., 1991.

dynamics, although it remains possible that estrogens influence the growth of neural processes.

In quail, as in roosters, crowing is an androgen-dependent behavior, and crowing and copulation occur in close temporal proximity. If a castrated male or a female quail is injected with testosterone, crowing will ensue. Testosterone must be converted to DHT in order to have a behavioral effect on crowing (Adkins and Pniewski, 1978), but, as stated above, testosterone must be converted to estrogen prior to eliciting a copulatory behavioral response. The latter conversion appears to occur at a much lower rate in the brains of males than in those of females; this may explain at least in part why females injected with testosterone never show the mounting behaviors associated with male copulatory behavior (Adkins and Adler, 1972). Copulation is integrated in the POM, and as described above, aromatase, which converts testosterone to estrogen, is present in that brain region. Vocalizations during courtship are probably regulated in the nucleus intercollicularis (ICo), a mesencephalic brain nucleus. This area of the brain contains the enzyme 5α-reductase, which converts testosterone into 5α-DHT. Thus, the enzymes that produce the behaviorally active metabolites of testosterone responsible for copulatory behavior and for crowing are located in the brain regions that underlie these two separate, but linked, reproductive behaviors (Balthazart and Ball, 1993).

One report has suggested that castration does not affect copulatory behavior in white-crowned sparrows (Moore and Kranz, 1983). These birds breed in the spring when the photoperiod exceeds 12 hours of light per day (see Chapter 10). If prepubescent male white-crowned sparrows are injected with testosterone while maintained under short-day light conditions, then they will not mate with receptive females. Exposure to long days appears to activate copulatory behavior in these sparrows permanently. If prepubertal males are castrated and then exposed to long days, they will continue to mate despite a complete lack of assayable testosterone in their blood (Moore and Kranz, 1983). The extent to which male mating behavior in other avian species may be divorced from hormonal control requires further investigation. Other species of sparrows appear to require androgens to elicit copulatory behavior (Balthazart, 1989). The extent to which species or individuals vary in their dependence on hormones for the maintenance of mating behavior does not seem to have any evident systematic organization.

Male Reproductive Behavior in Reptiles

Like birds, most male reptiles stop mating after castration (Crews and Silver, 1985). One notable exception is the red-sided garter snake (*Thamnophis sirtalis parietalis*), found on the midwestern Canadian prairies (Mason, 1987). Males of this species emerge from their winter hibernacula several weeks earlier than females, then attempt to mate with females as they emerge. Many males court a single female as she moves out of the hibernaculum (Figure 5.38). Females are much larger than males in this species. The male has two hemipenes one of which is inserted into one of the female's two vaginae. Other males stop courting the female as soon as one

5.38 A "mating ball" of male red-sided garter snakes forms when great numbers of males attempt to court and mate with females, which emerge from the winter hibernaculum several weeks after the males. As soon as one male successfully intromits, the female expresses a chemosignal that disperses the unmated suitors. Courtesy of David Crews.

male achieves an intromission. Apparently, a chemosensory signal is expressed by the copulating female that inhibits courtship behavior in the other males (Mason et al., 1989); the exact chemical identity of this substance is currently unknown.

Most male animals have an **associated reproductive pattern**. That is, during the breeding season, mating behavior coincides with maximal testis size, as well as maximal androgen concentrations and sperm production. The red-sided garter snake, however, exhibits a **dissociated reproductive pattern** (Crews, 1984, 1991). When the snakes are mating, the testes remain regressed; no sperm or androgens are being produced. The males inseminate females in the spring with sperm stored from the previous autumn. The testes begin to develop, produce androgens, and make sperm in midsummer, after all mating behavior has stopped (Figure 5.39). The testes then regress again, and the animal enters hibernation. The sperm are stored in a special organ until the following spring. Thus, copulation normally occurs in the complete absence of circulating steroids. Predictably, castration has no effect on subsequent mating behavior in this species. Removal of the adrenals, another potential source of androgens, or even the pituitary gland also has no effect on mating behavior in these snakes (Camazine et al., 1980; Crews et al., 1984). Neither treatment with steroid hormones nor treatment with gonadotropins affects copulation in this species (Crews, 1984). However, destruction of the MPOA (Krohmer and Crews, 1987) or removal of the pineal gland (Crews et al., 1988b; Nelson et al., 1987) eliminates mating behavior, probably because these two brain regions integrate the environmental information that allows the snakes to ascertain the appropriate season of the year for mating (see Chapter 10). The finding

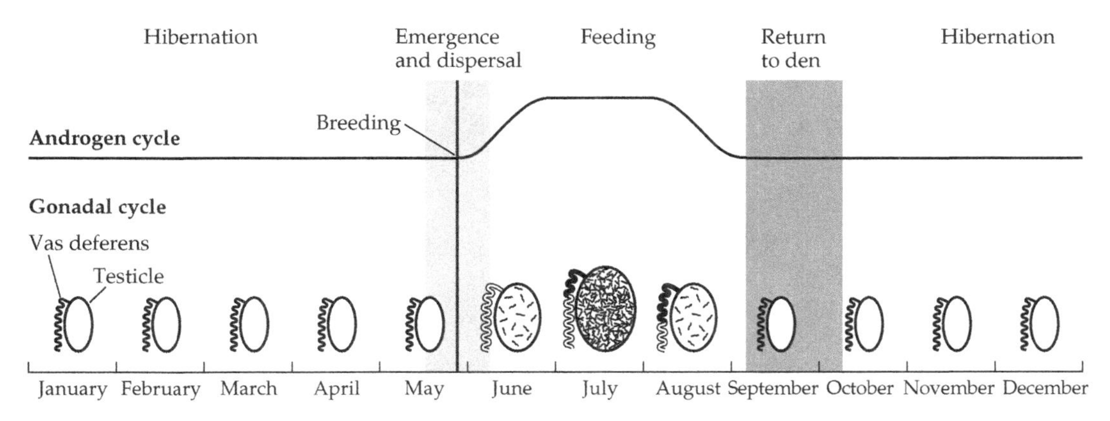

5.39 Red-sided garter snakes display a dissociated reproductive pattern. Male red-sided garter snakes emerge from hibernation and mate with females in the spring; however, the testes and vas deferens are regressed during the mating period, and the testes are not producing sperm or androgens at that time. In midsummer, the testes develop and begin producing androgens and sperm. The gonads regress prior to hibernation in the fall, and sperm produced in summer is stored for the mating period in the following spring. After Crews and Gartska, 1982;

that mating behavior is divorced from hormonal control among these snakes shows that position on the phylogenetic scale is not a perfect indicator of the independence of sexual behavior from direct hormonal control (see also Beach, 1947).

Male whiptail lizards of the genus *Cnemidophorus* rely on androgens to regulate their reproductive behavior. Courtship is highly ritualized in this species. The male approaches the female from the rear, climbs onto her back, bites her back or foreleg, then brings his tail underneath her to bring their respective cloacal areas together, forming a so-called "doughnut" posture (Figure 5.40A; Crews and Fitzgerald, 1980). The male dismounts after about 10 minutes and leaves the scene.

Some species within the genus *Cnemidophorus* are parthenogenetic; that is, all the members of these species are triploid females, which lay unfertilized eggs that all develop into daughters. These species evolved from hybrids of two sexually reproducing species. Interestingly, the occurrence of "male-like" behavior persists in at least five all-female *Cnemidophorus* species. Captive female *C. uniparens* alternate during the breeding season between displaying male-like "pseudosexual" behavior and female behavior. Female "receptive" behavior is observed before ovulation, when estrogen concentrations are relatively high (see Chapter 6); at this time, females allow other females to mount them. Male-like behavior is observed after ovulation, when blood concentrations of progesterone are elevated (Figure 5.41; Crews, 1987); postovulatory females court, mount, and form the doughnut mating posture with periovulatory females (Figure 5.40B). Ovariectomized females injected with progesterone show male-like pseudosexual behavior, and those injected with estradiol show female pseudosexual behavior. Thus, progesterone, rather than androgens, apparently mediates male-like behavior in these species.

5.40 Copulatory behavior in sexual and asexual species of whiptail lizards is similar. (A) The copulatory sequence (top to bottom) in *Cnemidophorus inornatus*, a sexually reproducing whiptail lizard. The male mounts the female, tucks his hindquarters under the female, and finally bites the female's back and intromits, forming the so-called doughnut posture. (B) Pseudosexual behavior in *C. uniparens*, a parthenogenetic whiptail lizard species that developed from hybrids of species that reproduce sexually. Although both participants are females, the copulatory sequence closely resembles that of the sexual whiptail species, including the performance of the doughnut posture. This behavior has been found to be necessary to maximize ovulation. From Crews and Fitzgerald, 1980.

Why do these parthenogenetic lizards engage in pseudosexual behavior at all (Crews, 1997)? Functionally, females that undergo mounting and pseudosexual mating behavior produce more offspring than females that do not engage in this behavior. Apparently, the act of pseudocopulation stimulates the release of additional ova. The reciprocal alternation between male-like and female behavior facilitates breed-

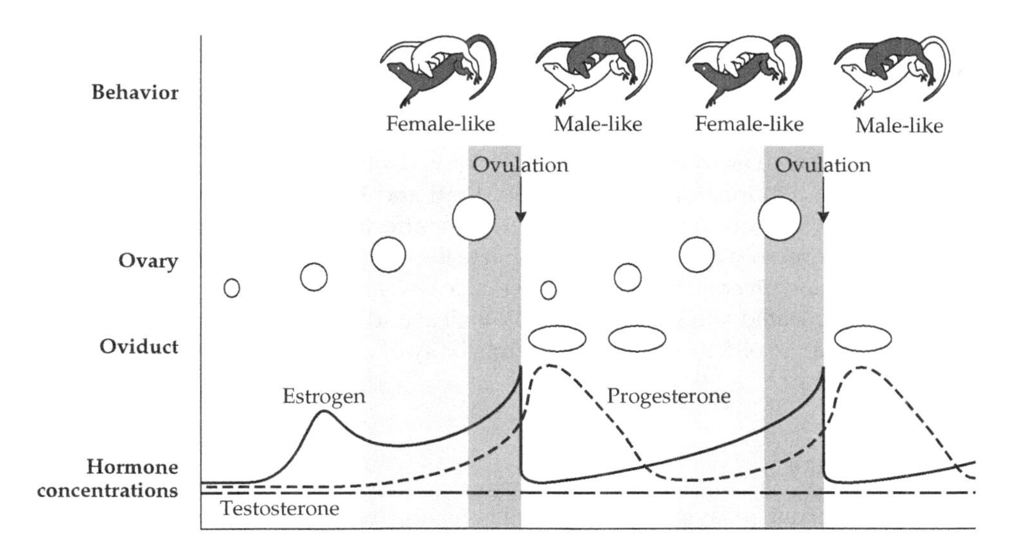

5.41 Hormones mediate pseudocopulation in parthenogenetic whiptail lizards. Females in parthenogenetic whiptail species alternate in performing female and male roles during pseudocopulation, and this alternation has clear hormonal correlates. The bottom portion of the figure shows hormonal cycles and events in the ovary and oviduct in relation to a hypothetical female's alternation between female and male-like behavior (top; black lizard). In the preovulatory state, when estrogen levels are increasing, the female behaves in a female-typical fashion. After ovulation, when her oviducts are full of eggs, blood progesterone levels increase, estrogen levels decrease, and she exhibits male-like pseudosexual behavior. This cycle repeats itself throughout the breeding season. Note that these behavioral shifts have no relation to testosterone levels, which remain constant throughout the cycle, despite the importance of this steroid hormone in mediating male behavior in sexually reproducing whiptail species. The neural substrates of male-like courtship and copulatory behavior have apparently been retained in the parthenogenetic species, but are mediated by progesterone. After Crews, 1987.

ing in these species (Crews, 1987). This seemingly altruistic behavior is to be expected when we recall that asexually reproducing animals share virtually all of their genes in common, so that an individual who helps other females to increase their production of offspring also enhances her own inclusive fitness. These unusual lizards provide a wonderful demonstration that the structures of the hormones involved in hormone–behavior interactions do not evolve very much; instead, new functions for those hormones evolve (Crews, 1997; Godwin and Crews, 2002).

Conclusions

How do hormones affect male sexual behavior? Based on our discussions of different species, a few generalizations can be made. Sex steroid hormones clearly do not act as a "switch" to activate sexual behavior. The presence of testosterone does not automatically stimulate mating behavior; rather, testosterone, or its metabolites, increases the probability that a sexual behavior will occur in the presence of appropriate stimuli. In both mammals and birds, sex steroid hormones appear to affect sensory input and central nervous system processing as well as behavioral output. Androgens, for example, amplify chemosensory information associated with estrous females in the male rodent's brain, allowing further processing of that incoming information (Pfaff and Pfaffmann, 1969; Stern, 1990; Wood, 1997). Steroid hormones are, in turn, affected by environmental stimuli. Peptide hormones also appear to modulate the effects of steroid hormones on sexual behavior. Overall, hormones affect male sexual behavior in many interacting and complicated ways. Several physiological and behavioral systems integrate information in order to enhance the probability of an individual male's reproductive success.

Summary

- Male sexual behavior consists of all behaviors necessary and sufficient to deliver male gametes to female gametes. Male sexual behavior can be divided into two phases: the appetitive phase (courtship, sex drive) and the consummatory phase (copulation, performance).
- Sex drive, the motivation to seek sexual contact, becomes extremely powerful and overtly expressed in males after puberty, when the testes begin to secrete androgens. Sexual performance and copulatory ability increase after puberty as well.
- Castration reduces sexual behavior in some proportion of individuals of virtually all vertebrate species examined. However, there is great individual and species variation in the extent to which sexual behavior is regulated by hormones. Testosterone treatment generally reverses the decrease in frequency of male sexual behavior observed after castration. More testosterone is necessary to restore sexual behavior in castrated males after it has ceased than is required to maintain male sexual behavior.
- There are three components of male rodent sexual behavior: mounting, intromission, and ejaculation. The corresponding penile responses—erections, flips, and cups—can be elicited from animals independent from higher brain influences. Testosterone does not seem to regulate sexual behavior or penile reflexes directly, but rather acts as a prohormone; metabolic products of testosterone, namely, estrogens and dihydrotestosterone (DHT), appear to be important in mediating copulatory behaviors and penile sensitivity, respectively.
- An intact chemosensory system is necessary for the proper function of sexual behavior in rodents. Lesions along the neural circuit that mediates olfactory information, from the olfactory bulbs to the amygdala to the medial preoptic area (MPOA), result in diminished sexual behavior. Electrical stimulation along this circuit increases the frequency of sexual behavior in male rodents.

- The MPOA appears to be crucial for integrating sensory and internal stimuli in order for normal copulation to occur. Destruction of the MPOA eliminates copulatory behaviors in virtually all vertebrate males. Implantation of minute amounts of testosterone into the MPOA restores mating behavior in castrated males. Testosterone must be converted into estrogen in order to have a behavioral effect; DHT implanted into the brain does not stimulate copulation. Mapping the distribution of steroid hormone receptors and aromatase activity has helped to pinpoint exactly where in the brain hormones act to regulate male sexual behavior. The use of mice with targeted deletions of specific genes has yielded some novel, and in some cases contradictory, information about hormone–behavior interactions underlying male sexual behavior.
- Females can affect hormone concentrations in males, and they can also affect the frequency and timing of their copulatory behaviors. The Coolidge effect is one well-known example in which novel females have substantial effects on the reproductive behavior of males. Plasma concentrations of LH and testosterone increase in males after exposure to females. This elevation in hormone concentrations can be classically conditioned.
- Individual differences exist in the frequency of sexual behavior exhibited by males. This difference in sexual activity does not appear to correlate with blood androgen concentrations.
- The copulatory behaviors of men cannot be classified according to muscular reflexes, as in other animals. Other male primates exhibit stereotyped mating positions. Androgens are important in maintaining nonhuman male primate mating behavior.
- The sexual behavior of men does not seem to be as dependent on blood androgen concentrations as that of some other taxa. However, there is great variation in the frequency of post-castration sexual activities among men. Hypogonadal men treated with a long-lasting androgen showed increases in several sexual behaviors and thoughts, indicating that androgens can affect sexual behavior in humans.
- The same brain regions underlying rodent sexual behavior appear to be involved in the hormonal mediation of primate sexual behavior, with the exception of the olfactory system.
- Female primates, like female rodents, affect sexual behavior and hormone concentrations in males, but results of studies comparable to the experiments with rodents are, for the most part, inconsistent.
- Opioids appear to inhibit sexual behavior, especially with chronic exposure, whereas treatment with opioid antagonists appears to facilitate the expression of sexual behavior.
- Birds, like rodents, appear to require androgens to maintain sexual behavior. Androgens are converted into estrogens in the preoptic medial nucleus (POM) to mediate copulatory behavior, or to DHT in other regions to mediate vocalizations.

- Most male reptiles stop breeding after castration, and mating behavior is restored by androgen replacement therapy. Red-sided garter snakes, however, engage in copulatory behavior when the testes are regressed and no circulating androgens can be detected. The independence of male copulatory behavior from hormone levels in these snakes provides strong evidence that not only animals with a highly developed brain can copulate without the presence of androgens.
- Individuals of female parthenogenetic lizard species engage in male-like behavior in order to increase the number of ova shed during ovulation. Although male copulatory behaviors in related sexual species are regulated by androgens, male-like behaviors in the all-female species are mediated by progesterone.

Questions for Discussion

1. Androgens cause sexual behavior in males. Is this statement really true? Defend your answer.
2. Discuss the following assertion: "Hormones play a lesser role in the mediation of male sexual behavior in primates than they do in rodents."
3. How do testosterone and its metabolites (DHT and estradiol) influence male sexual performance and motivation? What are the similarities and differences between the influences of steroid hormones on these two aspects of sexual behavior?
4. If you were put in charge of a large study to examine hormonal influences on the sexual behavior of men, how would you design your study? What would you measure?
5. What is sexual behavior? What behaviors do you think can reasonably be called sexual behaviors in the following species: male rats, male ring doves, and men?

Refer to the accompanying Student CD for additional resources, including Web links, videos, animations, and additional photos.

Suggested Readings

Adkins-Regan, E. 1996. Neuroanatomy of sexual behavior in the male Japanese quail from top to bottom. *Poultry and Avian Biology Reviews*, 7: 193–204.

Ball, G. F., and Balthazart, J. 2002. Neuroendocrine mechanisms regulating reproductive cycles and reproductive behavior in birds. In D. W. Pfaff, A. P.

Arnold, A. M. Etgen, S. E. Fahrbach, and R. T. Rubin (eds.), *Hormones, Brain and Behavior*, Volume 2, pp. 649–798. Academic Press, New York.
Godwin, J. and Crews, D. P. 2002. Hormones, brain and behavior in reptiles. In D. W. Pfaff, A. P. Arnold, A. M. Etgen, S. E. Fahrbach, and R. T. Rubin (eds.), *Hormones, Brain and Behavior*, Volume 2, pp. 545–586. Academic Press, New York.
Hull, E. M., Meisel, R. L., and Sachs, B. D. 2002. Male sexual behavior. In D. W. Pfaff, A. P. Arnold, A. M. Etgen, S. E. Fahrbach, and R. T. Rubin (eds.), *Hormones, Brain and Behavior*, Volume 1, pp. 1–138. Academic Press, New York.
Stone, C. P. 1922. The congenital sexual behavior of the young male albino rat. *Journal of Comparative Psychology*, 2: 95–153.

6

Female Reproductive Behavior

My wife's family owns a beautiful Chesapeake Bay retriever named Molly. Nowadays, Molly is an ideal pet; she is quiet, well-behaved, and obeys commands. But she had once exhibited behavior that was unusual—even somewhat frenzied. She routinely whimpered, barked, and howled. When she was inside the house, she frequently tried to bolt outside whenever a door was opened. If she was left in the yard unattended, she continually tried to leap over the fence. If she was in the yard with her owners and they opened the gate, Molly raced away, ignoring their pleas and commands. When Molly escaped the confines of her yard, she always headed west, toward the home of a male dog that lived in the neighborhood, instead of taking her preferred route toward the bay.

Molly's atypical behavior had been the result of her being in "heat," or **estrus**. The word *estrus* comes from the Latin word *oestrus*, which loosely translates to "in a frenzy, or possessed by the gadfly" (Feder, 1981). Many pet owners have experienced a dog or cat in "heat" and are often surprised by the intensity of effort displayed by their pets to mate (Figure 6.1). In a laboratory setting, an estrous rat will cross a highly charged electrified floor to gain physical access to a male (Warner, 1927); female rats in estrus will also diligently depress a lever or poke their noses through an opening many times in order to gain access to a male (Bermant, 1961; Matthews et al., 1997). Female rats (or dogs) that are not in estrus will not sustain pain or exert much effort to interact with males. The ovaries are involved in the

6.1 Estrous females are motivated to seek males. High circulating estrogen concentrations increase the motivation of females to engage in sexual behavior. This estrous Chesapeake Bay retriever is shown jumping a fence to visit a neighboring male dog; she does not behave in this manner when not in estrus. Courtesy of Brigid DeVries.

expression of estrous behavior in rats and dogs. Since she was spayed (ovariectomized) several years ago, Molly no longer displays estrous behavior.

The differences between estrous and **anestrous** (not in estrus) females are striking. First, a female in estrus will seek out males, initiate copulation, and prefer to remain in close proximity to males; the same female will not engage in these behaviors when not in estrus. Second, estrous females are more attractive to males than anestrous females; that is, conspecific males prefer to visit, and exert more effort to maintain close proximity to, estrous rather than anestrous females. Third, males mount estrous females preferentially as compared with anestrous females. Finally, only estrous females will permit mating to occur. Anestrous females will not tolerate male mounting behavior; in fact, anestrous females of many mammalian species will inflict serious injuries on persistent male suitors (Figure 6.2).

What physiological changes evoke these dramatic changes in a female's behavior as she cycles between an estrous and an anestrous state? Briefly, the hormones associated with the maturation of her ova (eggs) have also evolved to affect her nervous system in a number of ways. The resulting neural changes influence her behavior in such a way that the probability of mating, and ultimately of the successful production of offspring, is increased. Thus, mating behavior is tightly coupled in time with ovulation, and it occurs when successful fertilization of ova is most likely. Although analyses of the evolution of hormonal regulation of female sexual behavior are intriguing (see Wallen and Zehr, 2004; Thornton et al., 2003), the focus of this chapter will be on the proximate endocrine mechanisms underlying sexual behavior in females.

As noted in the previous chapter on males, reproductive hormones do not "turn on" sexual behavior per se; rather, hormones change the probability that specific stimuli will elicit particular behaviors that lead to successful copulation.

6.2 Nonestrous females are not motivated to mate. In female rodents, low circulating estrogen concentrations are not consistent with high motivation to mate. This female mouse is lying on her side showing disinterest by back-kicking a male that is attempting to mount her.

Hormones affect input systems: the acuity, sensitivity, and efficiency of the sensory systems are enhanced by reproductive hormones. Consequently, estrous females are better able than anestrous females to detect and respond to conspecific males. The endocrine changes associated with estrus also affect the central nervous system. Females' motivation, attention, and perception change as sex steroid concentrations fluctuate. Finally, because effectors, too, are affected by hormones, the way that a female moves and reacts to stimuli also changes. Her behavior, as well as the stimuli she emits (for example, chemosensory agents or auditory signals), affects the way males behave; the males' behavior may also feed back to alter the endocrine state of the female further.

As described previously, sexual behavior is behavior that has evolved to bring the two gamete types together. Female sexual behavior is defined as all behaviors necessary and sufficient to achieve fertilization of female gametes (ova) by male gametes (sperm). Sex drive provides a powerful motivational force urging females to seek sexual union, just as it does in males. Until recently, however, sexual motivation was not emphasized in studies of female mating behavior. Although male sexual behavior has been the subject of research for much longer than female sexual behavior, in some ways a more detailed understanding of the hormonal correlates and neural bases of sexual behavior has been achieved for females (Blaustein, 1996; Pfaff et al., 2000; Blaustein and Erskine, 2002). For example, most of the neural circuitry underlying one female reproductive behavior—namely, lordosis, the reflexive mating posture of female rats—as well as the specific effects of hormones on this circuit, have been elucidated (Pfaff et al., 2000; Micevych et al., 1997).

Female reproduction typically occurs in cycles. Mammalian female reproduction has six components: courtship, mating, ovulation, pregnancy, parturition, and lactation (Everett, 1961). This chapter will be limited to behavior associated with the first three components of female reproduction; the latter three components will be reviewed in Chapter 7, which discusses parental care. In common with the studies on males presented in Chapter 5, most of our knowledge about the physiological mechanisms underlying female sexual behavior has been gained through studies on nonhuman animals, especially rodents, and examples of such studies will be described below. But first, efforts to understand female sexual behavior will be presented from a historical perspective. Then female sexual behavior and reproductive cycles will be described. The last part of the chapter is devoted to experimental and correlational data relating hormones and female sexual behavior.

Early Discoveries about Female Sexual Behavior

Remarkably little progress was made in the study of female sexual behavior until the 20th century. It is difficult to discern whether this lack of progress was due to the inherent complexity of female sexual behavior or merely reflected a lack of interest on the part of a predominantly male research community that assumed that females are passive participants in copulation (but see Beach, 1976). A possible practical impediment to research is that ovariectomy is a difficult operation compared with removal of the testes.* In any event, prior to 1900, there were relatively few studies on the effects of female gonadectomy on behavior, in contrast to the wealth of information already available regarding the behavioral effects of male castration. However, in all the early studies, as well as in most subsequent ones, the results were quite consistent: in most adult female mammals, mating behaviors stop immediately after the removal of the ovaries (reviewed in Beach, 1948).† The development of ideas and methodologies associated with studies of hormonal correlates of female sexual behavior paralleled the studies on male sexual behavior presented in the previous chapter.

A major complicating factor in understanding the physiological mechanisms underlying female sexual behavior is the cyclic nature of that behavior. Adult males are generally able to mate any time they encounter an estrous female during the breeding season, but females of many mammalian species, especially those commonly used in laboratory research, enter and leave estrus on a regular basis. For instance, if a female Syrian hamster (*Mesocricetus auratus*) is placed in a cage with a vasectomized male for a short period each night, she will allow copulation every fourth night. Her **estrous cycle** is thus said to be 4 days in length. If the male were not vasectomized and copulation continued to ejaculation, her eggs would be fertilized, and she would not display estrous behavior again for 16 days, until immediately postpartum (that is, after giving birth). Sustained copulation, even with a vasectomized male, would activate the formation of specific ovarian structures that would also disrupt the display of 4-day estrous cycles. The nature of this disruption, termed *pseudopregnancy*, will be described later in this chapter.

The Development of the Vaginal Cytological Assay

Because estrous cycles stop after ovariectomy, it was surmised that the cycles in estrous behavior reflected cycles in ovarian function, but at the turn of the 20th century, it was difficult to discern these cyclic ovarian changes without surgical intervention. In most cases, after behavioral observations were made, the ovaries were removed and histologically fixed for microscopic examination, and the behavior was correlated with the presence or absence of different ovarian structures, such as

---

*In the context of females, the terms *ovariectomy*, *gonadectomy*, and *castration* all mean removal of the ovaries.

†Humans and rhesus monkeys are exceptions to this general rule, as we will see below.

follicles or corpora lutea (see Chapter 2). Only limited data could be obtained in this manner. Repeated behavioral tests could not be performed in the same animals because the behavior in question stopped after ovariectomy. Consequently, the first breakthrough in understanding the cyclicity of female sexual behavior was the technical development of noninvasive external markers that reliably reflected ovarian activity. The demonstration by Stockard and Papanicolaou (1917) that changes in vaginal cytology (cell types) could be closely correlated with changes in ovarian function in guinea pigs was an extremely important advance in the study of female reproductive behavior and physiology. This technique was soon extended to mice (Allen, 1922) and rats (Long and Evans, 1922), and eventually to hundreds of species (Asdell, 1964; Rowlands and Weir, 1984). This powerful tool allowed researchers to correlate mating behavior with ovarian function without direct surgical examination of the ovaries.

The technique for assessing changes in vaginal cytology is rapid and straightforward. A modified version of this histological test, in which cells from the cervix (instead of the vaginal lumen, or opening) are obtained and examined, is part of every woman's annual medical examination and is commonly called the Pap test, after Dr. Papanicolaou (Figure 6.3). In rodents, the procedure consists of swab-

6.3 Dr. George Nicholaus Papanicolaou was born in 1883 in Greece. He attained the M.D. degree from the University of Athens, and received a Ph.D. in Munich. Dr. Papanicolaou obtained a position in the Department of Anatomy at Cornell Medical College in 1913. His wife, Mary, also worked there as his technician. Dr. Papanicolaou remained at Cornell until a few months before his death in 1962. It was at Cornell that he began examining vaginal smears of guinea pigs to determine the external markers of the 16-day estrous cycle. Eventually, he began to study menstrual cycles in women. In 1933, he published a monograph titled "The Sexual Cycle of the Human Female as Revealed by the Vaginal Smear." It was while doing this work that he noticed cancer cells coming from the cervix. Dr. Papanicolaou's initial understanding of the significance of these cells as a diagnostic tool was not immediate. After an extensive collaboration with gynecologist Herbert Traut to validate the diagnostic potential of vaginal cells, he published the now famous monograph "Diagnosis of Uterine Cancer by the Vaginal Smear" in 1943. This diagnostic procedure was named the Pap test, and has been used to diagnose literally millions of cases of cervical cancer. Courtesy of National Library of Medicine.

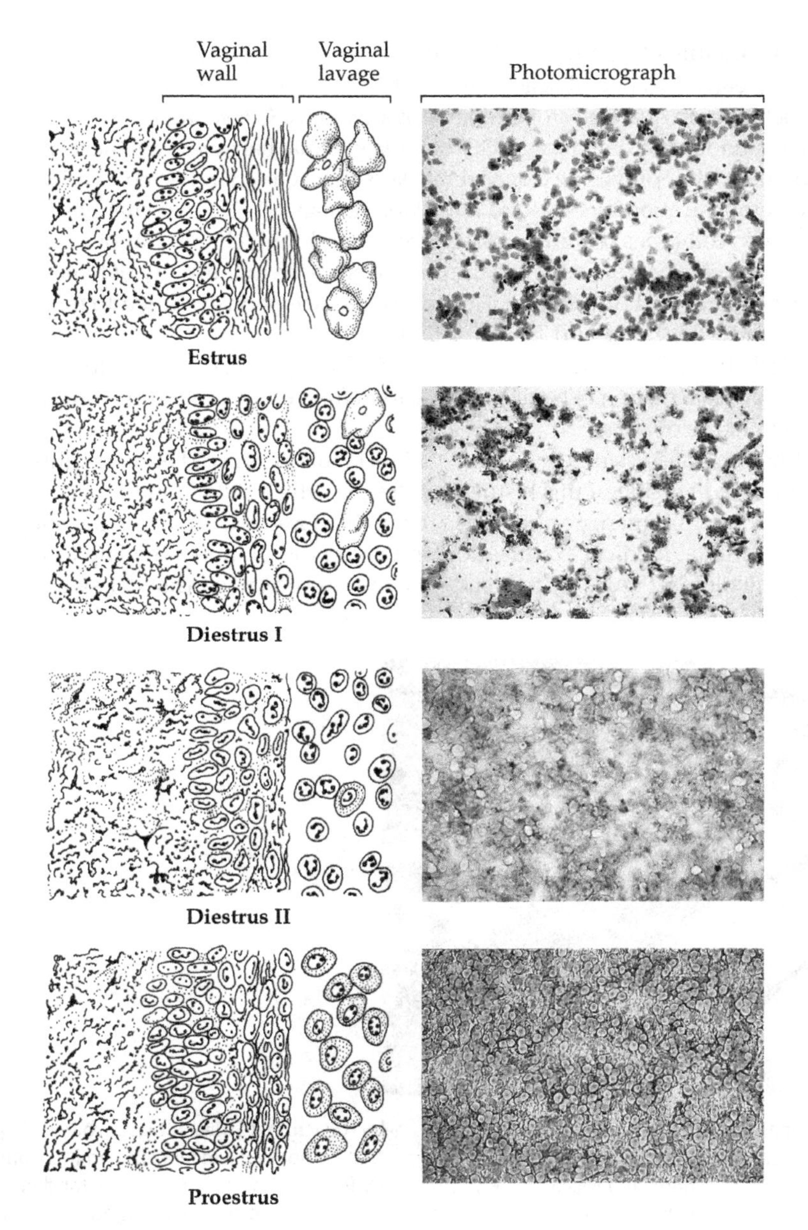

6.4 Cyclic changes in vaginal cell types in female rats can be observed in the vaginal wall and in cells obtained from daily vaginal lavage. During vaginal estrus, cornified epithelial cells are present for about 36 hours. Vaginal estrus is followed by vaginal diestrus, which lasts for about 48 hours in females displaying 4-day estrous cycles; this stage is marked by the appearance of leukocytes and a few nucleated epithelial cells, as well as a decrease in the number of cornified cells. The first day of diestrus is called diestrus I, and the second is called diestrus II. The final stage of the estrous cycle is called vaginal proestrus and lasts about 12 hours. Vaginal proestrus is characterized by an increase in the number of nucleated epithelial cells, as well as a sharp decrease in the number of leukocytes. After Wang, 1923; Feder, 1981; Ganong, 1995; Norris, 1996; photomicrographs courtesy of Deborah Drazen.

bing cells from the vaginal lumen after lavage (sterile wash), then examining those cells microscopically. When applied to rats, this method reveals changes in the cellular contents of the vaginal lumen that recur every 4 or 5 days (Figure 6.4; Long and Evans, 1922).* At one point in the cycle, "cornified" epithelial cells are present; these cells look like cornflakes under the microscope. This stage, which lasts about 36 hours, is arbitrarily considered the first stage of the cycle and is called **vaginal estrus**. Vaginal estrus is followed by a period during which cornified cells become reduced in number, and leukocytes (white blood cells), as well as a few nucleated epithelial cells, appear. This stage is called vaginal **diestrus** and has a duration of approximately 48 hours. The first day of diestrus is referred to as diestrus I, and the second day is often called diestrus II.† The next phase is characterized by the presence of many nucleated epithelial cells, as well as a dramatic reduction in the number of leukocytes. This final stage is called vaginal **proestrus** and lasts for about 12 hours. An obvious external marker of vaginal proestrus is that the thin vaginal membrane disappears, revealing a patent (open) vaginal canal. This external marker coincides with the onset of mating behavior.

These changes in vaginal cytology were soon correlated with ovarian changes (Figure 6.5; see also Figure 2.12). When researchers took vaginal smears from rats, then removed the ovaries from the same individuals, fixed them histologically, and examined them under a microscope, they found that the following correlations between changes in vaginal cytology and ovarian activity could be made (Boling et al., 1941; Feder, 1981):

1. Vaginal estrus is correlated with the presence of recently ruptured follicles following ovulation. In addition, tertiary follicles begin to develop from secondary follicles at this time. The next wave of ova released by the ovaries will eventually come from these tertiary follicles after they develop fully. The antrum of each tertiary follicle is small at this stage. The granulosa cells, the epithelial cells surrounding the follicle, induce the formation of a distinct layer of connective tissue called the theca interna.
2. As the vaginal smear becomes diestrous, the tertiary follicles become larger and the granulosa cells become more numerous. The antrum begins to fill with a clear fluid called the liquor folliculi. The rupture sites where previous ovulations occurred have been repaired, and endocrine luteal cells surround the fluid-filled cavities of the former follicles to form corpora lutea.
3. By diestrus II, the layer of connective tissue surrounding the theca interna differentiates into a layer of spindle-shaped cells called the theca externa. The enlarging tertiary follicles are now referred to as Graafian follicles. The corpora lutea are fully formed; the inner cavity has been filled with endocrine luteal cells and their supporting blood and connective tissues.

---

*For the sake of simplicity, only 4-day cycles will be considered here, but it is important to note that among rats there are either 4-day or 5-day cycles, but no 4.5-day cycles (the reason for this pattern will be explained below).

†Older studies sometimes refer to diestrus I as "metestrus." The 5-day rat estrous cycle includes an additional 24 hours of diestrus, called diestrus III.

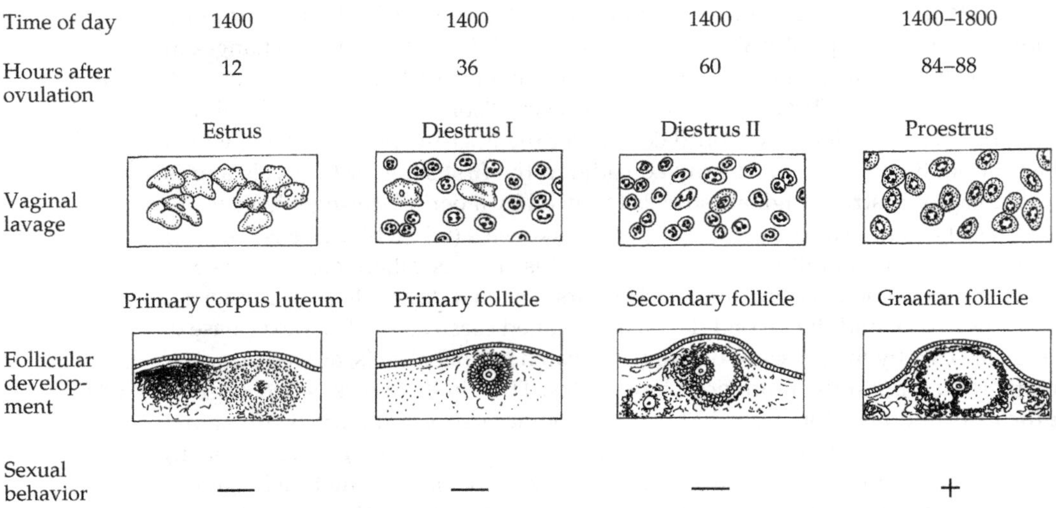

Time of day	1400	1400	1400	1400–1800
Hours after ovulation	12	36	60	84–88
	Estrus	Diestrus I	Diestrus II	Proestrus
Vaginal lavage				
	Primary corpus luteum	Primary follicle	Secondary follicle	Graafian follicle
Follicular develop-ment				
Sexual behavior	—	—	—	+

6.5 Cyclic changes in ovarian structure have been correlated with cyclic changes in vaginal cytology. Vaginal estrus occurs after ovulation and corresponds to the formation of corpora lutea. In the absence of stimuli associated with mating, luteal function wanes rapidly, and the corpora lutea regress. New follicles begin to develop during diestrus I and diestrus II. Vaginal proestrus coincides with maximal follicular development; behavioral estrus and mating occur during vaginal proestrus. After Feder, 1981.

4. As the vaginal cell cycle becomes proestrous, a growth spurt occurs in the Graafian follicles destined to ovulate (preovulatory swelling). Other follicles that are not going to ovulate regress. The corpora lutea also regress unless specific vaginal stimuli associated with intromission are experienced by the female.

When ovulation occurs, the cycle begins anew—unless the female mates. If she receives sufficient vaginal stimulation through copulation or simulated copulatory stimuli, then the corpora lutea do not regress. If the mating is fertile and the female becomes pregnant, then the corpora lutea remain large throughout most of the pregnancy. If the mating is sterile and the female does not become pregnant, then the corpora lutea remain large for approximately 14 days before regressing. Because it causes a number of physiological changes that resemble those of pregnancy, this state of sustained corpora luteal function is called *pseudopregnancy*. The estrous cycles are suspended during pregnancy or pseudopregnancy. Females also display seasonal anestrous, during which the ovaries are quiescent. Seasonal regulation of reproductive function is discussed in Chapter 10.

The linkage of vaginal cytology to ovarian structural changes enabled researchers to make a very good guess about ovarian function simply by observing cells obtained from the vaginal lumen. As these studies continued, other research was proceeding that linked mating behavior with vaginal cytological changes (see Figure 6.5). A classic monograph by Long and Evans (1922) provided an excellent description and partial quantification of the sexual behavior of female rats correlated with vaginal cytology. Long and Evans found that mating behavior is observed only when the vaginal smear shows many nucleated epithelial cells; in other words, as noted above, behavioral estrus coincides with vaginal proestrus. Female rats generally stop mating by the time vaginal estrus occurs. This unfortunate duplicity of terms has confused students for decades. Remember that behavioral estrus coincides with vaginal proestrus; vaginal estrus follows behavioral estrus.

As a result of these studies, vaginal cytological cycles could be correlated with cycles of estrous behavior, and thus, logically, estrous behavior could be correlated with ovarian function through observations of vaginal cytology. Because both the cycles of vaginal cytology and of estrous behavior stop after ovariectomy, researchers reasoned that the ovaries must produce a cyclic signal that drives the changes in both vaginal cell types and behavior. Because the cycles continue if the ovaries are denervated, it was reasoned that hormones must be mediating the cyclic signal. The next part of the puzzle involved understanding the endocrine products associated with the various ovarian structural changes. Relating these endocrine changes to behavioral alterations provided the framework for the modern era of sex research on females (Beach, 1981).

Research in the "Modern Era"

The two ovarian structures that were considered most likely to be producing hormones that influenced estrous behavior were the follicles and corpora lutea. Researchers tested these hypotheses by injecting laboratory animals with extracts from ovaries obtained from slaughterhouses. When chemical extracts derived from the follicles of hog ovaries were injected into ovariectomized mice, they caused four physiological changes that are normally observed in estrous female rodents: hyperemia (vascular development) of the reproductive tract, uterine growth, hypersecretion in the genital tract, and growth of the mammary glands (Allen and Doisy, 1923). Mating behavior was also observed in ovariectomized mice injected with the chemical extract from ovarian follicles. These four physiological changes, as well as the presence of mating behavior, served as important bioassays in the attempts to isolate and identify endocrine products from the ovarian follicles.

The substance in the follicles that produced these effects was called an *estrogen*, a generic term for "estrus-generating substances." Estrogens were the first steroid hormones to be chemically isolated when Doisy and co-workers (1929) and Butenandt (1929) independently succeeded in crystallizing the first known estrogenic material, now known as estrone, from the urine of pregnant women. Isolation of estrogens directly from ovarian tissue occurred a few years later (Doisy and MacCorquodale, 1936), using hormone isolation techniques that are quite crude by today's standards. For example, MacCorquodale obtained approximately 12 milligrams of relatively pure 17β-estradiol by processing over 3800 kg (4 tons) of sows' ovaries! However, the availability of even small amounts of these relatively pure hormones ushered in a new era for behavioral endocrinology and allowed the first direct correlations to be made between hormones and behavior.

Researchers soon established the role of estrogens in stimulating female sexual behavior. It was found that the fluid within the Graafian follicles is rich in estrogenic hormones, which are secreted into the general blood circulation. Rapidly increasing estrogen concentrations during the day of proestrus induce behavioral estrus, as well as cornification of vaginal epithelial cells. These findings were originally reported in mice (Allen and Doisy, 1923; Allen et al., 1924; Wiesner and Mirskaia, 1930), rats (Boling and Blandau, 1939), and guinea pigs (Dempsey et al., 1936). However, although estrogen injections induced estrous behavior in most ovariectomized rodents, in all these early studies a substantial minority (approximately 40%) of ovariectomized females failed to mate after estrogen treatment, regardless of dosage (Ring, 1944). It should be noted that the doses of estrogen provided in these studies were typically in a pharmacological, rather than a physiological, range. The observation that many animals did not mate even after receiving large doses of estrogen prompted several researchers to suggest that a supplemental factor might be working in concert with estrogen to induce behavioral estrus (Wiesner and Mirskaia, 1930).

When chemical extracts from corpora lutea, also obtained from slaughterhouses, were injected into ovariectomized rabbits, they effectively prepared the uterus for pregnancy and maintained pregnancy (Corner and Allen, 1929). The luteal steroid hormone responsible for these effects was called *progesterone*, so named because it produced a progestational condition (i.e., supportive of pregnancy) in the uterus. Progesterone was identified and isolated from ovarian tissue in the mid-1930s (Turner and Bagnara, 1971). It was not identified initially in urine because pregnanediol, the urinary metabolite of progesterone, is biologically inactive and thus difficult to quantify with bioassays.

Progesterone was discovered to be the supplemental factor that operates in concert with estrogen to induce behavioral estrus. In a series of papers, Young and his collaborators (for example, Dempsey et al., 1936; Young et al., 1935, 1938, 1939) demonstrated that estrogen alone was insufficient to produce behavioral estrus in guinea pigs. Because one of the ovarian changes associated with the onset of estrus is preovulatory swelling, and because injections of luteinizing hormone (LH) cause preovulatory swelling, Young and his associates reasoned initially that LH was the substance acting in concert with estrogen to cause estrus. However, subsequent experimentation revealed that LH does not act directly with estrogen to cause behavioral estrus. Rather, LH causes the formation of corpora lutea; in turn, the developing corpora lutea secrete progesterone, which acts synergistically with estrogen to cause behavioral estrus (Dempsey et al., 1936).

Progesterone exerts biphasic effects on sexual behaviors, and these biphasic effects appear to differ among species. Initially, studies reported that progesterone was important, although not required, for causing behavioral estrus in rats (Boling and Blandau, 1939) and a number of other species (Ring, 1944). Subsequently, progesterone has been discovered to be important in mediating estrous behavior during natural estrus in rats (Feder, 1981). However, Ball (1941) reported that progesterone administered to estrogen-primed, ovariectomized monkeys decreased sexual behavior. This counterintuitive finding was explained when the biphasic effects of progesterone on female mating behavior were eventually discovered. First, progesterone, in concert with estradiol, initiates female sexual behavior in rats. This preovulatory progesterone comes from the Graafian follicles. As the cycle continues after copulation and estrogen concentrations fall, the corpora lutea secrete larger amounts of progesterone, and elevated concentrations of progesterone in the blood inhibit female sexual behavior. The importance of progesterone in sexual behavior varies among species, as we will see below.

In rats and certain other rodent species, elevated estrogen secretion accompanies follicular development during diestrus II and proestrus, stimulating cornification of vaginal epithelial cells as well as estrous behavior. As just described, progesterone is also needed for the full expression of female sexual behavior in many species. In several rodent species, including rats, mice, guinea pigs, and hamsters, progesterone concentrations increase abruptly prior to ovulation to induce mating behavior. After ovulation, the corpora lutea secrete progesterone at high rates, which often serves to terminate estrous behavior. The neural and endocrine events that drive this cycle of estrogen and progesterone secretion, and thus drive the cycle of estrous behavior, will be described later in this chapter.

Mammalian Female Mating Behavior: A Description

Before we can understand the endocrine mechanisms underlying female mating behavior, we must first describe the behavior. Description always should precede analysis (Tinbergen, 1951).Thus, a description of female mating behavior, as observed in the laboratory, will be presented for several species.

Rodents

If an adult female rat is placed with an adult vasectomized male in a typical laboratory cage for 30 minutes every evening, a very consistent behavioral pattern will emerge (Long and Evans, 1922). Every fourth (or fifth) night, the female will be in behavioral estrus. The most prominent aspect of her estrous behavior is the assumption of the mating posture called **lordosis** (Figure 6.6), a name derived from a medical term that refers to curvature of the spine. When touched on or near the flanks, an estrous female rat will arch her back, deflect her tail, and stand completely immobile to aid the male's penile insertion. In the absence of lordosis, intromission and ejaculation are impossible (Figure 6.7; Diakow, 1974; Pfaff et al., 1978). In order to initiate or maintain mounting behavior by the male, she may approach him, then dart away. Hopping a short distance away, waiting, moving back, and wiggling the ears (actually, moving the head rapidly makes the ears appear to wiggle) are all behaviors that male rats find very attractive, and males are motivated to follow and usually mount a female performing these behaviors (Figure 6.8). After an intromission, the female will leave the mating posture, groom herself or the male, walk about the test arena, or perhaps rest.

Mice, hamsters, voles, lemmings, and many other rodent species display patterns of female copulatory behavior similar to those of rats. However, guinea pigs differ strikingly from other laboratory rodents in the latency to ejaculation (Young et al., 1938, 1939). Male guinea pigs mount females, intromit, and ejaculate immediately; there is no pelvic thrusting (see Chapter 5). Consequently, it is not neces-

6.6 Lordosis, the characteristic mating posture of estrous female rodents, as seen in (A) rats and (B) Syrian hamsters. The female arches her back and deflects her tail to allow a male to gain intromission. Lordosis occurs in response to tactile stimulation of the flanks and anogenital region, usually provided by a mounting male; in (B), the female exhibits lordosis in response to the male's investigation of her anogenital region and his previous tactile stimulation of her flanks. The female will maintain the lordosis posture for some time after a male has dismounted. (A) courtesy of Robert Meisel; (B) courtesy of Robert E. Johnston.

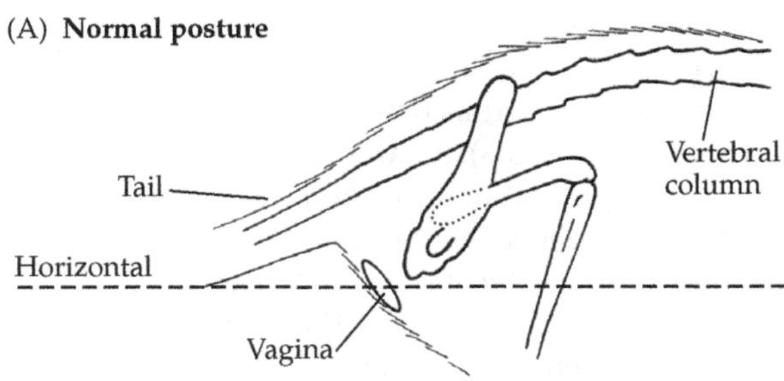

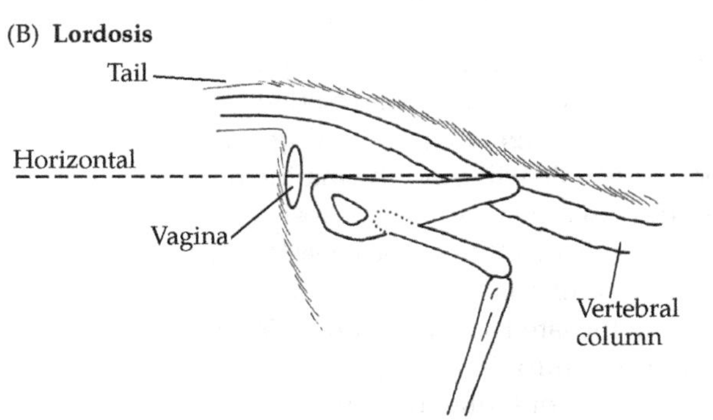

6.7 Lordosis makes successful copulation possible, as shown in these drawings based on X-ray images of female rats. (A) When the female is standing in her normal posture, it is impossible for the male to intromit. (B) During lordosis, the vaginal opening is accessible and the vagina is horizontal. Because rats have no manual dexterity, this precise postural adjustment on the part of the female is necessary for the male to intromit successfully. After Pfaff et al., 1978.

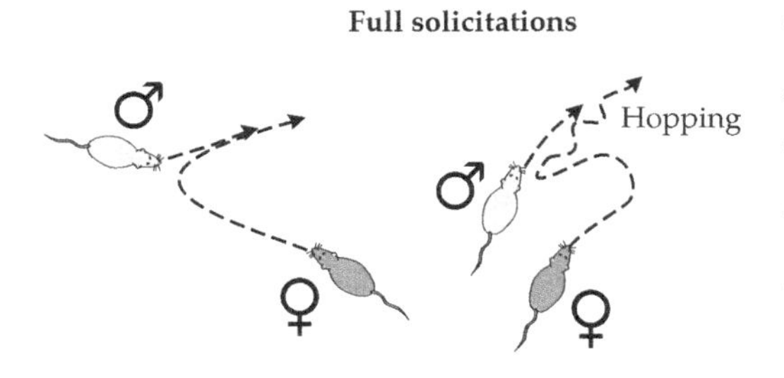

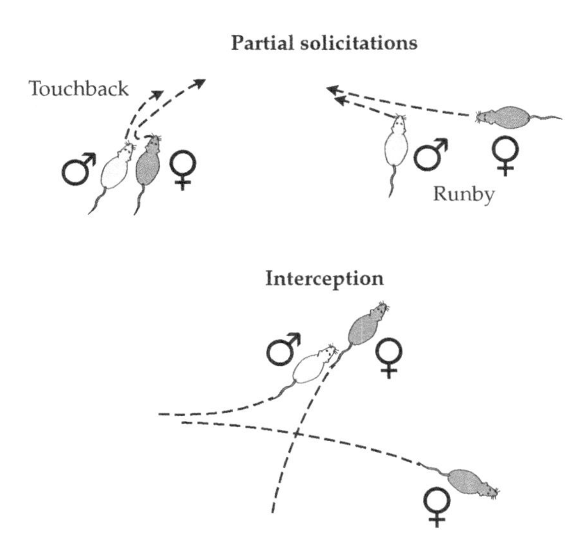

6.8 Female initiation of sexual interactions in rats often involves solicitational behaviors. In so-called full solicitations, a female rat will dart toward the male and run or hop away; the male often follows. During partial solicitations, the female may pause in front of the male (touchback) or run past (runby) the male. In mating situations involving multiple animals, interceptions may occur, whereby a female darts in front of a male that is following another female and distracts him. After McClintock, 1984.

sary for female guinea pigs to display lordosis for sustained periods of time, although they may maintain the lordosis posture for over 4 seconds in response to male mounting behavior in the laboratory, and if stimulated by manual palpation, may maintain lordosis for 10–20 seconds (Goy and Young, 1956/1957).

Canines

Female mammals in estrus have often been portrayed as "out of control" because they appear to be indiscriminate about their mating partners, but part of this portrayal results from the laboratory testing situations used, especially for rodents. Dogs, and many other mammalian species (especially primates), display substantial selectivity when in estrus (LeBoeuf, 1967). An estrous dog may absolutely refuse to mate with a male to whom she is not attracted, and there is substantial individual variation among female dogs in mate choice. Females may or may not choose their mates by rank in the pack; the alpha males of domestic canine packs do not appear to possess mating advantages over lower-ranked dogs (Beach and LeBoeuf, 1967).

An estrous dog is likely to seek out males to whom she is attracted. Male dogs are likely to seek out estrous females because a chemosensory cue in the urine of estrous females is very attractive to them (Beach and Gilmore, 1949).

When an estrous female and a male meet, she will orient her hindquarters toward his muzzle and deflect her tail. The male will lick her vulva, sometimes so vigorously that he lifts her hindquarters off the ground (Beach and LeBoeuf, 1967). The male then mounts the female and presses his erect penis into her vagina; the female may assist the male by backing into him as he thrusts. She may also compensate for missed intromissions by changing her position slightly. After a brief series of intromissions, the male ejaculates and dismounts. The two animals are locked together for some minutes as seminal fluid is transferred to the female (see Figure 5.8). An anestrous female dog will not allow a male to mount. She will sit down, try to move away, or growl and snap at the male.

Estrous occurs about once every 7–8 months, but recurs more frequently in unmated females (Asdell, 1964). Estrus persists for a week or 10 days. Termination of estrus is signaled when the female refuses to be mounted by males. Estrus is followed by about 2.5 months of diestrus, during which corpora lutea persist. After the corpora lutea regress, the female enters anestrus, which continues for about 4 or 5 months. The follicles begin to grow and secrete estrogens during proestrus, which lasts about 10 days. Vaginal bleeding is often observed during proestrus; the female will not mate during this time. The onset of estrus coincides with the female's acceptance of a male.

Primates

Early research conducted at the Yerkes Laboratories of Primate Biology in Orange Park, Florida and elsewhere, provided a very complete quantification of the sexual behavior of chimpanzees in a laboratory setting (e.g., Yerkes and Elder, 1936). This description of chimpanzee copulatory behavior was quite lively:

> What happens when a male and female meet, under the conditions of our experiment, has been found to depend upon individuality, physiological status, social relations, and environmental circumstance. ... [Various] postures and gestures [by the male] evidently are intended as appeals for sexual contact. If perchance the female does not respond to his solicitations, the male goes to her, usually examines her genitalia, and thereupon either attempts to copulate or turns away. Ordinarily, however, she responds instantly, when she sees the male ... by running quickly to him and crouching low or even flat upon the ground, with limbs flexed and genitalia directed toward him. To this female presentation [the male] bends over her back, with his hands on her shoulders or sides, or touching the floor. In this posture, he presses forward against her genital swelling and forces the long slender penis into the vagina. ... Insertion is followed by pelvic thrusts [which] may vary from as few as four or five to as many as twenty or thirty in our observation. Successful and mutually satisfactory copulation may be followed by manual, oral, or olfactory self- or mutual examination of the genitalia, and by grooming.

Obviously, some of the researchers' remarks, including "mutually satisfactory copulation," were inferences, but on the whole, their description is very accurate.

The mating posture of the female chimpanzee is rather stereotyped; there is little variation among females of this species. In contrast, as noted in Chapter 5, humans often seek variety in copulatory postures (Masters and Johnson, 1966). Variation in the sexual posture of some other nonhuman female primates has also been noted; for example, the ventral–ventral, or "face-to-face" position associated with humans has been observed in gorillas and orangutans (Nadler, 1976, 1988). Thus, among primates, there are species in which female sexual behavior is highly stereotyped (for example, chimpanzee and marmoset) and other species (human, gorilla, bonobo, and orangutan) that tend not to exhibit stereotyped female copulatory patterns.

Some primates, such as marmosets, have clearly defined estrous cycles, and mating behavior is limited to estrus. Other female primates, including rhesus monkeys, bonobos, and humans, do not limit sexual activity to a particular time and thus do not possess an estrous cycle per se (de Waal, 1995). There has been recurring controversy about the extent to which copulatory behavior in female humans and other anthropoid primates is regulated by ovarian hormones. Some investigators contend that because higher primates can copulate at any time during the ovarian cycle or even after ovariectomy, hormones are not involved in mediating female sexual behavior. Others have noted that under the appropriate circumstances, female primate sexual behavior shows a clear reliance on gonadal hormones. The controversy remains unresolved, although some consensus has emerged suggesting that sexual behavioral motivation in relation to fertility remains coupled to hormones, whereas sexual behavioral motivation in relation to social goals is uncoupled from hormones in female anthropoid primates (Wallen and Zehr, 2004).

Are Females Active Participants in Sexual Behavior?

Prior to the mid-1970s, the consensus among sex researchers, especially laboratory researchers, was that females were more or less passive recipients of male sexual attention. Although this perspective continues to dominate bird sexual research (see Cheng, 1992; van Tienhoven, 1983), Beach (1976) addressed this issue in mammals by noting that females often initiate sexual activities. We now know that females indeed take a very active role in initiating sexual activities, and in many species they virtually always act as the initiators of copulation (reviewed in Erskine, 1989; Wallen, 1990; Wallen and Zehr, 2004). The sex drive of females may equal or even exceed that of male conspecifics, but may be expressed only in very specific spatial, temporal, or social contexts. Recall that the number of offspring a female can produce in her lifetime is limited, as compared with males, who can potentially sire a large number of offspring. Thus, females generally have evolved to provide more resources to their offspring than males because they have a larger stake in the successful outcome of each breeding effort than males, who typically invest little in each reproductive effort (Clutton-Brock, 1991).

Consequently, females exhibit much greater selectivity than males in their choice of mating partners. This choosiness, which can easily be explained at an ultimate, evolutionary level of analysis, should not be invoked at the proximate level to suggest that females have less sex drive than males. Rather, the sex drive of females has evolved to maximize the reproductive success of each individual.

In retrospect, it may seem surprising that sex researchers "forgot" that *estrus* means "in a frenzy." And it may also seem surprising that they failed to notice the highly motivated behavior of pets like Molly in their attempts to copulate. Of course, the behavior of pets provides only anecdotal evidence at best, but there were many scientific reports, from field settings, of females that were highly motivated to engage in sexual behavior. For example, one study of a seminatural population of chimpanzees revealed that females risked physical attack whenever they approached males. Nevertheless, in most cases females were the initiators of copulation (Carpenter, 1942b; emphasis added):

> During her receptive period, her social status in the group shifts and she becomes a sexual incentive for the group's males. She actively approaches males and must overcome their *usual resistance* to close association; hence she becomes an object of attacks by them. Even other females attack her as a result of her shifted social status. ... Females 49, 105, 126, 144, "f.n." and 109 were all severely wounded during their estrous periods. Female 105 lost parts of both ears, was cut severely on the arm and received a network of wounds over her face and muzzle. Female 144 had a leg wound which compelled her to walk on three legs for several days. Female 145 was deeply cut on the thighs. Female "f.n." had a badly bruised nose while female 109 had a long, deep gash and her infant was wounded so severely that it died.

This account indicates that these females initiated copulation despite the threat of severe physical punishment and social disruption. Even the earliest laboratory accounts of primate mating (Ball and Hartman, 1935; Yerkes and Elder, 1936) indicated that females were highly motivated to copulate:

> Under the conditions which we have specified, copulation is determined and controlled almost entirely by the female. The male is suitor and servitor, not lord and dictator. There may be no suggestion of compulsion. The female ignores his solicitations if she sees fit; terminates the sexual union when she will—and that may be before orgasm and completion of ejaculation. (Yerkes and Elder, 1936)

What accounted for the failure of subsequent laboratory researchers to note the initiatives of females in copulation? As noted in Chapter 1, observer bias can adversely affect behavioral observations. Because feminine sexuality was widely assumed to be of a passive nature, it is possible that cultural constraints limited the interpretations of feminine sexual behavior observed in the laboratory. And in all fairness to those earlier sex researchers, the most striking aspect of female rodent sexual behavior in the laboratory test situation is the adoption of the rigid mating posture, lordosis. Maintaining immobility during copulation does indeed appear passive. It is also likely that sex researchers were victims of their own methods, born of their desire to control as many variables as possible during behavioral tests. Fieldwork produces highly valid results, but it is difficult, expensive, and occurs under uncontrolled conditions. Moving sex research into the laboratory permitted a high level of experimental control, but inadvertently removed important environmental and social variables that mediate female sexual behavior in nature. For example, pairing a single female with a single male has been the most common testing situation used in studies of rodents and primates, but it completely ignores the social reality of these animals. When a female is confined in a small space with a single male, her behavior may appear more passive than it would be if she could control the pacing of mating behavior or choose her mating partner, as is often the case in nature.

Components of Female Sexual Behavior

The knowledge that estrogens and progestins influence feminine sexual behavior is important, but is insufficient for students of behavioral endocrinology. In order to understand in greater detail how these hormones relate to behavior, researchers have been conceptually partitioning female sexual behavior into smaller and smaller component parts. Beach (1976) proposed the concept of dividing female sexual behavior into three components: (1) attractivity, (2) proceptivity, and (3) receptivity. **Attractivity** is the stimulus value of a female for a given male, a hypothetical construct that must be inferred by observation of a conspecific's behavior and must always be measured in relational terms. If a male will expend effort to sit next to a particular female, then she is more attractive than something else; perhaps she is more attractive to the male than being alone, an empty box, food, or another female. **Proceptivity** is the extent to which a female initiates copulation. It reflects her overt behavior, as well as her underlying motivational state. **Receptivity** reflects the stimulus value of the female for eliciting an intravaginal ejaculation from a male conspecific; in other words, receptivity is her state of responsiveness to the sexual initiation of another individual (Beach, 1976).

Proceptivity and receptivity overlap conceptually, as well as in practice. For example, a proceptive female rodent may initiate copulation by assuming the lordosis posture; this mating posture would also indicate her receptivity. The conceptual separation of proceptive behavior from receptive behavior was an important means of clarifying and operationalizing observations of female sexual behavior. Prior to Beach's classification scheme, hormonal effects on female "receptive" behavior, from assumption of the lordosis posture in rats to solicitation of copulation by rhesus monkeys, appeared to vary immensely. In retrospect, it is not surprising that few general principles arose relating hormonal influences to such poorly defined female sexual behavior. Beach's scheme of separating female sexual behavior into these three components has contributed to the acknowledgment of female sexual initiation, facilitated the study of female sexual behavior, provided a conceptual framework that equalizes the sexes in terms of sexual motivation and responsiveness, and thus facilitated the identification of hormone–behavior relationships underlying female sexual behavior. Sex steroid hormones, especially estrogens, affect all three components of female sexual behavior (Figure 6.9). Sorting out the degree to which hormones affect proceptiv-

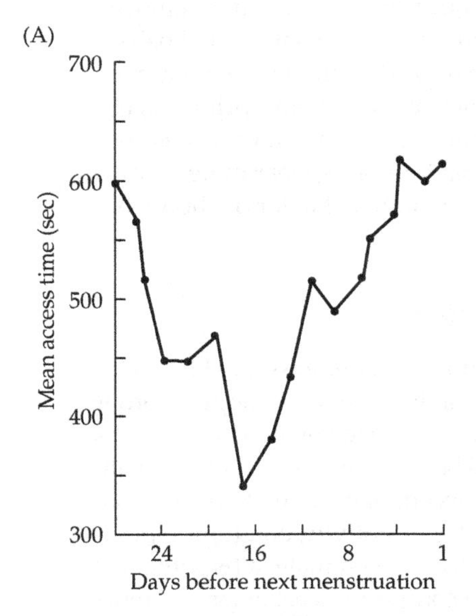

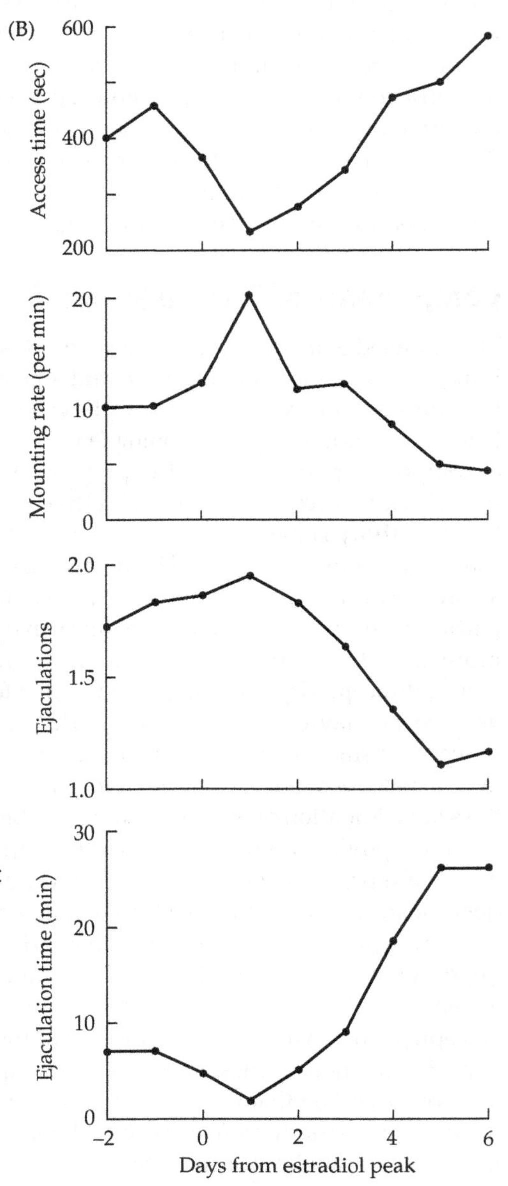

6.9 Estrogen mediates proceptivity, attractivity, and receptivity in rhesus monkeys. Female rhesus monkeys were taught to press a bar 250 times within 30 minutes to gain access to a male for a 1-hour behavioral test. (A) The average amount of time females required to press the bar 250 times is plotted across the menstrual cycle, showing a marked decrease around day 16. Ovulation would have occurred on about day 15 for these females. (B) Measures of proceptivity (mean access time for 250 bar presses), attractivity (male mounting rate), and receptivity (number of ejaculations and time to ejaculation) show that all these components of female sexual behavior are maximal when plasma estradiol concentrations peak near the time of ovulation. After Bonsall et al., 1978.

ity, attractivity, and receptivity has led to an increased understanding of the complex interactions between hormones and behavior and has also provided clues about the underlying physiological mechanisms.Other researchers have divided female sexual behavior into two phases: the precopulatory phase and the copulatory phase (Madlafousek and Hlinak, 1977). The precopulatory phase roughly corresponds to proceptive behavior in Beach's scheme. The precopulatory phase also corresponds to the appetitive phase described earlier for males, and may also be called courtship. The precopulatory phase involves all of the behaviors that allow a female to attract a male and initiate copulation. Behaviors as diverse as searching for males, emitting a chemical attractant, or assuming a mating posture and soliciting copulation may occur during courtship. Courtship functions as a communication opportunity during which information about species, readiness to mate, resources, and genetic endowment is shared. As with males, the precopulatory phase lasts much longer than the copulatory phase, which corresponds roughly to receptive behavior in Beach's scheme. Although this dichotomy, like any other classification scheme, may artificially divide a unified behavioral program, such partitioning can be valuable for elucidating the hormonal bases of female reproductive behavior, and possibly for discriminating the underlying neural mechanisms. As in males, sexual motivation, as represented by proceptive behaviors, may be regulated in females by different hormones—or to a different extent—than sexual performance, as represented by receptive behaviors.

A third classification system has been proposed for female sexual behavior. A strong argument has been made that the term "female sexual initiation" should be used for any feminine behavior that facilitates mating, and that "female sexual motivation" should be used for the state underlying sexual initiation or accommodation (Wallen, 1990; Zehr et al., 1998; Wallen and Zehr, 2004). Under this nomenclature, behaviors such as lordosis that facilitate copulation would be separated from solicitational behaviors that indicate sexual motivation. Hormones could affect these two components of female sexual behavior in the same way or differently. For example, female rodents differ from some female primates in their reliance on hormones for the expression of sexual behavior. Under laboratory conditions, female rodents appear to require ovarian hormones to maintain both facilitatory and motivational behaviors. In contrast, many primates appear to require hormones only for maintenance of female sexual motivation. Sex steroid hormones allow the expression of lordosis among female rodents in response to male tactile stimulation during mounting. As we have seen, lordosis is necessary for copulation because penile insertion is impossible unless the female stands immobile. Because of their manual dexterity, primates do not require the female to maintain a rigid mating posture to facilitate successful copulation. Female primates may permit copulation at all times during their ovarian cycle, but this copulation may not reflect their sexual motivation because female primates, especially those confined in the laboratory with a male, may copulate for a variety of reasons. Therefore, the only valid measure of primate sexual behavior would be female-initiated mating, because primate facilitatory behaviors are somewhat unnecessary.

Recently, another descriptive system has been proposed that parses female sexual behavior into (1) copulatory behaviors, (2) paracopulatory behaviors, and (3) progestative behaviors (Blaustein and Erskine, 2002). Copulatory behaviors are similar to receptive behaviors as described by Beach and include all behaviors that facilitate successful transfer of sperm to the ova. Paracopulatory behaviors are courtship-like behaviors that stimulate a male to mount and initiate copulation, and are similar to the proceptive behaviors previously described by Beach. Progestative behaviors are those behaviors that promote reproductive success, and include species-typical copulatory patterns and other activities that maximize fertilization (Blaustein and Erskine, 2002). Although many of the arguments for new terminology are valid and help to sharpen researchers' definitions and interpretations of behavior, especially the behavior of primates, Beach's nomenclature will be used in this chapter because of its enduring descriptive strengths.

Attractivity

Recall that bringing the two sexes together (attraction) when both are potentially fertile is the primary function of sexual behavior. Therefore, it should not be surprising that hormones associated with ovulation also mediate attractivity in females. In general, estrogens enhance attractivity. One way to test the attractiveness of females is to confine them separately and measure how much time a male spends with each of them. The assumption is that the duration of visitation by the male reflects the attractiveness of the female; the amount of time spent in close proximity to a female is called the PROX score (Phoenix, 1973). In this situation, males always spend more time in close proximity to estrous than anestrous females (reviewed in Beach, 1976). Ovariectomized females are rarely attractive to males. However, their attractivity is greatly enhanced by estradiol treatment. When spayed female dogs were injected with estradiol, the average duration of visits by males was increased sixfold (Beach and Merari, 1970). Similarly, ovariectomized female monkeys are rarely mounted by males (Figure 6.10). However, if

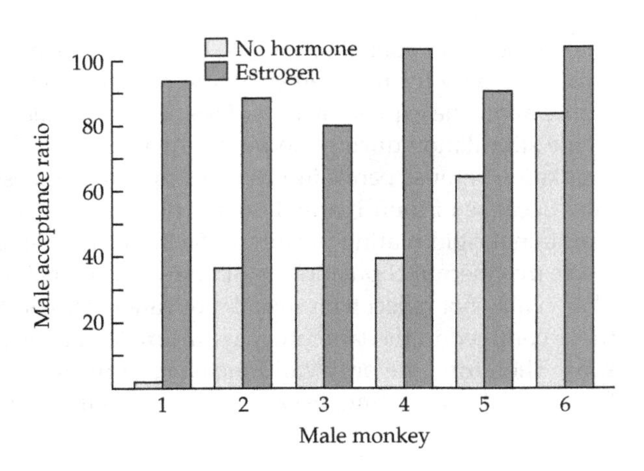

6.10 Estrogen increases attractivity in ovariectomized rhesus monkeys. The proportion of solicitations by a female monkey that results in male mounting behavior is called the male acceptance ratio. Males are more likely to mount an ovariectomized female after she is injected with estrogen (dark bars) than after she is injected with a biologically inert substance (light bars). Note the enormous individual differences in mounting behavior among males, and thus in the test female's attractivity to specific males. Male 1 rarely mounted the female in the absence of estrogen, but male 6 regularly mounted this female regardless of her hormonal condition. After Dixson et al., 1973.

ovariectomized monkeys are injected with estradiol, then they are mounted at a much higher frequency (Wallen, 1990).

In nature, high blood estradiol concentrations correspond to the time of maximal fertility. From an evolutionary perspective, it is predictable that females would be considered most attractive by male conspecifics when they are maximally fertile. Attracting males in order to mate is especially important for solitary animals such as cats or hamsters, which must signal their fertile condition to males over great distances. In a number of species, estradiol-induced attractivity is reduced or abolished by progesterone treatment, which is also predictable because high concentrations of progesterone are often associated with pregnancy or other nonfertile (from the male's perspective) states.

What exactly is attractivity? Both nonbehavioral and behavioral components sum together to form the stimulus bases of attractivity (Beach, 1976). Female attractivity is usually measured in terms of preference; that is, the extent to which a male prefers to be near one female as compared with others (Figure 6.11). As indicated in Chapter 5, the stimulus basis of male attractivity remains for the most part unspecified. As for females, nonbehavioral bases of attractivity include morphological changes that coincide with ovulation. For example, female primates in estrus experience a swelling of the perigenital skin. The visual stimulation provided by this conspicuous swelling attracts male attention (Carpenter, 1942b). Subordinate males in a troop of primates may not be permitted to get sufficiently close to females to have a real opportunity to mate, but even these males spend more time looking at estrous females than at anestrous females (Dixson et al., 1973).

Chemical cues are also important in forming the stimulus bases of female attractivity throughout the animal kingdom. For instance, chemosensory cues emanate from the urinary and vaginal secretions of estrous females (Box 6.1). Males sniff and lick females' external genitalia and vagina prior to copulation in many species of mammals, and males are much more likely to engage in this behavior with estrous than with anestrous females (Beach and Merari, 1968). Males of many mammalian species, especially carnivores, ungulates, and rodents, discriminate estrous from anestrous females on the basis of chemosensory signals

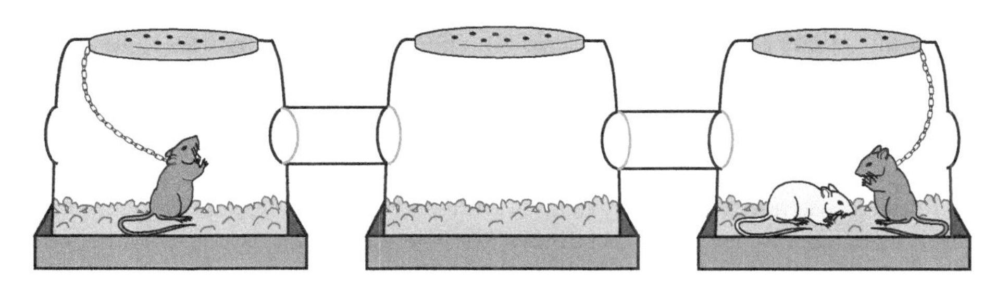

6.11 A three-chamber preference test is used to determine attractivity. All three chambers contain food and water. Females are tethered in the outer two chambers, and a male is free to move among all three chambers. He will spend the majority of the test period with the female to which he is most attracted.

BOX 6.1 Chemosignals and Courtship in the Red-Sided Garter Snake

An interesting twist to the story of chemical signals in courtship is provided by the red-sided garter snake (*Thamnophis sirtalis parietalis*) described in Chapter 5. Recall that the males of this species emerge from their winter hibernacula earlier than females, then attempt to mate with females as they emerge several weeks later. Females emerge one at a time, so many males court a single female as she leaves the hibernaculum, forming a large, writhing mating ball of snakes (see Figure 5.38).

How does a courting male identify the female in such a tangle of snakes? One possibility is that males discriminate by size; females are larger than males of this species. But size is not the primary cue male snakes use. They mainly rely on chemosensory information. Males produce one type of chemical signal, a squalene, that acts to identify them as males. Females produce another type of chemical signal, a methyl ketone, that males find very attractive. When males are courting, they flick their tongues along the scales of the female. As their sensory systems are stimulated by the female "attractive" cue, they begin to rub their chins on the female until one male gains intromission. As soon as an intromission occurs, the female expresses another chemical cue that immediately "turns off" the courtship behavior of the other males. Thus, female attractiveness is mediated directly by chemical factors: males are attracted to females emitting one chemosignal indicating readiness to mate and repelled by females emitting another chemical cue signaling that mating has already occurred (Mason et al., 1989). Estrogen is necessary for the production of the "attractive" chemical; however, injection of estrogen into adult males does not cause them to be courted by other males. In other words, estrogen alone is not the basis of female attractiveness in snakes.

In nature, however, about 15% of the males in a mating ball are courted by other males; that is,

that are processed through the vomeronasal organ (see Figure 5.14). Male ungulates commonly display a **flehmen response** when sexually excited, extending the neck and curling the upper lip to reveal the upper gums (Figure 6.12), which allows the female chemosignal to be delivered to the vomeronasal organ (Wysocki, 1979; Dulac and Torello, 2004). The vomeronasal organ is reduced or absent in anthropoid primates, and humans are notoriously poor among mammals in their olfactory abilities. For example, most farmers who raise cattle do so by artificial insemination: sperm is obtained from a bull by electrically induced ejaculation, and is introduced into estrous cows. However, discerning when a cow is in estrus has not been an easy task for cattle farmers, so dogs have been trained to discriminate estrous from anestrous cows. The dogs' chemosensory ability aids the farmers in knowing when artificial insemination will "take," although new electronic sensors have recently been developed to replace the dogs (Lane and Wathes, 1998).

In addition to physiological and morphological stimuli, various types of behavior may also increase the attractiveness of females. For example, females that actively solicit copulations have a higher stimulus value for males than

other males are observed slithering along their scales, tongue-flicking and chin-rubbing. What causes these males to be attractive objects of courtship? The courted males might have picked up the chemical attractant from a female that they were courting. But when researchers wiped off courted males with a chemical solvent and released them into a mating arena, this treatment did not reduce the incidence of courtship behavior directed at them (Mason, 1987). Chemical analyses of the skin secretions of courted males revealed that they produce the same attractant chemical that females do and none of the male-identifier chemosignal. These "she-males" have higher testosterone-to-estrogen ratios than typical males (blood plasma testosterone levels about 2.5 times higher than those of "he-males" and more than 3000 times higher than those of females) and also have very high concentrations of aromatase, the enzyme that converts testosterone into estradiol, in their skin. These differences may account for the she-males' producing the female attractant chemosignal.

But why have male snakes that are attractive to other males evolved, and how are they maintained in the population? The answer appears to be that these snakes are pursuing an alternative reproductive tactic. In mating competition tests, in which one she-male was placed with twenty he-males, the she-male obtained significantly more matings than the typical males (Mason et al., 1989). Apparently, the she-males confuse the other males, which waste time and energy courting the wrong animal in the mating ball. This provides the she-males with a competitive advantage during mating. Thus, in this species, some males exploit the "attractive" signals produced by females to gain a reproductive advantage over rival males.

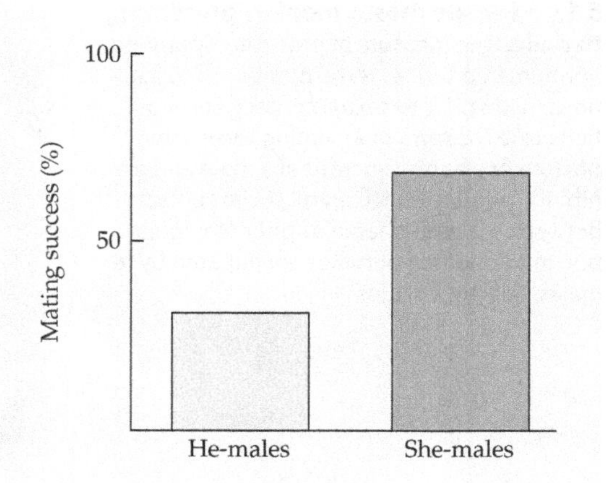

6.12 The flehmen response, seen here in a male red deer, is common in many male mammals, especially carnivores, ungulates, and rodents. In response to chemosignals obtained during anogenital investigation, the male extends the neck, curls the upper lip to reveal the gums, and usually presses the tongue against the roof of the mouth to force the chemosignals into the vomeronasal organ. Photo © Niall Benvie/naturepl.com.

6.13 Female rhesus monkey presenting to male. One measure of proceptivity among nonhuman primates is the number of solicitations made prior to sexual contact. Solicitations take the form of assuming the mating posture or standing in front of a male and exhibiting genitalia. Virtually all social contacts between males and females prior to copulation in nonhuman primates are initiated by females. Courtesy of Kim Wallen.

females that do not. As noted above, females of many nonhuman primates present their genitalia by backing their hindquarters toward a male (Figure 6.13); males spend more time with females that engage in this presentation behavior than with females that do not present (Dixson et al., 1973). Female rats also solicit mounts from males, as described above, by darting, hopping, and crouching behaviors and by running away from and returning to the male. As we will see below, in seminatural or natural settings, these female behaviors set the pace of the copulatory sequence, because virtually all male mounts follow a female solicitation (Erskine, 1989; Coopersmith et al., 1996).

All of these factors associated with attractivity, both behavioral and nonbehavioral, appear to involve sex steroid hormones. However, there are also nonhormonal factors that affect attractivity. Individual preferences exist among many species. For example, even when several ovariectomized female dogs are treated with equal amounts of estradiol, some consistently evoke more sexual responses from males than others (Beach and Merari, 1970; LeBoeuf, 1967). There is also sexual "favoritism" among male monkeys, apes, and humans (Herbert, 1970; Michael et al., 1972). A female may be very attractive to one male, but unattractive to another. This variation is consistently observed in the "Coolidge effect" (see Chapter 5), in which the stimulus value of a female is reduced for a male that has already copulated with her several times. She remains attractive to other males that have not just mated with her, but she has temporarily lost her stimulus value for the first male.

Understanding the stimuli underlying attractivity in various animal species is a daunting task. To get some idea of the complexity involved, think about the various components of individuals to whom you are attracted. Even though you can use language and can describe the features important in forming an attraction, it is very difficult to assign values to each feature. To some individuals, honesty may be a very important characteristic; to others, a clear complexion is important. How do you assign weight to each characteristic? Ask your friends to describe attractive people. How do you account for the vast differences among individuals in attraction? There is no reason to assume that the stimulus basis for attractivity is any less complicated in nonhuman animals (but see Grammer et al., 2003).

Proceptivity

Proceptive behavior comprises all of the appetitive activities shown by females. In other words, proceptivity is indicated by sexually solicitous behaviors that initiate sexual union, but not copulatory behavior per se. Proceptivity reflects a female's underlying motivational state in much the same way libido or sex drive reflects a male's motivational state (see Chapter 5). Molly's escapes from her yard and visits to her male neighbor indicated proceptivity. Estrous females, as well as being the most attractive to males, are also the most attracted to males.

Several behavioral measures have been used to assess proceptivity in females. The most common assay of proceptivity has been assessment of affiliative behaviors—efforts by females to establish and maintain proximity to males, which is a universal response of proceptive females. Female rhesus monkeys rarely interact with males except to initiate mating (Carpenter, 1942a,b). In one study, more than 80% of all proximate social interaction between male and female monkeys was initiated by females prior to mating (Cochran, 1979; Wallen et al., 1984). Another behavioral measure of proceptivity is the number of solicitations made before sexual contact, such as assumptions of the mating posture or presentations of the genitalia to the male (reviewed in Wallen, 1990) (see Figure 6.13). Such solicitations are commonly observed in primates, but are also observed in many disparate taxa, including birds, reptiles, and fishes. Vocalizations and head bobbing are also observed among monkeys that are soliciting copulations (Ball and Hartman, 1935; Carpenter, 1942a; Michael, 1972). A study of rhesus females was conducted during the nonbreeding season, when males are not responsive to females. Estradiol treatment during this time increased initiation of contact with males by ovariectomized female rhesus monkeys, despite the lack of behavioral feedback from the males (Zehr et al., 1998) (Figure 6.14).

Another female behavior that is highly proceptive is alternating approaches and withdrawals. This pattern is commonly observed in rodents and is also part of ungulate, canine, and primate copulation sequences. The female approaches the male, and if he follows her, she may pseudo-retreat, but eventually she allows him to investigate her. If the male does not follow her, then she approaches him again. This pattern of approach and withdrawal is very stimulating to most conspecific males and is in no way indicative of nonreceptivity or disinterest on the part of the female. Females may also initiate physical contact by investigating the male's anogenital region. Occasionally, mounting by highly proceptive females is observed (Beach, 1968). Females may mount males or other females, and these mounts are accompanied by thrusting pelvic movements (Beach, 1942d). In fact, ovulation may result from female–female mounting behavior among rabbits (Fee

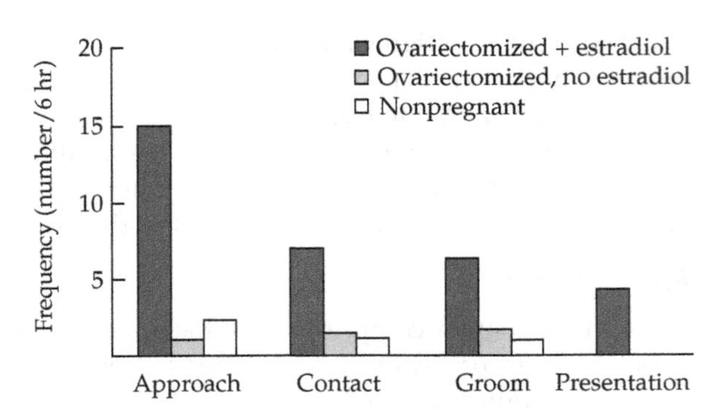

6.14 Estradiol enhances proceptivity in the absence of male interest. Estradiol treatment of ovariectomized rhesus monkeys during the nonbreeding season, when males are not responsive to females, increased the frequency of female initiation of contact with males. After Zehr et al., 1998.

and Parkes, 1930). After being mounted by a female several times, the male often mounts the female in turn.

A final determinant of female proceptivity is the attractiveness of the male mating partner. As noted above, females do not initiate copulation with males to whom they are not attracted. Castrated male rhesus monkeys, dogs, and hamsters elicit fewer approach responses by females than gonadally intact males in two-choice tests (reviewed in Beach, 1976). Female proceptivity is especially critical in species such as gorillas and rhesus monkeys because most copulation is a result of the females' initiative and mate selection (Cochran, 1979; Pomerantz and Goy, 1983; Wallen et al., 1984). Beach (1976) maintained that presentation to males by females is highest during midcycle, when ovulation occurs. This observation was based on laboratory, zoo, and natural populations of animals. In all species examined, high concentrations of estrogens facilitate proceptive behavior.

Receptivity

Receptivity, the consummatory phase of female mating behavior, can be defined as those female reactions that are necessary and sufficient for fertile copulation with a potent male. As Beach (1976) said, "Sexual receptivity is distinguished equally by the ubiquity of its usage and the infrequency of its definition." Prior to the development of Beach's classification scheme, behavioral measures of either receptivity or proceptivity were taken as evidence of receptivity.

Receptivity is indicated by a species-specific mating posture in all nonprimate mammals examined. For example, sows tread backward during copulation to help the boar gain intromission. Female rodents display the lordosis posture in response to tactile stimulation by the male. Females of several primate species back into males and literally seize the penis (reviewed in Wallen, 1990). Virtually all nonmammalian species also have a characteristic female mating posture (Crews and Silver, 1985). Most behavioral measures of receptivity are expressed in terms of ratios between a male's attempts to mate with a female and his success in doing so. This ratio is expressed in rats as a lordosis quotient (LQ), in dogs as a rejection coefficient, or in primates as an acceptance ratio (AR).

It has been proposed that female ring doves initiate their own receptive behavior by hearing themselves cooing. Cooing is usually performed by females in response to behavioral stimuli provided by males (Lehrman, 1961). It was first thought that the male induced receptivity in the female. However, the male actually induces the female to emit cooing vocalizations, and her behavior feeds back to affect her endocrine state and to initiate receptivity (Cheng, 1986; Cheng et al., 1998). If females are induced to coo in the absence of males, their follicles develop and secrete estrogens, and if females are devocalized by any of several means, including brain lesions, their ovaries do not develop in response to male stimulation. If the females' cooing vocalizations are recorded prior to devocalization and played back to them, even in the absence of males, the follicles develop and estrogens are secreted, inducing further receptive behavior (Cheng, 1992). Thus, ring doves present an interesting case of behavior in one individual inducing a behavioral change in the receiving animal that stimulates hormonal changes that cause further behavioral changes. In fact, part of the neural circuitry involved, located in the midbrain vocal nucleus, thalamus, and hypothalamus, has been identified (Cheng and Zuo, 1994). Specific hypothalamic neurons have been identified that respond to the female nesting coo and induce the secretion of LH (Cheng et al., 1998). Although the ring dove is the first species for which this sort of interactive feedback has been demonstrated, these kinds of complex social–behavioral–hormonal interactions are probably common and awaiting discovery.

There is abundant evidence that estrogen stimulation is very important in receptivity. For example, the display of the mating posture disappears in ovariectomized females of virtually all species examined. As noted above, lordosis can be induced in rodents by injections of estradiol and progesterone or sometimes estradiol alone (Feder, 1981). Rabbits, as well as rhesus monkeys and women, are maximally sensitive to tactile stimulation when the stimulation is concurrent with high blood concentrations of estrogen, but individuals of all these species will permit copulation at all stages of their ovarian cycle. These findings have led to confusion regarding the endocrine bases of female sexual receptivity among primates (Wallen, 1990).

FEMALE CONTROL OF COPULATION Much has been made of the so-called constant receptivity reported for humans and other primates. Their ability to copulate at any time during the ovarian cycle, or even in the complete absence of ovarian steroid hormones, caused many sex researchers to conclude that the sexual behavior of many primate species is not under hormonal control. However, as suggested above, the behavior of a female primate accepting copulation may not reflect the same motivational state as the behavior of a female rodent accepting copulation. A male rodent or canine cannot successfully copulate without female cooperation; the female must maintain a rigid posture. Primates do not exhibit this constraint, and copulation is mechanically possible without the female's cooperation. Although forced copulation is rare among most nonhuman primates, it has been consistently observed among orangutans (Mitani, 1985).

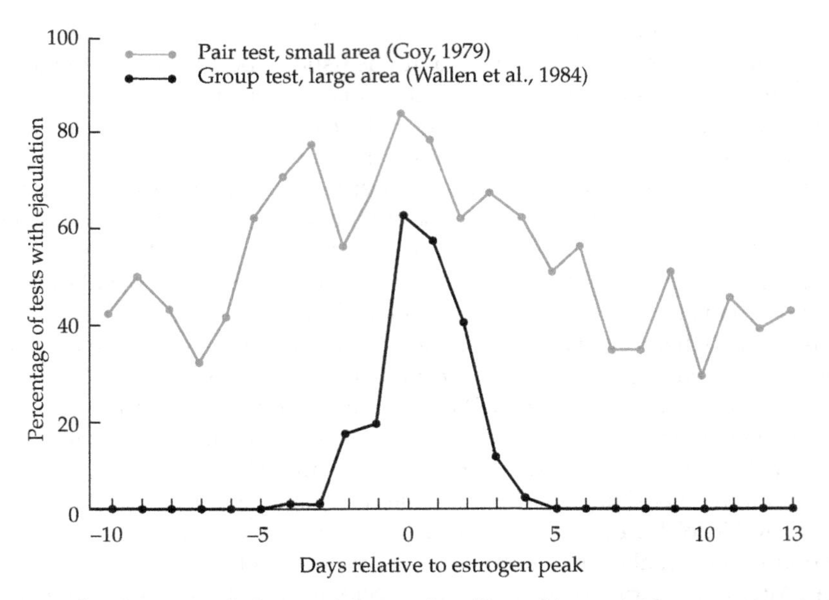

6.15 Endocrine control of receptivity can be affected by social factors. If female rhesus monkeys are tested with one male in a small area, there is no close relationship between estrogen concentration and receptivity (as measured by the number of tests with ejaculations). However, a clear relationship between receptivity and estrogen concentration emerges when females are tested in mixed social groups in a large area. After Wallen, 1990.

In a traditional laboratory test of paired partners, female primates require little sexual motivation to copulate. Frankly, there is little else to do. A male is provided in a small room, and there are no other animals with which to interact. Under these test conditions, a female primate's sexual behavior is unrelated to her ovarian cycle or endocrine state (Wallen, 1990). If animals are tested in more complex social conditions—that is, in the presence of multiple males and females—then females risk aggression from other females when they actively solicit males for sexual contact, especially if they are low-ranking females (Carpenter, 1942b; Wallen et al., 1984; Wallen and Zehr, 2004). Presumably, attacks by other females are aversive, and sexual motivation must be high to overcome this aversion. When female primates are tested in such group settings, copulatory behavior is tightly coupled to endocrine state (Figure 6.15; Wallen, 1990).

Is the sexual receptivity (or proceptivity) of women affected by hormones? This question has not been adequately answered, in part because the definitions of receptivity and proceptivity have become entangled and in part because social factors are not appropriately considered. Virtually all studies of human sexual behavior involve self-reports by couples in long-established, usually marriage, relationships. Under these circumstances there is little change in the frequency of copulation over the ovarian cycle (Adams et al., 1978; Udry and Morris, 1968), although some studies have revealed a peak in human female sexual activity around the time of ovulation. Ovulation is said to be "hidden" in women, by which it is meant that few women, and presumably fewer men, know precisely when ovulation occurs; however, the validity of this assumption remains unknown. Diaries of erotic thoughts or autosexual activities by women show a peak around the time of ovulation, suggesting an endocrine effect on sexual motivation (Adams et al., 1978).

Studies that report a peak in sexual behavior among married couples around the time of ovulation usually have not controlled for the possibility that males find periovulatory females more attractive, and hence, that men may initiate more sexual contacts around the time of ovulation. One study controlled for this possible confounding variable by examining the frequency of sexual activities among lesbian couples (Matteo and Rissman, 1984). This study reported a small peak in sexual activity around the time of ovulation, along with a secondary, perimenstrual peak. Other studies of heterosexual couples have also suggested a secondary perimenstrual peak in sexual behavior (for example, Sherwin, 1988a,b; Van Goozen et al., 1997; Harvey, 1987; Slob et al., 1991, 1996). A study of sexual arousability over the menstrual cycle used both subjective self-reports and objective (labium minor temperature) measures to discover cycles in erotic responsiveness (Slob et al., 1996). This study revealed that the order of testing (i.e., whether the test was first administered during the luteal or the follicular phase) could affect subjective reports of erotic arousal. Thus, caution must be exercised when gathering self-reports to make certain that multiple cycles are followed.

Part of the confusion about the role of hormones in the sexual behavior of women stems from imprecise assessments of ovulation. When the preovulatory surge of LH was pinpointed, it became clear that women were more sexually active on the days immediately before and on the day of the preovulatory LH surge (the time of maximal fertility). This pattern was pronounced when women initiated sexual contact, but not when sexual activity was initiated by their male partners, suggesting that women were more motivated to engage in sexual behavior, but not necessarily more attractive to their male partners, at the time of ovulation (Bullivant et al., 2004).

As indicated earlier in this chapter, female dogs will not accept males unless they are in estrus, or have been treated with estrogens. However, while sex steroid hormones are necessary for receptive behavior in dogs, they are not sufficient. For example, beagle females reject certain males, even when in estrus (LeBoeuf, 1967). Under natural conditions, nonanthropoid primate females are receptive only during the time of their cycle when estradiol blood concentrations are high, but even then they discriminate among males (Clutton-Brock, 1991). In virtually all nonmammalian species examined, females' blood concentrations of estrogens are elevated from some basal concentration at the time when they mate with males (reviewed in Crews and Silver, 1985).

FEMALE PACING OF COPULATION Sex steroid hormones are necessary and sufficient for normal sexual behavior among female rodents in the laboratory. There does not seem to be much selectivity of mating partners among estrous female rats

6.16 A breeding deme, the typical social organization of wild rats, consists of a small group of animals that includes a few adult males, several adult females, and many subadults. In this social setting, the females may undergo estrous cycles synchronously and mate with several males simultaneously. The burrows and runway systems in the rats' natural habitat promote a slower mating pace than is typical in a laboratory setting. From McClintock, 1987.

when they are tested in the typical laboratory apparatus with a single male. Under these conditions, an estrous rat will mate with virtually any adult male with which she is paired. But estrous rats do not normally mate with a single male in an enclosed space. Wild rats and mice typically live in small breeding units, called demes, consisting of one or two adult males, several females, and their offspring (Figure 6.16; see also Figure 5.22). The animals often mate in groups, and the pacing of mating behavior is usually under the control of the females. Group-living females often come into estrus together, as we will see below, and two or more estrous females may compete to copulate with two or more attendant males (Bronson, 1979; Calhoun, 1962a; McClintock and Adler, 1978). Because the first and last males to ejaculate with a female in a group mating situation sire proportionally more of her offspring than other ejaculating males, a female can to some degree choose the sires of her offspring by her choice of males for these roles.

When researchers tested groups of rats in a seminatural setting in which female rats could control the pace of mating behavior, the temporal pattern of rat copulatory behavior began to emerge (McClintock, 1987; McClintock and Adler, 1978). The natural pace of mating can also be simulated in a single-pair mating test by using a setting that allows the female to escape from the male between mating events (Erskine, 1985). Under conditions that allow females to control the pace of copulation, the intervals between intromissions are longer than those observed in more typical artificial settings (Erskine, 1985; McClintock and Adler, 1978). In group mating tests in a natural setting, the female retires to her burrow or moves behind some barrier for approximately 3 minutes after each intromission. In contrast, the average inter-intromission interval during standard paired rat mating tests is less than 1 minute. Females paired with a single male in a standard mating arena, but able to escape, also show different timing than females in nonpaced tests: females in paced paired tests stay away from the male for a longer time after an ejaculation than after an intromission, demonstrating that they are able to discriminate the type of vaginal stimulation they have received (Figure 6.17) (see Yang and Clemens, 1996). Anesthesia of the perianal region disrupts this discriminative ability (Bermant and Westbrook, 1966).

The larger size of the mating arena in natural (Calhoun, 1962a) or seminatural (McClintock and Adler, 1978) settings permits the full expression of female solicitation behaviors, including approach and withdrawal behavior, that are often missing or altered in standard mating test situations (Erskine, 1989). Approach and withdrawal behavior appears as the hopping and darting behaviors observed in typical laboratory settings. This behavior is an important component of normal rat sexual behavior and precedes 90% of intromissions in natural or seminatural settings. In contrast, only 3% of male-initiated contacts result in intromissions (McClintock, 1987).

The temporal pattern of copulatory events has an important role in rodent reproduction: namely, the induction of corpora luteal function. In Chapter 5, a vari-

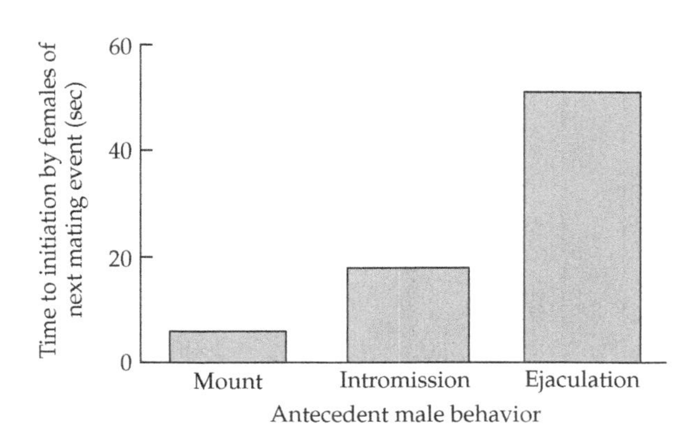

6.17 Female pacing of copulation. Female rats in paced paired mating tests, in which they are provided with an "escape" area away from the male, return to the male more slowly following an ejaculation than after an intromission or mount, indicating that they can discriminate among different types of vaginal stimulation. After Erskine, 1985.

ety of rodent copulatory patterns were presented. In some species, such as the guinea pig, males ejaculate on the first intromission, and there is no pelvic thrusting. In rats and mice, males mount a number of times, and there are multiple intromissions before ejaculation. Why have multiple intromissions? Physiologically, multiple intromissions stimulate sperm transport after ejaculation. But there are other important coadaptations between the physiology of females and the behavior of male conspecifics. If no mating stimuli, or insufficient stimuli, are received, a female undergoes recurrent estrous cycles of 4 or 5 days. The corpora lutea regress soon after being formed and do not secrete sufficient progesterone to build up the uterine endometrium to support pregnancy. Multiple intromissions maintain luteal function so that the fertilized ova can implant in the uterine wall. The sensory input from the multiple intromissions stimulates secretion of prolactin from the anterior pituitary in a characteristic twice-daily pulse. The secretion of prolactin in this manner supports the corpora lutea for about 10–12 days (Gunnet and Freeman, 1983). The corpora lutea then persist until the end of pregnancy without further prolactin support. Unlike a female guinea pig, a female rat cannot become pregnant with a single intromission and ejaculation—unless corpora luteal function has been activated by artificial vaginal stimulation.

The female's pacing of copulation may be a means of ensuring that she receives a pattern of stimulation as close as possible to the "vaginal code"—the so-called optimal pattern of stimulation for producing offspring—for her species (see Chapter 5) (Erskine et al., 2004; Lehmann and Erskine, 2004). In standard nonpaced mating tests, approximately ten intromissions are required to induce luteal function and ensure successful pregnancy. In a paced mating test in which the female controls the timing and duration of the copulatory activity, only five intromissions are necessary to induce corpora luteal function, and more females are impregnated after five intromissions than after ten intromissions (Figure 6.18). Males also need fewer intromissions to ejaculate in paced than in standard laboratory mating tests (Erskine, 1985). Presumably, his interaction with a female displaying the approach and withdrawal behavior pattern increases the male's arousal so that fewer intromissions are required

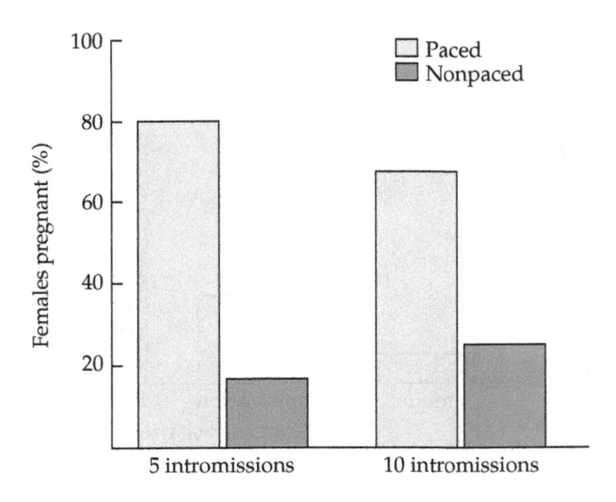

6.18 Paced mating enhances reproduction. The slower mating pace seen in female-paced mating tests optimizes reproduction: in this study, five male intromissions in a paced test resulted in more successful pregnancies than did ten intromissions in another paced test, and both paced tests resulted in more pregnancies than did the equivalent nonpaced tests. After Erskine, 1985.

prior to ejaculation (Erskine, 1989). It appears that the number, rather than the timing, of prior intromissions affects subsequent mating behavior and termination of estrus in rats (Coopersmith et al., 1996). Furthermore, paced mating appears to be more effective than nonpaced mating in inducing the expression of *c-fos* genes in neurons that are part of the neural circuitry underlying female sexual behavior (Erskine and Hanrahan, 1997). Although paced mating slightly changes pain thresholds in rats (Lee and Erskine, 2000), it does not seem to mediate mating-induced analgesia. A role of proper pacing of mating behavior in the reproduction of other species has been identified only in mice (Garey et al., 2002). Pacing is unnecessary in guinea pigs. Male guinea pigs ejaculate on the first intromission, but female guinea pigs have spontaneous corpora luteal activity, so multiple intromissions are unnecessary. Such coadaptation between male and female copulatory behavior and reproductive physiology is a fundamental characteristic defining a species.

Female Reproductive Cycles

The duration of estrus and the interval between estrous periods vary from species to species (Table 6.1). Rats, as we have seen, come into estrus every 4 or 5 days. For some species, like the golden-mantled ground squirrel (*Spermophilus lateralis*), the estrous period may last only a few hours per year during a single morning. In other species, such as rabbits, females will permit mating at any time during the breeding season. Female mammalian reproductive cycles can be classified into several types based on how and when ovulation and pseudopregnancy occur. Although these types will be presented here as if they were discrete categories, keep in mind that reproductive cycles actually vary along a continuum. The description of reproductive cycles that follows is largely based on the classification system put forth by Conaway (1971).

TABLE 6.1 Typical length of the estrous cycle of several common species when living under optimal conditions

Common name	Taxonomic name	Cycle length (days)
Laboratory rat	*Rattus norvegicus*	4–5
House mouse	*Mus musculus*	4–6
Guinea pig	*Cavia porcellus*	16
Golden hamster	*Mesocricetus auratus*	4
Deer mouse	*Peromyscus maniculatus*	4–5
Cotton rat	*Sigmodon hispidus*	4–20
Domestic cat	*Felis catus*	9–10, 2/year
Domestic dog	*Canis familiaris*	10, 2/year
Sheep	*Ovis aries*	16
Goat	*Capra hircus*	21
Cow	*Bos taurus*	21
Pig	*Sus scrofa*	22

Source: Asdell, 1964.

Types of Reproductive Cycles

The **follicular phase** of an ovarian cycle occurs prior to ovulation, when the follicles are developing. The **luteal phase** is that part of the cycle after ovulation during which the corpora lutea are active and producing progesterone. Successful fertilization usually occurs in the fallopian tubes, as the egg travels from the ovary to the uterus. Progesterone—the "progestational" hormone—normally induces the buildup of the uterine endometrium in preparation for the implantation of a **blastocyst**—the small mass of dividing cells that develops into an embryo—and supports the uterus during pregnancy. If a luteal phase occurs during a cycle when no mating, or a sterile mating, has taken place, then a **pseudopregnancy** will occur as the uterus is prepared for blastocyst implantation. Estrous cycles are suspended during pregnancy, pseudopregnancy, and lactation. Pseudopregnancy is considered *spontaneous* if the formation of functional corpora lutea always follows ovulation; if the formation of functional corpora lutea does not automatically follow ovulation, but requires some sort of additional stimulus, as in rats, pseudopregnancy is considered *induced*. In Conaway's classification system (see Table 6.2), pseudopregnancy is defined as "the occurrence of any functional luteal phase in a non-pregnant cycle" (Everett, 1961). Although the "luteal phase" of the ovarian cycle and "pseudopregnancy" are endocrinologically identical, the term *luteal phase* is reserved mainly for describing primate ovarian cycles. Exceptions include guinea pigs and sheep, which also both display spontaneous luteal phases.

The reproductive cycles of female anthropoid primates, including humans, are not technically estrous cycles because, as described above, mating behavior is not

TABLE 6.2 Types of female reproductive cycles in mammals

Cycle	Length	Features	Examples
Type 1			
Subtype 1.1.A	2–5 weeks	Spontaneous ovulation and pseudopregnancy; copulation limited to periovulatory period	Guinea pigs/other histricomorph rodents
Subtype 1.1.B	2–5 weeks	Spontaneous ovulation and pseudopregnancy; copulation may occur throughout cycle	Great apes, including humans
Subtype 1.2	>5 weeks	Spontaneous ovulation and pseudopregnancy; copulation limited to periovulatory period	Dogs and other canids
Type 2			
Subtype 2.1	3–5 weeks	Induced ovulation, but spontaneous pseudopregnancy	Rabbits and hares; lemmings and voles
Subtype 2.2	>5 weeks	Induced ovulation, but spontaneous pseudopregnancy	Cats, minks, ferrets, and skunks
Type 3	<1 week	Spontaneous ovulation, but induced pseudopregnancy	Rats, mice, and hamsters

Source: Conaway, 1971.

confined to a particular phase of the ovarian cycle. Women and females of a few other primate species may copulate on any day of the cycle (Wallen and Zehr, 2004). The ovarian cycle displayed by most primates is called the **menstrual** ("moon") **cycle** because of the recurring period of menstruation (sloughing off of the uterine endometrium) that typically occurs during each cycle, about once a month in humans (Figure 6.19). Dogs and some other mammals discharge blood from the vagina prior to estrus, but this discharge is fundamentally different from menstrual bleeding. Menstruation occurs when blood concentrations of estrogens and progesterone are basal. The endometrial layer of the uterus, no longer supported by the sex steroid hormones, is shed, and the corkscrew-shaped blood vessels that remain leak blood into the uterine lumen. The blood discharged by proestrous or estrous dogs, on the other hand, results from estrogen-induced stimulation of the uterine wall (hyperemia), which causes rapid growth of the endometrium and many tears in the supporting blood vessels. Thus, vaginal bleeding in canines results from steroid hormone stimulation, whereas menstrual bleeding in primates results from steroid hormone deprivation.

The Ecology of Reproductive Cycles

Table 6.2 describes the several different kinds of reproductive cycles in mammals. The type 3 reproductive cycle was once considered to be the pattern most typical of mammals because of the enormous amount of research that was conducted on species—typically rodents—that exhibit it. However, the actual number of mammalian species with type 3 reproductive cycles is probably quite low. Mammals with type 3 cycles do not have functional luteal phases if they do not mate; this ensures a rapid return to estrus. In nature, nonpregnant cycles are a rarity because most animals with type 3 cycles mate and become pregnant during a spontaneous postpartum estrus, and thus gestate a new litter while nursing the previous one.

Why are there so many different types of reproductive cycles? From an ecological perspective, each species' cyclic pattern can be seen as an adaptation for increasing its reproductive success. The cyclic pattern increases the probability that mating will occur when the female is fertile and that offspring will be produced and survive. Keep in mind that female mammals in the wild are typically pregnant, lactating, or in seasonal diestrus (anestrus); that is, reproductively quiescent. Repeated estrous cycles are laboratory artifacts that occur infrequently in free-ranging mammals, and sterile copulation is probably very rare in nature (Nalbandov, 1976). Repeated, sterile estrous cycles represent an abnormal state that females in the wild cannot afford (Conaway, 1971).

What are the relative advantages of the various reproductive cycle types? One correlation that is immediately apparent is that small, short-lived prey species (for example, rodents and lagomorphs [rabbits and hares]) display either type 2.1 or type 3 reproductive cycles, which minimize the time spent in a nonpregnant condition. Animals that live longer, such as predators, do not require as many adaptations to reduce nonpregnant intervals, and can focus more effort on rearing fewer offspring successfully. Another loose correlation occurs between induced ovulation and solitary living. Many of the solitary carnivores are induced ovulators; this adaptation ensures that ovulation occurs only when males are pres-

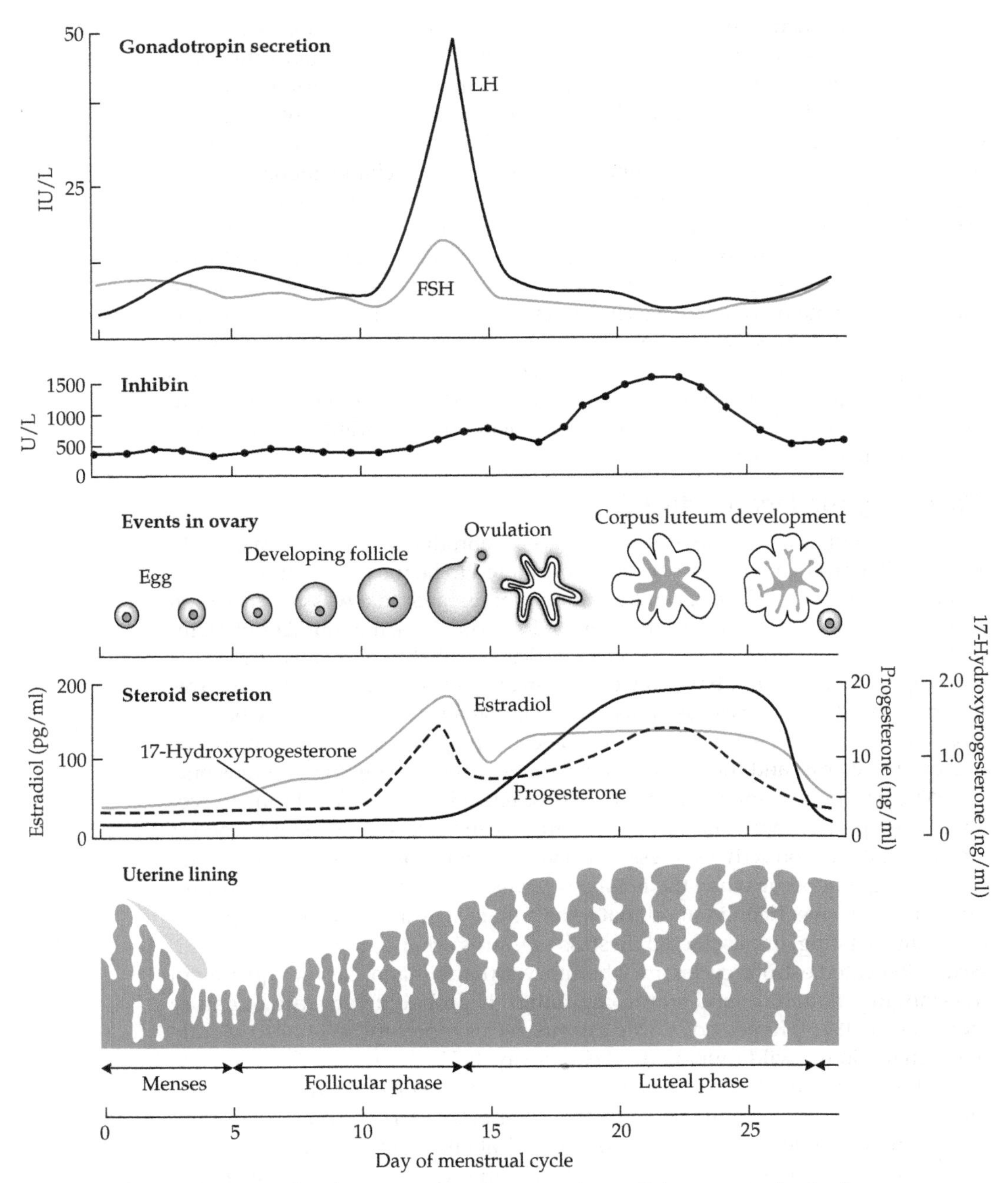

6.19 The human menstrual cycle. After menstrual bleeding has ceased, blood concentrations of LH and FSH increase gradually, stimulating the development of estrogen-secreting follicles. When estrogen concentrations are high, a positive feedback response occurs, causing a surge of gonadotropins that induces ovulation. After ovulation, negative feedback returns gonadotropin concentrations to baseline values, and estrogen concentrations wane. Corpora lutea are formed and secrete progesterone. If a woman is not pregnant, the corpora lutea regress, progesterone concentrations sharply decrease, and menstruation begins.

ent. In some species of induced ovulators, behavioral estrus is also induced by the presence of male conspecifics.

In many species of lagomorphs, behavioral estrus is induced by synchronized courtship displays by the males in the population (Conaway and Wight, 1962). In this case, group synchrony in mating probably confers individual advantages because births will be synchronized, and the chance of any one offspring becoming the victim of predation will be much less than if births were randomly timed throughout the breeding season. In contrast, many large group-living mammals are spontaneous ovulators. A social organization such as a herd probably guarantees that males will be present whenever a female comes into estrus.

Female prairie voles (*Microtus ochrogaster*) do not display regular estrous cycles, but are induced into behavioral estrus by the presence of a conspecific, fertile male or exposure to his urine (Richmond and Stehn, 1976). Undisturbed, isolated female prairie voles never display the cycles in vaginal cell types or vaginal opening observed in mice and rats. When a male is present, the female may ingest a few drops of his urine during mutual anogenital investigation. The urine is delivered to the vomeronasal organ (reviewed in Wysocki, 1979; Wysocki and Lepri, 1991). As described in Chapter 5, this chemosensory organ connects with the accessory olfactory bulb of the brain via the vomeronasal nerves. In many rodent species, the cell bodies of neurons that secrete GnRH are found in the olfactory bulb (Dluzen and Ramirez, 1983); these neurons project to the median eminence of the hypothalamus. The delivery of male urine to the vomeronasal organ of the female stimulates a cascade of endocrine events. Within an hour, GnRH is released from the olfactory bulb, which stimulates LH, and probably FSH, secretion from the anterior pituitary gland (Dluzen et al., 1981). The secretion of gonadotropins results in follicular development and subsequent production of estrogen. Elevated estrogen concentrations cause estrous behavior within 24 hours (Carter et al., 1986). Ovulation occurs approximately 12 hours after mating and is induced by copulation.

The behavior of a female prairie vole directly influences her endocrine state. If the female does not engage in anogenital investigation, she will not be induced into estrus. When females are housed with their fathers or brothers, they do not engage in anogenital investigation, are not induced into estrus, and thereby avoid incestuous mating (Carter et al., 1980). Because the ovaries are quiescent prior to the induction of estrus, proceptive behavior (anogenital investigation) is not mediated by ovarian steroid hormones in this species (Moffatt and Nelson, 1994).

In nature, the first estrus of the breeding season in prairie voles is induced by males, but subsequent matings occur during a spontaneous postpartum estrus (Nelson, 1985; Richmond and Conaway, 1969). Most pregnancies among natural populations of rodents are the result of mating during postpartum estrus. Rats ovulate about 3–6 hours after delivery of their young, and postpartum estrus occurs within 24 hours of the onset of parturition. The hormonal milieu of parturition is similar to the endocrine state associated with ovulation during an estrous cycle (Connor and Davis, 1980a,b). A female in postpartum estrus is attractive and signals males to mate with her (Greef and Merkx, 1982). Typically, she becomes pregnant again within a day after giving birth, and gestates a new litter while nursing the previous one. At about the time the current litter of pups is weaned, a new litter arrives, the female again enters estrus, and the sequence is repeated until the end of the breeding season (see Chapter 9). Thus, this reproductive pattern decreases the interval between successive litters and effectively doubles reproductive output. If a female fails to become pregnant during postpartum estrus, she enters a lactational diestrus and does not enter estrus again until her young are weaned 25–30 days later. This type of reproductive pattern is common among many mammalian groups, including marsupials, mustelids (mink, ferrets, skunks), and rodents. Despite the preponderance of pregnancies in the wild that arise from postpartum matings, little is known about the physiology of, or behavioral changes associated with, postpartum estrus (Blandau and Soderwall, 1941; Connor and Davis, 1980a,b; Everett, 1961; Greef and Merkx, 1982; Lu et al., 1976).

Why have estrous cycles evolved? Estrous cycles are probably an adaptation that allows a female to enter reproductive condition at the beginning of the breeding season or reenter estrus after a failed reproductive attempt. If nutritional availabil-

TABLE 6.3 Four related chemosensory-mediated effects that have been observed in the laboratory

	Effect	Physiology
Lee–Boot effect	When housed 4 per cage with no males present, female mice displayed longer estrous cycles due to lengthening of diestrous stage	Considered comparable to the pseudopregnancies induced by sterile matings because growth of the uterine lining was sometimes observed (decidual reaction)
Whitten effect		
(1) Estrus induction and synchronization	Presence of a male or his odors induced estrous behavior within 48 hours in group-housed female mice	Exposure to male urine induces GnRH release and estrus.
(2) Suppression of estrus	When female mice were housed >20 per cage with no males present, estrous cycles were suspended.	Exposure to female urine suppresses GnRH release and estrus
Bruce effect	Pregnant females aborted or resorbed their fetuses if exposed to male that was not the sire for >48 hours	Exposure to male urine induces GnRH release and estrus; this combination of endocrine events is incompatible with pregnancy
Vandenbergh effect		
(1) Acceleration of puberty	Female juvenile mice exposed to adult males matured earlier than those not exposed to adult males	Exposure to male urine induces GnRH release and puberty
(2) Delay of puberty	Female juvenile mice exposed to adult females matured later than those not exposed to adult females	Exposure to female urine inhibits GnRH release and puberty

ity wanes, or some other disruption of reproduction occurs, the estrous cycle returns the female to reproductive condition as soon as possible when conditions improve.

Social and Environmental Effects on Reproductive Cycles

Social conditions and environmental factors can dramatically affect female reproductive cycles, as several of the examples in the previous section demonstrate. Several effects of the social environment on reproductive cycles have been observed in laboratory studies. For example, female house mice (*Mus musculus*; type 3 cycle) that are housed together in groups of four or more often enter a period of anestrus. If a male conspecific, or his odor, is introduced into such a group, the females ovulate synchronously 3 or 4 days later (van der Lee and Boot, 1955; Whitten, 1956b). Thus, the organization of estrus among house mice can resemble that of prairie voles under the appropriate social conditions. These social effects on female reproductive cycles appear to be mediated by chemosensory cues from conspecifics. Four related chemosensory-mediated effects that have been observed in the laboratory are described in Table 6.3.

Cue	Reference
Chemosensory	van der Lee and Boot (1995)
Chemosensory (androgen-dependent substance in male urine)	Whitten (1956b, 1957)
Chemosensory (androgen-dependent substance in male urine)	
Chemosensory (androgen-dependent substance in male urine)	Bruce (1959, 1960; reviewed in Heske and Nelson, 1984)
Chemosensory (androgen-dependent substance in male urine)	Vandenbergh (1967, 1983, 1994)
Chemosensory	

THE ROLE OF PHEROMONES There appear to be two pheromones that are responsible for social effects on female reproductive cycles. One of these chemical signals comes from females and tends to suppress ovarian function. Depending on the reproductive condition of the recipient animal (prepubertal or cycling) and the strength of the stimuli, exposure to the female chemosignal either inhibits puberty or suspends estrus. This chemosignal probably acts by suppressing gonadotropin release from the anterior pituitary. The other chemical signal is emitted by males and, depending on the reproductive condition of the recipient female (prepubertal, suspended estrous cycling, or pregnant), accelerates puberty, induces estrus, or interrupts pregnancy. The male chemosignal appears to induce an abrupt release of LH, which stimulates follicular growth. Prolactin appears to interfere with the induction of LH release; for example, lactating females that become pregnant during postpartum estrus are protected against pregnancy block due to male chemosignals (Komisaruk et al., 1981).

From where do these chemical signals emanate? The entire body can, in theory, provide a multitude of chemical signals, but probably the most important source of chemical signals in rodents is the urine and feces. There is good evidence that the male primer pheromones are androgen-based components of urine; urine from castrated or prepubertal males is not very effective in producing estrus in females. The source of the pheromone emitted by females has not been identified, but there are strong indications that it, too, is excreted in urine.

The effects of social factors on ovarian cycles are not limited to rodents. Women who live together for extended periods of time eventually begin to synchronize their menstrual cycles (McClintock, 1971). Records of menstrual cycles obtained from undergraduate women attending Radcliffe College showed that menstrual synchronization was more likely among roommates than among women who lived in the same dormitory building, and that approximately 7 or 8 months of cohabitation is required before menstrual cycles become synchronized. This phenomenon may be mediated by chemosensory signals, but the experiments necessary to determine this cannot be conducted on humans. Analogous studies have, however, been conducted on rats, demonstrating that rats living in groups show a similar estrus-synchronizing phenomenon, with chemosensory cues providing the stimuli allowing synchronization of their estrous cycles (McClintock, 1978). Olfaction is probably a very important component of normal human sexual behavior, but it has rarely been studied until recently (but see Doty, 1986) (Box 6.2). Clinical reports on women with congenital absence of the olfactory bulbs revealed a markedly increased incidence of ovarian hypoplasia (nondevelopment) (Schwanzel-Fukuda et al., 1989), but this condition probably reflects the lack of GnRH neuronal migration rather than the loss of olfactory ability. It is likely that menstrual synchrony is part of a larger phenomenon of social influence on human ovulatory cycles (McClintock, 1998) similar to the chemosensory influences female mouse reproductive physiology and behavior.

ENVIRONMENTAL EFFECTS Environmental factors have an enormous influence on ovarian cycles. Lack of proper nutrition, stress, or lack of appropriate habitat can adversely affect ovarian cycles (Wade et al., 1997; Schneider et al., 1998). Stressful

BOX 6.2 Human Pheromones

Although there have been many claims of the existence of human pheromones—chemical signals that are released by an individual and travel through the air or water to affect the physiology or behavior of another individual—until recently there was no solid evidence of such substances. Nevertheless, one can buy perfumes, bearing such provocative names as Lure, Scent, and Instinct Pheromone, purported to contain human pheromones specifically formulated to attract the men or women of one's choice. The best part of these sales pitches (if you are selling these fragrances) is that the pheromones are essentially odorless (buyers can mix them into their favorite perfume or cologne) and affect the behavior of recipients without their conscious awareness. One such product claims to have 50 "known" human pheromones as ingredients. Despite these widespread claims for their existence, no human pheromones have been isolated and characterized, and until very recently, their existence was unproven (Stern and McClintock, 1998).

One study established the existence of two human pheromones in women, one produced prior to ovulation during the follicular phase, which shortens the ovarian cycle in recipients, and a second one produced around the time of ovulation, which lengthens the ovarian cycle in recipients (see the figure). These pheromones mediated a specific neuroendocrine response in women without the recipients' conscious awareness (Stern and McClintock, 1998). Women between the ages of 20 and 35 took part in the study. Nine of these women provided chemosensory samples produced during different times of their menstrual cycles by wearing absorbent pads under their arms for at least 8 hours daily. The pads were quartered and stored frozen until tested on other women. The pads were wiped under the noses of the recipient women. Pads obtained from the donor women during their follicular phase accelerated the periovulatory surge of LH and shortened the menstrual cycles of the recipient women by 1–14 days over two menstrual cycles (Stern and McClintock, 1998). Pads obtained from the donor women around the time of ovulation delayed the periovulatory LH surge and lengthened the menstrual cycle of recipient women by 1–12 days. Approximately 70% of the recipient women responded to the pheromones. Because the same donors provided all the chemosensory samples, this study, in addition to providing the first definitive evidence of human pheromones, demonstrated that the timing of ovulation in women could be manipulated. This work opens the door to a more "natural" way of manipulating the timing of ovulation, which could be helpful in birth control and fertility treatments.

A similar study used compounds collected from lactating women and their breast-feeding infants. These pheromones increased the sexual motivation of other women in terms of self-reported sexual desire and fantasies (Spencer et al., 2004). Interestingly, the increased sexual motivation took different forms, depending on whether or not the women had a regular sexual partner. Women with a regular sexual partner reported increased sexual desire, whereas women without a regular partner reported more sexual fantasies (Spencer et al., 2004). These results indicate that imperceptible chemical cues can affect hormone–behavior interactions in humans.

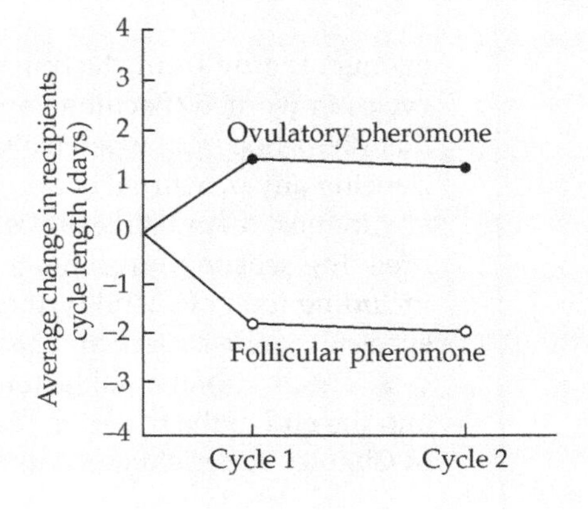

BOX 6.3 Illness Suppresses Female Sexual Behavior

Physical condition, including body mass and blood concentrations of specific metabolic fuels, has dramatic effects on female sexual physiology and behavior. Females with low energy, or otherwise in poor condition, are usually not receptive or proceptive toward males, and males are generally less attracted to females in poor condition than to females in good condition. Similarly, females that are ill are generally less receptive, proceptive, and attractive than females that are not ill.

The behavioral cues associated with illness are probably well known to you. An entire constellation of so-called "sickness behaviors," which occur in response to systemic diseases or localized infections, has been documented in humans and several other mammalian species. Obvious behavioral changes observed in sick individuals include lethargy, hypersomnia, malaise, anorexia, loss of interest, and reduction in goal-directed behaviors (Hart, 1988). To induce "sickness behaviors" experimentally, animals are treated with endotoxin (which consists of heat-inactivated shells of *E. coli* bacteria) or lipopolysaccharide (LPS, which is the major molecular component of the cell walls of *E. coli*); both treatments activate the immune system, but do not give the animals a replicating infection.

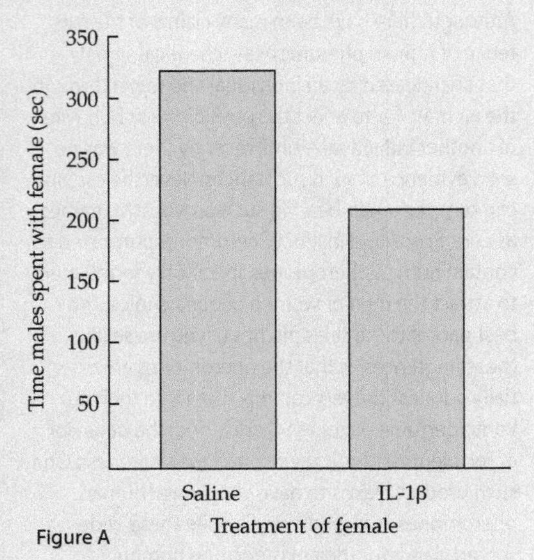

Figure A

Activation of the immune system is communicated to the neuroendocrine system by chemical messengers called cytokines. Cytokines are released by activated macrophages during immunological responses to infections. Several of these chemical messengers, including interleukin-1β (IL-1β), interleukin-6 (IL-6), and tumor necrosis factor-α (TNF-α), affect neuroendocrine processes in the hypothalamus and pituitary (Segreti et al., 1997).

events ranging from starting college to imprisonment can suspend menstrual cycles in women (Bachman and Kemmann, 1982; Bass, 1947). One study found that premenopausal women awaiting their execution on death row were not experiencing any menstrual cycles (Pettersson et al., 1973) (see Chapter 11).

In most mammals, estrous cycles occur only during the breeding season. The breeding season represents the "temporal fit" among many selective forces—including food availability, thermoregulatory pressures, climatic factors, and the species' mating system and gestation length—that best ensures reproductive success. The environmental factors that turn the ovaries on and off at the beginning and the end of the breeding season will be discussed in Chapter 10. As indicated in Chapter 3, perinatal events can also affect subsequent estrous cyclicity. Recall

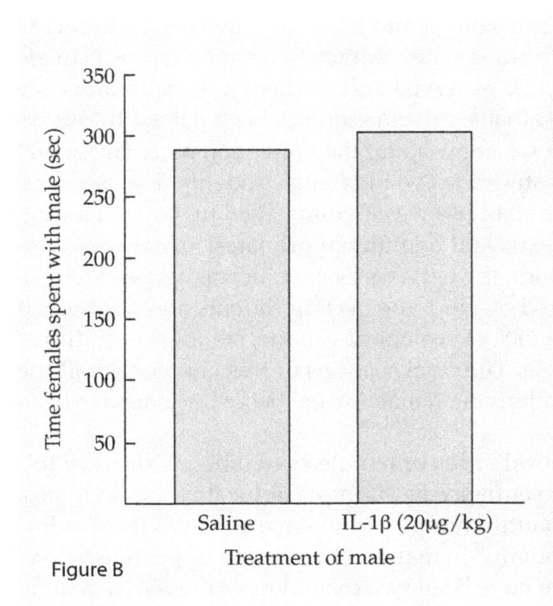

Figure B

Administration of IL-1β causes fever, hypersomnia, and slow-wave sleep, as well as reductions in locomotor activity, exploratory behaviors, food intake, and social contact (Yirmiya et al., 1995; Avitsur et al., 1997a,b). IL-1β treatment also inhibits GnRH gene expression in the hypothalamus, decreases GnRH release in rats on the afternoon of proestrus, and decreases circulating LH concentrations in ovariectomized rats and monkeys, as well as inhibiting steroid hormone synthesis in the ovaries and testes (Rivest and Rivier, 1993; Yirmiya et al., 1995).

Administration of IL-1β to female rats significantly reduced their sexual receptivity and proceptivity (Yirmiya et al., 1995). IL-1β–treated females also lost their preference for sexually active male rats. Surprisingly, IL-1β treatment of males did not diminish their sexual behavior. Both males and females displayed reduced locomotor activity after receiving IL-1β, which suggests that the sex difference in responsiveness to IL-1β is limited to sexual behavior. IL-1β also affected the attractivity of female, but not male, rats. When males had a choice between estrogen-treated, ovariectomized females injected either with IL-1β or with saline, males preferred saline-injected females (Avitsur et al., 1998); that is, IL-1β reduced attractiveness in females (Figure A). In contrast, when presented with a choice of males that had been injected either with IL-1β or with saline, females did not discriminate between these males except at high doses (Figure B).

Taken together, these studies suggest that cytokines have direct and dramatic effects on behavior. Furthermore, the effect of IL-1β on sexual behavior shows a clear sex difference. The ultimate cause of this sex difference probably reflects the differential effects of illness on the reproductive success of males and females. As these studies show, the interactions between the immune and endocrine systems that affect behavior are becoming a rapidly expanding area of research.

that proximity to brothers in the uterine horn during gestation increased estrous cycle length and reduced attractiveness in female rats and mice as compared with individuals gestated between two sisters (vom Saal and Bronson, 1980). Illness can also affect female sexual behavior (Box 6.3).

Experimental Analyses of Female Sexual Behavior

By now you have noted that a tremendous amount is known about estrous cycles. The research efforts aimed at understanding the temporal characteristics, physiological mechanisms, and adaptive functions of estrous cycles have been enormous. However, as the previous discussion has revealed, most laboratory studies examining female reproductive physiology and behavior have been conducted on females undergoing repeated estrous cycles. Although a strong case was made above that the chronic estrous cycles observed and studied in the laboratory are merely artifacts, a great deal of valuable information has been gained by studying those cycles: virtually all that we know about the regulation of female sexual behavior has been derived from studying cycling female rodents. The development of oral contraceptives for human use was accomplished by understanding the endocrinology of cycling rodents and nonhuman primates; all of our understanding of the relationships among the nervous system, hormones, and female sexual behavior has been achieved by studying cycling rodents and nonhuman primates. Therefore, despite their lack of ecological validity, laboratory studies of estrous cycles have been valuable. The final sections of this chapter detail the endocrine–neural interactions underlying female sexual behavior that have been revealed by such studies.

Attempts to localize the neural bases of female sexual behavior have followed, in many cases, the same sequence as attempts at localization in males. However, it is much easier to quantify receptivity in females in the form of lordosis than it is to quantify "receptivity" in males in the form of appetitive behavior. As is the case for males, there have been two major lines of research dealing with the interaction of female sexual behavior and hormones. The physiological line of research has attempted to determine the sources of the signals generating sexual behavior; that is, to discover the loci of the interaction between sex hormones and behavior (for example, Bard, 1936; Beach, 1944b; Brookhart et al., 1941). As with males, early research established that the interaction does not take place at the level of the gonads; consequently, much of the subsequent work has focused on the central nervous system. The second line of research has considered sexual motivation. The strength of the female sex drive has been tested and quantified using a variety of motivational tests (for example, Nissen, 1929), and in some cases, the neural bases of female sexual motivation have been discovered inadvertently during studies of receptivity. Pioneering sex researchers faced social pressures against positing a sex drive for females, as opposed to the "obvious" sex drive manifested by males. Then, as today, sexual behavior was a politically charged scientific topic. More recently, female sexual motivation has been quantified in terms of proceptive behavior (Erskine, 1989; Wallen, 1990; Pfaff, 1999).

This section will review the hormonal events associated with female ovarian cycles in rats and rhesus monkeys. The final section will present the neural bases of female sexual behavior.

Hormonal Correlates of Female Reproductive Cycles

Recall from Chapter 2 that there is a dynamic relationship among the gonadal sex steroid hormones, the gonadotropins, and the hypothalamic releasing hormones. Basically, this relationship is one of reciprocal inhibition. The hypothalamus secretes pulses of releasing hormones, which stimulate the release of gonadotropins from the anterior pituitary. Pulses of gonadotropins drive the release of sex steroid hormones by the gonads. Negative feedback works in the following way: as sex steroid

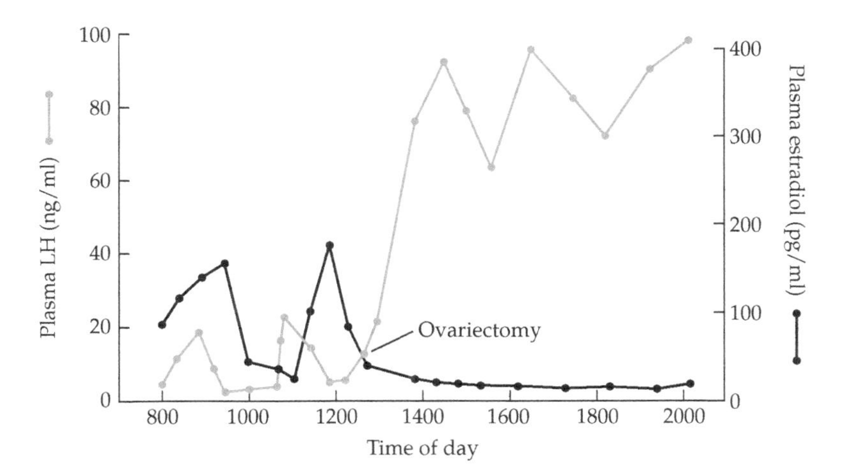

6.20 The castration response. Normally, the gonadotropins and sex steroid hormones exhibit a so-called seesaw profile. For example, increasing LH concentrations stimulate increased estradiol secretion until high concentrations of estradiol inhibit LH secretion through negative feedback, which in turn lowers estradiol concentrations. After ovariectomy, estradiol concentrations decline, and LH concentrations continue to rise, unchecked by negative feedback.

hormone concentrations increase, secretion of gonadotropins is reduced, which results in decreases in gonadotropin-stimulated steroid hormone synthesis (see Figure 2.36B). Thus, a simplified version of the relationship between gonadal sex steroid hormones and the gonadotropins is reflected in the metaphor of a seesaw (Moore and Price, 1932). Removal of the ovaries causes the well-known **castration response**: the removal of the source of steroid hormones removes the negative feedback that regulates gonadotropin secretion, resulting in sustained elevated gonadotropin concentrations, as well as elevated releasing hormone secretion (Figure 6.20).

THE OVARIAN CYCLE IN RODENTS Gonadotropin-releasing hormone (GnRH) is released by the hypothalamus in brief pulses. These pulses of GnRH stimulate cells in the anterior pituitary to release follicle-stimulating hormone (FSH) and luteinizing hormone (LH) (Figure 6.21). FSH, as its name implies, causes the follicles of the ovaries to grow. As the follicles develop, they produce estrogens and progestins. During the diestrous and proestrous periods of the rat estrous cycle, the follicles are stimulated to produce sex steroid hormones in increasing concentrations. On the afternoon of proestrus, the pulse amplitude and pulse frequency of GnRH release, and subsequently the pulse amplitude and pulse frequency of LH and FSH release, increase. Under these conditions, negative feedback fails momentarily, and the follicles release a large surge of estrogen and, soon thereafter, a smaller surge of progesterone. This preovulatory LH surge stimulates ovulation, reflecting the positive feedback feature of the hypothalamic–hypophyseal–gonadal control mechanism (see Figure 3.24). The elevated estrogen and progesterone blood concentrations

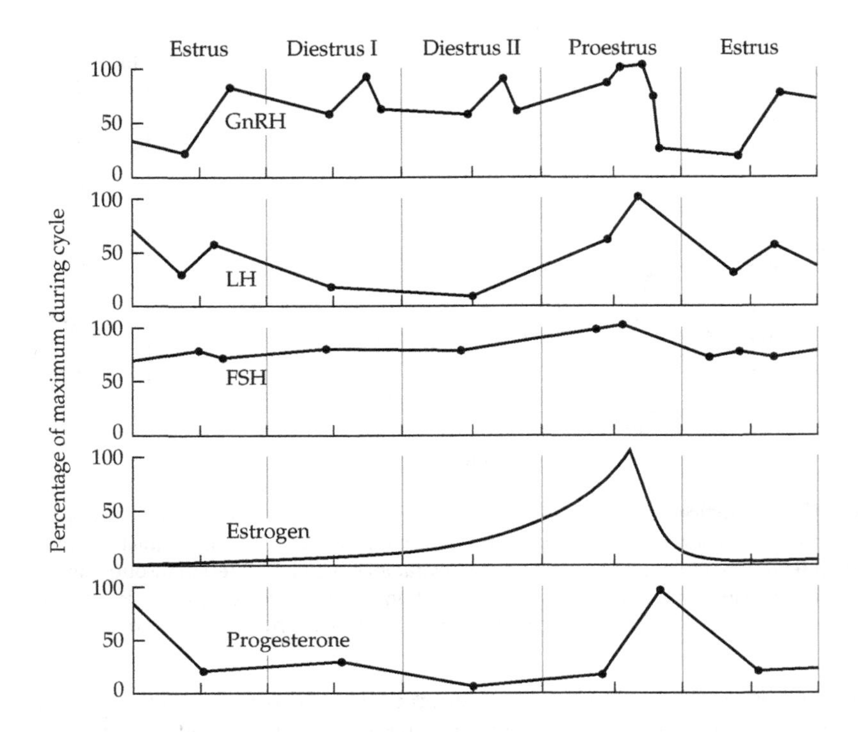

6.21 The ovarian cycle in rats. GnRH stimulates the secretion of LH and FSH, which cause a steady increase of estrogen until the afternoon of proestrus. Then the pulse amplitude and frequency of GnRH-release increase, causing a rapid increase in LH and FSH concentrations, and a corresponding surge of estrogen and progesterone that leads to ovulation and estrous behavior. Estrogen concentrations decline rapidly at vaginal estrus, but progesterone concentrations decline more gradually, especially if mating has occurred. After Ganong, 1995.

also stimulate estrous behavior. Estrogen concentrations fall back to baseline rapidly, but the blood progesterone concentration remains high for many hours if no mating occurs, or for many days if mating does occur. The elevated progesterone concentrations act eventually to terminate estrous behavior (Goy and Young, 1956/1957; Zucker, 1966, 1968).

The tonic release of GnRH throughout most of the estrous cycle is mediated by the medial basal hypothalamus, as it is in males (see Chapter 3). The release of increasingly higher pulses of GnRH is mediated by this so-called *pulse generator*. A neuronal pathway made up of neurons located in the anterior hypothalamus, preoptic area, and suprachiasmatic nucleus, as well as the arcuate nucleus, is involved in providing a daily signal for ovulation if the appropriate endocrine milieu is present. This *surge generator* normally transmits every afternoon; that is why there are 4- or 5-day estrous cycles in rats, but no 4.5-day cycles. Experimental disruption of the surge generator pathway or anesthesia during the afternoon of proestrus prevents transmission of the signal and delays estrus for another 24 hours. Only when the daily signal for a large surge of GnRH is coincident with previous estrogen priming will the positive feedback mechanism be engaged. Soon after ovulation, the negative feedback mechanism is restored, and low concentrations of GnRH, gonadotropins, and steroid hormones are secreted during early diestrus. The cycle resumes when pulses of GnRH stimulate brief pulses of gonadotropins, which stimulate follicular development anew. The 2 days of diestrus and 1 day of proestrus can be considered the follicular phase of the rat estrous cycle because follicular development occurs in anticipation of ovulation. Because the corpora lutea are not maintained after ovulation in unmated rats, there is no luteal phase. Thus, natural estrous behavior is the result of prolonged, high concentrations of estrogen followed by high concentrations of progesterone. Behavioral estrus can be induced in ovariectomized rats and guinea pigs with appropriately timed injections of estradiol and progesterone (Feder, 1981; Pfaff et al., 1994; Blaustein, 1996). In other species, such as Old World primates, a preovulatory GnRH and LH surge also occurs, but is not necessary for ovulation.

Although the sequential elevation of estradiol (and other estrogens) and progesterone is necessary to induce full estrous behavior in rats and guinea pigs, the sequential presence of progesterone and estradiol is necessary to induce estrus in sheep (Robinson, 1954). In other species, such as prairie voles (*Microtus ochrogaster*) and Djungarian hamsters (*Phodopus campbelli*), progesterone is unnecessary for the expression of estrous behaviors (Carter et al., 1987; Wynne-Edwards et al., 1987). Female musk shrews (*Suncus murinus*) secrete testosterone from their ovaries around the time of ovulation. The testosterone enters the circulation and is aromatized to estradiol in neurons located in the preoptic area and in the hypothalamus. The intracellular interaction of estradiol with its receptors produces behavioral estrus (Rissman, 1991). Testosterone and other androgens have been shown to be important in mediating female sexual behavior in several vertebrate species (Staub and De Beer, 1997). Thus, most female mammals require ovarian steroid hormones associated with ovulation to evoke behavioral estrus. However, the pattern of steroid hormone secretion is species-specific (Blaustein, 1996).

THE OVARIAN CYCLE IN PRIMATES The ovarian cycle of rhesus monkeys is virtually identical to that of humans (see Figure 6.19). In women and rhesus monkeys, only one follicle usually develops during each cycle. The corpus luteum is maintained after each ovulation (type 1.1B cycle) and persists for about 14 days in a nonpregnant cycle. The endometrial lining of the uterine wall develops in anticipation of blastocyst implantation. If the ovulated ovum is not fertilized in the fallopian tube and there is no blastocyst to implant, the endometrial lining is shed, and menstrual bleeding occurs. The follicular phase is more variable among individuals, ranging between 10 and 20 days. Variation in menstrual cycle length is almost always due to variation in the length of the follicular phase.

If a female rhesus monkey has a 28-day cycle, then the following endocrine profile occurs (Knobil and Hotchkiss, 1988): Blood plasma concentrations of FSH and LH slowly increase for approximately 10 days following the end of menstrual bleeding. Under the stimulation of FSH, the primary ovarian follicle secretes estrogens. Estrogen concentrations (primarily estradiol) increase gradually during the first week after the onset of menstrual bleeding, then increase sharply during the following week. Estrogen concentrations display a periovulatory surge, which stimulates a surge of LH release and a lesser surge of FSH release from the anterior pituitary. The gonadotropin surge stimulates ovulation of the ripe ovum. After ovulation, plasma estrogen concentrations plummet to basal levels, and gonadotropin concentrations also diminish. As the corpus luteum begins to function, plasma concentrations of progesterone increase. In a nonpregnant cycle, the corpus luteum begins to regress and progesterone concentrations fall back to baseline approximately 10–12 days after ovulation. The low progesterone concentrations evoke menstruation, and the cycle recurs. This cycle is similar to the human menstrual cycle (Table 6.4).

Androgens seem to be critical in mediating sexual behavior in women. The source of androgens in females appears to be the ovaries and adrenal glands. Blood concentrations of these androgens fluctuate during the menstrual cycle. In a study that assayed plasma concentrations of free estradiol, testosterone, and progesterone at weekly intervals, it was discovered that only free testosterone correlated positively with sexual desire, sexual thoughts, and anticipation of sexual activity (Alexander and Sherwin, 1993). Treatment of surgically induced menopausal (ovariectomized) women with several steroid hormones in double-blind studies revealed that androgen treatment was most effective in restoring sexual desire (Sherwin, 1988a; Sherwin and Gelfand, 1988; Sherwin et al., 1985). These results are reminiscent of the effects of androgens on sexual motivation among hypogonadal men (see Chapter 5; Davidson et al., 1979).

As in hypogonadal men, hormone concentrations above a certain threshold that maintains sexual behavior do not further increase the frequency of sexual behavior in ovariectomized women (Sherwin, 1988b). One study compared several aspects of sexual behavior in women using oral contraceptives and in nonusers (Alexander et al., 1990). Nonusers exhibited lower blood plasma concentrations of testosterone than oral contraceptive users, and they also displayed perimen-

TABLE 6.4 Daily production rates of sex steroid hormones in women at different phases of the menstrual cycle

	Menstrual phase		
Hormone	**Early follicular**	**Preovulatory**	**Midluteal**
Progesterone (mg)	1.0	4.0	25.0
17-Hydroxyprogesterone (mg)	0.5	4.0	4.0
Dehydroepiandrosterone (mg)	7.0	7.0	7.0
Androstenedione (mg)	2.6	4.7	3.4
Testosterone (μg)	144.0	171.0	126.0
Estrone (μg)	50.0	350.0	250.0
Estradiol (μg)	36.0	380.0	250.0

Source: After Yen and Jaffe, 1991.

strual decreases in plasma testosterone concentrations, which were associated with a reported drop in the level of sexual desire. There was no significant difference reported in the frequency of autosexual activities, but users of oral contraceptives reported more frequent and more satisfying sexual experiences than nonusers (Alexander et al., 1990). Of course, these results may reflect a relaxation of pregnancy fears rather than endocrine events. Taken together, these findings suggest that androgens induce sexual motivation, receptivity, and satisfaction in women (Bancroft et al., 1991a,b; Sherwin and Gelfand, 1988). Clinical trials with androgens have convincingly demonstrated that pharmacological doses of testosterone increase libido in postmenopausal women. The long-term safety of such doses is unclear (Seagraves, 2003). Androgens appear to enhance receptive and proceptive behaviors in female rhesus monkeys as well (e.g., Everitt and Herbert, 1971; Herbert and Trimble, 1967; Wallen and Goy, 1977).

The fluctuating concentrations of gonadal, pituitary, and hypothalamic hormones in female mammals complicate somewhat the understanding of hormone–behavior interactions. Generally, slowly rising concentrations of estrogen seem necessary to prime mammalian females for the subsequent elevated progesterone concentrations that induce behavioral estrus. As the ratio of estrogen to progesterone in the blood is reversed, progesterone often acts to stop estrous behavior. The hormonal correlates of the estrous cycle of rodents and of the menstrual cycle of primates differ in a few important ways. Ovulation and peak estrogen concentrations coincide in both rats and primates. However, rats display a periovulatory progesterone peak that is reduced or absent in primates. In rodents, and probably other nonprimate species, both sexual motivation and sexual performance are mediated by sex steroid hormones; copulation does not occur unless there are high blood concentrations of estrogen. In contrast, the ability to copulate is not linked to hormones in primates, but motivation to copulate in primates appears to be linked to a periovulatory peak in androgen concentrations. The following section reviews how these fluctuating hormones affect neurons directly to change female behavior.

Neural Mechanisms Mediating Female Sexual Behavior

Virtually everything that is known about the endocrine effects on neural tissue that mediate female sexual behavior has come from studies on the neural and endocrine control of lordosis in rats. As we have seen, lordosis is a receptive behavior, a sexual reflex observed in female rats. Estrogen and progesterone prime females for this behavior, and it occurs in response to tactile sensory information normally provided by a copulating male. Thus, sensory input is one aspect of the behavior on which hormones can be predicted to act. Relevant sensory input enters the female rat's nervous system during mating via cutaneous receptors and pressure-responsive sensory neurons on the flanks, rump, and perineum. Light touch or pressure on these areas between 40 and 450 millibars elicits the lordosis response. The receptive fields of sensory neurons on the flanks were found to increase in size in estrogen-treated females (Figure 6.22; Kow et al., 1979).

How does sensory information get to the hypothalamus? Information from the stimulated receptors enters the spinal cord, where the motoneurons control-

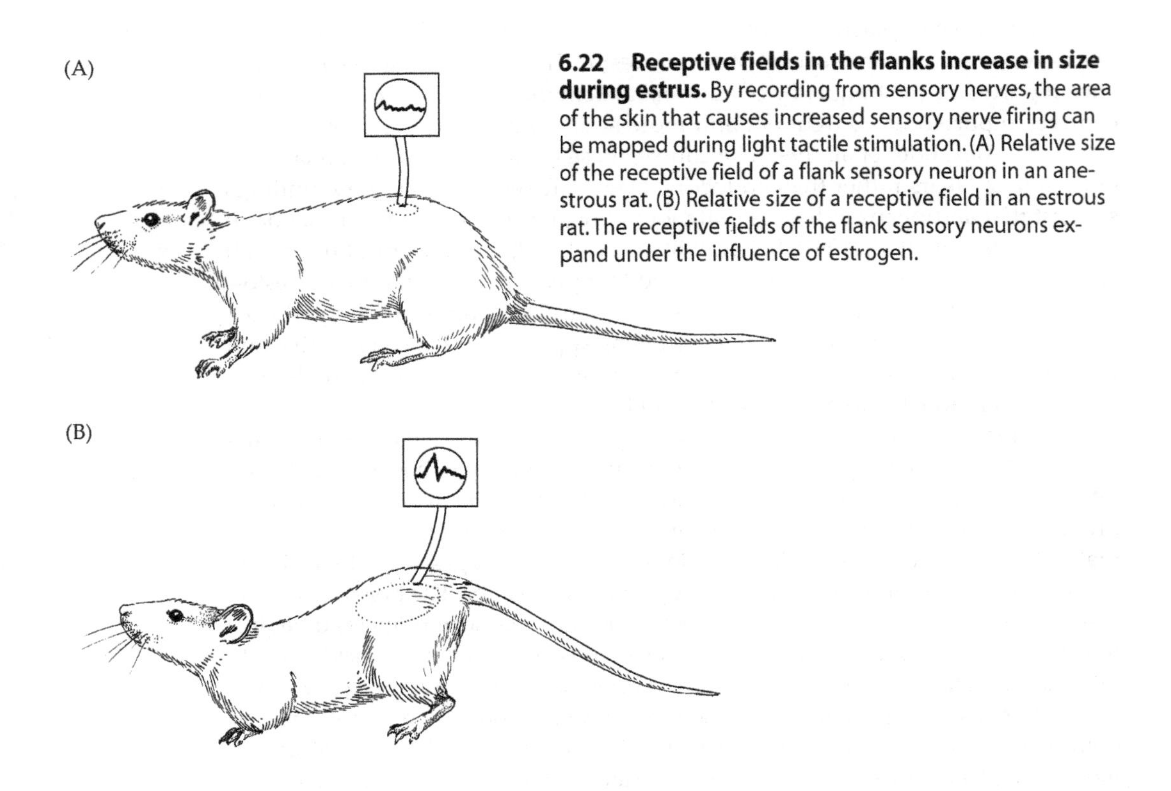

6.22 Receptive fields in the flanks increase in size during estrus. By recording from sensory nerves, the area of the skin that causes increased sensory nerve firing can be mapped during light tactile stimulation. (A) Relative size of the receptive field of a flank sensory neuron in an anestrous rat. (B) Relative size of a receptive field in an estrous rat. The receptive fields of the flank sensory neurons expand under the influence of estrogen.

ling the muscles involved in lordosis are located, and is sent to the medullary reticular formation (Figure 6.23A; Pfaff and Schwartz-Giblin, 1988). This pathway to the brain stem is necessary, but not sufficient, for lordosis to occur.

Hormones can also be predicted to act on the central brain mechanisms integrating the endocrine, social, and environmental stimuli coincident with mating. Several brain sites have been discovered to mediate lordosis (Figure 6.23B). Researchers found that lesions of the ventromedial nuclei of the hypothalamus (VMN) or destruction of their afferent and efferent fibers typically reduced the frequency of lordosis (Clark et al., 1981; Kennedy, 1964; Yamanouchi, 1980; but see Emery and Moss, 1984), demonstrating that the VMN are critical to the lordosis response. In order to map the neural circuit involved in lordosis, the incoming and outgoing fibers then had to be traced. Certain fibers leaving the VMN via a sweeping lateral–posterior pathway were found to be necessary for lordosis, whereas other exiting fibers were less critical (Pfaff et al., 1994). The essential axons descend to the midbrain central gray region; lesions of this region were found to reduce lordosis (Sakuma and Pfaff, 1979). Destruction of the midbrain ascending ventral noradrenergic bundle (VNAB) completely abolished lordosis (Hansen et al., 1980, 1981). Neurons in the midbrain central gray region project axons to the medullary reticular formation in the brain stem. This region of the

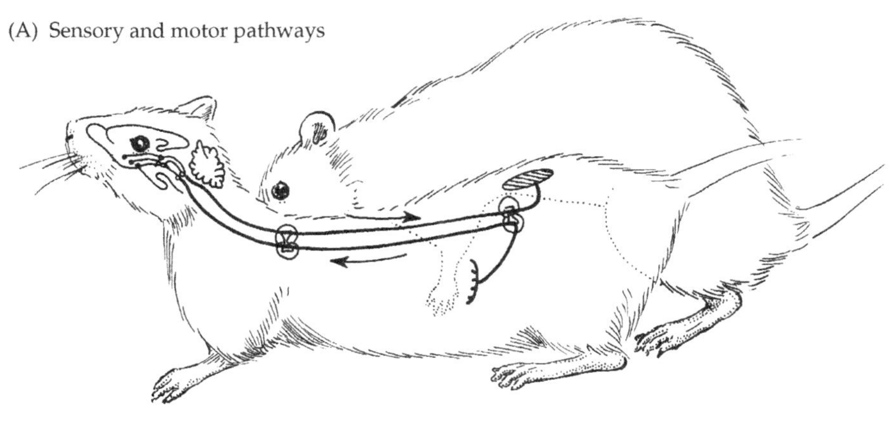

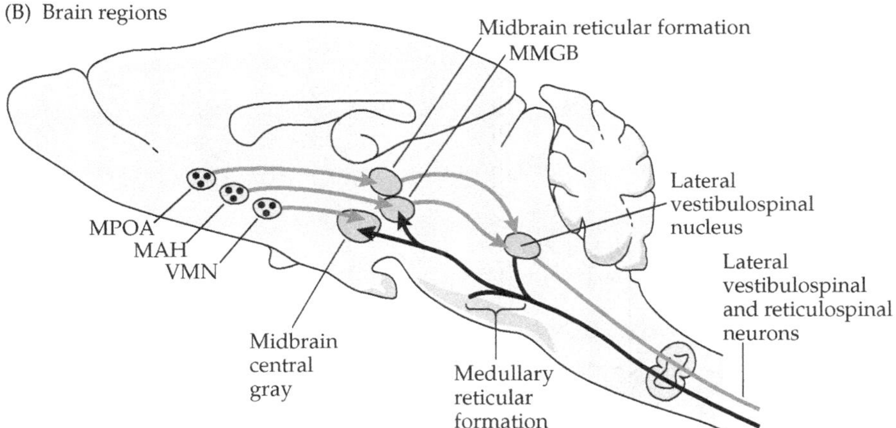

6.23 The neural basis of lordosis. (A) Male mounting behavior causes activation of pressure receptors in the female's flanks; axons from these receptors form a sensory nerve that projects to the dorsal root of the ganglion of the spinal cord. From the spinal cord, the sensory signals travel to the medullary reticular formation in the hindbrain and the midbrain central gray region (lower arrow). When estrogen concentrations are high, various brain regions activate the spinal motoneurons innervating the deep back muscles (upper arrow), resulting in the characteristic postural changes of lordosis. (B) Brain regions activated by estrogen and involved in lordosis include the ventromedial nucleus of the hypothalamus (VMN), the medial preoptic area (MPOA), and the medial anterior hypothalamus (MAH). Signals from these regions reach the midbrain central gray, midbrain reticular formation (MRF), and medial geniculate body (MMGB); the MRF and MMGB ultimately activate the spinal motoneurons innervating the back muscles involved in lordosis behavior. After Pfaff et al., 1994.

hindbrain controls motoneurons for axial muscles, especially the deep back muscles, which are critical for lordosis (see Figure 6.23A; Pfaff and Schwartz-Giblin, 1988). The connection of this brain stem region with the descending fibers from the midbrain central gray region permits lordosis only when sex steroid hormones

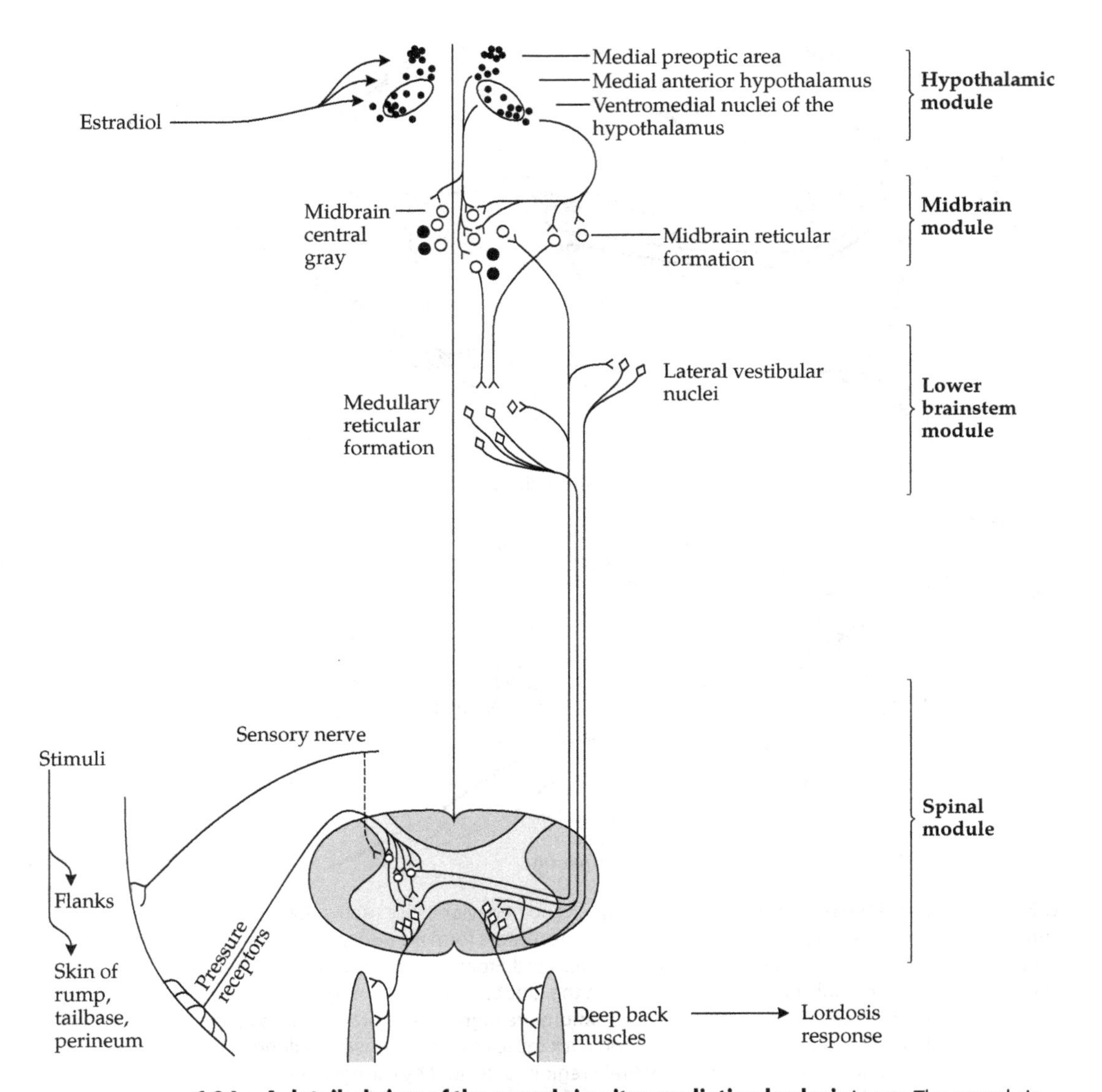

6.24 A detailed view of the neural circuitry mediating lordosis in rats. The neural circuits outlined in Figure 6.23 are separated into structural and functional neural modules. After Pfaff et al., 1994.

are available to neurons in the midbrain or to cells in the VMN. Figure 6.24 is a detailed diagram of these neural circuits.

What is the mechanism of action of estrogens and progestins on these brain regions? The steroid hormone receptors found in the brain are chemically similar to those found in the uterus. As described in Chapter 2, steroids can stimulate production of their own receptors as well as production of other steroid receptor types. Because all the steroid hormones are structurally similar, elevated estrogen concentrations can stimulate the production of both estrogen and progestin receptors in the cytosol of neurons throughout the nervous system. Thus, high concentrations of estradiol (injected by an experimenter, for example) can induce the production of progesterone receptors. This is the likely explanation for how estrogen alone induces estrous behavior in ovariectomized female rats. Anti-steroids—that is, substances that occupy steroid receptors without producing any biological effect—block the occurrence of lordosis (Pfaff, 1980; Pfaff et al., 1994; Delville and Blaustein, 1991; Blaustein and Olster, 1989).

Cells that concentrate estradiol and progesterone have been discovered throughout the vertebrate brain. These target cells show virtually the same distribution regardless of taxon, and they correspond closely with the distribution of receptors for androgens (see Figure 5.18). The areas with the highest density of estradiol- and progesterone-concentrating cells (i.e., highest number of receptors) are located in the forebrain, including the medial preoptic area, the anterior hypothalamus, and the ventromedial–ventrolateral hypothalamus, as well as the amygdala and midbrain central gray region (Figure 6.25) (Pfaff and Conrad, 1978; Blaustein, 1996). Many lesion studies and electrical stimulation studies have revealed that these regions are involved in the regulation of female sexual behavior.

There are hormone manipulations that can mimic the hormonal profile of natural estrus. One way to bring an ovariectomized mouse or hamster into estrus is to inject her with estradiol early one morning and about 48 hours later, inject her with progesterone. In about 6–8 hours, the female will display lordosis in response to the appropriate tactile stimulation. Behavioral estrus is observed approximately 4 hours after the progesterone injection. Another way to bring an ovariec-

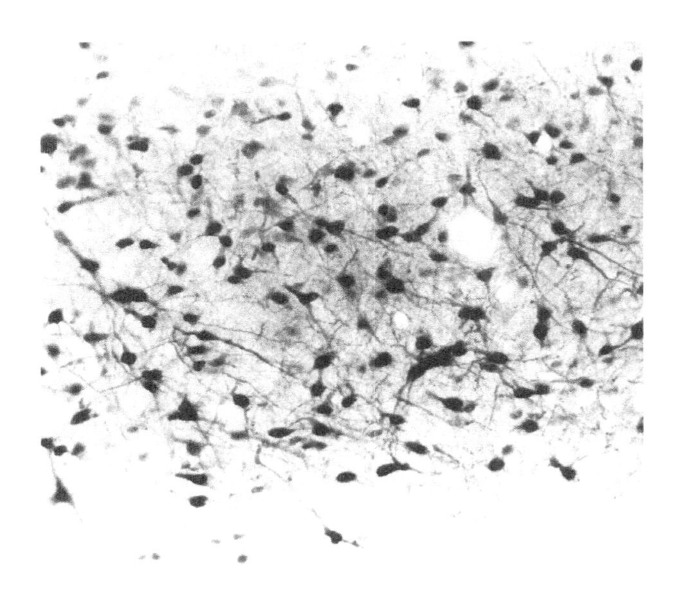

6.25 Estrogen receptor immunoreactivity shows that neurons in the hypothalamus have estrogen receptors (shown here with dark staining) in all parts of the cell (i.e., cytosol, nucleus, cellular processes). Courtesy of Jeffrey Blaustein.

tomized rat into estrus is to implant a Silastic capsule of estradiol under her skin, which will secrete estradiol at a constant rate that varies with the length of the capsule. If the female is then injected with progesterone about 4–8 hours prior to her pairing with a male, estrous behavior will ensue when the male is present.

Several experiments have demonstrated that estrogens must prime the central nervous system for further estrogen and progesterone exposure in order for estrous behavior to be exhibited. Implants of crystalline estradiol into the VMN induce lordosis in ovariectomized rats (Barfield and Chen, 1977; Lisk, 1962), but this lordotic behavior is relatively weak and infrequent compared with that observed among female rats in natural estrus. However, if crystalline estradiol is implanted into the VMN and the animal is then injected systemically with progesterone (or if crystalline progesterone is implanted into the VMN and the animal is then injected systemically with estradiol), virtually all females display lordosis that resembles that observed in females in natural estrus (Rubin and Barfield, 1983). A variety of techniques, including microimplants of protein synthesis inhibitors, transcription inhibitors, and anti-estrogens, have demonstrated that the VMN are critical for estradiol priming in rats (Blaustein and Olster, 1989). Thus, it can be concluded that estrogen treatment primes the central nervous system for subsequent estrogen or estrogen/progestin treatment (Parsons et al., 1979; Walker and Feder, 1977, 1979). Estrogen priming requires about 24 hours unless doses are very high, in which case the first lordosis response occurs in about 18–20 hours.

Progesterone facilitates estrus. Occupied progesterone receptors in relevant neurons mediate many of the behavioral effects of progesterone by serving as gene transcription factors (Blaustein, 1996) (see below). In many species, including rats and guinea pigs, progesterone initially facilitates estrous behavior, but eventually a refractory period occurs. In other species, especially reptiles, progesterone inhibits female receptive behavior (Godwin et al., 1996). The physiological explanation for the termination of estrous behavior is the down-regulation of progesterone receptors (see Chapter 2).

What is the mechanism of estrogen priming? One possibility is that receptor dynamics are affected by estrogen treatment; that is, estrogen secretion might induce the production of estrogen and progestin receptors in the central nervous system. This is a reasonable hypothesis, but in some studies no changes in steroid hormone receptor numbers were detected after estrogen priming (e.g., Parsons et al., 1979). Other studies have indicated that receptor numbers do change, but well after estrogen priming (reviewed in Pfaff et al., 1994). Interestingly, the behavioral changes first occur well after all bound estrogen has been metabolized by the liver, as shown by one study in which radioactive estradiol was given to ovariectomized rats. It was observed that peak binding of estrogen to nuclear receptors in the hypothalamus occurred 1 to 2 hours after injection; estrogen binding had returned to basal concentrations by 12 hours post-injection (McEwen et al., 1975).

Once estrogen binds to nuclear receptors in the hypothalamus and elsewhere, the estrogen–receptor complex attaches to the nuclear DNA and serves as a gene transcription factor. Thus gene transcription and translation are either activated or, in some cases, inhibited by estrogen treatment. Predictably, protein synthesis inhibitors block the effects of steroid hormones on estrous behavior. Over the course of a few hours after binding to its receptors, estrogen causes electrophysiological changes in the pattern and frequency of firing rates of neurons in the VMN, especially in slow-firing neurons in this region (Pfaff and Schwartz-Giblin, 1988). These changes alone cannot account for behavioral changes, because no lordosis is observed if only this part of the brain is exposed to estrogen priming. Electron microscopy studies have revealed that estrogen binding and subsequent DNA transcription promotes RNA synthesis. Growthlike processes appear in and around the VMN in response to estrogen treatment, which is consistent with the observed increases in RNA and structural protein synthesis after estrogen exposure. Conversely, treatment of the hypothalamus with protein synthesis inhibitors eliminates lordosis (Meisel and Pfaff, 1984). Thus, electrophysiological and structural changes occur in the VMN in response to estrogen priming, and continued exposure to high estradiol or progesterone concentrations induces the expression of lordosis in response to appropriate sensory input. Furthermore, intermittent exposure to estrogens is sufficient to prime ovariectomized rats and guinea pigs to respond behaviorally to progestins (Clark and Roy, 1983; Olster and Blaustein, 1988).

Many neurotransmitters have been implicated in the neural mediation of hormone-induced estrous behavior. For example, there is evidence that norepinephrine, dopamine, acetylcholine, serotonin, and GABA (each acting through their own specific receptor subtypes), as well as GnRH, prolactin, oxytocin, and substance P, facilitate estrous behavior often acting in the midbrain (Blaustein, 1996). Serotonin (acting through the 5-HT$_{1A}$ receptor), dopamine, β-endorphin, and CRH may inhibit lordosis. The effects of hormones probably affect these and other neurotransmitters. Estradiol mediates many neurotransmitters that facilitate female sexual behavior (McCarthy and Pfaus, 1996); for example, estradiol evokes oxytocin and enkephalin gene expression and increases the numbers of many types of postsynaptic neurotransmitter receptors (Flanagan-Cato and Fluharty, 1997). However, no clear pattern of neurotransmitter and steroid hormone receptor co-localization has been found in the brain. That is, knowing that a particular neuron has estrogen or progestin receptors does not tell us the type of neurotransmitter used by the neuron in question. However, recent work suggests that there can be "cross talk" between certain neurotransmitters (e.g., dopamine) and steroid hormone receptors (e.g., progestin). A dopamine D_1 receptor agonist, but not a dopamine D_2 receptor agonist, mimicked the effects of progesterone in facilitating sexual behavior in female rats (Mani et al., 1995). If the rats were treated with a progesterone antagonist, then the facilitatory effects of dopamine were blocked. These results suggest that neurotransmitters may affect in vivo gene expression and behavior by means of cross talk with steroid hormone receptors in the brain (Mani et al., 1995; Mani, 2001, 2003).

Two primary hypotheses regarding the role of estrogen in promoting lordosis have been proposed: the trigger hypothesis and the maintenance hypothesis (Pfaff et al., 1994). The trigger hypothesis states that one brief pulse of estrogen sets off a chain of events that, once triggered, continues to the end of the program with the display of lordosis. This hypothesis was generated in response to data indicating that a relatively brief exposure (approximately 30 minutes) to estrogen could result in lordosis (e.g., Bullock, 1970; Johnston and Davidson, 1979). However, the esterified estrogens used in these studies have relatively slow clearance rates and probably stayed in the blood and affected neural tissue for much longer than 30 minutes (Clemens and Weaver, 1985). The maintenance hypothesis contends that estrogen must be present continuously from the beginning of estrogen treatment throughout the behavioral test in order for lordosis to be displayed. However, tests using estradiol implants that could be inserted and removed rapidly revealed that two discontinuous exposures to estrogen could facilitate progesterone-evoked lordosis (Södersten et al., 1981). Researchers using a variety of temporal schedules of estradiol exposure reported that only 2 hours of total estrogen treatment were sufficient to increase progesterone receptor numbers in the hypothalamus, as well as to permit lordosis. The second estrogen treatment had to begin at least 4 hours after the end of the initial estrogen exposure, but not more than 13 hours later (Parsons et al., 1982).

More recently, a cascade hypothesis has been proposed to explain the endocrine events underlying lordosis. According to this hypothesis, specific and discrete events occur within the neurons of the VMN as a result of initial estrogen binding, and these events are required for later estrogen-dependent events to occur. This hypothesis differs from the maintenance hypothesis because continuous estrogen receptor occupation is not required; rather, estrogen must occupy receptors at specific, critical times in order for lordosis to be expressed (Pfaff and Schwartz-Giblin, 1988). When the two-pulse experiment described above was replicated, it was found that anesthesia (Roy et al., 1985) or protein synthesis inhibition (Meisel and Pfaff, 1984; Roy et al., 1985) prior to either pulse of estrogen (Parsons et al., 1982) or between the two pulses interfered with the facilitation of lordosis. These results suggest that electrophysiological events (suppressed by anesthesia), new protein synthesis, and incoming sensory information are all critical components of a cascade of events that begins with the initial exposure to estrogen and ends with the onset of the lordosis response.

The cascade hypothesis is reminiscent of the organizational/activational hypothesis presented in Chapter 3 to explain sex differences in neural function. Recall that sex steroid hormones organize neural structures perinatally in a masculine or feminine pattern, and that subsequent postpubertal hormonal stimulation activates these previously organized neural circuits, resulting in sexually differentiated behaviors. In the cascade of endocrine events leading to lordosis, there is a much shorter time course between the organizing effects of the initial estrogen exposure—late during diestrus II in the rat—and the activating effects of estrogen and progestin during late proestrus, but the principles are the same. The cascade hypothesis, along with the wonderfully detailed data on cellular and subcellular mechanisms now available (Pfaff et al., 2004), suggests that the organizational/activational principles proposed in 1959 may be the fundamental principles underlying all hormone–behavior interactions.

Gene Knockout Studies

As noted in previous chapters, studies of mice that have had specific genes deleted (knocked out) can be very useful in analyzing hormone–behavior interactions. In most cases studied, production of progesterone receptors is induced when estradiol interacts with estrogen receptors. Thus, many of the effects of progesterone on reproductive physiology and behavior could actually be due to a combination of progestins and estrogens. To separate out the effects of these two steroids, mice were created that lacked functional progesterone receptors ($PR^{-/-}$) (Lydon et al., 1996). Both male and female $PR^{-/-}$ mice develop to adulthood; males are fertile, but females are sterile. Female $PR^{-/-}$ mice display uterine hyperplasia and minimal mammary gland development, and they are anovulatory even when stimulated with exogenous gonadotropins (Lydon et al., 1996; Mani et al., 1997). When ovariectomized, neither $PR^{-/-}$ mice nor $PR^{+/+}$ mice displayed lordosis after treatment with estradiol. However, treatment with progesterone induced lordosis in virtually all of the estrogen-primed $PR^{+/+}$ mice tested, but in none of the $PR^{-/-}$ females (Figure 6.26) (Lydon et al., 1996). These results with PR knockout mice support previous work in which lordosis could not be elicited in estrogen-primed rats that received infusions of progesterone receptor antisense RNA into the VMN (Ogawa et al., 1994).

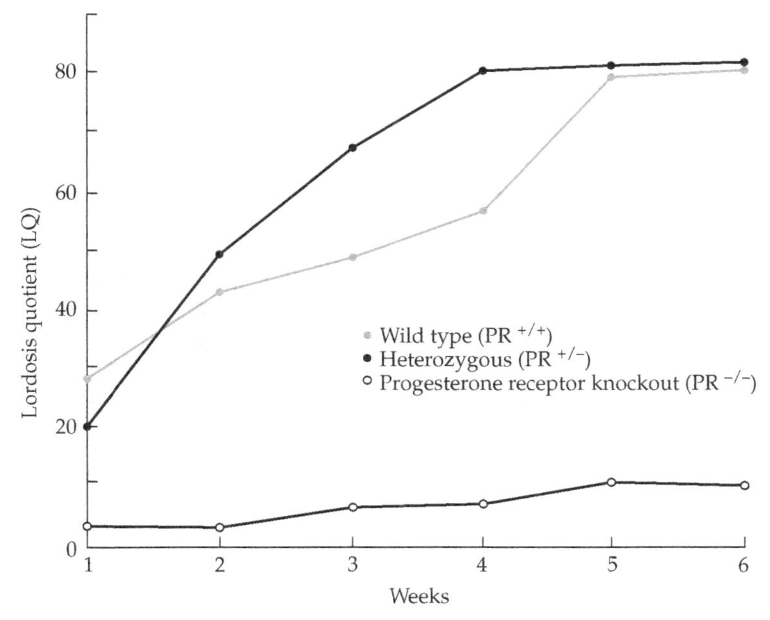

6.26 Lordosis does not occur in progesterone receptor knockout mice. Treatment with progesterone after estrogen priming induced lordosis in virtually all of the wild-type ($PR^{+/+}$) female mice tested, but in none of the progesterone receptor knockout ($PR^{-/-}$) females. After Lydon et al., 1996.

Knockout mice have also been developed for both the α and β subtypes of estrogen receptors (Lubahn et al., 1993; Ogawa et al., 1996a,b; Rissman et al., 1997a). When α estrogen receptor knockout mice (αERKO) and wild-type (WT) female mice were ovariectomized, then given equivalent doses of 17β-estradiol, only the WT mice displayed lordosis when tested with a wild-type male in a neutral arena (Rissman et al., 1997a). Progesterone treatment did not facilitate lordosis in αERKO females (Rissman et al., 1997b). When these αERKO females were given foster pups, induction of maternal behavior was also impaired (Ogawa et al., 1996a).

In a neutral mating arena, males mounted both αERKO and WT females, suggesting that mice of both genotypes were attractive to males. When tested in a three-chamber preference apparatus, males spent equivalent amounts of time with αERKO and WT females (Rissman et al., 1997b). When the conditions under which the female αERKO mice were tested were changed, however, a different pattern of results emerged (Ogawa et al., 1996a,b). When ovary-intact αERKO mice were placed in the males' home cages, some lordosis was observed. However, males often responded with aggressive attacks against the αERKO females (suggesting that they were not particularly attractive to them).

Taken together, the results of these behavioral analyses of steroid receptor knockout mice suggest that the presence of a functional estrogen receptor is necessary for female-typical reproductive behavior. Both estrogen receptor subtypes, ERα and ERβ, which bind to estradiol with similar affinity, have been identified in numerous sites in the brain (Shughrue et al., 1997; Mitra et al., 2003). For example, both receptor subtypes are present in the arcuate nucleus and the preoptic area (POA), whereas ERα is present in the VMN and ERβ in the paraventricular nuclei (PVN). These regions in and around the hypothalamus are important in sexual behavior, thermoregulation, and feeding behavior. Both estrogen receptor subtypes also are present in the amygdala and hippocampus, where they may mediate short-term memory and emotional responses. ERβ has also been identified in the cerebellum and in cortical regions (Mitra et al., 2003). The α estrogen receptor seems to be critical for mediating lordosis. Female mice missing both estrogen receptor subtypes (αβERKO mice), like αERKO female mice, failed to display lordosis after appropriate hormonal priming. Female βERKO mice, however, displayed lordosis to the same extent as female WT mice (Kudwa and Rissman, 2003).

A Neural Model of Lordosis

Pfaff and Schwartz-Giblin (1988) have modeled the neural regulation of lordosis. Their comprehensive and extraordinarily detailed model is based on the idea that several neural modules (specific subsections of the nervous system) function together to mediate the lordosis response. This model should be celebrated as the first complete circuitry map of a mammalian hormone–behavior interaction. Pfaff and colleagues mapped the sensory input, central integration, and effector pathways involved in lordosis and documented their interactions with hormones. Five modules are described by the model: the spinal cord module, lower brain stem module, midbrain module, hypothalamic module, and forebrain module. A brief synopsis of the role of each module is provided below (Figure 6.27; refer to Pfaff et al., 1994 for more details).

SPINAL CORD MODULE The spinal cord module receives the majority of somatosensory information during copulation. As mentioned above, lordosis is triggered by sensory input from the rump, flanks, and perianal region of the female. In addition to initially processing the sensory input, the spinal cord module generates the motoneuronal output that results in lordosis. Although the behavioral program of sensory input and motor output is located in the spinal cord, female rats differ from males in that the sexual response will not occur without input from the brain (Meisel and Sachs, 1994). Thus, female rats with transected spinal cords will not display lordosis.

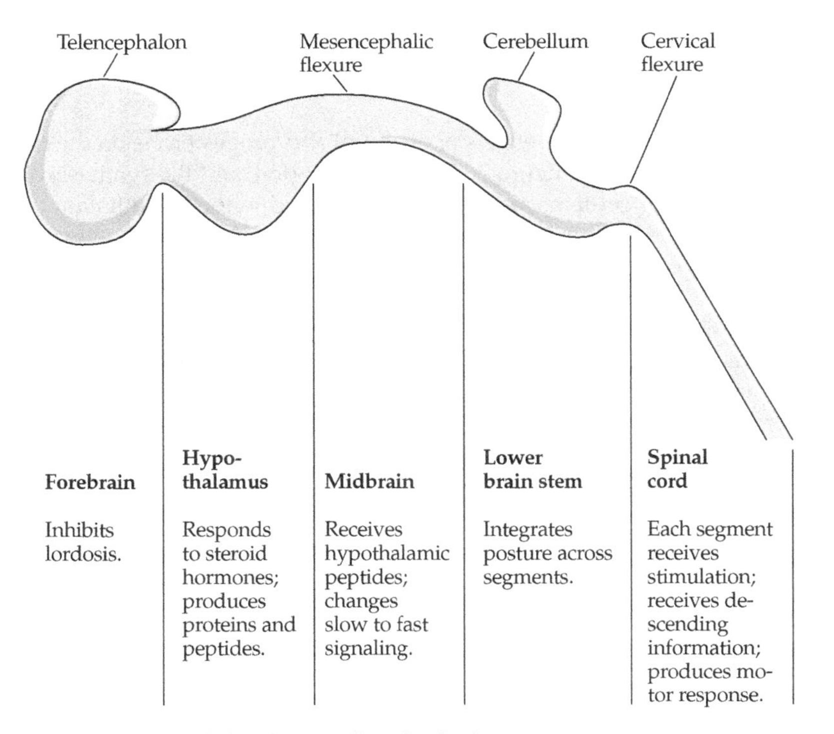

6.27 Five neural modules that mediate lordosis. The components of the nervous system that mediate lordosis in rats can be separated into five modules. The forebrain normally inhibits lordosis, but estrogen plus tactile stimulation disinhibits the behavior, as do lesions of the forebrain. Estradiol and progesterone affect the electrophysiological properties of neurons, RNA transcription, and the synthesis of proteins in the hypothalamic module. The hypothalamic module produces proteins and peptide hormones that interact with the midbrain, which serves to transduce these relatively slow endocrine changes into fast neural signals. The lower brain stem is critical for the postural changes involved in lordosis, and the spinal cord is responsible for moving sensory information associated with male mounting behavior to the brain and bringing motor signals from the brain to the deep back muscles that directly cause lordosis. After Pfaff et al., 1994.

LOWER BRAIN STEM MODULE Integration of information about posture and moment-to-moment corrections to posture are mediated in the lower brain stem module. Input from the vestibular organs and from proprioceptors throughout the body is necessary for maintaining a rigid posture and making corrections to maintain the weight of the mounting male.

MIDBRAIN MODULE The midbrain module receives input from the hypothalamus and elsewhere in the brain, and it translates and integrates these signals to mediate firing rates in the reticulospinal neurons in the lower brain stem module. Peptides are transmitted from the hypothalamus to the central gray region of the midbrain module, where these typically slow neuroendocrine signals (on the order of hours) are transduced to neural signals. If the midbrain module is not activated by peptidergic information from the hypothalamus and by steroid hormones, lordosis will not occur. Lordosis is steroid-dependent, and it is the midbrain module and, even more extensively, the hypothalamic module where the action of steroids is critical.

HYPOTHALAMIC MODULE The effects of estradiol and progesterone on the electrophysiological properties of neurons, RNA transcription, and the synthesis of new structural and other proteins are primarily mediated in the hypothalamic module. Components of this module are either inhibited (the MPOA) or activated (the VMN) by estrogens.

FOREBRAIN MODULE The forebrain module exerts primarily inhibitory effects on lordosis. Large frontal cortical lesions, as well as lesions to the olfactory bulbs or septum, facilitate lordosis. In the presence of estrogen, tactile stimulation disinhibits lordosis behavior.

According to this model, sex steroid hormones, as well as several peptide hormones, including GnRH, prolactin, β-endorphin, and substance P, modulate lordosis by acting on one or more of the neural modules during normal or experimentally induced estrus. When females with estrogen concentrations insufficient to evoke lordosis by themselves are primed with GnRH, prolactin, or substance P, lordosis results. GnRH is transported to the midbrain central gray region along axons from the medial basal hypothalamus, and in the presence of estrogen, stimulates excitatory responses in central gray neurons (Moss and McCann, 1973). Experimental treatment with GnRH antibody in the central gray region caused reductions in the lordosis response in rats (Sakuma and Pfaff, 1980). Application of CRH or β-endorphin to the central gray region also led to a decrease in the lordosis response (Pfaff and Schwartz-Giblin, 1988).

Neural Models of Proceptive Behaviors

In contrast to the comprehensive model that has been developed for receptive behavior, the neural loci mediating proceptive behavior have been only partially identified. Because lordosis is such a prominent feature of female rat mating behavior, the vast majority of brain lesion studies and hormone implant studies have used the presence or absence of lordosis as a behavioral assay, in much the same way that mounting and intromission have been employed in studies of the neural bases of male copulatory behavior. However, during the course of behavioral analyses of brain-altered females, changes in proceptive behaviors have occasionally been reported (Erskine, 1989).

As described above, implants of estradiol alone into the VMN induced lordosis in ovariectomized rats, but at a low frequency. Proceptive behaviors were never observed during mating tests under these conditions (Rubin and Barfield, 1983). However, a full complement of proceptive behaviors, including hopping, darting, and ear wiggling, was observed among estrogen-injected, ovariectomized female rats that received progesterone brain implants into the VMN, but not in those that received implants into the POA, midbrain central gray region, hippocampus, or medullary reticular formation (Rubin and Barfield, 1983). Infusions of progesterone into the VMN facilitated lordosis in less than 2 hours, suggesting that changes in electrophysiological parameters, rather than structural changes or protein synthesis, mediate the effects of progesterone (Glaser et al., 1983; McGinnis et al., 1981) (Box 6.4). The VMN are involved in the mediation of female sexual behavior across a wide range of taxa, including parthenogenetic whiptail lizards (*Cnemidophorus uniparens*) (see Chapter 5) (Kendrick et al., 1993).

Lesions of the VMN, or destruction of their afferent and efferent fibers, typically reduce the frequency of lordosis. Proceptive behaviors also appear to be affected by VMN lesions. As described above, projections from the VMN to the midbrain are important parts of the neural circuitry mediating lordosis (Kow and Pfaff, 1998). However, the midbrain appears to be less involved in the mediation of proceptive behaviors (Erskine, 1989). In fact, proceptive behaviors remain intact, or in some instances are enhanced, after destruction of the ascending ventral noradrenergic bundle (VNAB), a procedure that completely abolishes lordosis (Hansen et al., 1980, 1981). Thus, different brain regions may mediate receptive and proceptive behaviors. This possibility suggests that different neural substrates underlie sexual motivation and sexual performance in both males and females.

The brain mechanisms underlying primate sexual motivation and performance appear to be similar to those in rodents (Pfaus et al., 2003). For example, lesions of the anterior hypothalamus of estrogen-treated common marmosets (*Callithrix jacchus*), which extended to varying degrees into the medial hypothalamus, virtually abolished proceptive tongue-flicking and staring displays. Tongue-flicking during copulation also decreased, but the females did not increase the number of mounts that they refused or terminated, with the exception of one animal that had received more extensive damage to the medial hypothalamus (Kendrick and Dixson, 1986). These results suggest a neuroanatomical distinction between hypothalamic mechanisms that regulate proceptivity and receptivity in primates.

To determine the neural mechanisms underlying sexual behavior in female rhesus monkeys (*Macaca mulatto*), single-neuron activity in the ventromedial hypothalamus (VMH) and the medial preoptic area (MPOA) was recorded dur-

BOX 6.4 Nongenomic Behavioral Effects of Steroid Hormones

Traditionally, the behavioral effects of steroid hormones have been assumed to be mediated via intracellular receptor–ligand interactions that ultimately affect gene transcription (Flanagan-Cato and Fluharty, 1997). However, an alternative mechanism of steroid action has recently emerged (Moore and Orchinik, 1994). It was reported that corticosterone administration could rapidly interfere with mating behavior in a species of newt (*Taricha granulosa*) (Orchinik et al., 1991; Moore and Orchinik, 1994). Membrane-bound corticosterone receptors have been located in neuronal membrane fractions from these newts (Moore et al., 1995). The discovery of these membrane-bound steroid receptors and the rapidity of the response suggested that the genetic machinery could not have been engaged, and it was hypothesized that a nongenomic mechanism must mediate this rapid behavioral effect of steroid hormones.

Additional evidence for the existence of nongenomic effects of steroid hormones on behavior has been provided in mammals. For example, rapid onset of sexual receptivity can be induced in estrogen-primed female Syrian hamsters (*Mesocricetus auratus*) by providing progesterone that has been conjugated with a protein that presumably prevents the steroid from entering cells and interacting with intracellular receptors (DeBold and Frye, 1994; Frye and DeBold, 1993; Frye et al., 1992). Other research has demonstrated that estradiol evokes rapid electrophysiological effects in the CA1 neurons of the hippocampus, and that these rapid effects can be blocked by interfering with protein kinase A or with G protein receptors (Wong and Moss, 1992; Gu and Moss, 1996; Wong et al., 1996). More recent studies have shown that progesterone and neurosteroids modulate lordosis by acting in the ventral tegmentum and ventromedial hypothalamus via nongenomic effects (Frye, 2001a,b; Frye and Petralia, 2003).

Steroid hormones also bind to the GABA receptor–chloride ion channel complex (Majewska et al., 1986) as well as other membrane-bound receptors (Ke and Ramirez, 1990) to affect neurotransmission (Becker, 1990). Taken together, these findings suggest that steroid hormones can affect behavior both through traditional genomic actions and via alternative membrane-bound receptors. It appears that both mechanisms could act in concert to increase or decrease the likelihood of a particular behavior in a specific context (Frye et al., 1996; Frye, 2001a).

ing sexual interactions with a male partner (Aou et al., 1988). Proceptive presentation behavior with no mating evoked activity changes in about 40% of neurons tested in the VMH (mainly excitation) and MPOA (mainly inhibition). When the male's mating acts occurred, about 50% of VMH cells and about 90% of MPOA cells changed their firing rates during presentation behavior; MPOA cells significantly increased their excitatory firing rates. As copulation progressed, activity in about 40% of VMH and 70% of MPOA cells was observed. These findings suggest that the VMH and MPOA regulate primate sexual behavior in different ways: excitation of the VMH and inhibition of the MPOA are related to presentation behavior (proceptive behavior), whereas excitation of the MPOA is related to copulation with a male partner (receptive behavior). Furthermore, they suggest that the sexual behavior of a male partner modulates activity in both VMH and MPOA neurons of the female (Aou et al., 1988).

In women, recent work has examined the neural correlates of orgasm. As of yet, no definitive explanation for what triggers orgasm has emerged. The first brain imaging studies (PET, coupled with fMRI) during orgasm in women have recently been reported (Whipple and Komisaruk, 2002; Komisaruk et al., 2002). Increased brain activation during orgasm, compared with that during pre-orgasm sexual arousal, was noted in the following brain regions: the paraventricular nuclei (PVN) of the hypothalamus, periaqueductal gray region of the midbrain, hippocampus, and cerebellum (Whipple and Komisaruk, 2002). Further studies that compare brain areas activated by orgasm with those activated during sexual arousal without orgasm are needed to assess whether there are specific brain regions responsible for triggering orgasm in women (Meston et al., 2004).

In terms of sexual motivation, men are generally more interested in and responsive to visual sexually arousing stimuli than are women (Harmann et al., 2004). Functional magnetic resonance imaging (fMRI) revealed that the amygdala and hypothalamus were more strongly activated in men than in women viewing identical sexual stimuli (Figure 6.28). These results were the same even when women reported greater arousal (Harmann et al., 2004). Men and women showed similar activation patterns across several other brain regions, including ventral striatal regions involved in reward. These experimental results suggest that the

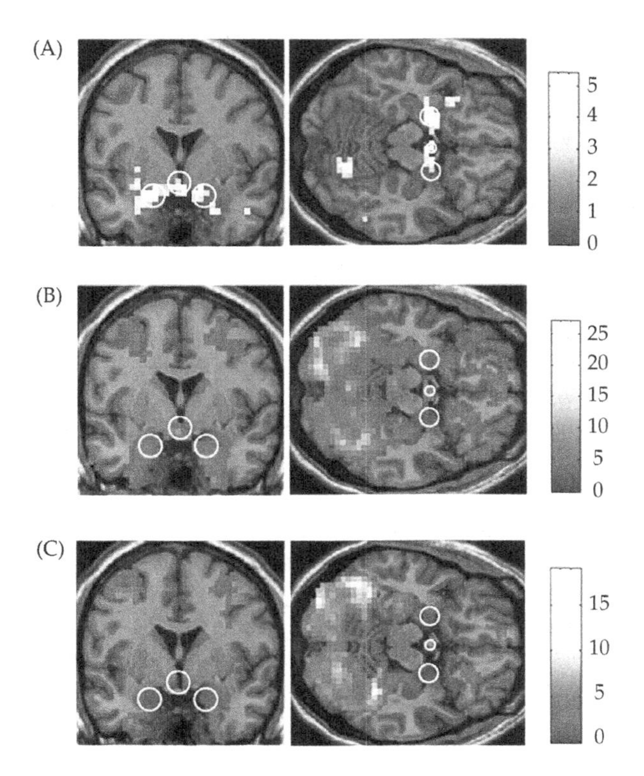

6.28 Brain imaging studies show differences in the responses of men and women to visual sexual stimuli. Functional MRI imaging reveals greater brain activation in the amygdala and hypothalamus of men than women. Regional activation maps indicate the brain activation contrasts between men and women viewing erotic pictures of couples as compared with men and women looking at a nonsexual point of reference on the computer screen (A–C). The white circles approximate the regions of interest. (A) Males versus females (left is coronal image; right is axial view showing additional right cerebellar activation). (B) Men at the same views. (C) Women at the same views. From Hamann et al., 2004.

amygdala mediates sex differences in responsiveness to appetitive and biologically salient stimuli; the human amygdala may also mediate the reportedly greater role of visual stimuli in male sexual behavior, paralleling prior animal findings.

To summarize, hormones affect female sexual behavior by affecting the input, central integration, and output functions of the central nervous system. Estrogens affect sensory input by increasing the receptive field size in sensory cells (Kow et al., 1979). Estrogens affect protein synthesis, the electrophysiological responses of neurons, and the appearance of growthlike processes on neurons (Pfaff and Schwartz-Giblin, 1988). Finally, estrogens affect the muscular output that results in lordosis, as well as the output of chemosensory stimuli important in attracting a mating partner (Beach and Gilmore, 1949).

Summary

- Copulatory behavior in females usually coincides with ovulation. Because females ovulate periodically, copulatory behavior is observed in cycles called estrous cycles. Female mammals are said to be in estrus when they permit copulation.
- Female sexual behavior has been the subject of formal study for much less time than male sexual behavior. Despite this disparity, much is known about the mechanisms underlying the hormone–behavior interactions involved in the regulation of female copulatory behavior. Ovariectomy consistently results in decreased sexual behavior in females from all vertebrate taxa.
- Cyclic changes in vaginal cytology have been correlated with changes in ovarian structure and subsequently with behavior. In rats, the vaginal estrous cycle consists of 2 or 3 days of diestrus, followed by a 12–18-hour proestrous phase, and then a 24–36-hour estrous period. Behavioral estrus and mating occur near the end of proestrus and end as the vaginal smear becomes estrous. Ovulation occurs near the beginning of vaginal estrus.
- Mating behavior coincides with the presence of a Graafian follicle; chemical extraction of this ovarian structure led to the discovery of the estrogen class of steroids. Mating behavior often stops with the onset of corpora luteal activity; chemical extraction of this ovarian structure led to the discovery of progesterone. Replacement studies using these steroids on ovariectomized animals demonstrated that estrogens and progesterone were required for mating behavior in guinea pigs, mice, rats, and many other species.
- Females of most vertebrate taxa display species-specific mating postures. In rodents, the characteristic mating posture is called lordosis. Females in lordosis arch their backs, deflect their tails, and remain immobile to allow male intromission. Female dogs deflect the tail and display a virtually rigid mating posture. Some female primates display stereotyped mating postures, but many do not.

- Females have historically been portrayed as passive recipients of male sexual attention. However, in many species, especially primates, females initiate virtually all sexual interactions. In an attempt to reduce variation, researchers have traditionally studied female copulatory behavior in single-pair tests. In this context, females appear rather passive. However, when they are tested in social groups that simulate natural conditions, females' initiation and control of copulatory activities become evident.
- Female sexual behavior can be divided into three components: (1) attractivity, (2) proceptivity, and (3) receptivity. Attractivity is the stimulus value of the female for a given male. Proceptivity is the extent to which females initiate copulation, and reflects overt behavior as well as the underlying motivational state. Receptivity reflects the stimulus value of the female for eliciting an intravaginal ejaculation from a male conspecific; in other words, receptivity is the state of responsiveness to the sexual behaviors of another individual. Proceptivity and receptivity overlap conceptually, as well as in practice. Generally, estrogens enhance attractivity, proceptivity, and receptivity, and progestins reduce these parameters.
- Female receptivity and control of copulation vary among species; some females copulate even when not in estrus, and others reject certain males even when in estrus. Pacing of copulatory behavior by female rats has important physiological consequences for induction of corpora luteal function and subsequent maintenance of pregnancy. Both sexual motivation and performance are mediated by sex steroid hormones in rodents, and probably in other nonprimate species as well; in these species, copulation does not occur in the absence of high blood concentrations of estrogen. In contrast, the ability to copulate is not linked to hormones among primates.
- Female reproductive cycles have been categorized into three basic types. In type 1 cycles, ovulation and pseudopregnancy are spontaneous. Humans and other primates display type 1 cycles. In type 2 cycles, ovulation is induced by copulation or other vaginal stimulation, but pseudopregnancy is spontaneous. Cats and ferrets display type 2 ovarian cycles. Both ovulation and corpora luteal formation are spontaneous in animals with type 3 cycles, but pseudopregnancy is induced via the release of prolactin following copulation. Mammals with type 3 cycles do not have functional luteal phases in nonpregnant cycles. Rats, mice, and hamsters have type 3 ovarian cycles.
- Reproductive cycles have evolved so as to maximize reproductive output, and thus vary with the ecology of the species. Repeated estrous cycles are laboratory artifacts and occur infrequently in free-ranging mammals. Females in nature are typically pregnant, lactating, or in seasonal diestrus. Most pregnancies in nature are the result of mating during postpartum estrus.
- Reproductive cycles can be influenced by a number of social and environmental factors. Rodents exert a number of effects, mediated by chemosensory fac-

tors, on the reproductive cycles of their conspecifics. Women who live together for extended periods of time may synchronize their menstrual cycles.

- Ovulation and peak estradiol concentrations coincide in both rats and primates. However, rats exhibit a periovulatory peak in progesterone that is reduced or absent in primates. Primates of many species, including humans, often display periovulatory peaks in androgen concentrations. Motivation to copulate appears to coincide with blood concentrations of androgens in primates.
- Several brain sites are necessary for lordosis. Lesions of the ventromedial nuclei of the hypothalamus (VMN) or destruction of their afferent and efferent fibers reduces the frequency of lordosis. Lesions of the central gray region also reduce lordosis behavior. Destruction of the midbrain ascending ventral noradrenergic bundle (VNAB) completely abolishes lordosis. The medullary reticular formation in the brain stem controls motoneurons innervating axial muscles, especially the deep back muscles that are critical for lordosis.
- Sensory input enters the female rat's nervous system during mating via cutaneous receptors on the flanks, rump, and perineum. Information from these skin receptors and from pressure-responsive neurons enters the spinal cord and is sent to the medullary reticular formation. The interaction of these ascending sensory messages with descending fibers from the midbrain central gray, which carries information from the VMN, permits lordosis only when sex steroid hormones are available to cells in the central gray and in the VMN.
- Estrogen promotes estrogen and progesterone receptor formation in the VMN, as well as stimulating RNA transcription, protein synthesis, electrophysiological changes in neurons, and delivery of peptides to the midbrain central gray region. According to the cascade hypothesis, estrogen induces these changes over several hours, or "primes the nervous system" for subsequent estrogenic facilitation of lordosis. The cascade hypothesis is similar in principle to the organizational/activational model of sexual differentiation: hormones affect subsequent behavior by causing structural changes prior to evoking electrophysiological or other fast-acting changes.
- A comprehensive model of the regulation of lordosis has been proposed, based on five neural modules (specific subsections of the nervous system)—namely, the spinal cord module, lower brain stem module, midbrain module, hypothalamic module, and forebrain module—that function together to mediate the lordosis response. Sex steroid hormones, as well as several peptide hormones, modulate lordosis behavior by acting on one or more of these neural modules.
- The neural and endocrine bases of proceptive behaviors are different from those of lordosis, a receptive behavior. Males and females are similar in that sexual motivation and sexual performance are organized separately in the nervous system.

Questions for Discussion

1. Discuss the proposition that a female mammal can be sexually receptive without being sexually attractive. Also provide examples of a female that is attractive, but not receptive. Would such observations negate a role for hormones in the mediation of attractivity and receptivity?
2. Document the following assertion: "Hormones play a lesser role in mediating sexual behavior in female primates than in female rodents." Is it reasonable to state that the sexual behavior of women is unaffected by sex steroid hormones?
3. Given the number of fundamental differences between rodents and primates, is it useful to study hormone–behavior interactions in rodents? Defend your answer.
4. Discuss the implications of the following quote in terms of the study of female sexual behavior: "It is an unfortunate accident that studies of reproductive physiology and behavior have been limited to a few domesticated species because our conceptual limits in the understanding of the regulation of female reproductive processes have become compressed and distorted."
5. Activation of female sexual behavior often requires both estrogen and progestins. If you discovered that a female rodent exhibited a type 2.1 reproductive cycle, would you predict that progesterone would be necessary for the expression of estrous behavior in this species? Why or why not?

Refer to the accompanying Student CD for additional resources, including Web links, videos, animations, and additional photos.

Suggested Readings

Anastasiadis, A. G., Davis, A. R., Salomon, L., Burchardt, M., and Shabsigh, R. 2002. Hormonal factors in female sexual dysfunction. *Current Opinions in Urology*, 12: 503–507.

Beach, F. A. 1976. Sexual attractivity, proceptivity, and receptivity in female mammals. *Hormones and Behavior*, 7: 105–138.

Blaustein, J. D., and Erskine, M. S. 2002. Feminine sexual behavior: Cellular integration of hormonal and afferent information in the rodent forebrain. In D. W. Pfaff, A. P. Arnold, A. M. Etgen, S. E. Fahrbach, and R. T. Rubin (eds.), *Hormones, Brain and Behavior*, Volume 1, pp. 139–214. Academic Press, New York.

Kow, L. M., and Pfaff, D. W. 1998. Mapping of neural and signal transduction pathways for lordosis in the search for estrogen actions on the central nervous system. *Behavioral Brain Research*, 92: 169–180.
Mani, S. K., Blaustein, J. D., and O'Malley, B. W. 1997. Progesterone receptor function from a behavioral perspective. *Hormones and Behavior*, 31: 244–255.
McMillan, H. J., and Wynne-Edwards, K. E. 1998. Evolutionary change in the endocrinology of behavioral receptivity: Divergent roles for progesterone and prolactin within the genus *Phodopus*. *Biology of Reproduction*, 59: 30–38.
Pfaff, D. W. 1999. *Drive: Neurobiological and Molecular Mechanisms of Sexual Motivation*. Cambridge: MIT Press.
Pfaus, J. G., Kippin, T. E., and Coria-Avila, G. 2003. What can animal models tell us about human sexual response? *Annual Review of Sex Research*, 14: 1–63.
Wallen, K., and Zehr, J. L. 2004. Hormones and history: The evolution and development of primate female sexuality. *Journal of Sexual Research*, 41: 101.

7

Parental Behavior

Thus far we have focused on the interactions between hormones and mating behavior. Mating behavior is critical for reproductive success because it brings the two sexes together in order to combine their genetic material. However, successful mating is insufficient if reproduction is to be judged successful. From an evolutionary perspective, the only currency of reproductive success is the production of successful offspring; that is, offspring that manage to survive and produce descendants. The offspring of many animal species require assistance from one or both parents in order to attain maturity and reproduce themselves.

The amount of assistance that parents provide varies widely both among and within species and reflects an optimal evolutionary strategy for maximizing fitness. The optimal strategy for each parent is to provide sufficient care, but no more than is absolutely necessary, to produce successful offspring. **Parental investment**, the extent to which parents compromise their ability to produce additional offspring in order to assist current offspring, may result in a conflict of interest between parents and offspring. As we saw in Chapter 3, parents usually share about 50% of their genes with each offspring, whereas each offspring is 100% related to itself, and acts accordingly! At one end of the parental care continuum are the numerous vertebrate species that provide absolutely no parental care. Females of many fish species, for example, simply release hundreds or even thousands of eggs from their bodies to be fertilized and leave them to face the vagaries of the cold, harsh world

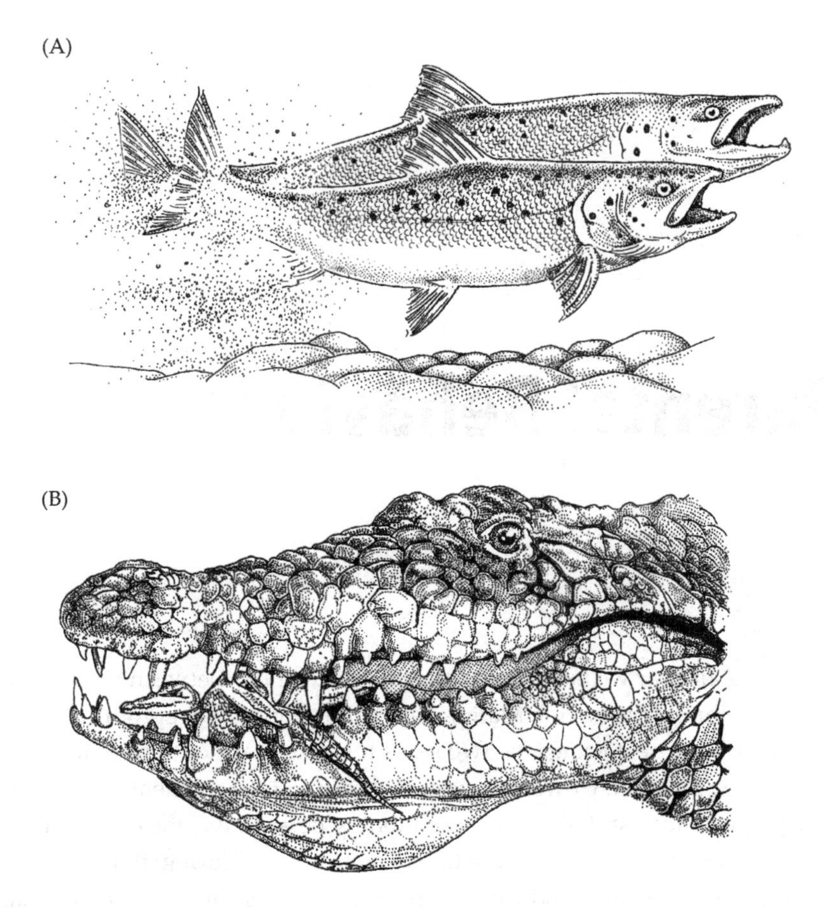

7.1 Some species provide little or no parental care. (A) Female salmon shed large numbers of eggs into the water to be fertilized by a courting male. The eggs remain where they fall, and no parental care is provided. The sheer number of eggs released ensures that some offspring will survive despite the absence of parental protection. (B) Female Nile crocodiles protect their newly hatched young by gently carrying them to the water in their mouths. Aside from nest building, this is the extent of parental care in this species; males make no parental contribution.

on their own (Figure 7.1A). Females of most reptilian species merely cover their newly deposited eggs with dirt and other debris, and provide no additional parental assistance. These young must fend for themselves immediately upon hatching. Parental care, ranging from rudimentary to complex, is observed among some invertebrate animals, mainly in the form of nest defense or provisioning the young with food (Box 7.1). Parents of many other vertebrate species go further, providing food, shelter, and protection from harm as their offspring mature (Figure 7.1B). Humans represent the other terminus of the parental care continuum, often providing substantial care and resources to their children for decades.

What Is Parental Behavior?

Parental behavior can be broadly defined as any behavior performed in relation to one's offspring (Rosenblatt et al., 1985), but more specifically, **parental behavior** is any behavior that contributes directly to the survival of fertilized eggs or offspring that have left the body of the female. If a father performs parental behavior, then a more specific term, **paternal behavior**, is typically used; similarly, parental behavior performed by mothers is called **maternal behavior**.

Why is parental care an important topic to study? Parental care is critical for infant survival among many species, including humans, and hence is critical for the reproductive success of the individual parent(s). Parental care plays an important role in the evolution of competition for mates because individuals may seek mates that can best provide care for their offspring; the intensity of mate competition within each sex influences the selection pressures operating on behavior, physiology, and morphology (Clutton-Brock, 1991). Parental care influences the course of physical and psychological development of the offspring, and it is also important among many species in the socialization of young. Individual humans display a wide range of parenting skills; it would be useful to identify any endocrine correlates of poor parenting because poor parenting is associated with numerous social problems for the offspring. The importance of mothers as nurturers of new life has been an inspiration to artists for eons (Figure 7.2).

As you will see, there is a wide diversity of parental care strategies among vertebrates. Some of this diversity is due to variation in the developmental maturity of the offspring when they are produced. Two broad categories of offspring development are found among vertebrates (Figure 7.3). In the first, females produce large numbers of immature and helpless, or **altricial**, young. Females that engage in this reproductive strategy may or may not display parental care. In the second, females produce a few **precocial** offspring that are well developed and may be able to survive with little or no parental intervention. There are trade-offs between these two strategies. If we consider the costs and benefits of producing precocial or altricial young for similar-sized individuals, precocial young require a greater initial investment per individual offspring, as the female must invest more energy in their development before birth or hatching, but less parental investment is usually required after birth. Altricial young require less prenatal investment, but if parental care is given, a greater amount is needed. Some species, rather than giving parental care to their altricial offspring, rely on the production of large num-

7.2 Women's ability to produce and nurture new life is accorded great respect in many African societies, as demonstrated by this mother and child wood carving from Mali, West Africa. The carving celebrates the nurturant power of women while depicting a nearly universal image of human maternal behavior. Photograph courtesy of the Kinsey Institute.

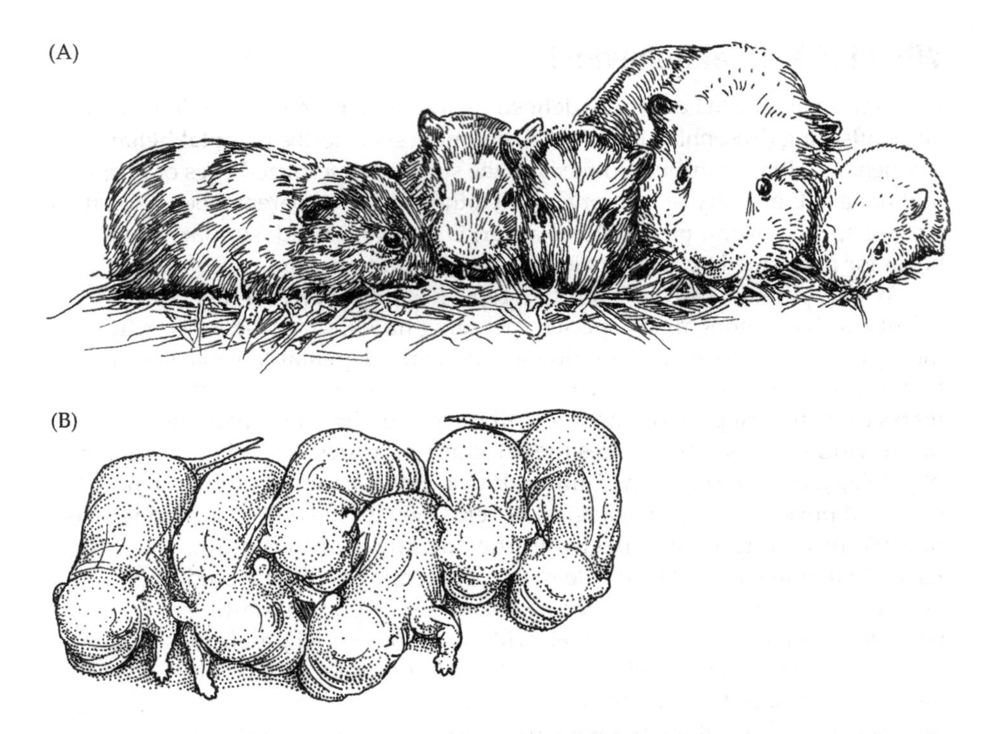

7.3 The extent and nature of parental care depends on the development of the offspring. There are vast differences in the developmental stages of offspring when produced, even among rodents. (A) Guinea pigs bear precocial young, which are born fully furred and mobile. (B) Rats bear altricial young, which are born without the ability to move or thermoregulate. Offspring born at these different stages of development obviously require different types of parental care.

bers of offspring to ensure that at least a few will survive. Other species, including humans, produce **semi-precocial** young that always demand significant parental care at birth, but can thermoregulate and cling to the mother.

In those animals that display parental care, the behaviors must be performed correctly, with little margin for error; they initially must be performed without previous experience; and they must usually begin immediately after the hatching or birth of the offspring. Given these constraints, the onset of parental behavior is often remarkable for its precision and suitability. The confidence with which first-time mother dogs attend to their newly arrived pups is impressive. A day or two prior to giving birth, a pregnant dog builds a nest (often using only the best shirts, towels, or sweaters available) into which her puppies will be delivered. The pups are born about 20 to 60 minutes apart. As each pup is born, the mother behaves solicitously toward it, licking off the amniotic fluid and membranes, and also vigorously licking the anogenital region to stimulate the elimination of wastes and other physiological processes. After consuming the placentas, she lies

7.4 Females of many species provide their offspring with food, shelter, and protection from harm. Mammals, such as this female dog, have evolved to provide a specialized secretion, milk, to feed the young. Because male mammals do not lactate, only females can meet the nutritional demands of the young. In some mammalian species, the male contributes by feeding the female while she nurses. In very rare cases, male mammals participate more fully in parental care.

on her side so the pups can attach to her nipples to nurse (Figure 7.4). The altricial puppies are blind and partially deaf, their coats are not completely established, and their thermoregulatory and locomotory abilities are not fully developed. During the first days of their lives, the mother continues to groom her pups as they nurse, and she also keeps them warm. She will retrieve pups by the scruff of the neck should they wander too far from the nest, and may move the entire litter if there are too many disturbances. If another dog, a well-known human acquaintance, or even the father of the litter comes too close to the pups, the mother may act very aggressively and can inflict serious damage on the intruder. The ferocity of this **maternal aggression** in previously docile animals can be surprising and dangerous to unsuspecting pet owners.

The parental competence exhibited by dogs and other nonhuman animals can lead to a powerful impression among human observers that nonhuman parents always perform parental behavior perfectly, and it is perhaps for this reason that comparatively few studies have examined the development of parental behavior. In all fairness to human parents, however, the "decisiveness" with which dogs and other nonhuman mothers seem to perform their parental behaviors is often an illusion. Upon close observation, it can be seen that errors are committed by inexperienced mothers and that trial-and-error learning often occurs

BOX 7.1 Parental Care among Insects

Compared with the enormous amount of research conducted on the ecology of invertebrate parental behavior, few studies have focused on the physiological bases of parental care among invertebrate animals. Although a wide range of parental behaviors are observed among invertebrates, most studies of the hormonal bases of invertebrate parental care have focused on insect species that carry young and build nests (Trumbo, 1996). The best examples of species that carry young are provided by cockroaches, which, depending on the species, are either oviparous (egg-laying), viviparous (give birth to live young), or ovoviviparous (eggs hatch internally). A rich "milk" is secreted from the walls of the brood sac in the viviparous species *Diploptera punctata*. Essentially, juvenile hormone (JH), the primary insect gonadotropin, which is secreted from the corpora allata, is inhibited throughout the cockroach's "pregnancy." With the onset of parturition and the termination of care, JH concentrations rise in preparation for the next brood (Rankin and Stay, 1985).

Many variations of the mother–offspring relationship exist among cockroaches. Displays of long-term association between mother and offspring occur in the Australian burrowing roach (*Macropanesthia*) and the woodroach (*Cryptocercus*) (e.g., Matsumoto, 1992; Ruegg and Rose, 1991). Young of the ovoviviparous cockroach *Leucophaea maderae* are reported to join in on their mother's

Photograph © Mark Moffett/Minden Pictures.

during the onset of maternal behavior. Parental "instinct" is not completely "hardwired," and one can observe variation from parent to parent within the same species. Although this species-specific parental variation probably provides the grist on which natural selection acts, and is thus important for the reproductive success of individuals, the behavioral patterns of parents within a species share much in common.

Ordinarily, female dogs do not act in a solicitous and protective manner in the presence of newborn puppies; however, female dogs that have just delivered their own pups virtually always behave maternally. What triggers the set of maternal behaviors in dogs? While one might suppose that the stimuli, neural or otherwise, associated with the birth process initiate the onset of parental behavior, foraging excursions soon after birth, as "Mom" presumably "shows them the ropes" (Trumbo, 1996). Little is known about the hormonal bases of these associations; clearly much more research needs to be conducted on these fascinating creatures.

Another parental behavior that has been studied among invertebrates is nest building. The proximate mechanisms underlying nest building have been well characterized in earwigs (*Dermaptera*). As in *Diploptera*, JH levels rise prior to egg laying, then remain low during most phases of parental care (Rankin et al., 1995). If the clutch of eggs is removed on the first day of the "parental phase," then JH concentrations increase, presumably to prepare for a new cycle of egg laying. If the eggs are returned within 24 hours, most females accept them. If mothers are separated from their eggs for 48 hours, then only 36% of mothers accept the eggs, and no mothers accept them after 72 hours of separation (Vancassel et al., 1984). These results suggest that high JH levels are incompatible with parental care in these insects.

A slight variation in JH profile is observed in the biparental burying beetle (*Nicrophorus orbicollis*, shown in the accompanying figure). These fascinating insects breed only when they locate a small vertebrate carcass, a somewhat rare resource that is necessary for the success of their offspring (Trumbo, 1997). Within 10 minutes of locating a carcass, JH concentrations double and ovarian mass increases dramatically in females (Trumbo, 1996). Neither feeding on the carcass nor mating per se induces the rapid increases in JH concentrations; rather, the beetle's evaluation of the carcass and sensory feedback seem to trigger the endocrine changes (Trumbo et al., 1995; Trumbo, 1997). The burying beetles inject antibiotic secretions into the carcass to delay decomposition during its preparation as a nest. After oviposition, JH concentrations decline, but remain elevated relative to those prior to the discovery of the carcass. When the young hatch out, about 5 days after the discovery of the carcass, a secondary peak of JH secretion is observed. At this time, the parents increase their locomotor activity and nest maintenance behavior, as well as regurgitating liquified carrion for the young (Trumbo, 1996). As the young become more independent, the parents' activities wane, and their primary parental duties consist of defense against predators. Tissue fluid JH concentrations decline during this latter part of the parental cycle, but rise again when the parents leave the nest to find a new carcass (Trumbo, 1996). All of these studies indicate that the hormone–behavior interactions underlying parental behavior are complex and environmentally mediated in insects, just as they are in vertebrate animals.

this hypothesis can be dismissed because dogs that give birth by cesarean section also behave maternally toward their puppies as soon as they recover from anesthesia. Indeed, many correlational studies in dogs, rabbits, and other mammals, especially rats, have demonstrated that it is the hormones associated with pregnancy and lactation that regulate the onset of mammalian parental behavior. In other vertebrate orders, hormones involved in egg production, nest building, egg laying, and incubation may activate parental behavior. In any case, hormones affect motivation to engage in parental care.

The hormones that trigger parental behavior, however, wane soon after the arrival of the young. What factors serve to maintain parental behavior? If you observe mother dogs as their puppies mature, you will notice that their maternal behaviors change over time. Nest building and pup retrieval behaviors wane, and the duration of nursing and play behaviors increases. Many maternal behaviors, including nursing, completely disappear after a few additional weeks. How are these changes in parental behavior coordinated with changes in the pups? What role, if any, do hormones play in the maintenance and termination of parental behavior? These questions will be answered in detail in this chapter.

Sex Differences in Parental Behavior

With the exception of most bird and some fish species, paternal behavior is rather rare in the animal kingdom; maternal care is much more common. Why does this sex difference exist? As described in Chapter 3, the two sexes produce different types of gametes. Females produce relatively few, large, immobile, resource-rich gametes (eggs), whereas males produce large numbers of small, mobile gametes containing few or no resources (sperm). One male can produce enough sperm to fertilize many more eggs than one female can produce. For example, a single ejaculation from a typical man (about 5 ml of semen) contains sufficient sperm, in theory, to fertilize all the women in North America. In contrast, a woman produces only a few hundred eggs in her lifetime. Because of this difference in reproductive potential, males and females differ fundamentally in how they can best achieve reproductive success.

Reproductive effort can be divided into mating effort and parental effort (Trivers, 1972). Males tend to concentrate their reproductive effort on mating because locating and fertilizing as many different females as possible is the best way for males to achieve maximal reproductive success. Females tend to put the majority of their reproductive effort into parental care because each offspring represents a substantial proportion of a female's lifetime investment of time and resources. A female can best increase her reproductive success by turning food into eggs or successful offspring at a faster rate. Mating with additional males typically does nothing to increase her reproductive success. Consider human females. Pregnancy lasts 9 months, and lactation-induced infertility may delay further reproductive efforts for another year or two. During this same time, a man could potentially fertilize hundreds of women.

The various factors limiting male and female reproductive success were first documented experimentally by Bateman (1948). He put equal numbers of male and female fruit flies in a large bottle and scored the number of matings and the number of offspring produced by each individual. Bateman was able to use specific genetic markers to assign parentage. He found that some males fertilized many females, while others never fathered offspring; however, all the females fared about equally well. The reproductive success of a male depended on the number of females he fertilized, whereas the females achieved maximal reproductive success, in most cases, with a single copulation.

A male forgoing additional mating opportunities in order to assist in the rearing of his offspring is typically at an evolutionary disadvantage relative to a nonparental conspecific male. The reproductive success of the nonparental male will be higher than that of the parental male—that is, unless the offspring cannot survive without parental assistance from two adults. Only in those situations in which two adults are required to guarantee the survival of the young, as is the case for many avian species, will parental males achieve higher reproductive success than nonparental males. Given these theoretical considerations, it is not surprising that paternal care is rare. Animals that have pursued the existing reproductive strategies have been more successful than animals that have not. It is not necessary for these strategies to be consciously directed, even among humans, to be successful. By analogy, birds do not require a conscious understanding of aerodynamic theories to fly. But birds that *behave as if* they do understand the theories of flight increase their reproductive success.

The rest of this chapter describes patterns of parental behavior in various species, the hormonal correlates of parental behavior, and the cues that are necessary and sufficient to elicit parental care. In addition, what is currently known about the neural foundations of parental behavior will be reviewed. As you will see, there is an enormous variety of parental behaviors among vertebrate taxa. But although parental care has been well studied in numerous species, little is known about its endocrine mechanisms in any but those few well-studied species. The vast majority of research addressing the hormonal correlates of parental behavior has been conducted on ring doves, sheep, and laboratory rats. Parental behaviors in birds are described first, followed by a description of the hormone–behavior interactions underlying them. Birds are interesting to study for a variety of reasons, including the fact that more is known about the endocrine correlates of paternal behavior in birds than in mammals. A discussion of parental behavior in mammals follows.

Parental Behavior in Birds

Birds show enormous diversity in parental behavior (Rosenblatt, 2003). Some avian species are so-called nest parasites that never engage in parental behavior. Female cuckoos and cowbirds deposit their eggs secretly into the nests of other birds, and the unknowing "adoptive" parents provide parental care for their "guests," incubating the eggs and feeding and protecting the hatchlings. In other avian species, such as chickens and ducks, only the female provides parental care. In some rare cases, only the male provides parental care. Mallee fowl and jacana males build and maintain nests and incubate the eggs deposited in them. The females compete with one another to lay eggs in different males' nests. Such instances of sex role reversal are interesting to study because they provide clues about the selective pressures driving the evolution of parental care. Biparental care, in which both male and female birds provide virtually equivalent care to their offspring, is the most common pattern of avian parental care (Figure 7.5). Nearly 90% of avian species are socially monogamous, and in contrast to mammals, paternal care is a common avian trait.

Avian Parental Behavior: A Description

Parental behavior in birds typically includes nest building, incubating the eggs, brooding the newly hatched nestlings, and taking care of the young until they are

7.5 Biparental care is common in birds. In most bird species, both parents make approximately equal parental investments in the form of nest building, incubation, feeding, and protection of the young. Immature birds require nearly continuous food intake to survive and develop successfully, and in most cases two parents are required to meet their needs.

ready to live independently. The extent to which the young are cared for depends on their developmental state at hatching. Altricial young are generally helpless after hatching and require substantial attention, feeding, brooding, and protection, whereas precocial young generally require only supervision. Many fowl, including chickens, pheasants, and ducks, produce precocial young. Maternal care in chickens involves nest building, incubation, and broody behavior, which consists of clucking and hovering over the chicks and nest. The hen may stimulate the chicks to feed by pecking at grains herself; she calls attention to the potential food by emitting a species-specific sound. But generally, the chicks are fully capable of feeding themselves after a day or two. If the chicks are threatened by an intruder, a broody hen will chase it with her wings extended and emit loud squawking sounds. Birds that produce altricial young, such as robins (*Turdus migratorius*) and starlings (*Sturnus vulgaris*), also build nests and incubate eggs, but they must provide their newly hatched young with food for several weeks. Regardless of whether altricial or precocial young are produced, many birds (sometimes both sexes) develop a *brood patch* on the breast, which loses feathers and becomes highly vascularized during incubation, facilitating heat transfer from the parent to the egg.

Males contribute some parental care in approximately 60% of avian subfamilies; in 20% of these subfamilies, the males provide virtually all of the parental care (Buntin, 1996). Males of some avian species, such as roosters, provide little or no paternal care. Males of other species, such as flycatchers, titmice, pigeons, and doves, are more or less equal partners in rearing their offspring. Male spotted sandpipers and Wilson's phalaropes, like male jacanas and mallee fowl, provide all the parental care to their young. Males may assist in one, several, or all of the parental tasks (Silver et al., 1985).

Why is paternal care common among birds? At an ultimate level of causation, as we have seen, males care for their offspring when they can better increase their own reproductive success by continuing to invest in those offspring than by seeking additional mates and fathering additional offspring. Most young birds are helpless at hatching and require constant food and warmth to develop sufficiently to leave the nest in just a few weeks. If males did not help to feed them, the hatchlings would die, and the males' fitness would suffer. In most cases, male birds are as capable as females of providing parental care in the form of nest construction, incubation of the eggs, and feeding of the young. This ability of avian fathers to feed their young—usually by regurgitating the results of recent foraging trips—contrasts sharply with most mammalian species, in which only the mother can meet the nutritional demands of the infants.

In some avian species, care of nestlings is provided by several male parental assistants, called **alloparents** or "helpers at the nest" (Emlen, 1978; Skutch, 1935), as well as by the parents. These helpers are usually elder brothers that are unable to set up their own breeding territories because of scarce resources, and they increase their reproductive fitness indirectly by helping their younger siblings.

Behavioral endocrinologists have focused their studies on two groups of birds—namely, the Galliformes (e.g., chickens) and the Columbiformes (e.g., pigeons and doves)—that are probably the two most atypical representatives of the class Aves (Buntin, 1996). Chickens have been studied because of their economic value, even though galliform birds are highly unusual in that only the mother provides any care. The pattern of avian parental behavior on which most endocrine investigations have focused, however, is one of the rarest: a unique mode of providing food to the young observed only among members of the family Columbidae. In this family, both sexes engage in full parental care. The male brings nest material to the female, who does most of the nest construction. Both parents incubate the eggs, and both help feed the young, which are called squab, after they hatch. So far, this pattern of behavior is not unusual among birds. What makes the pigeons and doves unique is that both parents produce *crop milk*, which is fed to the young. The crop milk, which resembles small-curd cottage cheese, is produced in a specialized exocrine gland called the crop sac, from which it is regurgitated to feed the squab (Figure 7.6) (Horseman and Buntin, 1995). Immediately after hatching, the squab are fed crop milk exclusively, but a mixture of seeds, insects, and crop milk is provided by the parents as the squab grow older. This unique avian adaptation of providing the young with "milk" is sufficiently similar to mammalian behavior to suggest to early endocrinologists that common underlying physiological mechanisms may be involved. Thus, pigeons and doves have proved to be attractive avian animal models in which to study hormonal effects on parental behavior.

7.6 A young wood pigeon feeds on crop milk, which is produced by both males and females in response to stimulation by prolactin. As the squab get older, a greater and greater proportion of the adult diet of seeds and insects is mixed with the crop milk until eventually only the adult diet is consumed, and parental crop milk production wanes.

Endocrine Correlates of Avian Parental Behavior

Parental behaviors, often performed by both mothers and fathers, are extensive among birds (Rosenblatt, 2002). The hormonal correlates of avian parental behaviors are described in the following sections.

AVIAN MATERNAL BEHAVIOR Which hormones mediate maternal behavior in birds? Hens exhibit maternal behavior, or broodiness, by making "clucking" vocalizations and by persistent incubation or "nesting" (Figure 7.7). The term *broodiness* can refer either to sitting on eggs in a nest or, more commonly, to protecting, covering, and warming the young under the wing. When hens become broody, they stop laying eggs, so there has been much research among poultry scientists aimed at preventing broodiness in order to increase egg production. Nearly 90 years ago, it was discovered that blood serum from a broody hen could induce a nonincubating hen to sit on a clutch of eggs (Leinhart, 1927). Oscar Riddle and his colleagues provided compelling evidence that prolactin induces broodiness in pigeons and chickens (Riddle et al., 1935a,b). Increased blood concentrations of prolactin are associated with broodiness in all female birds studied (Figure 7.8; see Goldsmith, 1983; Riddle et al., 1935b; Silverin and Goldsmith, 1983). Even in nest parasite species such as cowbirds, which never show broodiness or any other parental behavior, blood prolactin concentrations increase after egg laying (Rissman and Wingfield, 1984). Because the parental behavior of cowbirds differs

7.7 An extreme example of broodiness. After maternal behavior has been initiated by hormones, virtually any contact comfort, even when provided by the young of another species, is sufficient to maintain normal broody behavior in the hen.

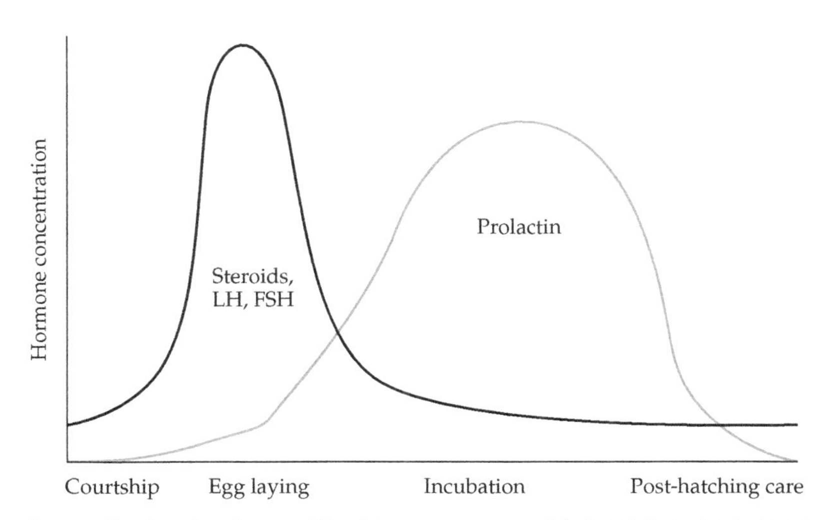

7.8 Generalized endocrine profile of temperate zone birds. While individual birds in a population may display different concentrations of reproductive hormones on any given sampling date, if these blood samples are arranged according to reproductive activity, a clear pattern of hormone concentration and reproductive function is observed. Sex steroid hormones usually increase coincident with the onset of courtship behavior, peak during the time of egg laying, and rapidly decrease to baseline levels prior to incubation. Prolactin concentrations begin to increase at the time of egg laying, remain high during incubation, then drop off gradually to baseline levels during post-hatching care. After Ball, 1991.

BOX 7.2 Daniel S. Lehrman

Daniel S. Lehrman made major contributions to behavioral endocrinology, particularly by untangling the complex interactions among social, behavioral, and hormonal stimuli that control courtship behavior in ring doves (e.g., Lehrman, 1965). An avid birdwatcher, Lehrman worked as a volunteer research assistant at the American Museum of Natural History for Dr. G. K. Noble while completing his undergraduate education at the City College of New York. At the museum, he was encouraged by Frank Beach and Ernst Mayr to study animal behavior. After completing Army duties, he returned to the museum in 1946 and met Theodore C. Schneirla, a well-known comparative psychologist, with whom he completed a doctoral degree in 1954. Lehrman joined the faculty at Rutgers University and founded the Institute of Animal Behavior in 1958; the institute has been an important research establishment and home to many prominent behavioral endocrinologists.

Despite his seminal work in behavioral endocrinology, Lehrman was probably most widely known for firing a major salvo at the European ethologists in 1953, in his "A Critique of Konrad Lorenz's Theory of Instinctive Behavior." Ethologists focused on "instinctual" behaviors, whereas comparative psychologists focused on learning. Although the publication of this biting attack on Lorenz's notions of "hard-wired" behavioral programs initially caused a number of caustic debates between North American comparative psychologists and European ethologists, the net effect of the paper was to bring the two schools together, setting up an eventual synthesis between the two scientific disciplines (Dewsbury, 1984).

Lehrman was a member of the National Academy of Sciences and is remembered by his students for his outstanding ability to communicate his ideas to scientists and nonscientists alike: he was famous for jumping onto tabletops and performing enthusiastic imitations of courting ring doves bowing and cooing.

Daniel S. Lehrman (1919–1972). Courtesy of Rae Silver.

substantially from that of other birds despite the similar hormonal profile, it is plausible to suggest that these behavioral differences are due to differences in the sensitivity of the central nervous system to hormones; experimental evidence supporting this proposition will be presented below.

The vast majority of studies on endocrine correlates of avian maternal behavior have been conducted on ring doves (*Streptopelia risoria*). Female ring doves, which were studied extensively by researcher Daniel S. Lehrman (Box 7.2), undergo a stereotyped sequence of mutually exclusive behaviors during their reproductive cycle (Lehrman, 1965). The cycle begins with courtship, which is followed by nest building, incubation, feeding the young, and finally resumption of courtship behavior. Changes in hormone concentrations have been found to correlate with the different stages of this cycle (see Figure 5.35). Incubation is initially evoked by progesterone (against a background of high estradiol concentrations), but around mid-incubation, it is sustained by prolactin, which is secreted in response to ventrum (belly) stimulation from contact with the eggs. Ovariectomy eliminates nest building and incubation behavior. Injections of either estradiol or progesterone fail to restore nest building or incubation behavior, but treatment with both steroid hormones restores these behaviors to normal values (Cheng and Silver, 1975). Prolactin is critical for stimulating brooding and development of the crop sac in ring doves (Figure 7.9) (Silver, 1978). Prolactin secretion wanes by the time the squab are 20 days old, at which time the mother no longer participates in the feed-

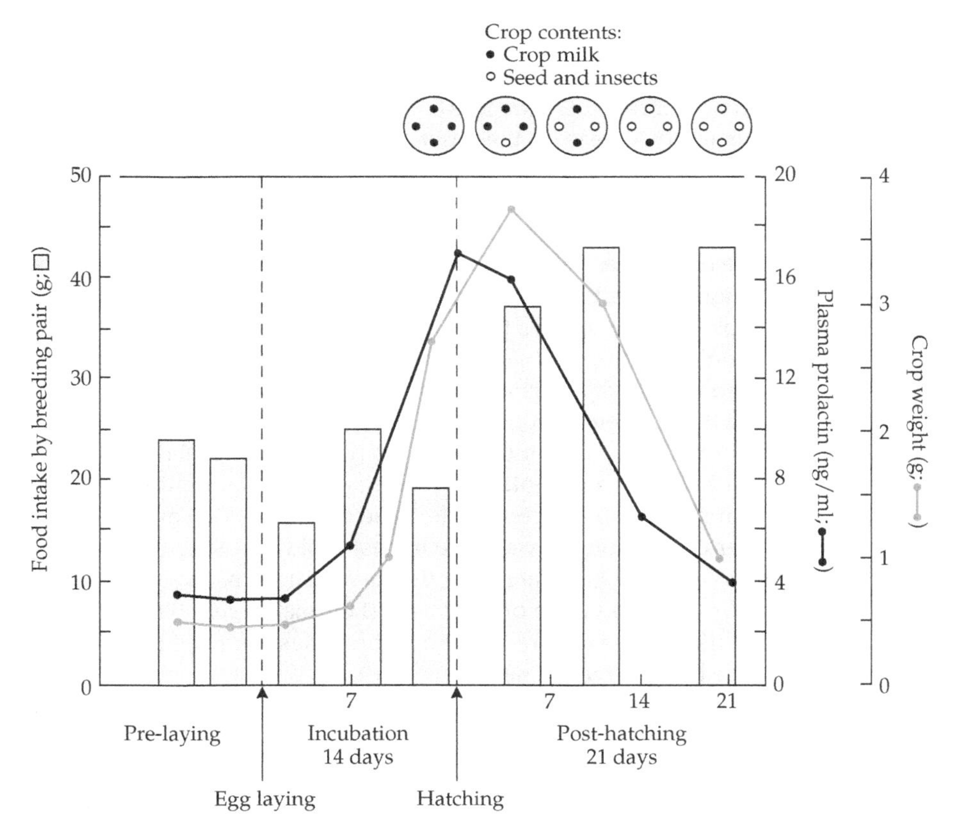

7.9 Prolactin concentrations, crop development, crop contents, and parental food intake are related in ring doves. Plasma prolactin concentrations increase post-hatching and are associated with elevated food consumption by the breeding pair, presumably to support foraging for insects for the squab. As prolactin begins to fall 10–14 days post-hatching, the contents of the crop shift from mainly crop milk to mainly seeds and insects. After Horseman and Buntin, 1995.

ing of her brood. Although the female and male begin to court and build a new nest at this time, the male continues to secrete prolactin and provide crop milk to the squab. Thus the onset, maintenance, and termination of parental behavior in pigeons and doves appear to depend primarily on prolactin concentrations.

AVIAN PATERNAL BEHAVIOR Despite the wide variety of parenting patterns found among male birds that provide parental care, there have been few studies of the endocrine correlates of avian paternal behavior in noncolumbiform birds. Early research on the endocrine bases of parental behavior focused on sex differences in the expression of parental care in species in which males normally do not provide any paternal assistance. Roosters typically do not exhibit paternal care, but a castrated rooster, or capon, becomes broody almost to the same extent as a hen when provided with foster chicks (Goodale, 1918). Tom turkeys also do not display paternal care, but males restrained so that they are forced to sit on a clutch of eggs eventually show incubation behavior (Taibell, 1928).

Generally, nest building and incubation behaviors by male passerine (i.e., perching) birds are observed after courtship and mating and correspond to a sharp decline in blood concentrations of androgens and sometimes progesterone. Prolactin concentrations generally increase at the onset of incubation behavior (Ball, 1991; Silverin, 1990). However, in field populations of sparrows, which may sire multiple broods, paternal behavior is observed while the birds maintain intermediate, and occasionally high, blood concentrations of testosterone (Wingfield and Moore, 1987). These elevated androgen concentrations are necessary to maintain the territorial defense behaviors required for successful rearing of the young. Males of species that display sex role reversal, such as the spotted sandpiper (*Actitis macularia*; see Box 3.1) and Wilson's phalarope (*Phalaropus tricolor*), exhibit the typical pattern of reduced steroid and increased prolactin concentrations; usually their prolactin concentrations exceed those of female conspecifics (Oring et al., 1986a,b).

As with maternal care, most of the work on hormonal correlates of avian paternal behavior has been accomplished on ring doves in laboratory settings (Buntin, 1996). During the reproductive cycle, male ring doves first exhibit courting behavior; after copulation, they engage in nest building. After the eggs are laid, males begin to incubate them, and after the squab hatch, males assist in feeding them for the first 3 weeks or so of their lives (Lehrman, 1965). These behaviors occur sequentially (see Figure 7.9), and hormones are important in the transition from one behavior to the next.

Although testosterone is required for courtship, this steroid hormone is not necessary for the onset of nest building behavior in male ring doves. Castrated males treated with daily injections of testosterone either courted females or built nests; which behavior they chose was solely dependent on the behavioral responses of the females (Silver, 1978). Thus, in contrast to that of females, maintenance of male nest building behavior is independent of hormonal status. This example makes the interesting point that virtually identical behaviors can have markedly different physiological bases in the two sexes. Apparently, stimuli from the female, rather than hormonal changes, also induce the male to begin incubation behavior (Silver, 1978). In both sexes, prolactin is required to stimulate broody behavior and production of crop milk (Lehrman and Brody, 1961). Testosterone, but not progesterone, appears to be necessary for maintaining incubation behavior in castrated male doves (Ramos and Silver, 1992; Lea et al., 1986).

The endocrine correlates of alloparenting have been studied in Florida scrub jays (Schoech, 2001). In male helpers, testosterone and prolactin concentrations were lower than those of fathers (Figure 7.10). Although the prolactin values of

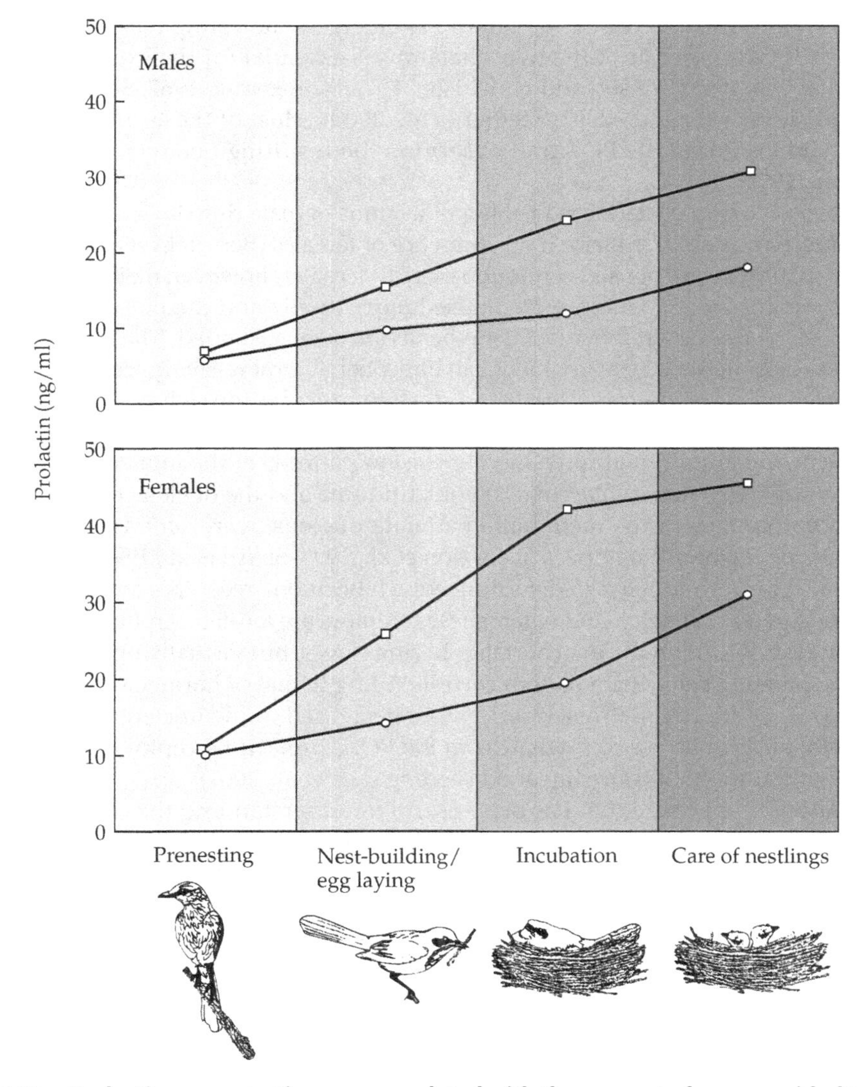

7.10 Prolactin concentrations are correlated with the amount of care provided to nestlings by Florida scrub jays. Breeding males and females (squares) have higher prolactin values than helpers (circles). After Schoech, 2001.

female helpers showed a pattern similar to those of male helpers, their blood estradiol patterns were more complicated. During the pre-nesting period, breeding females had higher estradiol concentrations than helper females, but female helpers' estradiol concentrations increased when the breeders were building nests and incubating. This increase in estradiol may prepare the helper females to strike out for their own breeding opportunities (Schoech, 2001).

NEURAL REGIONS ASSOCIATED WITH AVIAN PARENTAL BEHAVIOR Very few studies have investigated the neural substrates of avian parental behavior. However, in the few studies performed, a consistent pattern has emerged, indicating that the preoptic area (POA) anterior to the hypothalamus is essential for the expression of parental behavior (Erikson and Hutchison, 1977; Komisaruk, 1967; Slawski and Buntin, 1995; Lea et al., 2001; Ziegler et al., 2000). Most of these studies have involved lesioning the POA and observing the resulting behavioral deficits (Buntin, 1996).

Lesions of the posterior medial hypothalamus of male ring doves significantly reduced courtship behavior in the presence of females (Bernstein et al., 1993). If lesioned males were housed continuously with females, however, their courtship behaviors recovered; if they were housed individually and then introduced to females, the deficits in their courtship behavior were sustained. These findings suggest that the social environment can affect behavioral recovery from specific brain lesions. The neural circuitry underlying incubation behavior remains unknown (Buntin, 1996).

By using autoradiography and expression patterns of the immediate early gene *c-fos*, it was determined that the preoptic area and the nucleus tuberis, the avian homologue of the mammalian arcuate nucleus, were activated during parental behavior in ring doves (Georgiou et al., 1995; Sharp et al., 1996). Studies have also been conducted to determine which hormone receptors are activated during parental behavior, and where these receptors are located. Androgen receptor expression is high during courtship in ring doves, but virtually undetectable during parental behavior, a pattern corresponding to that of hormone concentrations (Lea et al., 2001). Prolactin receptors decreased in the nucleus tuberis in parental birds of both sexes, whereas in the POA, prolactin receptor expression was high throughout courtship and brooding (Lea et al., 2001).

Another approach that has been useful in understanding the neural substrates of avian parental behavior has employed comparisons of prolactin receptors in the brains of bird species that differ in their parental behavior. This comparative approach has yielded data suggesting that differences in parental care are mediated not by differences in hormones per se, but by differences in receptor numbers. For example, prolactin binding sites were compared in the brains of female cowbirds, red-winged blackbirds, and European starlings. Cowbirds, as you will recall, do not show parental care, but they exhibit prolactin profiles similar to those of other birds. In some brain areas, the pattern of prolactin binding was similar among all three species; however, the cowbirds displayed reduced prolactin binding in the POA compared with the species that exhibit parental care (Ball, 1991). Thus, it seems reasonable to conclude that species differences in parental behavior may be mediated by differences in target tissue sensitivity, rather than differences in blood concentrations of hormones.

In another study, male dark-eyed juncos (*Junco hyemalis*) were implanted with testosterone-filled or empty capsules, then evaluated for paternal behavior. Specific binding of radiolabeled prolactin was observed at brain sites previously implicated in the regulation of avian paternal behavior; namely, the POA, the ventromedial nuclei (VMN), and the paraventricular nuclei (PVN) (Schoech et al., 1998). Testosterone-treated males reduced their parental contributions to their offspring, but prolactin concentrations were not affected by the elevated testosterone concentrations. Furthermore, no differences were observed in the capacity for prolactin binding. These results indicate that testosterone does not block parental behavior by suppressing prolactin concentrations in the blood or by reducing prolactin receptor binding in brain regions that mediate paternal behavior (Schoech et al., 1998).

One important question regarding the hormone–brain relationship in avian parental behavior is how prolactin, a relatively large protein hormone, crosses the blood–brain barrier to interact with prolactin receptors on neurons. Short-term studies using intravenous injections of radiolabeled prolactin into the blood of ring doves revealed that prolactin does cross the blood–brain barrier and accumulates in brain neurons that have prolactin receptors (Buntin et al., 1993). Presumably, some sort of transport system exists to move prolactin into the brain, where it interacts with receptors on the surfaces of neurons in the hypothalamus and preoptic area (Buntin, 1996).

Parental Behavior in Mammals

Mammals have evolved specializations for taking care of their young (Bridges, 1990; Rosenblatt, 2003). The development of mammary glands, which provide nourishment to the young after they are born, makes mammals unique among vertebrates. Parental care—specifically, maternal care—is a particularly important mammalian characteristic (Numan and Insel, 2003).

Mammalian Maternal Behavior

Among the mammals, species exist that produce offspring at virtually every developmental stage. The **marsupials** have specialized in the production of altricial young. These mammals have a pouch that contains the mammary glands and also serves as a receptacle for the young. Most marsupials, including wallabies and kangaroos, have temporary or permanent pouches in which the embryonic young are carried for several months. When the young reach a certain species-specific age, they leave the pouch, but return to the mother to nurse. Other marsupial species, such as opossums (*Didelphis virginia*) and koalas (*Phascolarctos cinereus*), bear young that cling to the mother's fur outside of the pouch and are carried in this manner as the mother forages for food (Figure 7.11). These patterns of maternal care are adaptations based on the mothers' need to remain mobile, both to

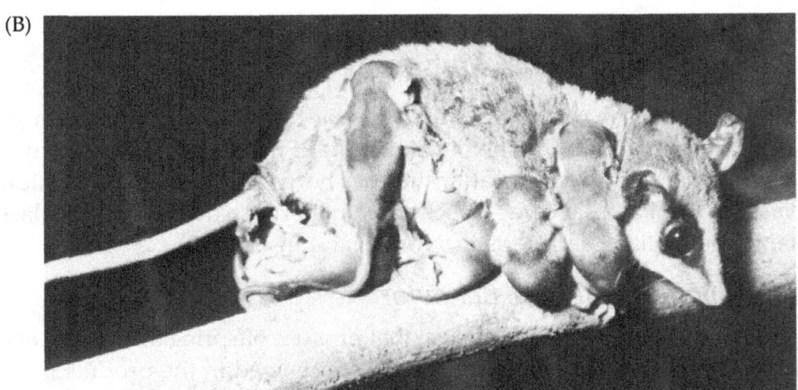

7.11 Marsupials have specialized in the production of altricial young, and have developed means for the mother to stay mobile and forage while the young are nursing. (A) Some marsupials, such as kangaroos, have a pouch in which the mammary glands are located, and which also serves as a carrier for the young. (B) In other species of marsupials, such as opossums and koalas, the young stay in the pouch only briefly and are carried by the mother, clinging to her fur, until they are weaned.

obtain food in their usually barren habitats and to escape predators (Sharman, 1970; Tyndale-Biscoe and Renfree, 1987). The young, in turn, are adapted to remain attached to the mother, either in the pouch or outside of it; should they become dislodged, the mother does not attempt to reattach them, and they invariably perish. Thus, marsupials do not invest in gestation, but instead invest in postpartum maternal care, which does not include extensive maternal *behavior*. Their unreliable habitat favors this arrangement because marsupial mothers can jettison a pregnancy, or even their young, if conditions turn bad. Because of the short gestation period, they can replace these young quickly when conditions improve.

Another group of mammals, the **monotremes**, produce young that are born at an even earlier developmental stage than those of the marsupials: as eggs. These odd creatures are considered to be modern representatives of very ancient forms of mammals that have retained the reptilian egg-laying trait (oviparity). In terms of maternal behavior, monotremes represent a transitional stage between reptiles, which lay eggs and may or may not bury them, and the much more solicitous modern mammalian species (Burghardt, 1988). As with the marsupials, there are two basic patterns of maternal care among the monotremes (Nowak and Paradiso, 1983; Rosenblatt et al., 1985). The first is evident in the duck-billed platypus (*Ornithorhynchus anatinus*) (Figure 7.12). The female of this species constructs a grass-lined burrow in which she lays her eggs. She then seals herself in the burrow with the eggs and incubates them until they hatch. She cares for the young and provides them with milk, which drips from the hair surrounding her mammary glands. Females of this species do not develop a pouch. The second pattern is seen in the spiny anteaters (*Tachyglossus aculeatus* and *Zaglossus bruijni*), which produce a single egg and carry it in a special pouch that develops at the time of egg delivery. The offspring, when it hatches, lives in the pouch and obtains milk from a milk-secreting gland that lacks a nipple.

The widest variety of maternal care patterns has evolved among the **eutherian** (*eu*, "true"; *therion*, "beast") mammals, which have placentas during pregnan-

7.12 Maternal care in monotremes. This duck-billed platypus is shown with her two altricial young in her burrow. She provides the young with milk, which drips from the hair surrounding her mammary glands. Courtesy of S. Jasper.

cy. There are three basic patterns of maternal care among eutherian mammals, initially described by Rosenblatt et al. (1985), which are related to several characteristics of the species, but mainly reflect the developmental status of the newborn. Eutherian mammals, in contrast to marsupials and monotremes, tend to live and reproduce in more stable and reliable habitats, and therefore have invested in relatively long gestations, which allow for greater development of the young before birth. Mechanisms have evolved in all mammalian species to provide parental care regardless of whether or not maternal behavior is present.

In the first pattern of maternal care, the mother provides her offspring with food, care, and shelter. This pattern is characteristic of dogs, as was described above (see Figure 7.4), as well as other carnivores, rats and most other rodent species, and insectivores (such as shrews) (Figure 7.13). Females exhibiting this pattern of mater-

7.13 Rat maternal care has three major components. Like mother dogs, rat mothers ("dams") must engage in three major behaviors in order for their altricial offspring to survive. First, they must lick their pups after birth to clean off the amniotic fluid and stimulate elimination of wastes. Second, they must adopt a nursing posture, huddling above the pups to allow access to the nipples and to provide warmth and protection. Finally, they must bring the pups back to the nest if they wander away. After Alberts and Gubernick, 1990.

nal care give birth to altricial young, which require a great deal of care. The mother must build a nest and deposit her young there, and she must visit her offspring frequently to feed them and protect them from predators.* Recall that a mother dog usually chooses a site for the nest and constructs it before the young are born.

After parturition, a mother dog eats the placenta (placentophagia also occurs in other mammalian species; see Figure 7.24) and licks the amniotic fluid off the puppies. In dogs, and probably in other mammals as well, this cleaning behavior is crucial for maternal acceptance of the newborns: if pups are removed from the nest immediately after birth, washed, and returned to the mother, she will reject them and will not provide any maternal care (Abitbol and Inglis, 1997). A mother dog also nurses her young and continues to lick them to stimulate various physiological processes.

In dogs and many other mammalian species, the mother will retrieve the young if they wander from the nest, but this is not always the case. Rabbit and hare mothers, for example, will not retrieve young that leave the nest. Mammalian mothers that retrieve their offspring often exhibit maternal aggression, attacking any animal that approaches their nest or young.

The many maternal behaviors described above can be classified according to whether or not a particular behavior is directed toward pups (Bridges, 1996). In this scheme, nest building, consumption of the placenta, and defense of pups are classified as non-pup-directed; pup-directed behaviors include grouping the pups together in the nest, huddling with them to provide warmth, retrieving them if they wander, licking their anogenital regions, and providing other forms of tactile stimulation. There may be different mechanisms underlying pup-directed versus non-pup-directed behaviors (González-Mariscal and Poindron, 2002).

In a second pattern of mammalian maternal care, females bear precocial young that are capable of a high degree of independent activity when they are born, though they may be confined to a nest at birth and for a few days afterward. Some hoofed mammals, such as white-tailed deer, cows, and Thomson's gazelles, give birth to one or two precocial offspring. These species are characterized as **hider-type** animals because the young remain hidden at the nest site for 7–10 days (Geist, 1971) (Figure 7.14A). The mother spends most of her time away from the nest foraging for food, but visits her offspring to nurse them. Hider-type offspring eventually grow mature enough to leave the nest and follow the mother about. Other hoofed and non-hoofed species, including sheep, goats, elephants, and many species of whales, have offspring that follow the mother about from birth; these species are characterized as **follower-type** animals (Figure 7.14B). The mother nurses her young and maintains constant vigilance, leading them away from danger and protecting them from predators.

In many species that bear precocial young, maternal care is associated with an exclusive bond that forms between the mother and her offspring (Gubernick and Klopfer, 1981). This bonding occurs soon after birth, which takes place in a

*Rabbits and hares show a similar pattern, except that they typically visit their litters only once daily to nurse the young; the rest of the day is spent foraging for food.

(A)

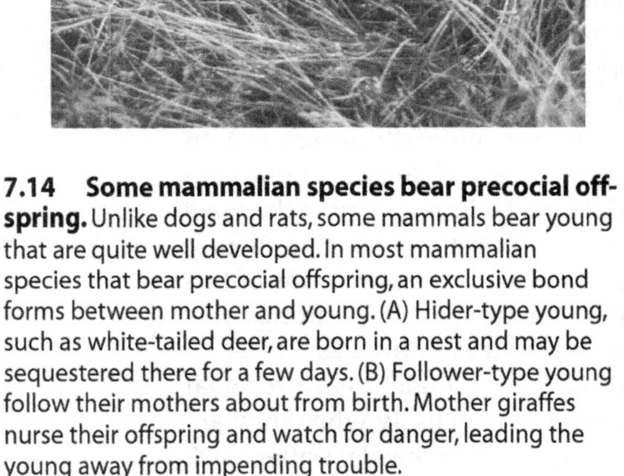

(B)

7.14 Some mammalian species bear precocial offspring. Unlike dogs and rats, some mammals bear young that are quite well developed. In most mammalian species that bear precocial offspring, an exclusive bond forms between mother and young. (A) Hider-type young, such as white-tailed deer, are born in a nest and may be sequestered there for a few days. (B) Follower-type young follow their mothers about from birth. Mother giraffes nurse their offspring and watch for danger, leading the young away from impending trouble.

secluded location away from the social group. Sheep and goat mothers that have just given birth emit characteristic low-pitched bleats in the presence of the newborn, which appear to attract the offspring to their mother. In these species, the mother initially licks her offspring and allows them access to her udders. If a mother sheep is separated from her newborn offspring, she does not form a maternal bond with that specific lamb. She continues to behave maternally for about 24 hours, but is nonexclusive about which lambs are permitted to nurse. Thus, there appear to be two distinct processes in sheep maternal behavior: (1) maternal responsiveness and (2) maternal selectivity (Lévy et al., 1996). Both processes are dependent on cues from the lamb; anosmia (inability to smell) interferes with the ewe's selective responses to her own lambs, but not with maternal responsiveness to lambs in general (Poindron and Le Neindre, 1980).

Many other mammalian species bear young that are neither precisely precocial nor precisely altricial; such offspring are usually described as semi-precocial. The offspring of humans and several other primate species are essentially helpless

7.15 Many primate species produce semi-precocial young. Female orangutans carry their infants with them continuously as they forage.

at birth, but they can cling to the mother in order to move about with her. Female orangutans carry their infants with them continuously for the first year of life (Figure 7.15). In other species, mothers alternate between carrying the young and placing them in a nest or crib; and in still other species, the mother may deposit her young in a tree nest, where they remain during their early development. Galagos hide their young while foraging during the day and return to carry them during the night.

Mammalian Paternal Behavior

Among mammals, paternal care is relatively uncommon, but it is observed in certain carnivore, rodent, and primate species (Kleiman and Malcolm, 1981). If it is assumed that paternal behavior occurs when it increases male reproductive success, then there should be obvious and compelling selective factors among the species displaying paternal care. Thus far, no such generalities have emerged, except the assumption that females would be unable to rear offspring successfully without the additional support of the male. Another supposition is that certainty of paternity must be very high in order for paternal care to occur (Werren et al., 1980). Males clearly would not increase their reproductive success by relinquishing the opportunity for additional matings in favor of rearing the offspring of other males. It could also be argued that continued proximity to the female after mating improves the male's chances of paternity and is a factor in fostering paternal care (Gowaty, 1996). Therefore, paternal behavior should be observed only in species that have a high degree of mating exclusivity between males and females.

Historically, paternal care was reported in fewer than 6% of rodent genera, and even in those relatively rare cases, the behavior was poorly documented with small sample sizes (for example, one or two individuals) in artificial laboratory environments (Gubernick and Alberts, 1987). These findings led some to question how often, if ever, paternal behavior is actually exhibited among mammals. Although it remains clear that paternal care is rare among mammals, recent studies have convincingly demonstrated paternal behavior in a number of species. Among rodents, paternal care has been documented in laboratory studies of several species of deer mice (*Peromyscus*) (Gubernick and Alberts, 1987; Hatton and Meyer, 1973; McCarty and Southwick, 1977), Syrian hamsters (*Mesocricetus aura-*

7.16 Male California mice display paternal behavior. With the exception of lactation, males of this species exhibit the same parental behaviors as females, and to the same extent. This species is reported to be monogamous in both the laboratory and the field. Courtesy of David Gubernick.

tus) (Marques and Valenstein, 1976), house mice (*Mus musculus*) (Priestnall and Young, 1978), grasshopper mice (*Onychomys torridus*), and two vole species (*Microtus ochrogaster, M. pinetorum*) (Gruder-Adams and Getz, 1985; Hartung and Dewsbury, 1979; Oliveras and Novak, 1986; Thomas and Birney, 1979; see Elwood, 1983 for review). California mice (*Peromyscus californicus*), for example, exhibit a biparental system of parental care. These animals form long-term pair bonds in the wild and in the laboratory and remain together throughout both the breeding and nonbreeding seasons (Ribble and Salvioni, 1990). Furthermore, these animals are exclusively monogamous (Ribble, 1990). Males and females exhibit the same parental behaviors to the same extent, with the obvious exception of lactation (Figure 7.16). Males spend as much time in the nest as females, help build nests, and carry young. Males also groom the pups and lick the anogenital regions of the pups to stimulate urination (Gubernick and Alberts, 1987).

Another well-studied biparental species is the common marmoset (*Callithrix jacchus*), a New World primate that also displays a monogamous mating system. The male assists during birth, chews food for the babies, and except during nursing sessions, he always carries the young (Hampton et al., 1966). Other nonhuman primates, including tamarins (*Leontocebus midas*), Japanese macaques (*Macaca fuscata*), gibbons (*Hylobates lar*), and siamangs (*Symphalangus syndactylus*), also show paternal involvement in the care of the young (Yogman, 1990; Zeigler et al., 2000). In humans, individual males vary from providing no parental care to providing substantial care to their young. In the limited confines of the laboratory setting, many species of rodents and primates exhibit paternal behavior even though they never display these behaviors in nature (Elwood, 1983; Redican and Taub, 1981; Suomi, 1977).

Endocrine Correlates of Mammalian Parental Behavior

The endocrine correlates of the onset, maintenance, and termination of maternal behavior in rats are probably the best understood of any species. Because the onset of maternal behavior coincides with the birth of the young, the hormones associated with pregnancy and lactation were historically considered likely candidates for causing maternal behavior. In many respects, lactation is the key to mammalian maternal behavior, but it is just one element in a complex system. Because prolactin and lactation are closely linked, prolactin was initially considered to be the critical hormone underlying maternal behavior (Riddle et al., 1935a), but a number of endocrine manipulations have since revealed that several additional hormones are involved; namely, oxytocin, β-endorphin, cholecystokinin, prostaglandins, relaxin, progesterone, and several estrogens (Bridges, 1996; Nelson and Panksepp, 1998). Virtually all mammalian species display elevated estrogen concentrations around the time of birth, and these hormones are important, if not critical, for the onset of maternal behavior (González-Mariscal and Poindron, 2002).

Pituitary prolactin and placental lactogen (a prolactin-like hormone) play leading roles in maternal behavior, but how they are related to other hormones has not yet been specified (Rosenblatt, 1990). Although some straightforward relationships between hormone concentrations and behavior have been reported, there are many subtle endocrine interactions as well. For example, although the generation of maternal behavior seems to be dependent on hormones, some components of maternal behavior are virtually independent of hormonal influences.

SENSITIZATION OR PUP INDUCTION An experimental paradigm often used to explore the occurrence and hormonal correlates of rat parental behavior is the presentation of foster pups. Researchers present pups of differing ages (typically 1–10 days of age) to adult females (and males) in different endocrine states and observe whether, and how quickly, the adults begin to behave parentally toward them. During the first day of exposure to pups, castrated or intact male rats normally ignore them, or in some cases attack them. Adult females that are not pregnant or pseudopregnant also ignore or attack the pups initially. Pregnant females, however, soon begin to behave maternally toward foster pups if they are first exposed to them during the last few days of their pregnancy (typically starting on day 16–18 of the 22-day pregnancy) (Figure 7.17). Pseudopregnant rats may also show a rapid onset of maternal behavior at the end of their 12–13-day pseudopregnancy. Once a female rat has weaned pups of her own, she retains an enhanced sensitivity to rat pups and will show a reduced latency to behave maternally in future tests (Bridges, 1990).

If foster pups are presented to an adult female rat that has never been pregnant (a nulliparous female) for several hours each day, then she begins to behave maternally after 5 or 6 days. This phenomenon is called **concaveation** (Rosenblatt, 1967; Wiesner and Sheard, 1933), **sensitization** (LeBlond, 1938), or pup-induced parental behavior (Fleming et al., 1996). All the behaviors, except maternal aggres-

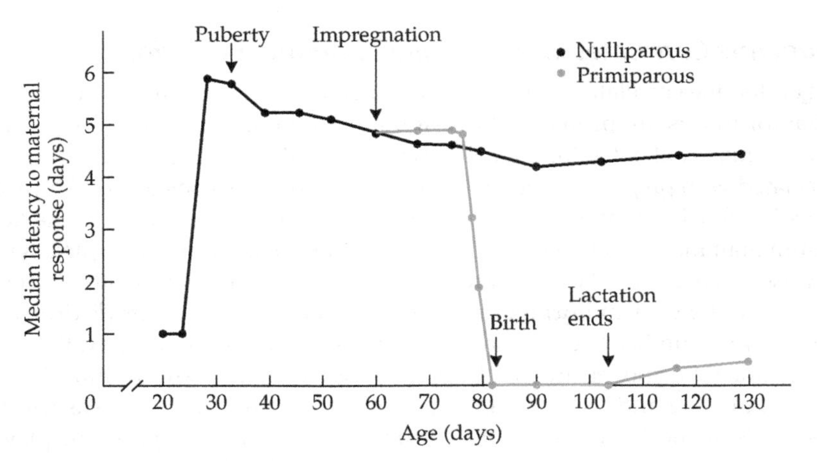

7.17 Latency to maternal behavior in rats. Prior to puberty, female rats behave maternally on the first day of exposure to foster pups. After puberty, 5–6 daily exposures to novel pups are required to elicit maternal behavior; that value gradually drops to about 4 days as rats age. Pregnant rats do not show maternal behavior during most of the gestation period, but during the last few days of pregnancy there is a rapid drop in the latency to display maternal behavior in response to pups. This sensitivity to pups remains high for some weeks after weaning. After Bridges, 1990.

sion in defense of the nest, that are typically observed in a rat that has just given birth are observed in these "sensitized" females; they even adopt a nursing posture, albeit an imperfect one in the absence of suckling by the pups (of course, they do not lactate without the proper hormonal priming). Remarkably, male rats also show this response to exposure to rat pups, although they take more time to achieve the level of maternal behavior observed in females, and the behavior is less consistent than that shown by females (Lubin et al., 1972; Mayer et al., 1979). Concaveation appears to have little or nothing to do with hormone concentrations, because the effect persists after the removal of the gonads, pituitary, or adrenal glands; the behavioral changes appear to be induced by the mere presence of the pups (Rosenblatt, 1967). An examination of the pattern of *fos* activation in the presence of pups revealed that maternal behavior induced by concaveation activated the same brain regions in virgin females as in mothers (Kalinichev et al., 2000).

Mother rats, or *dams*, on the other hand, behave maternally toward their own pups as soon as they are born, and this rapid onset of maternal behavior is hormonally mediated. Normally, adult rats are fearful of pups; they approach them very tentatively, sniff them, and withdraw quickly. One way in which the hormones associated with pregnancy and lactation predispose the new mother to behave maternally is by reducing the fear associated with the presence of pups (Fleming, 1986; Fleming and Luebke, 1981), as we will see below. Continued exposure to her pups also appears to allow the dam to overcome her initial fear of the newborns as she becomes accustomed to them. Consistent with the hypothesis that anxiety or fear blocks the onset of maternal behavior is the observation that the onset of maternal behavior is facilitated in female rats that are treated with tranquilizers such as benzodiazepines (Hansen et al., 1985).

Males may require longer exposure to pups than females to behave maternally in concaveation tests because, as open field tests and other evaluations have shown, males are generally more timid than females (Gray and Lalljee, 1974). As noted above for dogs, rats that give birth by cesarean section (C-section) typically behave maternally toward their pups as soon as they recover from anesthesia if they receive some time with the pups. However, if C-sectioned or normally parturient mother rats are not permitted this initial interaction with their pups, maternal responsiveness to pups wanes during the first week after birth (Bridges, 1977; Orpen and Fleming, 1987). By the tenth day postpartum, dams show the same latencies to respond to foster pups as nulliparous females.

MOTIVATION AND MATERNAL BEHAVIOR Like mating behavior, parental behavior can be divided into two components: motivation and performance. Hormones affect both the motivation to engage in parental care and the performance of specific actions required for parental care.

To untangle the effects of hormones and brain regions on maternal behavior, several studies have been conducted on rats. For example, female rats that were either postpartum or cycling (but had previous experience as mothers) were given the opportunity to press a bar to gain access to a pup (Lee et al., 2000). Females were also exposed to pups in their home cage. Different groups of these female rats had previously received lesions of the MPOA, the lateral amygdala, or the nucleus accumbens; other rats with sham lesions served as controls. Both postpartum and cycling females with MPOA lesions showed a reduced rate of bar-pressing for pups compared with control females (Lee et al., 2000). In postpartum females, amygdala lesions also reduced bar pressing. Rats in all the lesioned groups displayed decreased maternal behavior in the home cage. Taken together, these results indicate that the MPOA is part of a circuit that mediates both stereotyped maternal behaviors and motivation to perform them, but they also suggest a dissociation of the mechanisms mediating expression of the species-typical maternal behavior and those mediating motivation (Lee et al., 2000).

In another recent series of studies, the maternal motivation of rat dams was tested by determining whether they preferred a chamber previously associated with either cocaine or pups. This process is called *conditioned place preference* and is used commonly to understand the rewarding properties of various stimuli, including addictive drugs. During the early postpartum period (i.e., 8 days after birth), most dams preferred the chamber that had previously held their pups. In contrast, later in the postpartum period (i.e., 16 days after birth), most dams preferred the chamber in which they had received a shot of cocaine (Mattson et al., 2001). On postpartum day 10, however, the numbers of dams that preferred the cocaine-associated and the pup-associated chambers were nearly equal. Thus, postpartum day 10 appeared to be a midpoint in the shift in motivation of rat dams or in the reward properties that rat pups had for dams (Mattson et al., 2001). Regardless of their preference, the dams displayed equivalent levels of maternal behavior (Mattson et al., 2003). These studies indicate that differences in motivational state can be observed even while performance of maternal behavior remains the same. The factors, endocrine or otherwise, that underlie this distinction remain unspecified.

EXPERIENTIAL INFLUENCES ON MATERNAL BEHAVIOR The effects of various hormones and neuropeptides on maternal behavior change as a function of experience. In general, experienced mothers are "better" mothers. Females that have given birth previously display maternal care toward foster pups within 1 day, regardless of their hormonal status (Bridges, 1996). Furthermore, rats undergoing their second pregnancy respond to pups with maternal care within 1 day during mid-gestation, in contrast to primiparous (pregnant for the first time) rats, which require 7 to 8 days (Bridges, 1978). Even hypophysectomy appears to have little effect on the maintenance of maternal behavior (Erskine et al., 1980a). Thus, hormones appear to be important in priming first-time mothers to behave maternally at the end of their pregnancy. Subsequently, experiential factors appear to mediate maternal care for future offspring (Bridges, 1996).

It seems that a "maternal memory" that depends on experience with pups is normally formed immediately postpartum, and that this memory serves to reduce the latency to display maternal behavior (Bridges, 1975). Alison Fleming and her colleagues tested this hypothesis by administering cycloheximide, a protein synthesis inhibitor that interferes with memory formation in other contexts, to rat dams immediately before or after allowing them a 2-hour postnatal maternal experience with their pups. They found that these treatments blocked the expression of later maternal behavior toward foster pups (Fleming et al., 1990a). In order to specify the neural circuits involved in maternal memory, two subregions of the nucleus accumbens, the shell and the core, were lesioned. Lesions of the shell region, either before or shortly after a brief maternal experience, substantially disrupted maternal memory (Li and Fleming, 2003b). However, lesions in this brain area had no effect if the dam had more than 24 hours of experience with pups. Lesions of the core of the nucleus accumbens did not affect maternal memory (Li and Fleming, 2003b). Furthermore, when cycloheximide was infused into the shell region of the nucleus accumbens (but not elsewhere) of dams immediately after a single 1-hour interaction with their pups, maternal memory was disrupted. These results indicate that the shell, but not the core, of the nucleus accumbens is part of the brain circuitry mediating maternal memory (Li and Fleming, 2003b).

Cycloheximide apparently blocks the formation of a rat dam's memory of her early experiences with her pups by preventing the consolidation of new information (Davis and Squire, 1984). Precisely what new information is normally being consolidated is not known, although there are many hypotheses. Most likely, an association between stimuli correlated with the pups and the maternal response is stored during normal attachment. We know that protein synthesis is necessary for the formation of new memories. This protein synthesis may represent new growth of neural processes or an increase in neurotransmitter or receptor production. Thus, the protein synthesis inhibitor might prevent the consolidation of information by blocking either structural changes in or the enhancement of the efficacy of certain neurons. Based on chemical lesion studies, intracerebroventricular injections, microdialysis, and gene manipulations, researchers believe that the neurotransmitters involved in maternal behavior include dopamine, norepinephrine, and serotonin (Bridges, 1996; Thomas and Palmiter, 1997; Nelson and Panksepp, 1998).

The Onset of Maternal Behavior

The natural endocrine profile of late pregnancy and pseudopregnancy in rats includes a precipitous drop in blood plasma concentrations of progesterone after a dramatic increase throughout pregnancy; a steady, gradual increase in blood estradiol concentrations; and an increase in prolactin concentrations at the end of pregnancy and immediately postpartum (Figure 7.18A; Bridges, 1990). This endocrine pattern is also evident in near-term pregnant mice (McCormack and Greenwald, 1974), rabbits (Challis et al., 1973), and sheep (Chamley et al., 1973). In contrast, pregnancy in Old World primates, including humans, is characterized by high concentrations of both estrogens and progesterone throughout pregnancy, followed by a precipitous drop in the concentrations of both steroids at parturition (Figure 7.18B; Coe, 1990; Warren and Shortle, 1990). Oxytocin and endorphins increase around the time of primate parturition; oxytocin is important in the smooth muscle contractions necessary for giving birth (Bridges, 1996). The role of opioids remains controversial; they may reduce pain during childbirth, and they may be involved in the mediation of maternal behavior. Both stimulatory and inhibitory roles of β-endorphin in maternal behavior have been reported (reviewed in Bridges, 1996; González-Mariscal and Poindron, 2002). Primates also show increases in concentrations of ACTH and cortisol throughout gestation. The elevation of cortisol concentrations may serve to lower progesterone concentrations in addition to suppressing immune reactions of the mother toward her fetus (Coe, 1990).

Research on the topic of endocrine induction of maternal behavior in rats has fallen into two general categories. One approach has attempted to induce maternal behavior in nulliparous females via the administration or removal of hormones associated with pregnancy and lactation. Techniques used in such endocrine induction studies have included hormone replacement therapy and the transfusion of blood between pregnant and nonpregnant animals. This approach has also been used in other mammals. The other type of study has attempted to induce maternal behavior by terminating pregnancy at various stages. The logic behind these pregnancy termination studies is that (1) hormones change during pregnancy, (2) maternal behavior is caused by the hormones associated with pregnancy, and (3) because pregnancy termination leads to the rapid onset of maternal behavior before its normal onset at parturition, correlating the onset of maternal behavior with the coincident endocrine profile will reveal the hormonal cause of these behaviors.

About 80 years ago, Calvin P. Stone exchanged the blood of a female rat that was behaving maternally with the blood of one that was not by connecting their

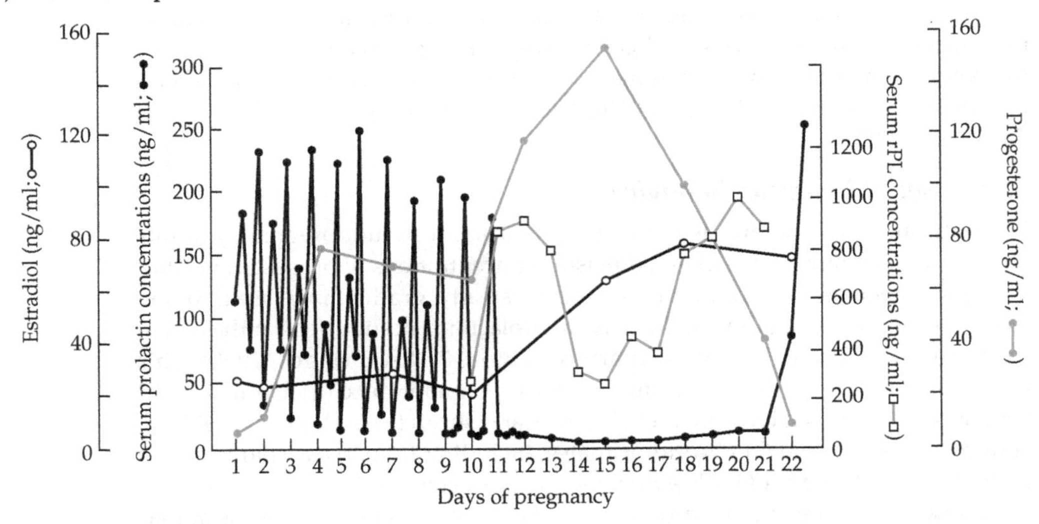

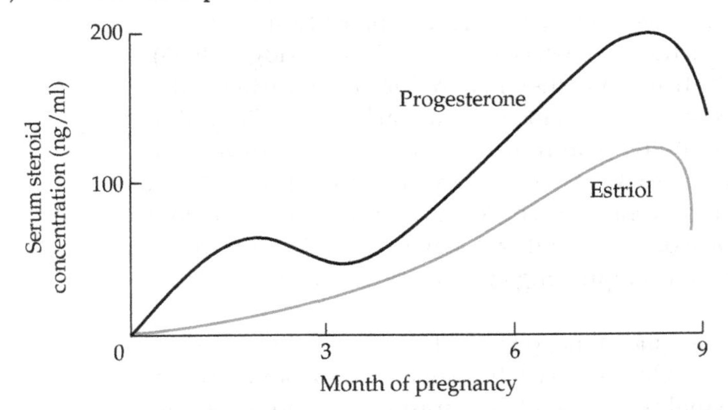

7.18 Hormone profile during pregnancy in rats and humans. (A) In rats, after blastocyst implantation, which occurs about 5 days after insemination, blood concentrations of progesterone begin to rise, peaking on days 15 and 16, then drop precipitously at the end of gestation. Blood estradiol concentrations remain stable during the first part of pregnancy, then rise dramatically at the end. Prolactin concentrations are high during the first half of pregnancy due to daily pulses of this hormone, but then decrease until the end of pregnancy, when they rise again; this hormone is necessary to support lactation and is important in mediating some maternal behaviors. Placental lactogen (rPL), secreted from the placenta, has actions similar to prolactin and growth hormone, supporting fetal growth and mammary gland development. (B) In humans, prolactin concentrations (not shown) increase gradually until parturition and stay elevated until nursing is completed. Concentrations of progesterone increase dramatically during the first trimester, then continue to increase until just prior to parturition. Estrogen (estriol) concentrations show a slow, continuous rise until the last few weeks of pregnancy, then quickly drop prior to parturition. After Rosenblatt et al., 1979.

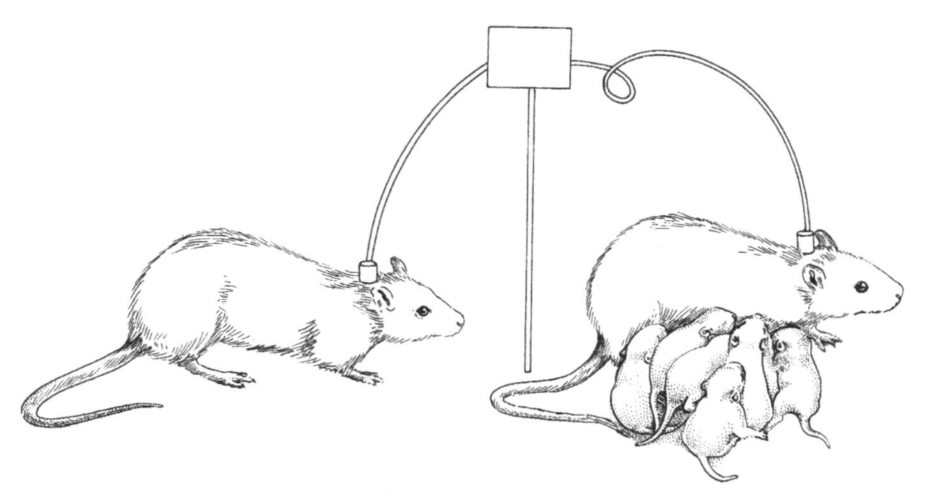

7.19 A blood-borne factor induces maternal behavior in nulliparous rats. When the blood of a new mother was transfused into a nulliparous rat, the recipient showed maternal behavior within 24 hours. Normally, 4–6 days of sensitization are required to induce maternal behavior in nulliparous females. After Terkel and Rosenblatt, 1968.

circulatory systems in what is called a parabiotic preparation. In this manner, Stone attempted to induce maternal behavior in the nonmaternal rat, but he was unsuccessful (Stone, 1925). Many years later, Terkel and Rosenblatt (1968, 1972) set up a transfusion system whereby the blood of one female could be exchanged with that of another while they were free to walk around and, apparently, behave normally (Figure 7.19). In one such experiment, the blood of a female that had just delivered a litter was exchanged with that of a nulliparous female. Virtually complete maternal behavior was observed in the nulliparous female within 24 hours of her receiving blood from the mother rat. This study demonstrated conclusively that a blood-borne factor is important in the induction of maternal behavior in rats.

The vast majority of work on the effects of pregnancy termination on the onset of maternal behavior has been conducted by Jay Rosenblatt and his colleagues (e.g., Bridges et al., 1978a; Rosenblatt and Siegel, 1975; Siegel and Rosenblatt, 1975, 1978). Female rats normally do not behave maternally toward foster pups at the beginning or middle of their pregnancies, but are intensely maternal at the end of pregnancy. The rationale behind these studies was to find out whether prematurely terminating pregnancy would induce the endocrine changes observed at birth and consequently induce maternal behavior. The researchers found that surgical removal of the uterus, fetuses, and placenta (hysterectomy) induces maternal behavior. Rats hysterectomized on day 16–19 of pregnancy respond rapidly with maternal behavior if they are presented with newborn foster pups 24–48 hours after surgery (Bridges et al., 1978b). Recall that nonpregnant female rats—with or without a uterus—require 5 or 6 days of periodic exposure to pups before they behave in a maternal manner.

Endocrinologically, the removal of the placenta is probably the main causative factor in the induction of maternal behavior in hysterectomized rats. The removal of the placenta eliminates the source of placental lactogen, which supports the corpora lutea during mid- to late pregnancy in rats. Consequently, removal of the placenta during hysterectomy leads to rapid regression of the corpora lutea and subsequent reductions of blood concentrations of progesterone. The rat placenta is also the source of a gonadotropin that typically suppresses ovarian estrogen production until very late in pregnancy. Removal of this source of estrogen inhibition results in a rapid pulse of estrogen secretion, which normally results in postpartum estrus. Thus, mid- to late pregnancy termination via hysterectomy results in an endocrine profile resembling that found in a normal rat immediately prior to parturition.

In other studies of the hormonal bases of maternal behavior, attempts have been made to mimic the endocrine patterns of pregnancy in ovariectomized females by injecting them with different hormones in various sequences (Moltz et al., 1970; Zarrow et al., 1971; Bridges, 1990, 1996; Rosenblatt et al., 1998; González-Mariscal, 2001). In one such study, nulliparous rabbits that were injected with estradiol for 18 days and received progesterone injections on days 2 through 15 exhibited nest building behavior (Zarrow et al., 1963; reviewed in González-Mariscal and Rosenblatt, 1996). Maternal behavior is relatively difficult to assess in rabbits because of its brevity. The energetic demands on a lactating rabbit doe are enormous, and foraging demands virtually all of her time. Consequently, rabbit mothers (called *does*) do not engage in as much maternal behavior as other species that produce altricial young; for instance, a doe nurses her litter for only about 3–4 minutes once per day (Ross et al., 1959; Zarrow et al., 1965). Thus, the maternal behavior of rabbits that is most commonly assayed in the laboratory is nest building. Near the time of birth, the pregnant doe builds a nest of straw and grass and lines it with her own fur, which loosens progressively throughout pregnancy so that it can be easily removed at this time. This fur loosening is dependent on hormonal changes. The increased estrogen-to-progesterone ratio at the end of pregnancy probably stimulates the release of prolactin from the anterior pituitary, as it does in rats and sheep. Treatment of pregnant rabbits with dopamine agonists (for example, ergot compounds), which block prolactin release, suppresses nest building; concurrent treatment with prolactin reverses the suppression (Zarrow et al., 1971). Thus, the maternal nest building observed at the end of pregnancy in rabbits correlates with an increased estrogen-to-progesterone ratio in the blood, as well as with elevated prolactin levels (González-Mariscal, 2001). In contrast to rabbits, house mice and Syrian hamsters build nests during early or mid-pregnancy. High concentrations of both estrogens and progestins induce maternal nest building in these species (Lisk, 1971; Richards, 1969; reviewed in González-Mariscal and Rosenblatt, 1996).

External factors interact with hormones in mediating maternal nest building in rabbits (González-Mariscal et al., 1998). The endocrine profile during pregnancy and parturition is very different from the endocrine profile during nursing. Yet, the quality (and perhaps quantity) of maternal care in rabbits remains constant.

As in other mammalian species, maintenance of rabbit maternal behavior appears to require sensory stimulation (e.g., olfactory, tactile, acoustic, and visual) that is provided by the offspring (reviewed in González-Mariscal and Poindron, 2002).

In another study, ovariectomized female rats were injected with estradiol for 11 consecutive days, progesterone on days 6 through 9, and prolactin on days 9 through 11 (Moltz et al., 1970). This pattern of hormone treatment caused full maternal behavior within 35 hours of exposure to pups. Although the induction of complex maternal behavior with a certain pattern of hormone injections is impressive, normal maternal behavior begins at or immediately before parturition, not 35 hours later. There must be other factors, endocrine as well as nonendocrine, involved in the induction of normal postpartum maternal behavior, or else the presentation of the hormones via systemic injections was insufficient to engage the neural machinery underlying this behavior. Infusions of prolactin directly into the MPOA of steroid-primed rats induce maternal behavior very quickly (Bridges and Freemark, 1995; Bridges et al., 1996; Bridges, 1996).

The role of progesterone in the induction of maternal behavior is not precisely understood (Numan and Insel, 2003). Progesterone appears to have two functions in rats. It probably acts with estradiol early in pregnancy to facilitate subsequent maternal behavior (both motivation and performance), but later in pregnancy it inhibits maternal behavior. That is, progesterone needs to be withdrawn before parturition if maternal behavior is to be displayed (Sheehan and Numan, 2002).

Recent studies suggest that placental lactogen may mediate the reduced latency to maternal behavior that occurs during pregnancy prior to parturition. Infusion of placental lactogen into the MPOA of steroid-primed rats was as effective as prolactin infusion in stimulating maternal behavior (Bridges et al., 1995, 1996). It appears that the conceptus (the fetus plus placenta) secretes placental lactogen, which can stimulate maternal behavior during and at the end of the pregnancy (Bridges et al., 1996). Thus, the endocrine communication between the developing conceptus and the mother ensures that maternal behavior will occur as soon as possible after birth (Bridges, 1996). This intriguing "manipulation" of maternal endocrine function is in the best interest of the fetus, but also benefits the mother's long-term fitness.

Other recent experiments have suggested a role for oxytocin in mediating the onset of rat maternal behavior. In all mammalian species that have been investigated, oxytocin concentrations increase during parturition (Fuchs and Dawood, 1980; Fuchs and Fuchs, 1984). Oxytocin causes the uterine and vaginal contractions of birth as well as milk letdown, processes that may induce other physiological changes leading to the onset of maternal behavior. If oxytocin is injected directly into the ventricles of the brain in estrogen-primed ovariectomized rats, maternal behavior is observed within 1 hour (Pedersen and Prange, 1979; C. A. Pedersen et al., 1982). However, rats are very sensitive to the testing environment, and differences in apparatus and test procedures may mask the effects of oxytocin on maternal behavior. Oxytocin appears to act as a neurotransmitter or neurohormone in the brain, because systemic injections of oxytocin into estrogen-primed ovariectomized rats do not induce maternal behavior (Insel, 1990a). Recall that peptides such as oxytocin do not cross the blood–brain barrier.

It is possible that vaginal stimulation during parturition causes oxytocin secretion, which in turn stimulates further contractions. To test this hypothesis, multiparous (having previously given birth several times), nonpregnant sheep were injected with a hormone regimen that included 12 days of progesterone followed by a large dose of estradiol, then were tested for maternal responsiveness to newborn lambs (Keverne et al., 1983). Half of the ewes received vaginocervical stimulation with a vibrator 5 minutes prior to the presentation of the lambs. Vaginocervical stimulation elevated oxytocin concentrations in the cerebrospinal fluid (Keverne and Kendrick, 1994). Only 20% of the nonstimulated ewes showed maternal behavior during a 1-hour test, but 80% of the stimulated ewes displayed maternal behavior. These results suggest that oxytocin is important in the induction of maternal behavior in sheep (Box 7.3). Recent evidence suggests that parturition activates the immediate early genes *fos* and *fosB* in neurons expressing oxytocin receptors, as well as in central oxytocinergic neurons in the medial preoptic area (MPOA), the piriform cortex, and the bed nucleus of the stria terminalis (BNST) (Lin et al., 2003). As we will see below, these neurons are critical for proper expression of maternal behavior.

In summary, a suite of hormones, neuropeptides, and neurotransmitters is responsible for the initiation and maintenance of maternal behavior in mammals. In rats, the factors that appear to have the greatest effects on maternal behavior are estrogens and lactogenic hormones (i.e., prolactin and placental lactogens) (Bridges, 1996). The supportive roles of prolactin, progesterone, oxytocin, cholecystokinin, and probably β-endorphins in maternal behavior are estrogen-dependent (Insel, 1990b; Bridges, 1996) (see Figure 7.18A). Several neurotransmitter systems are involved in the mediation of rat maternal behavior, but the dopaminergic system seems particularly important. The dopaminergic system can be stimulated by lactogenic hormones and inhibited by endogenous opioids (Bridges, 1996). Further research must be conducted to determine the extent to which the regulatory mechanisms underlying rat maternal behavior also mediate parental care in other mammals, including humans.

Maternal Aggression

Protection of offspring from predators or infanticidal conspecifics is an important component of parental care (Lonstein and Gammie, 2002). Maternal aggression is one mechanism by which female mammals protect their offspring. The onset of maternal aggression appears to be regulated by hormones. However, the hormonal control of maternal aggression appears to differ from the hormonal control of other components of maternal behavior. Maternal aggression is rarely observed in the absence of the offspring.

Most of the research on maternal aggression has been conducted on house mice—more specifically, on maternal aggression by house mice directed toward a male conspecific intruder. However, substantial work on the interaction of hormones and the somatosensory aspects of maternal aggression has been conducted on rats. Laboratory strains of female mice and rats are rather docile and do not usually attack conspecifics. However, beginning in the middle of gestation and

BOX 7.3 Maternal Behavior in Sheep

Research on sheep has been especially informative about the hormone–brain–behavior interactions associated with maternal behavior. Sheep are seasonal breeders, and most lambs are born early in the spring to coincide with the maximal availability of green vegetation for food. Sheep form feeding flocks and graze over a common home range (Lévy et al., 1996). When pregnant ewes are about to give birth, they seclude themselves away from the flock. The mother and her highly precocial lamb form a strong social bond within the first hour after birth. During this time, the mother learns to discriminate her lamb from the other lambs in the flock. This discrimination appears to be based on olfactory cues emitted by the lamb (Kendrick et al., 1992). Ewes normally permit only their own lambs to approach their udders and suckle. This discrimination is important from an evolutionary perspective because ewes that permitted other lambs to nurse would give up resources that could go to their offspring (Lévy et al., 1996). Such misdirected parental care would have strong negative consequences for fitness if there were insufficient milk for the ewes' own lambs. In fact, ewes are highly rejecting of approaching lambs with which they have not formed a social bond (Lévy et al., 1996). Thus, the establishment of an exclusive bond during the first hour is a critical component of sheep maternal care.

Importantly, establishment of the exclusive bond is separate from maternal responsiveness, a suite of behaviors that is directed toward any lamb immediately after parturition. Olfactory cues are also important in making lambs attractive stimuli to ewes. Both maternal responsiveness and establishment of the exclusive social bond characterize maternal behavior in sheep (Lévy et al., 1996).

As in other mammals, parturition in sheep is preceded by an increase in the ratio of estradiol to progesterone in the blood circulation. If nulliparous ewes are hormonally primed with estradiol and progesterone to mimic the conditions of the last week of pregnancy and presented with newborn lambs, the ewes are either indifferent or aggressive toward them. However, nulliparous ewes hormonally primed and given vaginocervical stimulation display a rapid onset of maternal behavior. The vaginocervical stimulation simulates parturition and evokes the release of oxytocin (Keverne and Kendrick, 1994). In vivo microdialysis studies of stimulated ewes revealed that oxytocin concentrations were elevated in the cerebrospinal fluid, as well as in the olfactory bulbs, MPOA, PVN, and substantia nigra (reviewed in Lévy et al., 1996), all brain areas that are important in sheep maternal behavior. When oxytocin is injected directly into the cerebrospinal fluid of nonpregnant ewes, the full complement of maternal behaviors is observed (Keverne and Kendrick, 1994). These results suggest that oxytocin is important for both maternal acceptance and other maternally motivated behaviors in sheep. The relevance of these findings to humans is potentially great.

peaking in intensity during the first week after birth, mouse dams react to an intruding male by displaying intense threat and biting behavior. Prior to parturition, progesterone appears to induce maternal aggression (Svare, 1990). Four converging lines of evidence point to a role for progesterone in mediating maternal aggression in pregnant mice. First, progesterone treatment elevates the rate of aggressive behavior in nulliparous mice (Mann et al., 1984). Second, pseudopregnant females become more aggressive as blood concentrations of progesterone increase (Barkley et al., 1979). Third, pregnant females begin to show signs of maternal aggression when peak concentrations of progesterone occur (Mann et al., 1984). Finally, surgical pregnancy termination, which dramatically reduces progesterone concentrations, eliminates maternal aggression when performed on day 15 of pregnancy, and progesterone replacement therapy partially restores maternal aggression in hysterectomized females (Svare et al., 1986). However, other hormones are also probably involved in maternal aggression, because progesterone treatment alone never induces the level of aggression in nulliparous females that is observed in newly parturient mice. Furthermore, maternal aggression is very high at the time of birth, when progesterone concentrations are relatively low, suggesting that other factors mediate aggressive behavior after the birth of the young.

The results of other studies suggest that in mice, and to a lesser extent in rats, suckling by the young stimulates postnatal maternal aggression (Svare and Gandelman, 1976). Intense maternal aggressive behaviors are observed in mice after 48 hours of suckling, and thelectomized (nipples surgically removed) mice display lower levels of maternal aggression than normal postpartum mice (Svare, 1990). Because suckling was found to be a critical stimulus for the expression of maternal aggression, it was logically suspected that prolactin is responsible for the induction of this behavior. However, this hypothesis was ruled out because neither postpartum hypophysectomy nor treatment with prolactin-inhibiting ergot compounds affects the occurrence of maternal aggression (Erskine et al., 1980b; Mann et al., 1980; Svare et al., 1982). Furthermore, no relationship has been found between blood plasma concentrations of prolactin and the initiation, maintenance, or decline of maternal aggression in house mice (Broida et al., 1981). However, prolactin does mediate maternal aggression in white-footed mice (*Peromyscus leucopus*) (Gleason et al., 1981) and Syrian hamsters (Wise and Pryor, 1977).

Thus, ovarian hormones present during pregnancy appear to regulate maternal aggression in mice and rats in two ways. First, ovarian steroids directly promote aggressive behavior during gestation. Second, these hormones indirectly induce aggressive behavior by stimulating nipple development for attachment and suckling by the young (Lonstein and Gammie, 2002). Whether ovarian hormones exert additional effects on postpartum aggressive behavior remains unresolved.

Nursing-induced postpartum aggression coincides with elevated levels of serotonin, and treatment with a serotonin antagonist reduces maternal aggression (Ieni and Thurmond, 1985). Endorphins, oxytocin, vasopressin, and nitric oxide also may play a role in maternal aggression (Lonstein and Gammie, 2002). Recently, it was reported that corticotropin-releasing hormone (CRH), an activator of fear and anxiety, regulates maternal aggression (Gammie et al., 2004). Intracerebroventricular injections of CRH inhibited maternal aggression, but not other maternal behaviors. These results suggest that decreased CRH is important for maternal aggression and may act by adjusting brain activity in response to an intruder (Gammie et al., 2004).

When the paraventricular nuclei of rat dams were destroyed by a chemical (ibotenic acid) lesion 2 days postpartum, maternal aggression increased at day 5 postpartum, a time when maternal aggression is normally high (Giovenardi et al., 1998). Recall from Chapter 2 that the neurons of the PVN are a source of oxytocin. To ascertain that oxytocin modulates maternal aggression, oxytocin antisense RNA was infused into the PVN on the fifth day postpartum, and maternal aggression was elevated (Giovenardi et al., 1998).

Maternal aggression per se does not seem to be a component of human maternal care, but direct tests are lacking. Experiments on humans are rarely conducted because of ethical considerations; consequently, correlational studies are usually preferred. In one study, women with high concentrations of circulating prolactin due to prolactin-secreting tumors reported higher levels of hostility in paper-and-pencil tests than normal controls; treatment with prolactin-inhibiting ergot compounds reduced these high hostility scores (Buckman and Kellner, 1985; Fava et al., 1981). In another study, new mothers rated themselves as more hostile on day 7 postpartum than did control females (hospital employees) (Mastrogiacomo et al., 1982, 1983).

One can imagine nonendocrine explanations for the increased hostility scores of newly parturient women, but studies on women differ fundamentally from the assessment of maternal aggression in other animals. Reliance on hostility scores, rather than on the aggressive behaviors themselves, reduces the likelihood that accurate information about endocrine correlates will be obtained. Paper-and-pencil tests of hostility are rarely validated to ascertain whether they truly correlate with the likelihood of agonistic behavior. As noted above, assessment of maternal aggression among nonhumans is always conducted in the presence of their offspring. Mouse dams, for instance, rarely display maternal aggression in the absence of their pups. Women, however, are generally separated from their babies during questionnaire testing. Direct tests of maternal aggressive behavior in the presence of their infants and concurrent measurements of hormone concentrations have not been performed on humans (but see Fleming, 1990).

Maintenance and Termination of Maternal Behavior

Hormones are clearly necessary to trigger maternal behavior when the offspring are first born. Yet maternal behavior persists after parturition, and long after the high hormone concentrations of pregnancy have returned to baseline levels. Are hormones involved in maintaining maternal behavior? Furthermore, mothers act less maternal as their offspring grow older (Figure 7.20). Are hormone concentrations changing, or are the young less effective in stimulating maternal care?

These questions have been addressed in rats, which display three stages of maternal behavior, but the answers remain speculative. It is important to emphasize that at each stage, the pups are trying to get food from their mother, and the pups' suckling behavior induces maternal nursing behaviors. During the first stage, the majority of maternal behaviors are initiated by the dam. She builds the nest,

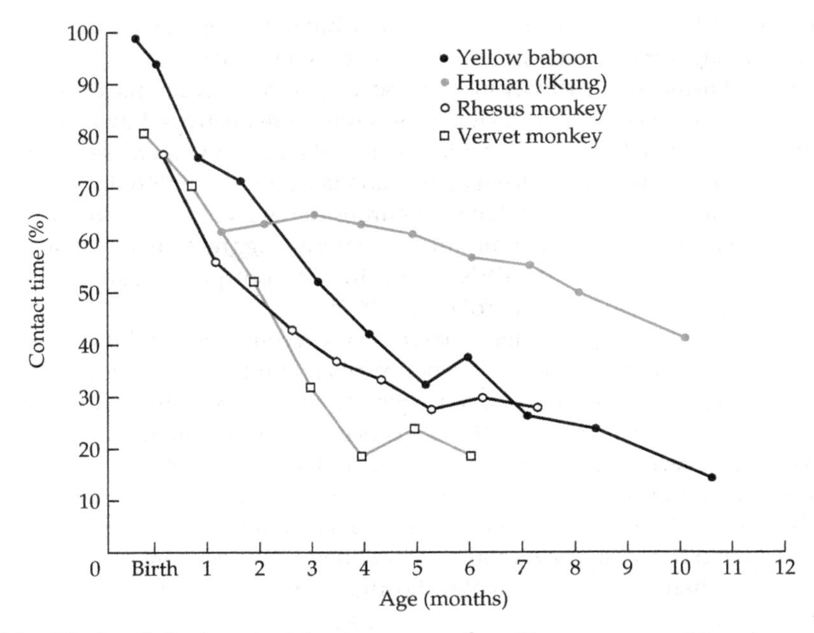

7.20 Mother–infant contact decreases over time. The percentage of time that mothers of four primate species and their young spent in close contact each day is plotted over the first year of the lives of the offspring. When the young are first born, virtually all of the mother's day is spent in contact with them, but in nonhuman primate species, this value drops to 10%–30% of the day after 6 months. At 10 months, approximately 40% of a human mother's time is spent in contact with her baby; this observation may reflect the slower maturation rate of humans or the increased solicitousness of human mothers. After Altmann, 1980.

delivers the young into it, and retrieves any pups that wander away from it. She huddles over her litter in the nursing posture. The pups must attach themselves to a nipple and stay attached, but the dam provides cues for the pups to promote the initial nipple attachment: she licks her vulva and ventrum during birth, spreading amniotic fluid across the front of her body, and the odor of the amniotic fluid attracts the newborn pups to the nipples. If the amniotic sacs of fetuses are injected with lemon oil during pregnancy, the pups are attracted to the odor of lemon oil after birth (Pedersen and Blass, 1982). As the pups age, the odor of their own saliva and their littermates' and that of their mother keeps attracting them back to the nipple. At an early age, nutritional state does not particularly influence the time it takes for pups to attach to a nipple, but as the pups age, the length of time since the last meal does influence the latency to reattach to a nipple.

In the second stage of rat maternal behavior, there is more mutually initiated contact. The dam may initiate care, but the pups often approach her for contact. Stimuli associated with the pups are critical for maintaining maternal care at this stage (Box 7.4). If a mother rat is provided with new 4- or 5-day-old foster pups every few days, she continues to build nests and behave maternally (Södersten and Eneroth, 1984). This behavior is maintained by tactile and other sensory stimuli from the young pups. In the normal situation, however, maternal care eventually wanes as the pups mature.

Maternal rejection of the pups and subsequent separation from them constitutes the third stage of maternal behavior in rats. The dam makes herself less available to the pups by "hiding" her ventrum from them, by rejecting them, or by remaining out of their reach (Reisbick et al., 1975). What are the signals that reduce maternal care as the pups grow older?

One such signal might be changes in the body temperature of the dam or the pups. As described above, rat pups are born blind, deaf, hairless, and unable to regulate their body temperatures. Rat pups are considered to be poikilothermic, like reptiles; the body temperature of an isolated rat pup is only slightly higher than the ambient temperature. However, although an individual pup cannot thermoregulate, a litter of pups can. After 4 hours of exposure to a 24°C ambient temperature, an individual pup possesses a body temperature of 26°C, while individuals in a huddle of four pups maintain a body temperature of 30°C (Alberts and Brunjes, 1978). Researchers examined the percentage of time that a rat pup spent on the outside of a huddle and found that there was considerable competition among littermates for inner and outer positions within the huddle. In one study, in which litters were culled to four, all four pups kept switching their positions within the litter. In another experiment (Alberts and Brunjes, 1978), one of the four pups in each litter was anesthetized, and the huddle was placed in either a hot or a cold environment (Figure 7.21). In the cold environment, the anes-

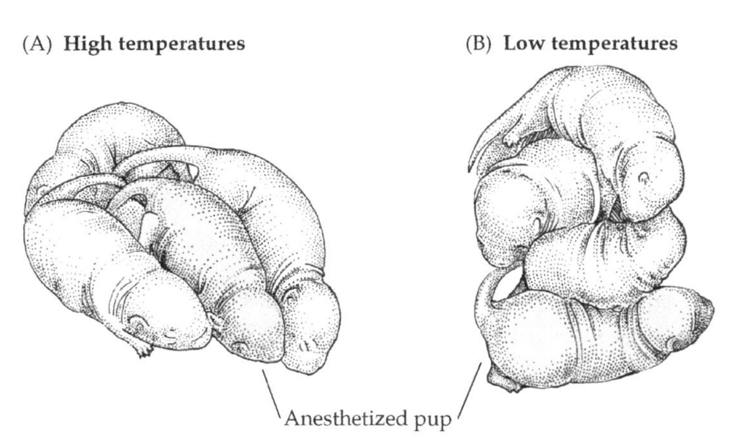

7.21 Thermoregulation in rat pups. An isolated rat pup's body temperature soon falls several degrees below the normal body temperature of 37°C, but pups in groups can thermoregulate because individuals move in and out of the huddle to maintain their body temperature in an ideal range. This behavioral thermoregulation can be demonstrated by anesthetizing one of the pups in the huddle. (A) At warm ambient temperatures, the pups maneuver so that the anesthetized pup is in the center of the huddle and becomes overheated. (B) At cool ambient temperatures, the anesthetized pup is shoved to the outside of the huddle. After Alberts and Brunjes, 1978.

BOX 7.4 Offspring Behavior and the Maintenance of Maternal Behavior

Research by Judith Stern has established the importance of somatosensory (tactile) feedback from pups for maintaining maternal care in rats (Stern, 1996). Most rat maternal behaviors, including nest construction, licking, and pup retrieval, involve the mouth. During the course of nuzzling or sniffing the pups, the female receives somatosensory input that stimulates further maternal behavior. Many maternal behaviors can be reduced or abolished by cutting or anesthetizing the nerves that innervate the area around the mouth (Stern, 1990). For example, a mother rat that has had her muzzle desensitized with a local anesthetic exhibits reductions in nest building, nest repair, pup retrieval, pup licking, and even biting of intruders.

Somatosensory information from the pups is also important for maintaining nursing behavior. Virtually all mammalian newborns arrive equipped with reflexes that allow them to find and attach to a nipple and ingest milk. Mothers' assistance of their offspring's nursing behavior ranges from the "passive tolerance" of marsupial mothers to the active participation in feeding seen in humans and other primate species (Stern and Johnson, 1990; Stern, 1990). In rats, ventral somatosensory information maintains the maternal nursing posture, an immobile, upright crouching posture in which the female stands over the pups. If the nipples are anesthetized or removed, then even an entire litter cannot stimulate the female to exhibit a normal nursing posture (Stern et al., 1992). Similarly, if the mouths of the pups are anesthetized so that they no longer root up against the mother trying to locate her nipples, an unanesthetized dam will fail to adopt the nursing posture. Thus, tactile stimuli from the pups are important in maintaining nursing behavior in mothers.

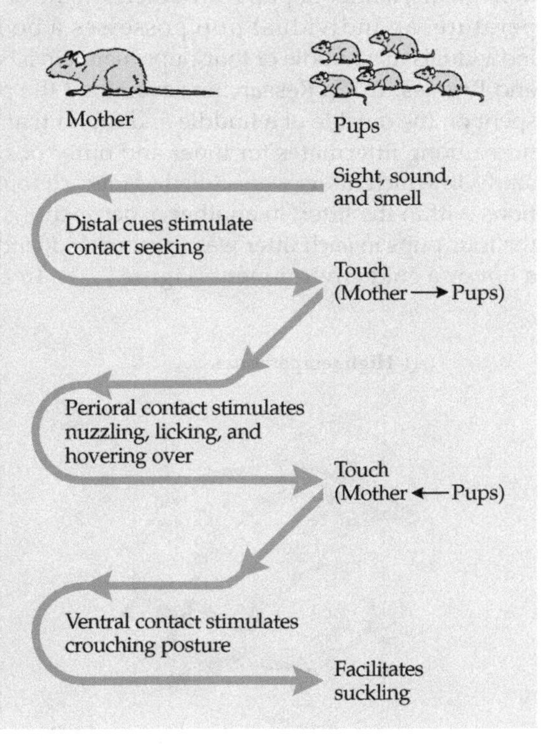

thetized pup spent up to 90% of the time exposed to the cold at the outer fringes of the huddle. In the hot environment, the anesthetized pup spent a considerable amount of time on the inside of the huddle.

Thermoregulation may play a major role in modulating mother–pup contact time during the third stage of rat maternal behavior, and hormones may underlie the physiological orchestration of changes in contact behavior. The amount of time a mother rat spends with her litter decreases as the pups get older. The dam initially spends about 75% of the day in contact with her litter. The bouts of time she spends on the nest decline progressively during the first 2 weeks postpartum; that is, the number of nest bouts remains constant, but the length of each nest bout decreases. Mother rats secrete progressively increasing amounts of prolactin during days 1–14 of lactation (Leon et al., 1990). When Adels and Leon (1986) treated mother rats with bromocryptine, an ergot compound that blocks prolactin production and secretion, they found that nest bout time did not decrease during the first 2 weeks of lactation, but remained high. Adels and Leon suspected that prolactin was acting through another hormone, and in a series of ablation experiments, they showed that removal of the adrenal glands was most effective in blocking the decrease in nest bout time. If the adrenal glands were removed, time on the nest did not show the normal decrease during days 1–14 of lactation, but removal of the adrenal medulla alone did not block the decline in nest bout time. Consequently, the researchers surmised that the decrease in nest bout time must be due to glucocorticoids (for example, corticosterone) rather than mineralocorticoids (for example, aldosterone). To discover which hormone would reverse the decline in nest bout time, they injected adrenalectomized dams with either corticosterone, aldosterone, progesterone, or a control treatment (the oil vehicle in which the injected steroids were dissolved). By this means, they eventually discovered that corticosterone was the critical hormone responsible for the decline in nest bout time. Normally, ACTH increases during the first 2 weeks of lactation and induces increased concentrations of corticosterone in the blood; the increased blood concentrations of corticosterone in turn lead to a reduction in nest bout time.

How do these hormones influence nest bout time? One possibility is that hormones could affect the efficiency of milk transfer. Mothers might regulate the amount of milk that their pups receive by utilizing cues associated with milk delivery to regulate nest bout time; the pups might become more efficient with age and remove a given amount of milk in less time. Faster transfer of nutrients would allow more time for the mother rat to forage for food. But this hypothesis can be dismissed, because the decline in nest bout time is also observed in dams with surgically closed nipples (Leon et al., 1978). Alternatively, the dam might monitor the pups' body temperatures and stay on the nest long enough to keep them warm. As the pups aged and began to thermoregulate on their own, she would need to spend less time with them. This hypothesis can also be rejected, because the body temperature of the pups does not directly affect nest bout time (Leon et al., 1978).

Another possibility is that mother rats monitor their own body temperature to decide when to leave the nest. Glucocorticoids increase the metabolic rate and thus chronically elevate the body temperatures of mother rats (Leon et al., 1990). They cannot dissipate their own body heat when crouching over a litter, so if they stay with the pups for too long, they are in danger of suffering hyperthermia and resultant brain damage. As the pups get older, they retain heat more efficiently, requiring dams to regulate their own body temperature by spending shorter bouts of time on the nest. When pups were experimentally heated or cooled, dams spent less time with hot pups than with cold pups (Woodside and Leon, 1980). When mother rats were shaved, they showed less of a decline in nest time than did furry mothers. Furthermore, direct heating of the MPOA of dams induced immediate termination of a nursing session (Woodside et al., 1980).

Thus, the researchers concluded that the suckling pups stimulate the secretion of prolactin by the dam, which in turn stimulates increased corticosterone production. Elevated glucocorticoid concentrations increase the mother's basal metabolic rate, which causes a rise in her body temperature. The dam's increased body temperature leads to a decline in time spent with the pups.

The results of some of these studies, however, are subject to alternative interpretations. It is possible that the overheated pups failed to induce nursing behavior in their mothers (Stern and Lonstein, 1996). It can be argued that the high temperatures used in these experiments, which rat mothers and pups do not normally experience in their natural habitats, had confounding effects. That is, hyperthermic pups usually do not suckle, and suckling is critical to inducing maternal behaviors. Because the studies were automated and assessed the time dams spent on the nest using photocells, no other behavioral changes were recorded. Another potential confounding variable is that the boxes in which the dams were tested at high temperatures were too small to allow them to adopt "heat-dissipating" nursing postures (Stern and Lonstein, 1996). Thus, the apparatus used in these studies might have altered the behavior under study in a manner reminiscent of way in which paired nonpaced mating tests altered female rat mating behavior (e.g., Erskine, 1989) (see Chapter 6).

The third stage of maternal care involves not only maternal rejection, but also separation initiated by the pups. At about 16 days of age, rat pups begin leaving the nest to feed on their own. How do they find their way back to the nest? They use odors. In home orientation tests using odor cues, pups orient toward the nest during days 12–14 or so, then start orienting toward their mother. This switch from nest to mother is an important change that frees the pups from the nest. Beginning on day 15 or 16 of lactation and continuing through day 27, mother rats emit a chemosensory signal that pups find very attractive. Leon and Moltz (1972) tested the attractiveness of odors from the nests of lactating dams compared with the nest odors of nulliparous females. They found that older pups have a stronger preference for the odors of lactating dams than do younger pups: 5-day-old pups showed no preference, but 15-day-old pups showed a strong preference. However, this preference is not specific to the odor of the pups' own mother (Leon, 1980).

What is the source of this attractive chemosensory signal? It is released in the anal excreta of mother rats—their urine does not attract pups. In addition to feces, rats excrete small pellets of partially digested food matter, called *caecotrophs*, which they usually reingest. Caecotrophs are distinguishable from feces in the caecum, and they are the source of the mother rat's attractive odor. Young rats consume their mother's caecotrophs as an energy-saving measure. In addition, a pup's digestive system is not very efficient, and it benefits from the bacteria found in caecotrophs in building its own internal gut fauna to aid in digestion, particularly digestion of cellulose (Moltz and Kilpatrick, 1978). Caecotrophs taken directly from the caecum of a lactating female attract young rats. Interestingly, caecotrophs taken from the caecum of a nulliparous female rat work almost as well as those of mothers in attracting pups. A nonlactating female, however, normally reingests all of her caecotrophs. A lactating female experiences an increase in appetite and food intake, stimulated by prolactin (Leon et al., 1990), and thus produces so many caecotrophs that some are left near the nest. Thus, it is not a specific chemical agent found only in lactating females' caecotrophs, but the emission rate, that is responsible for attracting pups. If prolactin secretion is blocked in the mother rat, the attractiveness of her nest odor is decreased, probably because she then consumes all the caecotrophs she produces, just like a nonlactating female. Pups lose interest in mothers' caecotrophs at about 28 days of age, when mothers reduce caecotroph production and the pups are about to be weaned.

If nulliparous female rats are sensitized to pups so that they behave maternally, and are prevented from ingesting their caecotrophs, then eventually the caecotrophs from these females are attractive to other rat pups (Leon, 1992). It requires about 14 days after concaveation for there to be sufficient chemical stimuli to attract pups. As in mothers, the attractiveness of their caecotrophs to pups increases as the pups age. Unlike those of lactating females, the blood prolactin concentrations of these rats do not change; again, it is the buildup of sufficient numbers of caecotrophs in the nest that makes the nest attractive to the pups, rather than a special chemical agent produced only by lactating females. The mechanism underlying this phenomenon in concaveated females is unknown.

Endocrine Correlates of Primate Maternal Behavior

The endocrine correlates of maternal behavior among primates generally, and humans specifically, remain largely unknown. The reasons for this are varied, but probably reflect an assumption by researchers that primate maternal care depends primarily on experiential, not hormonal, factors. First-time mothers are often poor mothers, whereas experienced mothers improve with practice (Figure 7.22). However, recent studies have indicated important roles for estradiol in nonhuman primate motivation to engage in maternal behavior, as well as for cortisol and the pattern of estradiol and progesterone secretion in mediating maternal behavior in women.

Maternal experience seems to play an important role in primate maternal behavior. In one study, adult female rhesus monkeys were presented with an infant for 1 hour daily on 5 consecutive days. Some of the females were nulliparous; others had previously given birth to several offspring. Some of the multiparous females were bilaterally ovariectomized, others were intact, and others were aged, postmenopausal animals. Essentially, the nulliparous females avoided the infant, whereas the multiparous females, regardless of endocrine status, immediately accepted the baby rhesus (Holman and Goy, 1980). One possible conclusion from this study is that hormones are involved in the onset of maternal behavior in first-time mother rhesus monkeys, but may not be necessary to stimulate maternal behavior in experienced females. Stimuli associated with primate

7.22 Maternal behavior in bonobos. Female bonobos engage in extended maternal care of their semi-precocial offspring. The experienced mother carries her infant and holds it to allow it to nurse. Photograph © Frans Lanting/Minden Pictures.

infants, primarily visual, can trigger appropriate maternal behavior in multiparous females (Coe, 1990).

However, hormones associated with pregnancy and parturition can affect motivation to behave maternally in nonhuman primates (Pryce, 1996). In test situations in which primigravid common marmosets were able to press a bar to receive sensory stimuli associated with infants, the rate of bar pressing increased at the end of pregnancy in association with increasing estradiol concentrations. In the same study, the pregnant marmosets pressed the bar at high rates to turn off the recorded cries of an infant (Pryce et al., 1993). A high rate of bar pressing continues postpartum even though estradiol values are decreasing (Figure 7.23). This correlational study suggests that estradiol primes females to behave in an appropriately maternal manner. Administration of progesterone and estradiol to marmosets increased their rates of bar pressing (Pryce, 1996).

Among human mothers, it is assumed that hormonal factors facilitate the initial postpartum expression of maternal care, but data to support this assumption are minimal. In fact, hormones are neither necessary nor sufficient for the appearance of basic parental behaviors (Fleming, 1990). Adoptive parents, siblings, grandparents, other relatives, and caregivers can grow attached to infants during a process that appears identical in qualitative and temporal character to the

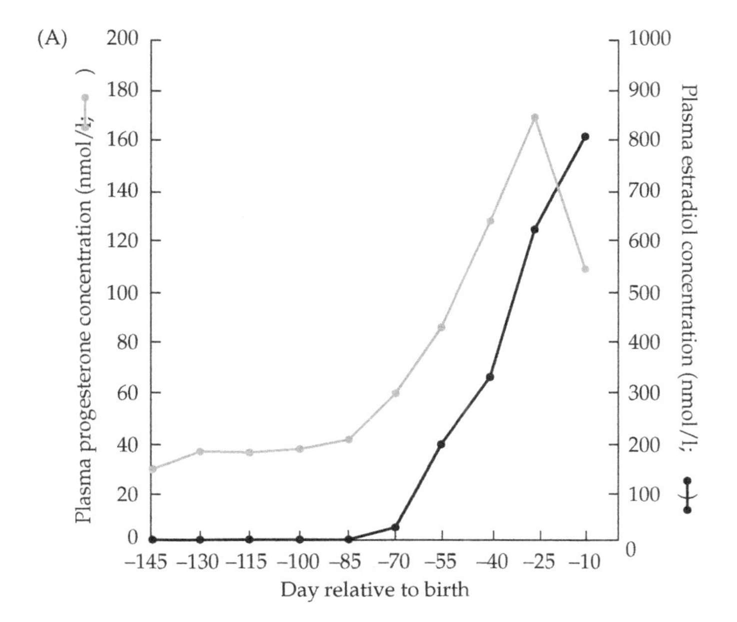

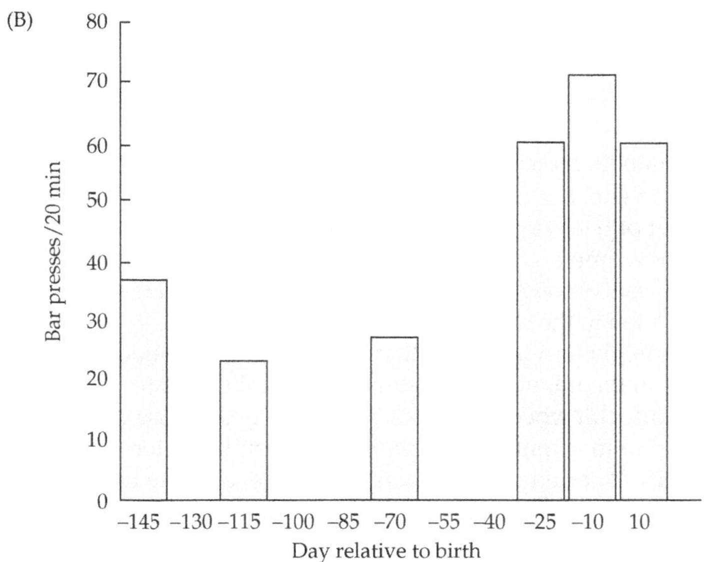

7.23 During pregnancy, female marmosets increase bar pressing for infant sensory reinforcement. (A) Plasma concentrations of estradiol and progesterone over the course of pregnancy in common marmosets. (B) The number of bar presses made by pregnant females to gain access to sensory stimuli associated with infants. After Pryce et al., 1993.

process observed in the biological mother. Obviously, the endocrine profiles of biological parents, adoptive parents, and other caregivers will vary in the presence of virtually identical behavioral patterns. There may be parallels between concaveation in rats and parental care in adoptive parents or caregivers.

Another issue in the study of endocrine correlates of human maternal behavior is that, in contrast to rats, humans have no clear-cut, universal set of maternal

7.24 Placentophagia is an important component of maternal care among nonhuman primates, such as this pigtailed macaque (*Macaca nemestrina*). Females consume the placenta and umbilical cord up to the ventrum (belly) of the infant. Consuming the placenta and umbilical cord rids the nest of a source of infection for the parents and offspring. Courtesy of Leonard Rosenblum.

behaviors. Human mothers typically provide most of the care for their infants, carry them, and nurse them, but this is by no means ubiquitous. There does not seem to be a single set of behaviors, or even attitudes, that characterize human maternal behavior. For example, consuming the placenta and umbilical cord after birth is virtually universal among nonhuman primates (Figure 7.24; Coe, 1990), but there is great variation in the occurrence of this behavior among human societies. (Placentophagia might be adaptive in that it reduces the likelihood of infection and hemorrhage in the infant.) Establishing hormonal correlates of human maternal behavior is difficult when the behavior itself eludes precise definition.

Information about human mothers' perceptions of their behaviors, attitudes, and feelings is typically collected on questionnaires; however, there has been no consensus about endocrine correlates of the attitudes observed in such studies (reviewed in Krasnegor and Bridges, 1990). Items on questionnaires are usually designed so that the answers are marked on a five- or seven-point scale reflecting a range of responses from "strongly agree" to "strongly disagree" (Fleming, 1990). Blood samples are sometimes obtained, or correlations between obvious endocrine status (for example, mid-pregnancy or menstruation) and self-reports are made. However, as we have seen, associations between self-reports of attitudes and actual behaviors are not always evident, and experiments to test behavior itself are difficult to perform on humans. These difficulties seriously impair the ability of researchers to ascertain the endocrine correlates of human maternal behavior.

Self-report studies have found differences in attitudes and feelings between new mothers and non-mothers. For example, the hedonic values of a variety of infant-associated (e.g., general body, urine, feces) and non-infant-associated (e.g., cheese, spices, and lotions) odors were rated by individuals in four distinct experimental groups: (1) new mothers, (2) mothers that were 1 month postpartum, (3) non-parent women, and (4) non-parent men (Fleming et al., 1993). The odors were rated on a scale that ranged from "extremely pleasant" to "extremely unpleasant," and all of the participants completed several attitude questionnaires as well. In common with the results of studies on nonhuman animals (e.g., Poindron and Lévy, 1990; Lévy et al., 1996), new mothers rated infant-associated odors as more pleasant than non-mothers. New mothers also reported more nurturant attitudes and feelings (Fleming et al., 1993). Similar studies indicate that several postpartum factors, including hormone concentrations and learning, are important in the establishment of a mother's attraction to her newborn infant's odors (Fleming et al., 1993; Fleming and Corter, 1995; Schaal et al., 1980).

In another series of studies, new mothers were asked to identify, in a two-choice test, a T-shirt that had been worn by their own infant (Fleming et al., 1995). This task was easily performed by virtually all new mothers. Similarly, new mothers can identify their infants based on their cries (Formby, 1967) and tactile features (Kaitz, 1992). The ability of new mothers to discriminate their infants from others appears to rely primarily on experience with the infants, although hormones may potentiate or enhance the effect (e.g., Fleming et al., 1995). Both mothers and fathers could discriminate between the odors of two samples of amniotic fluid—either parent could identify the amniotic fluid associated with their own infant (Schaal and Marlier, 1998).

A series of elegant experiments by Alison Fleming and her collaborators (e.g., Fleming, 1990; Fleming et al., 1987, 1990b) examined the endocrine correlates of the behavior of human mothers as well as the endocrine correlates of maternal attitudes as expressed in self-report questionnaires. In one such study, behavioral responses of new mothers to their 3–4-day-old infants were recorded. Responses such as patting, cuddling, or kissing the baby were called affectionate behaviors; talking, singing, or cooing to the baby were considered vocal behaviors. Both affectionate and vocal behaviors were considered approach behaviors (Fleming, 1990). Basic caregiving activities, such as changing diapers and burping the infants, were also recorded. In these studies, no relationship between hormone concentrations and maternal attitudes, as measured by the questionnaires, was found (Fleming, 1990). However, when behavior, rather than questionnaire responses, was compared with hormone concentrations, a different story emerged (Fleming et al., 1987). Blood plasma concentrations of cortisol were found to be positively associated with approach behaviors. In other words, women who had higher concentrations of blood cortisol, in samples obtained immediately before or after nursing, engaged in more physically affectionate behaviors and talked more often to their babies than mothers with lower concentrations of this steroid hormone. Additional analyses from this study revealed that the correlation was even greater for mothers that had reported positive maternal regard (feelings and

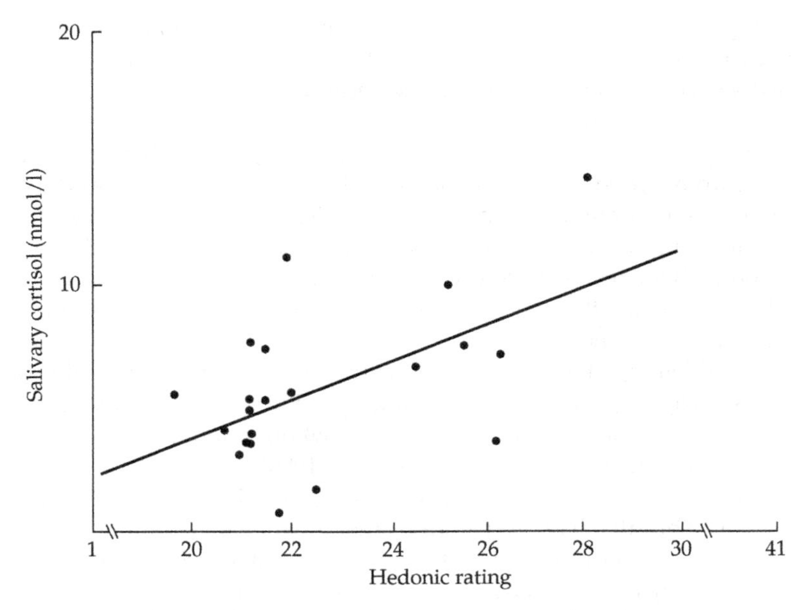

7.25 Hedonic ratings of their own infant's odors are positively correlated with mothers' cortisol levels. Saliva samples of cortisol from first-time mothers were compared with their hedonic ratings (from 1, "extremely unpleasant," to 41, "extremely pleasant") of their infant's T-shirt odors. Mothers with higher cortisol concentrations found their infant's odors more appealing than mothers with lower concentrations. After Fleming et al., 1997b.

attitudes) during pregnancy. In fact, nearly half of the variation in maternal behavior among the women could be accounted for by cortisol concentrations and positive maternal attitudes during pregnancy (Fleming et al., 1987).

Presumably, cortisol does not induce maternal behaviors directly, but it may act indirectly on the quality of maternal care by evoking an increase in the mother's general level of arousal (Mason, 1968), and thus increasing her responsiveness to infant-generated cues (Fleming, 1990). For example, new mothers with high cortisol concentrations were also more attracted to their infant's odors, were superior in identifying their infants, and generally found cues from infants highly appealing (Figure 7.25) (Fleming et al., 1997b). Although there have been a few examples of glucocorticoid involvement in the onset of maternal care in nonhuman animals (e.g., Keverne and Kendrick, 1992), it is possible that cortisol might simply reflect other endocrine changes involving oxytocin, CRH, and opioids, all of which affect maternal responsiveness in nonhuman animals (Fleming et al., 1997b), and all of which are secreted in concert with the glucocorticoids. Perhaps not surprisingly, experiential factors also affect maternal responsiveness in humans; experienced mothers find infants more attractive than do non-mothers and react to infant cues earlier in pregnancy than first-time mothers (Fleming et al., 1996). Moreover, the relation between cortisol and attraction to infant odors

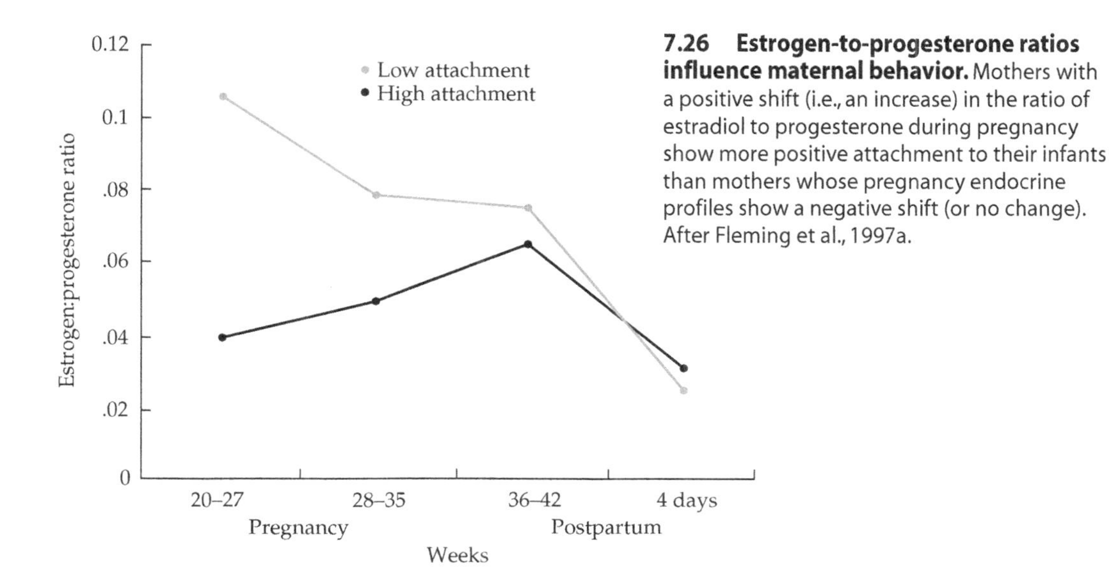

7.26 Estrogen-to-progesterone ratios influence maternal behavior. Mothers with a positive shift (i.e., an increase) in the ratio of estradiol to progesterone during pregnancy show more positive attachment to their infants than mothers whose pregnancy endocrine profiles show a negative shift (or no change). After Fleming et al., 1997a.

is found only in first-time mothers, in whom hormonal effects would presumably be most necessary.

Subsequent studies that have examined both human maternal behavior and hormone concentrations have revealed more surprises. Mothers who displayed a positive shift in the ratio of estradiol to progesterone during pregnancy were found to have more positive attachment to their infants than mothers whose endocrine profiles showed a negative shift (or no change) in the ratio of these two steroid hormones during pregnancy (Figure 7.26) (Fleming et al., 1997a). Taken together, the results of these studies on endocrine correlates of human maternal behavior suggest that a greater emphasis on observing behavior in human mothers is more likely to yield meaningful endocrine correlates than are scores from questionnaires (Fleming et al., 1996). These studies, in common with the studies described for sheep and rats, also emphasize the importance of early contact between mothers and infants at birth for optimal maternal behavior (Klaus and Kennel, 1976).

Endocrine Correlates of Paternal Behavior

Studies of the endocrine correlates of paternal behavior in mammals that display this behavior in nature have been limited to a few species, including California mice, dwarf hamsters, gerbils, tamarins, and marmosets. As described above, male California mice provide extensive paternal care to their offspring. In one study (Gubernick and Nelson, 1989), the parental behavior of fathers, expectant fathers (males living with their pregnant partners), and unmated males was assessed. The males were exposed to a novel 1–3-day-old pup during a 10-minute test. Relatively few unmated males displayed parental behavior (19%) as compared with fathers (80%) or expectant fathers (56%). Plasma concentrations of prolactin were higher in fathers than in expectant fathers or unmated males (Figure

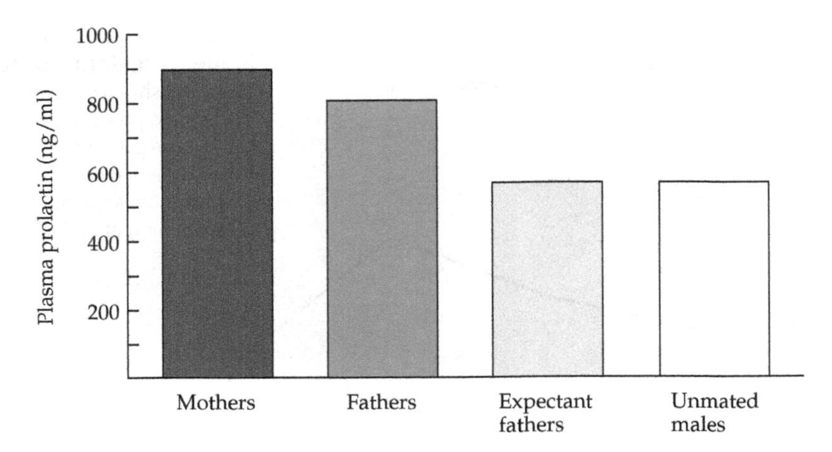

7.27 California mouse fathers show increased prolactin concentrations. Blood samples were obtained from California mouse mothers and fathers 2 days postpartum, from expectant fathers within 10 days of parturition, and from unmated males. Fathers showed concentrations of blood prolactin comparable to those of mothers. After Gubernick and Nelson, 1989.

7.27). Assessment of oxytocin levels revealed an increase immediately post-copulation, but there were no detectable changes in oxytocin levels after the pups were born, when the males behaved paternally (Gubernick et al., 1995). Testosterone concentrations did not differ among the three groups of males.

In contrast to birds and most mammals, testosterone appears necessary for paternal behavior in California mice. Castration reduced paternal behavior, whereas testosterone replacement maintained high levels of paternal behavior (Figure 7.28) (Trainor and Marler, 2001). Testosterone promotes paternal behavior in these mice through its conversion to estradiol (Trainor and Marler, 2002).

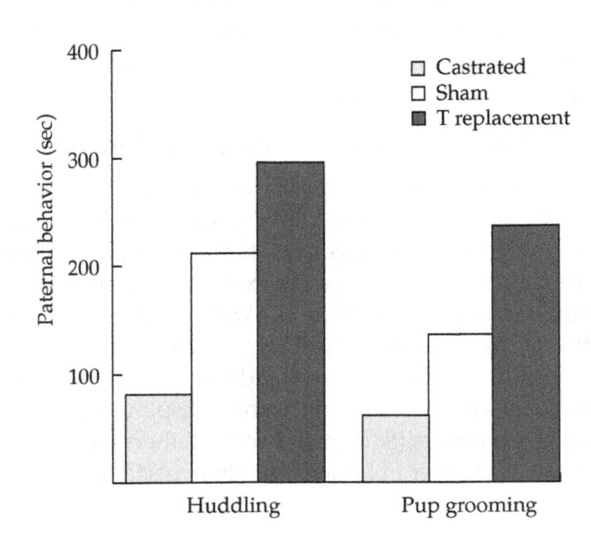

7.28 Testosterone is necessary for paternal care in California mice. Unlike most species of birds and mammals, male *Peromyscus californicus* require high testosterone concentrations to display paternal care. Castrated males displayed reduced huddling or pup grooming behaviors, but testosterone replacement restored this behavior in castrated males to the level of intact fathers or sham-castrated males. After Trainor and Marler, 2001.

California mouse fathers have more aromatase activity than non-fathers in the MPOA, a brain area known to regulate maternal care (Trainor et al., 2003).

Examination of plasma concentrations of prolactin in male rats, which do not display paternal behavior in nature, but were induced to show parental behaviors through concaveation, revealed no relationship between prolactin concentrations and paternal behavior (Samuels and Bridges, 1983; Södersten and Eneroth, 1984; Tate-Ostroff and Bridges, 1985). In male gerbils (*Meriones unguiculatus*) exposed to pregnant females or pups, prolactin values increased throughout pregnancy and remained elevated during the first 10 days after birth; plasma testosterone concentrations rose during pregnancy, then showed a steep decline in paternal males after their pups were born (Brown et al., 1995). However, there is substantial intraspecific variation in testosterone responses in gerbils; males seem to follow one of two reproductive tactics. Either testosterone remains elevated and males pursue additional copulations, or testosterone values decrease and males behave paternally (Clark et al., 1997). Studies on common marmosets have also found that males behaving paternally show a fivefold increase in blood concentrations of prolactin, and reductions in blood testosterone concentrations, compared with nonpaternal males (Dixson and George, 1982; reviewed in Zeigler, 2000). Similar findings have been reported for paternal dwarf hamsters (*Phodopus campbelli*) (Reburn and Wynne-Edwards, 1998; Wynne-Edwards, 1998).

The hormonal correlates of human paternal behavior resemble the factors associated with paternal behavior in other mammals; namely, testosterone, prolactin, and cortisol. Plasma prolactin concentrations were elevated in fathers just prior to parturition (Storey et al., 1998). Men who responded physiologically, with changes in heart rate, to the sounds of babies crying also displayed elevated prolactin concentrations compared with men who did not respond to infant cries (Elwood and Mason, 1994; Storey et al., 1998). The dynamics of cortisol secretion by men and women around the time of parturition are similar (Storey et al., 1998). Recall that elevated cortisol concentrations in new mothers are linked with the establishment of social bonding between mothers and infants (Fleming et al., 1997b). Cortisol concentrations were highly correlated between men and women partners during pregnancy and birth (Storey et al., 1998). In common with other paternal mammals, men displayed depressed testosterone values (–33%) immediately postpartum (Storey et al., 1998). Women experienced elevated testosterone levels near the end of pregnancy, but typically displayed a significant reduction in testosterone concentrations postpartum (Fleming et al., 1997a).

Another study observed endocrine changes in men who were becoming fathers for the first time (Berg and Wynne-Edwards, 2001). Men were recruited from prenatal classes to give saliva samples until the third month after their babies were born. Male volunteers from the general population were also chosen to provide samples, which were age-matched and obtained at the same time and season as those from the expectant fathers. Expectant fathers displayed lower testosterone and cortisol concentrations, and a higher proportion of samples with detectable estradiol concentrations, as well as higher estradiol values when detected, than control participants (Figure 7.29). The physiological importance of these hormon-

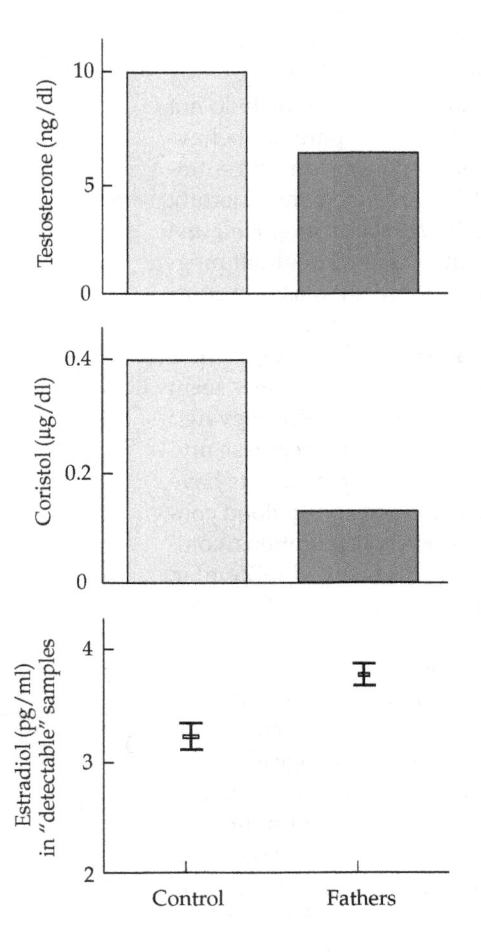

7.29 Human fathers display reduced testosterone and cortisol concentrations, as well as increased circulating estradiol concentrations. Concentrations of estradiol are shown only for those samples in which estradiol was detectable. After Wynne-Edwards, 2001.

al changes in men, if any, is not known, but these hormones have been shown to influence maternal behavior (Berg and Wynne-Edwards, 2001).

Before I became a father, I used to specifically request the "no baby" section of airplanes from my travel agent. To me, there was no more irritating sound than a crying infant. After I became a father, however, the sounds of crying infants evoked a much different response—one of trying to help or comfort the baby. Many of my male friends reported similar changes in their perceptions. Recent work has confirmed a potential hormonal basis for this anecdotal observation among new fathers (Fleming et al., 2002). New fathers and non-fathers were exposed to recorded infant cries and to control stimuli; heart-rate and endocrine responses, including salivary testosterone and cortisol, as well as blood prolactin concentrations prior to and after the stimuli, were assessed. Fleming and her co-workers reported that (1) fathers hearing the recorded crying felt more sympathetic and more alert than non-fathers, (2) fathers and non-fathers with low testosterone concentrations felt more sympathy and need to respond to the infant cries than fathers with high testosterone concentrations, and (3) fathers with high prolactin concentrations were also more alert and more responsive to the cries. The study also found endocrine differences between new and experienced fathers (Fleming et al., 2002). In common with previous experiments on biparental species, these studies of men indicate that human fathers are more responsive to infant cues than are non-fathers and that the responses of fathers to infant cues are associated with both hormone values and previous caregiving experience (Fleming et al., 2002).

Neural Changes Associated with Parental Behavior

The influences of hormones on the neural bases of parental care will be presented in this section. Again, most of the research on this topic has been done on maternal behavior in rats. There are very few studies addressing the neural control of parental behavior in nonmammalian species. Studies on rats indicate that estrogen promotes maternal behavior by enhancing pup-stimulated neural activity in the MPOA, the BNST, and the dorsal and intermediate lateral septum. These brain regions are considered part of a neural circuit underlying maternal behavior (Sheehan and Numan, 2002). In contrast, progesterone may inhibit maternal behavior in rats by inhibiting neural activity in parts of this brain circuitry (Sheehan et al., 2001; Sheehan and Numan, 2002).

Michael Numan, Alison Fleming, and Judith Stern have recently provided much insight into the interactions among the nervous system, hormones, and parental behavior. The first attempts to elucidate the neural bases of maternal behavior were made by Beach (1937) and Stone (1938c). Beach made multiple neocortical lesions in rats, and found that small lesions failed to disrupt the onset or maintenance of rat maternal behavior, but that larger lesions seriously interfered with the expression of most components of maternal care. Later, certain hypothalamic lesions were shown to seriously disrupt maternal behavior (Figure 7.30; Numan, 1990). Recently, several experiments have implicated the MPOA as a critical component of maternal behavior (Numan, 1988, 1990; Numan and Insel, 2003). Lesions of the MPOA eliminate performance of maternal behavior in rats (Numan, 1974; Numan and Sheehan, 1997).

Of course, the elimination of a behavior after the destruction of a brain region does not necessarily indicate that the destroyed brain area mediates the behavior in question. It is always possible that the destruction of axons of cell bodies originating outside of the lesioned area caused the change in observed behavior. One way to address this issue is to perform selective lesions of the neuronal cell bodies in a specific brain area. The excitotoxic *N*-methyl-DL-aspartic acid (NMA) is an amino acid that selectively kills neuronal cell bodies, but spares the axons that pass through them. When NMA was injected bilaterally into the MPOA of normally behaving mother rats 4 days postpartum, maternal behavior was severely curtailed. Bilateral NMA injections into the lateral preoptic area and adjoining regions, called the substantia innominata, also severely interfered with the display of maternal behavior (Numan et al., 1988). Estrogen implants into the MPOA stimulated maternal behavior in male rats (Rosenblatt et al., 1996; Rosenblatt and Ceus, 1998). Other brain areas are also involved in maternal behavior in rats; for instance, neurons in the lateral preoptic area contribute to pup retrieval, nest building, and nursing behaviors (Numan, 1990). There are also data suggesting that the preoptic area is involved in mediating parental care in hamsters (Miceli and Malsbury, 1982b), rabbits, (González-Mariscal, 2001), and ring doves (Komisaruk, 1967).

The nucleus accumbens appears to play a crucial modulatory role in the performance of maternal behaviors in rats (Li and Fleming, 2003a). Lesions of specific parts of the nucleus accumbens revealed that removing the shell, but not the core, disrupted pup retrieval. Females with lesions to the shell of the nucleus accumbens required more time to collect all their pups than non-lesioned rats. However, the latency to first pup retrieval was not affected by the lesions, suggesting a specific deficit in performance. No other component of maternal care was affected by these lesions. These results suggest that the shell of the nucleus accumbens may be required to maintain maternal attention or motivation (Li and Fleming, 2003a).

Additional evidence supporting the importance of specific brain regions in maternal behavior has been provided by immunocytochemical visualization of the nuclear protein product (c-Fos) of the immediate early gene *c-fos* as an indi-

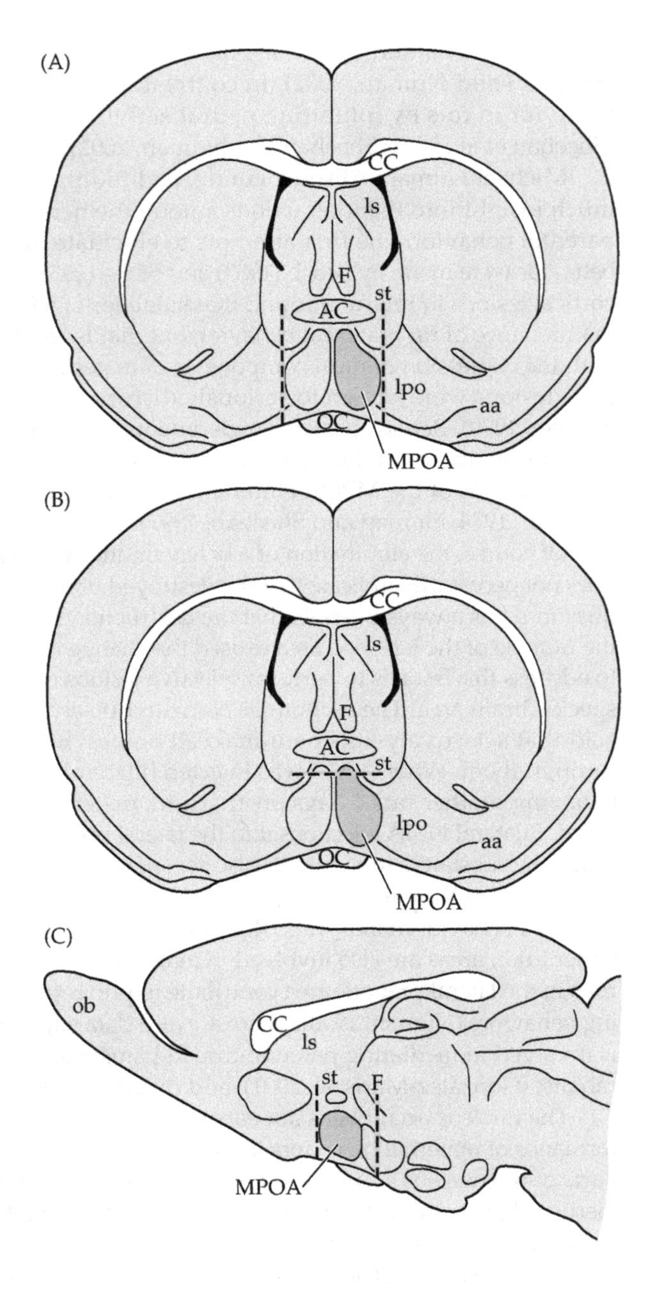

7.30 Lateral projections from the medial preoptic area (MPOA) are critical in rat maternal behavior. MPOA lesions severely disrupt the onset of maternal behavior in rats. (A) Coronal section of rat brain at the level of the MPOA, showing lesions (dashed lines) that sever the lateral connections of the MPOA to the rest of the brain. (B) Coronal section showing a lesion that severs the dorsal connections of the MPOA. (C) Sagittal section, showing lesions that cause anterior and posterior isolation. Only cuts severing the lateral connections interfere with maternal behavior. CC = corpus callosum; OC = optic chiasm; ob = olfactory bulb; MPOA = medial preoptic area; lpo = lateral preoptic area; aa = anterior amygdaloid nuclei; AC = anterior commissure; F = fornix; ls = lateral septum; st = bed nucleus of the stria terminalis. After Numan, 1990.

cation of neuronal activity. Several studies have reported an elevation in c-Fos production in the MPOA in rat dams (e.g., Fleming et al., 1994; Fleming et al., 1996; Lonstein et al., 1998; Numan and Numan, 1994; Numan et al., 1998). In one study, female rats with and without experience as mothers were allowed to interact with pups in a perforated box, then re-exposed to pups in the box, to the box alone, or left in the home cage without further stimulation (Fleming and Korsmit,

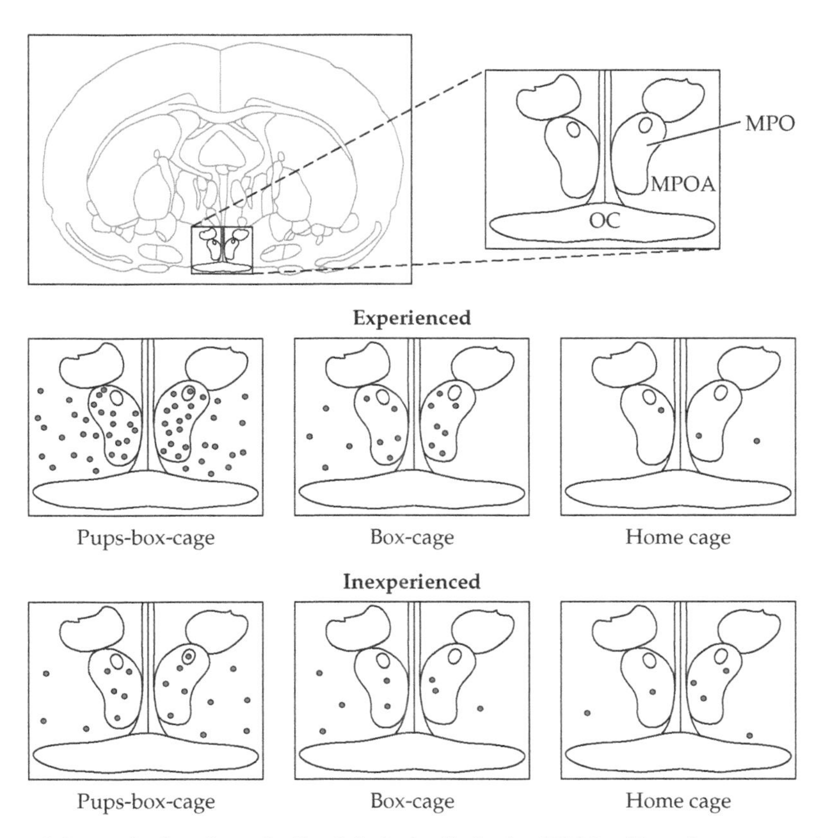

7.31 Schematic drawing of c-Fos-labeled cells in the MPOA of female rats. Female rats with and without experience as mothers were allowed to interact with pups in a perforated box, then re-exposed to pups in the box, to the box alone, or left alone in the home cage. Experienced females showed increased c-Fos concentrations after exposure to pups. Each black dot represents five labeled cells. MPOA = medial preoptic area; MPO = medial preoptic nucleus; OC =optic chiasm. After Fleming and Korsmit, 1996.

1996). Experienced rats showed increased c-Fos levels in the MPOA, the basolateral amygdala, the parietal cortex, and the prefrontal cortex (Fleming and Korsmit, 1996) in response to pup exposure (Figure 7.31). In another study using c-Fos as a marker for neural activity, dams were separated from their 5-day-old offspring for 48 hours to down-regulate the *c-fos* gene. The dams were then reunited with pups that were capable or incapable (due to anesthetized snouts) of suckling. In a third group, dams were presented with pups housed in a double wire mesh box so that they could see, smell, and hear the pups, but not touch them. Two additional control groups were also included in this study: one group of dams that was presented with the wire mesh box alone, and a group of females that had no further stimulation (Lonstein et al., 1998). Physical interaction with rat pups induced high c-Fos levels in the MPOA, regardless of whether or not the pups nursed (Lonstein et al., 1998). Expression of *c-fos* was relatively low in the three control groups. These results indicate that the MPOA is involved in maternal behavior, but not in response to neuroendocrine changes or sensory inputs.

As previously described, rats are quite neophobic—that is, they fear novel stimuli in their environment—and newly born pups are truly novel stimuli. Juvenile rats are often less neophobic than adults, and they readily perform parental tasks when presented with pups (Brunelli and Hofer, 1990). After an initial bout of sniffing the pups, adult nonpregnant female and male rats initially react to them in one of two ways. Most animals make a rapid retreat, then tend to ignore and deliberately avoid the squeaking pups. Some animals are apparently so frightened that they attack the pups and may even cannibalize them. As we saw above, newly parturient rat dams overcome their timidity and behave solicitously toward their offspring. One function of hormones in mediating the onset of maternal behavior might be reduction of the fear associated with the newly arrived pups. In fact, hormones not only reduce the negative valence associated with the pups, but also separately increase the positive valence of the pups, thus drawing the mother into close contact with them. In other words, the pups become attractive stimuli, a phenomenon new human parents frequently report about their new babies.

A major component of the avoidance of pups is a chemosensory-based repulsion (Rosenblatt, 1990). The repulsiveness of the pups' odors appears to be overcome by exposure to maternal hormones or by repeated experience with pups during the process of concaveation. This repulsion can also be overridden by making nulliparous females anosmic. As described in Chapter 5, rats have two sets of chemosensory receptors, the olfactory receptors and the vomeronasal organ. Chemosensory information from both inputs is relayed from the primary and accessory olfactory bulbs to the cortical and medial nuclei of the amygdala. The amygdala projects, via the stria terminalis, to both the BNST and the MPOA; the BNST also projects to the MPOA (Figure 7.32; Fleming and Rosenblatt, 1974).

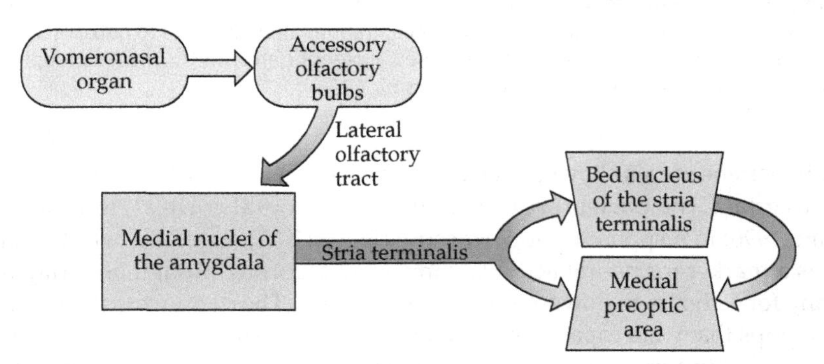

7.32 The vomeronasal organ–MPOA pathway. Information from the vomeronasal organ travels directly to the accessory olfactory bulbs, then to the medial nuclei of the amygdala via the lateral olfactory tract. Via the stria terminalis, information from the vomeronasal organ eventually reaches the BNST or the MPOA. Information from the BNST is also relayed to the MPOA. Lesions anywhere along this pathway interfere with chemosensory processing and hasten the onset of maternal behavior in rats. After Numan, 1990.

Lesioning both types of olfactory receptors with intranasal infusions of $ZnSO_4$ results in female nulliparous rats that do not avoid pups and which begin behaving maternally with a short latency.

The amygdala appears to inhibit maternal behavior in rats (Fleming et al., 1996; Sheehan and Numan, 2002). Lesions of the medial nuclei of the amygdala accelerate concaveation in nulliparous females (Fleming et al., 1980). Lesions anywhere along the pathway from the vomeronasal receptors to the amygdala also facilitate the onset of maternal behavior in nulliparous rats (Figure 7.33; Fleming and Rosenblatt, 1974; Fleming et al., 1979, 1980, 1983). Taken together, these results suggest that both olfactory and vomeronasal inputs to the MPOA normally inhibit maternal behavior in rats, and that maternal behavior occurs when the MPOA is released from this inhibition by hormones associated with late pregnancy. For example, implants of estradiol into the MPOA induces maternal behavior in rats. The medial amygdala projects to the anterior/ventromedial hypothalamic nuclei in a neural circuit that inhibits maternal behavior; the principal bed nucleus of the stria terminalis, ventral lateral septum, and dorsal premammillary nucleus also may be involved in this inhibitory circuit (Sheehan et al., 2001). Selective

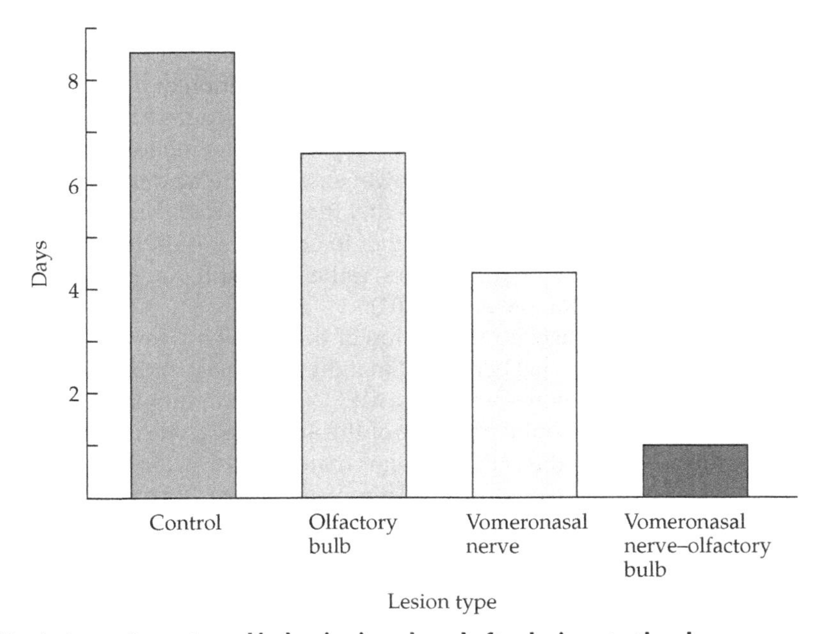

7.33 Latency to maternal behavior is reduced after lesions to the chemosensory organs. Adult female rats that cannot smell pups will behave maternally toward them almost as quickly as newly parturient mothers. Control rats began to behave maternally after about 8 days of exposure to pups. Lesions of the olfactory bulbs and vomeronasal nerve result in latencies to maternal behavior of about 7 and 4.5 days, respectively. If both the olfactory bulbs and vomeronasal nerve are damaged, female rats will behave maternally after only 1 day of exposure to pups. After Fleming et al., 1979.

lesions along the vomeronasal organ–amygdala pathway disinhibit the MPOA and allow the expression of maternal behavior in both virgin and pregnant rats.

The use of fMRI has revealed some important neural sites involved in maternal responsiveness among human mothers. In one early study, using only four participants, mothers listened to recorded infant cries and white noise control sounds while they underwent fMRI of the brain (Lorberbaum et al., 1999). Only the anterior cingulate and right medial prefrontal cortex showed significant changes in response to the infant cries as compared with white noise. More recently, fMRI was used to determine brain responses to infant crying and laughing in mothers and fathers of young children and in women and men without children (Seifritz et al., 2003). Regardless of parental status, women, but not men, displayed neural deactivation in the anterior cingulate cortex in response to both infant crying and laughing. Regardless of sex, parents showed increased activation in the amygdala and other limbic areas (Seifritz et al., 2003). Interestingly, parents (independent of sex) displayed more activation in response to infant crying, whereas non-parents displayed more activation in response to infant laughing! In another recent study, new mothers viewed photographs of their own baby, another baby, and adult faces while undergoing fMRI (Nitschke et al., 2004). Mothers displayed significant bilateral activation of the orbitofrontal cortex while viewing pictures of their own as compared with unfamiliar infants. While in the scanner, mothers rated their mood more positively when viewing pictures of their own infants than for unfamiliar infants, adults, or at baseline. Thus, activation of the orbitofrontal cortex correlated positively with pleasant mood ratings. Areas of the visual cortex also displayed differential activation in response to the mothers' own babies and unfamiliar infants; this activation, however, was unrelated to mood ratings (Nitschke et al., 2004). These data suggest that the orbitofrontal cortex activation represents the affective responses of a mother to her infant (Nitschke et al., 2004). Similar results were obtained when movies, rather than still photos, were shown to mothers during fMRI (Ranote et al., 2004).

But let us return to the central question of this chapter: How do hormones stimulate the onset of maternal behavior? In light of the results described above, one possibility is that hormones act on the MPOA to disinhibit maternal responsiveness. As noted above, in the presence of the appropriate hormonal priming, implants of estradiol into the MPOA trigger maternal behavior in rats. Similar implants into other neural regions do not stimulate maternal behavior (Fahrbach and Pfaff, 1986). As one might expect, the number of estrogen receptors in the rat MPOA (Figure 7.34) increases during pregnancy (Giordano and Rosenblatt, 1986). The protein products produced by cells in the MPOA in response to estrogen stimulation have yet to be fully characterized, but one of these products is oxytocin receptors (Champagne et al., 2001). In rats, central oxytocin receptor levels are functionally linked to behavioral differences in maternal care (Champagne et al., 2001; Francis et al., 1999; Insel and Shapiro, 1992).

As with human mothers, maternal care varies among rat mothers. Stable individual differences in pup licking and grooming emerge during the first week postpartum (Champagne et al., 2003). Such naturally occurring variations in maternal

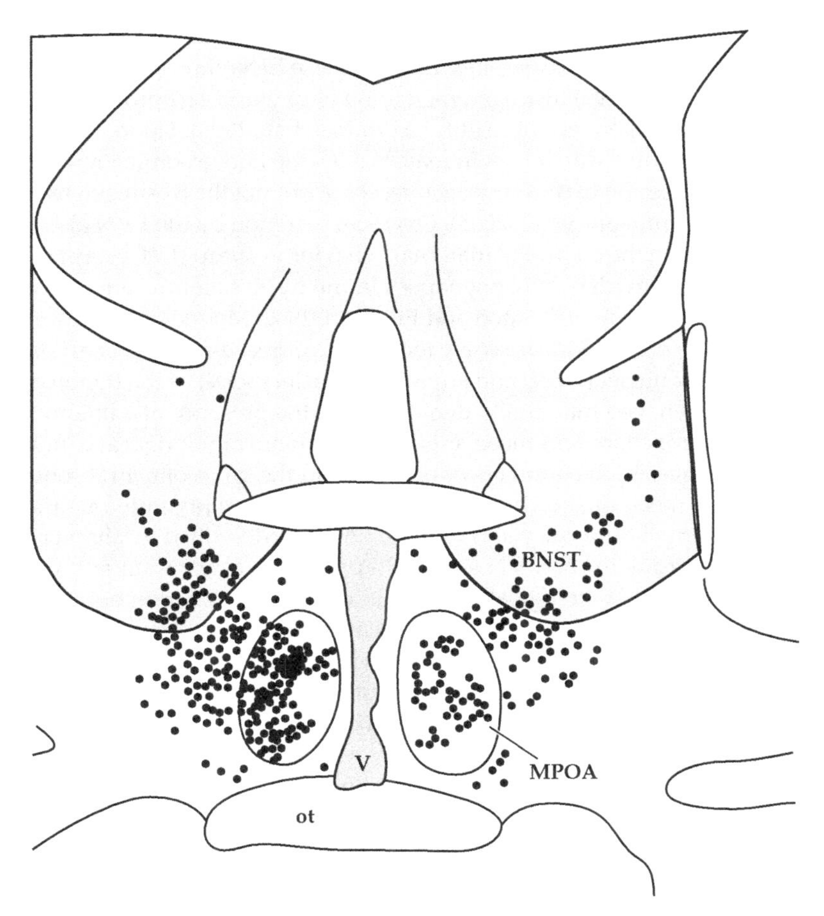

7.34 Estradiol receptors in the rat POA. In this schematic, magnified coronal section of a rat brain, each black circle represents a cell in the POA that binds estradiol and projects through or to the ventromedial midbrain. BNST = bed nucleus of the stria terminalis; MPOA = medial preoptic area; ot = optic tract; V = third ventricle. After Fahrbach et al., 1986.

behavior are associated with differences in estrogen-inducible oxytocin receptors in the MPOA (Champagne et al., 2001). Offspring of mothers that display high levels of pup licking and grooming display attenuated responses to stress and also display enhanced cognitive ability (Chapter 11) (Liu et al., 1997, 2000; Caldji et al., 1998; Francis et al., 1999). These individual differences in maternal care are transmitted across generations: offspring of mothers that engage in high rates of licking and grooming exhibit high rates of licking and grooming as mothers themselves, whereas the offspring of mothers expressing low rates of licking and grooming become mothers that display relatively low rates of licking and grooming their offspring (Francis et al., 1999). Oxytocin receptor binding in the MPOA is greater in lactating females that exhibit high rates of licking and grooming than in lactating females displaying low rates of these behaviors (Champagne et al., 2001; Francis et al., 1999). Estrogen regulation of oxytocin receptors in the MPOA requires the α estrogen receptor subtype (Young et al., 1998). Upon further examination, the individual differences in maternal licking and grooming reflected variation in the expression of the α estrogen receptor, but not the β estrogen receptor, in the MPOA (Champagne et al., 2003). Oxytocin secretion by the PVN also appears to be important in the control of maternal behavior in sheep (DaCosta et al., 1996).

There is also evidence that synapses in the hypothalamus are restructured during maternal behavior (Hatton and Ellisman, 1982; Modney and Hatton, 1990). Performing the electron microscopic technique of freeze-fracture analysis on the paraventricular nuclei (PVN) and supraoptic nuclei (SON) of the hypothalami of rats that had behaved maternally demonstrated the presence of a unique postsynaptic structural change in those animals. Restructuring of neural connections, as well as changes in the numbers of neurons and the types of neural connections and coupling mechanisms, have also been reported in juvenile rats that have behaved maternally (Modney and Hatton, 1990). As described in Chapter 2, magnocellular neurons in the PVN and SON produce vasopressin and oxytocin. Nearly all of the morphological changes in neural organization observed in rat dams are reversed approximately 1 month after their pups are weaned. It is likely that the secretion of oxytocin associated with parturition and lactation accounts for these changes in neuronal structure, but it remains possible that the change in postsynaptic specialization could reflect oxytocin-mediated effects on maternal behavior, which in turn causes the neural changes. Presumably, progesterone and prolactin also affect neuronal structure or function underlying rat maternal behavior. The ultrastructural changes that are observed in the PVN and SON of rat dams are also observed in nonlactating, nulliparous females that behave maternally after undergoing concaveation (Modney and Hatton, 1990). Other studies of neurons in the SON have demonstrated that oxytocinergic neurons show a reduction in dendritic length because of decreased branching (approximately 40% reduction) during lactation (Stern and Armstrong, 1998). In contrast, SON neurons containing vasopressin show an approximate 50% increase in dendritic length during lactation (Stern and Armstrong, 1998).

Thus, merely behaving maternally causes structural changes in the brain independently from hormonal influences. Do similar structural brain changes occur in males that behave in a paternal manner? This question is best addressed in a species that normally shows paternal behavior. Perinatal exposure to androgens suppresses parental responsiveness in many species of rodents (Kinsley, 1990). As we have seen, androgens organize sex differences in anatomical structures in several brain areas of rodents and primates, including the MPOA (see Chapter 4). Are these structural differences in the organization of the MPOA responsible, for instance, for the sex difference in propensity to behave parentally? There is a sex difference in the size of the MPOA in California mice; the volume of the MPOA in virgin males is larger than in nulliparous females. Paternal behavior in California mice is associated with a decrease in the number of neurons in

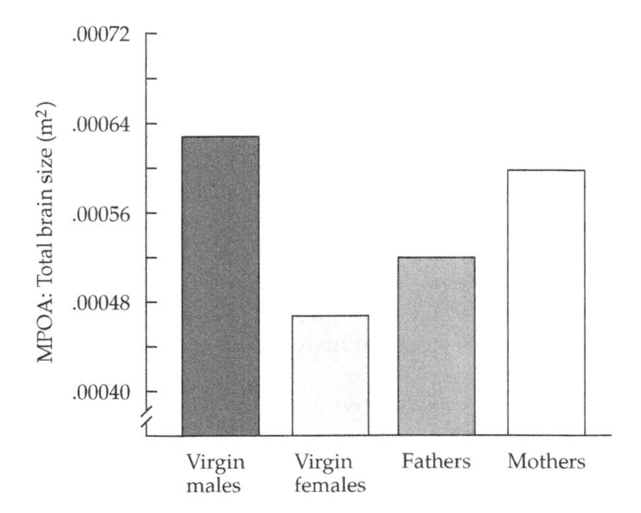

7.35 Effects of parental behavior on MPOA volume in the California mouse. There is a sex difference in the size of the MPOA in California mice: the volume of the MPOA of males without sexual experience is larger than that of nulliparous females. Paternal behavior caused the number of neurons in the MPOA to decrease and its overall size to diminish to approximately that found in females. The MPOA volume of females does not change significantly when they become mothers. After Alberts and Gubernick, 1990.

the MPOA (Alberts and Gubernick, 1990) and consequent shrinking of MPOA volume to approximately that of a female (Figure 7.35). There is no change in the size of the MPOA in females when they become mothers; this finding suggests that hormones act in females of this species, as in rats, to disinhibit maternal behaviors. The reduction in the number of neurons during paternal behavior suggests that the circuitry underlying this behavior is typically inhibited unless hormones activate some process leading to neuronal death, possibly by destroying the inhibitory neurons themselves. As noted above, the main endocrine correlates of paternal behavior thus far discovered are an increase in blood prolactin and a decrease in blood testosterone concentrations, but it has not been established whether prolactin or some other hormone causes the cell death associated with paternal behavior in this species.

You will recall from Chapter 1 that mice with a specific gene selectively disrupted are called knockout mice. Mice lacking the immediate early gene *fosB* display little maternal behavior (Brown et al., 1996). The lack of maternal behavior among *fosB* knockouts ($FosB^{-/-}$) does not correspond to a lack of fertility or the ability to lactate. $FosB^{-/-}$ mothers appear to have endocrine profiles similar to those of wild-type (WT) dams, but they fail to display nest building, cleaning and retrieving of pups, nursing, or protective crouching postures (Brown et al., 1996). It is not clear through what mechanism(s) the missing *fosB* gene affects maternal behavior, although as noted previously, expression of its protein products occurs in brain regions considered part of the maternal behavior neural circuit (Numan and Insel, 2003).

As we saw above, oxytocin is an important mediator of maternal behavior. It was therefore surprising that in two studies, female oxytocin (OT) knockout mice ($OT^{-/-}$) showed the full complement of normal maternal behaviors (Nishimori et al., 1996; DeVries et al., 1997). $OT^{-/-}$ females, however, failed to eject milk in response to suckling, so their pups had to be cross-fostered to WT mothers to survive. How do we interpret these data? Do we now conclude that oxytocin is *not* involved in maternal behavior? Not necessarily. There are alternative explanations for the results of both studies of $OT^{-/-}$ animals. It appears that maternal behavior is easily induced in the particular strain of mouse used in the first study simply by brief exposure to pups (Young et al., 1997). Thus, knocking out the OT gene in rats might be a better test of the hypothesis that OT is necessary for maternal behavior. The other $OT^{-/-}$ mouse study also had problems: the OT gene was only partially deactivated in this knockout. In addition, vasopressin concentrations were lower than normal because the neurophysin molecule was also damaged in this knockout (DeVries et al., 1997). Consequently, additional research must be conducted to clarify the role of oxytocin in maternal behavior (Winslow and Insel, 2002).

In contrast to the disquieting results associated with the $OT^{-/-}$ mice, mice with the prolactin receptor (PRLR) gene deleted display defective maternal behavior, as expected (Lucas et al., 1998). Specifically, pup-induced maternal behaviors were disrupted in both $PRLR^{-/-}$ and $PRLR^{+/-}$ females, and primiparous $PRL^{+/-}$ females displayed dramatic deficiencies in maternal care of foster pups (Lucas et al., 1998). These studies clearly indicate that the prolactin receptor is critical in mediating maternal behavior.

One final example from knockout mice adds an evolutionary twist to the role of genes in maternal care. *Mest* (also known as *Peg1*) is an imprinted gene, meaning that it is expressed only from the paternal allele during development (Keverne, 2001). *Mest*-deficient females displayed impaired maternal behavior and impaired placentophagia. These results provide evidence for the involvement of an imprinted gene from the father in the control of adult behavior of female offspring (Lefebvre et al., 1998).

These observations potentially open up a new area of inquiry into the mechanisms underlying hormonal mediation of maternal behavior. Behavioral studies of knockout mice may also evoke questions about established hormonal relationships. The missing genes may have subtle effects on maternal behavior; indeed, most knockout mice studied to date are successfully raised by their knockout mothers. The extent to which the altered behavior of knockout mice reflects atypical maternal behavior remains unknown (Caldji et al., 1998; Kinsley, 1994; Winslow and Insel, 2002). Cross-fostering studies comparing the behavior of knockout mice reared by WT dams with that of knockout mice reared by knockout mothers will be critical in the future to further untangle the roles of extrinsic (environmental) and intrinsic (hormonal) factors in maternal behavior.

Conclusions

A good understanding of the manner in which parental behavior is affected by hormones has recently been gained in a few species, but the parental behaviors shown by individuals in the various orders and families of birds and mammals are extremely diverse. Are the underlying endocrine mechanisms, and the hormones associated with them, equally varied, or do just a few select hormones produce this diversity of parental responses? Too few species have been studied to answer this question at this time. However, it is reasonable to propose that a combination of common hormonal signals is paired with a variety of extrinsic cues to elicit the species-appropriate constellation of parental behaviors.

As with other behaviors we have examined, hormones affect parental behavior at the level of input, integration, and output systems. Sensory inputs related to the perception of rat pups appear to be changed by hormones in parturient rat dams. Nonparental rats fear pups, but hormones alter mothers' perception of their pups to make them attractive stimuli. Sensory input associated with nursing probably is perceived as rewarding via the actions of oxytocin and endogenous opiates. The effects of hormones on central nervous system processing have been well documented in rat maternal behavior. Restructuring of neural connections, as well as changes in numbers of neurons and the types of neural connections and coupling mechanisms, has been reported in mother rats. Maternal behavior is integrated and consolidated in the MPOA and inhibited in the amygdala. Finally, hormones affect output systems to mediate maternal behavior. For instance, steroid hormones stimulate mammary gland development in mice; without mammary gland development, suckling stimuli could not trigger the onset of maternal aggression in this species. In birds, prolactin stimulates development of the brood patch, and feedback from this morphological feature encourages further incubation behavior.

Thus, parental behaviors are mediated by hormones via their effects on input, integration, and output systems, as well as on the interactions of those systems with experiential factors. Although much has been learned about this important topic, there is still a great deal left to learn. Parental care is fascinating to study because of its fundamental importance to the survival and reproductive success of individuals, and because of the role of hormones in the plasticity of brain and behavior. The diversity of parental care patterns found in nature provides an opportunity for comparative studies that will reveal how different animals have evolved different solutions to the common problem of increasing the odds of offspring survival and reproductive success.

Summary

- Parental behavior is any behavior that contributes directly to the survival of fertilized eggs or offspring that have left the body of the female.
- There are many patterns of mammalian parental care. The developmental status of the newborn is an important factor driving the type and quality of parental care in a species. Maternal care is much more common than paternal care.
- Birds show great diversity in parental care, ranging from nest parasite species that show absolutely no parental care to species in which both sexes build nests, incubate the eggs, and feed and protect the young, though most species provide parental care.

- Most studies on hormonal correlates of avian parental behavior have been performed on ring doves, which share the parental chores more or less equally and also produce a protein-rich crop milk that is regurgitated and fed to the squab. Ovariectomy eliminates nest building and incubation behaviors in female ring doves. Estradiol and progesterone replacement restores these behaviors. Prolactin mediates brooding and stimulates development of the crop sac in both sexes. Gonadectomy does not affect nest building or incubation behaviors in male ring doves; behavioral cues from the female elicit these behaviors in males.
- The vast majority of research on the hormonal correlates of mammalian parental behavior has been conducted on rats. Rats bear altricial young, and mothers perform a cluster of stereotyped maternal behaviors, including nest building, crouching over the pups to allow nursing and to provide warmth, pup retrieval, and increased aggression directed at intruders.
- Sensitization of adult nonparental rodents to pups by daily exposure causes them to behave maternally, but several days are required in order for maternal behavior to be observed. This process is called concaveation.
- The onset of maternal behavior in rats is mediated by hormones. Several methods of study, including pregnancy termination, hormone removal and replacement therapy, and parabiotic blood exchange, have been used to determine the hormonal correlates of rat maternal behavior. A precipitous decline in blood concentrations of progesterone in late pregnancy after sustained high concentrations of this hormone, in combination with high concentrations of estradiol, and probably prolactin and oxytocin, induces female rats to behave maternally in the presence of pups. This hormone profile at parturition overrides the usual fear response of adult rats toward pups and permits the onset of maternal behavior.
- Laboratory strains of mice and rats are usually docile, but mothers can be quite aggressive toward animals that venture too close to their offspring. Progesterone appears to be the primary hormone that induces this maternal aggression in rodents, but species differences exist. The role of maternal aggression in women's behavior has not been adequately described or tested.
- Female rats become less maternal as their pups mature, and spend less time with the pups during each nursing bout. One hypothetical mechanism underlying this phenomenon involves body heat regulation. According to this notion, prolactin stimulates corticosterone secretion, which increases metabolic rate and subsequently raises the mother's body temperature. As the pups grow larger and produce more heat, they block the mother's ability to dissipate her own body heat. When her brain temperature rises acutely during nursing, she leaves the litter to reduce her heat load and avoid hyperthermia.
- Two- to three-week-old rat pups are attracted to the nest odors of lactating dams. The excess production of caecotrophs by lactating dams attracts the young to the nest, or to their mother, until they are fully weaned.

- With the exception of cortisol, few clear-cut endocrine correlates of human or other primate maternal behavior have been discovered. Behavior per se is rarely tested in humans; rather, responses to questionnaires or surveys are used to study hormone–behavior interactions. A further confounding variable is a lack of consistent or universal definitions of human maternal care.
- Elevated prolactin concentrations appear to mediate paternal behavior in the two best-studied mammalian species, the California mouse and the common marmoset. Reduction in blood concentrations of testosterone is also associated with paternal behavior in marmosets and humans.
- The medial preoptic area is critical for the expression of rat maternal behavior. The amygdala appears to inhibit the expression of maternal behavior. Lesions of the amygdala or afferent sensory pathways from the vomeronasal organ to the amygdala disinhibit the expression of maternal behavior. Hormones or concaveation probably act to disinhibit the amygdala, thus permitting the occurrence of maternal behavior.
- There is evidence that structural changes in neural organization occur during natural maternal and paternal behavior. These changes are reversible and also occur in animals exhibiting parental behavior after concaveation.

Questions for Discussion

1. If hormones are important in the onset of rat maternal behavior, then why do hormone replacement regimes require over 24 hours to induce maternal behavior in the presence of pups, as compared with natural conditions, in which fully developed maternal behavior appears as soon as the pups are encountered?

2. A common endocrine correlate of the onset of paternal behavior is a reduction of blood concentrations of androgens. Why do you think this is true?

3. Discuss the interaction among hormones, maternal care, and stage of development of mammalian offspring at birth. Would you expect that hormones associated with parturition would be more or less important in mediation of maternal care of altricial offspring? Why or why not?

4. Are studies of the hormonal correlates of paternal care in birds and mammals useful for understanding the mechanisms of paternal care in humans? Why or why not? Does the phenomenon of adoptive human fathers behaving paternally provide any insights? How can we account for individual differences in paternal care among men?

5. What do concaveation studies tell us, if anything, about the neural circuits mediating maternal behavior in rats?

6. Defend or refute the following statement: "Differences in patterns of parental care reflect differences in receptor availability and distribution, rather than differences in circulating hormone concentrations."

Refer to the accompanying Student CD for additional resources, including Web links, videos, animations, and additional photos.

Suggested Readings

González-Mariscal, G., and Poindron, P. 2002. Parental care in mammals: Immediate internal and sensory factors of control. In D. W. Pfaff, A. P. Arnold, A. M. Etgen, S. E. Fahrbach, and R. T. Rubin (eds.), *Hormones, Brain and Behavior*, Volume 1, pp. 215–298. Academic Press, New York.

Lonstein, J. S., and Gammie, S. C. 2002. Sensory, hormonal, and neural control of maternal aggression in laboratory rodents. *Neuroscience and Biobehavioral Reviews*, 26: 869–888.

Numan, M., and Insel, T. R. 2003. *The Neurobiology of Parental Behavior*. Springer-Verlag, New York.

Rosenblatt, J. S., and Snowdon, C. T. (eds.). 1996. *Parental Care: Evolution, Mechanisms, and Adaptive Significance*. Academic Press, New York.

Silver, R. 1977. *Parental Behavior in Birds*. Dowden, Hutchinson and Ross, Stroudsburg, PA.

Stern, J. M. 1997. Offspring-induced nurturance: Animal-human parallels. *Developmental Psychobiology*, 31: 19–37.

Wiesner, B. P., and Sheard, N. M. 1933. *Maternal Behavior in Rats*. Oliver and Boyd, London.

8

Hormones and Social Behavior

Social behavior encompasses interactions between individuals from which one or more of the individuals benefit (see Wilson, 1975). This definition is purposely very broad, and it includes aggressive behavior, particularly in the context of territorial defense, as well as nonhostile interactions, including affiliation, courtship, and parental behaviors. The effects of hormones on courtship and parental behaviors were described in previous chapters; this chapter will emphasize the influence of hormones on affiliative and aggressive behaviors.

Animals display a wide range of social behaviors. Some animals are highly solitary and have few social interactions; in fact, after they have been weaned, their only social interactions may occur during mating. Other animals live together in large colonies where social interactions are frequent. Social behaviors that bring animals together are referred to as **affiliation** (Carter et al., 1997b). Social behaviors that keep animals apart are usually considered to be **territorial behavior** or **aggression**. Although these two types of social behaviors have the opposite effects on social interactions, the neural and endocrine mechanisms underlying affiliation and aggression do not have opposite effects on a single neural substrate; rather, they appear to involve different neural circuits. Whereas scientific interest in affiliation has blossomed beginning in the 1990s, formal studies of aggression have been common since the 1920s.

Affiliation

Affiliation is generally considered to have evolved from reproductive and parental behaviors; in other words, short-term associations have evolved into long-term social bonds (Crews, 1997). Both reproductive and parental bonds are critical for individual reproductive success. Not surprisingly, then, the hormones that are important in promoting affiliative behaviors appear to have been co-opted from their regulatory roles in reproductive and parental behaviors.

Imaging Studies of Humans

Parents of newborns or newly adopted children often report that they not only love their offspring "at first sight," but also "fall in love with them" over time (Leckman and Mayes, 1999). Such a statement is reminiscent of romantic love with a long-term partner. Indeed, the highly conserved behavioral and neural systems underlying romantic and parental love have been linked to both addiction (Insel, 2003) and obsessive–compulsive disorders (Leckman and Mayes, 1999). Given their commonalities, it is not surprising that the relationship between romantic and parental love has been investigated on a neurobiological level.

In one recent study, the brain activity of human participants who were "deeply in love" was monitored by fMRI while they looked at pictures of their love interests or pictures of friends of similar age, sex, and duration of relationship as their partners (Bartels and Zeki, 2000). Brain activation in response to the pictures of romantic partners occurred in the medial insular and anterior cingulate cortex, as well as in subcortical regions including the caudate nucleus and putamen. Significant deactivation was observed in the posterior cingulate cortex and amygdala. Because these brain regions are different from those associated with emotion in previous studies, it remains possible that they are part of the specialized circuitry underlying the affective state of "love" (Bartels and Zeki, 2000).

An additional study by the same research team compared the neural correlates of maternal and romantic love directly (Bartels and Zeki, 2004). In addition to monitoring women in love who looked at pictures of their partners and nonromantic friends with fMRI, as previously described, the researchers monitored new mothers who looked at pictures of their babies or of other infants with whom they were acquainted (Bartels and Zeki, 2004). Romantic and maternal love resulted in some overlapping areas of brain activation, including the putamen and caudate nucleus, as well as the medial insular and anterior cingulate cortex (Figure 8.1). Romantic love specifically activated the dentate gyrus/hippocampus, hypothalamus, and ventral tegmental area. Maternal love specifically activated the orbitofrontal cortex and the periaqueductal gray (PAG) region. Interestingly, the brain regions that showed the most activation are either part of the human brain's reward circuitry (Kelley and Berridge, 2002) or contain a high density of receptors for the nonapeptides oxytocin and vasopressin, which have been shown to be important in forming social attachments in nonhuman animals (Insel and Young, 2001; Kendrick, 2000; Pedersen, 1997; Bartels and Zeki, 2004). We will return to the role of oxytocin and vasopressin in regulating social bonds below.

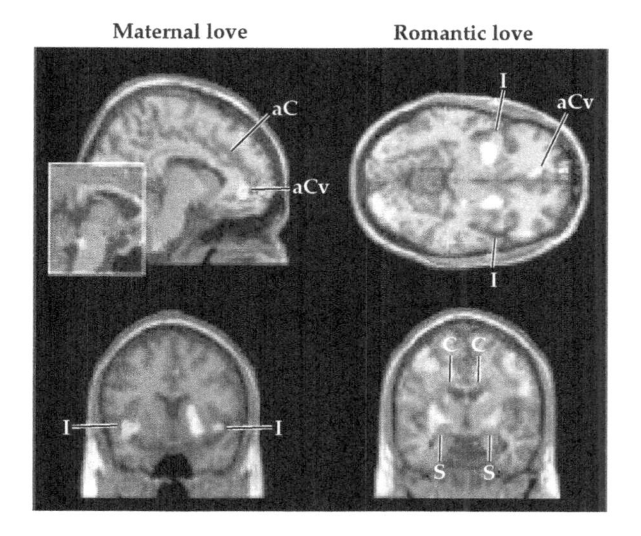

8.1 Romantic and maternal love evoke similar brain activity. The overlapping areas of activation include the anterior cingulate cortex (aC), the ventral anterior cingulate cortex (aCv), the medial insular cortex (I), the caudate nucleus (C), and the striatum (S). The darker gray reflects activation in the brains of women viewing pictures of their romantic partners as compared with pictures of a friend, whereas the whiter regions reflect activation in the brains of mothers viewing pictures of their infants as compared with pictures of other babies with whom they were acquainted. From Bartels and Zeki, 2004.

Adaptive Function of Affiliation

The evolutionary importance of the relationship between hormones and affiliation is suggested by considering a single hormone, testosterone. There are strong correlations among blood testosterone concentrations, testis size, sperm production, and social systems (i.e., monogamy versus polygamy) (Wingfield et al., 1997; Dixson, 1997). Males of monogamous species tend to have lower circulating testosterone concentrations, lower sperm numbers, and smaller testes than males of polygamous species. Formation of long-term pair bonds is critical for the reproductive success of some species, usually because both parents are required to rear the offspring. Because both parents have a 50% stake in the success of each offspring, they have evolved to cooperate because their offspring would die if either one abandoned the family. Hormones that evoke affiliation serve as the proximate means of bringing this about.

From the individual's perspective, the requirement for increased proximity during sexual reproduction may be potentially dangerous. Not only might the individual be attacked and maimed by a potential consort, but the potential for disease and parasite transmission increases during social gatherings (Crews, 1997). Mating behavior, as maternal behavior in rats, is a series of approach–avoidance behaviors. In order for laboratory rats to engage in successful maternal behavior, the inclination to flee from the frightening or aversive stimuli connected with pups must be suppressed, and the rats must approach the pups that previously elicited these aversive responses (Carter et al., 1997a). As noted in the previous chapter, neural–behavioral systems involving the medial amygdala have been implicated in avoidance behaviors, and neural–behavioral systems underlying approach behaviors include the medial preoptic area (MPOA) and the ventral portion of the bed nucleus of the stria terminalis (BNST) (Newman and Newman, 1997).

Voles have been important animal models in studies of affiliation. Voles are rodents, and although they are sometimes referred to as "field mice," they are

8.2 Prairie voles are socially monogamous rodents. Once a pair bond forms, the "couple" tends to stay together until death. Courtesy of Lowell L. Getz.

more closely related to lemmings than to laboratory mice (Figure 8.2). Voles inhabit much of the Northern Hemisphere, occupying a variety of habitat niches. Different species of voles display very different types of social organization. For example, prairie voles (*Microtus ochrogaster*), which inhabit grasslands throughout much of the midwestern United States, are socially monogamous. Meadow voles (*Microtus pennsylvanicus*), a closely related species, inhabit similar habitat throughout much of the eastern United States, but display a highly polygamous social system (reviewed in Klein and Nelson, 1998). Individuals of another closely related species, montane voles (*Microtus montanus*), inhabit grasslands in the West, and are also polygamous. Prairie voles are highly tolerant of other individuals, whereas meadow and montane voles are intolerant of other individuals during the breeding season, except to mate (Wolff, 1985). What are the endocrine bases of these differences in social organization?

Hormones and Affiliation

One obvious hormonal candidate is testosterone. Males of polygamous species generally have larger testes and higher testosterone concentrations than males of monogamous species (Dixson, 1997; Klein and Nelson, 1998). Plasma testosterone concentrations in male prairie voles are about half of those in meadow or montane voles (Klein and Nelson, 1998). However, supplemental testosterone treatment of male prairie voles does not make them polygamous, and castration of male meadow or montane voles does not make them monogamous (Roberts et al., 1996; Gaines et al., 1985).

Oxytocin has been implicated in several forms of affiliative behavior, including parental care, grooming, and sexual behavior (Carter et al., 1992, 1995). The locations of oxytocin receptors in the brains of monogamous prairie voles and polygamous montane voles have been characterized using in vitro receptor autoradiography (Insel and Shapiro, 1992). In prairie voles, oxytocin receptor densities were highest in the prelimbic cortex, BNST, nucleus accumbens, and the lateral aspects of the amygdala; oxytocin receptors were not evident in these brain regions in montane voles (Insel and Shapiro, 1992) (Figure 8.3). A similar pattern of social organization-dependent oxytocin receptor distribution was observed in

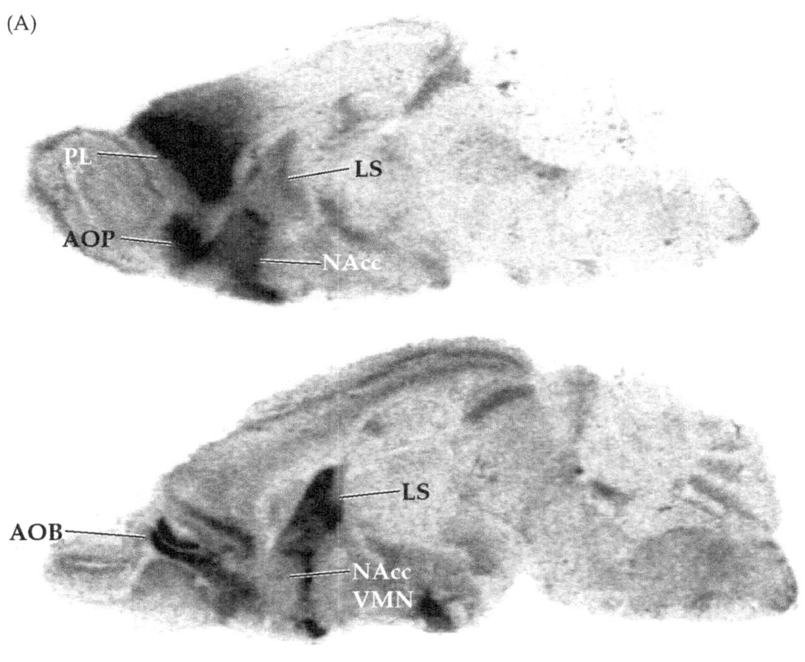

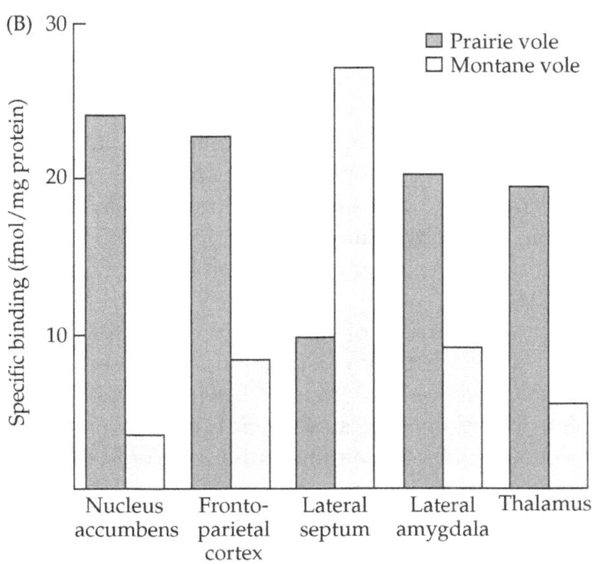

8.3 The pattern of oxytocin binding differs between monogamous and polygamous male voles. (A) In these autoradiograms, the dark regions represent ^{125}I-oxytocin antagonist binding in saggital sections of the brains of a monogamous prairie vole (top) and a polygamous montane vole (bottom). In the prairie vole, oxytocin receptor densities are highest in the prelimbic cortex (PL), nucleus accumbens (NAcc), and anterior olfactory nucleus (AOP); in the montane vole, receptors are not evident in these areas. LS = lateral septum; AOB = accessory olfactory bulb; VMN = ventromedial nucleus of the hypothalamus. (B) Levels of oxytocin binding in different brain regions of adult male prairie voles and montane voles. (A) from Insel, 1997; (B) after Insel 1997.

the brains of monogamous (*Peromyscus californicus*) and polygamous mice (*Peromyscus maniculatus*) (Insel et al., 1991), as well as in the brains of monogamous pine voles (*Microtus pinetorum*) and polygamous meadow voles (Insel and Shapiro, 1992). The differences in oxytocin receptor distribution appear around the time of weaning (Wang and Young, 1997). Differences in the patterns of vasopressin receptor distribution in the brains of prairie and montane voles have also been reported (Insel et al., 1994). Male prairie voles have higher numbers of the V1a subtype of vasopressin receptors in the ventral pallidum, medial amygdala, and mediodorsal thalamus than montane voles.

The difference in adult affiliative behaviors is reflected in the responses of infant prairie and montane voles to separation from their families (Shapiro and Insel, 1990). When prairie vole pups were isolated from their families for 5 minutes, they emitted over 300 distress vocalizations when they were 4–6 days old and approximately 600 distress calls at 8–10 days of age. Plasma corticosterone concentrations were elevated fourfold to sixfold in isolated prairie vole pups. When montane vole pups were socially isolated, they did not emit any distress vocalizations, and there were no changes in their blood corticosterone concentrations (Shapiro and Insel, 1990). If montane vole pups were stressed by tail suspension (being picked up and held by the tail) or halothane (a gaseous anesthetic) vapors, they emitted distress vocalizations and showed elevated corticosterone concentrations. This latter observation indicates that the montane vole pups were capable of responding to stress, but apparently did not find social isolation stressful (Shapiro and Insel, 1990).

Separation anxiety such as that displayed by prairie vole pups is part of the phenomenon of **attachment** (Bowlby, 1969; Ainsworth, 1972), in which one individual (usually an infant) strives to maintain proximity to a specific other individual (usually the primary caregiver). Generally, attached individuals display distress if separation or loss is experienced, and are motivated to reestablish contact after separation (Mendoza and Mason, 1997). The concept of attachment was developed for human infant–parent interactions, but it has been applied to nonhuman animals as well. Social bonding occurs in many species and is particularly salient in parent–offspring interactions (Pedersen, 1997) (see Chapter 7). Even when all their physiological requirements are met—for example, with full stomachs in a warm, dry environment—young animals of some species respond to social isolation promptly with characteristic distress vocalizations. In other words, babies cry when unattended (Kalin et al., 1988; Zimmerberg et al., 1994). The reaction of infants to social separation is fast and quite consistent across many disparate species (Panksepp et al., 1997).

A series of fascinating comparative studies of New World primates has provided insights into the hormonal bases of attachment behavior. This research has focused on squirrel monkeys (*Saimiri sciureus*) and titi monkeys (*Callicebus moloch*), two arboreal, omnivorous New World primate species. Squirrel monkeys are polygynous and live in large social groups containing individuals of both sexes and all ages. With the exception of mother–infant interactions, squirrel monkeys typically interact within same-sex cohorts (Mendoza et al., 1991; Mendoza and

Mason, 1997). Titi monkeys are socially monogamous; they live in small family groups, and the males provide substantial parental care (Mendoza and Mason, 1986). The infants spend more than 90% of their time riding on the backs of their fathers, transferring to their mothers only to nurse. Not surprisingly, titi infants form a very strong bond with their fathers (Figure 8.4).

If a squirrel monkey mother and her infant are separated for 30–60 minutes, both animals shows a robust increase in cortisol concentrations (Coe et al., 1978). Because separating a mother and infant causes substantial commotion and distress among their cagemates, a control procedure was conducted in which all of the steps involved in separating the mother and infant were taken, but the pair was reunited within seconds. In this case, there was no increase in cortisol in either the mother or infant (Mendoza and Mason, 1997). Separation of pairs of adult animals, regardless of their relationship within the social group, did not influence their cortisol values, although cortisol concentrations were markedly

(A)

(B)

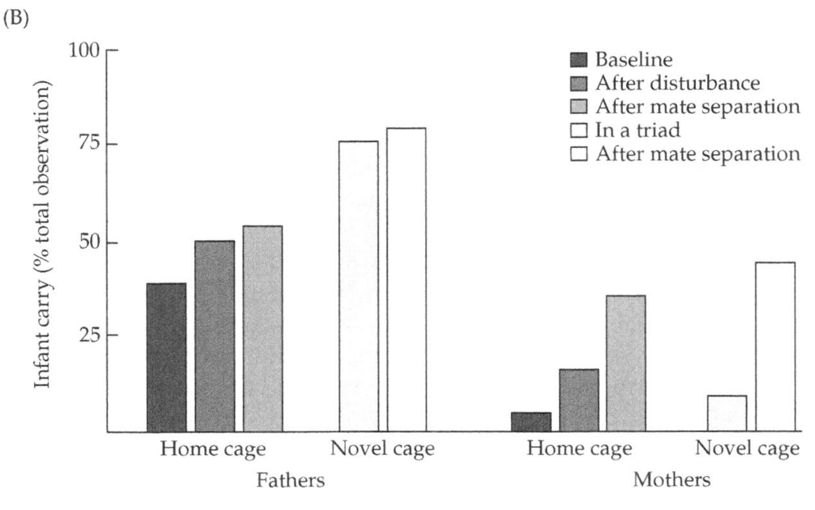

8.4 Infant titi monkeys receive most of their contact comfort from their fathers. (A) An infant titi monkey rides on the back of his father. (B) The graph shows the percentage of the observation time during which infants were carried by their fathers or their mothers under several conditions: at baseline, after a disturbance, and after mate separation in the home cage; and in a triad (infant, mother, and father) and after mate separation in a novel environment. Note that most of the time, fathers carried the offspring, even during times of stress. (A) photo courtesy of DeeAnn Reeder; (B) after Mendoza and Mason, 1986.

decreased during the initial formation of female, but not male, pairs (Saltzman et al., 1991).

A different pattern of results was obtained in similar studies of the socially monogamous titi monkey (Mendoza and Mason, 1997). Separation of titi infants from their fathers for 60 minutes caused a significant increase in their cortisol concentrations even if their mothers were present; separation from their mothers did not affect cortisol concentrations in the infant titi monkeys (Hoffman et al., 1996). Separation from the titi infant did not affect cortisol values in either parent. These different attachment patterns could contribute to the different social structures of these New World primate species (Mendoza and Mason, 1997), or they could be the result of these different social structures. These patterns may also serve to inform us about the different patterns of human social behavior (Keverne, 1992).

The formation of social bonds has also been studied in socially monogamous prairie voles (Carter et al., 1995; Carter and Keverne, 2002). Male and female prairie voles form long-term pair bonds, characterized by a social preference for a familiar partner, and in some cases selective aggression directed toward unfamiliar conspecifics (Getz and Carter, 1996; Winslow et al., 1993). Social preferences for a familiar partner are not restricted to heterosexual pairs, however; partner preferences can form between animals of the same sex and animals of the opposite sex following nonsexual cohabitation (DeVries et al., 1995a). In the laboratory, social preferences can develop when socially naive animals (animals that have been individually-housed since weaning) encounter and interact with strangers of the opposite sex (Carter et al., 1995). These social preferences can be determined in the three-chamber preference apparatus described in Chapter 6. Typically, when socially naive prairie voles meet, they engage in a brief session of olfactory investigation, followed by prolonged periods of sitting in close proximity (Gavish et al., 1983; Moffatt, 1994). Separation by a wire screen barrier prevents the development of social preferences, suggesting that physical contact or other forms of active interaction is critical for their development (Shapiro et al., 1986). It is important to note that social preferences remain stable during pair separation among monogamous species (Drickamer and Vessey, 1982; Wittenberger and Tilson, 1980). Partner preferences of heterosexual prairie vole pairs remain stable for up to 6 (males) or 8 (females) days (DeVries et al., 1995b).

Prairie voles have high basal corticosterone concentrations (600–1000 ng/ml) without any of the negative consequences normally associated with chronic glucocorticoid excess (Taymans et al., 1997). The resting corticosterone concentrations of prairie voles are 10–100 times higher than those of laboratory rats or mice (Taymans et al., 1997), and may be elevated even further in stressed animals. Corticosterone appears to modulate partner preferences in prairie voles, but in a sex-specific manner (DeVries et al., 1996). In socially naive males, injections of corticosterone, or the stress of being placed in a tank of water and thus forced to swim, facilitated the development of partner preferences (Table 8.1). Adrenalectomized males did not form partner preferences unless injected with corticosterone (DeVries et al., 1996). In socially naive female prairie voles, adrenalectomy facilitated partner preferences; partner preferences were formed within 1 hour of cohabitation among adrenalec-

TABLE 8.1 The effects of steroid hormones on prairie vole partner preference

Treatment	Males	Females
Gonadal steroid hormones	0	0
Corticosterone	+	–
Stress	+	–
Adrenalectomy	–	+

Source: Carter et al., 1997b
+ = increased partner preference; – = decreased partner preference;
0 = no change

tomized females (DeVries et al., 1995b). In contrast to males, corticosterone treatment or endogenous corticosterone elevation after swimming stress inhibited pair bond formation in females.

Corticosterone also appears to be important in the social behaviors of birds (Remage-Healey et al., 2003). Mate separation and reunion were studied in zebra finches (*Taeniopygia guttata*), a monogamous species. Plasma corticosterone concentrations and behavioral changes were recorded following an individual's separation from its mate, and again upon reintroduction of the mate or an opposite-sex cagemate. Corticosterone concentrations were elevated during separation from the mate (even in the presence of same-sex "friends"). Corticosterone concentrations returned to baseline levels upon reunion with the original mate, but not with the cagemate of the opposite sex (Remage-Healey et al., 2003). These results indicate that zebra finches display hormonal responses to separation and reunion specifically with a bonded mate and not with other familiar conspecifics. In addition, alterations in the birds' behavior during separation and reunion were consistent with monogamous pair bond maintenance behaviors observed in rodents (Remage-Healey et al., 2003).

In addition to glucocorticoids, nonapeptides appear to play an important role in partner preference formation. Sexual experience facilitates partner preference formation in both sexes of prairie voles (Carter et al., 1997a). Treatment with vasopressin, or vasopressin agonists, facilitates pair bonding in sexually experienced male voles; vasopressin also promotes the onset of aggression that is observed in male voles after mating (Winslow et al., 1993). Conversely, treatment of males with a vasopressin antagonist that binds to the V1a receptor (the most common type of vasopressin receptor) blocks the effects of vasopressin or sexual experience in facilitating pair bonding (Winslow et al., 1993). Unmated female prairie voles develop a partner preference following oxytocin infusions, but not after vasopressin or cerebrospinal fluid (a control procedure) brain infusions. Females treated with a selective oxytocin antagonist display normal mating behavior, but do not establish a pair bond (Insel and Hulihan, 1995; Williams et al., 1994). Treatment with a vasopressin antagonist does not block pair bond formation in sexually experienced females. These results suggest that oxytocin, released during mating, may be critical to the formation of a pair bond in female prairie voles; vasopressin appears to be more important for pair bonding in the male of this species (Young and Wang, 2004). As noted in the previous chapter, oxytocin is important in the formation of the exclusive bond between mothers and offspring in several species (Keverne and Kendrick, 1992; Kendrick, 2000). The interaction between nonapeptides and corticosterone in social bonding remains to be specified.

The mechanism underlying this sex difference in the actions of the nonapeptides in prairie voles is mysterious because the distribution of oxytocin and vasopressin receptors is similar in the two sexes. Some of the regions expressing receptors for the two nonapeptides appear to be involved directly in pair bond formation. For example, mating-induced partner preference in female prairie voles can be blocked by infusions of an oxytocin receptor antagonist into the prefrontal cortex and nucleus accumbens, but not the caudate putamen (Young et al., 2001) (Figure 8.5A). Blocking of the V1a receptors in the ventral pallidum, but not the medial amygdala or mediodorsal thalamus, also blocks partner preference formation in male prairie voles (Lim and Young, 2004) (Figure 8.5B). Taken together, these findings suggest that the prefrontal cortex, nucleus accumbens, and ventral pallidum are part of the pair bond neural circuitry. Probably not coincidentally, these regions are critical components of the brain's reward circuitry, suggesting that pair bond formation is rewarding (Young and Wang, 2004).

Dopamine, released in the medial nucleus accumbens, appears to mediate the rewarding properties of many addictive drugs, including nicotine and cocaine. This same reward mechanism might drive social contact in prairie voles (Cascio et al., 1998). Oxytocin receptors are located on the surfaces of some neurons in the medial nucleus accumbens. Brain injections of dopamine agonists facilitated pair bond formation in female voles, whereas brain injections of dopamine antagonists blocked pair bond formation. When the natural production of dopamine was measured by microdialysis in the brains of prairie voles during mating, a dramatic elevation in the concentrations of this neurotransmitter was observed (Gingrich et al., 1998). As noted in Chapter 5, mating evokes these changes in dopamine in male rodents such as rats and mice, which do not form pair bonds. What may differ between prairie voles and other rodents is the interaction among dopamine, oxytocin, and vasopressin in the reward circuitry (Young and Wang, 2004). Recall that meadow voles are closely related to prairie voles, but are polygamous. Although the distribution of dopamine receptors seems to be similar between the two species, the V1a receptor is expressed at higher levels in the ventral forebrain of the monogamous prairie voles than in the polygamous meadow voles (Insel et al., 1994). In a recent study, male meadow voles had a viral vector carrying a V1a receptor gene injected into the ventral forebrain, which caused them to overexpress the V1a receptor gene. These males spent a larger proportion of a 3-hour test period huddling and engaged in side-by-side contact with a partner (a female with which they had been housed previously for 24 hours) than did males that had received a viral vector carrying the *lacZ* gene (serving as a control for viral vector expression) or males that were inadvertently injected with the V1a receptor gene outside of the ventral pallidum (Lim et al., 2004). These results suggest that alterations in the expression of a single gene may be sufficient against the background of extant gene and neural circuitry to alter social behavior (Lim et al., 2004).

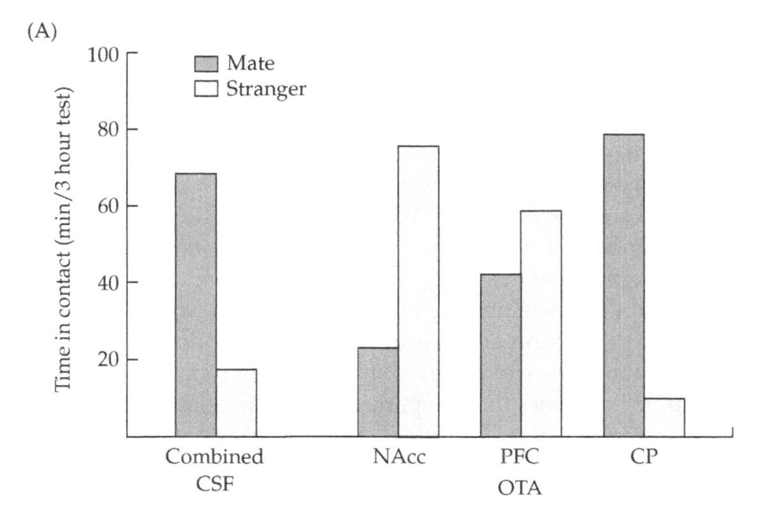

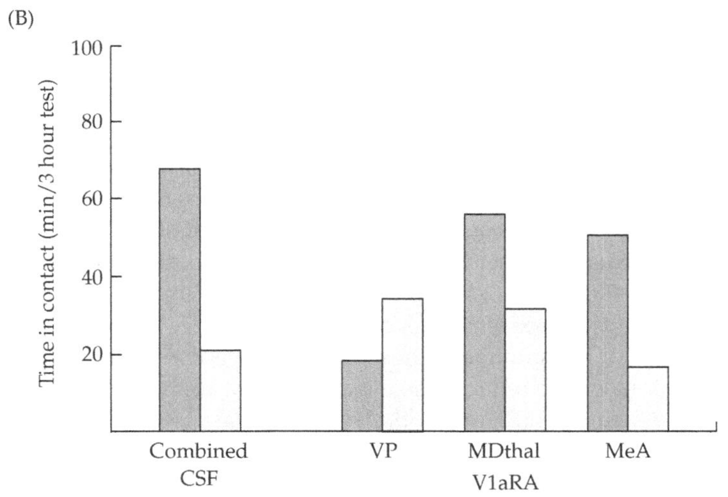

8.5 Social bonding is mediated by oxytocin and vasopressin in prairie voles. (A) When treated with cerebrospinal fluid (control treatment), female prairie voles prefer to spend more time with a social partner than with a stranger. This partner preference is abolished by infusion of a specific oxytocin antagonist (OTA) into the nucleus accumbens (NAcc) and prefrontal cortex (PFC), but not into the caudate putamen (CP). (B) Mating-induced partner preference in male prairie voles is disrupted by infusion of a specific V1a vasopressin receptor antagonist (V1aRA) into the ventral palladium (VP), but not into the mediodorsal thalamus (MDthal) or medial amygdala (MeA). After Young and Wang, 2004.

In addition to facilitating the rewarding properties of sexual behavior and the formation of pair bonds, oxytocin and vasopressin may be part of a social recognition system (Choleris et al., 2003; Young and Wang, 2004). In order to form social relationships, either positive or negative, it is critical to maintain social memory and social recognition. Female mice with the genes for oxytocin, oxytocin receptors, or the α or β estrogen receptor subtypes knocked out display social deficits (Choleris et al., 2003, 2004). These genes have been proposed to serve as a four-gene "micronet," an interacting small network of genes that serves to link the hypothalamic and limbic forebrain regions in the neural control of estrogen modulation of oxytocin evoked social recognition (Choleris et al., 2003, 2004). Male vasopressin (V1a) receptor knockout mice, but not V1b receptor knockout mice, display deficits in social recognition (Winslow and Insel, 2004). It is also possible that female V1b knockout mice have social deficits, but the data are too preliminary to say with certainty at this time. Whether these nonapeptides operate on sensory input, central processing functions such as memory, or other cognitive functions such as perception or attention remains to be determined. The role of these peptides in human social recognition and pair bonding remain unspecified.

Other peptide hormones that are part of the reward circuitry, particularly opioids, are involved in affiliative behaviors. These conclusions were the logical outcome of two observations; namely, that endogenous opioids exist in the brain, and that there are remarkable similarities between narcotic addiction and the comfort of social contact. Narcotic addiction and social contact both involve strong emotional attachments, and both processes have characteristic psychological and physiological symptoms during withdrawal. Physiological symptoms common to withdrawal from narcotics and removal of the attachment figure include crying, depression, inability to eat or sleep, and general irritability (Panksepp et al., 1980b). Endogenous opioids mediate isolation-induced distress vocalizations, and social contact increases endogenous opioid concentrations (Panksepp et al., 1997). These similarities suggest a common physiological mediation.

If opioids mediate contact comfort, then treatment with naloxone or naltrexone, both opioid antagonists, should block the onset of comfort. Young chickens display a characteristic sign of contact comfort when snuggled next to their mothers or cupped in a human hand: they rapidly close their eyes (Figure 8.6A). Treatment with naloxone reduces the effectiveness of contact comfort in causing chicks to close their eyes (Figure 8.6B; Panksepp et al., 1980a). In rodents, treatment with naltrexone blocks the effect of contact comfort in ameliorating the rate of distress vocalizations of pups (Panksepp et al., 1980b). Treatment with opioid antagonists does not appear to block the rewarding properties of social interactions in adult rats (Panksepp et al., 1997). It is possible, however, that social interactions among adult rats are not rewarding. Thus, it remains possible that opioids are important in the rewarding properties of development of social bonds in rats, but that some other system maintains social bonds in adulthood.

Opioids are important in the development of social bonds in rhesus monkeys (Kalin et al., 1988). Treatment with opioid antagonists decreases maternal bonding and play behavior, and increases distress vocalizations, in infant rhesus monkeys (Kalin et al., 1995).

Opioids also mediate social grooming in rodents and primates. Social grooming is important in maintaining social contact and serves to bond individuals. Treatment with opioid antagonists reduces the time primates spend in social

(A)

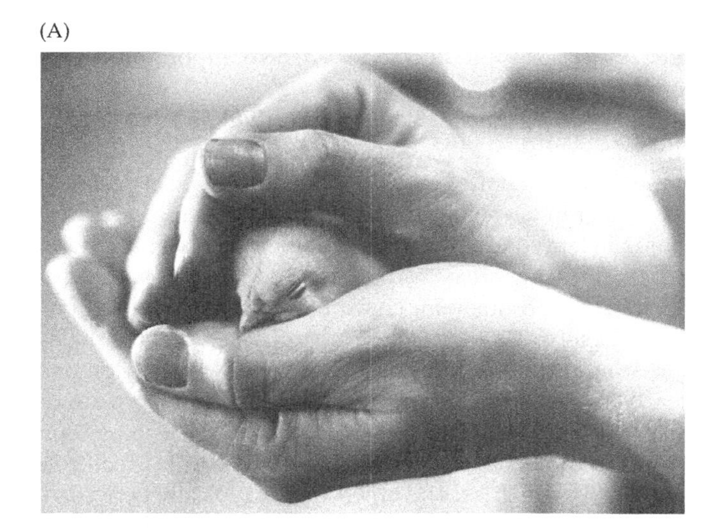

(B)

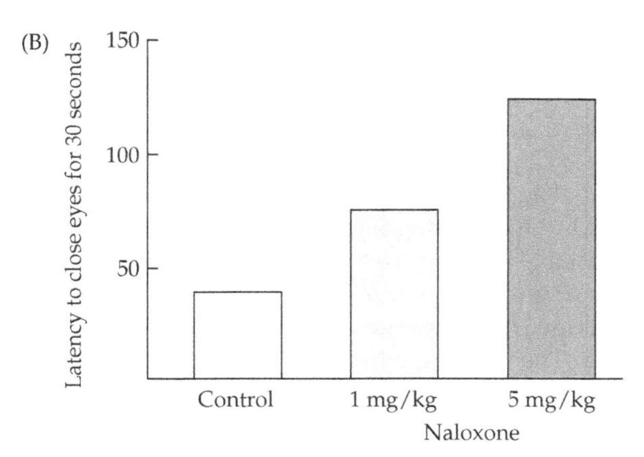

8.6 Contact comfort in chicks is mediated by opioids. (A) When nestled next to a hen or held in a person's cupped hands, a chick will rapidly close its eyes. (B) The eye-closing response can be blocked by administration of the opioid antagonist naloxone. (A) courtesy of Jack Panksepp; (B) after Panksepp et al., 1980a.

grooming (Benton and Brain, 1988; Keverne et al., 1989). Opioids are also critical in the formation of maternal bonds in sheep (Keverne and Carter, 2002).

In sum, the mechanisms underlying hormone–behavior interactions during affiliation are similar to those underlying mating and parental behavior. Many of the hormones involved appear to blunt the "stress" or fear of social contact—animals' aversive responses—and allow them to come together and engage in social behavior or mating behavior.

Aggression

Although aggression may seem superficially to be the opposite of affiliation, both types of behaviors serve to maximize the efficiency of social interactions. As mentioned above, affiliation and aggression are probably mediated by separate neural circuits. We have seen that hormones are important in affiliative behaviors. In the remainder of this chapter, we will see that hormones are also important mediators of aggressive behaviors.

A population of red deer (*Cervus elaphus*) has been studied on the island of Rhum, Scotland, since 1968 (Lincoln et al., 1972). The animals making up the main study population inhabit an area of about 12 square kilometers. During most of the year, the males live together in bachelor groups, grazing peacefully near the female herds. But beginning in late summer, as their antlers come into hard horn, the males move to traditional rutting areas, usually located in large patches of grasses and sedges. Males fight vigorously for control of these rutting locations throughout September and October. The battles are often fierce, with the males locking their antlers and attempting to inflict serious physical damage on one another (Figure 8.7). Nearly a quarter of the red deer stags on Rhum are wounded during the rutting season (Clutton-Brock et al., 1979); some animals may even die of their wounds. The social rank of a male and the size of the turf he controls are linked to his ability to win these battles. The oldest males, typically between 7 and 10 years old, are the most experienced fighters and sport fully developed antlers. Age and weaponry have their rewards: these senior males tend to hold the highest rank and win control of the largest rutting areas. After the rutting season has finished in late autumn, the males shed their antlers, their levels of aggression are substantially reduced, and they return to a relatively peaceful coexistence in their bachelor herds.

What is the goal of the stags' aggressive behavior? What makes this real estate so valuable that they risk life-threatening injuries? The ultimate explanation is probably somewhat predictable to you. Like the elephant seals described in Chapter 5, these males are fighting for the opportunity to mate. Each male is apparently trying to control a large tract of pastureland in order to attract and eventually to mate with females. The females, which are called hinds, gather in the victors' rutting territories to feed on the grasses there. Each territorial male keeps other males away from his "harem." Stags can maintain large harems only if they manage to win control of sufficiently large patches of green vegetation to support females and their subsequent offspring. Otherwise, the hinds stray from the harem in search of better food and subsequently mate with other males (Figure 8.8).

Males have been selected to maximize the number of females that graze on their hard-won territories because most of the females come into estrus, or rutting condition, for just 3 weeks or so during autumn (Guinness et al., 1978). It is relatively easy to explain, on an adaptive level, why males behave in an aggressive manner prior to the onset of the females' estrous season. A stag impregnates most of the hinds in his harem. A large territory can support more hinds than a small one, and the more hinds a stag has on his "property," the greater his reproductive fitness relative to males with access to fewer females (see Chapter 5). In many other cases, the adaptive, or ultimate, function of a seasonal cycle in aggressive behavior also appears to involve control of resources or mates to increase reproductive fitness. The remainder of this chapter, however, focuses on proxi-

8.7 Male red deer fight vigorously for control of rutting areas. Male red deer use their antlers to battle with one another during the rutting season (A), and attempt to wrestle their opponents to the ground (B). The loser of the aggressive encounter leaves the area (C), and the winner retains control of the rutting location and access to fertile females. Courtesy of Fiona Guinness.

8.8 A red deer stag with his harem. The victor in the rutting battles controls an area of grassland on which females graze. It is in the best interest of a male to control as large a parcel of grassland as possible because it can support a large number of females, and the male mates with all the females grazing on his territory. It is in a female's best interest to graze in a plot of grassland that is sufficient to provide good nutrition for her and her offspring. Courtesy of Fiona Guinness.

mate causes, such as the physiological mechanisms underlying this remarkable transformation in social organization and behavior. This part of the chapter will also explore some of the ways in which hormones influence, regulate, and mediate aggression and other social interactions.

There is little doubt that the red deer stags described above are behaving aggressively. But what do we really mean by this? Aggression is overt behavior with the intention of inflicting damage or other unpleasantness on another individual (Moyer, 1968, 1971). Intent is the key to differentiating aggressive behavior from other activities. For example, if you inadvertently kill hundreds of insects with your windshield while driving, your behavior could not be considered aggressive because your intent was merely to move at high speeds, and the death of the insects is an incidental occurrence. However, if you rampage through your apartment spraying insecticide after being bitten by a mosquito, your actions could be considered aggressive because of your intent to destroy insects. So different behaviors leading to the same outcome may or may not be considered aggressive, depending on the context in which the behaviors appear. Of course, it can be tricky to attribute intention to animals, but by careful observation of the circumstances and outcomes of behaviors, the goals of animals' behaviors can be ascertained.

Aggression has been divided into various types for ease of classification (Moyer, 1968, 1971). These different types of aggression appear to have different physiological causes and are expressed in different environmental and social contexts (Table 8.2). Maternal aggression and its underlying hormonal mediation have already been described in Chapter 7. Hormonal changes associated with

TABLE 8.2 **Types and tests of aggression**

Two schemes for classifying aggressive behavior	
Predatory	Predatory attack
Intermale	Self-defensive behavior
Fear-induced	Parental-defensive behavior
Irritable	Social conflict
Territorial	
Maternal	
Instrumental	
Common laboratory tests of aggression	
Muricide (mouse-killing) by cats or rats	
Shock-elicited fighting	
Isolation-induced aggression	
Resident-intruder aggression	

Source: Moyer, 1968, Brain et al., 1983

production of offspring—especially the relationship among blood concentrations of estrogens, progestins, and prolactin—are correlated with the onset of maternal aggression. Steroid hormones also underlie other types of aggressive behavior. Intermale aggression and territorial aggression, as well as sex-related and rank-related aggression, all appear to be mediated by androgens (Bouissou, 1983). Predatory aggression is another type of aggressive behavior that is performed in the context of obtaining food. Hormones may be involved in this type of aggression, but few studies have examined this possibility. Aggressive acts can also be classified as either offensive or defensive (Blanchard and Blanchard, 1984; 1988). Defensive aggression is used to protect against injury by others, whereas offensive aggression serves to obtain or retain resources important for survival and reproductive success. It has been proposed that each of these two categories of aggressive behaviors have similar endocrine and neural regulation in mammals (Blanchard and Blanchard, 1988).

Other types of aggression are most commonly evoked and studied in the laboratory. Learned aggression and irritable aggression are often studied in the form of restraint aggression, which results after an animal is held motionless. Another type of aggressive behavior commonly studied in the laboratory is fear-induced aggression. Many laboratory studies on aggression use albino house mice, which are normally quite docile. Consequently, the mice must be put into situations that promote aggression—for example, they are housed in isolation or given an electric shock. More commonly, mice are tested in a so-called resident–intruder test of aggression. Prior to the test, the resident mouse remains in his cage for 2 weeks or more. This causes the odors and other stimuli to become familiar to the resident mouse, which then defends its "territory" against the intruder—another mouse that is simply dropped into the resident's cage. Typically, the resident has a "home field advantage" and defeats the intruder mouse. These experiments are typically performed with males, but female mice housed in groups will also attack a novel lactating female. Examples of these types of experiments will be provided below, but it is important to keep in mind that the hormonal control of aggression in these contrived situations is likely to differ from the endocrine correlates of natural expressions of aggressive behavior.

The possibility for aggressive behavior exists whenever the interests of two or more individuals are in conflict (Svare, 1983). Conflicts are most likely to arise over limited resources such as territories, food, and mates. A social interaction decides which animal gains access to the contested resource. In many cases, a submissive posture or gesture on the part of one animal avoids the necessity of actual combat over a resource. Animals may also participate in threat displays or ritualized combat in which dominance is determined, but no physical damage is inflicted. In order to facilitate the study of aggressive interactions, the term **agonistic** was adopted to describe the entire behavioral repertoire of both aggressive and submissive actions within the context of a social interaction involving a conflict of interest (Scott and Fredericson, 1951). In the strict sense, however, submissive behaviors are not aggressive behaviors; consequently, hormonal mediation of the aggressive and the submissive components of an agonistic interaction may be different (Leshner and Moyer, 1975). Aggression and submission may represent the end points of a single behavioral continuum; alternatively, they may represent independent, but interacting, dimensions of the behaving individual. This is not a trivial semantic issue, but rather a conceptual issue that influences the manner in which the neural and endocrine bases of agonistic behavior are studied (Figure 8.9; Schlinger and Callard, 1990).

There is overwhelming circumstantial evidence that androgenic steroid hormones mediate aggressive behavior across many species. First, seasonal variations in blood plasma concentrations of testosterone and seasonal variations in aggression coincide. For instance, the incidence of aggressive behavior peaks for red deer stags in autumn, when their gonads are secreting large amounts of testosterone (Figure 8.10). Second, aggressive behaviors increase at the time of puberty, when blood concentrations of androgens rise. Juvenile red deer do not participate in the fighting during the rutting season. Third, in any given species,

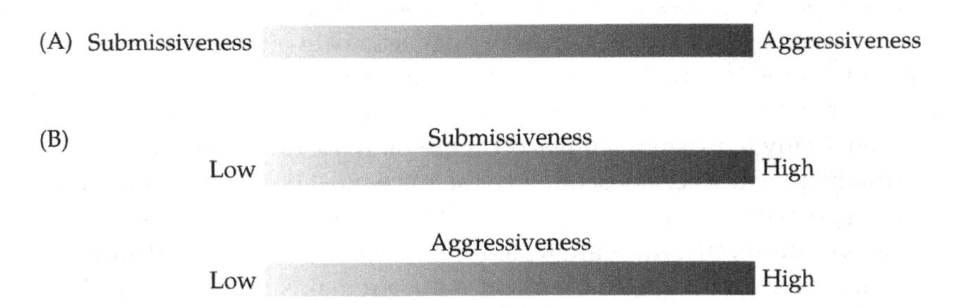

8.9 Two models of agonistic behavior. Aggressive and submissive behaviors can be seen as opposite ends of a single behavioral continuum (A), or they can be thought of as two separate aspects of an individual's behavior (B). Each model makes different predictions about how hormones might influence agonistic behavior.

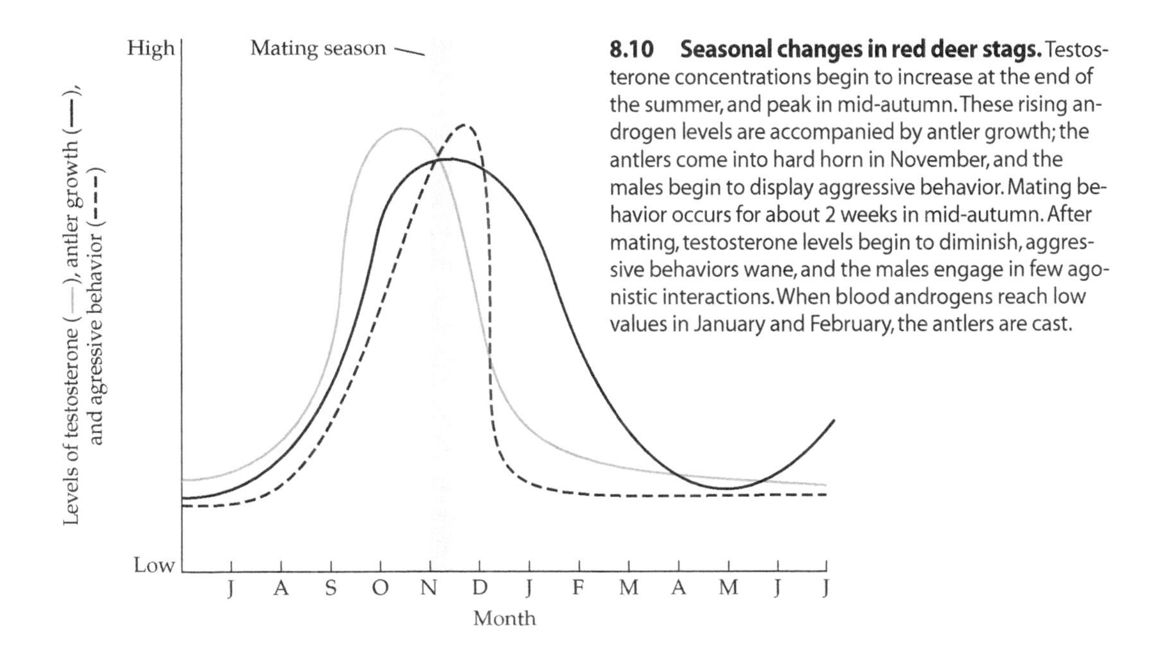

8.10 Seasonal changes in red deer stags. Testosterone concentrations begin to increase at the end of the summer, and peak in mid-autumn. These rising androgen levels are accompanied by antler growth; the antlers come into hard horn in November, and the males begin to display aggressive behavior. Mating behavior occurs for about 2 weeks in mid-autumn. After mating, testosterone levels begin to diminish, aggressive behaviors wane, and the males engage in few agonistic interactions. When blood androgens reach low values in January and February, the antlers are cast.

males are generally more aggressive than females. This is certainly true of red deer; relative to stags, red deer hinds rarely display aggressive behavior, and their rare aggressive acts are qualitatively different from the aggressive behavior of rutting males. Finally, castration typically reduces aggression in males, and testosterone replacement therapy restores aggression to pre-castration levels. There are some interesting exceptions to these general observations, however; such exceptions will be described later in this chapter.

Seasonal Changes in Social Behavior

Animals that are preparing to mate, or have recently mated, and require resources for their offspring are far more likely to behave aggressively than are nonbreeding individuals. Many animal species breed on a seasonal basis (see Chapter 10); consequently, a seasonal pattern of aggressive behavior is commonly observed. A number of experiments have been conducted on a variety of animal species to investigate seasonal changes in aggression and the role of androgens in this temporal variation.

Aggression and the Breeding Cycle: Red Deer

Red deer stags were the subject of one such experiment. Castration of stags during the winter caused them to plummet in social rank (Lincoln et al., 1972). Their reduction in social status mainly reflected the loss of their weapons: castrated stags promptly cast their antlers and acquired so-called velvet horn (see Figure 8.10). Normally, antlers develop throughout the summer. During the course of their development, the new antlers are covered by a mossy, soft, and highly vascularized tissue called velvet; thus, the deer are said to be in "velvet horn" during antler development. The velvet is shed at the end of the summer, when high blood testosterone concentrations destroy the blood supply to the velvet tissue and it dies. The antlers gain their familiar bony appearance as the velvet is shed, and the animals are said to be in "hard horn." After rutting, the testes regress during the winter, and low blood concentrations of testosterone cause the stags to cast their antlers in late winter. Thus, wintertime castration mimicked the normal reduction in testosterone and caused the stags to cast their antlers prematurely. In this study, the castrated males regained some of their position in the dominance hierarchy when the gonadally intact males also shed their antlers. However, they did not regain all of their former social rank.

Not only does the frequency of agonistic encounters change on a seasonal basis, but the quality of the encounters changes as well. Red deer stags use their antlers as both offensive and defensive weapons during the rutting season, but aggressive behavior among males lacking antlers is mainly expressed by kicking with their forelegs. Females typically use their forelegs in agonistic interactions throughout the year.

Some castrated stags in the study were implanted with slow-release testosterone capsules either in the winter or in the summer (Lincoln et al., 1972). Winter-implanted males retained their antlers, were more aggressive than other males, and climbed in social rank throughout the spring and summer. Without antlers, the normal stags were easily intimidated by the antler-bearing, testosterone-treated stags. The continuous supply of testosterone across seasons permitted the stags to maintain their antlers and evoked aggressive behavior. However, not all androgen-dependent behaviors were supported by the chronic androgen exposure. For instance, the testosterone-treated males showed mating behavior only during the appropriate autumnal breeding season, suggesting either that estrous females generate a cue that normally stimulates mating behavior in stags or that the neural substrate controlling reproductive behavior in this species is responsive to androgens only at a particular time of year.

Stags implanted with testosterone in the summer began to climb in social rank even before the androgen could stimulate antler development, indicating that their aggressive behavior increased in response to androgen treatment even without the external morphological changes. In other words, these males behaved aggressively even in the absence of weapons to fulfill their threats. However, the antlers are the main factor controlling social dominance. Amputating the antlers of gonadally intact stags caused an immediate change in their social status without any change in endocrine state (Bouissou, 1983). Thus, steroid hormones affect social status in red deer stags in two ways: first, by acting directly on the brain to promote aggressiveness, and second, by acting on the antlers, the effectors of aggressive behavior. Males of other ungulate species differ from red deer; for example, castrated horses (Tyler, 1972) and reindeer (Espmark, 1964) may still dominate other male conspecifics. It is not yet known what accounts for this variation among species.

Signals of Social Rank: Harris's Sparrows

There are other instances in which hormone concentrations appear to influence the perceived status of an individual through their effects on aggressiveness and morphological characters. Harris's sparrows (*Zonotrichia querula*) provide an elegant example of the endocrine mediation of social status. These birds set up individual territories and breed near the Arctic Circle during the summer. They migrate south to midwestern North America for the winter, where they join large mixed-sex and mixed-age, and even mixed-species, feeding flocks. The adaptive advantage of joining such a feeding flock probably arises from a "safety-in-numbers" strategy. Individuals compete for resources within the flock in the context of a social dominance hierarchy, or pecking order. The membership of each winter flock is variable and changes frequently; consequently, these birds require an easily identifiable signal of social status. Such a signal would be advantageous to both subordinate and dominant birds because of the savings in time and energy; subordinate birds would benefit by avoiding potentially lethal fighting over resources with dominant birds, and dominant birds would benefit by avoiding many needless agonistic interactions (Rohwer and Rohwer, 1978).

Harris's sparrows apparently do have such a signal. Individual sparrows exhibit tremendous variation in winter plumage coloration, mainly in the number of dark or light feathers on the breast and head (Figure 8.11; Rohwer, 1975). The appearance of the sparrows is fixed for the winter during the autumnal molt. Birds with darker coloration invariably win contests over lighter-colored birds. Direct confrontations are rare, however, because of the extent to which the plumage signal is respected. In most cases, lighter birds simply leave a feeding area upon the approach of a darker bird (Rohwer and Rohwer, 1978; Rohwer and Wingfield, 1981).

Harris's sparrows with darker winter coloration have higher testosterone concentrations than lighter-colored birds. Males and females have similar coloration during the breeding season, with relatively dark heads and breasts. When they molt at the end of the breeding season, the variation in coloration becomes apparent, with males generally darker than females. Possibly, the concentration of circulating androgens during the autumnal molt is reflected in the number of pigmented feathers (Rohwer and Wingfield, 1981); however, some females are darker than some males within a given flock.

An experiment was conducted to determine the degree to which coloration dictated the aggressive prowess of Harris's sparrows (Rohwer, 1977). The throat and crown feathers of some dark birds were bleached, and these same anatomical areas were dyed black on some lightly colored sparrows. Neither group of altered sparrows fared well with their new social identity. Newly bleached, formerly dominant birds frequently engaged in fierce fighting because subordinate birds attempted to displace them on the feeding areas and the dominant birds refused to yield. These formerly dominant, newly bleached birds became outrageous bullies, picking fights far in excess of what was necessary or reasonable for merely obtaining foraging positions. They appeared to be trying to regain their former

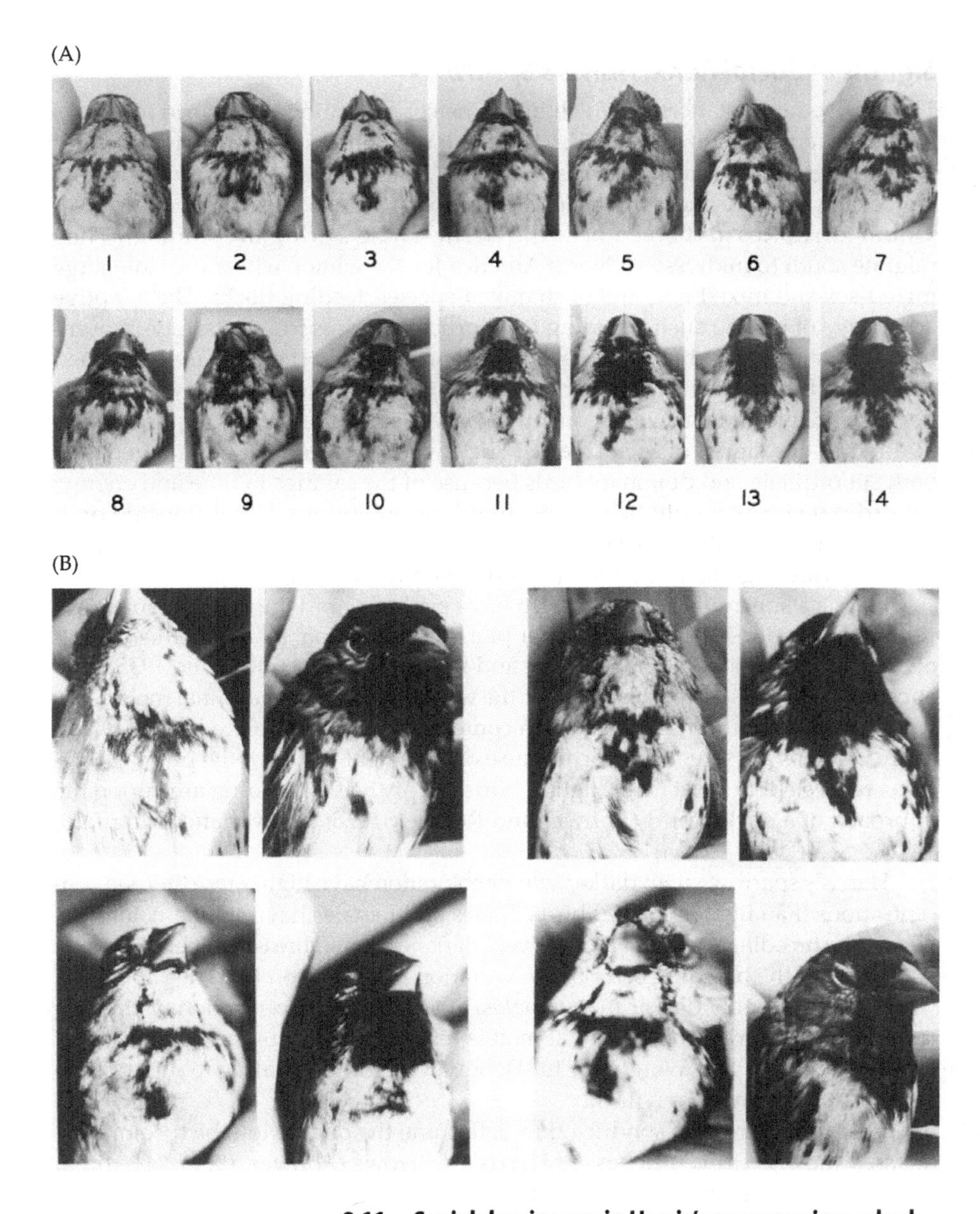

8.11 Social dominance in Harris's sparrows is marked by plumage coloration, with the darkest birds at the top and the lightest birds at the bottom of the "pecking order." (A) Fourteen sparrows arranged in a dominance scale, from submissive (1) to dominant (14). (B) Four individuals shown both with their natural coloration (left in each panel) and after their feathers were dyed (right). These birds' altered coloration did not allow them to rise in the social hierarchy. Courtesy of Sievert Rohwer.

social status. On the other hand, newly dyed, formerly subordinate birds did not fare much better in their new social station. These dyed sparrows were persecuted by the true dominant individuals, which resulted in a quadrupling of attacks aimed at them. Many of the newly dyed birds left the relative safety of the winter flocks and fed alone. Apparently, the behavior of these sparrows did not match their artificial status signals, and their flockmates retaliated against this blatant social cheating. If, however, individuals were both dyed and implanted with testosterone capsules, they showed dramatic increases in social status (Rohwer and Rohwer, 1978). These testosterone-treated birds apparently behaved in a manner appropriate to their new coloration.

The altered plumage coloration caused great difficulties for Harris's sparrows because of the confusion that ensued: behavior did not match social signals. For these birds, as well as for red deer, androgens appear to modify both behavior and morphology. In some cases, it is possible that behavior feeds back to elevate steroid hormone concentrations further and stimulate greater development of androgen-dependent secondary sex characters, including behavior, as we will see below. There was no correlation, however, between natural winter blood androgen concentrations and dominance status among wintering Harris's sparrows (Rohwer and Wingfield, 1981). Thus, the relationship between hormones and dominance is not apparent unless the hormone concentrations are monitored when the dominance hierarchy is being established.

Harris's sparrows, like many other migratory birds, must compete with resident birds for food and shelter in their winter habitat. Recent research with migratory redstarts (*Setophaga ruticilla*) demonstrated that the future reproductive success of migrants depends on the quality of their wintering habitat (Marra et al., 1998). At low latitudes, the resident birds are in reproductive condition during the winter and are extremely territorial. Although first reported as anomalous behavior, territoriality has recently been reported among wintering migrants and may be common among passerines (perching birds) that winter in the Neotropics (Greenberg, 1986; Marra et al., 1998). Wintering migrants are not in reproductive condition, and their gonads are not producing any measurable amounts of steroid hormones. The physiological mechanisms underlying territoriality or short-term site defense in these reproductively regressed nonresident migrants remain a mystery (Nelson et al., 1990). However, research on resident European robins (*Erithacus rubecula*) suggests that aggression during the winter may be testosterone-independent (Schwabl and Kriner, 1991), because when these birds are castrated during the winter, they continue to display aggressive behavior. Furthermore, treatment of free-living European robins in autumn and winter with the anti-androgen flutamide, which prevents androgens from binding to their receptors, failed to affect territorial aggression. These results support the hypothesis that androgen receptors are not primarily involved in the regulation of aggression outside of the breeding season (Schwabl and Kriner, 1991).

In European stonechats (*Saxicola torquata*), vernal (springtime) territorial aggression was reduced by using both flutamide and an aromatase inhibitor to block the conversion of testosterone to estrogen. However, the same treatment did not affect territorial behavior in winter, suggesting that in stonechats, territoriality is regulated by different hormonal mechanisms at different times of the year (Canoine and Gwinner, 2002). There appear to be substantial species differences in the endocrine regulation of aggressive behavior in birds during the nonbreeding season (Wingfield et al., 2005; Canoine and Gwinner, 2002; Moore et al., 2004). For example, in field studies of male song sparrows (*Melospiza melodia*), flutamide and an aromatase inhibitor significantly reduced territorial behavior in autumn and winter (Soma et al., 1999a,b; Soma and Wingfield, 1999). Similar results were obtained when an aromatase inhibitor (Fadrozole) was used alone (Soma et al., 2000a,b). Fadrozole treatment in combination with estradiol implants completely restored territorial aggression (Soma et al., 2000a). Considered together, these results suggest that estrogens, presumably aromatized from androgens in the brain, are involved in the regulation of territorial behavior outside the breeding season (Soma et al., 2000a,b). These results emphasize that estrogens, resulting from aromatization of androgens and acting through estrogen receptors, are important in the control of territorial aggression in autumn, whereas androgen receptor-mediated mechanisms appear to be less important at this time.

Given that androgen and estrogen concentrations both tend to be low in autumn and winter, what is the source of these steroids? It is possible that the steroids are produced de novo directly in the brain. Several studies have indicated that the avian brain can produce neurosteroids at high levels, so perhaps androgens are produced in the brain and converted by neurons into estrogens to regulate aggression outside of the breeding season (Baulieu, 1998; Saldanha et al., 1999). Alternatively, a biologically weak steroid or a steroid hormone precursor produced in the periphery could be converted to an active hormone in the brain. Of several possibilities, dehydroepiandrosterone (DHEA) seems the most likely candidate. DHEA is produced by the adrenal glands in significant amounts and also functions as a neurosteroid. The enzymatic machinery exists in the avian brain that could convert circulating DHEA to androstenedione, which in turn could be converted to testosterone or aromatized to estradiol (Labrie et al., 1995). DHEA appears to play an important role in territorial aggression during the nonbreeding season in spotted antbirds (*Hylophylax n. naevioides*) (Hau et al., 2004) and in song sparrows (Soma and Wingfield, 2001). DHEA concentrations were positively correlated with aggression in the spotted antbirds (Hau et al., 2004). Implants of DHEA into male song sparrows increased singing, one component of territorial aggression, and also resulted in growth of the HVC, a song control nucleus in the telencephalon (Soma et al., 2002) (see Chapter 4). It should be noted that the neurosteroid and circulating precursor hypotheses are not mutually exclusive, so both processes could be operating to provide steroids to regulate territorial behavior.

Aggression and Winter Survival: Rodents

Individuals of many rodent species shift from a highly territorial social strategy during the breeding season to a social, and highly interactive, existence during the winter. These species undergo reproductive regression at the end of the breeding season in response to short days (see Chapter 9); the resulting lower concentrations of androgens may enable this shift in social behavior. There are adaptive benefits that accrue from changing social systems on a seasonal basis. During the breeding season, animals benefit from controlling resources to promote their own survival and that of their offspring, and they defend those resources aggressively. However, rodents may benefit from group living during the winter because this strategy conserves energy and enhances survival during this time of low temperatures and reduced food availability. Many species of rodents conserve energy during the winter by forming aggregations of huddling animals (West and Dublin, 1984). These aggregations, like winter feeding flocks of birds, may consist of different sexes and different rodent species (Madison, 1984).

Even in the absence of huddling behavior, animals may tolerate one another at close quarters during the winter, whereas interspecific commingling is not tolerated during the breeding season. For example, meadow voles, as their name implies, occupy grasslands during their breeding season; the males are highly territorial at this time. Red-backed voles (*Clethrionomys gapperi*) prefer to breed in spruce forest habitats. But during the winter months, meadow voles move from open meadows into the forest habitats occupied by the red-backed voles, presumably to take advantage of the protective cover provided by the trees. In some cases, they occupy nests with other rodent species. Individual meadow voles, trapped during the winter and tested in paired encounters in a small neutral arena, exhibit less interspecific aggression than summer-trapped voles (Turner et al., 1975). The winter reduction in aggressiveness allows habitat sharing during harsh conditions. As the animals enter breeding condition in the spring, they once again establish mutually exclusive territories.

In another study, seasonal differences in social interactions between wild-caught adult and juvenile meadow voles were examined (Ferkin, 1988). Adults responded differently to juvenile male voles, but not to juvenile female voles, depending on the season. Adult male meadow voles were more aggressive toward juvenile males during the first part of the breeding season (May through August) than during the later months (Ferkin, 1988). In contrast, adult female voles were more aggressive toward juvenile males late in the breeding season than in the early portion of the breeding season. This seasonal shift in the aggressiveness of adult voles toward juvenile males may influence dispersal, and may also influence the formation of overwintering aggregations; overwintering groups of meadow voles typically include more females than males (Ferkin and Seamon, 1987).

Some individual male rodents do not undergo reproductive regression when exposed to short days; rather, these males maintain testicular function and produce sperm and androgens during these simulated winter conditions (Nelson, 1987; Prendergast et al., 2001). Males capable of continuous breeding would appear to be superior in fitness to individuals whose reproductive apparatus regresses, because they are capable of siring offspring at any time of year. The obvious advantage of continuous mating ability must incur some hidden costs, however, because only a minority of individuals adopt that strategy. If there were no costs to a strategy of continuous breeding, continuous breeders would eventually replace the seasonally breeding animals. One possible cost is that nonregressed males, with their elevated androgen concentrations, may remain too aggressive during the winter to benefit from communal huddling, and thus may give up substantial energy savings relative to reproductively regressed males in order to be reproductively competent all year long. The high behavioral and energetic costs associated with maintenance of the reproductive system would compromise survival, and may explain why nonregressive types do not normally predominate in temperate- or boreal-zone populations of rodents (Nelson, 1987; Nelson et al., 1989).

A field experiment on winter nesting behavior in a population of prairie voles supports this contention (McShea, 1990). The majority of voles in the population were reproductively inactive during the winter and formed groups of huddling individuals. However, there were two males in the population that remained in breeding condition and were never observed to huddle with other animals. In pairwise tests of aggression, these reproductively competent males were much more aggressive than reproductively quiescent individuals. Laboratory studies revealed that adult castration does not significantly affect aggression in this species (Demas et al., 1999). In another laboratory study, reproductive status influenced the odor preferences of meadow voles maintained in simulated winter day lengths (Gorman et al., 1993). Males that retained reproductive function under winter day-length conditions preferred the odors of females that also failed to inhibit reproduction during short days. This preference may facilitate the sporadic occurrences of winter breeding frequently reported for this species (reviewed in Nelson, 1987; Prendergast et al., 2001).

Although the results of most rodent studies to date indicate that low circulating testosterone concentrations reduce mating and aggressive behaviors (reviewed in Knol and Egberink-Alink, 1989; Rubinow and Schmidt, 1996), there are exceptions. In dusky-footed wood rats (*Neotoma fuscipes*), levels of intermale aggression, both in laboratory and in field encounters, rise dramatically during the breeding season, in a pattern that closely parallels the seasonal rise in testosterone concentrations (Caldwell et al., 1984). However, males castrated as adults also show the seasonal rise in aggression when tested in laboratory encounters with castrated opponents, and show no decrement in fighting ability when paired with intact opponents (Caldwell et al., 1984). These results demonstrate the independence of seasonal aggression from the proximate modulating effects of testosterone in this species. There is a growing list of other nondomesticated species that do not show reductions in aggression after castration. For example, castration does not decrease aggressiveness in male red-sided garter snakes (*Thamnophis sirtalis parietalis*) (Crews and Moore, 1986), European starlings (*Sturnus vulgaris*) (Pinxten et al., 2003), or Mongolian gerbils (*Meriones unguiculatus*) (Christianson et al., 1972). Both Syrian hamsters (*Mesocricetus auratus*) and Siberian hamsters (*Phodopus sungorus*) display increased aggressive behavior under short-day conditions despite decreased circulating gonadal steroids (Garrett and Campbell, 1980; Badura and Nunez, 1989; Jasnow et al., 2000) (Figure 8.12).

In addition to seasonal changes in androgen concentrations, seasonal changes in androgen receptor sensitivity occur in rodents. Consequently, it is not possible

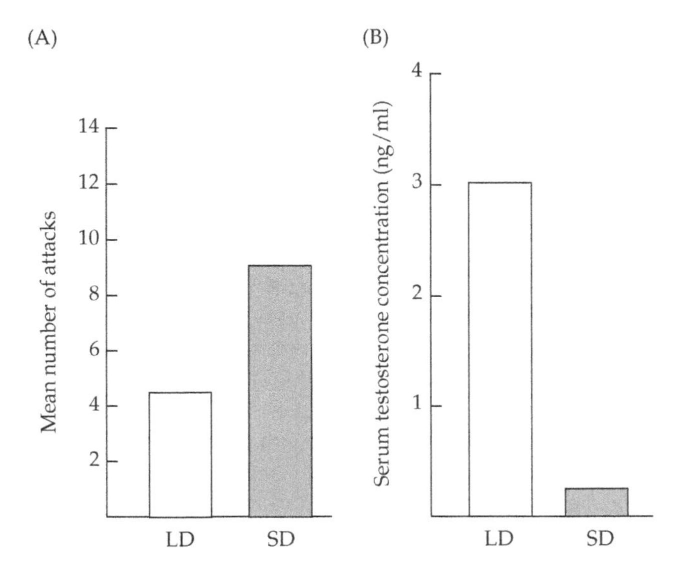

8.12 Aggression in Siberian hamsters is high when testosterone concentrations are low. (A) The number of attacks is higher among male Siberian hamsters housed under short-day conditions (SD) that simulate winter than among males housed under long-day conditions (LD). (B) Circulating blood testosterone concentrations, however, are very low under short-day conditions. After Jasnow et al., 2000.

to override wintertime reductions in aggressiveness merely by injecting testosterone. Injections of testosterone into reproductively regressed male deer mice (*Peromyscus maniculatus*) do not elevate the number of aggressive encounters to the level exhibited by hamsters in breeding condition. Presumably, the ineffectiveness of testosterone in stimulating aggressiveness reflects a lack of androgen receptors in the brain regions responsible for organizing aggressive behavior during the nonbreeding season. As with mating behavior, very high blood androgen concentrations are required to stimulate aggressive behavior immediately after the hamsters regrow their reproductive systems, as compared with those necessary to sustain aggressiveness in breeding animals (Berndtson and Desjardins, 1974). Desensitization of receptor tissues to steroid hormones is common after a prolonged absence of the hormone (Morin and Zucker, 1978; Powers et al., 1989), and this may be the mechanism by which seasonal changes in the sensitivity of central nervous system tissue are mediated (see Chapter 10). This last statement is somewhat speculative, however. Seasonal changes in receptors are only one of several factors, including changes in metabolizing enzymes, that may mediate seasonal changes in target tissue sensitivity.

Testosterone and the Energetic Costs of Aggression: Lizards

Elevated testosterone concentrations and aggression cause a decrease in survivorship in a number of species. A number of different factors, including injury, energetic costs, exposure to predators, or engaging in more "risky" behaviors, may mediate this effect. In male spiny lizards (*Sceloporus jarrovi*), seasonal changes in testosterone concentrations are tightly correlated with seasonal changes in the intensity of aggression, especially territorial aggression (Moore and Marler, 1987). During the autumnal breeding season, when territorial aggression is most fierce, testosterone concentrations are high; conversely, during the winter, when little territoriality is observed, testosterone concentrations are at their lowest. Interestingly, testosterone concentrations are highest during the fall, not during the summer breeding season, in these lizards. Artificial elevation of blood testosterone concentrations by implantation of Silastic capsules during the winter increased aggression above normal summer levels, but not to the level observed during the fall breeding season peak (Moore and Marler, 1987). Male lizards implanted with testosterone-filled capsules spent too much time engaged in territorial defense, aggressive displays, and other aggressive behaviors to forage adequately, and did not survive as long as males implanted with empty capsules. This decrease in survivorship probably reflects elevated energetic costs associated with aggressive behavior (Marler and Moore, 1988, 1991). Thus, males behaving aggressively during the winter incur significant costs even in the face of potential reproductive benefits. These costs include the energy expended in increased territorial aggression and displays, as well as longer daily activity periods, which could have exposed the males to greater predation risks (Marler and Moore, 1989; Marler et al., 1995). When oxygen consumption was measured directly, testosterone treatment did not increase resting, or basal, metabolic rates. The approximately 30% increase in metabolism evoked by testosterone treatment appeared to reflect primarily the energetic costs of territorial defense (Marler et al., 1995). Food supplements were able to compensate for the testosterone-induced costs of aggressive behavior (Marler and Moore, 1991). Other costs of testosterone include interference with paternal care, exposure to predators, increased risk of injury, loss of fat stores and possibly impaired immune system function, and oncogenic (cancer-causing) effects (Wingfield et al., 2001).

Do Seasonal Hormonal Changes in Primates Correlate with Aggression?

Seasonal breeding is observed among many primate species. Intermale aggression may increase during the breeding season in some species, including lemurs (*Lemur catta*) (Jolly, 1966) and rhesus monkeys (*Macaca mulatta*) (Michael and Zumpe, 1978, 1993). The frequency of aggressive acts may or may not increase during the breeding season in squirrel monkeys (*Saimiri sciureus*) (Baldwin, 1968; Mendoza et al., 1978). Correlations between periods of high androgen production and aggression are interesting, but do not by themselves demonstrate conclusively that elevated androgen concentrations cause aggression in primates. However, the results of several studies that have addressed the relation between decreases in blood plasma concentrations of androgens and receptor sensitivity to androgens have suggested that these correlations probably do represent causation, despite some notable exceptions. Evidence for and against the dominant view that androgens mediate aggressive behavior among primates will be reviewed later in this chapter.

Human males also show seasonal variation in blood testosterone concentrations (Dabbs, 1990; Smals et al., 1976). These seasonal changes have not been shown to cause aggressive behavior; however, this presumption has not been directly tested, despite an abundance of positive correlations. For example, peak incidences of violent urban crime in North America are associated with high ambient temperatures and peak androgen concentrations in some populations (for example, Smals et al., 1976). Similarly, inmates maintained in a Maryland prison showed higher levels of aggression, as assessed by rule infractions, during the summer than during the other three seasons (Haertzen et al., 1993). Of course, these correlations may simply represent increased contact between perpetrator and victim during the warmer weather of summer. However, an analysis of over 27,000 reports from women in 23 shelters scattered across five locations in the United States who had been abused by their live-in partners (both married and unmarried) revealed an annual rhythm of abuse with maximal levels occurring during the summer (Michael and Zumpe, 1986). These data suggest that violence by men against women increases during the summer independently of any major seasonal changes in contact between perpetrator and victim. Whatever the causal relationship, such correlations probably reflect factors other than hormone concentrations. As Joan Didion writes in *Slouching towards Bethlehem*, "In Switzerland the suicide rate goes up during the foehn [a strong, hot, dry wind], and in the courts of some Swiss cantons the wind is considered a mitigating circumstance for crime" (218–219).

Increases in Aggression at Puberty

There are countless reports of increased aggression in males undergoing puberty. During puberty, the testes grow larger, and under the influence of LH from the anterior pituitary, the Leydig cells secrete increasing amounts of androgens into the circulatory system. Plasma concentrations of androgens increase at the time of puberty and are associated with the elevation in aggressiveness. Intermale aggression in house mice, as well as isolation-induced aggression, is first observed at the time of puberty (Brain and Nowell, 1969; Levy and King, 1953).

As noted in Chapter 1, male livestock have been castrated prior to puberty since antiquity to make them more docile and manageable. Prepubertal castration is more effective in reducing all types of aggression in mice than castration in adulthood (Uhrich, 1938). This observation is not borne out, however, in all species. Prepubertal castration of dogs, for example, does not necessarily diminish their adult levels of aggression (LeBoeuf, 1970), and prepubertal castration does not preclude male primates from attaining a high social status. For instance, Narses (480–574 A.D.) was a famous Byzantine general and statesman who originally was a eunuch slave in Armenia. He had a reputation for ruthlessness and played an important role in suppressing the Nika riots. Narses served under the emperor Justinian I, who sent him to Italy in 538 A.D. to fight against the Goths. Eventually, Narses rose to become the exarch of Italy.

Although these are interesting exceptions, it is generally true that aggressiveness increases at puberty. The following sections review some specific examples of associations between hormone concentrations and aggressive behavior that coincide with puberty.

Social Influences on the Development of Aggressive Behavior

One important animal model in studies of social aggression has been Syrian hamsters. Prior to puberty (from 20–40 days of age), male Syrian hamsters engage in agonistic behavior mainly as a form of play, which is characterized by attacking the faces and cheeks of their opponents (Wommack et al., 2003). During puberty, the frequency of attacks diminishes, and "play fighting" morphs into the adult version of aggression (Delville et al., 2005). From about 40 to 50 days of age (mid-puberty), males undergo a transitional period during which attacks are aimed at the flanks. Finally, as they emerge from the pubertal period, males perform adult-like offensive aggressive behaviors, which are characterized by attacks to the underbelly and rear end (Wommack et al., 2003) (Figure 8.13). Exposure to social stress during puberty altered adult aggressive behavior in at least two ways. Repeated exposure of male peripubertal hamsters to aggressive adults hastened the onset of adultlike offensive, but not defensive, aggressive behaviors. If the exposure occurred early in puberty, the young hamsters were most likely to

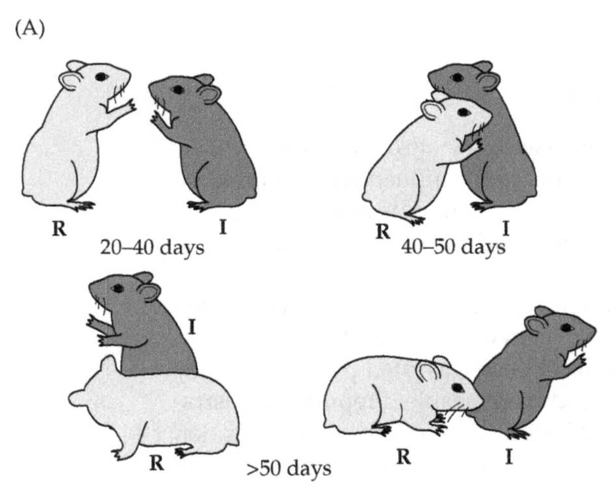

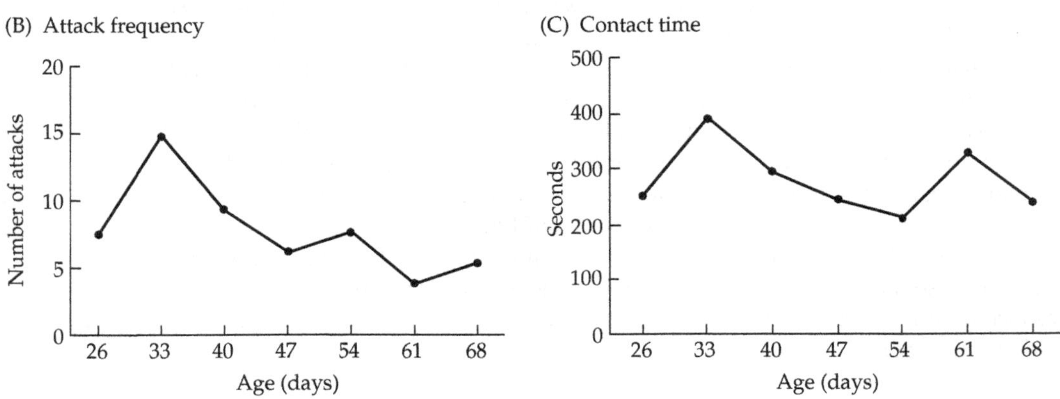

8.13 The type and amount of aggression varies during development in Syrian hamsters. (A) Different types of attacks are seen at different ages when resident (R) Syrian hamsters are paired with intruders (I). The different developmental stages are based on the targets of the attacks: face and cheeks before puberty, side and flanks during the transitional period, or lower belly or rump and rear in adulthood. (B) The number of attacks wane as the hamsters develop. (C) The amount of time that the resident maintains contact with the intruder gradually wanes as the hamsters grow older. After Wommack et al., 2003.

become aggressive, especially toward smaller opponents, post-puberty (Wommack et al., 2003). Furthermore, repeated social stress slowed the reduction in attack frequency normally seen during puberty. In contrast to males, female Syrian hamsters displayed a stable frequency of attacks through puberty and into adulthood, and there was no obvious mid-pubertal transitional period (Taravosh-Lahn and Delville, 2004).

In female hamsters, cortisol levels habituated to repeated exposures to aggressive adults (Taravosh-Lahn and Delville, 2004). Males do not appear to habituate to repeated exposure to social subjugation (Wommack and Delville, 2003). After 2 weeks of repeated exposure to aggressive adults, male hamsters persistently displayed elevated plasma cortisol concentrations, an endocrine marker of the maturation of agonistic behavior from play-fighting to adultlike attacks (Wommack et al., 2003; Wommack and Delville, 2003). Although castration has well-known effects on aggression in adults, castration of peripubertal hamsters did not affect the transition from play-fighting to adultlike fighting (Delville et al., 2005). The maturation of aggressive behavior is also correlated with the maturation of the hypothalamus, pituitary gland, and adrenal glands, or *adrenarche*, and activity of the hypothalamic–pituitary–adrenal (HPA) axis appears to control pre-adult agonistic behavior (Delville et al., 2005). As noted above, hamsters exposed to adult aggression begin to show adultlike aggressive behavior earlier in life. These animals undergo normal gonadal development, but they are also very aggressive toward smaller and younger individuals (Wommack et al., 2003). In effect, they become bullies.

What is the ecological significance of this accelerated development of agonistic behavior in Syrian hamsters? The consequences of this enhanced aggression may depend on population density (Delville et al., 2003). The relationship between high population density and elevated aggression is well established. In natural populations, juveniles would be most likely to be socially subjugated by adults when population levels were high or resources were low. To survive, they themselves would have to become more aggressive under these conditions, therefore increasing the level of aggression in the population (Delville et al., 2003). In contrast, social stress during late puberty does not affect the development of aggression. Indeed, hamsters experiencing repeated aggressive encounters at that time become submissive and may remain so for life (Huhman and Jasnow, 2005). Such animals would be unlikely to compete successfully for mates or resources.

Is It Adaptive for Rodents To Be Aggressive at Puberty?

In a study of prairie voles, a group of peripubertal males was implanted with slow-release capsules of testosterone, while another group received empty capsules. The testosterone-treated males exhibited a higher adult level of aggression in pairwise encounters than the control animals (Gaines et al., 1985), presumably because of the elevated concentrations of testosterone. How might such an increase in aggression at puberty be adaptive for these animals? Males of many rodent species, including voles, disperse at the time of puberty. Dispersal is a very dangerous time for animals; mortality rates are extremely high during this time.

Leaving home exposes the individual to many new threats, including predators and competing males. Dispersing males often find themselves encroaching on other males' territories as they learn to fend for themselves and try to find food, establish territories, and mate with females. Thus, dispersing males have evolved to be aggressive, which presumably increases their odds of survival and success in reproductive competition (Smale et al., 1997).

High blood concentrations of androgens are necessary to elicit dispersal in some mammalian species, but not in peripubertal male Belding's ground squirrels (*Spermophilus beldingii*) (Holekamp and Sherman, 1989; Smale et al., 1997). Male ground squirrels castrated at puberty exhibit dispersal behavior at the same rate and intensity as gonadally intact males. Interestingly, by the time these juvenile ground squirrels disperse, the mating season is over and adult males are no longer maintaining territories with the same intensity as they did at the onset of the breeding season.

The Timing of Puberty: Birds

In many avian species, reproductive and aggressive behaviors in males are closely linked. Many visual and vocal displays are used to attract females and repel other males simultaneously (Wingfield et al., 1994). Increasing vernal day lengths trigger reproductive maturation in many birds by stimulating gonadal growth and testosterone production (see Chapter 10). Peak concentrations of plasma testosterone correspond to peak levels of intermale territorial aggression in the late spring. At the end of the nesting period, the testes regress in response to the chronic long day lengths. Because long day lengths no longer stimulate the reproductive system, these birds are said to be refractory to photoperiod (day length), or *photorefractory*. Exposure to short days is then required to break photorefractoriness and allow the birds to respond to the stimulatory effect of long days the following spring.

Note that the photoperiodic control of gonadal function is fundamentally different in birds and rodents. Rodents are reproductively "on" unless exposed to short days, which turns them "off." Eventually, the reproductive systems of rodents break free of the inhibitory effects of chronic short days and regain their function. In other words, they turn "on" despite the short day lengths. The physiological mechanisms underlying this so-called "spontaneous" reproductive development are unknown.

The onset of puberty in most avian species is dependent on photoperiod. However, there is substantial variation in the timing of puberty; for example, some species reach puberty during their first year, others when they are 1 year old, and still others only after 2 or more years of age (Follett, 1991). This variation in the timing of puberty corresponds with variation among bird species in the timing of first territorial defense and mating. The majority of nontropical avian species hatch in the late spring and reach adult body size several weeks after hatching, but do not breed until the following spring. This is a reasonable strategy from an adaptive, functional perspective because young males probably could not secure suitable territories to induce a female to breed. From a physiological perspective, one might ask why long days do not stimulate the birds to undergo puberty in their first summer. There is increasing evidence that birds that normally breed at 1 year of age hatch in a photorefractory state (Follett, 1991). That is, these birds must experience short days in order to break photorefractoriness and be stimulated by long days. Thus, unless housed under artificial lighting conditions, these birds cannot undergo puberty until spring, when they first experience long days after a period of short days. Those birds, such as quail and chickens, that undergo puberty when less than 1 year of age are not photorefractory at hatching and can be stimulated by long days to undergo reproductive development during their first year.

When circulating testosterone concentrations increase, aggressive interactions among birds increase in frequency, especially during the establishment of territories or social dominance hierarchies (see below; Wingfield et al., 1987). Androgens are important in stimulating territorial behaviors, including singing. Experimental implants of testosterone capsules at the onset of territory establishment increased territory size or intermale aggression in several avian species, including the sharp-tailed grouse (*Tympanuchus phasianellus*) (Trobec and Oring, 1972), red-winged blackbird (*Agelaius phoenecius*) (Searcy and Wingfield, 1980), pied flycatcher (*Ficedula hypoleuca*) (Silverin, 1980), and white-crowned sparrow (*Zonotrichia leucophrys*) (M. C. Moore, 1984). Testosterone treatment after territories or social dominance hierarchies were already established, however, usually did not increase territory size or social status (see below; reviewed in Wingfield et al., 1990, but see Cawthorn et al., 1998).

Dispersal Strategies and Social Status in Primates

In contrast to many animals, the pubertal increase in blood concentrations of testosterone among primates is not always associated with elevated aggression. The onset of puberty in captive rhesus monkeys does not result in any discernible increase in aggressive behavior (Rose et al., 1978). Agonistic encounters between young captive male owl monkeys (*Aotus trivirgatus*) and their parents did not change in frequency at puberty, despite large increases in their plasma testosterone concentrations (Dixson, 1980).

As in many avian and mammalian species, puberty is associated with dispersal among many primate species (Carpenter, 1940; Charles-Dominique, 1977; Fossey, 1974). Male rhesus monkeys become targets of increasing aggressive outbursts by adult males as they approach puberty at 3 or 4 years of age. This chronic harassment appears to force the peripubertal males to leave the natal troop and attempt to join a new one. Essentially two strategies exist for joining a new troop: the newcomer may burst right into the group and seize membership status, or hang around the periphery and try to sneak into the group. Either strategy has costs, often fatal, for the newcomer. Attempts to join a new troop immediately are usually met with strong resistance by the resident males, and a great deal of aggression is aimed at the intruder. Peripheral animals are also in danger because they may starve to death or succumb to any one of a number of parasitic or stress-related illnesses.

Interestingly, several behavioral and hormonal factors predict which strategy an individual male will attempt in joining a new troop (Virgin and Sapolsky, 1997). Young males that are sons of high-ranking females often have "out-going, risk-taking" personalities (Suomi, 1991). These males venture farther from their mothers and engage in more rough-and-tumble play than do orphans or sons of low-ranking mothers. The sons of high-ranking females exhibit low vagal tone (a measure of autonomic reactivity; these individuals show a steady cardiac output and respiratory rate) and low plasma cortisol concentrations. In contrast, males that are not sons of high-ranking females have high vagal tone (heart rate, respiratory rate, and blood pressure are very reactive) and high blood concentrations of cortisol, suggesting that these males experience more stress in their lives. (In the laboratory, males with high vagal tone and high cortisol concentrations are also more likely to abuse alcohol than males with low vagal tone and low cortisol concentrations: Higley et al., 1991; Suomi, 1991, 1997.) In seminatural conditions, males with low vagal tone and low cortisol concentrations are more likely to employ the strategy of immediately establishing membership in a new troop. In contrast, males with high vagal tone and high cortisol concentrations are more likely to try to sneak into the troop after establishing familiarity from the periphery (Bolig et al., 1992).

In a study of male olive baboons (*Papio anubis*) living freely in stable hierarchies in Africa, some subordinate males displayed elevated basal glucocorticoid concentrations, and they showed a blunted glucocorticoid response and rapid suppression of testosterone concentrations during stress (Virgin and Sapolsky, 1997). These endocrine characteristics were initially interpreted as reflecting the chronic stress of their social position. However, long-term fieldwork revealed that these endocrine characters do not mark all subordinate individuals. Rather, endocrine profiles differed among subordinate males as a function of their individual "styles of social behavior." One subset of subordinate male baboons was identified that had significantly high rates of copulations, a behavior usually shown only by high-ranking males. Such behavior predicted the onset of a transition to dominance, as this subset of subordinate males was significantly more likely than other subordinates to move to the dominant half of the hierarchy over the subsequent 3 years (Virgin and Sapolsky, 1997). A second subset of subordinate males was the most likely to initiate fights and to displace aggression onto an uninvolved third party after losing a fight. Males in this second cohort generally had elevated testosterone and lower basal glucocorticoid concentrations compared with the remaining subordinate cohort. Taken together, these results suggest that variables other than rank alone may be associated with distinctive endocrine profiles, and that even in the face of a social stressor (such as social subordination), particular behavioral styles may attenuate the endocrine indices of stress (Virgin and Sapolsky, 1997; see Chapter 11).

Sex Differences in Social Behavior

Males are generally more aggressive than females. Certainly, human males are much more aggressive than females. Many more men than women are convicted of violent crimes in North America. The sex differences in human aggressiveness appear very early. At every age throughout the school years, many more boys than girls initiate physical assaults. Almost everyone will acknowledge the existence of this sex difference, but assigning a cause to behavioral sex differences in humans always elicits much debate (see Chapter 4). It is possible that boys are more aggressive than girls because androgens promote aggressive behavior and boys have higher blood concentrations of androgens than girls. It is possible that boys and girls differ in their aggressiveness because the brains of boys are exposed to androgens prenatally and the "wiring" of their brains is thus organized in a way that facilitates the expression of aggression. It is also possible that boys are encouraged and girls are discouraged by family, peers, and others from acting in an aggressive manner. These three hypotheses are not mutually exclusive, but it is extremely difficult to discriminate among them to account for sex differences in human aggressiveness.

What kinds of studies would be necessary to assess these hypotheses? It is usually difficult to separate out the influences of environment and physiology on the development of behavior in humans (see Chapter 4). For example, boys and girls differ in their rough-and-tumble play at a very young age, which suggests an early physiological influence on aggression. However, parents interact with their male and female offspring differently; they usually play more roughly with male infants than with females, which suggests that the sex difference in aggressiveness is partially learned (Smith and Lloyd, 1978). This difference in parental interaction style is evident by the first week of life. Because of these complexities in the factors influencing human behavior, the study of hormonal effects on sex-differentiated behavior has been pursued in nonhuman animals, for which environmental influences can be held relatively constant. Animal models for which sexual differentiation occurs postnatally are often used so that this process can be easily manipulated experimentally.

The Organization and Activation of Aggression: Mice

With the appropriate animal model, we can address the questions posed above: Is the sex difference in aggression due to higher adult blood concentrations of androgens in males than in females, or are males more aggressive than females because their brains are organized differently by perinatal hormones? Are males usually more aggressive than females because of an interaction of early and current blood androgen concentrations? As with humans, these possibilities are difficult to assess in red deer or Harris's sparrows, but relatively simple to examine in house mice, for which sexual differentiation of the brain occurs postnatally. If male mice are castrated prior to their 6th day of life, then treated with testosterone propionate in adulthood, they exhibit low levels of aggression. Similarly, females ovariectomized prior to day 6 of age but given androgens in adulthood do not express male-like levels of aggression. Treatment of perinatally gonadectomized males or females with testosterone prior to day 6 of age and also in adulthood results in a level of aggression similar to that observed in typical male mice (Edwards, 1969, 1970) (Table 8.3). Thus, in mice, the proclivity for males to act

TABLE 8.3 Effects of gonadectomy and androgen treatments on aggressive behavior in house mice

Treatment		
Before 6 days of age	In adulthood	Aggressive behavior in adulthood
Males		
None	None	+++
Castration	None	0
Castration	Androgen	+
Castration + androgen	Androgen	+++
Females		
None	None	0
Ovariectomy	None	0
Ovariectomy	Androgen	0
Ovariectomy + androgen	Androgen	+++

+++ = normal male levels of aggressive behavior; + = reduced levels; 0 = no aggressive behavior.

more aggressively than females is organized perinatally by androgens, but also requires the presence of androgens after puberty in order to be fully expressed. In other words, aggression in male mice is both organized and activated by androgens. The hormonal control of aggressive behavior in house mice is thus similar to the hormonal mediation of heterosexual male mating behavior in other rodent species (see Chapter 3). Aggressive behavior is both organized and activated by androgens in many rodent species, including rats, hamsters, and voles.

The world is much more complex than we often anticipate, however. Even when experimenters use a highly inbred animal model in constant laboratory conditions, it may be discovered that what appear to be constant conditions in fact contain many variables. Some of these hidden variables may profoundly affect the endocrine–behavior relationships under study. This problem arises in studies of aggression in house mice. In mice, as in other rodent species, fetuses are packed in the uterus like peas in a pod (Clark et al., 1992) (see Figure 3.28), and may thus be influenced by hormones produced by their developing siblings. A female mouse may be situated between two brothers (2M = 2 males), between a sister and brother (1M), or between two sisters (0M). In general, 2M females are more aggressive than their 0M sisters. Presumably, androgens produced by their brothers affect these females' nervous systems (see Chapter 3). This example should remind you that many subtle endocrine influences on behavior are possible.

Sex Differences in Dispersal: Ground Squirrels

Male Belding's ground squirrels leave home when they are about 50 days old. Dispersal is a gradual process taking 4–5 weeks to accomplish (Figure 8.14). It is also a robust sexually dimorphic trait: virtually all males disperse, whereas few

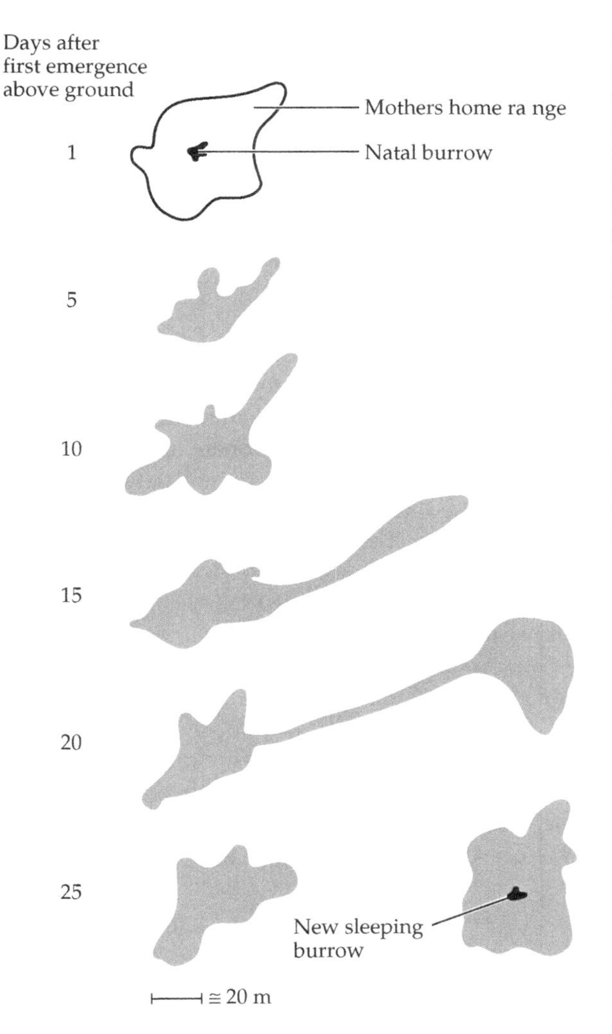

8.14 Dispersal in male Belding's ground squirrels. Although dispersal sounds like a sudden event, in ground squirrels it requires nearly a month to complete. At about 4 weeks of age, a young male emerges from his natal burrow, and his aboveground activity is limited to an area very close to the mouth of the burrow. After about 15 days above ground, the young male travels throughout his mother's home range, and also ventures well beyond it. Despite spending many hours away each day, he always returns to his mother's burrow at night until about 7–8 weeks of age (about 25 days above ground), when he establishes a new sleeping burrow and completes the dispersal process. When mapped, the dispersal process is visually reminiscent of an amoeba splitting during reproduction. After Holekamp and Sherman, 1989.

(<10%) females leave their natal region (Holekamp and Sherman, 1989). This sex difference in dispersal is common among mammals. A sex difference in dispersal also exists among avian species, but is reversed: female birds are far more likely to disperse from the natal nesting area than males.

As mentioned previously, peripubertal castration does not affect dispersal in male ground squirrels. If the activational effects of androgens are unimportant in sustaining this behavioral sex difference, then it is reasonable to suppose that it is mediated by early organizational hormonal events. In one test of this hypothesis, pregnant ground squirrels were trapped and caged at an outdoor camp near their home burrows (Holekamp, 1986). Their female offspring were injected with testosterone propionate dissolved in sesame seed oil, or with the oil vehicle alone, several days after birth. Immediately after the injections the families were returned to their home burrows, where the mothers successfully reared their offspring. Twelve of the androgen-treated females were re-trapped at 60 days of age or older. Seventy-five percent of them had dispersed from the mother's home burrow to a new nest site, following dispersal routes and traveling distances that were comparable to those of males. Approximately 70% of their male siblings had dispersed by this age. Only 8% of the females treated with the oil vehicle alone had left the natal area by 60 days of age. These results suggest an important role for early organizational effects of androgens on a complex behavior; namely, dispersal (Smale et al., 1997).

Hormones and Dominance Status: Canines

Hormonal mediation of sexual differentiation occurs both prenatally and postnatally in dogs (see Chapter 3). Female puppies can be masculinized in two steps, by treating their pregnant mothers with testosterone propionate during gestation and then injecting the newborn puppies with testosterone immediately after birth. Such dogs have highly masculinized external genitalia (Figure 8.15; see also Figure 4.13C) and exhibit male-like behavior. The social signals provided by these masculinized females elicit social confusion among beagles living in small, stable groups, just as the visually disparate signals of dyed Harris's sparrows do. The social confusion may result because the behavior of a masculinized female is somehow suspect, or she does not emit the appropriate odors or other sensory signals, or possibly some other feature about her is incongruent with her perceived social status.

The dominance hierarchy of a small group of beagles is relatively easy to determine with the right experimental tool. The right tool in this case is a large bone, preferably an oxtail bone, which is simply thrown into the middle of the dogs' pen when all the dogs are present and attending to the "bone provider,"

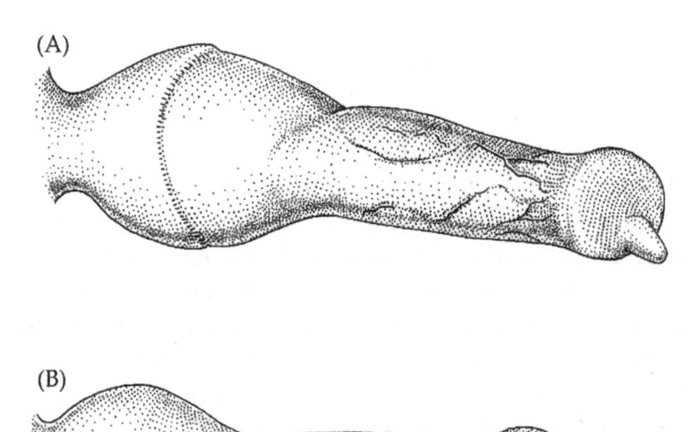

8.15 Perinatal exposure to androgens masculinizes female beagles. (A) The penis of a normal male beagle. (B) The pseudopenis of a female beagle exposed in utero and immediately after birth to androgens. If the penile (or pseudopenile) shaft is retracted, both males and masculinized females display comparable erections. Female dogs exposed to early androgen treatment also display very aggressive behavior and cause social disruption and confusion within small, stable groups. After Beach, 1984.

namely, the experimenter. During the normal course of events, the highest-ranking dog in a stable group gets the bone, usually with a minimum of conflict, growling threats, or any other hostile gestures. To discover which dog is the second in command, one simply removes the number one dog from the enclosure and repeats the bone competition test. In this way, and in pairwise tests between all of the dogs, the dominance hierarchy of a group can be determined (Beach et al., 1972). A variation of this technique, using chicken feed, was first used in classic studies on the pecking order of domestic fowl (Schjelderup-Ebbe, 1922).

Both dogs and chickens have relatively linear dominance hierarchies; that is, the highest-ranking individual, A, supplants all others, whereas the second-highest-ranking animal, B, supplants all others except A, and so on down the line. In some species, especially among primates, dominance status may be inherited and thereby relatively fixed over extended periods of time. In other cases, social organization may be quite plastic as low-ranking individuals continually test and overthrow the status quo. Other species do not have linear pecking orders. Dominance hierarchies among more complex animal societies, especially those of primates, may be circular; for example, A is dominant over B, who is dominant over C, who is dominant over A. Other dominance systems involve complex and dynamic cooperative coalitions; for example, A supplants B only with the aid of C (Bernstein, 1981). In general, dominance hierarchies represent a social adaptation that arose over evolutionary time owing to two factors: (1) not all individuals are equal in securing resources, and (2) it is a waste of time and energy to continually reestablish this inequality among group members (Wilson, 1975).

In one study of beagle social organization, the dominance hierarchies were separately determined for a group of five male beagles, a group of five female beagles, and a group of five masculinized female beagles injected with testosterone in adulthood (Beach et al., 1972). Dogs from the different groups were then paired, and their status was determined by which dog won two-dog bone competitions over several consecutive days of testing. In general, normal females, regardless of their rank within their own dominance hierarchy, gave way to males; however, the bottom one or two male dogs often lost to the top one or two female dogs. Female beagles that had not been exposed to testosterone also yielded to the masculinized females. Possession of the bone in virtually all of these interactions was determined without any overt aggressive behavior: if a female got the bone first, she would drop it when the male or masculinized female approached her. These social interactions became more interesting when males and masculinized females were paired in the bone competition. A masculinized female would not yield to a male if she obtained the bone first. The male behaved as if he expected the female to give up the prized bone; he would growl menacingly at her and eventually attack her when she refused to give up the bone. If the male obtained the bone first, the masculinized female would attack him and continue to harry him until he relinquished the bone or counterattacked. The social signals providing information about the masculinized females' status, or their understanding of canine social rules, whatever they may be, were somehow affected by the early hormonal treatment.

The inclination toward higher levels of aggression in male dogs, as in house mice, requires both early organizational exposure to androgens and activation in adulthood by these steroid hormones (see Chapter 3). Aggressive behavior in males of these species is thus organized and activated by androgens. Early androgen exposure alone elevates the incidence of aggressive behaviors in perinatally castrated males and females above the level observed in normal females, but does not increase aggression to the levels of normal males.

Sex Differences in Play Behavior: Primates

Although work with altricial rodents has provided remarkable insights into the process of sexual differentiation of behavior, the data generated with these rodents have some fundamental limitations in their application to understanding the process of sexual differentiation of human behavior. People undergo sexual differentiation in utero, not postpartum. Consequently, the organization of sexual differentiation in rodents, in which the hormonal effects occur postnatally, may differ fundamentally from the process in humans. Furthermore, masculinization in rodents depends on aromatization of androgens to estrogens. The aromatization of androgens does not appear to be important in the development and expression of masculine behaviors in primates (reviewed in Resko and Roselli, 1997; Balthazart and Ball, 1998). Although work with nonhuman primates is difficult and expensive, such work is more likely to provide insights into the sexual differentiation of humans than are studies of rodents.

Several fascinating studies that explored the role of gonadal steroid hormones in primate aggressive behavior have been reported. As is the case in humans, males of many nonhuman primate species are more aggressive than female conspecifics. This sex difference in behavior manifests itself early in development during the expression of play behavior. As noted previously, males engage in more rough-and-tumble play, more chasing, and more threatening behavior, and initiate play more often than females (see Chapter 3). The higher proclivity for aggression, either real or in the context of play behavior, in male than in female rhesus monkeys appears to require only prenatal exposure to androgens. Castrated male rhesus monkeys, for example, do not differ from intact males in the amount of threatening behavior, rough-and-tumble play, chasing, and play initiation they display (Goy and Phoenix, 1971). Many rodent species, in contrast, require androgens for both early organization and later activation of aggressive behavior. As described in Chapter 3, treatment of pregnant female rhesus monkeys with testosterone propionate produces pseudohermaphroditic female offspring. These androgen-treated females display threatening behavior, initiate play activities, and engage in rough-and-tumble play at frequencies intermediate between those of normal males and normal females (Goy, 1966, 1978; Phoenix, 1974; Phoenix et al., 1968). Thus, the expression of sex differences in play behavior requires the organizational effects of androgens, but not the activational effects.

The role of early androgen exposure in the mediation of sex differences in human aggressiveness has also been examined in "experiments of nature." As described in Chapter 3, girls who were exposed to high concentrations of androgens in utero because their adrenal glands malfunctioned were reported to engage in more rough-and-tumble play than their sisters who did not suffer from congenital adrenal hyperplasia (CAH) (Reinisch, 1974). Although these reports are interesting, observations of this type are less reliable in distinguishing cause–effect relationships than controlled experiments in which researchers are uninformed about the treatment condition of any individual. Recall that because the mothers were aware of their daughters' medical condition, it is possible that their ratings of their daughters' behavior may have been inadvertently biased (see Chapter 3).

Sex Role Reversals

In some species, females are more aggressive than males. It is interesting to examine these species' behaviors because studying them may shed light on aggressive behaviors among species with more typical sex roles. This section describes some species with reversed sex roles.

SANDPIPERS Some female sandpipers of the genera *Phalaropus* and *Actitis* exhibit brightly colored plumage, in contrast to the dull-colored males of their species (Tinbergen, 1935). Females compete among themselves for access to males, which are the predominant incubators, and are extremely aggressive toward other females (see Box 3.1). These reversals of the typical sex roles are not, however, accompanied by a reversal in the normal male:female ratio of blood androgen and estrogen concentrations. Androgen concentrations are 6 to 8 times higher in pre-incubating male than in female Wilson's phalaropes (*Phalaropus tricolor*) (Fivizzani et al., 1986) and spotted sandpipers (*Actitis macularia*) (Rissman and Wingfield, 1984).

Although androgen concentrations are lower in female than in male sandpipers at most times, androgens may still mediate aggressiveness among females. Before pair formation, female testosterone concentrations are quite low and are comparable to values reported for females of other avian species. Females sampled after pairing, however, displayed a temporary sevenfold elevation in blood testosterone concentrations (Fivizzani and Oring, 1986). Thus, female sandpiper territoriality and aggressiveness are not dependent on male-typical blood androgen concentrations, but may reflect increased sensitivity of brain structures to steroid hormones (Fivizzani and Oring, 1986). There is no sex difference in the amount of aromatase or 5α-reductase or in 5α-reductase enzyme activity in the brain, pituitary, or skin of the Wilson's phalarope, suggesting that neural sex steroid hormone receptors or changes in neural circuitry may underlie the sex role reversal observed in this species (Schlinger et al., 1989).

SYRIAN HAMSTERS Adult female Syrian hamsters are often more aggressive than males (Payne and Swanson, 1972). Aggression in these females can be moderated by social influences, particularly housing conditions. For example, ovariectomized females housed individually are more aggressive than intact females when they are brought into an arena with another hamster; however, group-

TABLE 8.4 **Effects of housing condition on aggressive behavior in Syrian hamsters**

Endocrine status	Aggressive behavior: Housed individually	Aggressive behavior: Group-housed
Females		
Ovariectomized	↑	↓↓
Ovariectomized + estradiol and progesterone	↑	↑↑
Males		
Castrated	↓	↓
Castrated + testosterone	↑	↑

↑ = increase; ↑↑ = substantial increase; ↓ = decrease; ↓↓ = substantial decrease

housed ovariectomized hamsters fight less than individually housed ovariectomized animals (Table 8.4). Treatment of ovariectomized females with estradiol and progesterone causes a greater increase in aggression in group-housed animals than in individually housed animals. Housing condition does not significantly affect the aggressive behavior of intact hamsters. As in most other species, castration of male hamsters reduces agonistic encounters, and testosterone replacement therapy ameliorates this reduction. It is interesting to note that male mice and rats have smaller adrenal glands than their female conspecifics, but that the reverse is true of Syrian hamsters (Chester-Jones, 1955).

Photoperiod can interact with gonadal steroid hormones to affect female hamster aggressive behavior. There was no difference in aggressiveness between ovariectomized females that were housed under short-day and long-day lighting conditions (Badura and Nunez, 1989; Elliott and Nunez, 1992). However, when treated with estradiol and progesterone, short-day hamsters were more aggressive than long-day females, especially when tested against males intruding into the females' home cage (Elliott and Nunez, 1992; Karp and Powers, 1993).

Agonistic behavior is not limited to fighting, but also includes scent marking in many species. Individuals of many mammalian species use chemosensory information such as scent marks to communicate social information (Doty, 1986; Johnston, 1985). In an agonistic context, scent marking has two social functions. The first is to advertise territoriality and thus prevent an aggressive interaction. The second kind of communication occurs after an aggressive encounter. In this case, the winner of the fight (i.e., the dominant individual) initiates scent marking as the loser retreats (Johnston, 1975, 1985). One type of scent-marking behavior commonly observed among mammals involves rubbing a part of the body, such as the chin or the flank, on a vertical substrate. Syrian hamsters engage in flank marking, a behavior that conveys information about dominance (Ferris et al., 1987) or mate choice (Huck et al., 1985). Males of other mammalian species engage in chin-marking behaviors to the same effect. Unlike most other mammals, in which males scent-mark more often than females, female Syrian hamsters flank-mark more often than male conspecifics (Hennessey et al., 1994).

Arginine vasopressin (AVP) plays a critical role in the expression of flank marking (Albers et al., 1992; Ferris, 1992). Microinjections of AVP into the MPOA–anterior hypothalamus, lateral septal nucleus, BNST, or PAG evoke increased flank marking in hamsters (Albers et al., 1986; Bamshad and Albers, 1996). Microinjections of an AVP antagonist into the MPOA–anterior hypothalamic regions result in a dramatic decline of flank-marking behavior (Albers et al., 1992; Ferris et al., 1993). Castration reduces the amount of flank marking in AVP-treated hamsters, and both testosterone and estradiol treatment rapidly restore flank marking in male and female hamsters, respectively (Huhman and Albers, 1993).

SPOTTED HYENAS As described in Chapter 3, spotted hyenas (*Crocuta crocuta*) are remarkable in the similarity of the external genitalia of the two sexes (Kruuk, 1972). As in beagles, there is a rigid dominance hierarchy within each sex (Frank, 1983), but in sharp contrast to beagles, female hyenas dominate males in most behavioral interactions. Do hormones regulate social interactions in this species? If so, are blood androgen concentrations higher in females than in males? These questions are unresolved; conflicting reports exist regarding blood androgen concentrations in this species. One study reported no difference in blood androgen concentrations between male and female spotted hyenas (Racey and Skinner, 1979), but the results of another study indicated the usual sex difference in blood androgen concentrations, with males possessing higher concentrations than females. Individual social status was a much better predictor of blood androgen level than sex; resident animals in general, and the top-ranking individuals of both sexes in particular, exhibited high blood serum androgen concentrations (Table 8.5; Frank et al., 1985; Yalcinkaya et al., 1993).

TABLE 8.5 Serum androgen concentrations in Kenyan spotted hyenas

Status	Androgen concentration (ng/ml ± SEM)
Males vs. females	
Adult males	1.93 ± 0.4
Prepubertal males	0.25 ± 0.1
Parous females	0.35 ± 0.1
Prepubertal females	0.18 ± 0.2
Residents vs. transients	
Residents	3.15 ± 0.6
Transients	0.81 ± 0.2

Source: Frank et al., 1985.

Individual Differences in Aggression

We all learn early in life that there are large individual differences in the expression of aggressive behavior. Some individuals resort to aggressive tactics with very little provocation, while others virtually never behave aggressively. You may have discovered this variation in preschool when you met your first bully, or you may have learned about it on your paper route when some dogs greeted you happily with wagging tails, whereas other dogs' raison d'être seemed to be to gnaw off your leg. What accounts for these differences in aggressiveness? Are individual differences in aggressive behavior explicable on the basis of blood plasma testosterone concentrations? In other words, are some animals more aggressive because they have higher blood concentrations of androgens, or do these aggressive animals have higher androgen concentrations because they are more aggressive?

Individual differences in aggressive behavior do not correlate with blood testosterone concentrations in mice (Barkley and Goldman, 1977; McKinney and Desjardins, 1973), although they do in rats (Schuurman, 1980). One experimental test of this question first rated male mice as aggressive or not aggressive. All the animals were then castrated, and they all became equally passive. The castrated mice were all provided with equal doses of testosterone and tested for aggressiveness again. Males that were previously aggressive became aggressive with testosterone replacement therapy, and previously nonaggressive males remained docile after testosterone treatment. This study indicates that high circulating concentrations of testosterone are necessary, but not sufficient, for stimulating aggressive behavior. The individual differences in aggressive behavior were probably due to differences in receptor sensitivity or in the ability of androgens to alter the perception of an aggression-evoking stimulus.

There are certainly strain differences among house mice in the extent to which aggressive behaviors are expressed and in the extent to which these aggressive behaviors are mediated by androgens. One systematic study in a Swiss strain of albino mice investigated the effects of castration on predatory, shock-induced, maternal, and isolation-induced aggression. Isolation-induced aggression was generally reduced after castration; post-gonadectomy treatment with testosterone, 5α-dihydrotestosterone (DHT), or estradiol restored this form of aggression (reviewed in Haug et al., 1986). In sharp contrast, castration increased territorial aggression when the intruder was a lactating female, and treatment with testosterone, DHT, or estradiol reversed the elevated rate of aggressive responses in this situation (Brain et al., 1983). These results strongly imply that steroid hormones do not merely "trigger" aggression, but act to affect the animal's perception of and response to aggression-provoking stimuli (Haug et al., 1986).

In another series of experiments (Whalen and Johnson, 1987), male mice were pitted against either lactating females or males from which the olfactory bulbs had been removed (reviewed by Johnson and Whalen, 1988). Bulbectomized males are used in these types of studies because they elicit, but do not initiate, aggressive behavior. Gonadally intact males and castrated males treated with testosterone attacked the bulbectomized males, but did not attack the lactating females. Untreated castrated males tended to display tremendous individual differences in aggressiveness, with some attacking either type of opponent, others attacking only one type of opponent, and others failing to attack any opponent. Because castration was associated with great individual variation in aggressive behavior, and because androgen treatment reduced that variation, Johnson and Whalen (1988) have proposed that testicular steroid hormones act to induce "behavioral homogenization": in other words, androgens reduce variability in male house mouse aggressiveness. This is an intriguing hypothesis to account for the disparate aggressive responses of males to different aggression-provoking stimuli, although further experiments are necessary to evaluate it fully.

Dominance in more complex social organizations may not be related to blood concentrations of testosterone, especially in stable groups. For example, dominant dogs or squirrel monkeys can be castrated without affecting their position in the dominance hierarchy (Dixson, 1980). Treatment of low-ranking individuals with androgens does not change their status in these species.

Experience is also important in the relationship between hormones and aggressive behavior (Miczek and Fish, 2005). Castration and hormone replacement studies of males representing several species of reptiles, fish, and birds clearly demonstrate reduced post-castration levels of aggression and restoration of aggression after testosterone treatment (e.g., Crews and Moore, 1986; Wingfield et al., 1987). In mammals, however, the effects of androgens in supporting aggressive behavior depend largely on experience. Castrated mice and rats without prior aggressive experience rarely fight when tested with a male conspecific (Christie and Barfield, 1979). If the animals are castrated after aggressive encounters have been experienced, however, aggressive behavior declines, but endures long after the surgery (e.g., Christie and Barfield, 1979; DeBold and Miczek, 1981, 1984). Rather than having an "obligatory" role in the regulation of aggression, as in fish, reptiles, and birds, androgens appear to exert a *modulatory* effect on mammalian aggressive behavior (Miczek and Fish, 2005; Johnson and Whalen, 1988).

Social Experience Feeds Back to Influence Hormone Concentrations

Hormones obviously affect behavior, but it should be emphasized that behavior, in turn, can feed back and affect hormone concentrations. Tests of male mice or Syrian hamsters in paired aggressive encounters revealed that their androgen levels were suppressed if they lost the fight (Lloyd, 1971; Huhman and Jasnow, 2005). This hormonal suppression lasted for many days after the defeat. Similarly, rhesus monkeys that were defeated by a higher-ranking male had profoundly reduced testosterone concentrations for weeks after the defeat. In contrast, winning males' circulating testosterone concentrations quadrupled within 24 hours of the victory (Bernstein et al., 1974).

Similar phenomena occur in humans. In sports, there is a so-called home field advantage. Home field advantage has been reported for several levels of competition, from elementary school to professional sports (Table 8.6) (Neave and

TABLE 8.6 **Home field advantage for English soccer leagues, 2000–2001**

Division (Number of teams)	Average home wins	Average away wins	Proportion of wins at home	Average home goals	Average away goals
Premiership (20)	9.20	4.75	0.66	29.35	20.25
Division 1 (24)	10.29	6.54	0.64	32.54	24.67
Division 2 (24)	10.08	6.62	0.61	35.08	26.00
Division 3 (24)	11.38	5.08	0.70	35.13	23.08

Source: Neave and Wolfson, 2003

Wolfson, 2003). There are several explanations for this advantage, including the home team's familiarity with the playing field, travel fatigue and disruption of routine for the visiting team, and home team crowd noises that inspire and encourage their team, as well as influence referee calls. Indeed, U.S. professional football teams with enclosed stadiums appear to have an enhanced home field advantage because their stadiums are noisier (Zeller and Jurkovac, 1989). All of these factors, as well as the possibility that humans defend a home "territory" in a sporting event in the same way many animals defend their home territory, may be mediated by hormones. In one series of studies, which assessed salivary testosterone concentrations in U.K. football (soccer) players, testosterone concentrations were significantly higher before home games than away games (Neave and Wolfson, 2003). The perceived rivalry with the opposing team influenced testosterone concentrations: testosterone was more elevated before playing a "challenging" rival than a more "moderate" rival, especially among defensive players such as the goalies (Neave and Wolfson, 2003). Because these games neither attracted large, noisy crowds nor required overnight travel or significant disruption of routine, it appears that testosterone is a primary mediator of the home field advantage, though much additional research on this question is required.

The Challenge Hypothesis: Birds

The influence of social experience on avian hormone concentrations has been studied in a number of different species (Wingfield et al., 1997). In one elegant field study, territorial male red-winged blackbirds were trapped under two different conditions. In one condition, a live decoy male was placed in a spring-loaded net trap located near the center of the target male's territory. A tape recorder repeatedly broadcast a specific male "advertising" song to ensure an aggressive response from the resident male. As you might imagine, the presence of a male intruder apparently singing with great abandon was certain to elicit a rapid behavioral response from the resident male. First, the resident increased his rate of singing behavior. Then he emitted the species-specific wingspread display, showing off his bright red epaulets (Figure 8.16). Because the tape continued to play and the intruder remained in place, the resident male approached and finally attacked the decoy, activating the net trap. A blood sample was immediately

8.16 Male red-winged blackbird in the wingspread display. This behavior, which shows off the bird's bright red epaulets, is an aggressive territorial signal to conspecific male intruders.

obtained from the resident. Other territorial males were trapped while foraging, without the use of a decoy. Blood testosterone concentrations were more variable in the birds caught during an escalating aggressive episode than in the foraging males, suggesting that the frequency of pulsatile androgen discharges in the aggressively behaving birds was increasing. There were no significant differences between the two groups in blood concentrations of corticoids (Harding and Follett, 1979). This study demonstrated that behavior or environmental stimuli can affect hormone concentrations.

It is difficult to discern whether aggressive behavior per se or simply the perception of aggression-provoking stimuli is critical in mediating hormonal responses. Another study on a different avian species begins to address this issue (Wingfield and Wada, 1989). Like red-winged blackbirds, male song sparrows exhibit endocrine changes in response to territorial disputes. Again, territorial intrusions were simulated by placing live, caged conspecifics in the center of a male's territory and broadcasting a tape of the conspecific song over a loudspeaker. Responding resident males were caught either 1–4 minutes, 5–10 minutes, or 10–60 minutes after the onset of aggression-provoking stimuli. Control birds were trapped while they were foraging and were not actively engaged in territorial disputes. Blood samples were obtained from all captured birds and later assayed for testosterone and LH concentrations. Neither LH nor testosterone concentrations differed significantly between foraging birds and birds involved in territorial disputes sampled 1–10 minutes after the onset of the agonistic interaction. Both testosterone and LH concentrations, however, increased in aggressively behav-

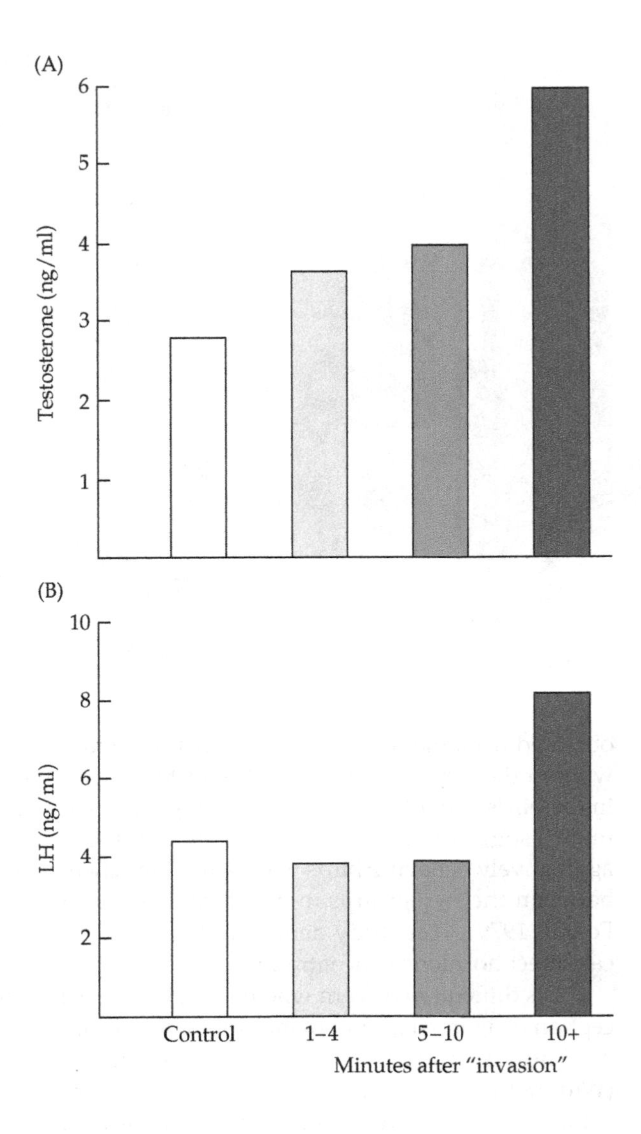

8.17 Effects of an intruder on resident song sparrows' hormone concentrations. After a simulated "invasion" of a male song sparrow's territory, plasma testosterone concentrations begin to rise (A). Note that the rise in testosterone levels 5–10 minutes after the detection of the intruder is not accompanied by a corresponding rise in LH (B), but that the sustained elevation in testosterone seen after 10 minutes appears to be mediated by LH. After Wingfield and Wada, 1989.

ing birds sampled 10–60 minutes after the agonistic encounter began (Figure 8.17). The initial pulse of testosterone in response to an intruder appeared to be independent from LH regulation, but persistently elevated testosterone concentrations appeared to be mediated by LH in the usual manner (see Chapter 3). Laboratory tests confirmed these field observations. Thus, in song sparrows, testosterone appears to increase the persistence of aggression following an intrusion rather than activating aggression per se. Such persistence would be highly adaptive in the breeding season, when reproductive success is at stake, but would not be adaptive in autumn, when other strategies (switch territories, float) are possible (Wingfield et al., 2005).

Additional laboratory studies examined the nature of the stimuli eliciting this behavioral–endocrine response in more detail (Wingfield and Wada, 1989). Simulated territorial intrusion by a heterospecific, a house sparrow (*Passer domesticus*), did not elicit any behavioral or hormonal response in resident song sparrows. Further experiments examined the contributions of various sensory inputs in affecting the endocrine system. Tactile cues did not influence hormone concentrations during aggressive encounters among male song sparrows. Song sparrows behaved similarly in response to a tape-recorded song (auditory channel), a voiceless conspecific male intruder (visual channel), or a combination of both stimuli. However, a tape recording alone or a devocalized male alone were less effective in stimulating elevations of LH and testosterone concentrations than a combination of both auditory and visual stimuli (Wingfield and Wada, 1989). In other words, although the males behaved similarly in response to various sensory stimuli that simulated a territorial intrusion, a combination of auditory and visual stimuli was necessary to provoke endocrine changes. Had these studies not been designed so well, it might have been concluded that behavior and hormones were not correlated.

A lack of association between hormone concentrations and aggressiveness in birds has been reported in a number of studies. This inconsistency may be explicable on the basis of timing. For example, captive groups of house sparrows were studied over a 2-week period, during which stable social dominance relationships were formed (Hegner and Wingfield, 1987). High-ranking birds were very aggressive during the first week, when the dominance hierarchy was being formed; high-ranking birds had higher blood testosterone concentrations, but not DHT or corticosterone concentrations, than low-ranking individuals during this time. High-ranking males continued to behave more aggressively than low-ranking males during the second week of the study, but their testosterone concentrations no longer differed from those of the low-ranking individuals. These results suggest that social rank (attained via agonistic interactions) and blood testosterone concentrations are correlated when social status is being established (Hegner and Wingfield, 1987) or is being actively challenged (for example, Harding and Follett, 1979; Wingfield and Wada, 1989). This may also be true of primates, as we will see below. Hormonal mediation of social status may wane after the initial relationships are formed and stabilized, and that status may then be maintained by nonendocrine factors. Thus, analyses of plasma hormone concentrations after dominance relationships have been established overlook the importance of hormones in mediating agonistic encounters during the formation of those relationships.

Field observations support these laboratory findings. A caged intruder will elicit ferocious fighting and huge surges in testosterone secretion in territorial males during the early part of the breeding season, when territorial boundaries are being established. An intruder may still elicit aggressive behavior later in the season, when territorial boundaries are more stable, but it is less intense and less sustained, and may not cause an increase in androgen production.

Taken together, these observations of birds have led to the formation of the so-called **challenge hypothesis** to explain the role of androgens in aggressive

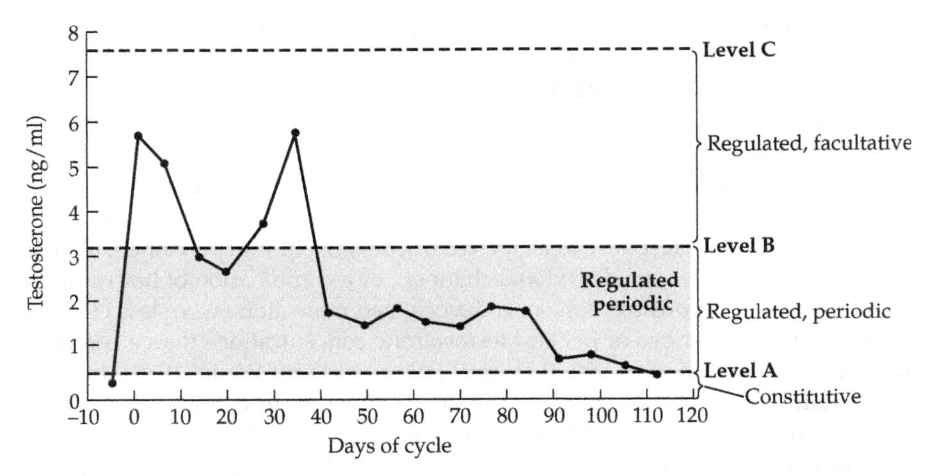

8.18 The challenge hypothesis predicts that testosterone will be elevated only during specific times of an individual's life history. Under challenging conditions, testosterone concentrations should be facultatively increased (from level B to level C) during specific events. In field studies of song sparrows (solid line), testosterone is raised to these high values only when setting up a territory and when initially guarding the first brood. During the rest of the breeding season, testosterone is regulated between levels A and B. After the breeding season ends, plasma testosterone concentrations fall into the constitutive range (below level A). After Wingfield et al., 2001.

behavior. Basically, the challenge hypothesis suggests that androgens are elevated and associated with aggressive behavior only when intermale competition is high; that is, androgen concentrations are elevated in a male by a challenge from another male (Wingfield, 1988; Wingfield et al., 1987). This pattern was observed in field studies of song sparrows, in which concentrations of testosterone were high only when males were setting up territories and guarding their first brood of offspring (Figure 8.18). In another example, laboratory tests of aggression in long-term competition between paired Japanese quail showed that testosterone concentrations were elevated in winners during the first 3 days after pairing, but by the 5th day, plasma concentrations of testosterone were indistinguishable between winners and losers (Ramenofsky, 1984).

Generation of the challenge hypothesis has come mainly from comparative studies of avian aggressiveness and hormone concentrations in field-trapped birds (Wingfield et al., 1997). If blood concentrations of androgens and LH are measured in wild male house sparrows, a seasonal change in these two hormones is observed in the population, with high concentrations from about March until late July (Wingfield et al., 1987). In house sparrows, however, as in many other species of birds, a clear relationship between hormone concentrations and aggressive behavior has been difficult to uncover because different birds in the population are engaged in different reproductive activities at any one time. House sparrows breed multiple times throughout the season, and they do not synchronize their

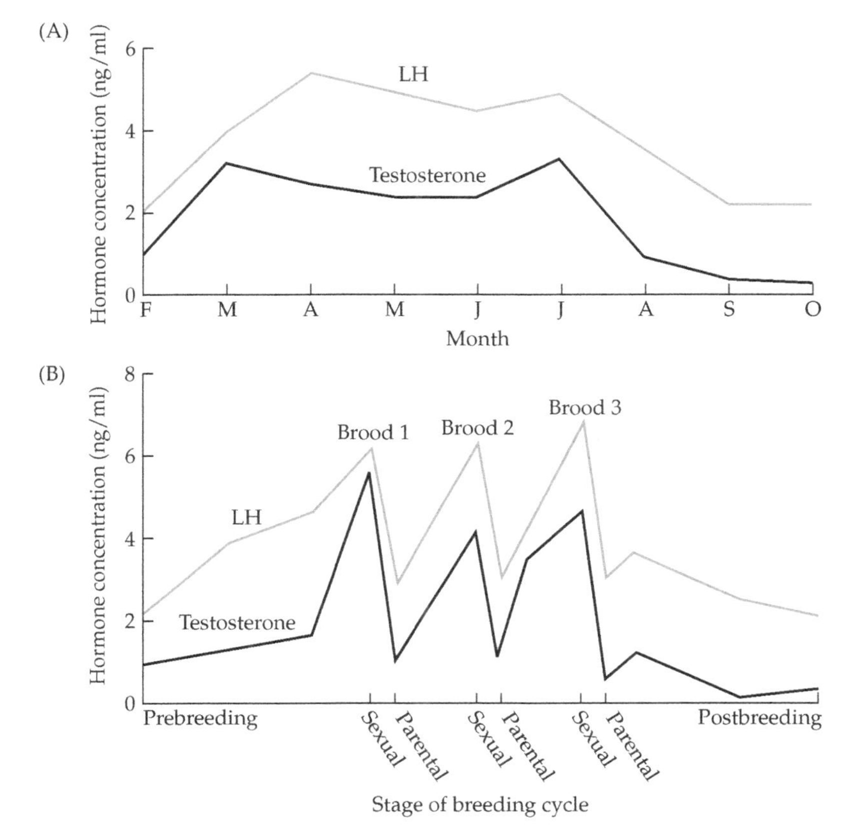

8.19 Seasonal changes in blood LH and testosterone concentrations in house sparrows. (A) When testosterone and LH concentrations from populations of free-living birds are plotted against a yearly calendar, a smooth population-wide seasonal pattern emerges, with a clear elevation in concentrations of these hormones from about March until late July. (B) When blood samples are linked to stages of the breeding cycle, a different pattern emerges: both hormones increase during the sexual stage of the breeding cycle and diminish prior to and during the parental stage of the breeding cycle. Thus, it might be expected that males are more aggressive when they are establishing territories and beginning a breeding effort than at other times. After Wingfield et al., 1987.

reproductive efforts. When blood samples are linked to the stage of the breeding cycle, a clear pattern emerges, showing that LH and testosterone concentrations are elevated during the sexual stage and diminished during the parental stage (Figure 8.19).

Figure 8.20 compares blood concentrations of testosterone for several avian species. Two peaks of testosterone are observed among song sparrows; the first corresponds to the establishment of the territory and the second corresponds to the mating period, during which the male guards his sexually receptive mate

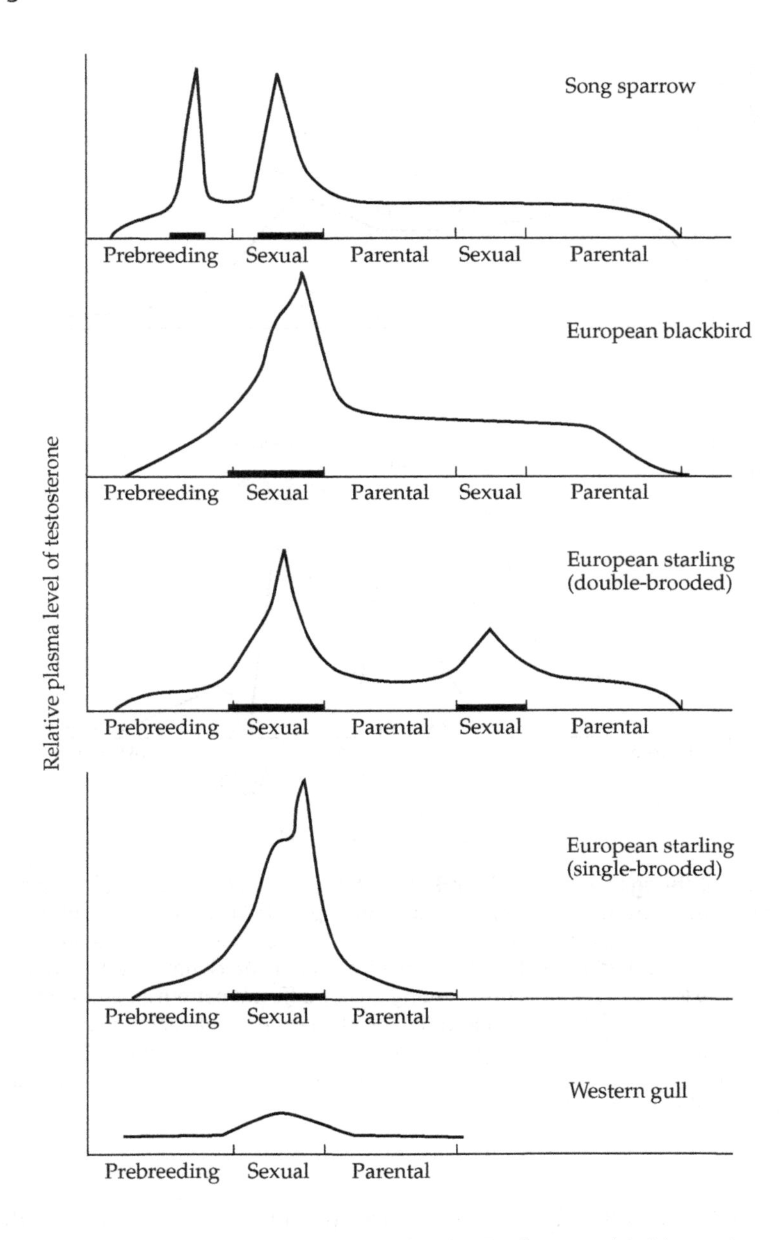

from other males. Male European blackbirds (*Turdus merula*), like male song sparrows, show no change in blood testosterone concentrations during the second mating episode of the season (Wingfield et al., 1987); this lack of androgen secretion has been hypothesized to reflect that little competition exists for nest sites in these species at this time. Other species, such as European starlings (*Sturnus vulgaris*) and house sparrows, nest in holes, which are scarce, and are extremely competitive; males of these hole-nesting species exhibit an increase in testosterone

◀ **8.20 Blood testosterone concentrations in birds may reflect competition.** The black bars correspond to periods when male–male aggressive interactions are frequent. During the song sparrow's breeding season, there are two peaks of testosterone; the first peak coincides with territory establishment and the second corresponds with the time of mating, when the male guards the female from other males. However, the second clutch of song sparrows and European blackbirds is not associated with a change in blood testosterone concentrations. Other species, such as the European starling, nest in holes, which are relatively scarce. Consequently, there is much competition over these nest sites, and blood testosterone concentrations increase during each egg-laying period. Colonial species, such as the western gull, display little male–male competition, and there is little change in androgen concentrations during the breeding season. After Wingfield et al., 1987.

secretion during each egg-laying period. In contrast, the western gull (*Larus occidentalis*), like many colonial species, is long-lived, and there is often a skewed sex ratio with excess females; thus, there is little intermale competition. These birds may form pair bonds that last 20 years. There is little seasonal change in the blood plasma testosterone concentrations of male western gulls.

The challenge hypothesis suggests something fundamental about how androgens mediate aggressive behavior, and it may account for much of the variation in the relationship between aggression and circulating androgen concentrations found in many species. Recent studies on additional species suggest that the challenge hypothesis may explain aggression in several species (e.g., Cardwell et al., 1996; Hirschenhauser et al., 2003; Thompson and Moore, 1992; Trainor et al., 2004).

Conditioned Social Defeat

As mentioned, in most encounters of matched combatants, the resident has a "home field" advantage. After defeat in the home cage of an aggressive conspecific, male Syrian hamsters will subsequently fail to defend their own home territory even if the intruder is a smaller, nonaggressive male (Huhman et al., 2003). This phenomenon, which has been called **conditioned defeat**, appears to evoke a stress response via fear conditioning (Huhman and Jasnow, 2005). The physiological effects of defeat include elevated HPA axis activity, such as increased plasma ACTH, β-endorphin, cortisol, and corticosterone concentrations, as well as decreased plasma testosterone and prolactin concentrations, hormones indicative of stress (Huhman et al., 1990, 1991). This endocrine profile is observed among previously defeated hamsters upon exposure to another male—even when the new opponent is blocked by a physical barrier (Huhman et al., 1992). This latter finding suggests that this endocrine response is a response to a psychological stressor, and not to the pain or anxiety of the combat itself. Social defeat also affects immune system responses (Fleshner et al., 1989; Jasnow et al., 2001). The physiological and behavioral consequences of conditioned social defeat persist for at least 33 days (Huhman et al., 2003), and perhaps throughout adulthood (Delville et al., 1998). Few female hamsters exhibit conditioned social defeat, although ACTH concentrations were reduced in those females that displayed low levels of submissive or defensive behavior (Huhman et al., 2003). In contrast to males, the conditioned defeat response did not persist beyond the first test among female hamsters. These results suggest that in male hamsters, conditioned defeat is a profound, persistent behavioral change characterized by a total absence of territorial aggression and by the frequent display of submissive and defensive behaviors (Huhman and Jasnow, 2005).

Hormones, Competition, and Violent Behavior: Humans

Social status and agonistic behavior may feed back to affect hormone concentrations in humans as well. The testosterone concentrations of male primates fluctuate as their social status changes; blood concentrations of testosterone tend to increase when dominance is achieved or defended, and decrease after defeat (Dixson, 1980). Several studies suggest that a similar effect exists among human males.

Participation in competitive sports is one way in which aggressive behavior is ritualized among humans. In one study, male graduate students participated in doubles tennis matches (Mazur and Lamb, 1980). The team winning the best of five sets split a $200 prize. Blood samples were obtained an hour prior to the match and within an hour of its conclusion. The majority of players whose teams had won decisively displayed post-game elevations in blood testosterone concentrations relative to the losers. Winners of close matches, in which the outcome was not assured until the end, did not differ from losers in their post-game plasma testosterone concentrations.

Male graduate students also participated in another study in which they had the opportunity to win money, but not through any effort on their own part. These individuals were paired, then each member of the pair was given a 50% chance of winning a $100 prize based on a lottery drawing; obviously the men had no influence on the outcome of this contest. A blood sample was obtained immediately prior to the lottery drawing and at four hourly intervals after the recipient of the prize had been determined. The act of becoming a "winner" did not affect blood testosterone concentrations in this situation; blood testosterone concentrations did not differ between winners and losers (Mazur and Lamb, 1980). These results suggest that blood androgen concentrations are more likely to be affected by winning a contest if its outcome is due to the man's own effort.

Consistent with this hypothesis are the intriguing results of a study in which the testosterone concentrations of new recipients of the M.D. degree were evaluated (Mazur and Lamb, 1980). Blood samples were obtained for several days before and after the graduation ceremony. The samples revealed an elevation of blood androgen concentrations 1 to 2 days after graduation. Whether this increase in steroid hormone concentrations reflected the graduates' change in social status or the disappearance of a stress-induced decline in gonadal function requires clarification. Certainly, the stress of intense physical competition can reduce androgen concentrations—even among winners. For example, blood concentrations of testosterone were reduced in both winners and losers of a 26-hour ice hockey cup tournament (Tegelman et al., 1988). In general, mood elevation was related to increased blood androgen concentrations in all of these studies.

The relationship among mood, competitive outcome, and androgen and cortisol concentrations was examined in college varsity tennis players. Saliva samples were obtained before and after six tennis matches, and the players' mood was evaluated prior to play as well as after the match. Most players experienced a pre-match anticipatory increase in salivary testosterone concentrations. Testosterone was further elevated in winners, especially those winners who reported positive moods, as compared with losers. Performance carried over to the next match in the sense that winners displayed higher and losers lower prematch testosterone concentrations as compared with their concentrations prior to the previous match. Winning or losing a match did not affect levels of cortisol, a hormone released during stress. Perhaps not surprisingly, top-seeded individuals had lower cortisol concentrations than low-ranking players; cortisol gradually decreased in all players as the season progressed (Booth et al., 1989).

Even the vicarious experience of winning or losing can affect testosterone concentrations among fans at sporting events, as we saw in Chapter 1: at a World Cup soccer match, mean testosterone concentrations increased among fans of the winning team and decreased significantly among fans of the losing team (Bernhardt et al., 1998) (see Figure 1.7). Thus, the performance of sports fans' favorite teams has the capacity to affect more than just civic pride; the testosterone concentrations of the fans can be affected as well. Indeed, spousal abuse by men increases in the Washington, D.C., area after the Washington Redskins win football games (White et al., 1992). It is possible that increased testosterone concentrations, evoked by the vicarious experience of winning, could trigger some men to cross the line into violence.

Studies like those just described that link hormones and behavior directly in humans are valuable because they are comparable to animal studies. However, the vast majority of research on hormones and social behavior in humans has relied on methods quite different from those used for nonhumans. This is not to diminish the value of these other types of correlational studies, but to remind you that conclusions about hormone–behavior interactions in humans are often derived from data that are fundamentally different from the data generated from nonhumans.

Although the organizational and activational roles of testosterone in aggression have been fairly well established in several species, such as mice and rats (Simon, 2002), the evidence is not compelling in other species, including humans (reviewed in Albers et al., 2002; Harris, 1999). Most studies of hormones and human aggression have been based on three sources of data; namely, hostility questionnaires, interviews, and criminal records. A blood sample may be analyzed and compared with test scores for aggressive tendencies or feelings. These procedures are very different from those used in research on nonhuman animals because the aggressive act is separated in time (by years in many prison studies) from the measurement of endocrine variables. Consequently, many contradictions have arisen. For example, meta-analyses of available studies about the role of hormones in human aggression indicate a positive, weak correlation between testosterone concentrations and aggression in humans (Archer, 1991; Book et al., 2001).

Self-reported measures of aggression in criminal and noncriminal populations, however, do not consistently correlate with testosterone concentrations (reviewed in Harris, 1999). Indeed, one study reported a negative correlation between aggressive behavior and testosterone in women (Gladue, 1991). Clinical, randomized, placebo-controlled studies have also yielded mixed results. In some studies, androgen administration in supra-physiological amounts to eugonadal men did not result in anger and aggression (O'Connor et al., 2002; Anderson et al., 1992; Tricker et al., 1996). In other studies, a positive association between androgen administration and aggressive behavior was observed (Kouri et al., 1995; Pope et al., 2000; Su et al., 1993).

In general, studies that have used psychological rating scales to quantify levels of aggressiveness or hostility have reported no relationship between blood or saliva androgen concentrations and aggressiveness (Doering et al., 1975; Monti et al., 1977; Persky et al., 1977). However, relationships between blood testosterone concentrations and behavior have been reported among aggressive, violent, and antisocial individuals, especially those incarcerated in prison (Ehrenkranz et al., 1974; Kreuz and Rose, 1972). Prison inmates exhibiting high circulating concentrations of testosterone, usually defined as the top 5% or 10% of the normal distribution, had committed violent crimes (Dabbs et al., 1987, 1989; Ehrenkranz et al., 1974), were more unruly in prison, and were judged more harshly by their parole boards (Dabbs et al., 1987, 1989). High testosterone concentrations have also been associated with male juvenile delinquency (Olweus, 1983). Although some studies of criminal populations show no association between plasma testosterone and violent behavior (for example, Matthews, 1979), the consensus is that violence among prison inmates and blood androgen concentrations are positively correlated. A similar relationship was observed among female prison inmates (Dabbs and Hargrove, 1997) (Figure 8.21).

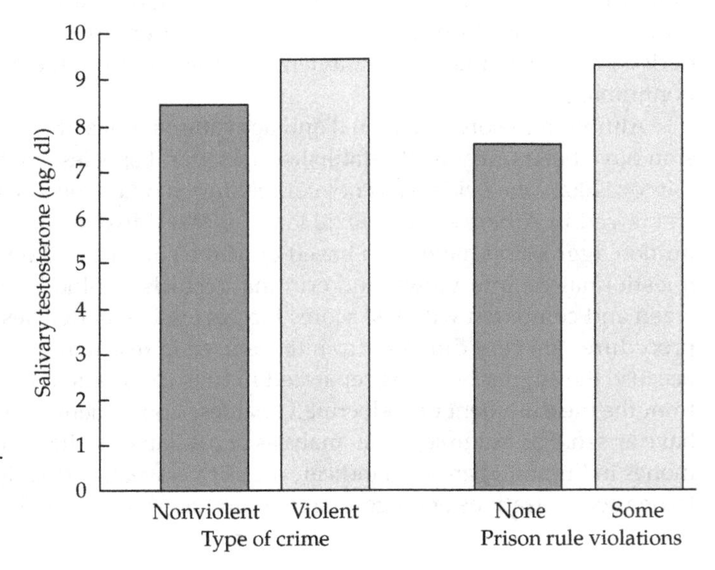

8.21 Relationship of testosterone concentrations in the saliva of female prisoners and their behavior. High salivary testosterone concentrations are linked to incarceration for violent crimes and to elevated prison rule violations. After Dabbs and Hargrove, 1997.

Two related hypotheses have been proposed to explain the association between high androgen concentrations and human antisocial behavior as observed in delinquent or criminal populations: first, that androgens directly mediate the antisocial activities, and second, that androgens promote a constellation of traits, including social dominance, competitiveness, and thrill-seeking, that may be expressed either as antisocial or as prosocial behavior depending on the individual's resources and background. For example, the same constellation of traits might be expressed as assaultive behavior in a gang fight among high-testosterone individuals living in an economically depressed neighborhood or as team-spirited competitive behavior among high-testosterone individuals on a college football team. To distinguish between these two possibilities, a large sample of 4462 United States military veterans was examined beginning in 1985. Psychological and medical evaluations and exhaustive laboratory tests, including measurements of blood androgen concentrations, were performed on these men. Analyses of their psychological profiles and saliva concentrations of testosterone indicated that androgens are associated with antisocial behavior in human males, although socioeconomic status has a small moderating effect (Dabbs and Morris, 1990).

Taken together, these studies suggest that androgens can affect human male aggressiveness and dominance. Some intriguing observations support this contention, albeit in a somewhat indirect manner. Satisfaction with family functioning at midlife, defined as 39–50 years of age in one study, is associated with nonaggressive tendencies among men. Certain psychological factors are also related to men's satisfaction with family life at midlife, including androgynous characteristics, marital satisfaction, parent–offspring communication, and emotional expressiveness. Multiple regression analysis revealed that adoption of androgynous behaviors—behaviors that are both masculine and feminine—was linked to low blood concentrations of testosterone. Similarly, marital satisfaction and the quality of the father–child relationship were negatively correlated with testosterone concentrations. In other words, low testosterone concentrations were associated with high marital and family satisfaction. According to this study's authors, dominance, independence, aggressiveness, and competitiveness may not be valuable male traits for maintaining stable family environments (Julian and McKenry, 1989). A study using similar methods examined testosterone concentrations in men with different occupations (Dabbs et al., 1990). Although the data are arranged from lowest to highest testosterone values, the only statistically significant difference was between NFL football players and ministers (Figure 8.22).

Few studies have addressed the role of androgens in aggressive behavior in women; no consistent correlation between androgen concentrations and aggressive behavior has been reported for women (Dabbs et al., 1989; Persky et al., 1982; Dabbs and Hargrove, 1997). However, subtle effects of androgens may influence aggression in women. Saliva testosterone concentrations did not differ between female prison inmates and female college students. But further analyses discovered that testosterone concentrations were highest in women prisoners convicted of unprovoked violent crimes and lowest in women convicted of "defensive" violent crimes, such as killing abusive husbands (Dabbs et al., 1989).

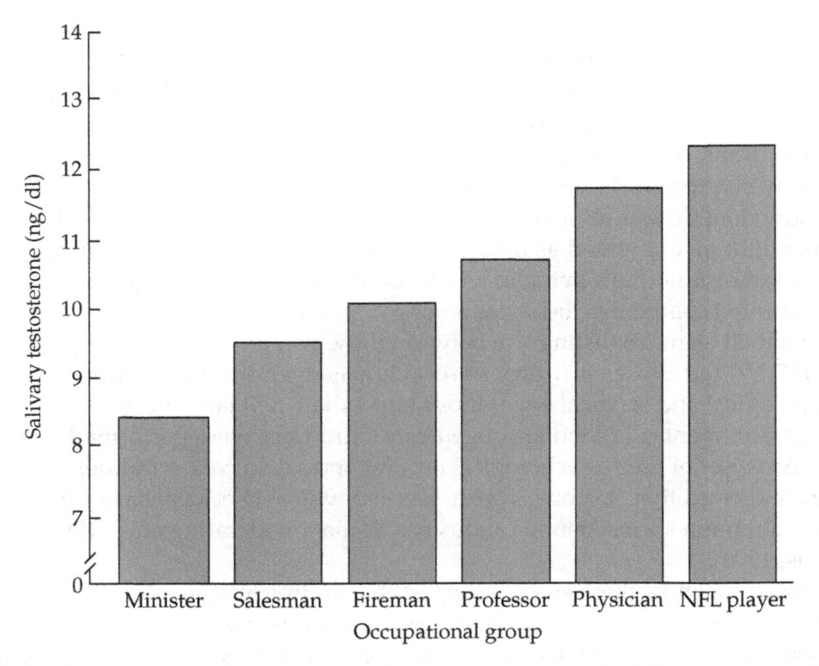

8.22 Salivary testosterone concentrations among men in six different occupations. After Dabbs et al., 1990.

Other data suggest that androgen concentrations in women, although much lower than those in men, may influence or be influenced by social status. For example, blood testosterone concentrations of female undergraduate students and professional, managerial, and technical workers are higher than those of clerical workers and housewives (Purifoy and Koopmans, 1979). As is the case in males, saliva testosterone concentrations are higher in female attorneys than in female teachers and nurses (Schindler, 1979).

Physiological Mechanisms Mediating Hormonal Effects on Aggressive Behavior

Brain Regions Associated with Aggression

A neural circuit comprising several regions of the prefrontal cortex, amygdala, hippocampus, MPOA, hypothalamus, anterior cingulate cortex, insular cortex, ventral striatum, and other interconnected structures has been implicated in the regulation of emotion, including impulsivity and aggression. Functional or structural abnormalities in one or more of these regions, or in the interconnections among them, can increase susceptibility to impulsive aggression and violence (Davidson, 2000). Although the brain mechanisms mediating aggression appear to be fairly constant among mammals, many details of the regulatory pathways involved are species-specific. For example, Syrian hamsters exhibit c-Fos immunoreactivity in the medial amygdala, BNST, ventrolateral hypothalamus,

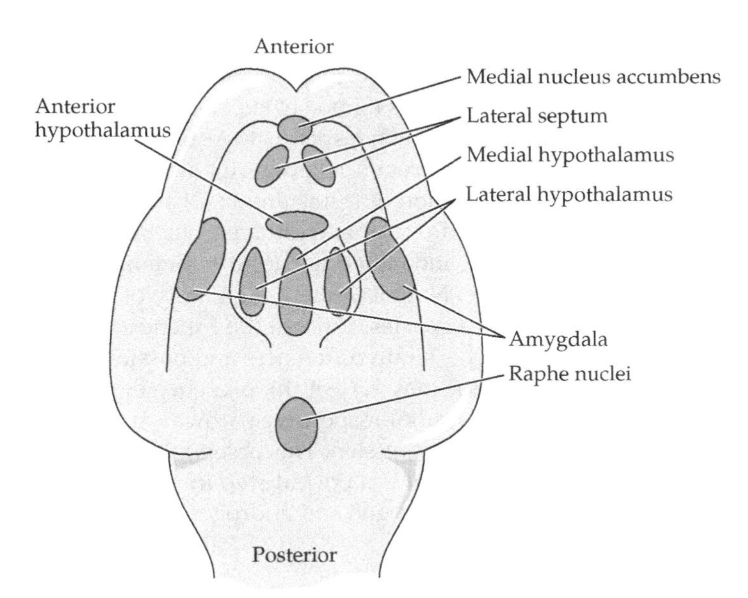

8.23 Components of the rat brain involved in aggression. This schematic cross section looking top-down, shows the major components known to be involved in defensive, predatory, and social aggression. After Albert and Walsh, 1984.

and dorsolateral part of the PAG after displaying offensive aggression toward an intruder (Delville et al., 2000). Because these brain regions have direct and indirect connections with the anterior hypothalamus, an integrated network centered on the anterior hypothalamus has been suggested to regulate offensive aggression (Delville et al., 2000). Lesions of the nucleus accumbens lessen testosterone-dependent aggression in male rats (Figure 8.23; Albert et al., 1990). Androgens also stimulate the hypothalamic tracts to promote aggressive behavior in primates (reviewed in Dixson, 1980).

A neural circuit involving the medial hypothalamus and PAG has been identified that mediates defensive rage behavior in cats (Gregg and Siegal, 2001). The hippocampus, amygdala, BNST, septal area, cingulate gyrus, and prefrontal cortex interconnect these structures and thus can modulate the intensity of attack and rage (Gregg and Siegal, 2001). In rats, attack behavior can be elicited by electrical stimulation of the intermediate hypothalamic area and the ventrolateral pole of the ventromedial nucleus of the hypothalamus, collectively termed the "attack area" (Kruk, 1991). Afferent and efferent connections to the attack area, including the amygdala, prefrontal cortex, septum, mediodorsal thalamic nucleus, ventral tegmentum, and PAG, are also involved in aggressive behavior (Gregg and Siegal, 2001). Importantly, neurons in these aggression-mediating areas are rich in both steroid hormone receptors and serotonin receptor subtypes 5-HT_{1A} and 5HT_{1B} (Simon et al., 1998). The neural circuitry for other types of aggression remains unspecified.

Brain Steroid Hormone Receptors

A series of studies that examined sex and strain differences in responses to steroid hormones, as well as studies that employed enzymatic inhibitors and steroid receptor blockers, revealed that testosterone and its major metabolites, estradiol and DHT, are important in regulating aggressive behavior (Simon, 2002). Four distinct regulatory pathways have been discovered through which testosterone promotes the display of aggression in adult males (Simon et al., 1996, 1998; Simon, 2002): (1) an androgen-responsive pathway, which responds to testosterone itself or its 5α-reduced metabolite, DHT; (2) an estrogen-responsive pathway, which responds to estradiol derived by aromatization of testosterone; (3) a synergistic or combined pathway, in which both the androgenic and estrogenic metabolites of testosterone facilitate aggressive behavior; and (4) a direct testosterone-mediated pathway, which utilizes only testosterone. Not all of these regulatory pathways are necessarily present in every male of a species. Indeed, the functional pathways appear to be determined by genotype; strain differences and possibly individual differences in aggressive behavior may reflect the presence of certain functional pathways (Simon, 2002). The estradiol-responsive pathway appears to be the one most commonly used to mediate aggression. This observation suggests that aromatization of testosterone to estradiol is a critical step in regulation of aggressive behavior. Not surprisingly, both estrogen and androgen receptors are important in the control of aggression.

ESTROGEN RECEPTORS The α estrogen receptor subtype (ERα) is important in mediating aggression in mice (Ogawa et al., 1997, 1998, 1999). Gonadally intact mice lacking the gene for ERα (αERKO knockout mice) displayed virtually no aggressive behavior (Ogawa et al., 1998). Even when castrated and given timed-release implants of testosterone, αERKO mice rarely displayed offensive aggression. This experiment demonstrated that these mice lacked androgen-responsive regulatory pathways mediating aggressive behavior (Ogawa et al., 1998). After the β estrogen receptor subtype was discovered, knockout mice were developed that lacked that receptor subtype. These βERKO mice displayed normal aggressive behavior, indicating that aggressive behavior is mediated by the α estrogen receptor subtype (Ogawa et al., 1999).

Although the β estrogen receptor subtype does not appear to be involved in murine aggression, it may be important in primate agonistic behavior. Recently, cynomolgus monkeys (*Macaca fascicularis*) were fed diets either high or low in soy phytoestrogens (Simon et al., 2004), which bind preferentially to ERβ. After 15 months, male monkeys fed the diet that was high in soy phytoestrogens were more aggressive than males fed the diet low in soy phytoestrogens (Simon et al., 2004). These behavioral results might represent decreased serotonin function in the dorsal raphe nucleus, where ERβ is the sole estrogen receptor subtype thus far detected in primates (Simon and Lu, 2005).

ANDROGEN RECEPTORS Using autoradiographic and immunocytochemical methods, the distribution of androgen receptors has been mapped in the brain (Simon and Lu, 2005). Androgen receptors are found in the BNST, MPOA, lateral septum, and medial amygdala of rodents (Simon, 2002). It is probably no coincidence that these same brain regions were identified as part of an "aggression brain circuit" in studies using lesions, hormone implants, and electrical stimulation (reviewed in Simon, 2002). It is still not known precisely how activation of androgen receptors contributes to aggressive behavioral output.

Brain Neurotransmitter Receptors

SEROTONIN Although nearly every neurotransmitter system has been reported to contribute in some way to aggressive behavior, serotonin (5-HT) is clearly the major regulator of aggressive behavior (Nelson and Chiavegatto, 2001). At least 14 different 5-HT receptor subtypes have been identified, but the 5-HT_{1A} and 5-HT_{1B} receptor subtypes appear to be critical mediators of aggression. Essentially, low 5-HT function or receptor activation is associated with high aggression, whereas high 5-HT function or receptor activation is associated with low aggression (reviewed in Albert and Walsh, 1984; Olivier et al., 1995; Olivier and Mos, 1992; Nelson and Chiavegatto, 2001; Simon and Lu, 2005). Depletion of 5-HT increases both offensive and defensive aggression (Vergnes et al., 1986). Brain 5-HT depletion increases behavioral responsiveness to sensory and painful stimuli (e.g., Lorens, 1978; Telner et al., 1979). Aggressive behavior in response to a wide variety of stimuli can be suppressed by 5-HT or by any treatment that elevates 5-HT (Ieni and Thurmond, 1985; Olivier and Mos, 1992). These relationships have been demonstrated in animals as diverse as lobsters, mice, hamsters, cats, monkeys, and even humans (Simon and Lu, 2005).

Testosterone and its metabolic by-products, estradiol and DHT, may support aggression by influencing serotonin function in one or several brain regions that are part of the brain circuitry underlying aggression (Simon, 2002). The distributions of steroid hormone receptors and serotonin receptors overlap in several distinct regions of the brain that are part of the neural circuit regulating intermale aggression, including the lateral septum, MPOA, medial amygdala, and BNST. Androgens and estrogens modulate 5-HT_{1A} and 5-HT_{1B} receptor function during offensive aggression (Cologer-Clifford et al., 1997, 1999). Serotonin agonists in the presence of androgen were much more effective in reducing aggression than were serotonin agonists plus estrogen treatment. Thus, serotonin appears to be able to modulate the ability of testosterone to support aggression. The relationship between serotonin and steroid hormones is bidirectional. Modulatory effects of neurochemicals on hormone-dependent behavior have been observed in studies of reproductive behavior (Etgen, 2002; Etgen et al., 1999; Hull et al., 1999) as well as anxiety and mood disorders (Bethea et al., 2000, 2002). Presumably, steroid hormones can affect serotonin receptor function by changing receptor gene expression, by exerting nongenomic effects on the receptors, or by indirectly influencing serotonin availability by acting on its synthesis, reuptake, or metabolism (Simon and Lu, 2005). Recent studies have also implicated nitric oxide (NO) (Box 8.1), GABA, and vasopressin as modulators of male aggression acting either directly on the aggression neural circuitry or indirectly by affecting serotonergic function (Nelson and Chiavegatto, 2001).

BOX 8.1 Nitric Oxide and Aggression

Nitric oxide (NO), first identified as an endogenous regulator of blood vessel tone (see Box 5.3), may also serve as a neurotransmitter (Nelson et al., 1997). NO is an endogenous gas that has several biochemical properties of a free radical. It is very labile, with a half-life of less than 5 seconds; consequently, many studies have manipulated NO indirectly by affecting the synthetic enzyme, nitric oxide synthase (NOS), that transforms arginine into NO and citrulline. Studies that have used this approach have determined that NO plays several critical physiological roles. Three distinct isoforms of NOS have been discovered: (1) a form found in the endothelial tissue of blood vessels (eNOS), (2) an inducible form found in macrophages (iNOS), and (3) a form found in neural tissue (nNOS). The neural isoform, nNOS, is localized at high densities within emotion-regulating brain regions.

Figure A. Photograph courtesy of Jay Van Rensselaer.

In order to examine the specific behavioral role of NO in neurons, knockout mice lacking the nNOS gene ($nNOS^{-/-}$) were created. Resident male $nNOS^{-/-}$ mice engaged in 3–4 times more aggressive encounters with intruders than wild-type (WT) resident mice. Against a WT intruder, nearly 90% of the aggressive encounters were initiated by the $nNOS^{-/-}$ animals (Figure A).

As noted in Chapter 1, behavioral studies of mice with targeted deletions of specific genes are subject to the criticism that the gene product is missing not only during the testing period, but also throughout ontogeny, when critical developmental processes, including activation of compensatory mechanisms, may be affected. To address this concern, WT mice were treated with 7-nitroindazole (7-NI), which specifically inhibits nNOS formation in vivo (Demas et al., 1997). Mice treated with 7-NI displayed substantially increased aggressive behavior in two different tests of aggressiveness as compared with control animals (Demas et al., 1997). Importantly, NOS activity in the brain was reduced more than 90% in the 7-NI-treated mice. Similarly, immunohistochemical staining for citrulline, a by-product of NO production, revealed a dramatic reduction in citrulline (and, presumably, NO, which cannot be measured directly in cells) in the 7-NI-

VASOPRESSIN Microinjections of arginine vasopressin into the brain of Syrian hamsters increased offensive aggression, but pre-treatment with a serotonin reuptake blocker (which increases the amount of serotonin in the synapses) blocked this elevation of aggression (Ferris et al., 1997). These results indicate that vasopressin and serotonin interact to influence aggressive behavior. Castration reduces vasopressin Type 1 receptor binding in the ventrolateral hypothalamus, a brain region involved in aggression; testosterone treatment reverses the effects of castration on vasopressin receptor binding. Similar effects have been observed in zebra finches and quail (Wingfield et al., 2005).

treated animals (Demas et al., 1997; Figure B). Taken together, these findings suggest that NO formation was virtually eliminated in these experimental animals. Because 7-NI treatment in WT mice caused aggression to be elevated to the levels displayed by nNOS$^{-/-}$ mice, it appears that nNOS is an important mediator of aggression. Interestingly, nNOS staining was reduced in the brains of Siberian hamsters housed in winter-like day lengths (Wen et al., 2005). Recall that hamsters are most aggressive when housed under short-day conditions (Jasnow et al., 2000).

As you now know, plasma androgen concentrations can affect the display of aggressive behavior. There were no differences between nNOS$^{-/-}$ and WT mice in blood testosterone concentrations either before or after agonistic encounters. However, testosterone is necessary to promote increased aggression in these mutants (Kriegsfeld et al., 1997). Castrated nNOS$^{-/-}$ mice displayed low levels of aggression that were equivalent to those observed among castrated WT males. Androgen replacement therapy restored the elevated levels of aggression in nNOS$^{-/-}$ mice. Importantly, inappropriate aggressiveness was never observed among female nNOS$^{-/-}$ mice in these test situations; however, when aggressive behavior was examined in female nNOS$^{-/-}$ mice in the context of maternal aggression, during which WT females are highly aggressive toward intruders, nNOS$^{-/-}$ dams were very docile (Gammie and Nelson, 1999)! No sensorimotor deficits were found among the mutant mice to account for these changes in aggressive behavior. However, these mice do show altered responses to stress (Bilbo et al., 2003).

Mice missing the gene for nNOS also display altered serotonin function (Chiavegatto et al., 2001). It appears that the excessive aggressiveness and impulsiveness of male nNOS knockout mice is related to selective decrements in serotonin (5-HT) turnover and deficient 5-HT1A and 5-HT1B receptor function in brain regions regulating emotion (Chiavegatto et al., 2001). These results indicate an important role for NO in normal brain 5-HT function and may have significant implications for the treatment of psychiatric disorders characterized by aggressiveness and impulsivity.

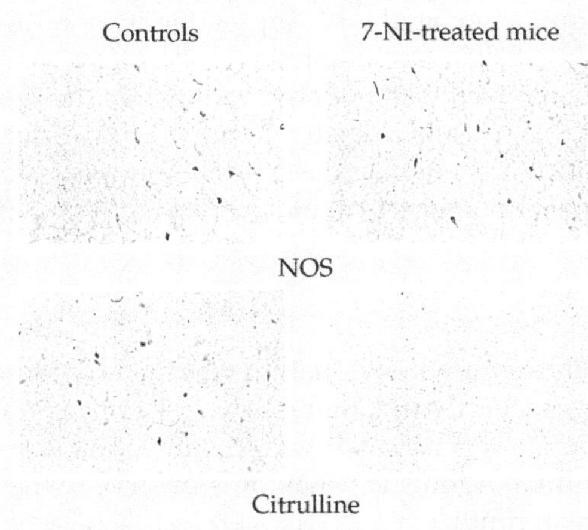

Figure B. From Demas et al., 1997.

In birds, both arginine vasotocin (AVT, the avian version of vasopressin) and vasoactive intestinal peptide (VIP) affect aggressive behavior in a context-dependent manner. Interestingly, the effects of these peptides differ depending on social structure (Goodson, 1998a,b). AVT increased male aggressive behavior in zebra finches, a colonial species (Goodson and Adkins-Regan, 1999), but decreased aggression in violet-eared waxbills (*Uraeginthus granatina*) and field sparrows (*Spizella pusilla*), both territorial species (Goodson, 1998a,b).

AVT may also mediate aggressive behaviors in a context-dependent manner in fishes (Wingfield et al., 2005). Bluehead wrasse (*Thalassoma bifasciata*), like the stoplight parrotfish described in Chapter 3, are female-to-male (protogynous) sex-changing fish, and males of this species display several distinct phenotypes. AVT reduced the number of aggressive behaviors emitted by large, colorful territorial males, whereas in nonterritorial males, AVT increased aggressive behaviors (Semsar et al., 2001). Thus, AVT evoked territorial aggression in nonterritorial morphs, but not in territorial terminal-phase males. The use of an AVT receptor antagonist had the opposite effects; that is, it caused territorial terminal-phase males to behave like nonterritorial males (Semsar et al., 2001). Thus, manipulation of the AVT system could shift social behavior from one state to another (Semsar and Godwin, 2004).

Conclusions

How do hormones affect social behavior? Hormones affect social behavior at the level of input systems; for example, androgens affect the perception of aggression-promoting stimuli associated with intermale agonistic interactions (Johnson and Whalen, 1988; Wingfield et al., 2005). Behavior can feed back, as in the case of territorial encounters between house sparrows, to affect hormonal and subsequent behavioral responses to intruders. Androgens, as well as peptide hormones, affect the central processing of stimuli associated with aggressive responding (Simon and Lu, 2005; Panksepp et al., 1980b). Steroid hormones also affect social behavior by influencing the effectors involved in aggressive behavior. In the case of red deer, androgens influence social interactions by acting on the antlers (Lincoln et al., 1972). Hormones, particularly endogenous opioids, may also provide reward properties to social contact or other social interactions.

Summary

- Social behavior involves interactions between individuals in which one or more of the animals benefits from the interaction. Affiliation refers to behaviors that bring animals together. Aggression is overt behavior with the intention of inflicting damage on another individual. Agonistic behavior is observed when the interests of two individuals are in conflict.
- Affiliative behaviors are affected by peptide hormones such as vasopressin and oxytocin as well as by glucocorticoids, all of which also influence parental behaviors. These hormones appear to function in affiliation by affecting the neurotransmitters dopamine and serotonin. Opioids also mediate affiliative behavior, especially in the context of social grooming and social contact.
- Androgens have been linked to aggressive behavior by several kinds of circumstantial evidence. Seasonal changes in blood androgen concentrations and aggression covary. Aggression is correlated with the onset of puberty, when androgen concentrations rise. Aggression is more common among males than females of most vertebrate species. Castration reduces aggression in males, and androgen treatment restores it.

- Seasonal changes in hormone concentrations and morphological signals affect aggressiveness in red deer and Harris's sparrows. The interaction between behavior, hormones, and self-perception of one's place in the social hierarchy is complex and subtle for some species. Among rodents, dramatic seasonal changes in social organization occur in response to blood concentrations of androgens.
- The onset of puberty heralds the initial display of aggression in males of many species. Pubertal aggression is linked to dispersal in many avian and mammalian species. Dispersing individuals must be aggressive at this time in order to compete within new social contexts.
- Males are generally more aggressive than females. In rodent species, the sex difference in aggressiveness is organized by androgens perinatally and also requires activational actions of androgens in adulthood for expression. The sex difference in aggressiveness among primate species appears to be organized prenatally and does not require androgen stimulation later in life for expression.
- Females of some species are more aggressive than males. Female sandpipers do not usually have higher blood concentrations of androgens than males; however, females show large elevations in blood concentrations of androgens after pairing, suggesting that testosterone may play a role in their aggressive behaviors. Female hyenas are also more aggressive than males; androgen concentrations in adults do not account fully for the behavioral sex reversal.
- There are great individual differences in aggressiveness. In some species, such as mice, aggressive behaviors do not correlate with blood testosterone concentrations, but in other species, such as rats and most mammalian vertebrates, the degree of aggressiveness and blood concentrations of testosterone are related.
- Social experience feeds back to influence androgen concentrations. Winning an agonistic encounter produces a sustained elevation in blood androgen concentrations in rodents and primates. In contrast, losing results in a chronic reduction of blood androgen concentrations.
- The act of fighting an intruder can elevate blood concentrations of testosterone in many species of birds. Obtaining blood samples during an aggressive encounter increases the likelihood that androgen concentrations will correlate positively with aggressive behavior, especially during the breeding season. The challenge hypothesis suggests that androgens mediate aggressive encounters only during the establishment of territories, or whenever males are highly competitive. Aggression at other times does not elicit the same kind of endocrine response and probably is not mediated by androgens.
- Social experience in competitive sports can feed back and affect human androgen concentrations. Elevation in social status is also associated with high blood androgen concentrations among humans.

- Assessments of human aggression are typically different from studies of aggression in nonhuman animals because the aggressive act is generally removed in time from the blood hormone assessment. Typically, aggression in humans is quantified on the basis of questionnaires or interviews. Nonetheless, prison inmates who have been convicted of violent crimes tend to have high concentrations of androgens.

Questions for Discussion

1. Suppose that you have obtained blood samples from matadors and bulls immediately after a number of bullfights. You notice that the matador's blood androgen concentration is elevated whenever he decisively defeats the bull, but that the bull's androgen concentration is elevated if it gores the matador. Is "winning" the critical factor in determining blood androgen concentrations? What other endocrine events might account for the data?
2. Design an experiment to test the proposition that professional football players are more aggressive than spectators because they have elevated thyroid hormone concentrations. Remember that hormones affect behavior and that behavior can feed back to affect blood hormone concentrations.
3. Johnson and Whalen (1988) propose that androgens cause a behavioral "homogenization" by reducing the variation in response. How might this work? Why is it that some castrated animals retain aggressive behaviors, whereas other conspecifics become docile after castration?
4. How might hormones affect the perception of a threat? How might hormones affect mood in such a way that stimuli become irritating? Design a test of your notions.
5. Compare and contrast the endocrine modulation of social affiliation and parental behaviors.

Refer to the accompanying Student CD for additional resources, including Web links, videos, animations, and additional photos.

Suggested Readings

Carter, C. S., and Keverne, E. B. 2002. The neurobiology of social affiliation and pair bonding. In D. W. Pfaff, A. P. Arnold, A. M. Etgen, S. E. Fahrbach, and R. T. Rubin (eds.), *Hormones, Brain and Behavior*, Volume 1, pp. 299–338. Academic Press, New York.

Insel, T. R., and Young, L. J. 2001. The neurobiology of attachment. *Nature Reviews: Neuroscience*, 2: 129–136.

Manuck, S. B., Kaplan, J. R., and Lotrich, F. E. 2005. Brain serotonin and aggressive disposition in humans and nonhuman primates. In R. J. Nelson (ed.), *Biology of Aggression*. Oxford University Press, New York. In press.

Nelson, R. J. (ed.). 2006. *Biology of Aggression*. Oxford University Press, New York. In press.

Nelson, R. J., and Chiavegatto, S. 2001. Molecular basis of aggression. *Trends in Neurosciences*, 24: 713–719.

Simon, N. 2002. Hormonal processes in the development and expression of aggressive behavior. In D. W. Pfaff, A. P. Arnold, A. M. Etgen, S. E. Fahrbach, and R. T. Rubin (eds.), *Hormones, Brain and Behavior*, Volume 1, pp. 339–392. Academic Press, New York.

Wingfield, J. C., Moore, I. T., Goymann, W., Wacker, D., and Sperry, T. 2005. Contexts and ethology of vertebrate aggression: Implications for the evolution of hormone–behavior interactions. In R. J. Nelson (ed.), *Biology of Aggression*. Oxford University Press, New York. In press.

9

Homeostasis and Behavior

In 1859, 12 pairs of European rabbits (*Oryctolagus cuniculus*) were released from a ranch in Victoria, the southeasternmost state in Australia, to provide a ready source of game for hunting. The rabbit population grew exponentially because the animals had no natural predators in Australia. By 1865, one hunting drive successfully killed over 20,000 rabbits. But the rabbits kept multiplying, and by 1900 they presented a serious problem for Australia, where the production of wool is an important economic activity. In addition to being very prolific, rabbits are extremely efficient foragers. Within 40 years after the initial 12 pairs were released, hundreds of millions of their descendants had spread throughout Australia and had destroyed valuable pastures and rangelands required for grazing sheep. The invading rabbits also destroyed many unique habitats, which led to the extinction and endangerment of several native marsupial species.

After several failed attempts by the government to reduce the number of rabbits by using poisons, predators, and other pest control procedures, the myxoma virus (which produces myxomatosis, a disease fatal to rabbits) was introduced in 1950. A myxomatosis epidemic broke out and killed over 99.9% of the rabbits. Only a few myxomatosis-resistant animals survived. However, these initially rare myxomatosis-resistant animals repopulated the continent once again. In some areas, only 2 years were necessary to restore the decimated populations. Recently, Australian government scientists were testing a new rabbit virus, called rabbit calicivirus disease (RCD), on nearby Wardang Island to make certain it would work properly.

However, before the testing was complete, the virus invaded mainland Australia and New Zealand. In some cases, rabbit mortality has been very high, but in other cases, only about half of the rabbits in a population have succumbed to RCD. A new concern in areas of Australia where RCD has drastically reduced the number of rabbits is that introduced foxes and feral cats will start killing native Australian species for food. An emergency program to kill foxes and feral cats has recently been instigated to reduce the damage from this new ecological imbalance. The "experts" hope that in the long run, everything will balance out, but there is a reasonably high chance that RCD-resistant rabbits will soon repopulate Australia and New Zealand.

In their colonization of Australia, the European rabbits were confronted with a wide variety of environmental conditions. The rapid spread of the animals throughout the Australian continent required them to adapt to markedly diverse habitats, including tropical forests, harsh deserts, fertile plains, and barren seacoasts. Besides undergoing the obvious climatic and reproductive adaptations, these pioneering rabbits had to cope with greatly varying levels of nutrient availability: for example, the prevalence of water and sodium differed widely across habitats and across seasons. Because physiological systems operate best within a very narrow range of conditions, rabbits living in different areas were forced to make very different biological adjustments depending on the habitat they encountered. For example, maintaining a certain level of sodium chloride, the salt that is common in your diet, is critical for sustaining life. Sodium chloride is scarce in some Australian habitats, and rabbits that colonized these areas were successful only to the extent that they coped with the sodium shortages. Rabbits that settled in sodium-rich habitats survived only if they could rid their bodies of excess sodium.

Individuals are motivated to maintain an optimal level of water, sodium, and nutrients in the body. Claude Bernard, the 19th-century French physiologist (Figure 9.1), was the first to describe the ability of animals to maintain a relatively constant internal environment, or *milieu intérieur*. Humans, for instance, maintain an optimal body temperature between 36° and 38°C; there are also optimal blood concentrations of sugars, proteins, sodium, potassium, and many other blood constituents, as well as an optimal blood pH. In 1929, Walter B. Cannon termed the process by which the body maintains this relatively constant internal milieu **homeostasis** (from the Greek, "standing the same").

Many homeostatic systems are completely physiological processes; however, many others—thirst, for example—also have one or more behavioral components. In still other cases, physiological mechanisms normally maintain homeostasis, but behavioral processes may be activated when the normal physiological regulatory systems are unable to restore equilibrium. If both adrenal glands are removed from a rat, for example, it will usually perish within a week. However, if salt water is made available, the adrenalectomized rat will drink it, and will survive as well as intact rats in the laboratory (Richter, 1936). Under normal conditions, when the adrenal glands are present, physiological homeostatic systems maintain sodium balance through the action of aldosterone; this steroid hormone, secreted from the adrenals, acts on the kidneys to conserve sodium (Denton, 1982). Without the adrenals, the animal cannot retain enough sodium to

9.1 Claude Bernard (1813–1878), a French physiologist who developed the concept of homeostasis.

sustain its life; however, it can compensate for the missing physiological sodium conservation mechanism by ingesting increased amounts of sodium. Prior to adrenalectomy, a rat, just like you, will avoid drinking seawater (about a 3.5%–7% sodium solution), but after adrenalectomy, a rat will avidly drink a saturated salt water solution. When physiological homeostatic systems fail, behavioral homeostatic processes are often engaged to sustain life.

Many anecdotal reports describe children, apparently suffering from some sort of nutritional deficiency, who consume seemingly bizarre items to maintain their health. For instance, a child suffering from a calcium deficiency might eat chalk or wallpaper paste in an attempt to ameliorate the calcium shortage. In the recent past, concerned parents admitted such children to the relatively sterile environment of a hospital, where they tragically died because they were no longer able to maintain homeostasis behaviorally (see, for example, Wilkins and Richter, 1940). Even in situations in which the physiological homeostatic systems are undeveloped, homeostasis can be maintained by behavior (recall the behavioral thermoregulation of rat pups described in Chapter 7).

Many homeostatic systems, such as those maintaining fluid and energy balance, normally require both physiological and behavioral mechanisms to orchestrate a relatively constant internal condition in the face of varying external conditions. Although the various homeostatic processes are presented here as isolated events for ease of explanation, it should be remembered that any physiological challenge is met with overlapping, redundant physiological responses. Concurrent physiological and behavioral responses are often mediated or signaled by changes in the release of multiple hormones acting in concert or in opposition. Individual hormones may have multiple physiological and behavioral effects.

This chapter will describe the hormonal and behavioral regulation of several homeostatic systems. Chapters 3–7 have emphasized the parsimonious relationship in which the hormones that mediate physiological reproductive processes have been co-opted during evolution to regulate mating and parental behaviors. Similarly, the hormones associated with physiological homeostatic mechanisms also appear to mediate behaviors critical for homeostatic maintenance. Different species use different hormonal signals to solve similar problems or challenges to their homeostatic processes. In other cases, different species rely on the same hormone to mediate very different solutions to similar problems. For example, female rats and hamsters have evolved two very different ways of coping with the energetic challenges of pregnancy. Rats increase their food intake and gain body fat during pregnancy, while hamsters do not change their food intake appreciably; rather, they utilize most of their established fat stores to meet the energetic demands of pregnancy. The same hormone, progesterone, elicits opposite, but similarly adaptive, responses in these two rodent species.

In this chapter, the conceptual bases of homeostasis will be presented. Then, the interaction among hormones, behavior, and sodium and water balance will be described as a model physiological homeostatic system. The effects of hormones and behavior on energy balance, eating, and body mass maintenance will also be discussed. Finally, the hormone–behavior interactions controlling specific hungers will be described.

Basic Concepts in Homeostasis

The archetypal homeostatic device is a thermostatically controlled heating and cooling system. The system in your home is probably controlled by a thermostat (Figure 9.2). A reduction in ambient temperature below the setting on the ther-

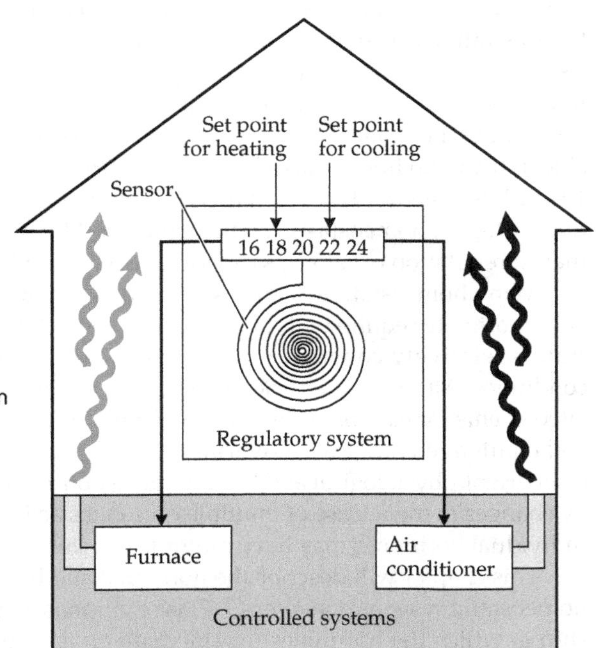

9.2 The thermostat is a common homeostatic device. A thermostat has a means to assign a set point and a detector mechanism to determine deviations from the set point and activate corrective measures. In this example, when the temperature decreases sufficiently from the set point, a signal is sent to activate the heating system; conversely, when the temperature increases sufficiently, the cooling system is engaged. Most physiological regulatory systems operate by similar negative feedback control loops.

mostat (21°C, for example) will activate the furnace, which in turn raises the temperature of the area controlled by the thermostat. When the ambient temperature increases to the "set" temperature, the thermostat shuts off the furnace (an example of negative feedback). If the ambient temperature rises above the set temperature, the cooling system is activated, shutting down if the area becomes too cool. The thermostat acts to keep the room temperature within a relatively narrow range, called the **set point**. The deviation in room temperature from the set point must be 1 or 2 degrees before the furnace or air conditioner is activated or turned off; otherwise, the system would be continuously engaging and turning off with only minor changes in temperature.

Similarly, rabbits, humans, and most other mammals maintain an internal body temperature of 37°C, but temperatures in a range between 35°C and 38°C are not life-threatening. However, a rapid elevation or reduction in body temperature of just 5°C would seriously endanger the individual. Mammals make physiological, morphological, and behavioral adjustments to maintain body temperature within this relatively narrow range. In common with a thermostat, the optimal set point of an animal can be changed depending on prevailing conditions. As you will learn in Chapter 10, there is a daily program that reduces body temperature significantly during an animal's inactive period, presumably by changing the set point.

When we are ill, our body temperature set point usually changes as well. We often develop a fever, which is an adaptive response because higher temperatures inactivate or kill the microbial invaders producing the illness. Iguanas provide a good example of this phenomenon. Lizards are *poikilothermic*, unable to maintain their body temperature physiologically. A terrarium can be set up in such a way that there is a warm end and a cool end with a relatively smooth temperature gradient in between (Figure 9.3). When healthy lizards were placed in such a terrarium, they settled along the temperature gradient so that their body temperature was 37°C (Kluger, 1978). If the reptiles were infected with bacteria, they positioned themselves in the terrarium so that their body temperature was elevated a few degrees. In other words, the infected lizards chose to be feverish: there was a behaviorally mediated change in the set point as a result of the infection. Again, the behavioral generation of a fever has important adaptive consequences. When infected lizards were prevented from elevating their temperature, their chances of surviving the infection were reduced (Kluger, 1986). This elegant interplay between physiology and behavior enhances survival.

The thermostat analogy helps us to understand the basic concepts behind homeostasis, but there are several aspects of the control of food and water balance, discussed later in the chapter, that cannot be described adequately by the thermostat analogy. For example, feedforward mechanisms, anticipation, and learning that occurs with ingestion may modify the basic thermostatic model.

What properties are needed for any system to maintain homeostasis? First, there has to be a *reference value* for the regulated variable (the set point). Second, there must be some sort of *detection mechanism* to detect any deviation from the reference value. Third, the homeostatic system must be able to *mobilize* the organism to make changes that will return the variable to the normal range when devi-

9.3 Behavioral thermoregulation in iguanas. (A) If an iguana is placed in a terrarium with a heat source at one end that produces a temperature gradient, it will tend to settle at that place along the temperature gradient where it can maintain a body temperature of about 37°C. (B) The behaviorally selected body temperatures of seven green iguanas placed in such a terrarium over the course of day 1 (open circles), on day 2, after each lizard received an injection of dead bacteria at 0900 hours (gray circles), and on day 3 (black circles). Note that the lizards' body temperatures were raised about 3 hours after the injection and that the rise persisted until about 1500 hours on day 3. (C) The percentage of iguanas injected with live bacteria and maintained at temperatures ranging from 34° to 42°C that survived the treatment; note that the number of animals surviving is highest at the highest ambient temperatures. After Kluger, 1978.

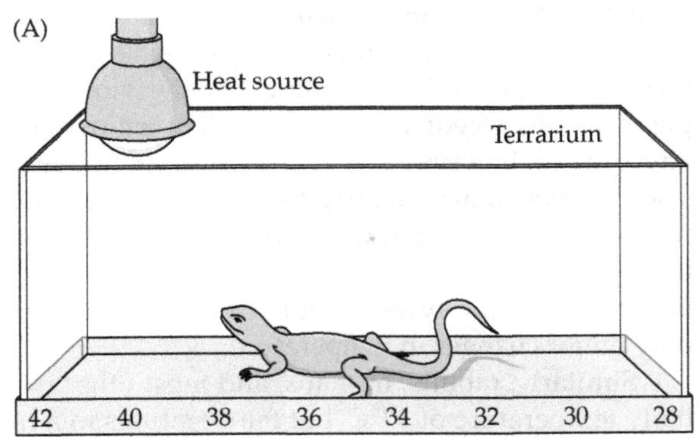

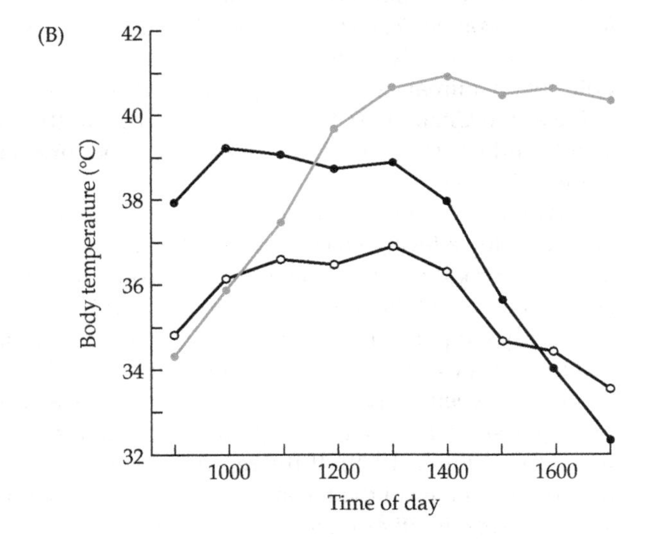

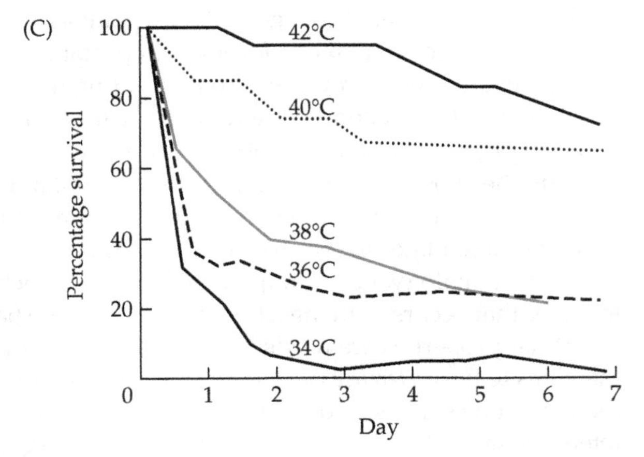

ation occurs. Finally, the detection mechanism must be able to *recognize* when the desired change occurs and shut down the mobilization process through a negative feedback mechanism. Homeostasis requires work, and work requires energy; thus, homeostasis requires energy.

Fluid Balance

Unicellular creatures living in the ocean rely on seawater to provide them with the nutrients, oxygen, water, and basic electrolytes necessary to sustain their life processes. These simple animals are at the mercy of the composition and temperature of the seawater; small deviations in water salinity, for instance, may be fatal. To some extent, the evolution of more complex multicellular organisms has required the compartmentalization of "seawater" within the body in the form of extracellular fluid. Essentially, we all carry within us a remnant of the seas in which unicellular organisms evolved—albeit a markedly diluted remnant—to bathe our cells. Homeostatic mechanisms have evolved to maintain the composition and temperature of this extracellular fluid at a relatively constant level. The processes regulating intake and excretion of water and sodium, the two main components of extracellular fluid, are closely linked in order to maintain them at ideal levels within relatively narrow ranges. The ability to maintain this relatively constant internal environment despite variable external conditions has allowed the radiation of multicellular animals into virtually every niche on the planet. The maintenance of a relatively constant internal environment liberated animals from the sea, or, as Claude Bernard stated, allowed *la vie en liberté*, the free life (Bernard, 1856).

Animals are watery creatures. By weight, mammals are approximately two-thirds water. The proportion of body water varies widely among individuals; in humans, body water content ranges from about 45% to 70% (Rolls and Rolls, 1982). The cells of animals require water for virtually all metabolic processes. Additionally, water serves as a solvent for sodium (Na^+), chloride (Cl^-), and potassium (K^+) ions, as well as sugars, amino acids, proteins, vitamins, and many other solutes, and it is therefore essential for the smooth functioning of the nervous system and for other physiological processes. Because water participates in so many processes, and because it is continuously lost through perspiration, respiration, urination, and defecation, it must be replaced periodically. In contrast to minerals or energy, very little extra water is stored in the body. When water use exceeds water intake, the body conserves water, mainly by reducing the amount of water excreted from the kidneys. Eventually, physiological water conservation can no longer compensate for water use and incidental water loss, and the animal searches for water and drinks. Animals appear to coordinate physiology and behavior so as to maintain body water concentration at some ideal range or set point.

The regulation of sodium intake and of water intake are closely linked to each other (Stricker and Verbalis, 1990b; Fluharty, 2002). Part of the reason for this is the way in which the kidney uses sodium to conserve water (Box 9.1). But sodium is also important in the movement of water between the two major fluid compartments in the body; namely, the extracellular and the intracellular compartments

BOX 9.1 Vertebrate Renal Function

The kidneys are remarkable organs. The human kidneys filter about 30 liters of blood each hour. About 1% of this filtrate is removed as waste and sent to the bladder for elimination. The rest of the blood plasma is reabsorbed in the kidney and returned to the circulation. The kidneys contribute to the maintenance of fluid balance by either conserving water or eliminating excess water, depending on the body's needs.

The functional unit of the kidney, the nephron, is where the reabsorption of water occurs (see the accompanying figure). There are about a million nephrons in each human kidney. As the filtrate from the blood passes into the nephron at high pressure, it enters a long, convoluted tubule that comprises several sections, each with a specific function: the proximal tubule; the loop of Henle, with its descending and ascending limbs; the distal tubule; and the collecting duct. Water flows passively out of the descending limb into the surrounding tissue, and sodium ions enter because there is a high concentration of sodium in the surrounding regions. The sodium concentration of the filtrate is greatest at the bottom of the loop of Henle; in humans that concentration is about 2%. As the filtrate moves out of the loop into the ascending limb, the sodium is actively pumped out into the surrounding tissue. The ascending limb is impervious to water, so water cannot passively follow the sodium. The filtrate is thus extremely dilute as it leaves the ascending limb of the loop of Henle. As it enters the distal tubule, it flows out of the tubule, into the surrounding tissue, and then into the capillaries by osmosis. The remaining filtrate flows into the collecting duct, and eventually moves to the bladder to be voided.

Vasopressin, or antidiuretic hormone (ADH), conserves water by acting on the distal tubules to increase their permeability to water and hence return more water to the circulation. Without ADH, the distal tubules become less permeable, resulting in diuresis, an increase in the amount of water eliminated in the urine.

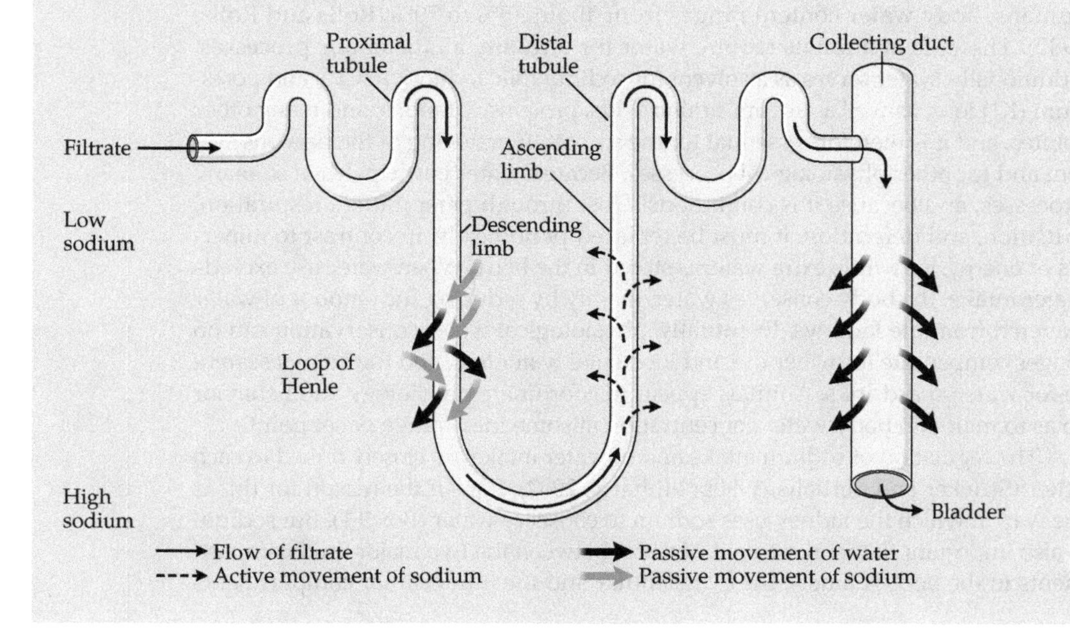

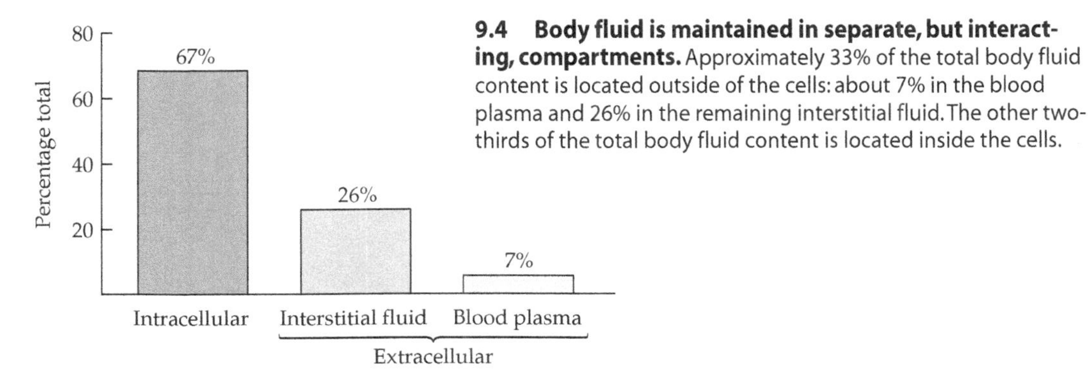

9.4 Body fluid is maintained in separate, but interacting, compartments. Approximately 33% of the total body fluid content is located outside of the cells: about 7% in the blood plasma and 26% in the remaining interstitial fluid. The other two-thirds of the total body fluid content is located inside the cells.

(Daniels and Fluharty, 2004). Approximately two-thirds of the total body fluid content is located in the intracellular compartment. The extracellular compartment comprises two separate but interacting parts: (1) the interstitial (between the cells) fluid and (2) the blood plasma. Of the one-third of the total body fluid content located in the extracellular compartment, about 7% is in the blood plasma and 26% is in the interstitial fluid (Figure 9.4). The distinction between the intracellular and extracellular compartments is not merely one of location. The fluids in the two compartments are fundamentally different in composition. Most of the body's sodium and chloride ions are located in the extracellular compartment, whereas most of the potassium ions are located in the intracellular compartment. These differences in fluid composition result from the properties of the cell membranes and blood vessel walls.

Mechanisms exist to balance both the intracellular and extracellular fluid levels. After mammals ingest water, most of it leaves the large intestine and eventually flows into one of the body's fluid compartments, which are separated by membranes. The blood vessel (capillary) walls act as a barrier between the blood plasma and the interstitial fluid, and permit the flow of all constituents of the plasma except proteins into the interstitial fluid. Cell membranes are the barriers between the extracellular and the intracellular fluid compartments. A number of physiological mechanisms are engaged to maintain the extracellular–intracellular differences in sodium and potassium concentrations.

In order to understand the dynamics of sodium and water and how they affect behavioral manifestations of thirst, it is necessary to review the concepts of osmolality, osmoregulation, and the regulation of blood plasma volume. Water can pass freely through semipermeable biological membranes, but many solutes (chemical substances dissolved in a solvent—in this case, water) cannot. When one compartment has a greater concentration of solutes than the other, water will tend to distribute itself so that the solute concentration on both sides of the membrane is equalized (Figure 9.5). In other words, water moves across the biological membrane to the compartment that has the higher concentration of solutes. The movement of water across a semipermeable membrane into a more concentrated solution is called **osmosis**. The concentration of solutes in a solution is its **osmolality**, and the control of this osmotic concentration is called **osmoregulation**.

9.5 Osmosis. The differences in composition of the different fluid compartments of the body are the result of properties of the cell membranes and blood vessel walls. If compartments containing a dilute and a concentrated solution are separated by a membrane that allows only water to pass through, water will flow into the compartment containing the concentrated solution until the osmolality of the compartments becomes equal.

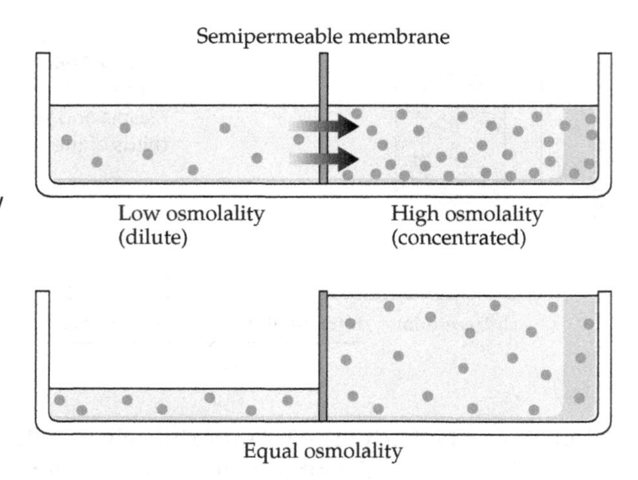

Table salt—sodium chloride—is an example of a solute that cannot pass easily through biological membranes. (Although chloride can move in and out of cells, it tends to stay outside of cells because it has a negative charge and is attracted to the positively charged sodium ions dissolved in the interstitial fluid.) In most cases, the body's extracellular fluid contains about 8.5 g of sodium chloride per liter, or stated differently, a 0.85% concentration of sodium chloride; this concentration is called physiological, isotonic, normal, or 0.9% saline. If an isotonic saline solution is injected into an animal, it has no effect, because it does not change the concentration of sodium chloride in the animal's interstitial fluid. However, an injection of hypertonic saline, a solution in which the concentration of sodium chloride exceeds 0.9%, will increase the sodium chloride concentration in the interstitial fluid. The resulting increase in interstitial osmolality draws water out of the cells, thus inducing cellular dehydration, a potent stimulus for thirst (Figure 9.6). This type of thirst is called **osmotic thirst**, and it is an experience that is well known to anyone who has consumed salty foods. Consumption of sugary foods also causes osmotic thirst because the excess glucose molecules in the interstitial fluid also pull water out of cells and induce cellular dehydration.

Vasopressin acts to conserve water as blood moves through the kidneys (see Box 9.1). If more water is consumed than needed, plasma osmolality is decreased. Reduced plasma osmolality inhibits thirst and suppresses the release of vasopressin from the posterior pituitary. Inhibition of vasopressin release causes *diuresis* in the kidney: water is removed from the blood plasma and sent to the bladder for elimination (Daniels and Fluharty, 2004). This function is the reason why vasopressin is also known as antidiuretic hormone (ADH).

In addition to the osmoregulatory mechanisms that regulate intracellular and interstitial fluid balance, a second important fluid regulatory system maintains blood plasma volume. Reduction of blood volume produces a potent stimulus for thirst. This type of thirst is called **hypovolemic thirst**, and it is manifested in the extreme by individuals experiencing hemorrhage; excessive perspiration, diar-

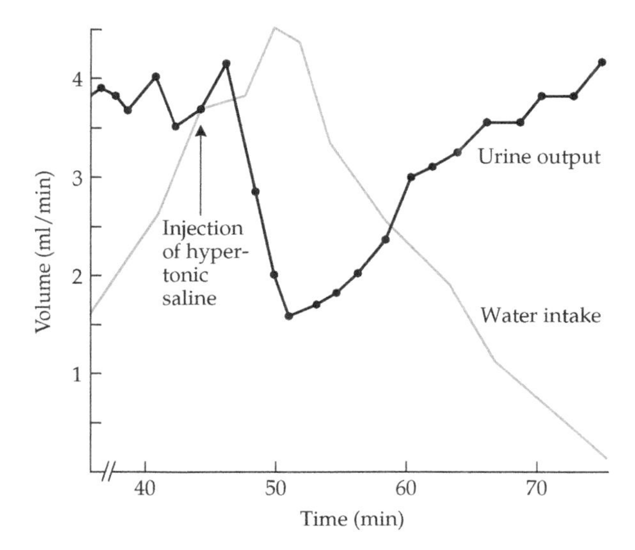

9.6 Osmotic thirst. An injection of hypertonic saline, a solution of saline in which the concentration of sodium chloride exceeds 0.9%, increases the sodium chloride concentration in the interstitial fluid. The resulting increase in osmolality draws water out of the cells, thus inducing cellular dehydration and osmotic thirst, and resulting in increased water intake. After Stricker and Verbalis, 1990b.

rhea, or heavy menstrual bleeding can also trigger hypovolemic thirst (Stricker et al., 1987). Hypovolemic and osmotic thirst differ in several important respects. During **hypovolemia**, or low blood volume, water, sodium, and other solutes are lost without necessarily pulling water out of the cells. Quenching osmotic thirst merely requires the ingestion of water, but alleviating hypovolemic thirst requires replacement of water, sodium, and other solutes (Figure 9.7).

There are two major ways to induce hypovolemic thirst experimentally. Inflicting a controlled hemorrhage has been the traditional method of causing hypovolemic thirst in animals; however, only acute behavioral changes can be studied with this method. In order to study gradual, long-term changes in behavior, another technique was developed that has been very useful in the study of hormonal effects on thirst and drinking behavior (Stricker, 1968): subcutaneous (under the skin) injection of polyethylene glycol (PEG). Because PEG is a relatively large colloidal molecule, it cannot cross the blood capillary membranes, and thus remains in the interstitial fluid. The colloidal particles cause an "equi-osmotic" sequestration of fluid; that is, both water and solutes are removed from the body fluids, reducing blood plasma volume without significant changes in blood solute content or induction of severe hypotension (i.e., low blood pressure). The degree of hypovolemia is directly related to the amount of colloidal material injected. When subcutaneous injections of PEG are given to rats, they begin drinking 1 or 2 hours later, when their plasma volume deficits approach 5%, and they continue to drink in short bouts for several hours thereafter (Stricker and Verbalis, 1988).

Hypovolemia compromises kidney function. The reduced blood volume and resulting low blood pressure prevent the kidneys from extracting water effectively. Consequently, hypovolemic rats consume water, but cannot completely correct their blood plasma volume because the ingested water enters all of their fluid

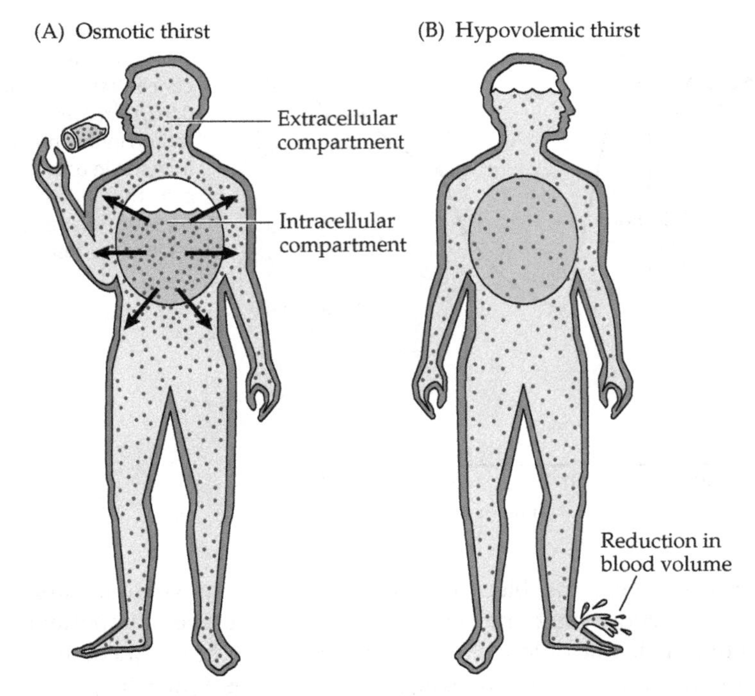

9.7 Two types of thirst result from different osmotic conditions. (A) Osmotic thirst is caused by cellular dehydration—which can occur after ingesting a salty snack or drink—when increased interstitial osmolality occurs and draws water out of the intracellular compartment. Water quenches osmotic thirst. (B) Hypovolemic thirst is caused by the loss of fluids and solutes—such as after hemorrhage. Replacement of both water and solutes is necessary to quench hypovolemic thirst.

compartments. The resulting combination of body fluid dilution and reduced water removal by the kidneys results in reduced blood plasma osmolality, a potent stimulus to stop drinking. Consequently, the rats stop drinking before attaining normal fluid balance. A hypovolemic individual requires salt to restore body fluid osmolality to normal levels. If given access to salt water or salty food, a hypovolemic rat will ingest the proper combination of water and salt to restore blood volume and osmolality to normal levels. Similarly, athletes often ingest salty beverages such as Gatorade after heavy perspiration when they are experiencing hypovolemic thirst. During the normal course of events—that is, during minor dehydration—the experience of thirst is a psychological manifestation of both osmotic and blood volume changes.

Thirst is defined here as a motivation to seek and ingest water (Stricker and Verbalis, 1988; Fitzsimons, 1998). Because thirst is a psychological hypothetical construct, we cannot measure it directly, but can infer it from behavior. In nonhuman animals, the degree of thirst can be operationally determined by the amount of effort an individual expends to drink. Individuals will work harder and endure greater noxious stimuli to obtain water as the length of water deprivation increases. Investigations of human thirst allow the added dimension of verbal reports: people can be asked to rate and describe their thirst (Figure 9.8). In humans, several parameters of thirst can be identified and described. For example, water-deprived humans rate the taste of water as more pleasant than do people who are not water-deprived (Rolls and Rolls, 1982).

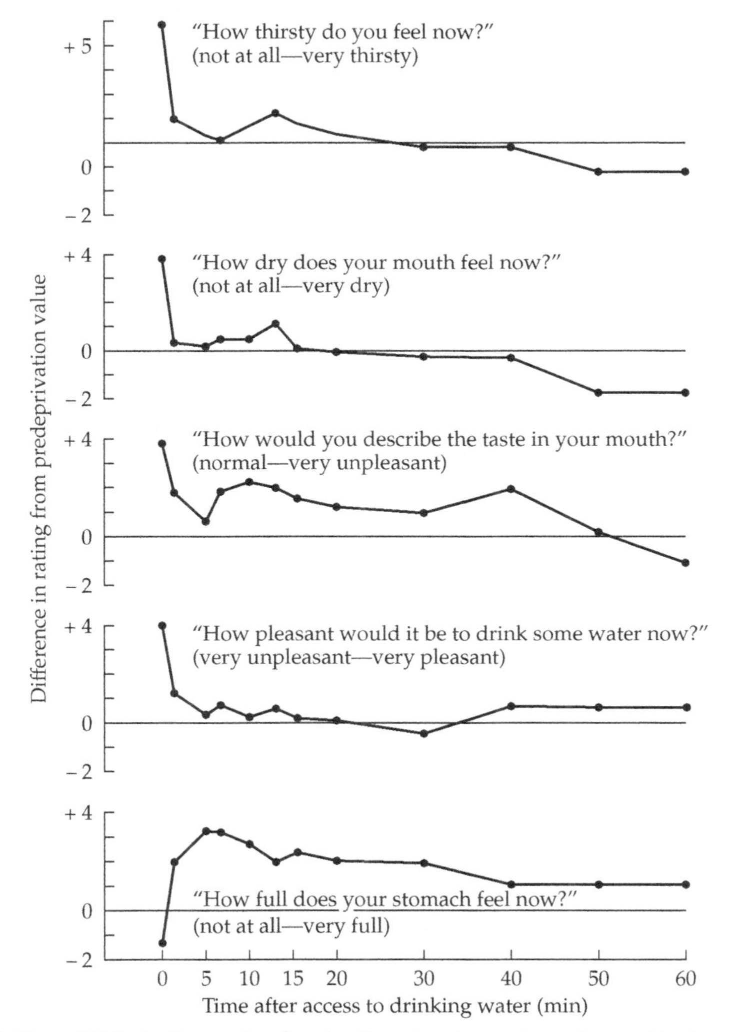

9.8 Rating of thirst after water deprivation. A rating scale can be used by human subjects to provide a measure of thirst and other sensations associated with fluid intake. Subjects were asked to rate their thirst, then deprived of water for 24 hours. They were then asked to rate their thirst again before and after being given access to drinking water, and their ratings were compared to their ratings on the same measures before they were deprived of water. Within 5 minutes of drinking, their thirst sensations were ameliorated, and the subjects reported that their stomachs felt full. After Rolls and Rolls, 1982.

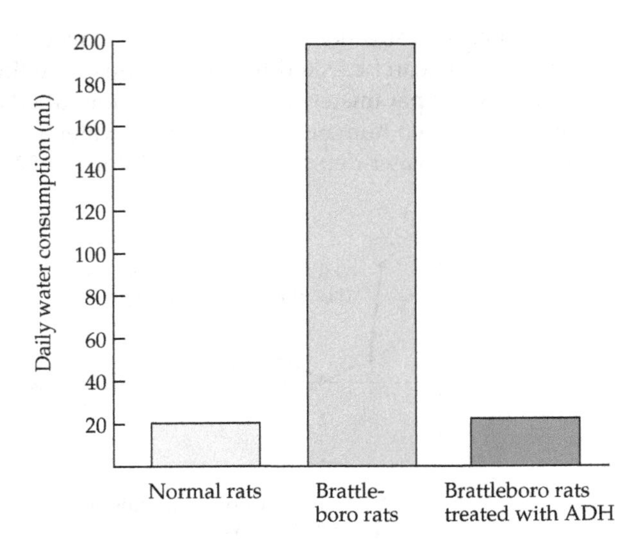

9.9 Vasopressin (ADH) mediates water consumption. Normal rats drink about 20 ml of water per day, but rats of the Brattleboro strain, which congenitally lack ADH, drink nearly 10 times as much water and also excrete about 10 times the volume of urine of normal rats. After ADH treatment, Brattleboro rats display normal levels of water consumption and urinary output.

Endocrine Regulation of Fluid Balance and Thirst

Vasopressin (ADH) acts on the tubules of the kidneys to retain water (see Box 9.1). Occasionally, a genetic error occurs in humans or rats so that ADH is not produced. Individuals lacking ADH suffer from diabetes insipidus (*diabetes*, "run through"; *insipidus*, "weak"), a condition that forces them to ingest copious amounts of water because they are almost constantly urinating. Essentially, all of their extracellular fluid is lost each day and must be replaced. Rats with this mutation, known as Brattleboro rats, urinate 200 ml/day and must drink about that amount daily to maintain their fluid balance. If these rats are injected with ADH, their water intake and urine output drop to the normal levels of about 20 and 15 ml/day, respectively (Figure 9.9).

Two types of stimuli associated with the need to balance body fluids normally evoke the release of ADH from the posterior pituitary. One of these stimuli is the cellular dehydration of cerebral osmoreceptors. Although virtually all cells in the body shrink in size as water moves into the interstitial space during osmotic dehydration, only these particular cells in the brain signal this condition to the paraventricular nuclei (PVN) and supraoptic nuclei (SON) of the hypothalamus, where ADH is made. These osmoreceptors are generally located in several brain structures located near the third ventricle, including the lateral preoptic area, the subfornical organ, and the organum vasculosum of the lamina terminalis (OVLT), as well as in the area postrema in the brain stem (Figure 9.10). Two different messages are sent from these central osmoreceptors in response to different levels of dehydration. A signal to release ADH from the posterior pituitary occurs in response to mild cellular dehydration. If dehydration persists after maximal reclamation of water has been achieved in the kidneys, then a second signal from the brain osmoreceptors stimulates drinking behavior. This dual control allows physiological water-saving systems to be engaged prior to behavioral responses and

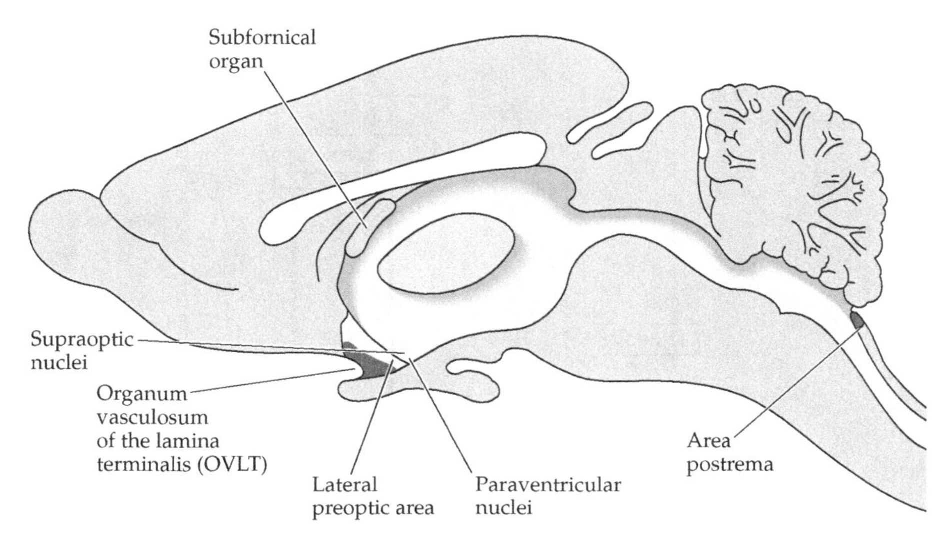

9.10 Brain structures involved in the mediation of thirst and water intake. Cerebral osmoreceptor cells respond to dehydration and shrinkage by signaling the paraventricular and supraoptic nuclei to synthesize vasopressin. These osmoreceptors are found in the circumventricular organs seen in this schematic drawing of a mid-sagittal section of the rat brain: the subfornical organ, the area postrema, the organum vasculosum of the lamina terminalis (OVLT), and the lateral preoptic area.

thus frees individuals from the need to drink frequently. ADH does not inhibit drinking behavior directly; rather, it causes the kidneys to retain water, which decreases blood plasma osmolality. Reduced plasma osmolality, in turn, inhibits drinking.

Another stimulus that triggers the release of vasopressin is a reduction of blood plasma volume. A loss of blood volume is detected by stretch receptors, called baroreceptors, in the walls of the cardiac blood vessels. These receptors signal the PVN and SON to release ADH, which acts as a vasoconstrictor (reduces the diameter of blood vessels) to increase blood pressure (Gauer and Henry, 1963). Hence, the same hormone has two distinct, but related, functions that work in concert when blood pressure drops, and possesses two common names: antidiuretic hormone (ADH), referring to its effect on the kidneys, and vasopressin, referring to its effect on the blood vessels. The cardiac baroreceptors also signal the brain directly via the vagus nerve* to stimulate thirst. A diagram of the regulation of fluid intake behavior is depicted in Figure 9.11.

Hypovolemia also stimulates the formation of another potent vasoconstrictor, angiotensin. Angiotensin is a hormone that comes from an unusual source:

*The vagus (Latin for "wandering") nerve is one of the cranial nerves that make up part of the peripheral nervous system (PNS). It extends far from the head and innervates the heart, lungs, digestive tract, and liver, carrying information between these organs and the central nervous system.

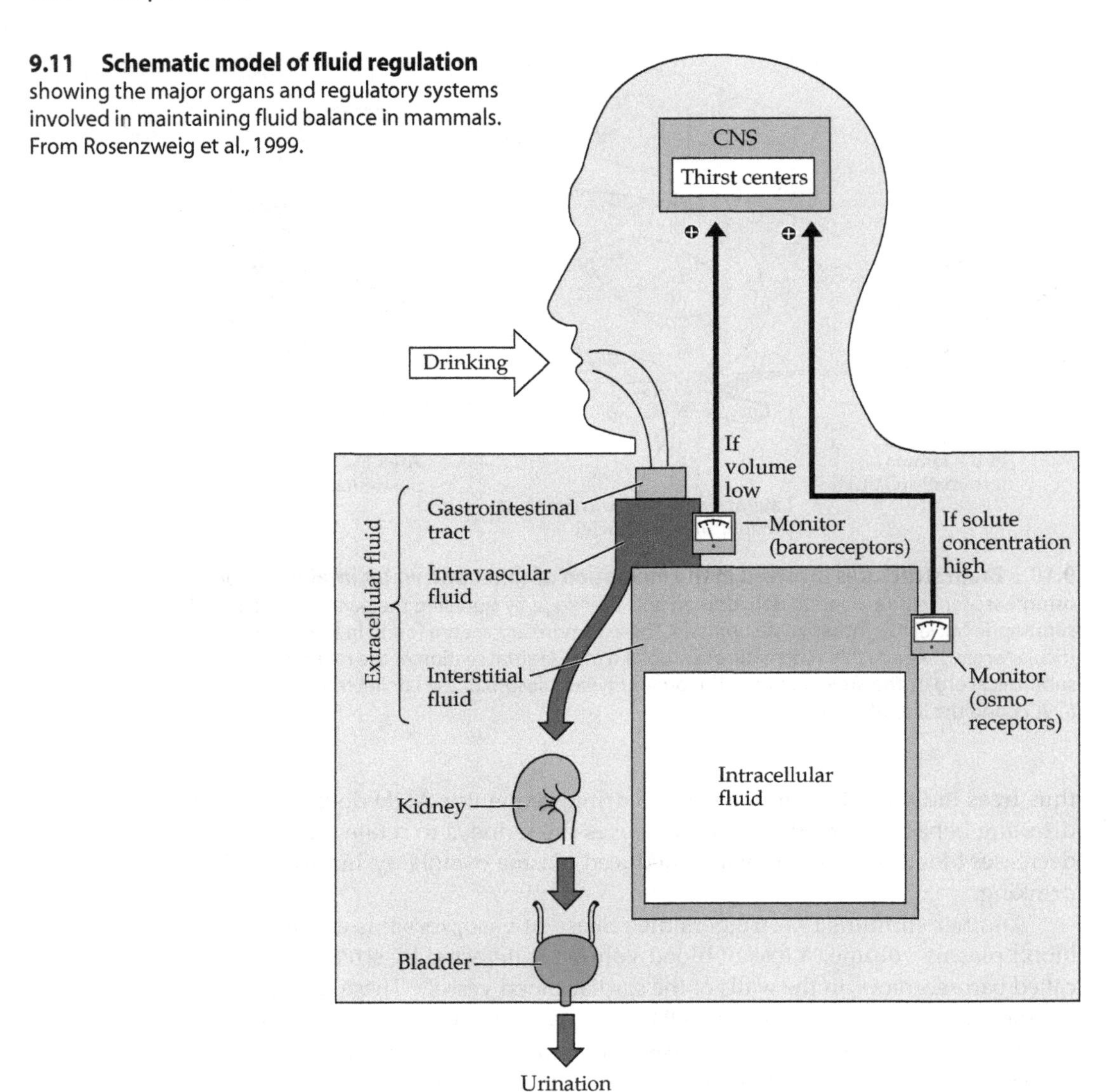

9.11 Schematic model of fluid regulation showing the major organs and regulatory systems involved in maintaining fluid balance in mammals. From Rosenzweig et al., 1999.

the blood. Angiotensin is not produced in any endocrine gland; however it is usually considered to be a hormone. In any case, a drop in blood pressure during hypovolemia stimulates the brain to signal the kidneys to make an enzyme called renin; the kidneys can also make renin in response to low blood pressure without any neural input. Renin converts a blood-borne substance called angiotensinogen into the various forms of angiotensin, including angiotensin II, the form involved in this response (Figure 9.12).

The role of angiotensin in drinking behavior is controversial. In rats, systemic injections of angiotensin (Fitzsimons and Simons, 1969), or injections directly into

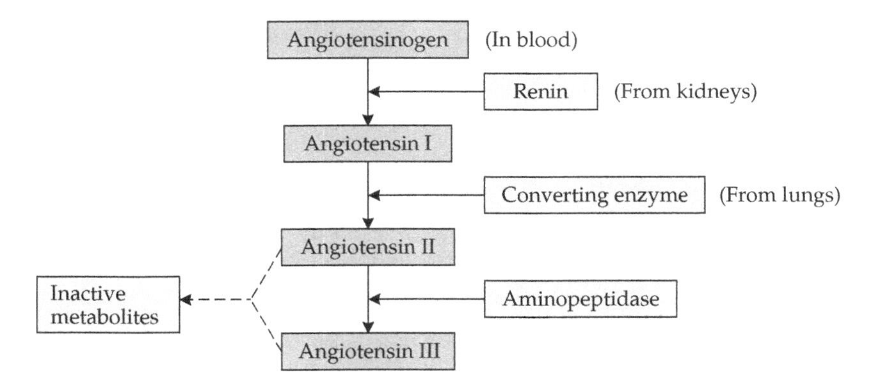

9.12 Synthesis of angiotensin in humans. The drop in blood pressure that occurs during hypovolemia initiates the formation of the potent vasoconstrictor angiotensin by stimulating the brain to signal the kidneys to make renin. Renin is an enzyme that converts the blood-borne substance angiotensinogen to three forms of angiotensin: angiotensin I, II, and III. Angiotensin II is the most biologically potent of the angiotensins, but all three exert some effects on water and sodium balance. The kidneys can also make renin in response to low blood pressure without any neural input. After Ganong, 1993.

the subfornical organ (Simpson et al., 1978), increase drinking behavior. However, the doses of angiotensin required to elicit drinking are much higher than the levels of this hormone produced naturally in response to hypovolemia (Stricker, 1978). Furthermore, removal of the kidneys, the source of renin, does not interfere with drinking behavior in response to hypovolemia in rats (Fitzsimons, 1961).

Despite the controversy surrounding the role of angiotensin in regulating thirst, it is well known that this parahormone stimulates the release of aldosterone from the zona glomerulosa of the adrenal gland. As we saw above, aldosterone is another hormone critical for maintaining fluid balance. Aldosterone promotes the retention of sodium in the kidney by stimulating sodium pumping in the ascending limb of the loop of Henle (see Box 9.1). The resulting retention of sodium at high concentrations in the kidneys causes water to be reabsorbed from the blood as it flows through the nephron, and thereby reduces the amount of water sent to the bladder for excretion. Without aldosterone, enormous amounts of water—approximately 30 liters per day in humans—and sodium—about 200 mmol/liter per day—would be lost in the urine.

Pharmacological tools have revealed two distinct types of angiotensin receptors in the brain, Type 1 and Type 2. Using specific receptor antagonists, it was discovered that only the Type 1, and not the Type 2, receptors are mediators of angiotensin II-related thirst and sodium intake (Fluharty and Sakai, 1995). Furthermore, intracerebroventricular administration of angiotensin Type 1 receptor antisense RNA diminished drinking behavior in rats (Sakai et al., 1995). Intracerebroventricular administration of mineralocorticoid receptor antisense RNA reduced sodium appetite in rats (Ma et al., 1997; Sakai et al., 1996) (Figure

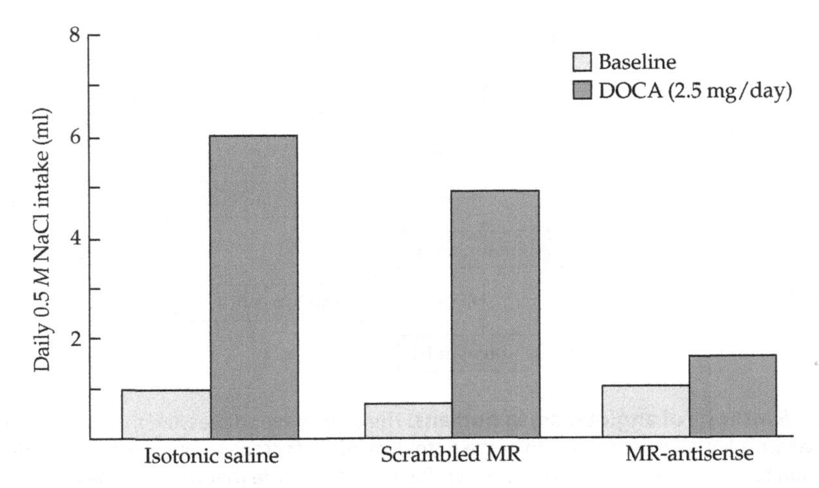

9.13 Antisense RNA for mineralocorticoid receptors blocks DOCA-induced elevation in saline intake. Deoxycorticosterone acetate (DOCA) (a synthetic mineralocorticoid with about 3% aldosterone activity) causes an increase in salt appetite when injected systemically, as shown by the increase in saline ingestion by animals treated with this substance. If animals treated with DOCA also have isotonic saline or scrambled (nonsense) mineralocorticoid receptor (MR) mRNA infused into the amygdala, their saline intake is unaffected. If DOCA-treated animals are treated with a MR antisense infusion, however, their saline intake is reduced. After Sakai et al., 1996.

9.13). These results of blocking the expression of either the angiotensin or aldosterone receptors demonstrate the role of angiotensin II and aldosterone in mediating water and sodium ingestion, respectively, as well as the importance of their actions in the central nervous system.

Sodium Balance

Humans that consume modern diets and animals maintained on commercial laboratory diets—both of which provide plenty of sodium—have a relatively large reservoir of sodium in the interstitial fluid (Daniels and Fluharty, 2004). This sodium reserve probably serves to buffer the brain from wide fluctuations in sodium availability during large variations in daily sodium intake and loss. When the sodium reservoir is depleted, individuals are motivated to seek and ingest sodium. Sustained sodium deprivation interferes with neural functioning and can rapidly lead to death. The detectors for blood levels of sodium appear to be located in the brain (Andersson et al., 1995; Stricker and Verbalis, 1990b; Weisinger et al., 1982).

Herbivores obtain all of their sodium and much of their water from the plants that they eat. Because there is a wide range of sodium content in plants, herbivores may be exposed to wide variations in sodium availability. Carnivores are not usually under the same sodium pressures as herbivores because the animals that they consume must maintain a relatively consistent sodium level. Essentially, carnivores have the problem of excreting sodium after a meal because their meals tend to be sodium-rich, whereas herbivores are adapted to retain sodium and excrete potassium after their meals.

Recall that European rabbits in Australia are confronted with environments that vary substantially in sodium availability. The rabbits must cope with these variations in order to survive and reproduce. Rabbits inhabiting the Snowy Mountains in Australia are exposed to an environment that is extremely low in sodium; analyses of the soil and plants in this alpine region revealed a profound lack of sodium (Blair-West et al., 1968). The rain is virtually free of sodium, and the spring snowmelt further reduces the availability of sodium by leaching it deeper into the ground. In contrast, the food plants of rabbits living in the harsh desert of central Australia are very high in sodium content. Similarly, rabbits inhabiting ocean coastal regions in Victoria are exposed to relatively high levels of sodium in plants because of the high concentrations of sodium in the coastal rainwater. Researchers have observed differences among the rabbits in these three regions in the quantities of water consumed and urine produced, as well as in other physiological parameters (Table 9.1).

The Australian rabbits occupying opposite ends of the sodium availability continuum are faced with a situation analogous to that of freshwater versus saltwater fish. Saltwater fish are essentially bags of fresh water surrounded by a concentrated saltwater solution. There is a high osmotic gradient favoring the movement of sodium ions into these fish, but they have evolved mechanisms to keep sodium out and to remove it from their bodies efficiently. Conversely, freshwater fish are essentially bags of saline surrounded by dilute water, which creates an osmotic gradient favoring the movement of sodium out of the fish. Freshwater fish have evolved effective sodium-retaining mechanisms to combat sodium depletion.

Modern bony fishes, called teleost fishes, are thought to have evolved in fresh water, then radiated to environments with vastly different degrees of salinity. Some fishes, termed *stenohaline* species, have adapted to specific saline environments and cannot tolerate salinities beyond very specific ranges. In contrast, *euryhaline* species can tolerate wide ranges of salinity. Euryhaline fishes in nature may be exclusively saltwater fishes (for example, the starry flounder, *Platichthes stellatus*), exclusively freshwater fishes (for example, the tilapia, *Sarotherodon mossambi-*

TABLE 9.1 **Comparison of sodium regulatory indices of rabbits living in three different habitats in Australia**

Habitat	Na^+ in urine (mmol/l)	Aldosterone (ng/100 ml)	Adrenal mass (g)	Zona glomerulosa size (as % of total adrenal cortex)
Snowy Mountains	0.53	69.0	0.12	22.6
Coast	18.00	21.0	0.11	15.6
Desert	139.00	9.0	0.13	15.3

Source: Blair-West et al., 1968.

TABLE 9.2 Seasonal changes in sodium regulatory indices of rabbits inhabiting the Snowy Mountains of Australia

Season	Na^+ in urine (mmol/l)	Aldosterone (ng/100 ml)	Adrenal mass (g)	Zona glomerulosa size (as % of total adrenal cortex)
Spring	0.59	130.0	0.18	34.4
Summer	0.53	69.0	0.12	22.6
Autumn	2.60	—	0.14	17.3
Winter	6.40	74.0	0.14	17.8

Source: Blair-West et al., 1968.

cus), or capable of exploiting their tolerance for a wide range of environmental salinities by migrating between salt water and fresh water (for example, Pacific salmon, *Oncorhynchus* sp., or Atlantic eels, *Anguilla* sp.). Those euryhaline fishes that move between fresh and salt water rely on hormones—most notably, prolactin, cortisol, and thyroid hormone—to adapt to their changing sodium requirements, and their ability to survive these changes involves a complex interaction among reproductive maturity, day length, and hormonal responses. The extent to which changing hormone levels themselves induce migration or are secondary to other maturational factors remains somewhat controversial.

The Australian rabbits are confronted with the same basic problem as fish; namely, how to survive in habitats with markedly different sodium availabilities. Several physiological adaptations are evident in rabbits living in sodium-poor habitats. Snowy Mountains rabbits excrete very little sodium in their urine, although there is a notable seasonal cycle of average urinary sodium excretion, which is greater in winter than in summer (Table 9.2). Average blood plasma concentrations of aldosterone mirror the seasonal urinary sodium excretion cycle, as does the average proportion of the adrenal gland made up by the zona glomerulosa, which varies nearly 50% in size. What is the reason for these seasonal variations? During spring and summer, the rapid growth of lush plants further reduces sodium concentrations per gram of plant material in the Snowy Mountains region. Herbivores living under these conditions must consume enormous quantities of food to maintain sodium at a level consistent with proper physiological functioning. Pregnancy and lactation increase this demand even further. During the spring and summer, rabbits living in the alpine regions of Australia display an avid salt appetite. When wooden pegs impregnated with various salts are made available to the rabbits, they clearly prefer sodium chloride (NaCl) and sodium bicarbonate ($NaHCO_3$), but they will also ingest magnesium chloride ($MgCl_2$) and potassium chloride (KCl) in smaller amounts (Figure 9.14; Myers, 1967). Thus, these animals possess physiological, morphological, and behavioral adaptations to obtain and to conserve sodium.

In contrast to Snowy Mountains rabbits, rabbits living in the desert inhabit a sodium-rich environment, and their physiological challenge is to limit sodium intake and reduce sodium levels in their bodies. These animals excrete consistent-

9.14 Rabbits inhabiting sodium-deficient habitats have voracious sodium appetites. This telescopic photograph shows rabbits living in sodium-deficient alpine regions of Australia devouring wooden pegs impregnated with sodium. Rabbits in sodium-rich habitats are never observed ingesting sodium-treated pegs. Photograph by E. Slater, CSIRO Wildlife Division, Canberra.

ly high levels of sodium throughout the year, and have concomitant low plasma aldosterone concentrations (see Table 9.1). The zona glomerulosa makes up only about 15% of the adrenal cortex. Rabbits living in desert or seashore habitats were never observed to ingest salt from salt licks provided by experimenters.

Regardless of sodium availability in the environment, each rabbit (and every other mammal, including you) must maintain the sodium levels in its body within precise limits. In rabbits and humans, sodium levels are maintained at 135–145 mmol/liter in the blood plasma. Obviously, animals living in sodium-rich habitats must excrete excess sodium to maintain blood sodium levels in the optimal range, whereas animals living in low-sodium environments must avoid losing precious sodium. Australian rabbits provide an example of how physiology, morphology, and behavior are linked to bring about fluid balance and maintain homeostasis.

Although many animals are generalists to some extent, most animals have evolved to inhabit specific environmental niches. Some animals have evolved in habitats that regularly experience sodium shortages, whereas others have evolved where sodium is plentiful. Like rats, Syrian hamsters (*Mesocricetus auratus*) will perish if subjected to bilateral adrenalectomy. Unlike adrenalectomized rats, however, adrenalectomized hamsters provided with salt water to drink will refuse it and die. Hamsters survive adrenalectomy only if provided with a solution of saline mixed with saccharin. Apparently, the sweet taste of saccharin masks the salty taste of the saline. When provided with a choice between a saccharin solution and a salt water solution, adrenalectomized hamsters will drink the saccharin solution exclusively and invariably perish. Do these facts indicate that hamsters are less intelligent than rats? No. These puzzling results are immediately reasonable in light of the fact that Syrian hamsters evolved in the desert, where sodium levels are high and water is scarce. They have evolved physiological and behavioral strategies to avoid sodium. It is unlikely that hamsters have ever had a need during their evolution to ingest sodium; consequently, behavioral homeostatic strategies in response to sodium loss from the kidneys are not available to this species. The difference in salt appetite between rats, rabbits, and humans on the one hand and hamsters on the other demonstrates the importance of understanding the evolutionary history of the animals in question when trying to understand the complex web of physiological and behavioral interactions mediating homeostatic processes.

Kangaroo rats (*Dipodomys merriami*) are desert rodents that rarely have access to water in their native habitat and probably never drink it if they do find it. Their water needs are met by chemically liberating water from the seeds they eat. Kangaroo rats have evolved to conserve water very effectively; they rarely urinate, and when they do, the urine is highly concentrated. These small rodents have the longest kidney tubules of any mammal, which accounts for their extreme efficiency in retaining water. Another mammal that has an unusual method of obtaining water is the elephant seal (*Mirounga angustirostris*). During their 4- to 5-month mating season, elephant seals engage in territorial and reproductive behaviors along the shoreline (see Figure 5.1), but do not eat or drink. All the water they need is liberated metabolically by hydrolysis of their substantial fat reserves. Elephant seals, like other marine mammals, have evolved to live in the sea, but their adaptation is limited in that they are no more capable of surviving on seawater than humans. During the nonbreeding season, they get all their water from the fish they consume (LeBoeuf, 1974). Kangaroo rats and elephant seals have adapted to their particular niches over the course of many eons of evolution. The rabbits invading Australia have not been there very long (in evolutionary time), and new mechanisms to balance water and sodium use have yet to evolve; consequently, behavioral adaptations, mediated in part by hormones, are critical for their continued survival.

How Do Hormones Regulate Drinking Behavior?

The hormones that maintain fluid balance interact with the sensory systems, or input systems, associated with water intake. In a typical study of these interactions, an animal is given a choice between two different drinking bottles (Figure 9.15). For example, a rat may be given a choice between tap water and an isotonic saline solution. The rat's fluid intake is measured daily, and a preference is calculated. For example, the rat might ingest 8 ml of water and 12 ml of saline during a single 24-hour period. In this case, we would say that saline is preferred over tap water by 67% ($8/12$). To examine the nature of this preference, the saline could be diluted by half to a 0.45% sodium concentration. If the rat still drinks more of the diluted saline than the tap water, an inference is made that the animal can

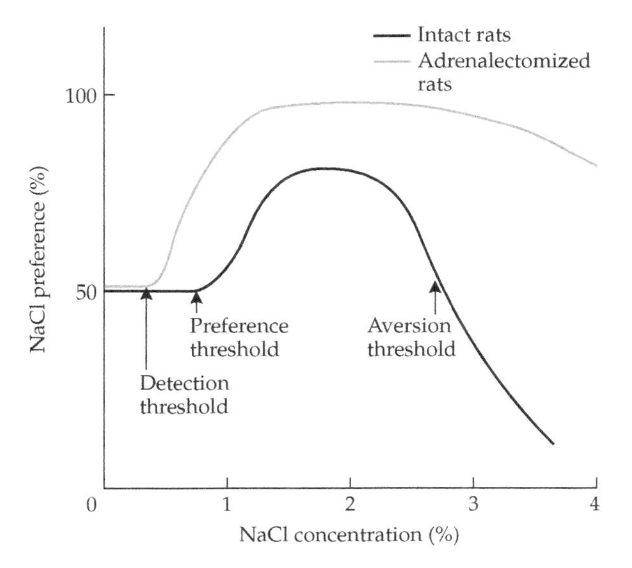

9.15 Adrenalectomy affects saline preference thresholds. The concentration at which an animal responds differently to different solutions is called the detection threshold. Normal rats (black line) prefer about a 1% solution of saline over water (the preference threshold), but as the concentration of sodium increases, the rats' preference changes to plain water (the aversion threshold). The gray line plots the saline preference pattern for adrenalectomized rats, which lack aldosterone. Note that adrenalectomized rats tolerate sodium solutions that normal rats or people would find extremely unpleasant; this change in sodium tolerance helps the adrenalectomized animal maintain its fluid balance.

detect the difference between the two and has a preference. In this manner, a **preference threshold** can be determined by serial dilutions. The preference threshold is the lowest concentration at which preference for the solution over water can be maintained. Obviously, animals will display different preference thresholds for different substances; for example, the preference threshold for sucrose solutions and for sodium solutions is different among most individuals. The **detection threshold** is the concentration at which the animal can tell the difference between solutions. These two parameters are not necessarily identical. Note in Figure 9.15 that the detection threshold for sodium is not different for adrenalectomized and intact animals, although the sodium-deficient animals drink more sodium at low and high concentrations. Adrenal-intact rats can discriminate among different low-sodium concentrations in a test in which they must do so to avoid electric shocks. Presumably, the differences among low-sodium solutions make a difference only to the adrenalectomized animals. Interestingly, the sensory neurons from the mouth of an adrenalectomized rat reduce their firing rate when sodium is applied to the tongue. Perhaps this is why adrenalectomized rats are able to ingest such high, normally aversive, concentrations of salt.

The hormones associated with fluid balance also mediate behavior by acting on the central processing systems. As described above, during mild cellular dehydration, ADH affects central drinking mechanisms indirectly by causing the kidneys to retain water, which results in reduced blood plasma osmolality; reduced plasma osmolality inhibits drinking. After persistent dehydration, ADH, aldosterone, and angiotensin II all increase drinking behavior by acting directly or indirectly on the central nervous system. Aldosterone stimulates sodium retention by the kidneys and thus maintains osmotic balance, but aldosterone affects drinking behavior only indirectly via its influence on osmotic thirst. Aldosterone may act directly on the central nervous system to regulate sodium appetite in conjunction with angiotensin. As we saw above, pharmacological doses of angiotensin elicit drinking behavior, presumably by acting on the subfornical organ in the brain (Simpson et al., 1978; Zigmond et al., 1991). The output systems—the effectors—of water ingestion are essentially unaffected by hormones. Hormones do not appear to increase the efficiency of water intake or otherwise affect the output systems of drinking behavior in any appreciable way.

Energy Balance

Animals eat food to meet the needs of the structural part of the body; that is, to obtain the raw materials to make bone, muscle, and other structures. Animals also eat food to obtain energy to fuel the body. Providing metabolic energy to the cells is absolutely critical to moment-to-moment survival. Consequently, homeostatic mechanisms have a predisposition toward storing metabolic fuels in the body, where they will be available to help the animal survive food shortages, rather than toward expending excess energy stored as fat.

All cellular processes require metabolic fuel to keep them going. The central nervous system is a major consumer of metabolic fuel. All the vital systems require energy to carry out their functions, and significant energy is necessary to maintain mammalian body temperatures at 37°C. Despite this persistent need for energy, there are fluctuations in the energy requirements of every individual, as the rate of energy utilization varies throughout the day as well as over the seasons. There are also fluctuations in energy acquisition. The balance between the amount of energy stored in the body, energy expenditure, and energy intake appears to be "regulated," but not in the same sense that water and sodium balance or body temperature are regulated. Whereas body temperature and sodium balance must be maintained within a relatively narrow window, the amount of energy stored in the body can vary widely among individuals of some species. In such species, individuals need a minimum of stored energy for survival and reproduction, but they may survive and reproduce especially well with surplus amounts of stored energy—that is, body fat. Over the course of evolution, this stored fat might have enhanced survival in environments in which energy availability fluctuated. Those individuals with large fat stores would have been more likely to survive harsh winters, droughts, or famine. Thus, it should not be surprising to find that in some species, there are fewer mechanisms to stop eating and weight gain than there are to promote eating and weight gain.

Eating tends to be episodic, even though the need for energy is more or less continuous (Stricker and Verbalis, 1988). Accordingly, energy acquisition and energy expenditure are never quite balanced. Individuals cannot eat continuously to meet moment-to-moment changes in energy demands; they must occasionally sleep, mate, or hunt for food. Furthermore, food availability is not constant; there are seasonal and daily cycles in food availability. All of this leads to a dynamic relationship between energy acquisition and energy expenditure. When more food is consumed than required, the excess energy is stored in the form of glycogen in the liver and

adipose tissue, or fat throughout the body. This stored energy is later tapped when the steady delivery of metabolic fuel from the intestines wanes after a meal.

Animals have homeostatic mechanisms that ensure long-term energy balance. These mechanisms function to keep body mass within a relatively fixed range over weeks, months, or even years. Other related mechanisms exist to regulate short-term energy balance, switching feeding behavior on or off. Multiple redundant mechanisms exist to control energy intake and expenditure, which makes the endocrine and neuroendocrine regulation of energy balance confusing to understand. Traditionally, experiments targeting a single hormonal or neuropeptide system ultimately fall short because another system kicks in to reverse the effects of the manipulated regulatory system. Another complicating factor in the study of energy balance is that although many different hormones may be directly involved in the control of food intake, many other substances may indirectly affect food intake, perhaps by increasing general arousal (orexin is an example because animals treated with this hormone that increase their food intake also sleep less) or by making the individual too sick to eat (e.g., CCK treatment appears to make individuals nauseated and thus reduces food intake). All of these caveats should be kept in mind as we review the interaction between hormones and energy balance.

In chronically well-fed animals, blood concentrations of insulin and leptin, and of metabolic fuels such as glucose and amino acids, are relatively high. Stores of body fat in adipose tissues and glycogen in the liver are high, as are concentrations of lipogenic (fat-synthesizing) enzymes. In contrast, well-fed animals have relatively low blood concentrations of glucagon, glucocorticoids, free fatty acids, lipolytic enzymes (which break down fat), and ketone bodies. Levels of certain neuropeptides in the brain, such as neuropeptide Y and CRH, are also low. Among fasting animals, the relative concentrations of all these substances are reversed.

As animals fluctuate between a well-fed and a fasting state, correlated changes occur in the secretion of a number of hormones, neurotransmitters, and neuromodulators. As noted in Chapter 1, an important test to determine the contribution of hormones in hormone–behavior interactions is the use of a specific hormone receptor antagonist to block the effects of the hormone on the behavior under study. Some hormones thought to be involved in energy balance are still awaiting this "acid test" (e.g., leptin).

Hormones influence the motivation to obtain food, as well as the consummatory behavior of eating (Schneider et al., 2002; Bartness et al., 2002). As with sexual behaviors, it is probably instructive to examine the effects of hormones on both appetitive and consummatory aspects of food intake. For example, in Syrian hamsters, food deprivation increases hoarding of food, but does not increase food intake. Leptin, which decreases food intake in rats and mice, appears to act on the motivation to procure food; leptin treatment of Syrian hamsters decreases food deprivation–induced food hoarding (Buckley and Schneider, 2003).

Metabolism during the Well-Fed State

After a meal, there are two phases of energy utilization and storage: (1) the *postprandial phase* and (2) the *postabsorptive phase*. The metabolic interactions that occur

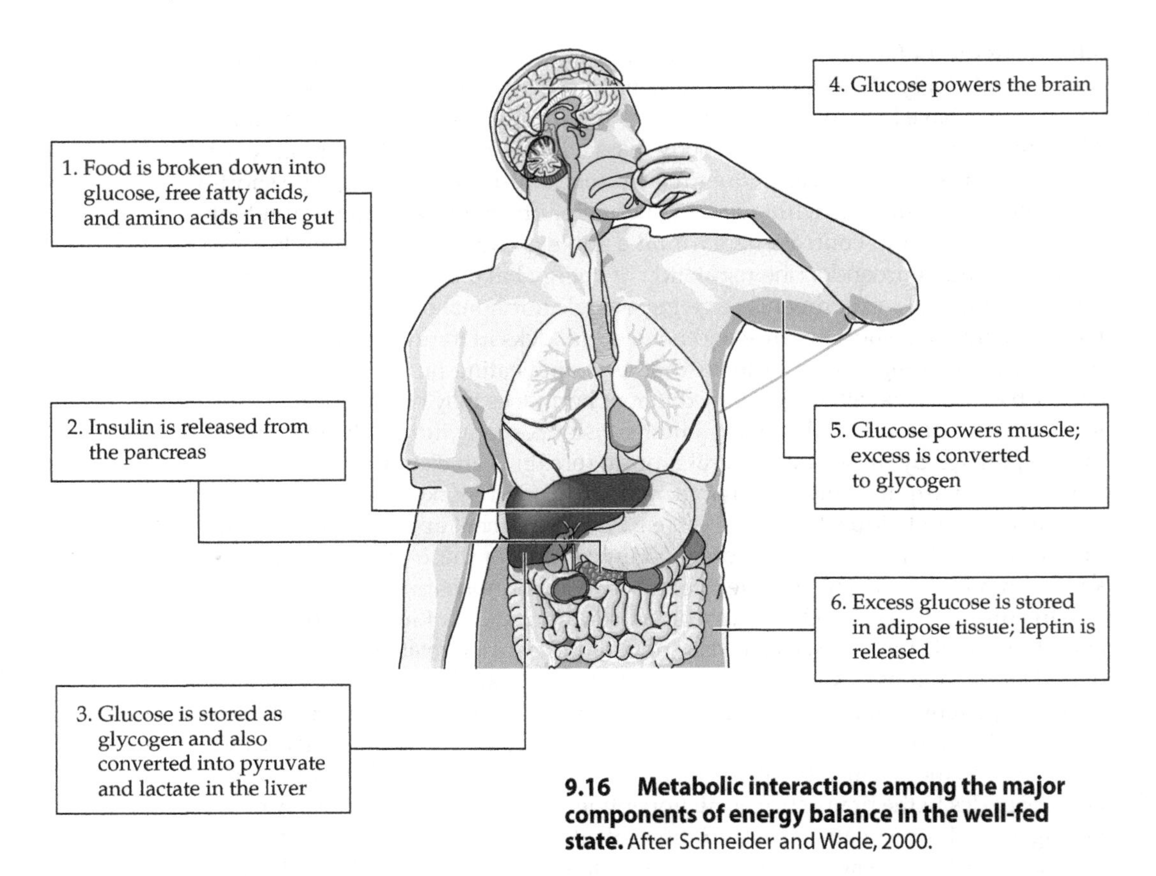

9.16 Metabolic interactions among the major components of energy balance in the well-fed state. After Schneider and Wade, 2000.

in a well-fed individual are outlined in Figure 9.16. The postprandial phase occurs immediately after the ingestion of food. When you eat a meal, a supply of **metabolic fuels**, in the form of simple sugars (e.g., glucose), fatty acids, and amino acids, enters the bloodstream almost immediately. Unless you are running a marathon immediately after eating, there tends to be an energy surplus after a meal. The excess energy is stored during the postabsorptive phase. The protein hormone insulin, secreted by the endocrine cells of the pancreas, acts to promote such storage. In doing so, insulin lowers the levels of metabolic fuels in the blood.

During the postabsorptive phase, insulin secretion rises while glucagon secretion falls. In the liver, insulin stimulates the conversion of glucose to glycogen, a stored form of sugar. Glycogen is stored in the liver and in muscle. Insulin also facilitates the transport of glucose into muscle and fat cells, as well as the transport of amino acids into muscle cells. In the liver, amino acids are converted into ketone bodies. In peripheral cells, insulin is necessary for glucose oxidation and lipogenesis, processes that result in the storage of fat in adipose tissue. Insulin also prevents the breakdown of glycogen in both muscle and liver cells, as well as the mobilization of metabolic fuels from adipose tissue. Although many hormones play a role in removing energy from storage, insulin is the sole hormone responsible for energy storage in vertebrates. The hormones involved in removing energy from storage include epinephrine, norepinephrine, glucocorticoids, thyroid hormones, growth hormone, somatomedin, and glucagon.

There are two phases of insulin release: the cephalic phase and the gastrointestinal (GI) phase. During the cephalic phase, a neurally triggered release of insulin from the pancreatic β-cells occurs as a result of the sensory stimuli associated with food intake (Steffens et al., 1990). So even before any new nutrients have arrived in the digestive system, an insulin-induced reduction in blood levels of metabolic fuels may be associated with an increase in hunger. You have probably experienced the cephalic phase of insulin release when the sights or smells of a favorite meal caused you to feel noticeably more hungry. The primary storage of excess nutrients taken in during a meal occurs in the GI phase, when insulin is released in response to the absorption of nutrients from the gut.

Metabolism during the Fasting State

Eventually, the influx of energy from the gut no longer exceeds the body's energy usage requirements. The body must then shift from putting energy into storage to getting it out of storage. This shift in energy balance occurs, for example, after strenuous exercise or after a prolonged fast. Most of us are confronted with this situation every morning when we awaken and need to break the nighttime fast, or *breakfast*. Energy reserves are mobilized from storage to meet your energy needs as you become active, even prior to eating your first food of the day.

Basically, the metabolic system is designed to provide sufficient levels of energy to the brain. The brain cannot utilize fatty acids, so it must receive a constant supply of glucose via the blood. During long-term food deprivation, the rest of the body is literally starved to feed the brain. Glucose is preferentially used by the brain, while cells in the periphery switch from metabolizing glucose to metabolizing free fatty acids mobilized from the lipids stored in adipose tissue. The switchover to using free fatty acids spares glucose for the brain. This mode of metabolism occurs not only when we are fasting, but when we are not consuming carbohydrates. Thus, the fasting metabolism is characteristic of individuals on low-carbohydrate diets.

There are several physiological methods of getting energy out of storage and into the bloodstream during a fast. The metabolic interactions that occur in a fasting individual are shown in Figure 9.17. First, **glycogenolysis**, the breakdown of stored glycogen, in the liver provides a rapid supply of glucose to the blood. Under extreme conditions, such as during strenuous exercise, glycogenolysis in muscle provides the extra energy for the physical work. Second, **lipolysis**, the breakdown of triglycerides stored in adipose tissue into free fatty acids and glycerol, provides oxidizable fuels for peripheral tissues. Glycerol can ultimately be converted into glucose in the liver and elsewhere. During fasting, the protein hormone glucagon is released from the α-cells of the pancreas, whereas the secretion of insulin from the pancreas is inhibited (Steffens et al., 1990). Glucagon induces both lipolysis and glycogenolysis.

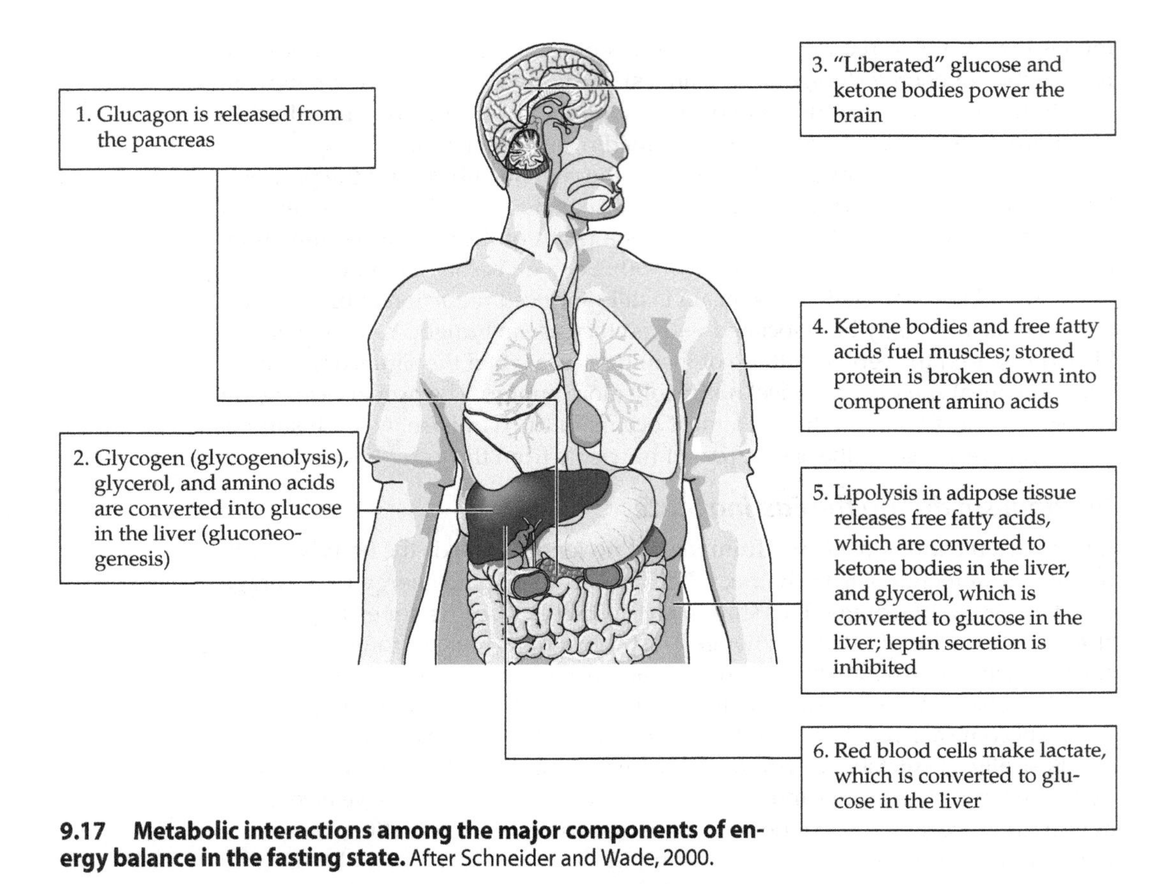

9.17 Metabolic interactions among the major components of energy balance in the fasting state. After Schneider and Wade, 2000.

Stored energy in the body can also be released via the sympathetic nervous system. The sympathetic system generally acts as an emergency system, although input to all organs can shift to favor the sympathetic or parasympathetic system in day-to-day regulation as well. Sympathetic nerves that innervate the fat cells can induce lipolysis (Steffens et al., 1990). The liver is also innervated with sympathetic nerves that may signal it to release stored glycogen (Sawchenko and Friedman, 1979). Another backup system that is used during fasting is a process called **gluconeogenesis**, the production of glucose from amino acids. Gluconeogenesis takes place entirely in the liver. It results in the excess production of ketone bodies, a side reaction of making sugar from amino acids; ketone production also results from the incomplete oxidation of free fatty acids (Friedman, 1990). Prolonged fasting, during which muscles are broken down for energy, produces toxic levels of ketone bodies, a condition called *ketosis*; the most serious effect of ketosis is the resulting change in blood pH, which has a deleterious effect on neural function. Diabetics with uncontrolled blood sugar levels can also enter ketosis when insufficient insulin is provided and may suffer serious consequences if blood pH fluctuations are prolonged. Similarly, people on long-term, very low carbohydrate diets can enter ketosis.

Glucagon release, gluconeogenesis, and stimulation of fat breakdown by the sympathetic nervous system are the most common mechanisms for raising blood glucose levels without eating, but there are additional methods of elevating blood glucose levels during emergencies to keep a relatively constant level of energy supplied to the brain. After prolonged fasting or very demanding exercise, growth hormone can trigger lipolysis in the fat cells. Furthermore, corticosterone or cortisol released during vigorous exercise can induce gluconeogenesis. During stress, epinephrine from the adrenal medulla acts to stimulate glycogenolysis in the liver.

Because the brain receives a relatively constant supply of glucose, you might guess that the detectors for changes in blood sugar levels might be found not in the brain, but rather in the periphery, where blood glucose changes are more pronounced. Logical as this assumption may be, it is not completely true. Most metabolic fuel detectors are located in the brain and liver, but fuel detectors are also located in muscle and fat cells.

Disordered Energy Metabolism

Problems with getting energy into storage are common. Again, insulin is the major actor in this physiological drama, and many metabolic difficulties occur if there are any problems with insulin secretion or insulin–receptor interactions. The best-known such disorders are the two types of diabetes mellitus. Type I diabetes is an autoimmune disorder in which the β-cells of the pancreas are destroyed by the immune system and an insulin deficiency results; it is also referred to as insulin-dependent diabetes mellitus. Type I diabetes usually has a rapid onset. It is most common among children and young adults and is commonly called "childhood diabetes." Type II diabetes usually develops more slowly in adults, and is usually evoked by obesity, stress, or menopause. Although the onset of Type II diabetes is usually after the age of 40, it is increasingly common among younger and younger obese individuals. Thus, the terms "childhood diabetes" and "adult-onset diabetes" no longer provide a useful distinction. Type II diabetes occurs because the cells of the body become resistant to insulin. Thus, glucose cannot enter cells, and glucose and insulin accumulate in the blood. During the early stages, this disorder can be controlled by diet; insulin treatment is usually not required. Indeed, most Type II diabetics have higher than average concentrations of circulating insulin because insulin receptors become insensitive to insulin. If the disorder is left uncontrolled, however, the pancreas can stop producing insulin, requiring the use of insulin treatment.

Individuals with either type of diabetes cannot move the surplus energy of a meal into storage; in other words, diabetics cannot make new fat. Consequently, extremely high levels of glucose remain in the blood, a condition that is commonly known as high blood sugar. Some of this excess glucose is excreted in the urine, giving the urine of diabetics a characteristic sweetness; hence the name of the disorder (*diabetes*, "run through"; *mellitus*, "sweet"). The brain does not require insulin to utilize glucose, and can function normally in the absence of insulin. All other cells in the body, however, require insulin to use glucose. Glucose dissolves easily in water, but not in oil; thus, it cannot easily move across cell membranes. Special protein receptors transport glucose into non-neural cells, and these receptors function only if the insulin receptors associated with them are activated. Brain cells have glucose transporters that do not require insulin. The liver metabolizes essentially no glucose, but instead metabolizes fats. The liver can also utilize fructose and alcohol as energy sources, but few of us can live on honey and beer (at least, not for long). Diabetic individuals become **hyperphagic**—that is, they eat more than normal individuals—because the rest of their cells are being starved. (This observation suggests that the detectors mediating hunger are not located *exclusively* in the brain, as we will see below.) Despite their enormous increase in food intake, Type I diabetics still cannot store fat in the absence of insulin. Most cells can use fatty acids to produce energy, although brain cells cannot. Thus, diabetic animals fed a very high-fat diet stop being hyperphagic because the energy requirements of their non-neural tissues can be met in this way (Friedman, 1978; Tepper and Friedman, 1991).

In contrast to diabetes mellitus, a condition exists in which individuals produce too much insulin, called hyperinsulinemia (Friedman, 1990). Hyperinsulinemia results in low blood sugar, and it usually leads to marked obesity because individuals with this condition are always hungry and eat frequently. Do these hyperinsulinemic individuals have high levels of insulin *because* they are obese and somewhat insulin-resistant, or are they obese because they have high levels of insulin? It should be emphasized here that not all obese people are hyperinsulinemic. There is no straightforward answer to this question in humans, and the issue remains unresolved. Certainly there is evidence that some people may be overly responsive to the cephalic phase of insulin release and become obese because they actually experience greater "hunger" at mealtime than individuals with normal cephalic insulin responsiveness. The appetite of such patients is huge, and they eat larger and longer meals than individuals with a normal cephalic insulin response. There also may be hyper-responsiveness to the GI phase of insulin release in some obese patients. Problems with the mobilization of fat stores, or with the monitoring of adiposity signals or blood glucose levels, may also be involved in obesity (Friedman, 1990; Seeley and Woods, 2003). The obese person or animal may not be able to mobilize normally the fat stores already present, so these individuals must eat frequently to meet their normal energy demands, and thus continue to deposit more and more fat.

Maintaining a Normal Energy Balance

Many hormones are involved in maintaining energy balance, and presumably, some of these hormones affect the behavioral systems that are important in maintaining energetic homeostasis. However, despite considerable progress during the past decade, the precise role that hormones play in mediating food intake and body mass has not been completely identified. Our inability to link hormones to ingestive behaviors is due to a lack of consensus regarding the stimuli eliciting hunger, as well as to the discovery of a seemingly endless supply of new peptides involved in the regulation of food intake.

All of the changes in metabolic fuels, hormones, and neuropeptides that occur as an animal shifts from a well-fed to a fasting state are correlated with one another. Which aspect(s) of these changes serve as signals to the neural mechanisms that control energy balance? Do one or all of these factors elicit hunger or control ingestion? Recent work has provided a detailed model of regulation of food intake in mammals, and has confirmed that several overlapping, redundant systems are involved in the regulation of food intake and energy balance.

One way to make sense of all this complexity is to organize the signals involved into two categories: primary sensory signals and secondary mediators or modulators. Primary sensory signals are those biochemical changes that result when fuels are metabolized to generate cellular energy. For example, changes in glucose metabolism might serve to generate the primary sensory stimulus that controls insulin secretion. Other metabolic sensory stimuli might include changes in the availability of free fatty acids. Some investigators believe that these primary sensory stimuli control hormone and neuropeptide secretion, and may also control food intake. Secondary modulators are the hormones and neuropeptides that affect food intake. Secondary modulators are usually considered to be "downstream" from the primary sensory stimuli. In other words, the secretion of hormones and neuropeptides is controlled by the primary sensory stimuli. Hormones and neuropeptides may have direct effects on mechanisms that control food intake, or they may have indirect effects on food intake via their effects on energy metabolism. Another way to organize the endocrine signals regulating food intake is to categorize them as either *peripheral* (originating outside the nervous system) or *central* (originating inside the nervous system).

Control of Food Intake

What regulates food intake? We eat because of a variety of external and internal cues, many of them learned. How much food, what kind of food, when we start a meal, and when we stop eating are all important questions when considering the control of food intake. Other extrinsic factors, such as who is eating with us, where we are eating, time of day, and stressors we are experiencing can also affect food intake.

Eating is an extraordinarily complex process that involves several intrinsic inputs, including the amount of fat stored in the body, the levels of glycogen stored in the liver, the biochemical qualities of the food being digested, neural and endocrine signals from the gut, and even the perceived pleasantness of the food (Morley et al., 1985a). Extrinsic factors, such as food availability and, particularly among humans, psychological and cultural influences, also regulate food intake. All of these intrinsic and extrinsic factors are integrated in the central nervous system to mediate feeding behavior.

The difficulty of inducing hunger experimentally stands in sharp contrast to the ease with which thirst can be induced, and has prompted the hypothesis that no single stimulus or set of stimuli exists for hunger (Stricker, 1984). **Hunger**, a strong motivation to seek out and ingest food, can thus be broadly defined as the psychological state experienced by individuals as satiety from a previous meal wanes. Somehow the body monitors long-term energy stores as well as energy utilization to regulate food intake. The endocrine system is central to this process.

Peripheral Signals

How does the brain maintain body mass and energy stores and monitor incoming and expended energy? Two main hypotheses emerged about 50 years ago. One suggested that the hypothalamus monitors the storage and use of triglycerides (the "lipostat" hypothesis) (Kennedy, 1953). The competing hypothesis suggested that the hypothalamus monitors the storage and use of glucose (the "glucostat" hypothesis) (Mayer, 1955). Considerable debate has ensued over the past half century over which model best explains the regulation of energy balance. As is often the case, neither hypothesis is completely correct or incorrect. Rather than arguing the merits of each hypothesis, it may be more useful to determine how components of each system (fat and glucose monitoring) are integrated to monitor energy resources and control ingestive behavior (Seeley and Woods, 2003).

Leptin, as you will recall from Chapter 1, is an **adipokine** hormone (produced by the adipose cells). It circulates in concentrations that are proportional to the total amount of fat in the body. When stored fat is being used for energy, blood levels of leptin fall faster than the levels of fat being metabolized; this rapid reduction in circulating leptin could signal a negative energy balance (Schwartz et al., 2000). Since the recognition that leptin was the protein encoded by the *ob* gene in the mid-1990s, rapid progress has been made in understanding the signals associated with energy balance.

Leptin receptors are located in several peripheral and brain regions; the arcuate nuclei of the hypothalamus, adjacent to the third ventricle, have the highest concentration of leptin receptors (Schwartz et al., 1996). Leptin is too large a peptide to cross the blood–brain barrier. It appears that an active transport mechanism ferries leptin into the brain. Thus, elevated leptin levels signal the hypothalamus that fat stores are increasing, which inhibits eating. In contrast, low leptin levels inform the hypothalamus of reduced fat stores, which stimulates eating (Seeley and Woods, 2003) (Figure 9.18).

Another potential adiposity signal is insulin, and recently, increased attention has focused on its role in the mediation of feeding behavior. There is an obvious association between insulin release and meal termination; hunger ensues when insulin levels drop at the end of the postabsorptive phase (Strubbe et al., 1977). Treatment of rats with streptozotocin, a drug that selectively destroys the pancreatic β-cells and induces diabetes, also causes long-term hyperphagia, demonstrating that lack of insulin can interfere with satiety. Presumably, insulin induces satiety indirectly, by its facilitation of the use and storage of ingested carbohydrates (Stricker, 1984), and via feedback mechanisms involving other hormones as well as the autonomic nervous system (Steffens et al., 1990). It is also possible that the oxidation of metabolic fuels itself generates some kind of signal that controls food intake; thus, the storage and mobilization of fat may indirectly affect food intake by changing the rate of oxidation of metabolic fuels. Levels of fatty acids and glucose in the blood are monitored by the liver; when insufficient levels of either fuel are available in the blood for oxidation, feeding might be induced (Friedman, 1990).

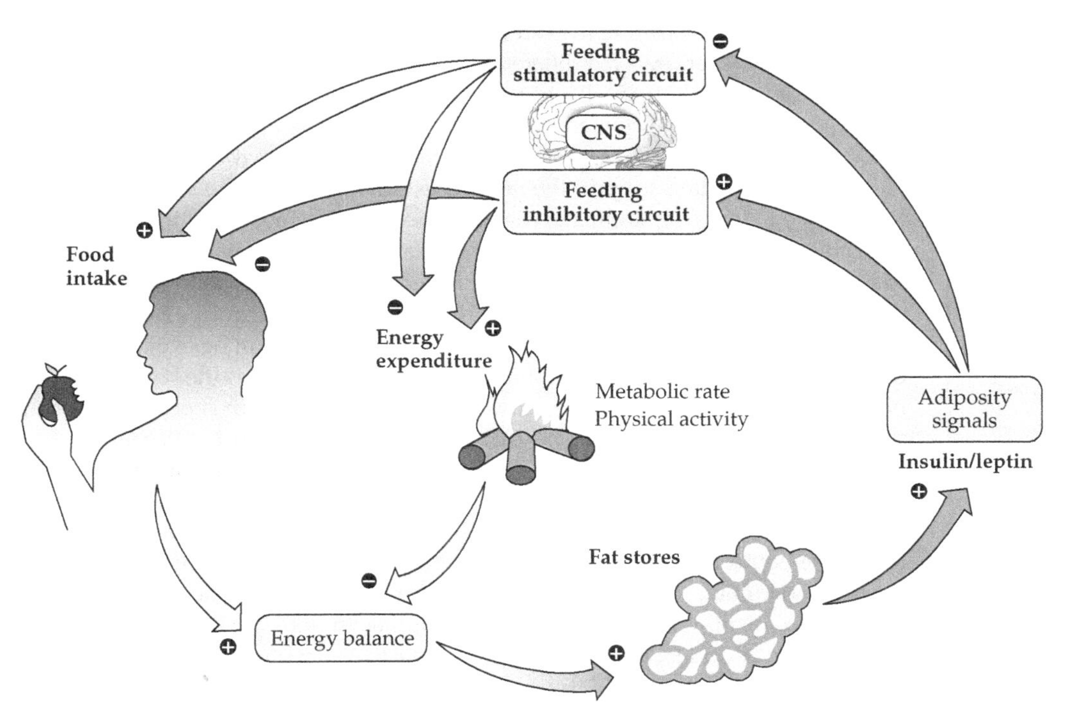

9.18 Leptin and insulin act as peripheral signals. (1) Leptin and insulin signals are secreted in proportion to the levels of fat stored in the body. (2) When elevated, these two hormones act on neuronal circuits in the hypothalamus. Leptin and insulin inhibit the feeding stimulatory circuits that curb energy use and stimulate eating, but activate the feeding inhibitory circuits that suppress eating and promote energy use. (3) Relatively low leptin and insulin levels during fasting have the opposite effect. (4) Eating generates neural and endocrine satiety signals in the hindbrain and gut. After Schwartz et al., 2000.

It has also been proposed that insulin signals the central nervous system directly about peripheral levels of metabolic fuels via the cerebrospinal fluid (Woods and Porte, 1983). Insulin levels increase in the cerebrospinal fluid of baboons immediately after a meal or an intravenous glucose injection (Woods et al., 1981), and intracerebroventricular infusions of insulin reduce food intake and body mass in baboons (Woods et al., 1979). Several researchers have suggested that tonic levels of central insulin (i.e., insulin in the brain) serve as an important signal to keep body mass and food intake within a healthy range. Thus, obesity might arise from impaired sensitivity to this central insulin signal (Schwartz et al., 1992; Woods et al., 1996). The evidence for this hypothesis is that plasma insulin concentrations are often correlated with body fat content.

It is difficult to determine whether insulin has direct effects on the mechanisms that control food intake or whether it changes food intake through indirect effects on metabolic fuel availability and oxidation. One way that this issue has been addressed is by dissociating the direct and indirect effects of insulin by measuring food intake in diabetic rats fed different diets (Friedman, 1978; Tepper and Friedman, 1991; Friedman et al., 1985). Diabetic rats, of course, have reduced plasma insulin concentrations, and they generally eat more food than nondiabetic rats. Untreated diabetic rats also tend to eat more food than diabetic rats treated with insulin (considered alone, these data support the notion that insulin acts in the brain to decrease food intake). However, untreated diabetic rats ate more food than nondiabetic rats when they were fed a low-fat, high-carbohydrate diet, but not when they were fed a high-fat, low-carbohydrate diet. Thus, insulin concentrations were not the critical determinant of food intake, because the diabetic rats in both groups had equivalent low levels of insulin. Their different levels of food intake can be explained by the differences in metabolic fuel availability in their diets. Insulin is the hormone that is necessary for peripheral glucose uptake, and therefore, insulin is necessary for the utilization of carbohydrate fuels. The diabetic rats could readily oxidize fats, but not carbohydrates. Consequently, the diabetic rats on the low-fat, high-carbohydrate diet may have been stimulated to overeat by their inability to oxidize the type of fuels they were ingesting. In contrast, the diabetic rats on a high-fat, low-carbohydrate diet were better able to oxidize the fuels they were ingesting, and thus did not overeat (Friedman, 1978; Friedman et al., 1985). It is true that when pharmacological doses of insulin are given, metabolic changes in both the body and the brain occur. However, antibodies to insulin (which inactivate insulin) infused into the brain cause animals to eat more with no effects on metabolism (McGowan et al., 1990). These results imply direct, rather than indirect, effects of insulin. Infusion of insulin directly into the cerebral ventricles of the brain decreases food intake and causes nonhuman animals to maintain their body mass at lower levels than before insulin treatment (Chavez et al., 1996).

Although insulin concentrations vary substantially before and after a meal, as well as throughout the day, insulin secretion is also affected by the amount of stored fat, as we saw above. That is, obese people and animals have relatively high circulating insulin levels, whereas lean people and animals have relatively low circulating insulin levels. Thus, insulin serves as an indicator of fat stores (Schwartz et al., 2000). Furthermore, insulin receptors have been located in the brain, especially in the arcuate nuclei (Baskin et al., 1994). Recall that the brain does not require insulin for glucose to enter its cells. Why, then, would insulin receptors be located in the arcuate nuclei? Apparently, these insulin receptors monitor metabolic fuels; namely, glucose. Leptin- and insulin-receptive neurons in the arcuate nuclei integrate this information to affect energy balance by acting on specific hypothalamic circuits (Seeley and Woods, 2003).

Central Signals: The Role of the Hypothalamus

How does activation of the insulin and leptin receptors in the arcuate nuclei affect food intake? It was believed for over two decades that mammalian feeding was regulated in a simple way, with two separate centers in the brain controlling hunger and satiety, respectively. A "hunger center" that facilitated eating was hypothesized to reside in the lateral hypothalamic area (LHA) because bilateral destruction of this brain region led to nearly complete **anorexia** (loss of appetite) and subsequent **aphagia** (absence of feeding behavior) in cats and rats (Anand and Brobeck, 1951). Similarly, a "satiety center" that inhibited feeding was hypothesized to be located in the ventromedial hypothalamus (VMH) because lesions of this brain area resulted in hyperphagia and enormous, rapid weight gain (Hetherington and Ranson, 1940). This dual-center theory of the control of feeding was very pervasive in shaping the research and theoretical framework in which experimental outcomes were interpreted (Stellar, 1954). The LHA and the VMH became the key components of putative regulatory systems that monitored blood glucose (the "glucostat" hypothesis) and triglyceride (the "lipostat" hypothesis) levels (Stricker, 1984).

Additional research displaced the dual-center theory of feeding regulation. Bilateral lesions of the LHA were shown to interfere with all motivated behaviors, not just feeding (Marshall et al., 1971). Lesions made anywhere along the dopaminergic tracts that pass through the LHA from the midbrain to the striatum while sparing the hypothalamic tissue produce behavioral deficits similar to those resulting from lesions of the LHA itself (Stricker et al., 1979). Other evidence suggests that the VMH is not a center for satiety, but part of a circuit including the PVN that regulates feeding behavior through the autonomic nervous system and its effects on energy metabolism, as well as mediating many other types of motivated behaviors. Lesions of the VMH suppress sympathetic nervous system activities while increasing parasympathetic activities, resulting in increased storage of fat while inhibiting lipolysis. This excessive storage of energy as fat leaves a shortage of energy to maintain the animal's daily metabolic processes. Consequently, VMH-lesioned animals develop voracious appetites, not because a satiety center has been damaged, but in order to maintain a steady delivery of metabolic fuels from the intestines. Eventually, as the animals become obese, the fat cells become insulin-resistant—that is, they no longer respond to insulin by taking up more glucose. Thus, food intake and body weight stabilize at a new set point. We now know that both of these lesions disrupt important neural circuits involved in food intake, which will be examined below.

As the dual-center control model has proved inadequate to explain feeding behavior, investigators have come to believe that the neuroendocrine system that controls food intake does so via both direct and indirect effects on energy balance. The current model of these direct effects is known as the *metabolic hypothesis*. Glucose, free fatty acids, and ketone bodies are metabolic fuels that can be oxidized to make usable energy for cellular processes. Food intake is responsive to changes in the oxidation of metabolic fuels. In order to study the metabolic signals that control ingestive behaviors, pharmacological agents have been given to animals to block oxidation of specific metabolic fuels. For example, 2-deoxy-D-glucose (2DG) or 5-thio-glucose (5TG) treatment blocks glucose oxidation, whereas methyl palmoxirate (MP) or mercaptoacetate (MA) treatment blocks free fatty acid oxidation. Treatment of satiated rats, dogs, cats, monkeys, and humans with these metabolic inhibitors causes rapid increases in food intake (Friedman and Tordoff, 1986; Ritter and Taylor, 1990; Ritter et al., 1992).

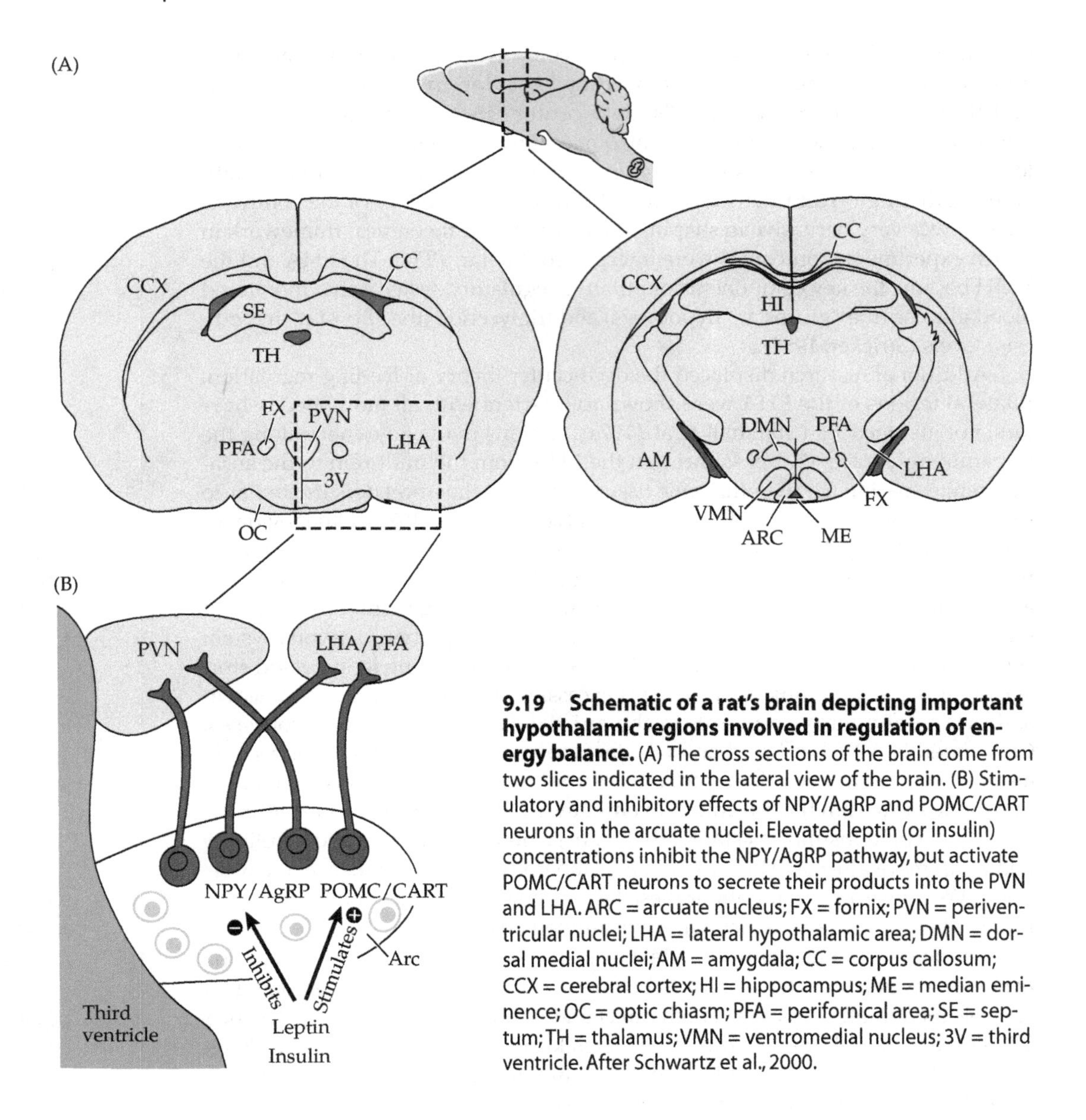

9.19 Schematic of a rat's brain depicting important hypothalamic regions involved in regulation of energy balance. (A) The cross sections of the brain come from two slices indicated in the lateral view of the brain. (B) Stimulatory and inhibitory effects of NPY/AgRP and POMC/CART neurons in the arcuate nuclei. Elevated leptin (or insulin) concentrations inhibit the NPY/AgRP pathway, but activate POMC/CART neurons to secrete their products into the PVN and LHA. ARC = arcuate nucleus; FX = fornix; PVN = periventricular nuclei; LHA = lateral hypothalamic area; DMN = dorsal medial nuclei; AM = amygdala; CC = corpus callosum; CCX = cerebral cortex; HI = hippocampus; ME = median eminence; OC = optic chiasm; PFA = perifornical area; SE = septum; TH = thalamus; VMN = ventromedial nucleus; 3V = third ventricle. After Schwartz et al., 2000.

Research on direct neuroendocrine control of food intake has focused on the hypothalamus (particularly the VMH and LHA) and the preoptic area for the past 50 years. Most of what is currently known about central nervous system control of food intake, however, comes from recent studies of the neuronal circuitry within the hypothalamus. The arcuate nucleui of the hypothalamus contain two opposing sets of neuronal circuitry: a feeding stimulatory circuit and a feeding inhibitory circuit (Figure 9.19). Both circuits send signals primarily to the PVN, but also to other nuclei of the hypothalamus, which then directly modulate feeding behavior. The feeding stimulatory and feeding inhibitory circuits are modulated by a number of peripheral hormonal signals that cross the blood–brain barrier. The feeding stimulatory circuit produces two neurotransmitters: neuropeptide Y (NPY) and agouti-related peptide (AgRP), both of which stimulate food intake (Schwartz et al., 2000). NPY directly signals the PVN to evoke feeding behavior, whereas AgRP indirectly promotes feeding by blocking appetite-inhibitory receptors in the PVN. The feeding inhibitory circuit also has two main signaling molecules: cocaine- and amphetamine-regulated transcript (CART) and pro-opiomelanocortin (POMC) and its products. During the well-fed state, elevated leptin (or insulin) concentrations from the periphery inhibit the NPY/AgRP pathway, and stimulate the POMC/CART neurons to secrete their products into the PVN and LHA (Schwartz et al., 2000). In contrast, during the fasting state, when leptin and insulin blood concentrations are relatively low the NPY/AgRP neurons are activated and the POMC/CART neurons are suppressed, leading to increased NPY and AgRP secretion and decreased POMC and CART secretion in the PVN, which increases food intake (Schwartz et al., 2000; Gale et al., 2004). This process can feed forward, resulting in obesity (Korner and Leibel, 2003).

Leptin and insulin both interact with receptors on neurons in the arcuate nuclei; thus, the arcuate nuclei are responsive to both circulating adiposity signals. Leptin infused into the arcuate nuclei acutely suppresses food intake (Satoh et al., 1997). Peptides that act as hormones elsewhere in the body serve as neuromodulators in the arcuate nuclei and have distinct effects on their target neurons. Some of these neuropeptides are *anabolic effectors*; that is, they increase appetite and food intake and suppress energy metabolism. Of course, other neuropeptides are *catabolic effectors*; they tend to decrease appetite and food intake and stimulate metabolism. A few of these anabolic and catabolic effector peptides are briefly described below (Table 9.3).

Central Anabolic Effectors: Peptides that Promote Food Intake (Orexigenic)

NEUROPEPTIDE Y Neuropeptide Y (NPY) is a potent activator of food intake in rats (Levine and Morley, 1984). This **orexigenic** peptide (from the Greek *orexis*, "appetite") appears to increase motivation to eat (Flood and Morley, 1991; Woods et al., 1998). Intracerebroventricular injections of NPY, or injections directly into the PVN, evoke marked food and water consumption in rats, both during the day (when these nocturnal animals rarely eat or drink) and at night (Kalra et al., 1991; Levine and Morley, 1984; Stanley and Leibowitz, 1985). NPY interacts with receptors in the PVN to cause hyperphagia. Targeted disruption of the NPY gene reduces hyperphagia and obesity in *ob/ob* mice (Erickson et al., 1996). NPY neurons project to the PVN, as well as to the LHA (see Figure 9.19). In the LHA, the NPY neurons synapse with neurons that secrete other peptides critical for body mass regulation: orexin and melanin-concentrating hormone (MCH) (Elias et al., 1998).

AGOUTI-RELATED PROTEIN The neurons that secrete NPY also secrete another orexigenic peptide with an odd name, agouti-related protein (AgRP). The path to the

TABLE 9.3 Neuropeptides involved in regulation of energy balance

Molecule	Regulation by adiposity signals
Orexigenic	
NPY	↓
AGRP	↓
MCH	↓
Orexin A and B (hypocretin 1 and 2)	↓
Galanin	?
Noradrenaline	?
Anorexigenic	
α-MSH	↑
CRH	↑
TRH	↑
CART	↑
IL-1β	↑
Urocortin	?
Glucagon-like peptide 1	?
Oxytocin	?
Neurotensin	?
Serotonin	?

Source: Schwartz et al., 2000.

discovery that this peptide regulates food intake was circuitous at best. It began with a mutant mouse strain that had a yellow (agouti) coat color and was obese. The yellow coat was discovered to be the result of the "agouti" protein, which, when cloned, was discovered to block MCH receptors in the skin. The blocked receptors led to less melanin and a lighter coat color. Obesity results when the agouti protein is overexpressed in the brain. As noted, this mutant protein led to the discovery of an endogenous agouti-related protein (AgRP) in the brain. AgRP is a potent stimulator of food intake; very small amounts infused into the ventricles of the brain stimulate food intake for up to 6 days (Lu et al., 2001). When circulating levels of leptin and insulin are low, the NPY/AgRP neurons in the arcuate nuclei are activated. Secretion of their peptide products activates the NPY/AgRP receptors in the PVN, which stimulate an increase in food intake (Figure 9.20).

MELANIN-CONCENTRATING HORMONE (MCH) Cell bodies of neurons that secrete MCH are located in the LHA. These neurons form synapses with several brain structures involved in motivation and movement, as well as with spinal neurons that regulate autonomic nervous system function. MCH, similar to MSH, received its name based on its actions on skin coloration in frogs. In the brains of mammals, it

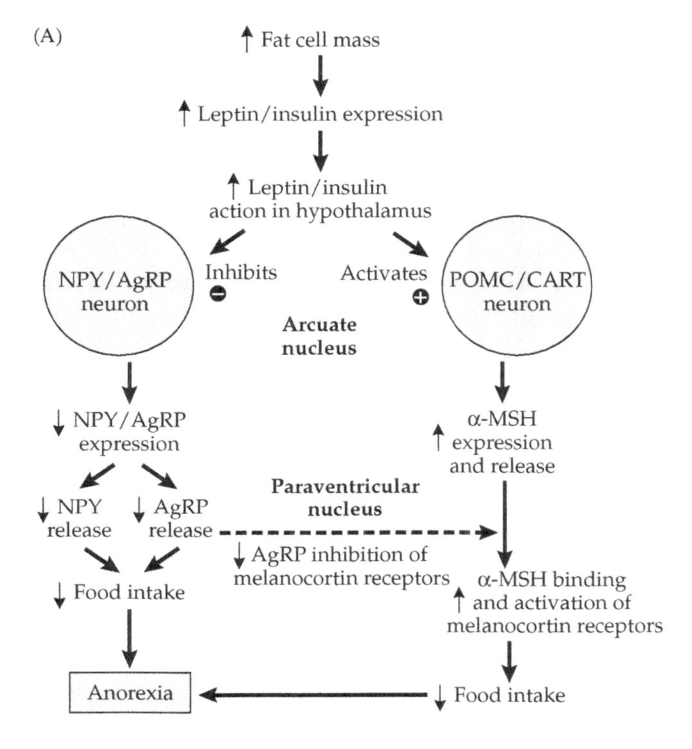

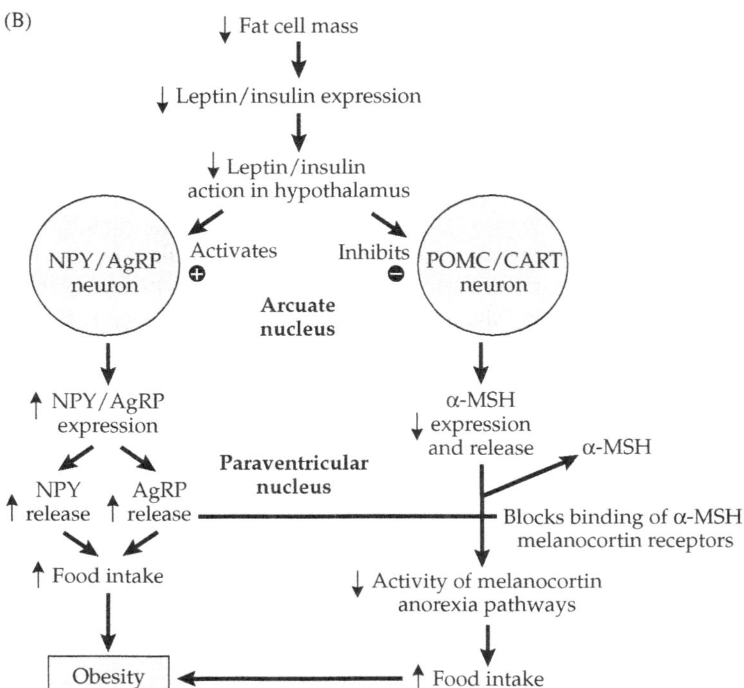

9.20 Current model of the role of the arcuate nuclei in monitoring metabolic fuels. During the well-fed state (A), the satiety signals leptin and insulin blood concentrations are relatively high. The concentrations activate the POMC/CART neurons and inhibit the NPY/AgRP neurons in the arcuate nucleus of the hypothalamus, leading to decreased NPY and AgRP secretion and increased POMC and CART secretion in the PVN, which in turn decrease food intake. Secretion of α-MSH (derived from POMC) in the PVN has an anorectic effect on food intake. During dieting and weight loss (B), the adiposity signals decrease. This stimulates NPY/AgRP neurons and inhibits POMC/CART neurons in the arcuate. NPY release from the PVN has an orexigenic effect, increasing food intake. Unchecked however, this process could feed forward, resulting in obesity. After Schwartz et al. 2000.

increases food intake and reduces metabolic rate. Injections of MCH into the lateral ventricles or various brain regions evoke feeding behavior (Dube et al., 1999).

OREXIN Another peptide produced by neurons whose cell bodies are located in the LHA is called orexin. There are two versions of orexin, A and B. Orexin A increases food intake, possibly by inhibiting sleep, whereas the physiological role of orexin B is not known at this time. Like MCH, orexin perfused into the lateral ventricles or other brain regions induces feeding behavior and increases metabolism (Dube et al., 1999). This neurohormone is called both orexin (by researchers studying its effects on food intake) and hypocretin (by sleep researchers). The axons of the orexin neurons project to most of the same targets as the MCH neurons (Seeley and Woods, 2003).

Central Catabolic Effectors: Peptides that Inhibit Food Intake (Anorexigenic)

MELANOCORTINS Recall from Chapter 2 that a large precursor molecule called proopiomelanocortin (POMC) is produced in the hypothalamus and elsewhere. This molecule can be cleaved to make several endocrine products that turn out to be important for regulation of food intake and metabolism: (1) the melanocortins, including α-MSH, β-MSH, and ACTH, and (2) the opioids. POMC-secreting neurons in the arcuate nuclei also express leptin and insulin receptors, and both leptin and insulin decrease POMC gene expression in those neurons (Cowley et al., 2001). Thus far, five melanocortin receptor subtypes have been identified; the melanocortin Type 3 and Type 4 receptors are common in the hypothalamus (Seeley and Woods, 2003). Targeted disruption of the gene for the MC4 receptor subtype results in obesity (Huszar et al., 1997). Thus, the actions of leptin and insulin to reduce food intake depend on their stimulation of POMC neurons to secrete α-MSH, which binds to MC4 receptors. Activation of MC4 receptors in the PVN has anorexic effects (Schwartz et al., 2000).

CART Individuals taking cocaine or methamphetamine become anorexic. It was discovered that a peptide termed cocaine- and amphetamine-regulated transcript (CART) is expressed in animals given these drugs (Douglass et al., 1995). Neurons in the arcuate nuclei that secrete POMC into the PVN and LHA also secrete CART. CART neurons also possess leptin receptors, and low circulating levels of leptin (and possibly insulin) reduce CART gene expression in the arcuate nuclei, whereas increasing leptin or insulin concentrations can increase CART gene expression. Axons from CART neurons also travel to other brain regions and to the regions of the spinal cord involved in regulation of the autonomic nervous system (Koylu et al., 1998). Activation of CART neurons increases metabolic rate. Thus, CART has a net catabolic effect on metabolism. Elevated circulating levels of leptin and insulin activate the POMC/CART neurons in the arcuate nuclei (see Figure 9.19). Secretion of these peptides activates the melanocortin receptors in the PVN, which decreases food intake.

Hindbrain and Brain Stem

Managing the intake and storage of energy is so critical to survival that several redundancies in these regulatory mechanisms have evolved. Although the necessary appetitive components of food intake are controlled in the hypothalamus, several parts of the system do not need the hypothalamus to maintain some components of energy balance. For example, brain regions in the posterior, or caudal, parts of the brain are able to act independently of the hypothalamus to control food consumption (Grill and Kaplan, 2002). The mechanisms that underlie feeding have been studied in animals that have undergone a surgical procedure to isolate the hypothalamus from the rest of the brain; these animals are termed *decerebrate* animals (Grill and Kaplan, 1990). Decerebrate rats display the ingestive motor program that underlies feeding behavior. These studies have demonstrated that even in the absence of information from the hypothalamus, decerebrate rats increase their food intake in response to 2DG and MP treatment, as well as in response to treatment with orexigenic peptides (Grill and Kaplan, 2002; Kaplan et al., 1993, 1998; Grill et al., 1997, 1998). These results show that the hypothalamus is not required for the control of feeding, and that metabolic and endocrine signals can act on the brain to control feeding.

Many data support the idea that areas in the caudal brain stem participate in the control of food intake (Grill et al., 1998; Grill and Kaplan, 2002). Treatment with metabolic inhibitors that affect food intake increases the activity of neurons in the area postrema and the nuclei of the solitary tract in the brain stem, as well as in forebrain regions such as the PVN (Ritter et al., 1992; Horn and Friedman, 1998). 2DG-induced increases in food intake are significantly attenuated in rats that receive varying amounts of damage to the area postrema and the medial nucleus of the solitary tract (Bird et al., 1983; Contreras et al., 1982; Hyde and Miselis, 1983; Ritter and Taylor, 1990). Infusions of 5TG into the fourth ventricle (in the brain stem), but not into the third ventricle (in the hypothalamic area), caused rats to increase their food consumption. Based on these findings, it has been hypothesized that detectors of glucose availability that control food intake may reside in the caudal brain stem.

However, there are also detectors of metabolic fuel availability and oxidation in peripheral tissues outside of the brain. Their presence was demonstrated by showing that metabolic inhibitors could increase food intake in animals in which the abdominal vagus nerve had been cut. Cutting the vagus nerve severs the connection between the brain and the internal organs. MP-induced, but not 2DG-induced, increases in food intake were abolished in vagotomized animals (Langhans and Scharrer, 1987; Ritter and Taylor, 1989, 1990). One neuroanatomical model of the pathways that regulate meal size is depicted in Figure 9.21.

Protein Hormones that Stop Food Intake

Insulin and leptin levels can regulate long-term patterns of food intake. Individuals can maintain their body mass within a narrow range for days, weeks,

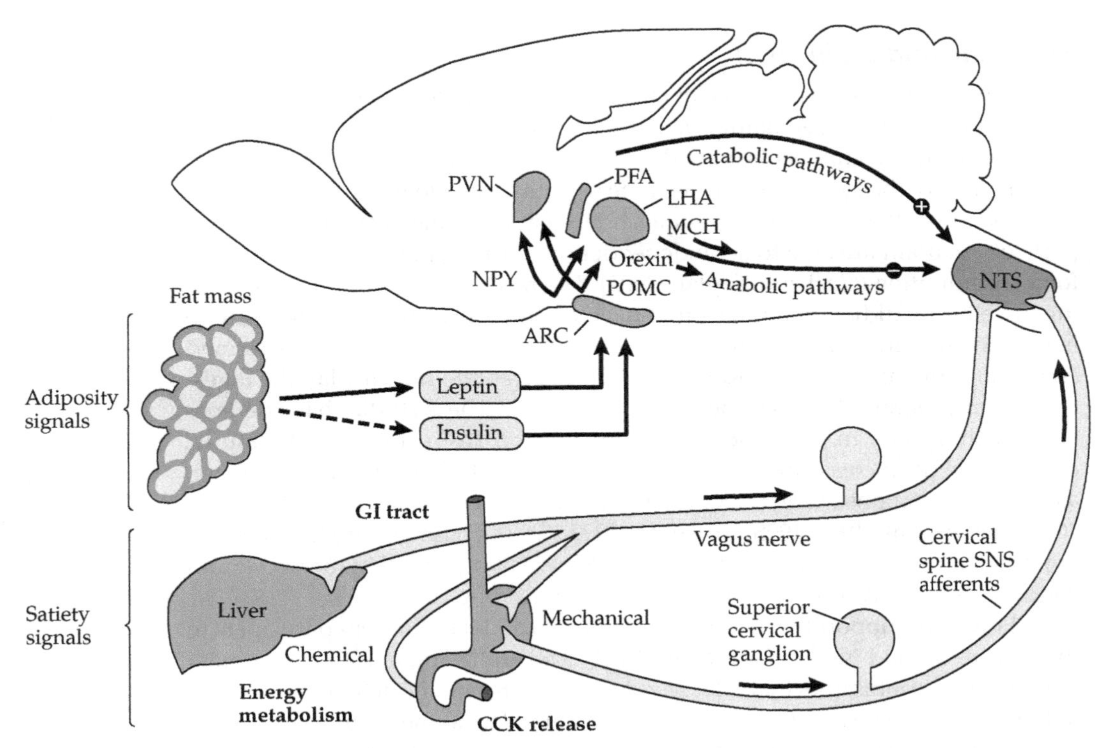

9.21 Neuroanatomical model showing the various pathways through which elevated adiposity signals (leptin and insulin) communicate with central autonomic circuits mediating food intake. Leptin and insulin appear to stimulate a catabolic pathway (i.e., POMC/CART neurons) and repress an anabolic pathway (i.e., NPY/AgRP neurons), both of which begin in the arcuate nuclei (ARC) and project to the PVN and LHA. Afferent information about satiety from the liver and from gut peptides such as CCK travels to the brain via the vagus nerve and sympathetic fibers of the nucleus of the solitary tract (NTS). Integration and consolidation of this information leads to termination of a meal. In contrast, reduced adiposity signals (e.g., during restricted caloric intake) may lead to increases in meal size by blocking brain stem responses to satiety signals. Leptin and insulin interact with hindbrain satiety circuits to regulate meal size. This phenomenon accounts for the common experience of following a large meal with a much smaller one to maintain energy balance. PFA = perifornical area. After Schwartz et al., 2000.

or even years. But what controls the size of individual meals? In general, a large meal is followed by a small meal, so there must be some sort of metabolic memory, but what factors regulate meal size? Because of the relatively long delay between when nutrients leave the gut and when they begin to be stored or used, factors other than insulin and leptin must be monitored to stop feeding behavior during a given meal.

Certainly neural signals of satiety exist. In the blowfly (*Phormia regina*), for example, feeding stops when sufficient food is ingested to overfill the crop sac. The overflow distends the foregut, which activates neural stretch receptors that signal the brain to stop feeding. Cutting the recurrent nerve between the foregut

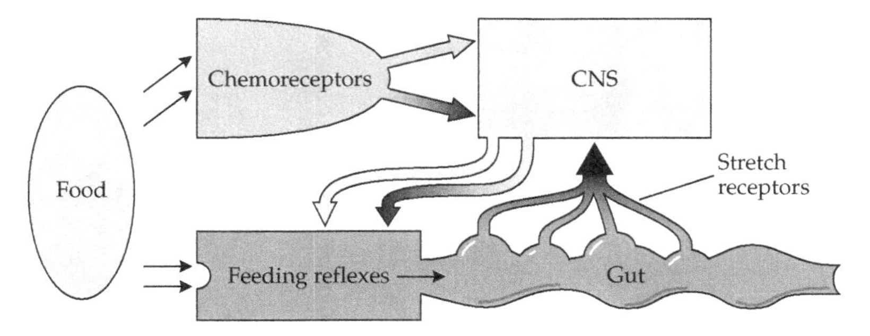

9.22 Neural control of feeding in the blowfly. Satiety in the blowfly is mediated via a simple negative feedback loop. When sufficient food is ingested to overfill the blowfly's crop sac, the overflow distends the foregut and activates neural stretch receptors that signal the brain to stop feeding. If the recurrent nerve between the foregut and the brain is severed, blowflies will continue to eat until they literally burst. After Rosenzweig and Leiman, 1989.

and the brain results in blowflies that continue to eat until they literally burst (Dethier, 1976). Satiety is thus mediated via a simple negative feedback loop (Figure 9.22). Similarly, neural signals of stomach distension in humans (Cannon, 1929; Thompson, 1980), other primates (Moran and McHugh, 1979, 1982), rats (Davis and Campbell, 1973), and dogs (Janowitz and Grossman, 1949) appear to inhibit feeding. The rate of gastric (stomach) emptying may also affect feeding behavior. A slow rate of gastric emptying would inhibit feeding for a longer time than a fast gastric contraction. In humans, a high-fat or high-protein meal leaves the stomach more slowly than a high-carbohydrate meal. Consequently, you actually may become hungry sooner after eating carbohydrate-rich Chinese food than after a rich, high-fat meal because of the faster rate of gastric emptying.

More to the point of this chapter, there are endocrine signals that stop feeding behavior. To demonstrate this point, an extra stomach and intestines can be transplanted into a rat. In the absence of neural connections, infusion of a liquid diet into the extra stomach results in a corresponding reduction in feeding behavior, even when the original stomach is empty. This observation suggests that some blood-borne product, possibly a hormone, is secreted in response to gastric distension by food (Koopmans, 1983). Filling the extra stomach with water does not reduce food intake. Further evidence of a blood-borne factor mediating stomach distension signals was obtained from hungry rats that ate much less after receiving blood transfusions from rats that had recently been fed (Davis et al., 1969). People who have had their stomachs surgically removed still report experiencing the sensation of hunger, but they usually reduce their caloric intake nonetheless (Mills and Stunkard, 1976).

Possibly the most salient example of the endocrine regulation of satiety is provided by a sea slug of the genus *Pleurobranchea* (Figure 9.23). This mollusk is a voracious cannibal that appears to live by one simple rule: Eat it. Anything less than about one-third its size that wanders too close to a *Pleurobranchea* is devoured.

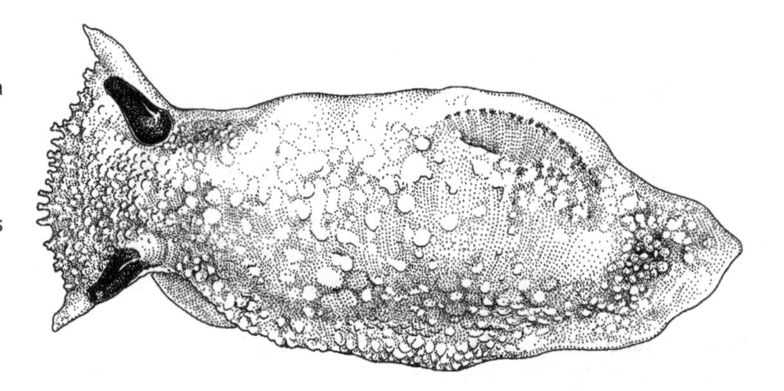

9.23 Feeding in *Pleurobranchea* is tightly regulated by hormones. Normally, this sea slug will eat anything less than about one-third of its size. However, *Pleurobranchea* avoids eating its own eggs because the peptide hormone that stimulates egg laying in this species also inhibits ingestive behavior.

Those of us who yearn for more simple lives can admire the straightforward rules by which this creature conducts itself, but simple lives can also become complicated. *Pleurobranchea* eggs are small and are therefore, by definition, food. Obviously, any species that incorporates its offspring into its menu will experience a decrease in reproductive success. However, the peptide hormone that stimulates egg laying in this species also stops feeding behavior (Mpitos and Davis, 1973). Treatment with the egg-laying hormone stops feeding behavior within 15 minutes and shortly thereafter causes oviposition.

Simple yet elegant regulatory systems such as that found in *Pleurobranchea* are uncommon among vertebrates. As we have seen, however, there are several complex, yet elegant, levels of endocrine mechanisms, usually involving peptide hormones, controlling feeding behavior in vertebrates. Similarly, there are endocrine mechanisms that signal individuals to stop eating.

CHOLECYSTOKININ Cholecystokinin (CCK) is one hormonal signal that invokes satiety. CCK is a gastrointestinal peptide hormone that is released during feeding to aid with digestion; it was named for its ability to promote the contraction of gallbladder muscle. Administration of CCK to hungry rats, mice, Syrian hamsters, chickens, rabbits, sheep, cats, dogs, or humans decreases their food intake (Morley et al., 1985b). CCK does not suppress water intake in thirsty animals, indicating that its satiating effect is specific to food intake. When CCK is released during a meal, it binds to peripheral receptors that signal the brain, via the vagus nerve, to stop feeding behavior. The effects of CCK on ingestive behaviors can be blocked by cutting the abdominal section of the vagus nerve (Bloom and Polak, 1981). Blood concentrations of CCK diminish only slightly as a meal is digested, suggesting that blood-borne CCK does not mediate satiety; however, CCK concentrations around the vagus nerve vary considerably during the course of a meal, suggesting that local release of CCK mediates satiety via the vagus nerve. It has been proposed that as these satiety signals stop, hunger is experienced again. Although this model is an extreme oversimplification of the endocrine control of vertebrate feeding, it suggests that the feedback principles displayed in the blowfly and *Pleurobranchea* could theoretically form the basis of satiety in vertebrates as well.

The exact nature of the inhibitory effects of CCK on feeding and the physiological role of CCK in normal satiety remain controversial. The satiety observed after CCK treatment has sometimes been attributed to the aversive or toxic effects of this compound: in other words, individuals stop eating after CCK treatment because it makes them sick. The aversive effects of CCK have been difficult to sort out. In order to determine whether CCK is aversive in humans, one can simply ask them. In some studies, CCK administration to people at concentrations sufficient to suppress eating caused mild to moderate nausea (Morley et al., 1985b). Nausea, abdominal bloating, cramping, and other symptoms of gastrointestinal distress were reported when CCK was given at doses higher than necessary to suppress food intake. In another study, however, CCK was reported to suppress hunger without nausea or other aversive effects (Stacher et al., 1978). The participants fasted overnight and were given either saline or one of several different doses of CCK the following morning. They were then seated in a dining room, and a large meal was prepared: "A piece of bread and butter was prepared in the subjects' presence. Fresh, good-smelling chives were cut and put on the bread. Following this, butter was melted in a pan, and bacon and two beaten eggs were added and fried. The aroma and sounds of the food being prepared was aimed to stimulate the subjects' appetite, and in fact, stimulated at least the experimenters' appetite to a rather large degree" (Stacher et al., 1978, p. 327). The participants were asked to rate their hunger throughout the preparation of the meal. Those who were injected with saline reported the most "voracious" appetites as the meal was being prepared. There was a dose-dependent relationship between CCK and reported intensity of hunger; higher doses of CCK were associated with lower levels of reported hunger. Note, however, that CCK was not paired with food intake in this study.

In the absence of language, it is even more difficult to ascertain whether CCK inhibits feeding via a true satiating mechanism or through secondary aversive effects. We can simply ask people if CCK makes them feel queasy, or observe them to see if they vomit after CCK treatment, to discover whether CCK causes nausea. But we cannot ask a rat if it is nauseated, and because rats do not vomit, it is difficult to discover whether something is aversive to them. However, learned aversions can be readily established by pairing flavors, such as saccharin, with substances known to produce illness, such as lithium chloride, and presenting the mixture to a rat. Upon subsequent presentations of a saccharin-flavored substance, the animal avoids it, demonstrating that it has learned that the sweet taste predicted the illness induced by lithium chloride. This learned aversion paradigm has been employed to explore the aversive effects of CCK. In some studies, CCK paired with saccharin did not cause rats to develop an aversion to saccharin (Gibbs et al., 1973). Other studies, however, have reported an aversive reaction to CCK in rats (Deutsch, 1982). Still other studies have shown that CCK antagonists increase the size of meals consumed after food deprivation, suggesting a physiological role for CCK in normal satiety (Moran et al., 1992). Peripheral treatment with CCK at doses sufficient to inhibit food intake caused synthesis of c-Fos, a marker of neuronal activation, in the brain stem, the nucleus of the solitary tract, and the dorsal vagal nucleus (Zittel et al., 1999). There are two CCK receptor subtypes, A and B. Rats lacking functional CCK-A receptors are hyperphagic, diabetic, and obese (Schwartz et al., 1999). This finding is species-specific, however, as CCK-A receptor knockout mice display normal body mass (Kopin et al., 1999). Further studies are required to understand the precise role of CCK in suppressing food intake.

BOMBESIN Another peptide that may mediate feeding behavior is bombesin, so named because it was originally isolated from the skin of the European frog *Bombina bombina* (Anastasi et al., 1971). This small protein, consisting of 14 amino acids, is distributed throughout the mammalian gastrointestinal and nervous systems (M. Brown et al., 1978), Experimental injections of bombesin reduce food intake in rats (Gibbs et al., 1979). The mechanism by which this peripheral bombesin treatment affects feeding is unknown; in contrast to CCK, cutting of the vagus nerve does not interfere with the satiating effects of bombesin (Smith et al., 1981a). The satiating effects of bombesin and CCK are additive, further suggesting that they have different mechanisms of action (Stein and Woods, 1981). Intracerebroventricular infusions of bombesin also reduce food intake in rats, indicating a central nervous system mechanism of action (Morley et al., 1985a). Although bombesin inhibits feeding, the extent to which it is important in the normal regulation of feeding remains to be determined (Hostetler et al., 1989).

CORTICOTROPIN-RELEASING HORMONE Corticotropin-releasing hormone (CRH) is another endocrine signal that stops feeding behavior. CRH rapidly reduces food intake and body mass after intracerebroventricular administration (Arase et al., 1988; Hotta et al., 1991; Rivest et al., 1989). A regulatory loop appears to exist between NPY, CRH, and the sympathetic nervous system. Discharges of sympathetic nerves to adipose tissues increased after intracerebroventricular injections of CRH and decreased after intracerebroventricular injections of NPY (Egawa et al., 1990), suggesting that these two neuromodulators have opposite effects on sympathetic activity (Heinrichs et al., 1992). The sympathetic nervous system could be a means of communication between fat depots and the brain; specific gut endocrine signals could influence the motivation to eat or stop eating, depending on energy stores, energy availability, and energy expenditures. Stressors also motivate eating in many individuals, and CRH dysregulation may be important in stress-evoked food intake (Box 9.2).

GLUCAGON-LIKE PEPTIDE 1 Glucagon-like peptide 1 (GLP-1) is produced both in the gut and in the brain. GLP-1 is released from the gut in response to food, and interacts with receptors in the brain stem, arcuate nuclei, and PVN (Small and Bloom, 2004). Peripheral administration of GLP-1 stimulates insulin secretion in humans. Administration of GLP-1 directly into the PVN of rats significantly suppresses food intake. In contrast, blocking GLP-1 receptors increases food intake (Turton et al., 1996).

Peripheral administration of GLP-1 reduces food intake in both humans and rats (Small and Bloom, 2004). In rats, GLP-1 treatment induces *c-fos* expression in

BOX 9.2 Comfort Food

When I am not feeling well, I crave a grilled cheese sandwich and tomato soup prepared in the specific manner that my mother used when I was a young boy. Food that makes people feel better is commonly called "comfort food." For example, many people eat ice cream or other high-fat/high-calorie foods, such as macaroni and cheese, when distressed. Recent work by Mary Dallman and her colleagues (Dallman et al., 2003; Pecoraro et al., 2004) suggest that when chronically stressed, rats (and perhaps people) crave high fat-food in an attempt to reduce anxiety. (Dallman et al., 2003). These mechanisms, the details of which have been worked out in rats, may explain some parts of the epidemic of obesity occurring in Western society.

This model proposes that glucocorticoids work differently in the long term than they do in the short term. Recall from Chapter 2 that stress causes CRH release from the hypothalamus, which stimulates ACTH release from the anterior pituitary, which stimulates the adrenal cortex to secrete glucocorticoids. Normally, high levels of glucocorticoids shut off CRH and ACTH through negative feedback, but when glucocorticoids are chronically present in the brain and body, those hormones maintain the stress response, instead of shutting it down, through a "feedforward" system (Dallman et al., 2003). CRH and glucocorticoids drive individuals to seek out pleasurable foods and direct the added calories to accumulate as abdominal fat.

In one study, Dallman and her colleagues simulated chronic stress by increasing the brain concentration of corticosterone. As corticosterone concentrations increased, the rats responded by drinking increasingly more sugar water, eating increasingly more lard, and gaining abdominal girth (Pecoraro et al., 2004). Although there was not a net increase in caloric intake, the types of calories consumed plus the high glucocorticoid values put the fat on the abdomen, where it increases the risk for cardiovascular disease and Type II diabetes in humans. Presumably, reducing stressors would alleviate the hypothalamic–pituitary–adrenal dysfunction of chronic stress and allow a better distribution of stored energy.

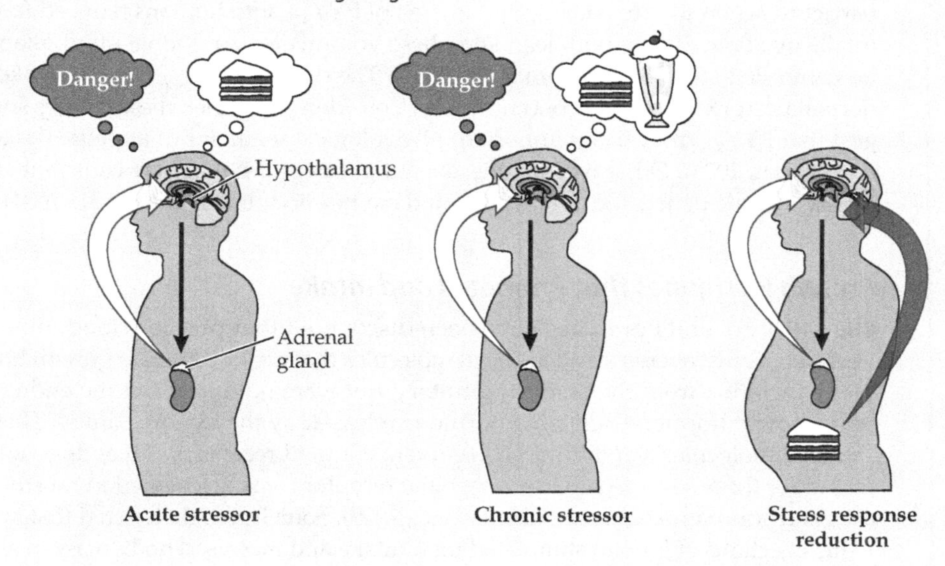

the brain stem. These findings, plus others, suggest that GLP-1 inhibits food intake in rodents by acting on the vagus nerve and the area postrema. The extent to which circulating GLP-1 mediates appetite in humans remains unspecified, although infusions of this peptide sufficient to attain normal postprandial concentrations reduce subsequent appetite and food intake (Flint et al., 2001). Additional research is required to see if drugs targeted for the GLP-1 receptors will have useful therapeutic value in controlling obesity.

ADIPONECTIN Adiponectin (ADP), like leptin, is a recently discovered adipokine hormone. ADP affects insulin actions in the periphery to facilitate glucose uptake in cells (Seeley et al., 2004). ADP enters the cerebrospinal fluid after intravenous injection, suggesting that it is actively transported into the brain (Qi et al., 2004). Intracerebroventricular infusion of ADP decreased body weight mainly by stimulating glucose use. ADP increased secretion of CRH by the hypothalamus, but did not act on other neuropeptide targets of leptin (Qi et al., 2004). ADP also induced distinct Fos immunoreactivity in brain regions associated with food intake. However, the mutant agouti mice described above did not respond to ADP or leptin, suggesting that melanocortin receptors may be a common target of these two peptides (Seeley et al., 2004). These results suggest that ADP has important central effects on energy balance (Qi et al., 2004).

PEPTIDE YY_{3-36} Peptide YY_{3-36} (PYY_{3-36}) is one of several recently discovered gut peptides that contribute to satiety (Wynne et al., 2004). In rodents, this peptide is released from the intestinal tract after the ingestion of food (Small and Bloom, 2004). Peptide YY_{3-36} decreases food intake by inhibiting hypothalamic NPY/AgRP neurons, which releases their inhibition of neighboring POMC/CART neurons (Gale et al., 2004). In common with leptin, this peptide crosses the blood–brain barrier to act on the arcuate nuclei. Infusion of PYY_{3-36} into humans reduced food intake by about 30% in both lean and obese volunteers in a double-blind, placebo-controlled study (Batterham et al., 2002). The dose of PYY_{3-36} infused evoked normal postprandial blood concentrations. Considered together, these studies suggest that PYY_{3-36} may be an important physiological regulator of appetite (Small and Bloom, 2004). Other recently discovered gut peptides that may contribute to satiety include pancreatic polypeptide and oxyntomodulin (Wynne et al., 2004).

Protein Hormones that Promote Food Intake

GHRELIN Several hormones have been discovered that promote food intake. Researchers discovered small artificial molecules that could stimulate growth hormone secretion from the anterior pituitary, but were not related to the endogenous growth hormone–releasing hormone released by the hypothalamus. These artificial molecules acted through G protein coupled receptors. An endogenous ligand for these so-called GH secretagogue receptors was discovered in rat stomachs and was named ghrelin (Kojima et al., 1999). Soon it was discovered that systemic injections of ghrelin stimulated food intake and increased body mass in rats

(Wren et al., 2001). Blood concentrations of ghrelin peak around the time of meal onset (Cummings et al., 2001). Intracerebroventricular injections of ghrelin strongly stimulated feeding in rats and increased body weight gain (Nakazato et al., 2001). Expression of ghrelin was discovered in a group of neurons adjacent to the third ventricle, between the dorsal, ventral, paraventricular, and arcuate hypothalamic nuclei (Cowley et al., 2001). These neurons project to key hypothalamic circuits, including those producing POMC, CRH, and especially NPY/AgRP. Ghrelin stimulated activity of NPY neurons in the arcuate nuclei and mimicked the effects of NPY in the PVN (Cowley et al., 2003).

As in Syrian hamsters, fasting increases appetitive behaviors such as foraging and hoarding, but not food intake, in Siberian hamsters (*Phodopus sungorus*) (Keen-Rhinehart and Bartness, 2005). Treatment with exogenous ghrelin mimics fasting to increase foraging and hoarding by Siberian hamsters, but unlike fasting, ghrelin also stimulates food intake in this species (Keen-Rhinehart and Bartness, 2005).

ENDORPHINS Endogenous opioids may also mediate food intake (Morley et al., 1983). Treatment of rats (Holtzman, 1975) and other species, including humans (Thompson et al., 1982), with the opioid antagonist naloxone reduces food intake. Apparently, naloxone reduces the hedonic value of food, making its consumption less "rewarding" (Morley et al., 1985a). Ingested sugars and oils are particularly salient cues for the release of endorphins and appear to provide calming and pain-reducing influences (Shide and Blass, 1989). Perhaps this is why we are more likely to reach for a candy bar than broccoli when we are stressed. Endogenous opioids may play a role in desire for "comfort food" (see Box 9.2).

The Role of the Liver

The liver may generate a signal for satiety. This organ is ideally situated to monitor changes in the concentrations of insulin and metabolic fuels (Granneman and Friedman, 1980). It is certainly important in the mediation of insulin-induced feeding, as elegantly demonstrated by Edward Stricker and his colleagues (Stricker et al., 1977). To understand their experiment fully, the use of metabolic fuels by the brain and liver will be briefly reviewed.

Two metabolic fuels can be used by the brain: glucose and ketone bodies. As described above, the brain primarily uses glucose, and insulin is not required to get energy into neurons. During prolonged starvation, the brain can function using ketone bodies as fuel. Fructose, like glucose, is a carbohydrate, but unlike glucose, it cannot easily cross the blood–brain barrier; consequently, fructose cannot be used as a fuel for the brain. The liver, on the other hand, can readily utilize fructose as a metabolic fuel, but it is unable to oxidize ketones and use them for energy.

Stricker and his co-workers exploited this disparity in fuel use between the brain and the liver to find out which organ mediated insulin-induced feeding. Rats were infused with (1) saline (controls), (2) fructose, to "feed" the liver but not the brain, or (3) β-hydroxybuterate, a ketone, to "feed" the brain but not the

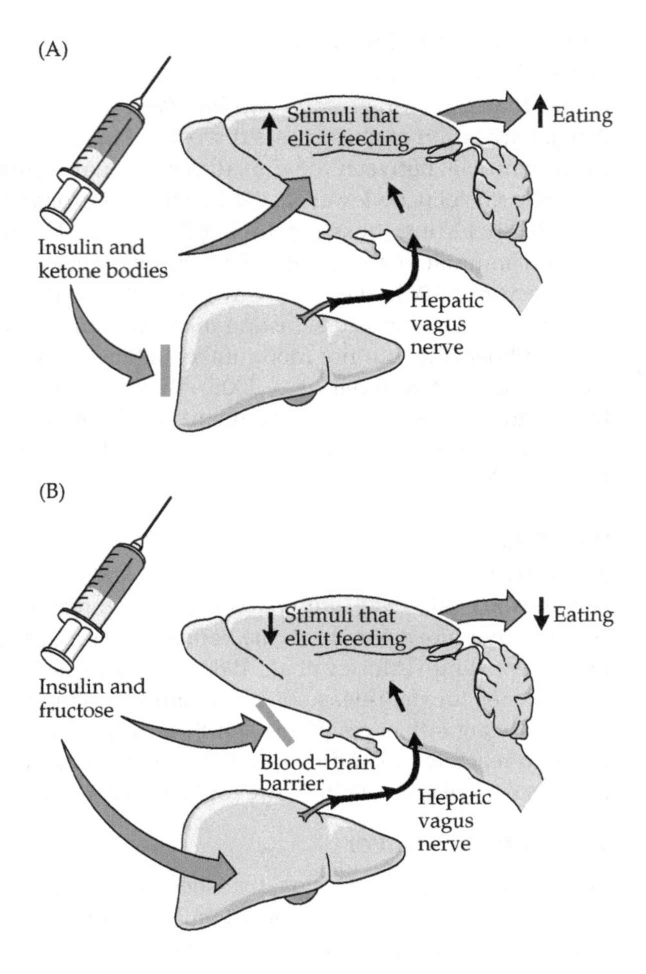

9.24 The role of the liver in insulin-induced feeding was determined by injecting animals with insulin, which lowers blood glucose levels and induces feeding, and then injecting them with either ketone bodies or fructose. (A) Ketones, which can be used as fuel by the brain but not by the liver, did not inhibit insulin-induced feeding. (B) Fructose, which can be used by the liver but not by the brain, inhibited insulin-induced feeding, indicating that the liver blocks insulin-induced feeding even if fuel levels in the brain are low. After Stricker et al., 1977.

liver (Figure 9.24; Friedman, 1990; Stricker et al., 1977). All three groups of rats were also treated with insulin. Insulin treatment of the control animals produced the expected reduction in blood levels of glucose and induced feeding behavior. Infusion of rats with β-hydroxybuterate did not affect insulin-induced feeding. However, infusions of fructose inhibited the onset of insulin-induced feeding, demonstrating that a "fed" liver, even in the presence of a "starved" brain, was sufficient to block insulin-induced feeding. These results indicate that the liver plays a role in mediating insulin-induced feeding, probably by monitoring its energy stores (glycogen) and communicating this information to the central nervous system via the hepatic branch of the vagus nerve (Friedman and Granneman, 1983; Sawchenko and Friedman, 1979; Steffens et al., 1990; Stricker et al., 1977).

Specific Hungers

What do we eat? In humans, much of what we consume is culturally determined: in other words, we learn what to eat. Many nonhuman animals also learn to eat foods that improve their health and to avoid foods that make them ill. For example, rats provided with a diet lacking thiamine (vitamin B_1) do not seek out foods rich in thiamine, but rather sample many foods, and over the course of several days begin to prefer foods containing thiamine. Humans are no better at detecting thiamine deficiencies than rats. In the 19th century, thousands of Asians died from beriberi, a degenerative disease of the nervous system resulting from thiamine deficiency. The cause of this outbreak of beriberi was the fashion of polishing rice before cooking. The hulls that were removed from the rice contained the thiamine. Once the practice of rice polishing was discontinued, beriberi became rare again. Because humans are unable to recognize thiamine in food, this vitamin has become a common food additive.

In contrast to learned preferences, some physiological requirements seem to elicit specific, unlearned hungers. For example, hungers for sodium and potassium appear to be innate (Milner and Zucker, 1965; Richter et al., 1938). However, even these hard-wired specific hungers can be modified by hormones. Most animals, even without prior experience, prefer sucrose-laced water over plain water. This preference is mediated by taste, because preferences also exist for water containing nonnutritive substances such as saccharin or Nutrasweet over plain water. An argument can be made that the intrinsic preference for sweet tastes is adaptive because it would increase intake of valuable energy resources. The yearning for chocolate, however, is not necessarily hard-wired at birth, but is an acquired taste.

Female mammals, including rats, mice, hamsters, monkeys, and human infants, consume greater quantities of sweet solutions than their male counterparts (Wade, 1976; Zucker et al., 1972). This sex difference is hormonally mediated; sex differences in taste preferences appear to be organized perinatally by androgens and also activated by sex steroid hormones in adulthood. Generally, estrogens stimulate sweet solution preferences; ovariectomy reduces the preference for sweet substances to the level of males (Wade, 1976). Castrated males show sweet solution preferences only slightly greater than those of gonadally intact males (Zucker, 1969).

Certainly, the taste for sodium and, to a lesser degree, potassium is innate in many species. Young animals that are sodium- or potassium-deprived seek out and prefer diets that contain these nutrients; these preferences are immediate and do not require feedback in terms of improvement of health, as in the case of thiamine. Several vertebrate species, including humans, show early and pervasive preferences for a slightly salty solution over plain water (Richter, 1936). One patient, a young boy of 3 or 4, was hospitalized because he was undergoing precocious puberty and experiencing development of secondary sex characters. He refused the bland hospital diet and died after a week in the hospital. An autopsy revealed that his adrenal glands were abnormal: the cells of the zona reticularis, which produce sex steroids, had invaded the zona glomerulosa, which normally produces aldosterone, the major sodium regulator. The parents reported that their son had consumed extremely salty foods and refused most foods that were not very salty at home. In fact, "salt" was one of the first words that he learned to say (Wilkins and Richter, 1940). Thus, failure of the adrenal glands to produce sufficient aldosterone can promote a powerful specific hunger for sodium.

Food Intake and Body Mass

As we have seen, the control of feeding, especially in humans, is a highly complex regulatory process involving both the endocrine and nervous systems. To complicate the mediation of food intake even further, gonadal steroid hormones also influence eating and subsequent body mass. Many animals also experience cyclic seasonal changes in body mass.

Estrogens and Progestins

After ovariectomy, a female rat increases her food intake and relatively quickly elevates her body mass about 20%–25% over that of a gonadally intact female. This hyperphagia eventually wanes, and her food intake stabilizes to maintain this new elevated body mass. Most of the increased body weight is the result of increased fat deposition (adiposity) (Leshner and Collier, 1973). Locomotor activity, as measured on a running wheel, is permanently suppressed by ovariectomy (Wang, 1923). In other words, ovariectomy leads to a fat rat. Ovariectomy also promotes obesity in other species, including cats, voles, and mice. Ovariectomy does not elevate food intake or body weight in other mammalian species; Syrian hamsters, for example, do not change body mass appreciably after ovariectomy. Bilateral ovariectomy in humans may or may not affect food intake and subsequent body mass. Menopause shifts fat distribution in humans from the female-like (pear-shaped) subcutaneous depots to a more male-like pattern (apple-shaped), with more fat deposited in the visceral depots.

Estrogens generally have catabolic effects, in contrast to androgens, which, as you know, are generally anabolic (see Chapter 2). Treatment of ovariectomized rats with estradiol benzoate reverses the increase in body mass by inducing a mild, transient hypophagia and a prolonged elevation in activity (Mook et al., 1972) (Figure 9.25). In contrast to the effects of estradiol, treatment of ovariectomized rats with progesterone does not prevent the increase in body mass; progesterone, in the absence of an ovary or estrogen, does not affect food intake, locomotor activity, body fat content, or body mass (Wade and Gray, 1979). In other words, ovariectomized female rats injected with only progesterone are as fat as untreated ovariectomized females. Treatment of an ovariectomized rat with both estradiol and progesterone also fails to prevent the increase in body mass. When gonadally intact females are given progesterone in relatively high doses (for example, 5 mg/day), they experience a weight gain resembling the weight gain observed after ovariectomy (Hervey and Hervey, 1967). Thus, it is reasonable to assume that progesterone blocks the catabolic effects of estradiol on food intake, adiposity, locomotor activity, and body mass. The effects of ovariectomy and progesterone treatment are virtually identical, but nonadditive; that is, both treatments cause the same physiological and behavioral outcomes.

If ovarian steroid hormones have these dynamic effects on food intake and body mass, then predictable changes in eating and body weight should accompany estrous cycles, pregnancy, and pseudopregnancy (Wade, 1986). Furthermore, females undergoing seasonal reproductive cycles should also display seasonal

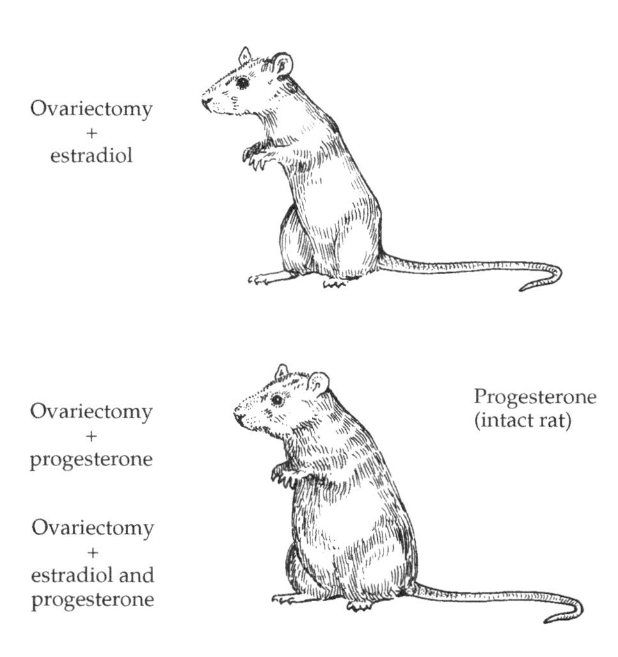

9.25 Effects of ovariectomy and progesterone treatment on body mass. Ovariectomized rats increase their food intake and increase their body mass by 20%–25%. Treatment with estradiol reverses these effects, but if estradiol is paired with progesterone, or if progesterone alone is administered, body mass is not reduced. Gonadally intact rats given progesterone increase their body mass in a manner similar to that seen in ovariectomized rats.

fluctuations in food intake and body mass. Energy balance does appear to oscillate throughout the estrous cycle. Recall from Chapter 6 that rat estrogen concentrations are highest during vaginal proestrus; eating and body mass are reduced immediately after proestrus (Wang, 1924). The low blood plasma estrogen concentrations during diestrus are associated with elevated food intake and body mass (Brobeck et al., 1947). Seasonally acyclic Syrian hamsters increase their body mass when days are short and blood levels of gonadal sex steroid hormones are low (Bartness and Wade, 1985); when the days grow longer and estrous cycles resume, hamsters often display fluctuations in body mass and locomotor activity, and perhaps food intake, similar to those of rats.

Correlations between ovarian steroid hormones and food intake have also been reported for primates. Food intake is higher during the luteal phase, when progesterone levels are high, than at other times during the menstrual cycle in both nonhuman primates (Czaja and Goy, 1975) and women (Dalvit, 1981). These changes in food intake are attributable to the fluctuations in blood levels of sex steroid hormones; ovariectomized rhesus monkeys show cyclic changes in food intake when injected with estradiol and progesterone in a pattern that mimics the hormonal profile of a normal menstrual cycle (Czaja and Goy, 1975). In women, food intake generally, and carbohydrate intake specifically, is higher during the 10-day period prior to menstruation (the luteal phase) than during the 10-day period immediately following the onset of menstrual bleeding (the follicular phase) (Dalvit-McPhillips, 1983; Wurtman and Wurtman, 1989). Carbohydrate metabolism may increase during the luteal phase, thus triggering an increase in physiological energy needs and subsequent elevated intake of this energy source (Pliner and Fleming, 1983; Wurtman and Wurtman, 1989).

As described in Chapter 6, blood plasma progesterone concentrations gradually increase throughout pregnancy; progesterone is also elevated during pseudopregnancy in rats. Food intake, fat storage, and overall body mass increase during both pregnancy and pseudopregnancy (Slonaker, 1924). Ewes entering their first estrous cycle of the breeding season do not display reproductive behavior; this first estrous cycle of the season is called a "silent heat." However, plasma levels of progesterone are high during the luteal phase of the silent heat, and food intake and body mass escalate accordingly during this time (Sadlier, 1969).

How might ovarian steroid hormones affect food intake and subsequent body mass? The most parsimonious possibility is that ovarian hormones affect the brain sites that control food intake and voluntary exercise. Changes in these behaviors would then lead to changes in adiposity and body mass. Alternatively, these steroids might have effects on body mass independent from food intake and exercise. One way to distinguish between these two possibilities is to "yoke" the amount of food available to one animal to the food intake of another. In this paradigm, the yoked animal receives the same amount of food per day as is consumed by a counterpart, usually a littermate. Ovariectomized rats still gain weight when they are fed exactly the same amount that a gonadally intact sister eats. In other words, ovariectomized rats appear almost destined to gain weight, even when provided with only sufficient calories to maintain a normal body mass in a gonadally intact animal. Are there changes in metabolic rate? The same result is obtained in ovariectomized rats yoked to ovariectomized females treated with estradiol. In both cases, the ovariectomized, untreated rats are less active than the intact or estrogen-treated females, and thus gain body mass.

Another way to examine the direct effects of hormones on body mass is to place hormone implants directly into the central nervous system. Implants of estradiol benzoate into the PVN or VMH of ovariectomized rats reduce food intake and body weight in a manner similar to systemic injections of estradiol (Butera and Beikirch, 1989; Wade and Zucker, 1970). Estrogen implants elsewhere in the diencephalon do not reliably affect food intake or body weight.

Adipose tissue metabolism is directly affected by steroid hormones. Alteration of the size of the fat deposits is the main effect of sex steroid hormones on body mass. Adipose tissue contains receptors for both estrogens and progestins (Gray and Wade, 1979; Wade and Gray, 1978). Estradiol and progesterone affect the activity of lipoprotein lipase (LPL), an enzyme found in fat cells that mediates the uptake of triglycerides from the bloodstream by the fat cells, as well as other coordinated changes in fat metabolism. LPL activity is high when fats are being stored; for example, after a meal. LPL activity is decreased by estradiol, but increased by progesterone. Consequently, fat cells cannot store triglycerides when exposed only to estradiol, but concurrent exposure to progesterone enhances fat storage by increasing the amount of triglycerides sequestered from the blood.

Why is the body mass effect of progesterone dependent on estradiol? Recall from Chapter 2 that estrogens turn on genes that code for, among other things, progestin receptors in the uterus and elsewhere. Normally, there are receptors for both estradiol and progesterone in the adipose tissue cytosol. After ovariectomy, progesterone receptors disappear from the fat cells. Within 6–12 hours of an estradiol injection, progesterone receptors again proliferate throughout the cytosol (Gray and Wade, 1979). In this case, the progesterone receptors in the fat cells are dependent on an anabolic effect of estrogens.

Clearly, ovarian steroid hormones can significantly affect food intake and body mass, as demonstrated by ovariectomy and hormone replacement studies. Does this effect of steroid hormones play any functional role in the regulation of food intake and body mass and in the maintenance of energy balance? The answer is probably yes. Pregnancy causes a great increase in food intake in many mammalian species. Pregnancy, and especially lactation, are very energy-demanding activities. The resting metabolic rate of a pregnant mammal more than triples as compared with that of a nonpregnant animal (Bronson, 1989). A large increase in food intake is required to maintain the fetuses. Fat deposition, and thus body mass, increases during pregnancy in most mammalian species, probably as a hedge against the coming energetic demands. New mother rats, for example, weigh more just after parturition than they did before becoming pregnant. If a pregnant female cannot find sufficient food, or lacks sufficient energy stores, the pregnancy is often aborted. Similarly, a lactating female with insufficient metabolic fuels may kill and eat some or all of her offspring to balance litter size with available metabolic energy supplies (Schneider and Wade, 1989b). Even estrous cycles are suspended if insufficient metabolic fuels are available to the female (Schneider and Wade, 1989a).

There are some interesting exceptions to the statement that pregnancy elevates food intake and increases body mass. As previously noted, Syrian hamsters do not increase their food intake during pregnancy; rather, they draw on their body fat reserves for energy (Wade et al., 1986). In Syrian and Siberian hamsters, progesterone appears to act in a manner opposite from its actions in rats; that is, progesterone causes a mobilization of fat and substantial decreases in body fat stores in pregnant hamsters (Schneider and Wade, 1987).

Fat deposition increases throughout the first two trimesters of pregnancy in humans. This observation makes sense when you consider the estrogen and progesterone levels found in the body during pregnancy. But why does fat accumulation virtually stop at the end of the second trimester? The answer to this question is another hormone: prolactin. High blood plasma levels of prolactin inhibit fat production. Prolactin has essentially the same effect as estradiol on fat cells; namely, it reduces LPL activity—except in mammary tissues. Consequently, fat continues to be deposited in the mammary glands through the end of pregnancy.

Androgens

Among vertebrates, males of many species are larger in size and eat more food than females. Much of this difference in size is organized perinatally by androgens, but the activational effects of anabolic steroids also are important in maintaining this sexual dimorphism. In most mammalian species, androgens promote elevated body mass and energy consumption (Wade, 1976). Males generally have more muscle mass than females, and the relatively higher concentrations of circulating androgens in males maintain this sex difference in body mass directly.

Castration decreases food intake and limits weight gain, including muscle mass gain, in rats (Mitchell and Keesey, 1974). The effects of castration on food consumption and body weight can be reversed by low-dose (50–200 μg/day) testosterone replacement therapy (Gentry and Wade, 1976). Bilateral implants of testosterone propionate into the VMH limit food intake in castrated male rats; this action of testosterone presumably reflects the effects of estrogen in this brain region after aromatization of the testosterone. Bilateral implants of dihydrotestosterone, an androgen that cannot be aromatized to an estrogen, into the VMH do not affect food intake (Nunez et al., 1980). These results suggest that androgens have both a central (CNS-induced change in food intake) and a peripheral (muscle mass) mechanism of action on food intake and body mass in male rats.

As is the case in many aspects of behavioral endocrinology, many of the general principles of androgen action have been generated from data obtained from rats. But many species-specific responses to these hormones have been reported that may require future modifications of the current general principles. For instance, a typical male meadow vole (*Microtus pennsylvanicus*) also eats less food and loses body mass after castration (Dark and Zucker, 1984), but individual voles vary markedly and predictably in their response to castration. Adult male meadow voles of identical age vary in body mass from about 35 to over 60 grams; all of this variation is due to differences in body fat stores. Voles heavier than 47 grams lost weight after castration, whereas voles lighter than 47 grams gained weight after castration. Males weighing about 47 grams at the time of castration did not display substantial body mass changes. Testosterone replacement reversed the effects of castration in the appropriate direction for the heavy and light males. The individual differences in body fat content did not reflect differences in blood plasma levels of testosterone, but rather appeared to reflect differences in target tissue sensitivity to androgens (Dark et al., 1987b).

As described in Chapter 6, female garter snakes (*Thamnophis sirtalis parietalis*) are about three times heavier than their male conspecifics. This difference in body size is apparent by 3 weeks of age, a time when plasma levels of androgens are high. Androgens appear to inhibit growth in male garter snakes. The sex difference in body size can be abolished by castration at any age; that is, castrated male snakes grow as large as females. Treatment with testosterone prevents the female-like body size in castrated males (Crews et al., 1985).

Seasonal Body Mass Cycles

Several species of rodents undergo seasonal shifts in body mass (Bartness et al., 2002). Many of us are familiar with sciurid rodents such as squirrels, chipmunks, and woodchucks, that undergo autumnal fattening prior to entering hibernation. As they hibernate, their fat stores are slowly oxidized to provide the energy they need to survive the winter. As will be described further in Chapter 10, these annual cycles of body mass are controlled by an endogenous circannual biological clock and persist for years, even when the animals are housed in laboratories under constant environmental conditions (Figure 9.26) (Barnes and York, 1990; Heller and Poulson, 1970; Mrosovsky, 1985).

Muroid rodents such as deer mice, hamsters, voles, and lemmings also undergo seasonal changes in body mass, primarily by altering the amount of body fat

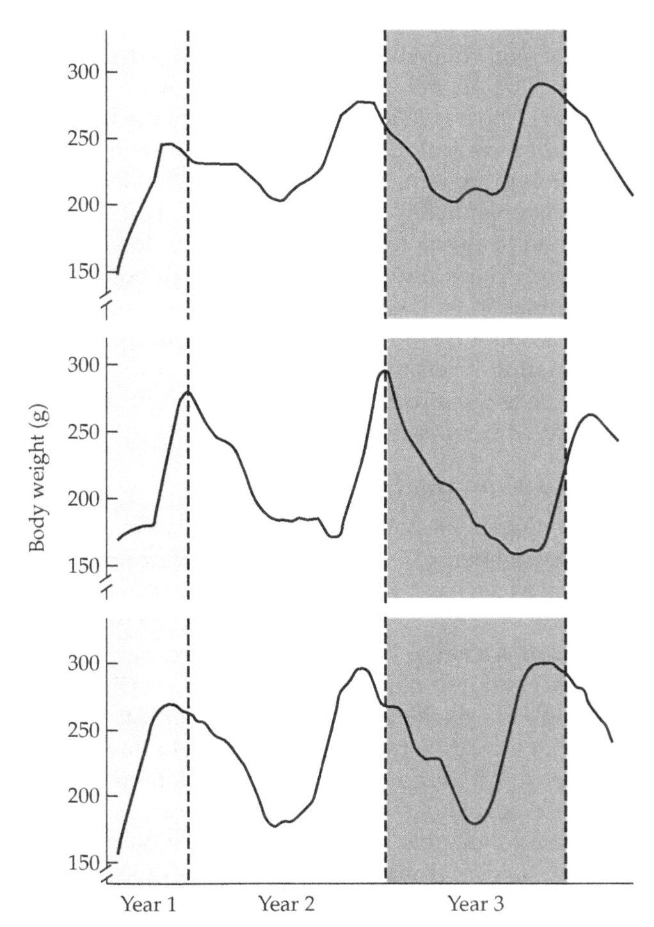

9.26 Annual body mass changes for three ground squirrels housed in a laboratory under constant conditions. The animals showed approximately one-year (circannual) fluctuations in weight gain and loss. These fluctuations correspond to seasonal body mass cycles that occur in nature. After Zucker, 1988.

stored. The seasonal cycles of body mass fluctuation in these animals are regulated, for the most part, by day length, or photoperiod. The onset of short days triggers an increase in body fat stores in Syrian hamsters. Food intake in this species, however, does not change in response to short day lengths. This observation suggests a direct effect of photoperiod on metabolic pathways. But that cannot be the whole story either, because the increase in body mass is somewhat independent of seasonal gonadal regression and the subsequent reductions in blood plasma concentrations of sex steroid hormones. As we will see in Chapter 10, day length is often encoded by melatonin secreted from the pineal gland; removal of the pineal gland therefore blocks reproductive regression in response to short photoperiods because animals cannot discern day length without the pineal melatonin rhythm. However, pinealectomy does not affect short day–induced fattening in Syrian hamsters (Bartness and Wade, 1984). The mechanism of action of this pineal-independent transducer of day length on body weight is unknown.

Other muroid species adopt an alternative strategy for winter survival. Instead of gaining weight, these animals lose body mass in anticipation of winter. A reduction in body mass results in an elevated surface-to-volume ratio, which permits increased heat loss to the environment and a consequent reduction in body fat, resulting in reduced insulation and energy reserves. However, these results of decreased body mass are apparently compensated for by the fact that a small individual has to ingest far fewer calories to maintain itself than a large individual. The resulting reduction in foraging time and concomitant exposure to unfavorable climatic conditions presumably enhances survival. Thus, meadow voles and other species respond to short days with a decrease in food intake and a subsequent reduction in body mass, although the decrease in body mass precedes the reduction in food intake in Siberian hamsters (Bartness and Wade, 1985; Dark and Zucker, 1986; Dark et al., 1983). Part of the weight loss can be attributed to reductions in circulating levels of gonadal steroid hormones, but most of the effect appears to be due to the direct influence of day length, mediated by melatonin, on metabolic pathways (Bartness and Wade, 1985).

Inhibition of Reproduction to Maintain Energy Balance

When food is abundant and the energetic demands of thermoregulation and foraging are minimal, there is plenty of energy available for the cellular processes necessary for day-to-day survival, as well as for growth and reproductive activities. However, when food is scarce or energetic demands are high, the processes that are necessary for survival take precedence. Thus, mammals have mechanisms to inhibit reproduction when they fall into negative energy balance (Bronson, 1999). Women who diet or engage in excessive exercise to the point that their body mass is less that 85% of what would be typical for their height have problems with fertility. Reports of menstrual irregularities, amenorrhea (cessation of menstrual cycles), and diminished sexual desire and sexual activity, as well as infertility, are common among fashion models, women with the eating disorder anorexia nervosa, and athletes or dancers who do not increase their food intake to compensate for their excessive energy expenditure (Wade and Jones, 2003;

Schneider, 2004; Schneider and Watts, 2002). The fact that reproductive inhibition occurs among females of all mammalian species and is entirely reversible upon refeeding or relaxation of energy demands should inform us that it is not a pathology or disease, but a biological adaptation for conserving energy in environments in which energy availability may fluctuate (see Chapter 10).

An individual that finds itself hungry in a low-temperature environment must conserve energy, and it must also find and eat food. Not surprisingly, the inhibition of reproduction in the service of energy balance is controlled by mechanisms that overlap with those that control food intake. Most of the peptide hormones, neuropeptides, and metabolic inhibitors that increase food intake also inhibit reproduction, and many of the same brain regions are involved in both processes (Wade and Jones, 2003). For example, in Syrian hamsters, estrous cycles can be interrupted by a 48-hour fast. Fat, but not lean, hamsters are protected from this fasting-induced anestrus. It was suggested that the fat hamsters were protected by their ability to use fatty acids as fuels by metabolizing lipids from their fat cells. To test this idea, very fat hamsters were either fasted or fed as much food as they wanted (ad libitum). Half of each group was treated with MP, an inhibitor of free fatty acid oxidation. Fat, fasted hamsters showed normal estrous cycles, but fat, fasted hamsters treated with MP did not. MP treatment had no effect on estrous cycles in hamsters fed ad libitum. Thus, the mechanisms that interrupt estrous cycles are sensitive to signals generated by the oxidation of metabolic fuels (Schneider and Wade, 1989a). Similarly, pulsatile LH secretion is affected by fasting, insulin, and metabolic inhibitors in rats, sheep, and monkeys (reviewed in Schneider, 2004).

Leptin, insulin, neuropeptide Y, CRH, CCK, galanin, nitric oxide synthase, and a variety of other substances are involved in the metabolic control of reproduction. As with the search for the signals that control food intake, the search for the signals that inhibit reproduction is complicated by the fact that hormones might have both direct and indirect influences on reproductive function. As discussed previously, leptin can increase free fatty acid oxidation and can prevent the storage of fat fuels in the form of triglycerides. Does leptin affect reproduction directly, through its effects on brain mechanisms that control estrous cycles, or indirectly, via its known effects on free fatty acid oxidation? To find out, Syrian hamsters were fasted and treated systemically with either leptin or leptin plus MP. Fasting-induced anestrus was prevented by treatment with high doses of leptin. However, leptin treatment did not prevent fasting-induced anestrus in hamsters pretreated with MP (Schneider et al., 1998). These results suggest that leptin affects reproduction via its indirect effects on fuel oxidation. It is important to note that there are currently no data to demonstrate that fluctuations in leptin concentrations within normal, healthy physiological ranges affect human reproduction (Schneider, 2004).

Regardless of whether leptin affects reproduction directly or indirectly (or both), the importance of normal daily fluctuations in leptin concentrations will have to be considered in relation to circadian variation in leptin secretion, as well as variation in the spacing of meals. There are some data that suggest a threshold effect of leptin on normal pubertal development (Cheung et al., 1997). More recently, ghrelin has been suggested to play a role in the link between fuel availability and reproduction (Barreiro and Tena-Sempere, 2004).

Final Thoughts

As already described, many hormonal and neural signals act in concert to balance energy intake and energy expenditure. It appears that hormones affect feeding behavior primarily by affecting the sensory systems related to feeding. CCK has been found to reduce neural activity in the brain regions where sensory information from the taste buds is received (Morin et al., 1983). Similarly, estradiol enhances the neural firing rates of afferent fibers from the tongue. Hormones also probably increase the sensory capabilities of animals hunting for food.

Hormones also have direct effects on the central processing mechanisms controlling food intake. CCK infusions into the brain appear to stop feeding behavior in rats. The hormone's mode of action, and how it would normally cross the blood–brain barrier, are unknown. CCK acts in the periphery by interacting with the vagus nerve (Weatherford et al., 1993). Insulin also exerts direct effects on the central nervous system to stop feeding behavior (Woods and Porte, 1983). Although insulin and leptin are too large to cross the blood–brain barrier, brain capillaries contain receptors for insulin and leptin that enable these hormones to be transported through the capillary walls into the brain interstitial fluid biologically intact (Schwartz et al., 1992). Other peptides, including leptin, ghrelin, GLP-1, PYY, and adiponectin have also been proposed to act centrally to modulate food intake (see Schneider and Watts, 2002, for review). Hormones have not been shown to increase the efficiency of effectors of food intake; that is, hormones do not increase chewing frequency or increase efficacy of the muscles involved in feeding.

Although we know a lot about the endocrine mediation of feeding behavior, particularly in rats, there are many different feeding strategies that probably have very different types of controlling mechanisms. For example, some snakes consume a large prey item and then do not eat again for many days as they digest their meal. Similarly, many cats will hunt prey and eat some of it, then leave the remains, or carry them to a resting spot. What cues tell snakes and cats that they are fed? What cues indicate hunger? The endocrine mediation of these signals may very well differ in many respects from that in rats and primates, or it may be identical. Many species go for weeks or even months without eating. Sometimes these fasting individuals are inactive, such as hibernating squirrels, but in other cases, the animals are engaged in other activities, as elephant seals are during the breeding season. The endocrine control of metabolism in these long-term fasting animals and the signals that stop them from feeding require further study. All in all, the study of hormones and homeostasis provides a unique opportunity to investigate the complex interactions among physiology, morphology, and behavior.

Summary

- Individuals are motivated to maintain specific endogenous levels of water, sodium, and nutrients. The process by which animals maintain a fairly stable internal environment is called homeostasis.
- Most homeostatic systems operate like a thermostatically controlled heating and cooling system. To maintain homeostasis, a system requires a detection mechanism to note any deviation from a reference value, or set point. The homeostatic system must also be able to mobilize the organism to make changes to return it to the reference value. Finally, the system must have some way to recognize when the desired change occurs and feed back to stop the mobilization process.
- In multicellular organisms, water and sodium levels are maintained within a narrow range by physiological and behavioral systems. Most water is located inside the cells; the water outside the cells is located both in blood vessels and in the interstitial fluid. Sodium is important in the movement of water between the extracellular and intracellular fluid compartments.
- Movement of water out of cells causes cellular dehydration, a potent stimulus for osmotic thirst. Consumption of salty or sugary foods induces osmotic thirst. Cellular dehydration of osmoreceptors in the brain causes ADH secretion from the posterior pituitary, which promotes water conservation by the kidneys. Drinking water alleviates osmotic thirst.
- Hypovolemic thirst is caused by reduced blood volume (hypovolemia). Quenching hypovolemic thirst requires ingestion of water, sodium, and other solutes.
- Hypovolemia stimulates the production of angiotensin II, a potent vasoconstrictor. Angiotensin II may also increase drinking behavior.
- Angiotensin II triggers the release of aldosterone from the adrenals; aldosterone promotes sodium retention and subsequent water conservation by the kidneys. Aldosterone affects drinking behavior indirectly by its influence on osmotic thirst.
- The need for metabolic fuels is continuous, but food intake is episodic; consequently, energy intake and expenditure are never quite balanced. After a meal, there is a surplus of energy that must be stored for later use. Insulin is the only hormone that promotes energy storage by cells.
- Although insulin is critical for energy storage, many other hormones, including epinephrine, norepinephrine, glucocorticoids, thyroid hormones, growth hormone, somatomedin, and glucagon, are involved in getting energy out of storage.
- The stimuli eliciting feeding are numerous and include physiological signals from fat stores, levels of glycogen stores, and the nature and quality of the food, as well as neural and hormonal signals from the gut.

- Insulin may signal the central nervous system about peripheral levels of metabolic fuels via the cerebrospinal fluid. Intracerebroventricular infusions of insulin reduce food intake.
- A complex network of neuropeptides and hormones acts to maintain energy balance. Several peptide hormones, including corticotropin-releasing hormone, leptin, neuropeptide Y, bombesin, and glucagon, influence food intake. Neuropeptide Y, AgRP, MCH, orexin A, and ghrelin are potent inducers of food intake.
- Cholecystokinin is released during ingestion and appears to provide an endocrine signal of satiety, acting to inhibit feeding via the vagus nerve.
- Endogenous opioids may alter the hedonic value of food. Treatment with opioid antagonists reduces food intake.
- Gonadal steroid hormones also act to affect food intake and body mass. Generally, estrogens reduce food intake and body mass, as well as increasing voluntary exercise. Progesterone blocks these effects of estrogen. Androgens generally increase food intake and body mass.
- Some mammalian species undergo seasonal changes in body mass. Individuals of some species elevate body fat levels to increase their chances of winter survival. An alternative strategy is to reduce body mass prior to winter so that less food must be consumed to survive.
- Many species must learn what foods to consume, but some physiological requirements elicit unlearned specific hungers. Specific hungers for sodium and potassium appear to be common in terrestrial vertebrates. Hormones mediate some specific hungers.
- Both leptin and insulin are peripheral adiposity signals that circulate in proportion to stored adipose tissue. Receptors for each hormone are found in the CNS, especially in the hypothalamus, and when activated appear to affect food intake. Impaired neuronal signal transduction by either hormone is associated with increased food intake and body fat stores.
- Insulin and leptin act in the arcuate nucleus of hypothalamus to suppress expression of neuropeptide Y and agouti-related peptide (NPY/AGRP) in neurons, which promotes an anabolic state, (increased food intake; decreased energy expenditure) while stimulating expression of pro-opiomelanocortin and cocaine and amphetamine-regulated transcript (POMC/CART) neurons that promote catabolic effects (anorexia; elevated energy expenditure).

Questions for Discussion

1. In what ways is a thermostat a good analogy, and in what ways is it a poor analogy, for the control of food intake and body mass?

2. Discuss the concept of "set point." In what ways is this concept limited in considering the annual changes in the body mass of squirrels or bears that gain weight prior to entering hibernation?
3. What are the advantages and disadvantages of "hard-wired" taste preferences? What are the advantages and disadvantages of hormonal mediation of taste preferences?
4. How are hormones that have central processing effects on feeding behavior different from or similar to neurotransmitters? Are the distinctions between hormones and neurotransmitters applicable when discussing chemical mediation of behavior in the brain?
5. Why are homeostatic processes controlling body fluid levels important? How would one go about studying the evolution of body fluid regulatory mechanisms?
6. Discuss the importance of the existence of neuropeptides that regulate energy homeostasis in both the brain and the gastrointestinal system. What are the advantages of having control both at the level of the brain and in the periphery?

Refer to the accompanying Student CD for additional resources, including Web links, videos, animations, and additional photos.

Suggested Readings

Bartness, T. J., Demas, G. E., and Song, C. K. 2002. Seasonal changes in adiposity: The roles of the photoperiod, melatonin and other hormones, and sympathetic nervous system. *Experimental Biology and Medicine (Maywood)*, 227: 363–376.

Fluharty, S. J. 2002. Neuroendocrinology of body fluid homeostasis. In D. W. Pfaff, A. P. Arnold, A. M. Etgen, S. E. Fahrbach, and R. T. Rubin (eds.), *Hormones, Brain and Behavior*, Volume 1, pp. 525–569. Academic Press, New York.

Grill, H. J., and Kaplan, J. M. 2002. The neuroanatomical axis for control of energy balance. *Frontiers in Neuroendocrinology*, 23: 2–40.

Schneider, J. E., and Watts, A. G. 2002. Energy balance, ingestive behavior, and reproductive success. In D. W. Pfaff, A. P. Arnold, A. M. Etgen, S. E. Fahrbach, and R. T. Rubin (eds.), *Hormones, Brain and Behavior*, Volume 1, pp. 435–523. Academic Press, New York.

Schwartz, M. W., Woods, S. C., Porte, D., Seeley, R. J., and Baskin, D. G. 2000. Central nervous system control of food intake. *Nature*, 406: 661–671.

Seeley, R. J., and Woods, S. C. 2003. Monitoring of stored and available fuel by the CNS: Implications for obesity. *Nature Reviews: Neuroscience*, 4: 901–909.

10

Biological Rhythms

When I was about 10 years old, I had a pet hamster. I do not recall his name, but vaguely recall an elaborate funeral upon his demise. What I remember best about this pet was that he had an exercise wheel inside his cage. Every night as I was trying to fall asleep, the hamster would begin to run in his wheel. Because the wheel made a squeaky noise with each revolution, it was very difficult to fall asleep. I tried oiling the wheel, but the nightly squeak continued. Finally, in desperation late one night, I put the hamster and his cage in my closet. As I was closing the closet door, I began to feel guilty, so I switched on the closet light for him. To my amazement, the hamster stopped running, at least until I fell asleep. I switched off the light the following morning. Over the course of the next several days, the hamster began to confine his wheel-running behavior to the period after the light was switched off each morning. Inadvertently, I had created a reverse light–dark cycle for my hamster. He now confined his running behavior to the daytime hours, when he was in my dark closet and I was in school. When I got home from school, I would play with the creature; before retiring, I would turn on the closet light and return the hamster to his cage, where he would curl into a ball and sleep through the night.

As you probably know, some animals (for example, cows, ground squirrels, and chickens) are **diurnal**, or active during the day, and some animals (for example, hamsters, raccoons, and owls) are **nocturnal**, or active at night. Other animals are **crepuscular** (for example, white-tailed deer), or primarily active at dawn and dusk.

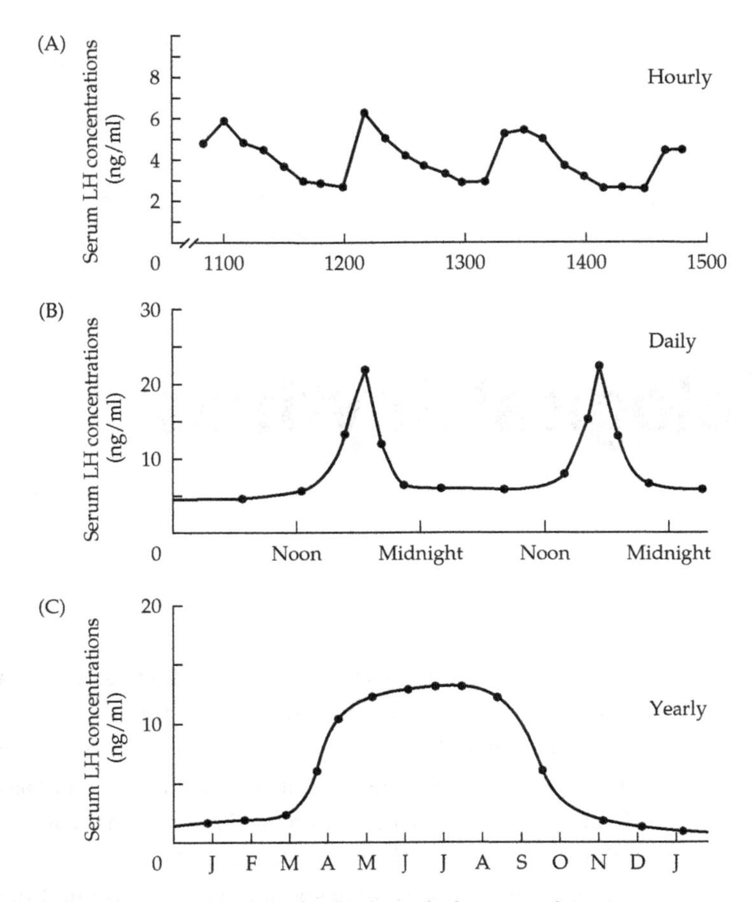

10.1 Endocrine function shows rhythmic variation over time. If LH concentrations in hamsters are observed over several different temporal scales, rhythmic release patterns emerge. (A) Assessments every 10 minutes reveal that LH is released in pulses that recur every 30–90 minutes. (B) Samples obtained every 6 hours will not detect 30–90-minute pulses, but reveal a daily rhythm of LH release. These daily surges consist of programmed increases in the 30–90-minute pulse frequency and pulse amplitude that result in elevated blood concentrations of LH. (C) Blood samples obtained once a month from hamsters housed in simulated winter conditions do not detect daily pulses, but an annual pattern of LH secretion emerges, showing that blood LH concentrations are generally higher during the breeding season in spring and summer than during autumn and winter.

Animals have evolved to restrict many of their behaviors, including feeding, drinking, mating, and locomotor activities, to specific temporal niches in response to a complex web of selective forces. This temporal variation in behavior presumably reflects temporal variation in underlying physiology. As we will discover, virtually all physiological processes vary over time. Specific to our interests here, endocrine secretion varies markedly over the course of minutes, over the course of a day, and over the course of a year (Figure 10.1). Because of this temporal variation in endocrine function, it should not be particularly surprising to you that behavior, especially behavior that is mediated by hormones, exhibits pronounced temporal variation. Much of this temporal variation is regulated by biological clocks.

The scientific study of biological clocks and their associated rhythms is called **chronobiology**. Although biological rhythms have been informally recognized by biologists for centuries, the formal study of biological timekeeping began about 50 years ago. One reason that chronobiology took so long to develop as a scientific discipline was that it had to counteract the dogma of homeostasis in the biological sciences and medicine. In Chapter 9 we learned that homeostatic processes work to maintain physiological parameters within specific ranges. Biologists and physicians often considered large fluctuations to be pathological, and many resisted the idea of the programmed changes in physiology and behavior that we now understand to underlie homeostatic processes. Many of the specific homeostatic ranges discussed in Chapter 9 are not static, but in fact change over time. To take these changes into account, the thermostat analogy that we used in that chapter must be updated to an automatic, electronic thermostat that can be programmed to limit heating or cooling to times when people are likely to be home. This temporal variation increases the efficiency of the home heating system. Similarly, biological clocks increase biological efficiency and thus increase fitness. This chapter will describe such temporal variations generally, and will focus specifically on the interaction among biological rhythms, hormones, and behavior.

Chronobiology has borrowed terms and concepts extensively from engineering disciplines to describe biological clocks and their associated rhythms. A **rhythm** is a recurrent event that is characterized by its period, frequency, amplitude, and phase (Figure 10.2; Aschoff, 1981). The **period** is the length of time required to complete one cycle of the rhythm in question; for instance, the amount of time required to go from peak to peak or trough to trough. **Frequency** is computed as the number of completed cycles per unit of time (for example, two cycles per day). **Amplitude** is the amount of change above and below the average value; that is,

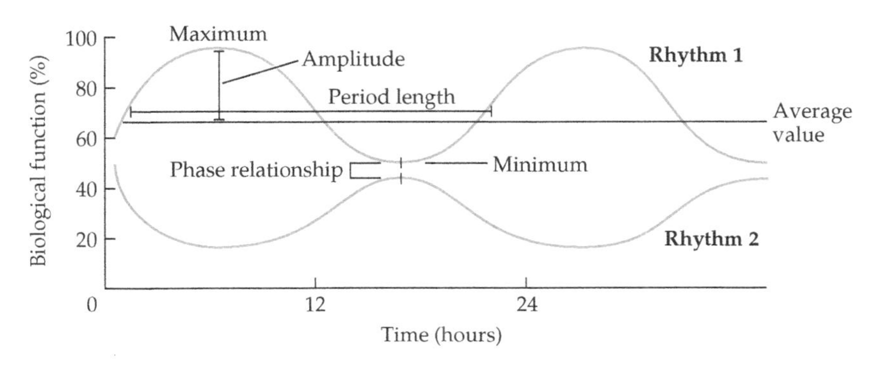

10.2 Components of biological rhythms. Rhythms can be analyzed in terms of amplitude, frequency, and period length. The relationship of one rhythm to another is expressed in terms of phase relationships. Both cycles have a period of 24 hours (frequency = 1/day).

the distance of the peak or nadir from the average. The **phase** represents a point in the rhythm relative to some objective time point during the cycle. For example, under normal conditions, the phase of onset of the activity portion of a hamster's activity–rest cycle corresponds closely with the onset of dark. Phase relations among various biological rhythms can also be described; for instance, the onset of the low body temperature phase of a hamster's daily body temperature cycle tightly corresponds with the onset of the sleep portion of its sleep–wake cycle.

Exogenous versus Endogenous Control of Biological Clocks

Some behavioral rhythms have been recognized since ancient times, but they have generally been attributed to **exogenous** (outside the organism) factors. For example, the activity–rest rhythms exhibited by many species, including my pet hamster, were believed to be caused by daily signals arising from the environment. Thus, one simplistic explanation of how the daily light–dark cycle might drive the activity–rest rhythm of hamsters might be that the light of day is too intense for their eyes, so they rest during the daylight hours and become active after dark. Of course, some rhythms do fade away in the absence of environmental cues. Although exogenous factors may serve a permissive role for biological rhythms, we now know that **endogenous** (inside the organism) timing mechanisms mediate many of the observed rhythms in physiology and behavior.

How is it established whether a rhythm is the result of exogenous factors or an endogenous clock? One type of convincing evidence is provided by isolation experiments. The persistence of a biological rhythm in the absence of environmental cues provides compelling evidence that the rhythm in question is generated within the animal, and not driven by the environment. If a biological rhythm disappears under constant conditions, then it is reasonable to suggest that some cyclic cue in the environment drives the biological rhythm. This logic was first applied to the study of biological rhythms in 1729, when the French astronomer Jean Jacques d'Ortous de Mairan found that the tension–relaxation pattern of a heliotropic plant—a plant with leaves that open during the day and close at night—persisted even when the plant was maintained in the total darkness of a cellar for several days (Figure 10.3; De Mairan, 1729). Augustin de Candolle, a French botanist, expanded on de Mairan's initial observation. Mimosa (*Mimosa pudica*) leaf movements also continued in constant dark conditions, but like most other internally generated biological rhythms, these rhythms displayed slight to moderate deviations from 24 hours in period length in the absence of environmental cues (de Candolle, 1832).

Examples of this sort were reported sporadically over the course of the next 200 years. Although such demonstrations were consistent with the existence of endogenously driven biological clocks, the scientific community largely maintained that these reported rhythms merely reflected rhythms in the environment. From the 1930s through the 1960s, biologists gradually began to accept the existence of endogenous biological clocks, although some scientists continued to argue that the persistent biological rhythms observed when all apparent environ-

10.3 De Mairan's experiment. Heliotropic plants isolated from sunlight continued to open and close in synchrony with the day–night cycle, suggesting that this rhythm has an endogenous, rather than an exogenous, source. De Mairan's report is the first recorded observation that biological rhythms can persist in the absence of environmental cues.

mental cues were absent could be explained away as reflecting subtle geophysical forces that simply had yet to be discovered (for example, Brown, 1972). However, recent evidence has convinced virtually everyone that the clocks driving physiology and behavior are inside the organism and are not driven by the environment. Several types of evidence support this conclusion, all of which will be more fully explored later in this chapter:

1. Animals maintained in constant conditions aboard a spacecraft orbiting far above the earth, presumably far away from subtle geophysical cues, display biological rhythms with periods similar to those observed on earth.
2. Animals maintained in adjacent, but separate, cages in the absence of environmental cues display biological rhythms with slightly different periods, suggesting that they are not being driven by the same subtle geophysical cues.
3. The period (and phase) of the biological rhythms of one individual can be transferred to another individual by means of tissue transplants.

Types of Biological Clocks and Rhythms

All animals, and virtually all plants, studied to date possess endogenous biological clocks that mediate biological rhythms. The periods of biological rhythms range from the 1-millisecond cycle of firing in some neurons to longer cycles such as the 16-day estrous cycles of guinea pigs, the 1-year cycle of ground squirrel

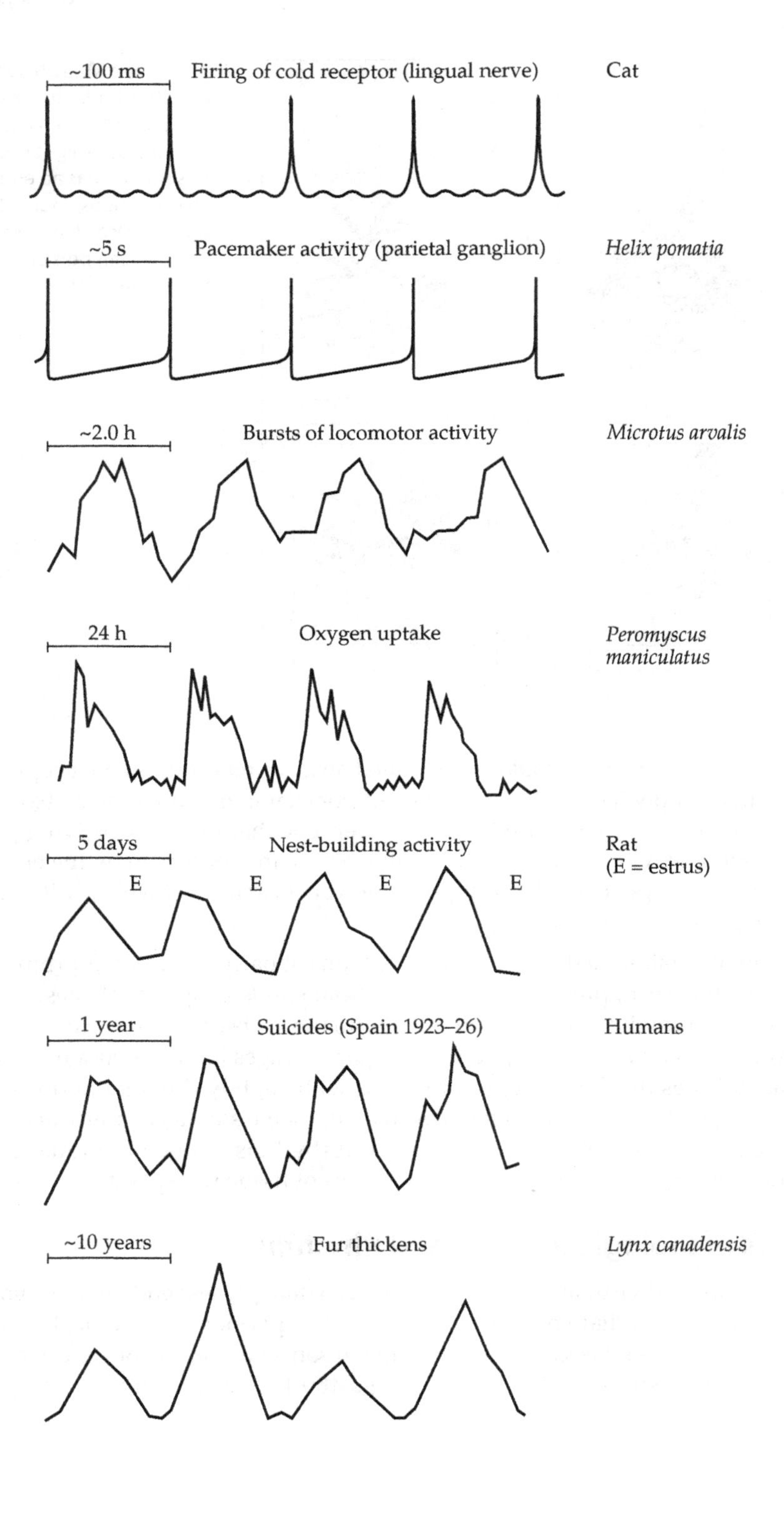
~100 ms
Firing of cold receptor (lingual nerve)
Cat
~5 s
Pacemaker activity (parietal ganglion)
Helix pomatia
~2.0 h
Bursts of locomotor activity
Microtus arvalis
24 h
Oxygen uptake
Peromyscus maniculatus
5 days
Nest-building activity
Rat
(E = estrus)
E
E
E
E
1 year
Suicides (Spain 1923–26)
Humans
~10 years
Fur thickens
Lynx canadensis

◀ **10.4 The diversity of biological rhythms.** The period lengths of biological rhythms range from milliseconds (top) to many years (bottom). The longest known period, the time that elapses between flowerings of certain bamboo species, lasts 100 years. After Aschoff, 1981.

hibernation, the 10-year cycle of the thickening of fur in the arctic lynx, or the 100-year cycle of the flowering of certain species of bamboo (Figure 10.4).

The periods of some biological rhythms, including most central nervous system, cardiovascular, and respiratory rhythms, vary widely within the same individual; for example, the period of heartbeats decreases (that is, the frequency increases) during exercise because heart rhythms are dependent on activity levels. The periods of other endogenous cycles, such as the ovarian cycle, are largely constant for the same individual, but there may be great interindividual or interspecific variation. However, there are four types of biological rhythms that are typically coupled with environmental cues, and the periods of these rhythms do not vary much under natural conditions. These relatively constant biological rhythms mimic the periods of the geophysical cycles of night and day (**circadian**), the tides (**circatidal**), the phases of the moon (**circalunar**), and the seasons of the year (**circannual**) (Table 10.1; Palmer, 1976). These rhythms persist when animals are isolated from the respective environmental cues, but in that case they only approximate the periods of the environmental cycles with which they are normally coupled, as we will see below. Thus, the terms for many biological rhythms use the prefix *circa*, from the Latin word meaning "about" (for example, *circa*, "about"; *dies*, "day"; *circadian*, "about a day") (Halberg, 1959). Although conducting an experiment that isolates animals from environmental cues is necessary to determine the extent of endogenous generation of any biological rhythm, it should be emphasized that the environment often exerts permissive effects on endogenously driven biological rhythms. For instance, chronic food restriction or maintenance in constant bright lights dampens the expression of estrous cycles in laboratory strains of rats.

An important experimental technique used in the study of biological rhythms involves keeping a log of physical activity over time. Basically, the monitoring of activity cycles is adapted to the species under investigation. Often small mammals are placed in a cage equipped with a running wheel, similar to the one in my pet

TABLE 10.1 **Comparison of biological rhythm types**

Rhythm type	Environmental cycle	Period length: Entrained	Period length: Free-running
Circadian	Revolution of earth	24 hours	22–26 hours
Circatidal	Tides	12.4 hours	11–14 hours
Circalunar	Phases of moon	29.5 days	26–32 days
Circannual	Seasons of the year	365.25 days	330–400 days

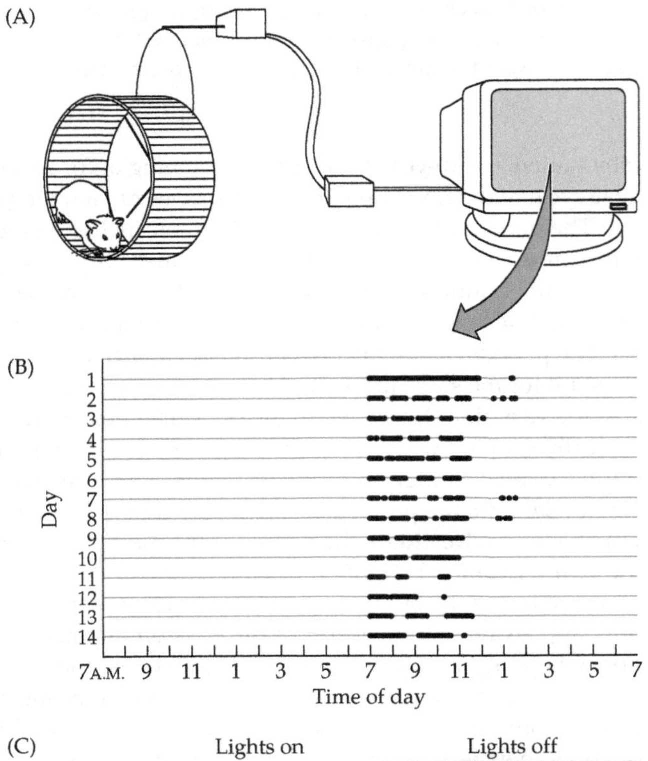

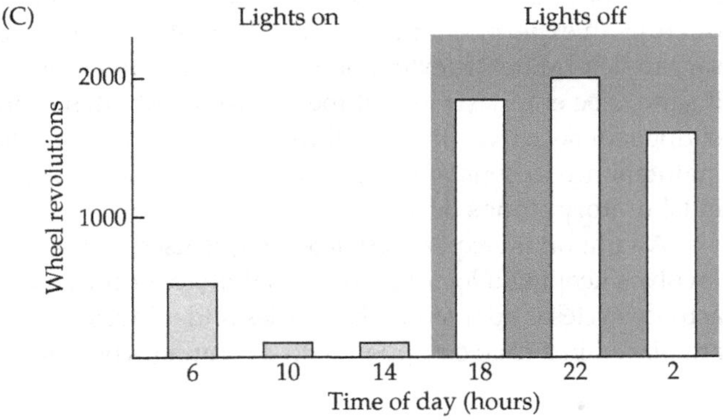

10.5 The collection of data on behavioral biological rhythms. (A) A common method of measuring rhythms in locomotor activity is a wheel-running apparatus that activates a pen with each revolution of the wheel. (B) Sustained wheel-running activity appears as a solid dark bar across the time line for each day. The activity record is read like sheet music, with each line representing 24 hours. The period of the daily biological rhythm can be measured from the onset of activity on one day to the onset of activity on the next. (C) The frequency of activity can be measured by counting the number of wheel revolutions each night. The amplitude is the distance between the peak and the trough of the number of wheel revolutions. Phase could be measured as the amount of time between the onset of running and the onset of darkness.

hamster's cage, connected to a counting device that automatically produces a continuous record of the animal's activity (Figure 10.5). The locomotor activity of small birds is often determined by monitoring perch-hopping activities around the clock. Humans can be equipped with electronic monitoring devices that transmit the movements of an individual to a central monitoring station.

Light serves as a potent environmental time cue, or **zeitgeber** (from the German, meaning "time giver"), for hamsters as well as for most other species (Pittendrigh and Daan, 1976). Temperature is an important zeitgeber for some poik-

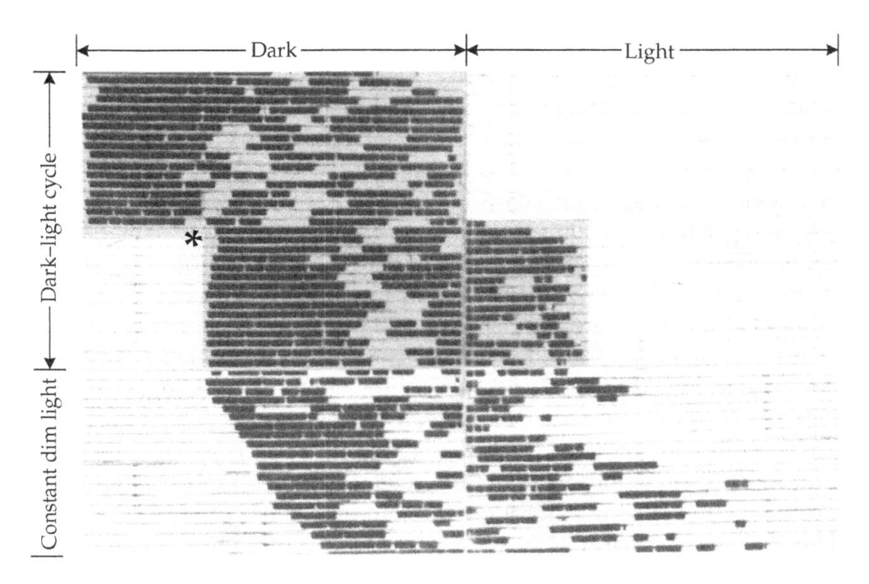

10.6 Wheel-running behavior of a hamster. When housed in a 12:12 dark–light cycle (lights off for 12 hours, then on for 12 hours), a hamster confines its wheel running to the dark part of the cycle (top). If the onset of darkness is shifted ahead by 4 hours (arrow), the hamster adjusts to the new cycle. If the dark–light cycle is suspended entirely, and the hamster is housed in constant dim light, it no longer expresses a 24-hour cycle of activity, but free-runs with a period greater than 24 hours; this free-running cycle represents the endogenous circadian rhythm of the individual. The asterisk indicates a 4 hour shift in light–dark cycle. From Zucker, 1980.

ilothermic animals, and possibly a secondary zeitgeber for birds and mammals, but light is the primary zeitgeber among homeothermic animals. If a hamster is placed in constant dim light so that there is no daily zeitgeber, the onset time of its locomotor activity begins to drift (Figure 10.6). Without the daily light–dark cycle providing a daily reset, the hamster's endogenous clock can only approximate 24-hour cycles. If a hamster usually begins to run in its wheel at 8:10 P.M., then in the absence of light–dark cycles, it may begin its wheel-running behavior 15 minutes later each day, with wheel-running activity onset at 8:25, 8:40, 8:55, and so on for subsequent days. Other hamsters may begin their wheel-running activity 10 minutes earlier each day. Biological rhythms such as these that are not synchronized with environmental cues are said to be **free-running**. Each individual displays its own free-running period. After a few weeks of constant lighting conditions, it is possible to observe different individuals housed in the same room running in their wheels at different times throughout the day. Thus, the free-running periods of biological clocks are precise, but are not exactly 24 hours. The observation that different hamsters display an array of different free-running periods when housed in the same room suggests that they are not synchronized by one another's behavior, and that subtle geophysical cues are not providing any temporal information.

Animals display species-specific times of locomotor activity onset that are often linked to the timing of food intake, water consumption, and reproductive behavior. In the case of the hamster—an experimental animal commonly used in studies of biological rhythms—locomotor activity begins right after lights are turned off each day. If lights are turned off at 8:00 P.M. each evening, hamsters begin to run in their wheels at about 8:05–8:15 P.M. each evening. More than 99% of their wheel-running activity is confined to the dark portion of the daily light–dark cycle. Hamsters also confine the majority of their eating, drinking, food hoarding, and sexual behaviors to the dark portion of the daily light–dark cycle. If the environmental light–dark cycles are *phase-shifted* (e.g., lights are turned off 4 hours later, as in Figure 10.6), hamsters adapt their activity rhythms to the new regimen in about four or five cycles. The circadian rhythms of other physiological processes, such as adrenocortical hormone release, body temperature, and blood plasma volume, may require longer to adapt to the new time. This process of adaptation to rapid phase shifts in environmental time cues results in what we commonly call "jet lag" when humans travel through different time zones (Box 10.1).

The process of synchronization of endogenous biological rhythms with a periodic cue in the environment—a zeitgeber—is called **entrainment**. We will focus on the role of light as a zeitgeber for daily activity rhythms. In order to understand entrainment, a basic understanding of phase-response curves is necessary. **Phase-response curves** are graphic representations of the differential effects that environmental light has on the timing of biological rhythms depending on the phase relationship of the light to the circadian organization of the animal in question (Figure 10.7). In other words, exposure to light at different times throughout the day does not result in uniform responses of free-running biological rhythms. If hamsters (or humans) are maintained in constant dark conditions, their biological rhythms begin to free-run, with the period of onset of locomotor activity varying slightly from 24 hours. The circadian period can thus be divided into a "subjective night" and a "subjective day." Because hamsters are nocturnal, the beginning of activity usually coincides with the onset of the creature's subjective night, whereas the rest period begins at the beginning of the animal's subjective day. For diurnal animals, the onset of activity coincides with the beginning of the animal's subjective day, and the rest period starts at the beginning of the animal's subjective night. If free-running hamsters that are housed in continuous darkness are exposed to an hour of light at any time throughout the middle of the subjective day, the time of onset of the next bout of locomotor activity is unaffected; that is, for a hamster with a free-running period of 24.25 hours, wheel-running still begins 24.25 hours after the start of the previous bout of activity. However, if the 1-hour pulse of light is given early in the subjective night (or late in the subjective day), then the hamster does not begin its activity 24.25 hours later, but rather delays its activity onset for 1–4 hours, depending on when exactly the pulse of light occurs. If the 1-hour pulse of light is given late in the subjective night (or early in the subjective day), again the hamster does not begin its activity 24.25 hours later, but rather advances its activity onset by 1–4 hours, depending on when exactly the pulse of light occurs. Thus, the largest phase delays can be produced early during the subjective night,

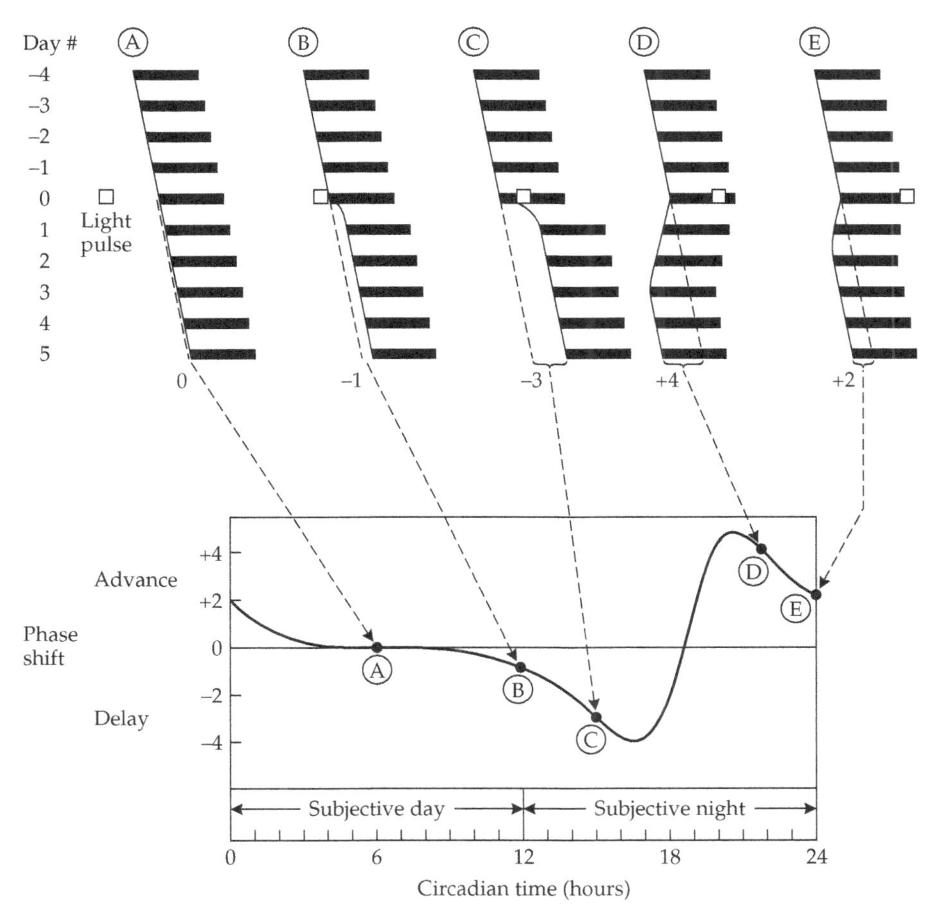

10.7 Phase-response curve. A free-running nocturnal individual maintained in constant dark conditions was exposed to a 1-hour light pulse at various times during its subjective day and night. The dark bars at the top of the figure are a schematic record of activity (wheel running), which occurred during the animal's subjective night. (A) When the light pulse was given in the middle of the subjective day, the subsequent activity onset time was unaffected; the middle of the subjective day is thus called the "dead zone." (B) A light pulse at the end of the subjective day phase-delayed activity the next day by about 1 hour. (C) With a light pulse at the start of the subjective night, a substantial phase delay (3 hours) was observed. (D) A light pulse given later during the subjective night caused a substantial advance (4 hours) of activity onset the next day. (E) Finally, a light pulse at the start of the subjective day caused a 2-hour phase advance in activity the following day. Light affects the endogenous circadian rhythms to entrain them to exactly 24 hours each day. After Moore-Ede et al., 1982.

when a nocturnal animal has just awakened and a diurnal animal has just retired, and the largest phase advances can be produced late during the subjective night; light appears to have little effect on the circadian system during the subjective day of either nocturnal or diurnal animals.

BOX 10.1 Jet Lag

The temporal niche to which humans have been exposed during the past million or so years has remained quite stable (Moore-Ede et al., 1982); it has been estimated that the time required for the earth to rotate once on its axis has slowed by only about 20 seconds during the last million years (Rosenberg and Runcorn, 1975). In contrast, the technological changes of the past century have had staggering and unprecedented effects on our temporal environment. The invention of electric lights has permitted round-the-clock shift work, and powerful security lights keep some environments brightly illuminated 24 hours a day. The development of jet travel has led to abrupt phase shifts not previously encountered by east–west travelers (most jet travel occurs on east–west, rather than north–south, routes because many commercial centers in Asia, Europe, and North America are at similar latitudes).

Jet lag, the set of physiological and behavioral responses to jet travel across time zones, involves phase shifts in all the zeitgebers at once. Symptoms include sleep disruption; disruption of digestive processes; impaired psychological processes, including attention, perception, and motivation; and a general feeling of malaise (Office of Technology Assessment, 1991). Most studies indicate that these symptoms are less pronounced when travelers phase-delay (travel west) than when they phase-advance (travel east). For example, one study discovered that resynchronization of psychomotor performance rhythms took longer for people on eastbound flights than for people on westbound flights, although eastbound travelers could hasten adaptation to local time by being outdoors (Klein and Wegmann, 1974; Moore-Ede et al., 1982). In general, the severity of jet lag symptoms also correlates with the number of time zones crossed. One study showed that physiological adaptation to west-bound travel (phase-delay) took less time than adaptation to east-bound travel (phase-advance) for travelers who crossed four or more time zones. Virtually everyone can easily adapt to 1-hour phase shifts and is adversely affected by 12-hour phase shifts, but there are tremendous individual differences in the effects of temporal phase shifts on performance, as well as in the speed of resynchronization following phase shifts (see the accompanying figure). Studies also show that elderly people have more difficulty with the effects of jet lag than younger people do. Appropriately timed melatonin treatment may provide some relief for the symptoms of jet lag (Arendt, 1998). Outdoor exercise has also been reported to ameliorate jet lag symptoms (Shiota et al., 1996).

Chronic jet lag can have serious effects on memory function. In one study, flight crews that worked on transmeridian flights were compared to ground crews (Cho et al 2000). The flight crews showed higher salivary cortisol values as compared to the ground crews. The flight crews also displayed memory deficits. A follow-up study examined two sets of flight crews: one flight crew group was designated as 'short-recovery' as they had less than 5 days to recover from an international flight (> 7 time zones), whereas the other flight crew group was designated as 'long-recovery' flight crew and had more than 15 days to recover from such flights. The flight crews were similar in most other ways. Ten female flight attendants were subjected to a functional MRI. Individuals working under the short-recovery schedule had smaller hip-

Although phase responses to light are best observed in free-running animals, these effects of light also operate on animals that are entrained to a light–dark cycle. Light impinges on the phase-advance and phase-delay portions of an animal's daily cycle of responsiveness to light to reset the biological rhythm to exactly

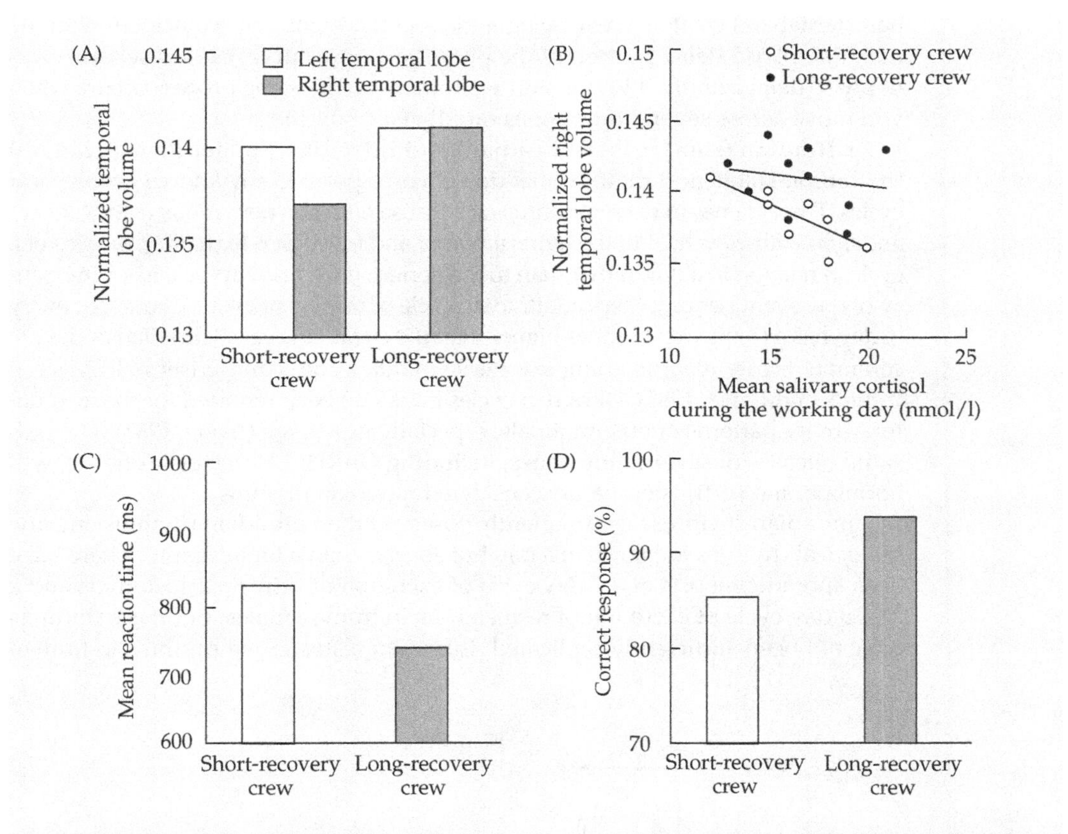

pocampal and right temporal lobe volumes than individuals working the long-recovery schedule (part A in above figure). These short-recovery flight attendants also displayed elevated cortisol (part B in above figure) and significant cognitive impairments (Cho, 2001). Performance on both reaction times and correct responses on a memory test were impared for the short-recovery crew (parts C and D in above figure). The hippocampus is where many memory processes occur. These studies indicate that chronic jet lag should be avoided.

However, even a 1-hour phase shift can be significant for some individuals. In many industrialized nations, the entire population undergoes a 1-hour phase shift twice a year with the change from standard time to daylight saving time. It usually requires several days for people to make the adjustment to this biennial phase shift in their circadian systems (Monk and Aplin, 1980). In the United States, the rate of traffic accidents increases significantly in the week following the time changes (Monk, 1980). Whether this statistic reflects an impairment of attention and psychomotor coordination because of alterations in our circadian systems, or missed appointments because of incorrectly set watches, is open to question.

24 hours—that is, to entrain the rhythm. In an analogous manner, if you owned a wristwatch that gained 15 minutes each day (period = 24.25 hours), you would be the zeitgeber that entrains the watch when you reset it to the correct time each morning. The phase-response curves give clues about what happened to my pet

hamster when I created a reverse light–dark cycle. Light, now coincident with the maximal phase-delay portion of the phase-response curve, caused several days of phase delays until a 12-hour shift was attained. The same process occurs when you move across several time zones rapidly (see Box 10.1).

Ultradian (shorter than circadian) and **infradian** (longer than circadian) rhythms are biological rhythms that do not correspond to any known geophysical cycles. These names may seem confusing because *ultra* is synonymous with "higher" and *infra* with "lower," but the terms *ultradian* and *infradian* refer to the *frequency* of a cycle in relation to a day, rather than to the period. If a circadian cycle has a frequency of one event per day, then an ultradian cycle of one event every 2 hours, or every 1/12 day, has a frequency 12 times higher than the circadian cycle. Ultradian cycles are common (Figure 10.8); an example is the 90-minute cycle characteristic of REM sleep (Schulz and Lavie, 1985). Ultradian cycles have also been reported for the locomotor activity patterns of other animals, especially newborns (Lavie, 1985). The pulsatile releases of several hormones, including GnRH, LH, testosterone, growth hormone, and corticosterone, are considered ultradian rhythms.

Infradian rhythms, less frequently observed than ultradian rhythms, include biological rhythms longer than a day but shorter than a lunar month. There have been sporadic reports of a 7-day cycle of excretion of urinary ketosteroids and a 21–28-day cycle of excretion of testosterone in human males, but these findings have not been sufficiently replicated. Infradian testosterone rhythms in human

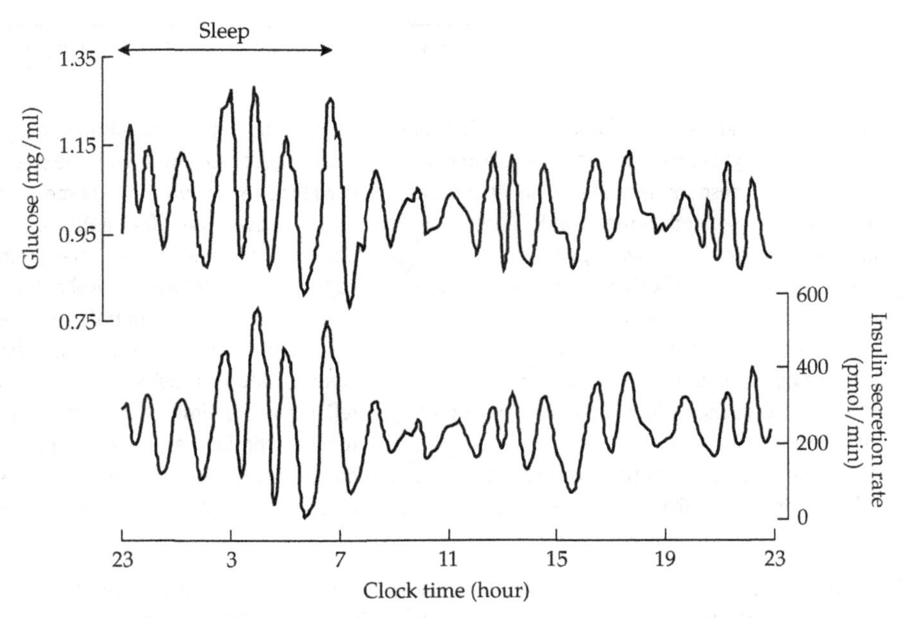

10.8 Ultradian rhythms in human plasma glucose level and insulin secretion rate over 24 hours. The mechanism(s) underlying these ultradian oscillations are unknown. After Simon and Brandenberger, 2002.

males are probably due to exogenous factors, such as pay schedules or exercise patterns, because these types of rhythms are rarely reported for men isolated from social and environmental cues. Generally, infradian rhythms in testicular function are rare among vertebrates. Dormice (*Glis glis*) have been reported to display infradian rhythms of about 60 days in body mass and body fat content (Grimes et al., 1981; Melnyk et al., 1983). The most common types of infradian rhythms are those associated with ovarian (estrous) cycles. Hamsters have very brief, 4-day, estrous cycles. Rats, as you may recall, have estrous cycles of 4 or 5 days. Guinea pigs and sheep have estrous cycles of approximately 16 days. Under the right conditions, these cycles are self-sustaining, endogenously generated biological rhythms that do not correspond to any known geophysical cue. Of course, human menstrual cycles roughly correspond to the period of one rotation of the moon around the earth—the lunar month of 29.5 days. However, the length of the human menstrual cycle seems to be only coincidentally similar to the length of the lunar month, rather than reflecting any adaptive link to the lunar cycle (Knobil and Hotchkiss, 1988).

Examples of Biological Rhythms in Behavior

We can illustrate these principles by looking at examples of the four different types of biological rhythms that occur in nature.

CIRCADIAN RHYTHMS Parental behavior in rabbits shows a distinct circadian rhythm. In the wild, rabbit pups are delivered in a special nursery den that is usually adjacent to the communal warren. The mother doe returns only once a day to nurse her pups, at intervals that at first seemed to field observers to be irregular—a strategy that could serve to deter predators. During the somewhat frenzied 3–5-minute nursing bout, the pups ingest an enormous quantity of milk, often exceeding a quarter of their body mass (Hudson and Distel, 1989; Zarrow et al., 1965). The doe does not interact with her pups during the rest of the day because she is foraging for food. However, the pups anticipate their mother's arrival; they emerge from the nesting material prior to the arrival of the doe in order to commence nursing immediately (Hudson and Distel, 1982; Jilge, 1993). What is the basis of this temporal synchrony? Obviously, the ability to anticipate the arrival of their mother indicates that a predictable nursing rhythm exists and that the pups have the ability to measure time.

In the laboratory, rabbits are born during the light portion of the day. Females return to the waiting litter every 23.15 hours, on average, for the first 10 days. The pups manifest anticipatory activity within the first few days of life (Figure 10.9). What is the zeitgeber for the pups? Remember, they are normally living in a constantly dark burrow, so the light–dark cycle is not available to them as a cue. The zeitgeber for this endogenous biological cycle of anticipatory activity must be the mother herself (Jilge, 1993), but how the mother might actually provide the temporal cue (for example, a nutritional, social, or physiological cue imparted to the infants via her milk), as well as the endocrine correlates of this phenomenon, remains unspecified.

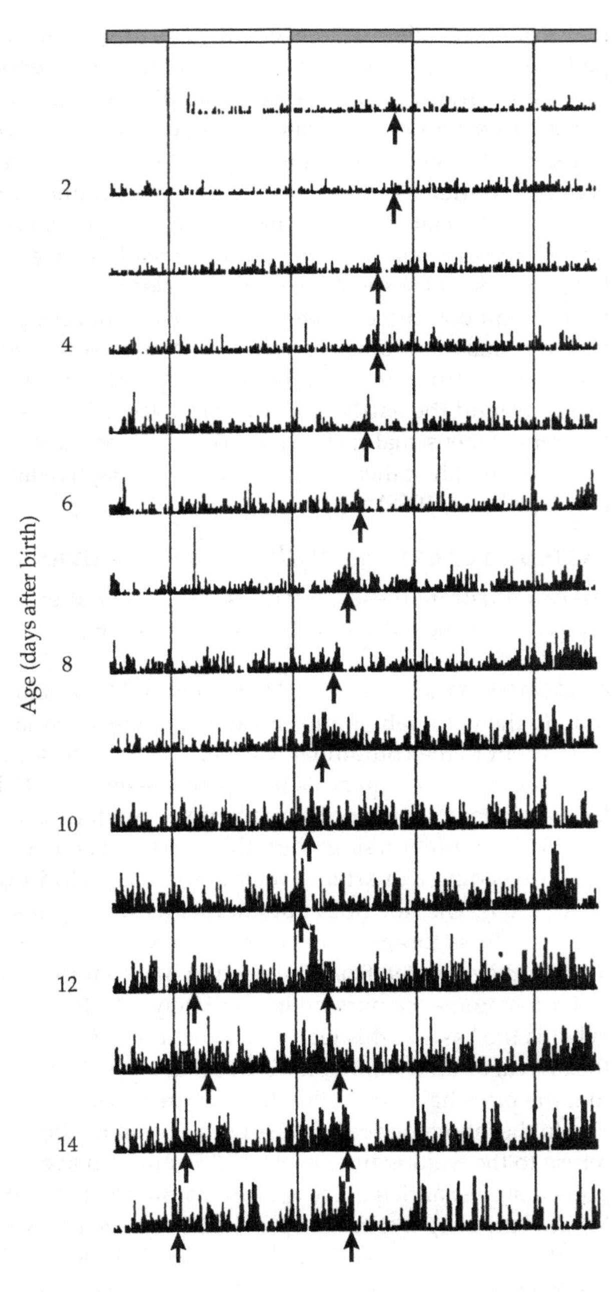

10.9 Development of circadian rhythms in rabbit pups. Rhythms of litter activity are double plotted (each line represents 48 hours, but the second 24 hour segment is repeated on the next day); the time of nursing is depicted by the arrow. Mother rabbits nurse their litters only once a day for an average of approximately 3.75 minutes per day. The dark and white horizontal bars at the top represent light and dark periods of each day. Note the anticipatory activity of the litter prior to the mother's return to the nursery cage. From Jilge, 1993.

CIRCATIDAL RHYTHMS An example of an endogenously generated circatidal rhythm can be seen in the locomotor activity pattern of fiddler crabs (*Uca pugnax*) (Palmer, 1990). These crabs are residents of the intertidal zone and can be observed moving along the marsh or beach during low tide, searching for food

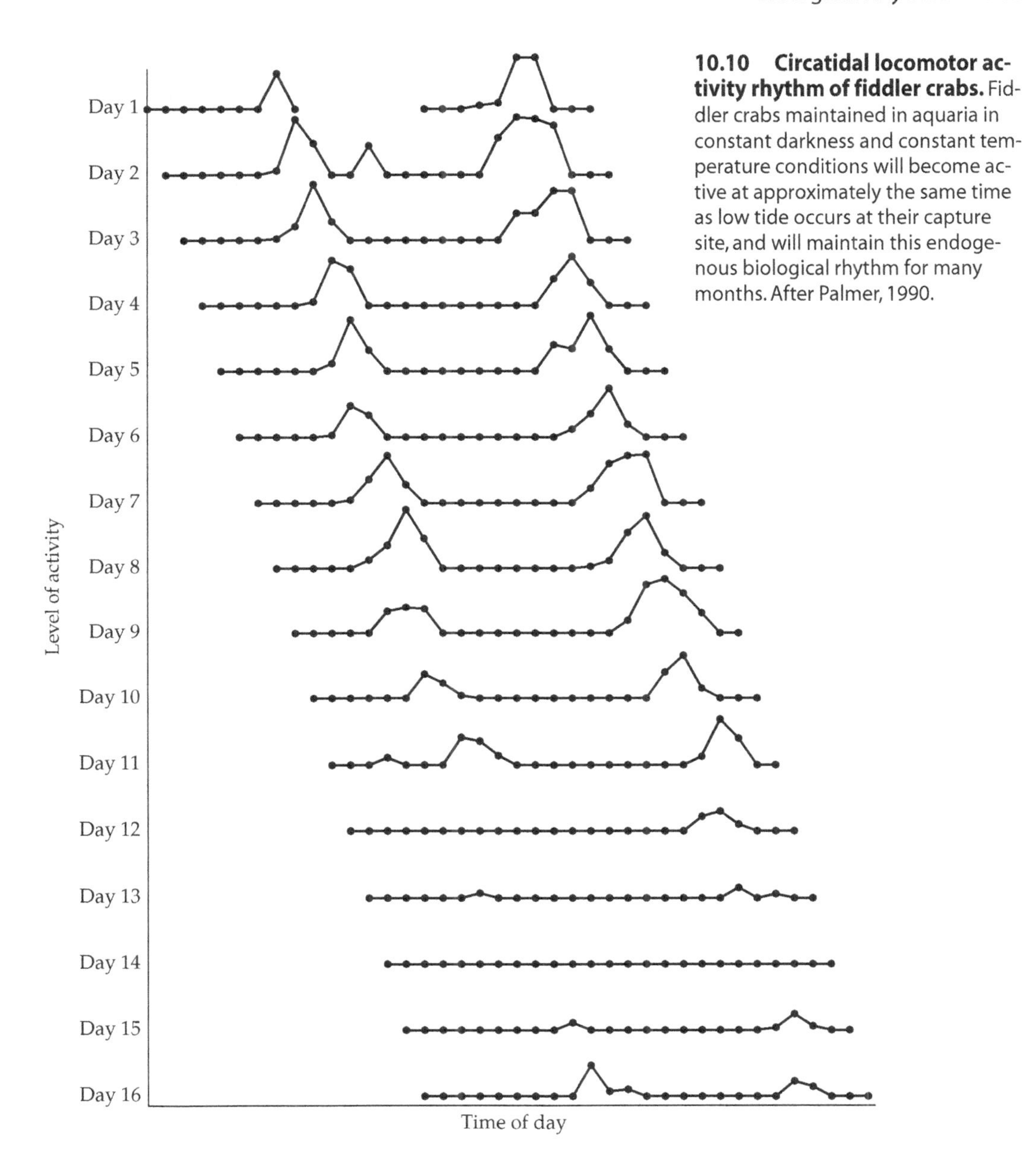

10.10 Circatidal locomotor activity rhythm of fiddler crabs. Fiddler crabs maintained in aquaria in constant darkness and constant temperature conditions will become active at approximately the same time as low tide occurs at their capture site, and will maintain this endogenous biological rhythm for many months. After Palmer, 1990.

and mates. The crabs return to their burrows prior to the onset of high tide, when their environment is flooded by the sea. If a fiddler crab is removed to an aquarium, it will retain its cycle of activity, with slight deviations of approximately 2%–3% from the 12.4-hour tidal cycle (Figure 10.10; Palmer, 1990). Its periods of locomotor activity will alternate with quiescence every 12.4 hours and will correspond to the occurrence of low tide in the area where it was collected.

CIRCALUNAR RHYTHMS A few animals have evolved biological rhythms that permit them to time their behavior within the 29.5-day lunar cycle (Richter, 1968). In general, nocturnal predators increase their evening activities around the time of a full moon, and decrease their activities around the time of a new moon. Prey species show the opposite pattern, avoiding nocturnal activities during evenings with bright moons. Insects may also show this pattern. For example, the ant lion (*Myrmeleon obscurus*) hunts by building a pit with steep walls of shifting sands (Figure 10.11A). The ant lion buries itself at the bottom and waits for an ant or other bug to slide into the pit. Its huge pincers pierce the body of its prey, and the ant lion sucks out the body fluids. Perhaps the ant lion could be called the vampire bug because it increases its activities during the full moon. Even in the con-

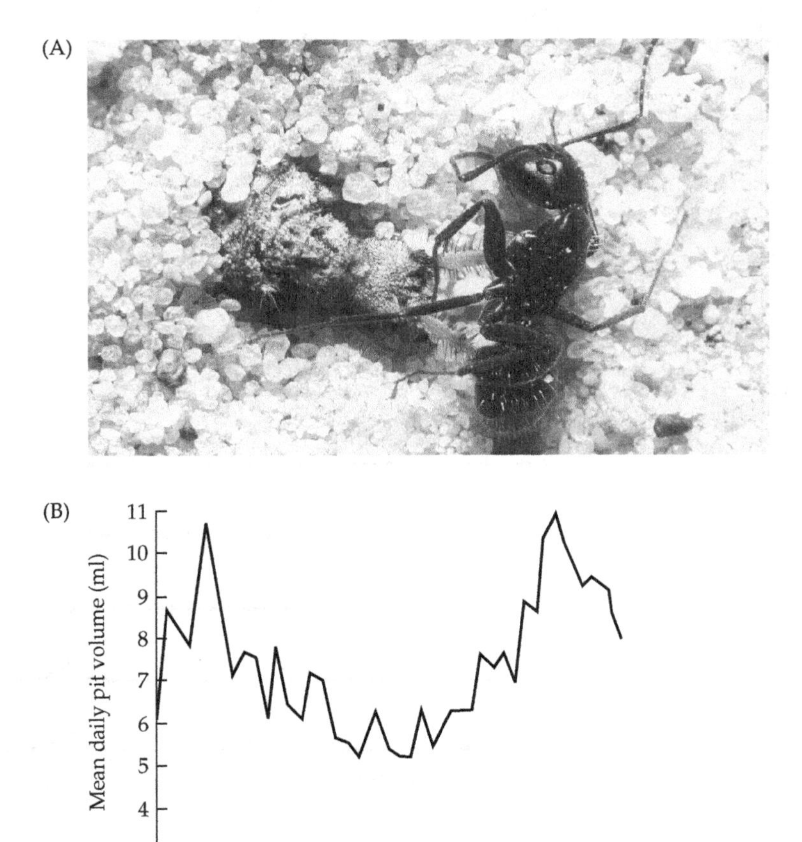

10.11 A circalunar behavioral rhythm. (A) The ant lion hunts by building a pit to catch passing ants or other insects. (B) 50 ant lions that were fed a single ant per day and maintained in constant conditions exhibited a circalunar rhythm in the average size of the pits they built; they dug larger pits when there was a full moon and smaller pits when there was a new moon. (A) from Robert Dunne/Bruce Colman; (B) after Youthed and Moran, 1969.

stant conditions of a laboratory, the ant lion builds a larger pit during full moons than during new moons (Figure 10.11B; Youthed and Moran, 1969).

CIRCANNUAL RHYTHMS Circannual behavioral rhythms include the migratory patterns of birds as well as the hibernation and seasonal changes in body mass observed in several mammalian species. Even when kept in a laboratory for years, birds show 10–12-month cycles of premigratory restlessness, weight gain, and reproductive competence (Gwinner, 1986). Mammals such as ground squirrels also show circannual cycles of weight gain, reproductive competence, and (if the temperature is sufficiently low) hibernation (see below). These annual rhythms usually reflect seasonal changes in energy availability in the environment.

There are also patterns of mixed or hybrid behavioral rhythms. The Pacific palolo worm (*Eunice viridis*) usually spends its time in burrows in the coral reefs off the coast of Fiji and Samoa (Burrows, 1945). But in the months of October, November, and December, on the last day of the last quarter of the moon, at sunrise, a 30-cm segment of each worm breaks off and swims to the surface, where it and other segments explode in unison, forming a slurry of eggs and sperm. The local epicures are aware of this cycle, which enables them to predict the event in advance and arrive at the coral reefs at dawn on the appropriate days to scoop up handfuls of this gastronomic treat. The palolo worms may be consumed raw or roasted, and I have been told that they taste like sushi stuffed with caviar when eaten raw (Figure 10.12).

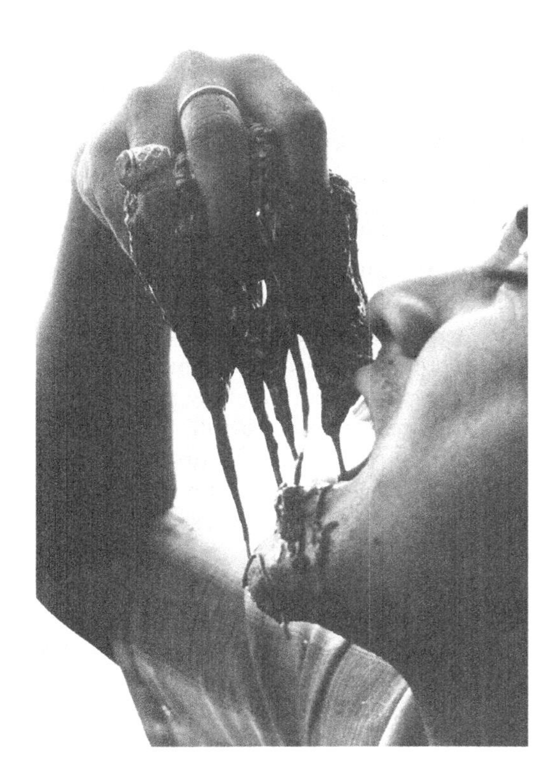

10.12 Palolo worms display a complex reproductive rhythm. These sea worms are native to the South Pacific, and move to the ocean surface to exchange sperm and eggs during the months of October, November, and December—but only at sunrise on the last day of the last quarter of the moon. Local residents consider the palolo a delicacy, and make use of their knowledge of its complex reproductive patterns in collecting the worms for eating.

Human knowledge of cyclic behavior, ranging from grunion "runs" on California beaches to the seasonal migrations of caribou in northern Canada, has allowed the exploitation of many animals for food and other resources.

Usefulness of Biological Clocks

Why have these elaborate timekeeping mechanisms evolved? In other words, what is the adaptive function of biological clocks? As the above examples indicate, biological clocks are useful for synchronizing the activities of animals with conditions in their environments, including their social environments, thus enabling them to prepare for predictable events (for example, night, winter, or the onset of reproductive function in potential mates); and for synchronizing the internal physiological and biochemical processes of animals in order to promote efficient functioning.

One of the most predictable features of life on earth is the regular pattern of environmental changes associated with the movement of our planet. Life evolved in a cyclic environment. As one chronobiologist wrote, "The rotation of the Earth on its polar axis gives rise to the dominant cycle of day and night; the revolutions of the Earth around the Sun give rise to the unfailing procession of the seasons; and the more complicated movements of the Moon in relation to the Earth and the Sun give rise to the lunar month and to the tidal cycles" (Saunders, 1977). Except for animals living at the bottom of the ocean or deep within caves, day follows night, and the seasons change. These orderly and predictable changes in the environment have existed since life first began to evolve on this planet, although the timing of these events may have changed over time. The rotation of the earth results in periodic exposure to the radiation of the sun, which causes predictable changes in light and ambient temperatures, as well as associated changes in the relative humidity of the air and in the oxygen levels of aqueous habitats. The biological clocks of animals and plants permit them to start or stop locomotor activities or activate photosynthetic machinery, respectively, in preparation for light (Figure 10.13).

The action of biological clocks in synchronizing the bodily functions of an individual has been compared to the role of the conductor in an orchestra. Internal processes may serve to prepare the body for certain activities to occur later; for example, the elevated adrenal secretions coinciding with the morning onset of activity prepare you for increased activity levels and for breaking your nightly fast. All eukaryotic and most prokaryotic organisms tested to date, from unicellular organisms to humans, display circadian rhythms. Of course, circadian rhythms have not evolved in organisms that live for less than 24 hours. Similarly, circannual rhythms have evolved only in animals that live for a year or more.

Physiological systems show a wide variety of rhythmic changes. For instance, dopamine turnover, body temperature, and blood plasma levels of potassium, sodium, cortisol, androgens, and growth hormone all show pronounced circadian rhythms (Figure 10.14; Halberg, 1977; Rusak, 1989). Some of these changes may be on the order of 100% to 200% from baseline values. Peak daily cortisol concentrations, for example, which usually occur just prior to or immediately after awak-

10.13 Linnaeus's "flower clock." Linnaeus exploited his knowledge of the activity patterns of heliotropic flowers to design this clock in 1751. He proposed that the clock would operate from 0600 to 1800 (6:00 A.M. to 6:00 P.M.), and that the time of day could be determined by noting which flowers were blooming.

ening, coincide with the onset of locomotor activity in the morning. This programmed elevation of cortisol concentrations increases blood pressure and cardiac output prior to the active phase of the day. We know that these increased cortisol concentrations are not driven by the increased activity levels themselves, because the same circadian rhythm is observed in bedridden patients under constant conditions (Aschoff, 1965). In some instances, the production and release of hormones corresponds to the timing of hormone receptor production. Neurotransmitter receptors are manufactured prior to a circadian-programmed increase in neurotransmitter production. Body temperature peaks in mid-afternoon, when people are most active, but again, muscular activity is not solely responsible for

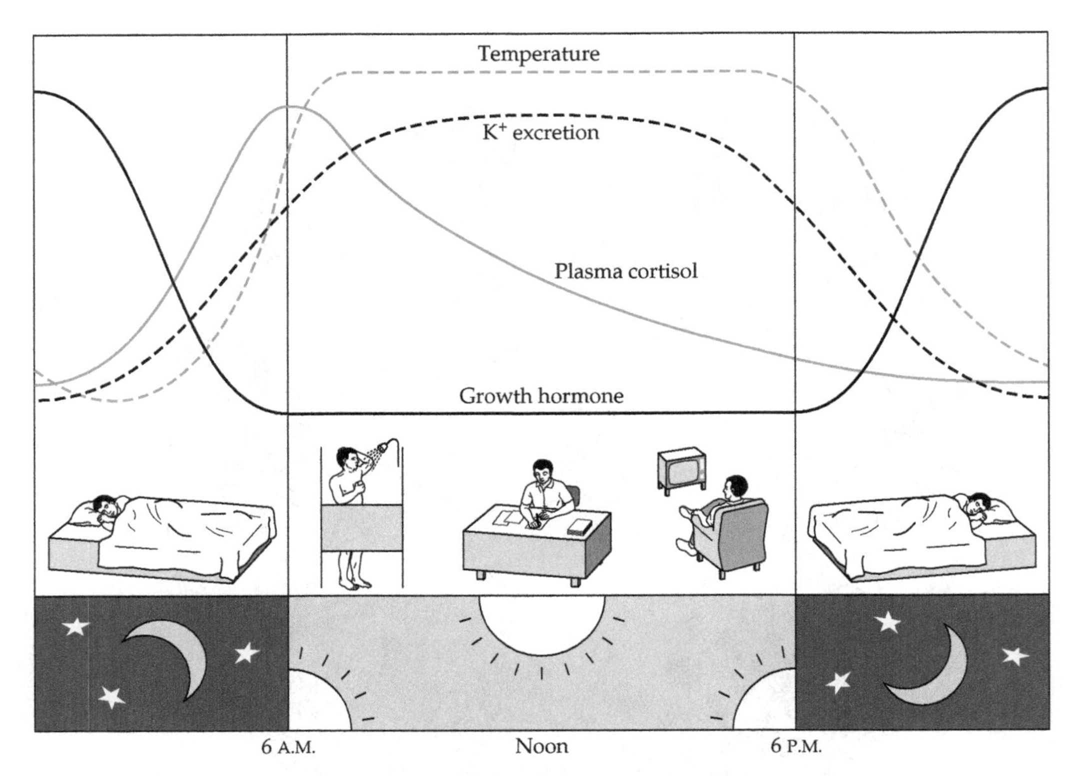

10.14 Circadian rhythms in human physiology and behavior. Activity levels, body temperature, urinary potassium excretion, plasma cortisol levels, and growth hormone secretion exhibit circadian rhythms that are closely coupled with the sleep–wake cycle. After Richardson and Martin, 1988.

this "heating." If you are inactive in bed under constant conditions, your peak body temperature still maintains a circadian rhythm (Moore-Ede et al., 1982).

Circadian Clocks

Circadian biological clocks have been the subject of the most study in chronobiology. Several generalizations can be made about circadian biological clocks on a phenomenological level: they are (1) inherited, (2) relatively independent of temperature, (3) relatively independent of chemical influence, (4) entrained (synchronized) only to limited cycle lengths, (5) relatively independent of behavioral feedback, and (6) found at every level of organization within an organism (Pittendrigh, 1981; Pittendrigh and Daan, 1976).

Localization and Characterization of Circadian Clocks

Where in the body are biological clocks located? Biological clocks are found in both plants and animals, so the nervous system cannot be the fundamental level of organization, although you will soon see that biological clocks in the nervous system play a critical role in the temporal organization of the physiology and behavior of birds and mammals. Because single-celled organisms exhibit circadian rhythms, it is reasonable to suggest that biological clocks are organized at the cellular level (Takahashi, 1991). There are two lines of research attempting to locate the fundamental cellular clock. One approach uses molecular biological techniques to discover the basic features of the clock. The second approach uses classic and molecular genetics to uncover the fundamental properties and location of the clock.

Much has been learned about the mechanisms of the cellular clock through the use of chemicals that promote or block specific cellular functions (Reppert and Weaver, 2002; Hastings et al., 2003). Typically, a chemical agent is given to an organism that is undergoing regular cycles under constant conditions. For example, providing deuterium oxide (heavy water), which affects cell membrane permeability to ions, to hamsters in constant dark conditions causes them to delay the onset of wheel-running behavior by several hours each night (Richter, 1977; Roberts, 1990). Changes in potassium movement across cell membranes alter some features of biological clocks; lithium, which can replace potassium in some physiological systems, also can affect biological rhythms (Klemfuss, 1992; Sweeney, 1976). Pharmacological studies such as these have indicated that important components of the biological clock include protein synthesis, membrane structures, mechanisms for the transport of calcium, potassium, and other ions, and activation of intracellular kinases (Hall and Rosbash, 1987; Takahashi et al., 1993; Reppert and Weaver, 2002; Hastings et al., 2003).

The second approach to understanding the mechanisms underlying cellular clocks has been to attempt to identify the genes that control the clock systems, then to identify the actions of those genes (Konopka, 1987; Takahashi, 1993; Reppert and Weaver, 2002). By studying mutants with unusual free-running circadian rhythms—that is, rhythms with periods substantially greater than or less than 24 hours—and back-crossing these animals with homozygous and heterozygous mates, insights into the locations of the genes that regulate biological clocks can be obtained. Clock mutants have been discovered in fruit flies (*Drosophila*), Syrian hamsters (*Mesocricetus auratus*) (Ralph and Menaker, 1988), and mice (*Mus musculus*) (Oishi et al., 2000), as well as in other simple plants and animals (Hall and Rosbash, 1987). In fruit flies, mutants that display no rhythmicity in their behavior, as well as flies that display 19-hour or 28-hour periods, have been discovered. Most of these mutations have been traced to a single gene, called *per*, which affects the period of the clock (see below) (Konopka and Benzer, 1971). The mechanisms for biological rhythms appear to be similar for all organisms; that is, they all use molecular feedback loops, although some of the specific genes involved differ (Takahashi et al., 1993; Reppert and Weaver, 2002).

On a physiological level, circadian oscillators and pacemakers have been discovered in several biological systems, mostly by damaging certain structures and observing that biological rhythms disappear after their destruction. Among vertebrates, the eyes of amphibians and the pineal glands of fishes, reptiles, and birds contain circadian oscillators (Cahill et al., 1991; Takahashi et al., 1989; Moore, 1997). Among mammals, the suprachiasmatic nuclei (SCN) of the anterior hypothalamus represent the master circadian clock of the body (Moore and Eichler, 1972; Stephan and Zucker, 1972).

Molecular Mechanisms of Circadian Clocks

Mutant fruit flies (*Drosophila melanogaster*) whose circadian rhythms show abnormal periodicities have provided important insights into the nature of the cellular biological clock. In a systematic study, fruit flies were exposed to a mutagenic compound, then observed for unusual biological rhythms (Konopka and Benzer, 1971). Observations of many mutant flies revealed three general types of mutations that affected biological rhythms: (1) flies with rhythms that had abnormally short periods (*per*^{s}), (2) flies with rhythms that had abnormally long periods (*per*^{l}), and arrhythmic flies (*per*^{0}). These flies differed in the period of their biological rhythms from normal, or wild-type, flies. Through selective mating experiments and the use of so-called visible traits (that is, traits whose genes and chromosomal locations are known), Konopka and Benzer were able to map these different mutations to the same location on the X chromosome. Through the use of what is termed a complementation test, it was discovered that all of the mutations occurred within a single section of the same gene on the X chromosome.

Follow-up studies attempted to isolate the gene, *per*, that was responsible for circadian rhythms in *Drosophila* and to elucidate the mechanism of action of this gene (Hardin et al., 1990; Zwiebel et al., 1991). The *per* gene codes for a protein, PER, that has many branched side chains of sugars (a proteoglycan). The PER protein may act in the cell nucleus to link short-term rhythms into a circadian oscillation (Hardin et al., 1990). Levels of *per* mRNA and the PER protein produced in response to the transcription of this gene exhibit a circadian rhythm: there is a 6–8-hour lag between the production of *per* mRNA and the production of the PER protein. The various mutant flies showed rhythms in *per* mRNA and PER protein synthesis with periods similar to the periods of their other abnormal biological rhythms.

Although much progress has been made, the precise mechanisms of the cellular clock remain unknown. It has been determined that just the 5′ end of the *per* gene is sufficient to cause the cycling of *per* mRNA expression, but is not sufficient to cause cycling in levels of the PER protein product. Thus, the PER protein rhythm does not simply reflect the *per* mRNA rhythm, as might be expected. Administering the wild-type PER protein to mutant flies results in normal (that is, wild-type) rhythms of the mutant mRNA in vivo, indicating that the PER protein feeds back to influence *per* gene transcription (Hardin et al., 1990). Other organisms, including humans, possess pieces of genes that resemble the *per* genes in *Drosophila*, and similar molecular mechanisms have been reported for *Neurospora* (a fungus) and for mice (Dunlap, 1998).

Another *Drosophila* mutant, called *timeless*, exhibits no circadian rhythms in constant darkness. These *timeless* mutants are missing the *tim* gene, and do not display translocation of the PER protein into the nucleus; perhaps the TIM protein is required for nuclear translocation of the PER protein.

Another protein that appears to be critical in flies (Allada et al., 1998; Gekakis et al., 1998) is called CLOCK. This protein is notable because it contains both a DNA-binding region and a site that binds to another protein (Dunlap, 1998). Another *Drosophila* protein was discovered, named CYCLE, that has proved to be an important component of the clock mechanism (Rutila et al., 1998). CYCLE is more commonly called BMAL now (or sometimes MOP). Flies with mutations of the *clock* or *BMAL* genes did not display any expression of the *per* and *tim* genes. It was discovered that CLOCK and BMAL bind together; then this so-called heterodimer binds to a specific region of DNA, called a promoter, that increases the transcription of the *per* and *tim* genes, but does not appear to affect other gene activities (Rutila et al., 1998; Allada et al., 1998).

Here's how the cycling works. As PER is produced, it is destroyed by a kinase, but over time, sufficient TIM builds up to bind to PER before the kinase can destroy it. TIM binds to any remaining PER protein and forms another heterodimer, which is resistant to the kinase. The TIM–PER heterodimer enters the nucleus, where it turns off the *clock* (negative feedback loop) and (possibly) the *BMAL* genes (Dunlap, 1998) (Figure 10.15). Over time, the TIM–PER heterodimers

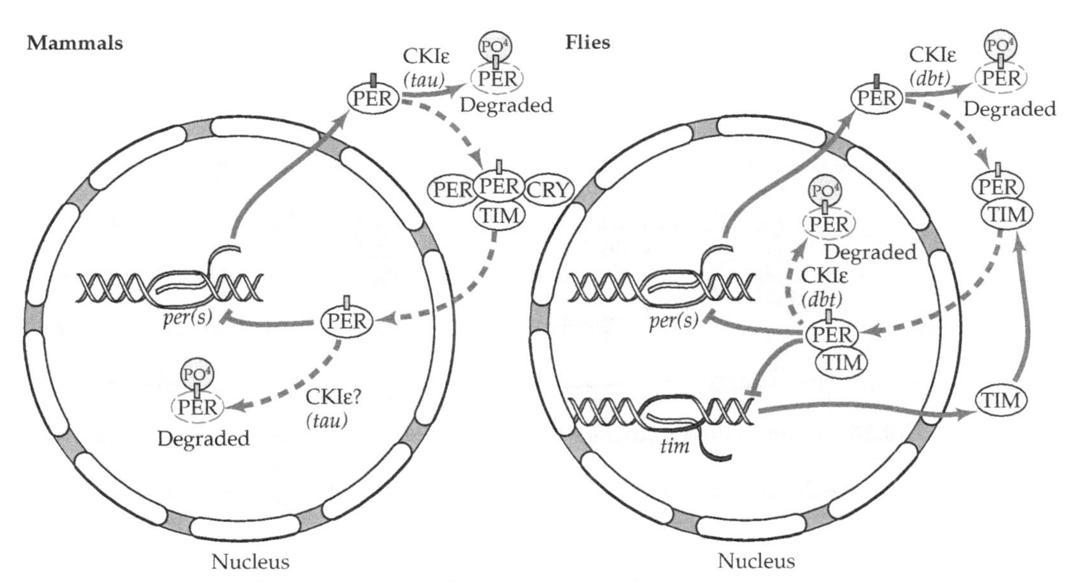

10.15 The genetics of biological clocks in mammals and fruit flies. In both groups, a set of three key genes produces proteins that interact to regulate the activity of certain other genes on a cycle lasting approximately 24 hours. One of the genes (*per*) codes for a protein (PER) that gradually builds up inside and outside the cell nucleus over time. Another key gene is *tau* in mammals and *dbt* in flies, which code for an enzyme that helps breakdown PER, slowing its rate of accumulation in the cell. But during peak periods of production of PER, more PER is available to bond with another protein (TIM) coded for by a third gene (*tim*). The PER-TIM complex does not degrade, so more PER re-enters the nucleus, where it blocks the activity of the very gene that produces it, though only temporarily. Then a new cycle of *per* gene activity and protein production begins. After Young, 2000.

are deactivated, and the CLOCK and BMAL proteins build up again to activate the *per* and *tim* genes (positive feedback loop) (Darlington et al., 1998). The temporal delay introduced by the kinase brings the oscillation to roughly 24 hours, although more regulatory loops and regulatory components may remain to be discovered.

Similar rhythmic patterns of gene expression provide the engine underlying cellular circadian clocks in other organisms, although the genes involved vary from species to species (Figure 10.16). Although these genes can be expressed in tissues throughout the body, their expression in the master clock, the SCN, is critical for integrated temporal coordination in mammals (Hastings et al., 2003). In mice, for example, essentially the same molecular feedback loops exist as in *Drosophila*, although the role of the *tim* gene appears to be performed by a gene called *cryp-*

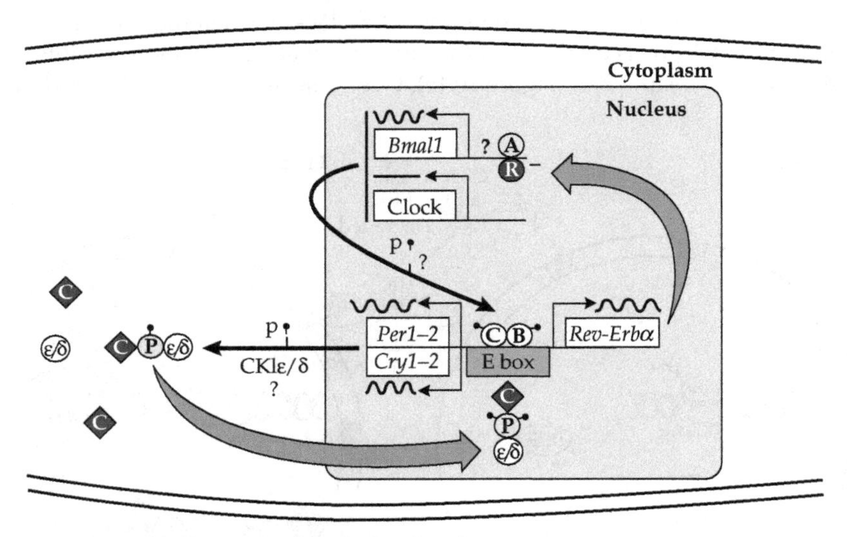

10.16 Genes regulate circadian timing in mice. Similar to flies, several genes have been identified that contribute to the circadian organization of cells. For example, *BMAL1* (Brain and Muscle ARNT-Like 1), *Clock*, *CK1*ε (caseine kinase 1 epsilon), *mPer* (murine period), and *mCry* (murine cryptochrome) all are circadian clock genes identified in mice. Some of these genes have several variants (e.g., *mPer* and *mCry* genes). In this schematic drawing, the clock mechanism comprises interactive positive (black) and negative (gray) feedback loops. CLOCK (C, circle) and BMAL1 (B, circle) proteins form heterodimers and activate transcription of the Per, Cry and Rev-Erb genes by binding to "molecular switches" (i.e., called E boxes) located in front of those genes. This is called a positive feedback loop. As Per and Cry (C, diamond) proteins accumulate, they form complexes with CKI /CKI (ε/δ, circle), that join with CLOCK–BMAL1 heterodimers to inhibit their own synthesis, and the protein levels decline. This part of the timekeeping process is a negative feedback loop. Thus, the BMAL1 and clock proteins promote activation of the Per and Cry genes, whereas Per proteins inhibit activation of those genes. The 24-hour cycling comes about as the BMAL1 and Clock proteins induce increased production of Per and Cry proteins. CK1ε protein also helps to regulate Clock protein levels by destabilizing Per protein.

tochrome (*Cry*) (Reppert and Weaver, 2002; Roenneberg and Merrow, 2003). TIM remains important in circadian timing in mice, however, as transgenic mice with low levels of TIM lose circadian activity in the SCN (Barnes et al., 2003). It is generally accepted that CLOCK and BMAL1 (also called MOP3) combine to form heterodimers that activate rhythmic transcription of three murine *period* genes (*mPer1*, *mPer2*, and *mPer3*), two *cryptochrome* genes (*mCry1* and *mCry2*), and at least one nuclear receptor gene (*Rev-Erbα*). *Per1* and *Per2* show rhythmic function, but *Per3* is only induced by light. The PER and CRY proteins move from the cytoplasm into the nucleus, and CRY protein interacts with the CLOCK–BMAL1 heterodimer to prevent further transcription, thereby closing the negative feedback loop. The positive feedback loop involves the rhythmic transcription of *Bmal1*, which peaks about 12 hours out of phase with the peak transcription of *mPer*, *mCry*, and *Rev-Erbα*. The Rev-Erbα protein suppresses *Bmal1* transcription, leading to the increase in PER and CRY levels coincident with the low levels of BMAL1. As CRY interacts with the CLOCK–BMAL1 heterodimer to inhibit transcription of *mPer*, *mCry*, and *Rev-Erbα*, the suppression of *BMAL1* transcription is released, and BMAL1 mRNA levels increase. Thus, both positive and negative feedback loops are regulated via the CLOCK–BMAL1 heterodimers through different protein products (Reppert and Weaver, 2002). Presumably, hormones that affect gene transcription could affect the expression of biological rhythms through this cycle.

Mutations of any of the genes encoding the clock proteins can affect the period of the clock, mainly by affecting protein stability. For example, phosphorylation of the PER2 protein is reduced in *tau* mutant hamsters (which exhibit abnormally short periods) because of a mutation in the casein kinase enzyme (Lowrey et al., 2000). Casein kinase normally "tags" the PER protein for degradation. In *tau* mutants, the inability of casein kinase to phosphorylate PER makes it more stable, so that it prematurely feeds back to the cell nucleus and evokes a shorter than normal period. Similarly, humans with familial advanced sleep phase syndrome (ASPS), an inherited sleep disorder in which the onset of sleep is advanced 2–4 hours each day, have mutations in the phospho-acceptor sites for casein kinase on the PER2 protein, which also leads to reduced phosphorylation and possibly increased protein stability (Toh et al., 2001). As the effects of mutated components of this molecular feedback system on physiology, behavior, and health are specified, an understanding of the functional roles of the proteins encoded by the clock genes will be gained.

The SCN as Master Circadian Clock

Several types of evidence suggest that the mammalian SCN contain circadian oscillators (Figure 10.17). First, bilateral lesions of the SCN eliminate circadian organization of physiology and behavior (Moore and Eichler, 1972; Stephan and Zucker, 1972). Second, thin slices of SCN tissue continue to display circadian rhythms of electrical activity even when maintained in a culture dish (Gillette, 1986; Shibata and Moore, 1988). Finally, if SCN tissue is transplanted into an SCN-lesioned recipient, the period or the phase of the recipient's rhythm will match that of the donor (Lehman et al., 1987) (Figure 10.18).

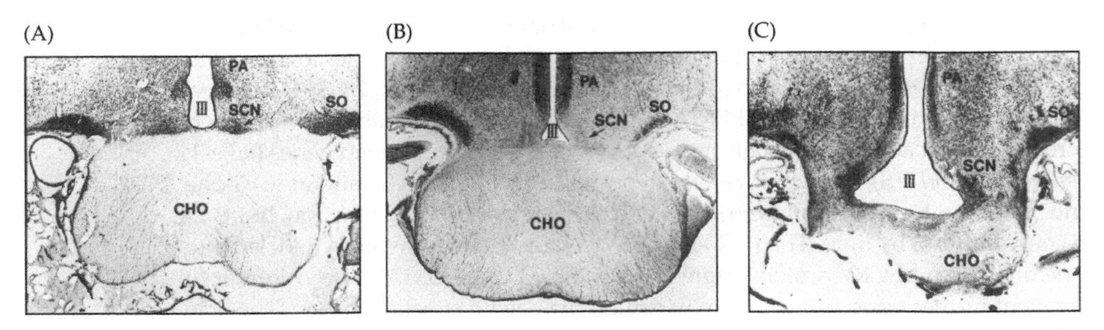

10.17 The SCN in mammals. Coronal sections through the anterior hypothalamus in (A) a squirrel monkey, (B) a rhesus monkey, and (C) a human, showing the location of the SCN bilateral to the third ventricle (III) and dorsal to the optic chiasm (CHO) in all three species. PA = paraventricular nuclei; SO = supraoptic nuclei. Courtesy of Ralph Lydic.

The SCN are paired clusters of approximately 8000–10,000 neurons, each located above where the optic nerves cross in the hypothalamus and below the third ventricle (see also Figure 3.25). Recordings from individual SCN neurons reveal that circadian timekeeping abilities are not emergent properties of the entire SCN, because each cell seems capable of generating its own specific circadian rhythm (Hastings et al., 2003). A daily rhythm of glucose utilization in the SCN was first observed in rats (a nocturnal species) (Schwartz and Gainer, 1977). Autoradiographic studies in which rats were injected with 2-deoxyglucose

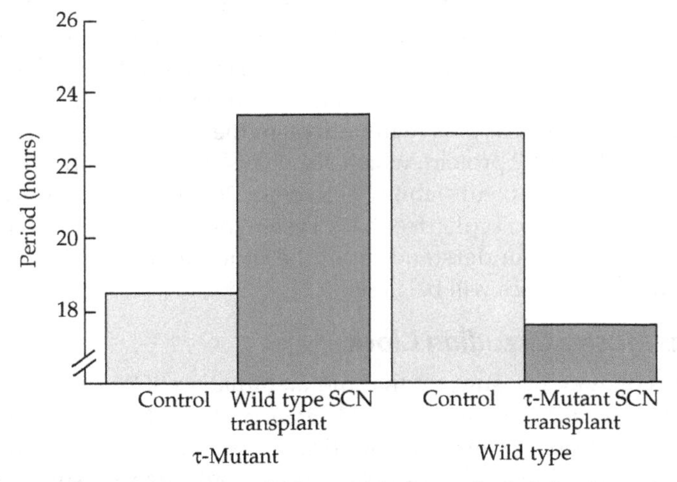

10.18 Restoration of circadian rhythms with the period of the donor by SCN transplants. Wild-type (i.e., approximately 24-hour free-running periods) or *tau* (τ)-mutant (i.e., approximately 20-hour free-running periods) hamsters received SCN lesions, then received SCN transplants from the other genotype. In all cases, the recipient hamsters took on the free-running periods of the SCN donors. After Ralph and Lehman, 1991.

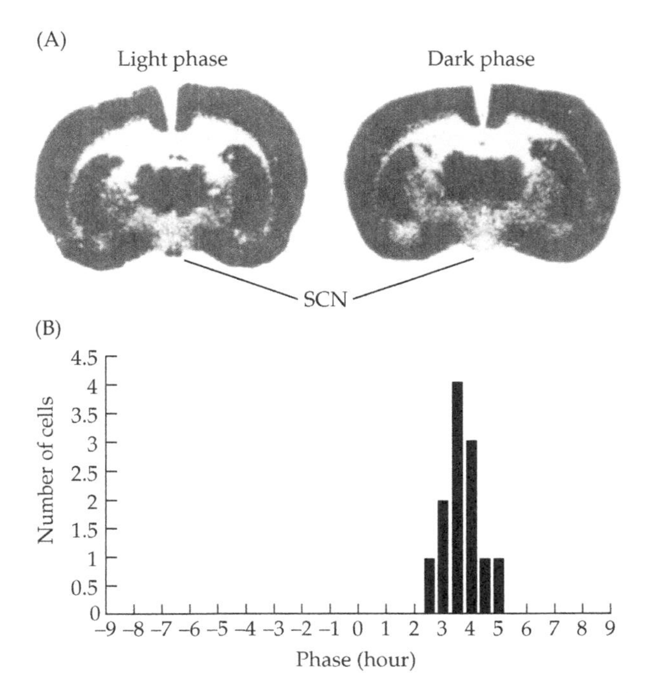

10.19 Activity of the SCN in mammals varies according to the time of day. (A) Rats utilize glucose during the day. Autoradiographs of coronal sections through the SCN of rats injected with radioactively-tagged glucose during the light or dark phase of the daily light–dark cycle show that energy use by the SCN in these nocturnal animals is higher during the day (dark staining) than at night. (B) Electrical activity in dispersed murine SCN neurons peaks around 4 hours after the lights are turned on (at time 0). (A) from Schwartz and Gainer, 1977; (B) after Aujard et al., 2001.

revealed that the SCN were metabolically active during the light phase of the daily light–dark cycle, but relatively inactive during the dark phase (Figure 10.19). Although some evidence suggests that other circadian pacemakers exist within the central nervous systems of mammals, the SCN clearly contain the "master clock" and are critical in the temporal organization of individuals. Bilateral lesions of the SCN provoke a loss of circadian organization of physiology and behavior (Figure 10.20). Indeed, SCN lesions of free-living ground squirrels increased the likelihood of their being killed by a predator, presumably because of inappropriate timing of their daily activities (Figure 10.21).

The central circadian oscillators have been located in several places in birds. The pineal gland is an important circadian clock among many species of passerine (perching) birds (Takahashi et al., 1989; Zimmerman and Menaker, 1979; Gwinner et al., 1997). Several types of evidence suggest that the avian pineal gland contains circadian oscillators. First, pineal tissue continues to display circadian rhythms of biosynthetic activity even when maintained in a culture dish (Takahashi et al., 1993). Second, in some species, if pineal gland tissue is transplanted into a pinealec-

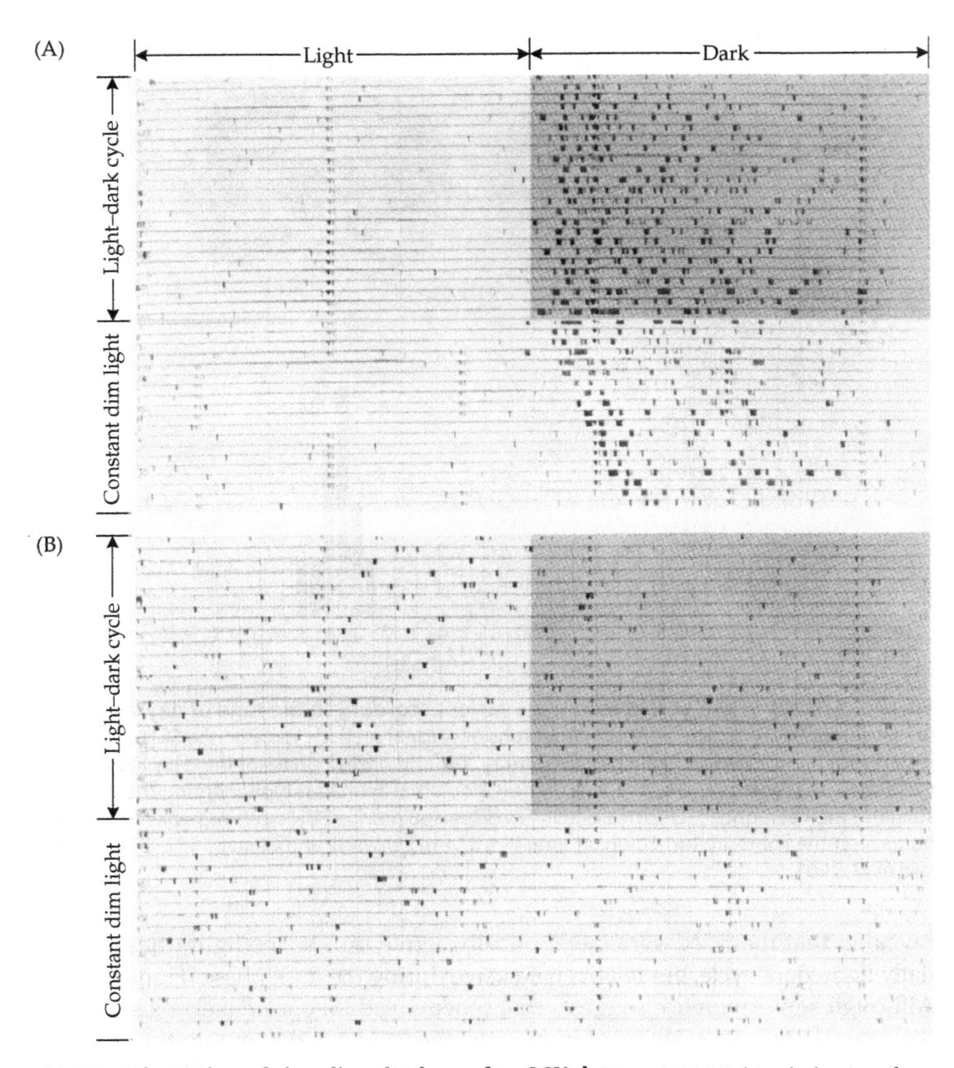

10.20 Disruption of circadian rhythms after SCN damage. (A) Drinking behavior of a normal hamster housed in a 12:12 light–dark cycle displays a 24-hour rhythm (top of panel); in constant dim light, a free-running circadian rhythm of 24.25 hours is displayed. (B) After lesions to both SCN, no temporal organization of the behavior is evident, regardless of lighting conditions. From Zucker, 1980.

tomized recipient, the period or the phase of the donor rhythm is imposed on the recipient (Gwinner et al., 1997). For example, a new circadian phase can be imposed in sparrows by implants of pineal glands from donor birds housed under different light–dark cycles than the recipient birds (Meijer and Rietveld, 1989). However, the pineal gland is not required for the expression of circadian rhythms in chickens and Japanese quail. In quail, the eyes control the circadian rhythm of melatonin secre-

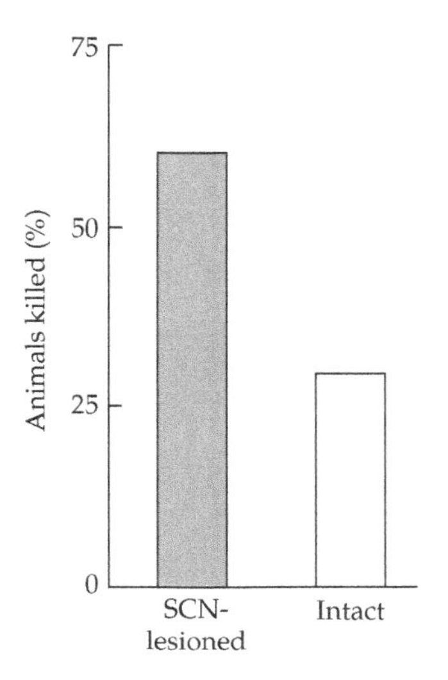

10.21 Antelope ground squirrels with SCN lesions were more likely to be killed by a predator than intact squirrels. Predation by a feral cat on free-living squirrels was monitored by infrared video camera. From DeCoursey et al., 1997.

tion, and they also contribute about 30% of the plasma melatonin. If the optic nerves of quail are cut, their activity rhythms are disrupted, but the ocular melatonin secretion rhythms remain intact (Takahashi, 1992). The role of the pineal gland in seasonal breeding is discussed below.

Although there have not been any clear demonstrations that the SCN mediate circadian organization in humans, several clinical case studies have reported that anterior hypothalamic damage in the vicinity of the optic chiasm was associated with disturbances in the daily sleep–wake or temperature cycle (reviewed in Cohen and Albers, 1991; Schwartz et al., 1986). In one case study of a 34-year-old female patient, A.H., a tumor was removed from the inferior region of the anterior hypothalamus that included the SCN (Cohen and Albers, 1991). A.H. displayed significant disruptions in her daily body temperature and sleep–wake cycles, as well as impairments in cognitive and behavioral function, after the surgery (Figure 10.22). The disruptions in cognition and behavior appeared to reflect an inability to maintain consistent levels of arousal and/or attention (Cohen and Albers, 1991). Arrhythmic circadian sleep–wake cycles have also been associated with schizophrenia (Wirz-Justice et al., 1997). Individuals with advanced sleep phase syndrome (ASPS), which affects about one-third of the elderly population, tend to fall asleep early in the evening, at about 7:00 P.M., and awaken spontaneously between 2:00 and 4:00 A.M. This type of ASPS can be treated with bright light. As noted above, people with the rare, inherited form of ASPS have a single base pair mutation in the human *per* gene (*hPer2*). This gene is located at the telomere of chromosome 2q and is homologous to *Drosophila* and mouse genes that shorten the period of circadian rhythms when mutated (Toh et al., 2001).

SCN Inputs and Outputs

In addition to a clock, any biological timekeeping system must have an input system (which usually brings light information to the clock) and an output system,

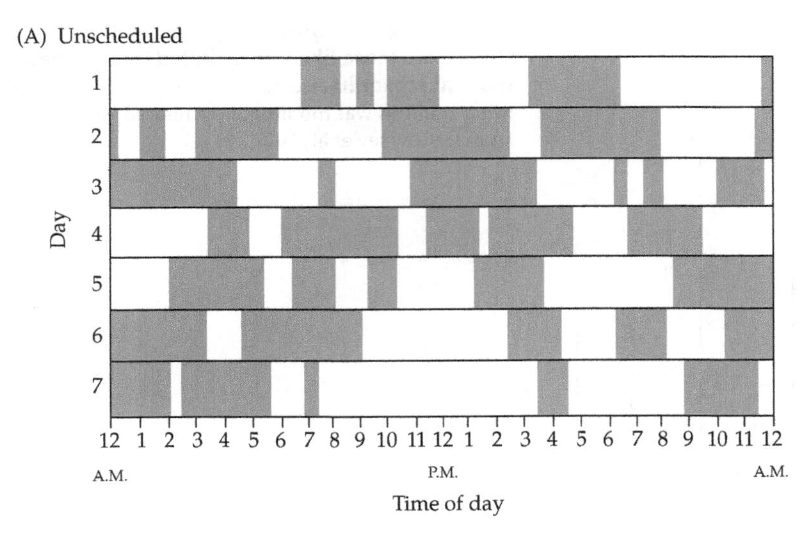

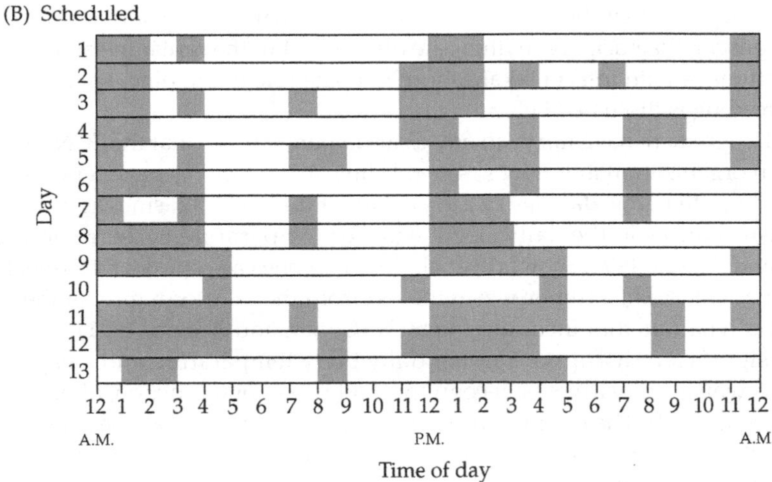

10.22 SCN damage disrupts the daily sleep–wake cycle in humans. (A) The sleep–wake cycle of A.H., a woman whose SCN was damaged during surgery, shows little daily temporal organization in an unscheduled environment (waking periods are shown in white; shaded areas represent time asleep). (B) A.H.'s sleep–wake cycle shows daily temporal organization if bedtime, waking time, and meals are scheduled by the nursing staff. Sleep–wake data were obtained every 15 minutes by the nursing staff. From Cohen and Albers, 1991.

which may be coupled to any number of species-specific physiological and behavioral systems (Takahashi et al., 1993; Abrahamson and Moore, 2001; Reppert and Weaver, 2002) (Figure 10.23). In some systems, these components may all occur in the same structure. For example, individual pineal cells from chickens possess

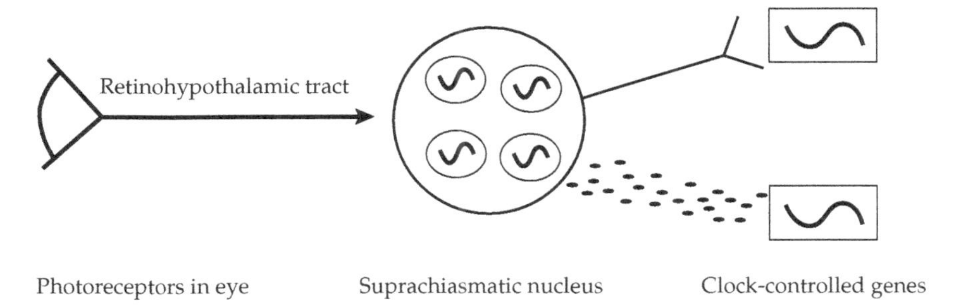

10.23 Schematic representation of the three main components of a mammalian biological clock system. Environmental light information is transduced from receptors in the eyes to the oscillators in the SCN. Temporal information leaves the SCN via neural and humoral signals to synchronize physiology and behavior. After Kriegsfeld et al., 2001.

a photoreceptor, a clock, and melatonin synthesis capabilities in vitro (Cassone, 1998; Nakahara et al., 1997). Similarly, a photoreceptor protein is located in the clock cells that are located throughout the head, thorax, and abdominal tissues (including the sensory bristles on the wings and legs) of *Drosophila* (Plautz et al., 1997). However, it is generally accepted that only the eyes contain photoreceptors that ultimately influence biological rhythms in mammals (Nelson and Zucker, 1981). The rod or cone photoreceptors traditionally associated with vision are sufficient, but not necessary, for entrainment (Berson, 2003; Bellingham and Foster, 2002). Mice that congenitally lack rods and cones entrain to light as well as sighted animals (Bellingham and Foster, 2002). Specialized retinal ganglion cells appear to convey photic information to the SCN. The photopigment used in the ganglion cells appears to be similar to a frog photopigment called melanopsin (Bellingham and Foster, 2002; Berson, 2003; Van Gelder, 2003). Targeted deletion of the gene for melanopsin in mice impairs circadian entrainment.

Environmental light entrains the oscillations of the SCN (Box 10.2). Light information received by photoreceptors in the eyes is transduced to the SCN via a glutamatergic pathway, called the retinohypothalamic tract (RHT), that is distinct from the classic visual system (Figure 10.24; Pickard, 1982; Rusak and Boulos, 1981). An indirect pathway from the intergeniculate leaflet of the thalamus, the geniculohypothalmic tract (GHT), also carries photic information to the SCN (Stopa et al., 1995). Neuropeptide Y appears to be the primary neurochemical signal from the GHT and may serve to phase-shift the endogenous circadian system (Gillespie et al., 1996; Huhman et al., 1996, 1997; Stopa et al., 1995). The SCN contain a heterogeneous population of neurons that express several neuropeptides, including neuropeptide Y, arginine vasopressin (AVP), vasoactive intestinal polypeptide (VIP), gastrin-releasing peptide, and somatostatin (Duncan, 1998; Hofman et al., 1996; Huhman et al., 1997; Stopa et al., 1995). Specifically, GABA-releasing neurons in the dorsal "shell" of the SCN express AVP, whereas neurons in the ventral "core" express VIP (Abrahamson and Moore, 2001; Hastings et al.,

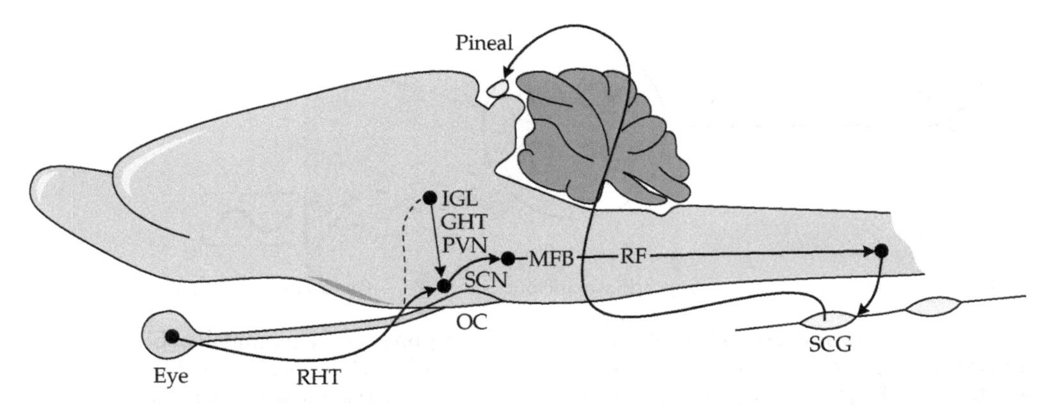

10.24 Input and output pathways to and from the SCN in mammals. Environmental light information enters the SCN from the eyes via the retinohypothalamic tract (RHT) or geniculohypothalamic tract (GHT) via the intergeniculate leaflet (IGL). One well-characterized output system involves projections from the SCN to the paraventricular nuclei (PVN), from the PVN to the medial forebrain bundle (MFB), and from the MFB to the superior cervical ganglion (SCG). Postganglionic noradrenergic fibers eventually project back into the brain and innervate the pineal gland, where neural information is transduced into a hormonal message. OC = optic chiasm; RF = reticular formation. After Klein et al., 1983.

2003). Both portions of the SCN innervate a wide range of targets. But it is not clear what is distinctive about the SCN that makes this section of the brain the primary oscillator, rather than some other brain tissue. Similar neurotransmitters are expressed in the SCN and elsewhere in the brain. Even the clock genes expressed in the SCN are expressed in other brain regions. Presumably, something about the unique neurotransmitter combination and connectivity within the SCN, as well as its constellation of inputs and outputs, make the SCN the central oscillator in mammals.

What are the SCN output pathways in mammals? A brain surgery can be performed that cuts all neural input to and output from the SCN. These isolated "islands" of brain tissue connected to the SCN continue to exhibit circadian rhythms in firing rates (Inouye and Kawamura, 1979). Because biological rhythms are so pervasive among physiological and behavioral systems, there are probably hundreds of output pathways leading from the SCN. Although endocrine rhythms are disrupted in animals with SCN islands, circadian locomotor rhythms persist or recover. This finding indicates that endocrine rhythms require a neural connection, whereas rhythms of locomotor activity do not. In addition to the SCN, mammalian clock genes are expressed rhythmically in several peripheral tissues, including the heart, lungs, adrenal glands, and liver, and these rhythms become disorganized, but continue, in SCN-lesioned animals, suggesting that the SCN communicates with these peripheral organs directly to synchronize gene expression (Hastings et al., 2003; Yoo et al., 2004). The autonomic nervous system is one potential route of communication between peripheral organs and the SCN, but additional research is necessary to demonstrate this functional connection.

BOX 10.2 Effects of Light on Gene Transcription

As we have seen, hamsters maintained in constant darkness exhibit free-running activity rhythms, but a pulse of light is sufficient to reset the phase of those rhythms (see Figure 10.7). Such light-induced phase shifts are caused by the activation of immediate early genes (IEGs) (Kornhauser et al., 1990) (see Chapter 5). To understand this process, it is useful to review the concept of gene transcription.

The process of gene transcription moves the protein-making instructions contained in the DNA molecule (which resides in the cell nucleus) to the protein-manufacturing apparatus, which lies outside the nucleus. The DNA unravels, and one of the exposed strands of DNA provides a template on which a messenger molecule, called messenger ribonucleic acid (mRNA), is formed. The newly transcribed mRNA leaves the cell nucleus and travels to the rough endoplasmic reticulum. After processing to remove extra bases that do not code for proteins (introns), the mRNA represents only the coding regions, or exons, of the gene. The mRNA then provides the instructions for production of chains of amino acids through a process called translation. If the resulting chain of amino acids is short, it is called a peptide; if it is long, it is called a protein.

Analyses of the regulation of gene transcription have revealed that several switches are involved in turning transcription on and off. In some cases, hormones or environmental input provide the stimulus to begin or end gene transcription. Each strand of the DNA molecule has a 3′ and a 5′ end. Transcription always proceeds from the 5′ end toward the 3′ end. Upstream from the start of the coding region of a gene are several short segments of the DNA molecule that do not code for proteins; rather, they powerfully affect the rate at which the coding region of the gene is transcribed into mRNA. This *enhancer* region acts like a rheostat switch to control the promoter region of the DNA. The *promoter* region is usually located about 30 bases upstream from the start of the coding sequence and is essential for the actual initiation of gene transcription. Steroid hormones and steroid hormone–receptor complexes may bind to specific receptor sequences within the enhancer region and act to repress or activate gene transcription. Approximately six different types of binding sequences have been located within enhancer regions of DNA, one of which is called the AP-1 complex.

The first genes activated when a cell receives an activation signal are the IEGs, which code for a variety of proteins in the Jun family, including Jun-B, c-Jun, and c-Fos. Jun-B and c-Fos bind together

Continued on following page

Maintained in darkness

After 1-hour light pulse

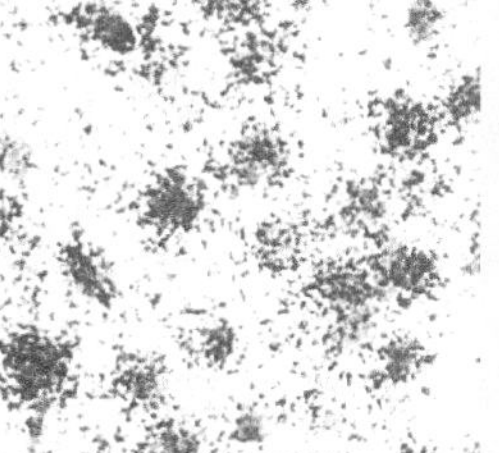

Figure A Photomicrographs courtesy of Jon Kornhauser.

BOX 10.2 *continued*

to form a dimer that binds to the AP-1 sequences in the enhancer regions of other genes to stimulate gene transcription. Thus, activation of IEGs is often the first indication that the genetic machinery within a cell has been activated by an external or hormonal stimulus. In addition to their effects on the enhancer region, IEG proteins can feed back to affect the promoter region, starting or stopping gene transcription.

Light appears to activate IEGs in the SCN of both nocturnal and diurnal animals (Katona et al., 1998). The photomicrographs shown in Figure A are in situ hybridization depictions (see Chapter 1) of different magnifications of the SCN from a hamster maintained in constant darkness (top), and from a hamster maintained in darkness and then exposed to a 1-hour pulse of light (bottom). Activation of the *c-Fos* gene is seen in the bottom left panel as light areas, and in the bottom right panel as dark spots lying over cells. Light also affects the expression of clock genes in the SCN. Mice receiving a 30-minute pulse of light showed a clock-gated up-regulation in the expression of *mPer1*. Activation of the *Per1* gene is plotted graphically in Figure B for various circadian times (CT) throughout the day.

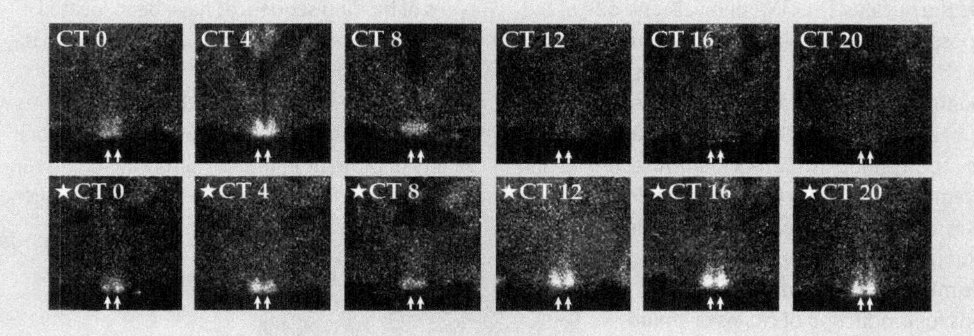

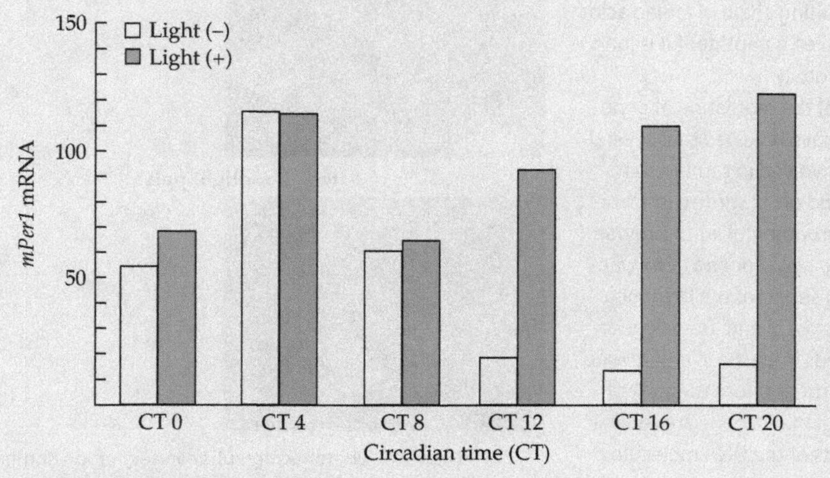

Figure B From Shigeyoshi et al., 1997.

It is generally accepted that two different types of output signals emerge from the SCN (see Figure 10.23). Neural efferents from the SCN appear to be necessary for maintaining most types of endocrine circadian rhythms. One of the largest efferent pathways from the SCN travels to a region below the paraventricular nuclei (PVN), called the subparaventricular zone (sPVZ). Other direct pathways from the SCN innervate hypothalamic areas that mediate neuroendocrine function, including the PVN (which mediate the release of corticotropin-releasing hormone, thyrotropin-releasing hormone, vasopressin, and oxytocin), the supraoptic nuclei (SON) (vasopressin and oxytocin), and the arcuate nuclei (thyrotropin-releasing hormone, corticotropin-releasing hormone, GnRH, and prolactin-inhibitory factor) (Kriegsfeld et al., 2002; 2004). Several other pathways directly and indirectly influence neuroendocrine function.

A second type of SCN output appears to be a humoral, diffusible signal. One candidate for such a signal is arginine vasopressin (AVP). The SCN display a circadian rhythm of AVP release and are responsible for a circadian rhythm of AVP in the cerebrospinal fluid (Majzoub et al., 1991). This AVP rhythm is lost after SCN lesion, but a dampened rhythm persists after transection of most of the SCN efferents (Reppert et al., 1987). However, the AVP rhythm cannot be responsible for mediating circadian locomotor activity rhythms, because Brattleboro rats, as you will recall from Chapter 9, lack AVP, and they display normal circadian organization (Reppert et al., 1987). The function of the circadian rhythm of AVP remains unspecified. Nitric oxide (NO) has been considered as another candidate for the diffusible SCN signal (Ding et al., 1994), although mutant mice that lack NO appear to have normal circadian locomotor rhythms (Kriegsfeld et al., 1999, 2001). Additional candidates include prokineticin-2 and transforming growth factor-α (Cheng et al., 2002; Jobst et al., 2004).

Some kind of diffusible factor must be responsible for maintaining the circadian rhythms of locomotor activity. Locomotor activity rhythms persist despite large SCN lesions as long as a group of cells that contain a specific calcium-binding protein (calbindin D28K) is spared (LeSauter and Silver, 1999; Kriegsfeld et al., 2004). If SCN are transplanted into the brains of arrhythmic, SCN-lesioned hamsters, circadian locomotor activity rhythms are restored (Lehman et al., 1987; Ralph et al., 1990) (see Figure 10.18). However, only transplants that contain the calbindin D28K cells restore circadian locomotor rhythms (LeSauter and Silver, 1999). Calbindin D28K is not the diffusible signal; instead, it probably defines a region of the SCN that may relay photic signals to those pacemaker cells in the dorsal SCN that do not receive information from the retina directly (Hamada et al., 2003). Additional evidence for a diffusible factor is the finding that if SCN implants are placed in microbaskets that block the formation of neural connections, circadian organization of locomotor activity is still restored (Silver et al., 1996). For example, if *tau* mutant hamsters, which exhibit short (about 20 hours) free-running periods, have their SCN destroyed, then receive encapsulated grafts of fetal SCN tissue from normal hamsters (Figure 10.25), they develop circadian rhythms that free-run with a period of approximately 24 hours. Similarly, if normal hamsters receive SCN lesions, then receive encapsulated SCN transplants

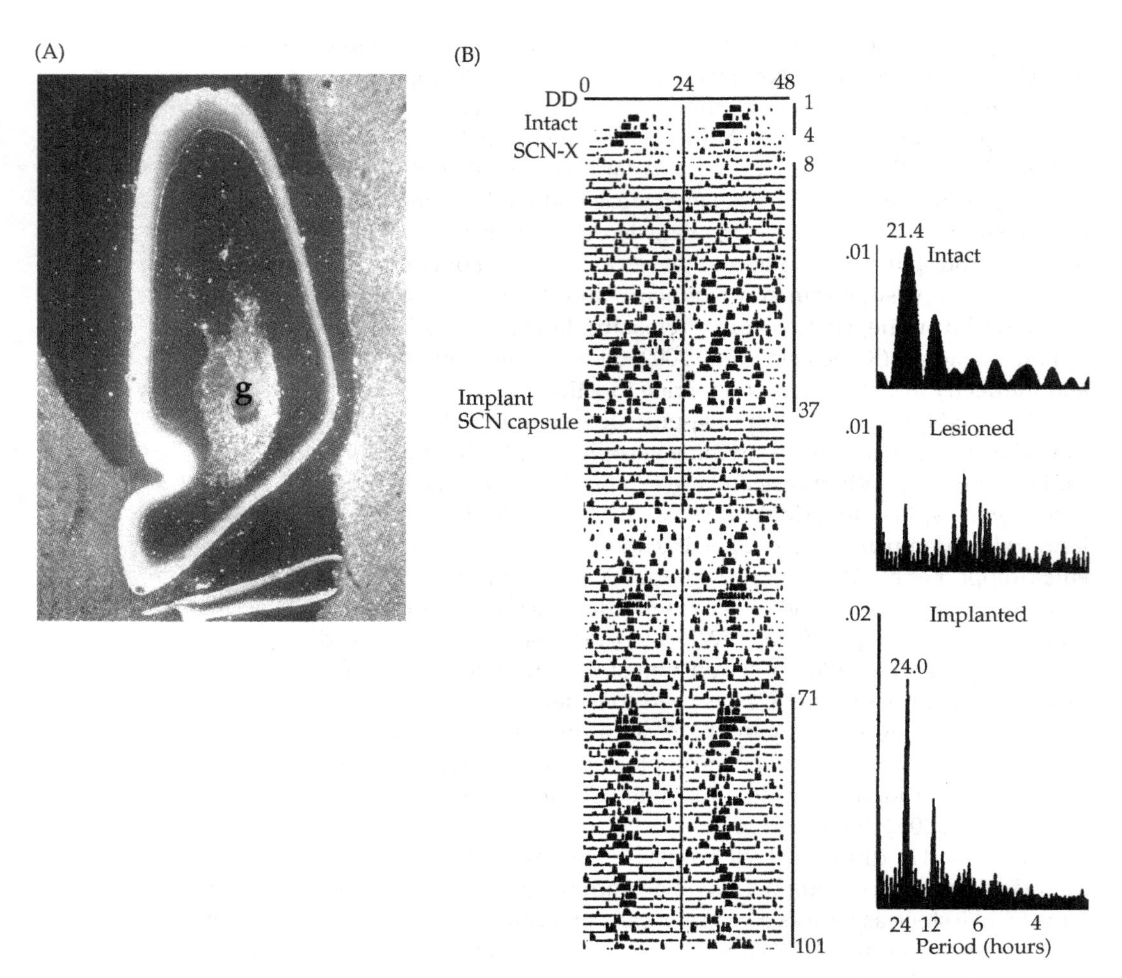

10.25 Photomicrograph of an encapsulated graft that restored circadian locomotor activity to a SCN-lesioned hamster. (A) The encapsulated graft (g) of SCN tissue lies in the third ventricle, and the basket encasing it prevents neural connectivity. (B) The wheel-running record of a *tau* mutant hamster that had a free-running period of 21.4 hours before the SCN were lesioned. The power spectral analysis on the far right of the actogram shows a peak at 21.4 hours prior to the SCN lesion (SCN-X), then random activity after the SCN lesion. After receiving an encapsulated graft from a normal donor, the *tau* mutant hamster exhibited a free-running period of 24.0 hours. From Silver et al., 1996.

from *tau* mutant hamsters displaying short free-running periods, the recipient hamsters develop the short free-running rhythms of the donor animal (Ralph and Lehman, 1991). The appearance of a donor rhythm in a transplant recipient indicates that the pacemaker cells that generate circadian rhythms exist in the SCN tissue. The phase relationship to the light–dark cycle can also be transplanted via donor SCN (LeSauter et al., 1997). In some way that is not completely understood, the SCN generate or coordinate overt circadian locomotor rhythms within individuals. SCN grafts do not restore circadian organization of the endocrine system (Lehman et al., 1991; Meyer-Bernstein et al., 1999), which appear to require the "hard wiring" of neural connectivity.

Perhaps the best-known output pathway from the SCN transduces light information into an endocrine message in the pineal gland, which secretes the hormone melatonin (see Figure 2.34). The SCN project to the PVN, which in turn project through the medial forebrain bundle (MFB) out of the central nervous system to the superior cervical ganglia (SCG), a part of the sympathetic nervous system. Neurons in the SCG project back into the CNS via the spinal cord to the pineal gland (see Figure 10.24). In the dark, these neurons secrete norepinephrine, which stimulates the pinealocytes to up-regulate their enzymatic activity to produce melatonin from serotonin. This multisynaptic pathway has been confirmed by injecting labeled pseudorabies, a transneuronal retrograde tracer, into the pineal gland (Card, 2000). The pattern of melatonin production and secretion is not a passive response to light and dark, but is a programmed rhythm that persists in constant conditions (although continuous bright light dampens the melatonin rhythm). Lesions of the SCN or any of its output tracts to the pineal gland eliminate circadian rhythms in melatonin production and secretion (Klein et al., 1983; Scott et al., 1995).

Among mammals, a retinal–SCN axis (with several output pathways, including the pineal gland) appears to underlie the circadian system. Indeed, this axis appears to exist among all vertebrate species, although there is great diversity among species in the extent to which each component contributes to circadian rhythms. There is greater complexity in the contributions of the component parts of the circadian systems of nonmammalian vertebrates than in mammals. For example, multiple input pathways and multiple oscillators have been discovered in different species of birds (Takahashi, 1991). The input pathways may involve receptors located in the retina, in the pineal gland, or dispersed throughout other brain regions. Extraretinal photoreception can be demonstrated by fitting a bird with opaque contact lenses; the bird will continue to be entrained to the daily light–dark cycle. However, if ink is injected under the scalp of the bird, it begins to free-run, because the light receptors necessary for entrainment are blocked. If the ink is removed, entrainment is observed again (Groos, 1982) (Figure 10.26).

Are there different clocks for rhythms of different lengths, or are longer rhythms simply the result of the multiplication of shorter rhythms? For example, is the 4-day estrous cycle of a hamster the result of a 4-day clock cycling once, or a 1-day clock cycling four times? Another possibility does not involve a clock at all, but rather assumes that estrous cycles simply reflect the cumulative amount of time required to carry out a certain sequence of physiological processes or stages. These hypotheses were evaluated in a study of free-running rhythms in female hamsters. Phase shifts of the circadian rhythm were accompanied by proportionate shifts in the 4-day (96-hour) estrous cycle (for example, if the daily light–dark cycle was changed to 25 hours, then the 4-day estrous cycle changed to 100 hours)

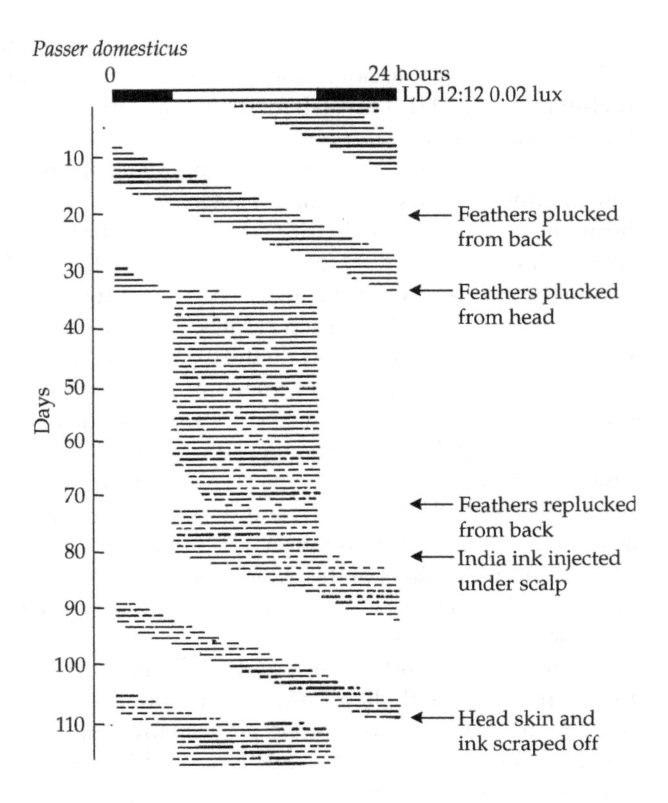

10.26 Circadian locomotor activity in birds is entrained by non-retinal photoreceptors in the brain. If a house sparrow (*Passer domesticus*) is fitted with opaque contact lenses, the bird will continue to be entrained to the daily light–dark cycle (even under dim illumination 0.02 lux) if the feathers on the head are plucked. As the feathers regrow, the bird will begin to free-run. However, if the feathers are plucked again, re-entrainment is attained. If ink is injected under the scalp of the bird, it begins to free-run again because the photoreceptors in the brain necessary for entrainment are blocked. If the ink is removed, entrainment is observed again. LD12:12 means lights are on for 12 hours and then it is dark for 12 hours. From Groos, 1982.

(Carmichael et al., 1981; Fitzgerald and Zucker, 1976). If the daily light–dark cycle was reduced to 20 hours, then the estrous cycle was shortened to 80 hours. These findings imply that the circadian rhythm may be, at least in part, the basis of the estrous cycle length. However, it appears that estrous cycle length is actually a combination of circadian organization and physiological limitations on stage length. Exposing ground squirrels chronically to 22-hour days significantly shortened the circannual rhythm of their reproductive function (Lee et al., 1986). However, in one study comparing normal hamsters (with free-running activity periods of about 24 hours) with *tau* mutant hamsters (with free-running activity periods of about 20 hours), it was reported that estrus recurred every 96 hours, rather than every 80 hours, in the mutants (Refinetti and Menaker, 1992). This observation indicates that the two rhythms are usually coupled, but that they have distinct underlying mechanisms. Incidentally, gestation required 16 days for both types of hamsters; the timing of gestation did not merely reflect 16 circadian cycles for the *tau* mutant hamsters.

Effects of Hormones on the SCN

Hormones may have many effects on daily locomotor activity cycles. As noted previously, hamsters in the laboratory become active 5 to 10 minutes after the lights

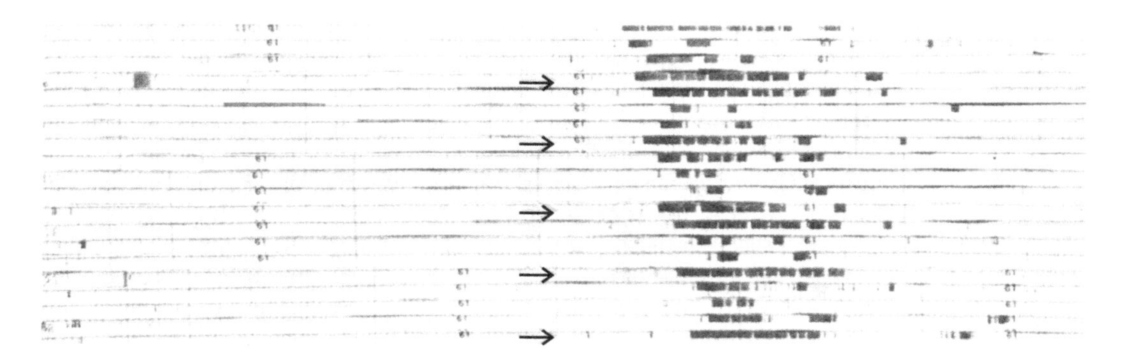

10.27 Estrogen-induced "scalloping" of activity patterns. Free-running female hamsters kept in chronic dark conditions begin locomotor activity at the expected time on most days, but on the day of estrus, the onset of their wheel-running behavior is phase-advanced (arrows) by rising levels of estrogens. From Zucker, 1980.

are turned off. This statement is true of males, but only partially true of females. Female hamsters display an interesting pattern of activity onset that has been termed "scalloping." Every fourth night, coincident with estrus, females show a spontaneous phase advance (Figure 10.27). Hamsters are solitary creatures, and it has been speculated that the earlier onset of activity during estrus increases the female's odds of locating a male. The scalloping pattern can be eliminated by ovariectomy. Furthermore, estradiol treatment of free-running, ovariectomized hamsters or rats reduces the period of locomotor activity onset, suggesting a direct effect of estrogens on the clock itself (Albers, 1981; Morin et al., 1977). Although there are low levels of estrogen receptors in the SCN (Gundlah et al., 2000), neurons in the preoptic area, amygdala, arcuate nuclei, and BNST possess the α estrogen receptor in relatively large numbers, and these structures communicate directly with the SCN (de la Iglesia et al., 1999). The connection from the SCN to the GnRH-secreting cells in the preoptic area is neural (de la Iglesia et al., 2003).

In addition to estradiol, other sex steroid hormones exhibit modulatory effects on biological oscillators. Progesterone lengthens the period of circadian rhythms in females, possibly by counteracting the effects of estradiol (Takahashi and Menaker, 1980). Sex steroid hormones can also affect daily activity rhythms in males. Castration lengthens and androgen replacement restores the free-running period of locomotor activity onset in male mice (Daan et al., 1975). Testosterone also seems to be important in consolidation of locomotor activity rhythms; castrated male rodents increase the duration of, and fluctuations in the onset of, the daily active period (Ellis and Turek, 1983; Morin and Cummings, 1981). Castrated rodents also reduce the number of daily wheel revolutions (Morin and Cummings, 1981). Testosterone replacement therapy reverses these effects of castration. Testosterone may have direct effects on the SCN, although androgen receptors have been identified in the SCN thus far only in ferrets (Kashon et al., 1996). Alternatively, in some cases, testosterone may be aromatized to estradiol, and estradiol could exert its effects either directly or on neurons that communicate with the SCN (de la Iglesia et al., 1999).

Removal of the pituitary gland (hypophysectomy) lengthens the free-running period of locomotor activity onset by about 12 minutes per day (Zucker et al., 1980). However, the endocrine sequelae of hypophysectomy are profound; in addition to disruptions in sex steroid hormone production, it alters many other endocrine functions. In order to separate the effects of hypophysectomy from other endocrine consequences of this surgery, endocrine manipulations of other systems mediated by the anterior pituitary have been attempted. Investigators have focused on the thyroid gland because of its obvious and direct effects on metabolic processes. Removal of the thyroid gland results in a shortening of the free-running period of locomotor activity onset in canaries, and thyroid hormone replacement therapy results in a corresponding lengthening of the period (Wahlstrom, 1965). Hypothyroidism induced by the drugs propylthiourea or propylthiouracil has been correlated with lengthened free-running periods of locomotor activity onset in hamsters (Beasley and Nelson, 1982; Morin et al., 1986). It remains unresolved whether alterations in thyroid hormone secretion per se or exposure to the anti-thyroid drugs accounts for the changes in period length (Morin, 1988). Injections of thyrotropin-releasing hormone directly into the SCN phase-advanced the onset of wheel-running behavior in hamsters (10 and 100 nM doses phase-advanced the behavior 18 and 35 minutes, respectively) (Gary et al., 1996). These hormonal influences on behavioral and physiological cyclic phenomena may eventually provide a key to understanding clock functions directly.

Effects of the SCN on Hormones

GLUCOCORTICOIDS A strong circadian rhythm of glucocorticoid secretion has been observed in many species of birds and mammals, including humans. As noted previously, cortisol secretion begins to rise during sleep, and the highest daily basal cortisol concentrations usually occur just prior to or immediately after awakening in humans, coincident with the onset of activity in the morning (see Figure 10.14). In rodents and other nocturnal animals, the corticosterone cycle is reversed, but the peak values coincide with the beginning of their active period as night begins (Albers et al., 1985) (Figure 10.28). These glucocorticoid rhythms persist in constant conditions in both rodents and primates (Czeisler et al., 1999; Moore and Eicler, 1972). Bilateral SCN lesions eliminate the corticosterone rhythms in rats (Moore and Eichler, 1972) and hamsters (Meyer-Bernstein et al., 1999). The timing of meals can significantly affect the timing of glucocorticoid rhythms (e.g., Wilkinson et al., 1979), and may therefore be another important factor in synchronizing biological rhythms, as we will see below.

GONADOTROPINS AND SEX STEROID HORMONES The SCN are important for the appropriate timing of gonadotropin secretion. The timing of gonadotropin secretion affects the timing of sex steroid secretion and the onset of reproductive behaviors,

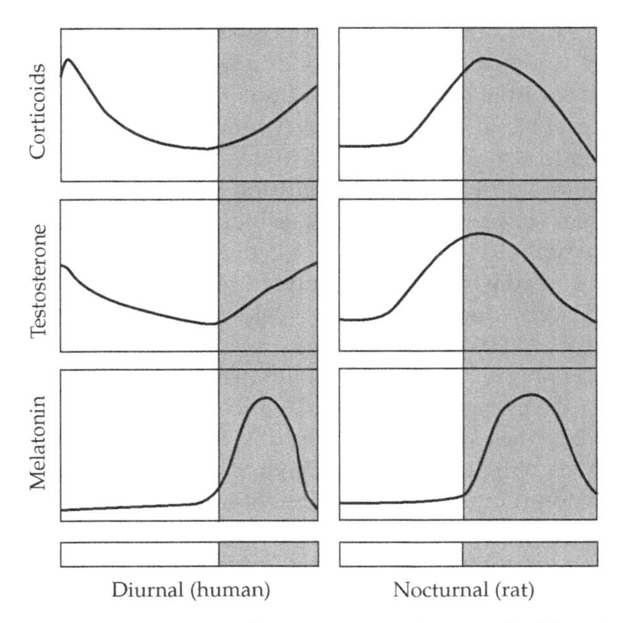

10.28 Daily circulation patterns of testosterone, glucocorticoid, and melatonin in humans and nocturnal rats. The shaded part of each graph represents night. After Kriegsfeld et al., 2001.

especially in females. For example, a neuronal pathway made up of neurons located in the anterior hypothalamus, preoptic area, and SCN, as well as the arcuate nuclei, is involved in providing a daily signal for ovulation if the appropriate endocrine milieu is present. This surge generator normally transmits every afternoon in rats (Gay et al., 1970); that is why there are 4- or 5-day estrous cycles in rats, but no 4.5-day cycles, as we saw in Chapter 6. Bilateral lesions of the SCN block the daily pulse of LH release (Gray et al., 1978). Despite the strong circadian organization of gonadotropin secretion in females, estrogen concentrations do not show consistent circadian rhythms. In contrast, testosterone concentrations display clear circadian rhythms in males that persist in constant conditions (Plant, 1981; Dibeu et al., 1983) despite weak or nonexistent circadian rhythms in gonadotropin concentrations in males (Velduis et al., 1986). In diurnal primates, blood testosterone concentrations begin to rise in the middle of the night and peak around dawn (Velduis et al., 1986) (see Figure 10.28).

MELATONIN A strong circadian pattern of melatonin secretion is observed in both nocturnal and diurnal animals. As we shall see below, day length (photoperiod) can be measured by the timing of melatonin secretion. Melatonin can phase-shift circadian locomotor activity rhythms in mice (Benloucif and Dubocovich, 1996; Arendt, 1998). Melatonin also affects entrainment in birds (Gwinner et al., 1997) and lizards (Bertolucci and Foa, 1998). Importantly, daily treatment with pharmacological doses of melatonin entrains the locomotor activity rhythms of nocturnal rodents such as rats and mice, but not diurnal chipmunks (Murakami et al., 1997). Despite the diurnal organization of humans, melatonin treatment may be useful in treating jet lag or sleep disorders (Waterhouse et al., 1998; Kripke et al., 1998). Melatonin has also proved effective in treating blind persons who are free-running. As you might imagine, this condition causes recurrent insomnia and daytime sleepiness because the endogenous biological rhythms drift in and out of phase with the typical 24-hour work day. In one study, subjects who were totally blind (i.e., lacked conscious perception of light) were given either 10 mg of melatonin or a placebo daily for several weeks at 1 hour before their preferred bedtime (Sack et al., 2000). When given the placebo, their sleep–wake cycles free-ran with a period of approximately 24.5 hours. The melatonin entrained the free-running sleep–wake cycle to 24 hours in about 86% of the subjects (Sack et al., 2000) (Figure 10.29). These researchers were able to entrain the sleep–wake cycles of blind people with a nightly dose of 0.05 mg of melatonin (Lewy et al., 2001). It is clear that melatonin influences entrainment in humans; however, the safety, toxicology, and teratogenic effects of melatonin have not been completely determined. Melatonin is sold over the counter in the United States as a nutritional supplement, but is strictly regulated as a neurohormone in Europe (Arendt, 1997; Guardiola-Lemaitre, 1997).

What is the zeitgeber synchronizing activity in humans? There are several possible candidates, including light–dark cycles, temperature rhythms, and social influences. Social cues do affect the timing of human circadian rhythms. For example, when a group of people is housed in an underground bunker with no time referents, the entire group may adopt a single free-running rhythm (Coleman, 1986). In some cases, the group members adopt the free-running rhythm of the individual with the most dominant personality (Palmer, 1976). In many other studies, the effects of social cues cannot be untangled from other potential zeitgebers, such as mealtimes or the timing of the sleep–wake cycle. However, as we saw with the studies of blind people, melatonin provides a potent zeitgeber to humans (Lewy et al., 2001).

10.29 Results of treatment of seven blind people with free-running circadian rhythms with melatonin. Each data point represents an assessment of circadian phase, as determined by the time that endogenous plasma melatonin concentrations rose above the threshold of 10 pg/ml. The slopes of the fitted regression lines are indicative of the subjects' circadian periods (shown in hours below the regression lines for the pre-treatment and placebo conditions). Treatment with melatonin or placebo began on day 1. In (A) and (B), the regression lines are arranged on a relative time scale in ascending order so that they can be easily compared. In (C), the time scale is absolute and shows the assessments of circadian phase and fitted regression lines for all seven subjects before (dashed lines) and after (solid lines) the melatonin trial. Treatment with melatonin resulted in entrainment (a circadian period of 24.0 hours) in all but one person (subject 7); on average, the rise in plasma melatonin after entrainment occurred at 24:18 (12:18 A.M). After Sack et al., 2000. ▶

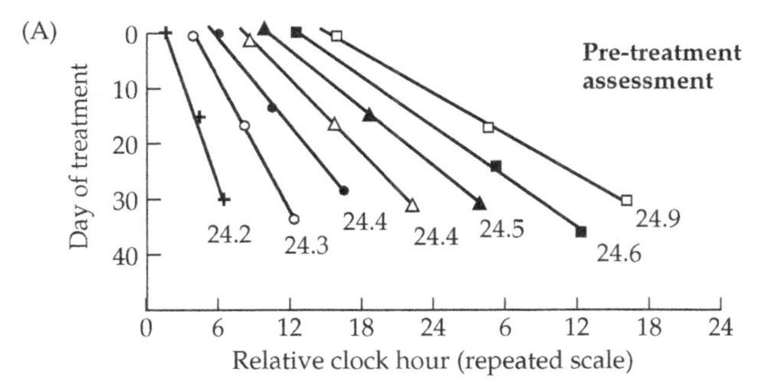
(A)
Pre-treatment assessment
Day of treatment
0
10
20
30
40
24.2
24.3
24.4
24.4
24.5
24.6
24.9
+ S1
○ S2
△ S3
▲ S4
• S5
■ S6
□ S7
0
6
12
18
24
6
12
18
24
Relative clock hour (repeated scale)

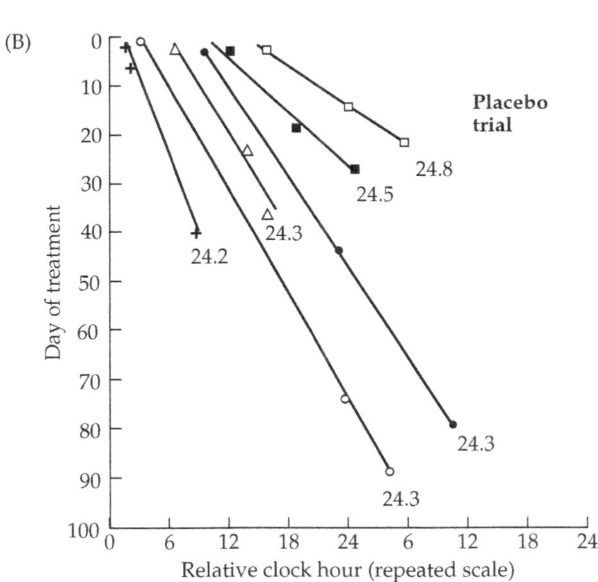
(B)
Placebo trial
Day of treatment
0
10
20
30
40
50
60
70
80
90
100
24.8
24.5
24.3
24.2
24.3
24.3
0
6
12
18
24
6
12
18
24
Relative clock hour (repeated scale)

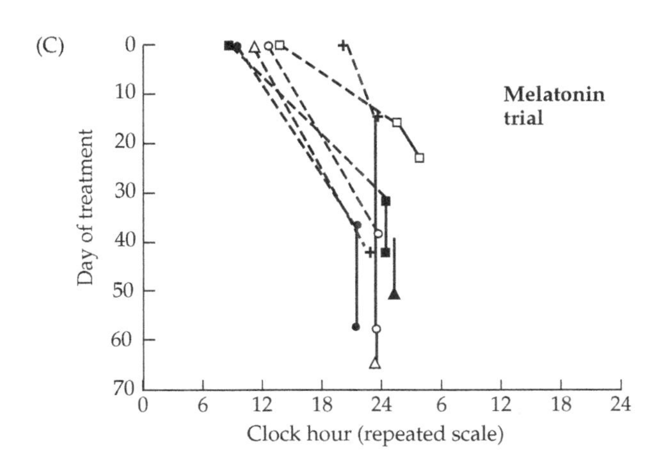
(C)
Melatonin trial
Day of treatment
0
10
20
30
40
50
60
70
0
6
12
18
24
6
12
18
24
Clock hour (repeated scale)

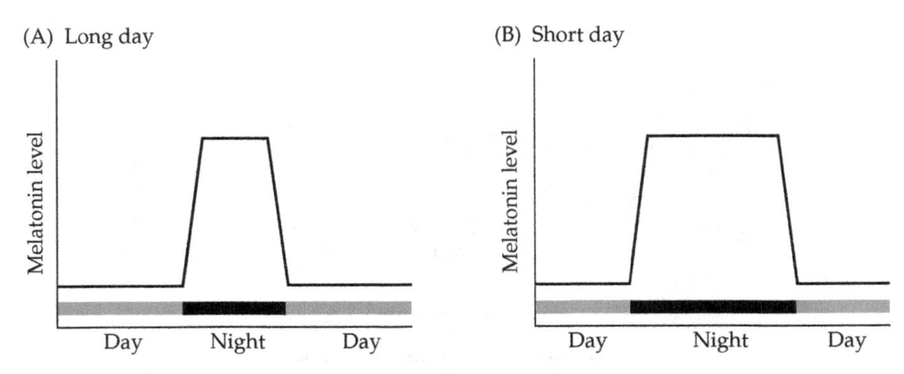

10.30 Melatonin measures night length. Melatonin is secreted only at night in both nocturnal and diurnal animals. Hamsters maintained under long-day (i.e., short-night) conditions display a relatively brief period of melatonin secretion, whereas hamsters maintained under short-day (i.e., long-night) conditions display an extended period of melatonin secretion. When the duration of melatonin secretion attains some threshold, a short-day response occurs. After Bartness and Goldman, 1989.

The importance of light as an entraining agent for human circadian rhythms has been established for rhythms of body temperature and the sleep–wake cycle (Czeisler et al., 1980; Dinges, 1984). The role of light in human circadian rhythms was emphasized by the discovery that bright light (>2500 lux) suppressed the production of pineal melatonin in humans (Lewy et al., 1980). In a number of mammalian species, including humans, melatonin is normally secreted in a circadian fashion, with an extended peak of secretion occurring at night (Figure 10.30). The duration of this sustained elevation of melatonin secretion varies with the length of the night (Bartness and Goldman, 1989; Illnerova et al., 1985); this variation provides a means of measuring day length and thus discerning the season of the year, as we will see below. Pineal melatonin secretion can be suppressed at night by light, and among many mammalian species, very little light indeed is necessary for this suppression. For example, exposure of hamsters to a very brief (1 msec), dim (0.1 lux—about the illumination provided by a half moon on a clear night) light pulse is sufficient to suppress melatonin production for the entire evening (Milette and Turek, 1986). Thus, exposure to only 1 msec of light per night, strategically placed to occur during the daily melatonin peak, will shorten the length of the nightly bout of melatonin secretion substantially and affect the animal's perception of seasonal cues.

In contrast to hamsters, the melatonin rhythms of humans are unperturbed by dim light. Over 2500 lux of illumination for about an hour is required to suppress melatonin secretion in humans (Lewy et al., 1980). In one study examining the effects of very bright light (7000–12,000 lux), typical room lighting (about 500 lux), and darkness, it was discovered that very bright light presented 3 hours prior to normal awakening was most effective in causing phase shifts in human body temperature (Czeisler et al., 1989). Daytime illumination levels approach

100,000 lux outdoors, and this is the level of light to which human biological clocks have evolved to respond (see Chapter 13). For people living at latitudes above 37°N, a typical December day may include commuting to work in the dark, 8 hours in an artificially illuminated office, a commute home in the dark, and 5 or 6 hours of low-level illumination at home. Individuals may not be exposed to 2500 lux of light for days or weeks at a time, and thus may be functionally considered to be living in a cave. Under these conditions, the biological clocks of some individuals do not receive sufficient light for entrainment, and begin to free-run. Under conditions that allow free-running, internal desynchronization—the dissociation of biological rhythms within an individual—may occur (Figure 10.31; Richardson and Martin, 1988). Thus, rhythms that are normally *phase-locked* (temporally linked so that all of the cycles show repeated phase relationships) when entrained may lose this relationship.

Because different rhythms may have different light intensity thresholds for entrainment, some rhythms may free-run in low-illumination conditions, whereas other rhythms may remain entrained. Desynchronization of these rhythms during the winter may account for a number of sleep, eating, and mood disorders, including jet lag. One well-known mood disorder, seasonal affective disorder (SAD), which is associated with dysfunctional biological timekeeping, is described

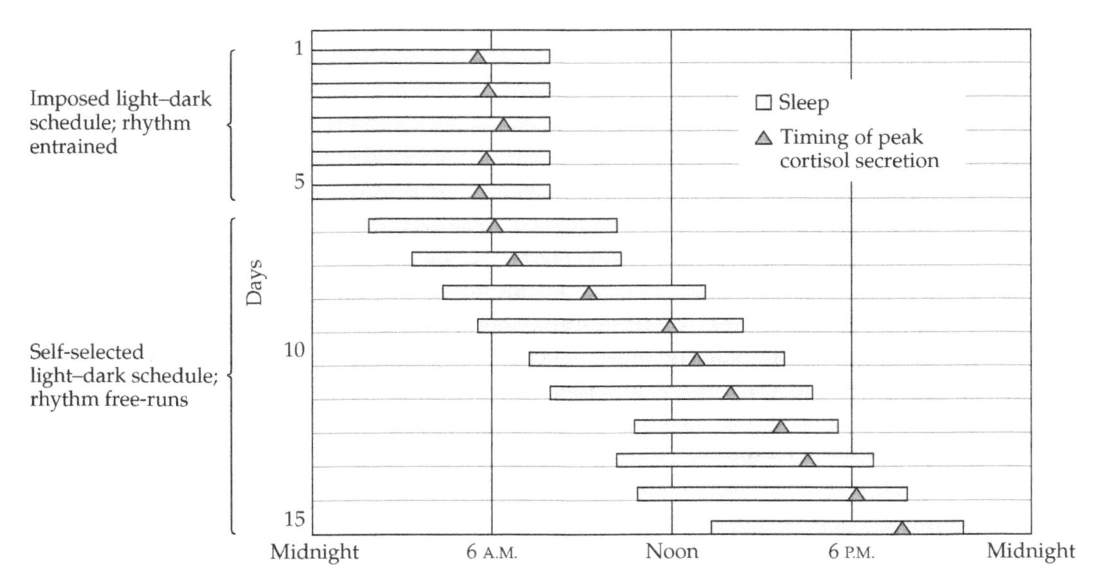

10.31 Internal desynchronization of circadian rhythms. Under normal conditions, or under an imposed light–dark cycle, the timing of peak cortisol secretion (triangles) is phase-locked to about 2 hours before awakening. When isolated individuals are allowed to select their own light–dark cycles with a light switch, free-running begins, and the rhythm of cortisol secretion and the sleep–wake cycle change at different rates, so that the peak of cortisol secretion occurs in the middle of the sleep period. Such internal desynchronization is thought to be the underlying cause of jet lag. After Richardson and Martin, 1988.

in detail in Chapter 13. In some cases, treatment with bright lights can ameliorate the affective symptoms associated with SAD.

Circadian Regulation of Food Intake

CIRCADIAN CONTROL OF FEEDING Animals that are nocturnal or diurnal confine most of their food intake to the dark or light part of the day, respectively. Some animals are flexible in the timing of their ingestive behaviors, and can eat throughout the day; you have probably had a midnight snack without ill effects. However, the timing of food intake is inflexible in other animals. Syrian hamsters, for example, are hard-wired to eat specific amounts during the light and dark phases, and they cannot compensate for shortages in one phase during the other. If hamsters eat a total of 10 g of food per day, they might eat 6 g at night and 4 g during the day. If their food is taken away during the day, then they eat only 6 g during the night and lose weight. If their food is restricted during the night, they eat only 4 g during the day, and lose weight to the point of starvation, rather than compensating with increased food intake during the day (Silverman and Zucker, 1976).

The regulation of food intake is complex and requires the coordination of several brain circuits and several orexigenic and anorexigenic hormones, as we saw in Chapter 9. If the SCN are lesioned in hamsters, their food intake is dispersed throughout the day and night. The SCN can presumably affect food intake by several pathways. For example, the SCN communicate directly with all of the hypothalamic nuclei that are involved in the regulation of food intake, including the arcuate nuclei, dorsal medial nuclei, ventromedial hypothalamus, and lateral hypothalamic areas (Kalra et al., 1999). SCN fibers innervate neuropeptide Y-, pro-opiomelanocortin-, and galanin-containing neurons in the arcuate nuclei, and lesions of the arcuate nuclei or PVN may disrupt the circadian pattern of orexigenic and anorexigenic hormone secretion and thus change the pattern of ingestive behaviors (Kriegsfeld et al., 2002). Neurotransmitters involved in food intake, such as serotonin, norepinephrine, and epinephrine, display circadian rhythms controlled by the SCN. For example, norepinephrine treatment appears to evoke food intake better at the onset of activity in nocturnal rodents than at other times (Currie and Wilson, 1993). Finally, the SCN can also affect the timing of food intake by affecting secretion of insulin and glucagon; bilateral lesions of the SCN block the circadian rhythm of insulin and glucagon concentrations in the blood (Yamamoto et al., 1987). It appears that the SCN may stimulate glucagon secretion and inhibit insulin secretion because glucagon concentrations are lower and insulin concentrations higher in SCN-lesioned rats than in SCN-intact animals (Yamamoto et al., 1987).

FOOD-ENTRAINABLE OSCILLATORS In some cases, nonphotic cues, such as scheduled handling or injection of saline, can entrain the free-running clocks of adult Syrian hamsters and mice (Hastings et al., 1998) (Table 10.2). Because many people who are totally blind (though not all, as we saw above) show entrainment of

TABLE 10.2 **Nonphotic zeitgebers**

Zeitgeber	References
Mealtimes	Stephan, 2002
Novel running wheels	Reebs and Mrosovsky, 1989
Foraging	Rusak et al., 1988
Social interactions	Mrosovsky, 1988
Dark pulses	Reebs et al., 1989; Van Reeth and Turek, 1989
Triazolam	Van Reeth and Turek, 1989; Mrosovsky and Salmon, 1990
Brotizolam	Yokota, 2000
Treadmill running	Marchant and Mistlberger, 1996
Limited access to home wheel	Edgar et al., 1991
Cold or refeeding	Mistlberger et al., 1996
Morphine	Marchant and Mistlberger, 1995

their biological rhythms, it is likely that humans also possess the capacity for nonphotic entrainment of their circadian clocks. One of the most potent nonphotic zeitgebers is scheduled feeding (reviewed in Stephan, 2002). Not surprisingly, limiting access to food to a specific time of day has dramatic effects on the temporal organization of behavior and physiology. The overt behavioral effect is meal anticipation. Pre-meal anticipatory behavior, elevation of core body temperature, and elevated serum corticosterone concentrations display dramatic circadian patterns in association with mealtime, and these patterns persist after SCN lesions. Although the SCN (and several other CNS sites, such as the amygdale, neocortex, hippocampus, and hypothalamus) have been ruled out as the integration site of food-entrainable oscillators (FEO) (e.g., Stephan, 1983), the search for the locus of a separate FEO has been unsuccessful (Marchant and Mistlberger, 1997).

The cloning of circadian clock genes and the discovery that these genes are expressed in many CNS structures outside the SCN and in peripheral tissues have led to new strategies for investigating potential loci of an FEO. Recently, clock gene expression in both the CNS and peripheral tissues has been observed to change in response to changes in periodic food availability. For example, food restriction to the light portion of the daily light–dark cycle (or to subjective day) shifted the expression of clock genes in the liver, but not in the SCN, of rats within 1 week (Damiola et al., 2000) (Figure 10.32). Because a secondary peak of glucocorticoids emerges at the time of anticipated feeding, glucocorticoids may, in part, mediate food-entrainable circadian rhythms. Restricted feeding alters micronutrient availability and consequent signaling pathways. Taken together, the evidence suggests that the daily light–dark cycle is the most potent zeitgeber for the SCN, whereas food intake appears to be the most important zeitgeber for peripheral oscillators, especially sites involved in metabolism (Kriegsfeld et al., 2002).

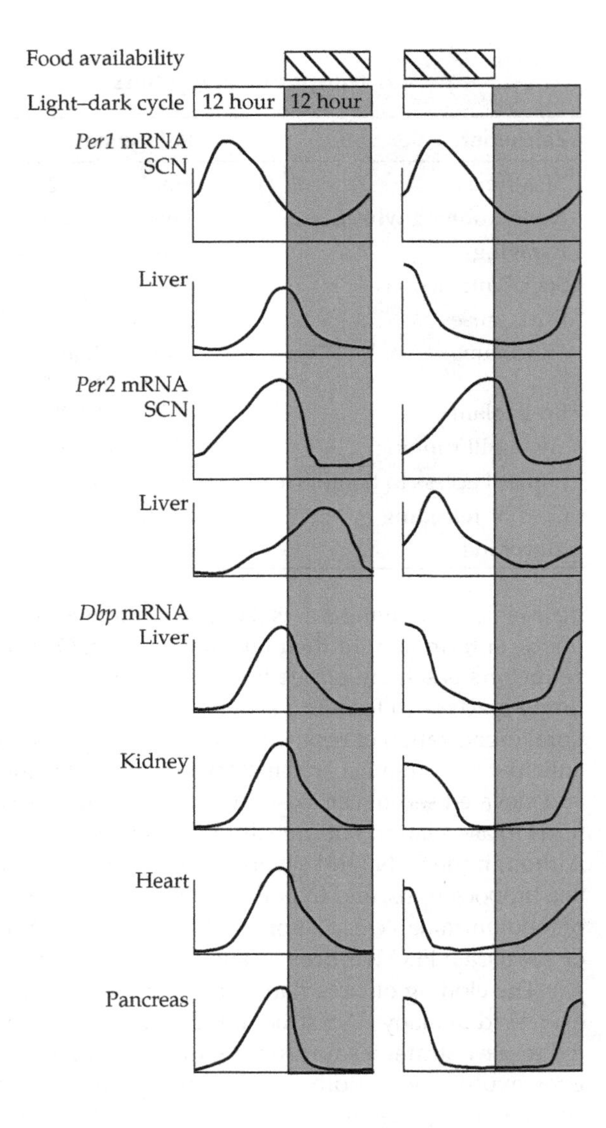

10.32 Shifting the time of food intake from night to day changes clock gene expression in the periphery, but not in the SCN. The left column of panels depict gene expression of *Per1* and *Per2* in the SCN and the circadian transcription factor, *Dbp*, in several peripheral organs when mice were given free access to food. Because mice are nocturnal, most of their food intake normally occurs at night. The right column of panels show the timing of expression of the same genes when food was restricted to the light portion of the day. After Damiola et al., 2000.

Circannual and Seasonal Rhythms

The effects of light on daily melatonin production are critical in mammalian **photoperiodism**, the use of day length to time annual cycles. A functional circadian clock is required to measure day lengths, and measurement of day lengths is necessary to make seasonal adjustments in behavior. The SCN can serve as both clock and calendar because its function changes in response to photoperiod. As we saw above, the SCN transduces light information from the eyes to the pineal gland, which secretes melatonin when it is dark. The duration of the nocturnal melatonin signal encodes day length, and many species use this information to coordinate physiological adaptations with the yearly climatic cycle.

The duration of the nightly sustained elevation of melatonin secretion varies with the length of the night; in other words, a relatively brief duration of elevated melatonin secretion codes for short nights, or long days, whereas a relatively long duration of elevated melatonin secretion codes for long nights, or short days (Bittman and Karsch, 1984; Carter and Goldman, 1983a) (see Figure 10.30). Hamsters experimentally exposed to a brief pulse of light each night, which shortens the duration of melatonin secretion, exhibit a short-night (long-day) hormone profile. They also exhibit physiological, morphological, and behavioral adaptations consistent with long days, such as maintaining complete reproductive function and their "summer" fur coat. Hamsters maintained in short-day light conditions exhibit longer bouts of elevated melatonin secretion, display reproductive regression, and develop a "winter" coat. Although daily melatonin rhythms are observed in birds and reptiles, they do not seem important for these animals.

There appear to be two different types of biological rhythms that account for seasonal behavioral cycles (Figure 10.33). One reflects a seasonal biological rhythm driven by environmental day length, but requires an endogenous circadian system to discern day length. The second type of seasonal biological rhythm comprises an endogenous organization and is commonly observed in animals that undergo seasonal hibernation in burrows and thus do not have access to environmental cues. There appear to be several instances in which both types of biological clocks interact to permit animals to discern the time of year.

Ultimate and Proximate Factors Underlying Seasonality

As noted above, animals have evolved to fill temporal as well as spatial niches. Many animals and plants are exposed to seasonal fluctuations in the quality of their environments. Most seasonal changes in behavior reflect strategies to manage an annual energy budget. Individuals generally restrict energetically expensive activities to a specific time of the year. Animals migrate or reduce their general levels of activity when food availability is low. Reproduction, preparation for migration, and other energetically demanding activities have evolved to coincide with abundant local food resources or other environmental conditions that promote survival. Precise timing of behavior is therefore a critical feature of individual reproductive success and subsequent fitness.

In some cases, physiological and behavioral changes that occur in direct response to environmental fluctuations have an obvious and immediate adaptive function. For example, reductions in the availability of food or water can inhibit reproductive function (Bronson, 1989; Nelson, 1987) (see Chapter 9). These types of environmental fluctuations have been termed the ultimate factors underlying seasonality (Baker, 1938). Many animals need to forecast the optimal time to breed so that spermatogenesis, territorial establishment, nest construction, or any other time-consuming preparations for reproduction will be complete at the start of the breeding season. Therefore, seasonally breeding animals frequently detect and

(A) **Type I rhythm**
Mixed: Driven by endogenous and exogenous signals (e.g., Syrian hamster reproduction)

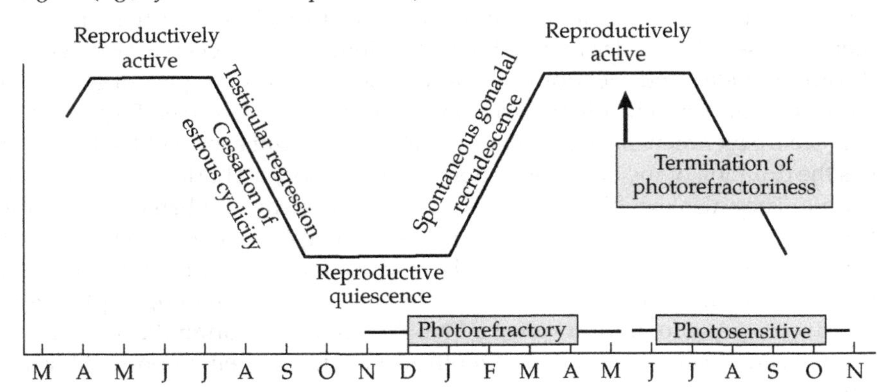

(B) **Type II rhythm**
Endogenous: Product of circannual clock(s) (e.g., ground squirrel body weight)

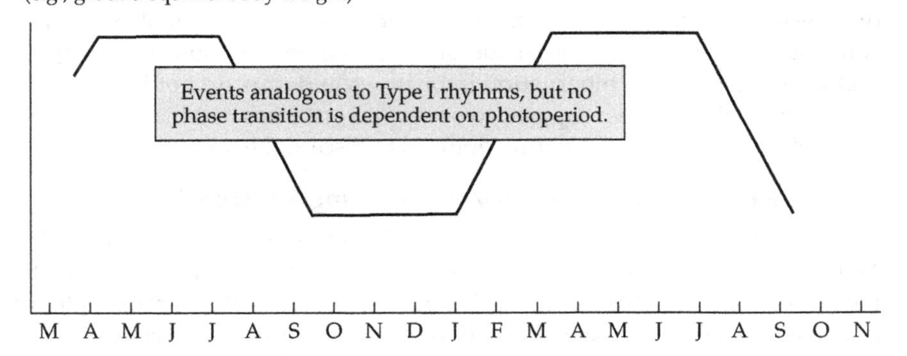

10.33 Schematic representation of two types of seasonal reproductive rhythms in rodents. (A) Type I (mixed) seasonal rhythms are mediated by both endogenous (interval timer) and exogenous (photoperiodic) signals. Decreasing photoperiods in autumn provoke gonadal regression and start an interval timer, which eventually (after several months) causes the animal to be nonresponsive to the inhibitory effects of short days and redevelop their reproductive systems. Thus, even animals inhabiting constantly dark hibernacula (e.g., burrows) emerge in the spring with full reproductive functions. Exposure to several weeks of long days is necessary to reset sensitivity to the inhibitory effects of short days. (B) Type II (circannual) rhythms are endogenous and have a period of 365 days or less. Photoperiodic information is necessary to entrain Type II rhythms to exactly 365 days. After Prendergast et al., 2002.

respond to environmental cues that accurately signal, well in advance, the arrival or departure of seasons favoring reproductive success. The cues used to predict environmental change may or may not have direct survival value, and can be referred to as proximate factors (Baker, 1938). The most notable example of such a proximate factor is day length, or **photoperiod**, a cue that can serve as a very precise reference for the time of year. Under some circumstances, proximate and ultimate factors may be identical (Negus and Berger, 1987). For example, some individuals may not begin breeding until food cues are detected (Bronson, 1989).

The most salient seasonal rhythm expressed by animals is the seasonal breeding cycle. Many animals, including small, short-lived mammals and birds, mate during the spring and rear their offspring during the summer when food is plentiful; these animals are called *long-day breeders*. Other animals, such as sheep, deer, and cattle, mate in autumn and have long gestation periods. These large mammals remain pregnant throughout the winter and, in common with small animals, give birth in the spring, when conditions are most suitable for survival. Animals that mate in autumn are called *short-day breeders*. Thus, many hormone–behavior interactions associated with the breeding season are mediated by biological clocks. Courtship, copulation, parental care, territorial defense, and aggression all exhibit pronounced seasonal fluctuations in occurrence that are linked directly to seasonal rhythms in reproductive function (Moffatt et al., 1993) (Figure 10.34A).

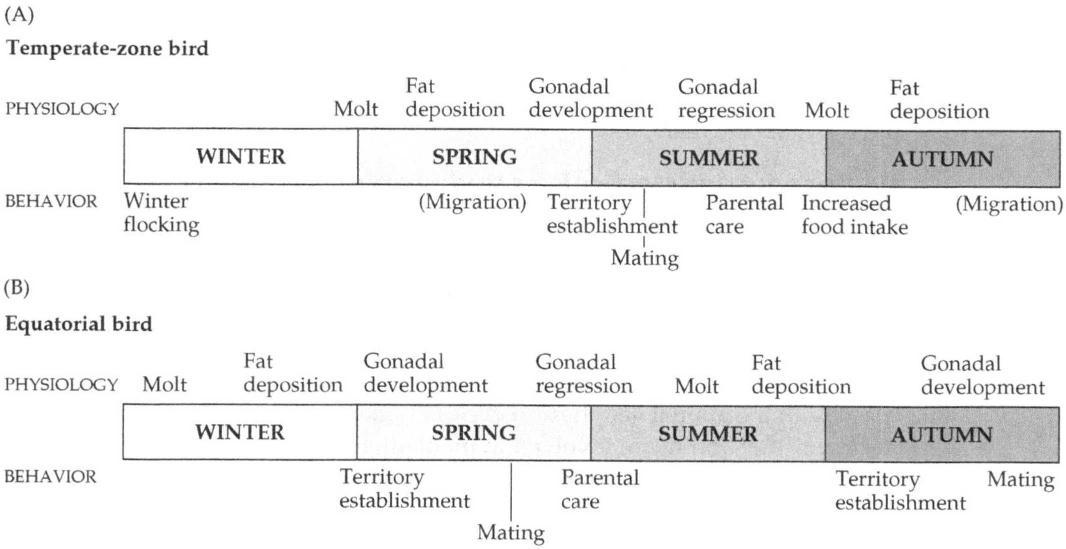

(A)

Temperate-zone bird

PHYSIOLOGY Molt | Fat deposition | Gonadal development | Gonadal regression | Molt | Fat deposition

WINTER | SPRING | SUMMER | AUTUMN

BEHAVIOR Winter flocking | (Migration) | Territory establishment | Mating | Parental care | Increased food intake | (Migration)

(B)

Equatorial bird

PHYSIOLOGY Molt | Fat deposition | Gonadal development | Gonadal regression | Molt | Fat deposition | Gonadal development

WINTER | SPRING | SUMMER | AUTUMN

BEHAVIOR Territory establishment | Mating | Parental care | Territory establishment | Mating

10.34 Annual breeding cycles in birds reflect energy constraints. (A) North temperate-zone birds may migrate north in the spring, then become reproductively active, establishing territories, courting, and rearing young. After breeding, the birds molt and increase their body fat stores in preparation for the autumn migration south. They feed all winter and undergo another molt prior to the next spring migration. Nonmigratory birds must bear the increased energetic costs associated with thermoregulation during the winter. Thus, molting, migration (or winter thermoregulation), and breeding are separated in time in temperate-zone birds so that they can cope with the enormous energetic demands of each activity. (B) Tropical birds may not have the same energetic constraints, so breeding and molting can occur at any time of year, and the frequency of the cycle between these energetic demands is limited only by the physiological capacity of the birds.

Seasonal changes in behavior are observed even among animals living in the tropics, where the annual cycle of changing day length is not as evident as it is at higher latitudes. Despite relatively constant photoperiodic and temperature conditions, seasonal variation in food availability is common for many tropical species. The irregularly timed onset of rain, or some coincident factor, induces tropical birds such as the red-billed quelea (*Quelea quelea*) to breed in East Africa (Disney et al., 1959); consequently, the onset of breeding in East African quelea varies from year to year (Murton and Westwood, 1980). In West Africa, where the onset of the rainy season is more consistent each year, the quelea display a predictable breeding season (Ward, 1965). In contrast, several species of oceanic seabirds inhabiting equatorial waters experience virtually no detectable seasonal variation in food supplies or other environmental factors. Reproductive activities and molting impose large, conflicting energetic demands on these birds, and so are separated in time (Figure 10.34B). However, in this relatively stable environment, breeding and molting may occur at any time of the year, and the frequency of the cycle between these two energetic demands is limited only by the physiological capacity of the birds, and varies among species. For example, breeding recurs every 8 months for the bridled tern (*Sterna anaethetus*) (Diamond, 1976), every 9 months for the brown pelican (*Pelecanus occidentalis*) (Harris, 1969) and Audubon's shearwater (*Puffinus lherminieri*) (Snow, 1965), and every 10 months for the swallow-tailed gull (*Creagrus fucatus*) of the Galápagos Islands (Snow and Snow, 1967). Thus, mating behavior in some tropical birds is distributed throughout the calendar year.

Neuroendocrine Mechanisms Underlying Seasonality

There is extensive literature on the mechanisms regulating seasonal breeding cycles. The principles of seasonality derived from this literature can serve as a basis for the examination of the sparser information base directly related to seasonal changes in behavior. The mechanisms that regulate seasonal reproductive changes may be classified under two categories. One set of mechanisms is directly responsible for timing seasonal rhythms and ensuring that they are synchronized with the annual geophysical cycles. In mammals, the pineal gland and its hormone, melatonin, are involved in mediating the effects of day length on the timing of a wide variety of seasonal changes in physiology and behavior (Goldman and Nelson, 1993). A second set of neuroendocrine mechanisms is directly responsible for regulating changes in the reproductive system. For example, changes in the rate or pattern of pituitary hormone secretion are important for driving changes in reproductive hormones. Seasonal changes in reproductive hormones, especially sex steroid hormones, result in a cascade of seasonal changes in steroid-dependent behaviors. These mechanisms will be referred to as *activational mechanisms* because they generally involve activational effects of hormones (see Chapter 3; Beach, 1975). In photoperiodic mammals, the various seasonal activational mechanisms are influenced by day length, and thus by pineal melatonin. For some species, including sheep, thyroid hormones play a permissive role in the termination of seasonal reproduction (Thrun et al., 1997).

Timing Mechanisms

Studies of seasonal rhythms in reproductive physiology have revealed a variety of timing mechanisms. Photoperiodism has been the most widely studied of these mechanisms. For the purposes of this discussion, photoperiodic mammals may be divided into two categories: (1) those that exhibit endogenous circannual cycles that are entrained, or synchronized, by photoperiodic cues; and (2) those that fail to exhibit endogenous circannual cycles in the absence of photoperiodic cues (see Figure 10.33) (Prendergast et al., 2002).

Individuals of many species of mammals fail to exhibit endogenous cycles when housed under a fixed day length. For example, a variety of relatively short-lived rodent species remain reproductively active so long as they are maintained under long photoperiods. Twelve hours of light per day is the critical photoperiod (i.e., the day length required to maintain reproductive function) for Syrian hamsters (Figure 10.35); other species, particularly those inhabiting high latitudes, have different critical photoperiods. Although these species require photoperiodic changes for the continuance of seasonal cycles, they still display a prominent element of endogenous seasonal timing. Short day lengths (<12.5 hours of light per day) induce reproductive regression, but after several months of exposure to short days, a "spontaneous" activation of the reproductive system occurs. This event seems to be triggered by an endogenous timing mechanism. Many of these species undergo various degrees of torpor in a dark burrow during winter, isolated from environmental cues; this mechanism allows them to prepare for the environmental changes that will take place in the spring (Elliott and Goldman, 1981; Reiter, 1970).

This "spontaneous" activation of reproductive function represents a phenomenon called **photorefractoriness**. In the case of Syrian hamsters, chronic maintenance under short day lengths causes the reproductive system to regress after about 10 weeks. After 20–30 weeks of continuous exposure to short days, however, the reproductive system is no longer inhibited by the short days, and is now

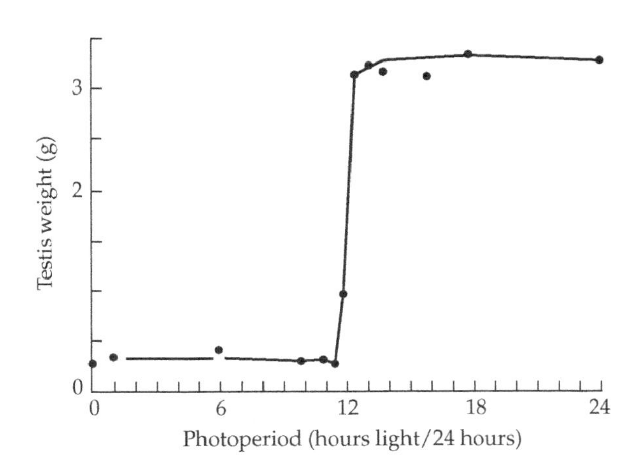

10.35 Critical day length for reproduction in male Syrian hamsters. Testicular responses of Syrian hamsters after exposure to various fixed photoperiods for about 3 months. Only day lengths of more than 12.5 hours maintained full gonadal size and function. Thus, the critical day length for this trait in this species is between 12 and 12.5 hours of light per day. From Elliott, 1976.

said to be photorefractory. After this time, the hamsters' reproductive systems will become active and remain totally functional, even if the hamsters are kept in total darkness for the rest of their lives. If photorefractory hamsters are switched to long day lengths (>12.5 hours of light per day), after 10 weeks they will be responsive again to the inhibitory effects of short day lengths on reproductive function. In other words, the long days eliminate photorefractoriness. Photorefractoriness is adaptive because it allows preparation of the reproductive system in advance of appropriate breeding conditions.

Deer, an example of the second type, exhibit circannual rhythms of reproductive activity that persist even when the animals are maintained under a constant day length, although these rhythms are not expressed under constant photoperiods of 12 hours or longer. However, changes in day length influence reproductive activity in deer, and natural photoperiodic changes are presumably largely responsible for establishing the seasonal pattern of reproduction (Goss, 1980, 1984; Goss and Rosen, 1973; Plotka et al., 1984). Sheep may also have the capacity to exhibit circannual cycles of reproductive activity in the absence of environmental cues, and these rhythms are clearly responsive to photoperiod under natural or artificial environmental conditions (Karsch et al., 1984; Sweeney et al., 1997) (Figure 10.36).

Perhaps the most striking examples of endogenous timing mechanisms in photoperiodic species are observed in several species of hibernating mammals. Most hibernators undergo gonadal regression before entering hibernation. The gonads shrink, and gamete and sex steroid hormone production stops. Yet when these animals emerge from their hibernacula in early spring, their reproductive systems have been activated, and they are fully capable of breeding. Because the process of spermatogenesis requires several weeks in mammals, reproductive activation in males must begin before emergence from hibernation.

HIBERNATION IN HAMSTERS Laboratory studies have demonstrated that pituitary gonadotropin secretion and testicular growth begin during the last few weeks of hibernation in Turkish hamsters (*Mesocricetus brandti*). Hibernation is terminated when testosterone levels exceed a certain threshold; this mechanism may serve to coordinate emergence from hibernation with testicular recrudescence (regrowth) (Hall and Goldman, 1980; Hall et al., 1982). The timing of testicular recrudescence in Turkish hamsters is probably accomplished by the same type of seasonal timing mechanism that operates in other photoperiodic rodents. Male Turkish hamsters that are exposed to short days in a warm environment do not hibernate, yet these animals exhibit a cycle of testicular regression and subsequent recrudescence very similar in timing to that observed in hibernating males (Darrow and Goldman, 1987). A similar mechanism for temporal coordination of the seasons of hibernation

10.36 Circannual LH rhythms in ewes under a simulated annual photoperiodic cycle. The sinusoidal line at the top of the figure indicates day length. LH rhythms (A) in intact ewes, (B) in pinealectomized ewes, and (C–F) in individual pinealectomized ewes treated with melatonin in winter, spring, summer, and autumn. Horizontal black bars indicate elevated LH concentrations. The shaded parts of the graphs indicate when intact ewes display elevated LH. After Woodfill et al., 1994.

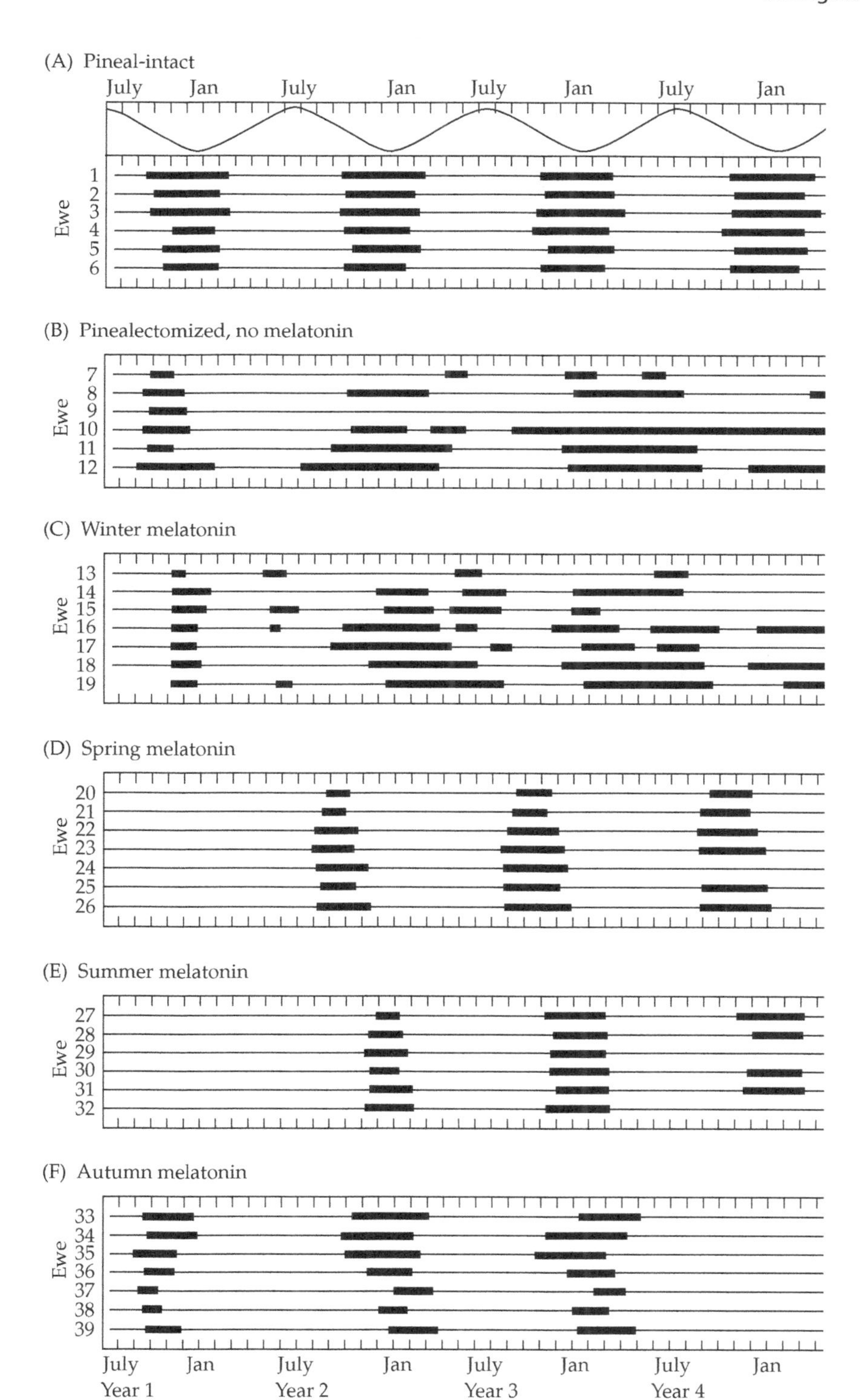
(A) Pineal-intact
July
Jan
July
Jan
July
Jan
July
Jan
Ewe
1
2
3
4
5
6
(B) Pinealectomized, no melatonin
7
8
9
10
11
12
(C) Winter melatonin
13
14
15
16
17
18
19
(D) Spring melatonin
20
21
22
23
24
25
26
(E) Summer melatonin
27
28
29
30
31
32
(F) Autumn melatonin
33
34
35
36
37
38
39
July
Year 1
Jan
July
Year 2
Jan
July
Year 3
Jan
July
Year 4
Jan

and reproduction appears to exist in European hamsters (*Cricetus cricetus*) (Darrow et al., 1988) and hedgehogs (*Erinacus europaeus*) (Saboureau, 1986).

Males that have just undergone testicular recrudescence produce more androgens and gonadotropins than animals whose reproductive function has been continuously maintained under long day lengths (Berndtson and Desjardins, 1974). This "overshoot" of endocrine activity may have functional behavioral consequences. Recall from Chapter 5 that long-castrated male rodents require higher androgen concentrations to maintain mating behavior than do recently castrated or intact animals (Damassa et al., 1977). The brain centers that control reproductive behavior may require sensitization by androgen exposure after prolonged gonadal quiescence in order to respond appropriately (Morin and Zucker, 1978).

HIBERNATION IN GROUND SQUIRRELS The seasonal organization of reproduction in golden-mantled ground squirrels (*Spermophilus lateralis*) has been the subject of many field and laboratory studies. These squirrels appear to possess an endogenous circannual clock (Berthold, 1837; Gwinner, 1986). When maintained under constant conditions in a laboratory setting, the squirrels display periods of reproductive function every 330–380 days. If the squirrels are exposed to a naturally changing photoperiod, they may entrain their endogenous cycles to exactly 365 days. A typical year in the lives of these squirrels living in northern California would begin with a summer of binge eating and deposition of body fat; body fat levels may double during the weight gain period of the summer. The squirrels then enter a burrow and go into deep torpor. During torpor, the body temperature approaches the ambient temperature of the hibernaculum, about 2°–4°C. Contrary to popular belief, the squirrels do not stay in deep hibernation throughout the winter, but arouse every few days. They usually do not eat during this time, and their bodies undergo a programmed utilization of fat stores (although other ground squirrel species may arouse and eat at this time) (Dark et al., 1989). The squirrels' reproductive systems become activated during hibernation. In males, blood concentrations of androgens rise and eventually become incompatible with torpor, inducing the final arousal in March. Thus, the onset of testicular androgen production in the spring both terminates hibernation and stimulates subsequent mating behavior. When males emerge from their hibernacula, they are fertile, and their high blood concentrations of androgens cause them to defend territories and court females, which emerge a few weeks later. Virtually all females are inseminated during their first day above ground in the spring, despite the fact that they possess very low fat reserves. A litter is born in late spring, and the young squirrels scramble to increase body mass during their first summer.

These seasonal changes in reproductive function, food intake, and body mass also occur in the laboratory when ambient conditions of temperature, day length, and food availability are maintained at constant levels (Zucker et al., 1991). There is a circannual clock in male and female ground squirrels that regulates gonadotropin-releasing factors (primarily GnRH) and, ultimately, steroid production. In males, the sensitivity of the hypothalamic–pituitary axis to the negative feedback effects of steroid hormones changes; during the end of the hibernation period, the negative feedback mechanisms become less sensitive to steroids, and gonadotropins are released in increasing amounts, stimulating greater and greater amounts of steroid production (Zucker, 1988a). In female ground squirrels, the feedback system appears to be switched on and off. Ovariectomy at the beginning of the breeding season results in a pronounced elevation of blood plasma levels of LH, indicating that negative feedback mechanisms are operational at this time. If ovariectomy is performed on females at the end of the breeding season, then there is no post-ovariectomy rise in plasma LH; the LH concentrations of ovariectomized squirrels remain basal until the following breeding season (Figure 10.37; Zucker, 1988a).

The annual cycle of body mass changes, with its corresponding changes in food intake, is also programmed by a circannual clock and persists under constant laboratory conditions. It appears that seasonal changes in food intake are

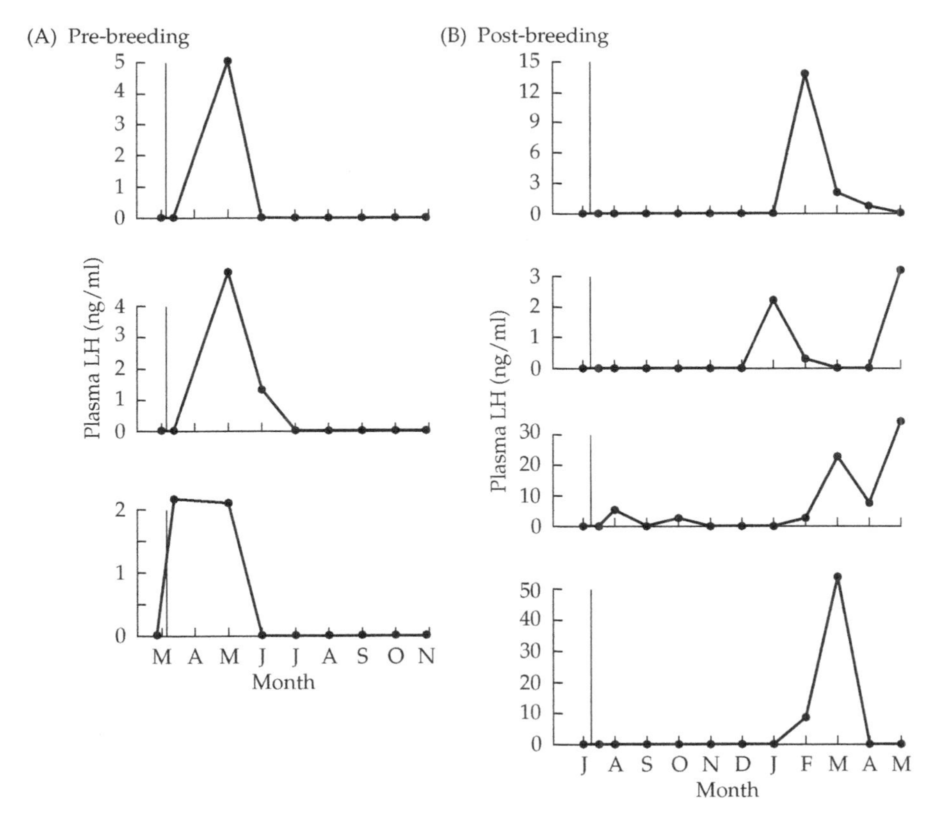

10.37 Circannual control of LH concentrations in ground squirrels. Female ground squirrels show regular annual cycles of breeding in nature and a circannual cycle of reproductive condition under constant conditions in the laboratory. (A) Ovariectomy (vertical line) of three female ground squirrels at the beginning of the breeding season led to the classic castration response of increased LH concentrations because it eliminated negative feedback control by steroid hormones. (B) When four other females were ovariectomized after the breeding season, no increase in plasma LH concentrations was observed until the next breeding season, suggesting that the steroid negative feedback mechanism is activated only at certain times of the year. After Zucker, 1988a.

secondary to seasonal changes in fat storage and utilization. If the squirrels are limited to levels of food availability comparable to what they were ingesting at the end of the hibernation phase, the animals still manage to gain amounts of body fat comparable to those of squirrels fed ad libitum (without restriction). They accomplish this feat by utilizing other components of the body and reducing activity. Even if the fat is removed by surgical aspiration, the animals make new fat cells and are able to catch up to intact control animals in body fat within a year (Dark et al., 1989). Thus, it appears that the circannual clock regulates fat storage and utilization, which causes changes in eating behavior. The exact physiological mechanisms underlying this phenomenon require clarification, but insulin and norepinephrine appear to be involved (Florant et al., 1989).

Thus, a continuum exists for annual cycles, from those that are completely autonomous and in which environmental cues play a limited role as entraining agents, to those that are driven wholly or in part by environmental factors. Ground squirrels are at one end of the continuum, with annual rhythms that are minimally influenced by environmental factors, except that day length entrains these rhythms. The annual cycles of Syrian hamsters, at the other end of the continuum, are driven by day length. However, during the winter, an endogenous timer appears to activate reproductive function prior to the mating season, which is coincident with, though not caused by, the long days of spring. Both ground squirrels and hamsters spend part of the year underground in relatively constant environments and require endogenous clocks and temporal processes to time part of their seasonal behavioral cycles.

BREEDING SEASONS IN BIRDS Nonmigrating temperate-zone birds are exposed to an annual photoperiodic cycle, and much of their seasonal cycle of behavior is driven by that cycle. Like rodents, nontropical birds mate and rear their young during the spring and summer, when days are increasing in length. In contrast to those of rodents, the reproductive systems of temperate-zone birds must be stimulated by long days to function, but it is probably more accurate to state that exposure to short days is necessary to "turn on" their reproductive systems. Breeding for most nontropical avian species is completed by the summer solstice. Some species breed more than once during the breeding season, but most undergo reproductive regression in response to the long days of summer. In other words, the previously stimulatory day lengths act to inhibit reproductive function; the birds are said to be photorefractory because long days no longer stimulate reproduction (Nicholls et al., 1988). In order to break photorefractoriness, the birds must experience a period of short days. After exposure to several weeks of short days, the birds are once again responsive to the stimulatory effects of long days on reproductive function (Nicholls et al., 1988).

Neural Mechanisms of Endogenous and Exogenous Seasonal Timekeeping

Virtually no information is available regarding the mechanisms of endogenous seasonal timekeeping in either birds or mammals. Attempts have been made to disrupt circannual rhythmicity by lesioning various brain regions. Neither lesions of the PVN, part of the output pathway from the SCN (see Figure 10.24), nor pinealectomy disrupt circannual cycles of body mass fluctuation in ground squirrels (Dark and Zucker, 1985). Lesions of the SCN disrupt other patterns of circannual rhythmicity in some individuals; however, most SCN-lesioned squirrels continue to display circannual rhythms, despite the fact that *circadian* rhythms are absent in these animals (Dark et al., 1985). This observation and others in both ground squirrels and birds suggest that circannual rhythms and circadian rhythms may not be regulated by the same neural substrate (Gwinner, 1986).

In contrast to our lack of understanding of the physiological substrate for endogenous seasonal timekeeping, much has been learned about the neuroendocrine basis of photoperiodism in mammals, especially over the past four decades. In virtually all species of mammals that have been carefully examined, pinealectomy severely interferes with most photoperiodic responses (Goldman, 1983). This was first demonstrated in Syrian hamsters, in which removal of the pineal gland prevented the inhibition of reproductive activity that typically occurs following exposure to day lengths of less than 12.5 hours (Hoffman and Reiter, 1965). This observation led to the common belief that the pineal gland exerts an inhibitory effect on the reproductive system (Reiter, 1970).

As we have seen, the pineal hormone, melatonin, has been shown to be the mediator of pineal effects on photoperiodic responses in a wide variety of mammals (Goldman and Nelson, 1993). Melatonin is secreted at night in both nocturnal and diurnal animals. The rhythm of pineal melatonin secretion is largely regulated by one or more circadian oscillators, probably in the SCN (Darrow and Goldman, 1986; Goldman and Darrow, 1983). The nocturnal pattern of melatonin synthesis and secretion is entrained, but not "driven," by the light–dark cycle. The pattern of nightly elevation of melatonin secretion persists in constant lighting conditions and corresponds roughly to the pattern of locomotor activity in nocturnal rodents (Goldman, 2001). In almost all mammals that have been examined, including species (such as the laboratory rat) whose reproductive cycles are generally considered to be nonresponsive to changes in day length, the duration of the nocturnal peak of melatonin secretion increases as the photoperiod decreases (Bartness et al., 1993; Darrow and Goldman, 1986; Illnerova et al., 1986; Karsch et al., 1984). In Siberian hamsters (*Phodopus sungorus*) and sheep, daily infusions of melatonin have been administered to pinealectomized animals and their reproductive responses measured. In both species, responses characteristic of animals exposed to long days (i.e., stimulation of reproduction for hamsters, inhibition for sheep) are elicited by daily melatonin infusions of short duration. Melatonin infusions of longer duration result in short-day responses (Bittman and Karsch, 1984; Carter and Goldman, 1983a,b). In Siberian hamsters, nonreproductive parameters, such as body mass and fat content, are also differentially affected by long-duration and short-duration infusions of melatonin (Bartness and Goldman, 1988a,b). The time of day at which the infusions are given does not appear to be critical in either sheep or hamsters (Bartness and Goldman, 1988b; Carter and Goldman, 1983a; Wayne et al., 1988). Based on these data, it has been proposed that changes in the duration of the nightly melatonin peak serve to convey a photoperiodic message to a variety of physiological systems (Bartness et al., 1993; Bittman, 1993; Goldman, 1983; Goldman and Elliott, 1988). Other species, including Syrian hamsters and white-footed mice (*Peromyscus leucopus*), respond to changes in the duration of the nightly melatonin peak in a manner similar to that reported for Siberian hamsters and sheep (Dowell and Lynch, 1987; Grosse et al., 1993).

In mammals, melatonin-binding sites have been located in several brain areas. In rats, Syrian hamsters, and Siberian hamsters, for example, the SCN, pars tuberalis, and median eminence display significant melatonin binding, as do several areas of the thalamus, hypothalamus, subiculum, and area postrema (Weaver et al., 1989). Interestingly, little melatonin binding was detected in the anterior pituitary gland of adult photoperiodic rodents, despite the profound influence of melatonin on the secretion of gonadotropins (Vanecek, 1988; Carlson et al., 1991). Among three suborders of Rodentia, considerable interspecific variation exists in the sites of high-affinity melatonin binding (Bittman et al., 1994). Among orders of mammals, variation in melatonin binding sites is also noteworthy; for example, numerous telencephalic, diencephalic, hypothalamic, and midbrain structures bind melatonin in sheep, whereas in ferrets only the pars tuberalis and pars distalis of the pituitary bind melatonin, despite the fact that both species respond reproductively to photoperiod (Bittman and Weaver, 1990; Weaver and Reppert, 1990). The pars tuberalis is the only structure that binds melatonin in all mammals and appears to feature prominently in the transduction of photoperiodic information for control of prolactin secretion (Morgan and Williams, 1989). Melatonin may have direct or indirect effects on the release of substances into the hypothalamic–pituitary portal system to regulate LH and FSH release from the anterior pituitary. Hamsters housed under short-day conditions exhibit reduced expression of LH mRNA, which may or may not be independent from steroidal activation (Bittman et al., 1992).

As previously described, the duration of the nightly peak of melatonin secretion is the critical parameter for transducing the effects of photoperiod on the hypothalamic–pituitary axis. A series of long-duration melatonin signals suppresses anterior pituitary gonadotropin secretion (Bartness et al., 1993; Goldman and Nelson, 1993; Prendergast et al., 2002). Melatonin must suppress GnRH secretion in the hypothalamus, attenuate its ability to stimulate pituitary FSH and LH release, or reduce gonadal responsiveness to gonadotropins (or any combination thereof). Neither in vivo nor in vitro studies provide consistent evidence that either day length or melatonin alter pituitary responsiveness to GnRH in rodents (e.g., Martin et al., 1977; Jetton et al., 1994), but photoperiod plays a significant modulatory role in ruminants (Fowler et al., 1992; Xu et al., 1992). Again, the reproductive effects of melatonin have been most extensively characterized in Syrian and Siberian hamsters. Among hypothalamic nuclei with high densities of melatonin binding, those in the mediobasal hypothalamus, specifically in the dorsomedial nuclei of the hypothalamus (DMN), appear to be essential for decoding photoperiod signals. Bilateral DMN lesions blocked the gonadal response to long-duration melatonin infusions in male Syrian hamsters (Maywood and Hastings, 1995; Maywood et al., 1996), and lesions of the adjacent ventromedial nuclei (VMN) evoked rapid redevelopment of testicular function in Syrian hamsters with regressed gonads (Bae et al., 1999). SCN lesions eliminated the antigonadal effects of long-duration melatonin signals in Siberian, but not Syrian, hamsters (Bartness et al., 1991; Bittman et al., 1979, 1989). Most mammals, studied to date, reduce prolactin concentrations in short days. DMN lesions spared the lactotropic prolactin response to short day lengths (or long melatonin signals) in Syrian hamsters, even though the responsiveness of gonadotropins to these signals was lost (Maywood and Hastings, 1995). This finding suggests that melatonin signals are transduced to the hypothalamic–pituitary axis via multiple parallel pathways that may be trait-specific (Maywood and Hastings, 1995; Lincoln, 1990, 1999). In support of this notion, SCN lesions that eliminated nocturnal melatonin secretion also abolished the gonadal response to short day lengths, although the prolactin response persisted (Bartness et al., 1991; Bittman et al., 1991). Microinfusions of melatonin into the SCN, reunions nuclei, or PVN each induced testicular regression in juvenile Siberian hamsters, but only infusions into the SCN yielded short-day-like prolactin concentrations (Badura and Goldman, 1992).

Melatonin may act on genes that provide day length information. In addition to the SCN, *Per1* mRNA is also expressed in other brain tissues, such as the pars tuberalis (PT) of the pituitary. The PT, as the SCN, has receptors for melatonin. As noted, the duration of photoperiod is transduced through the duration of melatonin secretion. Melatonin can affect the amplitude of *Per1* and inducible cAMP early repressor gene (ICER) expression in the PT. Both Syrian hamsters and Siberian hamsters display a robust and transient peak of *Per1* and *ICER* gene expression 3 hours after dawn in the PT, under both long photoperiods (16 hours of light per day [LD 16:8]) and short photoperiods (LD 8:16) (Messager et al., 1999, 2000). However, the amount of *Per1* and *ICER* gene expression is greatly attenuated in the PT, but not in the SCN, under short days or melatonin treatment in Siberian hamsters (Messager et al., 2000). These data show that the amount of gene expression for the *Per1* and *ICER* may be important to the long-term measurement of photoperiodic time intervals. The role of hormones affecting the expression of these genes remains to be determined.

Although seasonal changes in the pattern of melatonin secretion similar to those in mammals also occur in birds, birds do not appear to use the melatonin signal to time their reproductive effort (Chakraborty, 1995). Thus, the function of the annual fluctuation in the nocturnal melatonin signal in birds remains uncertain, but it has been implicated in the synchronization of circadian activity rhythms (Gwinner et al., 1997; Heigl and Gwinner, 1995; Menaker, 1968) and in seasonal changes in immune function (Bentley et al., 1998). In addition, it may be involved in seasonal changes in the brain areas underlying singing behavior (Bentley et al., 1999). Seasonal neuroplasticity has been documented in several species of songbirds within discrete telencephalic nuclei that are involved in song learning and production (Brenowitz et al., 1991, 1998; Nottebohm, 1981). Increases in the size of these song control nuclei depend largely on seasonal increases in circulating testosterone and its metabolites (Gulledge and Deviche, 1997; Nottebohm, 1980a; Smith et al., 1995), which are directly related to the annual reproductive cycles of these birds (Dawson, 1983). These seasonal changes are associated with changes in cell size and cell number in the various song control nuclei (Smith et al., 1997a) (see Chapter 4).

Despite the well-studied role of testosterone in these seasonal changes in bird brain morphology, there are both gonad- and testosterone-independent seasonal changes in the volumes of song control nuclei (Bernard et al., 1997; Smith et al., 1997b). Thus, the annual changes in melatonin secretion may be involved in the regulation of seasonal changes in the structure of the song control system. Presumably, melatonin exerts its effects directly on the neural substrate of the song control nuclei; melatonin-binding sites have been found in the song control systems of three songbird species (Gahr and Kosar, 1996; Whitfield-Rucker and Cassone, 1996). In starlings, the high vocal center (HVC), lateral magnocellular nucleus of the anterior neostriatum (lMAN), area X, and robust nucleus of the archistriatum (RA) all contain melatonin-binding sites (Gahr and Kosar, 1996; Whitfield-Rucker and Cassone, 1996; Aste et al., 2001). Taken together, this evidence suggests that although melatonin does not synchronize the reproductive activity of birds to a specific time of year, it is involved in seasonal modulation of aspects of avian physiology and behavior that have the potential to affect reproductive success.

Activational Mechanisms

Timing mechanisms interact with the neuroendocrine substrates of reproductive function to ensure appropriate timing of reproductive activities. In mammals, seasonal changes in reproductive activity are generally associated with changes in pituitary gonadotropin secretion. For example, Syrian hamsters, which are long-day breeders, exhibit decreased circulating concentrations of LH and FSH following exposure to simulated winter day lengths (Figure 10.38; Tamarkin et al., 1976). In sheep, which are short-day breeders, exposure to short days leads to increased LH secretion, manifested as an increase in the frequency of pulsatile LH release (Karsch et al., 1984). These types of changes in pituitary gonadotropin secretion lead to changes in gonadal growth and sex steroid hormone secretion (Berndston and Desjardins, 1974). Thus, the current understanding is that the exogenous factors that mediate seasonal changes in reproductive activity do so primarily via actions on the hypothalamic–pituitary axis. Photoperiod can also alter the ability of steroid hormones to activate sexual behavior.

STEROID-DEPENDENT REGULATION OF REPRODUCTION During periods of reproductive activity, one of the important mechanisms regulating pituitary secretion of FSH and LH is the gonadal hormone feedback system (see Chapter 3). In males, testicular androgens, especially testosterone, are capable of acting on the hypothalamic–pituitary axis to inhibit the secretion of both gonadotropins. In effect, this feedback system helps to maintain appropriate levels of gonadotropins; that is, the gonadotropins stimulate the biosynthesis and secretion of testicular androgens, and the negative feedback effect of the androgens prevents oversecretion of the gonadotropins. In females, a similar negative feedback system utilizes estro-

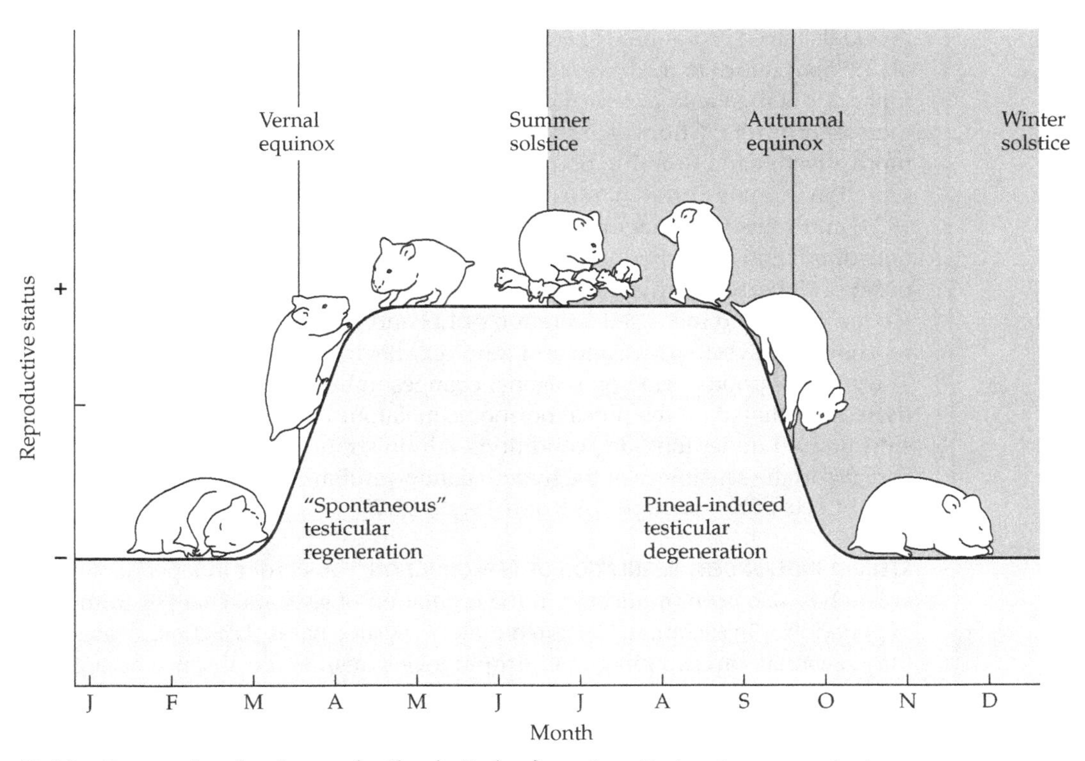

10.38 Seasonal cycle of reproduction in Syrian hamsters. During the summer, both male and female hamsters are reproductively competent and engage in breeding activities. When day lengths fall below some critical value in autumn (12.5 hours of light per day in the laboratory), the negative feedback mechanisms of the hypothalamus are enhanced, and become increasingly sensitive to the effects of steroid hormones. Thus, even low concentrations of steroid hormones completely inhibit secretion of GnRH, and consequently LH, which in turn suppresses secretion of steroid hormones by the gonads and leads to gonadal regression. Hamsters may spend the winter in torpor in a burrow. In the absence of light, the gonads regrow, and begin to produce steroid hormones and gametes again, in the early spring. Interestingly, the gonads can become functional again in spring when the day length is much less than what is required to maintain the reproductive systems in autumn. After Zucker, 1980.

gens and progestins to hold FSH and LH concentrations in check. This system is especially important for regulating the number of ovarian follicles that mature during each ovulatory cycle. As follicles become mature, they produce increasing amounts of estrogen, resulting in decreased levels of gonadotropins and a cessation of recruitment of new follicles (Bast and Greenwald, 1977; Bex and Goldman, 1975). In both sexes, a gonadal peptide hormone, called inhibin or folliculostatin, also serves to inhibit the secretion of FSH (Bernard et al., 2001).

One of the mechanisms that inhibits the secretion of pituitary LH and FSH during seasons of photoperiod-induced reproductive quiescence is an increased sensitivity of the hypothalamic–pituitary axis to the negative feedback effects of gonadal steroid hormones (Ellis and Turek, 1980a; Tamarkin et al., 1976; Turek et al., 1975). A return to the lower level of sensitivity in response to stimulatory photoperiods is then able to return the animal to a state of reproductive activity via increased pituitary hormone secretion. Low concentrations of testosterone are more effective in inhibiting post-castration increases in pituitary gonadotropin secretion in male Syrian hamsters (Tamarkin et al., 1976; Turek and Campbell, 1979) and rams (Pelletier and Ortavant, 1975) when the animals are exposed to nonstimulatory photoperiods. A similar phenomenon has been implicated in the onset of puberty, when a prepubertal decrease in sensitivity to steroid negative feedback leads to increased secretion of LH and FSH and thus activation of the reproductive system (McCann and Ramirez, 1964; Ramirez and McCann, 1963). The effect of photoperiod on seasonal changes in hypothalamic–pituitary sensitivity is mediated by the pineal hormone, melatonin. Thus, in male Syrian hamsters housed under long-day conditions, administration of melatonin induces an increase in the sensitivity of the hypothalamic–pituitary axis to the negative feedback effects of testosterone (Sisk and Turek, 1982).

STEROID-INDEPENDENT REGULATION OF REPRODUCTION A steroid-independent mechanism has also been implicated in the regulation of seasonal changes in the rate of gonadotropin secretion. Castrated male snowshoe hares (*Lepus americanus*) display seasonal variation in gonadotropin levels despite the absence of negative feedback from gonadal steroids (Davis and Meyer, 1973). Likewise, castration of male Syrian hamsters results in elevated blood concentrations of LH and FSH under both long-day and short-day conditions, although the post-castration gonadotropin levels are higher in animals housed under long-day conditions (Ellis and Turek, 1980b). A steroid-independent effect of photoperiod on gonadotropin secretion is particularly evident in female Syrian hamsters. Under long-day conditions, female hamsters exhibit an approximately eightfold to tenfold increase in baseline serum LH concentrations following ovariectomy, and LH concentrations can be returned to baseline by administration of estrogen (Yellon et al., 1989). After several weeks of exposure to short days, intact female Syrian hamsters become anovulatory, and their serum LH concentrations are very low during most of the day; however, these anovulatory females show large daily pulses of LH during the afternoon (Seegal and Goldman, 1975). This pattern of LH secretion continues following ovariectomy (Bridges and Goldman, 1975) or after combined ovariectomy and adrenalectomy (Bittman and Goldman, 1979). That is, removal of the source of steroid hormones does not result in any detectable increase in the baseline serum LH concentration in short-day female Syrian hamsters, and daily pulses of LH are still apparent in the steroid-deprived animals. These observations suggest that, in female Syrian hamsters, the effect of short days on LH secretion is mediated primarily via a steroid-independent mechanism. Seasonal variations in circulating and pituitary concentrations of gonadotropins have also been observed after ovariectomy in pony mares (Garcia and Ginther, 1976), ground squirrels (Zucker and Licht, 1983), and snowshoe hares (Davis and Meyer, 1973).

RELATIONSHIP OF BEHAVIOR TO REPRODUCTIVE STATE The mechanisms that mediate seasonal changes in reproductive state also affect behavior. A wide variety of behaviors vary in relation to an animal's reproductive state. Some of these, most notably mating behaviors, bear an obvious direct relationship to reproduction. In most vertebrates, mating occurs at about the same time as peak gamete production. Because gametogenesis is a steroid-dependent process in all vertebrates, steroid hormone regulation of mating behavior generally provides temporal coordination between gamete maturation and mating (although exceptions exist, as we will see below). Other behaviors, such as territorial behavior and migration, are less directly related to reproduction, but are frequently associated in an adaptive way with the reproductive process. It is probably because of the need for a close temporal association between fertility and behaviors that are directly or indirectly associated with reproduction that the regulation of many behaviors by the gonadal steroid hormones has evolved. As one would anticipate, the regulation of behavior by reproductive hormones is most evident for those behaviors most closely associated with reproduction; that is, mating behaviors.

The seasonal changes in the display of mating behaviors in many species are regulated primarily by seasonal changes in the amounts of circulating gonadal steroid hormones. In addition, seasonal fluctuations in behavioral sensitivity to steroid hormones have been observed. In castrated male Syrian hamsters, copulatory behavior can be restored by administration of testosterone; however, larger doses of the steroid are required to elicit the behavior in animals exposed to short days (Campbell et al., 1978; Morin and Zucker, 1978). The various components of masculine sexual responsiveness, including chemosensory behaviors, mounting, intromission, and ejaculation, are not all equally affected by exposure to short days. The effects of short-day exposure on these behaviors in male Syrian hamsters are prevented by pinealectomy (Miernicki et al., 1988). Female Syrian hamsters also exhibit decreased responsiveness to the activational effects of estrogen on lordosis behavior during exposure to short days (Badura et al., 1987). In contrast to males, this decrease in females' behavioral sensitivity to estrogen is not altered either by pinealectomy or by melatonin administration (Badura and Nunez, 1989). It is possible that photoperiod may influence female sexual behavior through a direct neural route. Neural input from the retina to the SCN is probably required for pineal-dependent responses to changes in day length. However, there are also direct retinal projections to the basal forebrain and hypothalamic regions outside the SCN (Pickard and Silverman, 1981; Youngstrom et al., 1987) that concentrate ovarian steroids (Fraile et al., 1987; Morrell and Pfaff, 1978) and may have a role in female sexual behavior.

DISSOCIATED REPRODUCTIVE PATTERNS Some notable exceptions to the usual association between reproductive hormones and sexual behavior have been reported (Crews, 1984). Some vertebrates exhibit a so-called *dissociated reproductive pattern*, whereby the production of gametes and mating do not occur at the same time of the annual cycle. In the red-sided garter snake (*Thamnophis sirtalis parietalis*), for example, sperm are produced during the summer and are stored in the male reproductive tract through the 8–9-month period of winter torpor (see Chapter 5); mating does not occur until the snake emerges from torpor in the spring. Experiments involving castration and treatment with androgens have revealed that the level of androgens present during the mating phase has no influence on the presence or intensity of mating behavior in these snakes (Crews, 1984). Rather, mating behavior, which persists for about 3 weeks, seems to occur only in snakes that have experienced a period of torpor. The pineal gland may be involved in determining when mating will occur, since removal of the pineal gland prior to entry into winter torpor prevents the display of mating behavior in the spring (Crews et al., 1988b; Nelson et al., 1987). A seasonal timing mechanism probably determines the time of mating in this species, and this mechanism may be partially or entirely independent of gonadal hormones. Clearly, for a species that has evolved a dissociated reproductive pattern, it is appropriate for reproductive behavior to be liberated from the influence of sex hormones. In this case, one must suppose that an alternative mechanism is employed to ensure that mating occurs at the proper time.

AGGRESSIVE BEHAVIOR Perhaps the most general case of seasonal regulation of a behavior by gonadal hormones is exemplified by the male mountain spiny lizard (*Sceloporus jarrovi*), in which a single behavior is largely regulated by testicular androgens during the breeding season in September and October, but is expressed independently of testicular hormones during another phase of the annual cycle. *S. jarrovi* begins to exhibit territorial behavior, expressed as intermale aggression, during the midsummer phase that precedes mating. At this time, castration does not result in a decrease in the level of intermale aggression (Moore and Marler, 1987). During the subsequent mating phase, the level of territorial behavior increases; this increase can be prevented by castration and reinstated by administration of androgens. Castrated mating-phase lizards do not stop showing territorial behavior altogether; rather, aggressiveness declines to a level similar to that exhibited during the earlier, pre-mating phase (Moore, 1987). Yet a third condition occurs in this species subsequent to the mating phase, in which territorial behavior is completely absent and the animals aggregate, tolerating close proximity and even physical contact by members of the same sex. During this phase, administration of androgens fails to stimulate aggressive behavior. It seems that testicular hormones act only to regulate the intensity of territorial behavior in *S. jarrovi* and that other, as yet unknown, mechanisms determine the overall annual pattern of territoriality.

There are many reports of seasonal changes in agonistic and territorial behavior among birds and mammals. For example, male starlings (*Sturnus vulgaris*) form rigorously defended territories during the breeding season (Feare, 1984). This territorial behavior is correlated with high circulating levels of androgens. At the end of the breeding season, blood androgen levels diminish, and territorial behaviors stop. The reduction in agonistic behavior allows the formation of so-called winter feeding flocks, which appear to confer advantages in predator avoidance and foraging success (Feare, 1984). Small rodents also display seasonal changes in aggressiveness and territorial behavior. Microtine rodents (lemmings and voles), and probably most rodent species in nontropical regions, form winter aggregations. Animals huddling together in the winter presumably benefit by reducing their energetic requirements. Presumably, the lack of circulating androgens permits the social tolerance necessary for this pattern of behavior to appear (see Chapter 8). Even in the absence of huddling behavior, animals may tolerate one another better in close quarters during the winter than during the breeding season. For example, male meadow voles (*M. pennsylvanicus*) are highly territorial in the spring and summer and occupy open meadows, whereas red-backed voles (*Clethrionomys gapperi*) breed in spruce forest habitats. During the winter months, meadow voles move into the forest habitats occupied by the red-backed voles, presumably to take advantage of the protective cover provided by the trees. In some cases, they share nests with other rodent species (Madison et al., 1984). Individual meadow voles trapped during the winter and tested in paired encounters in a small neutral arena exhibited less interspecific aggression than voles trapped in summer (Turner et al., 1975). This winter reduction in aggressiveness permits energy-saving habitat sharing. As the animals enter breeding condition in the spring, they reestablish mutually exclusive territories.

In wood rats (*Neotoma fuscipes*), seasonal changes in aggressive behavior are apparently independent of testicular hormones. The level of intermale aggression increases during the breeding season in this species, but this seasonal increase in aggressiveness is also observed in males that have been castrated postpubertally. Furthermore, the increased aggressiveness continues for some time after the breeding season ends. The independence of aggressive behavior from androgens in this species has been hypothesized to have evolved because the greatest threat to reproductive success from conspecific males comes during the breeding season, but the greatest need for nest defense comes later in the year, after the young have been weaned and begin seeking nests of their own (Caldwell et al., 1984). Syrian hamsters maintained in short-day conditions or given melatonin treatments to simulate short-day exposure display more aggression than long-day hamsters, despite their low testosterone concentrations (Jasnow et al., 2002). The independence of aggression from circulating testosterone concentrations seems to be more common than traditionally asserted (e.g., Demas et al., 1999).

DAILY ACTIVITY PATTERNS Field observations of several species of microtine rodents (e.g., *Microtus agrestis, M. oeconomus, M. montanus, Clethrionomys gapperi*, and *C. glareolus*) have indicated a seasonal shift in activity patterns (Erkinaro, 1961; Herman, 1977; Ostermann, 1956; Rowsemitt, 1986). These animals tend to be nocturnal during the summer and diurnal during the winter (Figure 10.39). The adaptive function of this seasonal shift in daily activity patterns probably involves energetic savings. By constraining the majority of its locomotor activity to the daylight hours during the winter, the animal avoids the coldest part of the day; likewise, bouts of activity during summer nights allow the animal to avoid thermal stress or dehydration (Rowsemitt et al., 1982; Rowsemitt, 1986). Predator avoidance may also contribute to the adaptive significance of this seasonal trait.

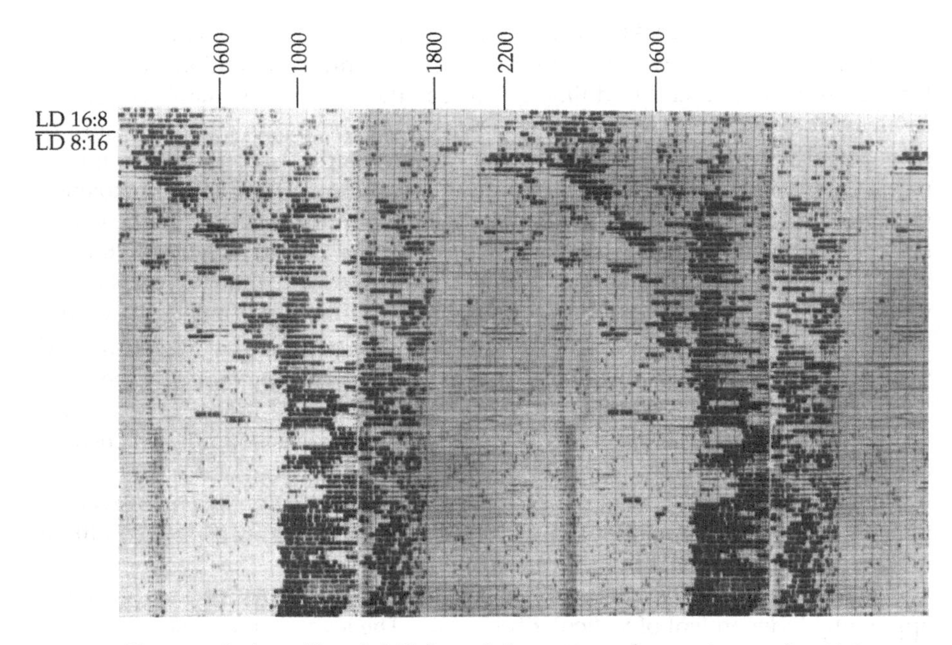

10.39 Photoperiod-mediated shift in activity patterns in montane voles. Male montane voles (*Microtus montanus*) maintained under long-day conditions (LD 16:8) change their locomotor activity from a nocturnal to a diurnal pattern if short-day (LD 8:16) conditions are imposed. From Rowsemitt et al., 1982.

Testosterone appears to mediate the seasonal shift in activity pattern in montane voles (*M. montanus*). Adult male voles were either castrated or left intact and maintained in long-day or short-day conditions. Testosterone replacement was given to some of the castrated animals via subcutaneously implanted Silastic capsules. Castrated montane voles showed increased diurnal and decreased nocturnal wheel-running activity compared with intact animals. Castrated voles implanted with testosterone increased their nocturnal activity relative to voles implanted with empty capsules. There was a great deal of individual variation among the experimental animals. Nevertheless, these results suggest that photoperiod primarily mediates this species' seasonal shift in activity patterns by affecting androgen production; that is, short-day animals tend to be diurnal and long-day animals tend to be nocturnal (Rowsemitt, 1986). Other environmental cues, such as temperature and food quality and quantity, may also affect activity patterns.

Although many subtle effects of steroids on the timing of activity have been reported in other rodent species (Ellis and Turek, 1983; Morin and Cummings, 1981; Morin et al., 1977), assessing the functional significance of these effects has been difficult. For example, the number of daily revolutions made in a running wheel significantly declines when Syrian hamsters are moved from long days to short days (Ellis and Turek, 1979). This decline in locomotor activity can be mimicked in long-day hamsters by castration and reversed by testosterone replacement (Ellis and Turek, 1983). Castrated male Syrian hamsters treated with testosterone but maintained under short day lengths do not increase their wheel-running behavior, suggesting that the neural tissues underlying this behavior become insensitive to steroids under short-day conditions (Ellis and Turek, 1983).

BRAIN SIZE AND LEARNING Seasonal changes in brain weight have been reported for several species of rodents and shrews (e.g., *C. glareolus*, *C. rutilus*, *M. oeconomus*, *M. gregalis*, *Sorex auraneus*, and *S. minutus*) (Bielak and Pucek, 1960; Pucek, 1965; Yaskin, 1984). Brain weights are greater in summer-captured than in winter-captured animals (Figure 10.40; Yaskin, 1984). The adaptive function of this seasonal variation in brain weight may involve energetic savings. Although the brain constitutes only 2%–3% of the total body mass in rodents and insectivores, it uses over 10% of the total energy expended by the animal. It has been suggested that minor reductions in brain mass could result in substantial energetic savings.

A significant part of this seasonal change in brain weight could be attributable to variation in water content; however, several parts of the brain—specifically, the neocortex and the basal portion of the brain (i.e., the corpus striatum)—show

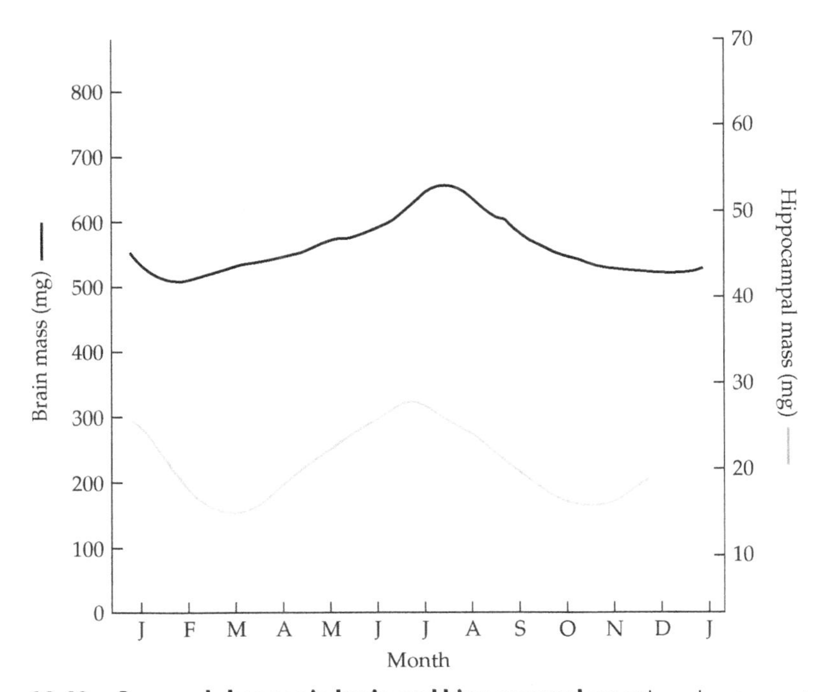

10.40 Seasonal changes in brain and hippocampal mass have been reported for several temperate-zone mammalian species, and are shown here for bank voles (*Clethrionomys glareolus*), corrected for changes in body mass. These changes in brain mass may account for seasonal variation in learning and memory, or they may be an energy-saving adaptation. After Yaskin, 1984.

seasonal cytoarchitectural changes in rodents and shrews. The relative weight of the forebrain and hippocampus declines during the winter, and the relative weight of the olfactory bulbs, myelencephalon, and cerebellum increases during the winter. A sex difference in brain weight is observed among bank voles (*C. glareolus*) only during the winter months; male brains are heavier than female brains at this time. The absolute and relative weight of the hippocampus is significantly higher in males than in females throughout the year, but the difference is most pronounced during the winter (Yaskin, 1984). Meadow voles also show seasonal changes in brain weight. Photoperiod appears to organize the seasonal fluctuation in brain weight in these animals (Dark et al., 1987b); males kept under short-day conditions have smaller brains than long-day males.

Despite the evidence for seasonal changes in brain weight in rodents, there has been relatively little research investigating seasonal changes in learning among mammalian species. There are a few studies addressing seasonal changes in learning and memory in fishes, reptiles, and birds. It is also known that winter-trapped voles make more errors and require longer to learn mazes than summer-captured voles (see Chapter 12); the extent to which this seasonal change is mediated by day length requires further study. Among reptiles, seasonal torpor appears to interfere with learning during the winter, but in studies of this phenomenon, it is not clear if learning or memory function is depressed because of cold exposure or if quiescent animals simply cannot make appropriate responses.

As noted above, song learning in birds also fluctuates on a seasonal basis, and the neural tissues underlying song learning change seasonally. Photoperiod is important in some avian species for the mediation of these changes, including recruitment of new neurons (Nottebohm, 1989). Testosterone, or its metabolites, appears to drive the seasonal changes in brain structure and bird song production (Nottebohm, 1981; Bottjer et al., 1986).

There are three types of evidence that support the role of testosterone in mediating the seasonal change in avian singing behavior. First, blood concentrations of testosterone are highest during the onset of the breeding season, and this peak coincides with maximal singing. Second, song production wanes after castration, and steroid replacement therapy (both androgens and estrogens) reinstates singing behavior (for example, Harding et al., 1983, 1988; Heid et al., 1985). Third, several of the song control nuclei (see Chapter 4) concentrate steroid hormones; for example, androgen receptors have been localized with autoradiographic and immunocytochemical techniques in the HVC, RA, lMAN, ICo, and nXIIts (Balthazart et al., 1992; Brenowitz and Arnold, 1992), and estrogen receptors have been localized in the HVC and ICo (Brenowitz and Arnold, 1989; Gahr et al., 1987, 1993).

In addition to the seasonal changes in singing behavior, there are substantial seasonal changes in the morphology of several song control nuclei. For example, the volume of the HVC and RA increased by 99% and 77%, respectively, among male canaries maintained under spring day lengths (>12 hours light/day) relative to birds housed under autumnal conditions (<12 hours light/day) (Nottebohm, 1981). Similar results have been reported for several other bird species, including red-winged blackbirds (*Agelaius phoeniceus*) (Kirn et al., 1989), rufous-sided towhees (*Pipilo erythrophthalmus*) (Brenowitz et al., 1991), orange bishop birds (*Euplectes franciscanus*) (Arai et al., 1989), and white-crowned sparrows (*Zonotrichia leucophrys*) (Smith et al., 1991). These seasonal changes in the size of certain brain structures are probably mediated by testosterone or its metabolites.

MATERNAL BEHAVIOR AND LITTER SEX RATIOS Seasonal fluctuations in the capacity to exhibit maternal behavior have not been examined in detail because it has been widely assumed that seasonally induced reproductive quiescence precludes the display of this behavior during part of the year. However, some environmental factors have been identified as having effects on maternal responses. Several avian species are known to adjust clutch size in response to changes in food availability (Lack, 1954). Syrian hamsters (Huck et al., 1986) and house mice (Marsteller and Lynch, 1987) display increased cannibalism of their young during periods of food restriction. The opportunity to hoard food reduces the incidence of, but does not abolish, cannibalism in Syrian hamsters (Miceli and Malsbury, 1982a). Interestingly, food restriction during development can have effects on second-generation offspring; litter survival and growth rate are reduced in hamsters born to a dam that had been food-restricted during development. In addition, the sex ratio of litters born to food-restricted female hamsters is skewed in favor of females (Huck et al., 1986). It is unclear whether this bias reflects a gestational event or an active culling of males via postpartum cannibalism. Montane voles display a similar sex ratio bias toward female offspring when the dam is given 6-methoxybenzoxazolinone (6-MBOA), a plant derivative that is made by young grasses when grazing voles injure the plants during the onset of the breeding season in this species (Berger et al., 1987). Ingestion of 6-MBOA appears to induce the birth of more females during the early portions of the breeding season.

The mechanisms of seasonal changes in maternal behavior may involve photoperiodic changes that modulate behavior through the endocrine system. Increased prolactin levels during pregnancy are required for the induction of the full maternal behavioral repertoire in rats (see Chapter 7; Loundes and Bridges, 1986). In hamsters, decreased prolactin levels induced by the administration of ergocornine or bromocriptine have been related to decreased maternal aggression toward male intruders, increased aggression toward pups, disruption of pup retrieval behavior, and an increased incidence of maternal cannibalism of pups (Wise and Pryor, 1977). Hamsters and several other species experience seasonal changes in circulating concentrations of prolactin that are largely under photoperiodic control (Blank and Desjardins, 1985; Goldman et al., 1981; Martinet et al., 1982; Smale et al., 1988; Worthy et al., 1985). It is not known whether day length also influences prolactin secretion during pregnancy or lactation. However, it has been reported that, while pinealectomized hamsters maintained under a natural photoperiod were able to bear litters during the winter, they displayed a high degree of cannibalism (Reiter, 1973/74). It is possible that the females in this study failed to secrete sufficient prolactin to support maternal behavior, because pinealectomy only partially prevents the effects of short-day exposure on prolactin production in female hamsters (Blask et al., 1986).

SEASONAL CHANGES IN COPING WITH STRESS The numbers and types of stressors animals encounter vary seasonally (Breuner et al., 1999). Many energetic adaptations have apparently evolved to attenuate the stress response during winter; at this time, energy shortages may limit animals' abilities to cope with stress. A comprehensive view of the mechanisms underlying the stress response (see Chapter 11) requires the consideration of its temporal organization. Circadian effects on adaptations to stressors have been reported; for example, there is a marked circadian rhythm in glucocorticoid secretion (e.g., Wetterberg, 1999; Chrousos, 1998; Dallman et al., 1993). Increased glucocorticoid release just prior to awakening each day elevates blood glucose concentrations in anticipation of the increased energy demands associated with wakefulness (see Figure 10.28). On this basis, we might expect increased glucocorticoid secretion during the energetically demanding phases of the annual cycle that encompass territorial defense, migration, low temperatures, or food scarcity. Low ambient temperatures and decreased food availability indeed evoke stress responses in mammals (reviewed in Nelson and Demas, 1996).

The primary endocrine components of the stress response are epinephrine and the glucocorticoids, hormones that suppress energy storage and promote energy use from adipose and liver stores (Sapolsky et al., 2000). Although a relatively steady supply of energy is required to sustain biological functions, daily and seasonal fluctuations in energy requirements occur in response to the challenges animals face. Most animals eat discontinuously, storing and accessing energy to maintain cellular function while engaged in nonfeeding activities. Eventually, however, the depletion of energy stores requires their replacement. In most habitats, food availability fluctuates on a daily and seasonal basis. For example, outside the tropics, food availability is generally low during the winter, when thermogenic energy demands are typically high. Consequently, energy intake and energy expenditures are often out of balance. A stress response, in the form of glucocorticoid secretion, occurs whenever a significant imbalance of energy is detected. The high energy demands associated with reproduction usually elicit stress responses (Nelson and Drazen, 1999). Thus, glucocorticoids are released during territorial defense or courtship behaviors. When energy is insufficient to support both reproduction and thermoregulation, a prolonged stress response can have pathological consequences (Sapolsky, 1998). Glucocorticoids released in response to stressful stimuli can compromise cellular and humoral immune function, reproduction, digestion, growth, and virtually any process that consumes significant energy (Sapolsky et al., 2000).

A direct link between melatonin and glucocorticoid biology has been established (Maestroni, 1993, 1995). Generally, melatonin and glucocorticosteroids enhance and compromise immune function, respectively (Aoyama et al., 1986, 1987). Melatonin can ameliorate the immunocompromising effects of glucocorticosteroids (Persengiev et al., 1991a,b), and glucocorticosteroids can reduce the immunoenhancing actions of melatonin (e.g., Poon et al., 1994). Environmental stressors such as low temperatures elevate blood glucocorticoid concentrations, which in turn suppress immune function (Ader and Cohen, 1993; Nelson et al., 2002). A positive balance between short-day enhanced immune status and glucocorticoid-induced immunosuppression may be essential for winter survival in small mammals (Nelson and Drazen, 1999; Sinclair and Lochmiller, 2000). Overcrowding, increased competition for scarce resources, low temperatures, reduced food availability, increased predator pressure, and lack of shelter may each contribute to immunosuppression. Each of these potential stressors may elevate blood concentrations of glucocorticoids. Winter breeding and the concomitant increase in sex steroid hormone secretion may also result in immunocompromise (Nelson et al., 2002). Presumably, winter breeding occurs when normal challenges from environmental stressors such as low temperature and reduced food availability are ameliorated. The advantages of winter reproduction must be balanced against the increased risks of autoimmune disease and susceptibility to opportunistic pathogens and parasites associated with winter steroidogenesis. Thus, reproductive and immune function seem intertwined.

In general, studies of captive animals reveal an inverse relation between dominance and glucocorticoid concentrations. However, this may not be the case in the wild. After capture, animals generally increase their secretion of epinephrine and glucocorticoids, and blood concentrations of steroids are generally much higher at this time than several hours later (e.g., Licht et al., 1983; Mahmoud and Licht, 1997; Mendonca and Licht, 1986; Orchinik et al., 1988). Assays of glucocorticoid by-products in the urine or feces of freely behaving dwarf mongooses (*Helogale parvula*) and African wild dogs (*Lycaon pictus*) suggest that high-ranking animals may be under high levels of social stress (Creel and Creel, 1996).

Seasonal variation in glucocorticoid concentrations is influenced by day length. Among Syrian hamsters, blood concentrations of glucocorticoids are lower in animals housed in short than in long photoperiods (De Souza and Meier, 1987; Ottenweiler et al., 1987). The decreased glucocorticoid concentrations in short-day hamsters could derive from the suppression of adrenal corticosteroid synthesis (e.g., Mehdi and Sandor, 1976) or from enhanced negative feedback inhibition of glucocorticoid secretion (Motta et al., 1967, 1969). Glucocorticoid receptors in the hippocampus, especially mineralocorticoid receptors (MRs) have been implicated in the negative feedback control of adrenal glucocorticoid secretion during stress responses (e.g., Sapolsky et al., 1984; Herman et al., 1993; Fischette et al., 1980). Changes in circulating glucocorticoid concentrations in short-day hamsters may be caused by alterations in hippocampal MR binding, glucocorticoid receptor mRNA expression, or both (Ronchi et al., 1998). Total hippocampal receptor binding of glucocorticoids was significantly higher in short-day hamsters than in long-day hamsters after 8 weeks of these light conditions, regardless of gonadal steroid status, primarily as a result of a significant increase in MR numbers in the former group. MR levels were also significantly elevated in the hypothalamus, but not in the brain, of short-day animals (Ronchi et al., 1998). Basal corticosterone and cortisol concentrations, gonadal mass, and testosterone concentrations did not differ between hamsters housed for 4 weeks in either long or short days (Ronchi et al., 1998). Corticosteroid concentrations were elevated to a similar degree in both long-day and short-day animals after 10 minutes of exposure to ether, but returned to baseline values after 60 minutes in short-day, but not in long-day, hamsters (Ronchi et al., 1998). Short days reduced the reactivity of the hypothalamic–pituitary–adrenal axis to the ether stress; this effect was evident after 2 months of short-day exposure (Ronchi et al., 1998). Short-day exposure likewise reduced the magnitude of the stress response elicited by exposure to low ambient temperatures (Demas and Nelson, 1996). Indeed, after only 18 days of exposure to short days, hippocampal MR mRNA expression was increased in short-day compared with long-day hamsters; this up-regulation of mRNA expression was associated with increased adrenal gland mass in short-day conditions (Lance et al., 1998). The HPA axis appears to respond relatively rapidly to changes in day length and may mediate short-day responses (Lance et al., 1998). Individual differences in stress responses are common, however, and whether individuals are stressed by a change in a specific environmental variable is probably a complex function of individual and species differences (Mason, 1975).

ODOR PREFERENCES Individual rodents have characteristic chemical signatures or odors. Generally, individuals prefer familiar, conspecific, heterosexual odors (Fortier et al., 1996; Sawry and Dewsbury, 1994) over unfamiliar, heterospecific, same-sex odors. In some species, reliance on chemosensory information for making reproductive decisions is marked (Carter et al., 1995).

Odor preferences vary on a seasonal basis. Aggressive behavior may depend on chemosensory stimuli in voles. Adult male meadow voles trapped in the field were more aggressive toward juvenile males during the first part of the breeding season (May through August) than during the later months; there were no seasonal differences in the males' responses to juvenile females (Ferkin, 1988). In contrast, adult female voles were more aggressive toward juvenile males during the late than during the early part of the breeding season. A seasonal shift in aggressiveness of adult voles toward juvenile males may facilitate the juveniles' dispersal as well as the formation of overwintering aggregations, which typically include more females than males (Ferkin and Seamon, 1987).

Seasonal shifts in odor preference are affected by photoperiod and by gonadal steroid hormones (Ferkin and Gorman, 1992; Ferkin et al., 1991, 1992). Male meadow voles housed under long-day conditions preferred female to male odors, whereas short-day males showed no preference (Ferkin and Gorman, 1992). The preference of long-day male voles for female odors was eliminated by castration and was restored by testosterone treatment; odor preference among short-day voles was unaffected by castration, but testosterone treatment induced a preference for female odors (Ferkin and Gorman, 1992). Thus, short-day conditions do not abolish responsiveness to artificially elevated blood testosterone for this behavioral trait. In contrast to the short-day-induced loss of responsiveness to testosterone in hamster reproductive neural circuitry, substrates that mediate odor preferences in male voles retain responsiveness to androgens out of season.

Photoperiod also affects odor preferences in female voles. In contrast to males, steroid treatments reverse odor preferences in gonadectomized females housed in long-day conditions, but are without effect on females housed in short-day conditions (Ferkin et al., 1991; Ferkin and Zucker, 1991). Prolactin and melatonin also influence the attractiveness of male odors to females (Ferkin and Kile, 1996; Ferkin et al., 1997; Leonard and Ferkin, 1999).

Photoperiod also affects odor perception. In contrast to more commonly studied rodents, female prairie voles do not undergo spontaneous estrous cycles; rather, they are induced into estrus by exposure to chemosignals expressed in conspecific male urine (Carter et al., 1995). Seasonal breeding among female prairie voles in the field may reflect photoperiod-mediated changes in the responsiveness of the chemosensory system to male urine (Moffatt et al., 1995). Responsiveness was assessed by localizing the product of the *c-fos* immediate early gene with an immunocytochemical procedure. Fos protein was revealed in a greater number of cells in the accessory olfactory system of female prairie voles, including the accessory olfactory bulbs, granule cell layer, and the medial and cortical divisions of the amygdala, 1 hour after exposure to a single drop of male urine than after exposure to skim milk. The number of cells producing Fos protein in the accessory olfactory system of females exposed to conspecific male urine was lower in short-day than in long-day females (Moffatt et al., 1995; Hairston et al., 2003).

SEASONAL CYCLES OF HUMAN REPRODUCTION Seasonal rhythms in rates of human conception, mortality, and suicide have been reported (Aschoff, 1981; Becker, 1981). In each case, it is generally necessary to sample a large population to obtain statistically significant data because the fluctuations from season to season are relatively small. Thus, these rhythms are quantitatively different from most of those discussed for other animals, which are far more obvious. Because the human data are derived from populations exposed to both natural and artificial environmental changes, it is clearly impossible to know the underlying causation of these rhythms.

It is interesting to consider the absence of major seasonal fluctuations in human reproductive activity in relation to the selective forces that presumably led to the evolution of reproductive seasonality in other species. Human reproduction is characterized by a relatively long gestation period and an extremely prolonged period of intensive parental care. These energy-demanding processes cannot be compressed into one portion of one year, as is typical for most seasonal species. There may be little selective advantage to beginning this lengthy process at any particular time of year.

With that stated, recent studies have reported a seasonal rhythm in human conceptions. These studies analyzed birth records from many countries from both the Northern and Southern Hemispheres covering over 3000 years of monthly data. At high latitudes, where changes in day length are most pronounced, the most conceptions (computed by back-dating birth records by 9 months) occur at about the time of the vernal equinox (Roenneberg and Aschoff, 1990a). The pattern of human conceptions is reversed 6 months in the Southern Hemisphere as compared with the Northern Hemisphere (Roenneberg and Aschoff, 1990b). Temperature has a moderating effect; conception rates are higher than the annual average between 5°C and 20°C and lower than the annual average at extreme temperatures. The influence of photoperiod on conception rates is more pronounced prior to 1930. The authors of the studies conclude that industrialization shielded humans from changes in photoperiod and temperature after that time through artificial lighting and temperature regulation (Roenneberg and Aschoff, 1990b). However, these studies can only be speculative in their conclusions because they are based on correlational data. There are many uncontrolled variables that could provide alternative explanations. However, the possibility of photoperiodic effects on human physiology and behavior remains (e.g., Reiter, 1998; Wehr, 1998).

There are no reports of the existence of endogenous circannual cycles in humans. Collecting such data would be problematic, because the studies necessary to test for the presence of circannual rhythms would require the isolation of individuals under constant environmental conditions for periods of more than a year. Clearly, there are major gaps in our knowledge because there are virtually no data that bear directly on the questions of whether humans are either photoperiodic or circannual. Nevertheless, the growing body of data on seasonal mammals may be useful in pointing the way to obtaining such information for humans. Future research into human seasonality might benefit from a consideration of its potential for contributing to human fitness. For example, seasonal changes in human immune function might be mediated by photoperiod, as they are in other mammalian species (Nelson and Demas, 1995).

Conclusions

An understanding of the rhythmic nature of behavior is important for behavioral scientists for two reasons. First, an awareness of the daily and annual variation in many behaviors may minimize any unintended influences of time of day or seasonality on experimental results. Care should be exercised in obtaining experimental animals, in maintaining appropriate lighting conditions, and in the timing of data collection. Second, seasonal and daily changes in many phenomena of interest to psychologists and biologists have been documented. Reliable daily and seasonal variation in learning and memory function, perception, communication, developmental rates, social behavior, parental behavior, and mating behavior has been reported for many species. Few data on the mechanisms underlying these seasonal changes in behavior are available. We do know that many hormone–behavior interactions appear to be linked to reproductive cycles, but virtually no data exist on the mechanisms underlying seasonal phenomena not linked to reproduction.

There is a lack of basic information about human seasonality. For example, it is not known if humans are photoperiodic or if they possess endogenous annual cycles. The possibility of seasonal variation in human developmental rates, learning, or perceptual abilities has not been well studied. The extent to which behaviors such as aggression in humans are seasonal, and if so, to what extent these seasonal fluctuations in behavior reflect seasonal changes in human hormone concentrations, also remains unknown. Studies on nonhuman mammals should be useful in obtaining information about the functions and mechanisms of seasonal cycles of behavior. In summary, biological clocks can no longer be ignored in describing normal behavior, treating abnormal conditions, or designing experiments.

Summary

- Endocrine function varies over a wide range of temporal scales. Many of these temporal changes are the result of biological rhythms generated by endogenous biological clocks. The study of biological clocks and their associated rhythms is called chronobiology. Many homeostatic processes show temporal fluctuations over the course of hours, days, months, or years.
- Some biological rhythms vary in period length, but four general classes of biological rhythms are normally synchronized to the geophysical cycles of day and night (circadian), the tides (circatidal), phases of the moon (circalunar), and the seasons of the year (circannual). These rhythms persist in the absence of the geophysical cues with which they are normally synchronized (entrained) with a period that approximates that of the geophysical cycle. Ultradian (shorter than circadian) and infradian (longer than circadian) rhythms are also observed, but do not correspond to any known geophysical cue.
- Biological rhythms have evolved to synchronize the activities of individuals with changes in their environment, allowing preparation for predictable events, and to synchronize the internal physiological and biochemical processes of individuals to permit efficient function.
- Biological clocks have several formal properties. Biological clocks are inherited, relatively independent of temperature, relatively independent of chemical influence, entrainable only to limited cycle lengths, relatively independent of behavioral feedback, and found at every level of organization within an individual plant or animal.
- Biological clocks are found in single cells. Within multicellular organisms, the various clocks in the body appear to be linked in a hierarchical organization that allows temporal coordination. The master clock appears to be located in the suprachiasmatic nuclei of the hypothalamus in mammals and in the pineal gland in some species of birds.
- Hormones can affect daily locomotor activity cycles. Estrogens accelerate clock function, and progesterone lengthens the period of activity cycles. Removal of the pituitary gland and castration also lengthen free-running activity rhythms.
- Light is important for entraining biological rhythms to geophysical cycles in virtually all species examined, including humans. Light affects the rhythm of melatonin secretion. Melatonin is produced by the pineal gland during the dark

phase of the day, resulting in a sustained elevation in blood levels of this hormone. Long nights (short days) are coded by relatively long periods of sustained elevated melatonin levels. Short nights (long days) are coded by relatively short periods of sustained elevated melatonin levels. Animals use the length of this nightly peak of melatonin secretion to assess the time of year.

- Many annual cycles of hormone–behavior interactions have been observed among animals. Most of these seasonal changes in behavior are correlated with seasonal changes in reproductive function. Thus, steroid-dependent behaviors wane when animals become reproductively quiescent and recur during the breeding season, when reproductive hormones are again secreted.
- Some annual cycles are the result of endogenous circannual clocks, whereas others are the result of environmental input; day length may be the most widely used cue in the regulation of seasonal rhythms. Seasonal rhythms in mating behavior, parental behavior, territoriality, aggression, and learning have been reported for a variety of species. Seasonal cycles in human conception have also been recently documented.

Questions for Discussion

1. Compare and contrast the concepts of "homeostasis" and "biological rhythms." Are their functions complementary or antagonistic?
2. Why is it important to have an understanding of biological clocks and rhythms in designing or interpreting behavioral endocrinology experiments?
3. Some have noted that males and females of the same species are quite different animals. Similarly, it can be said that animals examined during the summer are quite different animals than animals of the same species examined during the winter. Defend or refute this statement.
4. What adaptive advantages do biological clocks confer?
5. Postulate how disrupted biological rhythms might contribute to disease. Consider how artificial lighting might contribute to the problem. How might dysfunctional biological rhythms be repaired?

Refer to the accompanying Student CD for additional resources, including Web links, videos, animations, and additional photos.

Suggested Readings

Bronson, F. H., and Heideman, P. D. 1994. Seasonal regulation of reproduction in mammals. In E. Knobil and J. D. Neill (eds.), *The Physiology of Reproduction*, pp. 541–584. Raven Press, New York.

Hastings, M. H., Reddy, A. B., and Maywood, E. S. 2003. A clockwork web: Circadian timing in brain and periphery, in health and disease. *Nature Reviews: Neuroscience*, 4: 649–661.

Kriegsfeld, L. J., LeSauter, J., Hamada, T., Pitts, S. M., and Silver, R. 2002. Circadian rhythms in the endocrine system. In D. W. Pfaff, A. P. Arnold, A. M. Etgen, S. E. Fahrbach, and R. T. Rubin (eds.), *Hormones, Brain and Behavior*, Volume 2, pp. 33–91. Academic Press, New York.

Prendergast, B. J., Nelson, R. J., and Zucker, I. 2002. Mammalian seasonal rhythms: Behavior and neuroendocrine substrates. In D. W. Pfaff, A. P. Arnold, A. M. Etgen, S. E. Fahrbach, and R. T. Rubin (eds.), *Hormones, Brain and Behavior*, Volume 2, pp. 93–156. Academic Press, New York.

Reppert, S. M., and Weaver, D. R. 2002. Coordination of circadian timing in mammals. *Nature*, 418: 935–941

Rosenthal, N. E. 1993. *Winter Blues: Seasonal Affective Disorder*. Guilford Press, New York.

Takahashi, J., Turek, F. W., and Moore, R. Y. (eds.). 2001. *Handbook of Behavioral Neurobiology*. Volume 12, *Circadian Clocks*. Kluwer Academic/Plenum, New York.

11

Stress

It was a very sad day in my son's preschool. One of the three pet guinea pigs living in the classroom unexpectedly died. The teachers met with the students the following day to provide some comfort, answers, and closure. The teachers relayed to the students that the cause of death was probably unintentional "stress" brought on by the loud environment and intermittent "treats" provided by the students when the teachers were not watching. Because of the trauma, both to the rodents and to the children, schoolroom pets were thereafter limited to fish and insects.

Although the death of Snowball, the guinea pig, raised many theological questions in my preschooler during the next few evenings, he also asked several important biological questions: What is "stress"? Could "stress" really kill a guinea pig, and if so, could stress kill a person? Why did "stress" kill only one of the guinea pigs and not all three? Did the other two guinea pigs not notice the "stress"? Some version of these four basic questions forms the basis of this chapter: What is biological stress? Does stress impair life? Are there individual differences in the perception of and responses to stress, and if so, what are the physiological bases of these differences? Of course, we will also explore how hormones mediate the effects of stress.

There are many sources of stressors. Environmental factors, such as temperature extremes, are often perceived as stressors. In the case of the guinea pigs caged in a preschool classroom, the high levels of noise could have been a stressor. Stressors can also be physiological factors, such as insufficient food quality or quantity or water deprivation. Importantly, psychosocial factors, such as fighting, social subordination, or lack of control in a given situation, can be salient stressors. Psychological stressors for a pet guinea pig might include confinement, restraint when being held during petting, or even anticipation during the night of the onset of noise the following morning.

The goal of this chapter is to understand the concept of "stress," the hormonal correlates of stress, the adaptive versus maladaptive consequences of stress, and the effects of both short-term and long-term stress on behavior. Behaviors that are often influenced by hormones, including reproductive, parental, and social behaviors, are also affected by stress, as is drug abuse. All of these behaviors will be emphasized in this chapter.

Stress and Its Consequences

Life is challenging. The pressure of survival and reproduction takes its toll on every individual on the planet. The "wear and tear" of life eventually leads to death. All things being equal, animals that live the longest tend to leave the most offspring. In the Darwinian "game of life," individuals that leave the most successful offspring win. Although some of the variation in longevity among individuals of the same species merely reflects good fortune, a significant part of this variation reflects differences in the ability to cope with the demands of living, whether these demands occur during the daily commute, on the African Serengeti, or in a preschool classroom.

As we saw in Chapter 9, all living creatures are vessels of dynamic equilibrium, or homeostasis. Any perturbation to homeostasis requires the animal to expend energy to restore the original steady state. Among birds and mammals, an individual's total available energy is partitioned among many competing needs, such as growth, cellular maintenance, immune function, reproduction, and thermogenesis. During environmental energy shortages, processes that are not essential for immediate survival, such as growth and reproduction, are suppressed. If coping with homeostatic perturbations requires more energy than is readily available after these nonessential systems have been inhibited, then an individual's survival may be compromised.

All living organisms currently exist because they have evolved adaptations that allow individuals to cope with energetically demanding conditions. These demanding conditions range from final exam week for a well-fed college undergraduate in the United States to finding a meal on the African veldt for a wild dog. Surprisingly, the same neuroendocrine coping mechanisms are engaged in both of these cases, as well as in many other situations. Although stressors typically disrupt homeostasis, thereby affecting the brain and behavior, it is important to note that the brain itself can perceive psychological factors, such as taking an exam or giving a classroom presentation, as stressful and evoke a stress response that disrupts homeostasis. Because hormones are important mediators of stress, the stress response emphasizes the bidirectional relationship between hormones and behavior (and thus the brain).

11.1 A group of African wild dogs. Only the dominant alpha pair of each pack breeds. The other adult members of the pack bring food to the pups and otherwise assist in the breeding efforts of the alpha pair. In this species, the alpha animals have higher corticosterone concentrations than the subordinate animals of the pack. Courtesy of Scott Creel.

To illustrate the kinds of demanding conditions animals must face, let's look at the example of wild dogs more carefully. Despite their common name, African wild dogs (*Lycaon pictus*) are more closely related to jackals than to modern dogs (Figure 11.1). However, African wild dogs do possess several characteristics typical of domesticated dogs, including group hunting and the formation of strong social bonds. They live in packs of about 6 to 8 adults and several pups. Unlike other canine species, however, wild dog packs typically have only one breeding pair (called the alpha pair) (Creel et al., 1996, 1997). In one study, 82% of the alpha females within packs bred each year, and produced 76%–81% of the offspring; only 6%–17% of the subordinate females produced any offspring (Creel et al., 1997). Thus, the costs and benefits of defined social rank can be highly variable among the members of the pack. Not surprisingly, subordinate female African wild dogs tend to leave their natal packs and attempt to become the alpha female elsewhere, although some subordinate females stay in their natal packs and attempt to rise through the ranks to alpha status.

In general, dispersal is a demanding time for young animals (see Chapter 8). It requires moving to a new area, where they must learn where food and water can be found. Whatever protection they received from their natal group on their home territory is lost in the new neighborhood, and the dispersing individuals must establish new social relationships. A dispersing female wild dog faces many challenges when she leaves her natal pack. In order to become the reproductively successful alpha female in a new pack, a new female must attack and dominate the current alpha female of the pack and also defend against the other subordinate females of the pack vying for the top position. Thus, the intrusion of a newcomer results in increased social instability, and ultimately dominant dogs attempting to defend their status spend more time in aggressive interactions than subordinate animals do. Therefore, it is not surprising that dominant wild dogs have higher cortisol concentrations than low-ranked individuals.

The relationship between corticosteroid concentration and social dominance varies across species and may be influenced by the characteristics of the social hierarchy, including stability, number of aggressive encounters experienced by dominant versus subordinate individuals, and opportunity for affiliative interactions (reviewed in DeVries et al., 2003). Even among primates, there is no consistent relationship between the stress response and social rank (Box 11.1).

The Stress Response

When a stressor disrupts physiological homeostasis, an individual typically displays what is commonly referred to as a stress response. A **stress response** is a suite of physiological and behavioral responses that help to reestablish homeostasis. The stress response is relatively *nonspecific*; that is, many different stressors elicit a similar stress response (Selye, 1950). Two endocrine systems, one involving epinephrine (adrenaline) from the adrenal medulla and the other involving glucocorticoids from the adrenal cortex, constitute the major components of the stress response (Stratakis and Chrousos, 1995). Within seconds of perceiving a stressor, the sympathetic nervous system begins to secrete norepinephrine, and both adrenal medullae begin to secrete epinephrine. A few minutes later, the adrenal cortices begin to secrete glucocorticoids.

In 1915, Walter Cannon (see Figure 12.2) proposed his "emergency theory" of the adrenal glands, which suggested that the secretion of epinephrine from the adrenal medulla increases following an exposure to virtually any stressor, as a means of adapting to that stress. Many physiological studies conducted throughout the 1920s and 1930s demonstrated the stimulatory effects of epinephrine on the respiratory and cardiovascular systems. Epinephrine is usually the chemical message that acts first, because just a doubling of epinephrine from its resting values causes profound changes in respiration and cardiac tone (i.e., heart rate and blood pressure), whereas norepinephrine concentrations must increase fivefold to have similar effects (Ganong, 1997). This immediate, nonspecific component of the stress response was termed the **fight or flight response** because the physiological changes in cardiovascular tone, respiration rate, and blood flow to the muscles from the trunk could support either of those behavioral responses (Cannon, 1929). Importantly, the catecholamines (norepinephrine and epinephrine), working through a variety of mechanisms, also increase blood glucose levels. The elevated blood glucose fuels the fight or flight response. The catecholamines also increase alertness, and as we will see in Chapter 12, enhance learning and memory.

Modern perspectives on stress are less likely to focus on fight or flight than on a psychological feature of stress: namely, the degree of control the stressed individual has over the situation. In one study, for example, both epinephrine and

BOX 11.1 Stress and Social Dominance in Nonhuman Primates

In primates, the relationship between cortisol concentration and social dominance is influenced by the characteristics of the social hierarchy. For example, social instability alters the relationship between cortisol concentration and dominance rank in female rhesus monkeys. Low-ranking female rhesus monkeys in a stable, established group have higher cortisol concentrations than high-ranking females (Gust et al., 1993). In contrast, in a newly formed group with a relatively unstable hierarchy, there is no relationship between cortisol concentration and dominance rank. An interesting behavioral difference between established and newly formed groups is that the monkeys in newly formed groups engage in much more post-aggression reconciliatory behavior, such as grooming, which may relieve some of the stress associated with losing a skirmish and may ultimately result in low basal cortisol concentrations in both dominant and subordinate monkeys. Reconciliation following an aggressive interaction may be a behavioral strategy that aggressors use to ease tensions in unstable social groups in which strong social alliances have not yet been established (Gust et al., 1993). As the hierarchy becomes more stable, the expression of reconciliatory behaviors decreases, subordinate-directed aggression increases, and corticosterone concentrations increase among subordinate individuals.

In general, there is no consistent relationship between the stress response and social rank among nonhuman primates (Abbott et al., 2003). Rather, differences in stress responses probably reflect differences in social organization, stability, and specific behaviors. The accompanying table (see the next page) shows relative cortisol concentrations in subordinate animals of both sexes in several different species. The relative cortisol concentrations are basal cortisol concentrations in the subordinate animals expressed as a percentage of basal cortisol concentrations in dominant animals. The graphs below show the degree to which subordinate animals in each group are exposed to social stressors, and the degree to which they receive social support.

Continued on following page

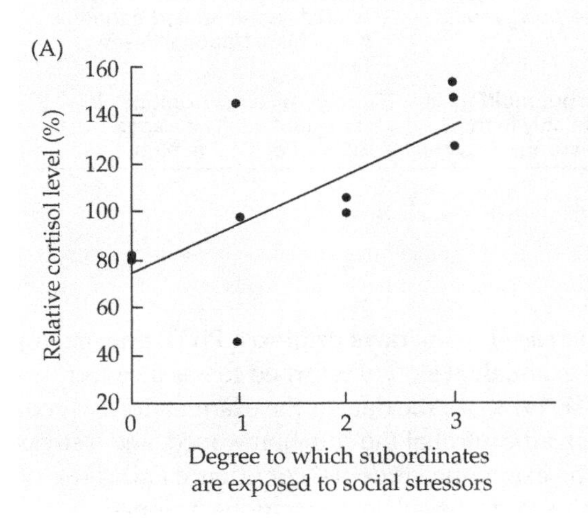

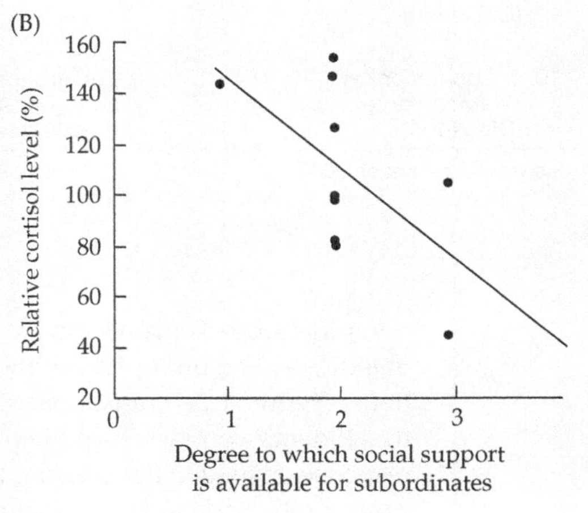

BOX 11.1 continued

Relative cortisol concentrations in subordinate animals of several primate species

Species	Relative cortisol (%)	Social/mating system	Group structure/ physical environment
1. Common marmoset (*Callithrix jacchus*), female [New World]	45	Small groups, singular cooperative breeders	Captive, mixed-sex groups of 3–6 unrelated adults in cages 0.6 × 0.9 m × 1.8 m high
2. Cotton-top tamarin (*Saguinus oedipus*), female [New World]	80	Small groups, singular cooperative breeders	Captive, families of parents and offspring in cages 0.8 × 1.5 m × 2.3 m high
3. Cotton-top tamarin (*S. oedipus*), male [New World]	82	Small groups, singular cooperative breeders	Captive, families of parents and offspring in cages 0.8 × 1.5 m × 2.3 m high
4. Squirrel monkey (*Saimiri scurieus*), female [New World]	98	Large groups, multi-male, multi-female, polygynous	Captive, single- or mixed-sex groups of 3–4 unrelated adults in cages 1.0 × 1.0 m × 2.3 m high
5. Rhesus monkey (*Macaca mulatta*), male [Old World]	99	Large groups, multi-male, multi-female, polygynous	Captive, troops of ~150 related and unrelated rhesus in 0.3-ha outdoor enclosure, Sebana Seca Field Station, Puerto Rico
6. Talapoin monkey (*Miopithecus talapoin*), female [Old World]	105	Large groups, but multi-male, multi-female only in the breeding season, polygynous	Captive, mixed-sex groups of 7–11 unrelated adults in cages 3.5 × 1.5 m × 1.7 m high
7. Cynomolgus monkey (*Macaca fascicularis*), female [Old World]	127	Large groups, multi-male, multi-female, polygynous	Captive, mixed-sex groups of 5–7 unrelated adults in outdoor pens 1.7 × 3.3 m
8. Squirrel monkey (*S. scurieus*), male [New World]	145	Large groups, multi-male, multi-female, polygynous	Captive, single- or mixed-sex groups of 3–4 unrelated adults in cages 1.0 × 1.0 m × 2.3 m high
9. Olive baboon (*Papio anubis*), male [Old World]	147	Large groups, multi-male, multi-female, polygynous	Free-ranging, troops of ~40–50 in related and unrelated baboons, Masai Mara National Reserve, Kenya
10. Talapoin monkey (*M. talapoin*), female [Old World]	154	Large groups, but multi-male, multi-female only in the breeding season, polygynous	Captive, mixed-sex groups of 7–11 unrelated adults in cages 3.5 × 1.5m × 1.7 m high

Source: Abbott et al., 2003

norepinephrine concentrations increased in the days prior to a Ph.D. final examination, peaked on the day of the exam, then slowly returned to basal concentrations (Figure 11.2) (Frankenhaeuser, 1978). As the date of the examination neared, the students' confidence and perceived control of the situation waned, and a stress response ensued. After passing the exam, the students' confidence and sense of control returned, and catecholamine and cortisol concentrations dropped.

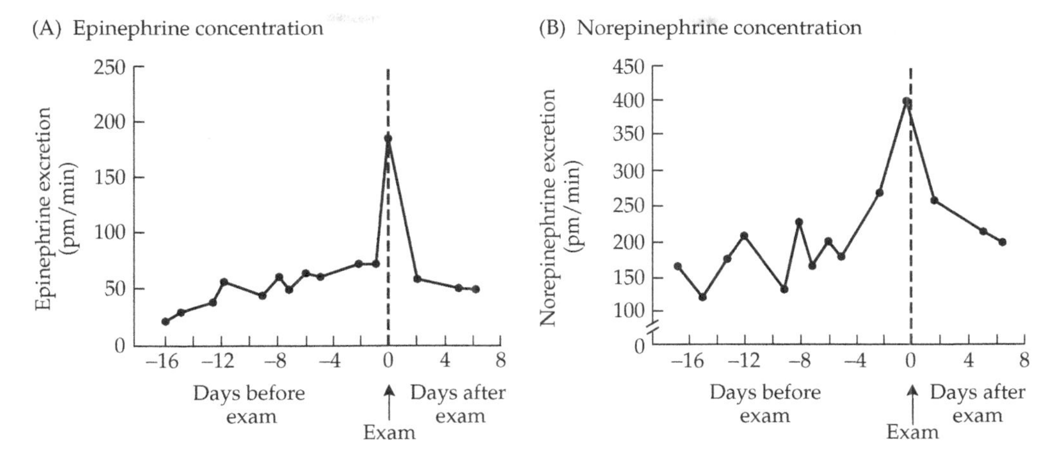

11.2 Catecholamine concentrations peaked on the day of a Ph.D. exam. Students showed a consistent elevation of (A) epinephrine and (B) norepinephrine concentrations in the days prior to the exam. Both catecholamines peaked on the day of the exam and returned to baseline after the students had coped successfully with the stressor (passing the exam). After Frankenhaeuser, 1978.

Other studies have shown that not just epinephrine and norepinephrine, but an entire suite of hormones known to be involved in the mediation of stress, change over the course of the stress response. The **hypothalamic–pituitary–adrenal axis (HPA axis)** is activated, so that corticotropin-releasing hormone (CRH), adrenocorticotropic hormone (ACTH), and glucocorticoids are released in response to stressors (Figure 11.3). In one classic study, blood samples were obtained from young military recruits preparing for their first parachute jump (Ursin et al., 1978). Basal blood hormone concentrations were determined prior to the jump. Hormone concentrations on the day of the jump, as well as on days of subsequent jumps, were also measured (Figure 11.4). Not surprisingly, catecholamine concentrations were high on the day of the first jump, although, as the recruits gained confidence, their stress responses were muted. Cortisol concentrations were also elevated on the first jump day, while testosterone concentrations were suppressed on the first jump day (Ursin et al., 1978).

Because epinephrine does not cross the blood–brain barrier, and because endocrine mediation of the stress response must affect the brain in order to affect behavior, much of this chapter will focus on the effects of glucocorticoids in the stress response. Glucocorticoids are good candidates for mediating the behavioral effects of stress because (1) these steroid hormones are released in response to numerous stressors, (2) steroid hormones can easily diffuse past the blood–brain barrier, and (3) there are glucocorticoids receptors in several brain regions.

Within minutes of the onset of a stressor, the adrenal cortex begins to secrete glucocorticoids (e.g., corticosterone in most rodents, birds, and reptiles; cortisol in most primates and carnivores) (Stratakis and Chrousos, 1995). Like epineph-

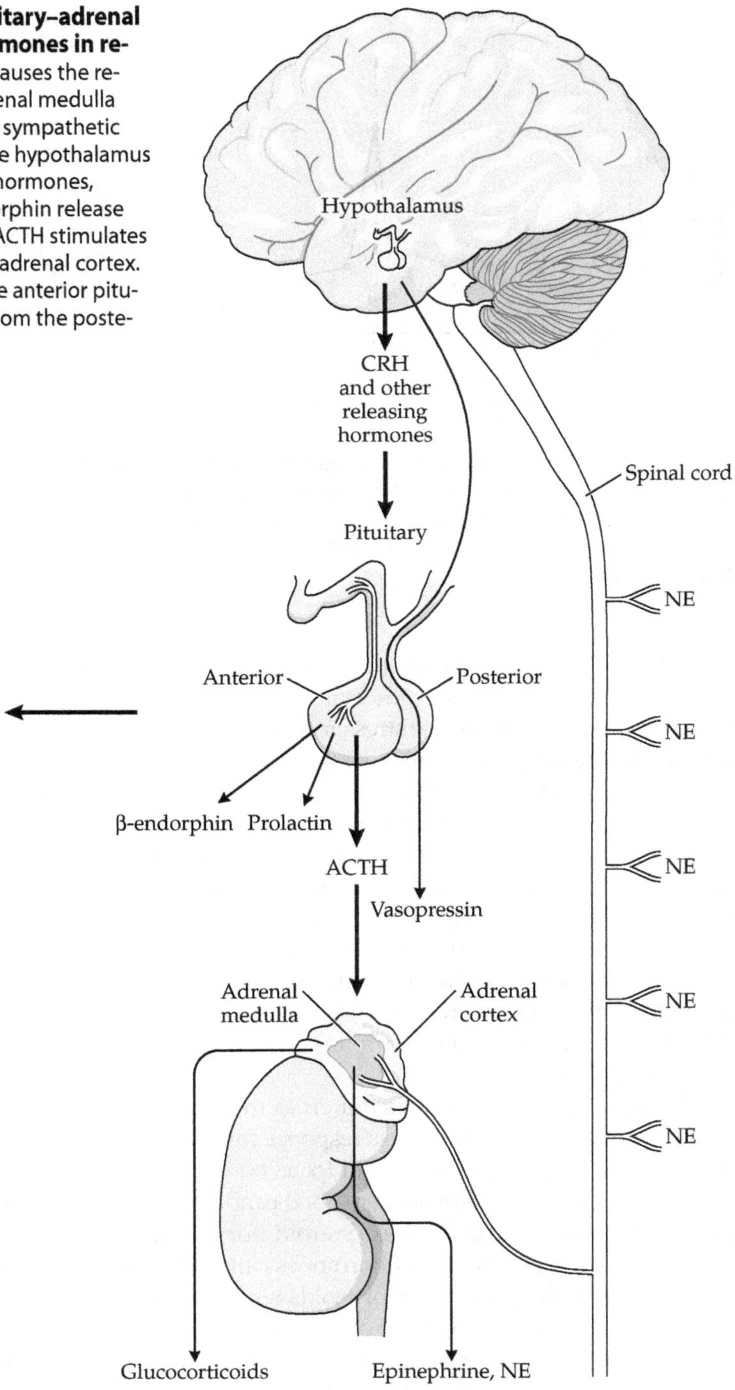

11.3 The hypothalamic–pituitary–adrenal (HPA) axis releases several hormones in response to stress. Initially, stress causes the release of epinephrine from the adrenal medulla and norepinephrine (NE) from the sympathetic nervous system. Moments later, the hypothalamus releases CRH and other releasing hormones, which stimulate ACTH and β-endorphin release from the anterior pituitary gland. ACTH stimulates corticosterone secretion from the adrenal cortex. Prolactin is often released from the anterior pituitary and vasopressin is released from the posterior pituitary during stress.

lic lecture or being confined in a tight space, can evoke a full physiological stress response; indeed, this psychological stress response actually causes homeostatic imbalance in an individual, rather than restoring it. Second, the homeostatic concept of stress does not account for individual variation in the perception of stressors. Most people would be terrified to jump out of an airplane with a parachute, and would report such an experience as stressful; however, some individuals seek out skydiving for pleasure. In order for a definition of stress to be useful, it must address how the same stimulus can be stressful to one individual and pleasurable to another. One variable to be considered in this regard is the extent to which the individual perceives the stressful situation as being under their control: individuals that have a sense of control feel less vulnerable to the effects of a stressful condition. Third, some definitions conflate "stress" with the "physiological response to stress"—that is, "stress" activates the adrenal glands to produce epinephrine and glucocorticoids. However, elevated glucocorticoid concentrations and activation of the sympathetic nervous system are caused by both stressful and pleasurable events.

Recently, Kim and Diamond proposed a three-part definition of stress that may prove useful in trying to unravel the interaction among stress, hormones, and behavior. Accordingly, stress is considered to be a condition in which individuals are aroused by aversive stimuli; the extent of the stress is determined by the individual's perception of control over the aversive stimuli (Kim and Diamond, 2002). First, stress provokes elevated arousal, which can be assessed by measuring locomotor activity, hormone concentrations (e.g., epinephrine or glucocorticoids), or electroencephalogram (EEG) activity. Second, because arousal can increase under both aversive and pleasurable conditions, for an event to be defined as stressful, the individual must perceive it as aversive (Kim and Diamond, 2002). Thus, jumping out of an airplane with a swatch of silk attached to a backpack at 2000 meters would provoke arousal in anyone; however, it would not be stressful to someone who enjoyed skydiving and did not find it aversive. In contrast, parachuting would be stressful for someone who was afraid of heights or for some other reason perceived it as aversive. The aversiveness of a stimulus can be judged by the extent to which an individual avoids the stimulus if given a choice. For example, a rat placed in a pool of water may be stressed initially, until it learns that there is a hidden platform just under the water surface to which it can escape. This final component of the definition of stress encompasses the concept of controllability. Numerous studies in both humans and nonhuman animals have indicated that the element of control in an aversive situation, as well as the related element of predictability, significantly ameliorates the long-term negative effects of the stressful experience, as we will see below.

Physiological Effects of the Stress Response

Almost immediately after exposure to a stressor, a physiological stress response with several components begins to unfold (see Figure 11.3). Within seconds, the sympathetic nervous system secretes norepinephrine, and the adrenal medulla secretes epinephrine. Stress also stimulates increased activity of the HPA axis, so that the hypothalamus releases CRH into the hypothalamic–hypophyseal portal system within seconds of perception of a stressor. The CRH travels to the anterior pituitary gland, where it stimulates the chromatophobes to secrete ACTH, which enters the general blood circulation and provokes the adrenal cortex to produce and secrete glucocorticoids. This cascade of endocrine events is often referred to as the *HPA response* and takes several minutes to be fully engaged. Several other hormones, including endorphins, prolactin, glucagon, thyroid hormones, vasopressin, and growth hormone may also be secreted from various endocrine tissues during an acute stress response.

The stress response has many adaptive effects:

1. Increased immediate availability of energy
2. Increased oxygen intake
3. Decreased blood flow to areas not necessary for movement
4. Inhibition of energetically expensive processes that are not related to immediate survival, such as digestion, growth, immune function, and reproduction
5. Decreased pain perception
6. Enhancement of sensory function and memory

Imagine that a rabbit is being chased by a fox. How does the stress response affect the rabbit in this situation? First, the rabbit needs a quick energy supply in order to sustain the sprint back to its warren. In such an emergency, the need for metabolic fuel is acute, large, and unpredictable (Sapolsky, 1992b, 1994). The immediate release of catecholamines raises the rabbit's respiration and cardiovascular rates within seconds; its body requires increased energy availability to sustain these high rates. The most obvious physiological change that occurs in response to a stressor is an immediate increase in available levels of glucose and oxygen in the blood. Epinephrine increases the delivery of oxygen to the tissues, and it raises sympathetic tone (i.e., the activation level of the sympathetic nervous system). Glucocorticoids, which are secreted within minutes, though probably not until the rabbit is back home, act on metabolic pathways to replenish the energy reserves used to escape the predator. Epinephrine causes blood flow to be temporarily rerouted to the muscles and away from the digestive system and other organs not critical for coping with the emergency situation. In other words, the rabbit's lunch can be digested later (if there is a "later"). Responses to injuries that might curtail movement (e.g., pain, inflammation) are inhibited by the stress-induced release of endorphins. If a rabbit that suffered scrapes and scratches as it darted into the brambles stopped to nurse its wounds, it would probably become the fox's dinner. Energetically expensive activities such as growth, reproduction, and some components of immune function are also suppressed until after the emergency has passed. Other components of immune function, such as trafficking of immune cells to the skin, where injuries might occur, are enhanced during stressful events (Dhabhar and McEwen, 1997, 1999) (Figure 11.6). Well after the rabbit is safe in its nest, the stress response subsides, parasympathetic tone increases, and metabolic rate returns to baseline.

Importantly, both the predator and the prey are experiencing similar acute stress responses during the chase, despite their disparate roles—another example

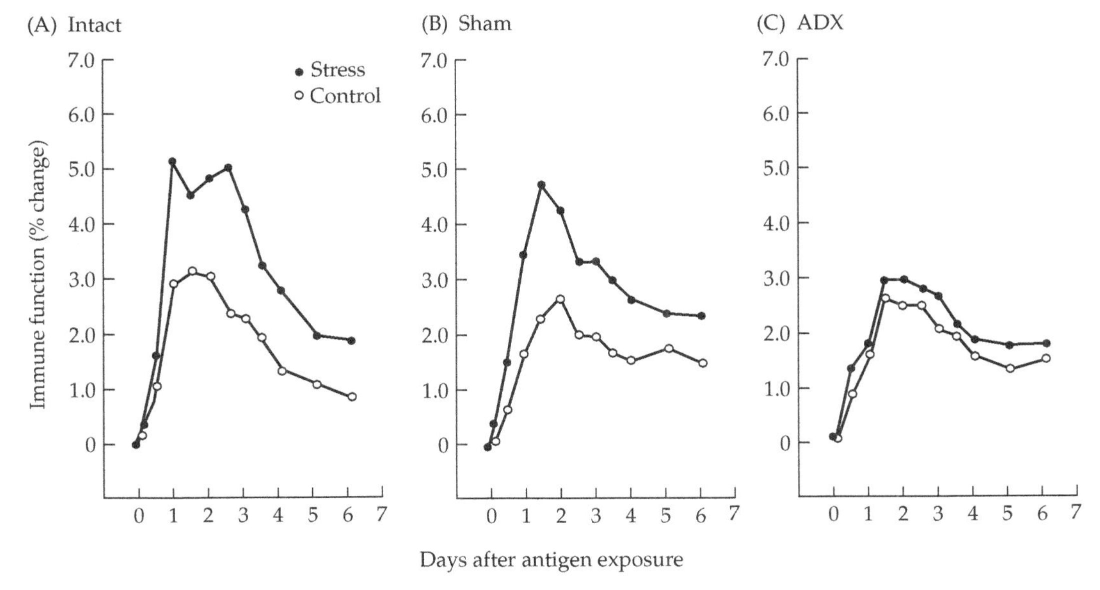

11.6 Acute stress can improve immune function. Restraint stress enhanced a type of immune response in adrenal-intact rodents (A) and in animals receiving a sham adrenalectomy (B). This effect was abolished by adrenalectomy, suggesting that it was mediated by glucocorticoids (C). After Dhabhar and McEwen, 1999.

of the nonspecificity of the stress response. If the attempts of the fox to obtain food are unsuccessful and she has not eaten in a long time, then she will experience a chronic stress response, the negative effects of which will be presented below.

Recently, it was discovered that in addition to provoking the release of ACTH and, subsequently, glucocorticoids, CRH itself mediates some aspects of the stress response, both adaptive and pathological. CRH appears to have a role in modulating anxiety under several circumstances (Coste et al., 2001). Transgenic mice that express high levels of CRH are more anxious in a novel environment than control mice. The same is true of wild-type mice and rats that are injected with exogenous CRH. Furthermore, CRH receptor antagonists decrease the anxiety normally associated with alcohol withdrawal and social defeat in rats and mice. Transgenic mice that express high levels of CRH also exhibit decreased female sexual receptivity, which is not reversed by adrenalectomy, and impaired spatial learning and memory.

Two CRH receptors with distinct binding patterns have been identified, and much of what we know about the influences of CRH on behavior has been gained through the use of transgenic mice that lack one or both of these receptor types. For example, CRH-R1 knockout mice, which lack functional CRH1 receptors, were less anxious, both at baseline and after alcohol withdrawal, than wild-type mice. CRH1 receptors also appear to mediate the hormonal, behavioral, and nociceptive (pain) responses to stress (Reul et al., 2002) and appear to be involved in negative feedback regulation of the HPA axis (Mueller et al., 2003). Mice with conditional CRH1 receptor gene inactivation (that is, the CRH receptor gene can be inactivated in adulthood) in neurons outside of the hypothalamus and pituitary showed a prolonged elevation in glucocorticoid concentrations following stressors. This finding supports the notion that the hippocampus and possibly the amygdala, where CRH1 receptors are located, and where they were presumably inactivated by the conditional gene inactivation (Mueller et al., 2003), provide negative feedback during stress responses. Under baseline conditions, the standard negative feedback regulatory mechanisms, including those of the hypothalamus and pituitary, are evoked as described in Chapter 2. However, during prolonged stress responses, the hippocampus and possibly the amygdala participate in terminating glucocorticoid secretion.

There are two types of corticosteroid receptors in the hippocampus, the type I mineralocorticoid receptors (MRs) and the type II glucocorticoid receptors (GRs). MRs have higher affinity for circulating glucocorticoids and are usually engaged under baseline conditions. Activation of these receptors is thought to modulate homeostatic balance. However, as glucocorticoids increase during a stress response, they bind to the low-affinity GRs; activation of these receptors influences negative feedback and brings the stress response back under control (Mayer and Fanselow, 2003). Chronic stress, genetic differences in corticosteroid receptor numbers or subtypes, or other individual differences in endocrine secretion, responsiveness, carrier proteins, or other factors can evoke dysregulation of the stress response and inappropriate or pathological stress responses, as we will see below.

Although ACTH is the pituitary hormone that is most often associated with stress, several other pituitary hormones, including vasopressin, prolactin, endorphins, and enkephalins, play important roles. For example, the stress response often includes an increase in blood concentrations of vasopressin, which acts to increase blood volume and blood pressure and makes the delivery of energy to the large muscles more efficient. Vasopressin also directly affects behavior; it has been shown to enhance memory consolidation and retrieval and to increase aggression in defense of a mate. Vasopressin also can work in concert with CRH to augment the release of ACTH from the anterior pituitary (de Kloet et al., 1991), as we will see below. The anterior pituitary also secretes prolactin, which functions in the stress response to suppress reproduction temporarily by acting at multiple sites within the hypothalamic–pituitary–gonadal (HPG) axis (Figure 11.7) (Van de Kar et al., 1991). The α-cells of the endocrine pancreas are stimulated to secrete glucagon. As you will recall from Chapter 2, glucagon serves to increase energy availability. The β-cells of the pancreas are inhibited by glucagon from secreting insulin. Endorphins and enkephalins are often released during stress to provide relief from pain; these hormones also suppress GnRH, thereby inhibiting reproductive function. Essentially, all of the physiological sequelae in the response to a stressor make more energy available for immediate use. Again, all processes that are involved in future survival or reproductive success (e.g., energy storage as fat, production of gametes, growth) are put on hold until a future is certain (Sapolsky, 1994).

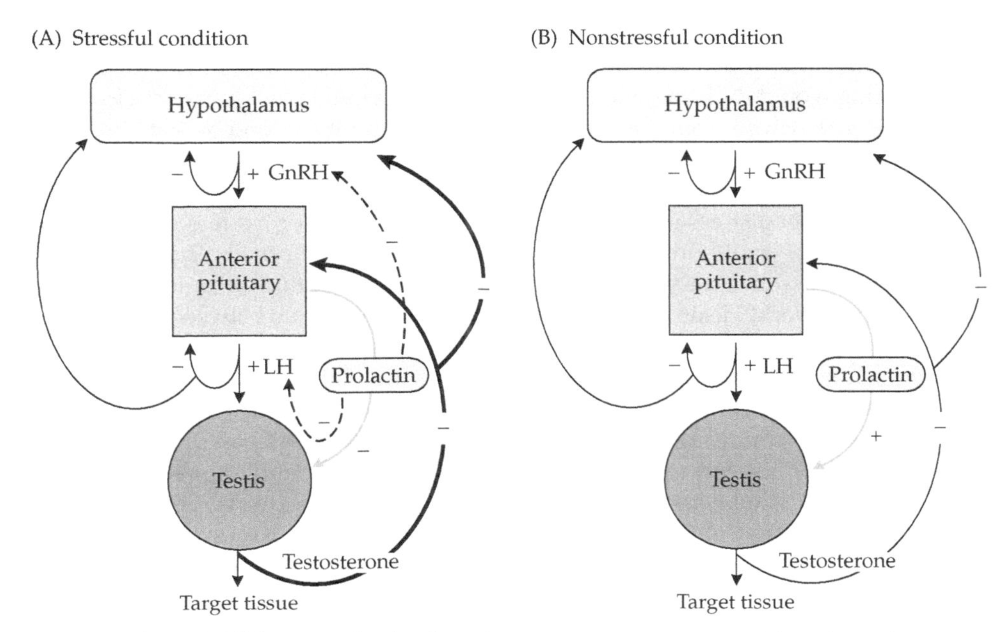

11.7 Prolactin inhibits reproductive function temporarily during stressful conditions by acting at multiple sites within the hypothalamic–pituitary–gonadal (HPG) axis. (A) Under the influence of prolactin, the Leydig cells become less responsive to LH and produce less testosterone, and the negative feedback mechanisms in the pituitary and hypothalamus are accentuated so that any testosterone produced causes further reductions in GnRH and gonadotropin secretion. (B) Normally, prolactin at low concentrations facilitates testosterone production.

Thus, when activated acutely, in response to a real threat, the stress response improves the chances of survival by orchestrating a shift in resources from nonessential, energetically expensive processes to those that promote immediate survival, such as the muscles and the brain. Under these circumstances, mounting a stress response is highly adaptive. However, if the stress response is not terminated efficiently, or is initiated too frequently or in response to inappropriate stimuli, then pathological consequences may ensue.

Pathological Effects of the Stress Response

In the short term, the stress response is adaptive, and for the most part, the neuroendocrine mechanisms underlying the stress response have been conserved across vertebrate phylogeny. Ideally, the stress response is initiated by stressful stimuli and serves to help the individual cope with the immediate energetic crisis; then the system is deactivated shortly after the crisis subsides. For example, if a female wild dog were trying to integrate herself into a new pack, she would probably be met by aggressive rebuffs from the females currently in the pack. Whether she decided to stay and fight or to run away, she would experience a short-term stress response that would increase her ability to mobilize and burn energy. If she were accepted into the pack, then a new social hierarchy would be established, fighting would decrease, and the stress response would be terminated. However, if she was driven from the pack, and she had to hunt for food or water alone in an unfamiliar territory, she would probably experience a long-term stress response, especially if her solo hunting success was low.

Prolonged stress shifts the useful, adaptive short-term stress response to a pathological condition that can jeopardize health and survival (Sapolsky, 1992b, 1994). Stress-related pathologies are a common contributing factor to human disease. The World Health Organization (WHO) estimates that between 5% and 15% of individuals living in industrialized societies suffer from stress-related disorders of mood and affect. Prolonged stressors are major contributors to disease onset and progression (Balbin et al., 1999; Chrousos, 2000; Yang and Glaser, 2000); such stressors induce or intensify many common human ailments, including cardiovascular disease, cancer, irritable bowel syndrome, and depression. Over 1 million U.S. employees are estimated to be absent on any given workday because of stress-related medical or psychological complaints (Goetzel et al., 1998). Furthermore, the direct medical expenses of individuals with self-reported chronic stress are elevated by nearly 50% (Goetzel et al., 1998).

Many of us perceive "foxes" (i.e., stressors) all around us—paying the bills, writing textbooks, public speaking, or dealing with our personal relationships may all evoke a full-blown stress response. Our bodies remain engaged for emergency action even though we may remain physically inactive. The consequences of this chronic stress are mainly negative. Some individuals may try to alter their perception of stress-evoking situations by ingesting alcohol or other drugs. As you will see below, stress accentuates the effects and the addictive properties of these substances. Some of us may use various psychological coping strategies, such as visualization, to alter our perception of stressors. Others will go through life seemingly unaware of stressors, and a few will embrace stressors, actually seeking out various experiences that engage the HPA axis, such as working in a big-city trauma center, skydiving, or riding roller coasters. Again, there is a high level of individual variation in perception of and response to stress.

The pathological effects of chronic stress involve cardiovascular, metabolic, reproductive, digestive, immune, anabolic, and psychological processes (Table 11.1) (Sapolsky, 1992a, 1994; Brown, 1994). For example, sympatho-adrenal activation, corticosteroid secretion, and increased cardiac tone may be useful for explosive activities such as pursuing prey or evading predators, but they are well-established contributors to the development of atherosclerotic plaques leading to heart disease, embolisms, and strokes when they are abnormally prolonged. There are several documented examples of natural disasters, such as earthquakes, precipitating heart attacks. Sometimes events that may seem trivial to most people, such as the 1996 elimination of the Dutch national football (soccer) team from the European Championships, are associated with a statistically significant increase in death from cardiovascular incidents among fans. However, it is much more common for stress to cause cumulative damage to cardiovascular health over years. For example, the

TABLE 11.1 **Pathological effects of long-term stress responses**

Acute stress response	Pathological state associated with chronic stress
Shift from energy storage to energy use	Fatigue, myopathy; steroid diabetes
Increased cardiovascular tone	Hypertension
Inhibited digestion	Peptic ulcers
Inhibited growth	Psychosocial dwarfism
Inhibited reproduction	Impotence; anovulation; loss of libido
Altered immune function and inflammatory responses	Impaired disease resistance; cancer
Enhanced cognition	Accelerated neural degeneration during aging
Enhanced analgesia	

Source: Sapolsky, 1992b

death of a child is associated with an increased incidence of myocardial infarction beginning 6 years after the loss (Jiong et al., 2002). Furthermore, the myocardial infarction rate was highest among parents whose children died unexpectedly (e.g., from SIDS).

Social status influences the development of arteriosclerosis among cynomolgus monkeys (Kaplan and Manuck, 1999). Dominant male monkeys developed more significant arteriosclerosis than lower-ranked individuals when housed in an unstable social group. Social status had no effect on arteriosclerotic progression, however, when these monkeys were housed in a stable social group (Kaplan and Manuck, 1999). In contrast, low-ranked female monkeys displayed more arteriosclerosis than high-ranked females regardless of social stability. These results emphasize the sex differences in perception of stress and its effects on disease progression.

With prolonged glucocorticoid secretion, myopathy (muscle loss) is inevitable, and in severe cases stress can induce the irreversible loss of muscle cells in the heart. The stress-induced breakdown of glycogen and lipids to elevate blood glucose concentrations cannot continue indefinitely. Reproductive function is also inhibited by high glucocorticoid concentrations. This effect has obvious negative consequences for the reproductive success of animals in the wild that are confronted by demanding environmental conditions. An important consequence of stress for reproductive function in humans is that infertile couples often experience an escalating effect of stress on fertility that interferes with the infertility treatment (Agarwal and Haney, 1994; Greenfield, 1997). In other words, the psychological stress of experiencing infertility, in addition to the stress of the treatments, causes more stress and consequently reinforces the infertility. Prolonged inhibition of digestion during stress can lead to ulceration and chronic bowel distress syndrome. Indeed, the appearance of gastric ulcers was one of the early indices of chronic stress in Selye's rats. Peptic ulcers can develop as a result of suppressed immune function that allows the proliferation of bacteria, especially *Helicobacter pylori*, in the stomach and digestive tract that can attack the lining of the stomach or duodenum. Chronic immunosuppression, as seen in AIDS patients, compromises long-term survival. Stress adds to AIDS-associated immunosuppression to reduce the chances of survival (Capitanio et al., 1998; Evans et al., 1997). Chronic stress also inhibits growth and repair processes. You might notice that your fingernails are more likely to be flimsy or brittle, or that your hair grows more slowly, during stressful periods in your life. Indeed, chronic stress delays cutaneous wound healing in humans and rodents, an effect that is mediated by elevated glucocorticoid concentrations and their effects on immune function at the wound site (Detillion et al., 2004). For example, postmenopausal women who served as caregivers for chronically ill spouses displayed higher cortisol values and slower healing from a biopsy wound on the arm than age-matched control women (Figure 11.8) (Kiecolt-Glaser et al., 1995). Chronic stress in children, usually as a result of parental deprivation and abuse, can inhibit growth hormone and result in a condition called psychosocial dwarfism, as we will see below.

The pathological long-term effects of chronic stress affect many hormone–behavior interactions. As we have seen, reproductive function is impaired in chronically stressed individuals (Welsh et al., 1999). Recall that only the alpha pair of each pack of African wild dogs breeds. A series of studies was conducted to discover why social subordinates do not breed, even though they are reproductively mature. It was hypothesized that subordinate wild dogs were reproductively suppressed by stress. These studies were conducted with the hope of reversing the recent population declines in this endangered species (Creel et al., 1996). By collecting feces from individual animals whose social status was known, the investigators were able to assay hormone concentrations (measured indirectly by assaying fecal metabolic breakdown products of the hormones of interest) without disturbing the animals

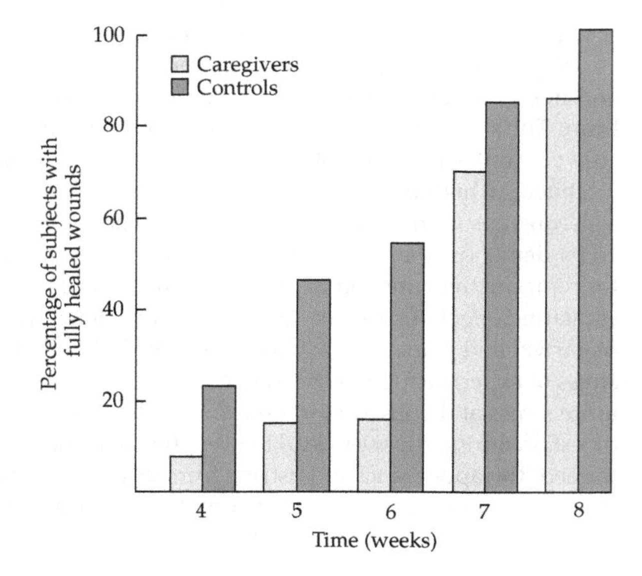

11.8 Wound healing is impaired in stressed caregivers. This graph depicts healing times for a standard-sized punch biopsy wound on the arm in postmenopausal women who served as caregivers for chronically ill spouses and control women of the same age. Slower healing increases the risk of infection. After Kiecolt-Glaser et al., 1995.

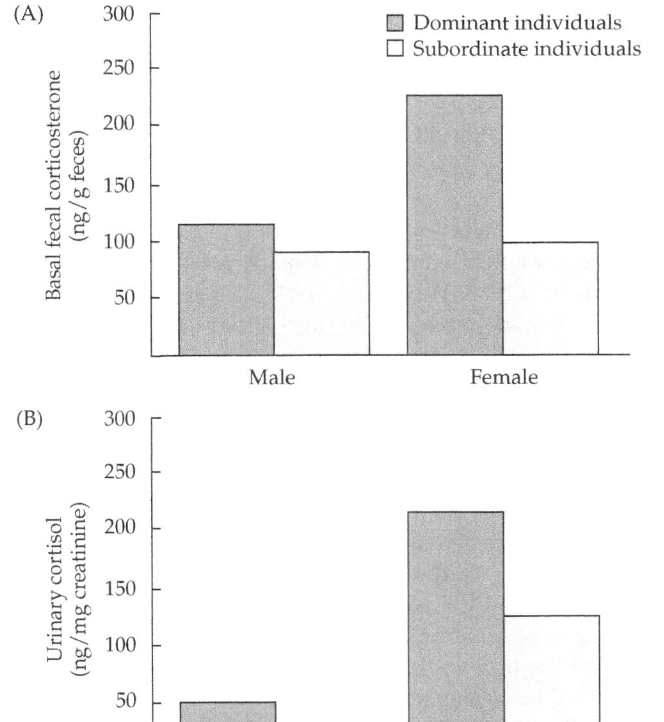

11.9 Corticosterone concentrations are elevated in dominant wild dogs and mongooses. In contrast to the findings in most laboratory animals, the alpha individuals in groups of (A) free-living African wild dogs and (B) captive mongooses have higher corticosterone values than subordinate animals. In other words, the dominant individuals appear to be more stressed than subordinate individuals. After Creel et al., 1997.

(Monfort et al., 1998). In contrast to the investigators' expectations, these studies, as well as studies of African dwarf mongooses (*Helogale parvula*), indicated that the dominant animals had high glucocorticoid concentrations (Figure 11.9) (Creel et al., 1996). It is possible that the dominant animals are exposed to the chronic psychological stressor of potential loss of their status. In any event, these results contrasted with the findings of previous laboratory studies that had reported that subordinate rats and mice were more highly stressed than the dominant members of their social group (Sapolsky, 1992b). As noted previously, if primates are representative (see Box 11.1), then perhaps social organization and social stability should be considered when assessing the effects of social status on stress hormones.

Factors that Affect Stress Responsiveness

Perinatal Stress

Thus far, we have considered only the activational effects of stress hormones. Glucocorticoids, like sex steroid hormones, can have organizational as well as activational effects on brain and behavior, although they are rarely conceptualized as such in the scientific literature. Thus, the effects of early stressors on brain function, hormone concentrations, and hormone receptors, as well as subsequent perception of and ability to cope with stressors, can be enduring and irreversible throughout life. The brains of stressed fetuses and infants are organized differently from the brains of individuals that do not experience stressors early in development, and they are differentially activated by glucocorticoids in adulthood.

PRENATAL STRESS Although we would all do well to try to limit the effects of chronic stress in our lives, sometimes they have already affected us before birth (Herrenkohl, 1986; Takahashi, 1998; Weinstock, 1997). A series of studies have examined the effects of prenatal stress on the physiology and behavior of rat pups. If pregnant dams are placed in clear plastic restraint devices under bright lights for 30–60 minutes each day during days 14–21 of their 21-day gestation, they respond with increased activation of the HPA axis. More importantly, their offspring display permanent changes in their brain morphology, physiology, and behavior. Maternal stress alters blood testosterone concentrations in fetal male rats, shifting the peak to 2 days earlier than that in nonstressed males (Ward and Weisz, 1980; Ward, 1984). When tested as adults, sons of stressed mothers often failed to ejaculate in mating tests (Ward et al., 1994), even though their adulthood concentrations of testosterone and LH were normal (I. L. Ward et al., 1996). The shift in the timing of the testosterone peak blocked the development of the normal sexual dimorphism of the brain, affecting the sexually dimorphic nuclei of the preoptic area (SDN-POA) as well as the spinal nucleus of the bulbocavernosus (SNB) and the dorsolateral nucleus of the spinal cord (Kerchner and Ward, 1992; Kerchner et al., 1995). The sex difference in the size of the rostral anterior commissure was also found to be altered by prenatal stress (Jones et al., 1997). Furthermore, when exposed to restraint stress as adults, both male and female rats whose mothers had experienced stress during gestation displayed an attenuated prolactin response (Kinsley et al., 1989), suggesting a permanent effect of prenatal stress on the HPA axis. Prenatally stressed female rats did not display any disturbances in normal estrous cycles, sexual behavior, pregnancy, parturition, pup survival, or maternal behavior when tested as adults (Beckhardt and Ward, 1983; but see Herrenkohl, 1986).

Recall from Chapter 3 that rat dams tend to spend more time licking the anogenital regions of male pups than female pups (Michel and Moore, 1995), and that this sex difference in eliciting maternal attention is mediated by testosterone. If dams are stressed while pregnant, they show no preference in grooming between their male and female offspring (Power and Moore, 1986). Presumably, the prenatal stress affects testosterone secretion so that the male and female offspring do not vary in the critical testosterone-dependent variable that dams normally use to discriminate male from female pups.

Prenatal stress impairs subsequent negative feedback within the HPA axis in adult rats, often resulting in elevated basal corticosterone blood concentrations (Henry et al., 1994). Subjecting pregnant rat dams to either stress or exogenous glucocorticoid administration decreases hippocampal glucocorticoid receptors

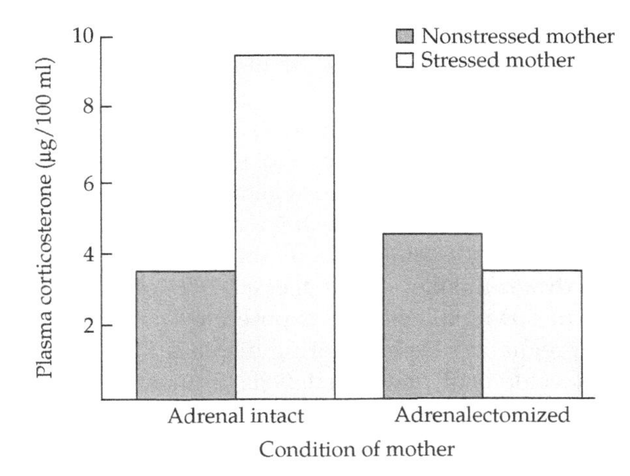

11.10 Adrenalectomy prevents the elevated stress response in rats stressed in utero. Corticosterone concentrations are elevated in response to restraint stress in adult rats whose mothers were stressed while pregnant. Adrenalectomy prevented this elevated stress response. After Barbazanges et al., 1996.

(GR) in their offspring; this decrease has been hypothesized to account for the impairment of negative feedback in the HPA axis in adulthood (Barbazanges et al., 1996). Damage to the hippocampal cells mediating negative feedback results in elevated glucocorticoid levels. If a pregnant female is adrenalectomized prior to stress or corticosterone treatment, the negative feedback mechanisms are not impaired in her adult offspring (Figure 11.10) (Barbazanges et al., 1996). The offspring of stressed rat dams are more likely to be anxious, and to self-administer drugs such as cocaine and amphetamines as adults, than their peers with nonstressed mothers (Deminiere et al., 1992; Diaz et al., 1995; see below). These results may have important implications for drug abuse in children of stressed parents.

Prenatally stressed human infants have reduced birth weights, experience developmental delays, and also display attentional deficits, hyperanxiety, impaired social behaviors, and possibly impaired strategies for coping with stressful conditions as adults (Weinstock, 1997). Some studies suggest that prenatal stress may increase the incidence of schizophrenia in adults. For example, the Netherlands National Psychiatric Registers provide evidence for an increased incidence of schizophrenia in individuals who were in utero during the 1940 invasion of The Netherlands by the German army (van Os and Selten, 1998). Furthermore, there was a twofold increased risk for schizophrenia among individuals conceived in the western cities of the Netherlands during the height of the Dutch Hunger Winter (1944/1945; Susser et al., 1996). It is also possible that the same kind of dysregulation of the HPA axis observed in rats exposed to prenatal stress serves as a precursor to the HPA disinhibition observed in many instances of human depression (Weinstock, 1997).

Like prenatal stress, exposure of humans to glucocorticoids in utero has a number of negative consequences, including low birth weight, which in turn is associated with adult cardiovascular and metabolic disorders such as hypertension, insulin resistance, hyperlipidemia, and ischemic heart and brain disease (Nyirenda and Seckl, 1998). There also is limited evidence that human babies exposed to exogenous corticosteroids late in gestation show behavioral problems in childhood and suffer from more cardiovascular disease than typical children (Doyle et al., 2000).

Human studies of prenatal stress, of course, are correlational; it would be unethical to manipulate stress levels in women in order to observe the effects on their babies. However, long-term studies of monkeys can be used to investigate the long-term consequences of prenatal stress for the behavior and reproductive success of adult primates. The effects of prenatal stress on the human HPA axis have been carefully modeled in rhesus monkeys (Clarke et al., 1994). To induce stress experimentally, pregnant rhesus monkeys were exposed to random loud noises throughout mid- to late pregnancy. Their offspring displayed low birth weights, impaired neuromotor development, attention deficits (Schneider, 1992b; Schneider and Coe, 1993), and impaired cognitive function (Schneider, 1992a) compared with their nonstressed peers. Additionally, enhanced stress responsiveness, higher frequencies of abnormal coping behaviors, and impaired social behaviors were observed in the offspring of female rhesus monkeys that were chronically stressed during pregnancy (Clarke and Schneider, 1993; Clarke et al., 1994).

NEONATAL STRESS Animal models have also been developed to determine the effects of stress in newborns. Several experimental paradigms have been examined to explain individual differences in coping with stress in adulthood. According to the leading hypothesis (Levine et al., 1967), early stressful experiences affect reactions to stress later in life. If a rat pup experiences certain types of mild stress, then it appears to be better able to cope with stress later in life. Human handling of pups or brief separation of pups from their mother evokes stress responses in the pups. However, upon the return of the pups to their mother, the rat dam spends extra time licking them. This gentle stimulation appears to reverse the negative effects of the stress on the pups and is responsible for the so-called "stress immunization" effect that allows the pups to cope with stress better as adults (Figure 11.11) (Liu et al., 1997). One way this is hypothesized to occur is by changing the pups' densities of hippocampal GRs, which are important in the negative feedback regulatory mechanisms of the HPA axis in adulthood (Liu et al., 1997), as we saw above.

A commonly used experimental paradigm for assessing separation stress involves the separation of infants from their mothers at different ages and for differing periods of time (Levine et al., 1991). Typically, rat pups are isolated in novel test arenas and receive a saline injection; this combined treatment of brief isolation (15–120 minutes) plus injection is considered to be a mild stressor (e.g., Suchecki et al., 1993). If 6-, 9-, or 12-day-old rat pups are isolated from their mother for 24 hours, their baseline ACTH concentrations are elevated. If these individuals are stressed, then the elevated ACTH concentrations are significantly higher than in nonstressed adult animals and the magnitude of the increase is higher as the age when the pups are stressed increases (Suchecki et al., 1993). Corticosterone concentrations show a similar effect, and the effect of mild stress on corticosterone

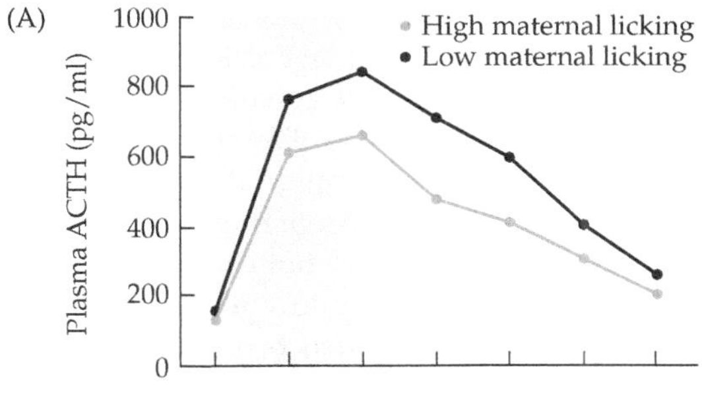

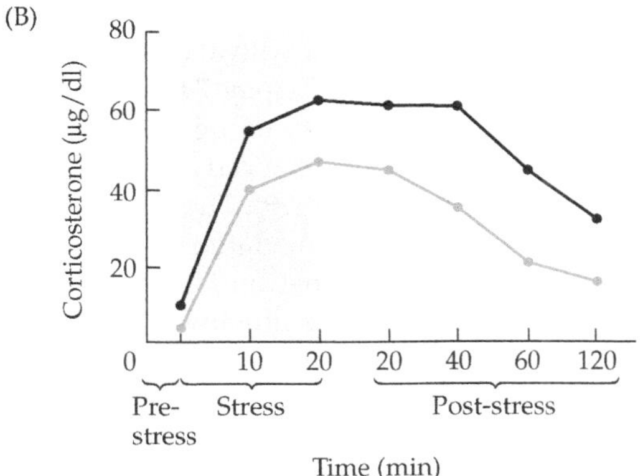

11.11 Stress responses are reduced in offspring that received maternal attention as pups. Both (A) ACTH concentrations and (B) corticosterone concentrations were lower before, during, and after a 20-minute restraint stress in rats (indicated by the shaded area) that received a high amount of maternal licking after separation from their mother as pups than in those that received a low amount. After Liu et al., 1997.

concentrations persists for at least for 4 days (Rosenfeld et al., 1992). Again, postnatal exposure to mild stressors blunts adult stress responsiveness (Levine et al., 1967; Sapolsky and Meaney, 1986). In contrast, elevated stress responses are observed throughout life in pups exposed to moderate or severe stressors (Rots et al., 1996). For example, rats that were severely stressed as pups may show long-term decreases in food intake and body mass in response to endogenous proinflammatory cytokines such as interleukin-1β (Kent et al., 1997).

Maternal tactile and feeding cues are sufficient to suppress the elevated concentrations of ACTH and corticosterone resulting from 24 hours of maternal deprivation (van Oers et al., 1998). Interaction with an anesthetized lactating female rat is sufficient to block the enhanced corticosterone stress response associated with isolation both during the isolation-induced stress and as adults (Stanton and Levine, 1990). Thus, it appears that the contact comfort of the mother is sufficient to prevent the stress response to isolation, although her grooming and licking behaviors are also important (Liu et al., 1997). Impairment of HPA function is more marked in female than in male rats exposed to isolation stress (McCormick et al., 1995).

Although brief maternal separation (e.g., 15 minutes, which is consistent with the amount of time the mother forages in the wild) appears to constrain stress responsiveness in adulthood, and 24 hours of maternal separation causes elevated responsiveness to stress in adulthood, as much as 3 hours of maternal separation has physiological effects that are not obvious by simply measuring circulating corticosterone values (Mirescu et al., 2004). Despite normal corticosterone concentrations, both basal and post-stress, rat pups that are separated from their mothers for 3 hours show impaired neurogenesis and cell proliferation in the dentate gyrus as adults. If rats that experienced 3 hours of maternal separation are adrenalectomized as adults, then the suppression of cell proliferation is reversed (Mirescu et al., 2004). These results indicate that prolonged, but not brief, periods of maternal separation result in enduring inhibition of neurogenesis and hippocampal plasticity as adults, caused not by elevated glucocorticoid concentrations, but by hypersensitivity to normal corticosterone concentrations (Mirescu et al., 2004).

How might early stressors organize the brain so that the activational effects of corticosterone in adulthood are less effective? Recall from Chapter 7 that offspring of mothers that display high levels of pup licking and grooming later display attenuated HPA axis responses to stress and also display enhanced cognitive ability (Liu et al., 1997, 2000; Caldji et al., 1998; Francis et al., 1999). The mechanisms underlying this brain organization have been recently elucidated in a series of elegant studies during the past few years by Michael Meaney and his colleagues.

High levels of pup licking, grooming, and arched-back nursing organize the brain of the recipient offspring so that they are resistant to stressors as adults, and female pups adopt the type of maternal behavior that they received as infants (Weaver et al., 2004). The immediate consequence of high levels of these maternal behaviors is an increase in GR gene expression in the hippocampus, which seems to be critical for the reduced stress effects observed later in adulthood (McCormick et al., 2000). But how can these transient effects on GR gene transcription have permanent organizational effects on brain and behavior? The answer appears to involve two epigenetic processes that affect gene expression: methylation and histones (Sapolsky, 2004) (Figure 11.12). Methylation is a process of long-term gene silencing that involves attaching a methyl group to the promoter region of a gene (Egger et al., 2004). Thus, demethylation often results in significant gene expression (Egger et al., 2004). The DNA of animals is wrapped around proteins called histones; DNA plus histones comprise the chromatin. The histones can bind tightly to the DNA so that transcription factors cannot access promoter sites. Specific types of maternal care evoked changes in methylation and chromatin structure, resulting in increased GR gene expression (Weaver et al., 2004). These changes emerged in early life, could be reversed by cross-fostering to mothers that displayed low levels of licking, grooming, and arched-backed nursing, and persisted into adulthood. Thus, these epigenetic influences of maternal care persist into adulthood and can be passed on to subsequent generations.

Could such organizational effects of glucocorticoids induced by style of maternal care apply to humans? The studies on isolation-induced stress in rodent pups are reminiscent of the famous biopsychological studies conducted on rhesus monkeys by Harry Harlow and colleagues, in which the factors that represent "motherness" were investigated. In these studies, infants were separated from their

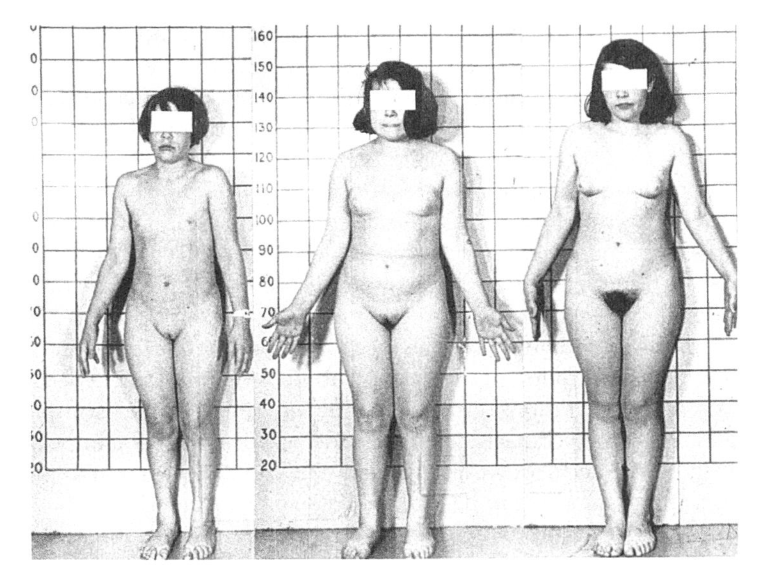

11.15 Psychosocial dwarfism is a rare occurrence seen most frequently in severe cases of child abuse, and is usually reversible after termination of the stressors. This girl was drinking from puddles in the street and eating from garbage cans before admission to the hospital at age 15.3 years (left panel). At admission, her height was in the normal range for 9-year-old children, and she displayed abnormal GH release. Both LH and FSH were nondetectable. After 1 month in the hospital, she grew 2.5 cm, and began to undergo puberty. At age 16 (center panel), she was living in foster care, and displaying significant GH release. Growth continued, and at 16.9 years (right panel), she had grown 17.5 cm. Courtesy of Dr. Claude Migeon.

The decrease in ODC concentrations in many different body sites suggests that either a general metabolic or an endocrine signal mediates the effects of tactile stimulation. Blood plasma concentrations of growth hormone (GH), but not of prolactin or thyroid-stimulating hormone, decreased in pups after maternal separation, whereas blood concentrations of corticosterone increased (Schanberg and

TABLE 11.2 Effect of stroking on ODC activity and serum GH concentrations in maternally deprived rat pups

	ODC serum concentration (% control ± SEM)			Serum GH concentration (% control ± SEM)
	Brain	Heart	Liver	
Control	100 ± 13	100 ± 6	100 ± 13	100 ± 20
Deprived[a]	68 ± 8	36 ± 7	32 ± 6	25 ± 7
Deprived and stroked	134 ± 15	93 ± 15	199 ± 41	82 ± 17

Source: Schanberg and Kuhn, 1985.

[a] $p < 0.05$ or better compared to controls for all values.

Kuhn, 1985). GH regulates ODC activity in the brain and other tissues. Stroking with a wet paintbrush also normalized GH concentrations in isolated pups. In an attempt to counteract the diminished concentrations of ODC after separation, isolated and unstroked pups were injected with growth hormone, but it was discovered that the tissues of these pups were unresponsive to the hormone. Stroking with a paintbrush maintained tissue responsiveness to GH.

After GH is secreted from the anterior pituitary gland, it binds, along with several types of somatomedins, to receptors on the surfaces of cells. Cells that bind GH are stimulated to grow and divide. The physiological mechanisms by which tactile stimuli increase GH secretion from the anterior pituitary, and how they induce tissue sensitivity to GH, are not yet known. But research into this fascinating area should provide more answers about how maternal behavior influences hormone concentrations, growth, and development in the offspring of rats and humans. This research has several practical applications; one important outcome has been the practice of stimulating growth and development in premature infants by increased handling in perinatal intensive care nurseries (Schanberg and Field, 1987). A low-tech improvement in premature infants' health, and a shorter hospital stay, are certainly impressive outcomes of these interesting studies in rats.

Glucocorticoids might also be involved in psychosocial dwarfism; however, cortisol concentrations are usually normal or in the low normal range in affected children. As mentioned above, glucocorticoid concentrations in young rats rise dramatically after maternal separation. In human children, only a doubling of glucocorticoid concentrations is sufficient to arrest growth (Sapolsky, 1994). Glucocorticoids can suppress growth by suppressing GH, inhibiting GH receptor sensitivity, and decreasing protein production and DNA synthesis. Whatever is actually suppressing the growth of afflicted children, the good news is that removing them from the stress-inducing environment allows them to catch up on their growth. In Table 11.3, the effects of emotional factors on growth are obvious in a single boy suffering from psychosocial dwarfism.

Reproductive Dysfunction

Every individual has limited energy resources, which must support maintenance of the body as well as reproduction. Other energetically expensive activities,

TABLE 11.3 Effects of external events on growth in a child with psychosocial dwarfism

Event	Plasma GH concentration (ng/ml)	Growth (cm/20 days)	Food intake (g/day)
Hospital admission	5.9	0.5	1663
100 days post-admission	13.0	1.7	1514
Favorite nurse on vacation	6.9	0.6	1504
Favorite nurse returns	15.0	1.5	1521

Source: Saenger et al., 1977.

including immune function and growth, must also be partitioned out of the total energy budget. Consequently, from an evolutionary perspective, there is a trade-off among competing energetic requirements, so that growth or immune function may be compromised to support reproduction when energy is somewhat limited (Ots and Horak, 1996). Alternatively, reproduction or immune function (or both, depending on the extent of the limitation) may be compromised when energy availability is significantly reduced.

Reproductive suppression in response to stress occurs on several physiological and behavioral levels and can have potentially severe effects in both humans and nonhuman animals (Welch et al., 1999). Stress is an important contributory factor in human sexual dysfunction and infertility. Stress-induced reproductive impairments in domestic animals may have a marked economic impact (Moberg, 1991): stress can reduce milk production in cows and egg production in chickens (Welsh et al., 1999). In sum, the scientific study of the effects of stress on reproductive function has important clinical and practical applications.

MALES A variety of stressors have been reported to suppress male reproductive function. Stress acts on the HPG axis in a number of ways to inhibit testosterone production, and low testosterone concentrations reduce both sexual motivation and performance (see Chapter 5). The release of CRH and endogenous opioids can directly suppress the release of GnRH (Hulse and Coleman, 1983; Jacobs and Lightman, 1980; Rasmussen et al., 1983; Rivier et al., 1986). CRH-containing neurons in the paraventricular nucleus (PVN) of the hypothalamus have been classified into two types: vasopressin-positive and vasopressin-negative neurons. Stress causes certain neurotransmitters to trigger the release of CRH from both types of neurons, and of vasopressin from the vasopressin-positive neurons. CRH travels to the anterior pituitary, where it binds to specific G protein-associated receptors in the ACTH-releasing cells. Activation of these CRH receptors stimulates the protein kinase A pathway which changes gene transcription or other cellular functions. CRH and its receptors have also been identified in the testes and ovaries, suggesting that this releasing hormone may also directly inhibit steroid production (Welsh et al., 1999). In the brain, vasopressin binds to its V1a receptors, which activate the protein kinase C pathway (see Chapter 2).

Both CRH and vasopressin can stimulate expression of the genes for pro-opiomelanocortin in the anterior pituitary. Recall from Chapter 2 that POMC can be processed by various enzymes to yield several peptides, including ACTH and β-endorphin. There is some evidence that different environmental stressors determine which enzymes are produced, and thus which peptides are cleaved from the POMC precursor molecule (DeWied, 1997). POMC-derived peptides may feed back to inhibit GnRH secretion (Welsh et al., 1999). Indeed, both endogenous and exogenous opiates have been found to inhibit GnRH and gonadotropin secretion (Figure 11.16) (Cameron, 1997; DeWied, 1997; Genazzani and Petraglia, 1989; Rivier and Rivest, 1991). Both sexual motivation and performance are impaired in male heroin addicts. More than half of the female heroin addicts interviewed in one study experienced menstrual abnormalities while taking heroin or

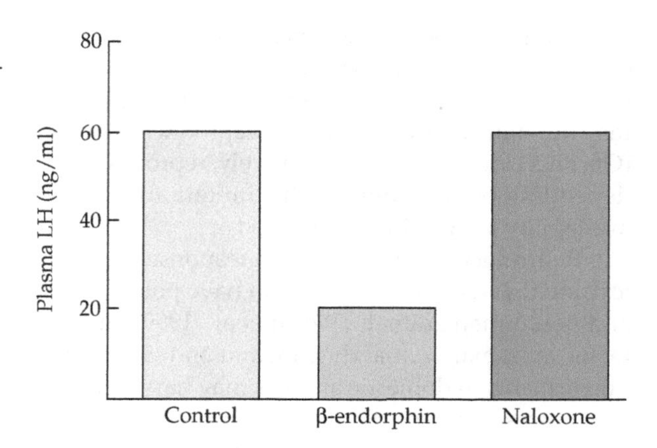

11.16 Opiates suppress gonadotropin production. Plasma LH concentrations are reduced after endorphin treatment; the reduction in LH by endorphin is reversed by treatment with the opiate antagonist naloxone. After Kalra et al., 1990.

methadone (Santen et al., 1975; Smith et al., 1982). Detailed endocrine studies revealed impaired gonadotropin secretion in most of these cases (Santen et al., 1975). Naloxone, an opioid antagonist, reverses reproductive suppression by opioids in humans and nonhuman animals (Genazzani and Petraglia, 1989; Gilbeau and Smith, 1985).

ACTH stimulates the production and secretion of glucocorticoids from the zona fasciculata and zona reticularis of the adrenal cortex. Glucocorticoids can inhibit reproduction in several ways. First, glucocorticoids at high concentrations can suppress GnRH and LH secretion (Welsh et al., 1999). Glucocorticoids have also been reported to inhibit the formation of proteins necessary for the production of hormone receptors, steroidogenic enzymes, and several intracellular signaling molecules (Rivier and Rivest, 1991). Cortisol inhibits testosterone secretion in men (Cummings et al., 1983) by acting on testicular LH receptors (Bambino and Hseuh, 1981). Glucocorticoids also suppress spermatogenesis more directly. The testosterone-producing Leydig cells have glucocorticoid receptors that appear to be involved in the normal process of cell growth, metabolism, and energy use. However, when glucocorticoid concentrations are elevated for long periods of time, especially in subordinate animals, an enzyme that normally neutralizes glucocorticoids at basal concentrations is overwhelmed, and testosterone production is curtailed (Ge et al., 1997). The resulting low testosterone concentrations may fail to support spermatogenesis, so that sperm counts fall and stressed individuals become infertile (Ge et al., 1997). In contrast, when dominant rats are housed under stressful conditions, their testosterone concentrations increase; it appears that dominant rats have higher levels of the enzyme that counteracts elevated glucocorticoid concentrations to ensure fertility (Monder et al., 1994). Thus, social status appears to affect testicular enzymes that, in turn, mediate androgen production and fertility (Figure 11.17).

A moderate amount of activity or exercise actually raises testosterone concentrations (Figure 11.18) (Elias, 1981; Grandi and Celani, 1990). However, as my grandmother used to say, "too much of a good thing is often a bad thing." Excessive

- Stress is a difficult concept to define precisely. For the purposes of this chapter, stress is defined as all the nonspecific effects of factors that can act on the body to increase energy consumption significantly above basal levels.
- The adaptive features of the stress response include an increase in the immediate availability of energy, an increase in oxygen uptake, a decrease in blood flow to areas not necessary for movement, and inhibition of digestion, growth, immune function, reproductive function, and pain perception, as well as enhancement of memory and perception.
- The pathological features of chronic stress include fatigue, myopathy, cardiovascular disease, gastric ulcers, psychosocial dwarfism, impotence, anovulation, compromised immune function, and potentially accelerated neural degeneration during aging.
- Prenatal stress can demasculinize male reproductive behaviors, but has little effect on female sexual behaviors. Mild stress during infancy can provide "stress immunization" and blunt stress responses later in life. Severe stress during infancy enhances stress responses.
- Psychosocial dwarfism is a rare syndrome found among human children that are raised in very stressful situations. GH is suppressed in these individuals; if they are removed from their stressful environments, they can catch up in their growth.
- Stress impairs male sexual behavior (both motivation and performance) by suppressing testosterone. Stress also affects female sexual behavior by interrupting the precise timing of neuroendocrine events necessary for successful ovulation and sexual behavior.
- Crowding evokes dramatic stress responses. In some social groups, it appears that the high-ranked individuals display pronounced stress responses, whereas in other social groups, low-ranked individuals display stress responses. Stress responses vary according to time of day and time of year.
- Individual variation in the stress response reflects variation in many physiological parameters, especially control, predictability, and outlets for dealing with frustration, but also reflects differences in the perception of stressors.
- Stress enhances and maintains substance abuse. Glucocorticoids function to accentuate the effects of dopamine to increase the rewarding properties of drugs of abuse.

Questions for Discussion

1. Discuss Hans Selye's quote, "Life is stress. Stress is life." Is stress truly a necessary component of life?
2. Why are psychological stressors as effective as physical stressors in evoking a physiological stress response? Is this an evolutionarily adaptive trait in nonhuman animals that has lost its adaptive value among modern humans?

3. Using thermometers and pH meters, infertile couples often try to maximize the odds that sperm and ova will meet in the oviduct(s). Given the effects of stress on reproductive function, how might you, as a clinician, advise (or treat) a couple undergoing fertility treatment?
4. Do you think that studies of the endocrinology of stress responses will help us understand or treat stress-related pathology? If so, how?
5. Given what you now know about the pathological effects of chronic stress, what sort of pharmacological interventions do you think would help to ameliorate the negative consequences of stress?

Refer to the accompanying Student CD for additional resources, including Web links, videos, animations, and additional photos.

Suggested Readings

Blanchard, D. C., McKittrick, C. R., Hardy, M. P., and Blanchard, R. J. 2002. Effects of social stress on hormones, brain, and behavior. In D. W. Pfaff, A. P. Arnold, A. M. Etgen, S. E. Fahrbach, and R. T. Rubin (eds.), *Hormones, Brain and Behavior*, Volume 1, pp. 735–772. Academic Press, New York.

Kim, J. J., and Diamond, D. M. 2002. The stressed hippocampus, synaptic plasticity and lost memories. *Nature Reviews: Neuroscience*, 3: 453–462.

Levine, S. 2002. Enduring effects of early experience on adult behavior. In D. W. Pfaff, A. P. Arnold, A. M. Etgen, S. E. Fahrbach, and R. T. Rubin (eds.), *Hormones, Brain and Behavior*, Volume 4, pp. 535–542. Academic Press, New York.

McEwen, B. S., and Wingfield, J. C. 2003. The concept of allostasis in biology and biomedicine. *Hormones and Behavior*, 43: 2–15.

Sapolsky, R. M. 1998. *Why Zebras Don't Get Ulcers: An Updated Guide to Stress, Stress-Related Diseases, and Coping*. W. H. Freeman, New York.

Tsigos, C., and Chrousos, G. P. 2002. Hypothalamic–pituitary–adrenal axis, neuroendocrine factors and stress. *Journal of Psychosomatic Research*, 53: 865–871.

Welsh, T. H., Kemper-Green, C. N., and Livingston, K. N. 1999. Stress and reproduction. In E. Knobil and J. D. Neill (eds.), *Encyclopedia of Reproduction*, 662–674. Academic Press, San Diego.

12

Learning and Memory

Many of us maintain vivid memories of frightening events. A colleague of mine recounts a terrifying experience that occurred during a boyhood game of kickball. He ran after a wayward ball, hoping to stop it before it rolled into the street. He made a valiant rescue attempt, but tripped on the curb and sprawled onto the pavement in front of an oncoming car. The driver slammed on the brakes and the car screeched to a halt, centimeters from him. Forty years later, my colleague insists that he can remember every detail of the incident, including the pattern of the front grille of the car. Now an entomologist, he claims that he can even identify, from memory, the genus of the dead moths that were stuck to the front of the car.

It is difficult to judge the accuracy of such accounts, but memories of stressful events in our lives seem to be particularly salient. In humans, the phenomenon of vivid memories of important, stressful events has been termed "flashbulb memory" (Brown and Kulik, 1977). Flashbulb memory is probably not a separate category of memory function; rather, it probably represents a class of memories that are more frequently rehearsed or more strongly encoded. Depending on their age, people are likely to have flashbulb memories of what they were doing when they heard about the terrorist attacks on New York and Washington, the explosion of the space shuttle *Challenger*, the suicide of Kurt Cobain, or the assassination of President Kennedy. Positive stressful events such as weddings, births of children, or winning lotteries are also remembered with great clarity (McCloskey et al., 1988). Other memories, however, are more fragile. For example, what you ate for dinner last night should be easily recalled, but you probably cannot remember what you ate for dinner a week ago, unless it made you ill.

Thus, some memories are relatively permanent whereas others are fleeting. Generally, important events are remembered better than unimportant events. How are memories tagged as important or trivial? My friend the entomologist cannot remember the front grille of every car that he has ever seen. The grille pattern of most cars is a trivial detail to him, and probably to you as well. Somehow, the brain sorts trivial and important events, and it has been proposed that hormones released during stress or high arousal may act as a physiological "marker" of importance. You would not want to remember every detail of your life. Unimportant details, such as the color of the shirt that your teacher wore on the forty-third class meeting of the third grade, are probably not useful to retain. Thus, as animals evolved, memory systems may have co-opted the endocrine signals of physiological stress to signal important events. At an ultimate level of causation, the marking of important events could enhance learning about adaptively significant aspects of the world. Presumably, the entomologist learned about the dangers of running into a street from just one experience, or learning "trial." The hormones secreted as he fell into the street and immediately thereafter notified his brain about the importance of the event. He can confidently claim that he learned from his experience because he never again ventured into the street without first checking for traffic.

Components of Learning and Memory

Many recent studies have demonstrated that the endocrine system has an important role in mediating learning and memory. In order to understand these hormonal effects, it is useful to understand several components of learning and memory. **Learning** can be defined as a process that expresses itself as an adaptive change in behavior that results from experience. The stages of learning include acquisition, consolidation, retrieval, and extinction. **Memory**, the encoding, storage, and retrieval (or forgetting) of information about past experience, is necessary if learning is to take place. Hormones could affect any one or all of these components of learning and memory.

Non-Associative Learning

All animals appear to be capable of changing their behavior as a result of experience. **Non-associative learning** occurs after repeated presentation of a single stimulus. **Sensitization** is one type of non-associative learning in which a stimulus that originally provoked little or no response begins to evoke stronger responses after several presentations, or a single intense presentation provokes stronger responses to other stimuli. For example, after an animal experiences a loud clanging noise, a slight noise to which no reaction would have occurred before might now elicit a startle response.

Another type of non-associative learning is **habituation**, a generalized and simple type of learning. Habituation involves learning *not* to respond after repeatative memory. **Procedural memory**, also termed "implicit" memory, is essentially memory for "knowing how." There are three types of procedural memory: skill learning, priming, and conditioning. Remembering how to play a song on a piano, ride a bicycle, or run a maze are types of procedural memory for *skill learning*. *Priming* involves a change in memory or processing of a stimulus as a result of previous experience. For example, if a person is exposed to the word *estrogen* on a list of words, then asked to fill in the blank after seeing "est_____," that person is more likely to say "estrogen" than someone who has not been primed. Finally, *conditioning* includes memory for both classical and operant conditioning, as previously described.

Declarative memory, also termed "explicit" memory, is memory for "knowing what"; that is, for knowing facts. Declarative memory can be divided into *semantic memory*, which is your general knowledge of facts and events, and *episodic memory*, which is your memory of personal events (episodes), such as a birthday party. Declarative memory has also been called *verbal memory* in humans, whereas procedural memories are generally nonverbal. Declarative memories are generally formed and forgotten relatively easily, whereas procedural memories require longer to form, and once learned, procedural memories, such as knowing how to swim, remain easier to retain.

Another way to categorize memory is by dividing it into working memory and reference memory. **Working memory** is similar to declarative and short-term memory in that it typically involves short-term memory for information that changes on a regular basis. It differs from **reference memory**, which generally refers to associations or discriminations requiring repetitious learning, as in learning the rules of a task or how to navigate around an environment such as a maze. Hormones have significant effects on spatial memory, which can involve both spatial working memory and spatial reference memory. **Spatial memory** can be defined as memory for the location of items or places in space. We usually think about navigation when we think of spatial memory, but spatial memory may include finding a specific reference in this textbook, finding your keys before you leave your apartment, or remembering the location of your classrooms on campus.

There are many other types of memory systems. A nervous system is not necessarily required for a memory system to be present. The immune system of vertebrates, for example, has a memory for microbes and responds vigorously to previously encountered agents. Likewise, calculators and computers have sophisticated memory systems. All memory systems share the ability to enter information into storage (also called encoding, acquisition, and consolidation), to retain stored information (storage), and to retrieve information from storage (retrieval). These three components of memory must operate or learning will not be observed. Hormones could affect any of these components of memory, or affect learning directly. It is rare to measure learning directly; only the results of learning can be measured and quantified. Consequently, statements about the effects of hormones on learning and memory are typically based on their effects on performance of a task, or observed behavior.

In addition to a functional memory system, successful learning involves several psychological components; namely, motivation, attention, and arousal. These components, like learning itself, are hypothetical psychological constructs designed to describe features of behavioral performance. These hypothetical constructs cannot be measured directly; only performance on a test designed to assess one of these constructs can be measured. For example, if your professor gave you an examination to test your learning of behavioral endocrinology, but asked only questions about geography, you would probably complain that the test was not a valid measure of your learning. Sometimes tests cannot discriminate one hypothetical construct from others. If you leave an answer blank on the exam, your professor does not know whether your failure to answer reflects a failure to learn the material, a failure to attend to the material when it was presented in class, or a failure to be sufficiently motivated to answer the question. Probably, your professor will assume that a failure in learning occurred and assign a grade accordingly. Learning, motivation, arousal, attention, and emotion interact and affect one another. In fact, one of the few laws in psychology states that learning is an inverted U-shaped function of arousal (Figure 12.1; Yerkes and Dodson, 1908). That is, learning does not occur in subjects that are too highly aroused (excited, agitated) or insufficiently aroused (asleep). Hormones are involved in arousal, motivation, and probably also in attention and emotion.

It is important to note that several studies have been conducted in which the influence of these so-called performance factors has been minimized—these studies are discussed later in this chapter. For example, post-training administration of hormones, particularly water-soluble steroids, eliminates the influence of exogenous hormones on performance factors. That is, the hormones are not available during the training or testing of an animal; they are present only during the memory consolidation processes that occur later in the home cage. Thus, careful

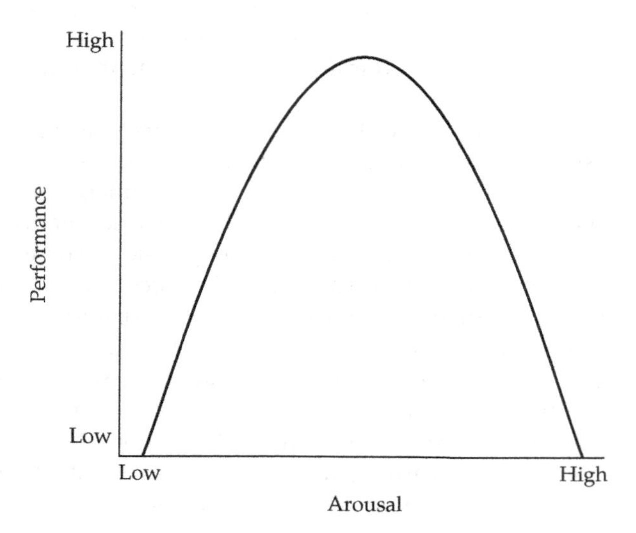

12.1 Arousal and learning. Optimal performance on learning tasks occurs at moderate levels of arousal. If arousal is either too low or too high, learning is adversely affected. Because hormones are involved in arousal, they can have effects on learned performance. After Yerkes and Dodson, 1908.

glands began releasing large quantities of epinephrine immediately after his learning "trial," and the elevated levels of epinephrine enhanced his memory to the extent that he vividly reports the incident 40 years later.

How does epinephrine enhance memory? Again, learning and memory involve encoding, storing, and retrieving information. Epinephrine, as well as other hormones, could facilitate any or all of these processes. The precise mechanisms of memory have yet to be elucidated, but neuroscientists certainly agree that the memory for active and passive avoidance tasks resides in the brain. Epinephrine is a molecule that is too large to cross the blood–brain barrier, although it is produced by a very few neurons in the brain as a neurotransmitter (Weil-Malharbe et al., 1959). We are faced with an apparent paradox: How can epinephrine secreted by the adrenal glands affect learning and memory processes if it cannot get to the neurons in the brain? Obviously, epinephrine must affect some process outside of the brain that subsequently influences the brain. There are currently two working hypotheses regarding how epinephrine affects memory: (1) epinephrine affects memory via its effect on blood glucose levels, and (2) epinephrine activates peripheral receptors that directly influence brain function.

THE GLUCOSE HYPOTHESIS One of the many physiological consequences of epinephrine secretion in a stressed animal is *hyperglycemia*, an increase in blood glucose levels. In general, glucose enhances memory for avoidance learning; the dose–response curve again resembles an inverted U (Figure 12.5; Gold, 1986; Parsons and Gold, 1992). The effects of glucose are time-dependent; injections of glucose delayed by 1 hour after training have no effect on retention and performance. These features are all common to glucose and epinephrine. The doses of glucose and of epinephrine that are most effective in enhancing memory both result in plasma glucose levels comparable to those found naturally in response to optimal training conditions. Glucose injected directly into the brain also enhances memory (Gold, 1987).

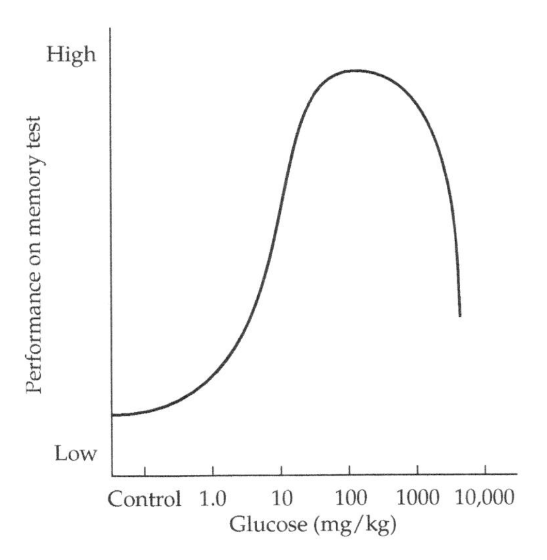

12.5 Effects of glucose on learning. Optimal performance on a memory test is attained with a dose of 100 mg/kg of glucose provided within 1 minute of training. Lower or higher doses, or doses provided later after training, do not enhance memory function. After Gold, 1987.

TABLE 12.2 Agents that enhance cognitive performance

Substance	Effect
Nootropics[a]: Aniracetam, piracetam, pramiracetam, oxiracetam	Elevation of glucose
Others: Amphetamine, epinephrine, glucose, hypertonic saline, painful stimuli, stress, vasopressin	Elevation of glucose
Choline, lecithin, physostigmine, phosphatidylersine	Elevation of synaptic ACh, but not glucose
Phlorizin[b]	Inhibition of glucose transport

[a]Pharmacological cognitive enhancers.
[b]Although phlorizin interferes with brain glucose utilization, it still enhances memory performance by some unknown means.

Additional evidence that epinephrine produces its effects on memory by raising blood glucose levels is based on negative findings. The memory-enhancing and memory-impairing effects of epinephrine treatment can be blocked by treating animals with adrenergic antagonists—drugs that block peripheral epinephrine receptors—but these blockers have no effect on memory enhancement produced by glucose treatment. These results are consistent with the notion that glucose release is a memory modulation step that occurs subsequent to the release of epinephrine. Because adrenergic antagonists block the receptors by which epinephrine acts, they prevent the effects of epinephrine, as well as those of treatments that act through epinephrine on memory. But the effects of glucose, which are "downstream" from the peripheral epinephrine receptors, remain intact. The agents that are most effective in enhancing human cognitive performance (i.e., improving learning and memory) share a common feature: they all elevate blood glucose levels (Table 12.2).

How does glucose produce its effects on memory? As we saw in Chapter 9, the brain requires a constant supply of glucose in order to function. Elevated blood glucose levels permit more glucose to enter neurons, which in turn stimulates an increase in the release of the neurotransmitter acetylcholine from neurons in the brain. Increases in acetylcholine levels in the brain synapses are characteristic of all known cognitive enhancers (see Table 12.2). Thus, cognitive enhancers must work by increasing glucose, which increases acetylcholine release, or by increasing acetylcholine levels directly. The neurobiological mechanism(s) by which glucose improves memory do appear to involve acetylcholine release. In a study of rats in a simple "T-maze," microdialysis revealed that acetylcholine levels were elevated in glucose-treated rats during memory testing (Ragozzino et al., 1996). The severe memory deficits observed in patients with advanced Alzheimer disease or AIDS are correlated with a marked reduction in neurons that secrete acetylcholine. Obviously, cognitive enhancers that increase the release of acetylcholine or prolong its half-life have great clinical significance.

The study of impaired learning and memory represents another approach to understanding the effects of hormones. Memory function in old rats diminishes in a fashion similar to the decline observed in normal elderly people. Epinephrine enhances the memory of very old rats; however, treatment of elderly human patients with epinephrine to enhance their memories would not be wise, considering the cardiovascular effects of this stimulatory hormone.

Unlike epinephrine, glucose is a relatively safe compound with which to study memory in humans. In a recent study, the effects of glucose on memory were tested in a group of healthy 70-year-old people. Individuals of this age typically have memory impairments even if they do not suffer from dementia or other aspects of Alzheimer disease. These 70-year-olds scored less well on a standard memory test than college students tested at the same time. Following the test of memory, the elderly subjects were asked to drink a glass of lemonade that had been prepared with either saccharine or glucose. Another memory test was then administered. Those individuals who had received glucose showed improved memory function relative to those who drank lemonade prepared with saccharine. It was noted that the glucose utilization efficiency of an individual predicted memory function among the aged, but not the young, subjects. Elderly people often have problems regulating their blood sugar levels. The normal decline in memory function during aging may thus reflect a diminishing ability to regulate blood levels of glucose. The extra glucose provided in the study probably allowed the brain cells to function better in the elderly subjects (Hall et al., 1989).

Subsequent studies have confirmed that glucose enhances both memory storage and retrieval in healthy elderly humans (Manning et al., 1998b). Glucose also improves learning and memory in people with Alzheimer disease, as well as in adults with Down syndrome (Korol and Gold, 1998; Manning et al., 1998a). Although earlier studies on college students failed to discover any effects of glucose on performance, more recent studies have revealed glucose enhancement of learning and memory in this age group when the tasks are challenging (Korol and Gold, 1998). In these challenging memory tests on college students, it was demonstrated that glucose augmented memory for material in a paragraph. Glucose also seemed to improve attentional skills in the students as well. However, neither face and word recognition nor working memory (defined here as the nonpermanent storage of information that is being processed in a cognitive task) were influenced by glucose treatment (Korol and Gold, 1998).

EFFECTS OF INSULIN If unregulated blood glucose levels interfere with learning, then diabetic animals and humans should exhibit learning difficulties. Surprisingly few studies, however, have directly addressed the relationship between diabetes and learning and memory in humans. Some studies have shown no strong correlation between learning and memory and insulin-dependent diabetes (Helkala et al., 1995; Crawford et al., 1995; Ryan and Williams, 1993; Wolters et al., 1996) or non-insulin-dependent diabetes (Worrall et al., 1996) in humans. However, most studies of insulin-dependent (Amiel and Gale, 1993; Langan et al., 1991; Lincoln et al., 1996) or non-insulin-dependent diabetes (Strachan et al., 1997) indicate a strong relationship between blood sugar and cognitive impairment. Verbal memory is the most commonly affected cognitive ability among diabetics. Although the effects of diabetes on cognitive function were most pronounced among aged people, cognitive impairments were observed among young adult diabetics as well (e.g., Dey et al., 1997). In insulin-dependent diabetes, chronic hyperglycemia and the recurrence of hypoglycemia appear to be associated with cognitive impairments. In non-insulin-dependent diabetes, the onset of cognitive impairments could reflect disruptions of blood glucose metabolism or a number of related problems, including hyperglycemia, changes in insulin concentration, hypertension, and changes in lipid levels (Kumari et al., 2000). These symptoms tend to co-occur and have been recently considered to be part of a syndrome associated with obesity and insulin resistance, termed "metabolic syndrome." Both insulin-dependent diabetes and non-insulin-dependent diabetes negatively affect measures of verbal and numerical reasoning, attention, concentration, verbal and visual memory, and verbal fluency (Kumari et al., 2000). Diabetes may also increase the risk of dementia (Park, 2001).

Insulin may be produced in the brain, and there are insulin receptors located in the brain, despite the observation that insulin is not necessary for neurons to take up glucose (see Chapter 9). However, activation of these insulin receptors in the brain may be important in associative learning (Zhao and Alkon, 2001). Genetic disruption of brain insulin receptors in mice impairs memory (Bruning et al., 2000). Furthermore, abnormal insulin levels and reduced numbers of brain insulin receptors are common among Alzheimer patients with severe memory impairments (Craft et al., 1998). Treatment of people with Alzheimer disease with insulin while keeping blood glucose levels constant significantly improved memory (Craft et al., 1996). These effects of insulin receptor activation in the brain may reflect improved neurotransmission (Zhao and Alkon, 2001).

To better understand the mechanisms of cognitive impairment in diabetic humans, several studies have been conducted on the effects of diabetes on the acquisition of passive avoidance tasks in nonhuman animals. Treatment of rodents with streptozotocin destroys the insulin-secreting β-cells of the pancreas, inducing diabetes (see Chapter 9). Streptozotocin-induced diabetic rodents display modest, but consistent, deficits in passive avoidance learning (Figure 12.6) (Leedom et al., 1987; Hoyer, 2003). Surprisingly, administration of phlorizin, a drug that inhibits glucose transport into cells, causes significant enhancement of memory performance in diabetic rats in comparison to control rats (Hall et al., 1992). The physiological mechanisms underlying these cognition-enhancing effects of phlorizin remain unspecified.

A common test of spatial learning is the Morris water maze (Figure 12.7). This "maze" consists of a round tank that is about 1.3 meters in diameter, filled with water; below the surface of the water is a small platform, which the animal must

12.7 Spatial memory can be assessed in the Morris water maze. (A) In the Morris water maze, rats must swim to find a submerged platform. The platform is visible here, but in testing conditions it is hidden below the surface of milky water. Finding the hidden platform requires relational learning because the only aids available to guide navigation are large extra-maze cues around the tank. (B) Representative swimming paths of rats in the Morris water maze. Rats learn to navigate to the platform from any position in the tank if they are normal or have had their neocortex lesioned. However, if the hippocampus is lesioned, the rat swims aimlessly until it randomly discovers the submerged platform. After Morris et al., 1982. ▶

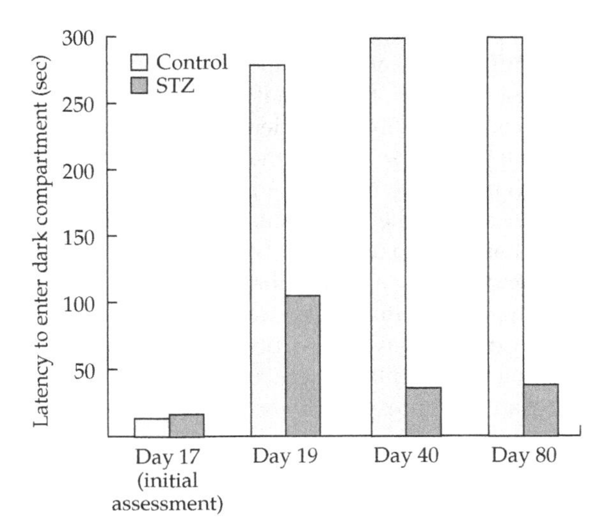

12.6 Streptozotocin impairs performance of passive avoidance tasks. Streptozotocin (STZ) was injected into rats for the first 18 days of the study. At the initial assessment on day 17, control animals and streptozotocin-treated animals showed little difference in the time they took to step through a door from a brightly lit compartment into a dark compartment. A foot-shock was applied on day 18 in the dark compartment. Latency to go from the lit compartment into the dark (preferred) compartment was assessed on days 19, 40, and 80. Note that the rats treated with streptozotocin showed a much shorter latency to enter the compartment where they previously received a shock. After Hoyer, 2003.

(A) The Morris water maze

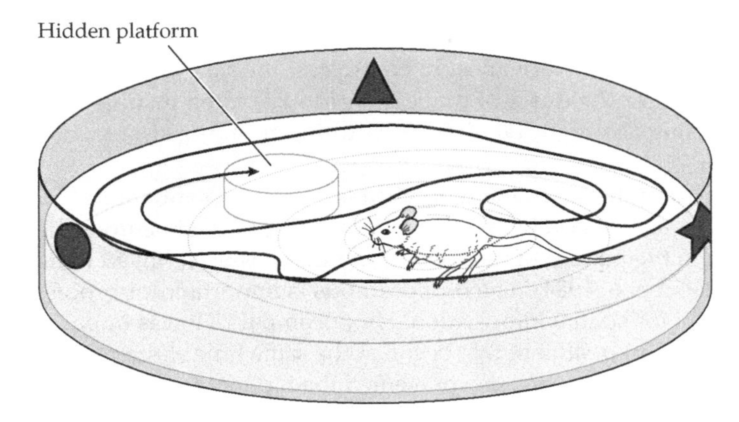

(B) Water maze trajectories

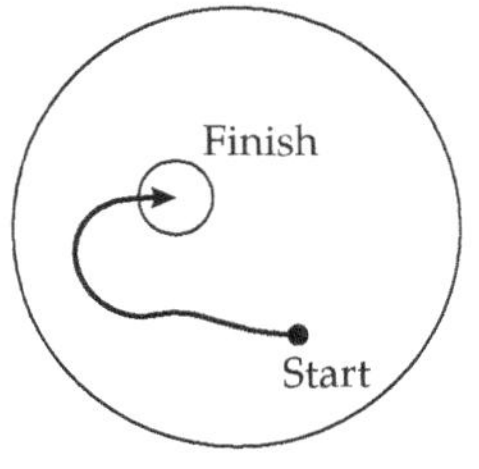

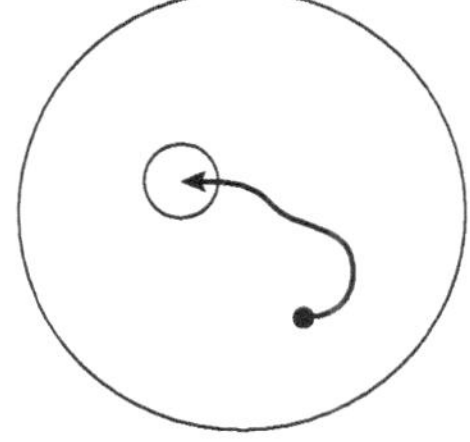

find and climb onto in order to escape the water (Morris et al., 1982). A rat or mouse is released into the water at different points along the edge of the tank. In order to find the platform on successive trials, the animal must depend on its memory to recall where the platform is located. In the hidden-platform version of the task, the platform is hidden by milky-white water, and the rat must use its spatial memory to locate the platform in relation to its current position with the aid of extra-maze visual cues. In a cued version of the task, animals are trained to find a visible platform whose location changes on successive trials; because spatial information is irrelevant to performance of this task, it is often used to control for those aspects of task performance that do not involve memory (e.g., swimming ability and motivation). An animal does not have to be food-restricted to learn either version of the task, which has its own built-in motivation—the water is at room temperature, and so is unpleasant, and presumably the rat does not want to drown.

Rats rendered diabetic with streptozotocin display deficits in spatial learning in the Morris water maze (Biessels et al., 1996; Lannert and Hoyer, 1998). Streptozotocin-treated rats were further tested in the water maze to determine whether their learning deficits could be prevented or reversed by insulin treatment (Biessels et al., 1998). Three experimental groups were evaluated: (1) rats that received streptozotocin, but no insulin, (2) rats that received streptozotocin, but also received insulin so that diabetes never developed, and (3) rats that received insulin 10 weeks after streptozotocin treatment began (i.e., after the onset of diabetes). In rats treated early, insulin prevented the streptozotocin-induced deficits in performance. In contrast, if the rats were not treated with insulin until after the onset of diabetes (group 3), then insulin was not effective in reversing the cognitive deficits resulting from the streptozotocin-induced diabetes.

In the next part of the experiment, in vitro long-term potentiation (LTP) in the hippocampus was assessed by monitoring the firing rates of neurons. LTP is an enhancement in the pattern of neuronal firing that has been linked to learning (Teyler and Discenna, 1984). The hippocampus is important for processing memory, especially for spatial information. Hippocampal LTP was impaired in the diabetic rats. Insulin treatment that began at the same time as streptozotocin treatment and prevented diabetes also prevented the hippocampal impairments in LTP; insulin treatment that began 10 weeks after the onset of diabetes only partially restored LTP (Biessels et al., 1998). Thus, it seems that insulin affects spatial memory in diabetic rats only if provided from the onset of diabetes.

Other studies have demonstrated that memory can also be enhanced by fructose, a sugar that is not metabolized by the neurons in the brain; other glucose analogs that are not well metabolized by the brain also improved retention (Messier and White, 1987). These latter studies also indicated that a much higher dose of glucose is optimal for memory enhancement than previously reported (e.g., 2.0 g/kg versus 0.1 g/kg in earlier studies) (Messier and White, 1987). The basis for this discrepancy in effective doses is not apparent. Perhaps at high doses these other sugars act at peripheral sites to activate membrane glucose transport mechanisms; at lower doses, the glucose may affect memory more directly via actions on neural function or processing.

THE PERIPHERAL RECEPTOR HYPOTHESIS The second working hypothesis for how epinephrine modulates memory is that it activates peripheral receptors that communicate with the central nervous system (McGaugh, 1989). In order to discover which neural receptors are involved in the memory effects of epinephrine, specific receptor agonists (mimics) and antagonists (blockers) have been employed.

The effects of epinephrine on memory can be blocked by drugs that block both the α- and β- epinephrine (adrenergic) receptors (Sternberg et al., 1985, 1986). In one study, rats were injected with an α- or a β-adrenergic antagonist—phenoxybenzamine or propranolol, respectively—30 minutes before they received either an injection of epinephrine or a foot-shock (which caused endogenous levels of epinephrine to rise). Glucose concentrations did not rise very much after epinephrine or foot-shock treatment in rats pre-treated with the β-blocker, propranolol; however, glucose concentrations increased after epinephrine treatment or foot-shock in rats pre-treated with the α-blocker, phenoxybenzamine (Manning et al., 1992). Furthermore, the mnemonic (memory-enhancing) effects of clenbuterol, an adrenergic agonist that crosses the blood–brain barrier, can be blocked only by centrally acting β-adrenergic antagonists (Introini-Collison and Baratti, 1986). Taken together, these data suggest that epinephrine acts on peripheral adrenergic receptors to initiate its effects on memory (McGaugh, 1989).

The amygdala has been considered for years to be involved in emotions (Figure 12.8). This almond-shaped part of the limbic forebrain (at the base of the temporal lobe) is also involved in learning and memory, both directly and indi-

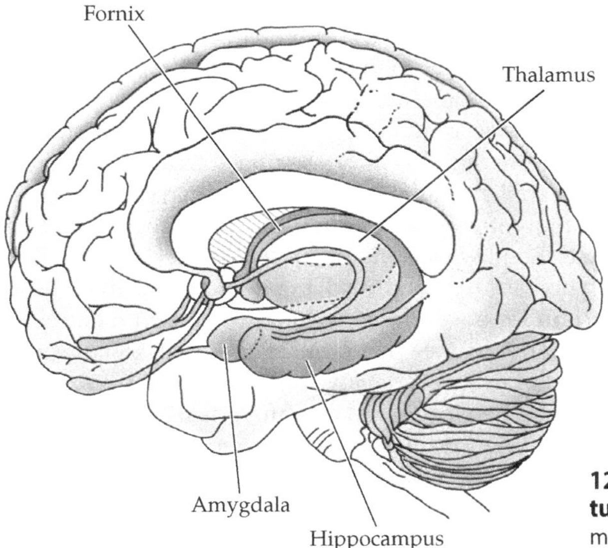

12.8 The amygdala and associated structures seen in a cutaway lateral view of the human brain.

rectly via its close association with the hippocampus. Electrical stimulation of the amygdala can increase memory retention (McGaugh and Gold, 1976). In an elegant series of experiments, it was demonstrated that memory for aversive conditioning can be modulated by post-training injections of epinephrine and norepinephrine directly into the amygdala. The increased retention induced by intra-amygdala injections of norepinephrine can be blocked by concurrent administration of β-adrenergic antagonists such as propranolol (Gallagher et al., 1981). The effect is observed only if the norepinephrine and the adrenergic antagonist are injected on the same side of the brain; in other words, the effect is *stereospecific*. Like that of epinephrine, this effect of norepinephrine is dose- and time-dependent. Recent evidence has indicated that peripheral epinephrine acts via β-noradrenergic receptors, which activate ascending neurons in the vagus nerve (Williams et al., 1998; Hoyer, 2003). These neurons travel to the nucleus of the solitary tract (NTS). The NTS projects noradrenergic fibers to the amygdala, especially the basolateral amygdala. The basolateral amygdala modulates the memory of emotional experiences by modulating memory consolidation via efferents to other brain regions, including the caudate nucleus, nucleus accumbens, and cortex (McGaugh, 2004). Blocking any part of this pathway prevents the memory-enhancing effects of epinephrine (Gold, 2003).

The degree of activation of the amygdala by emotionally arousing stimuli (both positive and negative) correlates directly with subsequent recall of those stimuli. In one fascinating study of humans, individuals read either an emotionally charged story or a similar story that had been judged by other people to be more emotionally neutral (Cahill et al., 1994). Some individuals were treated with propranolol, a potent β-adrenergic antagonist, which significantly impaired memory of the emotionally arousing story, but did not affect memory of the neutral story (Cahill et al., 1994). The researchers who conducted this study were able to rule out the possibility that the drug had nonspecific effects on attention or motivation. These results support the hypothesis that highly charged emotional memories require activation of β-adrenergic receptors (Cahill et al., 1994). Treatment with epinephrine or exposure to cold pressor stress (holding a hand in ice-cold water for 2 minutes, which evokes secretion of epinephrine) after participants viewed emotionally charged pictures enhanced the long-term recall of those pictures (Cahill and Alkire, 2003; Cahill et al., 2003). Similar effects have been reported in people after administration of yohimbine (an α-receptor blocker), which elevates norepinephrine levels (Southwick et al., 2002).

Recent imaging work on humans has confirmed and extended the hypothesis that the amygdala is important for encoding emotionally charged memories. Using either PET or fMRI, it was determined that recall of a series of scenes correlated highly with the activation of the amygdala (Cahill et al., 1996; Canli et al., 2000). Importantly, the association between recall and amygdala activation during learning was highest for the pictures judged as most emotionally intense (Figure 12.9) (Canli et al., 2000). Imaging studies of the contribution of emotions to memory consolidation have revealed an important sex difference (McGaugh, 2004). Women displayed enhanced activity in the left amygdala in relation to

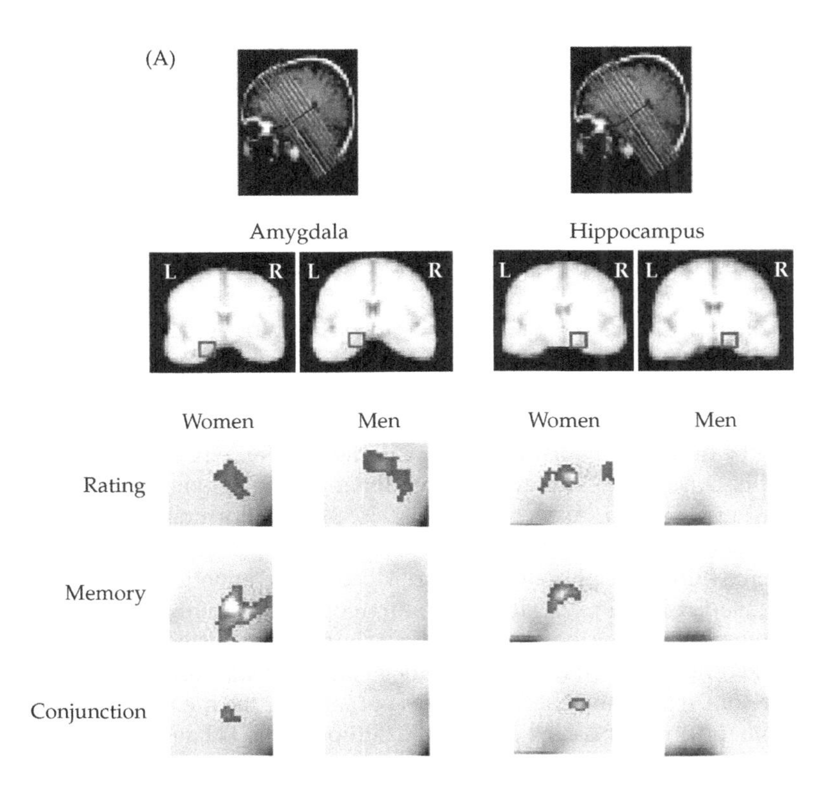

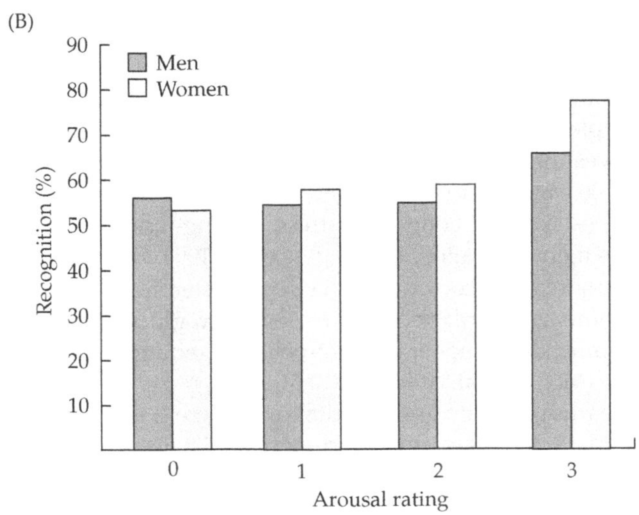

12.9 Colocalization of brain regions associated with emotional intensity ratings and recognition memory. (A) Areas of activation that correlated with highest emotional intensity ratings (Rating), those that correlated with the best subsequent memory (Memory), and results of a conjunction analysis of these two conditions (Conjunction) are shown for women and men. The images at the top show the slice containing the region of interest (light lines). The images below them show locations in the left amygdala and right hippocampus in which brain activation correlated with both rating and memory in women, but not men, in overlapping or adjacent brain regions. (B) Proportion of pictures rated neutral (0) to emotionally very arousing (3) that were recognized (summed percentage of familiar and remembered pictures) by men and women. From Canli et al. 2002.

improved memory recall, whereas men displayed enhanced activity in the right amygdala in the same relationship to memory (Cahill et al., 2001; Canli et al., 2002). Future work will have to sort out the importance of this structural sex difference in emotional processing.

The glucose and receptor hypotheses of how epinephrine affects memory are not incompatible. Epinephrine elevates blood glucose concentrations, which increases the amount of glucose that enters neurons in the brain. These neurons thus release higher concentrations of acetylcholine into the synapses. Epinephrine also appears to act directly on neurons to enhance their function. These direct, central effects of epinephrine may reflect endocrine activity, or epinephrine could affect memory directly (perhaps after entering the brain through the cerebrospinal fluid, thus circumventing the blood–brain barrier) by acting as a central neurotransmitter.

There may be other mechanisms underlying memory enhancement by epinephrine as well. During stressful events, additional adrenal hormones (i.e., the glucocorticoids) are secreted. As we shall see, these steroid hormones also have significant effects on learning and memory. In addition, the effects of epinephrine, administered either peripherally or centrally, could hypothetically be mediated via increased release rates of adrenocorticotropic hormone (ACTH). However, the memory-enhancing effect of post-training epinephrine administration is not blocked by dexamethasone, an artificial steroid that blocks the release of ACTH. Therefore, it appears that ACTH does not mediate the memory-enhancing effect of epinephrine (McGaugh et al., 1987), although ACTH and glucocorticoids have their own potent effects on memory.

Glucocorticoids

This chapter began by describing how memory is enhanced by stressful events. Acute stress appears to promote lasting memories, as does treatment with glucocorticoids. Treatment of people with glucocorticoids prior to learning words or pictures, for example, improves their recall on a subsequent memory test (Figure 12.10) (Abercrombie et al., 2003; Buchanan and Lovallo, 2001). Chronic stress, however, seems to have the opposite effect. Long-term stress, or long-term treatment with corticosterone, impairs memory (Luine, 1994; Luine et al., 1994; De Quervain et al., 1998). Thus, brief exposure to glucocorticoids (i.e., corticosterone or cortisol) enhances learning and memory, whereas chronic exposure to glucocorticoids appears to function as an **amnestic** (an agent that promotes forgetting) in most of the studies reported to date (McLay et al., 1998).

One common test of learning in rats uses a radial-arm maze (Figure 12.11). This task can be used to test both long-term and short-term spatial memory simultaneously. Typically, a radial-arm maze has eight runways, four of which are always stocked with a food treat (baited); the other four arms are always devoid of treats (unbaited). The solution to the maze involves making only one trip down each of the baited arms and avoiding the unbaited arms. This task, therefore, requires long-term, or reference, memory (to recall which of the eight arms are always baited) as well as short-term, or working, memory (to recall which of the

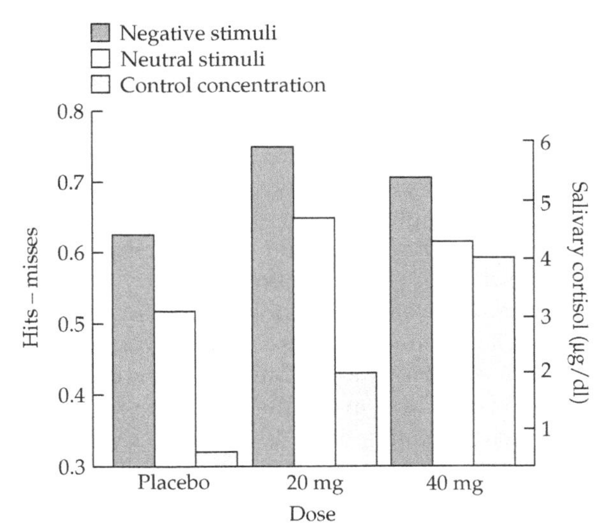

12.10 Recognition memory is related to cortisol concentrations. Treatment with cortisol increased subjects' recall of pictures; negative stimuli were more likely to be recalled than neutral stimuli. After Hoyer, 2003.

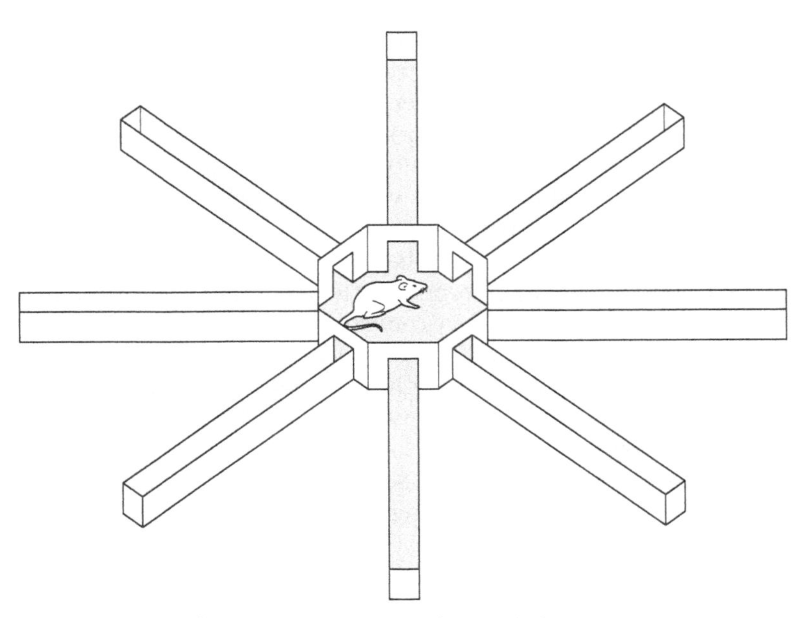

12.11 The radial-arm maze, popularized by David Olton, is a common apparatus used in assessments of spatial memory. The maze is open at the top and may have 8, 12, or even 36 arms; the greater the number of arms, the more difficult the maze. During training, the subject learns that only some of the arms are baited (have a food reward) at the end. To solve the maze successfully after training, the subject must make only one trip down each baited arm and avoid exploring the unbaited arms, a task that requires both long-term (reference) and short-term (working) memory. The animals navigate using extra-maze cues. After Olton and Samuelson, 1976.

eight baited arms have already been visited on a particular trial) (Olton and Papas, 1979; Olton and Samuelson, 1976).

In one study of spatial learning in the radial-arm maze, rats were chronically stressed by being placed in clear ventilated plastic containers, in which their movements were severely limited, for 6 hours per day for 3 weeks (Luine et al., 1994; Luine, 1994). After 21 days of this stress treatment, the rats were food-restricted, trained, and then tested on an eight-arm radial-arm maze. The performance of the restraint-stressed rats was impaired compared with that of rats that had not been restrained (Luine et al., 1994). In a follow-up study with the same design (Luine, 1994), blood samples were obtained, and blood concentrations of corticosterone were assayed. Rats with the highest corticosterone concentrations displayed the most errors in performance. Similarly, a study that correlated corticosterone consumption (provided in the rats' drinking water) and maze performance indicated that rats consuming the most corticosterone made the most errors in a radial-arm maze (Luine et al., 1993). Subsequent studies have shown that long-term corticosterone treatment impairs spatial learning in rats in a variety of testing situations (McLay et al., 1998). Treatment with RU38486 (a progestin and glucocorticoid receptor antagonist commonly called RU-486) infused directly into the dorsal hippocampus improved the performance of rats in the Morris water maze (Oitzl et al., 1998).

These studies on the effects of glucocorticoids on learning and memory were focused exclusively on acquisition and long-term storage. Acute stressors and corticosterone also appear to impair memory retrieval. Rats were trained to locate a submerged, hidden platform in a Morris water maze. After the rats were trained, they received a foot-shock (i.e., a mild stressor) either 2 minutes, 30 minutes, or 4 hours prior to testing. During the test, the platform was removed. In this situation, rats with good memories spend more time swimming over the area of the tank where the platform was previously located. Rats with poor memories swim around somewhat randomly, usually along the edge of the tank, suggesting that they do not recall where the escape platform ought to be. Rats shocked 30 minutes prior to memory testing performed poorly compared with rats tested 2 minutes or 4 hours after the foot-shock (Figure 12.12) (De Quervain et al., 1998). Because electric shocks result in elevated circulating glucocorticoid concentrations about 30 minutes later, these results suggest that elevated glucocorticoid concentrations at the time of memory assessment impair performance.

In contrast to the effects of chronic stress, several studies have confirmed that acute stress enhances performance on learning and memory tasks. For example, stress facilitates classical conditioning of the eyeblink response in rats (Shors et al., 1992). Stress-evoked facilitation occurs within minutes and persists for several days. Glucocorticoids appear to be involved in facilitation of both hippocampus-dependent and hippocampus-independent learning and memory. Adrenalectomy blocked the stress-evoked facilitation of learning (Beylin and Shors, 2003). Even when glucocorticoids were replaced at basal levels post-adrenalectomy, stress did not facilitate trace conditioning, a type of classical conditioning that is dependent on an intact hippocampus (Beylin and Shors, 2003). Adrenal demedullation, which

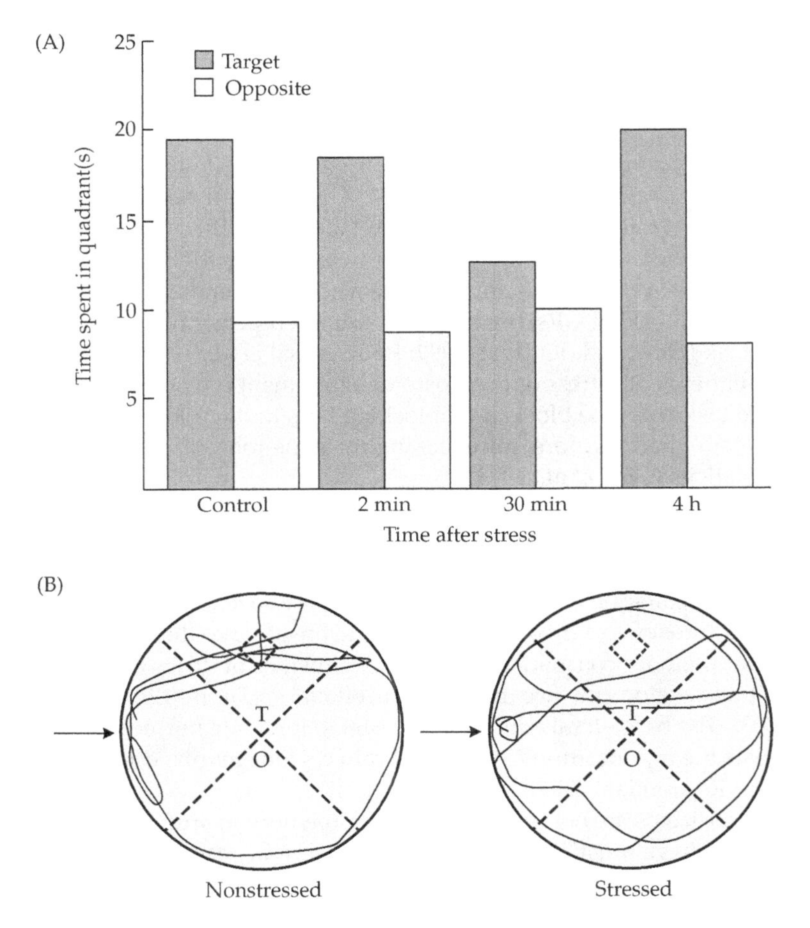

12.12 Retrieval of spatial learning is impaired by stress. (A) The amount of time spent in the target quadrant (T; where the platform was formerly located) as compared with the opposite quadrant (O) of the Morris water maze. Rats spent less time in the target quadrant 30 minutes after foot-shock, but this impairment was reversed 4 hours after foot-shock. (B) Representative swimming paths of a nonstressed control rat and a stressed rat. Note that while the nonstressed rat swam in the general area of the platform, the stressed rat appears to have swum across all areas of the maze randomly. The arrow indicates where rats entered the water. After De Quervain et al., 1998.

leaves the adrenal cortex intact, and glucocorticoid secretion responsive to stressful stimuli, facilitated stress-evoked learning. These experiments, as well as others, indicate that glucocorticoids are both necessary and sufficient for the temporary facilitation of learning caused by acute stressors, but cannot sustain these memories 24 or more hours later (Beylin and Shors, 2003).

Single injections of natural or synthetic (e.g., dexamethasone) glucocorticoids mimic acute stress and tend to facilitate memory consolidation (Roozendaal,

2000). Dexamethasone has high affinity for the glucocorticoid receptors (GRs or Type II receptors). In contrast, natural glucocorticoids, such as corticosterone, have high affinity for the mineralocorticoid receptors (MRs or Type I receptors). Typically, GRs are activated by corticosterone only when concentrations are elevated, such as during stressful events or during the circadian peak of glucocorticoid secretion (Reul and de Kloet, 1985). Therefore, it seems reasonable to hypothesize that GRs mediate the memory-facilitating effects of acute stress. To test this hypothesis, drugs that specifically blocked either GRs or MRs were infused into the brains of rats immediately before or immediately after learning a water maze. Blocking GRs, but not MRs, impaired performance on this spatial memory task (Oitzl and de Kloet, 1992; Roozendaal et al., 1996). The enhancing effects of either acute stress or corticosterone treatment on memory for a passive avoidance task could be blocked by blocking GRs in day-old chicks (Sandi and Rose, 1994a,b). Furthermore, mice lacking the gene for GRs display substantial memory deficits (Oitzl et al., 1998).

The memory-enhancing effects of glucocorticoids appear to involve the amygdala, which modulates the memory consolidation process that probably occurs elsewhere in the brain. Specific lesions of the basolateral amygdala block the memory-facilitating effects of glucocorticoids (Roozendaal, 2003; Roozendaal et al., 2002). Infusions of drugs that block GRs directly into the basolateral amygdala impair memory consolidation, whereas infusions of drugs that activate GRs into this brain region enhance memory consolidation (Figure 12.13) (Roozendaal, 2000, 2003). The basolateral amygdala appears to integrate hormonal information that signals the hippocampus and other brain regions involved in memory consolidation (Roozendaal, 2000).

Glucocorticoids are not only important for memory in artificial memory tasks, but appear to have functional significance for survival as well. Several species of birds, in common with many squirrels, hide caches of food in order to survive periods of time when food is scarce. These birds rely, at least in part, on spatial memory to find these previously hidden caches. It seems reasonable to assume that the ability to locate these hidden caches becomes critical in harsh habitats where the food supply is generally low or unpredictable. This assumption was recently tested in captive mountain chickadees (*Poecile gambeli*), half of which were maintained on a limited and unpredictable food supply while the rest were provisioned with unlimited food supplies for 60 days (Pravosudov and Clayton, 2001) (Figure 12.14). The birds maintained on an unpredictable food supply were more efficient at cache recovery and performed spatial (but not nonspatial) tasks more accurately than individuals from the same population that were fed ad libitum. The birds maintained on an unpredictable food supply displayed moderately elevated corticosterone concentrations compared with the birds provided with unlimited food (Pravosudov et al., 2001). Implanting corticosterone capsules (which increased blood concentrations of corticosterone to the level of birds receiving unpredictable food supplies) into mountain chickadees stimulated food intake and food caching as compared with birds implanted with an inert control pellet (Pravosudov, 2003). Chickadees implanted with corticosterone also displayed improved spatial (but

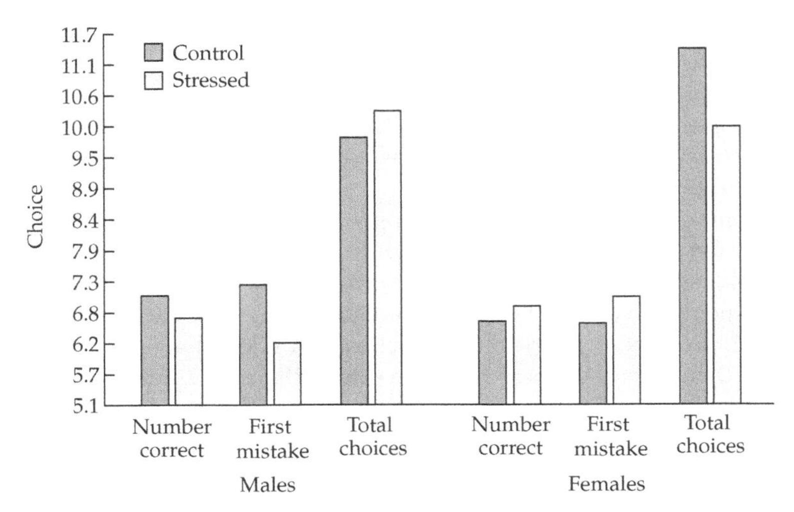

12.18 Sex differences in stress responsiveness. Chronic restraint stress impairs male performance and enhances female performance on the 8-arm radial-arm maze. Male performance is impaired by stress; they make fewer correct choices and make their first mistake earlier in their memory test than control rats do. Females show the opposite pattern. Courtesy of V. Luine.

Male rats exposed to recurring daily restraint (6 hours per day) in a plastic ventilated tube (a chronic psychological stressor) were unaffected in their performance on a radial-arm maze after a week of this treatment, but showed improvement in retention after 2 weeks of restraint as compared with unrestrained control rats (Luine et al., 1996). If the restraint continued for a third week, however, there was a reversal in performance; now the restrained rats displayed worse spatial memory than unrestrained controls (Luine et al., 1994). Restraint of female rats enhanced their performance in the radial-arm maze after 3 weeks (Bowman et al., 2001). A similar pattern of sexually dimorphic effects of chronic stress on performance has been reported for other spatial memory tasks and non-spatial visual (object) memory tasks (Beck and Luine, 2002).

It is also possible that chronic stress differentially affects neural structure or function. For example, after 21 days of restraint, apical dendritic branching and dendritic lengths in the CA3 region of the hippocampus of male rats were reduced. In contrast, female rats undergoing the same pattern of daily restraint did not display reductions in apical dendritic branching (Galea et al., 1997). This dendritic atrophy is mediated by corticosterone and can be prevented by blocking glucocorticoid release or actions.

Females in proestrus have a higher density of dendritic spines in the CA1 area of the hippocampus than males (Shors et al., 2001, 2004). In response to an acute stressor (intermittent tail-shocks), spine density was enhanced in the hippocampus of male rats, but reduced in the female hippocampus (Shors et al., 2001). The sex differences in hippocampal spine density were correlated with estradiol and testosterone concentrations; however, the stress-evoked spine density changes were not associated with circulating glucocorticoid concentrations. Thus, male and female rats display different densities of dendritic spines in the hippocampus under baseline conditions. Under stressful conditions, however, the neuronal structure of individuals of the two sexes responds in opposite directions to identical stressors (Shors et al., 2004).

In common with rats, humans display sex differences in the interactions between stress and memory. In young adults, stress-evoked elevation in cortisol levels was associated with impaired performance on memory tasks in men, but not in women (Wolf et al., 2001). Sex differences in responses to stress persist into old age. Elderly men experiencing psychosocial stress displayed higher HPA responses than elderly women (Kudielka et al., 1998). Sex differences in memory performance in response to stress might reflect sex differences in the activation of the amygdala during emotionally charged learning (Cahill et al., 2001). Importantly, there are several studies showing sex differences in spatial and nonspatial abilities in humans independent of stress. Several studies testing humans in computer "virtual" water mazes have shown a consistent male advantage (e.g., Astur et al., 1998, 2004). In contrast, women tend to outperform men on tasks using two- and three-dimensional object arrays in remembering the locations and identities of objects (e.g., Eals and Silverman, 1994). These results have been related to hunter–gatherer theories about why there should be sex differences in memory function among humans (Eals and Silverman, 1994).

In common with the behavioral effects of stress, the stress-evoked hippocampal structural changes are temporary. Good news for those of us who experience short-term, chronic psychological stressors: in rats, both behavioral impairments and hippocampal changes resolve about 5–10 days after the stressor stops (Conrad et al., 1999). Similar results have been reported in nonhuman primates. Psychosocial stressors reduce the numbers of hippocampal pyramidal neurons in subordinate males, but not in females (reviewed in McEwen, 2000).

Effects of Estrogens

The studies we have just discussed provide convincing evidence that estrogens are neuroprotective against stressors, and that they influence learning and memory. Estradiol appears to enhance spatial memory in a reliable, but subtle, manner (Luine, 1994; Daniel et al., 1997). Simple statements like this one, however, must be qualified. Estrogenic effects on learning and memory are complex and depend on many factors, including the timing of hormone administration and the gonadal state of the individual being assessed.

Estradiol seems to enhance memory when the task is difficult. In general, estrogens appear to enhance consolidation, and slightly enhance acquisition, of spatial reference memory tasks (Daniel et al., 1997; Luine et al., 1998; Fader et al., 1998) and of two-choice water-escape working memory tasks (a version of the Morris water maze) (O'Neil et al., 1996). Spatial memory in the Morris water maze is better during diestrus than during estrus (Frick and Berger-Sweeney, 2001). In contrast, several studies in rats and humans have demonstrated that spatial memory is impaired during the periovulatory portion of the ovarian cycle, when estradiol concentrations are normally high (e.g., Frye, 1995; Galea et al., 1995; Korol et al., 1994; Warren and Juraska, 1997).

In one of the first studies of the effects of estradiol on memory, castrated male rats were deprived of food (reduced to 85% of their baseline body mass to motivate them to find food in a maze) and trained on an eight-arm radial-arm maze. These males were then implanted with either empty Silastic capsules or capsules containing estradiol benzoate; males implanted with estradiol capsules had circulating estradiol concentrations of about 90 pg/ml of blood serum (Luine, 1994). Beginning 3 days after capsule implantation, the working memory of the males was assessed over twenty trials. There was no difference between estrogen-treated and control rats in performance (Figure 12.19) (Luine, 1994; Luine and Rodriguez, 1994). In the next eight trials, a 1-hour delay between the fourth and fifth arm choice was added. When the task was made more difficult by the 1-hour delay, the males treated with estrogen exhibited a small, but reliable, improvement in performance as compared with the control animals (Luine, 1994). In another study, both low (40 pg/ml) and high (200 pg/ml) doses of estradiol resulted in improved choice accuracy in a twelve-arm radial-arm maze by young and aged female rats (Williams, 1996). In yet another study, estrogen treatment of ovariectomized female rats augmented working memory, but not reference memory (Luine et al., 1998). Similarly, capsules containing estradiol enhanced the performance of ovariectomized female rats during acquisition of an eight-arm radial-arm maze when implanted for 30 days, then removed prior to training (Daniel et al., 1997). These results suggest that chronic treatment with estradiol produces an effect on memory, and that estradiol may induce changes in neuronal form or function that persist well after the hormone capsule is removed.

The effects of estrogens on spatial memory are probably mediated by the hippocampus. Infusion of a water-soluble form of estradiol directly into the hippocam-

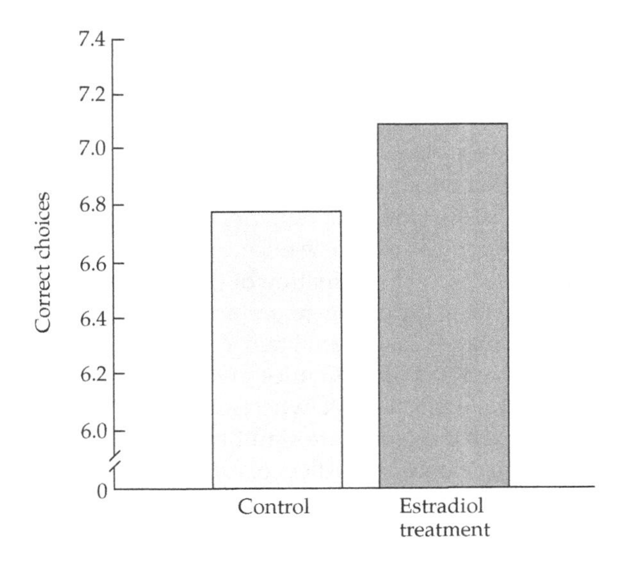

12.19 Estrogen improves spatial memory. Male rats were implanted with either an empty Silastic capsule (control) or with a capsule containing estradiol. The bars represent the average number of correct choices on an eight-arm radial-arm maze after a 1-hour delay between the fourth and fifth visits across several trials. After Luine and Rodriguez, 1994.

pus enhances the memory of ovariectomized rats for the Morris water maze, but only if given immediately after training (Packard et al., 1996; Packard and Teather, 1997). These results and others indicate that estradiol can also have acute effects on memory. Estradiol treatment 2 hours post-training did not affect retention (Packard and Teather, 1997), suggesting a time-limited effect on memory consolidation. Other studies have also demonstrated the memory-enhancing effects of acute estrogen treatment (e.g., Sandstrom and Williams, 2001, 2004). Ovariectomized rats treated with 17α-estradiol, 17β-estradiol, or diethylstilbesterol (DES) 30 minutes prior to or immediately after learning trials displayed rapid enhancement of visual and place memory, a special version of spatial memory (Luine et al., 2003). When these estrogenic hormones were given 2 hours after learning trials, they were ineffective in improving memory when the animals were tested 4 hours later. These results suggest that estrogens affect learning and memory consolidation processes, but not performance processes. In addition, the rapid effects of estrogens and the nonspecificity of estrogen type suggest that nongenomic mechanisms may be involved (Luine et al., 2003). Estrogen treatment of ovariectomized female rats at supraphysiological doses also increased performance on the hippocampal-dependent task of eyeblink conditioning (Leuner et al., 2004).

Estradiol receptors are located within the hippocampus, especially in the CA1 region, but also in the CA3 region and dentate gyrus (Loy et al., 1988; Maggi et al., 1989; Gould et al., 1990) (see Figure 12.15). Estrogen treatment, like naturally high estrogen concentrations around the time of ovulation, is associated with an increase in the density of dendritic spines in the CA1 region (Gould et al., 1990; Woolley et al., 1990a; McEwen et al., 1995; reviewed in Desmond and Levy, 1997) (see Chapter 4). The results of studies that have attempted to correlate hormone-induced changes in hippocampal circuitry with spatial learning performance in rats have been equivocal (Woolley, 1998). However, hippocampal LTP varies across the estrous cycle in association with these changes in neuronal connectivity (Warren et al., 1995). Hippocampal LTP is facilitated by estrogen treatment in awake rats (Cordoba-Montoya and Carrer, 1997). Estrogens also affect basal forebrain cholinergic neurons that might be important in passive avoidance and attentional tasks as well as in spatial learning (Gibbs, 1997). Prenatal gonadal steroids also affect hippocampal cell morphology: males have larger CA1 and CA3 pyramidal cell field volumes and cell body sizes than females (Isgor and Sengelaub, 1998).

Interestingly, estradiol does not increase CA1 dendritic spine density in female mice, as it does in female rats (Li et al., 2004). However, estradiol does increase the growth and maturation of CA1 dendritic spines in female mice and improves their spatial memory (Figure 12.20) (Li et al., 2004). Examination of estrogen receptor knockout mice revealed that the β subtype of the estrogen receptor is necessary for optimal learning and memory performance in a spatial task (the Morris water maze) (Rissman et al., 2002). Ovariectomized βERKO mice given high doses of estrogen replacement therapy failed to learn the task, whereas ovariectomized βERKO mice treated with low doses of estrogens were significantly delayed in learning the water maze. Wild-type female mice, regardless of estrogen treatment, learned the task (Rissman et al., 2002). A further link between estrogen and spatial memory in mice comes from studies in the water maze showing that the onset of

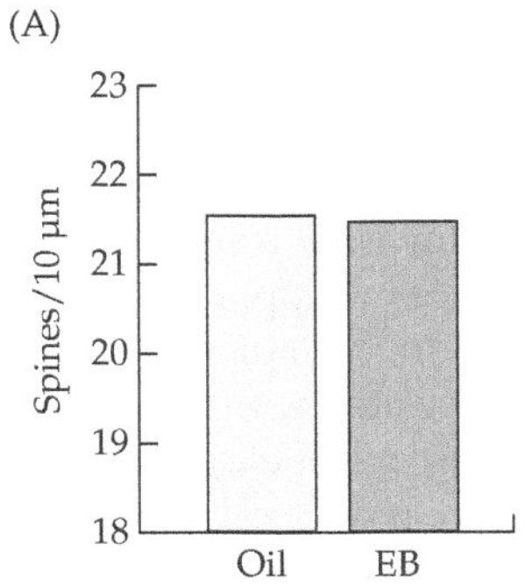

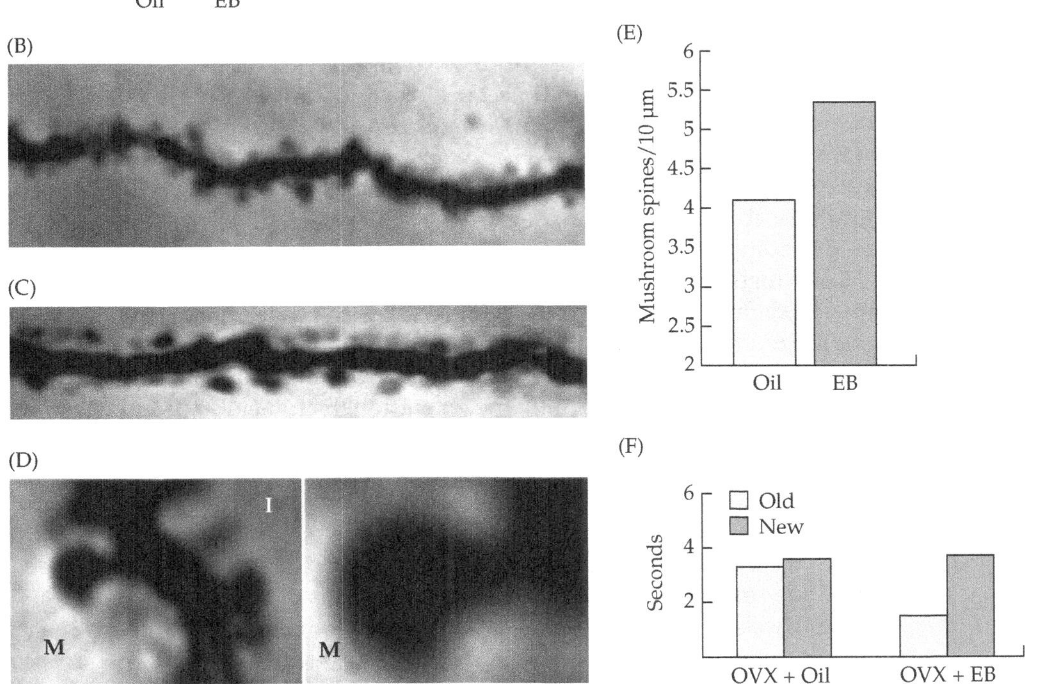

12.20 Estrogens influence CA1 dendritic spine function and spatial memory. Treatment with estradiol benzoate does not increase the overall density (A, B, C) of dendritic spines in mice. This treatment does facilitate spine maturation, however, as indicated by an increase in mushroom-shaped spines (part E, marked by M in part D). Spatial memory for object memory for new places in ovariectomized (OVX) mice improves with the same dose of estradiol benzoate (EB; part F). From Li et al., 2004.

the decline in spatial memory associated with aging occurs earlier in females than in males (and is associated with age-related hormone loss) (Frick et al., 2000), and that estradiol benzoate can significantly improve spatial memory and alter synaptic plasticity in aged female mice (Frick et al., 2002). These latter data, and similar findings, may have important implications for the design of estrogen replacement therapy.

Estrogen treatment of postmenopausal women has been reported to enhance verbal memory and maintain the ability to learn new material (Sherwin, 1996, 1997, 1998). In contrast, postmenopausal women participating in a short-term estrogen replacement therapy study with a randomized, double-blind crossover design showed no significant effects of estrogen treatment on memory (Polo-Kantola et al., 1998). The differences between the results of these studies may reflect differences in the memory tasks assessed. In another study of humans, male-to-female transsexuals receiving estrogen treatment prior to surgical sex reassignment did not differ from women on any tests of learning and memory (Miles et al., 1998). Few studies have examined the extent to which the decline in cognitive abilities and hormonal changes in aging men are related. However, what few data exist seem to support the idea that estrogens may play a role in supporting memory in aging men, as in women (Sherwin, 2003). Estrogen administration has been shown to enhance memory and reduce neuronal losses associated with Alzheimer disease in postmenopausal women in the short term (Simpkins et al., 1997), and to reduce the damage caused by blood reperfusion (restoration of blood flow) after cerebral ischemia (stroke) (Hurn et al., 1995). Estrogen replacement therapy in postmenopausal women, however, seems to provide inconsistent benefits to cognitive function (Maki and Hogervorst, 2003). Meta-analyses suggest that estrogen replacement protects against the onset of Alzheimer disease. However, studies of randomized trials of estrogen hormone replacement suggest that estrogens do not provide long-term benefits to women who are already stricken with Alzheimer disease (Maki and Hogervorst, 2003). There are not sufficient reliable data to provide advice about whether or not estrogen replacement therapy should be prescribed prophylactically to postmenopausal women.

Taken together, the evidence suggests that the effects of estrogens on memory may be slight and may affect only certain aspects of spatial memory. It has been hypothesized that the effects of estrogens on memory in females might reflect aspects of maternal behavior, rather than nonspecific effects on learning and memory (Woolley, 1998). Furthermore, the extent to which memory is enhanced by estrogen replacement therapy may depend on environmental factors such as cognitive and spatial complexity (Gresack and Frick, 2004). A study of ovariectomized female mice given estradiol or a control vehicle examined the effects of long-term exposure to an enriched (complex) environment. Mice were raised from 3 weeks to 6 months of age in standard (housed with cagemates but not exposed to enriching stimuli) or enriched (housed with cagemates but exposed to toys and running wheels for 3 hours per day) laboratory environments. At 6 months of age, estrogens improved spatial and nonspatial memory only in the mice exposed to the standard, impoverished environment; estrogens did not improve memory in the mice exposed to the enriched conditions (Gresack and Frick, 2004). The authors of this study propose that the changes induced by estrogens and by environmental enrichment may be associated with similar changes in hippocampal synaptic plasticity (Gresack and Frick, 2004).

Importantly, stress can interfere with the effects of estrogens. For instance, water maze training in female rats completely eliminated the typical estrogen-induced dendritic spine density increases observed in the CA1 region of the hippocampus (Frick et al., 2004).

Effects of Androgens

The effects of androgens on learning and memory have been studied for over 70 years (Tuttle and Dykshorn, 1928). In one of the best early experiments, male rats were castrated at 20, 50, 90, 130, or 170 days of age and later tested for maze-learning ability (Commins, 1932). Castrated and intact animals learned the mazes equally well. An extensive review of these early studies (Stone and Commins, 1936), including the reviewers' own work, in which castration failed to affect performance on a light discrimination task or on three different mazes, led to the conclusion in 1936 that castration does not affect learning in rats.

Many studies since then have investigated the role of testicular androgens in learning and memory, and the results of these studies generally agree with the early pronouncement that gonadal androgens do not affect learning and memory. This appears to be true of humans as well as nonhuman animals. For example, neither testosterone replacement therapy in hypogonadal men nor testosterone treatment of normal men resulted in changes in learning and memory performance (Alexander et al., 1998). In some cases, however, castration has small or subtle effects on learning, especially spatial learning or learning that affects anxiety levels. However, treatment of female rats with testosterone prior to 10 days of age masculinizes the hippocampus and improves spatial memory (Roof and Havens, 1992).

As we have seen, female rats generally outperform males in the acquisition of active avoidance tasks, although the difference is small (Haaren et al., 1990). In the laboratory strains of rats used in these studies, neither neonatal nor postpubertal castration affects performance. However, neonatal castration of males treated in utero with the anti-androgen cyproterone acetate raises their performance to levels typical of females, demonstrating that prepubertal organization by sex steroid hormones must be responsible for the difference in performance (Beatty, 1979). Similarly, it has been hypothesized that early organizational effects of androgens account for sexually dimorphic responses to classical conditioning of fearful stimuli (Anagnostaras et al., 1998).

Studies on seasonally breeding animals have documented a correlation between peak androgen concentrations and peak learning abilities. Goldfish trained in several tasks, including swimming with a tethered float, conditioned avoidance tasks, and maze learning, exhibit seasonal changes in learning ability (Agranoff and Davis, 1968; Shashoua, 1973). Maximal learning occurs between January and March, prior to spawning, when the gonads are developing and blood concentrations of steroid hormones are high. Poor learning is observed during the summer, after the spawning season. Photoperiod may regulate this seasonal cycle of learning ability in goldfish.

A sex difference in spatial learning performance has been reported for meadow voles (*Microtus pennsylvanicus*) and deer mice (*Peromyscus maniculatus*). During the breeding season, males of both species outperform females, but the sex difference disappears when the animals are not in breeding condition (Galea et al., 1996). In female voles, there was a significant correlation between plasma estradiol concentrations and retention, whereas there were no performance differences between males with low and high testosterone concentrations (Galea et al., 1995).

As described in Chapter 10, seasonal changes in brain size in several species of rodents are likely to be mediated by seasonal changes in androgen levels. Changes in learning and memory should correspond to these seasonal changes in brain mass, but this proposition essentially remains untested. In North America, over 95% of learning studies are performed on laboratory rats and mice (Burkhardt, 1987), which are not particularly seasonal.

There is a seasonal change in maze-learning ability in voles. Home range size shrinks during the winter in both meadow (*Microtus pennsylvanicus*) and prairie voles (*M. ochrogaster*). In one study, winter-trapped voles were compared with spring-trapped voles on their ability to learn mazes (Gaulin and FitzGerald, 1989). Males captured and tested during the winter made more errors while learning the mazes, and required more trials to negotiate the mazes without making any errors, than males tested during the spring. These results may reflect seasonal changes in brain size; brain weights are proportionally heavier in spring-trapped than in winter-trapped voles (Gaulin and FitzGerald, 1989; Yaskin, 1984). The difference in maze-learning abilities may also reflect differences in motivation between winter- and summer-trapped animals.

Testosterone does seem to have some positive reinforcing properties. A place preference develops in rats that have testosterone injected into the nucleus accumbens (Packard et al., 1997). If something "positive" occurs in a particular place, an individual will spend more time in that place in the future. For example, if a rat is given an injection of morphine in the right half of its cage, it develops a place preference and spends more time in the right half of the cage. Thus, one can determine whether a substance has positive reinforcing properties by observing whether or not a place preference develops. The rewarding properties of testosterone might help to explain some of the abuse associated with anabolic androgens (see Chapter 13).

The Effects of Peptide Hormones in Learning and Memory

There is convincing evidence that many peptides previously characterized as hormones also function as neurotransmitters in the central nervous system, including vasopressin, oxytocin, the opioids, and cholecystokinin (CCK). Of course, epinephrine and norepinephrine have been known for some time to function as both hormones and neurotransmitters. When there is evidence for a central site of action, as in the case of epinephrine, ACTH, vasopressin, the opioids, and CCK, one must consider the hypothesis that these chemical messengers exert their effects centrally by acting as neurotransmitters or neurohormones, rather than as hormones. This observation may account for the ability of these compounds, which cannot cross the blood–brain barrier, to act in the brain.

Adrenocorticotropic Hormone (ACTH)

In 1950, Hans Selye introduced the concept of the general adaptation syndrome as a model for the physiological coping strategies of an animal enduring stress (see Chapter 11). According to this model, stress causes the release of corticotropin-releasing hormone from the hypothalamus, which stimulates the secretion of ACTH from the anterior pituitary; ACTH, in turn, induces the production and release of glucocorticoids by the adrenal cortex. These hormones, along with epinephrine and norepinephrine, presumably prepare an animal to react to threatening stimuli—the so-called "fight or flight" response. In his early work, Selye assumed that the "stress" hormones had a relatively permanent effect on the central nervous system (i.e., learning), as well as affecting the immediate ability to respond (Selye, 1956). Many prominent learning theorists of the mid-1950s attempted to explain avoidance learning in terms of an anxiety reduction model. They hypothesized that an interruption in the "anxiety mechanisms" (i.e., the HPA axis) would disrupt the acquisition of an avoidance response.

ACTH affects memory independently from its effects on glucocorticoid secretion. In one early study, two groups of rats were subjected to a pole-jump test of active avoidance learning (Figure 12.21). The pituitary gland had been surgically removed in one group of rats; the control group had received a sham operation. (Recall that the pituitary is also called the hypophysis; the removal of this gland, therefore, is often called a **hypophysectomy**.) The control rats acquired the avoidance task much faster than the hypophysectomized animals (Applezweig and Baudry, 1955; Applezweig and Moeller, 1959). Injections of ACTH into the hypophysectomized animals restored their performance to the levels of the control animals. It is important to note that ACTH restores performance only in endocrine-deficient animals; ACTH treatment of intact animals does not enhance

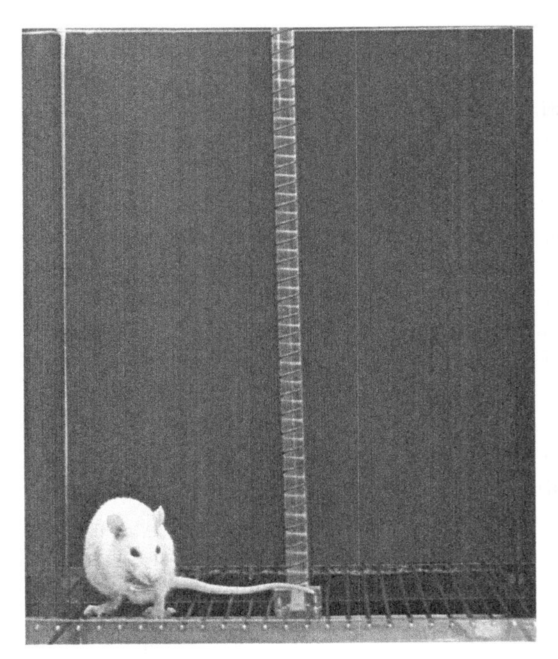

12.21 A test of active avoidance learning. In this "pole-jump" apparatus, a rat must jump on the pole to avoid a foot-shock when the light is illuminated. Courtesy of Organon International BV, Holland.

normal learning and memory performance (Murphy and Miller, 1955). Consequently, this peptide is not a cognitive enhancer (see Table 12.2).

In general, a deficit in ACTH leads to slower acquisition and poorer retention of information. Remember, however, that the surgical removal of the pituitary gland is not a minor physiological disruption. Hypophysectomy not only reduces ACTH levels, but affects many other endocrine and regulatory systems as well (see Chapter 2). DeWied (1964) tried to control for these problems by providing hormone "cocktails" to hypophysectomized rats. The addition of growth hormone and thyroid-stimulating hormone had restorative effects; that is, the animals were much healthier. Without the replacement of ACTH, however, the memory deficits were still observed.

Does ACTH act through the CNS or via the adrenal glands to affect memory? The best way to answer this question is to remove the adrenal glands. ACTH appears to act on the CNS through a nonadrenal route because it enhances acquisition of avoidance tasks even in rats that have been adrenalectomized. Because adrenalectomized animals lack feedback control, their ACTH levels rise. Furthermore, ACTH injections in rats lacking both pituitary and adrenal glands also restore learning. Therefore, the effect of ACTH on memory is independent of its classic endocrine effects (DeWied, 1974). It is unclear how ACTH reaches the brain to affect memory processes. ACTH consists of 39 amino acids and therefore, in common with epinephrine, is too large to cross the blood–brain barrier.

In addition to ACTH, small pieces of the ACTH molecule, devoid of any effect on the adrenal glands, also affect learning (for example, DeWied, 1974, 1977). The first hint of this possibility was evident in a number of studies in which the administration of α-MSH (melanocyte-stimulating hormone) and β-MSH restored avoidance responses in hypophysectomized rats (Kastin et al., 1975). Recall that ACTH is made from pro-opiomelanocortin (POMC), a giant molecule that is also the precursor of several other hormones, including α-MSH and β-MSH as well as the endorphins and enkephalins. The α-MSH and β-MSH molecules share the same sequence of amino acids with ACTH at positions 1–13 (see Figure 2.22). Treatment with smaller ACTH fragments, such as $ACTH_{1-10}$ or $ACTH_{4-10}$, also restored learning (DeWied, 1969, 1974). These fragments do not affect adrenocortical function. However, administration of $ACTH_{11-24}$ or $ACTH_{25-39}$ had no effect on avoidance learning (DeWied, 1974). In another series of studies, amino acids were systematically removed from both ends of the ACTH molecule; each ACTH fragment was then tested in vivo to determine its behavioral "potency" (DeWied, 1974, 1980). Through a long series of studies testing various bits and pieces of the ACTH molecule, it was determined that the behaviorally active portion of the ACTH molecule is $ACTH_{4-8}$, which resides at the amino end of the molecule. Artificial analogues of the ACTH molecule have also been produced and examined for behavioral effects. ACTH analogues with a D-isomer of phenylalanine at position 7 of the amino acid sequence exhibit behavioral effects that are opposite to those of comparable ACTH analogues with the normal L-isomer of phenylalanine at position 7: the D-isomer analogue facilitates extinction (i.e., decay of active avoidance responding), while the L-isomer form delays extinction (DeWied, 1980). This kind of dissection of a behavioral

non-opioid hormones. Rats also show place preferences in response to systemic injections of morphine (Amalric et al., 1987). This effect is also blocked by injections of naloxone.

Infant rats also experience the positively reinforcing properties of exogenous opioids. Young rats, like many animals, are *neophobic;* that is, they normally avoid novel experiences. In one experiment, 5-day-old rat pups were isolated from their mother and siblings, placed in the presence of a novel odor (orange extract) for 30 minutes, then injected either with a low dose of morphine (0.5 mg/kg) or with saline. For the next five days, these rats were tested to determine their responses to another presentation of the orange odor. The pups were placed in a small cage with plain bedding chips under one half and bedding chips mixed with orange extract under the other. Testing lasted 10 minutes, and the percentage of time spent in each half of the cage was recorded (Figure 12.23). Rat pups that had experienced morphine injections after exposure to the novel odor of orange extract spent 73% of their time over the orange-scented bedding. In contrast, control animals (injected with saline) spent only 34% of their time over the orange odor. Pre-treatment with naltrexone, another opioid antagonist, before conditioning blocked the positive association between the morphine state and the orange odor (Weller and Blass, 1988).

A similar paradigm has been used to test the effects of other substances on learning and memory. When opioid receptor agonists, including morphine, β-endorphin, and enkephalin, were given to adult rats immediately after training in low doses, their memory was impaired in a dose- and time-dependent manner (McGaugh, 1989). Opioid antagonists ameliorate the memory-impairing effects of opioids and their agonists. Studies have demonstrated that retention is enhanced by post-training administration of opioid antagonists, including naloxone, naltrexone, diprenorphine, levallorphan, nalmenfene, and β-funaltrexamine (McGaugh, 1989). The memory-enhancing effects of these opioid antagonists are also dose- and time-dependent, and have been found in studies using several types of training tasks, including passive avoidance, active avoidance, habituation, and appetitive spatial learning (McGaugh, 1989). Opioid antagonists also block the amnestic effects of electroconvulsive shock therapy (ECT) (Collier et al., 1987), but have had mixed results in human patients with memory disorders (McGaugh, 1989).

The amnestic effects of post-training, peripherally administered met-enkephalin can be blocked by the centrally acting opioid antagonists naloxone and naltrexone, as well as by the peripherally acting antagonist MR 2263 (Zhang et al., 1987). The memory-modulating effects of both central and peripheral injections of met-enkephalin (and peripheral naloxone) are attenuated in adrenal-denervated (or demedullated) animals (Conte et al., 1986). Denervating the adrenal glands prevents the release of epinephrine and norepinephrine. Of course, removal of the adrenal medulla also eliminates the release of many endogenous enkephalins and endorphins (see Chapter 2). Amnesia can be produced in adrenal-demedullated rats if epinephrine is administered prior to training (Conte et al., 1986). Taken together, these results suggest that peripheral injections of opioids affect memory via epinephrine. Predictably, injections of met-enkephalin ele-

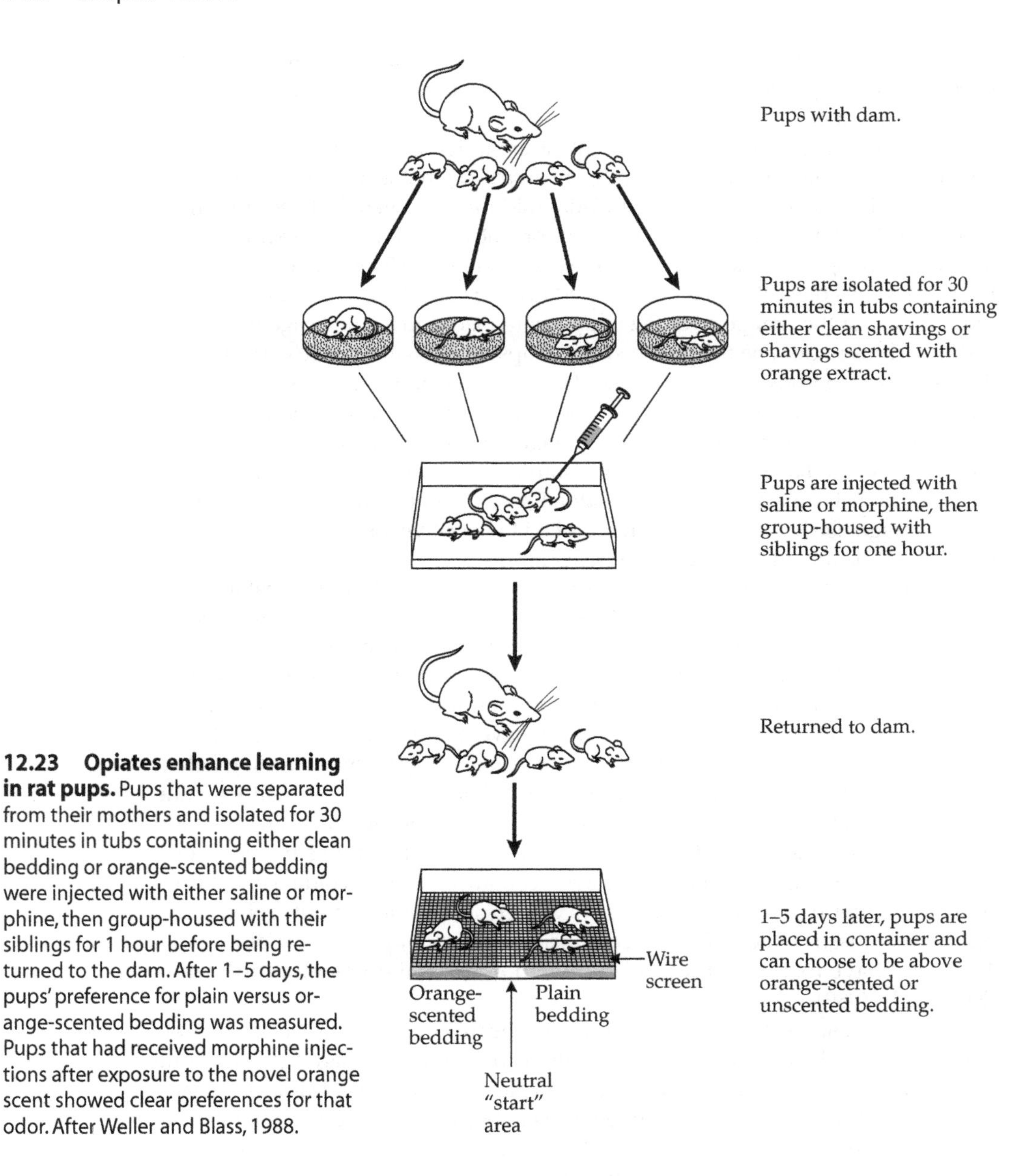

12.23 Opiates enhance learning in rat pups. Pups that were separated from their mothers and isolated for 30 minutes in tubs containing either clean bedding or orange-scented bedding were injected with either saline or morphine, then group-housed with their siblings for 1 hour before being returned to the dam. After 1–5 days, the pups' preference for plain versus orange-scented bedding was measured. Pups that had received morphine injections after exposure to the novel orange scent showed clear preferences for that odor. After Weller and Blass, 1988.

vate blood levels of glucose, probably by stimulating epinephrine secretion. This finding may seem contradictory to the results of previous studies indicating that epinephrine and glucose enhance memory. However, recall that the effects of epinephrine and glucose follow an inverted U-shaped curve; levels too high or too low reduce performance on learning and memory tasks. Only moderate levels of epinephrine and glucose enhance learning and memory.

Cholecystokinin

All animals need to find food in order to survive, so it should be advantageous for animals to remember where food has been found in the past. Therefore, one might suspect that hormones involved in feeding might modulate memory. At an ultimate level of explanation, it makes sense that hormones involved in feeding and digestion would be co-opted to enhance memory because the ability to remember the details of successful foraging behavior would have high survival value (Flood et al., 1987).

Cholecystokinin (CCK) is a gastrointestinal hormone that is released during feeding. CCK has been assigned several behavioral functions, including involvement with post-feeding satiety (Smith et al., 1981b). Injection of CCK inhibits feeding by rats via the vagus nerve (Bloom and Polak, 1981; Moran et al., 1992) (see Chapter 9). CCK has also been discovered in brain tissue.

CCK modulates learning and memory. In one study (Flood et al., 1987), two groups of hungry mice and one group of mice with free access to food were given a number of trials to learn to avoid an electric shock in one arm of a T-maze (Figure 12.24). The mice with free access to food and the mice in one of the food-restricted groups were given access to food immediately after training. The free-access mice tended to eat very little at this time, but the hungry mice feasted. Mice in the

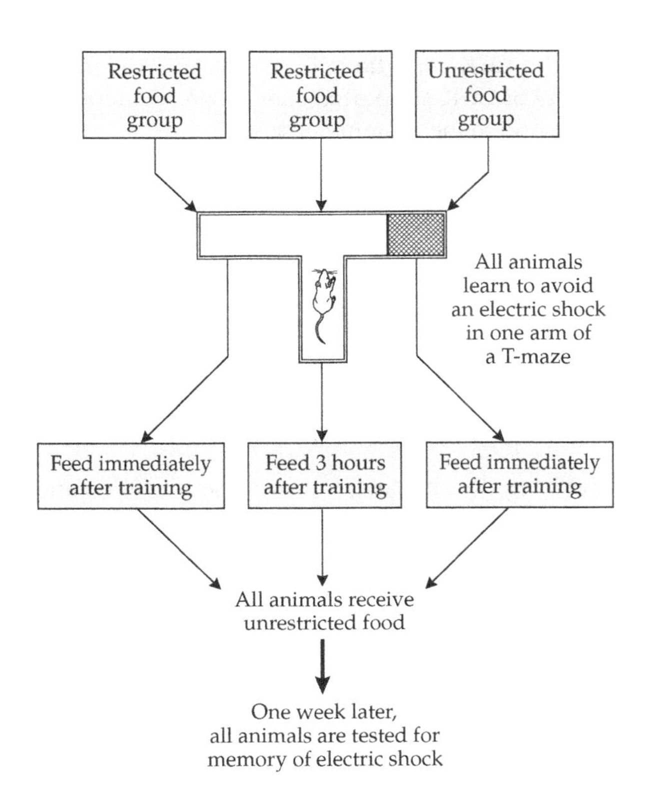

12.24 CCK enhances memory in mice. Mice from two food-restricted groups and from a group that had been fed ad libitum learned to avoid an electric shock in one arm of a T-maze. One of the food-restricted groups was fed immediately after training, and one was fed 3 hours after training; the freely fed group was fed immediately. When the animals were tested a week later, the food-restricted mice that were fed immediately after training showed the best retention, probably because the food caused these hungry mice to secrete CCK shortly after training. After Flood et al., 1987.

other food-restricted group received food 3 hours after training. During the next week, food was provided ad libitum to all the animals. At the end of the week, all of the mice were tested again in the T-maze. Access to food immediately after training enhanced memory retention for the aversive experience in the hungry mice, as compared with both the hungry mice for which feeding was delayed by 3 hours and the mice that received ad libitum. Presumably, the hungry mice that were fed immediately after training released CCK in response to their meal, which enhanced their memory of the shock. CCK injected intraperitoneally also enhances memory in mice. The memory-enhancing effects of CCK can be blocked by cutting the vagus nerve, which suggests that peripherally administered CCK produces its memory effects by activating ascending vagal fibers (Flood et al., 1987).

CCK also affects learning and memory functions in young rats. The normal aversion of young rats to a novel odor can be blunted by pairing such an odor (e.g., orange extract) with injections of CCK in a paradigm similar to that used to test the effects of morphine (see above; Weller and Blass, 1988). In one such study, rat pups preferred an orange-scented test arena 24 hours after the CCK injections, whereas pups injected with saline still avoided the orange scent. An antagonist (L-364,718) that selectively blocks peripheral CCK receptors completely eliminated CCK-induced conditioned odor preferences. These results strongly support the hypothesis that CCK affects learning in rats via a peripheral site of action. CCK "conditioning" follows an interesting developmental course: CCK-induced olfactory preferences were observed in rats that were 5, 11, and 22 days of age, but 28-day-old rats failed to exhibit CCK-conditioned preferences (Weller and Blass, 1990). Perhaps the ontogeny of CCK-supported conditioned preferences reflects natural events that correspond to the weaning process.

Considered together, the effects of hormones on learning and memory are usually subtle and often specific to a given situation or task. An understanding of the exact mechanisms by which hormones affect learning and memory function awaits additional advances in the understanding of the neurobiological mechanisms of learning and memory. Hormones may change structures in the brain, the activity of specific neurons, the uptake or release of neurotransmitters, or act in a coordinated way through all of these mechanisms.

Summary

- Several types of learning have been documented, including sensitization and habituation, as well as associative learning (classical and operant condition, appetitive and avoidance conditioning). There are also several types of memory including short-term memory (working memory) (spatial and nonspatial), long-term memory (reference memory) (spatial and nonspatial), procedural (implicit) memory including skill learning, priming, declarative (explicit), semantic (facts), and episodic (events). Hormones affect each type of learning and memory.

- The hormones that are associated with adaptation to stress enhance learning and memory.
- There are two hypotheses regarding how epinephrine affects memory: (1) it may affect memory via its effect on blood glucose levels, and (2) it may activate peripheral receptors that directly influence brain function.
- Elevated blood glucose levels facilitate the movement of glucose into neurons. More glucose entering neurons stimulates the release of more acetylcholine into the neuronal synapses.
- Elevated glucose levels may be a common pathway by which several hormones enhance memory. Hormones may increase blood sugar levels directly (e.g., epinephrine) or act through another agent (e.g., vasopressin) to increase blood sugar levels.
- Although acute glucocorticoids or stressors enhance learning and memory, chronically elevated glucocorticoids or chronic stressors impair spatial learning and memory. However, if glucocorticoid concentrations are too low, impairments in memory are also observed. Glucocorticoids appear to exert their effects on learning by affecting neuronal structure or function in the hippocampus or amygdala.
- Sex differences in learning and memory performance are common; however, these differences may be due to sex differences in spatial abilities, exploratory behaviors, or general anxiety levels. Acute stressors tend to improve memory in males, but impair memory in females. Chronic stress disrupts memory function in males, but the effects of chronic stress on female memory are not pronounced.
- Estrogens appear to enhance spatial learning in animals and other forms of memory in humans.
- In rats, androgens do not have major effects on learning and memory. In some species, enhanced learning and memory are reported during the breeding season, when blood concentrations of androgens are high.
- Many peptides previously characterized as hormones also function as neurotransmitters in the central nervous system, including vasopressin, oxytocin, the opioids, and cholecystokinin. Epinephrine and norepinephrine have been known for some time to function in both roles. When there is evidence for a central site of action of hormones on memory, as in the case of epinephrine, ACTH, vasopressin, the opioids, or CCK, one must consider the hypothesis that these chemical messengers exert their effects centrally by acting as neurotransmitters, rather than as hormones.
- The lack of ACTH impairs memory formation, and its presence can also protect against amnesia. A small piece of the ACTH molecule, devoid of endocrine activity, contains the features of the molecule that affect memory. The physiological mechanisms underlying the effects of ACTH on memory are unknown.

- Vasopressin also blocks amnesia and appears to act via its stimulatory effect on epinephrine secretion.
- Oxytocin may act as an amnestic agent. The mechanisms of its action are not known.
- Endogenous opioids seem to have amnestic properties in avoidance learning paradigms, but appear to enhance rewarding properties in trial-and-error learning tasks.
- CCK enhances memory, but requires an intact vagus nerve to act.

Questions for Discussion

1. Nearly all of the research on the effects of hormones on learning and memory has been conducted on laboratory rats and mice. Discuss the implications and limitations of this approach to studying learning.
2. In Chapter 10, it was established that hormone levels vary on a circadian basis. It was also documented that performance on a variety of cognitive skills varies as a function of time of day. In most studies on the effects of hormones, learning, and memory, rodents are tested during the experimenters' work day (the rodents' rest period). Discuss the proposition that hormones affect learning and memory by shifting circadian rhythms of performance.
3. Glucagon, secreted from the α-cells of the pancreas, acts in opposition to insulin and elevates blood sugar levels. Design a study to test the effects of glucagon on learning and memory. Discuss your predicted outcomes.
4. Opioids often act as reinforcers, or rewards, for behavior and consequently have marked effects on learning and memory. Discuss the possibility of other hormones acting as reinforcers and their potential influences on learning and memory. What are these hormones, and under what conditions would you expect them to exert effects on behavior?

> Refer to the accompanying Student CD for additional resources, including Web links, videos, animations, and additional photos.

Suggested Readings

Bowman, R. E., Beck, K. D., and Luine, V. N. 2003. Chronic stress effects on memory: Sex differences in performance and monoamines. *Hormones and Behavior*, 43: 48–59.

DeWied, D., Diamant, M., and Fodor, M. 1993. Central nervous system effects of the neurohypophyseal hormones and related peptides. *Frontiers in Neuroendocrinology*, 14: 251–302.

are legendary" (*Washingtonian*, April 1992, 67). Most North American adults understand this description to mean that the woman in question is possibly aggressive, certainly assertive, and probably cranky. These traits are often tolerated or even applauded in men. Does the assignment of these character traits to PMS trivialize the accomplishments of powerful women?

PMS must be considered within a social context because it affects only one sex and because its image is not positive. Feminist social scientists have noted that several cultural phenomena have interacted to establish PMS as a biologically based illness that affects only women and to make it a part of the social fabric (Brown-Parlee, 1990, 1991). For example, the increased awareness of PMS in North America has coincided with an increase of feminist ideas in public forums. A discussion of the meaning of feminism is beyond the scope of this book, but one of its central tenets is that equal opportunities for economic and political power should exist for women and men. This demand is reasonable in the North American political system because justice demands equal treatment for individuals who are equal. In theory, women should not have difficulty in making their case for equality. If, however, men and women are not biologically equivalent, then making an argument for equal economic and political opportunities for men and women is more difficult (Young, 1990). Some feminists have argued that the "pathologization" of normal menstruation is an attempt on the part of the male-dominated biomedical community to demonstrate that women and men are unequal in ways that disqualify women from holding positions of economic and political power.

Certainly, the biomedical community is not the objective, culture-free, unbiased fellowship of truth seekers that is often portrayed as its ideal. During the first feminist movement early in the 20th century, scientific evidence of the reproductive frailties of women provided by the medical community was used to support the view that women should not participate in higher education, vote, or drive automobiles. Others argue that many women actively participated in the "conspiracy" to medicalize PMS, apparently embracing it for the advantages of the "sick" role that accompanies it, especially the exemption from accountability for behaviors and feelings for which they would normally be held responsible. According to this argument, some women diagnosed with PMS have thereby been empowered by physicians to be "out of control" and have received subtle cultural acceptance for expressions of aggression and power that are typically unacceptable. The extreme version of this view is evident when PMS is used as a defense against criminal charges. In France, for instance, women defending themselves in court can plead a special form of temporary insanity if their crimes were committed during the late luteal phase of their menstrual cycle. In North America, PMS is not acceptable as a legal defense.

The feminist position does not question the reality of PMS, but it does question the existence of biological mechanisms underlying PMS, as well as emphasizing the potential limitations that the acceptance of PMS as a biologically based illness places on women, who are thus assumed to be subject to "raging hormones," "out of control," or "dysfunctional." From a social perspective, the important question is not whether women fluctuate in their moods or even in their levels of aggressive behavior, but rather whether their functioning is impaired by these monthly fluctuations. The vast majority of research from the social sciences has not detected any significant impairments of this sort (Ussher, 1989).

What Is PMS?

A consideration of PMS from a biomedical perspective reveals several problems with the characterization of this phenomenon. Consequently, understanding the relationship between hormones and the mood changes associated with PMS is difficult. The first issue that is often raised in biomedical discussions of PMS is whether PMS is a real phenomenon. Before this issue can be directly addressed, a distinction between *illness* and *disease* needs to be made. PMS is not a disease; it is a response to a natural cycle, and there is no known underlying pathology. However, a woman suffering from symptoms associated with menstruation can visit a physician, receive a diagnosis of PMS, get treatment, and report amelioration of the symptoms (Rapkin, 2003). This scenario suggests that PMS is real and should be considered an illness. One interesting issue, the significance of which remains unresolved, is the fact that approximately 50% of women that receive placebo treatments for their PMS symptoms report improvement in their symptoms. There are few other illnesses for which this is true. Of course, there is a placebo response for other symptoms not associated with PMS: approximately 30% of both men and women respond to placebo treatments for many symptoms. With PMS, there is reason to believe that the placebo effect wears off after 3 months; however, few clinical investigations of PMS have extended beyond 3 months. As you might expect, correlating hormonal changes with symptoms of PMS is challenging under these circumstances.

If the reality of PMS is accepted, several other problems remain to be addressed. First, estimates of the prevalence of PMS vary widely, despite the many cross-cultural studies that have indicated that some women in all societies examined report perimenstrual symptoms. The frequency of perimenstrual symptoms reported in different studies varies between 20% and 90% of North American and European women. A meta-analysis of all published reports yields a prevalence rate of about 45% of all women. Most of these women rate their symptoms as mild; however, some women are debilitated by PMS. The frequency of severe PMS symptoms ranges between 2% and 10% (Logue and Moos, 1986). It is generally accepted that about 3%–5% of women experience PMS symptoms to a degree that interferes with their usual functioning. This variation in prevalence rates is due in part to differences in the criteria used to define the syndrome (Endicott et al., 1981). The prevalence estimates reflect the types of symptoms included in each study (for example, psychological versus physiological) as well as the severity of symptoms included (Table 13.1).

A corollary of this problem is that all of the PMS data are based on self-reports. In some studies, the self-reports are retrospective; in others, they are prospective, or concurrent with the symptoms. In no other area of behavioral endocrinology is the self-report, or introspective, method the primary method used to generate behav-

TABLE 13.1 **Percentages of women who report perimenstrual symptoms in two tests designed to assess PMS**

Symptom	Test: Menstrual Distress Questionnaire[a]	Test: Premenstrual Assessment Form[b]
Breast pain	35	84
Weight gain	34	83
Swelling	36	77
Backaches	24	74
Sadness	43	76
Anxiety	30	70
Skin blemishes	34	69
Dizziness	5	25

Source: Logue and Moos, 1986.
Note: Tests were administered during the perimenstrual period.
[a]Severity ratings from "mild" to "severe"
[b]Severity ratings from "slight" to "extreme"

ioral data to correlate with hormonal data. Obviously, self-reports of PMS symptoms can be influenced by many factors; for example, simply being asked about the symptoms is likely to increase their perceived severity (Brown-Parlee, 1991). Women are also likely to rate their symptoms as more severe when asked about their perimenstrual period after the fact than during the experience.

A second problem with PMS is that neither the premenstrual nor menstrual period is well defined. When should symptoms be considered part of PMS? If they appear 3 days prior to menstruation? If they appear 5 days prior to menstruation? If they appear 10 days prior to menstruation? Furthermore, how long after the onset of menses should symptoms be included? Menstruation may continue for 3 days in some cases or 8 days in others. Is a menstrual period of 3 days equivalent to an 8-day period? Because there is no consensus about which data should be included, there is substantial difficulty in comparing PMS studies with one another.

A third problem with PMS is the lack of consensus concerning its symptoms. PMS was originally described in 1931 as "premenstrual tension." Symptoms of this condition included severe psychological tension, weight gain, headaches, and edema (swelling) that occurred 7–10 days prior to the onset of menses (Frank, 1931). Because these symptoms may also occur in postmenopausal women, as well as in men and children, diagnosis of PMS requires the additional criterion of regular recurrence. Thus, symptoms must manifest themselves on a cyclic basis and end with the onset or offset of menstruation. The current diagnostic criteria for the so-called Late Luteal Phase Dysphoric Disorder, as described in the Diagnostic and Statistical Manual of Mental Disorders, Fourth Edition (DSM-IV), published by the American Psychiatric Association in 1995, can be found in Box 13.1. At least five of these symptoms must be present for a diagnosis of this disor-

BOX 13.1 Diagnostic Criteria for Late Luteal Phase Dysphoric Disorder

A. In most menstrual cycles during the past year, symptoms in B occurred during the last week of the luteal phase and remitted within a few days after onset of the follicular phase. In menstruating females, these phases correspond to the week before, and a few days after, the onset of menses. (In nonmenstruating females who have had a hysterectomy, the timing of luteal and follicular phases may require measurement of circulating reproductive hormones.)

B. At least five of the following symptoms have been present for most of the time during each symptomatic late luteal phase, at least one of the symptoms being either (1), (2), (3), or (4):
   (1) marked affective lability, e.g., feeling suddenly sad, tearful, irritable, or angry
   (2) persistent and marked anger or irritability
   (3) marked anxiety, tension, feelings of being "keyed up," or "on edge"
   (4) markedly depressed mood, feelings of hopelessness, or self-deprecating thoughts
   (5) decreased interest in usual activities, e.g., work, friends, hobbies
   (6) easy fatigability or marked lack of energy
   (7) subjective sense of difficulty in concentrating
   (8) marked change in appetite, overeating, or specific food cravings
   (9) hypersomnia or insomnia
   (10) other physical symptoms, such as breast tenderness or swelling, headaches, joint or muscle pain, a sensation of "bloating," weight gain

C. The disturbance seriously interferes with work or with usual activities or relationships with others.

D. The disturbance is not merely an exacerbation of the symptoms of another disorder, such as Major Depression, Panic Disorder, Dysthymia, or a Personality Disorder (although it may be superimposed on any of these disorders).

E. Criteria A, B, C, and D are confirmed by prospective daily self-ratings during at least two symptomatic cycles. (The diagnosis may be made provisionally prior to this confirmation.)

Source: DSM-IV, 1995.

der (Freeman, 2003). With these strict criteria, the prevalence of the disorder drops to between 5% and 10% of women of reproductive age (Halbreich et al., 2003). The most recent version of the WHO International Classification of Diseases (1987) lists PMS as a gynecological, rather than a psychiatric, disorder. According to the WHO criteria, only one major distressing symptom must be present for a diagnosis of PMS (Freeman, 2003). According to the American College of Obstetricians and Gynecologists, a diagnosis of PMS requires at least one mood symptom and one physical symptom (Rapkin, 2003). Women with the more severe premenstrual dysphoric disorder tend to function worse in life than the community norm, and they show social dysfunction similar to that of women with major depression (Figure 13.1).

Within the past 15 years, the recognized symptoms of PMS have included anxiety, sadness, irritability, bloating, breast enlargement, dysmenorrhea (painful menstruation in the absence of identified pelvic pathology), increased appetite,

report an increase in sexual interest during the perimenstrual phase (Halbreich et al., 1982; Logue and Moos, 1988; Taylor, 1979). A heightened sex drive may represent reduced concern about unwanted pregnancies during the perimenstrual phase. However, women taking oral contraceptives also report an increase in sexual activity premenstrually, as well as during the time of ovulation (Englander-Golden et al., 1980; see Chapter 6).

Obviously, both endocrine and cognitive factors interact to influence PMS. As we have seen, the symptoms of PMS can be affected by social expectations, and can be improved by placebo treatments (Koeske, 1980; Parlee, 1982; Ruble, 1977). These facts, along with the lack of consistent endocrine correlates of PMS symptoms, have led some researchers to suggest that these symptoms are "all in the head" of the sufferers. However, the interaction of complex endocrine signals with cognitive factors is not unique to PMS. By analogy, recall that rat maternal behavior can be elicited after many days of exposure to pups, immediately after parturition, or after several days of specifically timed hormonal treatments involving at least three different hormones (see Chapter 7). In the latter case, the endocrine manipulations must still be combined with several days of exposure to pups prior to the onset of maternal behavior. Adults can overcome their fear of pups and begin acting maternally by means of frequent exposure to pups, the endocrine environment of pregnancy, or some combination of hormones and pup exposure. In other words, there is an interaction between cognitive (reduction in fear) and endocrine factors (the hormones that reduce the processing of fear-inducing chemical stimuli). Further research will have to focus on this interaction, keeping in mind that various subtypes of PMS are probably mediated by different underlying causes. A more holistic, integrative approach to the study of PMS may be required before the physiological, behavioral, and social causes of PMS are understood.

Hormones and Depression

The symptoms of **depression** include reduced mood (profound sadness); low self-esteem and feelings of worthlessness; general fatigue; feelings of guilt; disturbances (usually reductions) in sleep, sex drive, and food intake; anger; absence of pleasure; and agitated or retarded motor symptoms (DSM-IV, 1995). Agitated motor symptoms include pacing or hand-wringing; retarded motor symptoms include slow body movements or speech. In severe cases, delusions and hallucinations may be present.

Patients exhibiting the symptoms of depression are generally suffering from "clinical" depression. The term *major depression* describes a depressed state characterized by very severe symptoms that last chronically for at least 2 weeks. *Dysthymia* is characterized by less severe symptoms that last much longer, at least 2 years. Both of these types of depressive disorders occur in the absence of other psychological or physical problems, and are considered *primary depression*. *Secondary depression* refers to depression that results from some other physiological or psychological dysfunction. Individuals who are depressed for a transient

Depression	Melancholia	Normal	Hypomania	Mania

13.3 Continuum of mood from depression to mania. People may display unusual mood elevation or mood depression in response to environmental conditions. When these mood extremes persist and interfere with normal function, they become clinically significant.

period of time because of some clearly identifiable stressor or environmental stimulus (for example, grieving the loss of a friend or relative) are generally labeled "normal," "reactive," "neurotic," "exogenous," or "justified" depressives, or are said to be suffering from bereavement reactions (Corsini, 1984). The bouts of depression in these individuals are usually short-lived and exhibit a benign outcome; that is, these individuals recover fully on their own without treatment.

Another distinction made among types of clinical depression is between bipolar and unipolar depression. *Bipolar depression* is depression in the presence of at least one episode of mania, or elevated mood, which often involves inordinate feelings of confidence, power, and creative energies. *Unipolar depression* is depressed affect (mood) in the absence of manic episodes. Many of the studies on depression cited below used experimental groups that included both unipolar and bipolar depressives. Because the physiological mechanisms underlying these two subsets of depression may differ, the lumping of these individuals into one experimental group may mask hormonal correlates of depression, even though individuals in both groups often respond to the same antidepressant drugs.

Depression is typically considered to be at one end of a mood continuum with mania at the other end (Figure 13.3). Depressed individuals vary in both the severity and the duration of their symptoms. The prevalence of depressive symptoms is determined by means of psychological tests such as the Hamilton rating scale and the Zung self-rating scale (Hamilton, 1960). Studies of the endocrine correlates of depression use scores on these tests to measure the effects of hormone treatments. A number of endocrine correlates have been associated with depression (Rubin et al., 2002).

Endocrine Correlates of Depression

THYROID HORMONES The hormones of the hypothalamic–pituitary–thyroid axis have been implicated in depression (see Figure 11.3) (Musselman and Nemeroff, 1996; Sauvage et al., 1998). Administration of thyrotropin-releasing hormone (TRH) stimulates the release of thyroid-stimulating hormone (TSH) from the anterior pituitary gland and subsequent hormone production by the thyroid gland. Administration of TRH to depressed individuals has been attempted in several studies. In one such study, five patients that achieved a certain composite score on psychological tests were considered to be depressed (three were diagnosed as unipolar and two were bipolar) (Kastin et al., 1972a). Depression was reduced in four of these five patients by TRH treatment, as indicated by their improved scores on subsequent tests. Amelioration of the depressive symptoms varied in duration from 3 hours to several weeks, and the latency of action of TRH varied from 1 to 72 hours. No improvement of the depressive symptoms was reported when the patients were treated with saline. All of the depressed patients showed a smaller than normal TSH elevation in response to the TRH stimulation, which is unusual in the absence of thyroid dysfunction. Furthermore, although protein-bound iodine levels, basal metabolic rate, and rate of radioactive iodine uptake—all indicators of thyroid function—are usually within the normal range in depressed patients, the thyroid response to TSH is significantly lower in depressed patients than in nondepressed individuals (Ehrensing et al., 1974; Takahashi et al., 1974), suggesting some subclinical endocrine malfunction.

The depressive symptoms of PMS have also been associated with thyroid function. In one study, the responses of TSH and prolactin concentrations to TRH administration were examined in women who reported PMS symptoms and in women who did not. Previous studies on depressed patients had found both blunted and enhanced TSH secretion in response to TRH administration. TRH had also been found to stimulate prolactin release in nondepressed women, but did not affect prolactin levels in depressed women (Roy-Byrne et al., 1987). TRH was given during both the follicular and luteal phases. There were no significant differences in basal or maximal elevations of TSH or prolactin in response to the treatment between women with and without PMS symptoms, and neither TSH nor prolactin values differed between the luteal and follicular phases. However, the women with PMS showed much greater variation in TSH response to TRH treatment than the control women; that is, sometimes TSH levels were augmented, but at other times they were reduced (Roy-Byrne et al., 1987). Women without PMS showed stable responses of TSH to TRH. Variable TSH response to TRH could be present in a subgroup of women that suffer depression as part of their PMS symptoms.

In addition to changes in the response of TSH to TRH, recent studies have shown that depressed patients often exhibit (1) a very high level of antibodies against the thyroid gland, (2) high TRH concentrations in the cerebrospinal fluid, and (3) enhancement of antidepressant efficacy by co-treatment with triiodothyronine (T_3) (Musselman and Nemeroff, 1996; Sauvage et al., 1998).

GROWTH HORMONE AND PROLACTIN Basal growth hormone (GH) concentrations have been reported to be in the normal range in depressed patients. However, impaired GH responses to insulin-induced hypoglycemia in depressed patients have been reported by many investigators (for example, Sachar et al., 1973). Depressed patients also display inadequate GH responses to serotonin stimulation compared with nondepressed patients. TRH, which does not affect GH concentrations in normal individuals, evokes an abnormal increase in GH in depressed patients (Maeda et al., 1975). Elevated blood plasma prolactin concentrations have also been reported in depressed patients (Sachar et al., 1973; Nicholas et al., 1998). Although the results of correlative studies linking GH and prolactin to depression are somewhat contradictory (for example, TRH increased prolactin concentrations in depressed patients in one study [Maeda et al., 1975]

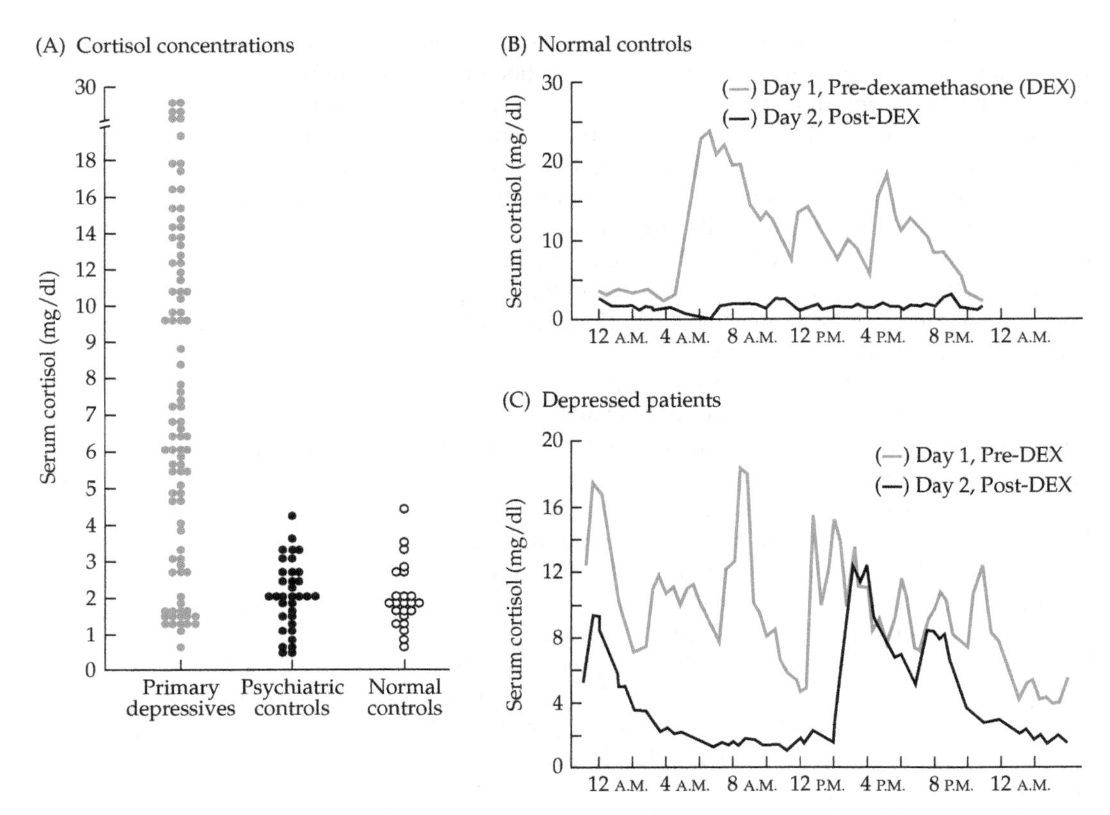

13.4 Cortisol secretion and depression. (A) Blood cortisol concentrations are often higher in people with primary depression than in nondepressed individuals or in people with other psychiatric disorders. (B) Dexamethasone inhibits the normal circadian pattern of cortisol secretion in nondepressed individuals. (C) Dexamethasone often fails to suppress cortisol secretion in depressed individuals. After Rosenzweig et al., 1999.

and decreased prolactin concentrations in another [Ehrensing et al., 1974]), they point to a fundamental difference in endocrine function between depressed patients and nondepressed individuals, and suggest differences in the physiological mechanisms underlying their endocrine feedback control systems (Dinan, 1998; Nicolas et al., 1998).

CORTISOL The negative feedback features of the hypothalamic–pituitary–adrenal (HPA) axis (see Figure 11.3) appear to be impaired in depressed patients. Excessive cortisol production has been reported in nearly 50% of depressed patients examined (Figure 13.4A) (Carroll, 1980). These increased serum cortisol concentrations do not appear to reflect the stress of coping with depression, because cortisol concentrations are at their highest 3–4 hours after sleep onset, when stress levels should be lowest, and decrease throughout the daylight hours, when stress levels are presumably highest (Carroll, 1980). Because cortisol is normally secreted in a pronounced circadian pattern, with peak concentrations measured in the early morning, this disturbance of the diurnal rhythm of cortisol secretion suggests an abnormal disinhibition of the neural centers regulating the release of adrenocorticotropic hormone (ACTH), the tropic hormone from the anterior pituitary gland that stimulates adrenal output.

The function of the HPA axis in depressed patients has also been assessed by means of the dexamethasone suppression test (Figure 13.4B,C). Administration of 1 to 2 milligrams of dexamethasone, an artificial steroid that mimics cortisol, at midnight normally suppresses blood plasma cortisol concentrations for 24 hours via a negative feedback mechanism (Cole et al., 2000). Dexamethasone failed to suppress cortisol production in 46% of depressed patients examined (Carroll et al., 1968). In most cases, plasma concentrations of cortisol were suppressed the morning after dexamethasone treatment, but increased shortly thereafter (Carroll et al., 1976). This inappropriate response to dexamethasone indicates a failure of the normal neural inhibiting mechanism for ACTH and cortisol secretion. It seems likely that a complex relation between the serotonergic system and the HPA axis (McAllister-Williams et al., 1998) underlies the hypersecretion of cortisol, the failure to respond normally to dexamethasone by suppressing cortisol production, and the disturbance of the circadian rhythm of cortisol secretion.

Cortisol normally counteracts the hypoglycemic effects of insulin; however, depressed patients do not show this response. High doses of insulin given during the diagnostic insulin tolerance test (ITT) suppress blood glucose levels, but glucose levels cannot be restored by cortisol in depressed patients (Carroll, 1980).

Taken together, findings in depressed patients of (1) high basal cortisol concentrations due to cortisol hypersecretion, (2) the failure to respond normally to dexamethasone by suppressing cortisol production, (3) high concentrations of corticotropin-releasing hormone (CRH) in the cerebrospinal fluid, (4) a blunted ACTH response to treatment with exogenous CRH, and (5) the disturbance of the circadian rhythm of cortisol secretion appear to indicate a central disinhibition of the HPA axis during depressive episodes (Heuser, 1998; Mitchell, 1998; Musselman and Nemeroff, 1996; Plotsky et al., 1998). These changes in the HPA axis appear to be state-dependent; that is, these factors normalize with the elevation of mood (Murphy, 1997; Plotsky et al., 1998).

In all of these examples of hormonal changes in depressed patients, it is unclear whether depression causes changes in hormone production or whether changes in hormone production cause depression. When depression is secondary to some endocrine dysfunction, then more definitive statements can be made about the direction of causality. For example, individuals suffering from primary hypothyroidism often present symptoms of depressed affect and intellect. However, patients suffering from primary adrenal disorders show conflicting mood responses to cortisol concentrations. For instance, patients with Cushing disease have adrenals that produce excessive cortisol, and depression is often a symptom of this disorder; however, patients with Addison disease have adrenal glands that produce insufficient cortisol, and depression is a defining symptom of this disease as well. Thus, an inverted U-shaped function exists for the effects of cortisol concentrations on mood (Figure 13.5).

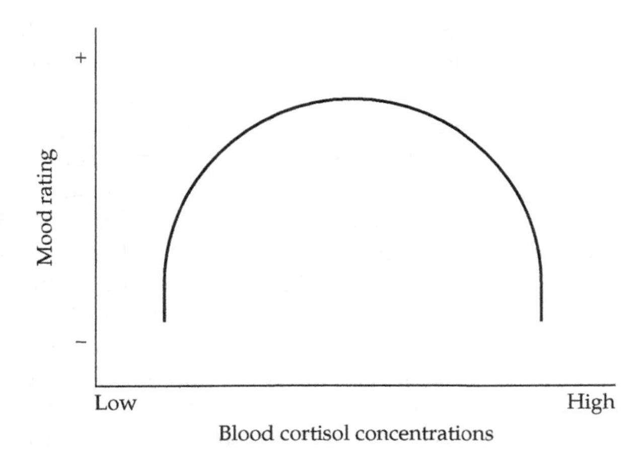

13.5 Cortisol and mood. Mood ratings in self-reports show an inverse U-shaped relationship with blood cortisol concentrations, with optimal mood ratings corresponding to moderate blood cortisol concentrations. If blood cortisol concentrations become too high or too low, as in Cushing disease or Addison disease, respectively, then mood ratings typically drop into the depressed category.

ESTROGEN Deficits in estrogen have been associated with depression (Fink et al., 1996; Halbreich, 1997; Zweifel and O'Brien, 1997). In one double-blind study, estrogen treatment was given to 40 women who were hospitalized with severe depression. None of the women who received placebo treatment showed changes in their affect; in fact, 47% of those patients deteriorated in their mood. However, over 90% of the depressed women treated with estrogen significantly improved their mood (Klaiber et al., 1979). It should be noted that substantial pharmacological doses (15–20 times the recommended therapeutic doses) of estrogen were administered in this study. Administration of estrogen in physiological doses improves mood in normal women (Sherwin and Gelfand, 1985; Sherwin and Suranyi-Cadotte, 1990), but not in clinically depressed women (Schneider et al., 1977). However, estrogen treatment seems to ameliorate depressed affect in postmenopausal women (Fink et al., 1996; Halbreich, 1997; Pearlstein et al., 1997; Rubinow et al., 1998; Zweifel and O'Brien, 1997). It is possible that estrogen withdrawal plays a role in postpartum depression.

Postpartum Depression

For most women, the postpartum period is a time of mother–infant bonding. However, a significant minority may suffer from **postpartum depression**. Like PMS, postpartum depression is a well-known outcome associated with the typical endocrine changes associated with reproductive function (Deakin, 1988; Hendrick et al., 1998; Llewellyn et al., 1997; Rubinow et al., 2002). Also in common with PMS, there is confusion regarding the precise definition of postpartum depression, as well as uncertainty that it is a unique disorder (Hamilton et al., 1988; Hopkins et al., 1984).

Many women experience mild to severe depression within a few days of giving birth (O'Hara et al., 1991; Susman, 1996). The mildest type of postpartum depression is called "maternity blues" or "baby blues" and typically lasts only 24–48 hours. A case of maternity blues is usually manifested by periods of crying and sadness; about 50% of women in North America display these symptoms after giving birth. Mild to moderate postpartum depression is experienced by about 15%–20% of women; this type of postpartum depression may last 4–8 weeks. Its symptoms include depressed affect, insomnia, crying, irritability, feelings of inadequacy, reduced coping ability, and fatigue. In very rare cases (fewer than 0.01% of women giving birth), women display a temporary, but severe, form of depression called postpartum depressive psychosis (Hopkins et al., 1984).

The exact precursors of postpartum mood disorders have not been specified, but a combination of biological, social, and psychological factors, including a personal or family history of depression, may be involved. Because dramatic changes in blood concentrations of estrogens, progesterone, and prolactin occur at the time of parturition, most studies of the hormonal correlates of postpartum mood changes have investigated the role of these hormones (O'Hara and Zekoski, 1988; Llewellyn et al., 1997). As in the case of PMS, no consistent correlation between mood changes and these hormones has been identified (George and Sandler, 1988; Gitlin and Pasnau, 1989; Kuevi et al., 1983; O'Hara et al., 1991).

A relationship between opioid peptides, specifically β-endorphin, and postpartum mood changes has been reported (Deakin, 1988; Ferin, 1984; George and Sandler, 1988). Blood plasma concentrations of β-endorphin change during pregnancy: concentrations are relatively constant during the first two trimesters, begin to rise during the end of pregnancy, peak during parturition, and then drop immediately afterward (Newnham et al., 1983, 1984; Smith et al., 1990). Because β-endorphin concentrations plummet within hours of parturition, it has been suggested that maternity blues, and perhaps other more severe postpartum mood disorders, may result from this "withdrawal" of endogenous opioids (Newnham et al., 1984). In support of this hypothesis, women who displayed the highest incidence of depressive symptoms from gestational week 38 through day 2 postpartum, as well as 3 months postpartum, had the greatest decreases in plasma β-endorphin concentrations after parturition (Smith et al., 1990). Increases in anxiety and tension are also associated with the postpartum decline in plasma β-endorphin levels (Brinsmead et al., 1985). Several correlative studies have established a relationship between opioids and postpartum mood; direct experimental evidence, however, remains lacking.

Although a fair amount of descriptive work has been done, only recently have data been collected that seem to suggest that endocrine changes may not necessarily be the primary cause of postpartum depression. There are several social changes that occur after the delivery of a baby that may lead to postpartum depression (Nonacs and Cohen, 1998; Hendrick et al., 1998). Hospitalization, for instance, may affect depressive symptoms in some women. The change in status to parent can be a sufficiently significant life event in itself to cause anxiety and depression (Richman et al., 1991). Women who remain at home with their children are more likely to display depressive symptoms than women who have careers outside the home (Gotlib et al., 1989). Not surprisingly, perhaps, women bringing unwanted pregnancies to term are more likely to become depressed than women with planned pregnancies.

In one study, repeated measurements of women's moods revealed that 25% exhibited increased depressive symptoms during pregnancy. However, only 10%

of the women in this study met the diagnostic criteria for depression, and only 7% could be diagnosed as depressed postpartum (Gotlib et al., 1989). These results indicate that mild depression may be common during and immediately after pregnancy, but that debilitating depression is rare. Men are nearly as likely as women to suffer postpartum depressive symptoms within 2 months of their child's birth (68% versus 82% respectively) (Richman et al., 1991). The similar incidence of depression seen in men, and in women both during and after pregnancy, despite very different hormone concentrations at those times, also suggests that endocrine changes may not be the primary cause of postpartum depression.

Seasonal Affective Disorder

Seasonal changes in behavior are legion in animals and humans (see Chapter 10). One seasonal rhythm in humans that has received much attention during the past several years is winter depression, or **seasonal affective disorder (SAD)** (Vessely and Lewy, 2002). SAD is characterized by depressed affect, lethargy, loss of libido, hypersomnia, excessive weight gain, carbohydrate cravings, anxiety, and inability to concentrate or focus attention, occurring during late autumn or winter (Rosenthal et al., 1988). In the Northern Hemisphere, symptoms usually begin between October and December and go into remission during March. These symptoms do not merely reflect the "holiday blues," because individuals suffering from SAD in the Southern Hemisphere display symptoms 6 months out of phase with Northern Hemisphere residents (Terman, 1988). With the onset of summer, SAD patients regain their energy and become active and elated, often to the point of hypomania or mania. Prevalence rates of SAD in the population range from 1% to 10%, with higher prevalence rates reported at higher latitudes (Rosenthal, 1993). Women seem to be affected by SAD more often than men (Kasper et al., 1989): the sex ratio of SAD prevalence in one epidemiological study was 3.5 women to 1 man. Over 80% of respondents to newspaper advertisements recruiting experimental subjects with SAD were women (Rosenthal and Wehr, 1987). Among menstruating women suffering from SAD, PMS is common during the fall and winter; often their PMS symptoms are reduced during the summer (Thase, 1989). Three features atypical of depression—hyperphagia, carbohydrate cravings, and hypersomnia—set SAD apart from nonseasonal depression. SAD is frequently diagnosed as "Bipolar II" depression or "Atypical Bipolar Disorder," particularly if hypomania or mania is present (DSM-IV).

Seasonal changes in mood and behavior may be closely related to alcoholism (Sher, 2004). A subset of individuals with alcoholism display seasonal patterns in their alcohol misuse. It is possible that such individuals are self-medicating an underlying seasonal affective disorder with alcohol or manifesting a seasonal pattern of alcohol-induced depression (Sher, 2004). Family and molecular genetic studies suggest the existence of a genetic link between SAD and alcoholism. In any case, the comorbidity of alcoholism and SAD suggests a link that could be causal in nature, and which should be considered by both mental health and drug and alcohol professionals when identifying, managing, and referring patients with comorbid alcoholism and SAD (Sher, 2004).

A potential diagnostic tool for SAD may be the presence of a reduced threshold for chemosensory detection (Postolache et al., 2002). People with SAD and control individuals were subjected to a detection threshold test in which phenyl ethyl alcohol was administered to each side of the nose in a counterbalanced order; the opposite nostril was occluded during the test (Postolache et al., 2002). Individuals with SAD were able to detect this odor at lower concentrations than people who did not have SAD. These results suggest that recurrent winter depression in humans may be associated with an enhanced olfactory ability (Postolache et al., 2002).

Improperly entrained circadian rhythms may be involved in SAD (Lewy et al., 1985, 1988; see Chapter 10). It has been hypothesized that changing the onset of sleep time resets biological clocks, resulting in amelioration of the depression (reviewed in Lewy et al., 1988). In one study, a depressed patient was phase-advanced in her sleep–wake cycle by 6 hours; her depression was temporarily ameliorated by this treatment (Wehr et al., 1979). Four of seven other patients who underwent spontaneous remission of depression displayed a spontaneous phase advance of their times of awakening (Wehr et al., 1979). Lithium, tricyclic antidepressants, and estrogen, all of which are used to treat depressive illnesses, also affect endogenous timekeeping mechanisms (Wehr et al., 1979). The fact that these pharmacological agents and sleep–wake cycle manipulations are both effective in ameliorating depression suggests that they work via similar mechanisms involving biological clocks.

More recently, bright lights have been used in place of sleep–wake therapy in the treatment of SAD. When patients are exposed to bright light, usually for a few hours in the morning, signs of remission of the SAD symptoms are often apparent within a few days (Figure 13.6A; Rosenthal et al., 1988). Phototherapy, like sleep–wake therapy, may work by phase-advancing biological rhythms. Bright light has been suggested to possess two antidepressant effects: (1) light treatment in the morning may ameliorate depression by realignment of inappropriately entrained circadian rhythms, and (2) light may also serve as a general "energizer ... of mood in a way that may be attributable wholly or in part to a placebo effect" (Lewy et al., 1988). Light treatment at different times during the day results in differing rates of mood improvement (Figure 13.6B; Lewy et al., 1987). Light treatment in the evening has no mood benefits. Phototherapy appears to shift circadian rhythms by shifting the entrainment of the nightly secretion of melatonin (Vessely and Lewy, 2002), as we will see shortly.

Serotonin may be involved in the symptoms of SAD (Skwerer et al., 1988; Wurtman and Wurtman, 1989). Tryptophan, an amino acid that normally circulates in the blood at low concentrations, is converted to serotonin in the brain, specifically in the raphe nuclei (Cooper et al., 1986). Diet affects this conversion process because carbohydrates stimulate pancreatic β-cells to secrete insulin, which in turn facilitates the uptake of sugars and non-tryptophan amino acids into peripheral cells. This action results in a relatively high ratio of tryptophan to other amino acids in the blood, and because tryptophan competes with the other amino acids for access to central nervous system tissue, carbohydrate ingestion results in more tryptophan crossing the blood–brain barrier, and thus higher production of serotonin

(A)

(B)

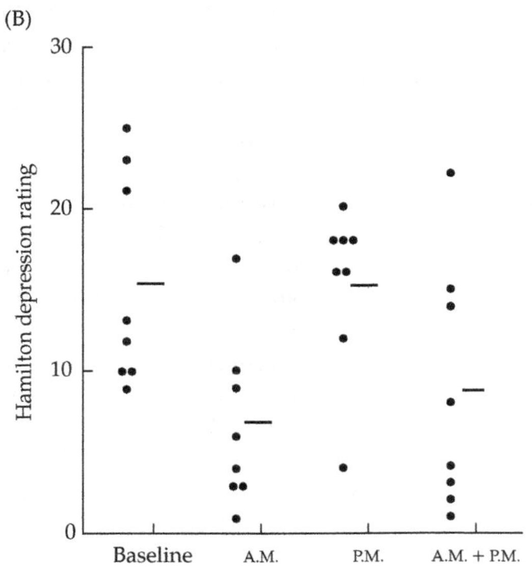

(Figure 13.7). Serotonin concentrations feed back to regulate the intake of carbohydrates. It is possible, therefore, that patients suffering from SAD have cyclic disruptions in their serotonin–carbohydrate regulating mechanisms (Wurtman and Wurtman, 1989). Serotonin is also involved in normal sleep onset, and faulty serotonin regulation may contribute to the hypersomnia reported in SAD patients. If it is true that symptoms of SAD result from faulty serotonin metabolism, then pharmacological interventions that elevate serotonin concentrations should be expected to reduce the severity of some SAD symptoms. Administration of the serotonin agonist δ-fenfluramine to patients with SAD reduces carbohydrate intake and the

Clearly, there are major gaps in our knowledge about SAD because there are virtually no consistent patterns that indicate whether humans are either photoperiodic or circannual, although it is possible that variation in human responsiveness to photoperiod may account for this inconsistency (Bronson, 2004). Nevertheless, the growing body of data from mammals with seasonal cycles may be useful in pointing the way to obtaining such information for humans and may eventually lead to effective pharmacological or behavioral treatments for SAD.

Hormones and Eating Disorders

Anorexia nervosa

Anorexia nervosa and bulimia are eating disorders that share several psychological features. **Anorexia nervosa** is most prevalent among adolescent women (Klein and Walsh, 2004), although a small, but increasing, number of boys are developing the disorder. If persistent, anorexia nervosa can be fatal. This eating disorder, first described in the 1870s, is characterized by greatly reduced food intake and body mass and a distorted body image, as well as suppressed or delayed onset of menstrual cycles. The DSM-IV describes two types of anorexia nervosa: (1) restricting (in which individuals maintain their low body mass mainly by severe dieting and excessive exercise), and (2) binge eating–purging (in which individuals maintain their low body mass by methods such as self-induced vomiting or overuse of laxatives, diuretics, and enemas). The prevalence of anorexia nervosa among women in North America is estimated at 0.5%. The cause of anorexia nervosa is unknown, but there is a strong genetic component, as well as other personality and situational factors that contribute to the disorder. Although this syndrome is most common in Western societies where being slender is valued, it has also been observed at similar population frequencies in non-Western societies in which being slender has reduced social value (e.g., Ung et al., 1997). The constellation of symptoms associated with anorexia nervosa appears to have existed as far back as the 1600s (Vanderreycken and Van Deth, 1994).

Anorexia nervosa is the only psychiatric disorder that requires an endocrine dysfunction (i.e., menstrual abnormalities) as a criterion for diagnosis (Negrão and Licinio, 2002). In addition to reproductive dysfunction, several additional endocrine disturbances have been reported in connection with anorexia nervosa (Table 13.2). Most characterizations of this disorder have been correlational; it is difficult to determine which endocrine relationships are unique to anorexia nervosa and which are the results of starvation. The studies required to answer this question would be unethical to perform, and even a natural occurrence of starvation requires immediate refeeding in any setting. There are no good animal models of anorexia nervosa or bulimia.

As mentioned, reproductive dysfunction is the most pronounced endocrine consequence of anorexia nervosa. Chronic malnutrition reduces body fat and available metabolic fuels, leading to regressed ovaries and disrupted ovulation, menstruation, fertility, and sexual behavior (Klein and Walsh, 2004). The reproductive hormone profile resembles that of prepubertal girls, with low concentrations of gonadotropins and sex steroid hormones (Boyar et al., 1974). GnRH secre-

TABLE 13.2 Endocrine alterations observed in anorexia nervosa and bulimia

Anorexia nervosa
Hypogonadism
Amenorrhea
Oligomenorrhea (irregular menstrual cycles)
Delayed puberty
Euthyroid sick syndrome
Hypercortisolism (elevated cortisol)
↑ CRH in central nervous system
↑ Basal/pulsatile GnRH secretion
↑ Osteoporosis
↓ Circulating leptin levels
↓ Insulin secretion
↑ Vasopressin in CNS
Altered melatonin levels
↑ Neuropeptide Y levels in cerebrospinal fluid
Bulimia
Amenorrhea
Oligomenorrhea
Anovulatoy cycles
Thyroid dysfunction
↑ Circulating GH
Normal bone density
Altered leptin
Normal insulin secretion and sensitivity
Altered melatonin
↓ Nighttime prolactin
↑ Levels of peptide YY_{3-36} in cerebrospinal fluid

Source: Negrão and Licinio, 2002

tion is impaired in anorexic women, but can be restored by normalization of body mass (Negrão and Licinio, 2002). Thyroid function is usually in the low normal range, but may be sufficiently low to result in the so-called euthyroid sick syndrome (Wartofsky and Burman, 1982). More recent studies have reported low circulating levels of thyroid hormone and TSH (Klein and Walsh, 2004). Other endocrine problems that are serious, but have less of a behavioral effect, include elevated cortisol concentrations (hypercortisolism; typically without Cushing disease), excessive GH concentrations with decreased concentrations of insulin-like growth factor (IGF), osteoporosis (probably due to the shifting GH/IGF ratio), and decreased insulin levels (Klein and Walsh, 2004).

Yesalis, 1989). After acute myocardial infarction, suicide is the most common cause of premature death among athletes suspected to have used anabolic androgen steroid hormones (Pässinen et al., 2000).

Adverse behavioral effects of anabolic steroid abuse have also been reported (Malone et al., 1995). The best-known of these effects is extremely aggressive behavior, also known as "roids rage." Although some of the adverse physiological effects of anabolic steroids may not be observed until many years after the cessation of use, the onset of adverse behavioral effects may be relatively rapid, even at therapeutic doses. Oxymetholone is a 17-α-methyl testosterone derivative that is a common, orally administered treatment for anemia. One 18-year-old male treated with oxymetholone for idiopathic aplastic anemia developed a temporary elevation in aggression (Barker, 1987). Prior to treatment, the patient had not displayed any notable aggressive behavior. During the steroid treatment, however, he broke a friend's nose and also destroyed some personal property. As noted above, anabolic steroid abusers are exposed to much higher doses than those prescribed therapeutically. These high doses have been associated with violent aggression, often accompanied by affective and psychotic symptoms. The behavioral pathology of androgens can, for the most part, be avoided by avoiding pharmacological self-administration of anabolic steroids. Most psychiatric symptoms subside when anabolic steroid use is discontinued. Unfortunately, no information is available on the long-term behavioral consequences of anabolic steroid abuse if the steroids were administered during puberty. Even the extent to which high endogenous androgen concentrations during adolescence evoke hyper-aggressiveness in human males remains unknown. In one well-known study, researchers contacted gymnasiums and offered members a cash payment to engage in a confidential interview about their steroid use. The respondents reported routinely using doses 10–100 times higher than recommended therapeutic doses. Fifteen of the 41 respondents regularly using anabolic steroids reported experiencing major psychiatric symptoms; 13 other individuals became manic or near manic. One user, convinced of his immortality, drove a car into a tree at 65 kilometers per hour while a friend videotaped him (Pope and Katz, 1988).

Anabolic steroid abuse has been associated with the perpetration of violent crimes (Canacher and Workman, 1989; Borowsky et al., 1997). A number of individuals have impulsively committed homicides while taking anabolic steroids. Carefully structured interviews with these perpetrators suggest strongly that steroid abuse was an important, if not the primary, factor in the manifestation of the extreme aggressive act in many such cases. Although the individuals interviewed may have emphasized the role of steroids in their violence to aid their legal positions, a consistent pattern of steroid-associated violence among previously nonaggressive individuals supports the possibility that these agents are involved in the mediation of violent behavior (Canacher and Workman, 1989; Orchard and Best, 1994).

Very few controlled studies of the effects of anabolic steroids on mood have been reported. In one study, 20 healthy male volunteers consecutively received a placebo, a low dose (40 mg) of methyltestosterone, a high dose (240 mg) of methyltestosterone, and another placebo over the course of 3 days (Su et al., 1993). A number of neuropsychiatric measures were obtained, which showed that the high testosterone dose caused positive changes in mood (e.g., euphoria, increased energy, and increased sexual arousal) as well as negative mood changes (e.g., irritability, mood swings, feelings of violence, and hostility) and cognitive impairments (e.g., distractibility, confusion, and impaired memory) (Su et al., 1993). One of the 20 (i.e., 5% incidence) study participants became acutely manic, and another individual became hypomanic (i.e., 5% incidence). The results of this study indicate that even brief anabolic steroid use can affect mood in an adverse manner (Su et al., 1993). Elevated mood was not observed in another study using lower doses of anabolic steroids (50, 100, and 200 mg) (Fingerhood et al., 1997).

Anabolic steroid abuse is also increasing among female athletes (Gruber and Pope, 2000). Interviews with women in gymnasiums revealed that approximately one-third had a history of anabolic androgen steroid abuse. Several behavioral problems were noted among these women, including chronic dissatisfaction and preoccupation with their bodies, rigid dietary practices, hypomania, and multiple substance dependencies (Gruber and Pope, 2000).

It has been proposed that a proportion of anabolic steroid abusers develop an addiction to steroids. This hypothesis is supported by the remarkable consistency with which anabolic steroid abusers fulfill the following commonly accepted criteria for psychoactive substance use disorder (Pope and Katz, 1990): (1) the steroids are used over longer periods than desired; (2) unsuccessful attempts to stop the steroid use occur; (3) substantial time is spent in procuring, using, or recovering from the effects of anabolic steroids; (4) use continues despite knowledge of the significant physical and behavioral problems it is causing; (5) characteristic withdrawal symptoms occur; and (6) more anabolic steroids are often taken to relieve the withdrawal symptoms (Kashkin and Kleber, 1989). These last two criteria in particular suggest that anabolic steroids might have direct rewarding properties similar to the reinforcing effects of drugs such as cocaine, amphetamines, morphine, and heroin (Shippenberg and Herz, 1987). Increasing reports of suicides in previously nondepressed young men who abruptly stopped using anabolic steroids have been noted. These tragic events appear to be linked with the constellation of symptoms that resemble withdrawal symptoms. Thus, the withdrawal symptoms manifested by anabolic steroid abusers and the symptoms of PMS or postpartum depression may result from a common underlying cause; namely, dependence on elevated steroid hormone concentrations. Additional research is required to ascertain the veracity of this hypothesis.

Research on rodents has demonstrated that androgens have rewarding properties (see also Chapter 12). For example, male Syrian hamsters preferred an aqueous solution of 200 μg/ml of testosterone over plain water in a two-bottle choice test (Johnson and Wood, 2001; Wood, 2002). When male rats or hamsters were catheterized to receive either an intravenous or an intracerebrovascular infusion of testosterone when they poked their noses into a hole, testosterone showed a modest (compared with other drugs of abuse) ability to serve as a reward (Wood et

Appendix: Hormones

Vertebrate steroid hormones

Hormone	Abbreviation	Source	Major biological action
Adrenal glands			
MINERALOCORTICOIDS			
Aldosterone		Zona glomerulosa of adrenal cortex	Sodium retention in kidney
11-Deoxycortico-sterone	DOC	Zona glomerulosa of adrenal cortex	Sodium retention in kidney
GLUCOCORTICOIDS			
Cortisol (Hydrocortisone)	F	Zona fasciculata and z. reticularis of adrenal cortex	Increases carbohydrate metabolism; antistress hormone
Corticosterone	B	Zona fasiculata and z. reticularis of adrenal cortex	Increases carbohydrate metabolism; antistress hormone
Dehydroepiandro-sterone	DHEA	Zona reticularis of adrenal cortex	Weak androgen; primary secretory product of fetal adrenal cortex
Ovaries			
Estradiol	E_2	Follicles	Uterine and other female tissue development
Estriol		Follicles	Uterine and mammary tissue development
Estrone		Follicles	Uterine and mammary tissue development

Vertebrate steroid hormones *continued*

Hormone	Abbreviation	Source	Major biological action
Ovaries			
Progesterone	P	Corpora lutea, placenta	Uterine development; mammary gland development; maintenance of pregnancy
Testes			
Androstenedione		Leydig cells	Male sex characters
Dihydrotestosterone	DHT	Seminiferous tubules and prostate	Male secondary sex characters
Testosterone	T	Leydig cells	Spermatogenesis; male secondary sex characters

Peptide and protein hormones

Hormone	Abbreviation	Source	Major biological action
Adipose Tissue			
Leptin (Ob protein)		Adipocytes	Regulation of energy balance
Adiponectin		Adipocytes	Modulates endothelial adhesion molecules
Plasminogen activator inhibitor-1	PAI-1	Adipocytes	Regulation of vascular hemostasis
Adrenal glands			
Met-enkephalin		Adrenal medulla	Analgesic actions in CNS
Leu-enkephalin		Adrenal medulla	Analgesic actions in CNS
Gut			
Bombesin		Neurons and endocrine cells of gut	Hypothermic hormone; increases gastrin secretion
Cholecystokinin (Pancreozymin)	CCK	Duodenum and CNS	Stimulates gallbladder contraction and bile flow; affects memory, eating behavior
Gastric inhibitory polypeptide	GIP	Duodenum	Inhibits gastric acid secretion
Gastrin		G-cells of midpyloric glands in stomach antrum	Increases secretion of gastric acid and pepsin
Gastrin releasing peptide	GRP	GI tract	Stimulates gastrin secretion
Ghrelin		Stomach mucosa/GI tract	Regulation of energy balance
Glucogon-like peptide-1	GLP-1	L cells of intestine	Regulates insulin secretion
Motilin		Duodenum, pineal gland	Alters motility of GI tract

Peptide and protein hormones *continued*

Hormone	Abbreviation	Source	Major biological action
Gut			
Secretin		Duodenum	Stimulates pancreatic acinar cells to release bicarbonate and water
Vasoactive intestinal polypeptide	VIP	GI tract, hypothalamus	Increases secretion of water and electrolytes from pancreas and gut; acts as neurotransmitter in autonomic nervous system
Peptide YY	PPY	GI tract	Regulation of energy balance/ food intake
Heart			
Atrial naturetic factor	ANF	Atrial myocytes	Regulation of urinary sodium excretion
Hypothalamus			
Agouti-related protein	AgRP	Arcuate nucleus	Regulation of energy balance
Arg-vasotocin	AVT	Hypothalamus and pineal gland	Regulates reproductive organs
Corticotropin-releasing hormone	CRH	Paraventricular nuclei, anterior periventricular nuclei	Stimulates release of ACTH and β-endorphin from anterior pituitary
Gonadotropin-releasing hormone (Luteinizing hormone–releasing hormone)	GnRH (LHRH)	Preoptic area; anterior hypothalamus; suprachiasmatic nuclei; medial basal hypothalamus (rodents and primates); arcuate nuclei (primates)	Stimulates release of FSH and LH from anterior pituitary
Gonadotropin-inhibiting hormone	GnIH	Species-dependent loci	Inhibits release of LH (in birds)
Somatostatin (Growth hormone–inhibiting hormone)		Anterior periventricular nuclei	Inhibits release of GH and TSH from anterior pituitary, inhibits release of insulin and glucagon from pancreas
Somatocrinin (Growth hormone–releasing hormone)	GHRH	Medial basal hypothalamus; arcuate nuclei	Stimulates release of GH from anterior pituitary
Melanotropin-release inhibitory factor (Dopamine)	MIF (DA)	Arcuate nuclei	Inhibits the release of MSH (no evidence of this peptide in humans)
Melanotropin-releasing hormone	MRH	Paraventricular nuclei	Stimulates the release of MSH from anterior pituitary (no evidence of this peptide in humans)
Neuropeptide Y	NPY	Arcuate nuclei	Regulation of energy balance
Neurotensin		Hypothalamus; intestinal mucosa	May act as a neurohormone

Peptide and protein hormones *continued*

Hormone	Abbreviation	Source	Major biological action
Hypothalamus			
Orexin A and B		Lateral hypothalamic area	Regulation of energy balance/food
Prolactin-inhibitory factor (Dopamine)	PIF (DA)	Arcuate nuclei	Inhibits PRL secretion
Prolactin-releasing hormone		Paraventricular nuclei	Stimulates release of PRL from anterior pituitary
Substance P	SP	Hypothalamus, CNS, intestine	Transmits pain; increases smooth muscle contractions of GI tract
Thyrotropin-releasing hormone	TRH	Paraventricular nuclei	Stimulates release of TSH and PRL from anterior pituitary
Urocortin		Lateral hypothalamus	CRH related peptide
Liver			
Somatomedins		Liver, kidney	Cartilage sulfation, somatic cell growth
Angiotensinogen		Liver, blood	Precursor of angiotensins, which affect blood pressure
Ovaries			
Relaxin		Corpora lutea	Permits relaxation of various ligaments during parturition
Inhibin		Follicles	Inhibits FSH secretion
Gonadotropin surge-attenuating factor	GnSAF	Follicles	Control of LH secretion during menstruation
Activin		Granulosa cells	Stimulates FSH secretion
Pancreas			
Glucagon		α-cells	Glycogenolysis in liver
Insulin		β-cells	Glucose uptake from blood; glycogen storage in liver
Somatostatin		δ-cells	Inhibits insulin and glucagon secretion
Pancreatic polypeptide	PP	Peripheral cells of pancreatic islets	Effects on gut in pharmacological doses
Pituitary			
Adrenocorticotropic hormone	ACTH	Anterior pituitary	Stimulates synthesis and release of glucocorticoids
Vasopressin (Antidiuretic) hormone)	ADH or AVP	Posterior pituitary	Increases water reabsorption in kidney
β-endorphin		Intermediate lobe of pituitary	Analgesic actions

Peptide and protein hormones *continued*

Hormone	Abbreviation	Source	Major biological action
Pituitary			
Follicle-stimulating hormone	FSH	Anterior pituitary	Stimulates development of ovarian follicles and secretion of estrogens; stimulates spermatogenesis
Growth hormone	GH	Anterior pituitary	Mediates somatic cell growth
Lipotropin	LPH	Anterior pituitary	Fat mobilization; precursor of opioids
Luteinizing hormone	LH	Anterior pituitary	Stimulates Leydig cell development and testosterone production in males; stimulates corpora lutea development and production of progesterone in females
Melanocyte-stimulating hormone	MSH	Anterior pituitary	Affects memory; affects skin color in amphibians
Oxytocin		Posterior pituitary	Stimulates milk letdown and uterine contractions during birth
Prolactin	PRL	Anterior pituitary	Many actions relating to reproduction, water balance, etc.
Thyroid-stimulating hormone (Thyrotropin)	TSH	Anterior pituitary	Stimulates thyroid hormone secretion
Placenta			
Chorionic gonadotropin	CG	Placenta	LH-like functions; maintains progesterone production during pregnancy
Chorionic somatomammotropin (Placental lactogen)	CS (PL)	Placenta	Acts like PRL and GH
Testes			
Müllerian inhibitory hormone	MIH	Fetal Sertoli cells of testes	Mediates regression of Müllerian duct system
Inhibin		Seminiferous tubules (and ovaries)	Inhibits FSH secretion
Activin		Sertoli cells	Stimulates FSH secretion
Thyroid/Parathyroid			
Calcitonin	CT	C-cells of thyroid	Lowers serum Ca^{2+} levels
Parathyroid hormone	PTH	Parathyroid gland	Stimulates bone resorption; increases serum Ca^{2+} levels
Thyroxine (Tetraiodothyronine)	T_4	Follicles of thyroid	Increases oxidation rates in tissue
Triiodothyronine	T_3	Follicles of thyroid	Increases oxidation rates in tissue

Peptide and protein hormones *continued*

Hormone	Abbreviation	Source	Major biological action
Thyroid/Parathyroid			
Parathyroid-related peptide	PTHrp	Parathyroid gland (and other tissues)	Regulation of bone/skin development
Thymus			
Thymosin		Thymocytes	Proliferation/differentiation of lymphocytes
Thymostatin		Thymocytes	Proliferation/differentiation of lymphocytes

Monoamine hormones

Hormone	Abbreviation	Source	Major biological action
Adrenal glands			
Epinephrine (Adrenaline)	EP	Adrenal medulla (and CNS)	Glycogenolysis in liver; increases blood pressure
Norepinephrine (Noradrenaline)	NE	Adrenal medulla (and CNS)	Increases blood pressure
Central nervous system			
Dopamine	DA	Arcuate nuclei of hypothalamus	Inhibits prolactin release (and other actions)
Serotonin	5-HT	CNS (also pineal)	Stimulates release of GH, TSH, ACTH; inhibits release of LH
Pineal gland			
Melatonin		Pineal gland	Affects reproductive functions

Lipid-based hormones

Hormone	Abbreviation	Source	Major biological action
Leukotrienes	LT	Lung	Long-acting broncho-constrictors
Prostaglandins E_1 and E_2	PGE_1 and PGE_2	Variety of cells	Stimulates cAMP
Prostaglandins $F_{1\alpha}$ and $F_{2\alpha}$	$PGF_{1\alpha}$ and $PGF_{2\alpha}$	Variety of cells	Active in dissolution of corpus luteum and in ovulation
Prostaglandin A_2	PGA_2	Kidney	Hypotensive effects
Prostacyclin I	PGI_2	Variety of cells	Increased second messenger formation
Thromboxane A_2	TX_2	Variety of cells	Increased second messenger formation

Glossary

Ablation Removal, especially by cutting.

Active avoidance A type of learning in which an individual must perform an action to avoid a noxious situation.

Activin Peptide hormone synthesized in the anterior pituitary gland and gonads that stimulates the secretion of follicle-stimulating hormone.

Adaptive function The role of any structural, physiological, or behavioral process that increases an individual's fitness to survive and reproduce as compared with other conspecifics.

Adiponectin A protein hormone secreted from adipose (fat) cells that affects several metabolic processes, including glucose regulation and fatty acid catabolism.

Adipose tissue Connective tissue in which fat is stored.

Adrenal cortex The outer layer(s) of the endocrine organ that sits above the kidneys in vertebrates and secretes steroid hormones.

Adrenal glands Paired, dual-compartment endocrine glands in vertebrates consisting of a medulla and a cortex. Epinephrine and norepinephrine are secreted from the medulla, and steroid hormones are released from the cortex.

Adrenal medulla The inner portion of the endocrine organ that sits above the kidneys in vertebrates and secretes epinephrine and norepinephrine.

Adrenocorticotrophic hormone (ACTH) A polypeptide hormone that is secreted by the anterior pituitary gland that stimulates the adrenal cortex to secrete corticosteroids, such as cortisol and corticosterone. ACTH is also known as corticotrophin.

Affiliation A form of social behavior that involves an individual's motivation to approach and remain in close proximity with a conspecific.

Aggression A form of social interaction that includes threat, attack, and fighting.

Agonist A chemical substance that binds to receptors for a hormone or neurotransmitter and causes a biological response that is indistinguishable from the response elicited by the natural hormone or neurotransmitter. Also called a mimic.

Agonistic Any behavior associated with fighting including aggression, submission, or retreat.

Aldosterone A mineralocorticoid that causes the kidneys to retain sodium.

Allele A specific version of a gene.

Alloparents Parental assistants; in birds, older male siblings that do not disperse, but remain at the nest site to assist the parents with the rearing of the current young.

Altricial Born or hatched at an early stage of development. Altricial offspring are generally quite helpless, and require substantial parental care to survive.

Amnestic A substance or event that causes forgetting.

Amplitude In biological rhythms, the amount of change in the rhythm above (to the peak) or below (to the nadir) the average value.

Amygdala An almond-shaped structure located near the base of each temporal lobe of the brain. The amygdala is critical for the integration of sensory information that is important in sexual behavior.

Anastomosis The joining together of two circulatory systems.

Androgen insensitivity *See* Testicular feminization mutation.

Androgens Any of the C_{19} class of steroid hormones, so named because of their andros- (male) generating effects. Androgens are the primary steroidal product secreted from the testes. Testosterone and dihydrotestosterone are biologically important androgens.

Androstenedione The primary sex hormone secreted by the human adrenal cortex.

Anestrus (n); anestrous (adj) The reproductive condition of a female mammal that is not in estrus, or mating condition.

Angiotensin II A parahormone that is produced in the blood by the actions of renin and angiotensinogen; it may be involved in the regulation of water balance and thirst.

Anlage The primordial substrate in a developing individual.

Antagonist A chemical substance that binds to receptors for a hormone or neurotransmitter, but does not cause a biological response. Also called a blocker.

Anterior pituitary Front part of the endocrine gland that extends from the base of the brain and secretes a number of tropic hormones in response to hormonal signals from the hypothalamus.

Antiamnestic A substance that protects against forgetting.

Antidiuretic hormone (ADH) A hormone secreted from the anterior pituitary that regulates removal of water from the blood via the kidneys. Also called vasopressin.

Apoptosis Programmed, orderly cell death that avoids immune system activation.

Appetitive learning Reinforcement of a behavior by a positive outcome.

Appetitive phase An ethological term, roughly equivalent to "courtship." All the behaviors an individual displays when attempting to gain access to an individual of the opposite sex for the purpose of mating.

Arcuate nucleus A hypothalamic nucleus located near the base of the hypothalamus that contains the cell bodies of many neurons that secrete hypothalamic releasing hormones.

Aromatase An enzyme that converts androgens into estrogens.

Aromatization The process of converting an androgen molecule to an estrogen molecule via the enzyme aromatase. The removal of the carbon at position 19 from the androgen precursors results in an estrogen with a phenolic A-ring, which is known as an aromatic compound.

Associated reproductive pattern The breeding pattern observed in most vertebrate species, in which reproductive behavior, maximal gonadal size and activity, high steroid concentrations, and gamete production coincide.

Attractivity The stimulus value of a female for a particular male. Attractivity is a hypothetical construct that must be inferred by observation of a conspecific's behavior.

Autocrine Pertaining to a signal secreted by a cell into the environment that affects the transmitting cell.

Autoradiography A technique used to detect a radiolabeled substance, such as a hormone, in a cell or organism, by placing a thin slice of the material in contact with a photographic emulsion, which displays darkened silver grains in response to the radioactive emissions.

Aversive learning A change in behavior to avoid some noxious outcome.

Basal metabolic rate (BMR) Rate of metabolism at rest.

Base A building block of DNA or RNA molecules. DNA contains four different bases (adenine, thymine, guanine, and cytosine), which form pairs that link the two strands of DNA together. The order of the bases provides the genetic information of the DNA molecule.

Behavioral sex The sex of an individual as discriminated on the basis of male-typical and female-typical behaviors.

β-endorphin An endogenous opioid. These peptide hormones are produced in the anterior pituitary gland and hypothalamus in vertebrates, and resemble opiates in their actions as "natural" pain killers.

Bioassay The use of a biological response to determine the presence or amount of a particular substance in a sample.

Biochronometry *See* Chronobiology.

Biological half-life The amount of time required to remove half of a hormone or other substance from the blood.

Blastocyst A fluid-filled sphere of cells that develops from a zygote. The embryo usually develops from the cluster of cells in the center of the blastocyst, whereas the external wall of the blastocyst develops into the placenta.

Blocker *See* Antagonist.

Blood–brain barrier (BBB) The mechanism of blood vessels in the brain that prevents toxic substances from diffusing into the brain.

Blot tests Techniques used to fractionate mixtures of proteins (Western), DNAs (Southern), or RNAs (Northern) so they can hybridize with markers that travel different distances in an electrophoretic gel based on their size.

Bruce effect The interruption of pregnancy caused by the odors of a strange male.

Calcitonin (CT) A polypeptide hormone that is secreted from the C cells associated with the thyroid gland which lowers blood calcium concentrations and affects blood phosphorus.

Cannulation A technique in which hollow electrodes or fine tubes (cannulas) are inserted into specific brain regions or into specific blood vessels, so that substances can be introduced precisely into a particular place or a blood sample obtained from a specific location.

Carrier protein Also called binding protein or transport protein; (1) one of several different plasma proteins that bind to hormones of low solubility (primarily thyroid and steroid hormones), providing a transport system for them. Some carrier proteins are specific for particular hormones, whereas others bind to many hormones. (2) One of several different types of trans-membrane proteins that function to transport small molecules across cell membranes.

Castration The surgical removal of the gonads. Synonymous with gonadectomy in both sexes, and with orchidectomy in males and ovariectomy or oophorectomy in females.

Catecholamines Hormones (e.g., epinephrine, norepinephrine) that are derived from tyrosine and secreted primarily from the adrenal medulla.

C-cells Thyroid cells found in the interstitial spaces between the thyroid follicle spheres that secrete calcitonin. C-cells are also called parafollicular cells.

Chimera An animal whose tissues are composed of two or more genetically distinct cell types; also called a mosaic.

Cholecystokinin (CCK) A hormone released by the lining of the small intestine that may be involved in satiation of food intake.

Cholesterol A white crystalline substance out of solution, and originally found in gallstones. It is also found in animal tissue, but typically is synthesized by the liver, and is an important part of cell membranes. A C-27 steroid, cholesterol is a precursor to steroid hormones.

Chromaffin cells Cells that make and store epinephrine secretory vesicles. Because these cells have an affinity for chromic salts, they were named chromaffin. These types of cells are common in the adrenal medulla.

Chromosomal sex The sex of an individual as determined by the sex chromosomes that an individual receives at fertilization.

Chronobiology The study of biological clocks and their associated rhythms. Also referred to as biochronometry.

Circadian rhythm A biological rhythm with a period of about 24 hours.

Circalunar rhythm A biological rhythm with a period of about 29.5 days that is closely tied to phases of the moon.

Circannual rhythm A biological rhythm with a period of about 12 months.

Circatidal rhythm A biological rhythm with a period of about 12.4 hours that is closely tied to changes in tides.

Cloaca The external sexual organ in birds. Males discharge sperm through the cloaca; female lay eggs through it.

Colloid A relatively large molecule that cannot pass through cell membranes. When injected under the skin or into the perineum, it draws in fluids by osmotic pressure.

Concaveation The process of becoming sensitized to newborn animals so that full maternal behavior is expressed. Also called pup induction or sensitization.

Congenital adrenal hyperplasia (CAH) A genetic deficiency that results in the overproduction of androgens by the adrenal glands. This syndrome has no reported ill effects in males, but causes various degrees of masculinization of the external genitalia in females, which may lead to erroneous assignment of sex at birth.

Consummatory phase An ethological term that encompasses the completion of a motivated behavior. In terms of sexual behavior, copulation represents the consummatory phase.

Corpus luteum An endocrine structure that forms from the remnants of the ovarian follicle after the egg is released. The corpus luteum secretes progestins, which support the uterine lining in preparation for blastocyst implantation.

Corticoids A class of C_{21} steroid hormones secreted primarily from the adrenal cortices. There are two main types of corticoids: glucocorticoids and mineralocorticoids.

Corticosterone Glucocorticoid produced in the adrenal cortices of most rodents and birds.

Corticotropin releasing hormone (CRH) A peptide hormone secreted by the hypothalamus that stimulates the release of ACTH (corticotrophin) by the anterior pituitary gland. CRH is also called corticotrophin releasing factor (CRF).

Courtship A system of communication that is used by one sex to induce mating in a receiving conspecific. Information about species, sex, physiological condition, and readiness to mate is exchanged during courtship.

Crepuscular Active at dawn and dusk.

Crop milk A prolactin-dependent substance discharged from the epithelial cells lining the crop in pigeons and doves, which is regurgitated to feed the young.

Cytokine A protein chemical messenger that evokes the proliferation of other cells, especially in the immune system. Examples of cytokines are interleukins and interferons.

Decidual reaction The formation of a deciduomata in the uterine wall. This process indicates that the corpora lutea are producing sufficient progestins to support implantation of blastocysts and it provides good evidence of pseudopregnancy.

Deciduomata A small vascular structure that forms in the uterine wall to support an implanting blastocyst.

Defeminization The removal of the potential for female traits.

Dehydroepiandrosterone (DHEA) A steroid hormone that is produced from cholesterol in the adrenal cortex, which is the primary precursor of natural estrogens, and is a weak androgen. DHEA is also called dehydroisoandrosterone or dehydroandrosterone.

Demasculinization The removal of the potential for male traits.

2-deoxyglucose (2-DG) A molecular mimic of glucose that provides no nutritional value. Cells that are actively working accumulate more 2-DG than inactive cells. If 2-DG is marked with a radioactive or immunoreactive tag, active neurons can be visualized.

Deoxyribonucleic acid (DNA) A nucleic acid that codes hereditary information.

Detection threshold The concentration at which an individual can tell the difference between two substances or stimuli.

Diabetes insipidus A condition characterized by the failure of the kidneys to retain water because of a lack of antidiuretic hormone. Symptoms include excessive urination, thirst, and water intake.

Diabetes mellitus A condition characterized by impaired glucose regulation because of reduced insulin secretion. Symptoms include high blood sugar levels and excessive urination, thirst, and water intake.

Dihydrotestosterone (DHT) A potent androgen that is derived from testosterone and binds more strongly to androgen receptors than testosterone. There are both 5-alpha and 5-beta forms of DHT.

Dimorphic Having two different forms; usually used regarding differences between the two sexes.

Dissociated reproductive pattern A breeding pattern observed in some vertebrate species in which reproductive behavior does not coincide with maximal gonadal size and activity. Instead, copulation occurs when steroid levels and gamete production are low.

Diurnal Active during the day.

Dopamine A neurotransmitter produced primarily in the forebrain and diencephalon that acts in the basal ganglia, olfactory system, and some parts of the cerebral cortex.

Down-regulation A process that is similar to negative feedback in which the overproduction of a hormone causes occupation of virtually all available receptors so that subsequent high levels of hormones cannot have a biological effect.

Ectocrine A parahormonal chemical substance that is secreted (usually by an invertebrate organism) into its immediate environment (air or water) which alters physiology or behavior of the recipient individual.

Ejaculation The forceful expulsion of semen from a male's body via the urethra.

Electrical stimulation Activation of nerve cells by electrical current.

Embryo A term used to describe the stage of development in mammals from implantation of the blastocyst until organ function begins to specialize. Embryonic development begins 1 week after conception in humans and continues until 9 weeks after conception.

Endocrine gland A ductless gland from which hormones are released into the blood system in response to specific physiological signals.

Endocrinology The scientific study of the endocrine glands and their hormones.

Endogenous Relating to a substance or process within the organism.

Endometrium The tissue lining the uterus into which a blastocyst implants.

Endorphin A hormone that acts as an endogenous opioid and binds the same receptors as morphine.

Entrainment The synchronization of biological rhythms to a periodic environmental cue.

Enzymoimmunoassay (EIA) An assay that uses the principle of competitive binding of an antibody to its antigen to determine the presence or quantity of a biological substance such as a hormone.

Epinephrine A catecholamine produced in the adrenal medulla that increases cardiac tone and glucose levels.

Estradiol The primary estrogenic steroid hormone produced by mammalian ovaries.

Estrogen Any of the C_{18} class of steroid hormones, so named because of their estrus-generating properties in female mammals. Biologically important estrogens include estradiol, estrone, and estriol.

Estrous cycle A recurrent cycle between mating and nonmating conditions.

Estrus The period during which female mammals will permit copulation.

Eutherian mammal The subclass of mammals that possess a placenta during pregnancy.

Evolutionary approach The perspective(s) adopted by biologists who assume that evolutionary processes are central to issues in ecology, systematics, and behavior.

Exocrine gland A gland that has a duct through which its product is secreted into adjacent organs or the environment. Examples of exocrine glands include the salivary, sweat, and mammary glands.

Exocytosis The extrusion or secretion of substances from a cell by the fusion of a vesicle membrane with the cell membrane. The vesicle contains the material to be extruded from the cell. This process occurs during hormone secretion and neurotransmitter release.

Exogenous Relating to a subject or process outside the organism.

Extirpation The complete surgical removal of a gland or other organ.

Fallopian tubes Tubes that connect the ovary to the uterus. Ova travel down the fallopian tubes after ovulation, and fertilization normally occurs in these tubes.

False negative A testing error in which an affected individual is reported to be unaffected.

False positive A testing error in which an unaffected individual is reported to be affected.

Feminization The induction of female traits.

Fertilization The penetration of an egg by a sperm with the subsequent combination of paternal and maternal DNA.

Fetus In humans, the stage of development that begins about 9 weeks after conception and continues until birth. The process of sexual differentiation begins in humans as embryos become fetuses.

Flehmen response A stereotyped behavior exhibited by males of many ungulate (hoofed) species when sexually excited. This response is observed after the male investigates a female's anogenital region and consists of extending the neck and curling the upper lip to reveal the upper gums, which allows the female chemosignal to be delivered to the vomeronasal organ.

Follicle An epithelial cell-lined sac or compartment of the thyroid gland, ovary, or other structure. Thyroid follicles are filled with colloidal material that contains thyroid hormones. In the ovary, each follicle contains an ovum. As the follicle and ovum mature, cells lining the follicle produce estrogens that regulate ovulation; after ovulation, the follicle develops into another endocrine structure called the corpus luteum.

Follicle-stimulating hormone (FSH) A gonadotropic hormone from the anterior pituitary that stimulates follicle development in females and sperm production in males.

Follicular phase The portion of the primate menstrual cycle that begins at the end of menstruation and ends at ovulation, characterized by high blood levels of estrogens and the development of follicles.

Freemartin A (usually) sterile female calf that is born as a twin to a male.

Free-running rhythm A biological rhythm that is not synchronized to its natural zeitgeber and expresses its own endogenous rhythm.

Frequency The number of completed cycles per unit of time; for example, two cycles per month.

G proteins A class of proteins located adjacent to the intracellular part of a hormone or neurotransmitter receptor that are activated when an appropriate ligand binds to the receptor.

Gametic sex The sex of an individual as determined by the production of ova by females and sperm by males.

Leydig cells The interstitial cells between the seminiferous tubules in the testes that produce androgens in response to luteinizing hormone from the anterior pituitary.

Ligand A substance that binds to a receptor molecule.

Lipid-based hormone A hormone derived from a fatty acid. Many lipid-based hormones are called eicosanoids, which are a class of oxygenated hydrophobic hormones that largely function as paracrine factors. All eicosanoids, including prostaglandins, leukotrienes, and thromboxanes, derive from arachidonic acid, which is a fatty acid derivative.

Lipolysis The breakdown of adipose tissue into free fatty acids.

Lordosis A female sexually receptive posture in which the hindquarters are raised and the tail is deflected to facilitate copulation.

Luteal phase The portion of the primate menstrual cycle that begins at ovulation and continues until the onset of menstruation, and is characterized by corpora luteal function and high blood levels of progesterone.

Luteinizing hormone (LH) A gonadotropin from the anterior pituitary that promotes formation of the corpora lutea in females and testosterone production in males.

Lux A unit of illumination equal to the direct illumination of a 1-meter surface by a uniform point source of 1 candle intensity.

Marsupial A mammal belonging to the subclass Metatheria that lacks a placenta, such as opossums and most Australian mammals. Most marsupials have a pouch (marsupium) in which the mammary glands are located and the young are transported.

Masculinization The induction of male traits.

Maternal aggression A type of aggressive behavior observed among new mothers when they fiercely defend their young from intruders.

Maternal behavior Parental behavior performed by the mother or another female.

Mating The behavioral process of bringing the two sex gamete types together for fertilization; i.e., copulation.

Medial preoptic area *See* Preoptic area.

Meiosis The process of gamete formation, in which each new haploid cell receives half of the chromosomes in the somatic cells.

Melanocyte-stimulating hormone (MSH) A peptide hormone secreted by the pituitary gland that regulates skin color in some vertebrates by stimulating melanin synthesis in melanocytes and melanin granule dispersal in melanophores. MSH is also called intermedin and melanotropin.

Melanotropin *See* melanocyte-stimulating hormone.

Melanotropin-inhibiting hormone (MIH) A peptide hormone that inhibits MSH secretion.

Melanotropin-releasing hormone (MRH) A hexapeptide that stimulates the secretion of melanotropin.

Melatonin An indole amine hormone released by the pineal gland.

Memory The encoding, storage, and retrieval of information about past experience.

Menopause The cessation of menstrual cycles in women following the depletion of eggs from the ovary.

Menstrual cycle The cyclic changes in hormone levels observed in women and other female primates that are associated with changes in ovarian follicular and luteal activity.

Menstruation The sloughing off of the endometrial lining of the uterus in women during a nonpregnant menstrual cycle. The corkscrew-shaped blood vessels that remain after the endometrial lining is shed leak blood into the uterine lumen.

Metabolic fuels The normal sources of energy for individuals, usually consisting of simple sugars (e.g., glucose), fatty acids, ketone bodies, and amino acids.

Mineralocorticoids One of the two types of corticoids secreted from the adrenal cortices. Aldosterone is the most important mineralocorticoid secreted by terrestrial vertebrates, and is important in ion exchange and water metabolism.

Mitosis The process of division of somatic cells that involves the duplication of DNA.

Monoamine A hormone or neurotransmitter that contains one amine group.

Monotreme A primitive egg-laying mammal, such as the duck-billed platypus and spiny anteater.

Morphological sex The sex of an individual as determined by body form.

Mounting A behavior observed among males of many species with internal fertilization in which the male assumes a copulatory position, but does not insert his penis (or other intromittive organ) into the female's vagina (or urogenital opening). This behavior is androgen-dependent.

Müllerian ducts A duct system present in both sexes during embryonic development that connects the gonads to the exterior. During normal development, the Müllerian duct system develops into the accessory sex organs in females and regresses in males.

Müllerian inhibitory hormone (MIH) A peptide hormone produced in the Sertoli cells in the developing testis that suppresses development of the Müllerian duct system, which prevents development of the uterus and cervix. Also called Müllerian inhibitory factor (MIF).

Negative feedback A regulatory system that tends to stabilize a process when its effects are pronounced by reducing its rate or output. A thermostat works on the principle of negative feedback to keep temperature within a certain range.

Neuroendocrinology The scientific study of the interaction between the nervous system and the endocrine system.

Neurohormone A hormone that is released into the blood from a neuron rather than from an endocrine gland.

Neurosecretory cells Cells in the central nervous system that secrete their products beyond the synapse to affect function in other cells.

Neurosteroid Steroids that are synthesized in the central nervous system (CNS) and the peripheral nervous systems (PNS), independently of the steroidogenic activity of the endocrine glands (e.g., gonads and adrenals).

Nocturnal Active at night.

Norepinephrine A substance that can act as either a hormone or neurotransmitter, which is secreted by the adrenal medulla and the nerve endings of the sympathetic nervous system. It is also called noradrenalin, and is chemically similar to epinephrine (adrenalin).

Nucleus (pl. nuclei) A collection of nerve cell bodies in the brain.

Ontogeny The development or course of development of an individual.

Opiates A class of compounds that exert opium-like effects, including pain insensitivity.

Opioids Peptide hormones located throughout the brain that bind to opiate receptors.

Orexigenic peptides Protein hormones found in the lateral hypothalamus that may trigger feeding.

Orexin *See* hypocretin.

Organizational/activational hypothesis The proposal that early androgen exposure permanently organizes the nervous system of mammals in a male-like manner. After pre- or perinatal organization by androgens, these hormones more readily activate male-typical postpubertal behaviors by acting upon the organized structures.

Oscillator A self-sustained biological timekeeper that functions in constant conditions.

Osmolality The concentration of solutes in a solution.

Osmoregulation The physiological and behavioral control of osmolality.

Osmosis Process of movement of a solvent through a semipermeable membrane (e.g., a living cell) into a solution of higher solute concentration that tends to equalize the concentrations of solute on the two sides of the membrane.

Osmotic thirst Motivation to consume water caused by increased osmolality in the brain.

Ovariectomy The surgical removal of the ovaries.

Ovary The female gonad, which produces estrogens, progestins, and ova.

Ovulation The release of a mature egg (ovum) from the ovary. Ovulation may occur spontaneously in response to a specific hormonal milieu or as a direct result of stimuli associated with copulation.

Ovum A haploid female gamete.

Oxytocin A peptide hormone secreted by the posterior pituitary that triggers milk letdown in lactating females and may be involved in other reproductive behaviors.

Pacemaker An oscillator that regulates some physiological or behavioral function.

Pancreas A composite vertebrate gland with both endocrine and exocrine functions. In humans, the pancreas is located within the curve of the duodenum behind the stomach and liver, and secretes digestive enzymes (exocrine function), insulin, glucagon, and somatostatin (endocrine function), as well as bicarbonate.

Paracrine Secretion of locally acting biological substances from cells.

Parathyroid gland In humans and most other eutherian mammals, four to eight islands of separate endocrine tissue associated with the thyroid gland. The parathyroid gland secretes parathyroid hormone, which increases blood calcium levels and decreases blood phosphate levels.

Parathyroid hormone (PTH) A protein hormone that is secreted by the parathyroid glands that regulates calcium and phosphate metabolism. PTH raises the extracellular calcium and blood plasma calcium concentrations.

Paraventricular nucleus A collection of cell bodies in the hypothalamus that produce vasopressin and oxytocin. These hormones are transported down the cells' axons for release from the posterior pituitary.

Parental behavior Behaviors performed in relation to one's offspring that contribute directly to the survival of fertilized eggs or offspring that have left the body of the female.

Parthenogenesis A type of asexual reproduction in which eggs can develop into offspring without fertilization.

Parturition The process of giving birth in mammals.

Passive avoidance A type of learning in which an individual must suppress some behavior that would otherwise be exhibited.

Paternal behavior Parental behavior performed by the father or another male.

Peptide hormone A class of hormones. Each consists of a relatively short chain of amino acids residues.

Perception The transduction of sensory information entering the nervous system into biologically useful information.

Perimenstrual syndrome (PMS) A constellation of symptoms, including anxiety, depression, moodiness, and fatigue, that recurs on a cyclical basis and is associated with menstruation.

Period The length of time required to complete one cycle of a rhythm, such as the amount of time between peaks in a cycle.

Permissive effect The process by which one hormone induces the production of receptors for a second hormone, or otherwise brings about the conditions necessary for the second hormone to be effective.

Phase A point in a rhythm relative to some objective time point during the cycle, or during the cycle of another rhythm.

Phase-response curve A graphic representation of the differential effects that a periodic environmental cue (usually light) has on the timing of biological rhythms.

Pheromone A chemical signal produced in one individual (the sender) that can alter the physiology and/or the behavior of another individual (the receiver).

Photoperiod Day length, or the amount of light per day.

Photoperiodism The use of photoperiod to time annual cycles.

Photorefractoriness The loss of responsiveness to changes in photoperiod.

Pineal gland An endocrine gland (also called the epiphysis), located in mammals between the telencephalon and diencephalon, that secretes melatonin, a hormone important in the regulation of daily and seasonal cycles.

Pinealocytes The primary cells of the pineal gland. They produce and secrete melatonin. Pinealocytes have a unique organelle called a *synaptic ribbon*, which serves as a specific marker for pinealocytes. Enzymes of the pinealocytes include 5-HT, N-acetyltransferase, and 5-hydroxyindole-O-methyltransferase which are used to convert serotonin to melatonin.

Pituitary gland An endocrine gland that has two distinct anatomical components in humans, the anterior pituitary and the posterior pituitary, which have different embryological origins and different functional roles in the endocrine system. The pituitary gland (also called the hypophysis) is connected to the median eminence of the hypothalamus by the pituitary stalk, or infundibulum.

Placenta A specialized organ produced by the mammalian embryo that is attached to the uterine wall and serves to provide nutrients, hormones, and energy to the fetus.

Plasma The clear, straw-colored fluid in which the blood cells are suspended.

Polypeptide An unbranched string of amino acids linked together by peptide bonds.

Portal system A special closed blood circuit in which two beds of capillaries are connected by a vein; thus, the flow of blood is in one direction only. The hypothalamus–pituitary portal system comprises the hypothalamic capillary bed and the anterior pituitary capillary bed, connected by a vein. The majority of blood flow, and hence endocrine information, is from the hypothalamus to the anterior pituitary.

Positive feedback A regulatory process that tends to accelerate an ongoing process by increasing production in response to the end product.

Posterior pituitary The rear part of the endocrine gland that extends from the base of the brain and stores and releases oxytocin and vasopressin (or some variant of these two nanapeptide hormones) which are produced in the hypothalamus.

Postpartum depression A decrease in mood that occurs within a few days of giving birth. Symptoms can range from mild (maternity blues) and brief (24–48 hours) to severe. Hormones associated with parturition may be involved in postpartum depression, but social and psychological factors may be as important as biological factors in the mediation of this syndrome.

Precocial Born or hatched at an advanced stage of development so that little or no parental intervention is required for survival.

Preference threshold The first detectable preference displayed by an individual for any substance or solution.

Pregnancy The state of a female mammal when she is carrying a developing fetus in her uterus.

Pregnenolone A C_{21} steroid prohormone that is the obligatory precursor for all other steroid hormones in vertebrates.

Preoptic area A region of the brain anterior to the hypothalamus. This region is usually divided into the lateral and medial preoptic areas.

Preoptic medial nucleus (POM) A collection of cell bodies in the songbird preoptic area that stains positively for estrogen receptors as well as for aromatase, and is thought to be critical in male avian copulatory behavior.

Preprohormone A sequence of amino acids that contains three different elements: (1) a signal sequence, (2) one or more copies of a peptide hormone, and (3) other peptide sequences that may or may not possess biological activity. Peptide hormones are first synthesized as preprohormones in the rough endoplasmic reticulum of endocrine cells.

Proceptivity The extent to which females initiate copulation.

Proestrus The vaginal cellular condition coincident with mating behavior (estrus) in female rodents.

Progestins A class of C_{21} steroid hormones, so named for their "progestational," or pregnancy-maintaining, effects in mammals. Progesterone is a common progestin.

Prohormone A molecule that can act as a hormone itself or can be converted into another hormone with different properties.

Prolactin A protein hormone that is highly conserved throughout vertebrate evolution and has many physiological functions, which can be broken down into five basic classes: (1) reproduction, (2) growth and development, (3) water and electrolyte balance, (4) maintenance of integumentary structures, and (5) actions on steroid-dependent target tissues or synergisms with steroid hormones to affect target tissues.

Prolactin inhibitory hormone (PIH) Dopamine inhibits prolactin secretion from the anterior pituitary.

Pro-opiomelanocortin (POMC) A precursor protein that consists of 241 amino acid residues. It is synthesized in the anterior and intermedate pituitary gland. Depending on the stimulus and site of production, POMC can be cleaved into a number of peptide hormones including ACTH, β-LPH, and met-enkephalin (in the anterior pituitary in response to CRH), or CLIP, γ-LPH, α-MSH, and β-endorphin (in the intermediate pituitary in response to dopamine).

Prostaglandins A family of lipid-based hormones that possess a basic 20-carbon fatty acid skeleton called prostanoic acid; prostaglandins are involved in several aspects of reproductive function.

Protandry A form of sequential hermaphroditism in which individuals begin life as males, then change into females.

Protein hormone A class of hormones. Each consists of a large chain of amino acids residues; also called a polypeptide.

Protein kinase An enzyme that transfers a phosphate group from adenine triphosphate (ATP) to the protein substrate of the enzyme. The activation of specific protein kinases represents the mechanism by which most protein hormones initially exert their effects on cellular function.

Protogyny A form of sequential hermaphroditism in which individuals begin life as females, then change into males.

Pseudohermaphrodite An individual, especially a human, born with ambiguous external genitalia.

Pseudopregnancy The luteal phase of the estrous cycle, or any period when there is a functional corpus luteum and buildup of the endometrial uterine layer in the absence of pregnancy.

Puberty The process of sexual maturation during which an individual becomes fertile.

Pulsatile secretion The episodic secretion of hormones in periodic bursts or spurts.

Radioimmunoassay (RIA) A technique used to measure hormones or other biological substances by using antibodies and purified radiolabeled ligands.

Receptivity The stimulus value of a female for eventually eliciting an intravaginal ejaculation from a male conspecific.

Receptor A chemical structure on the cell surface or inside the cell that has an affinity for a specific chemical configuration of a hormone, neurotransmitter, or other chemical compound.

5α-reductase An enzyme necessary to convert testosterone to 5α-dihydrotestosterone.

Relaxin A polypeptide hormone that is secreted by the corpus luteum during the last days of pregnancy in nonprimates; it relaxes the pelvic ligaments and prepares the uterus for labor.

Releasing hormone One of several polypeptides released from the hypothalamus that increase or decrease the release of hormones from the anterior pituitary gland.

Rhythm A recurrent event that is characterized by its period, frequency, amplitude, and phase.

Ribonucleic acid (RNA) A nucleic acid that implements the information stored in DNA. Two forms of RNA are transfer RNA and messenger RNA.

Scrotum An external pouch of skin that contains the testes.

Seasonal affective disorder (SAD) A psychological and physical syndrome, characterized by reduced mood, lethargy, loss of libido, hypersomnia, excessive weight gain, carbohydrate cravings, anxiety, and inability to concentrate or focus attention, that recurs during the late autumn and winter and subsides in the spring.

Second messenger A biological molecule released when a hormone binds to its receptor; the second messenger activates the cellular machinery of the target cell (the hormone is the first messenger).

Secretin A polypeptide hormone produced in the duodenum, in response to gastric acid secretion, to stimulate production of pancreatic secretions.

Seminiferous tubules The long, convoluted tubes in which spermatogenesis occurs.

Sensation The initial processing of sensory information as it enters the nervous system through sensory receptors.

Sensitization *See* concaveation.

Sequential hermaphrodite An animal that begins life as one sex, then changes to the other sex as an adult in response to environmental or genotypic factors.

Serotonin (5-HT) A neurotransmitter formed from tryptophan (5-hydroxytryptamine); 5-HT is also a vasoconstrictor and the precursor to melatonin formation in the pineal gland.

Sertoli cells Cells located along the basement membrane of the seminiferous tubules in which sperm cells are embedded while they mature.

Set point A reference value for a regulated physiological variable.

Sex determination The point at which an individual begins to develop as either a male or a female. In animals with sex chromosomes, sex determination occurs at fertilization; in animals that lack sex chromosomes, sex determination coincides with the presence of a specific environmental or social condition.

Sex drive The powerful motivational forces propelling individuals to seek copulation; in humans, often referred to as libido.

Sexual differentiation The process by which individuals develop the characteristics associated with being male or female. Among vertebrates, differential exposure to gonadal steroid hormones after sex determination causes sexual dimorphism in several structures, including the brain.

Sexual orientation The process of developing an erotic sexual attraction for other people. This term suggests that the process is mediated primarily by biological factors.

Sexual preference The process of developing an erotic sexual attraction for other people. This term suggests that the process is mediated primarily by learning and conscious choices.

Sexual selection A subset of natural selection that occurs when individuals within a population differ in their abilities to compete with members of the same sex for mates (intrasexual selection) or to attract mates of the opposite sex (intersexual selection).

Sexually dimorphic nucleus of the preoptic area (SDN-POA) A set of cell bodies anterior to the hypothalamus that is larger in male than in female humans and rodents. The functional significance of this brain dimorphism is unknown.

Signal transduction The intercellular or intracellular transfer of information (biological activation/inhibition) through a signal pathway which is mediated by a biologically active molecule (e.g., hormone, neurotransmitter) through the binding to a receptor/enzyme as part of a cascade of activation/inhibition of a second messenger system or an ion channel.

Simultaneous hermaphrodite An animal that possesses ovotestes that produce both eggs and sperm, and alternates between two behavioral roles in providing eggs or sperm during spawning.

Single-unit recording A technique that involves the placement of very small electrodes in or near one nerve cell to record changes in its neural activity before, during, and after some experimental treatment.

Somatocrinin *See* growth hormone-releasing hormone.

Somatomedins Insulin-like polypeptides (growth factors) produced in the liver and in some fibroblasts and released into the blood when stimulated by GH.

Somatostatin A polypeptide hormone that is mainly produced in the hypothalamus that inhibits the secretion of various other hormones, including somatotropin, glucagon, insulin, TSH, and gastrin.

References

Abbott, D. H., Keverne, E. B., Bercovitch, F. B., Shively, C. A., Mendoza, S. P., Saltzman, W., Snowdon, C. T., Ziegler, T. E., Banjevic, M., Garland, T. Jr., and Sapolsky, R. M. 2003. Are subordinates always stressed? A comparative analysis of rank differences in cortisol levels among primates. *Hormones and Behavior* 43:67–82.

Abbott, D. H., Saltzman, W., Schultz-Darken, N. J., Smith, T. E. 1997. Specific neuroendocrine mechanisms not involving generalized stress mediate social regulation of female reproduction in cooperatively breeding marmoset monkeys. *Ann. N. Y. Acad. Sci.*, 807:219–38

Abel, G. G., Barlow, D. H., Blanchard, E. B., and Arnold, D. 1977. The components of rapists' sexual arousal. *Arch. Gen. Psychiatry*, 34: 895–903.

Abercrombie, H. C., Kalin, N. H., Thurow, M. E., Rosenkranz, M. A., and Davidson, R. J. 2003. Cortisol variation in humans affects memory for emotionally laden and neutral information. *Behavioral Neuroscience*, 117: 506–516.

Abitbol, M. L. and Inglis, S. R. 1997. Role of amniotic fluid in newborn acceptance and bonding in canines. *J. Matern. Fetal Med.*, 6:49–52.

Abraham, G. E. 1980. The premenstrual tension syndromes. In L. K. McNall (ed.), *Contemp. Obstet. Gynecol. Nurs.*, Vol. 3, pp. 170–184. Mosby, St. Louis.

Abrahamson, E. E., and Moore, R. Y. 2001. Suprachiasmatic nucleus in the mouse: Retinal innervation, intrinsic organization and efferent projections. *Brain Research*, 916: 172–191.

Adams, D. B., Gold, A. R., and Burt, A. D. 1978. Rise in female-initiated sexual activity at ovulation and its suppression by oral contraceptives. *N. Engl. J. Med.*, 229:1145–1150.

Adan, R. A., Hillebrand, J. J., De Rijke, C., Nijenhuis, W., Vink, T., Garner, K. M., and Kas, M. J. 2003. Melanocortin system and eating disorders. *Annals of the New York Academy of Science*, 994: 267–274.

Adels, L. E., and Leon, M. 1986. Thermal control of mother–young contact in Norway rats: Factors mediating the chronic elevation of maternal temperature. *Physiol. Behav.*, 36:183–196.

Adkins, E. K., and Adler, N. T. 1972. Hormonal control of behavior in the Japanese quail. *J. Comp. Physiol. Psychol.*, 81:27–36.

Adkins, E. K., and Nock, B. L. 1976. The effects of the antiestrogens C1-628 on sexual behavior activated by androgen or estrogen in quail. *Horm. Behav.*, 7:417–429.

Adkins, E. K., and Pniewski, E. E. 1978. Control of reproductive behavior by sex steroids in male quail. *J. Comp. Physiol. Psychol.*, 92:1169–1178.

Adkins-Regan, E. 1987. Sexual differentiation in birds. *Trends Neurosci.*, 10:517–522.

Adkins-Regan, E. 1988. Sex hormones and sexual orientation in animals. *Psychobiology*, 16:335–347.

Adkins-Regan, E. 1996. Neuroanatomy of sexual behavior in the male Japanese quail from top to bottom. *Poultry Avian Biol. Rev.*, 7:193–204.

Adkins-Regan, E., Orgeur, P., and Signoret, J. P. 1989. Sexual differentiation of reproductive behavior in pigs: Defeminizing effects of prepubertal estradiol. *Horm. Behav.*, 23:290–303.

Agarwal, S. K., and Haney, A. F. 1994. Does recommending timed intercourse really help the infertile couple? *Obstet. Gynecol.*, 84:307–310.

Ågmo, A. 1997. Male rat sexual behavior. *Brain Res. Brain Res. Protoc.*, 1:203–209.

Ågmo, A., and Ellingsen, E. 2003. Relevance of non-human animal studies to the understanding of human sexuality. *Scandinavian Journal of Psychology*, 44: 293–301.

Agranoff, B. W., and Davis, R. E. 1968. The use of fishes in studies of memory formation. In D. Ingle (ed.), *The Central Nervous System and Fish Behavior*, pp. 193–202. University of Chicago Press, Chicago.

Aguzzi, A., Brandner, S., Sure, U., Ruedi, D., and Isenmann, S. 1994. Transgenic and knock-out mice: Models of neurological disease. *Brain Pathol.*, 4, 3–20.

Ahlenius, S., and Larsson, K. 1984. Apomorphine and haloperidol-induced effects on male sexual behavior: No evidence for actions due to stimulation of central dopamine autoreceptors. *Pharmacol. Biochem. Behav.*, 21:463–466.

Ainsworth, M. D. S. 1972. Attachment and dependency: A comparison. In J. L. Gerwirtz, *Attachment Dependency*, pp. 97–137. V. H. Winston, Washington, D.C.

Albers, H. E. 1981. Gonadal hormones organize and modulate the circadian system of the rat. *Am. J. Physiol.*, 241:R62–R66.

Albers, H. E., Hennessey, A. C., and Whitman, D. C. 1992. Vasopressin and the regulation of hamster social behavior. *Ann. N. Y. Acad. Sci.*, 652:227–242.

Albers, H. E., Pollock, J., Simmons, W. H., and Ferris, C. F. 1986. A V1-like receptor mediates vasopressin-induced flank marking behavior in hamster hypothalamus. *J. Neurosci.*, 6:2085–2089.

Albers, H. E., Huhman, K. L., and Meisel, R. L. 2002. Hormonal basis of social conflict and communication. In D. W. Pfaff, A. P. Arnold, A. M. Etgen, S. E. Fahrbach, and R. T. Rubin (eds.), *Hormones, Brain and Behavior*, Volume 1, pp. 393–434. Academic Press, New York.

Albert, D. J., and Walsh, M. L. 1984. Neural systems and the inhibitory modulation of agonistic behavior: A comparison of mammalian species. *Neurosci. Biobehav. Rev.*, 8:5–24.

Albert, D. J., Petrovic, D. M., Walsh, M. L., and Jonik, R. H. 1990. Medial accumbens lesions attenuate testosterone-dependent aggression in male rats. *Physiol. Behav.*, 46:625–631.

Alberts, J. R. 1974. Producing and interpreting experimental olfactory deficits. *Comp. Biochem. Physiol.*, 40A:971–974.

Alberts, J. R., and Brunjes, P. C. 1978. Ontogeny of thermal and olfactory determinants of huddling in the rat. *J. Comp. Physiol. Psychol.*, 92:897–906.

Alberts, J. R., and Galef, B. G. 1971. Acute anosmia in the rat: A behavioral test of a peripherally induced olfactory deficit. *Physiol. Behav.*, 12:657–670.

Alberts, J. R., and Gubernick, D. J. 1990. Functional organization of dyadic and triadic parent–offspring systems. In N. A. Krasnegor and R. S. Bridges (eds.), *Mammalian Parenting*, pp. 416–440. Oxford University Press, Oxford.

Alexander, G. M., and Hines, M. 2002. Sex differences in response to children's toys in nonhuman primates (*Cercopithecus aethiops sabaeus*). *Evolution and Human Behavior*, 23: 467–479.

Alexander, G. M., and Sherwin, B. B. 1993. Sex steroids, sexual behavior, and selection attention for erotic stimuli in women using oral contraceptives. *Psychoneuroendocrinology*,18:91–102.

Alexander, G. M., Sherwin, B. B., Bancroft, J., and Davidson, D. W. 1990. Testosterone and sexual behavior in oral contraceptive users and nonusers: A prospective study. *Horm. Behav.*, 24:388–402.

Alexander, G. M., Swerdloff, R. S., Wang, C., Davidson, T., McDonald, V., Steiner, B., and Hines, M. 1998. Androgen-behavior correlations in hypogonadal men and eugonadal men. II. Cognitive abilities. *Horm. Behav.*, 33:85–94.

Alexander, R. D., Hoogland, J. L., Howard, R. D., Noonan, K. M., and Sherman, P. W. 1979. Sexual dimorphisms and breeding systems in pinnipeds, ungulates, primates and humans. In N. A. Chagnon and W. Irons (eds.), *Evolutionary Biology and Human Social Behavior*, pp. 402–435. Duxbury Press, North Scituate, MA.

Allada, R., White, N. E., So, W. W., Hall, J. C., and Rosbash, M. 1998. A mutant *Drosophila* homolog of mammalian Clock disrupts circadian rhythms and transcription of period and timeless. *Cell*, 93:791–804.

Allen, E. 1922. The oestrus cycle in the mouse. *Am. J. Anat.*, 30:297–325.

Allen, E., and Doisy, E. A. 1923. An ovarian hormone: Preliminary report of its localization, extraction and partial purification, and action in test animals. *J.A.M.A.*, 81:819–821.

Allen, E., Francis, B. F., Robertson, L. L., Colgate, C. E., Johnston, C. G., Doisy, E. A., Kountz, W. B., and Gibson, H. V. 1924. The hormone of the ovarian follicle: Its localization and action in test animals, and additional points bearing upon the internal secretion of the ovary. *Am. J. Anat.*, 34:133–182.

Allsop, D. J., and West, S. A. 2003. Life history: Changing sex at the same relative body size. *Nature*, 425: 783–784.

Altemus, M., Hetherington, M., Kennedy, B., Licinio, J., and Gold, P. W. 1996. Thyroid function in bulimia nervosa. *Psychoneuroendocrinology*, 21: 249–261.

Altmann, J. 1980. *Baboon Mothers and Infants.* Harvard University Press, Cambridge, MA.

Altshuler, L. L., Hendrick, V., and Parry, B. 1995. Pharmacological management of premenstrual disorder. *Harv. Rev. Psychiatry*, 2:233–245.

Amalric, M., Cline, E. J., Martinez, J. L., Bloom, F. E., and Koob, G. F. 1987. Rewarding properties of β-endorphin as measured by conditioned place preference. *Psychopharmacology*, 91:14–19.

American Psychiatric Association. 1994. *Diagnostic and Statistical Manual of Mental Disorders: DSM-IV.* (4th ed.) American Psychiatric Association, Washington, D.C.

Anagnostaras, S. G., Maren, S., DeCola, J. P., Lane, N. I., Gale, G. D., Schlinger, B. A., and Fanselow, M. S. 1998. Testicular hormones do not regulate sexually dimorphic Pavlovian fear conditioning or perforant-path long-term potentiation in adult male rats. *Behav. Brain Res.*, 92:1–9.

Anand, B. K., and Brobeck, J. R. 1951. Localization of a feeding center in the hypothalamus of the rat. *Proc. Soc. Exp. Biol. Med.*, 77:323–324.

Anastasi, A., Erspamer, V., and Bucci, M. 1971. Isolation and structure of bombesin and alytesin, 2 analogous active peptides from the skin of the European amphibians Bombina and Alytes. *Experientia*, 27:123–140.

Anderson, R. A., Bancroft, J., and Wu, F. C. W. 1992. The effects of exogenous testosterone on sexuality and mood

Brantley, R. K., Wingfield, J., and Bass, A. H. 1993. Hormonal bases for male teleost dimorphisms: Sex steroid levels in *Porichthys notatus*, a fish with alternative reproductive tactics. *Horm. Behav.*, 27:332–347.

Breedlove, S. M. 1992. Sexual dimorphism in the vertebrate nervous system. *J. Neurosci.*, 12:4133–4142.

Bremner, J. D., Randall, P., Scott, T. M., Bronen, R. A., Seibyl, J. P., Southwick, S. M., Delaney, R. C., McCarthy, G., Charney, D. S., and Innis, R. B. 1995. MRI-based measurement of hippocampal volume in patients with combat-related posttraumatic stress disorder. *American Journal of Psychiatry*, 152: 973–981.

Bremner, J. D., Randall, P., Vermetten, E., Staib, L., Bronen, R. A., Mazure, C., Capelli, S., McCarthy, G., Innis, R. B., and Charney, D. S. 1997. Magnetic resonance imaging-based measurement of hippocampal volume in post-traumatic stress disorder related to childhood physical and sexual abuse: A preliminary report. *Biological Psychiatry*, 41: 23–32.

Brenowitz, E. A. 1991. Altered perception of species-specific song by female birds after lesions of a forebrain nucleus. *Science*, 251:303–305.

Brenowitz, E. A., and Arnold, A. P. 1989. Accumulation of estrogen in a vocal control brain region of a duetting song bird. *Brain Res.*, 359:364–367.

Brenowitz, E. A., and Arnold, A. P. 1992. Hormone accumulation in song regions of the canary brain. *J. Neurobiol.*, 23:871–880.

Brenowitz, E. A., Baptista, L., Lent, K., and Wingfield, J. 1998. Seasonal plasticity of the song control system in wild Nuttall's white-crowned sparrows. *J. Neurobiol.*, 34: 69–82.

Brenowitz, E. A., Nalls, B., Wingfield, J. C., and Kroodsma, D. E. 1991. Seasonal changes in avian song control nuclei without seasonal changes in song repertoire. *J. Neurosci.*, 11:1367–1374.

Brett, M., and Baxendale, S. 2001. Motherhood and memory: A review. *Psychoneuroendocrinology*, 26: 339–362.

Breuner, C. W., and Hahn, T. P. 2003. Integrating stress physiology, environmental change, and behavior in free-living sparrows. *Hormones and Behavior*, 43: 115–123.

Breuner, C. W., Wingfield, J. C., and Romero, L. M. 1999. Diel rhythms of basal and stress-induced corticosterone in a wild, seasonal vertebrate, Gambel's white-crowned sparrow. *Journal of Experimental Zoology*, 284: 334–342.

Bridges, R. S. 1975. Long-term effects of pregnancy and parturition upon maternal responsiveness in the rat. *Physiol. Behav.*, 14:245–249.

Bridges, R. S. 1977. Parturition: Its role in the long-term retention of maternal behavior in the rat. *Physiol. Behav.*,18:487–490.

Bridges, R. S. 1978. Retention of rapid onset of maternal behavior during pregnancy in primiparous rats. *Behav. Biology*, 24:113–117.

Bridges, R. S. 1990. Endocrine regulation of parental behavior in rodents. In N. A. Krasnegor and R. S. Bridges (eds.), *Mammalian Parenting*, pp. 93–117. Oxford University Press, Oxford.

Bridges, R. S. 1996. Biochemical basis of parental behavior in the rat. *Adv. Study Behav.*, 25:215–242.

Bridges, R. S., and Freemark, M. S. 1995. Human placental lactogen infusions into the medial preoptic area stimulate maternal behavior in steroid-primed, nulliparous female rats. *Horm. Behav.*, 29:216–226.

Bridges, R. S., and Goldman, B. D. 1975. Diurnal rhythms in gonadotropins and progesterone in lactating and photoperiod induced acyclic hamsters. *Biol. Reprod.*, 13:617–622.

Bridges, R. S., Robertson, M. C., Shiu, R. P., Friesen, H. G., Stuer, A. M., and Mann, P. E. 1996. Endocrine communication between conceptus and mother: Placental lactogen stimulation of maternal behavior. *Neuroendocrinology*, 64:57–64.

Bridges, R. S., Rosenblatt, J. S., and Feder, H. 1978a. Serum progesterone concentrations and maternal behavior in rats after pregnancy termination: Behavioral stimulation following progesterone withdrawal and inhibition by progesterone maintenance. *Endocrinology*, 102:258–267.

Bridges, R. S., Rosenblatt, J. S., and Feder, H. 1978b. Stimulation of maternal responsiveness after pregnancy termination in rats: Effect of time of onset of behavioral testing. *Horm. Behav.*, 10:235–245.

Brinkmann, A., Jenster, G., Ris-Stalpers, C., van der Korput, H., Bruggenwirth, H., Boehmer, A., and Trapman, J. 1996. Molecular basis of androgen insensitivity. *Steroids*, 61:172–5175.

Brinsmead, M., Smith, R., Singh, B., Lewin, T., and Owens, P. 1985. Peripartum concentrations of beta-endorphin and cortisol and maternal mood states. *Aust. N. Z. J. Obstet. Gynaecol.*, 12:194–197.

Brobeck, J. R., Wheatland, M., and Strominger, J. L. 1947. Variations in regulation of energy exchange associated with estrus, diestrus and pseudopregnancy in rats. *Endocrinology*, 40:65–72.

Broida, J., Michael, S., and Svare, B. 1981. Plasma prolactin is not related to the initiation, maintenance, and decline of maternal aggression in mice. *Behav. Neural Biol.*, 35:121–125.

Bronson, F. H. 1967. Effects of social stimulation on adrenal and reproductive physiology of rodents. In M. L. Conalty (ed.), *Husbandry of Laboratory Animals*. Academic Press, New York.

Bronson, F. H. 1979. The reproductive ecology of the house mouse. *Q. Rev. Biol.*, 54:265–299.

Bronson, F. H. 1988. Seasonal regulation of reproduction in mammals. In E. Knobil and J. D. Neill (eds.), *The Physiology of Reproduction*, Vol. 2, pp. 1831–1872. Raven Press, New York.

Bronson, F. H. 1989. *Mammalian Reproductive Biology*. University of Chicago Press, Chicago.

Bronson, F. H. 1996. Effects of prolonged exposure to anabolic steroids on the behavior of male and female mice. *Pharmacol. Biochem. Behav.* 53:329–334.

Bronson, F. H. 1999. Puberty and energy reserves: a walk on the wild side. In K. Wallen and J. E. Schneider (eds.), *Reproduction in Context*. MIT Press, Cambridge, MA.

Bronson, F. H. 2004. Are humans seasonally photoperiodic? *Journal of Biological Rhythms*, 19: 180–192.

Bronson, F. H. and Matherne, C. M. 1997. Exposure to anabolic-androgenic steroids shortens life span of male mice. *Med. Sci. Sports Exerc.*, 29:615–619.

Bronson, F. H., and Desjardins, C. 1982a. Endocrine responses to sexual arousal in male mice. *Endocrinology,* 111:1286–1291.

Bronson, F. H., and Desjardins, C. 1982b. Reproductive aging in male mice. In M. E. Reff (ed.), *Biological Markers of Aging,* pp. 87–93. NIH, Bethesda, MD. Publication no. 82-2221.

Brookhart, J. M., Dey, F. L., and Ranson, S. W. 1941. The abolition of mating behavior by hypothalamic lesions in guinea pigs. *Endocrinology,* 28:561–565.

Brooks-Gunn, J. 1986. Differentiating premenstrual syndromes. *Psychosomat. Med.,* 48:385–387.

Brosens, J. J., Tullet, J., Varshochi, R., and Lam, E. W. 2004. Steroid receptor action. *Best Practices in Research and Clinical Obstetrics and Gynaecology,* 18: 265–283.

Brown, C. S., Ling, F. W., Andersen, R. N., Farmer, R. G., and Arheart, K. L. 1994. Efficacy of depot leuprolide in premenstrual syndrome: Effect of symptom severity and type in a controlled trial. *Obstet. Gynecol.,* 84:77–786.

Brown, F. A. 1972. The clocks timing biological rhythms. *Am. Sci.,* 60:756–766.

Brown, J. R., Ye, H., Bronson, R. T., Dikkes, P., and Greenberg, M. E. 1996. A defect in nurturing in mice lacking the immediate early gene fosB. *Cell,* 86:297–309.

Brown, L. L., Siegel, H., and Etgen, A. M. 1996. Global sex differences in stress-induced activation of cerebral metabolism revealed by 2-deoxyglucose autoradiography. *Horm. Behav.,* 30:611–617.

Brown, M., Rivier, J., Kobayashi, R., and Vale, W. 1978. Cholecystokinin in feeding behavior. In S. R. Bloom (ed.), *Gut Hormones,* pp. 550–558. Raven Press, New York.

Brown, R. E. 1994. *An Introduction to Neuroendocrinology.* Cambridge University Press, Cambridge.

Brown, R. E., Murdoch, T., Murphy, P. R., and Moger, W. H. 1995. Hormonal responses of male gerbils to stimuli from their mate and pups. *Horm. Behav.,* 29:474–491.

Brown, R., and Kulik, J. 1977. Flashbulb memories. *Cognition,* 5:73–99.

Brown, W. A., Monti, P. M., and Corriveau, D. P. 1978. Serum testosterone and sexual activity and interest in men. *Arch. Sex. Behav.,* 7:97–103.

Brown-Parlee, M. 1990. Integrating biological and social scientific research on menopause. *Ann. N. Y. Acad. Sci.,* 592:379–389.

Brown-Parlee, M. 1991. The social construction of premenstrual syndrome: A case study of scientific discourse as cultural contestation. Working paper presented at: *The Good Body: Asceticism in Contemporary Culture.* Institute for the Medical Humanities, University of Texas, Galveston, TX.

Brown-Séquard, C. E. 1899. The effects produced on man by subcutaneous injections of a liquid obtained from the testicles of animals. *Lancet,* 105–106, reprinted in Carter, C. S. (ed.), *Hormones and Sexual Behavior.* Stroudsburg, Dowden, Hutchinson and Ross, Inc.

Brunelli, S. A., and Hofer, M. A. 1990. Parental behavior in juvenile rats: Environmental and biological determinants. In N. A. Krasnegor and R. S. Bridges (eds.), *Mammalian Parenting,* pp. 372–399. Oxford University Press, Oxford.

Bruning, J. C., Gautam, D., Burks, D. J., Gillette, J., Schubert, M., Orban, P. C., Klein, R., Krone, W., Muller-Wieland, D., and Kahn, C. R. 2000. Role of brain insulin receptor in control of body weight and reproduction. *Science,* 289: 2122–2125.

Bryden, A. A., Rothwell, P. J., and O'Reilly, P. H. 1995. Anabolic steroid abuse and renal-cell carcinoma. *Lancet,* 346: 1306–1307.

Buchanan, T. W., and Lovallo, W. R. 2001. Enhanced memory for emotional material following stress-level cortisol treatment in humans. *Psychoneuroendocrinology,* 26: 307–317.

Buckley, C. A., and Schneider, J. E. 2003. Food hoarding is increased by food deprivation and decreased by leptin treatment in Syrian hamsters. *American Journal of Physiology,* 285: R1021–R1029.

Bull, J. J. 1980. Sex determination in reptiles. *Q. Rev. Biol.,* 55:3–21.

Bull, J. J. 1983. *Evolution of Sex Determining Mechanisms.* Benjamin/Cummings, Menlo Park, CA.

Bullivant, S. B., Sellergren, S. A., Stern, K., Spencer, N. A., Jacob, S., Mennella, J. A., and McClintock, M. K. 2004. Women's sexual experience during the menstrual cycle: Identification of the sexual phase by noninvasive measurement of luteinizing hormone. *Journal of Sex Research,* 41: 82–93.

Bullock, D. W. 1970. Induction of heat in ovariectomized guinea pigs by brief exposure to estrogen and progesterone. *Horm. Behav.,* 1:137–143.

Bunnell, B. N., Boland, B. D., and Dewsbury, D. A. 1976. Copulatory behavior of the golden hamster (*Mescricetus auratus*). *Behaviour,* 61:180–206.

Buntin, J. D. 1996. Neural and hormonal control of parental behavior in birds. *Adv. Study Behav.,* 25:161–213.

Buntin, J. D., Ruzycki, E., and Witebsky, J. 1993. Prolactin receptors in dove brain: Autoradiographic analysis of binding characteristics in discrete brain regions and accessibility to blood-borne prolactin. *Neuroendocrinology,* 57:738–750.

Burgers, J. K., Nelson, R. J., Quinlan, D. M., and Walsh, P. C. 1991. Nerve growth factor, nerve grafts and amniotic membrane grafts restore erectile function in rats. *J. Urol.,* 146:463–468.

Burgess, L. H., and Handa, R. J. 1993. Hormonal regulation of androgen receptor mRNA in the brain and anterior pituitary gland of the male rat. *Molecular Brain Research,* 19: 31–38.

Burghardt, G. M. 1988. Precocity, play, and the ectotherm–endotherm transition: Profound reorganization or superficial adaptation? In E. M. Blass (ed.), *Handbook of Behavioral Neurobiology,* Vol. 9, pp. 107–148. Plenum, New York.

Burkhardt, R. W. 1987. The Journal of Animal Behavior and the early history of animal behavior in America. *J. Comp. Psychol.,* 101:223–230.

Burnett, A. L. 1995. Role of nitric oxide in the physiology of erection. *Biol. Reprod.,* 52:485–489.

Burnett, A. L., Johns, D. G., Kriegsfeld, L. J., Klein, S. L., Calvin, D. C., Demas, G. E., Schramm, L. P., Tonegawa, S., Nelson, R. J., Snyder, S. H., and Poss, K. D. 1998.

Ejaculatory abnormalities in mice with targeted disruption of the gene for heme oxygenase-2. *Nat. Med.*, 4:84–87.

Burrows, W. 1945. Periodic spawning of the Palolo worms in Pacific waters. *Nature*, 155:47–48.

Buston, P. 2003. Social hierarchies: Size and growth modification in clownfish. *Nature*, 424: 145–146.

Butenandt, A. 1929. Uber "progynon," ein kristall-isiertes weibliches Sexualhormone. *Naturwiessenschaften*, 17:879. Abstract translated by E. W. Henry in *Biol. Abstr.*, 5:85 (881).

Butera, P. C., and Beikirch, R. J. 1989. Central implants of diluted estradiol: Independent effects on ingestive and reproductive behaviors of ovariectomized rats. *Brain Res.*, 491:266–273.

Buvat, J., Lemaire, A., Buvat-Herbaut, M., Fourlinnie, J. C., Racadot, A., and Fossati, P. 1985. Hyperprolactinemia and sexual function in men. *Horm. Res.*, 22:196–203.

Byne, W., Tobet, S., Mattiace, L. A., Lasco, M. S., Kemether, E., Edgar, M. A., Morgello, S., Buchsbaum, M. S., and Jones, L. B. 2001. The interstitial nuclei of the human anterior hypothalamus: An investigation of variation with sex, sexual orientation, and HIV status. *Hormones and Behavior*, 40: 86–92.

Cahill, G. M., Grace, M. S., and Besharse, J. C. 1991. Rhythmic regulation of retinal melatonin: Metabolic pathways, neurochemical regulation and the ocular circadian clock. *Cell. Mol. Neurobiol.*, 11: 529–560.

Cahill, L., and Alkire, M. 2003. Epinephrine enhancement of human memory consolidation: Interaction with arousal at encoding. *Neurobiology of Learning and Memory*, 79: 194–198.

Cahill, L., Gorski, L., and Le, K. 2003. Enhanced human memory consolidation with post-learning stress: Interaction with the degree of arousal at encoding. *Learning and Memory*, 10: 270–274.

Cahill, L., Haier, R. J., Fallon, J., Alkire, M., Tang, C., Keator, D., Wu, J., and McGaugh, J. L. 1996. Amygdala activity at encoding correlated with long-term, free recall of emotional information *Proceedings of the National Academy of Sciences U.S.A.*, 93: 8016–8021.

Cahill, L., Haier, R. J., White, N. S., Fallon, J., Kilpatrick, L., Lawrence, C., Potkin, S. G., and Alkire, M. T. 2001. Sex-related difference in amygdala activity during emotionally influenced memory storage. *Neurobiology of Learning and Memory*, 75: 1–9.

Cahill, L., Prins, B., Weber, M., and McGaugh, J. L. 1994. Beta-adrenergic activation and memory for emotional events. *Nature*, 271:702–704.

Cain, D. P. 1974. The role of the olfactory bulbs in limbic mechanisms. *Psychol. Bull.*, 81:654–671.

Caldji, C., Tannenbaum, B., Sharma, S., Francis, D., Plotsky, P. M., and Meaney, M. J. 1998. Maternal care during infancy regulates the development of neural systems mediating the expression of fearfulness in the rat. *Proc. Natl. Acad. Sci. U.S.A.*, 95:5335–5340.

Caldwell, G. S., Glickman, S. E., and Smith, E. R. 1984. Seasonal aggression is independent of seasonal testosterone in wood rats. *Proc. Natl. Acad. Sci. U.S.A.*, 81:5255–5257.

Calhoun, J. B. 1961. Phenomena associated with population density. *Proc. Natl. Acad. Sci. U.S.A.*, 47:428–49.

Calhoun, J. B. 1962a. *The Ecology and Sociology of the Norway Rat.* U. S. Government Printing Office, Washington, D.C.

Calhoun, J. B. 1962b. Population density and social pathology. *Sci. Am.*, 206:139–148.

Camazine, B., Gartska, W., Tokarz, R., and Crews, D. 1980. Effects of castration and androgen replacement on male courtship behaviour in the red-sided garter snake (*Thamnophis sirtalis parietalis*). *Horm. Behav.*, 14:358–372.

Cameron, J. L. 1997. Stress and behaviorally-induced reproductive dysfunction in primates. *Sem. Reprod. Endocrinol.*, 15:37–45.

Cameron, J. L., Helmreich, D. L., and Schreihofer, D. A. 1993. Modulation of reproductive hormone secretion by nutritional intake: Stress signals versus metabolic signals. *Hum. Reprod.*, 8:162–167.

Campbell, C. S., Finkelstein, J. S., and Turek, F. W. 1978. The interaction of photoperiod and testosterone on the development of copulatory behavior in male hamsters. *Physiol. Behav.*, 21:409–415.

Campfield, L. A., Smith, F. J., and Burn, P. 1995. The OB protein (leptin) pathway—a link between adipose tissue mass and central neural networks. *Horm. Metab. Res.*, 28:619– 632.

Canacher, G. N., and Workman, D. G. 1989. Violent crime possibly associated with anabolic steroid use. *Am. J. Psychiatry*, 146:679.

Candolle, A. P. de 1832. *Physiologie Vegetale*, vol. 2. Bechet Jeune: Paris.

Canli, T., Desmond, J. E., Zhao, Z., and Gabrieli, J. D. E. 2002. Sex differences in the neural basis of emotional memories. *Proceedings of the National Academy of Sciences U.S.A.*, 99: 10789–10794.

Canli, T., Zhao, Z., Brewer, J., Gabrieli, J. D., and Cahill, L. 2000. Event-related activation in the human amygdala associates with later memory for individual emotional experience. *Journal of Neuroscience*, 20: RC99: 1–5.

Cannon, W. B. 1929. *Bodily Changes in Pain, Hunger, Fear and Rage.* Appleton, New York.

Canoine, V., and Gwinner, E. 2002. Seasonal differences in the hormonal control of territorial aggression in free-living European stonechats. *Hormones and Behavior*, 41, 1–8.

Capitanio, J. P., Mendoza, S. P., Lerche, N. W., and Mason, W. A. 1998. Social stress results in altered glucocorticoid regulation and shorter survival in simian acquired immune deficiency syndrome. *Proc. Natl. Acad. Sci. U.S.A.*, 95:471–4719.

Carani, C., Bancroft, J., DelRio, G., Granata, A. R. M., Faccinetti, F., and Marrama, P. 1990. The endocrine effects of visual erotic stimuli in normal men. *Psychoneuroendocrinology*, 15:207–216.

Carani, C., Rochira, V., Faustini-Fustini, M., Balestrieri, A., and Granata, A. 1999. Role of oestrogen in male sexual behaviour: Insights from the natural model of aromatase deficiency. *Clinical Endocrinology*, 51: 517–524.

Carboni, E., Imperato, A., Perezzani, L., and Di Chiara, G. 1989. Differential inhibitory effects of a 5-HT3 antagonist on drug-induced stimulation of dopamine release. *Eur. J. Pharmacol.*, 164:515–519.

Card, J. P. 2000. Pseudorabies virus and the functional architecture of the circadian timing system. *Journal of Biological Rhythms*, 15: 453–461.

Cardwell, J. R., and Liley, N. R. 1991a. Androgen control of social status in males of a wild population of stoplight parrot fish, *Sparisoma viride*. *Horm. Behav.*, 25:1–18.

Cardwell, J. R., and Liley, N. R. 1991b. Hormonal control of sex and color change in the stoplight parrot fish, *Sparisoma viride* (Scaridae). *J. Comp. Physiol.*, 115:299–317.

Cardwell, J. R., Sorensen, P. W., Van der Kraak, G. J., and Liley, N. R. 1996. Effect of dominance status on sex hormone levels in laboratory and wild-spawning male trout. *Gen. Comp. Endocrinol.*, 101:333–341.

Carlsen, E., Giwercman, A., Keiding, N., and Skakkebaek, N. E. 1992. Evidence for decreasing quality of semen during the past 50 years. *Br. Med. J.*, 305:609–613.

Carlson, M., and Earls, F. 1997. Psychological and neuroendocrinological sequelae of early social deprivation in institutionalized children in Romania. *Annals of the New York Academy of Sciences*, 807: 419–428.

Carmichael, M. S., Nelson, R. J., and Zucker, I. 1981. Hamster activity and estrous cycles: Control by a single versus multiple circadian oscillator(s). *Proc. Natl. Acad. Sci. U.S.A.*, 78:7830–7834.

Carpenter, C. R. 1940. A field study of the behavior and social relations of howling monkeys. *Comp. Psychol. Monogr.*, 16:1–21.

Carpenter, C. R. 1942a. Sexual behavior of free ranging rhesus monkeys (*Macaca mulatta*). I. Specimens, procedures, and behavioral characteristics of estrus. *J. Comp. Psychol.*, 33:113–142.

Carpenter, C. R. 1942b. Sexual behavior of free ranging rhesus monkeys (*Macaca mulatta*). II. Periodicity of estrus, homosexual, autoerotic and non-conformist behavior. *J. Comp. Psychol.*, 33:143–162.

Carroll, B. J. 1980. Dexamethasone suppression test in depression. *Lancet*, 8206:1249.

Carroll, B. J., Curtis, G. C., and Mendels, J. 1976. Neuroendocrine regulation in depression. I. Limbic system-adrenocortical dysfunction. *Arch. Gen. Psychiatry*, 33:1039.

Carroll, B. J., Martin, F. I. R., and Davies, B. 1968. Resistance to suppression by dexamethasone of plasma 11-OHCS levels in severe depressive illness. *Br. Med. J.*, 3:285–290.

Carter, C. S. (ed.). 1974. *Hormones and Sexual Behavior.* Stroudsburg, Dowden, Hutchinson and Ross, Inc.

Carter, C. S., and Keverne, E. B. 2002. The neurobiology of social affiliation and pair bonding. In D. W. Pfaff, A. P. Arnold, A. M. Etgen, S. E. Fahrbach, and R. T. Rubin (eds.), *Hormones, Brain and Behavior*, Volume 1, pp. 299–338. Academic Press, New York.

Carter, C. S., DeVries, A. C., and Getz, L. L. 1995. Physiological substrates of mammalian monogamy. *Neurosci. Biobehav. Rev.*, 19:303–314.

Carter, C. S., DeVries, A. C., Taymans, S. E., Roberts, R. L., Williams, J. R., and Getz, L. L. 1997a. Peptides, steroids, and pair bonding. *Ann. N. Y. Acad. Sci.*, 807:260–272.

Carter, C. S., Getz, L. L., and Cohen-Parsons, M. 1986. Relationship between social organization and behavioral endocrinology in a monogamous mammal. *Adv. Study Behav.*, 16:109–145.

Carter, C. S., Getz, L. L., Gavish, L., McDermott, J. L., and Arnold, P. 1980. Male-related pheromones and the activation of female reproduction in the prairie vole (*Microtus ochrogaster*). *Biol. Reprod.*, 23:1038–1045.

Carter, C. S., Lederhendler, I. I., and Kirkpatrick, B. 1997b. *Integrative Neurobiology of Affiliation*. Vol. 807. New York Academy of Sciences: New York.

Carter, C. S., Williams, J. R., Witt, D. M., and Insel, T. R. 1992. Oxytocin and social bonding. *Ann. N. Y. Acad. Sci.*, 652:204–211.

Carter, C. S., Witt, D. M., Auksi, T., and Casten, L. 1987. Estrogen and the induction of lordosis in female and male prairie voles (*Microtus ochrogaster*). *Horm. Behav.*, 21:65–73.

Carter, D. S., and Goldman, B. D. 1983a. Antigonadal effects of timed melatonin infusion in pinealectomized male Djungarian hamsters (*Phodopus sungorus sungorus*): Duration is the critical parameter. *Endocrinology*, 113:1261–1267.

Carter, D. S., and Goldman, B. D. 1983b. Progonadal role of the pineal in the Djungarian hamster (*Phodopus sungorus sungorus*): Mediation by melatonin. *Endocrinology*, 113:1268–1273.

Casanueva, F. F., and Dieguez, C. 2002. Ghrelin: The link connecting growth with metabolism and energy homeostasis. *Reviews in Endocrine and Metabolic Disorders*, 3: 325–338.

Cascio, C., Yu, G. Z., Insel, T. R., and Wang, Z. X. 1998. Dopamine D_2 receptor-mediated regulation of partner preferences in female prairie voles. *Soc. Neurosci. Abstr.*, 24:372.13.

Cassone, V. M. 1998. Melatonin's role in vertebrate circadian rhythms. *Chronobiol. Int.*, 15:457–473.

Catalano, S., Avila, D. M., Marsico, S., Wilson, J. D., Glickman, S. E., and McPhaul, M. J. 2002. Virilization of the female spotted hyena cannot be explained by alterations in the amino acid sequence of the androgen receptor (AR). *Molecular and Cellular Endocrinology*, 194: 85–94.

Cedrini, L., and Fasolo, A. 1970. Olfactory attractants in sex recognition of the crested newt: An electrophysiological research. *Monitore Zoology* (Italy), 5:223–229.

Chakraborty, S. 1995. Plasma prolactin and luteinizing hormone during termination and onset of photorefractoriness in intact and pinealectomized European starlings (*Sturnus vulgaris*). *Gen. Comp. Endocrinol.*, 99:185–91.

Challis, J. R. G., Davies, I. J., and Ryan, K. J. 1973. The concentrations of progesterone, estrone and estradiol-17β in the plasma of pregnant rabbits. *Endocrinology*, 93:971–976.

Chamley, W. A., Buckmaster, J. M., Cerini, M. R., Cumming, I. A., Goding, J. R., Obst, J. M., Williams, A., and Wingfield, C. 1973. Changes in the levels of progesterone, corticosteroids, estrone, estradiol-17β, luteinizing hormone, and prolactin in the peripheral plasma of the ewe during late pregnancy and at parturition. *Biol. Reprod.*, 9:30–35.

Champagne, F. A., Weaver, I. C., Diorio, J., Sharma, S., and Meaney, M. J. 2003. Natural variations in maternal care are associated with estrogen receptor alpha expression and estrogen sensitivity in the medial preoptic area. *Endocrinology*, 144: 4720–4724.

Champagne, F., Diorio, J., Sharma, S., and Meaney, M. J. 2001. Naturally occurring variations in maternal behavior in the rat are associated with differences in estrogen-inducible central oxytocin receptors. *Proceedings of the National Academy of Sciences U.S.A.*, 98: 12736–12741.

Chappel, S. C. 1985. Neuroendocrine regulation of luteinizing hormone and follicle stimulating hormone: A review. *Life Sci.*, 36:97–103.

Charles-Dominique, P. 1977. *Ecology and Behaviour of Nocturnal Primates.* Duckworth, London.

Chavez, M., Riedy, C. A., Van Dijk, G., and Woods, S. C. 1996. Central insulin and macronutrient intake in the rat. *Am. J. Physiol.*, 271:R727–R731.

Chehab, F. F., Lim, M. E., and Lu, R. 1996. Correction of the sterility defect in homozygous obese female mice by treatment with the human recombinant leptin. *Nat. Gen.*, 12:318–20.

Chen, G., Koyama, K., Yuan, X., Lee, Y., Zhou, Y.-T., O'Doherty, R., Newgard, C. B., and Unger, R. H. 1996. Disappearance of body fat in normal rats induced by adenovirus-mediated leptin gene therapy. *Proc. Natl. Acad. Sci. U.S.A.*, 93:14795–15799.

Cheng, M. Y., Bullock, C. M., Li, C., Lee, A. G., Bermak, J. C., Belluzzi, J., Weaver, D. R., Leslie, F. M., and Zhou, Q. Y. 2002. Prokineticin 2 transmits the behavioural circadian rhythm of the suprachiasmatic nucleus. *Nature*, 417: 405–410.

Cheng, M.-F. 1986. Individual behavioral response mediates endocrine changes induced by social interaction. In B. Komisaruk, H. Siegel, H. Feder, and M. F. Cheng (eds.), *Reproduction: A Behavioral and Neuroendocrine Perspective*, pp. 4–12. New York Academy of Sciences, New York.

Cheng, M.-F. 1992. For whom does the female dove coo? A case for the role of vocal self-stimulation. *Anim. Behav.*, 43:1035–1044.

Cheng, M.-F., and Lehrman, D. S. 1973. Relative effectiveness of diethylstilbestrol and estradiol benzoate in inducing female behavior patterns of ovariectomized ring doves (*Streptopelia risoria*). *Horm. Behav.*, 4:123–127.

Cheng, M.-F., and Silver, R. 1975. Estrogen–progesterone regulation of nest building behavior in ovariectomized ring doves (*Streptopelia risoria*). *J. Comp. Physiol. Psychol.*, 88:256–263.

Cheng, M.-F., and Zuo, M. 1994. Proposed pathways for vocal self-stimulation: met-enkephalinergic projections linking the midbrain vocal nucleus, auditory-responsive thalamic regions and neurosecretory hypothalamus. *J. Neurobiol.*, 25:361–379.

Cheng, M.-F., Peng, J. P., and Johnson, P. 1998. Hypothalamic neurons preferentially respond to female nest coo stimulation: demonstration of direct acoustic stimulation of luteinizing hormone release. *J. Neurosci.*, 18:5477–5489.

Cherry, J. A., Basham, M. E., Weaver, C. E., Krohmer, R. W., and Baum, M. J. 1990. Ontogeny of the sexually dimorphic male nucleus in the preoptic/anterior hypothalamus of ferrets and its manipulation by gonadal steroids. *J. Neurobiol.*, 21:844–857.

Chester-Jones, I. 1955. Role of the adrenal cortex in reproduction. *Br. Med. J.*, 11:156–160.

Cheung, C. C., Thornton, J. E., Kuijper, J. L., Weigle, D. S., Clifton, D. K., and Steiner, R. A. 1997. Leptin is a metabolic gate for the onset of puberty in the female rat. *Endocrinology*, 138:855–858.

Chevalier, G., and Deniau, J. M. 1990. Disinhibition as a basic process in the expression of striatal functions. *Trends Neurosci.*, 13:277–290.

Chiavegatto, S., Dawson, V. L., Mamounas, L. A., Koliatsos, V. E., Dawson, T. M., and Nelson, R. J. 2001. Brain serotonin dysfunction accounts for aggression in male mice lacking neuronal nitric oxide synthase. *Proceedings of the National Academy of Sciences U.S.A.*, 98: 1277–1281.

Childress, A. R., Ehrman, R., McLellan, A. T., MacRae, J., Natale, M., and O'Brien, C. P. 1994. Can induced moods trigger drug-related responses in opiate abuse patients? *J. Subst. Abuse Treat.*, 11:17–23.

Chitty, D. 1961. Variations in the weight of the adrenal glands of the field vole (*Microtus agrestis). J. Endocrinol.*, 22:287–293.

Cho, K. 2001. Chronic 'jet lag' produces temporal lobe atrophy and spatial cognitive deficits. *Nature Neuroscience*, 4:567–568.

Cho, K., Ennaceur, A., Cole, J.C., and Suh, C.K. 2000. Chronic jet lag produces cognitive deficits. *Journal of Neuroscience*, 20:RC66:1-5

Choleris, E., Gustafsson, J. A., Korach, K. S., Muglia, L. J., Pfaff, D. W., and Ogawa, S. 2003. An estrogen-dependent four-gene micronet regulating social recognition: A study with oxytocin and estrogen receptor-α and -β knockout mice. *Proceedings of the National Academy of Sciences U.S.A.*, 100: 6192–6197.

Choleris, E., Kavaliers, M., and Pfaff, D. W. 2004. Functional genomics of social recognition. *Journal of Neuroendocrinology*, 16: 383–389.

Christensen, L. W., and Clemens, L. G. 1975. Blockade of testosterone-induced mounting behavior in the male rat with intracranial application of the aromatization inhibitor, androst-1,4,6-triene-3,17-dione. *Endocrinology*, 97:1545–1551.

Christian, J. J. 1950. The adreno-pituitary system and population cycles in mammals. *J. Mammalogy*, 31:247–259.

Christian, J. J. 1959. The role of endocrine and behavioral factors in the growth of mammalian populations. In A. Gorbman (ed.), *Comparative Endocrinology*, pp. 71–97. Wiley, New York,

Christian, J. J., Lloyd, J. A., and Davis, D. E. 1965. The role of endocrines in the self-regulation of mammalian populations. *Rec. Prog. Horm. Res.*, 21:501–578.

Christianson, T., Wallen, K., Brown, B., and Glickman, S. E. 1972. Effects of castration, blindness and anosmia on social reactivity in the male Mongolian gerbil (*Meriones unguiculatus*). *Physiology and Behavior*, 10: 989–994.

Christie, M. H., and Barfield, R. J. 1979. Effects of castration and home cage residency on aggressive behavior in rats. *Hormones and Behavior*, 13: 85–91.

Chrousos, G. P. 1998. Stressors, stress, and neuroendocrine integration of the adaptive response. The 1997 Hans Selye Memorial Lecture. *Ann. N.Y. Acad. Sci.*, 85:311–335

Chrousos, G. P. 1998. Ultradian, circadian, and stress-related hypothalamic–pituitary–adrenal axis activity—a dynamic digital-to-analog modulation. *Endocrinology*, 139: 437–440.

Chrousos, G. P. 2000. Stress, chronic inflammation, and emotional and physical well-being: Concurrent effects and chronic sequelae. *Journal of Allergy and Clinical Immunology*, 106: S275–S291.

Chrousos, G. P., McCarty, R., Pacak, K., Cizza, G., Sternberg, E., Gold, P. W., and Kvetnansky, R. (eds.). 1995. *Stress: Basic Mechanisms Clinical Implications*. Vol. 771. New York Academy of Science, New York.

Chrousos, G. P., Torpy, D. J., and Gold, P. W. 1998. Interactions between the hypothalamic-pituitary-adrenal axis and the female reproductive system: Clinical implications. *Ann. Intern. Med.*, 129:229–240.

Chugani, H. T., Behen, M. E., Muzik, O., Juhasz, C., Nagy, F., and Chugani, D. C. 2001. Local brain functional activity following early deprivation: A study of postinstitutionalized Romanian orphans. *Neuroimage*, 14: 1290–1301.

Cinti, S., Frederich, R. C., Zingaretti, M. C., De Matteis, R., Flier, J. S., and Lowell, B. B. 1997. Immunohistichemical localization of leptin and uncoupling protein in white and brown adipose tissue. *Endocrinology*, 138:797–804.

Clancy, A. N., Zumpe, D., and Michael, R. P. 1995. Intracerebral infusion of an aromatase inhibitor, sexual behavior and brain estrogen receptor-like immunoreactivity in intact male rats. *Neuroendocrinology*, 61:98–111.

Clare, A. W. 1985. Hormones, behavior, and the menstrual cycle. *J. Psychosom. Res.*, 29:225–233.

Clark, A. S., and Roy, E. J. 1983. Behavioral and cellular responses to pulses of low doses of estradiol-17 beta. *Physiol. Behav.*, 30:561–565.

Clark, A. S., Davis, L. A., and Roy, E. J. 1985. A possible physiological basis for the dud-stud phenomenon. *Horm. Behav.*, 19:227–230.

Clark, A. S., Pfeifle, J. K., and Edwards, D. A. 1981. Ventromedial hypothalamic damage and sexual proceptivity in female rats. *Physiol. Behav.*, 27:597–602.

Clark, M. M., Desousa, D., Vonk, J., Galef, B. G. 1997. Parenting and potency: Alternative routes to reproductive success in male Mongolian gerbils. *Anim. Behav.*, 54:635–62.

Clark, M. M., vom Saal, F. S., and Galef, B. G. 1992. Intrauterine positions and testosterone levels of adult male gerbils are correlated. *Physiol. Behav.*, 51:957–960.

Clarke, A. S. and Schneider, M. L. 1993. Prenatal stress has long-term effects on behavioral responses to stress in juvenile rhesus monkeys. *Dev. Psychobiol.*, 26:293–304.

Clarke, A. S., Wittwer, D. J., Abbott, D. H., and Schneider, M. L. 1994. Long-term effects of prenatal stress on HPA axis activity in juvenile rhesus monkeys. *Dev. Psychobiol.*, 27:257–269.

Clemens, L. G., and Weaver, D. R. 1985. The role of gonadal hormones in the activation of feminine sexual behavior. In N. T. Adler, D. W. Pfaff, and R. W. Goy (eds.), *Handbook of Behavioral Neurobiology*, Vol. 7, pp. 183–227. Plenum Press, New York.

Clemens, L. G., Gladue, B. A., and Coniglio, L. P. 1978. Prenatal endogenous androgenic influences on masculine sexual behavior and genital morphology in male and female rats. *Horm. Behav.*, 10:40–53.

Clough, G. C. 1965. Lemmings and population problems. *Am. Sci.*, 53:99–112.

Clutton-Brock, T. H. 1991. *The Evolution of Parental Care.* Princeton University Press, Princeton.

Clutton-Brock, T. H., Albon, S. D., Gibson, R. J., and Guinness, F. E. 1979. The logical stag: Adaptive aspects of fighting in red deer (*Cervus elaphus L.*). *Anim. Behav.*, 27:211–225.

Cochran, C. G. 1979. Proceptive patterns of behavior throughout the menstrual cycle in female rhesus monkeys. *Behav. Neural Biol.*, 27:342–253.

Coe, C. L. 1990. Psychobiology of maternal behavior in nonhuman primates. In N. A. Krasnegor and R. S. Bridges (eds.), *Mammalian Parenting*, pp. 157–183. Oxford University Press, Oxford.

Coe, C. L., Mendoza, S. P., Somtherman, W. P., and Levine, S. 1978. Mother-infant attachment in the squirrel monkey: Adrenal response to separation. *Behav. Biology*, 22:256–263.

Cohen, I. T., Sherwin, B. B., and Fleming, A. S. 1987. Food cravings, mood, and the menstrual cycle. *Horm. Behav.*, 21:457–470.

Cohen, R. A., and Albers, H. E. 1991. Disruption of human circadian and cognitive regulation following a discrete hypothalamic lesion: A case study. *Neurology*, 41:726–729.

Cohen, S. L., Halaas, J. L., Friedman, J. M., Chait, B. T., Bennet, L., Chang, D., Hecht, R., and Collins, F. 1996. Human leptin characterization. *Nature*, 382:589.

Colborn, T., Myers, J. P., and Dumanoski, D. 1997. *Our Stolen Future: Are We Threatening Our Own Fertility, Intelligence, and Survival? A Scientific Detective Story.* E. F. Dutton, New York.

Cole, M. A., Kim, P. J., Kalman, B. A., and Spencer, R. L. 2000. Dexamethasone suppression of corticosteroid secretion: Evaluation of the site of action by receptor measures and functional studies. *Psychoneuroendocrinology*, 25: 151–167.

Coleman, R. M. 1986. *Awake at 3:00 A.M. By Choice or by Chance?* W. H. Freeman, New York.

Collier, T. J., Quirk, G. J., and Routtenberg, A. 1987. Separable roles of hippocampal granule cells in forgetting and pyramidal cells in remembering spatial information. *Brain Res.*, 409:316–328.

Cologer-Clifford, A., Simon, N. G., Lu, S. F., Smoluk, S. A. 1997. Serotonin agonist-induced decreases in intermale aggression are dependent on brain region and receptor subtype. *Pharmacol. Biochem. Behav.* 58(2): 425–430.

Cologer-Clifford, A., Simon, N. G., Richter, M. L., Smoluk, S. A., and Lu, S. 1999. Androgens and estrogens modulate 5-HT_{1A} and 5-HT_{1B} agonist effects on aggression. *Physiology and Behavior*, 65: 823–828.

Commins, D., and Yahr, P. 1984. Adult testosterone levels influence the morphology of a sexually dimorphic area in the Mongolian gerbil brain. *J. Comp. Neurol.*, 224:132–140.

Commins, D., and Yahr, P. 1985. Autoradiographic localization of estrogen and androgen receptors in the sexually dimorphic area and other regions of the gerbil brain. *J. Comp. Neurol.*, 231:473–489.

Commins, W. D. 1932. The effect of castration at various ages upon learning ability of male albino rats. *J. Comp. Psychol.*, 14:29–53.

Conaway, C. H. 1971. Ecological adaptation and mammalian reproduction. *Biol. Reprod.*, 1:239–247.

Conaway, C. H., and Wight, H. M. 1962. Onset of reproductive season and first pregnancy of the season in cottontails. *J. Wildlife Mgmt.*, 26:278–290.

Conn, P. M., and Freeman, M. E. (eds.). 2000. *Neuroendocrinology in Physiology and Medicine*. Humana Press, New York.

Connor, J. R., and Davis, H. N. 1980a. Postpartum estrus in Norway rats. I. Behavior. *Biol. Reprod.*, 23:994–999.

Connor, J. R., and Davis, H. N. 1980b. Postpartum estrus in Norway rats. II. Physiology. *Biol. Reprod.*, 23:1000–1006.

Conrad, C. D., LeDoux, J. E., Magarinos, A. M., and McEwen, B. S. 1999. Repeated restraint stress facilitates fear conditioning independently of causing hippocampal CA3 dendritic atrophy. *Behavioral Neuroscience*, 113: 902–913.

Conrad, C., and Roy, E. 1992. Selective loss of hippocampal granule cells following adrenalectomy: Implications for spatial memory. *J. Neurosci.*, 13:2582–2590.

Constantini, N. W., and Warren, M. P. 1994. Special problems of the female athlete. *Baillieres Clin. Rheumatol.*, 8:199–219.

Conte, C. O., Rosito, G. B. A., Palmini, A. L. F., Lucion, A. B., and de Almeida, A. M. R. 1986. Pre-training adrenaline recovers the amnestic effect of Met-enkephalin in demedullated rats. *Behav. Brain Res.*, 21:163–166.

Contreras, R. J., Fox, E., and Drugovich, M. L. 1982. Area postrema lesions produce feeding deficits in the rat: Effects of preoperative dieting and 2-deoxy-D-glucose. *Physiol. Behav.*, 29:875–884.

Coolen, L. M., Fitzgerald, M. E., Yu, L., and Lehman, M. N. 2004. Activation of mu opioid receptors in the medial preoptic area following copulation in male rats. *Neuroscience*, 124: 11–21.

Coolen, L. M., Veening, J. G., Petersen, D. W., and Shipley, M. T. 2003a. Parvocellular subparafascicular thalamic nucleus in the rat: Anatomical and functional compartmentalization. *Journal of Comparative Neurology*, 463: 117–131.

Coolen, L. M., Veening, J. G., Wells, A. B., and Shipley, M. T. 2003b. Afferent connections of the parvocellular subparafascicular thalamic nucleus in the rat: Evidence for functional subdivisions. *Journal of Comparative Neurology*, 463, 132–156.

Cooper, A. J. 1986. Progestins in the treatment of male sex offenders: A review. *Can. J. Psychiatry*, 31:73–79.

Cooper, J. R., Bloom, F. E, and Roth, R. H. 1986. *The Biochemical Basis of Neuropharmacology*. Oxford University Press, New York.

Coopersmith, C., Candurra, C., and Erskine, M. S. 1996. Effects of paced mating and intromissive stimulation on feminine sexual behavior and estrus termination in the cycling rat. *J. Comp. Psych.*, 110:176–186.

Cordoba-Montoya, D. A., and Carrer, H. F. 1997. Estrogen facilitates induction of long term potentiation in the hippocampus of awake rats. *Brain Res.*, 778:430–438.

Cormack, M., and Sheldrake, P. 1974. Menstrual cycle variations in cognitive ability: A preliminary report. *Int. J. Chronobiol.*, 2:53–55.

Corner, G. W., and Allen, W. M. 1929. Physiology of the corpus luteum. II. Production of a special uterine reaction (progestational proliferation) by extracts of the corpus luteum. *Am. J. Physiol.*, 88:326–339.

Corsini, R. J. 1984. *Encyclopaedia of Psychology*. Vol. 1. Wiley, New York.

Coste, S. C., Murray, S. E., and Stenzel-Poore, M. P. 2001. Animal models of CRH excess and CRH receptor deficiency display altered adaptations to stress. *Peptides*, 22: 733–741.

Cowley, M. A., Smart, J. L., Rubinstein, M., Cerdan, M. G., Diano, S., Horvath, T. L., Cone, R. D., and Low, M. J. 2001. Leptin activates anorexigenic POMC neurons through a neural network in the arcuate nucleus. *Nature*, 411: 480–484.

Cowley, M. A., Smith, R. G., Diano, S., Tschop, M., Pronchuk, N., Grove, K. L., Strasburger, C. J., et al. 2003. The distribution and mechanism of action of ghrelin in the CNS demonstrates a novel hypothalamic circuit regulating energy homeostasis. *Neuron*, 37: 649–661.

Craft, S., Newcomer, J., Kanne, S., Dagogo-Jack, S., Cryer, P., Sheline, Y., Luby, J., Dagogo-Jack, A., and Alderson, A. 1996. Memory improvement following induced hyperinsulinemia in Alzheimer's disease. *Neurobiology of Aging*, 17: 123–130.

Craft, S., Peskind, E., Schwartz, M. W., Schellenberg, G. D., Raskind, M., and Porte, D. 1998. Cerebrospinal fluid and plasma insulin levels in Alzheimer's disease: Relationship to severity of dementia and apolipoprotein E genotype. *Neurology*, 50: 164–168.

Crawford, S. G., Kaplan, B. J., and Field, L. L. 1995. Absence of an association between insulin-dependent diabetes mellitus and developmental learning difficulties. *Hereditas*, 122:73–78.

Creel, S., Creel, N. M., and Monfort, S. L. 1996. Social stress and dominance. *Nature*, 379:212.

Creel, S., Creel, N. M., Mills, M. G. L., and Monfort, S. L. 1997. Rank and reproduction in cooperatively breeding African wild dogs: Behavioral and endocrine correlates. *Behav. Ecol.*, 8:298–306.

Crews, D. 1984. Gamete production, sex hormone secretion, and mating behavior uncoupled. *Horm. Behav.*, 18:22–28.

Crews, D. 1987. Diversity and evolution of behavioral controlling mechanisms. In D. Crews (ed.), *Psychobiology of Reproductive Behavior*, pp. 88–119. Prentice Hall, Englewood Cliffs, NJ.

Crews, D. 1991. Trans-seasonal action of androgen in the control of spring courtship behavior in male red-sided garter snakes. *Proc. Natl. Acad. Sci. U.S.A.*, 88:3545–3548.

Crews, D. 1993. The organizational concept and vertebrates without sex chromosomes. *Brain, Behavior and Evolution*, 42: 202–214.

Crews, D. 1997. Species diversity and the evolution of behavioral controlling mechanisms. *Ann. N. Y. Acad. Sci.*, 807:1–21.

Crews, D. 2003a. The development of phenotypic plasticity: Where biology and psychology meet. *Developmental Psychobiology*, 43: 1–10.

Crews, D. 2003b. Sex determination: Where environment and genetics meet. *Evolution and Development*, 5: 50–55.

Crews, D., and Bergeron, J. M. 1994. Role of reductase and aromatase in sex determination in the red-eared slider (*Trachemys scripta*), a turtle with temperature-dependent sex determination. *J. Endocrinol.*, 143:279–289.

Crews, D., and Bull, J. J. 1987. Evolutionary insights from reptilian sexual differentiation. In F. P. Haseltine, M. E. McClure, and E. H. Goldberg (eds.), *Genetic Markers of Sex Differentiation*, pp. 11–26. Plenum, New York.

Crews, D., and Fitzgerald, K. 1980. "Sexual" behavior in parthenogenetic lizards (*Cnemidophorus*). *Proc. Natl. Acad. Sci. U.S.A.*, 77:499–502.

Crews, D., and Gartska, W. R. 1982. The ecological physiology of a garter snake. *Sci. Am.*, Nov., 158–168.

Crews, D., and Moore, M. C. 1986. Evolution of mechanisms controlling mating behavior. *Science*, 231: 121–125.

Crews, D., and Silver, R. 1985. Reproductive physiology and behavior interactions in nonmammalian vertebrates. In N. T. Adler, D. W. Pfaff, and R. W. Goy (eds.), *Handbook of Behavioral Neurobiology*, Vol. 7, Reproduction, pp. 101–182. Plenum Press, New York.

Crews, D., Bull, J. J., and Billy, A. J. 1988a. Sex determination and sexual differentiation in reptiles. In J. M. A. Sitsen (ed.), *Handbook of Sexology, Vol. 6: The Pharmacology and Endocrinology of Sexual Function*, pp. 98–121. Elsevier, New York.

Crews, D., Camazine, M., Diamond, R., Mason, R. T., Tokarz, R. R., and Garstka, W. R. 1984. Hormonal independence of courtship behavior in the male garter snake. *Horm. Behav.*, 18:29–41.

Crews, D., Diamond, M. A., Whittier, J., and Mason, R. 1985. Small male body size in garter snake depends on testes. *Am. J. Physiol.*, 249:R62–R66.

Crews, D., Hingorani, V., and Nelson, R. J. 1988b. Role of the pineal gland in the control of annual reproductive behavioral and physiological cycles in the red-sided garter snake (*Thamnophis sirtalis parietalis*). *J. Biol. Rhythms*, 3:293–302.

Critchlow, V., Liebelt, A., Bar-Sela, M., Mountcastle, W., and Lipscomb, H. S. 1963. Sex differences in resting pituitary-adrenal function in the rat. *Am. J. Physiol.*, 205:807–815.

Croiset, G., Marjoleen, J. M. A., and Kamphuis, P. J. 2000. Role of corticotrophin-releasing factor, vasopressin and the autonomic nervous system in learning and memory. *European Journal of Pharmacology*, 405: 225–234.

Cross, G. B., Marley, J., Miles, H., and Willson, K. 2001. Changes in nutrient intake during the menstrual cycle of overweight women with premenstrual syndrome. *British Journal of Nutrition*, 85: 475–482.

Cummings, D. C., Quigley, M. E., and Yen, S. S. 1983. Acute suppression of circulating testosterone levels by cortisol in men. *J. Clin. Endocrinol. Metab.*, 57:671–673.

Cummings, D. E., Purnell, J. Q., Frayo, R. S., Schmidova, K., Wisse, B. E., and Weigle, D. S. 2001. A preprandial rise in plasma ghrelin levels suggests a role in meal initiation in humans. *Diabetes*, 50: 1714–1719.

Cunningham, J. T. 1900. *Sexual Dimorphisms in the Animal Kingdom.* Black, London.

Cuthill, I. C., Hunt, S., Cleary, C., and Clark, C. 1997. Colour bands, dominance, and body mass regulation in male zebra finches (*Taeniopygia guttata*). *Proceedings of the Royal Society of London, Series B*, 264: 1093–1099.

Czaja, J. A. 1978. Ovarian influences on primate food intake: Assessment of progesterone actions. *Physiol. Behav.*, 21:923–928.

Czaja, J. A., and Goy, R. W. 1975. Ovarian hormones and food intake in female guinea pigs and rhesus monkeys. *Horm. Behav.*, 6:923–928.

Czeisler, C. A., Kronauer, R. E., Allan, J. S., et al. 1989. Bright light induction of strong (Type 0) resetting of the human circadian pacemaker. *Science*, 244:1328–1332.

Czeisler, C. A., Zimmerman, J. C., Ronda, J. M., et al. 1980. Timing of REM sleep is coupled to the circadian rhythm of body temperature in man. *Sleep*, 2:329–346.

Daan, S., Damassa, D., Pittendrigh, C. S., and Smith, E. R. 1975. An effect of castration and testosterone replacement on a circadian pacemaker in mice (*Mus musculus*). *Proc. Natl. Acad. Sci. U.S.A.*, 72:3744–3747.

Dabbs, J. M. 1990. Age and seasonal variation in serum testosterone concentrations among men. *Chronobiol. Int.*, 7:245–249.

Dabbs, J. M., and Hargrove, M. F. 1997. Age, testosterone, and behavior among female prison inmates. *Psychosom. Med.*, 59:477–480.

Dabbs, J. M., and Mohammed, S. 1992. Male and female salivary testosterone concentrations before and after sexual activity. *Physiol. Behav.*, 52:195–197.

Dabbs, J. M., and Morris, R. 1990. Testosterone, social class, and antisocial behavior in a sample of 4,462 men. *Psychol. Sci.*, 1:209–211.

Dabbs, J. M., Frady, R. L., Carr, T. S., and Besch, N. F. 1987. Saliva testosterone and criminal violence in young adult prison inmates. *Psychosom. Med.*, 49:174–182.

Dabbs, J. M., LaRue, D., and Williams, P. M. 1990. Testosterone and occupational choice: Actors, ministers, and other men. *J. Pers. Soc. Psychol.*, 59:1261–1265.

Dabbs, J. M., Ruback, R. B., Frady, R. L., Hopper, C. H., and Sgoutas, D. S. 1989. Saliva testosterone and criminal violence among women. *Pers. Indiv. Diff.*, 9:269–275.

DaCosta, A. P., Guevara-Guzman, R. G., Ohkura, S., Goode, J. A., and Kendrick, K. M. 1996. The role of oxytocin release in the paraventricular nucleus in the control of maternal behaviour in the sheep. *J. Neuroendocrinol.*, 8:163–177.

Dallman, M. F., Pecoraro, N., Akana, S. F., La Fleur, S. E., Gomez, F., Houshyar, H., Bell, M. E., Bhatnagar, S., Laugero, K. D., and Manalo, S. 2003. Chronic stress and obesity: A new view of "comfort food." *Proceedings of the National Academy of Sciences U.S.A.*, 100: 11696–11701.

Dallman, M. F., Strack, A. M., Akana, S. F., Bradbury, M. J., Hanson, E. S., Scribner, K. A., and Smith, M. 1993. Feast and famine: Critical role of glucocorticoids with insulin in daily energy flow. *Frontiers in Neuroendocrinology*, 14: 303–347.

Dalvit, S. P. 1981. The effect of the menstrual cycle on patterns of food intake. *Am. J. Clin. Nutr.*, 34:1811–1815.

Dalvit-McPhillips, S. P. 1983. The effect of the human menstrual cycle on nutrient intake. *Physiol. Behav.*, 31:209–212.

Daly, M., and Wilson, M. 1983. *Sex, Evolution, and Behavior.* Weber and Schmidt, Boston.

Damassa, D. A., Davidson, J. M., and Smith, E. R. 1977. The relationship between circulating testosterone levels and male sexual behavior in rats. *Horm. Behav.,* 8:275–286.

Damiola, F., Le Minh, N., Preitner, N., Kornmann, B., Fleury-Olela, F., and Schibler, U. 2000. Restricted feeding uncouples circadian oscillators in peripheral tissues from the central pacemaker in the suprachiasmatic nucleus. *Genes and Development,* 14: 2950–2961.

Daniel, J. M., Fader, A. J., Spencer, A. L., and Dohanich, G. P. 1997. Estrogen enhances performance of female rats during acquisition of a radial arm maze. *Horm. Behav.,* 32:217–225.

Daniels, D., and Fluharty, S. J. 2004. Salt appetite: A neurohormonal viewpoint. *Physiology and Behavior,* 81: 319–337.

Dark, J., and Zucker, I. 1984. Gonadal and photoperiodic control of seasonal body weight changes in male voles. *Am. J. Physiol.,* 247:R84–R88.

Dark, J., and Zucker, I. 1985. Circannual rhythms of ground squirrels: Role of the hypothalamic paraventricular nucleus. *J. Biol. Rhythms,* 1:17–23.

Dark, J., and Zucker, I. 1986. Photoperiodic regulation of body mass and fat reserves in the meadow vole. *Physiol. Behav.,* 38:851–854.

Dark, J., Pickard, G. E., and Zucker, I. 1985. Persistence of circannual rhythms in ground squirrels with lesions of the suprachiasmatic nuclei. *Brain Res.,* 332:201–07.

Dark, J., Stern, J., and Zucker, I. 1989. Adipose tissue dynamics during cyclic weight loss and weight gain of ground squirrels. *Am. J. Physiol.,* 256:R1286–1292.

Dark, J., Whaling, C. S., and Zucker, I. 1987b. Androgens exert opposite effects on body mass of heavy and light meadow voles. *Horm. Behav.,* 21:471–477.

Dark, J., Zucker, I., and Wade, G. N. 1983. Photoperiodic regulation of body mass, food intake, and reproduction in meadow voles. *Am. J. Physiol.,* 245:R334–R338.

Darlington, T. K., Wagner-Smith, K., Ceriani, M. F., Staknis, D., Gekakis, N., Steeves, T. D. L., Weitz, C. J., Takahashi, J. S., and Kay, S. A. 1998. Closing the circadian loop: CLOCK-induced transcription of its own inhibitors *per* and *tim. Science,* 280:1599–1603.

Darrow, J. M., and Goldman, B. D. 1986. Circadian regulation of pineal melatonin and reproduction in the Djungarian hamster. *J. Biol. Rhythms,* 1:39–53.

Darrow, J. M., Duncan, M. J., Bartke, A., Bona-Gallo., A., and Goldman, B. D. 1988. Influence of photoperiod and gonadal steroids on hibernation in the European hamster. *J. Comp. Physiol.,* 163:339–48.

Darrow, J. M., Yogev, L., and Goldman, B. D. 1987. Patterns of reproductive hormone secretion in hibernating Turkish hamsters. *Am. J. Physiol.,* 253:R329–R336.

Daugherty, J. E. 1998. Treatment strategies for premenstrual syndrome. *Am. Fam. Physician,* 58:197–198.

David, H. G., Green, J. T., Grant, A. J., and Wilson, C. A. 1995. Simultaneous bilateral quadriceps rupture: A complication of anabolic steroid abuse. *J. Bone Joint Surg. Br.,* 77:159–160.

David, K., Dingemanse, E., Freud, J., and Lanquer, E. 1935. Crystalline male hormone from testes (testosterone) are more active than androsterone prepared from urine or cholesterol. *J. Physiol. Chem.,* 233:281–282.

Davidson, J. M. 1966a. Activation of the male rat's sexual behavior by intracerebral implantation of androgen. *Endocrinology,* 84:1365–1372.

Davidson, J. M. 1966b. Characteristics of sex behaviour in male rats following castration. *Anim. Behav.,* 14:266–272.

Davidson, J. M. 1969. Effects of estrogen on the sexual behavior of male rats. *Endocrinology,* 84:1365–1372.

Davidson, J. M., Camargo, C. A., and Smith, E. R. 1979. Effects of androgen on sexual behavior in hypogonadal men. *J. Clin. Endocrinol. Metab.,* 48:955–958.

Davidson, J. M., Kwan, M., and Greenleaf, W. J. 1982. Hormonal replacement and sexuality in men. *J. Clin. Endocrinol. Metab.,* 11:599–623.

Davidson, J. M., Stefanick, M. L., Sachs, B. D., and Smith, E. R. 1978. Role of androgen in sexual reflexes of the male rat. *Physiol. Behav.,* 21:141–146.

Davidson, R. J. 2000. Dysfunction in the neural circuitry of emotion regulation: A possible prelude to violence. *Science,* 289, 591–594.

Davis, G. J., and Meyer, R. K. 1973. Seasonal variation in LH and FSH of bilaterally castrated snowshoe hares. *Gen. Comp. Endocrinol.,* 20:61–68.

Davis, H. P., and Squire, L. R. 1984. Protein synthesis and memory: A review. *Psychol. Bull.,* 96:518–559.

Davis, J. D., and Campbell, C. S. 1973. Peripheral control of meal size in the rat: Effect of sham feeding on meal size and drinking rate. *J. Comp. Physiol. Psychol.,* 83:379–387.

Davis, J. D., Gallagher, R. J., Ladove, R. F., and Turavasky, A. J. 1969. Inhibition of food intake by a humoral factor. *J. Comp. Physiol. Psychol.,* 67:407–417.

Davis, P. G., and Barfield, R. J. 1979. Activation of masculine sexual behavior by intracranial estradiol benzoate implants in male rats. *Neuroendocrinology,* 28:217–227.

Dawson, A. 1983. Plasma gonadal steroid levels in wild starlings (*Sturnus vulgaris*) during the annual cycle and in relation to the stages of breeding. *Gen. Comp. Endocrinol.,* 49:286–94.

Dawson, T. M., and Snyder, S. H. 1994. Gases as biological messengers: nitric oxide and carbon monoxide in the brain. *J. Neurosci.,* 14:5147–5159.

de Kloet, E. R., Joels, M., Oitzl, M., and Sutanto, W. 1991. Implication of brain corticosteroid receptor diversity for the adaptation syndrome concept. In G. Jasmin and M. Cantin (eds.), *Stress Revisited: Neuroendocrinology of Stress. Methods and Achievements in Experimental Pathology,* Vol 14:104–132. Karger, Basel.

de la Iglesia, H. O., Meyer, J., and Schwartz, W. J. 2003. Lateralization of circadian pacemaker output: Activation of left- and right-sided luteinizing hormone-releasing hormone neurons involves a neural rather than a humoral pathway. *Journal of Neuroscience,* 23: 7412–7414.

De Mairan, J. J. 1729. *Observation Botanique.* L'Academie Royale des Sciences Paris, 35–36.

De Quervain, D. J., Roozendaal, B., and McGaugh, J. L. 1998. Stress and glucocorticoids impair retrieval of long-term spatial memory. *Nature*, 394:787–790.

de Waal, F. B. M. 1987. Tension regulation and nonreproductive functions of sex among captive bonobos (*Pan paniscus*). *Natl. Geogr. Res.*, 3:318–335.

de Waal, F. B. M. 1995. Bonobo sex and society. *Sci. Am.*, 272:82–88.

de Waal, F. B. M., and Lanting, F. 1997. *Bonobo: The Forgotten Ape*, University of California Press, Berkeley.

Deakin, J. F. W. 1988. Relevance of hormone–CNS interactions to psychological changes in the puerperium. In R. Kumar and I. F. Brockington (eds.), *Motherhood and Mental Illness 2: Causes and Consequences*, pp. 113–132. Wright, Boston.

DeBold, J. F., and Frye, C. A. 1994. Progesterone and the neural mechanisms of hamster sexual behavior. *Psychoneuroendocrinology*, 19:563–579.

DeBold, J. F., and Miczek, K. A. 1981. Sexual dimorphism in the hormonal control of aggressive behavior of rats. *Pharmacology Biochemistry and Behavior*, 14, suppl. 1: 89–93.

DeBold, J. F., and Miczek, K. A. 1984. Aggression persists after ovariectomy in female rats. *Hormones and Behavior*, 18, 177–190.

Delville, Y., and Blaustein, J. D. 1991. A site for estradiol priming of progesterone-facilitated sexual receptivity in the ventrolateral hypothalamus of female guinea pigs. *Brain Res.*, 559(2):191–199.

Delville, Y., David, J. T., Taravosh-Lahn, K., and Wommack, J. C. 2003. Stress and the development of agonistic behavior in golden hamsters. *Hormones and Behavior*, 44: 263–270.

Delville, Y., De Vries, G. J., and Ferris, C. F. 2000. Neural connections of the anterior hypothalamus and agonistic behavior in golden hamsters. *Brain, Behavior and Evolution*, 55: 53–76.

Delville, Y., Melloni, R. H., and Ferris, C. F. 1998. Behavioral and neurobiological consequences of social subjugation during puberty in golden hamsters. *J. Neurosci.*, 18:2667–2672.

Delville, Y., Newman, M. L., Wommack, J. C., Taravosh-Lahn, K., and Cervantes, M. C., 2005. Development of aggression. In R. J. Nelson (ed.), *Biology of Aggression*. Oxford University Press, New York. In press.

Demas, G. E., Eliasson, M. J. L., Dawson, T. M., Dawson, V. L., Kriegsfeld, L. J., Nelson, R. J. and Snyder, S. H. 1997. Inhibition of neuronal nitric oxide synthase increases aggressive behavior in mice. *Mol. Med.*, 3: 611–617.

Demas, G. E., Moffatt, C. A., Drazen, D. L., and Nelson, R. J. 1999. Castration does not inhibit aggressive behavior in castrated adult male prairie voles (*Microtus ochrogaster*). *Physiology and Behavior*, 66: 59–62.

Demas, G. E., Polacek, K. M., Durazzo, A., and Jasnow, A. M. 2004. Adrenal hormones mediate melatonin-induced increases in aggression in male Siberian hamsters (*Phodopus sungorus*). *Hormones and Behavior*, 46: 582–691.

Deminiere, J. M., Piazza, P. V., Guegan, G., Abrous, N., Maccari, S., Le Moal, M., and Simon, H. 1992. Increased locomotor response to novelty and propensity to intravenous amphetamine self-administration in adult offspring of stressed mothers. *Brain Res.*, 586:135–139.

Dempsey, E. W. 1968. William Caldwell Young. An appreciation. In M. Diamond (ed.), *Perspectives in Reproduction and Sexual Behavior*, pp. 453–458. University of Indiana Press, Bloomington.

Dempsey, E. W., Hertz, R., and Young, W. C. 1936. The experimental induction of oestrus (sexual receptivity) in the normal and ovariectomized guinea pig. *Am. J. Physiol.*, 116:201–209.

Demski, L. S. 1987. Diversity in reproductive patterns and behavior in teleost fishes. In D. Crews (ed.), *Psychobiology of Reproductive Behavior*, pp. 2–27. Prentice Hall, Englewood Cliffs, NJ.

Dennis, R. 1992. Cultural change and the reproductive cycle. *Social Sci. Med.*, 34:485–489.

Denton, D. 1982. *The Hunger for Salt*. Springer-Verlag, Berlin.

Desmond, N. L., and Levy, W. B. 1997. Ovarian steroid control of connectivity in the female hipppocampus: An overview of recent experimental findings and speculations on its functional consequences. *Hippocampus*, 7:239–245.

Dessypris, A., Kuoppasalmi, K., and Adlercreutz, H. 1976. Plasma cortisol, testosterone, androstenedione and luteinizing hormone (LH) in a non-competitive marathon run. *J. Steroid Biochem.*, 7:33–37.

Dethier, V. G. 1976. *The Hungry Fly: A Physiological Study of the Behavior Associated with Feeding*. Harvard University Press, Cambridge, MA.

Detillion, C. E., Craft, T. K. S., Glasper, E. R., Prendergast, B., and DeVries, A. C. 2004. Social facilitation of wound healing. *Psychoneuroendocrinology*, 29: 1004–1011.

Dettling, A. C., Gunnar, M. R., and Donzella, B. 1999. Cortisol levels of young children in full-day childcare centers: Relations with age and temperament. *Psychoneuroendocrinology*, 24: 519–536.

Deutch, A. Y., and Roth, R. H. 1990. The determinants of stress-induced activation of the prefrontal cortical dopamine system. *Progr. Brain Res.*, 85:367–402.

Deutsch, J. A. 1982. Controversies in food intake regulation. In B. G. Hoebel and D. Novin (eds.), *The Neural Basis of Feeding and Reward*, pp. 137–148. Haer Institute, Brunswick, ME.

DeVries, A. C., DeVries, M. B., Taymans, S. E., and Carter, C. S. 1995a. The effects of social stress on social preferences are sexually dimorphic in prairie voles. *Proc. Natl. Acad. Sci. U.S.A.*, 93:11980–11984.

DeVries, A. C., DeVries, M. B., Taymans, S. E., and Carter, C. S. 1995b. Modulation of pair bonding by corticosterone in female prairie voles (*Microtus ochrogaster*). *Proc. Natl. Acad. Sci. U.S.A.*, 92:7744–7748.

DeVries, A. C., Glasper, E. R., and Detillion, C. E. 2003. Social modulation of stress responses. *Physiology and Behavior* 79: 399–407.

DeVries, A. C., Johnson, C.L., and Carter, C. S. 1996. Characterization of partner preference in male and female prairie voles (*Microtus ochrogaster*). *Can. J. Zool.*, 75:295–301.

DeVries, A. C., Taymans, S. E., Sundstrom, J. M., and Pert, A. 1998. Conditioned release of corticosterone by contextual stimuli associated with cocaine is mediated by corticotropin-releasing factor. *Brain Res.*, 786:39–46.

DeVries, A. C., Young, W. S., Nelson, R. J. 1997. Reduced aggressiveness in mice with targeted disruption of the gene for oxytocin. *J. Neuroendocrinol.*, 9:363–368.

DeVries, G. J., and Simerly, R. B. 2002. Anatomy, development, and function of sexually dimorphic neural circuits in the mammalian brain. In D. W. Pfaff, A. P. Arnold, A. M. Etgen, S. E. Fahrbach, and R. T. Rubin (eds.), *Hormones, Brain and Behavior*, Volume 4, pp. 137–191. Academic Press, New York.

DeVries, G. J., Buijs, R. M., Van Leeuwen, F. W., Caffe, A. R., and Swaab, D. F. 1985. The vasopressinergic innervation of the brain in normal and castrated rats. *J. Comp. Neurol.*, 233:236–254.

DeVries, G. J., Rissman, E. F., Simerly, R. B., Yang, L. Y., Scordalakes, E. M., Auger, C. J., Swain, A., Lovell-Badge, R., Burgoyne, P. S., and Arnold, A. P. 2002. A model system for the study of sex chromosome effects on sexually dimorphic neural and behavioral traits. *Journal of Neuroscience*, 22: 9005–9014.

DeWied, D. 1964. Influence of anterior pituitary on avoidance learning and escape behavior. *Am. J. Physiol.*, 207:255–259.

DeWied, D. 1969. Effects of peptide hormones on behavior. In W. F. Ganong and L. Martini (eds.), *Front. Neuroendocrinol.*, pp. 97–140. Oxford Press, New York.

DeWied, D. 1974. Pituitary–adrenal system hormones and behavior. In F. O. Schmidt and F. G. Worden (eds.), *The Neurosciences: Third Study Program*, pp. 653–666. MIT Press, Cambridge, MA.

DeWied, D. 1977. Peptides and behavior. *Life Sci.*, 20:195–204.

DeWied, D. 1980. Hormonal influences on motivation, learning, memory, and psychosis. In D. T. Krieger and J. C. Hughes (eds.), *Neuroendo-crinology*, pp. 194–204. Sinauer Associates, Sunderland, MA.

DeWied, D. 1984. Neurohypophyseal hormone influences on learning and memory processes. In G. Lynch, J. L. McGaugh, and N. M. Weinberger (eds.), *Neurobiology of Learning and Memory*, pp. 289–312. Guilford, New York.

DeWied, D. 1997. The neuropeptide story. *Front. Neuroendocrinol.*, 18:101–118.

Dewing, P., Shi, T., Horvath, S., and Vilain, E. 2003. Sexually dimorphic gene expression in mouse brain precedes gonadal differentiation. *Molecular Brain Research*, 118: 82–90.

Dewsbury, D. A. 1972. Patterns of copulatory behavior in male mammals. *Q. Rev. Biol.*, 47:1–33.

Dewsbury, D. A. 1979. *Comparative Animal Behavior.* McGraw-Hill, New York.

Dewsbury, D. A. 1984. *Comparative Psychology in the Twentieth Century.* Hutchinson Ross, Stroudsburg, PA.

Dey, J., Misra, A., Desai, N. G., Mahapatra, A. K., and Padma, M. V. 1997. Cognitive function in younger type II diabetes. *Diabetes Care*, 20: 32–35.

Dhabhar, F. S., and McEwen, B. S. 1997. Acute stress enhances while chronic stress suppresses immune function in vivo: A role for leukocyte trafficking. *Brain, Behavior, and Immunology*, 11: 286–306.

Dhabhar, F. S., and McEwen, B. S. 1999. Enhancing versus suppressive effects of stress hormones on skin immune function. *Proceedings of the National Academy of Sciences U.S.A.*, 96: 1059–1064.

Diakow, C. 1974. Motion picture analysis of rat mating behavior. *J. Comp. Physiol. Psychol.*, 88:704–712.

Diamond, A. W. 1976. Subannual breeding and moult cycles in the bridled tern *Sterna anaethetus* in the Seychelles. *Ibis*, 114:395–398.

Diamond, M. 1970. Intromission pattern and species vaginal code in relation to induction of pseudopregnancy. *Science*, 169:995–997.

Diamond, M. 1984. *Sex Watching*. Macdonald and Company, London.

Diaz, R., Ogren, S. O., Blum, M., and Fuxe, K. 1995. Prenatal corticosterone increases spontaneous and α-amphetamine induced locomotor activity and brain dopamine metabolism in prepubertal male and female rats. *Neuroscience*, 66:467–473.

Dickerman, R. D., Schaller, F., Prather, I., and McConathy, W. J. 1995. Sudden cardiac death in a 20-year-old bodybuilder using anabolic steroids. *Cardiology*, 86:172–173.

Dickinson, R. L. 1949. *Atlas of Human Sex Anatomy.* Williams & Wilkins, Baltimore.

Dietrich, A., and Allen, J. D. 1997a. Vasopressin and memory. I. The vasopressin analogue AVP4–9 enhances working memory as well as reference memory in the radial arm maze. *Behav. Brain Res.*, 87:197–200.

Dietrich, A., and Allen, J. D. 1997b. Vasopressin and memory. II. Lesions to the hippocampus block the memory enhancing effects of a AVP4–9 in the radial maze. *Behav. Brain Res.*, 87:201–208.

Dinan, T. G. 1998. Psychoneuroendocrinology of depression. Growth hormone. *Psychiatr. Clin. North Am.*, 21:325–339.

Ding, J. M., Chen, D., Weber, E. T., Faiman, L. E., Rea, M. A., and Gillette, M. U. 1994. Resetting the biological clock: Mediation of nocturnal circadian shifts by glutamate and NO. *Science*, 266: 1713–1717.

Dinges, D. F. 1984. The nature and timing of sleep. *Trans. Stud. Coll. Physicians Phila.* 6:177–206.

Disney, H. J., Lofts, B., and Murton, A. J. 1959. Duration of the regeneration period of the internal reproductive rhythm in a xerophilous equatorial bird, *Quelea quelea. Nature*, 184:1659–1660.

Dixson, A. F. 1980. Androgens and aggressive behavior in primates: A review. *Aggressive Behav.*, 6:37–67.

Dixson, A. F. 1997. Evolutionary perspectives on primae mating systems and behavior. *Ann. N. Y. Acad. Sci.*, 807:42–61.

Dixson, A. F., and George, L. 1982. Prolactin and parental behaviour in a male New World primate. *Nature*, 299:551–553.

Dixson, A. F., Everitt, G. J., Herbert, J., Rugman, S. J., and Scruton, D. M. 1973. Hormonal and other determinants of sexual attractiveness and receptivity in rhesus and talapoin monkeys. IVth International Congress of Primatology. Vol. 2. *Primate Reproductive Behavior*, pp. 36–63. Karger, Basel.

Dluzen, D. E., and Ramirez, V. D. 1983. Localized and discrete changes in neuropeptide (LHRH and TRH) and neurotransmitter (NE and DA) concentrations within the olfactory bulbs of male mice as a function of social interaction. *Horm. Behav.*, 17:139–145.

Dluzen, D. E., Ramirez, V. D., Carter, C. S., and Getz, L. L. 1981. Male vole urine changes luteinizing hormone-releasing hormone and norepinephrine in female olfactory bulb. *Science*, 212:573–575.

Dodson, R. E., and Gorski, R. A. 1993. Testosterone propionate administration prevents the loss of neurons within the central part of the medial preoptic nucleus. *Journal of Neurobiology*, 24: 80–88.

Doering, C. H., Brodie, H. K. H., Kraemer, H. C., Moos, R. H., Becker, H. B., and Hamburg, D. A. 1975. Negative affect and plasma testosterone: A longitudinal human study. *Psychosom. Med.*, 37:484–491.

Dohanich, G. 2002. Gonadal steroids, learning, and memory. In D. W. Pfaff, A. P. Arnold, A. M. Etgen, S. E. Fahrbach, and R. T. Rubin (eds.), *Hormones, Brain and Behavior*, Volume 2, pp. 265–327. Academic Press, New York.

Döhler, K. D., and Hancke, J. L. 1978. Thoughts on the mechanism of sexual brain differentiation. In K. D. Döhler and M. Kawakami (eds.), *Hormones and Brain Development*, pp. 153–157. Elsevier, Amsterdam.

Döhler, K. D., Hines, M., Coquelin, A., Davis, F., Shryne, J. E., and Gorski, R. A. 1982. Pre- and postnatal influence of diethylstilbestrol on differentiation of the sexually dimorphic nucleus in the preoptic area of the female rat brain. *Neuroendocrinol. Lett.*, 4:361–365.

Doisy, E. A., and MacCorquodale, D. W. 1936. The hormones. *Annu. Rev. Biochem.*, 5:315–354.

Doisy, E. A., Veler, C. D., and Thayer, S. 1929. Folluculin from urine of pregnant women. *Am. J. Physiol.*, 90:329–330.

Dominguez, J. M., Muschamp, J. W., Schmich, J. M., and Hull, E. M. 2004. Nitric oxide mediates glutamate-evoked dopamine release in the medial preoptic area. *Neuroscience*, 125: 203–210.

Doty, R. L. 1986. Odor-guided behaviour in mammals. *Experientia*, 42:257–271.

Doty, R. L. 1997. Studies of human olfaction from the University of Pennsylvania Smell and Taste Center. *Chemical Senses*, 22: 565–586.

Doty, R. L., and Anisko, J. J. 1973. Procaine hydrochloride olfactory block eliminated mounting in the golden hamster. *Physiol. Behav.*, 10:395–397.

Douglass, J., McKinzie, A. A., and Couceyro, P. 1995. PCR differential display identifies a rat brain mRNA that is transcriptionally regulated by cocaine and amphetamine. *Journal of Neuroscience*, 15: 2471–2481.

Dowell, S. F., and Lynch, G. R. 1987. Duration of the melatonin pulse in the hypothalamus controls testicular function in pinealectomized mice (*Peromyscus leucopus*). *Biol. Reprod.*, 36:1095–1101.

Doyle, L. W., Ford, G. W., Davis, N. M., and Callanan, C. 2000. Antenatal corticosteroid therapy and blood pressure at 14 years of age in preterm children. *Clinical Sciences* (London), 98: 137–142.

Drea, C. M., Weldele, M. L., Forger, N. G., Coscia, E. M., Frank, L. G., Licht, P., and Glickman, S. E. 1998. Androgens and masculinization of genitalia in the spotted hyaena (*Crocuta crocuta*). 2. Effects of prenatal anti-androgens. *Journal of Reproduction and Fertility*, 113: 117–127.

Drickamer, L. C. and Vessey S. H. 1982. *Animal Behavior: Concepts, Processes, Methods*. PWS, Boston.

Du, J., Lorrain, D. S., and Hull, E. M. 1998. Castration decreases extracellular, but increases intracellular, dopamine in medial preoptic area of male rats. *Brain Res.*, 782:11–17.

Dube, M. G., Kalra, S. P., and Kalra. P. S. 1999. Food intake elicited by central administration orexins/hypocretins: Identification of hypothalamic sites of action. *Brain Research*, 842: 473–477.

Dubocovich, M. L., Rivera-Bermudez, M. A., Gerdin, M. J., and Masana, M. I. 2003. Molecular pharmacology, regulation and function of mammalian melatonin receptors. *Frontiers in Bioscience*, 8: 1093–1108.

Dulac, C., and Torello, A. T. 2003. Molecular detection of pheromone signals in mammals: From genes to behavior. *Nature Reviews Neuroscience*, 4: 551–562.

Dunbar, I. 1975. Behaviour of castrated animals. *Vet. Rec.*, 96:92–93.

Duncan, M. J. 1998. Photoperiodic regulation of hypothalamic neuropeptide messenger RNA expression: Effect of pinealectomy and neuroanatomical location. *Mol. Brain Res.*, 57:142–148.

Dunlap, J. 1998. Circadian rhythms: An end in the beginning. *Science*, 280:1548–1549.

Dunn, A. J., Steelman, S., and Delanoy, R. 1980. Intraventricular ACTH and vasopressin cause regionally specific changes in cerebral deoxyglucose uptake. *J. Neurosci. Res.*, 5:485–495.

Dworkin, S. I., Goeders, N. E., Grabowski, J., and Smith, J. E. 1987. The effects of 12-hour limited access to cocaine: reduction in drug intake and mortality. *NIDA Res. Monogr.*, 76:221–225.

Eakin, R. M. 1973. *The Third Eye*. University of California Press, Berkeley.

Ebbinghaus, H. 1908. *Psychology: An Elementary Textbook*. D. C. Heath, Boston.

Eberhart, J. A. 1988. Neural and hormonal correlates of primate sexual behavior. *Comp. Primate Biol. Neurosci.*, 4:675–705.

Eckhardt, C. 1863. Untersuchungen uber die erektion des penis beim hund. *Beitr. Anat. Phisiol.*, 3:123–147.

Eddy, E. M., Washburn, T. F., Bunch, D. O., Goulding, E. H., Gladen, B. C., Lubahn, D. B., and Korach, K. S. 1996. Targeted disruption of the estrogen receptor gene in male mice causes alternation of spermatogenesis and infertility. *Endocrinology*, 137:4796–4805.

Edgar, D. M., Kilduff, T. S., Martin, C. E., and Dement, W. C. 1991. Influence of running wheel activity on free-running sleep/wake and drinking circadian rhythms in mice. *Physiology and Behavior*, 50: 373–378.

Edwards, D. A. 1969. Early androgen stimulation and aggressive behavior in male and female mice. *Physiol. Behav.*, 4:333–338.

Edwards, D. A. 1970. Post-neonatal androgenization and adult aggressive behavior in female mice. *Physiol. Behav.*, 5:465–467.

Egawa, M., Yoshimatsu, H., and Bray, G. A. 1990. Effect of corticotropin releasing hormone and neuropeptide Y on electrophysiological activity of sympathetic nerves to interscapular brown brown adipose tissue. *Neuroscience*, 34:771–775.

Egger, G., Liang, G., Aparicio, A., and Jones, P. A. 2004. Epigenetics in human disease and prospects for epigenetic therapy. *Nature*, 429: 457–463.

Ehrenkranz, J., Bliss, E., and Sheard, M. H. 1974. Plasma testosterone: Correlation with aggressive behavior and social dominance in man. *Psychosom. Med.*, 36:469–475.

Ehrensing, R. H., Kastin, A. J., Schalch, D. S., Friesen, H. G., Vargas, J. R., and Schally, A. V. 1974. Affective state and thyrotropin and prolactin responses after repeated injections of thyrotropin-releasing hormone in depressed patients. *Am. J. Psychiatry*, 131:714–718.

Eising, C. M., Eikenaar, C., Schwabl, H., and Groothuis, T. G. 2001. Maternal androgens in black-headed gull (*Larus ridibundus*) eggs: Consequences for chick development. *Proceedings of the Royal Society of London, Series B*, 268: 839–846.

Elias, C. F., Saper, C. B., Marotos-Flier, E., Tritos, N. A., Lee, C., Kelly, J., Tatro, J. B., et al. 1998. Chemically defined projections linking the mediobasal hypothalamus and the lateral hypothalamic area. *Journal of Comparative Neurology*, 402: 442–459.

Elias, M. 1981. Cortisol, testosterone and testosterone-binding globulin responses to competitive fighting in human males. *Aggressive Behav.*, 7:215–219.

Elliott, A. S., and Nunez, A. A. 1992. Photoperiod modulates the effects of steroids on sociosexual behaviors of hamsters. *Physiol. Behav.*, 51:1189–1193.

Elliott, J. A., and Goldman, B. D. 1981. Seasonal reproduction: Photoperiodism and biological clocks. In N. T. Adler (ed.), *Neuroendocrinology of Reproduction*, pp. 377–423. Plenum Press, New York.

Ellis, A. 1945. The sexual psychology of human hermaphrodites. *Psychosom. Med.*, 7:108–125.

Ellis, G. B., and Turek, F. W. 1979. Changes in locomotor activity associated with the photoperiodic response of testes in male golden hamsters. *J. Comp. Physiol.*, 132:277–284.

Ellis, G. B., and Turek, F. W. 1980a. Photoperiod-induced change in responsiveness of the hypothalamic-pituitary axis to exogenous 5-alpha-dihydrotestosterone and 17β-estradiol in castrated male hamsters. *Neuroendocrinology*, 31:205–209.

Ellis, G. B., and Turek, F. W. 1980b. Photoperiodic regulation of serum luteinizing hormone and follicle-stimulating hormone in castrated and castrated-adrenalectomized male hamsters. *Endocrinology*, 106:1338–1344.

Ellis, G. B., and Turek, F. W. 1983. Testosterone and photoperiod interact to regulate locomotor activity in male hamsters. *Horm. Behav.*, 17:66–75.

Elwood, R. W. 1983. Paternal care in rodents. In R. W. Elwood (ed.), *Parental Behaviour in Rodents*, pp. 235–257. Wiley, New York.

Elwood, R. W., and Mason, C. 1994. The couvade and the onset of paternal care: A biological perspective. *Ethol. Sociobiol.*, 15:145–156.

Emery, D. E., and Moss, R. L. 1984. Lesions confined to the ventromedial hypothalamus decrease the frequency of coital contacts in female rats. *Horm. Behav.*, 18:313–329.

Emlen, S. T. 1978. The evolution of cooperative breeding in birds. In J. R. Krebs and N. B. Davies (eds.), *Behavioural Ecology: An Evolutionary Approach*, pp. 245–281. Blackwell Scientific Publications, Oxford.

Endicott, J., and Halbreich, U. 1982. Retrospective report of premenstrual depressive changes: Factors affecting confirmation by daily ratings. *Psychopharm. Bull.*, 18:109–112.

Endicott, J., Halbreich, U., Schacht, S., and Nee, J. 1981. Premenstrual changes and affective disorders. *Psychosom. Med.*, 43:519–529.

Engelmann, M., Ebner, K., Wotjak, C. T., and Landgraf, R. 1998. Endogenous oxytocin is involved in short-term olfactory memory in female rats. *Behav. Brain Res.*, 90:89–94.

Engelmann, M., Wotjak, C. T., Neumann, I., Ludwig, M., and Landgraf, R. 1996. Behavioral consequences of intracerebral vasopressin and oxytocin: Focus on learning and memory. *Neurosci. Biobehav. Rev.*, 20:341–358.

Englander-Golden, P., Change, H., Whitmore, M. R., and Dienstbier, R. A. 1980. Female sexual arousal and the menstrual cycle. *J. Hum. Stress*, 6:42–48.

Erickson, J. C., Ahima, R. S., Hollopeter, G., Flier, J. S., and Palmiter, R. D. 1997. Endocrine function of neuropeptide Y knockout mice. *Regul. Pept.*, 70:199–202.

Erickson, J. C., Hollopeter, G., and Palmiter, R. D. 1996. Attenuation of the obesity syndrome of *ob/ob* mice by the loss of neuropeptide Y. *Science*, 274: 1704–1707.

Erikson, C. J., and Hutchison, J. B. 1977. Induction of nest-material collecting in male Barbary Doves by intracerebral androgen. *J. Reprod. Fertil.*, 50:9–16.

Erkinaro, E. 1961. The seasonal change of the activity of Microtus agrestis. *Oikos*, 12:157–63

Erskine, M. S. 1985. Effects of paced coital stimulation on estrus duration in intact cycling rats and ovariectomized and ovariectomized-adrenalectomized hormone-primed rats. *Behav. Neurosci.*, 99:151–161.

Erskine, M. S. 1989. Solicitation behavior in the estrous female rat: A review. *Horm. Behav.*, 23:473–502.

Erskine, M. S., and Hanrahan, S. B. 1997. Effects of paced mating on c-fos gene expression in the female rat brain. *J. Neuroendocrinol.*, 9:903–912.

Erskine, M. S., Barfield, R. J., and Goldman, B. D. 1980a. Postpartum aggression in rats: Effects of hypophysectomy. *J. Comp. Physiol. Psychol.*, 94:484–494.

Erskine, M. S., Barfield, R. J., and Goldman, B. D. 1980b. Postpartum aggression in rats: II. Dependence on maternal sensitivity to young and effects of experience with pregnancy and parturition. *J. Comp. Physiol. Psychol.*, 94:495–505.

Erskine, M. S., Lehmann, M. L., Cameron, N. M., and Polston, E. K. 2004. Co-regulation of female sexual behavior and pregnancy induction: An exploratory synthesis. *Behavioral Brain Research*, 153: 295–315.

Espmark, Y. 1964. Rutting behaviour in reindeer (*Rangifer tarandus L.*). *Anim. Behav.*, 12:159–163.

Etgen, A. 2002. Estrogen regulation of neurotransmitter and growth factor signaling in the brain. In D. W. Pfaff, A. P. Arnold, A. M. Etgen, S. E. Fahrbach, and R. T. Rubin (eds.), *Hormones, Brain and Behavior*, Volume 3, pp. 381–440. Academic Press, New York.

Etgen, A. M., Chu, H. P., Fiber, J. M., Karkanias, G. B., and Morales, J. M. 1999. Hormonal integration of neurochemical and sensory signals governing female reproductive behavior. *Behavior and Brain Research*, 105: 93–103.

Evans, D. L., Leserman, J., Perkins, D. O., Stern, R. A., Murphy, C., Zheng, B., Gettes, D., Longmate, J. A., Silva, S. G., van der Horst, C. M., Hall, C. D., Folds, J. D., Golden, R. N., and Petitto, J. M. 1997. Severe life stress as a predictor of early disease progression in HIV infection. *Am. J. Psychiatry*, 154:630–634.

Everett, J. W. 1961. The mammalian female reproductive cycle and its controlling mechanisms. In W. C. Young (ed.), *Sex and Internal Secretions*, Vol. 1, pp. 497–555. Williams & Wilkins, Baltimore.

Everitt, B. J. 1990. Sexual motivation: A neural and behavioral analysis of male rats. *Neurosci. Biobehav. Rev.*14:217–232.

Everitt, B. J., and Herbert, J. 1971. The effects of dexamethasone and androgens on sexual receptivity of female rhesus monkeys. *J. Endocrinol.*, 51:575–588.

Ewald, P. W. 1994. *Evolution of Infectious Diseases.* Oxford University Press, Oxford.

Ewer, J., Gammie, S. C., and Truman, J. W. 1997. Control of insect ecdysis by a positive-feedback endocrine system: Roles of eclosion hormone and ecdysis triggering hormone. *J. Exp. Biol.*, 200:869–881.

Fader, A. J., Hendricson, A. W., and Dohanich, G. P. 1998. Estrogen improves performance of reinforced T-maze alteration and prevents the amnestic effects of scopolamine administered systemically or intrahippocampally. *Neurobiol. Learn. Mem.*, 69:225–240.

Fahrbach, S. E., and Pfaff, D. W. 1986. Effects of preoptic region implants of dilute estradiol on the maternal behavior of ovariectomized, nulliparous rats. *Horm. Behav.*, 20:957–959.

Fahrbach, S. E., Morrell, J. I., and Pfaff, D. W. 1986. Effect of varying the duration of pre-test cage habituation on oxytocin induction of short-latency maternal behavior. *Physiol. Behav.*, 37:135–139.

Farrell, S. F., and McGinnis, M. Y. 2003. Effects of pubertal anabolic-androgenic steroid (AAS) administration on reproductive and aggressive behaviors in male rats. *Behavioral Neuroscience*, 117: 904–911.

Fava, G. A., Fava, M., Kellner, R., Serafini, E., and Mastrogiacomo, I. 1981. Depression, hostility and anxiety in hyperprolactinemic amenorrhea. *Psychother. Psychosom.*, 36:122–128.

Feare, C. 1984. *The Starling.* Oxford University Press, New York.

Feder, H. 1981. Perinatal hormones and their role in the development of sexually dimorphic behaviors. In N. T. Adler (ed.), *Neuroendocrinology of Reproduction*, 127–158. Plenum Press, New York.

Feder, H. H. 1971. The comparative action of testosterone propionate and 5α-androstan-17β-ol-3-one propionate on the reproductive behaviour, physiology and morphology of male rats. *J. Endocrinol.*, 51:242–252.

Feder, H. H. 1981. Estrous cyclicity in mammals. In N. T. Adler (ed.), *Neuroendocrinology of Reproduction*, pp. 279–348. Plenum Press, New York.

Feder, H. H., and Whalen, R. E. 1965. Feminine behavior in neonatally castrated and estrogen-treated male rats. *Science*, 147:306–307.

Feder, H. H., Naftolin, F., and Ryan, K. J. 1974. Male and female sexual responses in male rats given estradiol benzoate and 5α-androstan-17β-ol-3-one propionate. *Endocrinology*, 94:136–141.

Fee, A. R., and Parkes, A. S. 1930. Studies on ovulation. III. Effect of vaginal anaesthesia on ovulation in the rabbit. *J. Physiol. (Lond.)*, 70:385–397.

Ferin, M. 1984. Endogenous opioid peptides and the menstrual cycle. *Trends Neurosci.*, 7:194–196.

Ferkin, M. H. 1988. Seasonal differences in social behavior among adult and juvenile meadow voles, *Microtus pennsylvanicus. Ethology*, 79:116–125.

Ferkin, M. H., and Seamon, J. O. 1987. Odor preference and social behavior in meadow voles, *Microtus pennsylvanicus:* Seasonal differences. *Can. J. Zool.*, 65:2931–2937.

Fernald, R. D. 2002. Social regulation of the brain: Status, sex and size. In D. W. Pfaff, A. P. Arnold, A. M. Etgen, S. E. Fahrbach, and R. T. Rubin (eds.), *Hormones, Brain and Behavior*, Volume 2, pp. 435–444. Academic Press, New York.

Fernandez-Guasti, A., and Rodriguez-Manzo, G. 2003. Pharmacological and physiological aspects of sexual exhaustion in male rats. *Scandinavian Journal of Psychology*, 44: 257–263.

Ferris, C. F. 1992. Role of vasopressin in aggressive and dominant/subordinate behaviors. *Ann. N. Y. Acad. Sci.*, 652:212–226.

Ferris, C. F., Axelson, J. F., Shinto, L. H., and Albers, H. E. 1987. Scent marking and the maintenance of dominant/subordinate status in male golden hamsters. *Physiol. Behav.*, 40:661–664.

Ferris, C. F., Delville, Y., Grzonka, Z., Luber-Narod, J., and Insel, T. R. 1993. An iodinated vasopressin (V1) antagonist blocks flank marking and selectively labels neural binding sites in golden hamsters. *Physiol. Behav.*, 54:737–747.

Ferris, C. F., Melloni, R. H., Koppel, G., Perry, K. W., Fuller, R. W., and Delville, Y. 1997. Vasopressin/serotonin interactions in the anterior hypothalamus control aggressive behavior in golden hamsters. *J. Neurosci.*, 17:4331–4340.

Ferris, C. F., Snowdon, C. T., King, J. A., Duong, T. Q., Ziegler, T. E., Ugurbil, K., Ludwig, R., et al. 2001. Functional imaging of brain activity in conscious monkeys responding to sexually arousing cues. *Neuroreport*, 12: 2231–2236.

Ferris, C. F., Snowdon, C. T., King, J. A., Sullivan, J. M., Ziegler, T. E., Olson, D. P., Schultz-Darken, N. J., et al. 2004. Activation of neural pathways associated with sexual arousal in non-human primates. *Journal of Magnetic Resonance Imaging*, 19: 168–175.

Ferron, F., Considine, R. V., Peino, R., Lado, I. G., Dieguez, C., and Casanueva, F. F. 1997. Serum leptin concentrations in patients with anorexia nervosa, bulimia nervosa and non-specific eating disorders correlate with the body mass index but are independent of the respective disease. *Clinical Endocrinology*, 46: 289–293.

Fingerhood, M. I., Sullivan, J. T., Testa, M., and Jasinski, D. R. 1997. Abuse liability of testosterone. *J. Psychopharmacol.*, 11:59–63.

Fink, G., Sumner, B. E., Rosie, R., Grace, O., and Quinn, J. P. 1996. Estrogen control of central neurotransmission: Effect of mood, mental state, and memory. *Cellular Molecular Neurobiology*, 16:325–344.

Fiorino, D. F., Coury, A., and Phillips, A. G. 1997. Dynamic changes in nucleus accumbens dopamine efflux during the Coolidge effect in male rats. *J. Neurosci.*, 17: 4849–4855

Fitzgerald, K. M., and Zucker, I. 1976. Circadian organization of the estrous cycle of the golden hamster. *Proc. Natl. Acad. Sci. U.S.A.*, 73:2923–2927.

Fitzsimons, J. T. 1961. Drinking by rats depleted of body fluid without increase in osmotic pressure. *J. Physiol. (Lond.)*, 159:297–309.

Fitzsimons, J. T. 1998. Angiotensin, thirst, and sodium appetite. *Physiol. Rev.*, 78:583–686.

Fitzsimons, J. T., and Simons, B. J. 1969. The effect on drinking in the rat of intravenous infusion of angiotensin, given alone or in combination with other stimuli of thirst. *J. Physiol.*, 203:45–57.

Fivizzani, A. J., and Oring, L. W. 1986. Plasma steroid hormones in relation to behavioral sex role reversal in the spotted sandpiper, Actitis macularia. *Biol. Reprod.*, 35:1195–1201.

Fivizzani, A. J., Colwell, M. A., and Oring, L. W. 1986. Plasma steroid hormone levels in free-living Wilson's Phalaropes (*Phalaropus tricolor*). *Gen. Comp. Endocrinol.*, 62:137–144.

Flanagan-Cato, L. M., and Fluharty, S. J. 1997. Emerging mechanisms of the behavioral effects of steroids. *Curr. Opin. Neurobiol.*, 7:844–848.

Fleming, A. S. 1986. Psychobiology of rat maternal behavior. How and where hormones act to promote maternal behavior at parturition. *Ann. N. Y. Acad. Sci.*, 474:234–251.

Fleming, A. S. 1990. Hormonal and experimental correlates of maternal responsiveness in human mothers. In N. A. Krasnegor and R. S. Bridges (eds.), *Mammalian Parenting*, pp. 184–208. Oxford University Press, Oxford.

Fleming, A. S., and Corter, C. M. 1995. Psychobiology of maternal behavior in nonhuman mammals. In M. H. Bornstein (ed.), *Handbook of Parenting*, Vol. 2, pp. 59–85. Erlbaum Associates.

Fleming, A. S., and Korsmit, M. 1996. Plasticity inthe maternal circuit: Effects of maternal experience on Fos-Lir in hypothalamic, limbic, and cortical structures in the postpartum rat. *Behav. Neurosci.*, 110:567–582.

Fleming, A. S., and Luebke, C. 1981. Timidity prevents the nulliparous female from being a good mother. *Physiol. Behav.*, 27:863–868.

Fleming, A. S., and Pliner, P. 1983. Food intake, body weight, and sweetness preferences over the menstrual cycle in humans. *Physiol. Behav.*, 30:663–666.

Fleming, A. S., and Rosenblatt, J. S. 1974. Olfactory regulation of maternal behavior in rats: II. Effects of peripherally induced anosmia and lesions of the lateral olfactory tract in pup-induced virgins. *J. Comp. Physiol. Psychol.*, 86:233–246.

Fleming, A. S., Cheung, U. S., and Barry, M. 1990a. Cyclohexamide blocks the retention of maternal experience in postpartum rats. *Behav. Neural Biol.*, 53:64–73.

Fleming, A. S., Corter, C., Franks, P., Surbey, M., Schneider, B., and Steiner, M. 1993. Postpartum factors related to mother's attraction to newborn infant odors. *Dev. Psychobiol.*, 26:115–132.

Fleming, A. S., Corter, C., Stallings, J., and Steiner, M. 2002. Testosterone and prolactin are associated with emotional responses to infant cries in new fathers. *Hormones and Behavior*, 42: 399–413.

Fleming, A. S., Corter, C., Surbey, M., Franks, P., and Steiner, M. 1995. Postpartum factors related to mother's recognition of newborn infant odours. *J. Reprod. Infant Psychol.*, 13:197–210.

Fleming, A. S., Krieger, H., and Wong, P. Y. 1990b. Affect and nurturance in first-time mothers: Role of psychobiological influences. In B. Lerer and S. Gershon (eds.), *New Directions in Affective Disorders*, pp. 388–392, Springer-Verlag, New York.

Fleming, A. S., Miceli, M., and Morretto, D. 1983. Lesions of the medial preoptic area prevent the facilitation of maternal behavior produced by amygdaloid lesions. *Physiol. Behav.*, 31:502–510.

Fleming, A. S., Morgan, H. D., and Walsh, C. 1996. Experiential factors in postpartum regulation of maternal care. In J. S. Rosenblatt and C. T. Snowdon (eds.), *Parental Care: Evolution, Mechanisms, Adaptive Significance*, pp. 295–332. Academic Press, San Diego.

Fleming, A. S., Ruble, D. N., Krieger, H., and Wong, P. 1997a. Hormonal and experiential correlates of maternal responsiveness during pregnancy and the puerperium in human mothers. *Horm. Behav.*, 31:145–158.

Fleming, A. S., Steiner, M., and Anderson, V. 1987. Hormonal and attitudinal correlates of maternal behavior during the early postpartum period in first-time mothers. *J. Reprod. Infant Psychol.*, 5:193–205.

Fleming, A. S., Steiner, M., and Corter, C. 1997b. Cortisol, hedonics, and maternal responsiveness in human mothers. *Horm. Behav.*, 32:85–98.

Fleming, A. S., Suh, E. J., Korsmit, M., and Rusak, B. 1994. Activation of Fos-like immunoreactivity in the medial preoptic area and limbic structures by maternal and social interactions in rats. *Behav. Neurosci.*, 108:724–734.

Fleming, A. S., Vaccarino, F., and Luebke, C. 1980. Amygdaloid inhibition of maternal behavior in the nulliparous rat. *Physiol. Behav.*, 25:731–743.

Fleming, A. S., Vaccarino, F., Tambosso, L., and Chee, P. 1979. Vomeronasal and olfactory system modulation of maternal behavior in the rat. *Science*, 203:372–374.

Fleshner, M., Laudenslager, M. L., Simons, L., and Maier, S. F. 1989. Reduced serum antibodies associated with social defeat in rats. *Physiology and Behavior*, 45: 1183–1187.

Flint, A., Raben, A., Ersbøll, A. K., Holst, J. J., and Astrup, A. 2001. The effect of physiological levels of glucagon-like peptide-1 on appetite, gastric emptying, energy and substrate metabolism in obesity. *International Journal of Obesity*, 25: 781–792.

Flood, J. F., and Morley, J. E. 1991. Increased food intake by neuropeptide Y is due to an increased motivation to eat. *Peptides*, 12:1329–1332.

Flood, J. F., Smith, G. E., and Morley, J. E. 1987. Modulation of memory processing by cholecystokinin: Dependence on the vagus nerve. *Science*, 236:832–834.

Florant, G. L., Tokuyama, K., and Rintoul, D. A. 1989. Carbohydrate and lipid utilization in hibernators. In A. Malan and B. Canguilhem (eds.), *Living in the Cold*, pp. 137–145. John Libbey, London.

Fluharty, S. J., 2002. Neuroendocrinology of body fluid homeostasis. In D. W. Pfaff, A. P. Arnold, A. M. Etgen, S. E. Fahrbach, and R. T. Rubin (eds.), *Hormones, Brain and Behavior*, Volume 1, pp. 525–569. Academic Press, New York.

Fluharty, S. J., and Sakai, R. R. 1995. Behavioral and cellular analysis of adrenal steroid and angiotensin interactions mediating salt appetite. *Progr. Psychobiol. Physiol. Psychol.*, 16:177–212.

Follett, B. K. 1991. The physiology of puberty in seasonally breeding birds. In M. Hunzicker-Dunn and N. B. Schwartz (eds.), *Follicle Stimulating Hormone: Regulation of Secretion and Molecular Mechanisms of Action*, pp. 54–65. Springer-Verlag, New York.

Folman, Y., and Drori, D. 1966. Effects of social isolation and of female odours on the reproductive system, kidneys and adrenals of unmated male rats. *J. Reprod. Fertil.*, 11:43–50.

Forbes, T. R. 1949. A. A. Berthold and the first endocrine experiment: Some speculations as to its origin. *Bull. Hist. Med.*, 23:263–267.

Ford, C. S., and Beach, F. A. 1951. *Patterns of Sexual Behavior.* Harper and Row, New York.

Forest, M. G., and Cathiard, A. M. 1975. Pattern of plasma testosterone and androstenedione in normal newborns: Evidence for testicular activity at birth. *J. Clin. Endocrinol. Metab.*, 41:977–980.

Formby, D. 1967. Maternal recognition of infant's cry. *Dev. Med. Child Neurol.*, 9:293–298.

Fossey, D. 1974. Observations on the home range of one group of mountain gorillas (*Gorilla gorilla beringei*). *Anim. Behav.*, 22:568–581.

Fox, C. A., Ismail, A. A., Love, D. N., Kirkham, K. E., and Loraine, J. A. 1972. Studies on the relationship between plasma testosterone levels and human sexual activity. *J. Endocrinol.*, 52:51–58.

Fraile, I. G., Pfaff, D. W., and McEwen, B. S. 1987. Progestin receptors with and without estrogen induction in male and female hamster brain. *Neuroendocrinology*, 45:487–491.

Francis, D. D., Diorio, J., Liu, D., and Meaney, M. J. 1999. Non-genomic transmission across generations of maternal behavior and stress responses in the rat. *Science*, 286: 1155–1158.

Frank, L. G. 1986. Social organisation of the spotted hyaena: II. Dominance and reproduction. *Anim. Behav.*, 35:1510–1527.

Frank, L. G., Davidson, J. M., and Smith, E. R. 1985. Androgen levels in the spotted hyena *Crocuta crocuta*: The influence of social factors. *J. Zool. (Lond.)*, 206:525–531.

Frank, R. T. 1931. The hormonal causes of premenstrual tension. *Arch. Neurol. Psychiatry*, 26:1053–1057.

Frankenhaeuser, M. 1978. Psychoneuroendocrine approaches to the study of emotion as related to stress and coping. *Neb. Symp. Motiv.*, 26:123–162.

Freeman, E. W. 2003. Premenstrual syndrome and premenstrual dysphoric disorder: Definitions and diagnosis. *Psychoneuroendocrinology*, 28: 25–37.

Frick, K. M., and Berger-Sweeney, J. 2001. Spatial reference memory and neocortical neurochemistry vary with the estrous cycle in C57BL/6 mice. *Behavioral Neuroscience*, 115: 229–237.

Frick, K. M., Burlingame, L. A., Arters, J. A., and Berger-Sweeney, J. 2000. Reference memory, anxiety and estrous cyclicity in C57BL/6NIA mice are affected by age and sex. *Neuroscience*, 95: 293–307.

Frick, K. M., Fernandez, S. M., and Bulinski, S. C. 2002. Estrogen replacement improves spatial reference memory and increases hippocampal synaptophysin in aged female mice. *Neuroscience*, 115: 547–558.

Frick, K. M., Fernandez, S. M., Bennett, J. C., Prange-Kiel, J., MacLusky, N. J., and Leranth, C. 2004. Behavioral training interferes with the ability of gonadal hormones to increase CA1 spine synapse density in ovariectomized female rats. *European Journal of Neuroscience*, 19: 3026–3032.

Friedl, K. E., and Yesalis, C. E. 1989. Self-treatment of gynecomastia in bodybuilders who use anabolic steroids. *Physicians Sportsmedicine*, 17: 67–79.

Friedman, M. I. 1978. Hyperphagia in rats with experimental diabetes mellitus: A response to a decreased supply of utilizable fuels. *J. Comp. Physiol. Psychol.*, 92:109–117.

Friedman, M. I. 1990. Body fat and the metabolic control of food intake. *Int. J. Obesity*, 14:53–66.

Friedman, M. I., and Granneman, J. 1983. Food intake and peripheral factors after recovery from insulin-induced hypoglycemia *Am. J. Physiol.*, 244:R374–R382.

Friedman, M. I., and Tordoff, MG. 1986. Fatty acid oxidation and glucose utilization interact to control food intake in rats. *Am. J. Physiol.*, 251:R840–845.

Friedman, M. I., Ramirez, I., Edens, N. K., and Granneman, J. 1985. Food intake in diabetic rats: Isolation of primary metabolic effects of rat feeding. *Am. J. Physiol.*, 249:R44–R51.

Fry, D. M., and Toone, C. K. 1981. DDT-induced feminization of gull embryos. *Science*, 213:922–924.

Frye, C. A. 1995. Estrus-associated decrements in a water maze task are limited to acquisition. *Physiol. Behav.*, 57:5–14.

Frye, C. A. 2001. The role of neurosteroids and non-genomic effects of progestins and androgens in mediating sexual receptivity of rodents. *Brain Research: Brain Research Reviews*, 37: 201–222.

Frye, C. A. 2001a. The role of neurosteroids and non-genomic effects of progestins and androgens in mediating sexual receptivity of rodents. *Brain Research/Brain Research Reviews*, 37: 201–222.

Frye, C. A. 2001b. The role of neurosteroids and nongenomic effects of progestins in the ventral tegmental area in mediating sexual receptivity of rodents. *Hormones and Behavior*, 40: 226–233.

Frye, C. A., and DeBold, J. F. 1993. 3 alpha-OH-DHP and 5 alpha-THDOC implants to the ventral tegmental area facilitate sexual receptivity in hamsters after progesterone priming to the ventral medial hypothalamus. *Brain Res.*, 612:130–137.

Frye, C. A., and Petralia, S. M. 2003. Lordosis of rats is modified by neurosteroidogenic effects of membrane benzodiazepine receptors in the ventral tegmental area. *Neuroendocrinology*, 77: 71–82.

Frye, C. A., Mermelstein, P. G., and DeBold, J. F. 1992. Evidence for a non-genomic action of progestins on sexual receptivity in hamster ventral tegmental area but not hypothalamus. *Brain Res.*, 578:87–93.

Fuchs, A.-R., and Dawood, M. Y. 1980. Oxytocin release and uterine activation during parturition in rabbits. *Endocrinology*, 107:1117–1126.

Fuchs, A.-R., and Fuchs, F. 1984. Endocrinology of human parturition: A review. *Br. J. Obstet. Gynaecol.*, 91:948–967.

Gahr, M., and Kosar, E. 1996. Identification, distribution, and developmental changes of a melatonin binding site in the song control system of the zebra finch. *J. Comp. Neurol.*, 367:308–18.

Gahr, M., Flügge, G., and Güttinger, H.-R. 1987. Immunocytochemical localization of estrogen-binding neurons in the songbird brain. *Brain Res.*, 402:173–177.

Gahr, M., Güttinger, H.-R., and Kroodsma, D. A. 1993. Estrogen receptors in the avian brain: Survey reveals general distribution and forebrain areas unique to songbirds. *J. Comp. Neurol.*, 327:112–122.

Gaines, M. S., Fugate, C. L., Johnson, M. L., Johnson, D. C., Hisey, J. R., and Quadagno, D. M. 1985. Manipulation of aggressive behavior in male prairie voles (*Microtus ochrogaster*) implanted with testosterone in Silastic tubing. *Can. J. Zool.*, 63:2525–2528.

Gale, S. M., Castracane, V. D., and Mantzoros, C. S. 2004. Energy homeostasis, obesity and eating disorders: Recent advances in endocrinology. *Journal of Nutrition*, 134: 295–298.

Galea, L. A., Kavaliers, M., and Ossenkopp, K. P. 1996. Sexually dimorphic spatial learning in meadow voles (*Microtus pennsylvanicus*) and deer mice (*Peromyscus maniculatus*). *J. Exp. Biol.*, 199:195–200.

Galea, L. A., Kavaliers, M., Ossenkopp, K. P., and Hampson, E. 1995. Gonadal hormone levels and spatial learning performance in the Morris water maze in male and female meadow voles, *Microtus pennsylvanicus. Horm. Behav.*, 29:106–125.

Galea, L. A., McEwen, B. S., Tanapat, P., Deak, T., Spencer, R. L., and Dhabhar, F. S. 1997. Sex differences in dendritic atrophy of CA3 pyramidal neurons in response to chronic stress. *Neuroscience*, 81: 689–697.

Gallagher, M., Kapp, B. S., Pascoe, J. P., and Rapp, P. R. 1981. A neuropharmacology of amygdaloid systems which contribute to learning and memory. In I. Ben Ari (ed.), *The Amygdaloid Complex*, pp. 343–354. Elsevier, Amsterdam.

Galli-Taliadoros, L. S., Sedgwick, J. D., Wood, S. A., and Korner, H. 1995. Gene knock-out technology: A methodological overview for the interested novice. *J. Immunol. Meth*, 181, 1–15.

Gammie, S. C., and Truman, J. W. 1997. An endogenous elevation of cGMP increases the excitability of identified insect neurosecretory cells. *J. Comp. Physiol. A*: 180:329–338.

Gammie, S. C., Negron, A., Newman, S. M., and Rhodes, J. S. 2004. Corticotropin-releasing factor inhibits maternal aggression in mice. *Behavioral Neuroscience*, 118: 805–814.

Ganong, W. F. 1993. *Review of Medical Physiology*. (16th ed.). Appleton & Lange, Norwalk, CT.

Ganong, W. F. 1995. Review of Medical Physiology. (17th ed.). Appleton & Lange, Norwalk, CT.

Ganong, W. F. 1997. *Review of Medical Physiology*. (18th ed.). Lange, Los Altos, CA.

Ganong, W. F. 2003. *Review of Medical Physiology*. 21st edition. Lange, Los Altos, CA.

Garcia, M. C., and Ginther, O. J. 1976. Effects of ovariectomy and season on plasma luteinizing hormone in mares. *Endocrinology*, 98:958–962.

Garey, J., Goodwillie, A., Frohlich, J., Morgan, M., Gustafsson, J. A., Smithies, O., Korach, K. S., Ogawa, S., and Pfaff, D. W. 2003. Genetic contributions to generalized arousal of brain and behavior. *Proceedings of the National Academy of Sciences U.S.A.*, 100: 11019–11022.

Garey, J., Kow, L. M., Huynh, W., Ogawa, S., and Pfaff, D. W. 2002. Temporal and spatial quantitation of nesting and mating behaviors among mice housed in a seminatural environment. *Hormones and Behavior*, 42: 294–306.

Garrett, J. W., and Campbell, C. S. 1980. Changes in social behavior of the male golden hamster accompanying photoperiodic changes in reproduction. *Hormones and Behavior*, 14: 303–318.

Gary, K. A., Sollars, P. J., Lexow, N., Winokur, A., and Pickard, G. E. 1996. Thyrotropin-releasing hormone phase shifts circadian rhythms in hamsters. *Neuroreport*, 7:1631–1634.

Gatchel, R. J., Baum, A., and Krantz, D. S. 1989. *An Introduction to Health Psychology*. (2nd ed.). Newberry, New York.

Gauer, O. H., and Henry, J. P. 1963. Circulatory basis of fluid volume control. *Physiol. Rev.*, 43:423–481.

Gaulin, S. J. C., and FitzGerald, R. W. 1986. Sex differences in spatial ability: An evolutionary hypothesis and test. *Am. Nat.*, 127:74–88.

Gaulin, S. J. C., and FitzGerald, R. W. 1989. Sexual selection for spatial-learning ability. *Anim. Behav.*, 37:322–331.

Gavish, L., Carter, C. S., and Getz, L. L. 1983. Male-female interactions in prairie voles. *Anim. Behav.*, 31:511–517.

Gay, V. L., Midgley, A. R. Jr., and Niswender, G. D. 1970. Patterns of gonadotrophin secretion associated with ovulation. *Federation Proceedings*, 29: 1880–1887.

Ge, R. S., Hardy, D. O., Catterall, J. F., and Hardy, M. P. 1997. Developmental changes in glucocorticoid receptor and 11-hydroxysteroid dehydrogenase oxidative and reductive activities in rat Leydig cells. *Endocrinology*, 138:5089–5095.

Geist, V. 1971. *Mountain Sheep*. University of Chicago Press, Chicago.

Gekakis, N., Staknis, D., Nguyen, H. B., Davis, F. C., Wilsbacher, L. D., King, D. P., Takahashi, J. S., and Weitz, C. J. 1998. Role of the CLOCK protein in the mammalian circadian mechanism. *Science*, 280:1564–1569.

Genazzani, A. R., and Petraglia, F. 1989. Opioid control of luteinizing hormone secretion in humans. *J. Steroid Biochem.*, 33:751–755.

Gentry, R. T., and Wade, G. N. 1976. Androgenic control of food intake and body weight in male rats. *J. Comp. Physiol. Psychol.*, 90:18–25.

George, A. J., and Sandler, M. 1988. Endocrine and biochemical studies in puerperal mental disorders. In R. Kumar and I. F. Brockington (eds.), *Motherhood and Mental Illness 2: Causes and Consequences*, pp. 78–112. Wright, Boston.

Georgiou, G. C., Sharp, P. J., and Lea, R. W. 1995. [^{14}C]-2-deoxyglucose uptake in the brain of the ring dove (*Streptopelia risoria*): II. Differential uptake at the onset of incubation. *Brain Research*, 700: 137–141.

Gerall, A. A., and Givon, L. 1992. Early androgen and age-related modifications in female rat reproduction. In A. A. Gerall, H. Moltz, and I. L. Ward (eds.), *Sexual Differentiation*. Vol. 11. *Handbook of Behavioral Neurobiology*, pp. 313–354. Plenum Press, New York.

Gessa, G. L., Paglietti, E., and Quarantotti, B. P. 1979. Induction of copulatory behavior in sexually inactive rats by naloxone. *Science*, 204:203–205.

Getz, L. L., and Carter, C. S. 1996. Prairie vole partnerships. *Am. Sci.*, 84:56–62.

Getz, L. L., Hofmann, J. E., and Carter, C. S. 1987. Mating system and population fluctuation of the prairie vole (*Microtus ochrogaster*). *Am. Zool.*, 27:909–920.

Giantonio, G. W., Lund, N. L., and Gerall, A. A. 1970. Effect of diencephalic and rhinencephalic lesions on the male rat's sexual behavior. *J. Comp. Physiol. Psychol.*, 73:38–46.

Gibbs, J., Fauser, D. J., Rowe, E. A., Rolls, B. J., Rolls, E. T., and Maddison, S. P. 1979. Bombesin suppresses feeding in rats. *Nature*, 282:208–210.

Gibbs, J., Young, R. C., and Smith, G. P. 1973. Cholecystokinin elicits satiety in rats with open gastric fistulas. *Nature*, 245:323–325.

Gibbs, R. B. 1997. Effects of estrogen on basal forebrain cholinergic neurons vary as a function of dose and duration of treatment. *Brain Res.*, 757:10–16.

Gil, D., Graves, J., Hazon, N., and Wells, A. 1999. Male attractiveness and differential testosterone investment in zebra finch eggs. *Science*, 286: 126–128.

Gilbeau, P. M., and Smith, C. G. 1985. Naloxone reversal of stress-induced reproductive effects in the male rhesus monkey. *Neuropeptides*, 5:335–338.

Gilbertson, M. W., Shenton, M. E., Ciszewski, A., Kasai, K., Lasko, N. B., Orr, S. P., and Pitman, R. K. 2002. Smaller hippocampal volume predicts pathologic vulnerability to psychological trauma. *Nature Neuroscience*, 5: 1242–1247.

Gillespie, C. F., Huhman, K. L., Babagbemi, T. O., and Albers, H. E. 1996. Bicuculline increases and muscimol reduces the phase-delaying effects of light and VIP/PHI/GRP in the suprachiasmatic region. *J. Biol. Rhythms*, 11:137–144.

Gillette, M. U. 1986. The suprachiasmatic nuclei: Circadian phase-shifts induced at the time of hypothalamic slice preparation are preserved in vitro. *Brain Research*, 379: 176–181.

Gingrich, B. S., Cascio, C., Young, L. J., Liu, Y.,Wang, Z. X., and Insel, T. R. 1998. Oxytocin, dopamine, and enkephalin in the nucleus accumbens: A neurochemical cascade for pair bonding. *Soc. Neurosci. Abstr.*, 24:372.14.

Ginsberg, B., and Allee, W. C. 1942. Some effects of conditioning of social dominance and subordination in inbred strains of mice. *Physiol. Zool.*, 15:485–506.

Ginton, A., and Merari, A. 1977. Long range effects of MPOA lesions on mating behavior in the male rat. *Brain Res.*, 120:158–163.

Giordano, A. L., and Rosenblatt, J. S. 1986. Relationship between maternal behavior and nuclear estrogen receptor binding in preoptic area and hypothalamus during pregnancy in rats. *Soc. Neurosci. Abstr.*, p. 17.

Giovenardi, M., Padoin, M. J.,Cadore, L. P., and Lucion, A. B. 1998. Hypothalamic paraventricular nucleus modulates maternal aggression in rats: Effects of ibotenic acid lesion and oxytocin antisense. *Physiol. Behav.*, 63:351–359.

Giros, B., Jaber, M., Jones, S. R., Wightman, R. M., and Caron, M. G. 1996. Hyperlocomotion and indifference to cocaine and amphetamine in mice lacking the dopamine transporter. *Nature*. 379:606–612.

Gitlin, M. J., and Pasnau, R. O. 1989. Psychiatric syndromes linked to reproductive function in women: A review of current knowledge. *Am. J. Psychiatry*, 146:1414–1422.

Giwercman, A., and Skakkebaek, N. E. 1992. The human testis—an organ at risk? *Int. J. Androl.* 15:373–375.

Gladue, B. A. 1991. Aggressive behavioral characteristics, hormones, and sexual orientation in men and women. *Aggressive Behavior*, 17: 313–326.

Glaser, J. H., Rubin, B. S., and Barfield, R. J. 1983. Onset of the receptive and proceptive components of feminine sexual behavior in rats following the intravenous administration of progesterone. *Horm. Behav.*, 17:18–27.

Gleason, P. E., Michael, S. D., and Christian, J. J. 1981. Prolactin-induced aggression in female Peromyscus leucopus. *Behav. Neural Biol.*, 33:243–248.

Glickman, S. E., Frank, L. G., Davidson, J. M., Smith, E. R., and Siiteri, P. K. 1987. Androstenedione may organize or activate sex-reversed traits in female spotted hyenas. *Proc. Natl. Acad. Sci. U.S.A.*, 84:3444–3447.

Glickman, S. E., Frank, L. G., Licht, P., Yalcinkaya, T., Siiteri, P. K., and Davidson, J. 1992. Sexual differentiation of the female spotted hyena. One of nature's experiments. *Ann. N. Y. Acad. Sci.*, 662:135–159.

Go, V. L. E., DiMagno, E. P., Gardner, J. D., Lebenthal, E., Reber, H. A., and Scheely, G. A. 1993. *The Pancreas: Biology, Pathobiology and Disease*. Raven Press, New York.

Godwin, J., and Crews, D. P. 2002. Hormones, brain and behavior in reptiles. In D. W. Pfaff, A. P. Arnold, A. M. Etgen, S. E. Fahrbach, and R. T. Rubin (eds.), *Hormones,*

Brain and Behavior, Volume 2, pp. 545–586. Academic Press, New York.

Godwin, J., Hartman, V., Grammer, M., and Crews, D. 1996. Progesterone inhibits female-typical receptive behavior and decreases hypothalamic estrogen and progesterone receptor messenger ribonucleic acid levels in whiptail lizards (genus *Cnemidophorus*). *Horm. Behav.*, 30:138–144.

Goeders, N. E. 1997. A neuroendocrine role in cocaine reinforcement. *Psychoneuroendocrinology*, 22:237–259.

Goeders, N. E., and Guerin, G. F. 1996. Effects of surgical and pharmacological adrenalectomy on the initiation and maintenance of intravenous cocaine self-administration in rats. *Brain Res.*, 722:145–152.

Goeders, N. E., and Guerin, G. F. 1996a. Role of corticosterone in intravenous cocaine self-administration in rats. *Neuroendocrinology*, 64:337–348.

Goeders, N. E., and Guerin, G. F. 1997.Tolerance and sensitization to the behavioral effects of cocaine in rats: relationship to benzodiazepine receptors. *Pharmacol. Biochem. Behav.*, 57:43–56.

Goetzel, R. Z., Anderson, D. R., Whitmer, R. W., Ozminkowski, R. J., Dunn, R. L., and Wasserman, J., 1998. The relationship between modifiable health risks and health care expenditures: An analysis of the multi-employer HERO health risk and cost database. The Health Enhancement Research Organization (HERO) Research Committee. *Journal of Occupational and Environmental Medicine*, 40: 843–854.

Gold, P. E. 1986. Glucose modulation of memory storage processing. *Behav. Neural Biol.*, 45:342–349.

Gold, P. E. 1987. Sweet memories. *Am. Sci.*, 75:151–155.

Gold, P. E. 2003. Acetylcholine modulation of neural systems involved in learning and memory. *Neurobiology of Learning and Memory*, 80: 194–210.

Gold, P. E., and Van Buskirk, R. B. 1975. Facilitation of time-dependent memory processes with post-trial epinephrine injections. *Behav. Biol.*, 13:145–153.

Goldfoot, D. A., Wiegand, S. J., and Scheffler, G. 1978. Continued copulation in ovariectomized adrenal-suppressed stumptail macaques (*Macaca artoides*). *Horm. Behav.*, 11:89–99.

Goldman, B. D. 1983. The physiology of melatonin in mammals. In R. J. Reiter (ed.), *Pineal Research Reviews*, pp. 145–182. Alan R. Liss, New York.

Goldman, B. D., and Darrow, J. M. 1983. The pineal gland and mammalian photoperiodism. *Neuroendocrinology*, 37:386–396.

Goldman, B. D., and Elliott, J. A. 1988. Photoperiodism and seasonality in hamsters: Role of the pineal gland. In M. H. Stetson (ed.), *Processing of Environmental Information in Vertebrates*, pp. 203–218. Springer-Verlag, New York.

Goldman, B. D., and Nelson, R. J. 1993. Melatonin and seasonality in mammals. In H. S. Yu and R. J. Reiter (eds.), *Melatonin: Biosynthesis, Physiological Effects, and Clinical Applications*, pp. 225–252. CRC Press, Boca Raton, FL.

Goldman, B. D., Matt, K. S., Roychoudhury, P., and Stetson, M. H. 1981. Prolactin release in golden hamsters: Photoperiod and gonadal influences. *Biol. Reprod.*, 24:287–292.

Goldsmith, A. R. 1983. Prolactin in avian reproductive cycles. In J. Balthazart, J. E. Prove, and R. Gilles (eds.), *Hormones and Behaviour in Higher Vertebrates*, pp. 375–387. Springer-Verlag, Berlin.

Golier, J. A., Yehuda, R., Lupien, S. J., Harvey, P. D., Grossman, R., and Elkin, A. 2002. Memory performance in Holocaust survivors with posttraumatic stress disorder. *American Journal of Psychiatry*, 159: 1682–1688.

González-Mariscal, G. 2001. Neuroendocrinology of maternal behavior in the rabbit. *Hormones and Behavior*, 40: 125–132.

González-Mariscal, G., and Poindron, P. 2002. Parental care in mammals: Immediate internal and sensory factors of control. In D. W. Pfaff, A. P. Arnold, A. M. Etgen, S. E. Fahrbach, and R. T. Rubin (eds.), *Hormones, Brain and Behavior*, Volume 1, pp. 215–298. Academic Press, New York.

Gonzàlez-Mariscal, G., and Rosenblatt, J. S. 1996. Maternal behavior in rabbits: A historical and multidisciplinary perspective. *Adv. Study Behav.*, 25:333–360.

Gonzàlez-Mariscal, G., Cuamatzi, E., and Rosenblatt, J. S. 1998. Hormones and external factors: Are they "on/off" signals for maternal nest-building in rabbits? *Horm. Behav.*, 33:1–8.

Goodale, H. D. 1918. Feminized male birds. *Genetics*, 3:276–299.

Goodfellow, P. N., and Lovell-Badge, R. 1993. SRY and sex determination in mammals. *Annu. Rev. Genet.*, 27:71–92.

Goodson, J. L. 1998a. Territorial aggression and dawn song are modulad by septal vasotocin and vasoactive intestinal polypeptide in male field sparrows (*Spizella pusilla*). *Horm. Behav.*, 34:67–77.

Goodson, J. L. 1998b. Vasotocin and vasoactive intestinal polypeptide modulate aggression in a territorial songbird, the violet-eared waxbill (Estrildidae: *Uraeginthus granatina*). *General and Comparative Endocrinology*, 111: 233–244.

Goodson, J. L., and Adkins-Regan, E. 1999. Effect of intraseptal vasotocin and vasoactive intestinal polypeptide infusions on courtship song and aggression in the male zebra finch (*Taeniopygia guttata*). *Journal of Neuroendocrinology*, 11: 19–25.

Gooren, L. J. 1985. Human male sexual functions do not require aromatization of testosterone: A study using tamoxifern, testolactone, and dihydrotestosterone. *Arch. Sex Behav.*, 14:539–548.

Gordon, G. G., Southren, A. L., Tochimoto, S., Olivo, J., Altman, K., Rand, J., and Lemberger, L. 1970. Effect of medroxyprogesterone acetate (Provera) on the metabolism and biological activity of testosterone. *J. Clin. Endocrinol. Metab.*, 30:449–456.

Gorman, M. R., Ferkin, M. H., Nelson, R. J., and Zucker, I. 1993. Reproductive status influences odor preferences of the meadow vole, Microtus pennsylvanicus, in winter day lengths. *Can. J. Zool.*, 71:1748–1754.

Gorski, R. A. 1993. Editorial: Estradiol acts via the estrogen receptor in the sexual differentiation of the rat brain, but what does this complex do? *Endocrinology*, 133:431–432.

Gorwood, P. 2004. Eating disorders, serotonin transporter polymorphisms and potential treatment response. *American Journal of Pharmacogenomics*, 4: 9–17.

Goss, R. J. 1980. Photoperiodic control of antler cycles in deer. *J. Exp. Zool.*, 211:101–105.

Goss, R. J. 1984. Photoperiodic control of antler cycles in deer. VI. Circannual rhythms on altered day lengths. *J. Exp. Zool.*, 230:265–271.

Goss, R. J., and Rosen, J. K. 1973. The effects of latitude and photoperiod on the growth of antlers. *J. Reprod. Fertil. (Suppl.)*, 19:111–118.

Gotlib, I. H., Whiffen, V. E., Mount, J. H., Milne, K., and Cordy, N. I. 1989. Prevalence rates and demographic characteristics associated with depression in pregnancy and the postpartum. *J. Consult. Clin. Psychol.*, 57:269–274.

Gould, E., Tanapat, P., McEwen, B. S., Flugge, G., and Fuchs, E. 1998. Proliferation of granule cell precursors in the dentate gyrus of adult monkeys is diminished by stress. *Proc. Natl. Acad. Sci. U.S.A.*, 95:3168–3171.

Gould, E., Woolley, C. S., and McEwen, B. S. 1991. The hippocampal formation: Morphological changes induced by thyroid, gonadal, and adrenal hormones. *Psychoneuroendocrinology*, 16:67–84.

Gould, E., Woolley, C. S., Frankfurt, M., and McEwen, B. S. 1990. Gonadal steroids regulate dendritic spine density in hippocampal pyramidal cells in adulthood. *J. Neurosci.*, 10:1286–1291.

Gowaty, P. A. 1996. Field studies of parental care in birds: New data focus questions on variation among females. *Adv. Study Behav.*, 25:477–532.

Goy, R. W. 1966. Role of androgens in the establishment and regulation of behavioral sex differences in mammals. *J. Anim. Sci.*, 25:21–31.

Goy, R. W. 1967. William Caldwell Young. *Anat. Rec.*, 157:4–11.

Goy, R. W. 1978. Development of play and mounting behaviour in female rhesus virilized prenatally with esters of testosterone or dihydrotestosterone. In D. J. Chivers and J. Herbert (eds.), *Recent Advances in Primatology*, Vol. 1, pp. 449–462. Academic Press, London.

Goy, R. W., and Phoenix, C. H. 1971. Gonadal hormones and behavior of normal and pseudohermaphroditic female primates. In C. H. Sawyer and R. S. Gorski (eds.), *Steroids, Hormones and Brain Function*, pp. 193–202. University of California Press, Berkeley.

Goy, R. W., and Phoenix, C. H. 1972. The effects of testosterone propionate administered before birth on the development of behavior in genetic female rhesus monkeys. In C. Sawyer and R. Gorski (eds.), *Steroid Hormones and Brain Function*, pp. 193–201. University of California Press, Berkeley.

Goy, R. W., and Resko, J. A. 1972. Gonadal hormones and behavior of normal and pseudohermaphroditic nonhuman female primates. *Rec. Prog. Horm. Res.*, 28:707–733.

Goy, R. W., and Young, W. C. 1956/1957. Strain differences in the behavioral responses of female guinea pigs to alpha-estradiol benzoate and progesterone. *Behaviour*, 10:340–355.

Goy, R. W., Bridson, W. E., and Young, W. C. 1964. The period of maximal susceptibility of the prenatal female guinea pig to masculinizing actions of testosterone propionate. *J. Comp. Physiol. Psychol.*, 57:166–174.

Grady, K. L., Phoenix, C. H., and Young, W. C. 1965. Role of the developing rat testis in differentiation of the neural tissues mediating mating behavior. *J. Comp. Physiol. Psychol.*, 59:176–182.

Graham, J. M., and Desjardins, C. 1980. Classical conditioning: Induction of luteinizing hormone and testosterone secretion in anticipation of sexual activity. *Science*, 210:1039–1041.

Grammer, K., Fink, B., Møller, A. P., and Thornhill, R. 2003. Darwinian aesthetics: Sexual selection and the biology of beauty. *Biological Reviews of the Cambridge Philosophical Society*, 78: 385–407.

Grandi, M., and Celani, M. F. 1990. Effects of football on the pituitary-testicular axis (PTA): Differences between professional and non-professional soccer players. *Exp. Clin. Endocrinol.*, 96:253–259.

Granneman, J., and Friedman, M. I. 1980. Hepatic modulation of insulin-induced gastric acid secretion and EMB activity in rats. *Am. J. Physiol.*, 238:R346–R352.

Gray, G. D., and Dewsbury, D. A. 1973. A quantitative description of copulatory behavior in prairie voles (*Microtus ochrogaster*). *Brain Behav. Ecol.*, 8:837–852.

Gray, G. D., Smith, E. R., and Davidson, J. M. 1980. Hormonal regulation of penile erection in castrated male rats. *Physiol. Behav.*, 24:463–468.

Gray, G. D., Soderstein, P., Tallentire, D., and Davidson, J. M. 1978. Effects of lesions in various structures of the suprachiasmatic-preoptic region on LH regulation and sexual behavior in female rats. *Neuroendocrinology*, 25: 174–191.

Gray, J. A., and Lalljee, B. 1974. Sex differences in emotional behavior in the rat: Correlation between open field, defecation, and active avoidance. *Anim. Behav.*, 22:856–861.

Gray, J. M., and Wade, G. N. 1979. Cytoplasmic progestin binding in rat adipose tissues. *Endocrinology*, 104:1377–1382.

Gray, L. E., Ostby, J., Furr, J., Price, M., Veeramachaneni, D. N., and Parks, L. 2000. Perinatal exposure to the phthalates DEHP, BBP, and DINP, but not DEP, DMP, or DOTP, alters sexual differentiation of the male rat. *Toxicology Sciences*, 58: 350–365.

Greef, W. J. de, and Merkx, J. A. M. 1982. Receptivity, proceptivity and attractivity of pregnant and pseudopregnant rats. *Behav. Brain Res.*, 4:203–208.

Greenberg, R. 1986. Competition in migrant birds in the nonbreeding season. In R. J. Johnson (ed.), *Curr. Ornithol.*, pp. 281–307. Plenum, New York.

Greenfield, D. A. 1997. Does psychological support and counseling reduce the stress experienced by couples involved in assisted reproductive technology? *J. Assist. Reprod. Gen.*, 14:186–188.

Gregg, T. R., and Siegel, A. 2001. Brain structures and neurotransmitters regulating aggression in cats: Implications for human aggression. *Progress in Neuro-Psychopharmacological and Biological Psychiatry*, 25: 91–140.

Gresack, J. E., and Frick, K. M. 2004. Environmental enrichment reduces the mnemonic and neural benefits of estrogen. *Neuroscience*, 128: 459–471.

Griffin, J. E., and Ojeda, S. R. 1988. *Textbook of Endocrine Physiology.* Oxford University Press, New York.

Grill, H. J., and Kaplan, J. M. 1990. Caudal brainstem participates in the distributed neuronal control of feeding. In E. M. Stricker (ed.), *Handbook of Behav. Neurobiology: Neurobiology of Food Fluid Intake*, pp. 125–150. Plenum Press, New York.

Grill, H. J., and Kaplan, J. M. 2002. The neuroanatomical axis for control of energy balance. *Frontiers in Neuroendocrinology*, 23: 2–40.

Grill, H. J., Donahey, J. C. K., King, L, and Kaplan, J. M. 1997. Contribution of caudal brainstem to δ-fenfluramine anorexia. *Psychopharmacology*, 130:375–381.

Grill, H. J., Ginsberg, A. B., Seeley, R. J., and Kaplan, J. M. 1998. Brainstem application of melanocortin receptor ligands produces long-lasting effects on feeding and body weight. *J. Neurosci.*, 18:10128–10135.

Grimes, L. J., Melnyk, R. B., Martin, J. M., and Mrosovsky, N. 1981. Infradian cycles in glucose utilization and lipogenic enzyme activity in dormouse (*Glis glis*) adipocytes. *Gen. Comp. Endocrinol.*, 45:21–25.

Grisham, W., Kerchner, M., and Ward, I. L. 1991. Prenatal stress alters sexually dimorphic nuclei in the spinal cord of male rats. *Brain Res.*, 551:126–131.

Grön, G., Wunderlich, A. P., Spitzer, M., Tomczak, R., and Riepe, M. W. 2000. Brain activation during human navigation: Gender-different neural networks as substrate of performance. *Nature Neuroscience*, 3: 404–408.

Groos, G. 1982. The comparative physiology of extraocular photoreception. *Experientia*, 38:989–1128.

Grosse, J., Maywood, E. S., Ebling, F. J., and Hastings, M. H. 1993. Testicular regression in pinealectomized Syrian hamsters following infusions of melatonin delivered on non-circadian schedules. *Biol. Reprod.*, 49:666–674.

Gruber, A. J., and Pope, H. G. 2000. Psychiatric and medical effects of anabolic-androgenic steroid use in women. *Psychotherapy and Psychosomatics*, 69: 19–26.

Gruder-Adams, S., and Getz, L. L. 1985. Comparison of the mating system and paternal behavior in *Microtus ochrogaster* and *M. pennsylvanicus*. *J. Mammalogy*, 66:165–167.

Grunt, J. A., and Young, W. C. 1952. Differential reactivity of individuals and the response of the male guinea pig to testosterone propionate. *Endocrinology*, 51:237–248.

Grunt, J. A., and Young, W. C. 1953. Consistency of sexual behavior patterns in individual male guinea pigs following castration and androgen therapy. *J. Comp. Physiol. Psychol.*, 46:138–144.

Gu, Q., and Moss, R. L. 1996. 17 beta-Estradiol potentiates kainate- induced currents via activation of the cAMP cascade. *J. Neurosci.*, 16:3620–3629.

Guardiola-Lemaitre, B. 1997. Toxicology of melatonin. *J. Biol. Rhythms*, 12:697–706.

Gubernick, D. J., and Alberts, J. R. 1987. The biparental care system of the California mouse, *Peromyscus californicus*. *J. Comp. Psychol.*, 101:169–177.

Gubernick, D. J., and Klopfer, P. H. (eds.). 1981. *Parental Care in Mammals.* Plenum Press, New York.

Gubernick, D. J., and Nelson, R. J. 1989. Prolactin and paternal behavior in the biparental California mouse, *Peromyscus californicus*. *Horm. Behav.*, 23:203–210.

Gubernick, D. J., Winslow, J. T., Jensen, P., Jeanotte, L., and Bowen, J. 1995. Oxytocin changes in males over the reproductive cycle in the monogamous, biparental California mouse, *Peromyscus californicus*. *Horm. Behav.*, 29:5/1–/23.

Guillamón, A., and Segovia, S. 1997. Sex differences in the vomeronasal system. *Brain Res. Bull.*, 44:377–382.

Guinness, F. E., Albon, S. D., and Clutton-Brock, T. H. 1978. Factors affecting reproduction in red deer *(Cervus elaphus L.)*. *J. Reprod. Fertil.*, 54:325–334.

Gulledge, C., and Deviche, P. 1997. Androgen control of vocal control region volumes in a wild migratory songbird (*Junco hyemalis*) is region and possibly age dependent. *J. Neurobiol.*, 32:391–402.

Gunnar, M. R. 1998. Quality of early care and buffering of neuroendocrine stress reactions: Potential effects on the developing human brain. *Preventative Medicine*, 27: 208–211.

Gunnar, M. R., Morison, S. J., Chisholm, K., and Schuder, M. 2001. Salivary cortisol levels in children adopted from Romanian orphanages. *Development and Psychopathology*, 13: 611–628.

Gunnet, J. W., and Freeman, M. E. 1983. The mating-induced release of prolactin: A unique neuroendocrine response. *Endocrinol. Rev.*, 4:44–61.

Gust, D., Gordon, T. P., and Hambright, M. K. 1993. Relationship between social factors and pituitary–adrenocortical activity in female rhesus monkeys (*Macaca mulatta*). *Hormones and Behavior*, 27: 318–331.

Gutzke, W. H. N., and Crews, D. 1988. Embryonic temperature determines adult sexuality in a reptile. *Nature*, 332:832–834.

Gutzke, W. H. N., and Paukstis, G. L. 1983. Influence of the hydric environment on sexual differentiation of turtles. *J. Exp. Zool.*, 226:467–469.

Gwinner, E. 1986. *Circannual Rhythms.* Springer-Verlag, Berlin.

Gwinner, E., Hau, M., and Heigl, S. 1997. Melatonin: Generation and modulation of avian circadian rhythms. *Brain Res. Bull.*, 44:439–444.

Haaren, F. van, Hest, A., and Heinsbroek, R. P. W. 1990. Behavioral differences between male and female rats: Effects of gonadal hormones on learning and memory. *Neurosci. Biobehav. Rev.*, 14: 23–33.

Hadley, M. E. 1996. *Endocrinology.* (4th ed.). Prentice Hall, Englewood Cliffs, NJ.

Haertzen, C., Buxton, K., Covi, L., and Richards, H. 1993. Seasonal changes in rule infractions among prisoners: A preliminary test of the temperature-aggression hypothesis. *Psychological Reports*, 72: 195–200.

Haffen, K., and Wolff, E. 1977. Natural and experimental modification of ovarian development. In L. Z. Zuckerman and B. J. Wier (eds.), *The Ovary*, pp. 393–423. Academic Press, New York.

Halberg, F. 1959. Physiologic 24-hour periodicity in human beings and mice, the lighting regimen and daily routine. In R. B. Withrow (ed.), *Photoperiodism and Related Phenomena in Plants and Animals*, pp. 803–878. American Association for the Advancement of Science, Washington, D.C.

Halberg, F. 1977. Implications of biological rhythms for clinical practice. *Hosp. Prac.*, 12:139–149.

Halbreich, U. 1997. Role of estrogen in postmenopausal depression. *Neurology*, 48:S16-S19.

Halbreich, U. 2003. The etiology, biology, and evolving pathology of premenstrual syndromes. *Psychoneuroendocrinology*, 28: 55–99.

Halbreich, U., Borenstein, J., Pearlstein, T., and Kahn, L. S. 2003. The prevalence, impairment, impact, and burden of premenstrual dysphoric disorder (PMS/PMDD). *Psychoneuroendocrinology*, 28: 1–23.

Halbreich, U., Endicott, J., Schacts, D., and Nee, J. 1982. The diversity of premenstrual changes as reflected in the premenstrual assessment form. *Acta Psychiatrica Scand.*, 65:46–65.

Hall, J. C., and Rosbash, M. 1987. Genetic and molecular analysis of biological rhythms. *J. Biol. Rhythms*, 2:153–178.

Hall, J. L, Gonder-Frederick, L. A., Chewning, W. W., Silveira, J., and Gold, P. E. 1989. Glucose enhancement of performance on memory tests in young and aged humans. *Neuropsychologia*, 27:1129–1138.

Hall, J. L, Reilly, R. T., Cottrill, K. L, Stone, W. S., and Gold, P. E. 1992. Phlorizin enhancement of memory in rats and mice. *Pharmacol. Biochem. Behav.*, 41:295–299.

Hall, R. C., Popkin, M. K., Stickney, S. K., and Gardner, E. R. 1979. Presentation of steroid the psychoses. *J. Nerv. Ment. Dis.*, 167:229–236.

Hall, V. D., and Goldman, B. D. 1980. Effects of gonadal steroid hormones on hibernation in the Turkish hamster (*Mesocricetus brandti*). *J. Comp. Physiol.*, 135:107–114.

Hall, V. D., Bartke, A., and Goldman, B. D. 1982. Role of the testes in regulating the duration of hibernation in the Turkish hamster (*Mesocricetus brandti*). *Biol. Reprod.*, 27:802–810.

Hamada, T., LeSauter, J., Lokshin, M., Romero, M. T., Yan, L., Venuti, J. M., and Silver, R. 2003. Calbindin influences response to photic input in suprachiasmatic nucleus. *Journal of Neuroscience*, 23: 8820–8826.

Hamann, S., Herman, R. A., Nolan, C. L., and Wallen, K. 2004. Men and women differ in amygdale response to visual sexual stimuli. *Nature Neuroscience*, 7(4): 411–416.

Hamilton, J. A., Parry, B. L., and Blumenthal, S. J. 1988. The menstrual cycle in context. I. Affective syndromes associated with reproductive hormonal changes. *J. Clin. Psychiatry*, 49:474–480.

Hamilton, J. B., and Gardner, W. U. 1937. Effects in female young born of pregnant rats injected with androgens. *Proc. Soc. Exp. Biol. Med.*, 37:570–572.

Hamilton, M. 1960. A rating scale for depression. *J. Neurol. Neurosurg. Psychiatry*, 23: 56–61.

Hampton, J. K., Hampton, S. H., and Landwehr, B. T. 1966. Observations on a successful breeding colony of the marmoset, *Oedipomidas oedipus*. *Folia Primatol.*, 4:265–287.

Handa, R. J., Burgess, L. H., Kerr, J. E., and O'Keefe, J. A. 1994. Gonadal steroid hormone receptors and sex differences in the hypothalamic–pituitary–adrenal axis. *Hormones and Behavior*, 28: 464–476.

Handa, R. J., Burgess, L. H., Kerr, J. E., and O'Keefe, J. A. 1994. Gonadal steroid hormone receptors and sex differences in the hypothalamo-pituitary-adrenal axis. *Horm. Behav.*, 28:464–476.

Hansen, J. T., and Karasek, M. 1982. Neuron or endocrine cell? The pinealocyte as a paraneuron. In R. J. Reiter (ed.), *The Pineal and Its Hormones*, pp. 1–9. Alan R. Liss, New York.

Hansen, S., Ferreira, A., and Selart, M. E. 1985. Behavioural similarities between mother rats and benzodiazepine-treated non-maternal animals. *Psychopharmacology*, 86:344–347.

Hansen, S., Köhler, C., and Ross, S. B. 1982. On the role of the dorsal mesencephalic tegmentum in the control of masculine sexual behavior in the rat: Effects of electrolytic lesions, ibotenic acid and DSP4. *Brain Res.*, 240:311–320.

Hansen, S., Stanfield, E. J., and Everitt, B. J. 1980. The role of the ventral bundle noradrenergic neurones in sensory components of sexual behaviour and coitus-induced pseudopregnancy. *Nature*, 286:152–154.

Hansen, S., Stanfield, E. J., and Everitt, B. J. 1981. The effects of lesions of lateral tegmental noradrenergic neurons on components of sexual behavior and pseudopregnancy in female rats. *Neuroscience*, 6:1105–1117.

Hardie, L. J., Rayner, D. V., Holmes, S., and Trayhurn, P. 1996. Circulating leptin levels are modulated by fasting, cold exposure and insulin administration in lean, but not Zucker (*fa/fa*) rats as measured by ELISA. *Biochem. Biophys. Res. Commun.*, 223:660–665.

Hardin, P. E., Hall, J. C., and Rosbash, M. 1990. Feedback of the Drosophila period gene product on circadian cycling of its messenger RNA levels. *Nature*, 343:536–540.

Harding, C. F., and Follett, B. K. 1979. Hormone changes triggered by aggression in a natural population of blackbirds. *Science*, 203:918–920.

Harding, C. F., Sheridan, K., and Walters, M. J. 1983. Hormonal specificity and activation of sexual behavior in male zebra finches. *Horm. Behav.*, 17:111–113.

Harding, C. F., Walters, M. J., Collado, M., and Sheridan, K. 1988. Hormonal specificity and activation of social behavior in male red-winged blackbirds. *Horm. Behav.*, 22:402–418.

Hardy, D. F., and DeBold, J. F. 1971. Effects of mounts without intromission upon the behavior of female rats during the onset of estrogen-induced heat. *Physiol. Behav.*, 7:643–645.

Harley, V. R., and Goodfellow, P. N. 1995. The biochemical role of SRY in sex determination. *Mol. Reprod. Dev.*, 39:184–193.

Harlow, H. F. 1959. Love in infant monkeys. *Sci. Am.*, (June), 100–106

Harlow, H. F., and Harlow, M. K. 1965. The affectional systems. In A. N. Schrier, H. F. Harlow, and F. Stollnitz (eds.), *Behavior of Nonhuman Primates*. Vol. 2, pp. 287–334. Academic Press, New York.

Harlow, H. F., Harlow, M. K., and Suomi, S. 1971. From thought to therapy: Lessons from a primate laboratory. *Am. Sci.*, 59:538–549.

Harris, G. W. 1937. The induction of ovulation in the rabbit by electrical stimulation of the hypothalamo-hypophysial mechanism. *Proc. R. Soc. Lond.*, B122:374–394.

Harris, G. W. 1948. Electrical stimulation of the hypothalamus and the mechanism of neural control of the adenohypophysis. *J. Physiol.*, 107:418–429.

Harris, G. W. 1955. *Neural Control of the Pituitary Gland.* E. Arnold, London.

Harris, G. W., and Jacobsohn, D. 1952. Functional grafts of the anterior pituitary gland. *Proc. R. Soc. Lond. B*, 139:263–276.

Harris, J. A. 1999. Review and methodological considerations in research on testosterone and aggression. *Aggressive and Violent Behavior*, 4: 273–291.

Harris, M. P. 1969. Breeding seasons of sea-birds in the Galapagos Islands. *J. Zool. (Lond.)*, 159:145–165.

Hart, B. L. 1974. Gonadal androgen and sociosexual behavior of male mammals: A comparative analysis. *Psychol. Bull.*, 7:383–400.

Hart, B. L. 1979. Activation of sexual reflexes of male rats by dihydrotestosterone but not estrogen. *Physiol. Behav.*, 23:107–109.

Hart, B. L. 1988. Biological basis of the behavior of sick animals. *Neurosci. Biobehav. Rev.*, 12:123–137.

Hart, B. L., Wallach, S. J. R., and Melese-d'Hospital, P. Y. 1983. Differences in responsiveness to testosterone of penile reflexes and copulatory behavior of male rats. *Horm. Behav.*, 17:274–283.

Hartgens, F., and Kuipers, H. 2004. Effects of androgenic-anabolic steroids in athletes. *Sports Medicine*, 34: 513–554.

Hartung, T. G., and Dewsbury, D. A. 1979. Paternal behaviour in six species of muroid rodents. *Behav. Neural Biol.*, 26:466–478.

Harvey, S. M. 1987. Female sexual behavior: fluctuations during the menstrual cycle. *J. Psychosom. Res.*, 31:101–110.

Hastings, M. H., Duffield, G. E., Ebling, F. J., Kidd, A., Maywood, E. S., Schurov, I. 1998. Non-photic signalling in the suprachiasmatic nucleus. *Biol. Cell*, 89:495–503.

Hastings, M. H., Reddy, A. B., and Maywood, E. S. 2003. A clockwork web: Circadian timing in brain and periphery, in health and disease. *Nature Reviews: Neuroscience*, 4: 649–661.

Hatton, D.C., and Meyer, M. E. 1973. Paternal behavior in cactus mice *(Peromyscus eremicus). Bull. Psychonom. Soc.*, 2:330.

Hatton, J. D., and Ellisman, M. H. 1982. A restructuring of hypothalamic synapses is associated with motherhood. *J. Neurosci.*, 2:704–707.

Hau, M., Stoddard, S. T., and Soma, K. K. 2004. Territorial aggression and hormones during the non-breeding season in a tropical bird. *Hormones and Behavior*, 45: 40–49.

Haug, M., Brain, P. F., and Kamis, A. B. 1986. A brief review comparing the effects of sex steroids on two forms of aggression in laboratory mice. *Neurosci. Biobehav. Rev.*, 10:463–468.

Haupt, H. A., and Rovere, G. D. 1984. Anabolic steroids: A review of the literature. *Am. J. Sports Med.*, 12:469–484.

Hawkins, M. B., Thornton, J. W., Crews, D., Skipper, J. K., Dotte, A., and Thomas, P. 2000. Identification of a third distinct estrogen receptor and reclassification of estrogen receptors in teleosts. *Proceedings of the National Academy of Sciences U.S.A.*, 97: 10751–10756.

Hayes, T. B., Collins, A., Lee, M., Mendoza, M., Noriega, N., Stuart, A. A., and Vonk, A. 2002a. Hermaphroditic, demasculinized frogs after exposure to the herbicide atrazine at low ecologically relevant doses. *Proceedings of the National Academy of Sciences U.S.A.*, 99: 5476–5480.

Hayes, T., Haston, K., Tsui, M., Hoang, A., Haeffele, C., and Vonk, A. 2002b. Herbicides: Feminization of male frogs in the wild. *Nature*, 419: 895–896.

Hayes, T., Haston, K., Tsui, M., Hoang, A., Haeffele, C., and Vonk, A. 2003. Atrazine-induced hermaphroditism at 0.1 ppb in American leopard frogs (*Rana pipiens*): Laboratory and field evidence. *Environmental Health Perspectives*, 111: 568–575.

Hebebrand, J., and Remschmidt, H. 1995. Anorexia nervosa viewed as an extreme weight condition: Genetic implications. *Human Genetics*, 95: 1–11.

Hebebrand, J., Blum, W. F., Barth, N., Coners, H., Englaro, P., Juul, A., Ziegler, A., Warnke, A., Rascher, W., and Remschmidt, H. 1997. Leptin levels in patients with anorexia nervosa are reduced in the acute stage and elevated upon short-term weight restoration. *Molecular Psychiatry*, 2: 330–334.

Hegner, R. E., and Wingfield, J. C. 1987. Social status and circulating levels of hormones in flocks of house sparrows, *Passer domesticus. Ethology*, 76:1–14.

Heid, P., Güttinger, H.-R., and Prove, E. 1985. The influence of castration and testosterone replacement on the song architecture of canaries (*Serinus canarius*). *Z. Tierpsychol.*, 69:224–236.

Heigl, S., and Gwinner, E. 1995. Synchronization of circadian rhythms of house sparrows by oral melatonin: effects of changing period. *J. Biol. Rhythms*, 10:225–33.

Heim, N., and Hursch, C. J. 1979. Castration for sex offenders: Treatment or punishment? A review and critique of recent European literature. *Arch. Sex Behav.*, 8:281–304.

Heimer, L., and Larsson, K. 1966. Impairment of mating behavior in male rats following lesions in the preoptic-anterior hypothalamic continuum. *Brain Res.*, 3:248–263.

Heinrichs, S. C., Cole, B. J., Pich, E. M., Menzaghi, F., Koob, G. F., and Hauger, R. L. 1992. Endogenous corticotropin-releasing factor modulates feeding induced by neuropeptide Y or a tail-pinch stressor. *Peptides*, 13:879–884.

Heinsbroek, R. P., van Haaren, F, and van de Poll, N. E. 1988. Sex differences in passive avoidance behavior in rats: Sex-dependent susceptibility to shock-induced behavioral depression. *Physiol. Behav.*, 43:201–206.

Helkala, E. L.,Niskanen, L., Viinamaki, H., Partanen, J., and Uusitupa, M. 1995. Short-term and long-term memory in elderly patients with NIDDM. *Diabetes Care*, 18:681–685.

Heller, H. C., and Poulson, T. L. 1970. Circannian rhythms: II. Endogenous and exogenous factor controlling reproduction and hibernation in chipmunks (*Eutamias*) and ground squirrels (*Spermophilus*). *Comp. Biochem. Physiol.*, 33:357–383.

Helvacioglu, A., Yeoman, R. R., Hazelton, J. M., and Aksel, S. 1993. Premenstrual syndrome and related hormonal changes. Long-lasting gonadotropin releasing hormone agonist treatment. *J. Reprod. Med.*, 38:864–870.

Hendrick, V., Altshuler, L. L., and Suri, R. 1998. Hormonal changes in the postpartum and implications for postpartum depression. *Psychosomatics*, 39:93–101.

Henn, W., and Zang, K. D. 1997. Mosaicism in Turner's syndrome. *Nature*, 390:569.

Hennessey, A. C., Huhman, K. L., and Albers, H. E. 1994. Vasopressin and sex differences in hamster flank marking. *Physiol. Behav.*, 55:905–911.

Henry, C., Kabbaj, M., Simon, H., Le Moal, M., and Maccari, S. 1994. Prenatal stress increases the hypothalamo-pituitary-adrenal axis response in young and adult rats. *J. Neuroendocrinol.*, 6:341–345.

Herbert, J. 1970. Hormones and reproductive behavior in rhesus and talapoin monkeys. *J. Reprod. Fertil., (Suppl.)*, 11:119–140.

Herbert, J., and Trimble, M. R. 1967. Effect of oestradiol and testosterone on the sexual receptivity and attractiveness of the female rhesus monkey. *Nature*, 216:165–166.

Heriot, A. 1974. *The Castrati in Opera*. DaCapo, New York.

Herman, T. B. 1977. Activity patterns and movements of subarctic voles. *Oikos*, 29:434–444.

Herrenkohl, L. R. 1986. Prenatal stress disrupts reproductive behavior and physiology in offspring. *Ann. N. Y. Acad. Sci.*,474:120–128.

Hervey, E., and Hervey, G. R. 1967. The effects of progesterone on body weight and composition in the rat. *J. Endocrinol.*, 37:361–384.

Hetherington, A. W., and Ranson, S. W. 1940. Hypothalamic lesions and adiposity in the rat. *Anat. Rec.*, 78:149–172.

Heuser, I. 1998. Anna-Monika-Prize paper. The hypothalamus-pituitary-adrenal system in depression. *Pharmacopsychiatry*, 31:10–13.

Hews, D. K., Thompson, C. W., Moore, I. T., and Moore, M. C. 1997. Population frequencies of alternative male phenotypes in tree lizards: Microgeographic variation and common-garden rearing studies. *Behav. Ecol. Sociobiol.*, 41:371–380.

Higley, J. D., Hasert, M. F., Suomi, S. J., and Linnoila, M. 1991. Nonhuman primate model of alcohol abuse: Effects of early experience, personality, and stress on alcohol consumption. *Proc. Natl. Acad. Sci. U.S.A.*, 88:7261–7265.

Hinde, R. A. 1970. *Animal Behaviour: A Synthesis of Ethology and Comparative Psychology*. McGraw-Hill, New York.

Hines, M. 1982. Prenatal gonadal hormones and sex differences in human behavior. *Psychol. Bull.*, 92:56–80.

Hines, M. 2004. *Brain Gender*. Oxford University Press, Oxford.

Hines, M., and Kaufman, F. R. 1994. Androgen and the development of human sex-typical behavior: Rough-and-tumble play and sex of preferred playmates in children with congenital adrenal hyperplasia (CAH). *Child Development*, 65: 1042–1053.

Hines, M., Fane, B. A., Pasterski, V. L., Mathews, G. A., Conway, G. S., and Brook, C. 2003. Spatial abilities following prenatal androgen abnormality: Targeting and mental rotations performance in individuals with congenital adrenal hyperplasia. *Psychoneuroendocrinology*, 28: 1010–1026.

Hirschenhauser, K., Winkler, H., and Oliveira, R. F. 2003. Comparative analysis of male androgen responsiveness to social environment in birds: The effects of mating system and paternal incubation. *Hormones and Behavior*, 43: 508–519.

Hodges, J. K., and Hearne, J. P. 1978. A positive feedback effect of oestradiol on LH release in the male marmoset monkey. *J. Reprod. Fertil.*, 52:83–86.

Hoffman, K. A., Mendoza, S. P., Hennessy, M. B., and Mason, W. A. 1996. Responses of infant titi monkeys, *Callicebus moloch*, to removal of one or both parents: Evidence for paternal attachment. *Dev. Psychobiol.*, 28:399–407.

Hoffman, R. A., and Reiter, R. J. 1965. Pineal gland: Influence on gonads of male hamsters. *Science*, 148:1609–1615.

Hoffmann, K., Illnerova, H., and Vanecek, J. 1981. Effect of photoperiod and of one minute light at night-time on the pineal rhythm on N-acetyltransferase activity in the Djungarian hamster, Phodopus sungorus. *Biol. Reprod.*, 24:551–556.

Hofman, M. A., Zhou, J. N., and Swaab, D. F. 1996. Suprachiasmatic nucleus of the human brain: An immunocytochemical and morphometric analysis. *Anatomical Record*, 244:552–562.

Hoksbergen, R. A., ter Laak, J., van Dijkum, C., Rijk, S., Rijk, K., and Stoutjesdijk, F. 2003. Posttraumatic stress disorder in adopted children from Romania. *American Journal of Orthopsychiatry*, 73: 255–265.

Holekamp, K. E. 1986. Proximal causes of natal dispersal in Belding's ground squirrels (*Spermophilus beldingii*). *Ecol. Monogr.*, 56:365–391.

Holekamp, K. E., and Sherman, P. W. 1989. Why male ground squirrels disperse. *Am. Sci.*, 77:232–239.

Holman, S. D., and Goy, R. W. 1980. Behavioral and mammary responses of adult female rhesus to strange infants. *Horm. Behav.*, 14:348–357.

Holstege, G., Georgiadis, J. R., Paans, A. M. J., Meiners, L. C., van der Graaf, F. H. C. E., and Reinders, A. A. T. S. 2003. Brain activation during human male ejaculation. *Journal of Neuroscience*, 23: 9185–9193.

Holtzman, S. G. 1975. Effects of narcotic antagonists on fluid intake in the rat. *Life Sci.*, 16:1465–1470.

Honda, S., Harada, N., Ito, S., Takagi, Y., and Maeda, S. 1998. Disruption of sexual behavior in male aromatase-deficient mice lacking exons 1 and 2 of the *cyp19* gene. *Biochemical and Biophysics Research Communications*, 252: 445–449.

Hook, E. B. 1973. Behavioral implications of the human XYY genotype. *Science*, 179:139–149.

Hopkins, J., Marcues, M., and Campbell, S. B. 1984. Postpartum depression: A critical review. *Psychol. Bull.*, 95:498–515.

Horn, C. C., and Friedman, M. I. 1998. Metabolic inhibition increases feeding and brain Fos-like immunoreactivity as a function of diet. *Am. J. Physiol.*, 275: R448–R459.

Horseman, N. D., and Buntin, J. D. 1995. Regulation of pigeon cropmilk secretion and parental behaviors by prolactin. *Annu. Rev. Nutr.*, 15:213–238.

Horvath, T. L., Garcia-Segura, L. M., and Naftolin, F. 1997. Lack of gonadotropin-positive feedback in the male rat is associated with lack of estrogen-induced synaptic plasticity in the arcuate nucleus. *Neuroendocrinology*, 65: 136–140.

Hostetler, A. M., McHugh, P. R., and Moran, T. H. 1989. Bombesin affects feeding independent of a gastric mechanism or site of action. *Am. J. Physiol.*, 257:R1219–R1224.

Hotta, M., Shibasaki, T., Yamauchi, N., Ohno, H., Benoit, R., Ling, N., and Demura, H. 1991. The effects of chronic central administration of corticotropin-releasing factor on food intake, body weight, and hypothalamic-pituitary-adrenocortical hormones. *Life Sci.*, 48:1483–1491.

Howard, R. A., and Lively, C. M. 1994. Parasitism, mutation accumulation, and the maintenance of sex. *Nature*, 367:554–557.

Howdeshell, K. L., Hotchkiss, A. K., Thayer, K. A., Vandenbergh, J. G., and vom Saal, F. S. 1999. Exposure to bisphenol A advances puberty. *Nature*, 401: 763–764.

Hoyer, S. 2003. Memory function and brain glucose metabolism. *Pharmacopsychiatry*, 36: S62–S67.

Huck, U. M., Lisk, R. D., and Gore, A. C. 1985. Scent marking and mate choice in the golden hamster. *Physiol. Behav.*, 35:389–393.

Huck, U. W., Labov, J. B., and Lisk, R. D. 1986. Food restricting young hamsters (*Mesocricetus auratus*) alters sex ratio and growth of subsequent offspring. *Biol. Reprod.*, 36:592–598.

Hudson, R., and Distel, H. 1982. The pattern of behaviour of rabbit pups in the nest. *Behaviour*, 79:255–271.

Hudson, R., and Distel, H. 1989. Temporal pattern of suckling in rabbit pups: A model of circadian synchrony between mother and young. In S. M. Reppert (ed.), *Research in Perinatal Medicine*, Vol. 9, *Development of Circadian Rhythmicity and Photoperiodism in Mammals*, pp. 83–102. Perinatology Press, Ithaca.

Hughes, T. K., Fulep, E., Juelich, T., Smith, E. M., and Stanton, G. J. 1995. Modulation of immune responses by anabolic androgenic steroids. *Int. J. Immunopharmacol.*, 17:857–863.

Hughes, T. K., Rady, P. L., and Smith, E. M. 1998. Potential for the effects of anabolic steroid abuse in the immune and neuroendocrine axis. *J. Neuroimmunol.*, 83:162- 167.

Huhman, K. L., and Albers, H. E. 1993. Estradiol increases the behavioral response to arginine vasopressin (AVP) in the medial preoptic-anterior hypothalamus. *Peptides*, 14:1049–1054.

Huhman, K. L., and Jasnow, A. M. 2005. Conditioned defeat. In R. J. Nelson (ed.), *Biology of Aggression*. Oxford University Press, New York. In press.

Huhman, K. L., Bunnell, B. N., Mougey, E. H., and Meyerhoff, J. L. 1990. Effects of social conflict on POMC-derived peptides and glucocorticoids in male golden hamsters. *Physiology and Behavior*, 47: 949–956.

Huhman, K. L., Gillespie, C. F., Marvel, C. L., and Albers, H. E. 1996. Neuropeptide Y phase shifts circadian rhythms in vivo via a Y_2 receptor. *NeuroReport*, 7:1249–1252.

Huhman, K. L., Gillespie, C. F., Marvel, C. L., and Albers, H. E. 1997. Peptidergic mechanisms of action in the suprachiasmatic nucleus. *Ann. N. Y. Acad. Sci.*, 814:300–304.

Huhman, K. L., Moore, T. O., Ferris, C. F., Mougey, E. H., and Meyerhoff, J. L. 1991. Acute and repeated exposure to social conflict in male golden hamsters: Increases in plasma POMC-peptides and cortisol and decreases in plasma testosterone. *Hormones and Behavior*, 25: 206–216.

Huhman, K. L., Moore, T. O., Mougey, E. H., and Meyerhoff, J. L. 1992. Hormonal responses to fighting in hamsters: Separation of physical and psychological causes. *Physiology and Behavior*, 51: 1083–1086.

Huhman, K. L., Solomon, M. B., Janicki, M., Harmon, A. C., Lin, S. M., Israel, J. E., and Jasnow, A. M. 2003. Conditioned defeat in male and female Syrian hamsters. *Hormones and Behavior*, 44: 293–299.

Huie, M. J. 1994. An acute myocardial infarction occurring in an anabolic steroid user. *Med. Sci. Sports Exer.*, 26:408–413.

Hull, E. M., Du, J., Lorrain, D. S., and Matuszewich, L. 1995. Extracellular dopamine in the medial preoptic area: Implications for sexual motivation and hormonal control of copulation. *J. Neurosci.*, 15:7465–7471.

Hull, E. M., Du, J., Lorrain, D. S., and Matuszewich, L. 1997. Testosterone, preoptic dopamine, and copulation in male rats. *Brain Res. Bull.*, 44:327–334.

Hull, E. M., Lorrain, D. S., Du, J., Matuszewich, L., Lumley, L. A., Putnam, S. K., and Moses, J. 1999. Hormone-neurotransmitter interactions in the control of sexual behavior. *Behavior and Brain Research*, 105: 105–116.

Hull, E. M., Meisel, R. L., and Sachs, B. D. 2002. Male sexual behavior. In D. W. Pfaff, A. P. Arnold, A. T. Etgen, S. E. Fahrbach, and R. T. Rubin (eds.), *Hormones, Brain and Behavior*, Volume 1, pp. 1–138. Academic Press, New York.

Hulse, G. K., and Coleman, G. J. 1983. The role of endogenous opioids in the blockade of reproductive function in the rat following exposure to acute stress. *Pharmacol. Biochem. Behav.*, 19:795–799.

Hurn, P. D., Littleton-Kearny, M. T., Kirsch, J. R., Dharmarajan, A. M., and Traystman, R. J. 1995. Postischemic cerebral blood flow recovery in the female: Effect of 17β-estradiol. *J. Cereb. Blood Flow Metab.*,15:666–672.

Huszar, D., Lynch, C. A., Fairchild-Huntress, V., Dunmore, J. H., Fang, Q., Berkemeier, l. R., Gu, W. et al. 1997. Targeted disruption of the melanocortin-4 receptor results in obesity. *Cell*, 88: 131–141.

Hutchison, J. B. 1970. Differential effects of testosterone and oestradiol on male courtship in Barbary doves (*Streptopelia risoria*). *Anim. Behav.*, 18:41–51.

Hutchison, J. B., and Steimer, T. 1983. Hormone-mediated behavioural transitions: A role for brain aromatase. In J. Balthazart, E. Prove, and R. Gilles (eds.), *Hormones and Behaviour in Higher Vertebrates,* pp. 261–274. Springer Verlag, Berlin.

Hyde, T. M., and Miselis, R. R. 1983. Effects of area postrema/caudal medial nucleus of solitary tract lesions on food intake and body weight. *Am. J. Physiol.,* 244:R577–R587.

Ieni, J. R., and Thurmond, J. B. 1985. Maternal aggression in mice: Effects of treatments with PCPA, 5-HTP and 5-HT receptor antagonists. *Eur. J. Pharmacol.,* 111:211–220.

Iijima, M., Arisaka, O., Minamoto, F., and Arai, Y. 2001. Differences in children's free drawings: A study on girls with congenital adrenal hyperplasia. *Hormones and Behavior,* 40: 99–104.

Illnerova, H., and Vanecek, J. 1984. Circadian rhythm in inducibility of rat pineal N-acetyltransferase after brief light pulses at night: Control by a morning oscillator. *J. Comp. Physiol. A,* 154:739–744.

Illnerova, H., Hoffmann, K., and Vanecek, J. 1986. Adjustments of the rat pineal N-acetyltransferase rhythm to change from long to short photoperiod depends on the direction of the extension of the dark period. *Brain Res.,* 362:403–408.

Illnerova, H., Zvolsky, P., and Vanecek, J. 1985. The circadian rhythm in plasma melatonin concentration of the urbanized man: The effect of summer and winter time. *Brain Res.,* 328:186–192.

Inouye, S. T., and Kawamura, H. 1979. Persistence of circadian rhythmicity in a mammalian hypothalamic "island" containing the suprachiasmatic nucleus. *Proc. Natl. Acad. Sci. U.S.A.,* 76:5962–5966.

Insel, T. R. 1990a. Oxytocin and maternal behavior. In N. A. Krasnegor and R. S. Bridges (eds.), *Mammalian Parenting,* pp. 260–280. Oxford University Press, Oxford.

Insel, T. R. 1990a. Regional changes in brain oxytocin receptors postpartum: Time-course and relationship to maternal behaviour. *J. Neuroendocrinol.,* 2:539–545.

Insel, T. R. 1990b. Regional changes in brain oxytocin receptors postpartum: Time-course and relationship to maternal behaviour. *J. Neuroendocrinol.,* 2:539–545.

Insel, T. R. 1997. A neurobiological basis of social attachment. *Am. J. Psychiatry,* 154:726–735.

Insel, T. R. 2003. Is social attachment an addictive disorder? *Physiology and Behavior,* 79: 351–357.

Insel, T. R., and Fernald, R. D. 2004. How the brain processes social information: Searching for the social brain. *Annual Review of Neuroscience,* 27: 697–722.

Insel, T. R., and Hulihan, T. J. 1995. A gender-specific mechanism for pair bonding: oxytocin and partner preference formation in monogamous voles. *Behav. Neurosci.,* 109: 782–789.

Insel, T. R., and Shapiro, L. E. 1992. Oxytocin receptor distribution reflects social organization in monogamous and polygamous voles. *Proc. Natl. Acad. Sci. U.S.A.,* 89:5981–5985.

Insel, T. R., and Young, L. J. 2001. The neurobiology of attachment. *Nature Reviews: Neuroscience,* 2: 129–136.

Insel, T. R., Gilhard, R., and Shapiro, L. E. 1991. The comparative distribution of forebrain receptors for neurohypophyseal peptides in monogamous and polygamous mice. *Neuroscience,* 43:623–630.

Insel, T. R., Wang, Z. X., and Ferris, C. F. 1994. Patterns of brain vasopressin receptor distribution associated with social organization in microtine rodents. *J. Neurosci.,* 14:5381–5392.

Introini-Collison, I. B., and Baratti, C. M. 1986. Opioid peptidergic systems modulate the activity of beta-adrenergic mechanisms during memory consolidation processes. *Behav. Neural Biol.,* 46:227–241.

Isgor, C., and Sengelaub, D. R. 1998. Prenatal gonadal steroids affect adult spatial behavior, CA1 and CA3 pyramidal cell morphology in rats. *Horm. Behav.,* 34:183–198.

Jacob, S., Garcia, S., Hayreh, D., and McClintock, M. K. 2002. Psychological effects of musky compounds: Comparison of androstadienone with androstenol and muscone. *Hormones and Behavior,* 42: 274–283.

Jacob, S., Hayreh, D. J., and McClintock, M. K. 2001. Context-dependent effects of steroid chemosignals on human physiology and mood. *Physiology and Behavior,* 74: 15–27.

Jacobs, M., and Lightman, S. L. 1980. Studies in the opioid control of anterior pituitary hormones. *J. Physiol. (Lond.),* 300:53–60.

Jacobs, P. A., Brunton, M., Melville, M. M., Brittain, R. P., and McClemont, W. F. 1965. Aggressive behavior, mental subnormality and the XYY male. *Nature,* 208:1351–1352.

James, V. H. T. 1992. *The Adrenal Gland.* Raven Press, New York.

Janowitz, H. D., and Grossman, M. I. 1949. Some factors affecting the food intake of normal dogs and dogs with esophagostomy and gastric fistula. *Am. J. Physiol.,* 159:143–148.

Janowsky, D. S., Berens, S. C., and Davis, J. M. 1973. Correlations between mood, weight, and electrolytes during the menstrual cycle: A renin-angiotensin-aldosterone hypothesis of premenstrual tension. *Psychosom. Med.,* 35:143–154.

Jasnow, A. M., Drazen, D. L., Huhman, K. L., Nelson, R. J., and Demas, G. E. 2001. Acute and chronic social defeat suppresses humoral immunity of male Syrian hamsters (*Mesocricetus auratus*). *Hormones and Behavior,* 40: 428–433.

Jasnow, A. M., Huhman, K. L., Bartness, T. J., and Demas, G. E. 2002. Short days and exogenous melatonin increase aggression of male Syrian hamsters (*Mesocricetus auratus*). *Hormones and Behavior,* 42: 13–20.

Jasnow, A. M., Huhman, K. L., Bartness, T. J., and Demas, G. E. 2000. Short-day increases in aggression are inversely related to circulating testosterone concentrations in male Siberian hamsters (*Phodopus sungorus*). *Hormones and Behavior,* 38: 102–110.

Jenkins, R. L., Wilson, E. M., Angus, R. A., Howell, W. M., and Kirk, M. 2003. Androstenedione and progesterone in the sediment of a river receiving paper mill effluent. *Toxicological Sciences,* 73: 53–59.

Jenkins, R., Angus, R. A., McNatt, H., Howell, W. M., Kemppainen, J. A., Kirk, M., and Wilson, E. M. 2001. Identification of androstenedione in a river containing paper mill effluent. *Environmental Toxicology and Chemistry*, 20: 1325–1331.

Jensen, B. K. 1982. Menstrual cycle effects on task performance examined in the context of stress research. *Acta Psychol.*, 50:159–178.

Jensen, T. S., Genefke, I. K., Hyldebrandt, N., Pedersen, H., Petersen, H. D., Weile, B.1982. Cerebral atrophy in young torture victims. *N. Engl. J. Med.*, 307:1341.

Jilge, B. 1993. The ontogeny of circadian rhythms in the rabbit. *J. Biol. Rhythms*, 8:247–260.

Jobst, E. E., Robinson, D. W., and Allen, C. N. 2004. Potential pathways for intercellular communication within the calbindin subnucleus of the hamster suprachiasmatic nucleus. *Neuroscience*, 123: 87–99.

Johanson, C. E., Balster, R. L., and Bonese, K. 1976. Self-administration of psychomotor stimulant drugs: the effects of unlimited access. *Pharmacol. Biochem. Behav.*, 4:45–51.

Johnson, F., and Whalen, R. E. 1988. Testicular hormones reduce individual differences in the aggressive behavior of male mice: A theory of hormone action. *Neurosci. Biobehav. Rev.*, 12:93–99.

Johnson, L. R., and Wood, R. I. 2001. Oral testosterone self-administration in male hamsters. *Neuroendocrinology*, 73: 285–292.

Johnston, A. L., and File, S. E. 1991. Sex differences in animal tests of anxiety. *Physiol. Behav.*, 49:245–250.

Johnston, P. G., and Davidson, J. M. 1979. Priming action of estrogen: Minimum duration of exposure for feedback and behavioral effects. *Neuroendocrinology*, 28:155–159.

Johnston, R. E. 1985. Communication. In H. I. Siegel (ed.), *The Hamster: Reproduction Behavior*, pp. 121–154. Plenum Press, New York.

Jolly, A. 1966. *Lemur Behavior*. University of Chicago Press, Chicago.

Jones, H. E., Ruscio, M. A., Keyser, L. A., Gonzalez, C., Billack, B., Rowe, R., Hancock, C., Lambert, K. G., and Kinsley, C. H. 1997. Prenatal stress alters the size the rostral anterior commissure in rats. *Brain Res. Bull.*, 42:341–346.

Jost, A. 1979. Basic sexual trends in the development of vertebrates. In *Sex, Hormones and Behavior*. CIBA Foundation Symposium, No. 62. Elsevier, Amsterdam.

Judson, O. P. 1997. A model of asexuality and clonal diversity: Cloning the red queen. *J. Theoret. Biol.*, 186:33–40.

Julian, T., and McKenry, P. C. 1989. Relationship of testosterone to men's family functioning at mid-life: A research note. *Aggressive Behav.*, 15:281–289.

Kaitz, M. 1992. Recognition of familiar individuals by touch. *Physiol. Behav.*, 52:565–567.

Kalin, N. H., Shelton, S. E., and Barksdale, C. M. 1988. Opiate modulation of separation-induced distress in nonhuman primates. *Brain Research*, 440: 285–292.

Kalin, N. H., Shelton, S. E., and Barksdale, C. M. 1988. Opiate modulation of separation-induced distress in non-human primates. *Brain Res.*, 440:285–292.

Kalin, N. H., Shelton, S. E., and Lynn, D. E. 1995. Opiate systems in mother and infant primates coordinate intimate contact during reunion. *Psychoneuroendocrinology*, 7: 735–742.

Kalinichev, M., Rosenblatt, J. S., Nakabeppu, Y., and Morrell, J. I. 2000. Induction of *c-fos*-like and *fosB*-like immunoreactivity reveals forebrain neuronal populations involved differentially in pup-mediated maternal behavior in juvenile and adult rats. *Journal of Comparative Neurology*, 416: 45–78.

Kalra, P. S., Fuentes, M., Sahu, A., and Kalra, S. P. 1990. Endogenous opioid peptides mediate the interleukin-1-induced inhibition of the release of luteinizing hormone (LH)-releasing hormone and LH. *Endocrinology*, 127:2381–2386.

Kalra, S. P., Dube, M. G., Sahu, A., Phelps, C. P., and Kalra, P. S. 1991. Neuropeptide Y secretion increases in the paraventricular nucleus in association with increased appetite for food. *Proc. Natl. Acad. Sci. U.S.A.*, 88:10931–10935.

Kannan, C. R. 1987. *The Pituitary Gland*. Plenum Medical Book, New York.

Kano, T. 1992. *The Last Ape: Pygmy Chimpanzee Behav. Ecology*. Stanford University Press. Stanford, CA.

Kaplan, J. M., Seeley, R. J., and Grill, H. J. 1993. Daily caloric intake in intact and chronic decerebrate rats. *Behav. Neurosci.*, 107:876–881.

Kaplan, J. M., Song, S., and Grill, H. J. 1998. Serotonin receptors in the caudal brainstem are necessary and sufficient for the anorectic effect of peripherally administered mCPP. *Psychopharmacology*, 137:43–49.

Kaplan, J. R., and Manuck, S. B. 1999. Status, stress, and atherosclerosis: The role of environment and individual behavior. *Annals of the New York Academy of Sciences*, 896: 145–161.

Karp, J. D., and Powers, J. B. 1993. Photoperiodic and pineal influences on estrogen-stimulated behaviors in female Syrian hamsters. *Physiol. Behav.*, 54:19–28.

Karsch, F. J. 1987. Central actions of ovarian steroids in the feedback regulation of pulsatile secretion of luteinizing hormone. *Annu. Rev. Physiol.*, 49:365–382.

Karsch, F. J., Bittman, E. L., Foster, D. L., Goodman, R. L., Legan, S. J., and Robinson, J. E. 1984. Neuroendocrine basis of seasonal reproduction. *Rec. Prog. Horm. Res.*, 40:185–232.

Karsch, F. J., Dierschke, D. J., and Knobil, E. 1973. Sexual differentiation of pituitary function: Apparent difference between primates and rodents. *Science*, 179:484–486.

Kashkin, K. B., and Kleber, H. D. 1989. Hooked on hormones? An anabolic steroid addiction hypothesis. *J. A. M. A.*, 262:3166–3170.

Kasper, S., Wehr, T., Bartko, J., Gaist, P., and Rosenthal, N. 1989. Epidemiological findings of seasonal changes in mood and behavior. A telephone survey of Montgomery County, Maryland. *Arch. Gen. Psychiatry*, 46:823–833.

Kastin, A. J., Ehrensing, R. H., Schalch, D. S., and Anderson, M. S. 1972a. Improvement in mental depression with decreased thyrotropin response after administration of thyrotropin-releasing hormone. *Lancet*, 780:740–742.

Kastin, A. J., Sandman, C. A., Stratton, L. O., Schally, A. V., and Miller, L. H. 1975. Behavioral and electrographic changes in rat and man after MSH. *Prog. Brain Res.*, 42:143–150.

Katona, C., Rose, S., and Smale, L. 1998. The expression of Fos with the suprachiasmatic nucleus of the diurnal rodent *Arvicanthis niloticus*. *Brain Res.*, 791:27–34.

Ke, F. C., and Ramirez, V. D. 1990. Binding of progesterone to nerve cell membranes of rat brain using progesterone conjugated to 125I-bovine serum albumin as a ligand. *J. Neurochem.*, 54:467–472.

Keen-Rhinehart, E., and Bartness, T. J. 2005. Peripheral ghrelin injections stimulate food intake, foraging and food hoarding in Siberian hamsters. *American Journal of Physiology*. In press.

Kelley, A. E., and Berridge, K. C. 2002. The neuroscience of natural rewards: Relevance to addictive drugs. *Journal of Neuroscience*, 22: 3306–3311.

Kendrick, K. M. 2000. Oxytocin, motherhood and bonding. *Experimental Physiology*, 85: 111S–124S.

Kendrick, K. M., and Dixson, A. F. 1986. Anteromedial hypothalamic lesions block proceptivity but not receptivity in the female common marmoset (*Callithrix jacchus*). *Brain Research*, 375: 221–229.

Kendrick, K. M., Fabre-Nys, C., Blache, D., Goode, J. A., and Broad, K. D. 1993. The role of oxytocin release in the mediobasal hypothalamus of the sheep in relation to female sexual receptivity. *J. Neuroendocrinol.*, 5:13–21.

Kendrick, K. M., Levy, F., and Keverne, E. B. 1992. Changes in the sensory processing of olfactory signals induced by birth in sleep. *Science*, 256:833–836.

Kennedy, G. C. 1953. The role of depot fat in the hypothalamic control of food intake in the rat. *Proceedings of the Royal Society of London, Series B*, 140: 579–592.

Kennedy, G. C. 1964. Hypothalamic control of the endocrine and behavioural changes associated with oestrus in the rat. *J. Physiol. (Lond)*, 166:395–407.

Kent, S., Tom, C., and Levine, S. 1997. Effect of interleukin-1 on pituitary-adrenal responses and body weight in neonatal rats: Interaction with maternal deprivation. *Stress*, 1:213–230.

Kerchner, M., and Ward, I. L. 1992. SDN-MPOA volume in male rats is decreased by prenatal stress, but is not related to ejaculatory behavior. *Brain Res.*, 581:244–251.

Kerchner, M., Malsbury, C. W., Ward, O. B., and Ward, I. L. 1995. Sexually dimorphic areas in the rat medial amygdala: Resistance to the demasculinizing effect of prenatal stress. *Brain Res.*, 20:251–260.

Keverne, E. B. 1992. Primate social relationships: Their determinants and consequences. *Advances in the Study of Behavior*, 21: 1–37.

Keverne, E. B. 2001. Genomic imprinting, maternal care, and brain evolution. *Hormones and Behavior*, 40: 146–155.

Keverne, E. B., and Kendrick, K. M. 1992. Oxytocin facilitation of maternal behavior in sheep. *Ann. N. Y. Acad. Sci.*, 652:83–101.

Keverne, E. B., and Kendrick, K. M. 1994. Maternal behaviour in sheep and its neuroendocrine regulation. *Acta Paediatr.* 397:47–56.

Keverne, E. B., Levy, F., Poindron, P., and Lindsay, D. R. 1983. Vaginal stimulation: An important determinant of maternal bonding in sheep. *Science*, 219:81–83.

Kibble, M. W., and Ross, M. B. 1987. Adversive effects of anabolic steroids in athletes. *Clin. Pharm.*, 6:686–692.

Kiecolt-Glaser, J. K., Marucha, P. T., Malarkey, W. B., Mercado, A. M., and Glaser, R. 1995. Slowing of wound healing by psychological stress. *Lancet*, 346: 1194–1196.

Kikuyama, S., Toyoda, F., Ohmiya, Y., Matsuda, K., Tanaka, S., Hayashi, H. 1995. Sodefrin: A female-attracting peptide pheromone in newt cloacal glands. *Science*, 267:1643–1645.

Kikuyama, S., Toyoda, F., Yamamoto, K., Tanaka, S., and Hayashi, H. 1997. Female-attracting pheromone in newt cloacal glands. *Brain Res. Bull.*, 44:415–422.

Kim, J. J., and Diamond, D. M. 2002. The stressed hippocampus, synaptic plasticity and lost memories. *Nature Reviews: Neuroscience*, 3: 453–462.

King, J. M. 1979. Effects of lesions of the amygdala, preoptic area, and hypothalamus on estradiol-induced activity in the female rat. *J. Comp. Physiol. Psychol.*, 93:360–367.

Kinsey, A. C., Pomeroy, W. B., and Martin, C. E. 1948. *Sexual Behavior in the Human Male*. W. B. Saunders, Philadelphia.

Kinsey, A. C., Pomeroy, W. B., Martin, C. E., and Gebhard, P. 1953. *Sexual Behavior in the Human Female*. Saunders, Philadelphia.

Kinsley, C. H. 1990. Prenatal and postnatal influences on parental behavior in rodents. In N. A. Krasnegor and R. S. Bridges (eds.), *Mammalian Parenting*, pp. 347–371. Oxford University Press, Oxford.

Kinsley, C. H. 1994. Developmental psychobiological influences on rodent parental behavior. *Neurosci. Biobehav. Rev.*18:269–280.

Kinsley, C. H., Mann, P. E., and Bridges, R. S. 1989. Alterations in stress-induced prolactin release in adult female and male rats exposed to stress, *in utero*. *Physiol. Behav.*, 45:1073–1076.

Kirn, J. R., Clower, R. P., Kroodsma, D. E., and DeVoogd, T. J. 1989. Song-related brain regions in the red-winged blackbird are affected by sex and season but not repertoire size. *J. Neurobiol.*, 11:139–163.

Kitay, J. I. 1963. Pituitary-adrenal function in the rat after gonadectomy and gonadal hormone replacement. *Endocrinology*, 73:253–260.

Klaiber, E. L., Broverman, D. M., Vogel, W., and Kobayski, V. 1979. Estrogen therapy for severe persistent depressions in women. *Arch. Gen. Psychiatry*, 36:550–554.

Klaus, M. H., and Kennell, J. H. 1976. *Maternal-Infant Bonding*. Mosby: St. Louis.

Kleiman, D. G., and Malcolm, J. R. 1981. The evolution of male parental investment. In D. J. Gubernick and P. H. Klopfer (eds.), *Parental Care in Mammals*, pp. 347–387. Plenum, New York.

Klein, D. A., and Walsh, B. T. 2003. Eating disorders. *International Review of Psychiatry*, 15: 205–216.

Klein, D. A., and Walsh, B. T. 2004. Eating disorders: Clinical features and pathophysiology. *Physiology and Behavior*, 81: 359–374.

Klein, D.C., Smoot, R., Weller, J. L., Higa, S., Markey, S. P., Creed, G. H., and Jacobowitz, D. M. 1983. Lesions of the paraventricular nucleus area of the hypothalamus disrupt the suprachiasmatic–spinal cord circuit in the melatonin rhythm generating system. *Brain Res. Bull.*, 10:647–652.

Klein, K. E., and Wegmann, H. M. 1974. The resychronization of human circadian rhythms after transmeridian flights as a result of flight direction and mode of activity. In L. E. Scheving, F. Halberg, and J. E. Pauley (eds.), *Chronobiology*, pp. 564–570. Igaku, Tokyo.

Klein, S. L., and Nelson, R. J. 1998. Adaptive immune responses are linked to the mating systems of arvicoline rodents. *Am. Nat.*, 151:59–67.

Klemfuss, H. 1992. Rhythms and the pharmacology of lithium. *Pharmacol. Ther.*, 56:53–78.

Kluger, M. J. 1978. The evolution and adaptive value of fever. *Am. Sci.*, 66:38–43.

Kluger, M. J. 1986. Is fever beneficial? *Yale J. Biol. Med.*, 59:89–95.

Kluver, H. N., and Tinbergen, N. 1953. Territory and the regulation of density of titmice. *Arch. Neerlandaises de Zool.*, 10:265–289.

Klüver, H., and Bucy, P. C. 1939. Preliminary analysis of functions of the temporal lobes in monkeys. *Arch. Neurol. Psychiatry*, 42:979–1000.

Knobil, E., and Hotchkiss, J. 1988. The menstrual cycle and its neuroendocrine control. In E. Knobil and, J. D. Neill (eds.), *The Physiology of Reproduction*, Vol. 2, pp. 1971–1994. Raven, New York.

Knobil, E., and Neill, J. D. 1994. *The Physiology of Reproduction.* (2nd ed.). Raven Press, New York.

Koeske, R. K. D. 1980. Theoretical perspectives on menstrual cycle research: The relevance of attributional approaches for the perception and explanation of premenstrual emotionality. In A. J. Dan, E. A. Graham, and C. P. Beecher (eds.), *The Menstrual Cycle*, pp. 161–181. Springer Verlag, New York.

Kojima, M., Hosoda, H., Date, Y., Nakazato, M., Matsuo, H., and Kangawa, K. 1999. Ghrelin is a growth-hormone-releasing acylated peptide from stomach. *Nature*, 402: 656–660.

Komisaruk, B. R. 1967. Effects of local brain implants of progesterone on reproductive behavior in ring doves. *J. Comp. Physiol. Psychol.*, 64:219–224.

Komisaruk, B. R., Terasawa, E., and Rodriguez-Sierra, J. F. 1981. How the brain mediates ovarian responses to environmental stimuli. In N. T. Adler (ed.), *Neuroendocrinology of Reproduction*, pp. 349–376. Plenum Press, New York.

Komisaruk, B. R., Whipple, B., Crawford, A., Grimes, S., Kalnin, A. J., Mosier, K., Liu, W. C., and Harkness, B. 2002. Brain activity (fMRI and PET) during orgasm in women, in response to vaginocervical self-stimulation. *Abstracts of Society for Neuroscience*, 841: 17.

Konopka, R. J. 1987. Genetics of biological rhythms in *Drosophila*. *Annu. Rev. Genet.*, 21:227–236.

Konopka, R. J., and Benzer, S. 1971. Clock mutants of *Drosophila melanogaster*. *Proc. Natl. Acad. Sci. U.S.A.*, 68:2112–2116.

Koob, G. F. 1992. Drugs of abuse:anatomy, pharmacology and function of reward pathways. *Trends Pharmacol. Sci.*, 13:177–184.

Koob, G. F., and Bloom, F. E. 1988. Cellular and molecular mechanisms of drug dependence. *Science*, 242:715–723.

Koopmans, H. S. 1983. A stomach hormone that inhibits food intake. *J. Autonomic Nerv. Syst.*, 9:157–171.

Kopin, A. S., Foulds-Mathes, W., McBride, E. W., Nguyen, M., Al-Haider, W., Schmitz, F., Bonner-Weir, S., Kanarek, R., and Beinborn, M. 1999. The cholecystokinin-A receptor mediates inhibition of food intake yet is not essential for the maintenance of body weight. *Journal of Clinical Investigations*, 103: 383–391.

Kopin, I. J. 1995. Definitions of stress and sympathetic neuronal responses. *Ann. N. Y. Acad. Sci.*,771:19–30.

Korach, K. S. 1994. Insights from the study of animals lacking functional estrogen receptor. *Science*, 266:1524–1527.

Kornhauser, J. M., Nelson, D. E., Mayo, K. E., and Takahashi, J. S. 1990. Photic and circadian regulation of c-fos gene expression in the hamster suprachiasmatic nucleus. *Neuron*, 5:127–134.

Korol, D. L., and Gold, P. E. 1998. Glucose, memory, and aging. *Am. J. Clin. Nutr.*, 67:764S–771S.

Korol, D. L., Unick, K., Goosens, K., Crane, C., Gold, P. E., and Foster, T. C. 1994. Estrogen effects on spatial performance and hippocampal physiology in female rats. *Soc. Neurosci. Abstr.*, 20:1436.

Korpelainen, H. 1990. Sex ratios and conditions required for environmental sex determination in animals. *Biol. Rev.*, 65:147–184.

Kosman, M. E., and Gerard, R. W. 1955. The effect of adrenaline on a conditioned avoidance response. *J. Comp. Physiol. Psychol.*, 48:506–508.

Kotrschal, K., Hirschenhauser, K., and Möstl, E. 1998. The relationship between social stress and dominance is seasonal in greylag geese. *Anim. Behav.*, 55:171–176.

Kow, L.-M., and Pfaff, D. W. 1998. Mapping of neural and signal transduction pathways for lordosis in the search for estrogen actions on the central nervous system. *Behav. Brain Res.*, 92:169–180.

Kow, L.-M., Montgomery, M. O., and Pfaff, D. W. 1979. Triggering of lordosis reflex in female rats with somatosensory stimulation: Quantitative determination of stimulus parameters. *J. Neurophysiol.*, 42:195–202.

Koylu, E. O., Couceyro, P. R., Lambert, P. D., and Kuhar, M. J. 1998. Cocaine- and amphetamine-regulated transcript peptide immunohistochemical localization in the rat brain. *Journal of Comparative Neurology*, 391: 115–132.

Krasnegor, N. A., and Bridges, R. S. (eds.). 1990. *Mammalian Parenting*. Oxford University Press, Oxford.

Kreuz, L. E., and Rose, R. M. 1972. Assessment of aggressive behavior and plasma testosterone in a young criminal population. *Psychosom. Med.*, 34:321–332.

Kriegsfeld, L. J., Dawson, T. M., Dawson, V. L., Nelson, R. J. and Snyder, S. H. 1997. Aggressive behavior in male mice lacking the gene for nNOS is testosterone-dependent. *Brain Res.*, 769:66–70.

Kriegsfeld, L. J., Demas, G. E., Dawson, T. M., Dawson, V. L., Lee, S., and Nelson, R. J. 1999. Circadian behavior in mice with targeted disruption of the gene for the neuronal isoform of nitric oxide synthase. *Journal of Biological Rhythms*, 14: 20–27.

Kriegsfeld, L. J., Drazen, D. L., and Nelson, R. J. 2001. Circadian organization in male mice lacking the gene for endothelial nitric oxide synthase (eNOS$^{-/-}$). *Journal of Biological Rhythms*, 16: 142–148.

Kriegsfeld, L. J., LeSauter, J., and Silver, R. 2004. Targeted microlesions reveal novel organization of the hamster suprachiasmatic nucleus. *Journal of Neuroscience*, 24: 2449–2457.

Kriegsfeld, L. J., LeSauter, J., Hamada, T., Pitts, S. M., and Silver, R. 2002. Circadian rhythms in the endocrine system. In D. W. Pfaff, A. P. Arnold, A. M. Etgen, S. E. Fahrbach, and R. T. Rubin (eds.), *Hormones, Brain and Behavior*, Volume 2, pp. 33–91. Academic Press, New York.

Kripke, D. F. 1981. Photoperiodic mechanisms for depression and its treatment. In C. Perris, G. Struwe, and B. Jansson (eds.), *Biological Psychiatry*, pp. 1249–1252. Elsevier, Amsterdam.

Kripke, D. F., Elliott, J. A., Youngstedt, S. D., and Smith, J. S. 1998. Melatonin: Marvel or marker? *Ann. Med.*, 30:81–87.

Krohmer, R. W., and Crews, D. 1987. Temperature activation of courtship behavior in the male red-sided garter snake (*Thamnophis sirtalis parietalis*): Role of the anterior hypothalamus-preoptic area. *Behav. Neurosci.*, 101:228–236.

Kruk, M. R. 1991. Ethology and pharmacology of hypothalamic aggression in the rat. *Neuroscience and Biobehavioral Review*, 15: 527–538.

Kruuk, H. 1972. *The Spotted Hyena: A Study of Predation and Social Behavior.* University of Chicago Press, Chicago.

Kudielka, B. M., Hellhammer, J., Hellhammer, D. H., Wolf, O. T., Pirke, K. M., Varadi, E., Pilz, J., and Kirschbaum, C. 1998. Sex differences in endocrine and psychological responses to psychosocial stress in healthy elderly subjects and the impact of a 2-week dehydroepiandrosterone treatment. *Journal of Clinical Endocrinology and Metabolism*, 83: 1756–1761.

Kudwa, A. E., and Rissman, E. F. 2003. Double oestrogen receptor α and β knockout mice reveal differences in neural oestrogen-mediated progestin receptor induction and female sexual behaviour. *Journal of Neuroendocrinology*, 15: 978–983

Kuederling, I., Evans, C. S., Abbott, D. H., Pryce, C. R., and Epple, G. 1995. Differential excretion of urinary oestrogen by breeding females and daughters in the red-bellied tamarin (*Saguinus labiatus*). *Fol. Primatol.*, 64:140–145.

Kuevi, V., Causon, R., Dixson, A. F., Everard, D. M., Hall., J. M., Hole, D., Whitehead, S. A., Wilson, C. A., and Wise, J. C. M. 1983. Plasma amine and hormone changes in 'post-partum' blues. *Clin. Endocrinol.*, 19:39–46.

Kuga, M., Ikeda, M., Suzuki, K., and Takeuchi, S. 2002. Changes in gustatory sense during pregnancy. *Acta Otolaryngologia* Supplement, 546: 146–153.

Kuiper, G. C., Enmark, E., Pelto-Huikko, M., Nilsson, S., and Gustafsson, J. A. 1996. Cloning of a novel estrogen receptor expressed in rat prostate and ovary. *Proc. Natl. Acad. Sci. U.S.A.*, 93:5925–5930.

Kumari, M., Brunner, E., and Fuhrer, R. 2000. Mechanisms by which the metabolic syndrome and diabetes impair memory. *Journals of Gerontology, Series A: Biological Sciences and Medical Sciences*, 55: B228–B232.

Kurzer, M. S. 1997. Women, food, and mood. *Nutr. Rev.*, 55:268–276.

Labrie, F., Belanger, A., Simard, J., Luu-The, V., and Labrie, C. 1995. DHEA and peripheral androgen and estrogen formation: Intracrinology. *Annals of the New York Academy of Science*, 774: 16–28.

Lacey, E. A., and Sherman, P. W. 1991. Social organization of naked mole-rat colonies: Evidence for division of labor. In P. W. Sherman, J. U. M. Jarvis, and R. D. Alexander (eds.), *The Biology of the Naked Mole-Rat*, 275–336. Princeton University Press, Princeton, NJ.

Lack, D. 1954. *The Natural Regulation of Animal Numbers.* Clarendon Press, Oxford.

Ladle, R. J., Johnstone, R. A., and Judson, O. P. 1993. Coevolutionary dynamics of sex in a metapopulation-Escaping the red queen. *Proc. R. Soc. Lond. B.*, 253:155–160.

LaFerla, J. J., Anderson, D. L., and Schalch, D. S. 1978. Psychoendocrine response to sexual arousal in human males. *Psychosom. Med.*, 40:166–172.

Lahr, G., Maxson, S., Mayer, A., Just, W., Pilgrim, C., and Reisert, I. 1995. Transcription of the Y chromosomal gene, Sry, in adult mouse brain. *Mol. Brain Res.*, 33:179–182.

Lal, S., Ackman, D., Thavundayil, J. X., Kiely, M. E., and Etienne, P. 1984. Effect of apomorphine, a dopamine receptor agonist, on penile tumescence in normal subjects. *Prog. Neuropsychopharmacol. Biol. Psychiatry*, 8:695–699.

Lal, S., Laryea, E., Thavundayil, J. X., Nair, N. P. V., Negrete, J., Ackman, D., Blundell, P., and Gardiner, R. J. 1987. Apomorphine-induced penile tumescence in impotent patients—preliminary findings. *Prog. Neuropsychopharmacol. Biol. Psychiatry*, 11:235–242.

Lance, S. L., and Wells, K. D. 1993. Are spring peeper satellite males physiologically inferior to calling males? *Copeia*, 1993:1162–1166.

Lane, A. J. P., and Wathes, D.C. 1998. An electronic nose to detect changes in perineal odors associated with estrus in the cow. *J. Dairy Sci.*, 81:2145–2150.

Langan, S. J., Deary, I. J., Hepburn, D. A., and Frier, B. M. 1991. Cumulative cognitive impairment following recurrent severe hypoglycaemia in adult patients with insulin-treated diabetes mellitus. *Diabetologia*, 34: 337–344.

Langevin, R., Paitich, D., Hucker, S., Newman, S., Ramsay, G., Pope, S., Geller, G., and Anderson, C. 1979. The effect of assertiveness training, Provera and sex of therapist in the treatment of genital exhibitionism. *J. Behav. Ther. Exp. Psychiatry*, 10:275–282.

Langhans, W., and Scharrer, E. J. 1987. Evidence for a vagally mediated satiety signal derived from hepatic fatty acid oxidation. *Autonomic Nerv. Sys.*, 18:13–18.

Lank, D. B., Smith, C. M., Hanotte, O., Burke, T., and Cooke, F. 1995. Genetic polymorphism for alternative mating behaviour in lekking male ruff *Philomachus pugnax*. *Nature*, 378:59–62.

Lannert, H., and Hoyer, S. 1998. Intracerebroventricular administration of streptozotocin causes long-term diminutions in learning and memory abilities and in cerebral energy metabolism in adult rats. *Behavioral Neuroscience*, 112: 1199–1208.

Larkin, G. 1991. Carcinoma of the prostate. *New England Journal of Medicine*, 324: 1892–1893.

Larsen, P. R., Wilson, J. D., Schlomo, M., Foster, D. W., and Kronenberg, H. M. 2003. *Williams Textbook of Endocrinology*. Saunders, Philadelphia.

Larsson, K. 1962. Mating behavior in male rats after cerebral cortex ablation. I. Effects of lesions in the dorsolateral and the median cortex. *J. Exp. Zool.*, 151:167–176.

Larsson, K. 1966. Individual differences in reactivity to androgen in male rats. *Physiol. Behav.*, 1:255–258.

Larsson, K. 1969. Failure of gonadal and gonadotrophic hormones to compensate for an impaired sexual function in anosmic male rats. *Physiol. Behav.*, 4:733–737.

Larsson, K. 1979. Features of the neuroendocrine regulation of masculine sexual behavior. In C. Beyer (ed.), *Endocrine Control of Sexual Behavior*, pp. 77–163. Raven Press, New York.

Larsson, K. 2003. My way to biological psychology. *Scandinavian Journal of Psychology*, 44: 173–187.

Lashley, K. S. 1938. Experimental analysis of instinctive behavior. *Psychol. Rev.*, 45:445–471.

Lavie, P. 1985. Ultradian rhythms: Gates of sleep and wakefulness. In H. Schulz and P. Lavie (eds.), *Ultradian Rhythms in Physiology and Behavior*, pp. 148–164. Springer-Verlag, Berlin.

Le Mouellic, H., Lallemand, Y., and Brulet, P. 1990. Targeted replacement of the homeobox gene Hox-3.1 by the *Escherischia coli* lacZ in mouse chimeric embryos. *Proc. Natl. Acad. Sci. U.S.A.*, 87, 4712–4716.

Lea, R. W., Clark, J. A., and Tsutsui, K. 2001. Changes in central steroid receptor expression, steroid synthesis, and dopaminergic activity related to the reproductive cycle of the ring dove. *Microscopy Research and Technique*, 55: 12–26.

Lea, R. W., Vowels, D. M., and Dick, H. R. 1986. Factors affecting prolactin secretion during the breeding cycle of the ring dove (*Streptopelia risoria*) and its possible role in incubation. *J. Endocrinol.*, 110: 447–458.

LeBlond, C. P. 1938. Extra-hormonal factors in maternal behavior. *Proc. Soc. Exp. Biol.*, 38:66–70.

LeBoeuf, B. J. 1967. Interindividual associations in dogs. *Behaviour*, 29:268–295.

LeBoeuf, B. J. 1970. Copulatory and aggressive behavior in the prepubertally castrated male dog. *Horm. Behav.*, 1:127–136.

LeBoeuf, B. J. 1974. Male-male competition and reproductive success in elephant seals. *Am. Zool.*, 14:163–176.

Leckman, J. F., and Mayes, L. C. 1999. Preoccupations and behaviors associated with romantic and parental love: Perspectives on the origin of obsessive–compulsive disorders. *Child and Adolescent Psychiatric Clinics of North America*, 8: 635–665.

Lee, A., Clancy, S., and Fleming, A. S. 2000. Mother rats bar-press for pups: Effects of lesions of the MPOA and limbic sites on maternal behavior and operant responding for pup-reinforcement. *Behavioural Brain Research*, 108: 215–231.

Lee, R., Jaffe, R., and Midgley, A. 1974. Lack of alteration of serum gonadotropins in men and women following sexual intercourse. *Am. J. Obstet. Gynecol.*, 120:985–987.

Lee, T. M., Carmichael, M. S., and Zucker, I. 1986. Circannual variations in circadian rhythms of ground squirrels. *Am. J. Physiol.*, 244:R607–R610.

Leedom, L. J., Meehan, W. P., and Zeidler, A. 1987. Avoidance responding in mice with diabetes mellitus. *Physiol. Behav.*, 40:447–451.

Lefebvre, L., Viville, S., Barton, S. C., Ishino, F., Keverne, E. B., and Surani, M. A. 1998. Abnormal maternal behaviour and growth retardation associated with loss of the imprinted gene Mest. *Nat. Genet.*, 20:163–169.

Lehman, M. N., and Winans, S. S. 1982. Vomeronasal and olfactory pathways to the amygdala controlling male hamster sexual behavior: Autoradiographic and behavioral analyses. *Brain Res.*, 240:27–41.

Lehman, M. N., Silver, R., Gladstone, W. R., Kahn, R. M., Gibson, M., and Bittman, E. L. 1987. Circadian rhythmicity restored by neural transplant. Immunocytochemical characterization of the graft and its integration with the host brain. *J. Neurosci.*, 7:1626–1638.

Lehmann, M. L., and Erskine, M. S. 2004. Induction of pseudopregnancy using artificial VCS: Importance of lordosis intensity and prestimulus estrous cycle length. *Hormones and Behavior*, 45: 75–83.

Lehne, G. K. 1988. Treatment of sex offenders with medroxyprogesterone acetate. In J. B. A. Sitsen (ed.), *Handbook of Sexology*, Vol. 6, *The Pharmacology and Endocrinology of Sexual Function*, pp. 516–525. Elsevier Science Publishers, Amsterdam.

Lehrman, D. S. 1961. Hormonal regulation of parental behavior in birds and infrahuman mammals. In W. C. Young (ed.), *Sex and Internal Secretions*, Vol. 2, pp. 1268–1382. Williams & Wilkins, Baltimore.

Lehrman, D. S. 1965. Interaction between internal and external environments in the regulation of the reproductive cycle of the ring dove. In F. A. Beach (ed.), *Sex and Behavior*, pp. 335–380. Wiley, New York.

Lehrman, D. S., and Brody, P. 1961. Does prolactin induce incubation behavior in the ring dove? *J. Endocrinol.*, 22:369–375.

Leinhart, R. 1927. Contribution a l'étude de l'incubation. *C. R. Soc. Biol. (Paris)*, 97:1296–1297.

Leitner, S., Voigt, C., Garcia-Seguar, L.-M., Van't Hof, T., and Gahr, M. 2001. Seasonal activation and inactivation of song motor memories in wild canaries is not reflected in neuroanatomical changes of forebrain song areas. *Hormones and Behavior*, 40: 160–168.

Leon, M. 1980. Development of olfactory attraction by young Norway rats. In D. Müller-Schwarze and R. M. Silverstein (eds.), *Chemical Signals*, pp. 193–209. Plenum, New York.

Leon, M. 1992. Neuroethology of olfactory preference development. *J. Neurobiol.*, 23:1557–1573.

Leon, M., and Moltz, H. 1972. The development of the pheromonal bond in the albino rat. *Physiol. Behav.,* 8:683–686.

Leon, M., Coopersmith, R., Beasley, L. J., and Sullivan, R. M. 1990. Thermal aspects of parenting. In N. A. Krasnegor and R. S. Bridges (eds.), *Mammalian Parenting,* pp. 400–415. Oxford University Press, Oxford.

Leon, M., Croskerry, P. G., and Smith, G. K. 1978. Thermal control of mother-young contact in rats. *Physiol. Behav.,* 21:793–811.

Lepri, J. J., and Wysocki, C. J. 1987. Removal of the vomeronasal organ disrupts the activation of reproduction in female voles. *Physiol. Behav.,* 40:349–355.

Lepri, J. J., Wysocki, C. J., and Vandenbergh, J. G. 1985. Mouse vomeronasal organ: Effects on chemosignal production and maternal behavior. *Physiol. Behav.,* 35:809–814.

LeSauter, J., Romero, P., Cascio, M., and Silver, R. 1997. Attachment site of grafted SCN influences precision of restored circadian rhythm. *J. Biol. Rhythms,* 12:327–338.

Leshner, A. I., and Collier, G. 1973. The effects of gonadectomy on the sex differences in dietary self-selection patterns and carcass compositions of rats. *Physiol. Behav.,* 11:671–676.

Leshner, A. I., and Moyer, J. A. 1975. Androgens and agonistic behavior in mice: Relevance to aggression and irrelevance to avoidance-of-attack. *Physiol. Behav.,* 15:695–699.

Leuner, B., Mendolia-Loffredo, S., and Shors, T. J. 2004. High levels of estrogen enhance associative memory formation in ovariectomized females. *Psychoneuroendocrinology,* 29: 883–890.

Levernz, J. B., Wilkinson, C. W., Wamble, M., Corbin, S., Grabber, J. E., Raskind, M. A., and Peskind, E. R. 1999. Effect of chronic high-dose exogenous cortisol on hippocampal neuronal number in aged nonhuman primates. *Journal of Neuroscience,* 19: 2356–2361.

Levine, A. S., and Morley, J. E. 1984. Neuropeptide Y: A potent inducer of consummatory behavior in rats. *Peptides,* 5:1025–1029.

Levine, S. 2002. Enduring effects of early experience on adult behavior. In D. W. Pfaff, A. P. Arnold, A. M. Etgen, S. E. Fahrbach, and R. T. Rubin (eds.), *Hormones, Brain and Behavior,* Volume 4, pp. 535–542. Academic Press, New York.

Levine, S., Haltmeyer, G. C., and Karas, G. G. 1967. Physiological and behavioral effects of infantile stimulation. *Physiol. Behav.,* 2:55–59.

Levine, S., Huchton, D. M., Wiener, S. G., and Rosenfeld, P. 1991. Time course of the effect of maternal deprivation on the hypothalamic-pituitary-adrenal axis in the infant rat. *Dev. Psychobiol.,* 24:547–558.

Levy, A. D., Baumann, M. H., and Van De Kar, L. D. 1994. Monoaminergic regulation of neuroendocrine function and its modification by cocaine. *Front. Neuroendocrinol.,* 15:85–156.

Levy, A., and Lightman, S. L. 1997. *Endocrinology.* Oxford University Press, Oxford.

Lévy, F., Porter, R. H., Kendrick, K. M., Keverne, E. B., and Romeyer, A. 1996. Physiological, sensory, and experiential factors of parental care in sheep. *Adv. Study Behav.,* 25:385–422.

Levy, J. V., and King, J. A. 1953. The effects of testosterone propionate on fighting behaviour in young male C57BL/10 mice. *Anat. Rec.,* 117:562–563.

Lewy, A. J., Ahmed, S., and Sack, R. L. 1996. Phase shifting the human circadian clock using melatonin. *Behav. Brain Res.,* 73:131–134.

Lewy, A. J., Bauer, V. K., Ahmed, S., Thomas, K. H., Cutler, N. L., Singer, C. M., Moffitt, M. T., and Sack, R. L. 1998. The human phase response curve is about 12 hours out of phase with the PRC to light. *Chronobiology International,* 15: 71–83.

Lewy, A. J., Bauer, V. K., Cutler, N. L., Sack, R. L., Ahmed, S., Thomas, K. H., Blood, J. L., and Jackson, J. M. 1998. Morning vs evening light treatment of patients with winter depression. *Arch. Gen. Psychiatry,* 55:890–896.

Lewy, A. J., Bauer, V. K., Hasler, B. P., Kendall, A. R., Pires, M. L., and Sack, R. L. 2001. Capturing the circadian rhythms of free-running blind people with 0.5 mg melatonin. *Brain Research,* 918: 96–100.

Lewy, A. J., Emens, J., Sack, R. L., Hasler, B. P., and Bernert, R. A. 2003. Zeitgeber hierarchy in humans: Resetting the circadian phase positions of blind people using melatonin. *Chronobiology International,* 20: 837–852.

Lewy, A. J., Sack, R. L., and Singer, C. M. 1985. Melatonin, light and chronobiological disorders. In D. Evered and S. Clark. (eds.), *Photoperiodism, Melatonin and the Pineal.* Ciba Foundation Symposium, 117, pp. 231–252. Pitman, London.

Lewy, A. J., Sack, R. L., Miller, S., et al. 1987. Antidepressant and circadian phase-shifting effects of light. *Science,* 235:352–354.

Lewy, A. J., Sack, R. L., Singer, C. M., White, D. M., and Hoban, T. M. 1988. Winter depression and the phase-shift hypothesis for bright light's therapeutic effects: History, theory, and experimental evidence. *J. Biol. Rhythms,* 3:121–134.

Lewy, A. J., Wehr, T. A., Goodwin, F. K., Newsome, D. A., and Markey, S. P. 1980. Light suppresses melatonin secretion in humans. *Science,* 210:1267–1269.

Li, C., Brake, W. G., Romeo, R. D., Dunlop, J. C., Gordon, M., Buzescu, R., Magarinos, A. M., Allen, P. B., Greengard, P., Luine, V. N., and McEwen, B. S. 2004. Estrogen alters hippocampal dendritic spine shape and enhances synaptic protein immunoreactivity and spatial memory in female mice. *Proceedings of the National Academy of Sciences U.S.A.,* 101: 2185–2190.

Li, M., and Fleming, A. S. 2003a. The nucleus accumbens shell is critical for normal expression of pup-retrieval in postpartum female rats. *Behavioural Brain Research,* 145: 99–111.

Li, M., and Fleming, A. S. 2003b. Differential involvement of nucleus accumbens shell and core subregions in maternal memory in postpartum female rats. *Behavioral Neuroscience,* 117: 426–445.

Liberzon, I., Trujillo, K. A., Akil, H., and Young, E. A. 1997. Motivational properties of oxytocin in the conditioned place preference paradigm. *Neuropsychopharmacology,* 17:353–359.

senger RNA is tissue specific: Differential regulation of variant transcripts by early-life events. *Molecular Endocrinology*, 14: 506–517.

McElreavey, K., Barbaux, S., Ion, A., and Fellous, M. 1995. The genetic basis of murine and human sex determination: A review. *Heredity*, 75:599–611.

McEwen, B. S. 2000. The neurobiology of stress: From serendipity to clinical relevance. *Brain Research*, 886: 172–189.

McEwen, B. S. 2001. Estrogen effects on the brain: Multiple sites and molecular mechanisms. *Journal of Applied Physiology*, 91: 2785–2801.

McEwen, B. S., and Wingfield, J. C. 2003. The concept of allostasis in biology and biomedicine. *Hormones and Behavior*, 43: 2–15.

McEwen, B. S., Gould, E., Orchinik, M., Weiland, N. G., and Woolley, C. S. 1995. Oestrogens and the structural and functional plasticity of neurons: Implications for memory, ageing and neurodegenerative processes. *Ciba Found. Symp.*, 191:52–66.

McEwen, B. S., Pfaff, D. W., Chaptal, C., and Luine, V. N. 1975. Brain cell nuclear retention of [3H]estradiol doses able to promote lordosis: Temporal and regional aspects. *Brain Res.*, 86:155–161.

McFadden, D. 2000. Masculinizing effects on otoacoustic emissions and auditory evoked potentials in women using oral contraceptives. *Hearing Research*, 142: 23–33.

McFadden, D. 1999. Intersex infants and otoacoustic emissions. *Urology*, 53: 240.

McFadden, D. 2002. Masculinizing effects in the auditory system. *Archives of Sexual Behavior*, 31: 91–111.

McGaugh, J. L. 1983. Hormonal influences on memory. *Annu. Rev. Psychol.*, 34:297–323.

McGaugh, J. L. 1989. Involvement of hormonal and neuromodulatory systems in the regulation of memory storage. *Annu. Rev. Neurosci.*, 12:255–287.

McGaugh, J. L. 2004. The amygdala modulates the consolidation of memories of emotionally arousing experiences. *Annual Review of Neuroscience*, 27: 1–28.

McGaugh, J. L., and Gold, P. E. 1976. Modulation of memory by electrical stimulation of the brain. In M. R. Rosenzweig and E. L. Bennett (eds.), *Neural Mechanisms of Learning and Memory*, pp. 549–560. MIT Press, Cambridge, MA.

McGaugh, J. L., and Roozendaal, B. 2002. Role of adrenal stress hormones in forming lasting memories in the brain. *Current Opinions in Neurobiology*, 12: 205–210.

McGaugh, J. L., Bennett, M. C., Liang, K. C., Juler, R. G., and Tam, D. 1987. Memory-enhancing effect of post-training epinephrine is not blocked by dexa-methasone. *Psychobiology*, 15:343–344.

McGee, L. C., Juhn, M., and Domm, L. V. 1928. The development of secondary sex characters by injection of bull testes. *Am. J. Physiol.*, 87:406–435.

McGill, T. E. 1962. Sexual behavior in three inbred strains of mice. *Behaviour*, 19:341–350.

McGill, T. E. 1977. Reproductive isolation, behavioral genetics, and function of sexual behavior in rodents. In J. S. Rosenblatt and B. R. Komisaruk (eds.), *Reproductive Behavior and Evolution*, pp. 73–109. Plenum Press, New York.

McGinnis, M. Y., Lumia, A. R., and Possidente, B. P. 2002. Effects of withdrawal from anabolic androgenic steroids on aggression in adult male rats. *Physiology and Behavior*, 75: 541–549.

McGinnis, M. Y., Parsons, B., Rainbow, T. C., Krey, L. C., and McEwen, B. S. 1981. Temporal relationship between cell nuclear progestin receptor levels and sexual receptivity following intravenous progesterone administration. *Brain Res.*, 218:365–371.

McGinnis, M. Y., Williams, G. W., and Lumia, A. R. 1996. Inhibition of male sex behavior by androgen receptor blockade in preoptic area or hypothalamus, but not amygdala or septum. *Physiol. Behav.*, 60:783–789.

McGowan, M. K., Andrews, K. M., Kelly, J., and Grossman, S. P. 1990. Effects of chronic intrahypothalamic infusion of insulin on food intake and diurnal meal patterning in the rat. *Behav. Neurosci.*, 104:373–385.

McGregor, G. P., Desaga, J. F., Ehlenz, K., Fischer, A., Heese, F., Hegele, A., Lammer, C., Peiser, C., and Lang, R. E. 1996. Radioimmunological measurement of leptin in plasma of obese and diabetic human subjects. *Endocrinology*, 137:1501–1504.

McKinney, T. D., and Desjardins, C. 1973. Intermale stimuli and testicular function in adult and immature house mice. *Biol. Reprod.*, 9:370–378.

McLay, R. N., Freeman, S. M., and Zadina, J. E. 1998. Chronic corticosterone impairs memory performance in the Barnes maze. *Physiol. Behav.*, 63:933–937.

McNeilly, A. S., Sharpe, R. M., and Fraser, H. M. 1983. Increased sensitivity to the negative feedback effect of testosterone induced by hyperprolactinemia in the adult male rat. *Endocrinology*, 112:22–28.

McShea, W. J. 1990. Social tolerance and proximate mechanisms of dispersal among winter groups of meadow voles (*Microtus pennsylvanicus*). *Anim. Behav.*, 39:346–351.

Meijer, J. H., and Rietveld, W. J. 1989. Neurophysiology of the suprachiasmatic circadian pacemaker in rodents. *Physiol. Rev.*, 69:671–707.

Meisel, R. L., and Sachs, B. D. 1994. The physiology of male sexual behavior. In E. Knobil and J. D. Neill (eds.), *The Physiology of Reproduction*, Vol. 2, pp. 3–105. Raven Press, New York.

Meisel, R. L., Lumia, A. R., and Sachs, B. D. 1980. Effects of olfactory bulb removal and flank shock on copulation in male rats. *Physiol. Behav.*, 25:383–387.

Meisel, R. L., O'Hanlon, J. K., and Sachs, B. D. 1984. Differential maintenance of penile responses and copulatory behavior by gonadal hormones in castrated male rats. *Horm. Behav.*, 18:56–64.

Meisel, R., and Pfaff, D. W. 1984. RNA and protein synthesis inhibitors: Effects on sexual behavior in female rats. *Brain Res. Bull.*, 12:187–193.

Melis, M. R., and Argiolas, A. 1995. Dopamine and sexual behavior. *Neurosci. Biobehav. Rev.*19:19–38.

Melloni, R. H., Connor, D. F., Hang, P. T., Harrison, R. J., and Ferris, C. F. 1997. Anabolic-androgenic steroid exposure during adolescence and aggressive behavior in golden hamsters. *Physiol. Behav.*, 61:359–364.

Melnyk, R. B., Mrosovsky, N., and Martin, J. M. 1983. Spontaneous obesity and weight loss: insulin action in the dormouse. *Am. J. Physiol.*, 245:R396–402.

Menaker, M. 1968. Extraretinal light perception in the sparrow. I. Entrainment of the biological clock. *Proc. Natl. Acad. Sci. U.S.A.*, 59: 414–21.

Mendelson, J. H., and Mello, N. K. 1982. Hormones and psycho-sexual development in young men following chronic heroin use. *Neurobehav. Toxicol. Teratol.*, 4:441–445.

Mendonca, M. T., and Licht, P. 1986. Seasonal cycles in gonadal activity and plasma gonadotropin in the musk turtle, *Sternotherus odoratus*. *Gen. Comp. Endocrinol.*, 62:459–469.

Mendoza, S. P., and Mason, W. A. 1986. Parental division of labour and differentiation of attachments in a monogamous primate (*Callicebus moloch*). *Anim. Behav.*, 34:1336–1347.

Mendoza, S. P., and Mason, W. A. 1997. Attachment relationships in New World primates. *Ann. N. Y. Acad. Sci.*,807:203–209.

Mendoza, S. P., Lowe, E. L., and Resko, J. A. 1978. Seasonal variations in gonadal hormones and social behavior in squirrel monkeys. *Physiol. Behav.*, 20:515–522.

Mendoza, S. P., Lyons, D. M., and Saltzman, W. 1991. Sociophysiology of squirrel monkeys. *Am. J. Primatol.*, 23:37–54.

Mercer, J. G., Hoggard, N., Williams, L. M., Lawrence, C. B., Hannah, L. T., and Trayhurn, P. 1996. Localization of leptin receptor mRNA and the long form splice variant (Ob-Rb) in mouse hypothalamus and adjacent brain regions by in situ hybridization. *FEBS Lett.*, 387:113–116.

Messager, S., Hazlerigg, D. G., Mercer, J. G., and Morgan, P. J. 2000. Photoperiod differentially regulates the expression of *Per1* and ICER in the pars tuberalis and the suprachiasmatic nucleus of the Siberian hamster. *European Journal of Neuroscience*, 12: 2865–2870.

Messager, S., Ross, A. W., Barrett, P., and Morgan, P. J. 1999. Decoding photoperiodic time through *Per1* and ICER gene amplitude. *Proceedings of the National Academy of Sciences U.S.A.*, 96: 9938–9943.

Messier, C., and White, N. M. 1987. Memory improvement by glucose, fructose, and two glucose analogs: A possible effect on peripheral glucose transport. *Behav. Neural Biol.*, 48:104–127.

Meston, C. M., Hull, E., Levin, R. J., and Sipski, M. 2004. Disorders of orgasm in women. *Journal of Sexual Medicine*, 1: 66–68.

Mewis, C., Spyridopoulos, I., Kuhlkamp, V., and Seipel, L. 1996. Manifestation of severe coronary heart disease after anabolic drug abuse. *Clin. Cardiol.*, 19:153–155.

Meyer, W. J., Walker, P. A., Emory, L. E., and Smith, E. R. 1985. Physical, metabolic, and hormonal effects on men of long-term therapy with medroxyprogesterone acetate. *Fertil. Steril.*, 43:102–106.

Meyer-Bernstein, E. L., Jetton, A. E., Matsumoto, S. I., Markuns, J. F., Lehman, M. N., and Bittman, E. L. 1999. Effects of suprachiasmatic transplants on circadian rhythms of neuroendocrine function in golden hamsters. *Endocrinology*, 140: 207–218.

Miceli, M. O., and Malsbury, C. W. 1982a. Availability of a food hoard facilitates maternal behaviour in virgin female hamsters. *Physiol. Behav.*, 28:855–856.

Miceli, M. O., and Malsbury, C. W. 1982b. Sagittal knife cuts in the near and far lateral preoptic area–hypothalamus disrupt maternal behavior in female hamsters. *Physiol. Behav.*, 28:857–867.

Micevych, P. E., Eckersell, C. B., Brecha, N., and Holland, K. L. 1997. Estrogen modulation of opioid and cholecystokinin systems in the limbic-hypothalamic circuit. *Brain Res. Bull.*, 44:335–343.

Michael, R. P. 1972. Determinants of primate reproductive behaviour. In E. Diczfalusy and C. C. Standley (eds.), *The Use of Non-human Primates in Research on Human Reproduction*. WHO Symposium, Karolinska Institute, Stockholm.

Michael, R. P., and Zumpe, D. 1978. Annual cycles of aggression and plasma testosterone in captive male rhesus monkeys. *Psychoneuroendocrinology*, 3:217–220.

Michael, R. P., and Zumpe, D. 1986. An annual rhythm in the battering of women. *American Journal of Psychiatry*, 143: 637–640.

Michael, R. P., and Zumpe, D. 1996. Biological factors in the organization and expression of sexual behaviour. In I. Rosen (ed.), *Sexual Deviation*, pp. 452–287. Oxford University Press, Oxford.

Michael, R. P., Clancy, A. N., and Zumpe, D. 1995. Distribution of androgen receptor-like immunoreactivity in the brains of Cynomolgus monkeys. *J. Neuroendocrinol.*, 7:713–719.

Michael, R. P., Zumpe, D., and Bonsall, R. W. 1992. The interaction of testosterone with the brain of the orchidectomized primate fetus. *Brain Res.*, 570:68–74.

Michael, R. P., Zumpe, D., Kerverne, E. B., and Bonsall, R. W. 1972. Neuroendocrine factors in the control of primate behavior. *Rec. Prog. Horm. Res.*, 28:665–706.

Michel, G. F., and Moore, C. L. 1995. *Developmental Psychobiology: An Interdisciplinary Science*. MIT Press: Cambridge, MA.

Michiels, N. K., and Newman, L. J. 1998. Sex and violence in hermaphrodites. *Nature*, 391:647.

Miczek, K. A., and Fish, E. W. 2005. Monoamines, GABA, glutamate and aggression. In R. J. Nelson (ed.), *Biology of Aggression*. Oxford University Press, New York. In press.

Middleman, A. B. and DuRant, R. H. 1996. Anabolic steroid use and associated health risk behaviours. *Sports Med.*, 21:251–225.

Miernicki, M., Pospichal, M., Karg, J., and Powers, J. B. 1988. Photoperiodic effects on male sexual behavior. *Conference on Reproductive Behavior*, pp. 64, Omaha, NE.

Migeon, C. J., and Wisniewski, A. B. 1998. Sexual differentiation: from genes to gender. *Horm. Res.*, 50:245–251.

Migeon, C. J., Wisniewski, A. B., Gearhart, J. P., Meyer-Bahlburg, H. F., Rock, J. A., Brown, T. R., Casella, S. J., Maret, A., Ngai, K. M., Money, J., and Berkovitz, G. D. 2002. Ambiguous genitalia with perineoscrotal hypospadias in 46,XY individuals: Long-term medical, surgical, and psychosexual outcome. *Pediatrics*, 110: e31.

Mikhelashvili-Browner, N., Yousem, D. M., Wu, C., Kraut, M. A., Vaughan, C. L., Oguz, K. K., and Calhoun, V. D.

2003. Lack of sex effect on brain activity during a visuomotor response task: Functional MR imaging study. *American Journal of Neuroradiology*, 24: 488–494.

Miles, C., Green, R., Sanders, G., and Hines, M. 1998. Estrogen and memory in a transsexual population. *Horm. Behav.*, 34:199–208.

Milette, J. J., and Turek, F. W. 1986. Circadian and photoperiodic effects of brief light pulses in male Djungarian hamsters. *Biol. Reprod.*, 35:327–335.

Miller, W. L., Baxter, J. D., and Eberhardt, N. L. 1983. Peptide hormone genes: Structure and evolution. In D. T. Krieger, M. J. Brownstein, and J. B. Martin (eds.), *Brain Peptides*, pp. 16–78. Wiley-Interscience, New York.

Mills, M. J., and Stunkard, A. J. 1976. Behavioral changes following surgery for obesity. *Am. J. Psychiatry*, 133:527–531.

Milner, P., and Zucker, I. 1965. Specific hunger for potassium in the rat. *Psychonom. Sci.*, 2:17–18.

Mirescu, C., Peters, J. D., and Gould, E. 2004. Early life experience alters response of adult neurogenesis to stress. *Nature Neuroscience*, 7: 841–846.

Mistlberger, R. E., Marchant, E. G., and Sinclair, S. V. 1996. Nonphotic phase-shifting and the motivation to run: Cold exposure reexamined. *Journal of Biological Rhythms*, 11: 208–215.

Mitani, J. C. 1985. Mating behavior of male orangutans in the Kutai Game Reserve, Indonesia. *Anim. Behav.*, 33:392–402.

Mitchell, A. J. 1998. The role of corticotropin releasing factor in depressive illness: A critical review. *Neurosci. Biobehav. Rev.*, 22:635–651.

Mitchell, J. S., and Keesey, R. E. 1974. The effects of lateral hypothalamic lesions and castration upon the body weight and composition of male rats. *Behav. Biol.*, 11:69–82.

Mitra, S. W., Hoskin, E., Yudkovitz, J., Pear, L., Wilkinson, H. A., Hayashi, S., Pfaff, D. W., et al. 2003. Immunolocalization of estrogen receptor beta in the mouse brain: Comparison with estrogen receptor alpha. *Endocrinology*, 144: 2055–2067.

Moberg, G. P. 1991. How behavioral stress disrupts the endocrine control of reproduction in domestic animals. *J. Dairy Sci.*, 74:304–311.

Modney, B. K., and Hatton, G. I. 1990. Motherhood modifies magnocellular neuronal interrelationships in functionally meaningful ways. In N. A. Krasnegor and R. S. Bridges (eds.), *Mammalian Parenting*, pp. 305–323. Oxford University Press, Oxford.

Moffatt, C. A. 1994. Seasonal and Social Regulation of Reproductive Behavior in Female Prairie Voles, *Microtus ochrogaster*. Dissertation, Johns Hopkins University, Baltimore, MD.

Moffatt, C. A., and Nelson, R. J. 1993. Day length influences proceptive behavior of female prairie voles (*Microtus ochrogaster*). *Physiol. Behav.*, 55:1163–1165.

Moffatt, C. A., DeVries, A. C., and Nelson, R. J. 1993. Winter adaptations of male deer mice (*Peromyscus maniculatus*) and prairie voles (*Microtus ochrogaster*) that vary in reproductive responsiveness to photoperiod. *J. Biol. Rhythms*, 8:221–232.

Mogenson, G. J., Jones, D. L., and Yim, C. Y. 1980. From motivation to action: Functional interface between the limbic system and the motor system. *Prog. Neurobiol.*, 14:69–97.

Moltz, H., and Kilpatrick, S. J. 1978. Response to the maternal pheromone in the rat as protection against necrotizing enterocolitis. *Neurosci. Biobehav. Rev.*, 2:277–280.

Moltz, H., Lubin, M., Leon, M., and Numan, M. 1970. Hormonal induction of maternal behavior in the ovariectomized nulliparous rat. *Physiol. Behav.*, 5:1373–1377.

Monder, C., Hardy, M. P., Blanchard, R. J., and Blanchard, D.C. 1994. Comparative aspects of 11-hydroxysteroid dehydrogenase. Testicular 11-hydroxysteroid dehydrogenase: Development of a model for the mediation of Leydig cell function by corticosteroids. *Steroids*, 59:69–73.

Money, J. 1961. Sex hormones and other variables in human eroticism. In W. C. Young (ed.), *Sex and Internal Secretions*, pp. 1383–1400. Williams & Wilkins, Baltimore.

Money, J. 1988. The ethics of pornography in the era of AIDS. *J. Sex Marital Ther.*, 14:177–183.

Money, J., and Bennett, R. G. 1981. Postadolescent paraphilic sex offenders: Antiandrogenic and counseling therapy follow-up. *Int. J. Ment. Health*, 10:122–133.

Monk, T. H. 1980. Traffic accident increases as a possible indicant of desynchronosis. *Chronobiologia*, 7:527–529.

Monk, T. H., and Aplin, L. C. 1980. Spring and autumn Daylight Savings Time changes: Studies of adjustment in sleep timings, mood, and efficiency. *Ergonomics*, 23:167–178.

Montague, C. T., Farooqi, I. S., Whitehead, J. P., Soos, M. A., Rau, H., Wareham, N. J., Sewter, C. P., Digby, J. E., Mohammed, S. N., Hurst, J. A., Cheetham, C. H., Earley, A. R., Barnett, A. H., Prins, J. B., and O'Rahilly, S. 1997. Congenital leptin deficiency is associated with severe early-onset obesity in humans. *Nature*, 387:903–908.

Monti, P. M., Brown, W. A., and Corriveau, D. P. 1977. Testosterone and components of aggressive and sexual behavior in man. *Am. J. Psychiatry*, 134:692–694.

Mook, D. G., Kenney, N. J., Roberts, S., Nussbaum, A. I., and Rodier, W. I. 1972. Ovarian-adrenal interactions in regulation of body weight in female rats. *J. Comp. Physiol. Psychol.*, 81:198–211.

Mooney, R. 1999. Sensitive periods and circuits for learned birdsong. *Current Opinions in Neurobiology*, 9: 121–127.

Moore, C. L. 1984. Maternal contributions to the development of masculine sexual behavior in laboratory rats. *Dev. Psychobiol.*, 17:347–356.

Moore, C. L. 1986. Interaction of species-typical environmental and hormonal factors in sexual differentiation of behavior. *Ann. N. Y. Acad. Sci.*, 474:108–119.

Moore, C. L., and Morelli, G. A. 1979. Mother rats interact differently with male and female offspring. *J. Comp. Physiol. Psychol.*, 93:677–684.

Moore, I. T., Walker, B. G., and Wingfield, J. C. 2004. The effects of combined aromatase inhibitor and anti-androgen on male territorial aggression in a tropical population of rufous-collared sparrows, *Zonotrichia capensis. General and Comparative Endocrinology*, 135: 223–229.

Moore, M. C. 1984. Changes in territorial defense produced by changes in circulating levels of testosterone: A possible hormonal basis for mate-guarding in white-crowned sparrows. *Behaviour*, 88:215–226.

Moore, M. C. 1987. Castration affects territorial and sexual behavior of free-living male lizards, *Sceloporus jarrovi. Anim. Behav.*, 35:1193–1199.

Moore, M. C. 1991. Application of organization-activation theory to alternative male reproductive strategies: A review. *Horm. Behav.*, 25:154–179.

Moore, M. C., and Marler, C. A. 1987. Effects of testosterone manipulations on nonbreeding season territorial aggression in free-living male lizards, *Sceloporus jarrovi. Gen. Comp. Endocrinol.*, 65:225–232.

Moore, M. C., Hews, D. K., and Knapp, R. 1998. Hormonal control and evolution of alternative male phenotypes: Generalizations of models for sexual differentiation. *Am. Zool.*, 38:133–151.

Moore, R. M., and Eichler, V. B. 1972. Loss of a circadian adrenal corticosterone rhythm following suprachiasmatic lesions in the rat. *Brain Res.*, 42:201–206.

Moore, R. Y. 1997. Circadian rhythms: Basic neurobiology and clinical applications. *Annu. Rev. Med.*, 48:253–266.

Moore, W. V. 1988. Anabolic steroid use in adolescence. *J.A.M.A.*, 260:3484–3486.

Moore-Ede, M. C., Sulzman, F. M., and Fuller, C. A. 1982. *The Clocks That Time Us.* Harvard University Press, Cambridge.

Morali, G., Asuncion, M., Soto, P., Contreras, J. L., Arteaga, M., Gonzalez-Vidal, D. M., and Beyer, C. 2003. Detailed analysis of the male copulatory motor pattern in mammals: Hormonal bases. *Scandinavian Journal of Psychology*, 44: 279–288.

Moran, T. H., Ameglio, P. J., Schwartz, G. J., and McHugh, P. R. 1992. Blockade of type A, not type B, CCK receptors attenuates satiety actions of exogenous and endogenous CCK. *Am. J. Physiol.*, 262:R46–R50.

Moran, T. H., and McHugh, P. R. 1979. Cholecystokinin: Gastric emptying and feeding in *Macaca mulatta. Fed. Proc.*, 38:1131.

Morin, L. P. 1988. Propylthiouracil, but not other antithyroid treatments, lengthens hamster circadian period. *Am. J. Physiol.*, 255:R1–R5.

Morin, L. P., and Cummings, L. A. 1981. Effect of surgical or photoperiodic castration, testosterone replacement or pinealectomy on male hamster running rhythmicity. *Physiol. Behav.*, 26:825–38.

Morin, L. P., and Zucker, I. 1978. Photoperiodic regulation of copulatory behavior in the male hamster. *J. Endocrinol.*, 77:249–258.

Morin, L. P., Fitzgerald, K. M., and Zucker, I. 1977. Estradiol shortens the period of hamster circadian rhythms. *Science*, 196:305–307.

Morin, L. P., Gavin, M. L., and Ottenweller, J. E. 1986. Propylthiouracil causes phase delays and circadian period lengthening in male and female hamsters. *Am. J. Physiol.*, 250:R151–R160.

Morin, M. P., DeMarchi, P., Champagnat, J., Vanderhaeghen, J. J., Rossier, J., and Denavit-Staubie, M. 1983. Inhibitory effect of cholecystokinin octapeptide on neurons in the nucleus tractus solitarius. *Brain Res.*, 265:333–338.

Morley, J. E., Bartness, T. J., Gosnell, B. A., and Levine, A. S. 1985a. Peptidergic regulation of feeding. *Int. Rev. Neurobiol.*, 27:207–298.

Morley, J. E., Levine, A. S., Bartness, T. J., Nizielski, S. E., Shaw, M. J., and Hughes, J. J. 1985b. Species differences in the response to cholecystokinin. *Ann. N. Y. Acad. Sci.*, 448:413–416.

Morley, J. E., Levine, A. S., Yim, G. K. W., and Lowy, M. T. 1983. Opioid modulation of appetite. *Neurosci. Biobehav. Rev.*, 7:281–305.

Morokoff, P. J., Baum, A., McKinnon, W. R., and Gillilland, R. 1987. Effects of chronic unemployment and acute psychological stress on sexual arousal in men. *Health Psychol.*, 6:545–60.

Morrell, J. I., and Pfaff, D. W. 1978. A neuroendocrine approach to brain function: Localization of sex-steroid concentrating cells in vertebrate brains. *Am. Zool.*, 18:447–460.

Morris, R. G. M., Garrud, P., Rawlins, JNP., and O'Keefe, J. 1982. Place navigation impaired in rats with hippocampal lesions. *Nature*, 297:681–683.

Morris, R., and Nostren-Bertrand, M. 1996. NOS and aggression. *Trends Neurosci.*, 19:277–278.

Mortola, J. F. 1997. From GnRH to SSRIs and beyond: Weighing the options for drug therapy in premenstrual syndrome. *Medscape Womens Health*, 2:3–8.

Moss, H. B., Panzak, G. L., and Tarter, R. E. 1993. Sexual functioning of male anabolic steroid abusers. *Arch. Sex. Behav.*, 22;1–12.

Moss, R. L., and McCann, S. M. 1973. Induction of mating behavior in rats by luteinizing hormone-releasing factor. *Science*, 181:177–179.

Moss, R. L., McCann, S. M., and Dudley, C. A. 1975. Releasing hormones and sexual behavior. *Prog. Brain Res.*, 42:37–46.

Moulton, D. G. 1967. Olfaction in mammals. *Am. Zool.*, 7:421–429.

Mounzih, K., Lu, R., and Chehab, F. F. 1997. Leptin treatment rescues the sterility of genetically obese *ob/ob* males. *Endocrinology*, 138:1190–1193.

Moyer, K. E. 1968. Kinds of aggression and their physiological basis. *Commun. Behav. Biol.*, 2A:65–87.

Moyer, K. E. 1971. *The Physiology of Hostility.* Markham, Chicago.

Moyer, K. E. 1976. *The Psychobiology of Aggression.* Harper and Row, New York.

Mpitos, G. J., and Davis, W. J. 1973. Learning: Classical and avoidance conditioning in the mollusk, *Pleurobranchea. Science*, 180:317–320.

Mrosovsky, N. 1985. Cyclical obesity in hibernators: The search for the adjustable regulator. In J. Hirsch and T. B. Van Itallie (eds.), *Recent Advances in Obesity Research: IV*, pp. 45–56. J. Libbey, London.

Mrosovsky, N. 1988. Phase response curves for social entrainment. *Journal of Comparative Physiology A*, 162: 35–46.

hamster (*Mesocricetus auratus* Waterhouse). *Physiol. Behav.*, 8:687–691.

Pearlstein, T., Rosen, K., and Stone, A. B. 1997. Mood disorders and menopause. *Endocrinol. Metab. Clin. North Am.*, 26:279–294.

Pecoraro, N., Reyes, F., Gomez, F., Bhargava, A., and Dallman, M. F. 2004. Chronic stress promotes palatable feeding, which reduces signs of stress: Feedforward and feedback effects of chronic stress. *Endocrinology*, 145: 3754–3762.

Pedersen, C. A., and Prange, A. J. 1979. Induction of maternal behavior in virgin rats after intracerebroventricular administration of oxytocin. *Proc. Natl. Acad. Sci. U.S.A.*, 76:6661–6665.

Pedersen, C. A., Ascher, J. A., Monroe, Y. L., and Prange, A. J. 1982. Oxytocin induces maternal behavior in virgin female rats. *Science*, 216:648–649.

Pedersen, P. E., and Blass, E. M. 1982. Prenatal and postnatal determinants of the first suckling episode in albino rats. *Dev. Psychobiol.*, 15:349–355.

Pederson, C. A. 1997. Oxytocin control of maternal behavior: Regulation by sex steroids and offspring stimuli. *Ann. N. Y. Acad. Sci.*, 807:126–145.

Peele, D. B., and Vincent, A. 1989. Strategies for assessing learning and memory, 1978–1987: A comparison of behavioral toxicology, psychopharmacology, and neurobiology. *Neurosci. Biobehav. Rev.*, 13:33–38.

Pelletier, J., and Ortavant, R. 1975. Photoperiodic control of LH release in the ram. *Acta Endocrinol.*, 78:442–450.

Perachio, A. A. 1978. Hypothalamic regulation of behavioural and hormonal aspects of aggression and sexual performance. In D. J. Chivers and J. Herbert (eds.), *Recent Advances in Primatology*, Vol. 1, pp. 549–565. Academic Press, New York.

Perachio, A. A., Alexander, M., and Marr, L. D. 1973. Hormonal and social factors affecting evoked sexual behavior in rhesus monkeys. *Am. J. Phys. Anthropol.*, 38:227–232.

Perachio, A. A., Marr, L. D., and Alexander, M. 1979. Sexual behavior in male rhesus monkeys elicited by electrical stimulation of preoptic and hypothalamic areas. *Brain Res.*, 177:127–144.

Perkins, M. S., Perkins, M. N., and Hitt, J. C. 1980. Effects of stimulus female on sexual behavior of male rats given olfactory tubercle and corticomedial amygdaloid lesions. *Physiol. Behav.*, 25:495–500.

Peroulakis, M. E., Goldman, B. D., and Forger, N. G. 2002. Perineal muscles and motoneurons are sexually monomorphic in the naked mole-rat (*Heterocephalus glaber*). *Journal of Neurobiology*, 51: 33–42.

Perras, B., Droste, C., Born, J., Fehm, H. L., and Pietrowsky, R. 1997. Verbal memory after three months of intranasal vasopressin in healthy old humans. *Psychoneuroendocrinology*, 22:387–396.

Perry, A. N., and Grober, M. S. 2003. A model for social control of sex change: Interactions of behavior, neuropeptides, glucocorticoids, and sex steroids. *Hormones and Behavior*, 43: 31–38.

Perry, P. J., Andersen, K. H., and Yates, W. R. 1990. Illicit anabolic steroid use in athletes. A case series analysis. *Am. J. Sports Med.*, 18:422–428.

Persengiev, S., Marinova, C., and Patchev, V. 1991a. Steroid hormone receptors in the thymus: A site of immunomodulatory action of melatonin. *International Journal of Biochemistry*, 23: 1483–1485.

Persengiev, S., Patchev, V., and Velev, B. 1991b. Melatonin effects on thymus steroid receptors in the course of primary antibody responses: Significance of circulating glucocorticoid levels. *International Journal of Biochemistry*, 23: 1487–1489.

Persky, H. Dreisbach, L., Miller, W. R., O'Brien, C. P., Khan, M. A., Lief, H. I., Charney, N., and Straus, D. 1982. The relation of plasma androgen levels to sexual behaviors and attitudes of women. *Psychosom. Med.*, 44:305–319.

Persky, H., O'Brien, C. P., Fine, E., Howard, W. J., Khan, M. A., and Beck, R. W. 1977. The effect of alcohol and smoking on testosterone function and aggression in chronic alcoholics. *Am. J. Psychiatry*, 134:621–625.

Petitti, D. B., and Porterfield, D. 1992. Worldwide variations in the lifetime probability of reproductive cancer in women: Implications of best-case, worst-case, and likely-case assumptions about the effect of oral contraceptive use. *Contraception*, 45: 93–104.

Petrie, M., Schwabl, H., Brande-Lavridsen, N., and Burke, T. 2001. Maternal investment: Sex differences in avian yolk hormone levels. *Nature*, 412: 498.

Pettersson, F., Fries, H., and Nillius, S. J. 1973. Epidemiology of secondary amenorrhea. I. Incidence and prevalence rates. *Am. J. Obstet. Gynecol.*, 117:80–86.

Pettit, H. O., and Justice, J. B. 1991. Effect of dose on cocaine self-administration and dopamine levels in the nucleus accumbens. *Brain Res.*, 539:94–102.

Pfaff, D. W. 1980. *Estrogens and Brain Function.* Springer-Verlag, New York.

Pfaff, D. W., and Conrad, L. C. A. 1978. Hypothalamic neuroanatomy: Steroid hormone binding and patterns of axonal projections. In G. Bourne (ed.), *International Review of Cytology*, Vol. 54, pp. 245–265. Academic Press, New York.

Pfaff, D. W., and Pfaffmann, C. 1969. Olfactory and hormonal influences on the basal forebrain of the male rat. *Brain Res.*, 15:137–156.

Pfaff, D. W., and Schwartz-Giblin, S. 1988. Cellular mechanisms of female reproductive behaviors. In E. Knobil and J. Neill (eds.), *The Physiology of Reproduction*, pp. 1487–1568. Raven Press, New York.

Pfaff, D. W., Diakow, C., Montgomery, M., Jenkins, F. A. 1978. X-ray cinematographic analysis of lordosis in female rats. *J. Comp. Physiol. Psychol.*, 92:937–941.

Pfaff, D. W., Schwartz-Giblin, S., McCarthy, M. M., and Kow, L.-M. 1994. Cellular and molecular mechanisms of female reproductive behaviors. In E. Knobil and J. Neill (eds.), *The Physiology of Reproduction.* Vol. 2, pp. 107–220. Raven Press, New York.

Pfaff, D. W., Vasudevan, N., Kia, H. K., Zhu, Y. S., Chan, J., Garey, J., Morgan, M., and Ogawa, S. 2000. Estrogens, brain and behavior: Studies in fundamental neurobiology and observations related to women's health. *Journal of Steroid Biochemistry and Molecular Biology*, 74: 365–373.

Pfaus, J. G. 1996. Homologies of animal and human sexual behaviors. *Horm. Behav.*, 30: 187–200.

Pfaus, J. G., and Gorzalka, B. B. 1987. Opioids and sexual behavior. *Neurosci. Biobehav. Rev.* 11:1–34.

Pfaus, J. G., and Heeb, M. M. 1997. Implications of immediate-early gene induction in the brain following sexual stimulation of female and male rodents. *Brain Res. Bull.*, 44:397–407.

Pfaus, J. G., and Phillips, A. G. 1991. Role of dopamine in anticipatory and consummatory aspects of sexual behavior in the male rat. *Behav. Neurosci.*, 105:727–743.

Pfaus, J. G., Kippin, T. E., and Coria-Avila, G. 2003. What can animal models tell us about human sexual response? *Annual Review of Sex Research*, 14: 1–63.

Pfeiffer, C. A. 1935. Origin of functional differences between male and female hypophyses. *Proc. Soc. Exp. Biol. Med.*, 32:603–605.

Pfeiffer, C. A. 1936. Sexual differences of the hypophysis and their determination by the gonads. *Am. J. Anat.*, 58:195–225.

Phoenix, C. H. 1973. Sexual behavior in rhesus monkeys after vasectomy. *Science*, 179:493–494.

Phoenix, C. H. 1974. Effect of dihydrotestosterone on sexual behavior of castrated male rhesus monkeys. *Physiol. Behav.*, 12:1045–1055.

Phoenix, C. H., Goy, R. W., and Resko, J. A. 1968. Psychosexual differentiation as a function of androgenic stimulation. In M. Diamond (ed.), *Perspectives in Reproduction and Sexual Behavior*, pp. 215–246. Indiana University Press, Bloomington, IN.

Phoenix, C. H., Goy, R. W., Gerall, A. A., and Young, W. C. 1959. Organizing action of prenatally administered testosterone propionate on the tissues mediating mating behavior in the female guinea pig. *Endocrinology*, 65:369–382.

Phoenix, C. H., Slob, A. K., and Goy, R. W. 1973. Effects of castration and replacement therapy on sexual behavior of adult male rhesuses. *J. Comp. Physiol. Psychol.*, 81:472–481.

Piazza, P. V., Deminiere, J. M., Maccari, S., Mormede, P., Le Moal, M., and Simon, H. 1990. Individual reactivity to novelty predicts probability of amphetamine self-administration. *Behavioral Pharmacology*, 1: 339–345.

Piazza, P. V., Deroche, V., Deminiere, J. M., Maccari, S., Le Moal, M., and Simon, H. 1993. Corticosterone in the range of stress-induced levels possesses reinforcing properties: implications for sensation-seeking behaviors. *Proc. Natl. Acad. Sci. U. S. A.*, 90: 11738–11742.

Piazza, P. V., Maccari, S., Deminiere, J. M., Le Moal, M., Mormede, P., and Simon, H. 1991. Corticosterone levels determine individual vulnerability to amphetamine self-administration. *Neurobiology*, 88:2088–2092.

Pickard, G. E. 1982. The afferent connections of the suprachiasmatic nucleus of the golden hamster with emphasis on the retinohypothalamic projection. *J. Comp. Neurol.*, 211:65–83.

Pickard, G. E., and Silverman, A. J. 1981. Direct retinal projections to the hypothalamus, piriform cortex and accessory optic nuclei in the golden hamster as demonstrated by a sensitive anterograde horseradish peroxidase technique. *J. Comp. Neurol.*, 196:155–172.

Pierce, J. G. 1988. Gonadotropins: Chemistry and biosynthesis. In E. Knobil and J. D. Neill (eds.), *The Physiology of Reproduction*, pp. 1335–1348. Raven Press, New York.

Pike, T. W., and Petrie, M. 2003. Potential mechanisms of avian sex manipulation. *Biological Reviews*,78: 553–574.

Pinckard, K. L., Stellflug, J., Resko, J. A., Roselli, C. E., and Stormshak, F. 2000. Review: Brain aromatization and other factors affecting male reproductive behavior with emphasis on the sexual orientation of rams. *Domesticated Animal Endocrinology*, 18: 83–96.

Pinxten, R., De Ridder, E., De Cock, M., and Eens, M. 2003. Castration does not decrease nonreproductive aggression in yearling male European starlings (*Sturnus vulgaris*). *Hormones and Behavior*, 43: 394–401.

Pittendrigh, C. S. 1981. Circadian systems: Entrainment. *Handbook of Behavioral Neurobiology*, Vol. 4, pp. 95–124.

Pittendrigh, C. S., and Daan, S. 1976. A functional analysis of circadian pacemakers in nocturnal rodents. IV. Entrainment: Pacemaker as clock. *J. Comp. Physiol.*, 106:291–331.

Platt, J. R. 1964. Strong inference. *Science*, 146:347–353.

Plautz, J. D., Kaneko, M., Hall, J. C., and Kay, S. A. 1997. Independent photoreceptive circadian clocks throughout *Drosophila. Science*, 278:1632–1635.

Pliner, P., and Fleming, A. S. 1983. Food intake, body weight, and sweetness preferences over the menstrual cycle in humans. *Physiol. Behav.*, 30:663–666.

Plotka, E. D., Seal, U. S., Letellier, M. A., Verme, L. J., and Ozoga, J. J. 1984. Early effects of pinealectomy on LH and testosterone secretion in white-tailed deer. *J. Endocrinol.*, 103:1–7.

Plotsky, P. M., Owens, M. J., and Nemeroff, C. B. 1998. Psychoneuroendocrinology of depression. Hypothalamic-pituitary-adrenal axis. *Psychiatr. Clin. North Am.*, 21:293–307.

Poindron, P., and Le Neindre, P. 1980. Endocrine and sensory regulation of maternal behavior in the ewe. *Adv. Study Behav.*, 11:75–119.

Poindron, P., and Levy, F. 1990. Physiological, sensory and experiential determinants of maternal behavior in sheep. In N. A. Krasnegor and R. S. Bridges (eds.). *Mammalian Parenting: Biochemical, Neurobiological, and Behavioral Determinants*, pp. 133–156. Oxford University Press, New York.

Polak, J. M., and Van Noorden, S. 1997. *Introduction to Immunocytochemistry*. Microscopy Handbook Series, 37, Springer-Verlag: New York. 144 pp.

Polo-Kantola, P., Portin, R., Polo, O., Helenius, H., Irjala, K., and Erkkola, R. 1998. The effect of short-term estrogen replacement therapy on cognition: A randomized double-blind, cross-over trial in postmenopausal women. *Obstet. Gynecol.*, 91:459–466.

Pomerantz, S. M. 1990. Apomorphine facilitates male sexual behavior of rhesus monkeys. *Pharmacol. Biochem. Behav.*, 35:659–664.

Pomerantz, S. M., and Goy, R. W. 1983. Proceptive behavior of female rhesus monkeys during tests with tethered males. *Horm. Behav.*, 17:237–248.

Pope, H. G., and Katz, D. L. 1989. Homicide and near-homicide by anabolic steroid users. *J. Clin. Psychiatry*, 51:28–31.

Pope, H. G., Kouri, E. M., and Hudson, J. I. 2000. Effects of supraphysiologic doses of testosterone on mood and aggression in normal men: A randomized controlled trial. *Archives of General Psychiatry*, 57: 133–140.

Postolache, T. T., Wehr, T. A., Doty, R. L., Sher, L., Turner, E. H., Bartko, J. J., and Rosenthal, N. E. 2002. Patients with seasonal affective disorder have lower odor detection thresholds than control subjects. *Archives in General Psychiatry*, 59: 1119–1122.

Power, K. L., and Moore, C. L. 1986. Prenatal stress eliminates differential maternal attention to male offspring in Norway rats. *Physiol. Behav.*, 38:667–671.

Powers, J. B., and Winans, S. S. 1975. Vomeronasal organ: Critical role in mediating sexual behavior of the male hamster. *Science*, 187:961–963.

Powers, J. B., Steel, E. A., Hutchison, J. B., Hastings, M. H., Herbert, J., and Walker, A. P. 1989. Photoperiodic influences on sexual behavior in male Syrian hamsters. *J. Biol. Rhythms*, 4:61–78.

Prat, J., Gray, G. F., Stolley, P. D., and Coleman, W. 1977. Wilms tumor in an adult associated with androgen abuse. *Journal of the American Medical Association*, 237: 2322–2323.

Pravosudov, V. V. 2003. Long-term moderate elevation of corticosterone facilitates avian food-caching behaviour and enhances spatial memory. *Proceedings of the Royal Society of London, Series B*, 270: 2599–2604.

Pravosudov, V. V. 2005. Corticosterone and memory in birds. In A. Dawson and P. J. Sharp (eds.), *Functional Avian Endocrinology*. Narosa Publishing House, New Delhi, India. In Press.

Pravosudov, V. V., and Clayton, N. S. 2001. Effects of demanding foraging conditions on cache retrieval accuracy in food-caching mountain chickadees (*Poecile gambeli*). *Proceedings of the Royal Society of London, Series B*, 268: 363–368.

Pravosudov, V. V., and Omanska, A. 2005. Dominance-related changes in spatial memory are associated with changes in hippocampal cell proliferation rates in mountain chickadees. *Journal of Neurobiology*, 62: 31–41.

Pravosudov, V. V., Kitaysky, A. S., Wingfield, J. C., and Clayton, N. S. 2001. Long-term unpredictable foraging conditions and physiological stress response in mountain chickadees (*Poecile gambeli*). *General and Comparative Endocrinology*, 123: 324–331.

Pravosudov, V. V., Lavenex, P., and Clayton, N. S. 2002. Changes in spatial memory mediated by experimental variation in food supply do not affect hippocampal anatomy in mountain chickadees (*Poecile gambeli*). *Journal of Neurobiology*, 51: 142–148.

Pravosudov, V. V., Mendoza, S. P., and Clayton, N. S. 2003. The relationship between dominance, corticosterone, memory, and food caching in mountain chickadees (*Poecile gambeli*). *Hormones and Behavior*, 44: 93–102.

Prendergast, B. J., Kriegsfeld, L. J., and Nelson, R. J. 2001. Photoperiodic polyphenisms in rodents: Neuroendocrine mechanisms, costs, and functions. *Quarterly Review of Biology*, 76: 293–325.

Priestnall, R., and Young, S. 1978. An observational study of caretaking behavior of male and female mice housed together. *Dev. Psychobiol.*, 11:23–30.

Pryce, C. R. 1996. Socialization, hormones, and the regulation of maternal behavior in nonhuman simian primates. *Adv. Study Behav.*, 25:423–473.

Pryce, C. R., Döbeli, M., and Martin, R. D. 1993. Effects of sex steroids on maternal motivation in the common marmoset (*Callithrix jacchus*): Development and application of an operant system with maternal reinforcement. *J. Comp. Psychol.*, 107:99–115.

Pucek, M. 1965. Water contents and seasonal changes of the brain weight in shrews. *Acta Theriol.*, 10:353–367.

Purifoy, F. E., and Koopmans, L. H. 1979. Androstenedione, testosterone, and free testosterone concentration in women of various occupations. *Soc. Biol.*, 26:179–188.

Purvis, K., and Haynes, N. B. 1974. Short-term effects of copulation, human chorionic gonadotrophin injection and non-tactile association with a female on testosterone levels in the male rat. *J. Endocrinol.*, 60:429–439.

Purvis, K., Landgren, B., Cekan, Z., and Diczfalusy, E. 1976. Endocrine effects of masturbation in men. *J. Endocrinol.*, 70:439–444.

Putnam, S. K., Du, J., Sato, S., and Hull, E. M. 2001. Testosterone restoration of copulatory behavior correlates with medial preoptic dopamine release in castrated male rats. *Hormones and Behavior*, 39: 216–224.

Putnam, S. K., Sato, S., and Hull, E. M. 2003. Effects of testosterone metabolites on copulation and medial preoptic dopamine release in castrated male rats. *Hormones and Behavior*, 44: 419–426.

Qi, Y., Takahashi, N., Hileman, S. M., Patel, H. R., Berg, A. H., Pajvani, U. B., Scherer, P. E., and Ahima, R. S. 2004. Adiponectin acts in the brain to decrease body weight. *Nature Medicine*, 10: 524–529.

Quartermain, D. 1976. The influence of drugs on learning and memory. In M. R. Rosenzweig and E. L. Bennett (eds.), *Neural Mechanisms of Learning and Memory*, pp. 508–520. MIT Press, Cambridge, MA.

Quétel, C. 1990. *History of Syphilis*. Johns Hopkins University Press, Baltimore.

Quinlan, D. M., Nelson, R. J., Partin, A. W., Mostwin, J. L., and Walsh, P. C. 1989. The rat as a model for the study of penile erection. *J. Urol.*, 141:656–661.

Racey, P. A., and Skinner, J. D. 1979. Endocrine aspects of sexual mimicry in spotted hyaenas, *Crocuta crocuta*. *J. Zool. (Lond.)*, 187:315–326.

Ragozzino, M. E., Unick, K. E., and Gold, P. E. 1996. Hippocampal acetylcholine release during memory testing in rats: Augmentation by glucose. *Proc. Natl. Acad. Sci. U.S.A.*, 93:4693–4698.

Raine, A. 2002. Biosocial studies of antisocial and violent behavior in children and adults: A review. *Journal of Abnormal Child Psychology*, 30: 311–326.

Raloff, J. 1994a. The gender benders. *Sci. News*, 145:24–27.

Raloff, J. 1994b. That feminine touch. *Sci. News*, 145:56–59.

Ralph, M. R., and Lehman, M. N. 1991. Transplantation: A new tool in the analysis of the mammalian hypothalamic circadian pacemaker. *Trends Neurosci.*, 14:362–366.

Ralph, M. R., and Menaker, M. 1988. A mutation in the circadian system in the golden hamster. *Science*, 241:1225–1227.

Ralph, M. R., Foster, R. G., Davis, F. C., and Menaker, M. 1990. Transplanted suprachiasmatic nucleus determines circadian period. *Science*, 247:975–978.

Ramenofsky, M. 1984. Agonistic behavior and endogenous plasma hormones in male Japanese quail. *Anim. Behav.*, 32:698–708.

Ramirez, V. D., and McCann, S. M. 1963. Comparisons of the regulation of luteinizing hormone (LH) secretion in immature and adult rats. *Endocrinology*, 72:452–464.

Ramos, C., and Silver, R. 1992. Gonadal hormones determine sex differences in timing of incubation by doves. *Horm. Behav.*, 26:586–601.

Rankin, S. M., and Stay, B. 1985. Ovarian inhibition of juvenile hormone synthesis in the viviparous cockroach, *Diploptera punctata. Gen. Comp. Endocrinol.*, 59:230–237.

Rankin, S. M., Palmer, J. O., Yagi, K. J., Scott, G. L., and Tobe, S. S. 1995. Biosynthesis and release of JH during the reproductive cycle of the ring-legged earwig. *Comp. Biochem. Physiol. C*, 110:241–251.

Ranote, S., Elliott, R., Abel, K. M., Mitchell, R., Deakin, J. F., and Appleby, L. 2004. The neural basis of maternal responsiveness to infants: An fMRI study. *Neuroreport*, 15: 1825–1829.

Rapkin, A. 2003. A review of treatment of premenstrual syndrome and premenstrual dysphoric disorder. *Psychoneuroendocrinology*, 28: 39–53.

Rasmussen, D. D., Liu, J. H., Wolf, P. L., and Yen, S. S. 1983. Endogenous opioid regulation of gonadotropin-releasing hormone release from the human fetal hypothalamus in vitro. *J. Clin. Endocrinol. Metab.*, 57:881–884.

Reburn, C. J., and Wynne-Edwards, K. E. 1999. Hormonal changes in males of a naturally biparental and a uniparental mammal. *Horm. Behav.* 35:163–167.

Reddy, V. V. R., Naftolin, F., and Ryan, K. J. 1974. Conversion of androstenedione to estrone by neural tissues from fetal and neonatal rat. *Endocrinology*, 94:117–121.

Redican, W. K., and Taub, D. M. 1981. Male parental care in monkeys and apes. In M. E. Lamb (ed.), *The Role of the Father in Child Development*, pp. 345–385. Wiley, New York.

Reebs, S. G., and Mrosovsky, N. 1989. Effects of induced wheel running on the circadian activity rhythms of Syrian hamsters: Entrainment and phase response curve. *Journal of Biological Rhythms*, 4: 39–48.

Reebs, S. G., Lavery, R. J., and Mrosovsky, N. 1989. Running activity mediates the phase-advancing effects of dark pulses on hamster circadian rhythms. *Journal of Comparative Physiology A*, 165: 811–818.

Reed, C. A. 1946. The copulatory behavior of small mammals. *J. Comp. Psychol.*, 39:185–206.

Refinetti, R., and Menaker, M. 1992. Evidence for separate control of estrous and circadian periodicity in the golden hamster. *Behav. Neural Biol.*, 58:27–36.

Reid, R. L., and Yen, S. S. C. 1981. Premenstrual syndrome. *Am. J. Obstet. Gynecol.*, 139:85–97.

Reijmers, L. G., van Ree, J. M., Spruijt, B. M., Burbach, J. P., and DeWied, D. 1998. Vasopressin metabolites: A link between vasopressin and memory. *Progress in Brain Research*, 119: 523–535.

Reinisch, J. M. 1974. Fetal hormones, the brain and human sex differences: A heuristic integrative review of the recent literature. *Arch. Sex. Behav.*, 3:51–90.

Reisbick, S., Rosenblatt, J. S., and Mayer, A. D. 1975. Decline of maternal behavior in the virgin and lactating rat. *J. Comp. Physiol. Psychol.*, 89:722–732.

Reiter, R. J. 1970. Endocrine rhythms associated with pineal gland function. In L. W. Hedlund, J. M. Franz, and A. D. Kenny (eds.), *Biological Rhythms and Endocrine Function*, pp. 43–78. Plenum, New York.

Reiter, R. J. 1973. Comparative physiology: Pineal gland. *Annu. Rev. Physiol.*, 35:305–328.

Reiter, R. J. 1973/74. Influence of pinealectomy on the breeding capability of hamsters maintained under natural photoperiod and temperature conditions. *Neuroendocrinology*, 13:366–370.

Reiter, R. J. 1982. *The Pineal and Its Hormones.* Alan R. Liss, New York.

Reiter, R. J. 1998. Melatonin and human reproduction. *Ann. Med.*, 30:103–108.

Remage-Healey, L., Adkins-Regan, E., and Romero, L. M. 2003. Behavioral and adrenocortical responses to mate separation and reunion in the zebra finch. *Hormones and Behavior*, 43: 108–114.

Remage-Healey, L., and Bass, A. H. 2004. Rapid, hierarchical modulation of vocal patterning by steroid hormones. *Journal of Neuroscience*, 24: 5892–5900.

Reppert, S. M., and Weaver, D. R. 2002. Coordination of circadian timing in mammals. *Nature*, 418: 935–941.

Resko, J. A., and Roselli, C. E. 1997. Prenatal hormones organize sex differences of the neuroendocrine reproductive system: observations on guinea pigs and nonhuman primates. *Cell. Molec. Neurobiol.*, 17:627–648.

Reul, J. M. H. M., and de Kloet, E. R. 1985. Two receptor systems for corticosterone in the rat brain: Microdistribution and differential occupation. *Endocrinology*, 117: 2505–2512.

Reyes, R. J., Zicchi, S., Hamed, H., Chaudary, M. A., and Fentiman, I. S. 1995. Surgical correction of gynaecomastia in bodybuilders. *Br. J. Clin. Pract.*, 49:177–179.

Rhen, T., and Lang, J. W. 1994. Temperature-dependent sex determination in the snapping turtle: Manipulation of the embryonic sex steroid environment. General and Comparative *Endocrinology*, 96:243–254.

Ribble, D. O. 1990. Population and social dynamics of the California mouse (*Peromyscus californicus*). Ph.D. Dissertation, University of California, Berkeley.

Ribble, D. O., and Salvioni, M. 1990. Social organization and nest co-occupancy in *Peromyscus californicus*, a monogamous rodent. *Behav. Ecol. Sociobiol.*, 26: 9–15.

Richards, M. P. M. 1969. Effects of oestrogen and progesterone on nest building in the golden hamster. *Anim. Behav.*, 17:356–361.

Richardson, G. S., and Martin, J. B. 1988. Circadian rhythms in neuroendocrinology and immunology: Influence of aging. *Prog. Neuroendocrinol. Immunol.*, 1:16–20.

Richman, J. A., Raskin, V. D., and Gaines, C. 1991. Gender roles, social support, and postpartum depressive symptomatology: The benefits of caring. *J. Nerv. Ment. Dis.*, 179:139–147.

Richmond, M. E., and Stehn, R. A. 1976. Olfaction and reproductive behavior in microtine rodents. In R. L. Doty (ed.), *Mammalian Olfaction, Reproductive Processes and Behavior*, pp. 197–217. Academic Press, New York.

Richmond, M., and Conaway, C. H. 1969. Induced ovulation and oestrus in *Microtus ochrogaster. J. Reprod. Fertil. (Suppl.)*, 6:357–376.

Richter, C. P. 1936. Increased salt appetite in adrenalectomized rats. *Am. J. Physiol.*, 115:155–161.

Richter, C. P. 1968. Inherent twenty-four hour and lunar clocks of a primate: The squirrel monkey. *Commun. Behav. Biol.*, 1:305–332.

Richter, C. P. 1977. Heavy water as a tool for study of the forces that control length of period of the 24-hour clock of the hamster. *Proc. Natl. Acad. Sci. U.S.A.*, 74:1295–1299.

Richter, C. P., Holt, L. E., and Barelare, B. 1938. Nutritional requirements for normal growth and reproduction in rats studied by the self selection method. *Am. J. Physiol.*, 122:734–744.

Riddle, O. 1924/25. Birds without gonads: Their origin, behavior, and bearing on the theory of the internal secretion of the testis. *Br. J. Exp. Biol.*, 2:211–246.

Riddle, O., Bates, R. W., and Lahr, E. L. 1935a. Maternal behavior induced in virgin rats by prolactin. *Proc. Soc. Exp. Biol. Med.*, 32:730–734.

Riddle, O., Bates, R. W., and Lahr, E. L. 1935b. Prolactin induces broodiness in fowl. *Am. J. Physiol.*, 111:352–360.

Ring, J. R. 1944. The estrogen-progesterone induction of sexual receptivity in the spayed female mouse. *Endocrinology*, 34:269–275.

Rissman, E. F. 1991. Frank A. Beach Award. Behavioral endocrinology of the female musk shrew. *Horm. Behav.*, 25:125–127.

Rissman, E. F., and Wingfield, J. C. 1984. Hormonal correlates of polyandry in the spotted sandpiper, *Actitis macularia. Gen. Comp. Endocrinol.*, 56:401–405.

Rissman, E. F., Early, A. H., Taylor, J. A., Korach, K. S., and Lubahn, D. B. 1997b. Estrogen receptors are essential for female sexual receptivity. *Endocrinology*, 138:507–510.

Rissman, E. F., Heck, A. L., Leonard, J. E., Shupnik, M. A., and Gustafsson, J. A. 2002. Disruption of estrogen receptor β gene impairs spatial learning in female mice. *Proceedings of the National Academy of Sciences U.S.A.*, 99: 3996–4001.

Rissman, E. F., Wersinger, S. R., Taylor, J. A., and Lubahn, D. B. 1997a. Estrogen receptor function as revealed by knockout studies: Neuroendocrine and behavioral aspects. *Horm. Behav.*, 31:232–243.

Ritter, S., and Taylor, J. S. 1989. Capsaicin abolishes lipoprivic but not glucoprivic feeding in rats. *Am. J. Physiol.*, 256:R1232–1239.

Ritter, S., and Taylor, J. S. 1990. Vagal sensory neurons are required for lipoprivic but not glucoprivic feeding in rats. *Am. J. Physiol.*, 258:R1395–1401.

Ritter, S., Calingasan, N. Y., Hutton, B., and Dinh, T. T. 1992. Cooperation of vagal and central neural systems in monitoring metabolic events controlling feeding behavior. In S. Ritter, R. C. Ritter, and C. D. Barnes (eds.). *Neuroanatomy Physiology of Abdominal Vagal Afferents*, pp. 240–277. CRC Press, Boca Raton, FL.

Rivest, S., and Rivier, C. 1993. Interleukin-1 beta inhibits the endogenous expression of the early gene c-fos located within the nucleus of LH-RH neurons and interferes with hypothalamic LH-RH release during proestrus in the rat. *Brain Res.*, 613:132–142.

Rivest, S., Deshaies, Y., and Richard, D. 1989. Effects of corticotropin-releasing factor on energy balance in rats are sex dependent. *Am. J. Physiol.*, 257:R1417–R1422.

Rivier, C., and Rivest, S. 1991. Effects of stress on the activity of the hypothalamic-pituitary-gonadal axis: Peripheral and central mechanisms. *Biol. Reprod.*, 45:523–532.

Rivier, C., Rivier, J., and Vale, W. 1986. Stress-induced inhibition of reproductive functions: Role of endogenous corticotropin-releasing factor. *Science*, 231:607–609.

Robbins, T. W. and Everitt, B. J. 1992. Functions of dopamine in the dorsal and ventral striatum. *Sem. Neurosci.*, 4:119–128.

Roberts, J. T., and Essenhigh, D. M. 1986. Adenocarcinoma of prostate in 40-year-old bodybuilder. *Lancet*, 2(8509): 742.

Roberts, M. H. 1990. Commentary: The properties of cell water relative to the temperature compensation of circadian rhythms. *J. Biol. Rhythms*, 5:175–176.

Roberts, R. L., Zullo, A., Gustafson, E. A., and Carter, C. S. 1996. Perinatal steroid treatments alter alloparental and affiliative behavior in prairie voles. *Horm. Behav.*, 30:576–582.

Robinson, T. J. 1954. The necessity for progesterone with estrogen for the induction of recurrent estrus in the ovariectomized ewe. *Endocrinology*, 55:403–408.

Roenneberg, T. and Aschoff, J. 1990a. Annual rhythm of human reproduction: I. Biology, sociology, or both? *J. Biol. Rhythms*, 5:195–216.

Roenneberg, T., and Aschoff, J. 1990b. Annual rhythm of human reproduction: II. Environmental correlations. *J. Biol. Rhythms*, 5:217–239.

Roenneberg, T., and Merrow, M. 2003. The network of time: Understanding the molecular circadian system. *Current Biology*, 13: R198–R207.

Rohwer, S. 1975. The social significance of avian winter plumage variability. *Evolution*, 29:593–610.

Rohwer, S. 1977. Status signaling in Harris sparrows: Some experiments in deception. *Behaviour*, 61:107–129.

Rohwer, S., and Rohwer, F. C. 1978. Status signalling in Harris' sparrows: Experimental deceptions achieved. *Anim. Behav.*, 26:1012–1022.

Rohwer, S., and Wingfield, J. C. 1981. A field study of social dominance, plasma levels of luteinizing hormone and steroid hormones in wintering Harris' Sparrows. *Z. Tierpsychol.*, 57:173–183.

Rolls, B. J., and Rolls, E. T. 1982. *Thirst*. University of Cambridge Press, Cambridge.

Romero, L. M. 2002. Seasonal changes in plasma glucocorticoid concentrations in free-living vertebrates. *General and Comparative Endocrinology*, 128: 1–24.

Romero, L. M., Ramenofsky, M., and Wingfield, J. C. 1997. Season and migration alters the corticosterone response to capture and handling in an arctic migrant, the white-crowned sparrow (*Zonotrichia leucophrys gambelii*). *Comp. Biochem. Physiol. C.*, 116:171–177.

Romero, L. M., Soma, K. K., and Wingfield, J. C. 1998b. Hypothalamic-pituitary-adrenal axis changes allow seasonal modulation of corticosterone in a bird. *Am. J. Physiol.*, 274:R1338–1344.

Romero, L. M., Soma, K. K., and Wingfield, J. C. 1998c. The hypothalamus and adrenal regulate modulation of corticosterone release in redpolls (*Carduelis flammea*): An arctic-breeding song bird. *Gen. Comp. Endocrinol.*, 109:347–355.

Romero, L. M., Soma, K. K., and Wingfield, J. C. 1998d. Changes in pituitary and adrenal sensitivities allow the snow bunting (*Plectrophenax nivalis*), an arctic-breeding song bird, to modulate corticosterone release seasonally. *J. Comp. Physiol. B*, 168:353–358.

Romero, L. M., Soma, K. K., O'Reilly, K. M., Suydam, R.,and Wingfield, J. C. 1998a. Hormones and territorial behavior during breeding in snow buntings (*Plectrophenax nivalis*): An arctic-breeding songbird. *Horm. Behav.*, 33:40–47.

Ronchi, E., Spencer, R. L., Krey, L. C., and McEwen, B. S. 1998. Effects of photoperiod on brain corticosteroid receptors and the stress response in the golden hamster (*Mesocricetus auratus*). *Brain Research*, 780: 348–351.

Roof, R. L., and Havens, M. D. 1992. Testosterone improves maze performance and induces development of a male hippocampus in females. *Brain Research*, 572: 310–313.

Roozendaal, B. 2000. Glucocorticoids and the regulation of memory consolidation. *Psychoneuroendocrinology*, 25: 213–238.

Roozendaal, B. 2003. Systems mediating acute glucocorticoid effects on memory consolidation and retrieval. *Progress in Neuropsychopharmacology and Biological Psychiatry*. 27: 1213–1223.

Roozendaal, B., Brunson, K. L., Holloway, B. L., McGaugh, J. L., and Baram, T. Z. 2002. Involvement of stress-released corticotropin-releasing hormone in the basolateral amygdala in regulating memory consolidation. *Proceedings of the National Academy of Sciences U.S.A.*, 99: 13908–13913.

Roozendaal, B., Portillo-Marquez, G., and McGaugh, J. L. 1996. Basolateral amygdala lesions block glucocorticoid-induced modulation of memory for spatial learning. *Behavioral Neuroscience*, 110: 1074–1083.

Rose, R. M., Bernstein, I. S., Gordon, T. P., and Lindsley, J. G. 1978. Changes in testosterone and behavior during adolescence in the male rhesus monkey. *Psychosom. Med.*, 40:60–70.

Rose, R. M., Berstein, I. S., and Holaday, J. W. 1971. Plasma testosterone, dominance rank and aggressive behavior in a group of male rhesus monkeys. *Nature*, 231:366–368.

Roselli, C. E., Abdelgadir, S. E., and Resko, J. A. 1997. Regulation of aromatase gene expression in the adult rat brain. *Brain Res. Bull.*, 44:351–358.

Rosenberg, G. D., and Runcorn, S. K. 1975. Conclusions. In G. D. Rosenberg and S. K. Runcorn (eds.), *Growth Rhythms and the History of the Earth's Rotation*, pp. 535–538. John Wiley and Sons, New York.

Rosenblatt, J. S. 1967. Nonhormonal basis of maternal behavior in the rat. *Science*, 156:1512–1514.

Rosenblatt, J. S. 1990. Landmarks in the physiological study of maternal behavior with special reference to the rat. In N. A. Krasnegor and R. S. Bridges (eds.), *Mammalian Parenting*, pp. 40–60. Oxford University Press, Oxford.

Rosenblatt, J. S. 2002. Hormonal basis of parenting in mammals. In M. H. Bornstein (ed.), *Handbook of Parenting*, 2nd edition, Volume 2, pp. 31–60. Lawrence Erlbaum, Mahwah, NJ.

Rosenblatt, J. S. 2003. Outline of the evolution of behavioral and nonbehavioral patterns of parental care among the vertebrates: Critical characteristics of mammalian and avian parental behavior. *Scandinavian Journal of Psychology*, 44: 265–271.

Rosenblatt, J. S., and Aronson, L. R. 1958. The decline of sexual behaviour in male cats after castration with special reference to the role of prior sexual experience. *Behaviour*, 12:285–338.

Rosenblatt, J. S., and Ceus, K. 1998. Estrogen implants in the medial preoptic area stimulate maternal behavior in male rats. *Horm. Behav.*, 33:23–30.

Rosenblatt, J. S., and Siegel, H. I. 1975. Hysterectomy-induced maternal behavior during pregnancy in the rat. *J. Comp. Physiol. Psychol.*, 89:685–700.

Rosenblatt, J. S., Hazelwood, S., and Poole, J. 1996. Maternal behavior in male rats: Effects of medial preoptic area lesions and presence of maternal aggression. *Horm. Behav.*, 30:01–215.

Rosenblatt, J. S., Mayer, A. D., and Siegel, H. I. 1985. Maternal behavior among the nonprimate mammals. In N. T. Adler, D. Pfaff, and R. W. Goy (eds.), *Handbook of Behavioral Neurobiology*, pp. 229–298. Plenum, New York.

Rosenblatt, J. S., Olufowobi, A., and Siegel, H. I. 1998. Effects of pregnancy hormones on maternal responsiveness, responsiveness to estrogen stimulation of maternal behavior, and the lordosis response to estrogen stimulation. *Horm. Behav.*, 33:104–114.

Rosenblatt, J. S., Siegel, H. I., and Mayer, A. D. 1979. Blood levels of progesterone, estradiol and prolactin in pregnant rats. *Adv. Study Behav.*, 10:225–311.

Rosenfeld, P., Wetmore, J. B., and Levine, S. 1992. Effects of repeated maternal separations on the adrenocortical response to stress of preweanling rats. *Physiol. Behav.*, 52:787–791.

Rosenthal, N. E. 1993. *Winter Blues: Seasonal Affective Disorder.* Guilford Press, New York.

Rosenthal, N. E., and Wehr, T. A. 1987. Seasonal affective disorders. *Psychiatric Ann.*, 17:670–674.

Rosenthal, N. E., Sack, D. A., Skwerer, R. G., Jacobsen, F. M., and Wehr, T. A. 1988. Phototherapy for seasonal affective disorder. *J. Biol. Rhythms*, 3:101–120.

Rosenzweig, M. R., and Leiman, A. L. 1989. *Physiological Psychology.* (2nd ed.). Random House, New York.

Rosenzweig, M. R., Leiman, A. L., and Breedlove, S. M. 1999. *Biological Psychology: An Introduction to Behavioral, Cognitive, and Clinical Neuroscience.* (2nd ed.). Sinauer Associates, Sunderland, MA.

Ross, S., Denenberg, V. H., Frommer, G. P., and Sawin, P. B. 1959. Genetic, physiological, and behavioral background of reproduction in the rabbit. V. Nonretrieving of neonates. *J. Mammal.*, 40:91–96.

Rothschild, A. J., Langlais, P. J., Schatzberg, A. F., Miller ,M. M., Saloman, M. S., Lerbinger, J. E., Cole, J. O., and Bird, E. D. 1985. The effects of a single acute dose of dexamethasone on monoamine and metabolite levels in rat brain. *Life Sci.*, 36:2491–2501.

Rots, N. Y., du Jong, J., Workel, J. O., Levine, S., Cools, A. R., and De Kloet, E. R. 1996. Neonatal maternally deprived rats have as adults elevated basal pituitary-adrenal activity and enhanced susceptibility to apomorphine. *J. Neuroendocrinol.*, 8:501–506.

Rous, S. N. 1996. *Urology: A Core Textbook*, Blackwell Science Press:Cambridge, MA.

Rowe, F. A., and Smith, W. E. 1972. Effects of peripherally induced anosmia on mating behavior of male mice. *Psychonom. Sci.*, 27:33–34.

Rowlands, I. W., and Weir, B. J. 1984. Mammals: Non-primate eutherians. In G. E. Lamming (ed.), *Marshall's Physiology of Reproduction.* Vol. 1, *Reproductive Cycles of Vertebrates*, pp. 455–658. Churchill Livingstone, Edinburgh.

Roy, E. J., Lynn, D. M., and Clark, A. S. 1985. Inhibition of sexual receptivity by anesthesia during estrogen priming. *Brain Res.*, 337:163–166.

Roy-Byrne, P. P., Rubinow, D. R., Hoban, M. C., Grover, G. N., Blank, D. 1987. TSH and prolactin responses to TRH in patients with premenstrual syndrome. *Am. J. Psychiatry*, 144:480–484.

Rubin, B. S., and Barfield, R. J. 1983. Progesterone in the ventromedial hypothalamus facilitates estrous behavior in ovariectomized, estrogen-primed rats. *Endocrinology*, 113:797–804.

Rubin, R. T., Dinan, T. G., and Scott, L. V. 2002. The neuroendocrinology of affective disorders. In D. W. Pfaff, A. P. Arnold, A. M. Etgen, S. E. Fahrbach, and R. T. Rubin (eds.), *Hormones, Brain and Behavior*, Volume 5, pp. 467–514. Academic Press, New York.

Rubinow, D. R., and Schmidt, P. J. 1996. Androgens, brain, and behavior. *Am. J. Psychiatry*, 153:974–984.

Rubinow, D. R., Hoban, M. C., Grover, G. N., Galloway, D. S., Roy-Byrne, R., Andersen, R., and Merriam, G. R. 1988. Changes in plasma hormones across the menstrual cycle in patients with menstrually related mood disorder and in control subjects. *Am. J. Obstet. Gynecol.*, 158:5–11.

Rubinow, D. R., Schmidt, P. J., and Roca, C. A. 1998. Estrogen-serotonin interactions: Implications for affective regulation. *Biol. Psychiatry*, 44:839–850.

Rubinow, D. R., Schmidt, P. J., Roca, C. A., and Daly, R. C. 2002. Gonadal hormones and behavior in women: Concentrations versus context. In D. W. Pfaff, A. P. Arnold, A. M. Etgen, S. E. Fahrbach, and R. T. Rubin (eds.), *Hormones, Brain and Behavior*, Volume 5, pp. 37–73. Academic Press, New York.

Ruble, D. N. 1977. Premenstrual symptoms: A reinterpretation. *Science*, 187:291–292.

Ruble, D. N., and Martin, C. L. 1998. Gender development. In W. Damon and N. Eisenberg (eds.). *Handbook of Child Psychology.* Vol. 3. , pp. 933–1016. Wiley, New York.

Ruegg, D., and Rose, H. A. 1991. Biology of *Macropanesthia rhinoceros* Saussure (Dictoptera: Balberidae). *Ann. Ent. Soc. Am.*, 84:575–582.

Rusak, B. 1989. The mammalian circadian system: models and physiology. *J. Biol. Rhythms*, 4:121–134.

Rusak, B., and Boulos, Z. 1981. Pathways for photic entrainment of mammalian circadian rhythms. *Photochem. Photobiol.*, 34:267–273.

Rusak, B., Mistlberger, R. E., Losier, B., and Jones, C. H. 1988. Daily hoarding opportunity entrains the pacemaker for hamster activity rhythms. *Journal of Comparative Physiology A*, 164: 165–171.

Rushen, J. 1986. Some problems with the physiological concept of "stress". *Aust. Vet. J.*, 63:359–361.

Rutila, J. E., Suri, V., Le, M., So, W. V., Rosbash, M., and Hall, J. C. 1998. CYCLE is a second bHLH-PAS clock protein essential for circadian rhythmicity and transcription of *Drosophila* period and timeless. *Cell*, 93:805–814.

Ryan, C. M., and Williams, T. M. 1993. Effects of insulin-dependent diabetes on learning and memory efficiency in adults. *J. Clin. Exp. Neuropsychol.*, 15:685–700.

Sabatinelli, D., Flaisch, T., Bradley, M. M., Fitzsimmons, J. R., and Lang, P. J. 2004. Affective picture perception: Gender differences in visual cortex? *Neuroreport*, 15: 1109–1112.

Saboureau, M. 1986. Hibernation in the hedgehog: Influence of external and internal factors. In H. C. Heller, X. J. Musacchia, and L. C. H. Wang (eds.), *Living in the Cold: Physiological and Biochemical Adaptations*, pp. 253–263, Elsevier, New York.

Sachar, E. O., Hellman, L., Roffwang, H. Halpern, F., Fukushima, D., and Gallager, T. 1973. Disrupted 24-hour patterns of cortisol secretion in psychotic depression. *Arch. Gen. Psychiatry*, 28:19–24.

Sachs, B. D. 1995a. Context-sensitive variation in the regulation of erection. In J. Bancroft (ed), *The Pharmacology of Sexual Function Dysfunction*, pp. 97–108. Elsevier Science, Amsterdam.

Sachs, B. D. 1995b. Placing erection in context: The reflexogenic-psychogenic dichotomy reconsidered. *Neurosci. Biobehav. Rev.*, 19:211–224.

Sachs, B. D., Akasofu, K., Citron, J. H., Daniels, S. B., and Natoli, J. H. 1994. Noncontact stimulation from estrous females evokes penile erection in rats. *Physiol. Behav.*, 55:1073–1079.

Sachs, B. D., and Bitran, D. 1990. Spinal block reveals roles for brain and spinal cord in the mediation of reflexive penile erections in rats. *Brain Res.*, 528:99–108.

Sachs, B. D., and Meisel, R. L. 1979. Pubertal development of penile reflexes and copulation in male rats. *Psychoneuroendocrinology*, 4:287–296.

Sachs, B. D., and Meisel, R. L. 1988. The physiology of male sexual behavior. In E. Knobil and J. Neill (eds.), *The Physiology of Reproduction*, pp. 1393–1485. Raven Press, New York.

Sack, R. L., and Lewy, A. J. 2001. Circadian rhythm sleep disorders: Lessons from the blind. *Sleep Medicine Reviews*, 5: 189–206.

Sack, R. L., Brandes, R. W., Kendall, A. R., and Lewy, A. J. 2000. Entrainment of free-running circadian rhythms by melatonin in blind people. *New England Journal of Medicine*, 343: 1070–1077.

Sadlier, R. M. F. S. 1969. *The Ecology of Reproduction in Wild and Domestic Mammals,* Methuen, London.

Sakai, R. R., Ma, L. Y., He, P. F., and Fluharty, S. J. 1995. Intracerebroventricular administration of angiotensin type 1 (AT1) receptor antisense oligonucleotides attenuate thirst in the rat. *Regul. Pept.,* 59:183–92.

Sakai, R. R., Ma, L-Y., Zhang, D. M., McEwen, B. S., and Fluharty, S. J. 1996. Intracerebroventricular administration of mineralocorticoid receptor antisense oligonucleotides attenuates adrenal steroid-induced salt appetite in rats. *Neuroendocrinology,* 64:425–529.

Sakata, J. T., and Crews, D. 2004. Developmental sculpting of social phenotype and plasticity. *Neuroscience and Biobehavioral Reviews,* 28: 95–112.

Sakuma, Y., and Pfaff, D. W. 1979. Mesencephalic mechanisms for integration of female reproductive behavior in the rat. *Am. J. Physiol.,* 237:R285–R290.

Sakuma, Y., and Pfaff, D. W. 1980. Lh-RH in the mesencephalic central gray can potentiate lordosis reflex of female rats. *Nature,* 283:566–567.

Saldanha, C. J., Clayton, N. S., and Schlinger, B. A. 1999. Androgen metabolism in the juvenile oscine forebrain: A cross-species analysis at neural sites implicated in memory function. *Journal of Neurobiology,* 40: 397–406.

Saltiel, A. R., and Kahn, C. R. 2001. Insulin signaling and the regulation of glucose and lipid metabolism. *Nature,* 414: 799–806.

Saltzman, W., Mendoza, S. P., and Mason, W. A. 1991. Sociophysiology of relationships in squirrel monkeys: I. Formation of female dyads. *Physiol. Behav.,* 50: 271–280.

Saltzman, W., Schultz-Darken, N. J., and Abbott, D. H. 1996. Behavioural and endocrine predictors of dominance and tolerance in female common marmosets, *Callithrix jacchus. Anim. Behav.,* 51: 657–674.

Samuels, M. H., and Bridges, R. S. 1983. Plasma prolactin concentrations in parental male and female rats: Effects of exposure to rat young. *Endocrinology,* 113:1647–1654.

Sanders, D., Warner, P., Backstrom, T., and Bancroft, J. 1983. Mood, sexuality, hormones, and the menstrual cycle: Changes in mood and physical state. *Psychosom. Med.,* 45: 487–500.

Sandi, C., and Rose, S. P. R. 1994a. Corticosterone enhances long-term memory in one-day-old chicks trained in a weak passive avoidance paradigm. *Brain Research,* 647: 106–112.

Sandi, C., and Rose, S. P. R. 1994b. Corticosteroid receptor antagonists are amnestic for passive avoidance learning in day-old chicks. *European Journal of Neuroscience,* 6: 1292–1297.

Sandstrom, N. J., and Williams, C. L. 2001. Memory retention is modulated by acute estradiol and progesterone replacement. *Behavioral Neuroscience,* 115: 384–393.

Sandstrom, N. J., and Williams, C. L. 2004. Spatial memory retention is enhanced by acute and continuous estradiol replacement. *Hormones and Behavior,* 45: 128–135.

Santen, F. J., Sofsky, J., Bilic, N., and Lippert, R. 1975. Mechanism of action of narcotics in the production of menstrual dysfunction in women. *Fertil. Steril.,* 26: 538–548.

Sapolsky, R. M. 1992a. Neuroendocrinology of the stress-response. In by J. B. Becker, S. M. Breedlove, and D. Crews (eds.). *Behavioral Endocrinology,* pp. 287–324. MIT Press, Cambridge, MA.

Sapolsky, R. M. 1992b. *Stress, the Aging Brain, and the Mechanisms of Neuron Death.* MIT Press, Cambridge, MA.

Sapolsky, R. M. 1994. *Why Zebras Don't Get Ulcers: A Guide to Stress, Stress-Related Diseases, and Coping.* W. H. Freeman, New York.

Sapolsky, R. M. 2004. Mothering style and methylation. *Nature Neuroscience,* 7: 791–792.

Sapolsky, R. M., and Meaney, M. J. 1986. Maturation of adrenocortical stress response: Neuroendocrine control mechanisms and the stress hyporesponsive period. *Brain Res. Rev.,* 11: 65–76.

Sapolsky, R. M., Krey, L. C., and McEwen, B. S. 1984. Glucocorticoid-sensitive hippocampal neurons are involved in terminating the adrenocortical stress response. *Proceedings of the National Academy of Sciences U.S.A.,* 81: 6174–6177.

Sapolsky, R. M., Krey, L., and McEwen, B. S. 1985. Prolonged glucocorticoid exposure reduces hippocampal neuron numbeer: Implications for aging. *J. Neurosci.,* 5: 1221–1228.

Sapolsky, R. M., Romero, L. M., and Munck, A. U. 2000. How do glucocorticoids influence stress responses? Integrating permissive, suppressive, stimulatory, and preparative actions. *Endocrine Reviews,* 21: 55–89.

Sapolsky, R.M. 2002. Chickens, eggs and hippocampal atrophy. *Nature Neuroscience,* 5: 1111–1113.

Sapolsky, R.M. 2004. Mothering style and methylation. *Nature Neuroscience,* 7: 791–792.

Sar, M., and Stumpf, W. E. 1977. Distribution of androgen target cells in rat forebrain and pituitary after [3H]-dihydrotestosterone administration. *J. Steroid Biochem.,* 8: 1131–1135.

Satinoff, E., and Prosser, R. A. 1988. Suprachiasmatic nuclear lesions eliminate circadian rhythms of drinking and activity, but not of body temperature, in male rats. *J. Biol. Rhythms,* 3: 1–22.

Sato, T., Matsumoto, T., Kawano, H., Watanabe, T., Uematsu, Y., Sekine, K., Fukuda, T., et al. 2004. Brain masculinization requires androgen receptor function. *Proceedings of the National Academy of Sciences U.S.A.,* 101: 1673–1678.

Saunders, D. S. 1977. *An Introduction to Biological Rhythms.* Wiley, New York.

Sauvage, M. F., Marquet, P., Rousseau, A., Raby, C., Buxeraud, J., and Lachatre, G. 1998. *Toxicol. Appl. Pharmacol.,* 149: 127–135.

Sawchenko, P. E., and Friedman, M. I. 1979. Sensory functions of the liver: A review. *Am. J. Physiol.,* 236: R5–R20.

Scalia, F., and Winans, S. S. 1976. New perspectives on the morphology of the olfactory system: Olfactory and vomeronasal pathways in mammals. In R. L. Doty (ed.), *Mammalian Olfaction, Reproductive Processes, and Behavior,* pp. 7–28. Academic Press, New York.

Schaal, B., and Marlier, L. 1998. Maternal and paternal perception of individual odor signatures in human amniotic fluid-potential role in early bonding? *Biol. Neonate*, 74: 266–273.

Schaal, B., Montagner, H., Hertling, E., Bolzoni, D., Moyse, A., and Quichon, R. 1980. Olfactory stimulation in the relationship between child and mother. *Reprod. Nutr. Dev.*, 20: 843–858.

Schaefer, G. J. 1988. Opiate antagonists and rewarding brain stimulation. *Neurosci. Biobehav. Rev.*, 12: 1–17.

Schanberg, S. M., and Field, T. M. 1987. Sensory deprivation stress and supplemental stimulation in the rat pup and preterm human neonate. *Child Dev.*, 58: 1431–1447.

Schanberg, S. M., and Kuhn, C. M. 1985. The biochemical effects of tactile deprivation in neonatal rats. *Perspect. Behav. Med.*, 2: 133–148.

Scheider, J.E. and Wade, G.N. 2000. Inhibition of reproduction in service of energy balance. In: *Reproduction in Context: Social and Environmental Influences on Reproduction*. K. Wallen and J.E. Schneider. MIT Press: Cambridge. pp 35-82.

Schildkraut, J. M., Batos, E., and Berchuck, A. 1997. Relationship between lifetime ovulatory cycles and overexpression of mutant p53 in epithelial ovarian cancer. *J. Natl. Cancer Inst.*, 89: 932–938.

Schindler, G. L. 1979. Testosterone concentration, personality patterns, and occupational choice in women. Ph.D. dissertation, University of Houston.

Schjelderup-Ebbe, T. 1922. Beiträge zür Socialpsychologie des haushuhns. *Z. Psychol.*, 88: 225–252.

Schlinger, B. A. 1998. Sexual differentiation of avian brain and behavior: Current views on gonadal hormone-dependent and independent mechanisms. *Annual Review of Physiology*, 60: 407–429.

Schlinger, B. A., and Callard, G. V. 1990. Aggressive behavior in birds: An experimental model for studies of brain–steroid interactions. *Comp. Biochem. Physiol.*, 97A: 307–316.

Schlinger, B. A., Fivizzani, A. J., and Callard, G. V. 1989. Aromatase, 5α- and 5β-reductase in brain, pituitary and skin of the sex-role reversed Wilson's phalarope. *J. Endocrinol.*, 122: 573–581.

Schlinger, B. A., Soma, K. K., and London, S. E. 2001. Neurosteroids and brain sexual differentiation. *Trends in Neuroscience*, 24: 429–431.

Schlotts, F. 1988. *The PMS Book.* Ivory Tower Publications, Watertown, MA.

Schmidt, P. J., Nieman, L. K., Danaceau, M. A., Adams, L. F., and Rubinow, D. R. 1998. Differential behavioral effects of gonadal steroids in women with and those without premenstrual syndrome. *N. Engl. J. Med.*, 338: 209–216.

Schmidt, P. J., Nieman, L. K., Grover, G. N., Muller, K. L., Merriam, G. R., and Rubinow, D. R. 1991. Lack of effect of induced menses on symptoms in women with premenstrual syndrome. *N. Engl. J. Med.*, 324:1174–1179.

Schneider, J. E. 2004. Energy balance and reproduction. *Physiology and Behavior*, 81: 289–317.

Schneider, J. E., and Wade, G. N. 1987. Body composition, food intake, and brown fat thermogenesis in pregnant Djungarian hamsters. *Am. J. Physiol.*, 253: R314–320.

Schneider, J. E., and Wade, G. N. 1989a. Availability of metabolic fuels controls estrous cyclicity of Syrian hamsters. *Science*, 244:1326–1328.

Schneider, J. E., and Wade, G. N. 1989b. Effects of maternal diet, body weight and body composition on infanticide in Syrian hamsters. *Physiol. Behav.*, 46: 815–821.

Schneider, J. E., and Wade, G. N., 2000. Inhibition of reproduction in service of energy balance. In: *Reproduction in Context: Social and Environmental Influences on Reproduction* pp. 35–82.. K. Wallen and J. E. Schneider. MIT Press, Cambridge.

Schneider, J. E., and Watts, A. G. 2002. Energy balance, ingestive behavior, and reproductive success. In D. W. Pfaff, A. P. Arnold, A. M. Etgen, S. E. Fahrbach, and R. T. Rubin (eds.), *Hormones, Brain and Behavior*, Volume 1, pp. 435–523. Academic Press, New York.

Schneider, J. E., Buckley, C. A., Blum, R. M., Zhou, D., Szymanski, L., Day, D. E., and Bartness, T. J. 2002. Metabolic signals, hormones and neuropeptides involved in control of energy balance and reproductive success in hamsters. *European Journal of Neuroscience*, 16: 377–379.

Schneider, J. E., Goldman, M. D., Tang, S., Bean, B., Ji, H., and Friedman, M. I. 1998. Leptin indirectly affects estrous cycles by increasing metabolic fuel oxidation. *Horm. Behav.*, 33: 217–228.

Schneider, M. A., Brotherton, P. L., and Hailes, J. 1977. The effect of exogenous oestrogens on depression in menopausal women. *Med. J. Aust.*, 2: 162–163.

Schneider, M. L. 1992a. The effect of mild stress during pregnancy on birth weight and neuromotor maturation in rhesus monkey infants (*Mucaca mulatta*). *Infant Behav. Dev.*, 15: 389–403.

Schneider, M. L. 1992b. Prenatal stress exposure alters postnatal behavioral expression under conditions of novelty challenge in rhesus monkey infants. *Dev. Psychobiol.*, 25: 529–540.

Schneider, M. L., and Coe, C. L. 1993. Repeated social stress during pregnancy impairs neuromotor development of the primate infant. *Dev. Behav. Pediatr.*, 14: 81–87.

Schoech, S. J. 2001. Physiology of helping in Florida scrub-jays. In P. W. Sherman and J. Alcock (eds.), *Exploring Animal Behavior: Readings from American Scientist*, 3rd ed. Sinauer Associates, Inc, Sunderland, MA.

Schoech, S. J., Ketterson, E. D., Nolan, V., Sharp, P. J., and Buntin, J. D. 1998. The effect of exogenous testosterone on parental behavior, plasma prolactin, and prolactin binding sites in dark-eyed juncos. *Hormones and Behavior*, 34: 1–10.

Schulz, H., and Lavie, P. 1985. *Ultradian Rhythms in Physiology and Behavior.* Springer-Verlag, Berlin.

Schuurman, T. 1980. Hormonal correlates of agonistic behavior in adult male rats. *Prog. Brain Res.*, 53: 415–420.

Schwabl, H. 1993. Yolk is a source of maternal testosterone for developing birds. *Proceedings of the National Academy of Sciences U.S.A.*, 90: 11446–11450.

Schwabl, H. 1996. Maternal testosterone in the avian egg enhances postnatal growth. *Comparative Biochemistry and Physiology A*, 114: 271–276.

Schwabl, H., and Kriner, E. 1991. Territorial aggression and song of male European robins (*Erithacus rubecula*) in autumn and spring: Effects of antiandrogen treatment. *Hormones and Behavior*, 25: 180–194.

Schwabl, H., and Kriner, E. 1991. Territorial aggression and song of male European robins (*Erithacus rubecula*) in autumn and spring: Effects of antiandrogen treatment. *Horm. Behav.*, 25:180–194.

Schwanzel-Fukuda, M., Bick, D., and Pfaff, D. W. 1989. Luteinizing hormone-releasing hormone (LHRH)-expressing cells do not migrate normally in an inherited hypogonadal (Kallmann) syndrome. *Mol. Brain Res.*, 6: 311–326.

Schwartz, G. J., Whitney, A., Skoglund, C., Castonguay, T. W., and Moran, T. H. 1999. Decreased responsiveness to dietary fat in Otsuka Long-Evans Tokushima fatty rats lacking CCK-A receptors. *American Journal of Physiology*, 277: R1144–R1151.

Schwartz, M. W., Figlewicz, D. P., Baskin, D. G., Woods, S. C., and Porte, D. Jr. 1992. Insulin in the brain: a hormonal regulator of energy balance. *Endocr. Rev.*, 13: 387–414.

Schwartz, M. W., Seeley, R. J., Campfield, L. A., Burn, P., and Baskin, D. G., 1996. Identification of hypothalamic targets of leptin action. *Journal of Clinical Investigations*, 98: 1101–1106.

Schwartz, M. W., Woods, S. C., Porte, D., Seeley, R. J., and Baskin, D. G. 2000. Central nervous system control of food intake. *Nature*, 406: 661–671.

Schwartz, W. J., and Gainer, H. 1977. Suprachiasmatic nucleus: Use of ^{14}C-labeled deoxyglucose uptake as a functional marker. *Science*, 197: 1089–1091.

Schwartz, W. J., Bosis, N. A., and Hedley-Whyte, E. T. 1986. A discrete lesion of ventral hypothalamus and optic chiasm that disturbed the daily temperature rhythm. *J. Neurol.*, 233: 1–4.

Scott, C. J., Jansen, H. T., Kao, C. C., Kuehl, D. E., and Jackson, G. L. 1995. Disruption of reproductive rhythms and patterns of melatonin and prolactin secretion following bilateral lesions of the suprachiasmatic nuclei in the ewe. *Journal of Neuroendocrinology*, 7: 429–443.

Scott, J. P., and Fredericson, E. 1951. The causes of fighting in mice and rats. *Physiol. Zool.*, 24: 273–309.

Searcy, W. A., and Wingfield, J. C. 1980. The effects of androgen and anti-androgen on dominance and aggressiveness in male red-winged blackbirds. *Horm. Behav.*, 14: 126–135.

Sedivy, J. M., and Sharp, P. A. 1989. Positive genetic selection for gene disruption in mammalian cells by homologous recombination. *Proc. Nat. Acad. Sci. U.S.A.*, 86, 227–231.

Seegal, R. F., and Goldman, B. D. 1975. Effects of photoperiod on cyclicity and serum gonadotropins in the Syrian hamster. *Biol. Reprod.*, 12: 223–231.

Seeley, R. J., and Woods, S. C. 2003. Monitoring of stored and available fuel by the CNS: Implications for obesity. *Nature Reviews: Neuroscience*, 4: 901–909.

Seeley, R. J., D'Alessio, D. A., and Woods, S. C. 2004. Fat hormones pull their weight in the CNS. *Nature Medicine*, 10: 454–456.

Segraves, R. T. 2003. Emerging therapies for female sexual dysfunction. *Expert Opinion on Emerging Drugs*, 8: 515–522.

Segreti, J., Gheusi, G., Dantzer, R., Kelley, K. W., and Johnson, R. W. 1997. Defect in interleukin-1beta secretion prevents sickness behavior in C3H/HeJ mice. *Physiol. Behav.*, 61: 873–878.

Seifritz, E., Esposito, F., Neuhoff, J. G., Luthi, A., Mustovic, H., Dammann, G., von Bardeleben, U., Radue, E. W., Cirillo, S., Tedeschi, G., and Di Salle, F. 2003. Differential sex-independent amygdala response to infant crying and laughing in parents versus nonparents. *Biological Psychiatry*, 54: 1367–1375.

Selye, H. 1936. A syndrome produced by diverse nocuous agents. *Nature*, 138: 32–35.

Selye, H. 1937a. Studies on adaptation. *Endocrinology*, 21: 169–188.

Selye, H. 1937b. The significance of the adrenals for adaptation. *Science*, 85: 247–248.

Selye, H. 1950. *Stress*. Acta, Inc., Montreal.

Selye, H. 1956. *The Stress of Life*. McGraw-Hill, New York.

Semsar, K., and Godwin, J. 2004. Multiple mechanisms of phenotype development in the bluehead wrasse. *Hormones and Behavior*, 45: 345–353.

Semsar, K., Kandel, F. L., and Godwin, J. 2001. Manipulations of the AVT system shift social status and related courtship and aggressive behavior in the bluehead wrasse. *Hormones and Behavior*, 40: 21–31.

Serra, G. B. 1983. *The Ovary*. Raven Press, New York.

Severino, S. K., and Moline, M. L. 1989. *Premenstrual Syndrome: A Clinician's Manual*. Guilford, New York.

Shaham, Y. and Stewart, J. 1994. Exposure to mild stress enhances the reinforcing efficacy of intravenous heroin self-administration in rats. *Psychopharmacology*, 114: 523–527.

Shaham, Y., and Stewart, J. 1995. Stress reinstates heroin-seeking in drug-free animals: an effect mimicking heroin, not withdrawal. *Psychopharmacology*,119: 334–341.

Shapiro, H. A. 1937. Effect of testosterone propionate on mating. *Nature*, 139: 588–589.

Shapiro, L. E., and Insel, T. R. 1990. Infant's response to social separation reflects adult differences in affiliative behavior: A comparative developmental study in prairie and montane voles. *Dev. Psychobiol.*, 23: 375–393.

Shapiro, L. E., Austin, D., Ward, S. E., and Dewsbury, D. A. 1986. Familiarity and female mate choice in two species of voles (*Microtus ochrogaster* and *Microtus montanus*). *Anim. Behav.*, 34: 90–97.

Shapiro, S., Schlesinger, E. R., and Nesbitt, R. E. L. 1968. *Infant, Perinatal, Maternal, and Childhood Mortality in the United States*. Harvard University Press, Cambridge, MA.

Sharman, G. B. 1970. Reproductive physiology of marsupials. *Science*, 167: 1221–1228.

Sharp, P. J., Li, Q., Georgiou, G. C., and Lea, R. W. 1996. Expression of *fos*-like immunoreactivity in the hypothalamus of the ring dove (*Streptopelia risoria*) at the onset of

Stone, C. P. 1938a. Loss and restoration of copulatory activity in adult male rats following castration and subsequent injections of testosterone propionate. *Endocrinology*, 23: 529.

Stone, C. P. 1938b. Activation of impotent male rats by injections of testosterone propionate. *J. Comp. Psychol.*, 25: 445–450.

Stone, C. P. 1938c. Effects of cortical destruction on reproductive behavior and maze learning in albino rats. *J. Comp. Psychol.*, 26: 217–236.

Stone, C. P. 1939. Copulatory activity in adult male rats following castration and injections of testosterone propionate. *Endocrinology*, 24: 165–174.

Stone, C. P., and Commins, W. D. 1936. The effect of castration at various ages upon the learning ability of male albino rats: II. Relearning after an interval of one year. *J. Genet. Psychol.*, 48: 20–28.

Stone, C. P., Barker, R. G., and Tomlin, M. I. 1935. Sexual drive in potent and impotent male rats as measured by the Columbia obstruction apparatus. *J. Genet. Psychol.*, 47: 33–48.

Stopa, E. G., Johnson, J. K., Friedman, D. I., Ryer, H. I., Reidy, J., Kuo-LeBlanc, V., and Albers, H. E. 1995. Neuropeptide Y receptor distribution and regulationin the suprachiasmatic nucleus of the Syrian hamster *(Mesocricetus auratus)*. *Peptide Res.*, 8: 95–100.

Storey, A. E., Courage, C., and Wynne-Edwards, K. 199. Of mice and men: Why is there variation in the social cues that trigger mammalian paternal care? Paper presented at the Animal Behaviour Society meeting, June 1999, College Park, MD.

Stowers, L., Holy, T. E., Meister, M., Dulac, C., and Koentges, G. 2002. Loss of sex discrimination and male-male aggression in mice deficient for TRP2. *Science*, 295: 1493–1500.

Strachan, M. W. J., Deary, I. J., Ewing, F. M. E., and Frier, B. M. 1997. Is type II diabetes associated with an increased risk of cognitive dysfunction? *Diabetes Care*, 20: 438–445.

Stratakis, C. A. and Chrousos, G. P. 1995. Neuroendocrinology and pathophysiology of the stress system. *Ann. N. Y. Acad. Sci.*, 771: 1–18.

Strauss, R. H., Liggett, M. T., and Lanese, R. R. 1985. Anabolic steroid use and perceived effects in ten weight-trained women athletes. *J. A. M. A.*, 253: 2871–2873.

Stricker, E. M. 1968. Some physiological and motivational properties of the hypovolemic stimulus for thirst. *Physiol. Behav.*, 3: 379–385.

Stricker, E. M. 1978. The renin-angiotensin system and thirst: Some unanswered questions. *Fed. Proc.*, 37: 2704–2710.

Stricker, E. M. 1984. Biological bases of hunger and satiety: Therapeutic implications. *Nutr. Rev.*, 42: 333–340.

Stricker, E. M., and Verbalis, J. G. 1988. Hormones and behavior: The biology of thirst and sodium appetite. *Am. Sci.*, 76: 261–267.

Stricker, E. M., and Verbalis, J. G. 1990a. Control of appetite and satiety: Insight from biologic and behavioral studies. *Nutr. Rev.*, 48: 49–56.

Stricker, E. M., and Verbalis, J. G. 1990b. Sodium appetite. In E. M. Stricker (ed.), *Handbook of Behavioral Neurobiology*, Vol. 10, pp. 387–419. Plenum, New York.

Stricker, E. M., Cooper, P. H., Marshall, J. F., and Zigmond, M. J. 1979. Acute homeostatic imbalances reinstate sensorimotor dysfunctions in rats with lateral hypothalamic lesions. *J. Comp. Physiol. Psychol.*, 93: 512–521.

Stricker, E. M., Hosutt, J. A., and Verbalis, J. G. 1987. Neurohypophyseal secretion in hypovolemic rats: Inverse relation to sodium appetite. *Am. J. Physiol.*, 252: R889–R896.

Stricker, E. M., Rowland, N., Saller, C. F., and Friedman, M. I. 1977. Homeostasis during hypoglycemia: Central control of adrenal secretion and peripheral control of feeding. *Science*,196: 79–81.

Strobel, A., Issad, T., Camoin, L., Ozata, M., and Strosberg, A. D. 1998. A leptin missense mutation associated with hypogonadism and morbid obesity. *Nat. Genet.*, 18: 213–215.

Strubbe, J. H., Steffens, A. B., and De Ruiter, L. 1977. Plasma insulin and the time pattern of feeding in the rat. *Physiol. Behav.*, 18: 81–86.

Su, T. P., Pagliaro, M., Schmidt, P. J., Pickar, D., Wolkowitz, O., and Rubinow, D. R. 1993. Neuropsychiatric effects of anabolic steroids in male normal volunteers. *J. A. M. A.*, 269: 2760–2764.

Suchecki, D., Mozaffarian, D., Gross, G., Rosenfeld, P., and Levine, S. 1993. Effects of maternal deprivation on the ACTH stress response in the infant rat. *Neuroendocrinology*, 57: 204–212.

Sundstrom, I, and Backstrom, T. 1998. Patients with premenstrual syndrome have decreased saccadic eye velocity compared to control subjects. *Biological Psychiatry*, 44: 755–764.

Sundstrom, I., Nyberg, S., and Backstrom, T. 1997. Patients with premenstrual syndrome have reduced sensitivity to midazolam compared to control subjects. *Neuropsychopharmacology*, 17: 370–381.

Suomi, S. J. 1977. Adult male-infant interactions among monkeys living in nuclear families. *Child Dev.*, 48: 1215–1270.

Suomi, S. J. 1991. Early stress and adult emotional reactivity in rhesus monkeys. *Ciba Found. Symp.*, 156: 171–183.

Suomi, S. J. 1997. Early determinants of behaviour: Evidence from primate studies. *British Medical Bulletin*, 53: 170–184.

Susman, J. L. 1996. Postpartum depressive disorders. *J. Fam. Pract.*, 43: S17–24.

Susser, E., Neugebauer, R., Hoek, H. W., Brown, A. S., Lin, S., Labovitz, D., and Gorman, J. M. 1996. Schizophrenia after prenatal famine: Further evidence. *Archives of General Psychiatry*, 53: 25–31.

Svare, B. 1983. *Hormones and Aggressive Behavior.* Plenum, New York.

Svare, B. 1990. Maternal aggression: Hormonal, genetic, and developmental determinants. In N. A. Krasnegor and R. S. Bridges (eds.), *Mammalian Parenting*, pp. 118–132. Oxford University Press, Oxford.

Svare, B., and Gandelman, R. 1976. Postpartum aggression in mice: Experiential and environmental factors. *Horm. Behav.*, 7: 407–416.

Svare, B., Mann, M. A., Broida, J., and Michael, S. 1982. Maternal aggression exhibited by hypophysectomized parturient mice. *Horm. Behav.*, 16: 455–461.

Svare, B., Miele, J., and Kinsley, C. 1986. Mice: Progesterone stimulates aggression in pregnancy-terminated females. *Horm. Behav.*, 20: 194–200.

Swaab, D. F., Chung, W. C. J., Kruijver, F. P. M., Hofman, M. A., and Ishunina, T. A. 2001. Structural and functional sex differences in the human hypothalamus. *Hormones and Behavior*, 40: 93–98.

Swann, J. M., Wang, J., and Govek, E. K. 2003. The MPN mag: Introducing a critical area mediating pheromonal and hormonal regulation of male sexual behavior. *Annals of the New York Academy of Science*, 1007: 199–210.

Swann, J., and Fiber, J. M. 1997. Sex differences in function of a pheromonally stimulated pathway: role of steroids and the main olfactory system. *Brain Res. Bull.*, 44: 409–413.

Sweeney, B. M. 1976. Evidence that membranes are components of circadian oscillators. In J. W. Hastings and H. G. Schweiger (eds.), *Molecular Basis of Circadian Rhythms*, pp. 267–281. Abakon Verlagsgesellschaft, Berlin.

Sweeney, T., Donovan, A., Karsch, F. J., Roche, J. F., and O'Callaghan, D. 1997. Influence of previous photoperiodic exposure on the reproductive response to a specific photoperiod signal in ewes. *Biol. Reprod.*, 56: 916–920.

Szymusiak, R., and Satinoff, E. 1982. Acute thermoregulatory effects of unilateral electrolytic lesions of the medial and lateral preoptic area in rats. *Physiol. Behav.*, 28: 161–170.

Taibell, A. 1928. Riseglio artificiale di istinti tipicamenta feminili nei maschi di faluni uccelli. *Atti. Soc. Nat. Mat. Modena*, 59: 93–102.

Takahashi, J. S. 1991. Circadian rhythms: From gene expression to behavior. *Curr. Opinion Neurobiol.*, 1: 556–561.

Takahashi, J. S. 1992. Circadian clock genes are ticking. *Science*, 258: 238–240.

Takahashi, J. S. 1993. Circadian-clock regulation of gene expression. *Curr. Opinion Genet. Devel.*, 3: 301–309.

Takahashi, J. S., and Menaker, M. 1980. Interaction of estradiol and progesterone: Effects on circadian locomotor rhythms in female golden hamsters. *Am. J. Physiol.*, 239: R497–R504.

Takahashi, J. S., Kondo, H., Yoshimura, M., and Ochi, Y. 1974. Thyrotropin responses to TRH in depressive illness: Relation to clinical subtypes and prolonged duration of depressive episode. *Folia Psychiatrica Neurol. Japonica*, 28: 355–365.

Takahashi, J. S., Kornhauser, J. M., Koumenis, C., and Eskin, A. 1993. Molecular approaches to understanding circadian oscillations. *Annu. Rev. Physiol.*, 55: 729–753.

Takahashi, J. S., Murakami, N., and Nikaido, S. S. 1989. The avian pineal, a vertebrate model system of the circadian oscillator: Cellular regulation of circadian rhythms by light, second messengers, and macromolecular synthesis. *Rec. Prog. Horm. Res.*, 45: 279–352.

Takahashi, J. S., Pinto, L. H., and Vitaterna, M. H. 1994. Forward and reverse genetic approaches to behavior in the mouse. *Science*, 264, 1724–1733.

Takahashi, L. K. 1998. Prenatal stress: Consequences of glucocorticoids on hippocampal development and function. *Int. J. Dev. Neurosci.*, 16: 199–207.

Tamarkin, L., Hutchison, J. S., and Goldman, B. D. 1976. Regulation of serum gonadotropins by photoperiod and testicular hormone in the Syrian hamster. *Endocrinology*, 99: 1528–1533.

Tanagho, E. A., Lue, T. F., and McClure, R. D. 1988. *Contemporary Management of Impotence Infertility*, Williams & Wilkins, Baltimore.

Tanner, J. M. 1962. *Growth at Adolescence.* Blackwell, Oxford.

Taravosh-Lahn, K., and Delville, Y. 2004. Aggressive behavior in female golden hamsters: Development and the effect of repeated social stress. *Hormones and Behavior*, 46: 428–435.

Tartaglia, L. A., Dembski, M., Weng, X., Deng, N., Culpepper, J., Devos, R., Richards, G. J., Campfield, L. A., Clark, F. T., Deeds, J., Muir, C., Sanker, S., Moriarty, A., Moore, K. J., Smutko, J. S., Mays, G. G., Woolf, E. A., Monroe, C. A., and Tepper, R. I. 1995. Identification and expression cloning of a leptin receptor, OB-R. *Cell*, 83: 1263–1271.

Tate-Ostroff B. and Bridges, R. S. 1985. Plasma prolactin levels in parental male rats: effects of increased pup stimuli. *Horm. Behav.*, 19: 220–226.

Taylor, J. W. 1979. The timing of menstruation-related symptoms assessed by a daily symptoms rating scale. *Acta Psychiatrica Scand.*, 60: 87–105.

Taylor, W. N., and Black, A. B. 1987. Pervasive anabolic steroid use among health club athletes. *Ann. Sports Med.*, 3: 155–159.

Taymans, S. E., DeVries, A. C., DeVries, DeVries, M. B., Nelson, R. J., Friedman, T. C., Castro, M., Detera-Wadleigh, S., Carter, C. S., and Chrousos, G. P. 1997. The hypothalamic-pituitary-adrenal axis of prairie voles (*Microtus ochrogaster*): Evidence for target tissue glucocorticoid resistance. *Gen. Comp. Endocrinol.*, 106: 48–61.

Tegelman, R., Carlström, K., and Pousette, A. 1988. Hormone levels in male ice hockey players during a 26-hour cup tournament. *Int. J. Androl.*, 11: 361–368.

Telner, J., Lepore, F., and Guillemot, J. P. 1979. Effects of seratonin content on pain sensitivity in the rat. *Pharmacol. Biochem. Behav.*,10: 657–661.

Tepper, B. J., and Friedman, M. I. 1991. Altered acceptability of and preference for sugar solutions by diabetic rats is normalized by high-fat diet. *Appetite*, 16: 25–38.

Terkel, J., and Rosenblatt, J. S. 1968. Maternal behavior induced by maternal blood plasma injected into virgin rats. *J. Comp. Physiol. Psychol.*, 65: 479–482.

Terkel, J., and Rosenblatt, J. S. 1972. Humoral factors underlying maternal behavior at parturition. *J. Comp. Physiol. Psychol.*, 80: 365–371.

Terman, M. 1988. On the question of mechanism in phototherapy for seasonal affective disorder: Considerations for clinical efficacy and epidemiology. *J. Biol. Rhythms*, 3: 155–172.

Teyler, T. J. and Discenna, P. 1984. Long-term potentiation as a candidate mnemonic device. *Brain Res.*, 319: 15–28.

Thase, M. E. 1989. Comparison between seasonal affective disorder and other forms of recurrent depression. In N. E. Rosenthal and M. C. Blehar (eds.), *Seasonal Affective Disorders and Phototherapy*, pp. 64–78. Guilford Press, New York.

Thiessen, D. D., and Rodgers, D. A. 1961. Population density and endocrine function. *Psychol. Bull.*, 58: 441–451.

Thomas, J. A., and Birney, E. C. 1979. Parental care and mating system of the prairie vole, *Microtus ochrogaster. Behav. Ecol. Sociobiol.*, 5: 171–186.

Thomas, S. A., and Palmiter, R. D. 1997. Impaired maternal behavior in mice lacking norepinephrine and epinephrine. *Cell*, 91: 583–592.

Thompson, C. I. 1980. *Controls of Eating*. Jamaica, New York, Spectrum.

Thompson, C. W., and Moore, M. C. 1992. Behavioral and hormonal correlates of alternative reproductive strategies in a polygynous lizard: tests of the relative plasticity and challenge hypotheses. *Horm. Behav.*, 26: 568–585.

Thompson, C., Stinson, D., and Smith, A. 1990. Seasonal affective disorder and season-dependent abnormalities of melatonin suppression of light. *Lancet*, 336: 703–706.

Thompson, D. A., Welle, S. L., Lilavivat, U., Pericaud, L., and Campbell, R. G. 1982. Opiate receptor blockade in man reduces 2-Deoxy-D-glucose induced food intake, but not hunger, thirst, and hypothermia. *Life Sci.*, 31: 847–852.

Thornton, J. W., Need, E., and Crews, D. 2003. Resurrecting the ancestral steroid receptor: Ancient origin of estrogen signaling. *Science*, 301: 1714–1717.

Thrun, L. A., Dahl, G. E., Evans, N. P., and Karsch, F. J. 1997. Effect of thyroidectomy on maintenance of seasonal reproductive suppression in the ewe. *Biol. Reprod.*, 56: 1035–1040.

Thys-Jacobs, S. and Alvir, M. J. 1995. Calcium-regulating hormones across the menstrual cycle: Evidence of a secondary hyperparathyroidism in women with PMS. *J. Clin. Endocrinol. Metab.*, 80: 2227–2232.

Thys-Jacobs, S., Starkey, P., Bernstein, D., and Tian, J. 1998. Calcium carbonate and the premenstrual syndrome: Effects on premenstrual and menstrual symptoms. Premenstrual Syndrome Study Group. *Am. J. Obstet. Gynecol.*, 179: 444–452.

Tian, M., Broxmeyer, H. E., Fan, Y., Lai, Z., Zhang, S., Aronica, S., Cooper, S., Bigsby, R. M., Steinmetz, R., Engle, S. J., Mestek, A., Pollock, J. D., Lehman, M. N., Jansen, H. T., Ying, M., Stambrook, P. J., Tischfield, J. A., and Yu, L. 1997. Altered hematopoiesis, behavior, and sexual function in μ-opioid receptor-deficient mice. *J. Exp. Med.*, 185: 1517–1522.

Tiemstra, J. D. and Patel, K. 1998. Hormonal therapy in the management of premenstrual syndrome. *J. Am. Fam. Pract.*, 11: 378–381.

Tinbergen, N. 1935. The behavior of the red-necked Phalarope in spring. *Ardea*, 24: 1–42.

Tinbergen, N. 1951. *The Study of Instinct.* Oxford University Press, Oxford.

Toates, F. 1995. *Stress: Conceptual Biological Aspects.* Wiley: New York.

Toh, K. L., Jones, C. R., He, Y., Eide, E. J., Hinz, W. A., Virshup, D. M., Ptaček, L. J., and Fu, Y.-H. 2001. An *hPer2* phosphorylation site mutation in familial advanced sleep phase syndrome. *Science*, 291: 1040–1043.

Tonegawa, S. 1995. Mammalian learning and memory studied by gene targeting. *Ann. N. Y. Acad. Sci.*, 758, 213–217.

Toran-Allerand, C. D. 1984. On the genesis of sexual differentiation of the central nervous system: Morphogenetic consequences of steroidal exposure and possible role of a-fetoprotein. In G. J. DeVries, J. P. C. DeBruin, H. B. M. Uylings, M. A. Corner (eds.), *Sex Differences in the Brain*, pp. 63–98. Elsevier, Amsterdam.

Toyoda, F., Tanaka, S., Matsuda, K., and Kikuyama, S. 1994. Hormonal control of response to and secretion of sex attractants in Japanese newts. *Physiol. Behav.*, 55: 569–576.

Toyoda, F., Yamamoto, K., Iwata, T., Hasunuma, I., Cardinali, M., Mosconi, G., Polzonetti-Magni, A. M., and Kikuyama, S. 2004. Peptide pheromones in newts. *Peptides*, 25: 1531–1536.

Trainor, B. C., and Marler, C. A. 2001. Testosterone, paternal behavior, and aggression in the California mouse, *Peromyscus californicus. Hormones and Behavior*, 40, 32–42.

Trainor, B. C., and Marler, C. A. 2002. Testosterone promotes paternal behaviour in a monogamous mammal via conversion to oestrogen. *Proceedings of the Royal Society of London, Series B*, 269: 823–829.

Trainor, B. C., Bird, I. M., Alday, N. A., Schlinger, B. A., and Marler, C. A. 2003. Variation in aromatase activity in the medial preoptic area and plasma progesterone is associated with the onset of paternal behavior. *Neuroendocrinology*, 78, 36–44.

Trainor, B. C., Bird, I. M., and Marler, C. A. 2004. Opposing hormonal mechanisms of aggression revealed through short-lived testosterone manipulations and multiple winning experiences. *Hormones and Behavior*, 45: 115–121.

Tramontin, A. D., and Brenowitz, E. A. 2000. Seasonal plasticity in the adult brain. *Trends in Neuroscience*, 23: 251–258.

Tricker, R., Casaburi, R., Storer, T. W., Clevenger, B., Berman, N., Shirazi, A., and Bhasin, S. 1996. The effect of supraphysiological doses of testosterone on angry behavior in healthy eugonadal men: A clinical research center study. *Journal of Clinical Endocrinology and Metabolism*, 81: 3754–3758.

Triemstra, J. L., and Wood, R. I. 2004. Testosterone self-administration in female hamsters. *Behavioral Brain Research*, 154: 221–229.

Trivers, R. L. 1972. Parental investment and sexual selection. In B. G. Campbell (ed.), *Sexual Selection and the Descent of Man*. 1871–1971, pp. 136–179. Aldine, Chicago.

Trivers, R. L. 1974. Parent-offspring conflict. *American Zoologist* 14: 249–264.

Trobec, R. J., and Oring, L. W. 1972. Effects of testosterone propionate implantation on lek behavior of sharp-tailed grouse. *Am. Midland Nat.*, 87: 531–536.

Truitt, W. A., and Coolen, L. M. 2002. Identification of a potential ejaculation generator in the spinal cord. *Science*, 297: 1566–1569.

Truitt, W. A., Shipley, M. T., Veening, J. G., and Coolen, L. M. 2003. Activation of a subset of lumbar spinothalamic neurons after copulatory behavior in male but not female rats. *Journal of Neuroscience*, 23: 325–331.Alexander, G. M. 2003. An evolutionary perspective of sex-typed toy preferences: Pink, blue, and the brain. *Archives of Sexual Behavior*, 32: 7–14.

Truman, J. W. 1992. The eclosion hormone system of insects. *Progr. Brain Res.*, 92: 361–374.

Trumbo, S. T. 1996. Parental care in invertebrates. *Adv. Study Behav.*, 25: 3–51.

Trumbo, S. T. 1997. Juvenile hormone-mediated reproduction in burying beetles: From behavior to physiology. *Arch. Ins. Biochem. Physiol.*, 35: 479–490.

Trumbo, S. T., Borst, D. W., and Robinson, G. E. 1995. Rapid elevation of juvenile hormone titer during behavioral assessment of the breeding resource by the burying beetle, *Nicrophorus orbicollis*. *J. Insect Physiol.*, 41: 535–543.

Tsai, C. 1925. The relative strength of sex and hunger motives in the albino rat. *J. Comp. Psychol.*, 5: 407.

Tserotas, K., and Merino, G. 1998. Andropause and the aging male. *Arch. Androl.*, 40: 87–93.

Tsigos, C., and Chrousos, G. P. 2002. Hypothalamic–pituitary–adrenal axis, neuroendocrine factors and stress. *Journal of Psychosomatic Research*, 53: 865–871.

Turek, F. W., and Campbell, C. S. 1979. Photoperiodic regulation of neuroendocrine-gonadal activity. *Biol. Reprod.*, 20: 32–50.

Turek, F. W., Elliott, J. A., Alvis, J. D., and Menaker, M. 1975. The interaction of castration and photoperiod in the regulation of hypophyseal and serum gonadotropin levels in male golden hamsters. *Endocrinology*, 96: 854–860.

Turner, B. N., Perrin, M. R., and Iverson, S. L. 1975. Winter coexistence of voles in spruce forest: Relevance of seasonal changes in aggression. *Can. J. Zool.*, 53: 1004–1011.

Turner, C. D., and Bagnara, J. T. 1971. *General Endocrinology*. Saunders, Philadelphia.

Turton, M. D., O'Shea, M., Gunn, I., Beak, S. A., Edwards, C. M. B., Meeran, K., Choi, S. J. et al. 1996. A role for glucagon-like peptide-1 in the central regulation of feeding. *Nature*, 379: 69–72.

Tuttle, W. W., and Dykshorn, S. 1928. The effect of castration and ovariectomy on spontaneous activity and ability to learn. *Proc. Soc. Exp. Biol. Med.*, 25: 569–570.

Twiggs, D. G., Popolow, H. B., and Gerall, A. A. 1978. Medial preoptic lesions and male sexual behavior: Age and environmental interactions. *Science*, 200: 1414–1415.

Tyler, S. J. 1972. The behaviour and social organization of the New Forest ponies. *Anim. Behav. Monogr.*, 5: 87–196.

Tyndale-Biscoe, H., and Renfree, M. 1987. *Reproductive Physiology of Marsupials*. Cambridge University Press, Cambridge.

Tyrrell, C. L., and Cree, A. 1998. Relationships between corticosterone concentration and season, time of day and confinement in a wild reptile *(Tuatara, Sphenodon punctatus)*. *Gen. Comp. Endocrinol.*, 110: 97–108.

Udry, J. R., and Morris, N. M. 1968. Distribution of coitus in the menstrual cycle. *Nature*, 220: 593–595.

Uhrich, J. 1938. The social hierarchy in albino mice. *J. Comp. Psychol.*, 23: 373–413.

Ung, E. K., Lee, S., and Kua, E. H. 1997. Anorexia nervosa and bulimia—a Singapore perspective. *Singapore Medical Journal*, 38: 332–335.

Uno, H., Tarara, R., Else, J. G., Suleman, M. A., and Sapolsky, R. M. 1989. Hippocampal damage associated with prolonged and fatal stress in primates. *J. Neurosci.*, 9: 1705–1711.

Ursin, H., Baade, E., and Levine, S. 1978. *Psychobiology of Stress: A Study of Coping Men*. Academic Press: London.

Ussher, J. 1989. *The Psychology of the Female Body*. Routledge, New York.

Valenstein, E. S., and Young, W. C. 1955. An experiential factor influencing the effectiveness of testosterone proprionate in eliciting sexual behavior in male guinea pigs. *Endocrinology*, 56: 173–185.

Van de Kar, L. D., Richardson-Morton, K. D., and Rittenhouse, P. A. 1991. Stress: Neuroendocrine and pharmacological mechanisms. In G. Jasmin and M. Cantin (eds.). *Stress Revisited: Neuroendocrinology of Stress. Methods and Achievements in Experimental Pathology*, Vol 14: 133–173. Karger, Basel.

van der Lee, S., and Boot, L. M. 1955. Spontaneous pseudopregnancy in mice. *Acta Physiol. Pharmacol. Neer.*, 4: 442–443.

Van Dis, H., and Larsson, K. 1971. Induction of sexual arousal in the castrated male rat by intracranial stimulation. *Physiol. Behav.*, 6: 85–86.

Van Dyck, C. H., Malison, R. T., Staley, J. K., Jacobsen, L. K., Seibyl, J. P., Laruelle, M., Baldwin, R. M., Innis, R. B., and Gelernter, J. 2004. Central serotonin transporter availability measured with [^{123}I]β-CIT SPECT in relation to serotonin transporter genotype. *American Journal of Psychiatry*, 161: 525–531.

van Furth, W. R., Wolterink, G., and van Ree, J. M. 1995. Regulation of masculine sexual behavior: Involvement of brain opioids and dopamine. *Brain Res. Rev.*, 21: 162–184.

Van Gelder, R. N. 2003. Making (a) sense of non-visual ocular photoreception. *Trends in Neurosciences*, 26: 458–461.

Van Goozen, S. H., Wiegant, V. M., Endert, E., Helmond, F. A., and Van de Poll, N. E. 1997. Psychoendocrinological assessment of the menstrual cycle: the relationship between hormones, sexuality, and mood. *Arch. of Sex. Behav.*, 26: 359–382.

Van Oers, H. J., de Kloet, E. R., Whelan, T., and Levine, S. 1998. Maternal deprivation effect on the infant's neural stress markers is reversed by tactile stimulation and feeding but not by suppressing corticosterone. *J. Neurosci.*, 18: 10171–10179.

van Os, J., and Selten, J. P. 1998. Prenatal exposure to maternal stress and subsequent schizophrenia: The May 1940 invasion of The Netherlands. *British Journal of Psychiatry*, 172: 324–326.

Van Reeth, O., and Turek, F. W. 1989. Stimulated activity mediates phase shifts in the hamster circadian clock induced by dark pulses or benzodiazepines. *Nature*, 339: 49–51.

van Tienhoven, A. 1983. *Reproductive Physiology of Vertebrates.* Cornell University Press, Ithaca.

Vancassel, M., Foraste, M., Strambi, A., and Strambi, C. 1984. Normal and experimentally induced changes in hormonal hemolymph titres during parental behavior of the earwig, *Labidura riparia. Gen. Comp. Endocrinol.*, 56: 444–456.

Vandenbergh, J. G. 1969. Endocrine coordination in monkeys: Male sexual responses to the female. *Physiol. Behav.*, 4: 261–264.

Vanderreycken, W., and Van Deth, R. 1994. *From Fasting Saints to Anorexic Girls.* New York University Press, New York.

Vasey, P. L. 2002. Same-sex sexual partner preference in hormonally and neurologically unmanipulated animals. *Annual Review of Sex Research*, 13: 141–179.

Veldhuis, J. D., Evans, W. S., Demers, L. M., Thorner, M. O., Wakat, D., and Rogol, A. D. 1985. Altered neuroendocrine regulation of gonadotropin secretion in women distance runners. *J. Clin. Endocrinol. Metab.*, 61: 557–563.

Vergnes, M., Depaulis, A., and Boehrer, A. 1986. Parachlorophenylalanine-induced seratonin depletion increases offensive but not defensive aggression in male rats. *Physiol. Behav.*, 36(4): 653–658.

Vermeulen, A., Rubens, R., and Verdonck, L. 1972. Testosterone secretion and metabolism in male senescence. *J. Clin. Endocrinol. Metab.*, 34: 730–735.

Vessely, L. H., and Lewy, A. J. 2002. Melatonin as a hormone and as a marker for circadian phase position in humans. In D. W. Pfaff, A. P. Arnold, A. M. Etgen, S. E. Fahrbach, and R. T. Rubin (eds.), *Hormones, Brain and Behavior*, Volume 5, pp. 121–141. Academic Press, New York.

Viau, V., and Meaney, M. J. 1991. Variations in the hypothalamic-pituitary-adrenal response to stress during the estrous cycle of the adult rat. *J. Neurosci.*, 12: 2503–2511.

Virgin, C. E., and Sapolsky, R. M. 1997. Styles of male social behavior and their endocrine correlates among low-ranking baboons. *Am. J. Primatol.*, 42: 25–39.

vom Saal, F. S. 1979. Prenatal exposure to androgen influences morphology and aggressive behavior of male and female mice. *Horm. Behav.*, 12: 1–11.

vom Saal, F. S., and Bronson, F. H. 1980. Variation in length of the estrous cycle in mice due to former uterine proximity to male fetuses. *Biol. Reprod.*, 22: 777–780.

vom Saal, F. S., and Finch, C. E. 1988. Reproductive senescence: Phenomena and mechanisms in mammals and selected vertebrates. In E. Knobil and J. Neill (eds.), *Physiology of Reproduction*, pp. 2351– 2413. Raven Press, New York.

Waber, D. 1976. Sex differences in cognition: A function of maturation rate. *Science*, 192: 572–574.

Wade, G. N. 1972. Gonadal hormones and behavioral regulation of body weight. *Physiol. Behav.*, 8: 523–534.

Wade, G. N. 1976. Sex hormones, regulatory behaviors and body weight. *Adv. Study Behav.*, 6: 201–279.

Wade, G. N. 1986. Sex steroids and energy balance: Sites and mechanisms of action. *Ann. N. Y. Acad. Sci.*, 474: 389–399.

Wade, G. N., and Gray, J. M. 1978. Cytoplasmic 17β-[3H]estradiol binding in rat adipose tissues. *Endocrinology*, 103: 1695–1701.

Wade, G. N., and Gray, J. M. 1979. Gonadal effects on food intake and adiposity: A metabolic hypothesis. *Physiol. Behav.*, 22: 583–593.

Wade, G. N., and Jones, J. E. 2003. Lessons from experimental disruption of estrous cycles and behaviors. *Medicine and Science in Sports and Exercise*, 35: 1573–1580.

Wade, G. N., and Zucker, I. 1970. Modulation of food intake and locomotor activity in female rats by diencephalic hormone implants. *J. Comp. Physiol. Psychol.*, 72: 328–336.

Wade, G. N., Jennings, G., and Trayhurn, P. 1986. Energy balance and brown adipose tissue thermogenesis during pregnancy in Syrian hamsters. *Am. J. Physiol.*, 250: R845–R850.

Wade, G. N., Lempicki, R. L., Panicker, A. K., Frisbee, R. M., and Blaustein, J. D. 1997. Leptin facilitates and inhibits sexual behavior in female hamsters. *Am. J. Physiol.*, 272: R1354–R1358.

Wade, J. 2001. Zebra finch sexual differentiation: The aromatization hypothesis revisited. *Microscopy Research and Technique*, 54: 354–363.

Wade, J., and Arnold, A. P. 1996. Functional testicular tissue does not masculinize development of the zebra finch song system. *Proc. Natl. Acad. Sci. U.S.A.*, 93: 5264–5268.

Wahlstrom, G. 1965. *Circadian Clocks.* North-Holland, Amsterdam.

Walker, W. A., and Feder, H. H. 1977. Inhibitory and facilitory effects of various anti-estrogens on the induction of female sexual behavior by estradiol benzoate in guinea pigs. *Brain Res.*, 134: 455–465.

Wallen, K. 1990. Desire and ability: Hormones and the regulation of female sexual behavior. *Neurosci. Biobehav. Rev.*, 14: 233–241.

Wallen, K., and Baum, M. J. 2002. Masculinization and defeminization in altricial and precocial mammals: Comparative aspects of steroid hormone action. In D. W. Pfaff, A. Arnold, A. Etgen, S. Fahrbach, and R. Rubin (eds.), *Hormones, Brain, and Behavior*, Volume 4, pp. 385–423. Academic Press, New York.

Wallen, K., and Goy, R. W. 1977. Effects of estradiol benzoate, estrone, and propionates of testosterone or dihydrotestosterone on sexual and related behaviors of ovariectomized rhesus monkeys. *Horm. Behav.*, 9: 228–248.

Wallen, K., and Zehr, J. L. 2004. Hormones and history: The evolution and development of primate female sexuality. *Journal of Sexual Research*, 41: 101–112.

Wallen, K., Winston, L. A., Gaventa, S., Davis-DaSilva, M., and Collins, D.C. 1984. Periovulatory changes in female sexual behavior in group-living rhesus monkeys. *Horm. Behav.*, 18: 431–450.

Wang, C., and Swerdloff, R. S. 1997. Androgen replacement therapy. *Ann. Med.*, 29: 365–370.

Wang, G. H. 1923. The relation between 'spontaneous' activity and oestrous cycle in the white rat. *Comp. Psychol. Monogr.*, 2: 1–27.

Wang, G. H. 1924. The changes in the amount of daily food-intake of the albino rat during pregnancy and lactation. *Am. J. Physiol.*, 71: 735–741.

Wang, Z., and Young, L. J. 1997. Ontogeny of oxytocin and vasopressin receptor binding in the lateral septum in prairie and montane voles. *Brain Res.: Dev. Brain Res.*, 104: 191–195.

Ward, I. L. 1984. The prenatal stress syndrome: Current status. *Psychoneuroendocrinology*, 9: 3–11.

Ward, I. L. 1992. Sexual behavior: The product of perinatal hormonal and prepubertal social factors. In *Sexual Differentiation*, Vol. 11 of *Handbook of Behavioral Neurobiology*, edited by A. A. Gerall, H. Moltz, and I. L. Ward, Plenum Press: New York.

Ward, I. L., and Reed, J. 1985. Prenatal stress and prepubertal social rearing conditions interact to determine sexual behavior in male rats. *Behav. Neurosci.*, 99: 301–309.

Ward, I. L., and Stehm, K. E. 1991. Prenatal stress feminizes juvenile play patterns in male rats. *Physiol. Behav.*, 50: 601–605.

Ward, I. L., and Weisz, J. 1980. Maternal stress alters plasma testosterone in fetal males. *Science*, 207: 328–329.

Ward, I. L., Ward, O. B., French, J. A., Hendricks, S. E., Mehan, D., and Winn, R. J. 1996. Prenatal alcohol and stress interact to attenuate ejaculatory behavior, but not serum testosterone or LH in adult male rats. *Behav. Neurosci.*, 110: 1469–1477.

Ward, I. L., Ward, O. B., Winn, R. J., and Bielawski, D. 1994. Male and female sexual behavior potential of male rats prenatally exposed to the influence of alcohol, stress, or both factors. *Behav. Neurosci.*, 108: 1188–1195.

Ward, M. W., and Holimon, T. D. 1999. Calcium treatment for premenstrual syndrome. *Annals in Pharmacotherapy*, 33: 1356–1358.

Ward, O. B., Wexler, A. M., Carlucci, J. R., Eckert, M. A., and Ward, I. L. 1996. Critical periods of sensitivity of sexually dimorphic spinal nuclei to prenatal testosterone exposure in female rats. *Horm. Behav.*, 30: 407–415.

Ward, P. 1965. Seasonal changes in the sex ratio of Quelea quelea. *Ibis*, 107: 397–399.

Warden, C. J., Jenkins, T. N., and Warner, L. H. 1935. *Comparative Psychology: A Comprehensive Treatise: Principles and Methods.* Vol. 1. Ronald Press: New York, NY.

Warner, L. H. 1927. A study of sex behavior in the white rat by means of the obstruction method. *Comp. Psychol. Monogr.*, 4:1–68.

Warren, M. P., and Shortle, B. 1990. Endocrine correlates of human parenting. In N. A. Krasnegor and R. S. Bridges (eds.), *Mammalian Parenting*, pp. 209–230. Oxford University Press, Oxford.

Warren, S. G., and Juraska, J. M. 1997. Spatial and nonspatial learning across the rat estrous cycle. *Behav. Neurosci.*, 111: 259–266.

Warren, S. G., Humphreys, A. G., Juraska, J. M., and Greenough, W. T. 1995. LTP varies across the estrous cycle: Enhanced synaptic plasticity in proestrus rats. *Brain Res.*, 703: 26–30.

Wartofsky, L., and Burman, K. D. 1982. Alterations in thyroid function in patients with systemic illness: The "euthyroid sick syndrome." *Endocrine Reviews*, 3: 164–217.

Watanabe, Y., Gould, E., and McEwen, B. S. 1992. Stress induces atrophy of apical dendrites of hippocampal CA3 pyramidal neurons. *Brain Res.*, 588: 341–345.

Waterhouse, J., Reilly, T., and Atkinson, G. 1998. Melatonin and jet lag. *Br. J. Sports Med.*, 32: 98–99.

Watson, J. S. 1969. Operant conditioning of visual fixation in infants under visual and auditory reinforcement. *Dev. Psychol.*, 1: 508–516.

Watson, J. T., and Adkins-Regan, E. 1989. Testosterone implanted in the preoptic area of male Japanese quail must be aromatized to activate copulation. *Horm. Behav.*, 23: 432–447.

Wayne, N. L., Malpaux, B., and Karsch, F. J. 1988. How does melatonin code for day length in the ewe: Duration of nocturnal melatonin release or coincidence of melatonin with a light-entrained sensitive period. *Biol. Reprod.*, 39: 66–75.

Weatherford, S. C., Laughton, W. B., Salabarria, J., Danho, W., Tilley, J. W., Netterville, L. A., Schwartz, G. J, and Moran, T. H. 1993. CCK satiety is differentially mediated by high- and low-affinity CCK receptors in mice and rats. *Am. J. Physiol.*, 264: R244–R249.

Weaver, D. R., Rivkees, S. A., and Reppert, S. M. 1989. Localization and characterization of melatonin receptors in rodent brain by in vitro autoradiography. *J. Neurosci.*, 9: 2581–2590.

Weaver, I. C. G., Cervoni, N., Champagne, F. A., D'Alessio, A. C., Sharma, S., Seckl, J. R., Dymov, S., Szyf, M., and Meaney, M. J. 2004. Epigenetic programming by maternal behavior. *Nature Neuroscience*, 7: 847–854.

Wehr, T. A. 1998. Effect of seasonal changes in daylength on human neuroendocrine functions. *Horm. Res.*, 49: 118–124.

Wehr, T. A., Wirz-Justice, A., Goodwin, F. K., Duncan, W., and Gillin, J. C. 1979. Phase advance of the circadian sleep-wake cycle as an antidepressant. *Science*, 206: 710–713.

Weil-Malharbe, H., Axelrod, J., and Tomchick, R. 1959. Blood-brain barrier for adrenaline. *Science*, 129: 1226–1228.

Weinstock, M. 1997. Does prenatal stress impair coping and regulation of hypothalamic–pituitary–adrenal axis? *Neuroscience and Biobehavioral Reviews*, 21: 1–10.

Weinstock, M. 1997. Does prenatal stress impair coping and regulation of hypothalmic-pituitary-adrenal axis. *Neurosci. Biobehav. Rev.*, 21: 1–10.

Weisinger, R. S., Considine, P., Denton, D. A., Leksell, L, McKinley, M. J., Mouw, D. R., Muller, A. F., and Tarjan, E. 1982. Role of sodium concentration of the cerebrospinal fluid in the salt appetite of sheep. *Am. J. Physiol.*, 242: R51–R63.

Weiss, J. M. 1968. Effects of coping response on stress. *J. Comp. Physiol. Psychol.*, 65: 251–260.

Weiss, J. M. 1972. Biological factors in stress and disease. *Sci. Am.*, 226: 104–113.

Weller, A., and Blass, E. M. 1988. Behavioral evidence for cholecystokinin-opiate interactions in neonatal rats. *Am. J. Physiol.*, 255: R901–907.

Weller, A., and Blass, E. M. 1990. Cholecystokinin conditioning in rats: Ontogenesis determinants. *Behav. Neurosci.*, 104: 199–206.

Welsh, T. H., Kemper-Green, C. N., and Livingston, K. N. 1999. Stress and reproduction. In E. Knobil and J. D. Neill (eds.). *Encyclopedia of Reproduction*, pp. 662–674. Academic Press, San Diego.

Wemyss-Holden, S. A., Hamdy, F. C., and Hastie, K. J. 1994. Steroid abuse in athletes, prostatic enlargement and bladder outflow obstruction: Is there a relationship? *British Journal of Urology*, 74: 476–478.

Wen, J. C., Hotchkiss, A. K., Demas, G. E., and Nelson, R. J. 2005. Photoperiod affects neuronal nitric oxide synthase and aggressive behavior in male Siberian hamsters (*Phodopus sungorus*). *Journal of Neuroendocrinology*. 16: 916–921.

Werren, J. H., Gross, M. R., and Shine, R. 1980. Paternity and the evolution of male parental care. *J. Theor. Biol.*, 82: 619–631.

West, S. D., and Dublin, H. T. 1984. Behavioral strategies of small mammals under winter conditions: Solitary or social? *Bull. Carnegie Museum Nat. Hist.*, 10: 293–300.

Wetterberg, L. 1999. Melatonin and clinical application. *Reproduction, Nutrition, and Development*, 39: 367–382.

Wever, R. A. 1989. Light effects on human circadian rhythms: A review of recent Andechs experiments. *J. Biol. Rhythms*, 4: 161–185.

Whalen, R. E., and Johnson, F. 1987. Individual differences in the attack behavior of male mice: A function of attack stimulus and hormonal state. *Horm. Behav.*, 21: 223–233.

Whipple, B., and Komisaruk, B. R. 2002. Brain (PET) responses to vaginal-cervical self-stimulation in women with complete spinal cord injury: Preliminary findings. *Journal of Sex and Marital Therapy*, 28: 79–86.

White, G. F., Katz, J., and Scarborough, K. E. 1992. The impact of professional football games upon violent assaults on women. *Violence Victims*, 7: 157–171.

White, S. A., and Fernald, R. D. 1997. Changing through doing: Behavioral influences on the brain. *Progress in Hormone Research*, 52: 455–474.

Whitfield-Rucker, M., and Cassone, V. 1996. Melatonin binding in the house sparrow song control system: sexual dimorphism and the effect of photoperiod. *Horm. Behav.*, 30: 528–37.

Whitten, W. K. 1956a. The effect of removal of the olfactory bulbs on the gonads of mice. *J. Endocrinol.*, 14: 160–163.

Whitten, W. K. 1956b. Modification of the oestrus cycle of the mouse by external stimuli associated with the male. *J. Endocrinol.*, 13: 399–404.

Whitten, W. K. 1957. Effect of exteroceptive factors on the oestrus cycle of mice. *Nature*, 180: 1456–1457.

Whittenberger, J. R., and Tilson, R. L. 1980. The evolution of monogamy: Hypotheses and evidence. *Annu. Rev. Ecol. Syst.*, 11: 197–232.

Widdowson, E. 1951. Mental contentment and physical growth. *Lancet*, (13 June), 1316–1328.

Widowski, T. M., Porter, T. A., Ziegler, T. E., and Snowdon, C. T. 1992. The stimulatory effect of males on the initiation, but not the maintenance of ovarian cycling in cotton-top tamarins (*Saguinus oedipus*). *Am. J. Primatol.*, 26: 97–108.

Wiesner, B. P., and Mirskaia, L. 1930. *Q. J. Exp. Physiol.*, 20: 273–294.

Wiesner, B. P., and Sheard, N. M. 1933. *Maternal Behavior in Rats.* Oliver and Boyd, London.

Wilkins, L., and Richter, C. P. 1940. A great craving for salt by a child with cortico-adrenal insufficiency. *J. A. M. A.*, 114: 866–868.

Wilkinson, C. W., Shinsako, J., and Dallman, M. F. 1979. Daily rhythms in adrenal responsiveness to adrenocorticotropin are determined primarily by the time of feeding in the rat. *Endocrinology*, 104: 350–359.

Williams, C. L. 1996. Short-term but not long-term estradiol replacement improvs radial arm maze performance of young and aging rats. *Soc. Neurosci. Abstr.*, 22: 1164.

Williams, C. L., and Meck, W. H. 1991. The organizational effects of gonadal steroids on sexually dimorphic spatial ability. *Psychoneuroendocrinology*, 16: 155–176.

Williams, C. L., Barnett, A. M., Meck, W. H. 1990. Organizational effects of early gonadal secretions on sexual differentiation in spatial memory. *Behav. Neurosci.*, 104: 84–97.

Williams, C. L., Men, D., Clayton, E. C., and Gold, P. E. 1998. Norepinephrine release in the amygdala after systemic injection of epinephrine or escapable footshock: Contribution of the nucleus of the solitary tract. *Behavioral Neuroscience*, 112: 1414–1422.

Williams, J. R., Catania, K. C., and Carter, C. S. 1992. Development of partner preferences in female prairie voles (*Microtus ochrogaster*): The role of social and sexual experience. *Horm. Behav.*, 26: 339–349.

Williams, J. R., Insel, T. R., Harbaugh, C. R., and Carter, C. S. 1994. Oxytocin centrally administered facilitates formation of a partner preference in female prairie voles (*Microtus ochrogaster*). *J. Neuroendocrinol.*, 6: 247–250.

Willingham, E., Baldwin, R., Skipper, J. K., and Crews, D. 2000. Aromatase activity during embryogenesis in the brain and adrenal-kidney-gonad of the red-eared slider turtle, a species with temperature-dependent sex determination. *General and Comparative Endocrinology*, 119: 202–207.

Wilson, A. P., and Vessey, S. H. 1968. Behavior of free-ranging castrated rhesus monkeys. *Folia Primatol.*, 9: 1–14.

Wilson, E. O. 1975. *Sociobiology.* Belknap Press of Harvard University Press, Cambridge, MA.

Wilson, J. D., George, F. W., and Griffin, J. E. 1981. The hormonal control of sexual development. *Science*, 211: 1278–1284.

Wilson, J. R., Adler, N., and LeBoeuf, B. 1965. The effects of intromission frequency on successful pregnancy in the female rat. *Proc. Natl. Acad. Sci. U.S.A.*, 53: 1392–1395.

Winans, S. S., and Powers, J. B. 1974. Neonatal and two-stage olfactory bulbectomy: Effects on male hamster sexual behavior. *Behav. Biol.*, 10: 461–471.

Wingfield, J. C. 1988. Changes in reproductive function of free-living birds in direct response to environmental perturbations. In M. H. Stetson (ed.), *Processing of Environmental Information in Vertebrates*, pp. 121–148. Springer-Verlag, Berlin.

Wingfield, J. C., and Farner, D. S. 1980. Control of seasonal reproduction in temperate-zone birds. *Prog. Reprod. Biol.*, 5: 62–101.

Wingfield, J. C., and Moore, M. C. 1987. Hormonal, social, and environmental factors in the reproductive biology of free-living male birds. In D. Crews (ed.), *Psychobiology of Reproductive Behavior*, pp. 148–175. Prentice Hall, Englewood Cliffs, NJ.

Wingfield, J. C., and Wada, M. 1989. Changes in plasma levels of testosterone during male–male interactions in the song sparrow, Melospiza melodia: Time course and specificity of response. *J. Comp. Physiol. A*, 166: 189–194.

Wingfield, J. C., Ball, G. F., Dufty, A. M., Hegner, R. E., and Ramenofsky, M. 1987. Testosterone and aggression in birds. *Am. Sci.*, 75: 602–608.

Wingfield, J. C., Breuner, C., Jacobs, J. D., Lynn, S., Maney, D., Ramenofsky, M., and Richardson, R. 1998. Ecological basis of hormone–behavior interactions: The "emergency life history stage." *American Zoologist*, 38: 191–206.

Wingfield, J. C., Deviche, P., Sharbaugh, S., Astheimer, L. B., Holberton, R., Suydam, R., and Hunt, K. 1994. Seasonal-chanages of the adrenocortical responses to stress in redpolls, *Acanthis flammea*, in Alaska. *J. Exp. Zool.*, 270: 372–380.

Wingfield, J. C., Hegner, R. E., Dufty, A. M., and Ball, G. F. 1990. The "challenge hypothesis": Theoretical implications for patterns of testosterone secretion, mating systems, and breeding strategies. *Am. Nat.*, 136: 829–846.

Wingfield, J. C., Jacobs, J., and Hillgarth, N. 1997. Ecological constraints and the evolution of hormone-behavior interrelationships. *Ann. N. Y. Acad. Sci.*, 807: 22–41.

Wingfield, J. C., Lynn, S., and Soma, K. K. 2001. Avoiding the "costs" of testosterone: Ecological bases of hormone-behavior interactions. *Brain, Behavior and Evolution*, 57: 239–251.

Wingfield, J. C., Moore, I. T., Goymann, W., Wacker, D., and Sperry, T. 2005. Contexts and ethology of vertebrate aggression: Implications for the evolution of hormone-behavior interactions. In R. J. Nelson (ed.), *Biology of Aggression*. Oxford University Press, New York. In press.

Wingfield, J. C., Whaling, C. S., and Marler, P. 1994. Communication in vertebrate aggression and reproduction: The role of hormones. In E. Knobil and J. D. Neill (eds.), *The Physiology of Reproduction* (2nd ed.), pp. 303–342. Raven Press, New York.

Winslow, J. T., and Insel, T. R. 2002. The social deficits of the oxytocin knockout mouse. *Neuropeptides*, 36: 221–229.

Winslow, J. T., and Insel, T. R. 2004. Neuroendocrine basis of social recognition. *Current Opinion in Neurobiology*, 14: 248–253.

Winslow, J. T., Hastings, N., Carter, C. S., Harbaugh, C. R., and Insel, T. R. 1993. A role for central vasopressin in pair bonding monogamous prairie voles. *Nature*, 365: 545–548.

Wirz-Justice, A., Bucheli, C., Graw, P., Kielholz, P., Fisch, H. U., and Woggon, B. 1986. Light treatment of seasonal affective disorder in Switzerland. *Acta Psychiatrica Scand.*, 74: 193–204.

Wirz-Justice, A., Cajochen, C., and Nussbaum, P. 1997. A schizophrenic patient with an arrhythmic circadian rest-activity cycle. *Psychiatry Res.*, 14: 83–90.

Wise, D. A., and Pryor, T. L. 1977. Effects of ergocornine and prolactin on aggression in the postpartum golden hamster. *Horm. Behav.*, 8: 30–39.

Wise, R. A. 1996. Neurobiology of addiction. *Curr. Opin. Neurobiol.*, 6: 243–251.

Wise, R. A., and Rompre, P. P. 1989. Brain dopamine and reward. *Annu. Rev. Psychol.*, 40: 191–225.

Wisniewski, A. B., Migeon, C. J., Gearhart, J. P., Rock, J. A., Berkovitz, G. D., Plotnick, L. P., Meyer-Bahlburg, H. F. L., and Money, J. 2001. Congenital micropenis: Long-term medical, surgical and psychosexual follow-up of individuals raised male or female. *Hormone Research*, 56: 3–11.

Wisniewski, A. B., Migeon, C. J., Malouf, M. A., and Gearhart, J. P. 2004. Psychosexual outcome in women affected by congenital adrenal hyperplasia due to 21-hydroxylase deficiency. *Journal of Urology*, 171: 2497–2501.

Wisniewski, A. B., Migeon, C. J., Meyer-Bahlburg, H. F. L., Gearhart, J. P., Berkovitz, G. D., Brown, T. R., and Money, J. 2000. Complete androgen insensitivity syndrome: Long-term medical, surgical and psychosexual outcome. *Journal of Clinical Endocrinology and Metabolism*: 85: 2664–2669.

Wisniewski, A. B., Nguyen, T. T., and Dobs, A. S. 2002. Evaluation of high-dose estrogen and high-dose estrogen plus methyltestosterone treatment on cognitive task performance in postmenopausal women. *Hormone Research*, 58: 150–155.

Witkin, H. A., Mednick, S. A., Schulsinger, F., Bakkestrom, E., Christiansen, K. O., Goodenough, D. R., Hirschhorn, K., Lundsteen, C., Owens, D. R., Philip, J. et al., 1976. Criminality in XYY and XXY men. *Science*, 193:547–555.

Wolf, O. T., Schommer, N. C., Hellhammer, D. H., McEwen, B. S., and Kirschbaum, C. 2001. The relationship between stress induced cortisol levels and memory differs between men and women. *Psychoneuroendocrinology*, 26: 711–720.

Wolff, E., and Wolff, E. 1951. The effects of castration on bird embryos. *J. Exp. Zool.*, 116: 59–97.

Wolff, J. O. 1985. Behavior. In R. H. Tamarkin (ed.). *Biology of New World* Microtus. American Society of Mammalogists, Shippensberg, PA.

Wolters, C. A., Yu, S. L., Hagen, J. W., and Kail, R. 1996. Short-term memory and strategy use in children with insulin-dependent diabetes mellitus. *J. Consult. Clin. Psychol.*, 64: 1397–1405.

Wommack, J. C., and Delville, Y. 2003. Repeated social stress and the development of agonistic behavior: Individual differences in coping responses in male golden hamsters. *Physiology and Behavior*, 80: 303–308.

Wommack, J. C., Taravosh-Lahn, K., David, J. T., and Delville, Y. 2003. Repeated exposure to social stress alters the development of agonistic behavior in male golden hamsters. *Hormones and Behavior*, 43: 229–236.

Wong, M., and Moss, R. L. 1992. Long-term and short-term electrophysiological effects of estrogen on the synaptic properties of hippocampal CA1 neurons. *J. Neurosci.*, 12: 3217–3225.

Wong, M., Thompson, T. L., and Moss, R. L. 1996. Nongenomic actions of estrogen in the brain: physiological significance and cellular mechanisms. *Crit. Rev. Neurobiol.*, 10: 189–203.

Wood, D. L., Sheps, S. G., Elveback, L. R., and Schirder, A. 1984. Cold pressor test as a predictor of hypertension. *Hypertension*, 6: 301–306.

Wood, G. E., and Shors, T. J. 1998. Stress facilitates classical conditioning in males, but impairs classical conditioning in females through activational effects of ovarian hormones. *Proc. Natl. Acad. Sci. U.S.A.*, 95: 4066–4071.

Wood, G. E., Beylin, A. V., and Shors, T. J. 2001. The contribution of adrenal and reproductive hormones to the opposing effects of stress on trace conditioning in males versus females. *Behavioral Neuroscience*, 115: 175–187.

Wood, R. I. 1997. Thinking about networks in the control of male hamster sexual behavior. *Horm. Behav.*, 32: 40–45.

Wood, R. I. 2002. Oral testosterone self-administration in male hamsters: Dose-response, voluntary exercise, and individual differences. *Hormones and Behavior*, 41: 247–258.

Wood, R. I., Johnson, L. R., Chu, L., Schad, C., and Self, D. W. 2004. Testosterone reinforcement: Intravenous and intracerebroventricular self-administration in male rats and hamsters. *Psychopharmacology*, 171: 298–305.

Woods, S. C., and Porte, D. 1983. The role of insulin as a satiety factor in the central nervous system. *Adv. Metab. Disord.*, 10: 457–468.

Woods, S. C., Chavez, M., Park, C. R., Riedy, C., Kaiyala, K., Richardson, R. D., Figlewicz, D. P., Schwartz, M. W., Porte, D. Jr, and Seeley, R. J. 1996. The evaluation of insulin as a metabolic signal influencing behavior via the brain. *Neurosci. Biobehav. Rev.*, 20: 139–144.

Woods, S. C., Figlewicz, D. P., Madden, L., Porte, D., Sipols, A. J., and Seeley, R. J. 1998. NPY and food intake: Discrepancies in the model. *Regul. Pept.*, 75: 403–408.

Woods, S. C., Lotter, E. C., McKay, L. D., and Porte, D. 1979. Chronic intracerebroventricular infusion of insulin reduces food intake and body weight of baboons. *Nature*, 282: 503–505.

Woods, S. C., West, D. B., Stein, L. J., McKay, L. D., Lotter, E. C., Porte, S. G., Kenney, N. J., and Porte, D., Jr. 1981. Peptides and the control of meal size. *Diabetologia*, 20: 305–313.

Woodside, B., and Leon, M. 1980. Thermoendocrine influences on maternal nesting behavior in rats. *J. Comp. Physiol. Psychol.*, 94: 41–60.

Woodside, B., Pelchat, R., and Leon, M. 1980. Acute elevation of the heat load of mother rats curtails maternal nest bouts. *J. Comp. Physiol. Psychol.*, 94: 61–68.

Woolley, C. S. 1998. Estrogen-mediated structural and functional synaptic plasticity in the female rat hippocampus. *Horm. Behav.*, 34: 140–148.

Woolley, C. S., Gould, E., and McEwen, B. S. 1990a. Exposure to excess glucocorticoids alters dendritic morphology of adult hippocampal pyramidal neurons. *Brain Res.*, 531: 225–231.

Woolley, C. S., Gould, E., and McEwen, B. S. 1990b. Naturally occurring fluctuations in the dendritic spine density in adult hippocampal neurons. *J. Neurosci.*, 10: 4035–4039.

Worrall, G. J., Chaulk, P. C., and Moulton, N. 1996. Cognitive function and glycosylated hemoglobin in older patients with type II diabetes. *J. Diabetes Complications*, 10: 320—24.

Worthy, K., Haresign, W., Dodson, S., McLeod, B. J., Foxcroft, G. R., and Haynes, N. B. 1985. Evidence that the onset of the breeding season in the ewe may be independent of decreasing plasma prolactin concentrations. *J. Reprod. Fertil.*, 75: 237–246.

Wrase, J., Klein, S., Gruesser, S. M., Hermann, D., Flor, H., Mann, K., Braus, D. F., and Heinz, A. 2003. Gender differences in the processing of standardized emotional visual stimuli in humans: A functional magnetic resonance imaging study. *Neuroscience Letters* 348: 41–45.

Wren, A. M., Small, C. J., Abbott, C. R., Dhillo, W. S., Seal, L. J., Cohen, M. A., Batterham, R. L. et al. 2001. Ghrelin causes hyperphagia and obesity in rats. *Diabetes*, 50: 2540–2547.

Wu, F. C. 1997. Endocrine aspects of anabolic steroids. *Clin. Chem.*, 43: 1289–1292.

Wurtman, J. J., Wurtman, R. J., Mark, S., Tsay, R., Gilbert, W., and Growdon, J. 1985. D-Fenfluramine selectively suppresses carbohydrate snacking by obese subjects. *Int. J. Eat. Disord.*, 4: 89–99.

Wurtman, R. J. 1975. The effects of light on man and other animals. *Annu. Rev. Physiol.*, 37: 467–83.

Wurtman, R. J., and Wurtman, J. J. 1989. Carbohydrates and Depression. *Sci. Am.*, 68: 68–75.

Wuttke, W., Arnold, P., Becker, D., Creutzfeldt, O., Langenstein, S., and Tirsch, W. 1975. Circulating hormones, EEG and performance in psychological tests of women with and without oral contraceptives. *Psychoneuroendocrinology*, 1: 141–151.

Wynne, K., Stanley, S., and Bloom, S. 2004. The gut and regulation of body weight. *Journal of Clinical Endocrinology and Metabolism*, 89: 2576–2582.

Wynne-Edwards, K. E. 1998. Evolution of parental care in *Phodopus*: Conflict between adaptations for survival and adaptations for rapid reproduction. *Am. Zool.*, 38: 238–250

Wynne-Edwards, K. E. 2001. Hormonal changes in mammalian fathers. *Hormones and Behavior*, 40: 139–145.

Wynne-Edwards, K. E., Terranova, P. F., and Lisk, R. D. 1987. Cyclic Djungarian hamsters, Phodopus campbelli, lack the progesterone surge normally associated with ovulation and behavioral receptivity. *Endocrinology*, 120: 1308–1316.

Wynne-Edwards, V. C. 1965. Self-regulating systems in populations of animals. *Science*, 147: 1543–1548.

Wysocki, C. J. 1979. Neurobehavioral evidence for the involvement of the vomeronasal system in mammalian reproduction. *Neurosci. Biobehav. Rev.*, 3: 301–341.

Wysocki, C. J., and Lepri, J. J. 1991. Consequences of removing the vomeronasal organ. *J. Steroid Biochem. Mol. Biol.*,39: 661–669.

Wysocki, C. J., Katz, Y., and Bernhard, R. 1983. Male vomeronasal organ mediates female-induced testosterone surges in mice. *Biol. Reprod.*, 28: 917–922.

Xiao, E., and Ferin, M. 1997. Stress-related disturbances of the menstrual cycle. *Ann. Med.*, 29: 215–219.

Yalcinkaya, T. M., Siiteri, P. K., Vigne, J. L., Licht, P., Pavgi, S., Frank, L. G., and Glickman, S. E. 1993. A mechanism for virilization of female spotted hyenas in utero. *Science*, 25;260: 1929–1931.

Yamanouchi, K. 1980. Inhibitory and facilitatory neural mechanisms involved in the regulation of lordosis behavior in female rats: effects of dual cuts in the preoptic area and hypothalamus. *Physiol. Behav.*, 25: 721–725.

Yamauchi, T., Kamon, J., Ito, Y., Tsuchida, A., Yokomizo, T., Kita, S., Sugiyama, T., et al. 2003. Cloning of adiponectin receptors that mediate antidiabetic metabolic effects. *Nature*, 423: 762–769.

Yang, E. V., and Glaser, R. 2000. Stress-induced immunomodulation: Impact on immune defenses against infectious disease. *Biomedicine and Pharmacotherapeutics*, 54: 245–250.

Yang, L. Y., and Clemens, L. G. 1996. Relation of intromissions to the female's postejaculatory refractory period in rats. *Physiol. Behav.*, 60: 1505–1511.

Yarbrough, W. G., Quarmby, V. E., Simental, J. A., Joseph, D. R., Sar, M., Lubahn, D. B., Olsen, K. L., French, F. S., and Wilson, E. M. 1990. A single base mutation in the androgen receptor gene causes insensitivity in the testicular feminized rat. *J. Biol. Chem.*, 265: 8893–8900.

Yaskin, V. A. 1984. Seasonal changes in brain morphology in small mammals. In J. F. Merrit (ed.), *Winter Ecology of Small Mammals*, pp. 183–189. Special Publication of the Carnegie Museum of Natural History, No. 10.

Yehuda, R., Giller, E. L., and Mason, J. W. 1993. Psychoneuroendocrine assessment of posttraumatic stress disorder: Current progress and new directions. *Prog. Neuropsychopharmacol. Biol. Psychiatry*, 17: 541–550.

Yehuda, R., Kahana, B., Binder-Brynes, K., Southwick, S. M., Mason, J. W., and Giller, E. L. 1995. Low urinary cortisol excretion in Holocaust survivors with posttraumatic stress disorder. *Am. J. Psychiatry*, 152: 982–986.

Yellon, S. M., Hutchison, J. S., and Goldman, B. D. 1989. Sexual differentiation of the steroid feedback mechanism regulating follicle-stimulating hormone secretion in the Syrian hamster. *Biol. Reprod.*, 41: 7–14.

Yen, S. S., Morales, A. J., and Khorram, O. 1995. Replacement of DHEA in aging men and women. Potential remedial effects. *Ann. N. Y. Acad. Sci.* , 774: 128–142.

Yerkes, R. M., and Dodson, J. D. 1908. The relation of strength of stimulus to rapidity of habit formation. *J. Comp. Neurol. Psychol.*, 18: 458–482.

Yerkes, R. M., and Elder, J. H. 1936. Oestrus, receptivity, and mating in chimpanzee. *Comp. Psychol. Monogr.*, 13: 1–39.

Yesalis, C. E., Barsukiewicz, C. K., Kopstein, A. N., and Bahrke, M. S. 1997. Trends in anabolic-androgenic steroid use among adolescents. *Arch. Pediatr. Adolesc. Med.*, 151: 1197–1206.

Yirmiya, R., Avitsur, R., Donchin, O., and Cohen, E. 1995. Interleukin-1 inhibits sexual behavior in female but not in male rats. *Brain Behav. Immunity*, 9: 220–233.

Yogman, M. W. 1990. Male parental behavior in humans and nonhuman primates. In N. A. Krasnegor and R. S. Bridges (eds.), *Mammalian Parenting*, pp. 461–481. Oxford University Press, Oxford.

Yokota, S. I., Horikawa, K., Akiyama, M., Moriya, T., Ebihara, S., Komuro, G., Ohta, T., and Shibata, S. 2000. Inhibitory action of brotizolam on circadian and light-induced *per1* and *per2* expression in the hamster suprachiasmatic nucleus. *British Journal of Pharmacology*, 131: 1739–1747.

Yonkers, K. A. 1997. Antidepressants in the treatment of premenstrual dysphoric disorder. *J. Clin. Psychiatry*, 58: 4–10.

Yoo, S. H., Yamazaki, S., Lowrey, P. L., Shimomura, K., Ko, C. H., Buhr, E. D., Siepka, S. M., et al. 2004. PERIOD2::LUCIFERASE real-time reporting of circadian dynamics reveals persistent circadian oscillations in mouse peripheral tissues. *Proceedings of the National Academy of Sciences U.S.A.*, 101: 5339–5346.

Young, E. and Korszun, A. 1998. Psychoneuroendocrinology of depression. Hypothalamic-pituitary-gonadal axis. *Psychiatry Clin. North Am.*, 21: 309–323.

Young, I. M. 1990. *Justice and the Politics of Difference.* Princeton University Press, Princeton.

Young, L. J., and Wang, Z. 2004. The neurobiology of pair bonding. *Nature Neuroscience*, 7: 1048–1054.

Young, L. J., Lim, M., Gingrich, B., and Insel, T. R. 2001. Cellular mechanisms of social attachment. *Hormones and Behavior*, 40: 133–148.

Young, L. J., Wang, Z., Donaldson, R., and Rissman, E. F. 1998. Estrogen receptor is essential for induction of oxytocin receptor by estrogen. *Neuroreport*, 9: 933–936.

Young, L. J., Winslow, J. T., Wang, Z., Gingrich, B., Guo, Q., Matzuk, M. M., and Insel, T. R. 1997. Gene targeting approaches to neuroendocrinology: Oxytocin, maternal behavior, and affiliation. *Horm. Behav.*, 31: 221–231.

Young, M. W. 2000. Marking time for a kingdom. *Science*, 288: 451–453.

Young, W. C. 1937. The vaginal smear picture, sexual receptivity, and the time of ovulation in the guinea pig. *Anat. Rec.*, 67: 305–539.

Young, W. C. 1961. The hormones and mating behavior. In W. C. Young (ed.), *Sex and Internal Secretions*, Vol. 2, pp. 1173–1239. Williams & Wilkins, Baltimore.

Young, W. C., and Plough, H. H. 1926. On the sterilization of *Drosophila* by high temperature. *Biol. Bull.*, 51: 189–198.

Young, W. C., Dempsey, E. W., and Myers, H. I. 1935. Cyclic reproductive behavior in the female guinea pig. *J. Comp. Psychol.*, 19: 313–335.

Young, W. C., Dempsey, E. W., Hagquist, C. W., and Boling, J. L. 1939. Sexual behavior and sexual receptivity in the female guinea pig. *J. Comp. Psychol.*, 27: 49–68.

Young, W. C., Dempsey, E. W., Myers, H. I., and Hagquist, C. W. 1938. The ovarian condition and sexual behavior in the female guinea pig. *Am. J. Anat.*, 63: 457–487.

Youngstrom, T. G., Weiss, M. L., and Nunez, A. A. 1987. A retinal projection to the paraventricular nuclei of the hypothalamus in the Syrian hamster (*Mesocricetus auratus*). *Brain Res. Bull.*, 19: 747–750.

Yousem, D. M., Maldjian, J. A., Siddiqi, F., Hummel, T., Alsop, D. C., Geckle, R. J., Bilker, W. B., and Doty, R. L. 1999. Gender effects on odor-stimulated functional magnetic resonance imaging. *Brain Research*, 818: 480–487.

Youthed, G. J., and Moran, R. C. 1969. The lunar day activity rhythm of myrmeleontid larvae. *J. Insect Physiol.*, 15: 1259–1271.

Zahavi, A., and Zahavi, A. 1997. *The Handicap Principle.* Oxford University Press: New York.

Zamorano, P. L., Mahesh, V. B., DeSevilla, L. M., horich, L. P., Bhat, G. K., and Brann, D. W. 1997. Expression and localization of the leptin receptor in endocrine and neuroendocrine tissues of the rat. *Neuroendocrinology*, 65: 223–228.

Zarrow, M. X., Denenberg, V. H., and Anderson, C. O. 1965. Rabbit: Frequency of suckling in the pup. *Science*, 150: 1835–1836.

Zarrow, M. X., Farooq, A., Denenberg, V. H., Sawin, P. B., and Ross, S. 1963. Maternal behaviour in the rabbit: Endocrine control of maternal-nest building. *J. Reprod. Fertil.*, 6: 375–383.

Zarrow, M. X., Gandelman, R., and Denenberg, V. H. 1971. Prolactin: Is it an essential hormone for maternal behavior in the mammal? *Horm. Behav.*, 2: 343–354.

Zehr, J. L., Maestripieri, D., and Wallen, K. 1998. Estradiol increases female sexual initiation independent of male responsiveness in rhesus monkeys. *Horm. Behav.*, 33: 95–103.

Zeller, R. A., and Jurkovac, T. 1989. A dome stadium: Does it help the home team in the National Football League? *Sport Place International*, 3: 37–39.

Zhang, S., McGaugh, J. L., Juler, R. G., and Introini-Collison, I. B. 1987. Naloxone and [Met5]-enkephalin effects on retention: Attenuation by adrenal denervation. *Eur. J. Pharmacol.*, 138: 37–44.

Zhang, Y., Proenca, R., Maffei, M., Barone, M., Leopold, L., and Friedman, J. M. 1994. Positional cloning of the mouse obese gene and its human homologue. *Nature*, 372: 425–432.

Zhao, W.-Q., and Alkon, D. L. 2001. Role of insulin and insulin receptor in learning and memory. *Molecular and Cellular Endocrinology*, 177: 125–134.

Ziegler, T. E. 2000. Hormones associated with non-maternal infant care: A review of mammalian and avian studies. *Folia Primatologia*, 71: 6–21.

Ziegler, T. E., Wegner, F. H., Carlson, A. A., Lazaro-Perea, C., and Snowdon, C. T. 2000. Prolactin levels during the periparturitional period in the biparental cotton-top tamarin (*Saguinus oedipus*): Interactions with gender, androgen levels, and parenting. *Hormones and Behavior*, 38: 111–122.

Zigmond, M. J., Abercrombie, E. D., Berger, T. W., Grace, A. A., and Stricker, E. M. 1991. Compensations after lesions of central dopaminergic neurons: Some clinical and basic implications. *Trends Neurosci.*, 13: 290–296.

Zimmerberg, B., Brunelli, S. A., and Hofer, M. A. 1994. Reduction of rat pup ultrasonic vocalizations by the neuroactive steroid allopregnanolone. *Pharmacol. Biochem. Behav.*, 47: 735–738.

Zimmerman, N. H., and Menaker, M. 1979. The pineal gland: The pacemaker within the circadian system of the house sparrow. *Proc. Natl. Acad. Sci. U.S.A.*, 76: 999–1003.

Zinn, A. R. 1993. Turner Syndrome: The case of the missing sex chromosome. *Trends Genet.*, 9(3): 90–93.

Zittel, T. T., Glatzle, J., Kreis, M. E., Starlinger, M., Eichner, M., Raybould, H. E., Becker, H. D., and Jehle, E. C. 1999. C-fos protein expression in the nucleus of the solitary tract correlates with cholecystokinin dose injected and food intake in rats. *Brain Research*, 846: 1–11.

Zucker, I. 1966. Facilitory and inhibitory effects of progesterone on sexual responses of spayed guinea pigs. *J. Comp. Physiol. Psychol.*, 62: 376–381.

Zucker, I. 1968. Biphasic effects of progesterone on sexual receptivity in the female guinea pig. *J. Comp. Physiol. Psychol.*, 65, 472–478.

Zucker, I. 1969. Hormonal determinants of sex differences in saccharine preferences, food intake, and body weight. *Physiol. Behav.*, 4: 595–602.

Zucker, I. 1980. Light, behavior, and biologic rhythms. In D. T. Krieger and J. C. Hughes (eds.), *Neuroendocrinology*, pp. 93–101. Sinauer Associates, Sunderland, MA.

Zucker, I. 1988a. Neuroendocrine substrates of circannual rhythms. In D. J. Kupfer, T. H. Monk, and J. D. Barchas (eds.), *Biological Rhythms and Mental Disorders*, pp. 219–251. Guilford Press, New York.

Zucker, I. 1988b. Seasonal affective disorders: Animal models Non fingo. *J. Biol. Rhythms*, 3: 209–223.

Zucker, I., and Licht, P. 1983. Circannual and seasonal variations in plasma luteinizing hormone levels of ovariectomized ground squirrels (*Spermophilus lateralis*). *Biol. Reprod.*, 28: 178–85.

Zucker, I., Cramer, C. P., and Bittman, E. L. 1980. Regulation by the pituitary gland of circadian rhythms in the hamster. *J. Endocrinol.*, 85: 17–25.

Zucker, I., Lee, T. M., and Dark, J. 1991. The suprachiasmatic nucleus and annual rhythms in mammals. In D.C. Klein, S. M. Reppert, and R. Y. Moore (eds.), *The Suprachiasmatic Nucleus: The Mind's Clock*, pp. 246–260. Elsevier, New York.

Zucker, I., Wade, G. N., and Ziegler, R. 1972. Sexual and hormonal influences on eating, taste preferences, and body weight of hamsters. *Physiol. Behav.*, 8: 101–111.

Zucker, K. J. 1996. Commentary on Diamond's "Prenatal predisposition and the clinical management of some pediatric conditions. *J. Sex Marital Ther.*, 22: 148– 160.

Zweifel, J. E. and O'Brian, W. H. 1997. A meta-analysis of the effect of hormone replacement therapy upon depressed mood. *Psychoneuroendocrinology*, 22: 189–212.

Index

B

N

Q

R

T

U

V

About the Book

Editor: Graig Donini
Production Editor: Sydney Carroll
Copy Editor: Norma Roche
Production Manager: Christopher Small
Book Production: Joan Gemme
Book Design: Jefferson Johnson
Cover Design: Joan Gemme
Book and Cover Manufacturer: Courier Companies, Inc.